REINFORCED CONCRETE STRUCTURES

ANALYSIS, DESIGN AND DETAILS IN SI UNITS

By the Same Author

1. **Analysis of Structure Vol. I**
2. **Analysis of Structure Vol. II**
3. **Steel Structures (Including Timber & Plastic)**
4. **Advanced Theory of Structures and Matrix Method**
5. **Design of Concrete Bridges**
6. **Handbook of Civil Engineering**
7. **Limit State Design of Reinforced Concrete Structures**

Analysis, Theory & Details

LIMIT STATE DESIGN OF REINFORCED CONCRETE STRUCTURES

Prof. V.N.Vazirani & H.Mehra

CONTENTS

DESIGN OF REINFORCED
CONCRETE STRUCTURES

Prof. V.N. VAZIRANI *M.Sc. (Structures)*
(Formerly, Deptt. of Civil Engg.
Delhi College of Engineering, Delhi)

and

Dr. M.M. RATWANI *Ph.D. (Structures)*
(Formerly, Deptt. of Civil Engg.,
Indian Institute of Technology, Delhi)

KHANNA PUBLISHERS®
Investing in Learning®
Operational Office :
4575/15, Onkar House, Opp. Happy School,
Ground Floor, Daryaganj, New Delhi 110 002
Phones : 011-45033819 • *Mob.* 09811541460
email : contactus@khannapublishers.in

Published by :

Romesh Chander Khanna & Vineet Khanna
for **KHANNA PUBLISHERS**
2-B, Nath Market, Nai Sarak
Delhi- 110 006 (India)

Website : www.khannapublishers.in

Hologram & Description

To all readers of our books, from yourself if being defrauded by pirates to prevent, please make sure that there is an Hologram on the cover of our books with the below specifications. If you find any book without Hologram and Description, please mail us at **contactus@khannapublishers.in**

Thanking you

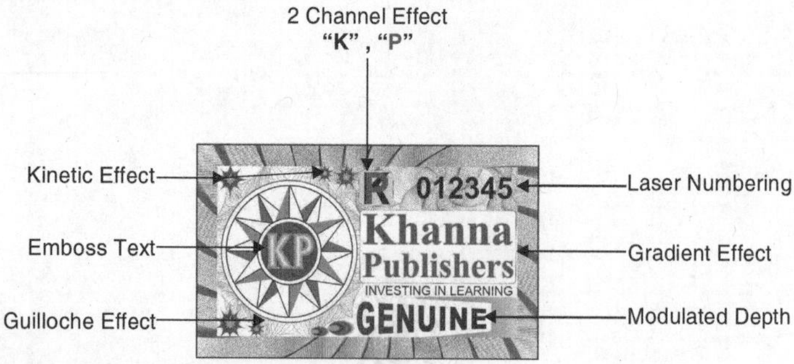

ISBN No. 978-81-7409-232-8

First Edition : 1977
Sixteenth Edition
Seventh Reprint : 2025

***Price :* ₹ 925.00**

PREFACE

The authors have attempted to present in this volume detailed and simple treatment of the thoery and design of R.C.C. stuctures. Complete designs of various structures have been worked out with detailed drawings. Besides giving the basic principles for design of prestressed concrete, theory alongwith detailed drawings & charts as applied to concrete structures has been given which will be of imense use for engineers for design and reference purpose.

While the work done by previous authors in the field of R.C.C. structures has greatly influenced the subject matter of this book, its speciality lies not only in entirely fresh and analytical approach to the problem of design with certain basically new conception but also in the more comprehensive treatment which makes the design of structures fuller in all respects, even to the extent of minor details.

The revised edition features, some advanced topics on Shell Construction and Folded Plate Construction with emphasis on design. Another feature of this edition is two chapters on Design Problems, and Graphs for Compression with Bending, which will prove to imense help to the readers.

The authors are grateful to the readers for sending suggestions which have enabled us to make changes and improve the text of the book and will welcome them in future improvement of the text.

Finally the authors would like to express their sense of gratitude to Shri Romesh Chander Khanna for giving personal attention and expediting the publication of the book.

—V.N. Vazirani
—M.M. Ratwani

CONTENTS

1

Introduction

1·1. Concrete, Reinforced Concrete and Prestressed Concrete

Concrete is a stonelike material which is obtained by carefully mixing in proportion cement, sand and gravel or other aggregate, and water to harden in forms of the shape and dimensions of the desired structures. The bulk of the material consists of fine and coarse aggregate. Cement and water interact chemically to bind the aggregate particles into a solid stonelike material. Additional water, over and above that is needed for chemical reaction, is necessary to give the mixture workability which enables it to fill the forms and surround the embedded reinforcing steel before hardening. Concrete of various strength properties can be obtained by appropriate proportions of the constituent materials. Special cements such as high-early strength cements, and special aggregates such as light weight or heavy weight aggregates and special curing methods such as steam curing permit an even wider variety of properties to be obtained. Before being placed in form this material is referred to as concrete mix. Concrete mixed in a concrete factory *i.e.* central mixing plant and supplied to the user ready for use is called ready-mixed concrete.

Production takes place by mixing the materials in concrete mixers, which ensure homogeneity of the product on completion of mixing the ingredients measured in definite quantities for the given concrete composition.

The properties of concrete depend to a very large extent on the proportions of the mix, on the thoroughness with which the various constituents are intermixed, and on the conditions of humidity and temperature in which the mix is maintained from the moment it is placed in the forms untill it is fully hardened The process of controlling these conditions is known as curing.

(*a*) **Cement**. According to the purpose for which the concrete is required, the mix may contain cement from one of the following groups ; all kinds of ordinary Portland cement, Portland blast-furnace cement, pozzolanic cement, and sand-Portland-cement, high alumina cement, expanding cement and non-shrinking cement, super-sulphate cement or slag sulphate cement.

(*b*) **Aggregates.** In ordinary structural concretes the aggregates occupy about 70 to 75 percent of the volume of the hardened mass. The remainder consists of hardened cement paste, uncombined water and air voids. The uncombined water and air voids do not contribute to the strength of concrete. The more densely the aggregate can be packed, the better the strength, weather resistance and enconomy of the concrete. The gradation of the particle sizes in the aggregates to produce dense concrete is very important. It is also important that the aggregate should have good strength, durability and weather resistance. It should be free from impurities such as loam, silt and organic matter which may weaken the bond with cement and have unfavourable chemical reaction with cement.

Natural aggregates are generally classified as fine and coarse. The size of a fine aggregate shall not exceed 4·75 mm gauge. Aggregates of bigger sizes are called coarse aggregates. When favourable gradation is desired, aggregates are separated by sieving into two or three size groups of sand and several size groups of coarse aggregate. These are then combined to get a densely packed aggregate. The maximum size of coarse aggregate in reinforced concrete is governed by the requirement that it shall easily fit into the forms and between the reinforcing bars. For reinforced concrete work, aggregates having a normal size of 20 mm are generally considered satisfactory. For heavily reinforced concrete members as in the case of ribs of main members, the nominal maximum size of the aggregate should usually be restricted to 5 mm less than the minimum clear distance between the main bars or 5 mm less than the minimum cover to the reinforcement whichever is smaller.

Crushed rock or gravel in the form of a mixture of several fractions can be used as coarse aggregate. The fine aggregate *i.e.* sand is used as far as possible in two fractions. For preparing the concrete mix the fractions of the aggregates are measured separately.

Plums above 160 mm upto any reasonable size may be used in plain concrete work upto a maximum limit of 20 percent by volume of concrete. The plums shall be distributed evenly and shall be not closer than 150 mm from the surface.

(*c*) **Concrete.** The unit weight of stone concrete *i.e.* concrete with natural stone aggregates varies from 23,000 N/m^3 to 24000 N/m^3.

For special purposes lightweight concretes and heavy concretes are being used. A variety of lightweight aggregates consist of pumice or cinders used for insulating concretes, processed aggregates like expanded shales, clays, slates, slags or pelletized fly ash for structural light weight concrete. These are light in weight due to the individual aggregate particles are porous or cellular which is achieved by using gas or steam in processing the aggregates in rotary kilns at as high temperature as 1100 °C.

Three classes of lightweight concretes are

(*i*) low density concretes which are employed for insulation with unit weight not exceeding 8000 N/m³.

(*ii*) Moderate strength concretes with unit weights from 9500 N/m³, to 1300 N/m³, having compressive strength of 7 N/mm² to 15 N/mm² chiefly used as fill.

(*iii*) Structural concretes with unit weight of 14,000 N/m³ to 19,000 N/m³ and compressive strenghts equal to those of stone concretes.

Heavy weight concrete is required for shielding gamma and X-radiation in nuclear reactor and similar installations, for protective structures, as counterweight of lift bridges. Heavy aggregates such as heavy iron ores or barium sulfate rock crushed to suitable sizes. Steel in the form of scrap, punchings or fines is also used. Unit weights of heavy-weight concretes with natural heavy rock aggregates range from 30,000 N/m³ to 32,000 N/m³. If iron punchings are added to high density ores unit weight can be as high as 38,000 N/m³. The weight may be as high as 52,000 N/m³ if ores are used for the fines and steel for the coarse aggregate.

(*d*) **Water-cement ratio.** Water used for mixing and curing shall be clean and free from injurious amounts of oils, acids, alkalis, salts, sugar, organic materials or other substances that may be deleterious to concrete or steel. Potable water is generally considered satisfactory for mixing concrete.

For complete hydration of a given amount of cement, an amount of water equal to about 25 percent of that cement by weight, is needed for chemical reaction. An additional 10 to 15 percent is added to provide mobility for the water in the cement paste during the hydration process so that it can reach the cement particles. This makes for a total minimum water-cement ratio of 0·35 to 0·40 by weight. Water-cement ratios in concretes are generally considerably larger than this minimum, to provide the necessary workability of the concrete mix. Any amount of water above the 25 percent consumed in the chemical reaction produces pores in the cement paste. Since only the solids and not the voids resist stress, strength increases directly as the fraction of the total volume occupied by the solids increases. Thus the strength of cement paste decreases directly with increasing water-cement ratio.

The chemical process involved in the setting and hardening liberates heat, known as heat of hydration. In large concrete masses, such as dams, this heat gets dissipated very slowly and therefore there will be rise in temperature consequently increase in volume of concrete during hydration and subsequently on cooling there will be contraction and cracking. To avoid the serious cracking and weakening, special measures must be taken and low heat Portland cement be used.

(*e*) **Workability**. The concrete mix proportions chosen should be such that the concrete is of adequate workability for the placing conditions of the concrete and can be properly compacted with the means available.

The workability is measured by two methods—

(*i*) Slump test (*ii*) Compaction factor test.

In slump test, a slump test cone is filled one fourth at a time with concrete and compaction is done with a 16 mm rod by stroking evenly with 25 strokes. After the whole cone is filled and compacted, the cone is removed without disturbing the concrete. The vertical settlement of the concrete is called the slump. The slump required for various works is—

(*a*) Concrete for road work......25 mm to 50 mm

(*b*) Concrete for beams and slabs......50 mm to 100 mm

(*c*) Concrete for walls, thin vertical members and columns......75 mm to 125 mm

(*d*) Vibrated concrete......15 mm to 25 mm

(*e*) Mass concrete......25 mm to 50 mm.

In compacting factor test, concrete is poured into top hopper and then the door is opened at the bottom so as concrete to fall into the second hopper, kept at some distance from first hopper. The door of second hopper is opened so as the concrete to fall into a cylinder kept at distance from second hopper. Extra concrete in the cylinder is scraped off. The weight of concrete in cylinder is ascertained and let it be W_1. Then the same cylinder is emptied and filled with concrete mix and compacted so as there are no voids and the weight of voidless concrete is ascertained and let it be W_2. The ratio W_1/W_2 is called compaction factor. A compaction factor of 0·95 represents good workability and factor 0·92 represents medium workability and factor 0·85 represents low workability. The compaction factor required for various placing conditions are as given below:

	Compaction factor
(*a*) Concreting of shallow sections with vibration	...0·75 to 0·8
(*b*) Concreting of lightly reinforced sections with vibration	...0·80 to 0·90
(*c*) Concreting of lightly reinforced section without vibration, or heavily reinforced section with vibrations	...0·85 to 0·92
(*d*) Concreting of heavily reinforced section without vibration	...Above 0·92

(*f*) **Durability.** The durability of concrete depends on the resistance to deterioration and the environment in which it is placed. The resistance of concrete to weathering, chemical attack, abrasion, frost and fire depends upon quality of constituent materials.

One of the main characteristics influencing the durability of concrete is its permeability. With strong, dense aggregates, a suitably low permeability is achieved by having a sufficiently low water-cement ratio by ensuring as thorough compaction of the concrete as possible, and by ensuring sufficient hydration of cement through proper curing methods.

Therefore for given aggregates, the cement content should be sufficient to provide adequate workability with a low water-cement ratio so that concrete can be completely compacted with the means available.

(*g*) **Transporting, Placing, Compacting and Curing.** Concrete shall be transported from the mixer to the form work as rapidly as possible by methods which will prevent the segregation or loss of any of the ingredients and maintaining the required workability.

The individual components of concrete tend to segregate because of their dissimilarity. If the concrete is very wet, standing in containers or forms, the heavier gravel components tend to settle and the lighter materials particularly water tend to rise. Lateral movement such as flow in the forms, tends to separate the coarse gravel from the finer components of the mix. The previously common methods of conveying such as chutes and conveyer belts have been discarded because they cause segregation of components Now a days concrete is conveyed by bottom dump buckets or in wheelbarrows or pumping through steel pipelines.

Placing is the process of transferring the fresh concrete from the conveying device to its final place in the forms. The concrete shall be deposited as nearly as practicable in its final position to avoid rehandling. Prior to placing, loose rust must be removed from the reinforcement, forms must be cleaned and hardened surfaces of previous concrete lifts must be cleaned and treated appropriately. Placing and compacting are critical in their effect on the final quality of the concrete. Proper placement must avoid segregation, displacement of forms or of reinforcement in the forms, and poor bond between successive layers of concrete.

Immediately after placing, the concrete should be compacted by means of hand tools or vibrators. The concrete should be worked around the reinforcement, around embedded fixtures and into corners of the formwork. Such compacting prevents honey combing, assures close contact with forms and reinforcement and serves as a partial remedy to possible prior segregation. Compacting is achieved by hand tamping with a variety of special tools, but now more commonly and successfully with high frequency power driven vibrators. The vibrators may be internal type which are immersed in concrete or the external type which are attached to the forms.

Over-vibration or vibration of very wet mixes is harmful and should be avioded. Under-vibration is also harmful.

Whenever vibration has to be applied externally, the design of formwork and the disposition of vibrators should receive special consideration to ensure efficient compaction and to avoid surface blemishes.

The internal vibrators are preferable but must be supplemented by the external vibrators where narrow forms or other obstactles make immersion impossible.

Fresh concrete gains strength most rapidly during the first few days and weeks. Structural design is generally based on the 28 days strength, about 70 percent of which is reached at the end of the first week after placing. The final strength depends greatly on the conditions of moisture and temperature during this initial period. The maintenance of proper conditions during this time is known as curing. Thirty percent or more of the strength can be last by premature drying out of the concerte. Similarly strength may be lost to the extent of thirty percent if the outside temperature falls to 2°C or lower during the first few days unless the concrete is maintained continuously moist for a long time thereafter. Freezing of fresh concrete may reduce its strength by as much as 50 percent.

To prevent such damage, the concrete should be protected from loss of moisture for at least 7 days and in case of more sensitive work upto 14 days. When high-early-strength cements are used, curing periods can be cut in half.

Curing can be achieved by keeping exposed surfaces continuously wet through sprinkling, ponding, covering with wet burlap, or the like. Recent methods include the use of sealing compounds, which when properly used, form evaporation reta ding membranes, and water proof papers. Besides strength, proper moist curing provides better shrinkage control.

To protect the concrete against low temperature during cold wauther, the mixing water and occasionally, the aggregates are heated, temperature insulation is used where possible and special admixturctemperature.rly calcium chloride are used. When air temperatures are very low, external heat may have to be supplied in additon to insulation

(*h*) **Under-water Concreting.** When it is necessary to deposit concrete under water, the following points should be noted :

(*i*) Such concrete should not be considered as 'Design mix concrete'.

(*ii*) The concrete shall contain atleast 10 percent more cement than that required for the same mix placed in the dry, the quantity of extra cement varying with conditions of placing. The volume of mass of the coarse aggregate shall be not less than one and a half times, not more than twice that of the fine aggregate. The materials shall be so proportioned as to produce a concrete having a slump of not less than 100 mm and not more than 180 mm.

(*iii*) Cofferdams or forms shall be sufficiently tight to ensure still water if practicable, and in any case to reduce the flow of water to less than 3 metres per minute through the space into which concrete is to be deposited. Cofferdams or forms in still water shall be sufficiently tight to prevent loss of mortar through the walls. Dewatering by pumping shall not be done when concrete is being placed or until 24 hours thereafter.

(*iv*) Cencrete shall be deposited continuously until it is brought to the required height. While depositing the top surface shall be kept as nearly level as possible and the formation of seams avoided. The methods to be used for depositing concrete under water shall be one of the following :

(*I*) *Tremie*. When concrete is to be deposited under water by means of tremie, the top section of the tremie shall be a hopper, large enough to hold one entire batch of the mix or the entire contents of the transporting bucket if any. The tremie pipe shall be not less than 200 mm in diameter and shall be large enough to allow a free flow of concrete and strong enough to withstand the external pressure of the water in which it is suspended, even if a partial vacuum develops inside the pipe. Preferably flanged steel pipe of adequate strength for the job should be used. A separate lifting device shall be provided for each tremie pipe with its hopper at the upper end Unless the lower end of the pipe is equiped with an approved automatic check valve, the upper end of the pipe shall be plugged with a wadding of the gunny sacking or other approved material before delivering the concrete to the tremie pipe through the hopper, so that when the concrete is forced down from the hopper to the pipe, it will force the plug along with any water in the pipe, down the pipe and out of the bottom end, thus establishing continuous stream of concrete.

It will be necessary to raise slowly the tremie in order to cause a uniform flow of the concrete, but the tremie shall not be emptied so that water enters the pipe At all times after the placing of concrete is started and until all the concrete is placed, the lower end of the tremie pipe shall be below the top surface of the plastic concrete. This will cause the concrete to build up from below instead of flowing out over the surface, and thus avoid formation of laitance layer. If the charge in the tremie is lost while depositing, the tremie shall be raised above the concrete surface, and unless sealed by a check valve it shall be plugged at the top end, as at the beginning, before refilling for depositing concrete.

(*II*) *Drop bottom bucket*. The top of the bucket shall be covered with a canvass flap. The bottom doors shall open freely downward and outward when tripped. The bucket shall be filled completely and lowered slowly to avoid backwash. The bottom doors shall not be opened until the bucket rests on the surface upon which the concrete is to be deposited and when discharged, shall be with drawn slowely until well above the concrete.

(*III*) *Bags*. Bags of atleast 0·028 m^3 capacity of jute or other coarse cloth shall be filled about two thirds full of concrete, the spare end turned under so that bag is square ended and securely tied. They shall be placed carefully in header and stretcher courses so that the whole mass is interlocked. Bags used for this purpose shall be free from deleterious materials.

(*IV*) *Grouting*. A series of round cages made from 50 mm mesh of 6 mm steel and extending over the full height to be concreted shall be prepared and laid vertically over the area to be concreted so that the distance between centres of the cages and also to the faces of the concrete shall not exceed one metre. Stone aggregate of not less than 50 mm nor more than 200 mm size shall be deposited outside the steel cages over the full area and height to be concreted with due care to prevent displacement of the cages.

A stable 1·2 cement sand grout with a water-cement ratio of not less than 0·6 and not more than 0·8 shall be prepared in a mechanical mixer and sent down under pressure (about 0·2 N/mm^2) through 38 to 50 mm diameter pipes terminating into steel cages, about 50 mm above the bottom of the concrete. As the grouting proceeds, the pipe shall be raised gradually upto a height of not more than 600 mm above its starting level after which it may be withdrawn and placed into the next cage for further grouting by the same procedure.

After grouting the whole area for a height of 600 mm, the same operation shall be repeated, for the next layer of 600 mm and so on.

The amount of grout to be sent down shall be sufficient to fill all the voids which may be either ascertained or assumed as 55 percent of the volume to be concreted.

1·2. Design Considerations

In past four decades or so reinforced cement concrete (RCC) has established itself as a very satisfactory construction material. Reinforced concrete has reached very advanced stage both in design and construction techniques. Reinforced concrete has been success fully used in construction of buildings, bridges, dams, bunkers, silos, water tanks and many other structures.

The first basic step in the design of any structure is layout. The layout of a structure is governed by the functions to be performed by the structure.

The design of RCC structures will depend on the layout. The layout depends on the functions to be performed by the structure. Planning and proportioning a structure is called structural design. Some of the factors which will govern the design are :

(*i*) Adequate strength and rigidity.

(*ii*) Should not interfere with the functions for which the structure is intended.

(*iii*) Be economical in first cost and maintenance.

(*iv*) Should be strong enough to last for the design service life.

(*v*) Be readily adaptable to future extension.

The design will consist of :

(*i*) Working out the forces and loads which the structure has to carry.

(*ii*) Arrangement of various beams, columns etc. to support the structure.

(*iii*) Computation of stresses, shears, moments etc. in the members of the structure.

(*iv*) Providing adequate sections to resist the stresses in the members.

1·3. Loads

For the purpose of computing the maximum stresses in any structure or member of structure the following loads should be taken into account :

(*i*) *Dead loads.* The weight of the RCC and all the material supported by it permanently.

(*ii*) *Live loads.* These are the loads which temporarily rest at one place such as—

(*a*) The goods stored on a warehouse floor.

(*b*) The furniture and occupants of an office.

(*c*) The snow load on the roof or deck of a bridge.

(*d*) Material stored in bunkers and silos.

(*e*) Moving loads on bridges.

(*iii*) *Dynamic effects.* Live loads which produce vibrations by impact such as moving trains, trucks and cranes, produce greater effects than would be produced by the same loads if stationary. The additional effect is called the dynamic effect.

(*iv*) *Wind loads.*

(*v*) *Seismic loads.*

(*vi*) *Erection loads.* Storage of construction material and erection equipment including all loads due to operation of such equipment.

(*vii*) *Temperature effects.* The provisions for maximum expansion and contraction has to be allowed in the structure.

1·4. Concrete and Reinforced Concrete

Concrete. Concrete is a material obtained by cementing together inert materials like sand, gravel broken stone or some other suitable materials. Cementing material generally used is portland cement. Cement and water react chemically to bind sand, known as fine aggregate and gravel or broken stone known as coarse aggregate. The resulting product is like an artificial stone. When the aggregates, cement and water are mixed in proper proportions, the resulting product after some time is hard, strong in compression and shear, brittle and weak in tension. These properties of concrete are affected by several factors like proportions of cement, coarse aggregate and fine aggregate ; amount of water, temperature at the time of mixing, humidity at the time of moulding of concrete in forms and temperature and humidity maintained subsequently. This process of maintaining desired temperature and humidity is known as curing of concrete.

The plain concrete, being strong in compression is suitable for massive construction, where it is subjected to compressive stresses. It is not suitable at places where it is subjected to tensile stresses.

Reinforced concrete. As discussed above, concrete is strong in compression and weak in tension, so steel reinforcement is used to take up tensile stresses at places where section is subjected to tensile stress. Such a concrete is known as reinforced concrete. Cement concrete shrinks a little and grips very fast on the steel reinforcement. This produces bond between cement concrete and steel, and thus tensile stresses are transferred to steel reinforcement.

Concrete has good resistance to damage by fire and protects steel bars from buckling and twisting at high temperature. Also concrete offers very good resistance to atmospheric adverse conditions and thus stops steel reinforcement from rusting.

1·5. Advantages of RCC Construction

Following are the advantages of RCC construction :

1. Concrete can be easily moulded to any desired shape.
2. The materials for RCC are easily available.
3. It is easy to make.
4. It is durable.
5. By proper proportioning of mix, concrete can be made water-tight.

6. Its monolithic character gives it more rigidity.
7. It is fire resisting.
8. Its maintenance cost is practically nil.

1·6. Steel as Reinforcement

Steel is used as reinforcement to take up tensile stresses in RCC structures because of the following reasons :

(1) Its tensile strength is high.

(2) It can develop good bond with concrete.

(3) Its coefficient of expansion is nearly same as that for concrete.

(4) It is easily available.

The steel bars used for reinforcing concrete are generally of plain round mild steel with diameters between 5 mm and 50 mm. It is only in very heavy structures that bars of diameter larger than 40 mm are used. There is a growing practice to use high yield point steel bars which are usually mild steel bars that have been old drawn to increase their strength : but the difficulty of placing and bending these bars may outweigh the initial saving in the weight of steel required.

Mesh fabric is used frequently as reinforcement in concrete roads, walls and floor slabs. It can also be used in column footings. It is used in place of bar reinforcement only when there are definite advantages in cost and placing. The common types of mesh fabrics in use are welded fabric, expanded metal and ribbed metal lathing. The fabric is made in standard size lengths and widths.

Deformed bars conforming to IS : 1130—1966 and cold-twisted steel conforming to IS : 1566—1967 are being widely used.

All reinforcement in RCC construction shall be clean and free from loose mill scales, dust, loose rust and coats of paints, oil or other coatings which may destory or reduce bond.

1·7. S.I. Units

S.I. units have been used in revised code of practice for plane and reinforced concrete IS 456—1978. The stresses are given in N/mm², *i.e.* Newtons per millimetre square.

$$1\ \text{N/mm}^2 = \frac{100}{9{\cdot}81}\ \text{kg/cm}^2$$

$$= 10{\cdot}19\ \text{kgf/cm}^2 \qquad [1\ \text{kgf} = 9{\cdot}81\ \text{N}]$$

Newton is defined as a force to produce acceleration of 1 metre/sec² in a mass of 1 kg. Unit of mass is kg.

By this revised code the stresses are taken as :

(*i*) *The modulus of elasticity of steel* shall be taken as 200 kN/mm².

(*ii*) *The characteristic strength of concrete* is defined as the strength of materirl below which not more than 25 percent of the least results are expected to fall.

Table 1·1

Grade designation	*Specified charaeteristic compressive strength of 28 days f_{ck} (N/mm²)*
M 10	10
M 15	15
M 20	20
M 25	25
M 30	30
M 35	35
M 40	40

Letter 'M' refers to the mix and the number to the specified characteristic compressive strength of 15 cm cube at 28 days, expressed in N/mm².

Grades of concrete less than M 15 shall not be used in reinforced concrete.

M 5 and M 7·5 grades of concrete may be used for lean concrete bases and simple foundations for masonry walls.

(*iii*) *Increase in strength of concrete with age.* Where it can be shown that a member will not receive its full load within a period of 28 days after the casting of the member, the characteristic compressive strength given in Table 1·1 may be increased by factor as given in Table 1·2.

Table 1·2

Minimum age of member when full design load is expected months	*Age factor*
1	1·0
3	1·10
6	1·15
12	1·20

(*iv*) *Tensile strength of concrete.* The following formula should be used to arrive at the tensile strength from the compressive strength.

Flexural strength,

$$F_{cr}=\sqrt{f_{ck}}\ \text{N/mm}^2$$

where f_{ck} is the characteristic compressive strength of concrete.

(*v*) *The modulus of elasticity for structural concrete* may be assumed as

$$F_c=5700\sqrt{f_{ck}}\ \text{N/mm}^2$$

where F_c is the short term static modulus of elasticity.

(*vi*) *Shrinkage strain of concrete* may be taken as 0·0003.

(*vii*) *Creep of concrete.* Creep of concrete depends on stress in the concrete, age at loading and the duration of loading besides water content at the time of mixing and cement content. The creep coefficient defined as ultimate creep strain by elastic strain at the age of loading may be taken as given in Table 1·3

Table 1·3

Age of loading	*Creep coefficient*
7 days	2·2
28 days	1·6
1 year	1·1

(*viii*) *Thermal expansion,* The coefficient of thermal expansion depends upon the aggregate used and the nature of cement, cement content the relative humidity and the size of sections. Table 1·4 gives thermal coefficient for various aggregates.

Table 1·4

Type of aggregate	*Coefficient of thermal expansion for concrete/°C*
Quatzite	$1{\cdot}2$ to $1{\cdot}3\times10^{-5}$
Sandstone	$0{\cdot}9$ to $1{\cdot}2\times10^{-5}$
Granite	$0{\cdot}7$ to $0{\cdot}95\times10^{-5}$
Basalt	$0{\cdot}8$ to $0{\cdot}95\times10^{-5}$
Limestone	$0{\cdot}6$ to $0{\cdot}9\times10^{-5}$

(*ix*) *Design of concrete mix.* The mix shall be designed to produce that grade of concrete having the required workability and a characteristic strength not less than appropriate value as given in Table 1·2.

As long as the quality of materials does not change, a mix design done earlier may be considered adequate for later work.

Nominal concrete mix may be used for concrete of grades M 5, M 7·5, M 10, M 15 and M 20. The properties of materials for nominal concrete mix shall be in accordance with Table 1·5.

Table 1·5

Grade of concrete	*Total quantity of dry aggregate by mass per 50 kg of cement to be taken as the sum of individual mass of fine and coarse aggregate mix (kg)*	*Proportion of fine aggregate to coarse aggregate (by mass)*	*Quantity of water per kg of cement max litres*
M 5	800	Generally 1 : 2	60
M 7·5	625	but subject to an	45
M 10	480	upper limit of	34
M 15	350	1 : 1½ and a lower	32
M 20	250	limit of 1 : 2½	30

The cement content of the mix specified in Table 1·5 for any nominal mix shall be proportionately increased if the quantity of water in a mix has to be increased to overcome the difficulties of placement and compaction, so that the water cement ratio as specified is not exceeded.

In the case of vibrated concrete, the limits specified may be suitably reduced to avoid segregation.

The quantity of water used in concrete mix, for reinforced concrete work should be sufficient, but not more than necessary to produce a large concrete of adequate workability for its purpose, which will surround and properly grip all the reinforcement. Workability of the concrete should be controlled by maintaining a water content that is found to give concrete which is just sufficiently wet to be placed and compacted without difficulty with the means available.

(*x*) *Modular ratio* $m=\frac{280}{3\sigma_{cbc}}$

where σ_{cbc} is permissible compressive stress due to bending in concrete in N/mm^2 as specified in Table 1·6.

(*xi*) *Permissible stress in concrete.* Permissible stresses for the various grades of concrete σ_{cbc} shall be taken as given in Table 1·6.

Table 1·6

Grade of concrete	Permissible stress in compression: Bending σ_{cbc}	Permissible stress in compression: Direct σ_{cdc}	Permissible stress in bond (average) for plain bars in tension τ_{bd}	m
M 10	3·0	2·5	—	—
M 15	5·0	4·0	0·6	18
M 20	7·0	5·0	0·8	13
M 25	8·5	6·0	0·9	11
M 30	10·0	8·0	1·0	9
M 35	11·5	9·0	1·1	8
M 40	13·0	10·0	1·2	7

(*xii*) *Permissible stresses in concrete in direct tension.* For members in direct tension, when full tension is taken by the reinforcement alone, the tensile stress shall be not greater than the values given in Table 1·7. The tensile stress shall be calculated as $\frac{F_t}{A_c+mA_{st}}$.

Table 1·7

Grade of concrete	M 10	M 15	M 20	M 25	M 30	M 35	M 40
Tensile stress N/mm²	1·2	2·0	2·8	3·2	3·6	4·0	4·4

where F_t=Total tension on the member minus pretension in steel if any, before concreting

A_c=Cross-sectional area of concrete excluding any finishing material and reinforcing steel

m=Modular ratio

A_{st}=Cross-sectional area of reinforcing steel in tension.

(*xiii*) *Bond stress for deformed bars.* In the case of deformed bars, the bond stresses as given in Table 1·7 may be increased by 40 per cent.

(*xiv*) *Permissible stresses in steel reinforcement.* Permissible stresses in steel reinforcement shall not exceed the values given in Table 1·8.

Table 1·8. Permissible stresses in N/mm²

Type of stress in steel reinforcement	*Mild steel bars conforming to grade I of IS : 432 or deformed mild steel bars IS : 1139*	*Medium tensile steel conforming to IS : 432, or deformed medium tensile steel bars IS : 1139*	*High yield strength deformed bars IS : 1139 or IS : 1786*
Tension σ_{st}		Half the guaranteed yield stress subject to max of 190	
(*a*) Upto 20 mm	140		230
(*b*) Over 20 mm	130		230
Compression in columns σ_{sc}	130	130	190
Compression in bars in a slab beam, resistances of concrete taken into account	The calculated compressive stress in the surrounding concrete multiplied by 4·5 times the modular ratio or σ_{sc} whichever is lower.		
Compression in bars, resistances of concrete not taken		Half the guaranteed yield stress subject to a maximum	
(*a*) Upto 20 mm	140	190	190
(*b*) Over 20 mm	130		190

(*xv*) *Permissible stress in long columns.* The maximum permissible stress in a reinforced concrete column or part thereof having a ratio of effective column length l_{ef} to least lateral radius of gyration i_{min}, above 40 shall not exceed the appropriate permissible stresses multiplied by C_r given by

$$C_r = 1{\cdot}25 - \frac{l_{ef}}{160\, i_{min}}$$

For approximate value, $C_r = 1{\cdot}25 - \dfrac{l_{ef}}{48\, b}$

where b is least dimension, for $\dfrac{l_{ef}}{b} > 12$.

(*xvi*) *Density of concrete.* The unit weight of plain concrete and reinforced concrete made with sand and gravel and crushed stone aggregate may be taken as 24,000 N/m³ and 25,000 N/m³ respectively.

(*xvii*) *Design shear strength of concrete.* The permissible shear stress in concrete in beams without shear reinforcement is given in Table 1 9.

Table 1·9

$\frac{100\ A_s}{bd}$	Permissible shear stress in concrete τ_c N/mm^2 for grade of concrete					
	M 15	*M 20*	*M 25*	*M 30*	*M 35*	*M 40*
0·25	0·22	0·22	0·23	0·23	0·23	0·23
0·50	0·29	0·30	0·31	0·31	0·31	0·32
0·75	0·34	0·35	0·36	0·37	0·37	0·38
1·00	0·37	0·39	0·40	0·41	0·42	0·42
1·25	0·40	0·42	0·44	0·45	0·45	0·46
1·50	0·42	0·45	0·46	0·48	0·49	0·49
1·75	0·44	0·47	0·49	0·50	0·52	0·52
2·00	0·44	0·49	0·51	0·53	0·54	0·55
2·25	0·44	0·51	0·53	0·55	0·56	0·57
2·50	0·44	0·51	0·55	0·57	0·58	0·60
2·75	0·44	0·51	0·56	0·58	0·60	0·62
3·00	0·44	0·51	0·57	0·60	0·62	0·63

A_s is that area of longitudinal reinforcement which continues at least one effective depth beyond the section being considered except at supports where the full area of tension reinforcement may be used.

(*xviii*) For solid slabs (excluding flat slabs) the permissible shear stress in concrete shall be $k\tau_c$, where k has the values as given in Table 1·10.

Table 1·10

Overall depth of slab in mm	300 or more	275	250	225	200	175	150 or less
k	1·00	1·05	1·10	1·15	1·20	1·25	1·30

(*xix*) *Permissible shear stress in flat slabs*, When shear reinforcement is not provided, the calculated shear stress at the critical section shall not exceed $k_s\ \tau_c$.

where $k_s=(0{\cdot}5+\beta_c)$ but not greater than 1, β_c being the ratio of short side to long side of the column a capital.

and $\tau_c=0{\cdot}25\sqrt{f_{ck}}$ in limit state method of design and

$0{\cdot}16\sqrt{f_{ck}}$ in working stress method of design.

When the shear stress exceeds $k_s\ \tau_c$ shear reinforcement is provided In case actual shear stress is greater than $1{\cdot}5\ \tau_c$, the flat slab shall be redesigned.

(xx) *Permissible loads in compression members.* The axial load P for pedestal or short column reinforced with longitudinal bars and lateral ties shall not exceed that given by the following formula :

$$P=\sigma_{cc}A_c+\sigma_{sc}A_{sc}$$

where σ_{cc} = Permissible stress in concrete in direct compression

A_c = Cross-sectional area of concrete excluding any finishing material and reinforcing steel

σ_{sc} = Permissible compressive stress for column bars

A_{sc} = Cross-sectional area of the longitudinal steel.

Short-column with helical reinforcement such as the ratio of the volume of helical reinforcement to the volume of the core shall not be less than $0{\cdot}36\left(\frac{A_g}{A_c}-1\right)\frac{f_{ck}}{f_y}$

where A_g = Gross area of the section

A_c = Area of core of the helically reinforced column measured to the outside diameter of the helix

f_{ck} = Characteristic compressive strength of the concrete

f_y = Characteristic strength of the helical reinforcement but not exceeding 415 N/mm^2.

The permissible load in short column satisfying above helical reinforcement requirements shall be 1·05 times the permissible load for similar member with lateral ties or rings.

1·8. Formwork

Formwork is a temporary structure in which the wet concrete is placed to harden till it reaches required strength. It may be constructed of timber steel, asbestos or plastic. Formwork is one of the major cost items in reinforced concrete structures.

The following types of formwork are in common use :

1. *Timber.* This is most commonly used material in the construction of formwork. It can easily be put into required shape and can easily be fixed and dismantled. Mostly commercial soft woods can be used.

2. *Plywood.* Resin bonded plywood panels are sometimes used instead of timber planks, especially where large exposed areas of concrete are to be constructed *e.g.* faces of retaining walls and the under sides of floor slabs. Compared to timber, shuttering of this type reduces labour costs in fixing and dismantling because of use of large panels. This type eliminates joint marks and with careful fixing of panels, the hardened concrete surface will not require any finishing treatment.

3. *Steel.* Steel formwork is in the form of panels made by welding steel sheets to angle framing. Steel shuttering is initially

more expensive than timber formwork but because it can be continually re-used, it is eventually more economical. This type is most suitable for curved construction as circular chimneys and columns. This type of shuttering is extensively used in precast concrete industry because it can be constructed with a high degree of accuracy.

4. *Moulds.* This type of formwork is made in steel, asbestos or glass fibre. They are precisely made and, therefore, produce a very smooth surface to the hardened concrete. This type of shuttering is mostly used to form the underside or ribbed floor slabs.

5. *Permanent formwork.* The moulds described above are sometimes left in place after the concrete has hardened and such a formwork is called as permanent formwork. Any shuttering that is left in place after concrete has hardened is called permanent formwork.

6. *Travelling formwork.* This is an economical way to forming certain types of structures. This type of formwork travels along the path of the structure to be formed. The type of structures where this type of formwork is used are tall chimneys, silos and bunkers where plan shape is constant throughout the height of the structure.

2

Singly Reinforced Beams

2·1. A beam is a structural member which is primarily designed to resist bending stresses. A beam reinforced only on one side of the section to resist tensile stresses is known as singly reinforced beam.

The simple theory of bending is applicable to homogeneous sections, *i.e.* sections of same material throughout and uniform in quality. The relation between the internal stresses of homogeneous section and the external bending moment is given by

$$f = \frac{M.y}{I}$$

where f = internal stress at distance y from neutral axis.

M = external bending momentum at the section.

I = moment of inertia of cross-section about N.A.

The reinforced concrete beams are non-homogeneous. So, it is rather difficult to evolve exact theory for behaviour of reinforced concrete sections. However, with certain assumptions, simple theory of bending with certain modifications can be applied to reinforced concrete sections. The assumptions made are—

(*i*) Plane cross-sections before loading remain plane after loading.

(*ii*) Stress is proportional to deformation, *i.e.* modulus of elasticity is constant at all stresses. It is also assumed that modulus of elasticity is not function of duration of stress.

(*iii*) Concrete area on tension side is assumed to be ineffective.

(*iv*) Tension is entirely taken by steel due to perfect bond between steel and concrete.

(*v*) The steel area is assumed to be concentrated at centroid of steel area.

First assumption is the basis of theory of bending. The strain at any section is proportional to its distance from neutral axis.

Second assumption implies that material is perfectly elastic and stress strain curve is a straight line nearly upto yield line. For concrete, relation between stress and strain is nearly a straight line over a small range of stress. At higher stresses straight line relationship between stress and strain does not hold good (Fig. 2·1).

Young's modulus for concrete is generally found by secant method from stress-strain curve.

The modulus of elasticity of concrete is not constant, but is a function of age, quality and duration of stress. The extact formula derived taking into consideration the above factors will be too cumbersome to be of any practical use. Therefore, modulus of elasticity is assumed to be constant.

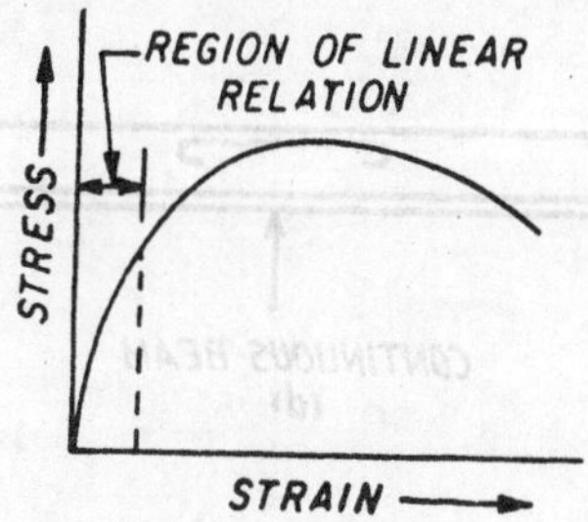

Fig. 2·1

Below neutral axis where tensile strength of concrete is exceeded, it cracks. Only small portion of concrete near neutral axis in tension zone does not crack and resists tensile forces. As this quantity is very small its effect is neglected.

The design of reinforced concrete sections is based on the assumption that concrete in tension zone cracks and steel takes entire tensile force, which is transferred to it by bond between steel and concrete.

Ex. 2·1. *Determine the moment taken by 200 mm×400 mm concrete beam if the working stress of concrete in tension is 0·6 N/mm².*

Sol. $M_r = fZ = 0{\cdot}6 \times 1/6 \times bd^2$

$= 0{\cdot}6 \times 1/6 \times 200 \times 400 \times 400$

$= 3{,}200{,}000$ N-mm.

2·2. Placing of Steel in Reinforced Concrete Sections

In RCC structure all the tension is to be taken by steel. From the B.M. diagram the position of tensile zones of beam is known which determines the position of tensile steel. For sagging bending moment tension will be below neutral axis and for hogging moments the tension will be above neutral axis. Fig. 2·2 shows the position of steel reinforcement in different types of beams.

A beam carrying vertical loads is made symmetrical about vertical centre line, so the vertical loads cause uniplanar bending of the beam and no torsional effects are produced.

All concrete area in tension is neglected for design purposes and tension is entirely taken by steel and so the depth of the beam from compression edge to centre of tensile reinforcement is effective and it is called 'effective depth of the beam'.

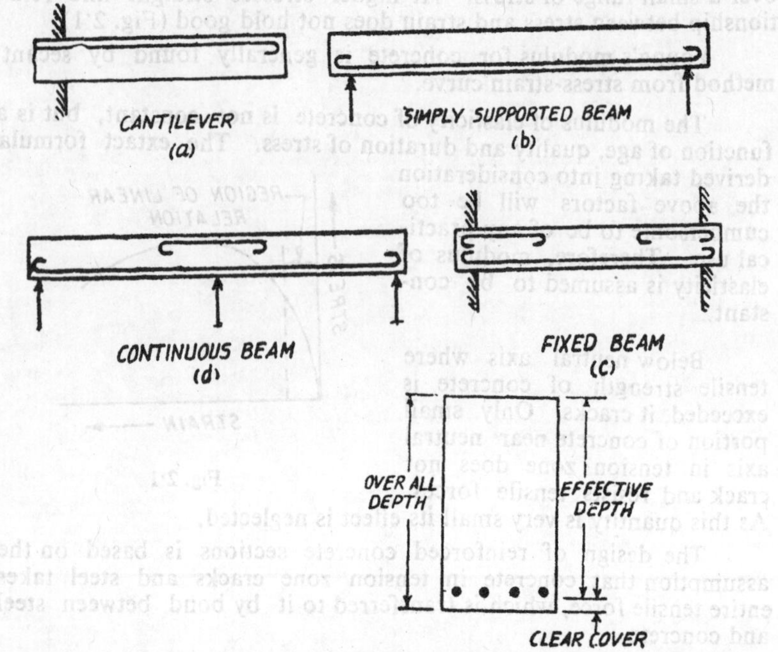

Fig. 2·2

2·3. Behaviour of Beam Under Gradually Increasing Load

When beams are subjected to load there are changes in stress and deformation. Tests have shown that the beams fail either along a vertical (normal) plane or a diagonal plane as shown in Fig. 2·3.

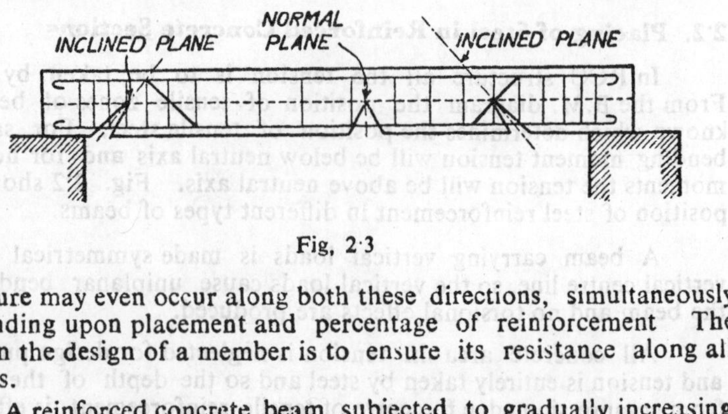

Fig. 2·3

Rupture may even occur along both these directions, simultaneously depending upon placement and percentage of reinforcement The aim in the design of a member is to ensure its resistance along all planes.

A reinforced concrete beam subjected to gradually increasing load till failure undergoes three stages of deformations.

Stage I. When the beam is loaded in initial stages, the stresses in compression and tension zones of concrete are within elastic limits and therefore, deformation are mainly elastic. Stress deformation diagram is a straight line. The normal stress diagram is triangular and the steel reinforcement jointly with the concrete take tension shown in Fig. 2·4.

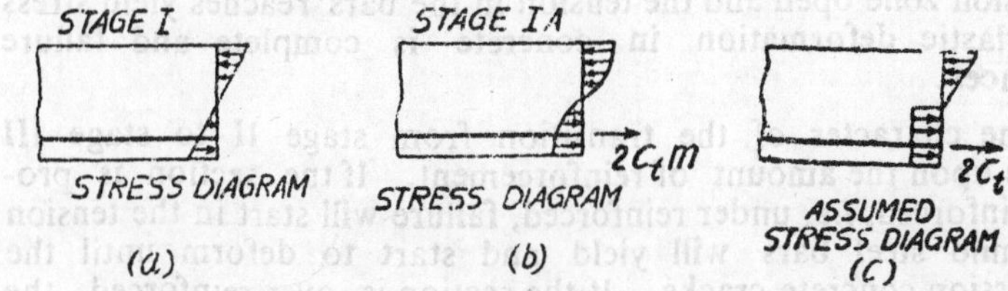

Fig. 2·4

As the load is further increased, stresses in the section increase. In the tension zone deformation develops rapidly till ultimate resistance is reached in the extreme fibres. Plastic deformation becomes noticeable in the compression zone. The condition of stress deformation is shown at stage IA in Fig. 2·4 (*b*) and used for calculating fissure width. The normal stress diagram is assumed rectangular for tension zone and triangular for compression zone as shown in Fig. 2·4. The stress in steel will be $2c_t\, m$ as Young's modulus in concrete at ultimate tension is 0·5 E_c.

The ultimate deformation of the concrete in tension before cracks develop will be $e_c = \dfrac{2c_t}{E_c}$ and the deformation in steel will be same,

$$e = \frac{2c_t}{E_c}$$

$$\text{Stress in steel} = \frac{2c_t}{E_c} \times E_s = 2c_t\, m.$$

Stage II. As the load is increased further, stresses in the tension zone become more complicated. The steel grips the concrete

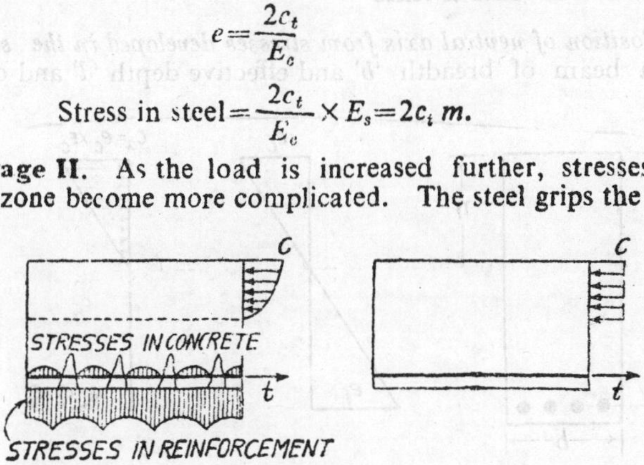

Fig. 2·5

between the cracks. The tension in concrete is more away from the cracks, while tension in steel is maximum at the crack. Stresses increase in the compression zone, plastic deformation sets in and normal stress diagram becomes curved. The stage serves for calculating bending deformation and fissure widening. A rectangular diagram for normal stresses in compression is assumed for concrete as shown in Fig. 2·5.

Stage III. With a further increase in the load, the cracks in the tension zone open and the tension in the bars reaches yield stress σ_{sy}. Plastic deformation in concrete is complete and failure commences.

The character of the transition from stage II to stage III depends upon the amount of reinforcement. If the section is properly reinforced, *i.e.* under reinforced, failure will start in the tension zone, mild steel bars will yield and start to deform until the compression concrete cracks If the section is over-reinforced, the compression zone concrete will fail before the bars reach their stress limit. Over reinforced beams are never used as they are also uneconomical

Stage III serves as basis for computing carrying capacity. Stress diagram for compression zone is adopted as rectangular as shown in Fig. 2·6.

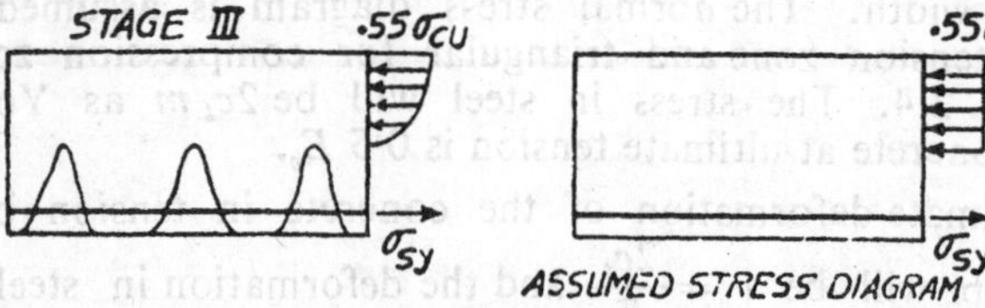

Fig. 2·6

2·4. Position of Neutral Axis

(*a*) *Position of neutral axis from stresses developed in the section.* Consider a beam of breadth '*b*' and effective depth '*d*' and overall depth '*D*'.

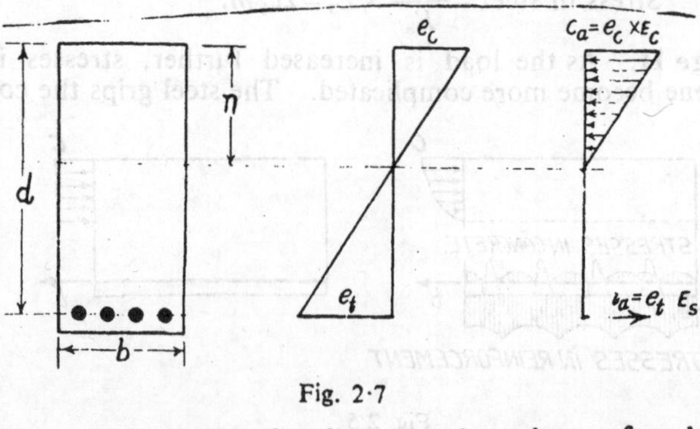

Fig. 2·7

A plane section before bending remains plane after bending, the strain diagram is a straight line, the neutral fibre being at distance n from compressive face on the beam. The stress at any point will be equal to strain at that point multiplied by Young's modulus. Let e_c and e_t be the maximum strain in concrete and strain at the centre of gravity of steel respectively Let c_a and t_a be the maximum compressive stress in concrete and stress in steel respectively.

$$c_a = e_c \times E_c$$
$$t_a = e_t \times E_s$$

Now $$\frac{e_c}{n} = \frac{e_t}{d-n}$$

$$\therefore \quad \frac{c_a/E_c}{n} = \frac{t_a/E_s}{d-n}$$

$$\therefore \quad \frac{d-n}{n} = \frac{t_c}{c_a} \times \frac{E_c}{E_s} = \frac{t_a}{c_a} \times \frac{1}{m}$$

where, $$m = \frac{E_s}{E_c} = \text{modular ratio}$$

$$\frac{d}{n} - 1 = \frac{t_a}{mc_a}$$

$$\frac{d}{n} = 1 + \frac{t_a}{mc_a}$$

$$n = d.\frac{1}{1+\dfrac{t_a}{mc_a}} = k'd$$

where k' is called neutral axis constant and is given by $\dfrac{1}{1+\dfrac{t_a}{mc_a}}$.

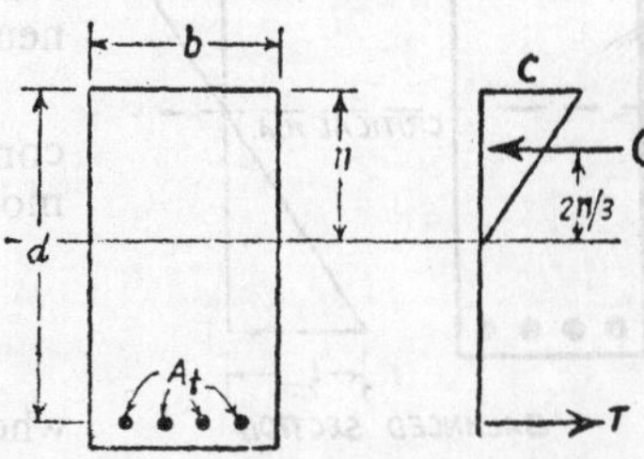

Fig. 2·8

(b) *Position of neutral axis when section is given.* The resisting moment is supplied by the compression in concrete and tension in steel. The small tension taken by concrete is neglected. Neutral axis will be centre of gravity of composite section consisting of concrete area in compression and area of tensile steel.

Let A_t be area of tensile steel.

Total compressive force = Total tensile force.

$$\therefore \quad b \times n \times \frac{c_a}{2} = t_a \times A_t$$

$$b \times n \times \frac{c_a}{2} = m \times c_a \times \frac{d \times n}{n} \times A_t$$

$$\therefore \quad b \times n \times \frac{n}{2} = m \times A_t(d-n)$$

$$\therefore \quad \frac{bn^2}{2} + mA_t \times n - mA_t d = 0$$

$$\therefore \quad n = \frac{-mA_t + \sqrt{(mA_t)^2 + 2bmA_t d}}{b}$$

$m \times A_t$ is called equivalent concrete area.

As $b \times n \times \frac{n}{2}$ is the moment of concrete area about N.A. and $m \times A_t \times (d-n)$ is the moment of m times steel area about N.A., therefore, neutral axis is the centre of gravity of concrete area in compression and m times steel area concentrated at the centroid of steel.

The neutral axis constant $k' = n/d$ for different percentages of steel for various mixes and is given in graph 1 in Appendix.

2·5. (a) Critical and Balanced Section

When the maximum stresses in steel and concrete simultaneously reach allowable values, the section is called balanced or critical section. For particular values of allowable stresses in concrete and steel and modular ratio m the depth of neutral axis is fixed.

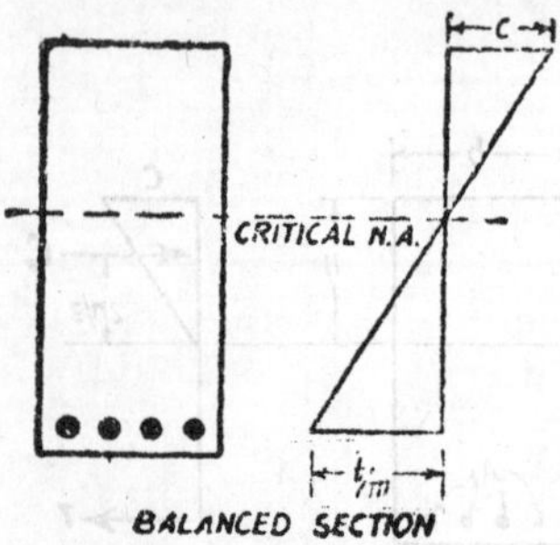

Fig.2·9

If c and t are allowable stresses in concrete and steel respectively and m is modular also

Depth of critical neutral axis,

$$n_c = \frac{1}{1 + t/cm} \times d = kd$$

where k is critical neutral axis depth constant and is given by $\frac{1}{1 + t/cm}$.

The values of m for different mixes are such that the product of m and c is a constant. Therefore, for particular grade of steel t/mc will be constant. Therefore, k will have the same value for all mixes. For $t = 140$ N/mm^2, $k = 0·4$.

The compressive force $= b \times n_c \times \frac{c}{2}$

Total tensile force $= t \times A_t$

$$\text{Lever arm} = a = d - \frac{n_c}{3} = d - \frac{kd}{3} = d(1 - k/3) = jd$$

where j is known as lever arm constant for balanced section and is equal to $1 - k/3$. The level arm constant will be the same for all mixes for $t = 140$ N/mm^2, $j = 0·87$.

Moment of the resistance

$$= M = b \times n_c \times \frac{c}{2}\left(d - \frac{n_c}{3}\right) = b \times kd \times \frac{c}{2} \times jd$$

$$= \tfrac{1}{2}\, c \times k \times j \times bd^2.$$

$\frac{1}{2}c \times k \times j$ is constant for particular values of allowable stresses and is denoted by Q.

$$M = Qbd^2$$

$$M = t \times A_t \times jd$$

$$A_t = \frac{Qbd^2}{t \times jd}$$

$$\frac{A_t}{bd} = \frac{Q}{t \times j}$$

$$p = \% \text{ steel area} = \frac{A_t}{bd} \times 100$$

p is constant for particular values of stresses.

Therefore, balanced section has definite percentage of steel.

(*b*) **Under reinforced section.** When the percentage of steel in a section is less than that required for a balanced section, the section is called under-reinforced section. In this case concrete stress will not reach maximum allowable value. The neutral axis depth will be smaller than that in the balanced section as shown in Fig. 2·10. The moment of resistance of such a section will be governed by allowable tensile stress in steel.

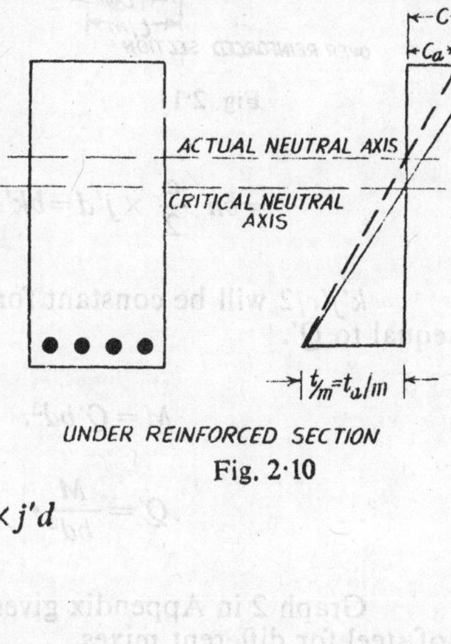

UNDER REINFORCED SECTION

Fig. 2·10

Moment of resistance

$$= t \times A_t \times (d - n/3)$$

$$= t \times A_t \times \left(d - \frac{k'd}{3}\right)$$

$$M = t \times A_t \times j'd$$

where j' is lever arm constant.

$$A_t = \frac{pbd}{100}$$

$$M = \frac{t \times p}{100} \times j'bd^2$$

For particular % of steel, $\frac{tpj}{100}$ will be a constant say Q

$$M = Q'bd^2$$

$$\therefore \quad Q' = \frac{M}{bd^2}.$$

Graph 2 in Appendix gives values of j for various percentages of steel for different mixes.

Graph 3 in Appendix gives values of Q for various percentages of steel for different mixes.

(*c*) **Over reinforced section.** When the percentage of steel in section is more than that required for a balanced section, the section is called over-reinforced section. In this case steel stress will not reach maximum value. The neutral axis depth will be greater than that in the balanced section as shown in Fig. 2·11. Moment of resistance of such a section will be governed by compressive stress in concrete.

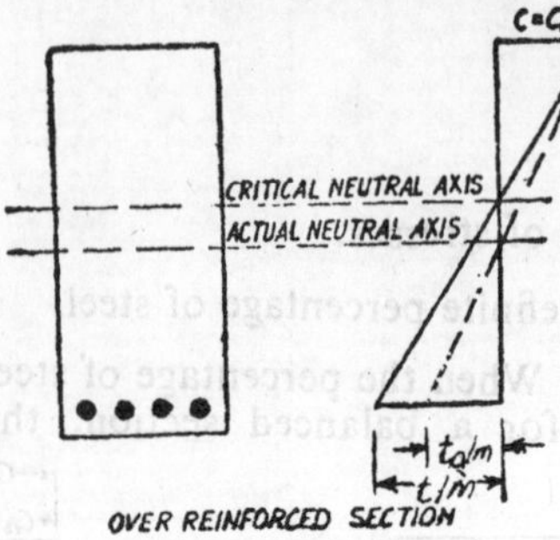

Fig. 2·11

Moment of resistance

$$=bn\,\frac{c}{2}\times(d-n/3)$$

$$=bn\,\frac{c}{2}\times j'd=bk'd\,\frac{c}{2}\times j'd=\frac{k'j'c}{2}\times bd^2.$$

$k'j'c/2$ will be constant for particular percentage for steel, say equal to Q'.

$$M=Q'bd^2.$$

$$\therefore \qquad Q'=\frac{M}{bd^2}.$$

Graph 2 in Appendix gives values of j' for various percentages of steel for different mixes.

Graph 3 in Appendix gives values of Q' for various percentages of steel for different mixes.

2·6. Effect of Varying Percentage of Steel on Moment of Resistance

The moment of resistance of a section with varying percentage of steel is given in Table 2·1. The stresses taken are

$$c=5\ \text{N/mm}^2,\ t=140\ \text{N/mm}^2 \text{ and } m=18.$$

Table 2·1

Percentage Reinforce-ment	*n, depth of N.A.*	*Lever arm*	*Moment of resistance $=Q'\ bd^2$*	*Remarks*
0·1	0·17 *d*	0·943 *d*	0·1320 bd^2	Under reinforced
0·2	0·235 *d*	0·92 *d*	0·2576 bd^2	Under reinforced
0·4	0·314 *d*	0·895 *d*	0·501 bd^2	Under reinforced
0·6	0·369 *d*	0·877 *d*	0·736 bd^2	Under reinforced
0·714	0·40 *d*	0·87 *d*	0·870 bd^2	Balanced
0·8	0·412 *d*	0·866 *d*	0·891 bd^2	Over reinforced
1	0·446 *d*	0·851	0·949 bd^2	Over reinforced
2	0·562 *d*	0·813	1·142 bd^2	Over reinforced
3	0·631 *d*	0·790 *a*	1·246 bd^2	Over reinforced

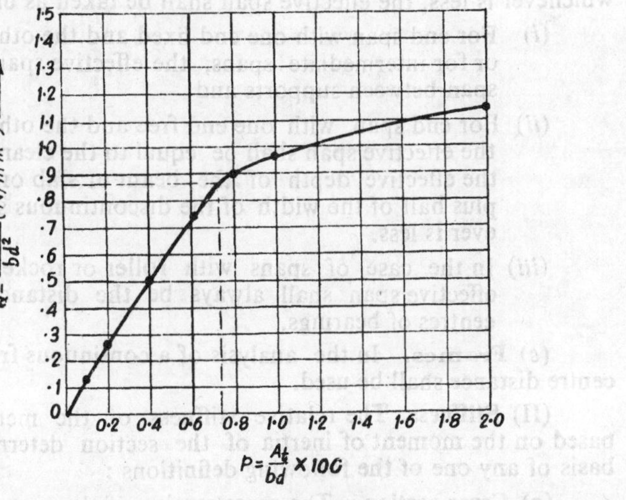

Fig. 2·12

The values of moment of resistance constant Q' are plotted against percentage reinforcement in Fig. 2·12. The moment of resistance increases rapidly with percentage of steel till the section is balanced. With further increase of steel the increase in moment of resistance is small. Hence balanced section is a critical section.

2·7. General Notes on Design of Beams and Slabs (I.S.I.)

All structures may be analysed by the linear elastic theory to calculate internal actions produced by design loads.

(I) Effective Span. The effective span of a member shall be as follows :

(*a*) **Simply supported beam or slab.** The effective span for a member that is not built integrally with its supports shall be taken as clear span plus the effective depth of slab or beam or centre to centre of supports, whichever is less.

(*b*) **Continuous beam or slab.** In the case of continuous beam or slab if the width of the support is less than 1/12th of the clear span, the effective span shall be as for simply supported. If the supports are wider than 1/12th of the clear span or 600 mm whichever is less, the effective span shall be taken as under :

(*i*) For end span with one end fixed and the other continuous, or for intermediate spans, the effective span shall be clear span between supports and

(*ii*) For end span with one end free and the other continuous, the effective span shall be equal to the clear span plus half the effective depth of the beam or slab or the clear span plus half of the width of the discontinuous support whichever is less.

(*iii*) In the case of spans with roller or rocker bearings, the effective span shall always be the distance between the centres of bearings.

(*c*) **Frames.** In the analysis of a continuous frame, centre to centre distance shall be used.

(II) **Stiffness.** The relative stiffness of the member may be based on the moment of inertia of the section determined on the basis of any one of the following definitions :

(*a*) **Cross-section.** The cross-section of the members ignoring reinforcement.

(*b*) **Transformed section.** The concrete cross-section plus the area of reinforcement transformed on the basis of modular ratio.

(*c*) **Cracked section.** The area of concrete in compression plus the area of reinforcement transformed on the basis of modular ratio.

The assumptions made shall be consistent for all members of the structure throughout any analysis.

For deflection calculations, moment of inertia as specified below should be used.

(A) **Short-term deflection.** The short-term deflection may be calculated by the usual methods for elastic deflection using the short-term modulus of elasticity of concrete, E_c and an effective moment of inertia I_{eff} as given below:

$$I_{eff}=\frac{I_r}{1\cdot 2-\frac{M_r}{M}\times\frac{a}{d}\left(1-\frac{n}{d}\right)\frac{b_w}{b}}$$

but $$I_r \leq I_{eff} \leq I_{gr}$$

where, I_r=Moment of inertia of the cracked section,

M_r=Cracking moment, equal to $\frac{f_{cr} I_{gr}}{y_t}$ where f_{cr} is the modulus of rupture of concrete, I_{gr} is the moment of inertia of the gross section about the centroidal axis, neglecting the reinforcement, and y_t is the distance from centroid axis of gross section, neglecting the reinforcement, to extreme fibre in tension.

M=maximum moment under service loads

a=lever arm

n=depth of neutral axis

d=effective depth

b_w=breadth of web

b=breadth of compression face.

For continuous beams, deflection shall be calculated using the values of I_r, I_{gr} and M_r modified by the following equation :

$$X_e=k_1\left[\frac{X_1+X_2}{2}\right]+(1-k_1)X_0$$

where, X_e=modified value of X (*i.e.* I_r, I_{gr} or M_r)

X_1, X_2=values of X at the supports

k_1=coefficient given in Table 2·2

X_0=value of I_r, I_{gr} or M_r as appropriate.

Table 2·2. Values of Coefficient k_1

k_2	0·500 less	0·6	0·7	0·8	0·9	1·0	1·1	1·2	1·3	1·4
k_1	0	0·03	0·08	0·16	0·30	0·50	0·73	0·91	0·97	1·0

$$k_2=\frac{M_1+M_2}{M_{f1}+M_{f2}}$$

where, M_1, M_2=Supports moments and

M_{f1}, M_{f2}=Fixed end moments.

Deflection due to temperature difference between top surface and bottom surface of $t=\frac{kt\alpha L^2}{D}$. For cantilever, $k=\frac{1}{2}$. Simply supported $k=1/8$, where α = coefficient of expansion, L=span. D=depth of the section.

(B) Deflection due to shrinkage. The deflection due to shrinkage a_{cs} may be computed from the following equation :

$$a_{cs}=k_3\psi_{cs}l^2$$

where, k_3 is a constant depending upon the support conditions .

0·5 for cantilevers

0·125 for simply supported members

0·086 for members continuous at one end and

0·063 for fully continuous members.

ψ_{cs} is shrinkage curvature equal to $k_4\frac{\epsilon_{cs}}{D}$

where, ϵ_{cs} is the ultimate shrinkage strain of concrete which may be estimated from the following values of creep coefficient that is ultimate creep strain/elastic strain at the age of loading.

Age of loading	*Creep coefficient*
7 days	2·2
28 days	1·6
1 year	1·1

The ultimate creep strain, estimated as described above does not include the elastic strain.

$$k_4=0{\cdot}72\times\frac{p_t-p_c}{\sqrt{p_t}}<1{\cdot}0 \text{ for } 0{\cdot}25<p_t-p_t<1{\cdot}0$$

$$k_4=0{\cdot}65\times\frac{p_t-p_c}{\sqrt{p_t}}<1{\cdot}0 \text{ for } p_t-p_c\geqslant1{\cdot}0$$

where $p_t=\frac{100\,A_{st}}{bd}$ and $p_c=\frac{100A_{cs}}{bd}$ and D is the total depth of the section and l is the length of span.

(C) Deflection due to creep. The creep deflection due to permanent load $a_{cc(perm)}$ may be obtained from the following equation :

$$a_{cc(perm)}=a_{cc(perm)}-a_{1(perm)}$$

where, $a_{1cc(perm)}$=initial plus creep deflection due to permanent loads obtained using an elastic analysis with an effective modulus elasticity $E_{cc}=\frac{E_c}{1+\theta}$, θ being the creep coefficient.

and $a_{1(perm)}$=short term deflection due to permanent load using E_c.

(III) Control of deflection. The deflection of a structure or part thereof shall not adversely effect the appearance of

efficiency of the structure of finishes or partitions. The deflection generally should be limited to the following :

(*i*) The final deflection due to all loads including the effects of temperature, creep and shrinkage and measured from the cast level of the supports of floors, roofs and other horizontal members, should not normally exceed span/250.

(*ii*) The deflection including the effects of temperature creep and shrinkage occurring after erection of partitions and the application of finishes should not normally exceed span/350 or 20 mm whichever is less.

For beams and slabs, the vertical deflection limits may generally be assumed to be satisfied provided that the span to depth ratios are not greater than the values obtained as below :

(*a*) Basic values of span to effective depth ratios for spans upto 10 m.

Cantilever	...	7
Simply supported	...	20
Continuous	...	26

(*b*) For spans above 10 m, the values in (*a*) may be multiplied by 10/span in metres, except for cantilever in which case deflection calculations should be done.

(*c*) Depending on the area of the type of steel for tension reinforcement, the values in (*a*) and (*b*) shall be modified as per Fig. 2·13.

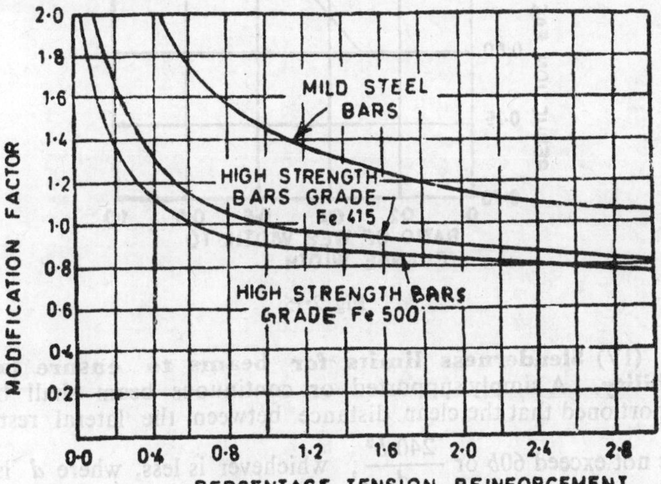

Fig. 2·13

(*d*) Depending on the area of compression reinforcement, the value of span to depth ratio be further modified as per Fig. 2·14.

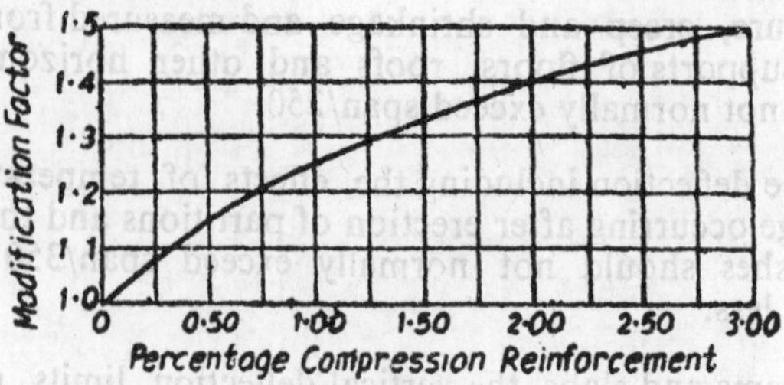

Fig. 2·14

(*e*) For flanged beams, the values of (*a*) and (*b*) modified as per Fig. 2 15 and the reinforcement percentage for use in Figs. 2·13 and 2·14 should be based on area of section equal to $b_f d$, where b_f is effective flange width.

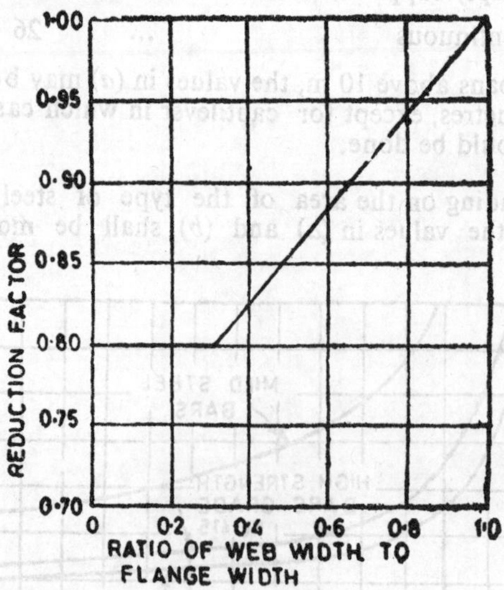

Fig. 2·15

(IV) **Slenderness limits for beams to ensure lateral stability.** A simply supported or continuous beam shall be so proportioned that the clear distance between the lateral restraints does not exceed $60b$ or $\dfrac{240\,b^2}{d}$, whichever is less, where d is the effective depth of the beam and b the breadth of the compression face midway between the lateral restraints.

For a cantilever, the clear distance from the free end of the cantilever to the lateral restraint shall not exceed $25b$ or $\frac{100\,b^2}{d}$ whichever is less.

(V) **Requirement Governing Reinforcement.** Reinforcing of same type and grade shall be used as main reinforcement in a structural member. However, simultaneous use of two different types or grades of steel for main and secondary reinforcement respectively is permissible.

Bars may be arranged singly, or in pairs in contact, or in groups of three or four bars bundled in contact. Bundles shall not be used in a member without stirrups. Bundled bars shall be tied together to ensure the bars remain together. Bars larger than 36 mm diameter should not be bundled.

For curtailment of tension reinforcement in flexural members, the reinforcement shall extend beyond the point at which it is no longer required to resist flexure (a point at which reinforcement is no longer required to resist flexure is where the resistance moment of the section, considering only the continuing bars, is equal to the design moment) for a distance equal to the effective depth of the member or 12 times the bar diameter, whichever is greater except at simple support or end of cantilever.

Flexural reinforcement shall not be terminated in tension zone unless any one of the following conditions is satisfied :

(*a*) The shear at the cut-off point does not exceed two-thirds that permitted, including the shear strength of web reinforcement provided.

(*b*) Stirrup area in excess of that required for shear and torsion is provided along each terminated bar over a distance from the cut-off point equal to three-fourths the effective depth of the member. The excess stirrup area shall not be less than $0{\cdot}4\ bsF_y$, where b is the breadth of beams, s is the spacing and F_y is the characteristic strength of reinforcement in N/mm². The resulting spacing shall not exceed $d/8\beta_b$, where β_b is the ratio of the area of bars cut-off to the total area of bars at the section, and d is the effective depth.

For 36 mm and smaller bars, the continuing bars provide double the area required for flexture at the cut-off point and the shear does not exceed three-fourths that permitted.

Positive Moment Reinforcement

(*a*) At least one-third the positive moment reinforcement in simple members and one-fourth the positive moment reinforcement in continuous members shall extend along the same face of the member into the support, to a length equal to $L_d/3$ where L_d is the development length.

(*b*) When a flexural member is a part of the primary lateral load resisting system, the positive reinforcement required to be extended into the support as described in (*a*) above shall be anchored to develop its design stress in tension at the face of support.

(*c*) At simple supports and at points of inflection, positive moment tension reinforcement shall be limited to a diameter such that L_d computed for f_d, *i.e.* $L_d=\frac{\phi f_d}{4\tau_{bd}}$ does not exceed $\frac{M_1}{V}+L_0$ where,

M_1=moment of resistance of the section assuming all reinforcement at the section to be stressed to f_d.

f_d=0·87 f_y in the case of limit state design and the permissible stress σ_{st} in the case of working stress design.

V=shear force at the section due to design loads.

L_0=sum of the anchorage beyond the centre of the support and the equivalent anchorage value of any hook or mechanical anchorage at simple support, and at a point of inflection, L_0 is limited to the effective depth of the member or 12ϕ, whichever is greater and, ϕ=diameter of bar.

The value of M_1/V in the above expression may be increased by 30 per cent when the ends of the reinforcement are confined by a compressive reaction.

(*i*) **Negative moment reinforcement.** At least one-third of the total reinforcement provided for negative moment at the support shall extend beyond the point of inflection for a distance not less than the effective depth of the member or 12ϕ or one-sixteenth of the clear span whichever is greater.

(*ii*) **Curtailment of bundled bars.** Bars in a bundle shall terminate at different points spaced apart at not less than 40 times the bar diameter except for bundles stopping at a support.

(*iii*) **Special members.** Adequate end anchorage shall be provided for tension reinforcement in flexural members, where reinforcement stress is not directly proportional to moment, such as sloped, stepped, or tapered footings, brackets, deep beams, and members in which the tension reinforcement is not parallel to compression face.

(*iv*) **Reinforcing splicing.** Where splices are provided in the reinforcing bars, they shall as far as possible be away from the sections of maximum stress and be staggered. It is recommended that splices in flexural members should not be at sections where the bending moment is more than 50 per cent of the moment of resistance, and not more than half the bars shall be spliced at a section.

Where more than one-half of the bars are spliced at a section or where splices are made at points of maximum stress, special precautions shall be taken, such as increasing the length of lap and/or using spirals or closely-spaced stirrups around the length of the splice.

(*v*) **Lap splices.** Lap splices shall not be used for bars larger than 36 mm, larger diameters bars may be welded. But in all cases of important connections, tests shall be conducted to prove that the joints have the full strength of bars connected. In case where welding is not practical, lapping of bars larger than 36 mm may be permitted, in which case additional spirals should be provided around the lapped bars.

Lap splices shall be considered as staggered if the centre to centre distance of splices is not less than 1·3 times the lap length calculated as follows :

Lap-length including anchorage value of hooks in fluxural tension shall be L_d or 20ϕ whichever is greater and for direct tension $2L_d$ or 30ϕ whichever is greater. The straight length of the lap shall not be less than 15ϕ or 20 cm, where L_d is the development length.

When bars of two different diameters are to be spliced, the lap length shall be calculated on the basis of the smaller bar.

When splicing of welded wire fabric is to be carried out lap splices of wires shall be made so that overlap measured between the extreme cross wires shall be not less than the spacing of cross wires plus 10 cm.

In case of bundled bars, lapped splice of bundled bars shall be made by splicing one bar at a time, such individual splices within a bundle shall be staggered.

(*vi*) **Spacing of Reinforcement.** The diameter of a round bar shall be taken as its nominal diameter and in the case of bars which are not round or in the case of deformed bars or crimped bars, the diameter shall be taken as the diameter of a circle giving an equivalent effective area. Where spacing limitations and minimum concrete cover are based on bar diameter, a group of bars bundled in contact shall be treated as a single bar of diameter derived from the total equivalent area. The minimum spacing of bars shall be as follows :

(*a*) The horizontal distance between two parallel main reinforcing bars shall usually be not less than the greatest of the following :

(*b*) The diameter of the bar if the diameters are equal.

(2) The diameter of the larger bar if the diameters are unequal.

(3) 5 mm more than the nominal maximum size of coarse aggregate.

This does not preclude the use of larger size of aggregates beyond the congested reinforcement in the same member, the size of aggregates may be reduced around congested reinforcement.

(*b*) Greater horizontal distance than the minimum specified in (*a*) should be provided whenever possible. However when needle vibrators are used the horizontal distance between bars of a group may be reduced to two-thirds the nominal maximum size of the coarse aggregate, provided that sufficient space is left between groups of bars to enable the vibrator to be immersed.

(*c*) Where there are two or more rows of bars, the bars shall be 15 mm, two-thirds the nominal maximum size of aggregate or the maximum size of bar, whichever is the greatest.

Maximum distance between bars in tension shall be as follows :

(*a*) **Beams.** The horizontal distance between parallel reinforcement bars of groups, near the tension of a beam shall not be greater than the value given in Table 2·3, depending on the amount of redistribution carried out in analysis and the characteristic strength of the reinforcement.

Table 2·3. Maximum Clear Distance Between Bars

f_y N/mm²	*Percentage Redistribution to or from section considered*				
	−30	−15	0	+15	+30
	Clear distance between bars mm				
250	215	260	300	300	300
415	125	155	180	210	235
500	105	130	150	175	195

Note. The spacings given in the table are not applicable to members subjected to particularly aggressive environments unless in the calculations of the moment of resistance, f_y has been limited to 300 N/mm² in limit state design and σ_{st} limited to 165 N/mm² in working stress design.

(*vii*) **Cover to Reinforcement.** Reinforcement shall have concrete cover and thickness of such cover (exclusive of plaster or other decorative finish) shall be as follows :

(*a*) At each end of reinforcing bar not less than 25 mm, nor less than twice the diameter of such bar.

(*b*) For a longitudinal reinforcing bar in a column, not less than 40 mm, nor less than diameter of such bars. In the case of column of minimum dimension 200 mm or under, whose reinforcing bars do not exceed 12 mm, a cover of 25 mm may be used.

(*c*) For longitudinal reinforcing bar in a beam, not less than 25 mm, nor less than the diameter of such bar.

(*d*) For tensile, compressive, shear or any other reinforcement in a slab, not less than 15 mm, nor less than the diameter of such bar.

(*e*) For any other reinforcement, not less than 15 mm, nor less than the diameter of such bar.

Increased cover thickness may be provided when surfaces of concrete members are exposed to the action of harmful chemicals (as in the case of concrete in contact with earth faces contaminated with such chemicals) acid, vapour, saline atmosphere, sulphurous smoke (as in the case of steam-operated railways) etc. and such increase of cover may be between 15 mm and 50 mm beyond the figures given in (*vii*) as may be specified by the engineer-in-charge.

For reinforced concrete members totally immersed in sea water, the cover shall be 40 mm more than specified earlier.

For reinforced concrete members, periodically immersed in sea water or subjected to sea spray, the cover of concrete shall be 50 mm more than that specified earlier.

(*viii*) Requirement of reinforcement for structural members—

(*a*) **Tension reinforcement for beams.** The minimum area of tension reinforcement shall not be less than that given by equation

$$\frac{A_s}{bd} = \frac{0{\cdot}85}{F_y}$$

where, A_s = minimum area of tension reinforcement.

b=breadth of the beam or the breadth of the web of T-beam.

d=effective depth.

F_y=characteristic strength of reinforcement in N/mm^2.

Maximum reinforcement area of tension reinforcement shall not exceed 0·04 bD, where D is total depth.

(*b*) **Compression reinforcement in beams.** The maximum area of compression reinforcement shall not exceed 0·04 bD. Compression reinforcement in beams shall be enclosed by stirrups for effective lateral restraint.

(*c*) **Side face reinforcement.** Where the depth of the web in a beam exceed 750 mm, side face reinforcement shall be provided along the two faces. The total area of such reinforcement shall be not less than 0·1 per cent of the web area and shall be distributed equally on two faces at a spacing not exceeding 300 mm or web thickness whichever is less.

(*d*) **Transverse reinforcement in beams for shear and tension.** The transverse reinforcement in beams shall be taken around the outer-most tension and compression bars. In T-beams and I-beams such reinforcement shall pass around longitudinal bars located close to the outer face of the flange.

Maximum spacing of shear reinforcement measured along the axis of the member shall not exceed 0·15 d for vertical stirrups and d for inclined stirrups at 45°, where d is the effective depth of the section under consideration. In no case shall the spacing exceed 450 mm.

Maximum shear reinforcement in the form of stirrups shall be provided such that—

$$\frac{A_{sv}}{b \times s_v} \geqslant \frac{0{\cdot}85}{f_y}$$

where, A_{sv}=total cross-sectional area of stirrup legs effective in shear

s_v=stirrups spacing along the length of the member

b=breadth of the beam or breadth of the web of flanged beam

f_y=characteristic strength of the stirrups reinforcement in N/mm^2 which shall not be taken greater than 415 N/mm^2.

However, in members of minor structural importance such as lintels, or where the maximum shear calculated is less than half the permissible value, this provision need not be complied with.

(*e*) **Distribution of torsion reinforcement.** When a member is designed for torsion, torsion reinforcement shall be provided as under :

(1) The transverse reinforcement for torsion shall be rectangular closed stirrups placed perpendicular to the axis of the member. The spacing of the stirrups shall not exceed any of x_1, $\frac{x_1+y_1}{4}$ and 300 mm, where x_1 and y_1 are respectively the short and long dimensions of the stirrups.

(2) Longitudinal reinforcement shall be placed as close as is practicable to the corners of the cross-section and in all cases, there shall be atleast one longitudinal bar in each corner of the ties. When the cross-sectional dimension of the member exceeds 450 mm, additional bars shall be provided to satisfy the requirements of minimum reinforcement and spacing.

(*ix*) **Moment and shear coefficients for continuous beams and slabs.** Unless more exact estimates are made, for beams of uniform cross-section which support substantially uniformly distributed loads over three or more spans which do not differ by more than 15 per cent of the longest, the bending moments and shear forces used in design may be obtained using the coefficients given in Table 2·4 and 2·5 respectively.

For moments at supports where two unequal spans meet or in case where the spans are not equally loaded, the average of the two values for the negative moment at the support may be taken for design.

When coefficient given in Table 2·4 are used for calculation of bending moments, redistribution of moments shall not be permitted.

Table 2·4. Bending Moment Coefficients

Type of Load	*Span moments*		*Support moments*	
	Near middle of end span	*At middle of interior span*	*At support next to the end support*	*At other interior support*
Dead load and imposed load fixed	$+\frac{1}{12}$	$+\frac{1}{24}$	$-\frac{1}{10}$	$-\frac{1}{12}$
Imposed load not fixed	$+\frac{1}{10}$	$+\frac{1}{12}$	$-\frac{1}{9}$	$-\frac{1}{9}$

Note. For obtaining the bending moment, the coefficients shall be multiplied by the total design load and effective span.

Table 2·5. Shear Force Coefficients

Type of load	*At end support*	*At support next to the end support*		*At all other interior supports*
		Outer side	*Inner side*	
Dead load and imposed load fixed	0·4	0·6	0·55	0·5
Imposed load not fixed	0·45	0·6	0·6	0·6

Note. For obtaining the shear force, the coefficient shall be multiplied by the total design load.

Where a member is built into a masonry wall which develops only partial restraint, the member shall be designed to resist a negative moment at the face of the support of $Wl/24$ where W is the total design load and l is the effective span, or such other restraining moment as may be shown to be applicable. For such a condition shear coefficient given in Table 2·5 at the end support may be increased by 0·05.

Point of centres flexture may be assumed as 1/7 span from bearing centre line and 1/5 span from centre line of intermediate support.

(*xi*) **Critical sections for moment and shear.** For monolithic construction, the moments computed at the face of the supports shall be used in the design of the members at those sections.

The shears computed at the face of the support shall be used in the design of the member at section except as given below :

When the reaction in the direction of the applied shear introduces compression into the end region of the member, sections located at distance less than d from the face of the support may be designed for the same shear as that computed at support (Fig. 2·16).

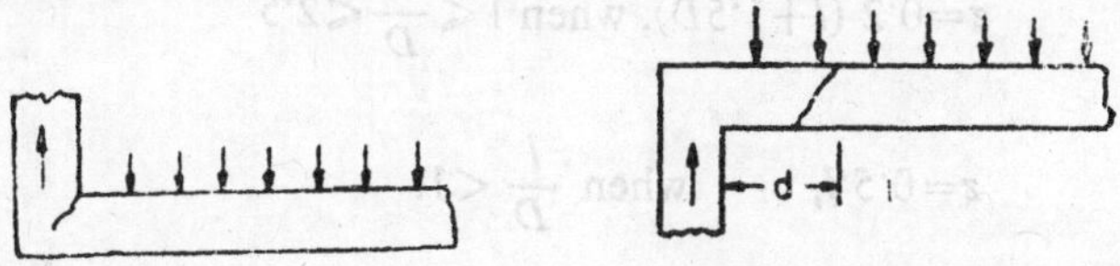

Critical section at the face of supports

Critical section at distance '*d*' from the face of support

Fig. 2·16

(*xii*) **Redistribution of moments.** Redistribution of moments may be done as follows :

Except where the simplified analysis using coefficients is used, the moments over the supports for assumed arrangement for loading including the dead load moments may each be increased or decreased by not more than 15 per cent, provided that these modified moments over the supports are used for the calculation of the corresponding moments in the spans.

(*xiii*) **Effective depth of beams.** Effective depth of beam is the distance between the centroid of the area of tension reinforcement and the maximum compression face excluding the thickness of finishing material not placed monolithically with the member and the thickness of any concrete provided to allow for wear. This will not apply to deep beams.

(*xiv*) **Deep beams.** A beam shall be deemed to be a deep beam when the ratio of effective span to overall depth, l/D is less than –

(*i*) 2·0 for a simply supported beam.

(*ii*) 2·5 for a continuous beam.

The lever arm z for a deep beam shall be determined as below :

(*a*) For simply support beams :

$$z=0{\cdot}2\ (l+2D), \text{ when } 1\leqslant\frac{l}{D}\leqslant 2$$

or $$z=0{\cdot}6\ l, \quad \text{when } \frac{l}{D}<1.$$

(*b*) For continuous beams :

$$z=0{\cdot}2\ (l+1{\cdot}5D), \text{ when } 1\leqslant\frac{l}{D}\leqslant 2{\cdot}5$$

or $$z=0{\cdot}5\ l, \quad \text{when } \frac{l}{D}<1.$$

where l is the effective span taken as centre to centre distance between supports or 1·15 times the clear span, whichever is smaller, and D is the overall depth.

Positive tensile reinforcement required to resist positive bending moment in any span of a deep beam shall—

(*a*) extend without curtailment between supports ;

(*b*) be embedded beyond the face of each support, so that at the face of the support it shall have a development length not less than 0·8 L_d where L_d, is the development length of ordinary beams, for the design stress in the reinforcement ;

(*c*) be placed within a zone of depth equal to $0{\cdot}25\ D-0{\cdot}05\ l$ adjacent to the tension face of the beam where D is the overall depth and l is effective span.

For tensile reinforcement required to resist negative bending moment over a support of a deep beam —

(*a*) It shall be permissible to terminate not more than half of the reinforcement at a distance of 0·5 D from the face of the support.

(*b*) The remainder shall extend over the full span.

When ratio of clear span to overall depth is in the range 1·0 to 2·5, tensile reinforcement over a support of a deep beam shall be placed in two zones comprising—

(*a*) a zone of 0·2 *D*, adjacent to the tension face, which shall contain a proportion of the tension steel, given by

$$0{\cdot}5\left(\frac{l}{D}-0{\cdot}5\right)$$

where, l=clear span and D=overall depth.

(*b*) a zone measuring 0·3 *D* in either side of the mid-depth of the beam, which shall contain the remainder of the tensile steel, evenly distributed.

For span to depth ratios less than unity, the steel shall be evently distributed over a depth of 0·8 *D* measured from the tension face.

If forces are applied to a deep beam in such a way that hanging action is required, bars or suspension stirrups shall be provided to carry all the forces concerned.

Side face reinforcement shall comply with requirements of minimum reinforcement of walls *i.e.* the minimum ratio of vertical reinforcement to gross concrete area shall be—

(*a*) 0·0012 for deformed bars not larger than 16 mm diameter and with a characteristic strength of 415 N/mm² or greater.

(*b*) 0·0015 for other types of steel.

(*c*) 0·0015 for welded wire fabric not larger than 16 mm in diameter.

Vertical reinforcement shall be spaced not farther than 40 mm.

The minimum ratio of horizontal reinforcement to gross concrete area shall be—

(*a*) 0·0020 for deformed bars not larger than 16 mm in diameter.

(*b*) 0·0025 for other types of bars.

(*c*) 0·0020 for welded wire fabric not larger than 16 mm in diameter.

The minimum reinforcement may not always be sufficient to provide adequate resistance to the effects of shrinkage and temperature.

2·8. Types of Problems

In design of reinforced concrete beams four types of problems are met with.

Type 1. *To design beam to resist given bending moment.* In this, design is worked out as a balanced section.

(*a*) Work out the constant—k, j and Q for given allowable stresses.

(*b*) In most of the cases breadth b of the beam is governed by the size of supporting column or aesthetic sense. The usual ratio of breadth to depth is between 1/2 to 1/4. From the equation $M=Qdb^2$, depth of beam is fixed.

(*c*) Calculate the amount of tensile steel from equation

$$A_t=\frac{M}{tjd}.$$

(*d*) Adequate cover is provided to steel and overall depth D of the beam is found out.

Type 2. *To determine the permissible B.M. or loading on beams.* Given the dimensions of sections and maximum allowable stresses.

(*a*) Find the position of neutral axis. The neutral axis will be centroid of concrete area in compression and equivalent area of tensile steel.

(*b*) For the given maximum stresses determine the depth of critical neutral axis n_c.

(*c*) If the depth of critical N.A. is more than the depth of actual N.A. the section is under-reinforced and the stress in steel will reach the maximum value. Moment of resistance will be $t\times A_t\times(d-n/3)$.

If the depth of critical N.A. is less than the depth of the actual N.A. the section is over-reinforced and the stress in concrete will reach the maximum value and moment of resistance will be

$$b\times n\times\frac{c}{2}\times(d-n/3).$$

If the span and nature of the loading is given the permissible loading on the beam can be worked out.

Type 3. *To check the design of beam for stresses.* Given the section, area of steel and modular ratio m, to determine the stresses developed in concrete and steel when the section is subjected to given B.M.

(*a*) Determine the position of N.A.

(*b*) Determine level arm $a=d-n/3$.

(*c*) Moment of resistance $M=t_a\times A_t\times a=\dfrac{b\times n\times c_a}{2}\times a$

where c_a and t_a are actual maximum stresses developed in concrete and steel respectively.

Type 4. *To determine the area of tensile reinforcement.* Given the dimension of section of beam, B.M. to be resisted and maximum allowable stresses.

(*a*) For the given stresses determine constants k, j and Q.

(*b*) For the given dimensions of the section, find the moment of resistance of critical section, $M=Qbd^2$.

(*c*) (*i*) If the actual B.M. is less than the moment of resistance of critical section the section is to be designed as under-reinforced.

$$\text{M.R.}=t\times A_t(d-n/3) \qquad ...(1)$$

Taking moments about N.A.

$$b\times n\times\frac{n}{2}=mA_t(d-n) \qquad ...(2)$$

$$A_t=\frac{bn^2}{2m(d-n)}$$

Substituting in (1)

$$\text{M.R.}=\frac{t\times bn^2}{2m(d-n)}\times(d-n/3).$$

This can be solved for n. Knowing the value of n, A_t can be calculated.

(*ii*) If the actual B.M. is more than the moment of resistance of critical section, the section is to be designed as over-reinforced.

$$\text{M.R.}=bn\,\frac{c}{2}\,(d-n/3).$$

From this equation value of n can be found out.

Area of steel can be found out from the equation

$$bn\times\frac{n}{2}=mA_t\times(d-n).$$

Alternately, putting $M=Q'bd^2$, value of Q' can be evaluated. From the graph 3 in Appendix for given stresses percentage p of steel can be found out. Area of steel will be $\dfrac{p\times b\times d}{100}$.

Ex. 2·1. *Design an RCC rectangular beam to resist a B.M. of 2×10^8 N-mm. The width of the beam is to be 300 mm, $c=5$ N/mm², $t=140$ N-mm², $m=18$.*

Sol. $\text{M.R.}=Qbd^2$

$$2\times10^8=0{\cdot}87\times300\times d^2$$

$$d^2=\frac{2\times10^8}{0{\cdot}87\times300}$$

$$d=875{\cdot}4 \text{ mm}$$

$$A_t=\frac{M}{tjd}=\frac{2\times10^8}{140\times0{\cdot}87\times875{\cdot}4}=1885 \text{ mm}^2$$

Provide 6 bars of 20 mm ϕ, $A_t=1885$ mm².

Overall depth of the beam will be equal to effective depth plus half the diameter of the bars used plus 25 mm.

$$D=875{\cdot}4+10+25=910{\cdot}4 \text{ mm}.$$

From practical consideration depth of beam is not specified in fractions. Overall depth will be fixed at 920 mm. Effective depth will be 920−25−10=885 mm.

From Table 2·1 it is clear that for slightly under-reinforced beam lever arm will be slightly greater than that for critical section. Therefore, area of steel based on increased effective depth and lever arm constant as for critical section, will be slightly greater and section will be safe.

$$A_t \text{ required}=\frac{M}{tjd}$$

$$=\frac{1\times10^8}{140\times0{\cdot}87\times885{\cdot}4}$$

$$=1855 \text{ mm}^2$$

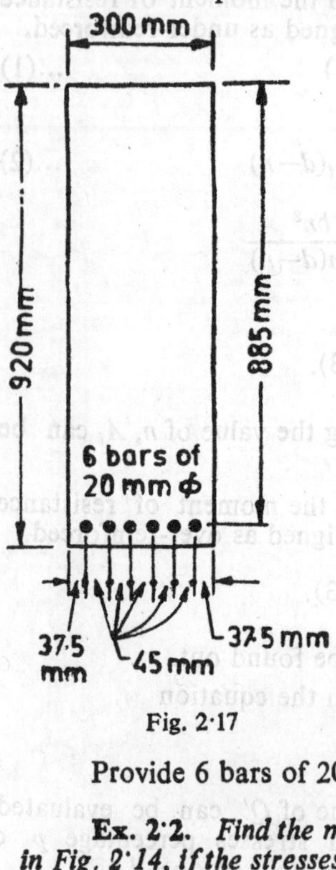

Fig. 2·17

Provide 6 bars of 20 mm ϕ, $A_t=1885$ mm².

Ex. 2·2. *Find the moment of resistance of the RCC beam shown in Fig. 2·14, if the stresses in steel and concrete are not to exceed 140 N/mm² and 5·0 N/mm², respectively, m=18.*

If the span of the beam is 4 m, find the maximum superimposed distribution load the beam can carry.

Sol. Let n be the depth of N.A.

Taking moments about N.A.

$$b \times n \times \frac{n}{2} = mA_t(d-n)$$

$$\therefore \quad 220 \times n \times \frac{n}{2} = 18 \times 4 \times 314(400-n)$$

$$n = 201{\cdot}1 \text{ mm}$$

Depth of critical N.A.

$$= kd = 0{\cdot}4 \times 400 = 160 \text{ mm.}$$

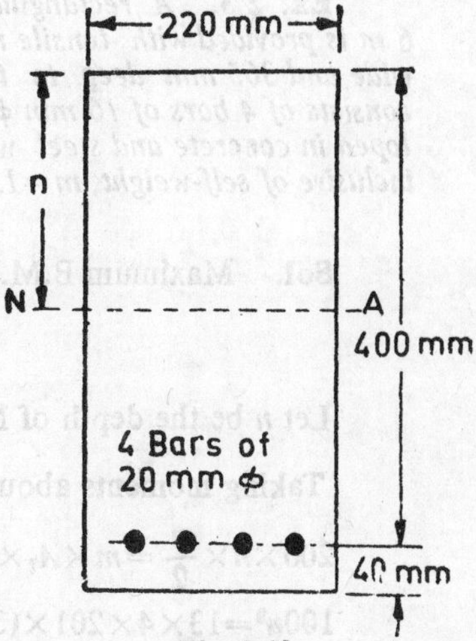

Fig. 2·18

Depth of actual N.A. is more than that of critical N.A. Therefore, section is over-reinforced. Concrete reaches maximum stress.

$$\text{M.R.} = bn \times \frac{c}{2}(d - n/3)$$

$$= 220 \times 201{\cdot}1 \times \frac{5{\cdot}0}{2} \times \left(400 - \frac{201{\cdot}1}{3}\right)$$

$$= 36{,}830{,}000 \text{ N-mm}$$

If w is the total uniformly distributed load per metre.

$$\text{Maximum B.M.} = \frac{wl^2}{8}$$

$$= \frac{w \times 4 \times 4}{8} \times 1000 = 36{,}830{,}000$$

$$\therefore \quad w = 18{,}415 \text{ N/m.}$$

Self-weight per metre

$$= \frac{440}{1000} \times \frac{220}{1000} \times 1 \times 25{,}000 = 2420 \text{ N}$$

Superimposed load $= 18{,}415 - 2420 = 15{,}995$ N/m.

Moment of resistance from graph 3

$$\text{Percentage of steel } p = \frac{1256}{220 \times 400} \times 100 = 1{\cdot}427\%$$

Q' from graph is 1·045

$$\text{M.R.}=1{\cdot}045\times220(400)^2=36{,}784{,}000 \text{ N-mm.}$$

Ex. 2·3. *A rectangular beam simply supported over a span of 6 m is provided with tensile reinforcement only. The beam is 200 mm wide and 365 mm deep to the centre of tensile reinforcement, which consists of 4 bars of 16 mm ϕ. Calculate the maximum stresses developed in concrete and steel when the beam carries a load of 8000 N/m inclusive of self-weight, m=13.*

Sol. Maximum B.M. $=\dfrac{wl^2}{8}=\dfrac{8000\times6\times6}{8}\times1000$ N-mm

$$=36{,}000{,}000 \text{ N-mm}$$

Let n be the depth of N.A.

Taking moments about N.A.

$$200\times n\times\frac{n}{2}=m\times A_t\times(d-n)$$

$$100n^2=13\times4\times201\times(365-n)$$

$$n=149{\cdot}8 \text{ mm}$$

Alternative from graph 1, for

$$p=\frac{804}{200\times365}\times100$$

$$=1{\cdot}1$$

$$k'd=0{\cdot}412\times365$$

$$=150{\cdot}4 \text{ mm}$$

$$a=d-\frac{n}{3}$$

$$=365-\frac{149{\cdot}8}{3}$$

$$=315{\cdot}1 \text{ mm}$$

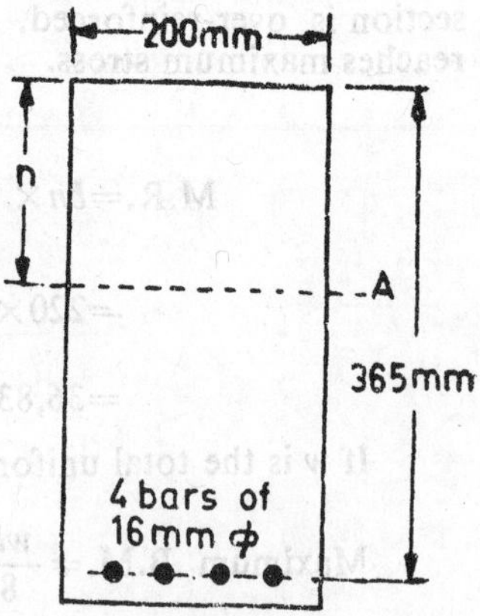

Fig. 2·19

$$\text{M.R.}=\frac{b\times n\times c_a}{2}\times a$$

$$36{,}000{,}000=200\times\frac{149{\cdot}8}{2}\times c_a\times315{\cdot}1$$

$$c_a=\frac{36{,}000{,}000}{100\times149{\cdot}8\times315{\cdot}1}=7{\cdot}626 \text{ N/mm}^2$$

Also $\qquad \text{M.R.}=A_t\times t_a\times a$

$$36{,}000{,}000=804\times t_a\times 315{\cdot}1$$

$$t_a=\frac{36{,}000{,}000}{804\times 315{\cdot}1}=142{\cdot}1\ \text{N/mm}^2.$$

Ex. 2·4. *A rectangular beam 200 mm wide has effective depth of 350 mm. It is subjected to a bending moment of 20,000 N-m. Find the area of tensile reinforcement if the permissible stresses are c=5·0 N/mm², t=140 N/mm², m=18.*

Sol. (*i*) For critical section,

M.R.$=Qbd^2$

$=0{\cdot}87\times 200\times 350\times 350$

$=21{,}315{,}000$ N-mm

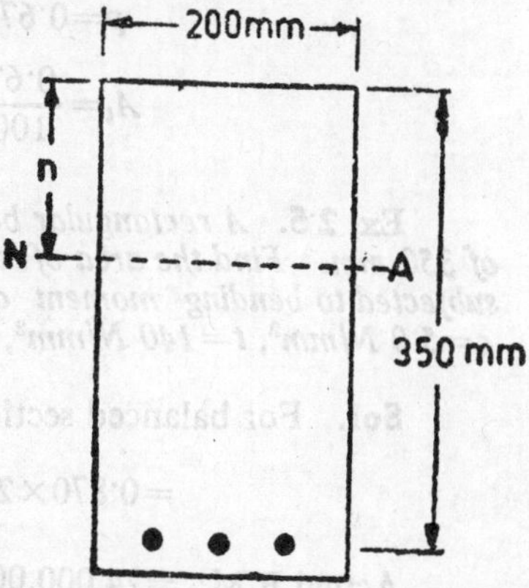

Actual B.M.

$=20{,}000{,}000$ N-mm.

Therefore, beam is to be designed as under-reinforced.

Let n be the depth of N.A.

Taking moments about N.A.

$$200\times n\times\frac{n}{2}=18\times A_t(350-n)$$

$$A_t=\frac{100n^2}{18(350-n)} \qquad ...(a)$$

Moment of resistance$=t\times A_t\times(d-n/3)$

$$20{,}000{,}000=140\times\frac{100n^2}{18(350-n)}\times\left(350-\frac{n}{3}\right)$$

Solving this equation by trial and error $n=135$ mm.

$$20{,}000{,}000=140\times A_t\times\left(350-\frac{135}{3}\right)$$

$$A_t=\frac{20{,}000{,}000}{140\times 305}=468\ \text{mm}^2. \qquad ...(i)$$

(*ii*) *Area of steel assuming lever arm as for balanced action.*

$$A_t=\frac{M}{tjd}=\frac{20{,}000{,}000}{148\times 0{\cdot}87\times 350}$$

$$=471\ \text{mm}^2 \ ...(ii)$$

Comparing (*i*) and (*ii*) it is seen that area of steel found by assuming lever arm as for balanced section does not vary much from that found by accurate method. Thus it is evident that for slightly under reinforced sections area of steel can be calculated on the basis of lever arm for balanced section.

(*iii*) *Use of graph 3.*

$$Q'=\frac{20{,}000{,}000}{200\times(350)^2}=0{\cdot}816.$$ From graph for mix M 15.

$$p=0{\cdot}67\%$$

$$A_t=\frac{0{\cdot}67}{100}\times200\times350=469 \text{ mm}^2.$$

Ex. 2·5. *A rectangular beam 200 mm wide has an effective depth of 350 mm. Find the area of tensile reinforcement required when it is subjected to bending moment of 24,000 Nm. Permissible stresses are* $c=5{\cdot}0$ *N/mm²*, $t=140$ *N/mm²*, $m=18$.

Sol. For balanced section $\text{M.R.}=Qbd^2$

$$=0{\cdot}870\times200\times(350)^2=21{,}315{,}000 \text{ N-mm}$$

Actual B.M. =24,000,000 N-mm

This section is to be designed as over-reinforced.

$$\text{M.R.}=b\times n\times\frac{c}{2}\left(d-\frac{n}{3}\right)$$

$$24{,}000{,}000=200\times n\times\frac{5}{2}\left(350-\frac{n}{3}\right)$$

$$n=162{\cdot}3 \text{ mm.}$$

Taking moments about N.A.

$$200\times n\times\frac{n}{2}=18\times A_t\,(d-n)$$

$$200\times162{\cdot}3\times\frac{162{\cdot}3}{2}=18\times A_t\,(350-162{\cdot}3)$$

$$A_t=\frac{100\times162{\cdot}3\times162{\cdot}3}{18\times187{\cdot}7}=778 \text{ mm}^2.$$

Percentage of steel from graph 3.

$$Q'=\frac{24{,}000\times 1000}{200\times(350)^2}=0{\cdot}0979.$$

From graph, percentage of steel corresponding to mix M 15 is 1·11.

$$A_t=\frac{1{\cdot}11\times 200\times 350}{100}=777 \text{ mm}^2.$$

Ex. 2·6. *Find the moment of resistance of an RCC beam having trapezoidal section as shown in Fig. 2·21. The beam is reinforced with 4 bars 16 mm ϕ.*

$$c=5{\cdot}0 \text{ N/mm}^2,\ t=140 \text{ N/mm}^2,\ m=18.$$

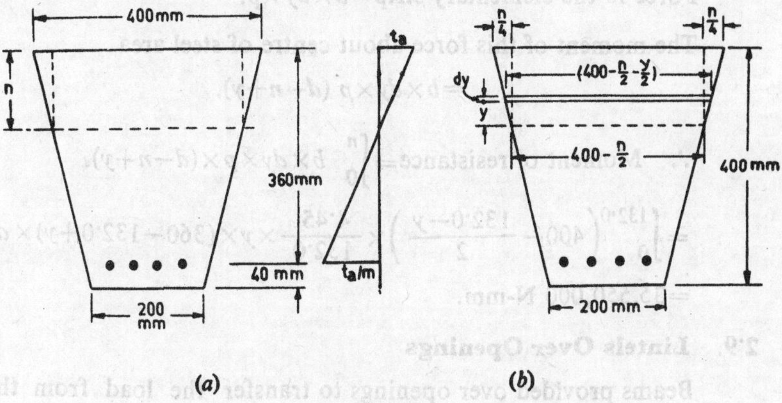

Fig. 2·21

Sol. Let n be the depth of N.A.

Taking moments about N.A.

$$\left(400-\frac{n}{2}\right)\times n\times\frac{n}{2}+\frac{1}{2}\times n\times\frac{n}{2}\times\frac{2}{3}\,n$$
$$=18\times 4\times 201\times(360-n)$$

Solving the above equation by trial and error

$$n=132 \text{ mm}.$$

Depth of critical N.A. $=0{\cdot}4\times 360$

$$=144 \text{ mm}.$$

∴ Section is under-reinforced. Stress in steel will reach the maximum value.

$$c_a = \frac{t}{m} \times \frac{n}{d-n}$$

$$= \frac{140}{18} \times \frac{132{\cdot}0}{350-132{\cdot}0}$$

$$= 4{\cdot}5 \text{ N/mm}^2.$$

Consider an elementary strip of area $b \times dy$ at y distance from N.A. Let p be the stress at the level of strip.

Breadth of strip $b = 400 - \frac{n-y}{2}$ and $p = c_a \times \frac{y}{n}$.

Force in the elementary strip $= b \times dy \times p$.

The moment of this force about centre of steel area

$$= b \times dy \times p\ (d-n+y).$$

∴ Moment of resistance $= \int_0^n b \times dy \times p \times (d-n+y)$.

$$= \int_0^{132{\cdot}0} \left(400 - \frac{132{\cdot}0 - y}{2}\right) \times \frac{4{\cdot}45}{132{\cdot}0} \times y \times (360 - 132{\cdot}0 + y) \times dy$$

$= 35{,}550{,}000$ N-mm.

2·9. Lintels Over Openings

Beams provided over openings to transfer the load from the wall above to the sides are called lintels.

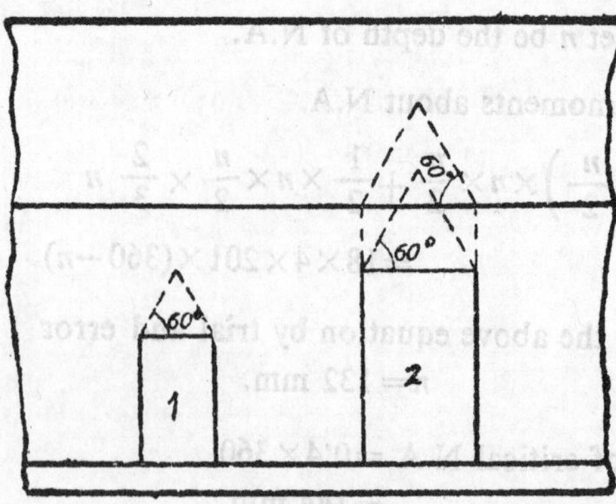

Fig. 2·22

As the mansonry is well-bonded, there is arch action over the lintel and only some portion of the wall load above the opening will come on the beam. Generally the load of equilateral triangle formed by drawing lines at 60° from the end of the lintel is taken to act on the lintel. If the height of the wall above the lintel is not sufficient to form equilateral triangle, the whole of the rectangular load above the lintel is taken to act on the lintel. In case a slab comes within that triangular area, the load from masonry above the lintel and the load from the slab plus the triangular load above the slab is taken to act on the lintel as shown in Fig. 2·22.

Ex. 2·7. *A lintel is to be provided over an opening 2 m wide. The height of opening is 2·5 m and the floor height is 5 m. The wall is 40 cm thick. Assume density of masonry as 19,200 N/m³. Allowable stresses are c=5 N/mm², t=140 N/mm² and m=18. Design a suitable section for the lintel.*

Sol. Assume effective depth of lintel as 140 mm and overall depth as 180 mm.

Span of lintel $=2+0{\cdot}14$

$=2{\cdot}14$ m.

Load from wall W_1

$$=\frac{2{\cdot}14\times1{\cdot}07\sqrt{3}}{2}\times0{\cdot}4\times19{,}200$$

$=15{,}230$ N.

Self-weight of lintel W_2.

$$=2{\cdot}14\times\frac{40}{100}\times\frac{180}{1000}\times25{,}000=3852\text{ N}$$

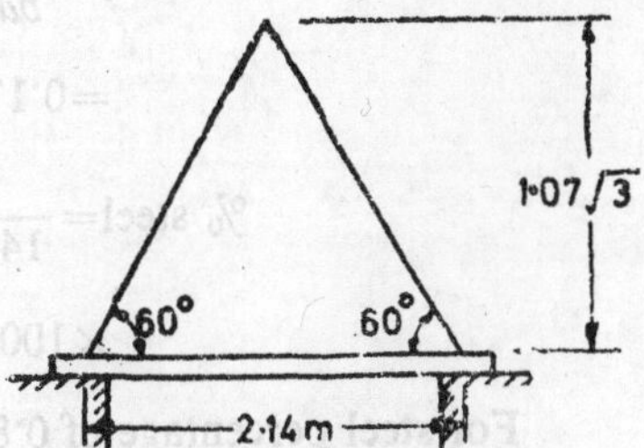

Fig. 2·23

$$\text{Max. B.M.}=\frac{W_1 l}{6}+\frac{W_2 l}{8}$$

$$=\frac{15{,}230\times2{\cdot}14}{6}+\frac{3852\times2{\cdot}14}{8}$$

$=5432+1030$ N-m

$=6{,}462$ N-m

$=6{,}462{,}000$ N-mm.

$$d=\sqrt{\frac{M}{Qb}}$$

$$=\sqrt{\frac{6,462,000}{0{\cdot}87\times400}}$$

$$=136{\cdot}3 \text{ mm.}$$

Pr ovide overall depth of 180 mm and effective depth of 140 mm,

$$A_t=\frac{6,462,000}{140\times0{\cdot}87\times140}$$

$$=379 \text{ mm}^2$$

Provide 4 bars of 12 mm ϕ.

Area provided=452 mm²

$$\text{Maximum S.F.}=\frac{15230}{2}+\frac{3852}{2}$$

$$=9541 \text{ N}$$

Shear stress $\tau_v=\frac{F}{bd}=\frac{9541}{400\times140}$

$$=0{\cdot}17 \text{ N/mm}^2.$$

$$\%\text{ steel}=\frac{452}{140\times400}\times100=0{\cdot}81$$

For steel percentage of 0·81, from Table 1·9, $\tau_c=0{\cdot}35$ N/mm². Use minimum steel for stirrups. For mild steel $f_y=250$ N/mm².

$$\frac{A_{sv}}{b\times s_v}\geqslant\frac{0{\cdot}4}{f_y}$$

$$A_{sv}\geqslant\frac{b\times s_v\times0{\cdot}4}{f_y}$$

Maximum spacing $=0{\cdot}75\,d=0{\cdot}75\times140=105$ mm

Use 6 mm ϕ,

$$2\times28\geqslant\frac{400\times s_v\times0{\cdot}4}{250}$$

$$\therefore \qquad s_v<\frac{2\times28\times250}{400\times0{\cdot}4}\leqslant87{\cdot}5$$

Use 6 mm ϕ stirrups at 85 mm centre to centre.

2·10. Balanced Section of New Concrete Mixes (M, 15, M 20, M 25, M 30, M 35, M 40)

The permissible stresses for new concrete mixes are given in Table 1·6 in N/mm² (Newtons per millimetre square). The permissible stresses in steel are given in Table 1·8.

From Article 2·5—

Critical neutral axis depth constant is given by

$$\eta_o=\frac{1}{1+t/cm}\times d=kd$$

Lever arm constant $j=1-k/3$

Moment of resistance constant

$$Q=\tfrac{1}{2}\times c\times k\times j$$

Area of steel for balanced section

$$A_t=\frac{Qbd^2}{tjd}$$

For a balanced section percentage of steel p is given by

$$p=\frac{A_t}{bd}\times 100=\frac{Q}{tj}\times 100$$

The values of m, k, j, Q, p are given in Table 2·6.

Table 2·6. Constants for Critical Section

Grade of concrete	*Permissible stresses bending σ_{bc} or c in (N/mm²)*	*m*	*k*	*j*	*Q*	*p (percentage)*
M 10	3·0	31	0·40	0·87	0·52	0·43
M 15	5·0	18	0·40	0·87	0·87	0·72
M 20	7·0	13	0·40	0·87	1·21	0·99
M 25	8·5	11	0·40	0·87	1·41	1·21
M 30	10·5	9	0·40	0·87	1·74	1·43
M 35	11·5	8	0·40	0·87	2·00	1·65
M 40	13·0	7	0·40	0·87	2·26	1·85

2·11. Balanced Section for Deformed Bars

$$mc=\frac{280}{3}$$

$$t=\frac{mc(d-n)}{n};\quad t=\frac{280\,(d-n)}{3n}$$

$$\therefore\quad \frac{t\times 3}{280}=\frac{d}{n}-1$$

$$\therefore\quad \frac{3t}{280}+1=\frac{d}{n}$$

$$\therefore\quad \frac{3t+280}{280}=\frac{d}{n}$$

$$n=\frac{280d}{3t+280}$$

or
$$t=190,\ n=\frac{28d}{85}$$

$$a=d-\frac{n}{3}=d-0{\cdot}11\,d=0{\cdot}89\,d$$

$$\text{M.R.}=\frac{c}{2}\,bn\left(d-\frac{n}{3}\right)$$

$$=\frac{cb}{2}\times\frac{28d}{85}\left(d-\frac{28d}{3\times 85}\right)$$

$$=\frac{cbd^2\times 28}{2\times 85}\left(\frac{255-28}{255}\right)$$

$$=\frac{cbd^2\times 28\times 227}{2\times 85\times 255}=0{\cdot}1466\,cbd^2$$

M 15, M.R. $=0{\cdot}733\,bd^2$

M 20, M.R. $=1{\cdot}0262\,bd^2$

M 25, M.R. $=1{\cdot}2461\,bd^2$

M 30, M.R. $=1{\cdot}466\,bd^2$

$$mA_t\,(d-n)=\frac{bn^2}{2}$$

$$\therefore\quad A_t=\frac{b}{2}\left(\frac{28d}{85}\right)^2\times\frac{1}{m\left(d-\frac{28d}{85}\right)}$$

$$=\frac{b}{2}\left(\frac{28d}{85}\right)^2\times\frac{85}{md\times 57}$$

$$=\frac{bd\;28\times 14}{85\times 57\;m}$$

$$=\frac{bd}{m}\times 0{\cdot}809$$

$$p=\frac{A_t}{bd}\times 100=\frac{8{\cdot}09}{m}$$

M 15, $p=0{\cdot}449\%$

M 20, $p=0{\cdot}622\%$

M 25, $p=0{\cdot}735\%$

M 30, $p=0{\cdot}889\%$

The results have been tabulated in Table 2·7.

Table 2·7

Constant for Balanced Section, Deformed Bars
σ_{st}=190 N/mm²

Grade of concrete	*Permissible stress c N/mm²*	*m*	*k*	*j*	*Q*	*p*
M 15	5·0	18	0·33	0·89	0·7330	0·449
M 20	7·0	13	0·33	0·89	1·0262	0·622
M 25	8·5	11	0·33	0·89	1·2461	0·735
M 30	10·0	9	0·33	0·89	1·4660	0·899

For $t=230$,
$$n=\frac{280d}{690+280}=\frac{28d}{97}=2{\cdot}289d$$

$$a=d-\frac{n}{3}=d-\frac{28d}{291}=\frac{263d}{291}=0{\cdot}904d$$

$$\text{B.M.}=\frac{c}{2}bn\left(d-\frac{n}{3}\right)=\frac{cb}{2}\times\frac{28d}{97}\left(d-\frac{28d}{291}\right)$$

$$=\frac{cbd^2}{2}\times\frac{28}{97}\times\frac{263}{291}=0{\cdot}1304cbd^2$$

$$mA_t(d-n)=\frac{bn^2}{2}$$

$$\therefore \quad A_t=\frac{b\left(\frac{28d}{29}\right)^2}{2m\left(d-\frac{28d}{97}\right)}=\frac{b\times(28)^2d}{2\times97\times69m}$$

$$\therefore \quad p=\frac{A_t}{bd}\times100=\frac{(28)^2\times100}{2\times97\times69m}$$

$$=\frac{5{\cdot}857}{m}$$

for $t=210$, $$n=\frac{280d}{630+280}=\frac{28d}{91}=0{\cdot}308d$$

$$a=d-\frac{n}{3}=d-\frac{28d}{273}\;\frac{245d}{273}$$

$$=0{\cdot}897\,d$$

$$\text{M.R.}=\frac{c}{2}\,bn\left(d-\frac{n}{3}\right)$$

$$=\frac{cb}{2}\times\frac{28d}{91}\left(d-\frac{28d}{273}\right)$$

$$=\frac{cbd^2}{2}\times\frac{28}{91}\times\frac{245}{273}$$

$$=0{\cdot}1381\;cbd^2$$

$$mA_t(d-n)=\frac{bn^2}{2}.$$

$$\therefore \quad A_t=\frac{b\left(\frac{28d}{91}\right)^2}{2m\left(d-\frac{28d}{91}\right)}=\frac{bd(28)^2}{2\times91\times63m}$$

$$p=\frac{A_t}{bd}\times100$$

$$=\frac{(28)^2\times100}{2\times91\times63m}=\frac{6{\cdot}838}{m}$$

Tables 2·8 and 2·9 give constants for balanced section using deformed bars with $\sigma_{st}=230\ \text{N/mm}^2$ and $210\ \text{N/mm}^2$ respectively.

Table 2·8

Constant for Balanced, Section, Deformed Bars

σ_{st}=230 N/mm²

Grade of concrete	*Permissible stress c N/mm²*	*m*	*k*	*j*	*Q*	*p*
M 15	5·0	18	0·289	0·904	0·6520	0·326
M 20	7·0	13	0·289	0·904	0·9128	0·451
M 25	8·5	10	0·289	0·904	1·1084	0·532
M 30	10·0	9	0·289	0·904	1·3040	0·651

Table 2·9

Constant for Balanced Section, Deformed Bars,

σ_{st}=210 N/mm²

Grade of concrete	*Permissible stress c N/mm²*	*m*	*k*	*j*	*Q*	*p*
M 15	5·0	18	0·308	0·897	0·6905	0·325
M 20	7·0	13	0·308	0·897	0·9567	0·526
M 25	8·5	11	0·308	0·897	1·1739	0·622
M 30	10·0	9	0·308	0·897	1·3810	0·760

Table 2·10 gives values of M/b for various values of effective depths d for various mixes and σ_{st}.

Graphs 21, 22, 23 have been drawn for σ_{st}=190, 210, 230 and concrete mixes of M 15, M 20, M 25, M 30 giving moment of resistance of sla

Ex. 2·8. *Balanced Section,* $t=\dfrac{mc(d-n)}{n}$ *rmed bars are used take, c=5 N/mm², t=190 N/mm² and m=18.*

For deformed bars

Balanced Section, $t=\dfrac{mc(d-n)}{n}$

Table 2·10. M.R. of Rectangular Section with Deformed Bars

Concrete mix	σ_{st}	d mm	60	80	100	120	140	160	180	200	250	300	350
(1)	(2)	(3)	(4)	(5)	(6)	(7)	(8)	(9)	(10)	(11)	(12)	(13)	(14)
	190		26·988	46 912	73·300	105·55	143 65	187·65	237·49	293·20	458·13	659·70	897·93
M 15	210	$\frac{M}{b}$ N	29·858	44·192	69·050	99 433	135·34	176·77	723·72	276·20	431·56	621·45	845·86
	230		23·472	41·728	65·200	93·888	127 79	166·91	211·25	260·80	407·50	586·80	798·70
	190		36·943	65·677	110·262	147·77	201·14	262·71	332·46	410·48	641·38	923·58	1257·1
M 20	110	$\frac{M}{b}$ N	34·801	61·869	96·670	139·20	189 47	247·48	313·21	386·68	604·19	870·03	1184·2
	230		32·861	58·419	91·280	131·64	178 91	233·68	295·75	365·12	570·50	821·52	1182·2
	190		44·860	79·750	124·61	179·14	244·24	319·00	403·75	498·44	778·81	1121·5	1526·5
M 25	210	$\frac{M}{b}$ N	42·260	75·130	117·39	169·04	230·08	300·52	380·34	469·56	733 69	1056·5	1438·0
	230		39·902	70·938	110·84	159·61	217 25	283·75	359·12	443·36	692 75	9975·6	1357·8
	190		52·776	93·824	146·06	211·10	287·34	375·30	474·98	586·40	916·25	1319·4	1795·9
M 30	210	$\frac{M}{b}$ N	49·716	88·384	138·10	198·86	270·68	353·54	447·44	552·46	863·13	1242·9	1691·7
	230		46·944	83·456	130·40	187·78	255 58	333·82	422·50	521·60	815·00	1173·6	1597·4

400	450	500	500	600	650	700	750	800	850	900	950	1000	1050	1100	1150	1200
(15)	(16)	(17)	(18)	(19)	(20)	(21)	(22)	(23)	(24)	(25)	(26)	(27)	(28)	(29)	(30)	(31)
1172·8	1483·3	1830·5	2217·3	2638·8	3096·9	3391·7	4123·1	4691·2	5295·9	5937·3	6616·3	7330·0	8081·3	8869·3	9693·0	10555·2
1104·8	1308·3	1726·3	2088·8	3485·8	2917·4	3383·5	3884·1	4419·2	4988·9	5593·1	6231·8	6905·0	7612·8	8335·1	9131·9	9943·2
1043·2	1320·3	1630·0	1972·8	2347·2	2754·0	3149·8	3667·5	4172·8	4710·7	5281·2	5884·3	6520·0	7188·3	7889 2	8622·7	9388·8
1641·9	2078·1	2565·5	3104·3	3694·3	4335·7	5028·4	5572·4	6567·7	7414·3	8312·2	9261·5	10262 0	11313 0	12417·0	13571·5	14777·3
1546·7	1957·6	2416·8	2924·3	3480·3	4084·3	4736·8	5437·7	6186·9	6984·4	7830·3	8724·5	9667·9	10657·9	11697·1	12784·6	13920·5
1460·5	1848·4	2282·0	2761·2	3286·1	3856·6	4472·7	5134·5	5841·3	6595·0	7393·7	8238·0	9128·0	10063·6	11044·9	12071·8	13144·3
1993·8	2523·3	3115·3	3769·5	4486·0	5264·8	6105·0	7009·3	7975·0	9003·1	10093·4	11246·1	12461·0	13738·3	15077·8	16979·7	17943·8
1878·2	2377·1	2934·8	3551·0	4226·0	4959·7	5752·1	6603·2	7513·0	8482·4	9508·6	10594·4	11739·0	12942·2	14042·0	15524·8	16904·2
1773·4	2244·5	2717·7	3352·8	3990·2	4683·0	5431·2	6234·8	7093·8	8008·2	8978·0	10003·3	11084·0	12220·1	13411·6	14058·6	15961·0
2245·6	2968·7	3665·0	4434·7	5277·5	6193·9	7183·4	8246·3	9382·4	10591·0	11874·6	13230·7	14660·0	16162·7	17738·6	19387·9	21110·4
2209·6	2796·5	3452·5	4177·5	4971·7	5834·7	6766·9	7768·1	8838·4	9977·7	11186·1	12463·5	13810·0	15225·5	16710·1	18263·7	19886·4
2086·4	2640·6	3260·0	3944·6	4694·4	5506·4	6389·6	7335·0	8345·6	9421·4	10562·4	11768·6	13040·0	14376·6	15778·4	17245·4	18777·6

$$\therefore \quad 190 = 18 \times 5\left(\frac{d}{n} - 1\right)$$

$$\frac{19}{9} + 1 = \frac{d}{n}$$

$$\therefore \quad n = \frac{9d}{28} = 0{\cdot}32\,d$$

$$a = d - \frac{n}{3} = d - \frac{9d}{28 \times 3} = \frac{25d}{28} = 0{\cdot}89d$$

$$\text{M.R.} = \frac{c}{2}bn\left(d - \frac{n}{3}\right) = \frac{5{\cdot}0}{2} \times b \times \frac{9d}{28} \times \frac{25d}{28}$$

$$= 0{\cdot}7175\,bd^2$$

$$\text{B.M.} = 2 \times 10^8 = 0{\cdot}7175\,bd^2$$

$$b = 300 \text{ mm}$$

$$\therefore \quad d = \sqrt{\frac{2 \times 10^8}{0{\cdot}7175 \times 300}} = 964 \text{ mm.}$$

$$A_t \times t = \frac{c}{2}\,bn$$

$$A_t = \frac{5}{2}\,b \times \frac{9d}{28 \times 190} = 0{\cdot}00423\,bd$$

Percentage of steel $p = \frac{A_t \times 100}{bd} = 0{\cdot}423\%$

$$A_t = 0.00423 \times 300 \times 964 = 1223{\cdot}3 \text{ mm}^2.$$

Use 4 bars 20 mm ϕ, $A_t = 1256 \text{ mm}^2$

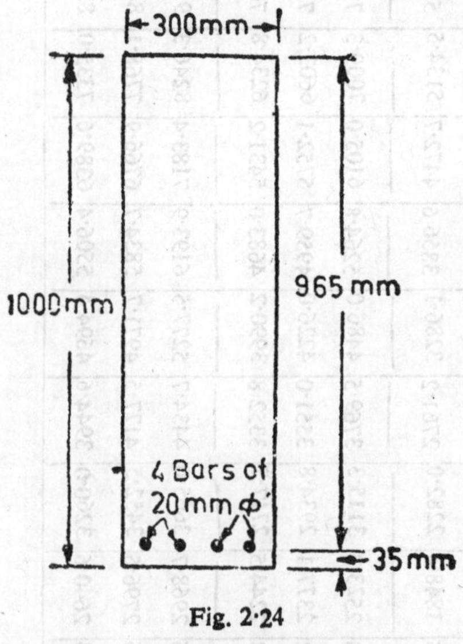

Fig. 2·24

Overall size of the beam
$= 964 + 25 + 10 = 999$ mm
say 1000 mm.

$\therefore$ Use 1000 mm $\times$ 300 mm rectangular beam.

Using Table 2·7

$Q = 0{\cdot}733$

$M = Q\,bd^2$

$$d = \sqrt{\frac{2 \times 10^8}{0{\cdot}733 \times 300}} = 954 \text{ mm}$$

Overall depth $= 1000$ mm

Effective depth $= 1000 - 25 - 10$

$= 965$ mm

$$A_t = \frac{2 \times 10^8}{0{\cdot}89 \times 965 \times 190}$$

$= 1226 \text{ mm}^2$

Use 4 bars 20 mm ϕ

$A_t = 1226 \text{ mm}^2$.

Ex. 2·9. *In Ex. 2·5, if deformed bars are used, find the tensile reinforcement. Take $c=5$ N/mm², $t=230$ N/mm² and $m=18$.*

Sol. $\text{M.R.}=Qbd^2$

$$=0{\cdot}6520\ bd^2$$
$$=0{\cdot}6520\times 200\times(350)^2$$
$$=15{,}974{,}000 \text{ N-mm}$$
$$<20{,}000{,}000 \text{ N-mm}$$

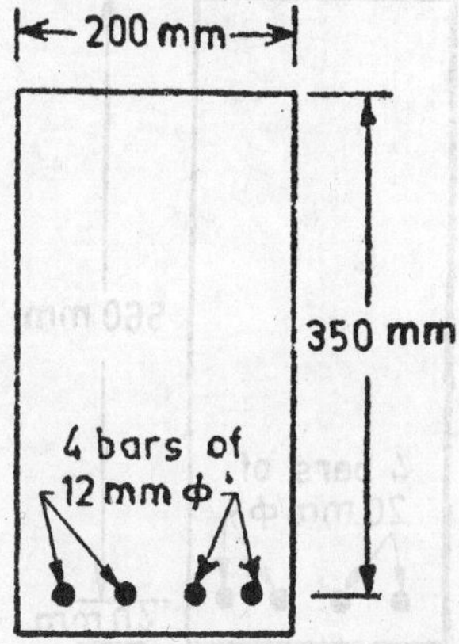

Fig. 2·25

Therefore, the beam is designed as over-reinforced. Let n be the neutral axis depth,

$$20{,}000{,}000=\frac{c}{2}bn\left(d-\frac{n}{3}\right)$$
$$=\frac{5}{2}\times 200\times n\left(350-\frac{n}{3}\right)$$

$n=130{\cdot}5$ mm

Take moments about N.A.

$$b\times\frac{n^2}{2}=A_t m\,(d-n)$$

$$200\times\frac{(130{\cdot}5)^2}{2}=A_t\ 18\ (350-130{\cdot}5)$$

$$A_t=\frac{100\times(130{\cdot}5)^2}{18\times 219{\cdot}5}$$

$$=431 \text{ mm}^2$$

Use 4 bars of 12 mm φ,

$$A_t=452 \text{ mm}^2.$$

PROBLEMS

1. Design an RCC rectangular beam to resist a B.M. of 8×10^8 N-mm. The width of the beam is to be kept 300 mm, $c=5{\cdot}0$ N/mm², $t=140$ N/mm², $m=18$.

2. Design an RCC rectangular beam having span of 6 m and carrying superimposed load of 33,000 N/m run. Use mix M 20.

3. Find the moment of resistance of the RCC rectangular beam shown in Fig. 2·26. Allowable stresses are

$$c=5{\cdot}0 \text{ N/mm}^2,\ t=140 \text{ mm}^2,\ m=18.$$

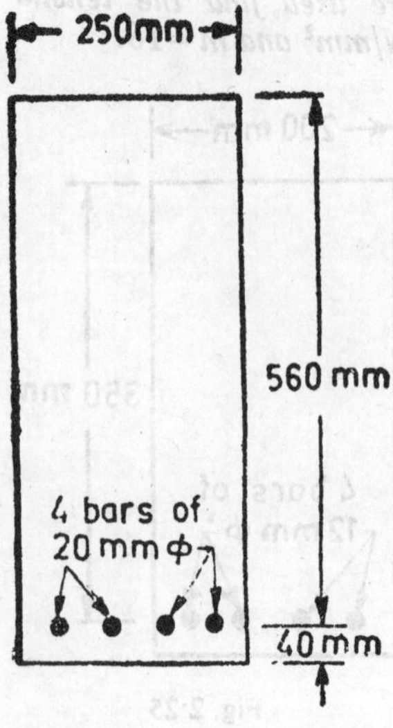

Fig. 2·26

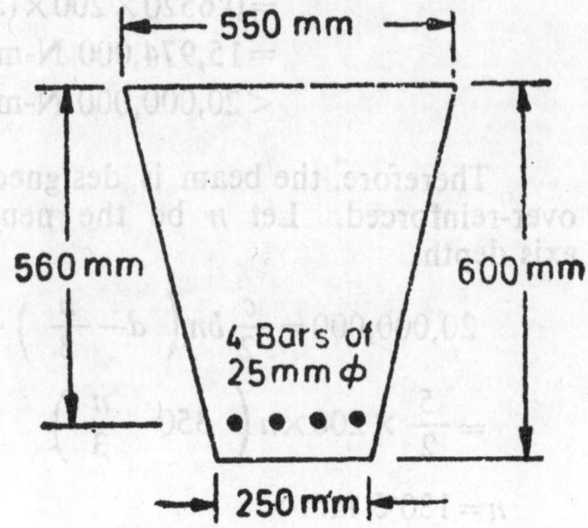

Fig. 2·27

If the span of the beam is 6 m, find the maximum superimposed distributed load that the beam can carry.

4. Find the moment of resistance of the trapezoidal section shown in Fig. 2·27.

5. Find the uniformly distributed load the beam section shown in Fig. 2·28, can safely carry over an effective span of 3 m, $c=5{\cdot}0$ N/mm², $t=140$ N/mm², $m=18$.

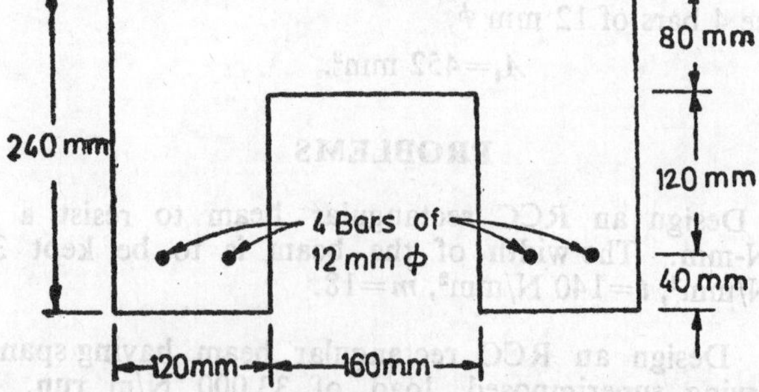

Fig. 2·28

6. A rectangular beam having breadth 250 mm and effective depth 460 mm is reinforced with 4 bars 90 mm ϕ on the tension side only. It is subjected to B.M. of 50,000,000 N-mm at a cross-section. Find the maximum stresses developed in concrete and steel, $m=18$.

7. A rectangular beam 300 mm wide has an effective depth of 882·5 mm. Find the area of tensile reinforced required if it is subjected to a B.M. of 180,000 Nm, $c=5·0$ N/mm², $t=140$ N/mm², $m=18$.

8. Find the area of tensile steel if the beam in Problem 7 is subjected to B.M. of 250,000 Nm.

9. A lintel is to be provided over an opening of 1·5 m width. The height of the opening is 2·5 and the floor height is 5 m. The wall is 400 mm thick. Design a suitable lintel for the opening.

10. A lintel is to be provided over an opening of 2 m wide. The height of opening is 2·5 m and the floor height is 3·5 m. The wall is 40 cm thick and the load from roof slab on wall is 10,000 N/m. Design suitable section for the lintel.

11. In problem 1, if deformed bars with $t=190$ N/mm² are used, design the section.

12. In problem 3, if deformed bars with $t=190$ N/mm² are used, find the moment of resistance.

13. In problem 7, if deformed bars with $t=210$ N/mm² are used, design the reinforcement.

3

Slabs Spanning in One Direction

3·1. Slabs are structural members whose depth *D* is small as compared to its span *l* and breadth *b*. The simplest form of slab is one supported on two supports, as shown in the Fig. 3·1.

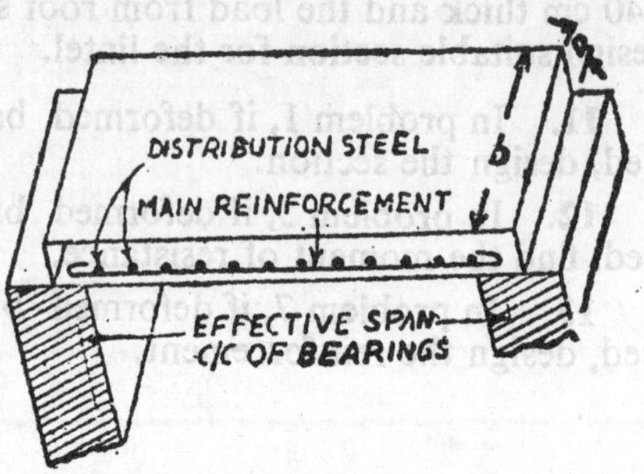

Fig. 3·1

The slabs are designed on assumption that they consist of a number of strips of span *l* and breadth one metre though in practice slab is cast in one piece and it is not composed of separate strips. The strip can be analysed by the same formulae which are used for design of rectangular beams in Chapter 2. The bending moments are calculated for a width of one metre. The load per square metre of the slab will be load per metre run of the assumed beam having width one metre and span *l*.

As the slab spans between two supports, the bars in the direction of span are said to be the main or longitudinal bars and, therefore, are positioned nearer to the bottom surface of the slab than the other bars. The bars in the direction at right angles to the span are called distribution steel. All slabs must have transverse steel to assist in distribution of stresses. These stresses are caused by the superimposed loading, temperature changes and shrinkage during drying process. As the distribution steel is wired to main steel it keeps main steel in position during the pouring of the wet concrete into the formwork.

3·2. General Notes on Design of Slabs (I.S.I.)

Use relative code requirements from Chapter 2 applicable to slabs.

(I) Effective span of slabs. The effective span shall be as follows :

(*a*) **Simply Supported.** The effective span shall be taken as clear span plus the effective depth of slab or centre to centre of supports whichever is less.

(*b*) **Continuous slab.** In the case of continuous slab, the width of the support is less than $\frac{1}{12}$th of the clear span, the effective span shall be as in (*a*). If the supports are wider than $\frac{1}{12}$th the clear span or 600 mm whichever is less, the effective span shall be taken as under :

(*i*) For end span with one end fixed and the other continuous or for intermediate spans, the effective span shall be the clear span between supports.

(*ii*) For end span with one end free and the other continuous, the effective span shall be equal to the clear span plus half the effective depth of the slab or the clear span plus half the width of the discontinuous support, whichever is less.

(II) **Control of deflection.** For slabs, the vertical deflection limits may generally be assumed to be satisfied provided that the span to depth ratios are not greater than the values obtained as below—

(*a*) Basic values of span to effective depth ratios for spans upto 10 m :

Cantilever	...	7
Simply supported	...	20
Continuous	...	26

For slabs spanning in two directions, the shorter of the two spans should be used for calculating the span to effective depth ratios.

For two way slabs of small spans (upto 3·5 m) with mild steel reinforcement, the span to overall depth ratios given below may generally be assumed to satisfy vertical deflection limits for loading class upto 3000 N/m² :

Simply supported slabs	35
Continuous slabs	40

(*b*) For spans above 10 m, the values in (*a*) may be multiplied by 10/span in metres, except the cantilever in which case deflection calculations should be made.

The deflections shall generally be limited to the following :

(*i*) The final deflections due to all loads including the effects of temperature, creep and shrinkage and measured from the as cast level should not normally exceed span/250.

(*ii*) The deflection including the effects of temperature, creep nd shrinkage occurring after erection of partitions and the appli-ation of finishes should not normally exceed span/350 or 20 mm vhichever is less.

(III) **Moments and shear coefficients for continuous labs.** The slabs continuous over supports shall be designed ccording to the provisions applicable to continuous beams.

Bending moments in slabs (except flat slabs) constructed nonolithically with the supports shall be calculated by taking such labs either as continuous over supports and capable of free rota-ion, or as members of a continuous framework with the supports, aking into account the stiffness of such supports. If such supports re formed due to beams which justify fixity at the support or labs, then the effects on the supporting beam, such as, bending of ne web in the transverse direction of the beam and tension in the ongitudinal direction of the beam, wherever, applicable, shall also e considered in the design of the beam.

For the purpose of calculation of moments in slabs in a mono-thic structure, it will generally be sufficiently accurate to assume nat members connected to the ends of such slabs are fixed in posi-on and direction at the ends remote from their connections with ne slabs.

(IV) **Slabs carrying concentrated load.** If a solid slab upported on two opposite edges, carries concentrated loads, the naximum bending moment caused by the concentrated loads shall e assumed to be resisted by an effective width of slab (measured arallel to the supporting edges) as follows :

(*a*) For a single concentrated load, the effective width shall be alculated in accordance with the following equation, provided that shall not exceed the actual width of the slab :

$$b_{ef}=kx\left(1-\frac{x}{l_{ef}}\right)+a$$

here b_{ef}=effective width of slab

k=constant having the values given in table 3·1 depending upon the ratio of the width of the slab (l') to the effective span l_{ef}

x=distance of the centroid of the concentrated load from the nearer support.

l_{ef}=effective span

a=width of the contact area of the concentrated load measured parallel to the supported edge.

And provided further that in case of a load near the nsupported edge of a slab, the effective width shall not exceed the bove value nor half the above value plus the distance of the load om the unsupported edge.

(*b*) For two or more concentrated loads placed in a line in the direction of the span, the bending moment per metre width of the slab shall be calculated separately for each load according to its appropriate effective width of slab calculated as in (*a*) above and added together for design calculations.

(*c*) For two or more loads not in a line in the direction of the span, if the effective width of slabs for one load does not overlap the effective width of slab for another load, both calculated as in (*a*) above, the slab for each load can be designed separately. If the effective width of slab for one load overlaps the effective width of slab for an adjacent load, the overlapping portion of the slab shall be designed for the combined effects to two loads.

Table 3·1

Values of k for Simply Supported and Continuous Slabs

l'/l_{ef}	*k for simply supported slabs*	*k for continuous slabs*
0·1	0·40	0·40
0·2	0·80	0·80
0·3	1·16	1·16
0·4	1·48	1·44
0·5	1·72	1·68
0·6	1·96	1·84
0·7	2·12	1·96
0·8	2·24	2·08
0·9	2·36	2·16
0·10 and above	2·48	2·24

(*d*) For cantilever solid slabs, the effective width shall be calculated in accordance with the following equation :

$$b_{ef}=1{\cdot}2\,a_1+a$$

where b_{ef}=effective width,

a_1=distance of the concentrated load from the face of the cantilever support, and

a=width of contact area of the concentrated load measured parallel to the supporting edge.

Provided that the effective width of the cantilever slab shall not exceed one-third of the length of the cantilever slab measured parallel to the fixed edge.

And provided further that when the concentrated load is placed near the extreme ends of the length of cantilever slab in the direction parallel to the fixed edge, the effective width shall not exceed the above value, nor shall it exceed half the above value plus the distance of the concentrated load from the extreme end measured in the direction parallel to the fixed edge.

For slabs other than solid slabs, the effective width shall depend on the ratio of the transverse and longitudinal flexural rigidities of the slab. When the ratio is one, that is, when the transverse and longitudinal flexural rigidities are approximately equal, the value of effective width as for solid slabs may be used. But as the ratio decreases, a proportionately smaller value should be taken.

Any other recognised method of analysis may be used for analysis of above cases of the slabs with the approval of engineer-in-charge.

Maximum distance between bars in tension :

(*i*) The horizontal distance between parallel main reinforcement bars shall not be more than three times the effective depth of a solid slab or 450 mm whichever is small.

(*ii*) The horizontal distance between parallel reinforcement bars provided against shrinkage and temperature shall not be more than five times the effective depth of a solid slab or 450 mm whichever is smaller.

Ex. 3·1. *Design a reinforced concrete slab to carry load of 6,000 N/m² inclusive of its own weight on an effective span of 4 m simply supported. The maximum stresses are not to exceed* $c=5{\cdot}0$ *N/mm² and* $t=140$ *N/mm²,* $m=18$.

Sol. Consider 1 m width of slab

$$\text{Maximum B.M.} = \frac{wl^2}{8} = \frac{6000\times4\times4}{8} = 12{,}000 \text{ N-m}$$

$$= 12{,}000{,}000 \text{ N-mm}$$

Moment of resistance $=Qbd^2$

$$12{,}000{,}000 = 0{\cdot}870\times1000\times d^2$$

$$d=\sqrt{\frac{12{,}000{,}000}{870}}=117{\cdot}5 \text{ mm}$$

Provide overall depth of 200 mm. (see below)

Minimum overall depth required

$$=\frac{l}{20}=\frac{4\times1000}{20}=200 \text{ mm}$$

$$\text{Effective depth} = 200-15-5=180 \text{ mm}$$

$$A_t=\frac{M}{tjd}=\frac{12{,}000{,}000}{140\times0{\cdot}87\times180}=547 \text{ mm}^2$$

Area of one 10 mm ϕ bar $= 79$ mm^2

$$\text{Spacing or bars} = \frac{1000 \times 79}{547} = 144 \text{ mm}$$

Provide 10 mm ϕ bars at 140 mm c/c.

Distribution steel $= 0{\cdot}15\%$ of gross cross-sectional area of concrete

$$= \frac{0{\cdot}15}{100} \times 1000 \times 200 = 300 \text{ mm}^2$$

Use 5 mm ϕ bars.

Area of one bar $= 20$ mm^2

$$\text{Spacing of bars} = \frac{1000 \times 20}{300} = 66{\cdot}7 \text{ mm}$$

Provide 5 mm bars at 60 mm c/c.

Ex. 3·2. *A reinforced concrete floor consists of a slab 140 mm thick with bars of 10 mm ϕ at 100 mm c/c, the centre of the bars being 20 mm from the underside of the slab. The span is 2·8 m and the slab is simply supported. The maximum allowable stresses are $c = 5{\cdot}0$ N/mm^2 in concrete and $t = 140$ N/mm^2 in steel and $m = 18$. Find the safe load on slab uniformly distributed including the weight of the slab.*

Sol. Consider one metre width of slab. In 1 m width of slab 10 bars are effective.

$$\text{Area of steel} = 79 \times 10$$
$$= 790 \text{ mm}^2$$

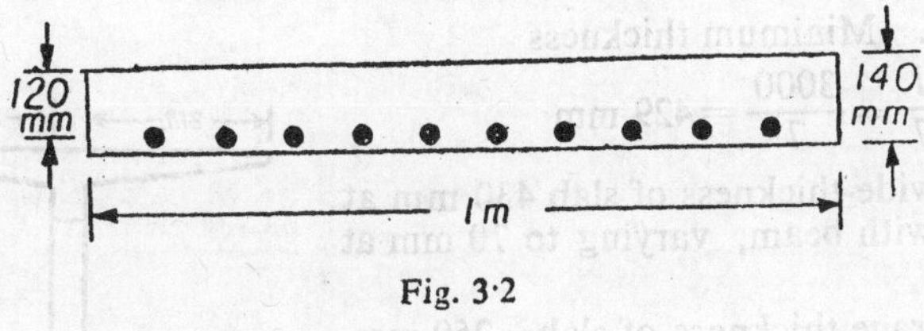

Fig. 3·2

Let n be the depth of N.A.

Taking moments about N.A.

$$1000 \times n \times \frac{n}{2} = 18 \times 790 \times (120 - n)$$

$$n = 45{\cdot}9 \text{ mm.}$$

Depth of critical N.A.

$$=kd=0{\cdot}4\times120=48 \text{ mm}$$

∴ Section is under-reinforced. Steel reaches maximum stress.

Moment of resistance $=t\times A_t\times(d-n/3)$

$$=140\times790\times\left(120-\frac{45{\cdot}90}{3}\right)$$

$$=140\times790\times104{\cdot}70$$

$$=11{,}580{,}000 \text{ N-mm}$$

$$=11{,}580 \text{ N-m}$$

Moment of resistance from graph 3.

$$\text{Percentage of steel } p=\frac{A_t}{bd}\times100=\frac{790}{1000\times120}\times100=0{\cdot}658$$

From graph 3 for mix M 15

$$Q'=0{\cdot}8$$

$$M=Q'bd^2=0{\cdot}8\times1000\times(120)^2$$

$$=11{,}520{,}000 \text{ N-mm or } 11{,}520 \text{ N-m}$$

$$\text{B.M.}=\frac{wl^2}{8}=11{,}580$$

$$w=\frac{11{,}580\times8}{(2{\cdot}8)^2}$$

$$11{,}816 \text{ N/m}^2$$

Permissible load is $11{,}816 \text{ N/m}^2$.

Ex. 3·3. *The roof of a railway station platform consists of RCC slab cantilevering 3 m on each side of RCC beam which is supported by columns 5 m apart.*

Design the slab taking live load on roof as 1500 N/m² and floor finish as 500 N/m².

Sol. Minimum thickness

$$=\frac{l}{7}=\frac{3000}{7}=429 \text{ mm}$$

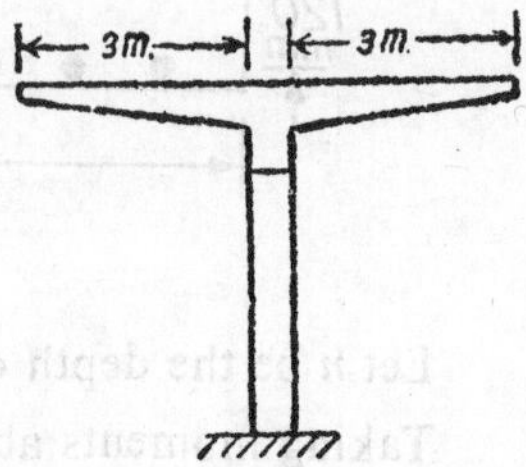

Fig. 3·3

Provide thickness of slab 430 mm at junction with beam, varying to 70 mm at edge.

Average thickness of slab = 250 mm.

Dead load per square metre based on average thickness

$$=\frac{250}{1000}\times25{,}000$$

$$=6250 \text{ N/m}^2$$

Total load per sq. m=1500+500+6250=8250 N

$$\text{Max. B.M.}=\frac{wl^2}{2}=8250\times\frac{3\times3}{2}=37{,}125 \text{ N-m}$$

$$=37{,}125{,}000 \text{ N-mm}$$

$$d=\sqrt{\frac{M}{Qb}}=\sqrt{\frac{37{,}125{,}000}{0{\cdot}870\times1000}}=206{\cdot}6 \text{ mm}$$

Overall depth provided is 430 mm.

Effective depth =410 mm

$$A_t=\frac{37{,}125\times1000}{140\times0{\cdot}87\times410}=743{\cdot}4 \text{ mm}^2.$$

Use 10 mm ϕ bars, area of one bar

$$=79 \text{ mm}^2$$

$$\text{Spacing}=\frac{1000\times79}{743{\cdot}4}=106{\cdot}3 \text{ mm}$$

Use 10 ϕ mm bars at 100 mm centres

$$\text{Distribution steel}=\frac{0{\cdot}15}{100}\times430\times1000=645 \text{ mm}^2$$

Using 8 mm ϕ bars,

$$\text{Spacing}=\frac{1000\times50}{645}=77{\cdot}5 \text{ mm.}$$

Use 8 mm ϕ bars at 75 mm c/c. Near the edge of cantilever pitch of distribution steel will be gradually increased to 250 mm.

Curtailment of main reinforcement. Half the bars will be curtailed at centre of overhang. Pitch of main reinforcement at this section will be 130 mm as shown in Fig. 3·4 (*a*).

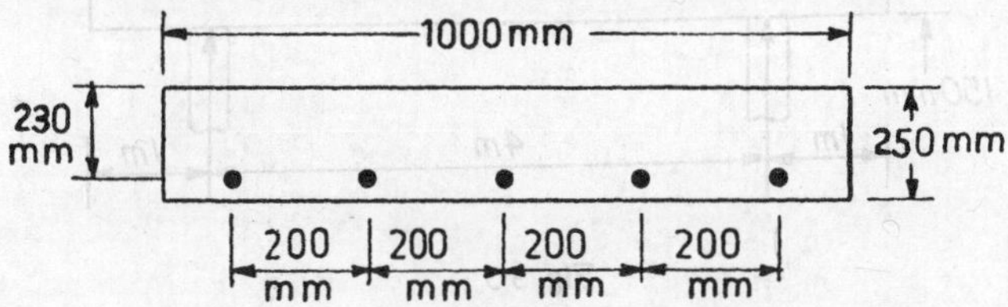

Fig. 3·4 (*a*)

Area of steel in 1 m width

$$=\frac{79\times1000}{200}=395 \text{ mm}^2.$$

$$\%\text{ steel}=\frac{395}{230\times1000}\times100=0{\cdot}178.$$

Percentage of steel for balanced section is 0·714. Therefore, section is under-reinforced.

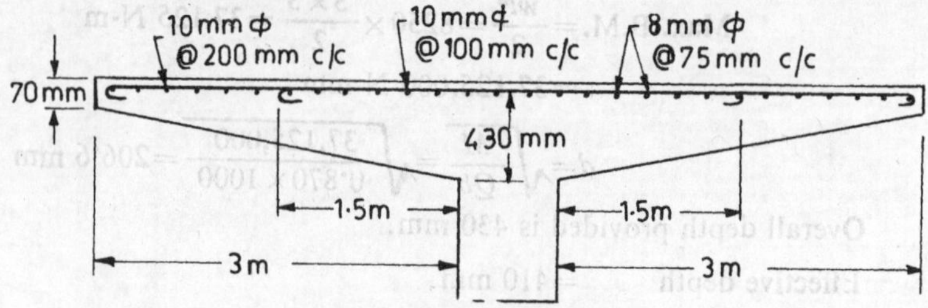

Fig. 3·4 (*b*)

Approximate moment of resistance $= t \times A_t \times jd$

$$= 140 \times 395 \times 0{\cdot}87 \times 230 = 11{,}065{,}530 \text{ N-mm}$$

$$\text{Actual B.M.} = \frac{8000 \times (1{\cdot}5)^2}{1} \times 1000 \text{ N-mm}$$

$$= 9{,}000{,}000 < 11{,}065{,}530$$

Section provided is adequate.

Ex. 3·4. *Fig. 3·5 shows the cross-section of a roof slab. The supports being 4 m apart. Design the slab. Live load=1500 N/m². Floor finish=500 N/mm², c=5·0 N/mm², t=140 N/mm², m=18.*

Sol. Minimum thickness

$$= \frac{\text{span}}{26} = \frac{4000}{26} = 153{\cdot}8$$

Assume thickness of slab as 155 mm throughout for calculation of dead load.

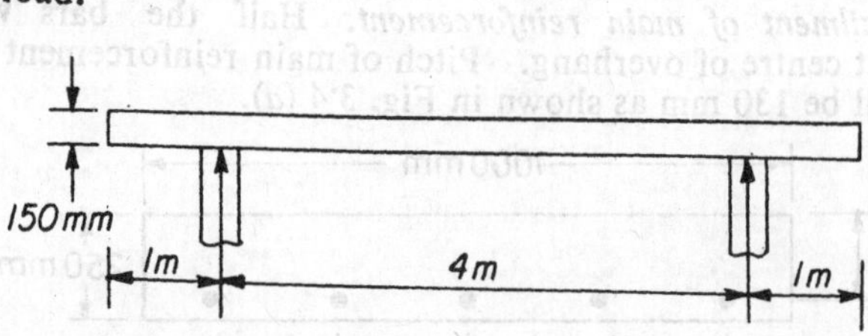

Fig. 3·5

Dead load of slab/sq. m.

$$= \frac{155}{1000} \times 25{,}000 = 3875 \text{ N}$$

Total dead load $= 3875 + 500 = 4375$ N/m²

Maximum+ve B.M. in supported span will occur when only supported span *BC* is loaded with live load. Load in supported span will be (4375+1500) N/m. Max. B.M. at centre *E*.

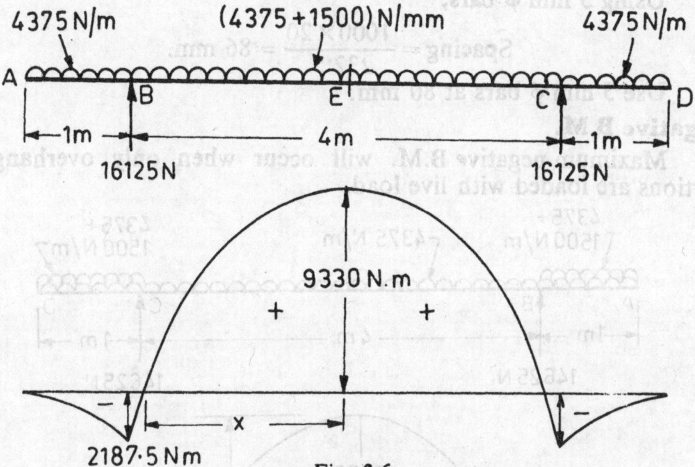

Fig. 3·6

$$=16{,}125\times2+4375\times2{\cdot}5-5875\times2\times1$$
$$=9562{\cdot}5 \text{ N-m}$$
$$=9{,}562{,}500 \text{ N-mm.}$$

Position of point of contra-flexure

B.M. at distance x from centre

$$=9562{\cdot}5-5875\times\frac{x^2}{2}=0$$

$$\therefore \qquad x=\sqrt{\frac{9562{\cdot}5}{5875}}=1{\cdot}804 \text{ m.}$$

Design of Section

Effective depth $\quad d=\sqrt{\dfrac{M}{Qb}}=\sqrt{\dfrac{9{,}562{,}500}{1000\times0{\cdot}870}}=104{\cdot}8$ mm.

Provide overall depth of 155 mm, Effective depth = 135 mm.

$$\left(\text{Minimum thickness required}=\frac{4000}{26}=153{\cdot}8 \text{ mm}\right)$$

$$A_t=\frac{9{,}562{,}500}{140\times0{\cdot}87\times135}=581{\cdot}6 \text{ mm}^2$$

Using 10 mm ϕ bars,

$$\text{Spacing}=\frac{1000\times79}{581{\cdot}6}=135{\cdot}8 \text{ mm}$$

Provide 10 m ϕ bars at 130 mm c/c

Distribution steel $\quad =\dfrac{0{\cdot}15}{100}\times1000\times155=232{\cdot}5$

Using 5 mm ϕ bars,

$$\text{Spacing} = \frac{1000 \times 20}{232 \cdot 5} = 86 \text{ mm.}$$

Use 5 mm ϕ bars at 80 mm.

Negative B.M.

Maximum negative B.M. will occur when only overhanging portions are loaded with live load.

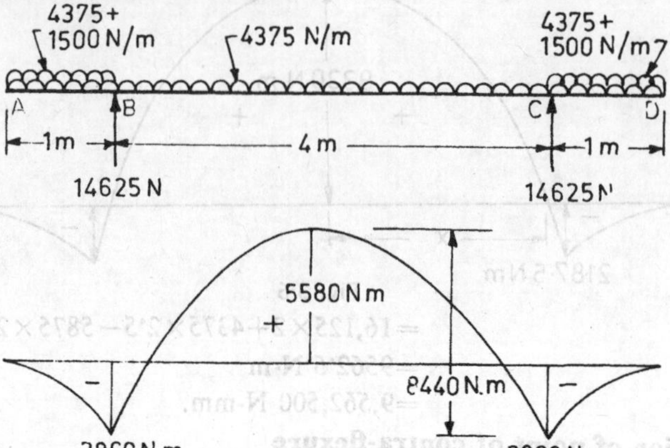

Fig. 3·7

Maximum −ve B.M.

$$=(4375+1500)\times 1 \times \tfrac{1}{2} = 2937{\cdot}5 \text{ N-m.}$$

Use same depth of 155 mm as for supported span.

The section will be designed as under-reinforced.

$$A_t = \frac{2937{\cdot}5 \times 1000}{140 \times 0{\cdot}87 \times 135} = 178{\cdot}6 \text{ mm}^2$$

Using 10 mm ϕ bars,

$$\text{Spacing} = \frac{1000 \times 79}{178{\cdot}6} = 442 \text{ mm.}$$

Use 10 mm ϕ bars 260 mm c/c. Alternate bars from supported span will be bent up. (Alternate bar bent up).

Position of point of contra-flexure

B.M. at centre of span

$$=5580 \text{ N-m}$$

If point of contra-flexure is at x distance from the centre of span,

$$M_x = 5580 - \frac{4220 \times x^2}{2} = 0$$

$$x = \sqrt{\frac{5580 \times 2}{4220}} = 1{\cdot}63 \text{ m}$$

Distance from support=0·4 m.

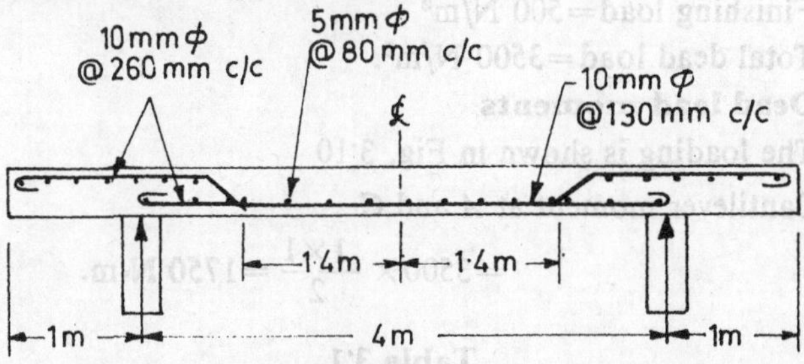

Fig. 3·8

Ex. 3·5. *An RCC bent for a city bus-stand is shown in Fig. 3·9. The total length of the bus-stand is 8 m. The bents are spaced at 3 m centres so that the slab overhangs 1 m on either end beyond the centres of bent ribs. Live load may be assumed at 2000 N/m². Design the roof slab for the stand if the ribs are to be provided above the roof. Show the arrangements of slab reinforcement and connection with the bent ribs.*

Sol. The slab is continuous over these supports with equal overhangs on either side of end supports.

The minimum overall depth required

$$=\frac{3000}{26}=115·4 \text{ mm}$$

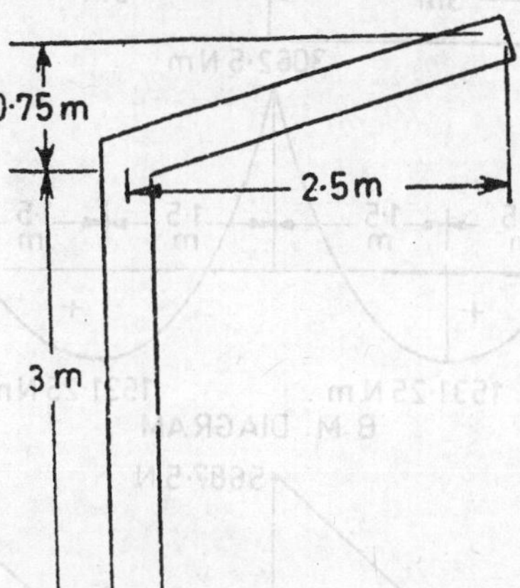

Fig. 3·9

Provide overall depth of 120 mm

Consider 1 m width of slab

$$\text{Dead load of slab}=\frac{120}{1000}\times1\times1\times25000$$

$$=3000 \text{ Nm}^2$$

Finishing load=500 N/m²

Total dead load=3500 N/m².

Dead load moments

The loading is shown in Fig. 3·10

Cantilever moment at *A* and *C*

$$=3500\times\frac{1\times1}{2}=1750 \text{ N-m.}$$

Table 3·1

A		½ *B* ½			*C*
+1750	−2625	+2625	−2625	+2625	−1750
	+ 875	+ 437·5	−437·5	− 875	
+1750	+1750	+3062·5	−3062·5	+1750	−1750

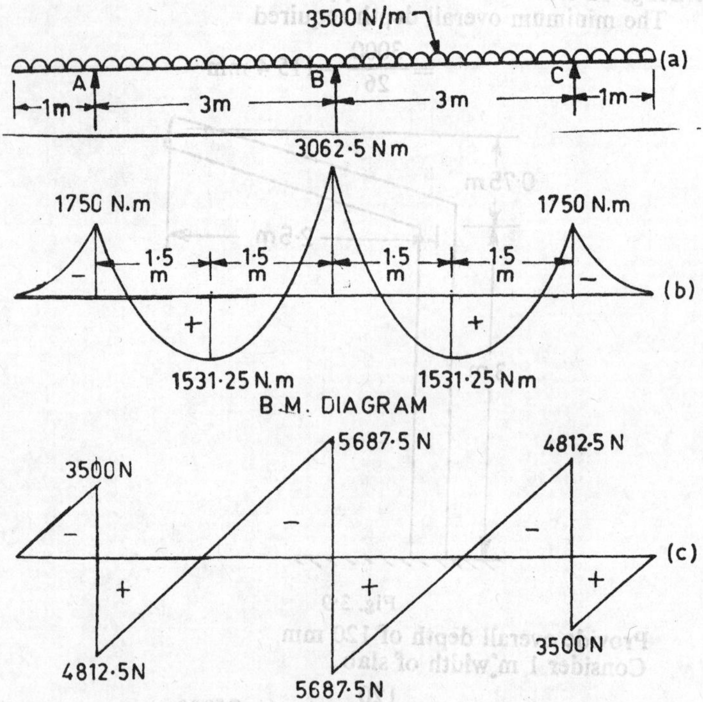

Fig. 3·10

Fixed end moment for span *AB* and *BC*

$$=\frac{wl^2}{12}=\frac{3500\times 3^2}{12}=2625 \text{ N-m.}$$

Analysis is done by moment distribution as shown in table 3·1.

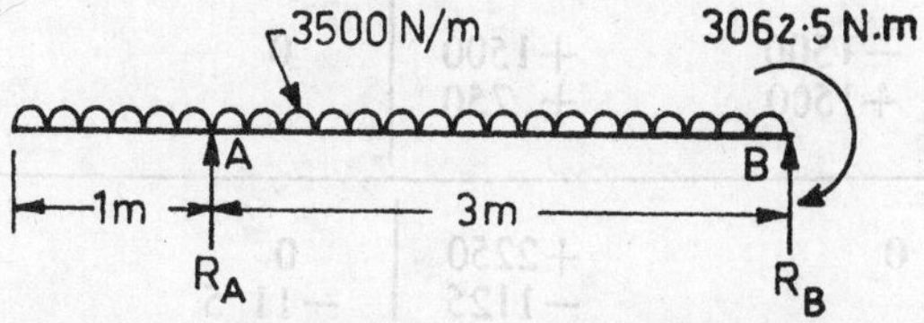

Fig. 3·11

$$R_A=\frac{3500\times 4\times 2-3062\text{·}5}{3}=8312\text{·}5$$

$R_B=3500\times 8-2\times 8312\text{·}5=11,375$ N

B.M. and shear force diagrams are shown in Fig. 3·10.

For maximum positive moment in span *AB*, live load is placed in span *AB* as shown in Fig. 3·12.

Analysis is shown in table 3·2.

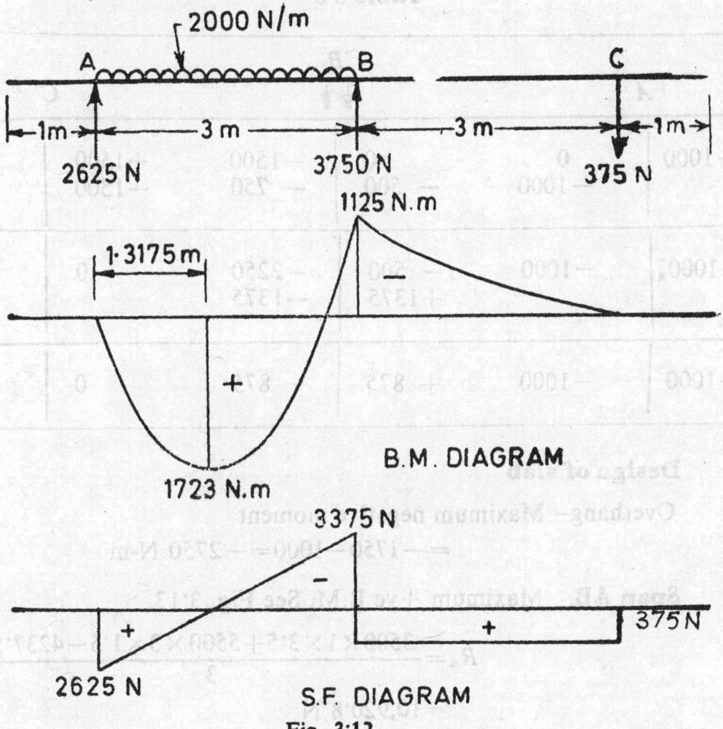

Fig. 3·12

Table 3·2

A		B $\frac{1}{2}$	$\frac{1}{2}$	C
	−1500 +1500	+1500 + 750	0	0
	0	+2250 −1125	0 −11 5	0 0
	0	+1125	−1125	0

For minimum moment in span *AB*, live load is placed in cantilever adjacent to *AB* and on span *BC*. Moment distribution is shown in Table 3·3.

Table 3·3

	A	B $\frac{1}{2}$	$\frac{1}{2}$	C
+1000	0 −1000	0 − 500	−1500 − 750	+1500 −1500
+1000	−1000	− 500 +1375	−2250 −1375	0
+1000	−1000	+ 875	− 875	0

Design of slab

Overhang– Maximum negative moment

$$=-1750-1000=-2750 \text{ N-m}$$

Span AB. Maximum +ve B.M. See Fig. 3·13.

$$R_A=\frac{3500\times1\times3{\cdot}5+5500\times3\times1{\cdot}5-4237{\cdot}5}{3}$$

$$=10{,}920{\cdot}8 \text{ N}$$

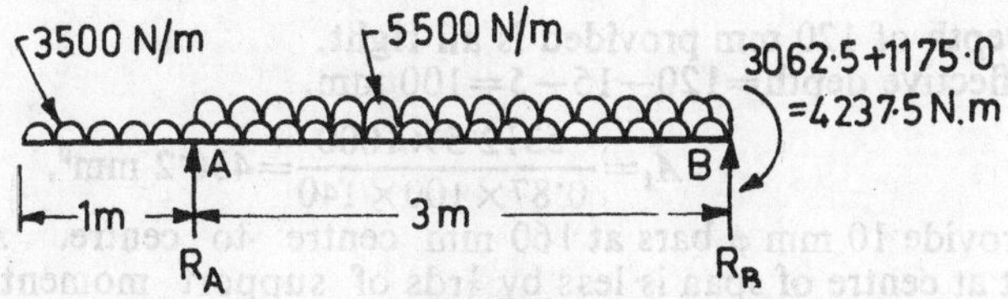

Fig. 3·13

For zero shear in AB, $10{,}920·8-3500-5500\ x=0$

$$x=\frac{7420·8}{5500}$$

$$=1·35 \text{ metres from } A.$$

$$\text{Max. +ve B.M.}=10{,}920·8\times1·35-3500\times1·85-5500\times\frac{(1·35)^2}{2}$$

$$=3256·2 \text{ N-m.}$$

Minimum moment in AB. For B.M. see Fig. 3·14.

$$R_A=\frac{5500\times1\times3·5+3500\times3\times1·5-3937·5}{3}$$

$$=10{,}354·2 \text{ N}$$

For zero shear in AB, $10{,}354·2-5500\times1-3500\ x=0$

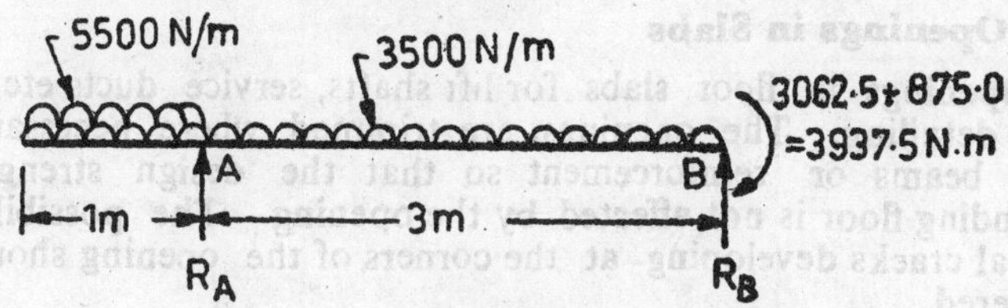

Fig. 3·14

$$x=\frac{4854·2}{3500}=1·39 \text{ m from } A.$$

$$\text{Minimum moment}=10{,}354·2\times1·39-5500\times1·89-3500\times\frac{(1·39)^2}{9}$$

$$=616·2 \text{ N-m.}$$

Maximum negative moment at support B when live load on both spans AB and $BC=-3062·5-1125-1225=-5312·5$ N-m.

Use M 15 mix.

Effective depth of slab

$$=\sqrt{\frac{5312·5\times1000}{0·87\times1000}}=78·1 \text{ mm}$$

Depth of 120 mm provided is all right.
Effective depth=120−15−5=100 mm.

$$A_t=\frac{5312{\cdot}5\times1000}{0{\cdot}87\times100\times140}=436{\cdot}2\ \text{mm}^2.$$

Provide 10 mm ϕ bars at 160 mm centre to centre. As the moment at centre of span is less by $\frac{1}{3}$rds of support moment every third bar will be stopped and remaining two taken in as shown in Fig. 3·15.

$$\text{Distribution steel}=\frac{0{\cdot}15}{100}\times1000\times120=180\ \text{mm}^2$$

Use 6 mm ϕ bars at 160 mm centre to centre.

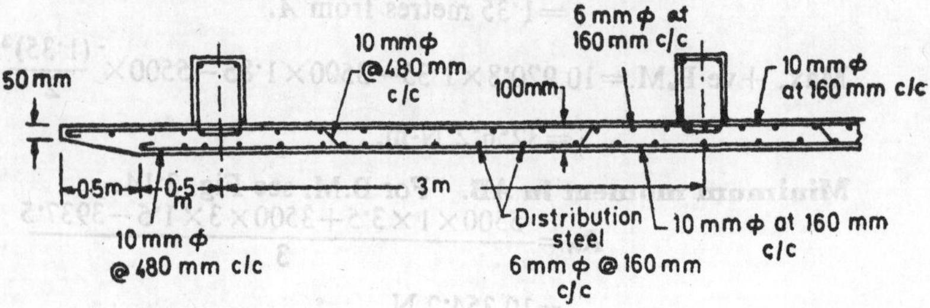

Fig. 3·15

The arrangement of steel reinforcement is shown in Fig. 3·15.

3·3. Openings in Slabs

Openings in floor slabs for lift shafts, service ducts etc. need special detailing. The openings are trimmed where necessary by special beams or reinforcement so that the design strength of surrounding floor is not affected by the opening. The possibility of diagonal cracks developing at the corners of the opening should be considered.

When the openings are small and the trimmer beams are not provided for strengthening the edges, the extra reinforcement to be provided is shown in Fig. 3·16. The normal reinforcement in both

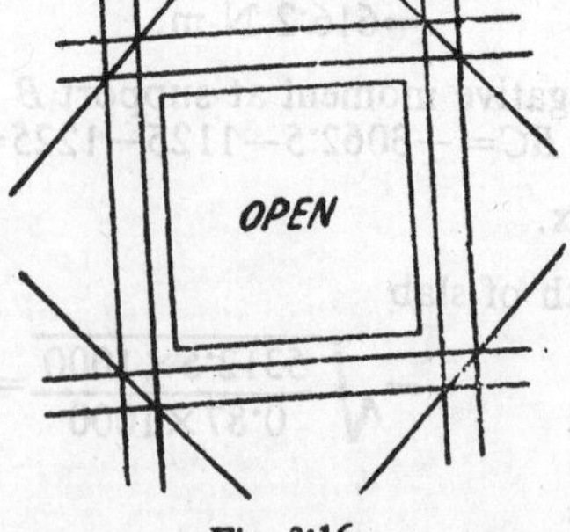

Fig. 3·16

directions of the slab is cut-off at the opening and extra-reinforcement of cross-sectional area not less than the area of reinforcing bars that would have crossed the opening in that direction is placed alongside each edge of the opening. The extra bars should extend 40 times the diameter of the bar on either side of the opening. Generally four extra bars of nominal diameter are provided diagonally across the corners of the opening as shown in Fig. 3·16.

Ex. 3·6. *In example 3·2, if deformed bars with $t=190$ N/mm² are used, find the moment of resistance and the safe load on slab uniformly distributed including the weight of the slab.*

Consider one metre width of slab. In 1 metre width of slab 10 bars are effective.

Sol. Area of steel $=79\times10=790$ mm²

Let n be the depth of N.A.

$$1000\times n\times\frac{n}{2}=18\times790\,(120-n)$$

$$n=45{\cdot}9\text{ mm.}$$

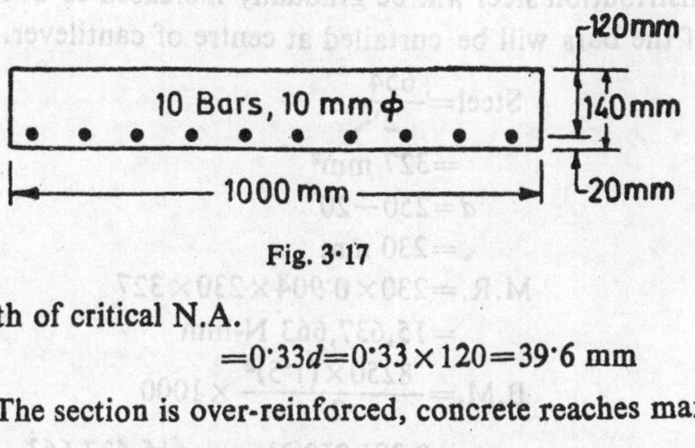

Fig. 3·17

Depth of critical N.A.

$$=0{\cdot}33d=0{\cdot}33\times120=39{\cdot}6\text{ mm}$$

∴ The section is over-reinforced, concrete reaches max. stress

$$\therefore \quad \text{M.R.}=\frac{c}{2}\,bn\left(d-\frac{n}{3}\right)$$

$$=\frac{5}{2}\times1000\times45{\cdot}9\times\left(120-\frac{45{\cdot}9}{3}\right)$$

$$=12{,}014{,}300\text{ N-mm}$$

$$\text{B.M.}=\frac{wl^2}{8}=\frac{w\times(2{\cdot}8)^2\times1000}{8}=980\,w$$

$$w=\frac{12{,}014{,}300}{980}=12{,}259\text{ N/m}^2.$$

Ex 3·7. *In Example 3·3 Using deformed bars with t=230 N/mm², design the slab.*

Sol. Keep the thickness of 430 mm at junction of beam varying to 70 mm at edge.

Max, B.M.=37,125,000 N-mm

$$d=\sqrt{\frac{M}{Qb}}=\sqrt{\frac{37{,}125{,}000}{0{\cdot}652\times1000}}=238{\cdot}6\text{ mm}$$

Overall depth provided is 430 mm

Effective depth=430−15−5=410 mm

$$A_t=\frac{37{,}125{,}000}{230\times0{\cdot}904\times410}$$

$$=435{\cdot}5\text{ mm}^2$$

Use 10 mm ϕ bars at 120 mm c/c (minimum steel).

$$A_t=654\text{ mm}^2$$

$$\text{Distribution steel}=\frac{0{\cdot}15}{100}\times400\times1000$$

$$=645\text{ mm}^2.$$

Use 8 mm ϕ bars at 70 mm c/c. Near the edge of cantilever pitch of distribution steel will be gradually increased to 250 mm.

Half the bars will be curtailed at centre of cantilever.

$$\text{Steel}=\frac{654}{2}$$

$$=327\text{ mm}^2$$

$$d=250-20$$

$$=230\text{ mm}$$

$$\text{M.R.}=230\times0{\cdot}904\times230\times327$$

$$=15{,}637{,}663\text{ N-mm}$$

$$\text{B.M.}=\frac{8250\times(1{\cdot}5)^2}{2}\times1000$$

$$=9{,}281{,}250\text{ N-mm}<15{,}637{,}663.$$

Reinforcement in the slab is shown in Fig. 3·18.

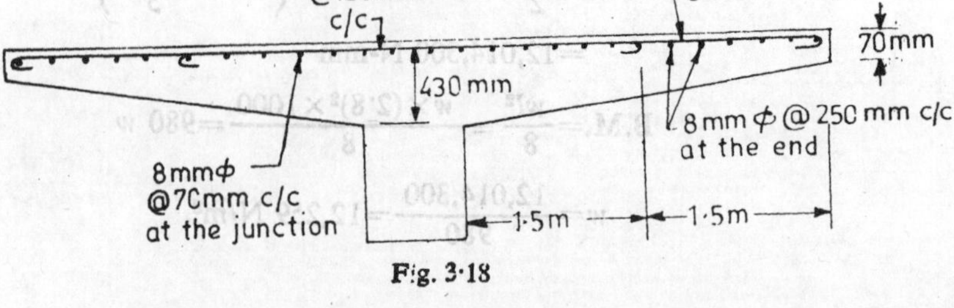

Fig. 3·18

Ex. 3·8. *Design a continuous floor slab for an office building 16 metres long and 6 metres wide, supported on 4 RCC intermediate beams. Take live load as 3000 N/m². Use M 15 concrete and mild steel bars.*

Sol. Provide 160 mm bearing at each end of slab. The length of slab centre to centre of bearings will be 16+0·16=16·16 metres.

The B.M. for continuous slab will be $wl^2/10$ for end span and $wl^2/12$ for intermediate spans. Therefore to make the B.M. in end span equal to B.M. in the intermediate spans, the end span is taken as 0·90 l to 0·94 l where 'l' is the length of intermediate span. There will be three intermediate spans and two end spans.

Take l=intermediate span=3·34 metres.

$$\text{End span } l' = \frac{16{\cdot}16 - 3\times 3{\cdot}34}{2}$$

$$= \frac{6{\cdot}24}{2} = 3{\cdot}07 \text{ metres}$$

$$\frac{l'}{l} = \frac{3{\cdot}07}{3{\cdot}34} = 0{\cdot}92$$

Minimum thickness of slab

$$= \frac{3{\cdot}34\times 1000}{26} = 128{\cdot}5 \text{ mm}$$

Assume thickness of slab as 130 mm

(*i*) Dead load of slab $= \dfrac{130\times 25{,}000}{1000} = 3250 \text{ N/m}^2$

(*ii*) Floor finish $= 1000 \text{ N/m}^2$

(*iii*) Live load $= 3000 \text{ N/m}^2$

$$w_D = 3250 + 1000 = 4250 \text{ N/m}^2$$

$$w_L = 3000 \text{ N/m}^2$$

Consider 1 metre width of continuous slab,

dead load per running metre $w_D = 4250 \text{ N/m}^2$

live load per running metre $w_L = 3000 \text{ N/m}^2$

(*a*) End span=3·07 metres

+B.M. at centre of span

$$= \frac{w_D l^2}{12} \times \frac{w_L l_1^2}{10}$$

$$= \frac{4250\times 3{\cdot}07^2}{12} + \frac{3000\times 3{\cdot}07^2}{10}$$

$$= 6165{\cdot}5 \text{ N-m.}$$

—B.M. at penultimate support

$$=-\left[\frac{w_D l_1^2}{10}+\frac{w_L l_1^2}{9}\right]$$

$$=-\left[\frac{4250\times 3{\cdot}07^2}{10}+\frac{3000\times 3{\cdot}07^2}{9}\right]$$

$$=-7147{\cdot}2 \text{ N-m}$$

(*b*) +B.M. at centre of intermediate span

$$=\frac{w_D l^2}{24}+\frac{w_L l^2}{12}$$

$$=\frac{4250\times 3{\cdot}34^2}{24}+\frac{300\times 3{\cdot}34^2}{12}$$

$$=4764{\cdot}4 \text{ N-m.}$$

—B.M. over intermediate supports

$$=-\left[\frac{w_D l^2}{12}+\frac{w_L l^2}{9}\right]$$

$$=-\left[\frac{4250\times 3{\cdot}34^2}{12}+\frac{3000\times 3{\cdot}34^2}{9}\right]$$

$$=-7669{\cdot}5 \text{ N-m}$$

Absolute max B.M.=7669·5 N-m

$$\therefore \qquad d=\sqrt{\frac{M}{Qb}}=\sqrt{\frac{7669{\cdot}5\times 1000}{0{\cdot}87\times 1000}}$$

$$=93{\cdot}9 \text{ mm}$$

Thickness provided is 130 mm

Using 10 mm ϕ bars and using cover as 15 mm.

Effective depth=130—15—5=110 mm

$$A_t=\frac{M}{at}=\frac{7669{\cdot}5\times 100}{0{\cdot}87\times 100\times 140}=629{\cdot}7 \text{ mm}^2$$

Use 10 mm ϕ bars at 120 mm centres

$$A_t=654 \text{ mm}^2$$

Distribution steel $=\frac{0{\cdot}15}{100}\times 1000\times 130=195 \text{ mm}^2$

Use 6 mm ϕ bars at 140 mm centres, $A_t=202$ mm²

Point of contra flexture=0·2 of span.

0·2×3·34×1000=668 mm, say 670 mm

0·2×3·02×1000=614 mm, say 615 mm

End span at support $=\frac{1}{7}\times 3{\cdot}07\times 1$ m=438·6, say 440 mm.

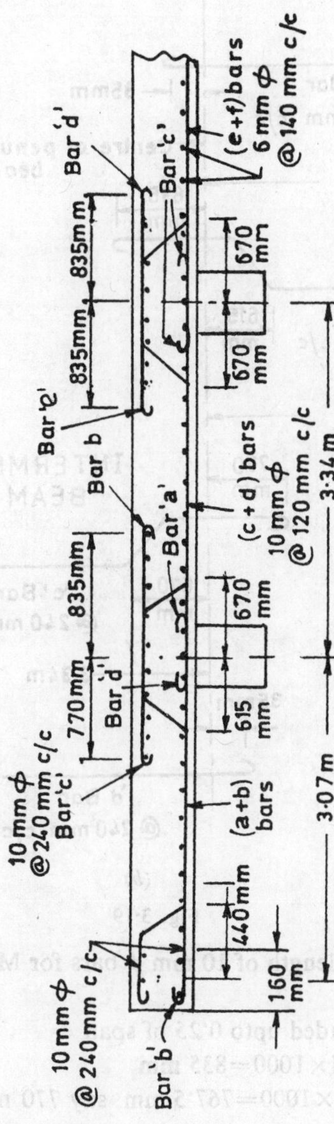

10 mm ϕ
@ 240 mm c/c
Bar 'b'
10 mm ϕ
@ 240 mm c/c
Bar 'c'
160 mm
440 mm
(a+b) bars
3·07 m
770 mm
Bar 'd'
615 mm
670 mm
835 mm
Bar 'a'
Bar b
(c+d) bars
10 mm ϕ
@ 120 mm c/c
3·34 m
Bar 'e'
835 mm
670 mm
835 mm
670 mm
Bar 'c'
Bar 'd'
(e+f) bars
6 mm ϕ
@ 140 mm c/c

Fig. 3·19 (a)

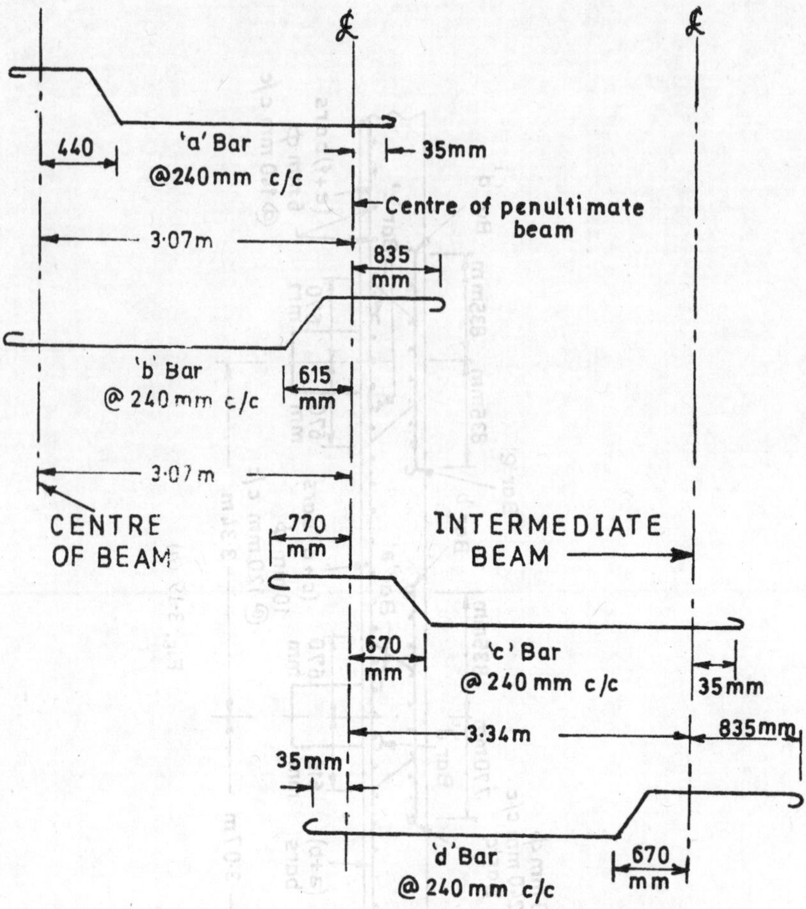

(*b*)

Fig. 3·19

Development length of 10 mm ϕ bars for M 15 (see table 4·8) is 583 mm.

Bars are extended upto 0·25 of span

$0{\cdot}25 \times 3{\cdot}34 \times 1000 = 835$ mm

$0{\cdot}25 \times 3{\cdot}07 \times 1000 = 767{\cdot}5$ mm, say 770 mm.

Bent-up bars provided at supports will be extended by 770 mm on end slab portion and 835 mm on intermediate spans. This will provide sufficient development length. The arrangement of reinforcement is shown in Fig. 3·19

PROBLEMS

1. Design a reinforced concrete slab to carry superimposed load of 4000 N/m² on an effective span of 3 m simply supported.

2. A reinforced concrete floor consists of slab 150 mm thick with bars 12 mm ϕ at 100 mm c/c, the centres of the bars being 20 mm from the underside of the slab. The span is 4 m and slab is simply supported. Find the uniformly distributed superimposed load that the slab can safely carry $c=5$ N/mm², $t=140$ N/mm², $m=18$.

3. Fig. 3·20 shows the cross-section of a roof slab. The supports being 4 m apart. Design the slab.

Live load =1500 N/m²

Floor finish =500 N/m².

Fig. 3·20

4. A 140 mm RCC slab has to carry load of 7000 N/m² inclusive of its own weight on an effective span of 4 m simply supported. The maximum stresses do not exceed $c=5$ N/mm² $t=140$ N/mm², $m=18$. Design the slab.

5. Design an RCC slab to carry load of 5000 N/m², over an effective span of 4 m, simply supported.

If the above slab has to carry an additional load of 1000 N/m², design the slab providing same depth, $c=5$ N/mm², $t=100$ N/mm², $m=18$.

4

Shear, Bond, Development Length and Anchorage

4·1. Shear in Homogeneous Beams

Shear in a beam is caused by the variation of bending moment along the span. A beam when loaded has different stresses at adjacent sections in the same fibres because of different bending moments. Fig. 4·1 shows portion of rectangular beam between two cross-sections XX and X_1X_1. The variation of stress intensity on

(a)

(b) Stress diagram. (c) Variation of shear stress

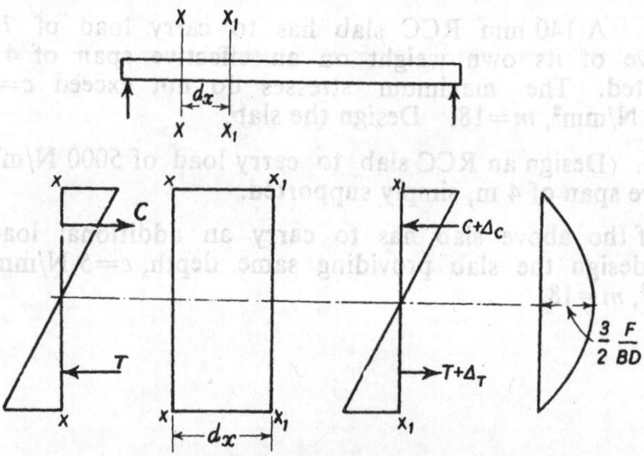

Fig. 4·1

each face of the portion separated is shown in Fig. 4·1 (b). The stresses on face X_1X_1 are shown greater as the B.M. at X_1X_1 is more. Because of unequal stresses at two cross-sections in the same fibre, there is tendency in each fibre to slide over adjacent fibre, which is prevented by horizontal shear stress. These horizontal shear stresses give rise to vertical stresses.

Shear stresses at any section is given by

$$q=\frac{FA\bar{y}}{bI}$$

where, F is S.F.

$A\bar{y}$ is moment of area above the section about N.A.

b is breadth of section.

I is moment of inertia.

The variation of shear stress across the cross-section is shown in Fig. 4·1 (*c*). The variation is parabolic with maximum ordinate at N.A., one and half times the value of average shear stress.

4·2. Shear in Reinforced Concrete Rectangular Beams

Consider short length of a reinforced concrete beam between two sections XX and X_1X_1 at distance dx apart as shown in Fig. 4·2 (*a*).

The forces acting on the element, consist of compression C and $C+\Delta C$, tension T and $T+\Delta T$, acting at their respective centriods, and total vertical shear F. It is assumed that cross-section X_1X_1 is subjected to greater B.M. than cross-section XX. It is further assumed that the section chosen are so close that the shear

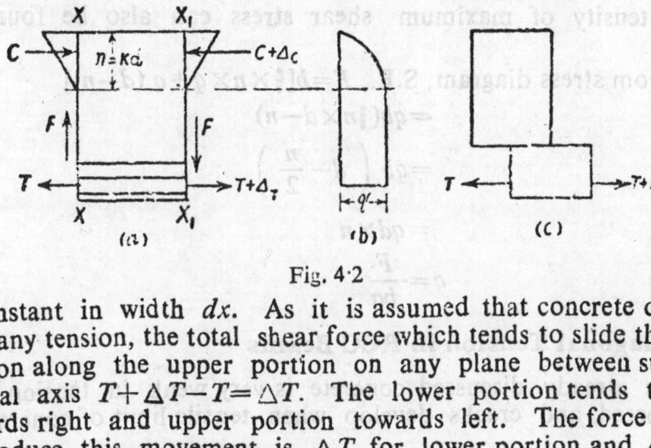

Fig. 4·2

is constant in width dx. As it is assumed that concrete does not take any tension, the total shear force which tends to slide the lower portion along the upper portion on any plane between steel and neutral axis $T+\Delta T-T=\Delta T$. The lower portion tends to slide towards right and upper portion towards left. The force tending to produce this movement is ΔT for lower portion and ΔC for upper portion and ΔC will be equal to ΔT. The shearing strength of horizontal plane should be sufficient to prevent sliding.

If q is the intensity of horizontal shear stress on such a plane,

$$q \times b \times dx = \Delta T$$

$$q = \frac{\Delta T}{b \times dx} \qquad \ldots(a)$$

The forces acting on the small length dx must be in equilibrium.

$$\therefore \quad \Sigma M = 0$$

$$\therefore \quad (T+\Delta T)\left(d - \frac{k'd}{3}\right) = T\left(d - \frac{k'd}{3}\right) + F \times dx$$

$$\therefore \quad \Delta T \times a\left(1-\frac{k'}{3}\right)=F \times dx$$

$$\Delta T=\frac{F \times dx}{a}$$

$$\left[\text{Lever arm}=a=d\left(1-\frac{k'}{3}\right)\right]$$

Substituting in (*a*)

$$q=\frac{F}{ab}.$$

This represents the value of shear stress along any plane between the steel and neutral axis. For the portion of section above neutral axis the distribution of the shear stress will be as for a homogeneous section. The distribution of shear stress will be parabolic.

Shear stress will be given by $\frac{FA\bar{y}}{bI}$

Intensity of maximum shear stress can also be found as follows :

From stress diagram, S.F., $F=b[\frac{2}{3}\times n\times q+q\,(d-n)]$

$$=qb(\tfrac{2}{3}n\times d-n)$$

$$=qb\left(d-\frac{n}{2}\right)$$

$$=qd\times n$$

$$q=\frac{F}{ba}$$

4·3. Diagonal Tension in RCC Beams

As already discussed concrete is very weak in tension when unreinforced, and cracks develop when tensile limit of concrete is

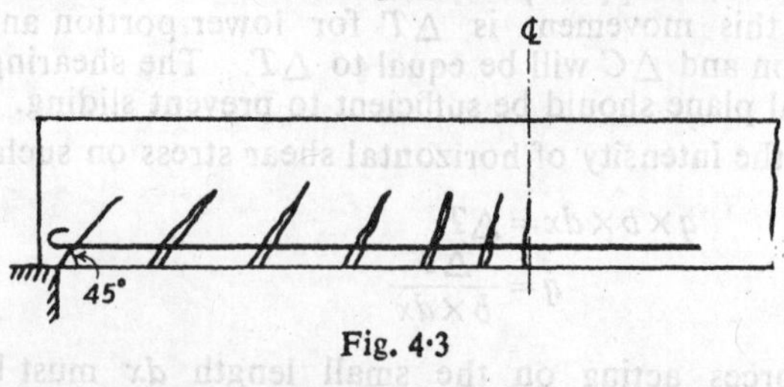

Fig. 4·3

exceeded. The pattern of cracking is shown in Fig. 4·3. In order to avoid failure, it is necessary to reinforce section. The purpose of providing reinforcement is not to prevent cracking as that is not possible, but to prevent widening of cracks and thus to avoid failure.

Consider a small cube removed from a beam parallel to the length of the beam and below N.A. It is subjected to shear stress of intensity q parallel to four sides and tensile stress f_x in horizontal directions as shown in Fig. 4·4.

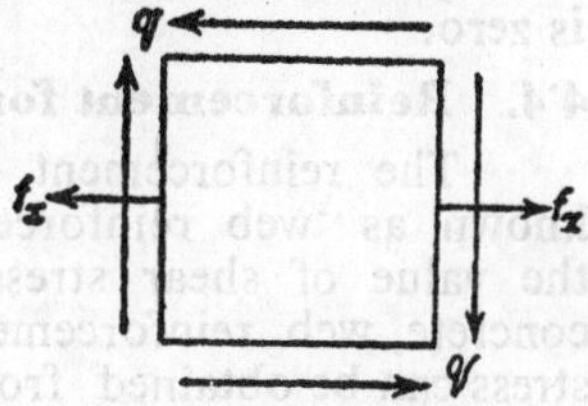

Fig. 4·4

Principal stresses on the cube are given by,

$$p=\frac{f_x}{2}\pm\sqrt{\left(\frac{f_x}{2}\right)^2+q}$$

and inclination of principal planes is given by

$$\tan 2\theta=\frac{-2q}{f_x}.$$

The maximum principal tensile stress will be

$$\frac{f_x}{2}+\sqrt{\left(\frac{f_x}{2}\right)^2+q^2}$$

and other principal stress will be compressive.

At a section, where B.M. is zero, $f_x=0$ and principal tension will be q.

$$\tan 2\theta=-x \quad \text{or } \theta=45^\circ \quad \text{or} \quad 135^\circ.$$

Principal tension acts along one diagonal of the cube. The principal tension acting along a diagonal is called diagonal tension.

At a section, where shear $q=0$, principal tension will be f_x, *i.e* tensile stress.

$$\tan 2\theta=0 \quad \therefore \quad \theta=0.$$

Principal tension has value of f_x and acts in a horizontal direction.

From the above discussion it is evident that at the mid-span of the beam where bending moment is maximum and shear is least the direction of diagonal tension will be horizontal and cracks will be vertical. As the supports are approached the S.F. increases and B.M. reduces for simply supported beam. The shear stress will increase and bending stress will reduce. Tan $2\theta=-2q/f_x$ numerically increases and at supports, where $f_x=0$, θ will be 45°. The cracks will start as 45° as shown in Fig. 4·3.

To avoid failure of beam against diagonal tension it is reinforced with requisite amount of steel. The reinforcement of diagonal tension is provided in two forms—vertical stirrups looped about the main reinforcement and bent up bars inclined at an angle across the region of diagonal tension. In the design of diagonal tension reinforcement assumption is made that diagonal tension is inclined at 45° and is equal to shear at that section. The assumption holds good only at a cross-section, where normal tension is zero.

4·4. Reinforcement for Diagonal Tension

The reinforcement provided to take up diagonal tension is known as 'web reinforcement or shear reinforcement. Whenever the value of shear stress exceeds the permissible shear stress of concrete, web reinforcement is to be provided. Safe value of shear stress can be obtained from Table 4·1. The web reinforcement is to be provided for entire amount of shear force. The section is to be redesigned if the shear stress exceeds four times the allowable shear stress.

Fig. 4·5 shows the section of an RCC beam near support, where shear is maximum and the diagonal tension stresses are inclined at 45° to horizontal.

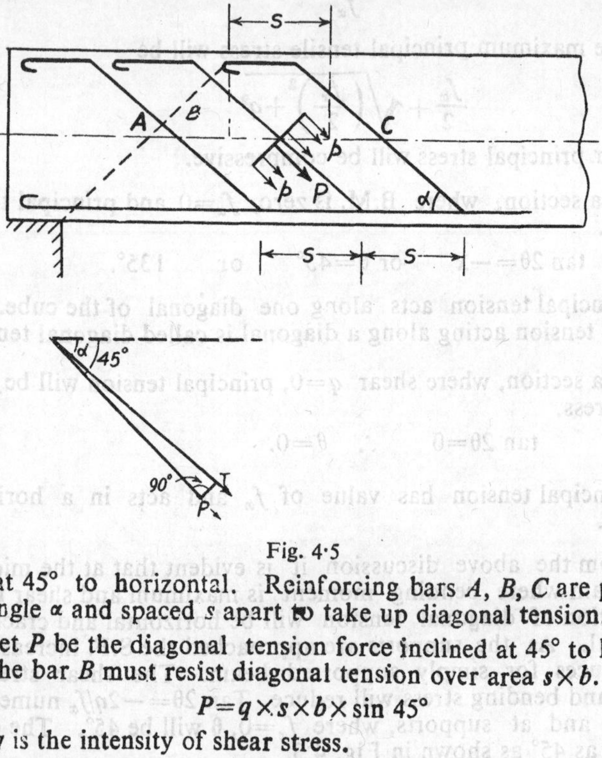

Fig. 4·5

clined at 45° to horizontal. Reinforcing bars *A*, *B*, *C* are provided at an angle α and spaced s apart to take up diagonal tension.

Let P be the diagonal tension force inclined at 45° to horizontal. The bar B must resist diagonal tension over area $s\times b$.

$$P=q\times s\times b\times \sin 45° \qquad \ldots(a)$$

where q is the intensity of shear stress.

Bar is inclined as an angle α', so the force in bar will be

$$T=\frac{P}{\cos(45°-\alpha)} \qquad \ldots(b)$$

Substituting P from (a)

$$T=\frac{q\times s\times b\times \sin 45°}{\cos(45°-\alpha)}$$

$$=\frac{qsb \sin 45^\circ}{\cos 45^\circ \cos \alpha+\sin 45^\circ \sin \alpha}$$

$$=\frac{qsb}{\cos \alpha+\sin \alpha} \quad ...(c)$$

But, $$q=\frac{F}{b\times a}$$

where F in S.F. at the section, b is breadth of section and a is a lever arm.

Substituting in (c),

$$T=\frac{F\times s}{a\,(\cos \alpha+\sin \alpha)} \quad ...(d)$$

Let A_s be the area of bent up bars and t be the allowable stress in the bar.

$$T=A_s\times t.$$

Substituting in (d),

$$A_s\times t=\frac{F\times s}{a\,(\cos \alpha+\sin \alpha)}$$

$$s=\frac{A_s.t_a}{F}(\cos \alpha+\sin \alpha)$$

If the bars are bent up at 45°

$$s=\frac{A_s.t_a}{F}\times\sqrt{2} \text{ or } \frac{A_s t\sqrt{2}}{qb}$$

If the vertical stirrups are provided, $\alpha=90°$

Pitch, $$s=\frac{A_s.t_a}{F} \text{ or } \frac{A_s t}{qb}.$$

Vertical stirrups with two legs are provided and their ends are properly anchored in compression zone by providing hooks.

4·5. Inclined Bars as Shear Reinforcement

When inclined bars are used as shear reinforcement, they usually consist of main tension steel bent at 45° at place where it is not required to resist any stresses due to bending moment. This is usually done near the supports.

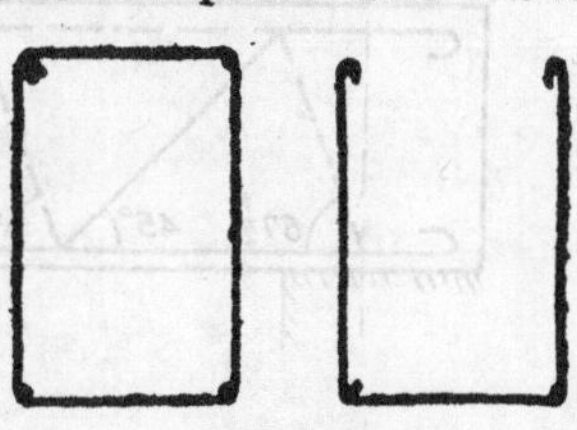

Fig. 4·6

The use of inclined bars as shear reinforcement to form stirrup system at pitch s discussed in last article is not always practicable. It is only in large beams *e.g.* bridge girders having considerable amount of tension reinforcement that sufficient bars can be bent up those enough to form a stirrup system. Generally a few bars are bent up near the support. The bars are generally bent up in pairs.

The action of one pair of bars bent up to prevent cracking is shown in Fig. 4·7. Vertical component of the force in the bar resists shear force.

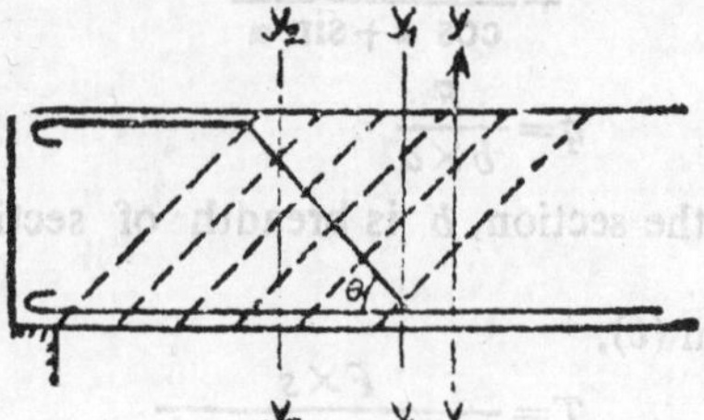

Fig. 4·7

Total force in inclined bars

$$= t \times A_s, \text{ where } A_s \text{ is area of bars bent up}$$

Vertical component of this force

$$= t \times A_s \times \sin \theta.$$

If the shear force F is less than $tA_s \sin \theta$, no shear reinforcement is required. If F is more than $tA_s \sin \theta$, vertical stirrups are to be provided for balance of shear $(F - tA_s \sin \theta)$.

This bar is effective only up to section Y_1Y_1. Beyond Y_1Y_1 if shear intensity is more than allowable value, reinforcement is to be provided.

In case number of bars can be bent up as shown in Fig. 4·7 then the resistance of shear can be found by assuming that the inclined bars from the web system of a lattice girder with line through centre of gravity of compression zone as axis of compression flange and horizontal tension bar as tension flange. Other web system of lattice girder assumed to be produced by concrete inclined at $67\frac{1}{2}°$ as shown in Fig. 4·8.

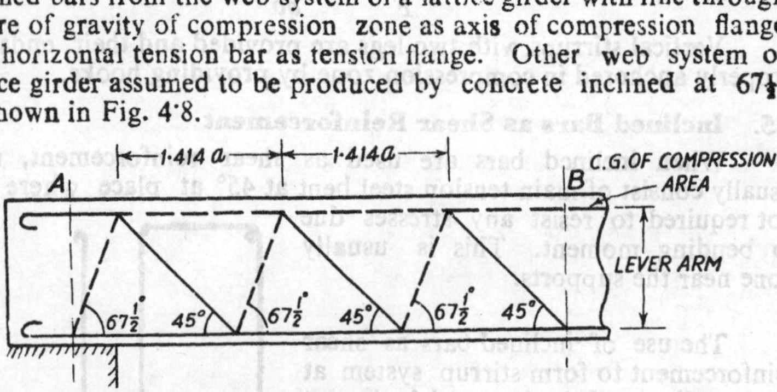

Fig. 4·8

Shear resistance at any section of beam between A and B is given by $A_s \times t \times \sin 45°$, where A_s is area of pair of bars bent up together.

If shear force F is greater than this resistance, stirrups are to be provided for balance of shear.

Beyond B if shear stress is greater than that permissible in concrete, stirrups are to be provided for this shear.

If the bars are bent up closer say at $0.707a$ as shown in Fig. 4·9, resistance to shear is to be taken twice as for any shear crack two sets of bars are effective.

Shear resistance $= 2A_s \times t \times \sin 45°$.

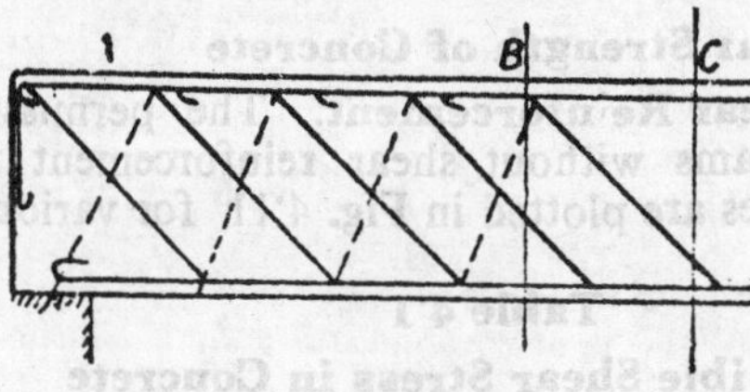

Fig. 4·9

Between B and C only one set of bars is effective and resistance to shear $= A_s \times t \times \sin 45°$.

If the bars are bent at distance $2a$ apart, the force in

bar $BC = \dfrac{t \times A_s \times \sin 45°}{\sin 90°}$

$= t \times A_s \times 0.707$

where t is stress in bar AB.

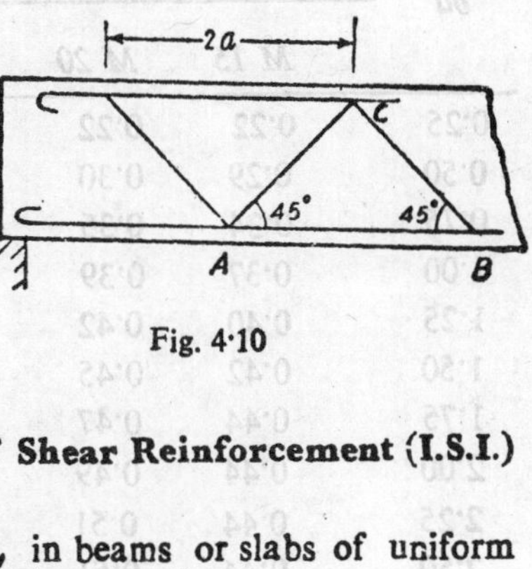

Fig. 4·10

Shear taken

$= t \times A_s \times 0.707 \times \sin 45°$

$= 0.5\, tA_s$.

4·6. General Rules on Design of Shear Reinforcement (I.S.I.)

(i) Nominal Shear Stress

(a) The nominal shear stress τ_v in beams or slabs of uniform depth shall be calculated by the following equations :

$$\tau_v = \frac{F}{bd}$$

where F = shear force due to design loads.

b = breadth of the member, which for flanged sections shall be taken as the breadth of the web, b_w, and

d = effective depth.

(b) **Beams of varying depth.** In the case of beams of varying depth, the equation shall be modified as :

$$\tau_v = \frac{F \pm \dfrac{M}{d} \tan \beta}{bd}$$

where M=bending moment at the section

β=angle between the top and the bottom edges of the beam.

The negative sign in the formula applies when the bending moment M increases numerically in the same direction as the effective depth d increases, and the positive sign when the moment decreases numerically in this direction.

(*ii*) **Design Shear Strength of Concrete**

(*a*) **Without Shear Reinforcement.** The permissible shear stress in concrete in beams without shear reinforcement is given in table 4 1. These values are plotted in Fig. 4·11 for various concrete mixes.

Table 4·1

Permissible Shear Stress in Concrete

$\frac{100\,A_s}{bd}$	*Permissible Shear Stress in Concrete* τ_c *N/mm*^{2} *for Grade of Concrete*					
	M 15	*M 20*	*M 25*	*M 30*	*M 35*	*M 40*
0·25	0·22	0·22	0·23	0·23	0·23	0·23
0·50	0·29	0·30	0·31	0·31	0·31	0·32
0·75	0·34	0·35	0·36	0·37	0·37	0·38
1·00	0·37	0·39	0·40	0·41	0·42	0·42
1·25	0·40	0·42	0·44	0·45	0·45	0·46
1·50	0·42	0·45	0·46	0·48	0·49	0·49
1·75	0·44	0·47	0·49	0·50	0·52	0·52
2·00	0·44	0·49	0·51	0·53	0·54	0·55
2·25	0·44	0·51	0·53	0·55	0·56	0 57
2·50	0·44	0·51	0·55	0·57	0·58	0·60
2·75	0·44	0·51	0·56	0·58	0·60	0·62
3·00 and above	0·44	0·51	0·57	0·60	0·62	0·63

Note. A_s is the area of longitudinal tension reinforcement which continues at least one effective depth beyond the section being considered except at supports where the full area of tension reinforcement may be used, provided the detailing conforms to proper anchoring.

(*iii*) For solid slabs the permissible shear stress in concrete shall be $k\tau_c$, where k has the value given in table 4·2.

Table 4·2
Values of k to be Used for Solid Slabs

Overall depth of slab in mm	300 or more	275	250	225	200	1	150 or less
k	1·00	1·05	1·10	1·15	1·20	1·25	1·30

Note. This does not apply to flat slabs.

(*iv*) **Shear Strength of Members under Axial Compression.** For members subjected to axial compression P, the permissible shear stress in concrete τ_c given in table 4·1, shall be multiplied by $\delta=\frac{5P}{A_g f_{ck}}$, but not exceeding 1·5, where P=axial compressive force in newtons, A_g=gross area of the concrete section in mm² and f_{ck}=characteristic compressive strength of concrete.

(*v*) **When Shear Reinforcement** is provided the nominal shear stress τ_v in beams shall not exceed $\tau_{c\,max}$ given in table 4·3.

For slabs, τ_v shall not exceed half the value of $\tau_{c\,max}$ given in table 4·3.

Table 4·3
Maximum Shear Stress $\tau_{c\,max}$ N/mm²

Concrete Grade	M 15	M 20	M 25	M 30	M 35	M 40
$\tau_{c\,max}$ *N/mm²*	1·6	1·8	1·9	2·2	2·2	2·5

When τ_v is less than τ_c given in table 4·1, minimum shear reinforcement according to formula :

$$\frac{A_{sv}}{b\times s_v} > \frac{0{\cdot}4}{f_y}$$

where, A_{sv}=total cross-sectional area of stirrup legs effective in shear

s_v=stirrup spacing along the length of the members

b=breath of the beam or breadth of the web of flanged beam

f_y=characteristic strength of the stirrup reinforcement in N/mm² which shall not be taken greater than 415 N/mm².

However in members of minor structural importance such as lintels or where the maximum shear stress calculated is less than half the permissible value, this provision need not be complied with.

The maximum spacing of shear reinforcement measured along the axis of the member shall not exceed 0·75*d* for vertical stirrups and *d* for inclined stirrups at 45°, where *d* is the effective depth of the section under consideration. In no case shall the spacing exceed 450 mm.

(*vi*) **Design of Shear Reinforcement.** When τ_v exceeds τ_c given in table 4·1, shear reinforcement shall be provided in any of the following forms :

(*a*) Vertical stirrups

(*b*) Bent-up bars, along with stirrups

(*c*) Inclined stirrups.

Where bent-up bars are provided their contribution towards shear resistance shall not be more than half that of the total shear reinforcement.

Shear reinforcement shall be provided to carry a shear equal to $F-\tau_c bd$. The strength of shear reinforcement V_s shall be calculated as given below :

(*a*) For vertical stirrups $V_s=\dfrac{\sigma_{sv}A_{sv}d}{s_v}$

(*b*) For inclined stirrups or a series of bars bent-up at different cross-sections, $V_s=\dfrac{\sigma_{sv}A_{sv}}{s_v}(\sin\alpha+\cos\alpha)$.

(*c*) For single bar or single group of parallel bars, all bent-up at the same cross-section, $V_s=\alpha_{sv}A_{sv}\sin\alpha$, where

A_{sv}=total cross-sectional area of stirrups legs or bent-up bars within a distance of s_v.

s_v=spacing of the stirrups or bent-up bars along the length of the member.

τ_c=design shear strength of the concrete.

b=breadth of the member which for flanged beams shall be taken as the breadth of the web b_w.

σ_{sv}=permissible tensile stress in shear reinforcement which shall not be taken greater than 230 N/mm².

α=angle between the inclined stirrup or bent-up bar and the axis of the member, not less than 45°.

d=effective depth.

Note. Where more than one type of shear reinforcement is used to reinforce the same portion of the beam, the total shear resistance shall be computed as the sum of the resistance for the various types separately. The area of the stirrups shall not be less than the minimum.

The values of vertical shear $\frac{V_s}{d}$ for various stirrup diameters and spacings are given in table 4·4. The shear taken by bent bars of various diameters is shown in Table 4·5.

Ex. 4·1. *A beam having 200 mm width and 450 mm effective depth supports a total uniformly distributed load of 200,000 N. Calculate the maximum shear stress and the spacing of 10 mm diameter stirrups near the supports. The concrete mix used is M 15.*

Sol. Max. S.F. $= \frac{200,000}{2} = 100,000$ N

Nominal shear stress

$$\tau_v = \frac{F}{bd} = \frac{100,000}{200 \times 450} = 1{\cdot}11 \text{ N/mm}^2$$

Area of steel for balanced section

$$= 0{\cdot}714 \text{ (percentage table 2·1)}$$

Allowable shear stress

$$\tau_c = 0{\cdot}34 \text{ N/mm}^2$$

Net shear for which stirrups are to be provided

$$= F - \tau_c bd$$
$$= 100,000 - 0{\cdot}34 \times 200 \times 450$$
$$= 1000,000 - 30,600 = 69,400 \text{ N}$$

$$V_s = \frac{\sigma_{sv} A_{sv} d}{s_v}.$$

Now $\sigma_{sv} = 140$ N/mm², Mild steel $A_{sv} = 2 \times 79$

$$\therefore \quad s_v = \frac{\sigma_{sv} A_{sv} d}{V_s} = \frac{140 \times 2 \times 79 \times 450}{69,400}$$
$$= 143{\cdot}4 \text{ mm c/c.}$$

Minimum shear reinforcement,

$$\frac{A_{sv}}{b \times s_v} \geq \frac{0{\cdot}4}{f_y}, \quad (f_y = 250 \text{ N/mm}^2)$$

$$\frac{2 \times 79}{200 \times s_v} \geq \frac{0{\cdot}4}{250} \therefore s_v \leq \frac{2 \times 79 \times 250}{200 \times 0{\cdot}4} \leq 493{\cdot}8 \text{ mm.}$$

Maximum spacing $= 0{\cdot}75\, d$ or 450 mm whichever is less.

$$0{\cdot}75d = 0{\cdot}75 \times 450 = 337{\cdot}5 \text{ mm.}$$

Use 10 mm stirrups at 140 mm centre to centre.

Ex. 4·2. *A simply supported beam 200 mm wide and 400 mm effective depth is simply supported over a span of 3 metres. It carries a uniformly distributed load, including its own weight of 30,000 N/m. The beam is reinforced with 6 bars of 16 mm φ. The concrete used is M 15. Determine the portions of the span over which it is necessary to provide shear reinforcement. If two bars can be bent up to take up* *shear determine whether it is necessary to provide stirrups and find spacing of stirrups.*

Sol. Maximum shear force

$$F=\frac{3\times 30{,}000}{2}=45{,}000 \text{ N}$$

$$\text{Shear stress } \tau_v=\frac{F}{bd}=\frac{45{,}000}{200\times 400}=0{\cdot}563 \text{ N/mm}^2$$

$$\text{Steel percentage}=\frac{6\times 201\times 100}{200\times 400}=1{\cdot}51\%$$

$$\tau_c=0{\cdot}42 \text{ N/mm}^2$$

Shear reinforcement is necessary

$$V_s=\text{Net shear}=45{,}000-0{\cdot}42\times 200\times 400=11{,}400 \text{ N}$$

Use 6 mm ϕ stirrups

$$s_v=\frac{\sigma_{sv}\times A_{sv}\times d}{V_s}=\frac{140\times 2\times 28\times 400}{11{,}400}$$

$$=275 \text{ mm.}$$

Distance

Let the distance upto which shear reinforcement is required

$$F=45{,}000-30{,}000\, x=0{\cdot}42\times 200\times 400=33{,}600$$

$$x=\frac{45{,}000-33{,}600}{30{,}000}=0{\cdot}38 \text{ metre.}$$

Shear reinforcement is required upto 0·38 metre from both ends.

Minimum shear reinforement

$$f_y=250 \text{ N/mm}^2$$

$$\frac{A_{sv}}{b\times s_v}\geqslant\frac{0{\cdot}4}{f_y}$$

$$\therefore \quad s_v\leqslant\frac{A_{sv}\times f_y}{0{\cdot}4b}=\frac{2\times 28\times 250}{0{\cdot}4\times 200}\leqslant 175 \text{ mm.}$$

Maximum spacing 0·75*d* or 450 mm whichever is less

$$0{\cdot}75d=0{\cdot}75\times 400=300 \text{ mm.}$$

Use 6 mm ϕ of stirrups at 175 mm centre to centre throughout the beam.

Two bars bent up

Shear taken by two bars $=\sigma_{sv}\times A_{sv}\sin\alpha$.

Taking $\alpha=45°$

Shear taken by two bars $=140\times 2\times 201\times\frac{1}{\sqrt{2}}=39{,}796$ N

$$\text{Steel percentage}=\frac{4\times 201\times 100}{200\times 400}=1{\cdot}005.$$

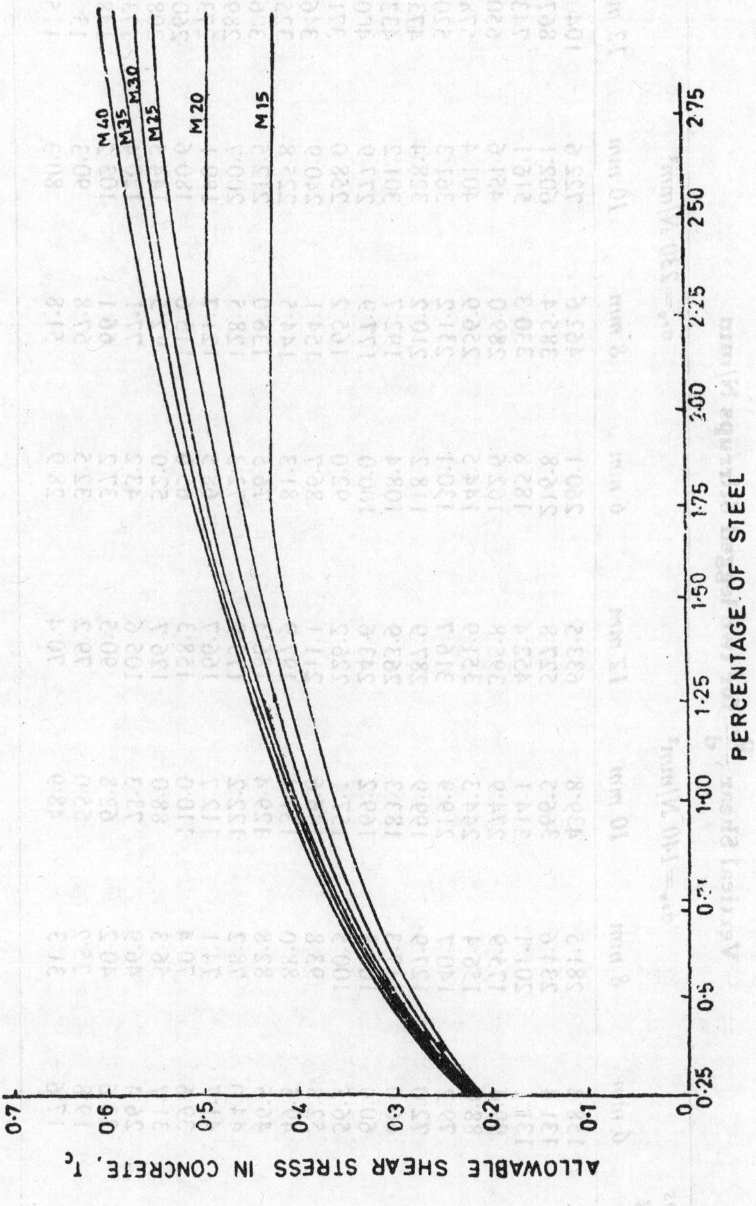
ALLOWABLE SHEAR STRESS IN CONCRETE, τ_c
0·7
0·6
0·5
0·4
0·3
0·2
0·1
0
0·25
0·5
1·00
1·25
1·50
1·75
2·00
2·25
2·50
2·75
PERCENTAGE OF STEEL
M40
M35
M30
M25
M20
M15

Fig. 4·11

Table 4·4

Vertical Shear $\frac{V_s}{d}$ for two legged Stirrups N/mm

Stirrups spacing in mm	$\sigma_{sv}=140\ N/mm^2$				$\sigma_{sv}=230\ N/mm^2$			
	6 mm	8 mm	10 mm	12 mm	6 mm	8 mm	10 mm	12 mm
50	158·3	281·5	439·8	633·5	260·1	462·6	722·6	1040·5
60	131·4	234·6	366·5	527·8	216·8	385·4	602·1	867·1
70	131·1	201·1	314·1	452·4	185·8	330·3	516·1	743·2
80	99·4	175·9	274·9	395·8	162·6	289·0	451·6	650·3
90	88·0	156·4	244·3	351·9	144·5	256·9	401·4	578·1
100	79·2	140·7	219·9	316·7	130·1	231·2	361·3	520·2
110	72·0	127·9	199·9	287·9	118·2	210·2	328·4	473·0
120	66·0	117·3	183·3	263·9	108·4	192·7	301·2	433·5
130	60·9	108·3	169·2	243·6	100·0	177·9	277·9	400·2
140	56·5	100·5	157·1	226·2	92·0	165·2	258·0	371·6
150	52·8	93·8	146·6	211·1	86·7	154·1	240·9	346·8
160	49·5	88·0	137·4	197·9	81·3	144·5	225·8	325·2
170	46·6	82·8	129·4	186·3	76·5	136·0	212·5	306·0
180	44·0	78·2	122·2	175·9	72·3	128·5	200·7	289·0
190	41·7	74·1	112·7	166·7	68·5	121·7	190·1	273·8
200	39·6	70·4	110·0	158·3	65·0	115·6	180·6	260·1
250	31·7	56·3	88·0	126·7	52·0	92·5	144·5	208·1
300	26·4	46·9	73·3	105·6	43·2	77·1	120·4	173·8
350	22·6	40·2	62·8	90·5	37·2	66·1	103·2	148·6
400	19·8	35·2	55·0	79·2	32·5	57·8	90·3	13 ·1
450	17·6	31·3	48·9	70·4	28·9	51·8	80·3	115·6

$$\tau_c = 0{\cdot}37$$

$$V_s = \text{Net shear}$$

$$= 45{,}000 - 39{,}796 - 0{\cdot}37 \times 200 \times 400$$

$$= 45{,}000 - 39{,}796 - 29{,}600 = -24{,}396.$$

Use minimum shear reinforcement 6 mm ϕ stirrups at 175 mm centres throughout the beam.

In Ex. 4·2, using Fig. 4·11, for M 15, $p = 1{\cdot}51$, $\tau_c = 0{\cdot}423$.

From table 4·4, for 6 mm ϕ stirrups, for

$$\frac{V_s}{d} = \frac{11{,}400}{400} = 28{\cdot}5$$

$$\text{Spacing of stirrups} = 250 + \frac{50 \times 3{\cdot}2}{5{\cdot}3}$$

$$= 250 + 30{\cdot}2 = 280{\cdot}2 \text{ mm.}$$

Table 4·5

Shear Bent-up Bars Values of V_s for Single Bar in N

Bar diameter mm	$\tau_{sv} = 140\ N/mm^2$ *upto 20 mm diameter* $\sigma_{sv} = 130\ N/mm^2$ *over 20 mm diameter*		$\sigma_{sv} = 230\ N/mm^2$	
	$\alpha = 45°$	$\alpha = 60$	$\alpha = 45$	$\alpha = 60$
10	7,780	9,520	12,770	15,640
12	11,200	13,710	18,390	22,530
16	19,900	24,380	32,700	40,050
18	25,190	30,860	41,390	50,690
20	31,100	38,090	51,090	62,580
22	34,940	42,800	61,820	75,720
25	45,120	55,260	79,830	97,770
28	56,600	69,320	100,140	122,650
32	73,930	90,540	130,800	160,190
36	93,750	114,600	165,540	202,750

In Ex. 2, using table 4·5,

Shear taken by 2 bars 16 mm ϕ, bent at 45°

$$= 19{,}900 \times 2 = 39{,}800 \text{ N.}$$

Allowable shear stress may be calculated from the following equation :

$$\tau_c = 0{\cdot}04\sqrt{f_{ck}} + 0{\cdot}19p$$

where f_{ck} = characteristic strength of concrete and p is the percentage of steel,

In Ex. 4·2, for M 15, $p=1{\cdot}51$, $\tau_c=0{\cdot}04\sqrt{15}+0{\cdot}19\times1{\cdot}51$
$=0{\cdot}155+0{\cdot}287=0{\cdot}442.$

Maximum value of τ_c is not to exceed $0{\cdot}114\sqrt{f_{ck}}$. For M 15. $\tau_c=0{\cdot}442$.

4·7. Bond

When a reinforcing bar is embedded in concrete, the concrete adheres to its surface and resists any force that tries to pull out or push the rod. This is called the bond between steel and concrete. The intensity of this adhesive force is called bond stress.

4·8. Development of Stress in Reinforcement and Development Length of Bars

The calculated tension or compression in any bar at any section shall be developed on each side of the section by an appropriate development length or end anchorage or by a combination thereof.

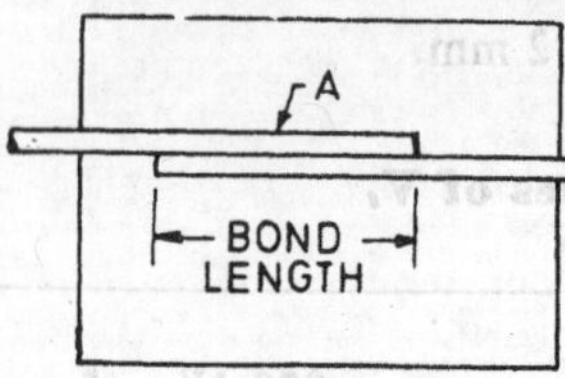

Fig. 4·12

Consider a bar A having stress σ_s. The bar has to transfer the stress to bar B.

Let l be the overlap length of the bar.

Total force in bar $A=\frac{\pi}{4}\phi^2\sigma_s$

This force is to be transferred by bond. If τ_{bd} is the design bond stress,

$$\frac{\pi}{4}\phi^2\sigma_s=\tau_{bd}\,(\pi\phi)l$$

$$l=\frac{\phi\sigma_s}{4\tau_{bd}}$$

where, ϕ = nominal diameter of the bar

σ_s = stress in bar at the section considered.

The development or anchorage length is the necessary length between the point of maximum stress is a bar and the end of the bar.

The development length L_d is given by $L_d=\frac{\phi\sigma_s}{4\tau_{bd}}$.

The development length includes anchorage values of hooks in tension reinforcement.

The bars of sections other than circular, the development length should be sufficient to develop the stress in the bar by bond. Permissible stress in bond (average) for plain bars in tension is given in table 4·6.

Table 4·6

Permissible Stress in Bond τ_{bd}

Grade of concrete	M15	M20	M25	M30	M35	M40
τ_{bd} *in N/mm²*	0·6	0·8	0·9	1·0	1·1	1·2

1. The values of bond stress shall be increased by 25 per cent for bars in compression.

2. In the case of deformed bars, the bond stress as given in table 4·6 may be increased by 40 per cent.

4·9. Bundled Bars in Contact

The development length of each bar of bundled bars shall be that for the individual bar, increased by 10 per cent for two bars in contact, 20 per cent for three bars in contact and 33 per cent for four bars in contact.

4·10. Anchoring Reinforcing bars

(*i*) **Anchoring bars in tension.** Deformed bars may be used without end anchorages provided development length is satisfied. Hooks as shown in Fig. 4·13 should normally be provided for plain bars in tension. The anchorage value of a standard U type hook shall be equal to 16 times the diameter of the bar.

The anchorage value of bend shall be taken as 4 times the diameter of the bar for each 45° bend subject to a maximum of 16 times the diameter of the bar.

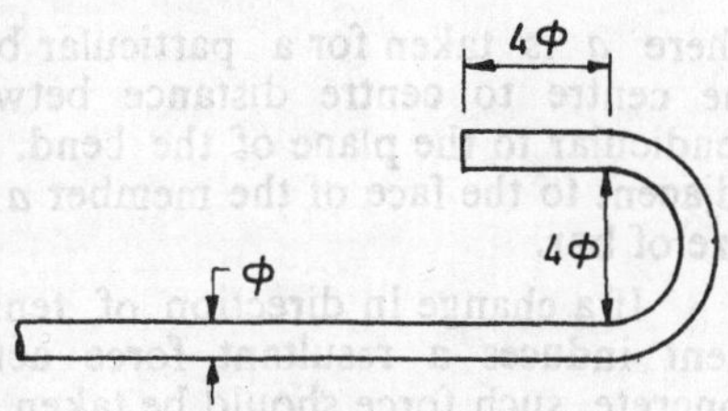

Fig. 4·13. U-type Hook.

(*ii*) **Anchoring bars in compression.** The anchorage length of straight bar in compression shall be equal to the development length of bars in compression. The projected length of hooks, bends and straight lengths beyond bends if provided for a bar in compression, shall be considered for development length.

(*iii*) **Mechanical devices for anchorage.** Any mechanical or other device capable of developing the strength of the bar without damage of concrete may be used as anchorage with the approval of the engineer-in-charge.

(*iv*) **Anchoring shearing reinforcement**

(*a*) *Inclined bars.* The development length sha'l be as for bars in tension. In tension zone the length shall be measured from the end of the sloping or inclined portion of the bar and in the compression zone, from the mid-depth of the beam.

(*b*) *Stirrups.* In case of secondary reinforcement such as stirrups and transverse ties, complete development lengths and anchorage shall be deemed to have been provided when the bar is bent through an angle of atleast 90° round a bar of atleast its own diameter and is continued beyond the end of the curve for a length of atleast eight diameters or when the bar is bent through an angle of 135° and continued beyond the end of the curve for a length of atleast six bar diameters or when the bar is bent through an angle 180° and is continued beyond the end of the curve for a length of atleast four bar diameters.

(*v*) **Bearing stresses at bends.** The bearing stress in concrete for standard bends and hooks need not be checked. In any other bend the bearing stress$=\frac{F_{bt}}{r\phi}$ where

F_{bt}=tensile force due to design loads or a bar in group of bars

r=internal radius of the bend and

ϕ=size of the bar or in bundle the size of bar of equivalent area.

For working stress the bearing stress shall not exceed $\frac{f_{ck}}{1+2d/a}$ where a is taken for a particular bar or group of bars in contact as the centre to centre distance between bars or group of bars perpendicular to the plane of the bend. For a bar or group of bars, adjacent to the face of the member a shall be taken as the cover plus size of bar.

If a change in direction of tension or compression reinforcement induces a resultant force acting outward tending to split the concrete, such force should be taken up by additional links or stirrups. Bent tension bar at a re-entrant angle should be avoided.

Anchorage value of hooks and bends, is given in Table 4·7 in cm.

Development lengths for various reinforcements are given in Tables 4·8, 4·9 and 4·10.

Table 4·7

Anchorage Value of Hooks and Bends in cm

Bar diameter mm	6	8	10	12	16	18	20	22	25	28	32	36
Anchorage value of hook	9·6	12·8	16·0	19 2	25 6	28·8	32·0	35·2	40·0	44·8	51·2	57 6
Anchorage value of 90° bend	4·8	6·4	8·0	9 6	12·8	14·4	16·0	17·6	20·0	22·4	25·6	78·8

Mild steel, $k=2$
Cold worked steel $k=4$

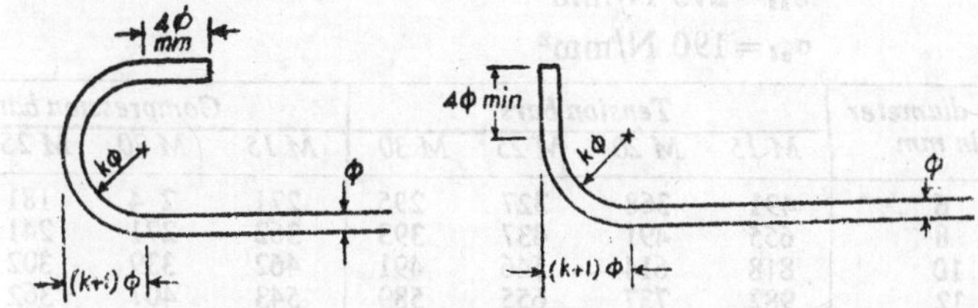

Fig. 4·14

Table 4·8

Development length for plain bars in mm.

$\sigma_{st}=140$ N/mm² for bars upto 20 mm diameter
$\sigma_{st}=130$ N/mm² for bars over 20 mm diameter
$\sigma_{sc}=130$ N/mm² for all bars.

Bar diameter in mm	*Tension bars*				*Compression bars*			
	M 15	*M 20*	*M 25*	*M 30*	*M 15*	*M 20*	*M 25*	*M 30*
6	350	263	233	210	260	195	173	156
8	467	350	311	280	347	260	231	208
10	583	438	389	350	433	325	289	260
12	700	525	467	420	520	390	347	312
16	933	700	622	560	693	520	462	416
18	1050	788	700	630	780	585	520	468
20	1167	875	778	700	867	650	578	520
22	1192	894	794	715	953	715	636	572
25	1354	1016	903	813	1083	813	722	650
28	1517	1138	1011	910	1213	910	809	728
32	1733	1300	1156	1040	1387	1040	924	832
36	1950	1463	1300	1170	1560	1170	1040	936

Table 4·9

Development length for deformed bars in mm.

$\sigma_{st}=230$ N/mm²
$\sigma_{st}=190$ N/mm²

Bar diameter in mm	*Tension bars*				*Compression bars*			
	M 15	*M 20*	*M 25*	*M 30*	*M 15*	*M 20*	*M 25*	*M 30*
6	411	308	274	246	271	204	181	163
8	548	411	365	329	362	271	241	217
10	685	513	456	411	452	339	302	271
12	821	616	548	493	543	407	362	326
16	1095	821	730	657	724	543	483	434
18	1232	924	821	739	814	611	543	489
20	1369	1027	913	821	905	679	603	543
22	1506	1129	1004	904	995	746	663	597
25	1711	1283	1141	1027	1131	848	754	679
28	1917	1438	1278	1150	1267	950	844	760
32	2190	1643	1460	1314	1448	1086	965	869
36	2464	1848	1643	1479	1629	1221	1086	977

Table 4·10

Development length for deformed bars in mm

$\sigma_{st}=275$ N/mm²

$\sigma_{st}=190$ N/mm²

Bar diameter in mm	*Tension bars*				*Compression bars*			
	M 15	*M 20*	*M 25*	*M 30*	*M 15*	*M 20*	*M 25*	*M 30*
6	491	368	327	295	271	2 4	181	163
8	655	491	437	393	362	271	241	217
10	818	614	546	491	462	339	302	271
12	982	737	655	589	543	407	362	326
16	1310	982	873	786	724	543	483	434
18	1473	1105	982	884	814	611	5-3	489
20	1637	1228	1091	982	905	679	603	543
22	1801	1350	1200	1080	995	746	663	597
25	2046	1535	1364	1228	1131	848	754	679
28	2292	1719	1528	1375	1267	950	844	760
32	2619	1964	1746	1571	1448	1086	965	869
36	2946	2210	1965	1768	1629	1221	1600	977

Ex. 4·3. *A hook 10 mm diameter is embedded in concrete for a distance of 80 mm as shown in Fig. 4·15. Calculate the maximum load that can be suspended from the hook if the bond stress is not to exceed 0·6 N/mm².*

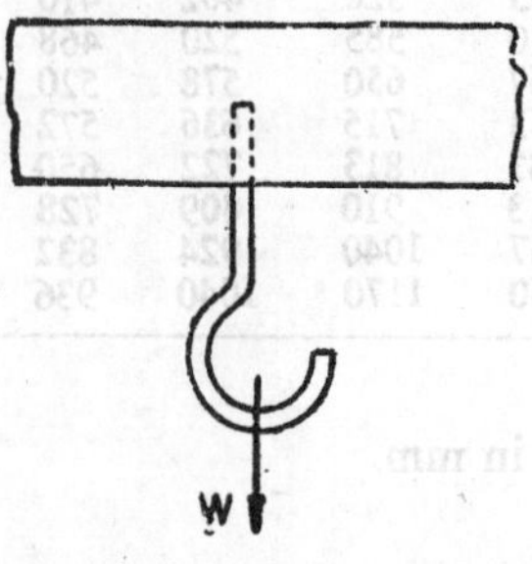

Fig. 4·15

Sol. Total force that can be resisted

$$=\pi dl\times\tau_{bd}$$

$$=\pi\times10\times80\times0{\cdot}6$$

$$=481\pi\ \text{N}=1517{\cdot}5\ \text{N}.$$

Load that can be suspended

$$=1507{\cdot}5\ \text{N}.$$

Ex. 4·4. *A reinforced concrete beam having 4·0 metres span is loaded with a uniformly distributed load. The beam is reinforced with 7 bars of 16 mm ϕ at the mid-span. Find the distance from the centre of the beam where the first bar can be curtailed as regards the tensile stresses. Also calculate the actual point of curtailment so that the bar is able to resist its full shear of tensile force at the mid span. Allowable bond stress = 0·6 N/mm². The steel is stressed to its full limit of 140 N/mm² at the mid-span.*

Sol. Area of steel at mid-span$=\dfrac{M}{t\times a}$. Therefore, area of steel A_t at a section is proportional to bending moment.

Let M be the B M. at centre of span, where number of bars is n. At a section x distance from mid-span let B.M. be reduced to M_x and number of bars curtailed be n_1.

$$\therefore \quad \frac{n-n_1}{n}=\frac{M_x}{M}$$

$$\therefore \quad \frac{n_1}{n}=\frac{M-M_x}{M}$$

B.M. at centre $M=\frac{wl^2}{8}$

B.M at distance x from centre $=\frac{wl^2}{8}-\frac{wx^2}{2}=M_x$

$$\therefore \quad M-M_x=\frac{wx^2}{2}$$

$$\therefore \quad \frac{n_1}{n}=\frac{wx^2/2}{wl^2/8}=\frac{4x^2}{l^2}$$

$$\therefore \quad x=\frac{1}{2}\sqrt{\frac{n_1}{n}}$$

If, $n_1=1$ and $n=7$, $x=\frac{4}{2}\sqrt{\frac{1}{7}}=0{\cdot}756$ m.

First bar can be bent at 0·756 m from centre of span as regards tensile stresses.

If the bar is to develop full tensile stress, its embedded length should be sufficient to provide bond strength to avoid slipping. Let l be length required to develop full stress,

$$\frac{\pi}{4}d^2\times 140=\pi dl\times \tau_{bd}$$

$$\therefore \quad l=\frac{140\times 16}{4\times 6}=933{\cdot}3 \text{ mm}=0{\cdot}9333 \text{ m}$$

Thus length is more than 0·756 m.

If the first bar is curtailed at 0·9333 m on either side of centre of beam, it will be able to develop full tensile stress.

Ex. 4·5. *In Ex. 4·1 if deformed bars with $\sigma_{st}=230$ N/mm² are used, find the pitch of 10 mm diameter stirrups.*

Allowable stress $\sigma_{sv}=230$ N/mm²

Max. shear $=\frac{200{,}000}{2}=100{,}000$ N.

Balanced steel $=0{\cdot}325$

Allowable shear stress $\tau_c=0{\cdot}29-(0{\cdot}07)\times\frac{0{\cdot}175}{0{\cdot}25}=0{\cdot}24$

Net shear force $V_s = 100{,}000 - 0{\cdot}24 \times 200 \times 450 = 78{,}400$

$$s_v = \frac{\sigma_{sv} A_{sv} d}{V_s} = \frac{230 \times 2 \times 79 \times 450}{78{,}400} = 208{\cdot}5 \text{ mm}$$

Minimum shear reinforcement,

$$\frac{A_{sv}}{b \times s_v} > \frac{0{\cdot}4}{f_y}$$

$f_y = 415$ N/mm²,

$$\therefore \qquad s_v < \frac{A_{sv} f_y}{0{\cdot}4 b} > \frac{2 \times 79 \times 415}{0{\cdot}4 \times 200} < 819{\cdot}6$$

Use spacing of 205 mm centre to centre.

Ex. 4·6. *In Ex. 4·2, if deformed bars with $\sigma_{st} = 230$ N/mm² are used, determine whether it is necessary to provide stirrups and find spacing of stirrups.*

$$\text{Max. S.F.} = \frac{3 \times 30{,}000}{2} = 45{,}000 \text{ N}$$

From Ex. 4·2, $\tau_c = 0{\cdot}42$ N/mm².

Net shear V_s for which reinforcement is required

$$= 45{,}000 - 0{\cdot}42 \times 200 \times 400 = 11{,}400 \text{ N.}$$

Use 6 mm ϕ stirrups,

$$s_v = \frac{\sigma_{sv} \times A_{sv} \times d}{V_s} = \frac{230 \times 2 \times 28 \times 200}{11{,}400} = 451{\cdot}9 \text{ mm}$$

Minimum shear reinforcement,

$f_y = 415$ N/mm²

$$s_v < \frac{A_{sv} \times f_y}{0{\cdot}4 b} = \frac{2 \times 28 \times 415}{0{\cdot}4 \times 200} < 290{\cdot}5$$

Maximum spacing 0·75 d or 450 mm whichever is less

$0{\cdot}75\, d = 0{\cdot}75 \times 400 = 300$ mm.

Use 6 mm φ stirrups at 290 mm centre to centre throughout the beam.

Ex. 4·7. *A reinforced concrete beam having 4·0 m span is loaded with a uniformly distributed load. The beam is reinforced with 7 bars of 16 mm ϕ at the mid span. Find the distance from the centre where the first bar can be curtailed as regards the tensile stress. Also calculate the actual point of curtailment, so that bar is able to resist its full share of tensile force at the mid span. Deformed bars are used with $\sigma_{st} = 230$ N/mm².*

As in Ex. 4·4, $\dfrac{n_1}{n} = \dfrac{M - M_r}{M}$

$$x = \frac{l}{2}\sqrt{\frac{n_1}{n}}$$

For bending the first bar, for $n=7$, $n_1=1$, $l=4$ m,

$$x=\frac{4}{2}\sqrt{\frac{1}{7}}=0{\cdot}756 \text{ m.}$$

If the bar is to develop full tensile stress, the embedded length should be sufficient to provide bond strength to avoid slipping. Let l be length required to develop full shear.

$$\frac{\pi}{4}\times d^2\times 320=\pi dl\times\tau_{bd}$$

Assuming M 15 concrete,

$$\tau_{bd}=0{\cdot}6\times1{\cdot}4=0{\cdot}84 \text{ N/mm}^2 \quad \text{[deformed bars]}$$

$$\frac{d\times 230}{4}=l\times 0{\cdot}84$$

Now $d=16$ mm, $l=\dfrac{16\times730}{47{\cdot}84}=1095$ mm$=1{\cdot}095$ m

The first bar should be curtailed at 1·1 m from centre.

PROBLEMS

1. A simply supported beam 300 mm wide and 882·5 mm effective depth is simply supported over a span of 4 m. It carries a uniformly distributed load including its own weight of 100,000 N/m. It is reinforced with 6 bars of 20 mm ϕ. If the maximum permissible shear stress in concrete is 5 N/mm², determine the parts of the span over which it is necessary to provide shear reinforcement. If 2 bars can be bent up to take shear, determine whether it is necessary to provide stirrups. Lever arm constant$=0{\cdot}87$.

2. A beam 250 mm wide and 400 mm effective depth supports a total uniformly distributed load of 200,000 N. Calculate the maximum shear stress and the spacing 10 mm diameter stirrups near the support. Lever arm constant $=0{\cdot}87$.

3. The lever arm of a beam is 400 mm and at a section 6 mm ϕ stirrups two legged, are provided at 160 mm centres. Calculate the shear resistance of the stirrups.

4. A rectangular beam is 300 mm wide and 600 mm effective depth, and, is simply supported at its ends. Calculate the total uniformly distributed load inclusive of self-weight, the beam can be permitted, to carry if the shear stress is not to exceed 0·5 N/mm². Assume $c=5$ N/mm², $t=140$ N/mm² and $m=18$, calculate the maximum span over which the beam can be allowed to support this load.

5. In problem 1, find the number of bars to be taken straight, if the bond stress is not to exceed 1 N/mm².

6. In problem 1, find the distance from the centres of beam when the first two bars can be curtailed as regards the tensile stresses. Also calculate the actual point of curtailment so that bar is able to resist its full share of tensile force at the mid span. Allowable bond stress 0·6 N/mm². Assume that the steel is stressed to its full limit 140 N/mm² at mid span.

7. A beam is 4 m in span simply supported at its ends and has to support a load of 10,000 N/m run inclusive of its own weight. Design the beam section. Calculate the shear and bond stresses at supports. Also design shear reinforcement.

5

Doubly Reinforced Beams

5·1. In RCC design it is economical to design a section such that all compression is carried by concrete and tension by steel and both the materials are stressed to maximum allowable stresses, thus resulting in balanced or critical section. However, it sometimes becomes necessary for a section to resist more bending moment than it can resist as a balanced section. This becomes necessary when overall size of a section is limited by structural, head room and architectural requirements. In such cases, the additional moment of resistance can be provided for by designing the section as over-reinforced section.

From Graph 3 in the Appendix, it is evident that the rate of increase in moment of resistance decreases as the reinforcement increases beyond that for balanced section From Graph 4 it is also seen that stress is steel decreases considerably when the section is over reinforced. When the tensile reinforcement provided is twice that required for balanced section the stress in steel reduces to about 65% only and the moment of resistance of such section is about 22% more than that of the balanced section.

A better method of designing a section, which has to resist more B.M. than it can resist as a balanced section, is by providing extra reinforcement in compression zone nearer to the outermost fibres and additional reinforcement on tension side more than required for balanced section. In this way considerable economy can be achieved. Such sections reinforced both in compression and tension sides are called doubly reinforced sections.

By putting total reinforcement in compression and tension zones equal to twice the reinforcement required for a balanced section, the moment of resistance increases by about 36%. Thus doubly reinforced beams will be more economical than over-reinforced beams.

5·2. Design of Doubly Reinforceed Sections

The design of doubly reinforced sections is based on the same assumptions as singly reinforced sections.

Notations

d=effective depth of beam.

b=breadth of the beam.

f=cover to centre of compressive steel.

M=total B.M. to be resisted by the section.

M_1=moment that can be developed by balanced section of concrete without compression reinforcement.

M_2=moment in excess of M_1 which is to be provided for by compression steel. $M_2=M-M_1$.

A_c=area of compressive steel.

A_t=total area of tensile steel$=A_{t_1}+A_{t_2}$.

A_{t_1}=area of tensile steel required to develop M_1.

A_{t_2}=area of additional tensile steel necessary to develop M_2.

t=stress in tensile steel.

t_c=stress in compressive steel.

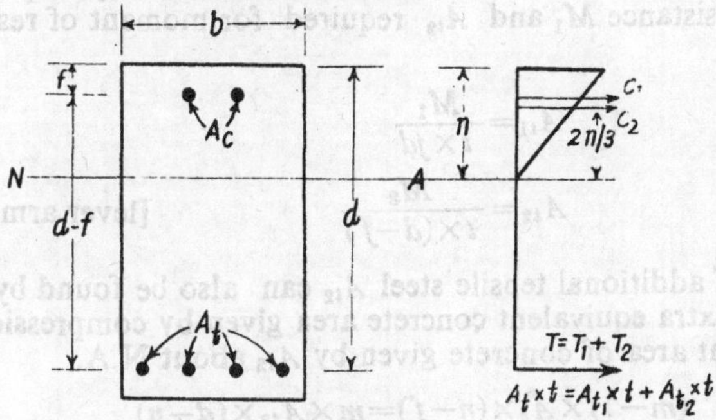

Fig. 5·1

Strains are proportional to the distance from N.A.

$$\frac{\text{Strain in top fibre of concrete}}{\text{Strain in compression steel}}=\frac{n}{n-f}$$

$$\frac{c/E_c}{t_c/E_s}=\frac{n}{n-f}$$

$$\frac{t}{t_c}\times\frac{E_s}{E_c}=\frac{n}{n-f}$$

$$t_c=c\times\frac{n-f}{n}\,m$$

$c\times\dfrac{n-f}{n}$ is the stress in concrete at level of compressive steel say c_1.

$$t_c=mc_1$$

Total equivalent concrete area resisting compression

$$=(nb-A_c)+mA_c$$
$$=nb+(m-1)A_c.$$

Thus steel provides extra equivalent concrete area of $(m-1)\ A_c$.

Taking moments about centre of tensile steel.

Moment of resistance $M=C_1\times(d-n/3)+C_2(d-f)$

$$=bn\ \frac{c}{2}\ (d-n/3)+(m-1)\ A_c\ c\times\frac{n-f}{n}(d-f)$$

$$=Qbd^2+(m-1)\ A_c\ c\times\frac{n-f}{n}\ (d-f)$$

$$=M_1+M_2.$$

Thus total moment of resistance consists of M_1 for balanced section and M_2 provided by compression steel.

Total area of tensile steel will be sum of A_{t1} required for moment of resistance M_1 and A_{t2} required for moment of resistance M_2.

$$A_{t1}=\frac{M_1}{t\times jd}$$

$$A_{t2}=\frac{M_2}{t\times(d-f)} \qquad \text{[lever arm}=d-f\text{]}$$

Area of additional tensile steel A_{t2} can also be found by taking moments of extra equivalent concrete area given by compression steel and equivalent area of concrete given by A_{t2} about N.A.

$$\therefore \qquad (m-1)\times A_c\times(n-f)=m\times A_{t2}\times(d-n)$$

$$A_{t2}=\frac{(m-1)}{m}\times A_c\times\frac{n-f}{d-n}.$$

5·3. General Notes on Design of Doubly Reinforced Beams (I.S.I.)

(*i*) The compression reinforcement shall be effectively anchored in two directions at right angles over the distance, where it is required to act in compression at points not further apart, centre to centre, than 16 times the diameter of the anchored bar or 48 times the diameter of the anchoring bar whichever is less. The secondary reinforcement used for this purpose should pass round or be hooked over, both the compressive and tensile reinforcement.

(*ii*) *Percentage of compression reinforcement*. The percentage of steel in compression should preferably not exceed 4 per cent ; but if it does, only 4 per cent should be allowed for in the calculation of the resisting moment of the beam. This percentage should be calculated on the basis of total cross-sectional area of the rectangle of depth and width equivalent to the maximum depth and width of the beam. In case of T and L-beams maximum width of the beam shall be the maximum width of the rib.

5·4. Type of Problems

Type 1. *Design of doubly reinforced beams.* Given maximum B.M. allowable stresses, *m*, and the dimensions to which size of the beam is to be restricted.

(*i*) For given stresses, values of n_c, *j*, *k* and *Q* are found.

(*ii*) Assume certain cover for compressive steel, *i.e.* value of *f*.

(*iii*) Moment of resistance of singly reinforced section

$$M_1 = Qbd^2.$$

Balance of moment for which compressive steel is to be provided

$$M_2 = M - M_1 = M - Qbd^2.$$

(*iv*) Moment resisted by compressive steel is given by

$$(m-1)\ A_c \times c \times \frac{n-f}{n} \times (d-f) = M - M_1.$$

For this, value of A_c can be found.

(*v*) Tensile reinforcement A_t will consist of A_{t1} and A_{t2},

where $$A_{t1} = \frac{M_1}{tjd}$$

and $$A_{t2} = \frac{M_1 - M_2}{t(d-f)}.$$

Alternatively $$A_{t2} = \frac{(m-1)}{m} \times A_c \times \frac{n-f}{d-n}.$$

The section can also be designed from graphs 6 to 11 in Appendix. For given B.M. and dimensions of section, find $Q' = \frac{M}{bd^2}$. Decide about value of *f*. Referring to the graph of particular mix and for particular value of *f*, find the percentages of steel reinforcement in compression and tension for the value of *Q* found above. Provide requisite reinforcement.

Type 2. *To determine the permissible loading or B.M. on beams.* Given the dimensions of section, area of steel reinforcement and maximum allowable stresses.

(*i*) Determine the depth of neutral axis by taking moment of con crete and equivalent steel area about N.A.

(*ii*) For given permissible stresses, determine depth of critical neutral axis n_c.

(*iii*) By comparing depth of actual neutral axis with critical N.A., determine whether concrete reaches maximum stress or steel reaches maximum stress.

If n is greater than n_c, the concrete reaches maximum allowable stress c first and moment of resistance can be found out by taking moments about tensile steel.

$$\text{M.R.}=b\times n\,\frac{c}{2}\times(d-n/3)+(m-1)\times A_c\times c\times\frac{n-f}{n}\times(d-f).$$

If n is less than n_c steel reaches maximum stress t first. Corresponding to maximum stress t in steel, stress in concrete will be $c_a=\frac{t}{m}\times\frac{n}{(d-n)}$.

Moment of resistance can be found taking moments about tensile steel.

$$\text{M.R.}=b\times n\times\frac{c_a}{2}\times\left(d-\frac{n}{3}\right)+(m-1)\times A_c\times c_a\times\frac{n-f}{n}\times(d-f)$$

Type 3. *To check design of a beam for stresses.* Given dimension of section, area of steel reinforcement and B.M. on section.

(*i*) Determine N.A. of the section by equating areas of concrete and equivalent areas of steel about N.A.

(*ii*) If c_a is compressive stress in top fibre of concrete, stress in compressive still will be $m\times c_a\times\frac{n-f}{n}$.

(*iii*) Taking moments of compressive forces about tensile steel and equating to external B.M.

$$M=(m-1)\,A_c\times\frac{n-f}{n}\,c_a\times(d-f)+bn\times\frac{c_a}{2}\times(d-n/3).$$

From this value of c_a, stress in top fibre of concrete can be found.

(*iv*) Stress in tensile steel will be $t_a=mc_a\times\frac{d-n}{n}$.

(*v*) Stress in compressive steel will be $mc_a\times\frac{n-f}{n}$.

Ex. 5·1. *A reinforced concrete beam is simply supported over a span of 4 m and carries a uniformly distributed load of 30,000 N/m inclusive of self-weight. If the size of the beam is to be limited to 250 mm×500 mm, determine the areas of tensile and compressive* *reinforcement required. Allowable stresses are c=5 N/mm², t=140 N/mm², m=18.*

Sol. Assume 40 mm cover to centre of tensile steel. Effective depth=460 mm.

$$\text{Max. B M., } M=\frac{wl^2}{8}\times\frac{30{,}000\times4\times4}{8}\times1000$$
$$=60{,}000{,}000 \text{ N-mm}$$
$$M_1=Qbd^2$$
$$=0{\cdot}87\times250\times(460)^2$$
$$=46{,}030{,}000 \text{ N-mm}$$
$$M_2=M-M_1$$
$$=60{,}000{,}000-46{,}030{,}000$$
$$=13{,}970{,}000 \text{ N-mm}$$

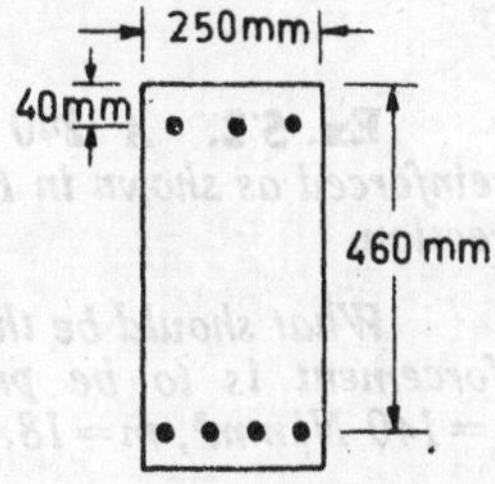

Fig. 5·2

Keep f, cover to centre of compression steel 40 mm.

$$n=kd=0{\cdot}4\times460=184 \text{ mm}$$

$$M_2=(m-1)\ A_c\times c\times\frac{n-f}{n}\times(d-f)$$

$$13{,}970{,}000=17\times A_c\times5\times\frac{184-40}{184}(460-40)$$

$$\therefore\quad A_c=\frac{13{,}970{,}000}{17\times5}\times\frac{184}{144\times420}=499{\cdot}9 \text{ mm}^2$$

Provide two bars of 16 mm ϕ and one bar of 12 mm ϕ.

$$A_c=2\times201+113=402+113=515 \text{ mm}^2$$

$$A_{t1}=\frac{M}{tjd}=\frac{46{,}030{,}000}{140\times0{\cdot}87\times460}$$
$$=821{\cdot}5 \text{ mm}^2$$

$$A_{t1}=\frac{M_2}{t(d-f)}=\frac{13{,}970{,}000}{140(460-40)}=237{\cdot}6 \text{ mm}^2$$

Total area of tensile steel

$$=821{\cdot}5+237{\cdot}6=1059{\cdot}1 \text{ mm}^2$$

Provide 4 bars of 20 mm ϕ

$$A_t=1256 \text{ mm}^2.$$

Design of section from graphs

$$Q'=\frac{M}{bd^2}=\frac{60{,}000{,}000}{250\times460\times460}=1{\cdot}164$$

$$f=\frac{40}{460}\times d=0{\cdot}087\ d.$$

Referring to graph 6 of mix M 15 and interpolating for $f=0{\cdot}087\ d$, values of reinforcements corresponding to value of $Q'=1{\cdot}134$, $p_c=0{\cdot}44\%$ and $p_t=0{\cdot}92\%$.

$$A_c=0{\cdot}44\times\frac{250\times460}{100}=506\ \text{mm}^2$$

$$A_t=0{\cdot}02\times\frac{250\times460}{100}=1058\ \text{mm}^2.$$

Ex. 5·2. *A 240 mm×520 mm reinforced concrete beam is reinforced as shown in Fig. 5·3. Find the moment of resistance of the section.*

What should be the depth of the beam if no compression reinforcement is to be provided? Allowable stresses are $c=5\ N/mm^2$, $t=140\ N/mm^2$, $m=18$.

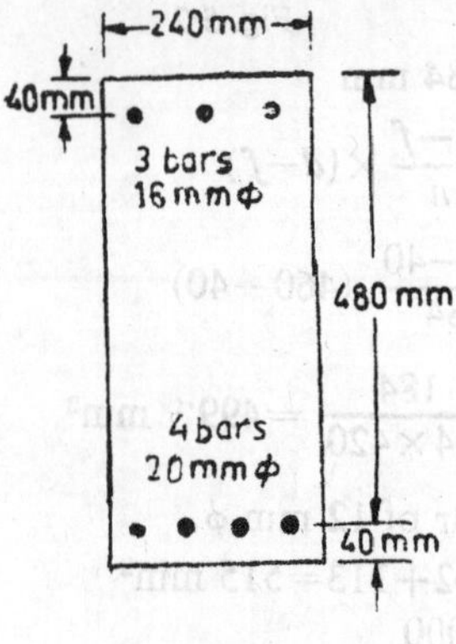

Fig. 5·3

Sol. Let n be the depth of N.A.

Taking moments about N.A.

$$bn\times\frac{n}{2}+(m-1)\ A_c\times(n-f)$$

$$=m\ A_t\times(d-n)$$

$$\therefore\quad 24\times\frac{n^2}{2}+17\times201\times3\times(n-40)$$

$$=18\times4\times314\times(480-n)$$

$$n=198{\cdot}4\ \text{mm}$$

Depth of critical N.A.

$$=0{\cdot}4\times480=191\ \text{mm}<198{\cdot}4$$

Therefore, section is over-reinforced.

Concrete reaches maximum allowable stress.

Moment of resistance

$$=bn\times\frac{cd}{2}\times(d-n/3)+(m-1)\ A_c\times\frac{n-f}{n}\times c\times(d-f)$$

$$=24\times\frac{198{\cdot}4\times5}{2}\times\left(480-\frac{198{\cdot}4}{3}\right)+17\times603$$

$$\times\frac{198{\cdot}4-40}{198{\cdot}4}\times5\times(480-40)$$

$$=600\times198{\cdot}4\times413{\cdot}9+17\times603\times\frac{158{\cdot}4}{198{\cdot}4}\times5\times440$$

$$=67{,}280{,}000\ \text{N-mm}.$$

If the beam is to be singly reinforced, for balanced design

$$M=Qbd^2$$

$$67{,}280{,}000 = 0{\cdot}87 \times 240 \times d^2$$

$$d^2 = \frac{67{,}280{,}000}{0{\cdot}87 \times 240}$$

$$d = 567{\cdot}6 \text{ mm}$$

Provide overall depth of 600 mm

Effective depth $= 575$ mm

$$A_t = \frac{M}{tjd} = \frac{67{,}280{,}000}{140 \times 0{\cdot}87 \times 575} = 960{\cdot}7 \text{ mm}^2.$$

Provide 5 bars of 16 mm ϕ, $\quad A_t = 1005 \text{ mm}^2$.

Ex. 5·3. *A reinforced concrete beam is 200 mm wide and 400 mm deep from the compression edge to the centre of tension steel. The compression reinforcement consists of 2 bars of 20 mm ϕ and tensile reinforcement 3 bars of 20 mm ϕ. The centre of compression steel is at distance of 40 mm from top edge of the beam. If $m=18$ and B.M. at the section is 40,000 N·m, calculate the maximum compressive stress in concrete and stresses in both compressive and tensile steels.*

Sol. Let n be the depth of N.A.

Taking moment of equivalent areas about N.A.

$$200 \times n \times \frac{n}{2} + (m-1)A_c \times (n-f)$$

$$= m \times A_t \times (d-n)$$

$$\therefore \quad 100n^2 + 17 \times 628 \times (n-40)$$

$$= 18 + 942 \times (400-n)$$

$$\therefore \quad n = 163{\cdot}9 \text{ mm}$$

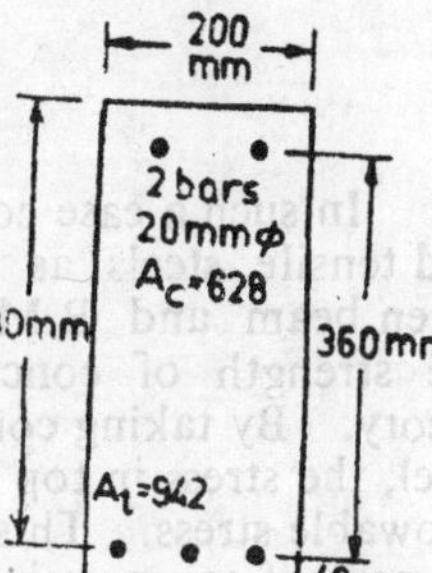

Fig. 5·4

Let c_a be the compressive stress in top fibre of concrete.

Stress in compressive steel

$$= c_a \times \frac{163{\cdot}9 - 40}{163{\cdot}9} = \frac{123{\cdot}9}{163{\cdot}9} c_a.$$

$$\text{Moment of resistance} = 200 \times 163{\cdot}9 \times \frac{c_a}{2}\left(400 - \frac{163{\cdot}9}{2}\right) + 17$$

$$\times 628 \times \frac{123{\cdot}9}{163{\cdot}9} \times c_a \times 360$$

$$= 40{,}000 \times 1000$$

$$c_a = 4{\cdot}67 \text{ N/mm}^2$$

Stress in compressive steel

$$=\frac{123{\cdot}9}{163{\cdot}9}\times 4{\cdot}670\times 18$$

$$=63{\cdot}6 \text{ N/mm}^2$$

Stress in tensile steel

$$=18\times 4{\cdot}670\times \frac{400-163{\cdot}9}{163{\cdot}9}$$

$$=18\times \frac{4{\cdot}670\times 236{\cdot}1}{163{\cdot}9}$$

$$=121{\cdot}5 \text{ N/mm}^2.$$

5·5. Steel Beam Theory

The moment of resistance of a doubly reinforced beam is sometimes calculated neglecting compressive strength of concrete completely. Equal reinforcement is provided both on tension and compression sides. The moment of resistance is taken as moment of steel couple only, the stress in compressive steel being working stress of 140 N/mm². Lever arm is taken as distance between centre of the tension and compression steel, *i.e.* $(d-f)$.

$$A_t=A_c=\frac{M}{140(d-f)}$$

In such a case concrete serves only as a web and compression and tensile steels as flanges of an imaginary steel joist. For a given beam and B.M. the areas of steel will be same whatever be the strength of concrete. Thus there is no justification for this theory. By taking compressive stress of 140 N/mm² in compression steel, the stress in top fibres in concrete will be much more than allowable stress. This method is not usually recommended. When the area of compressive steel comes more than that of tensile steel then steel beam theory may be applied.

Ex. 5·4. *A beam 250 mm×600 mm is to resist a B.M. of 17,500 N-m. Find the areas of compressive and tensile steels based on steel beam theory.*

Sol. Moment of resistance

$$=t\times A_t\times (d-f).$$

Assume 40 mm covers to centres of tensile and compression steels.

$$175{,}000\times 1000=140\times A_t(600-40-40)$$

$$A_t=\frac{17{,}500{,}000}{520\times 14}=2400 \text{ mm}^2.$$

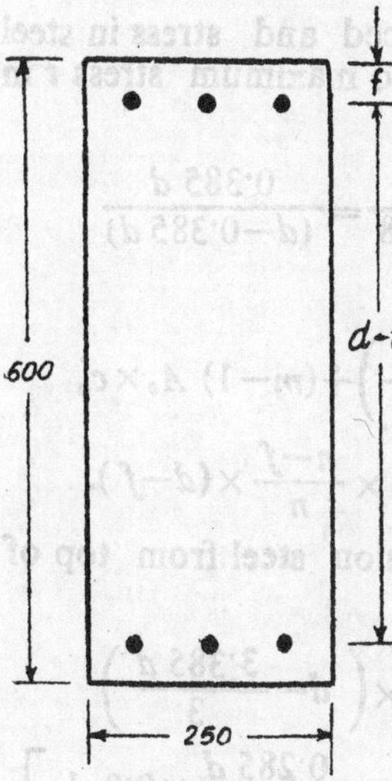

Fig. 5·5

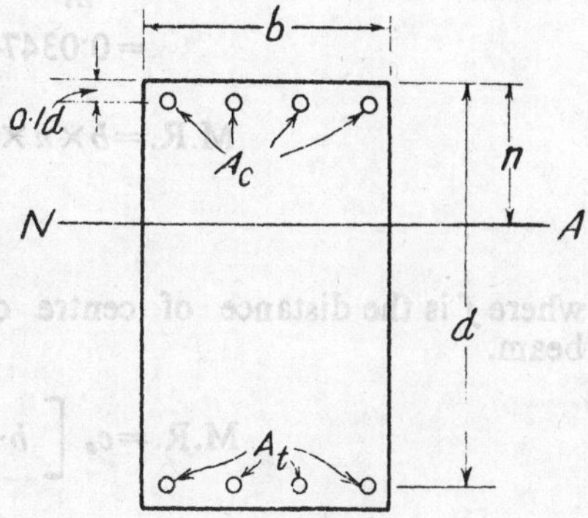

Fig. 5·6

Provide 4 bars of 28 mm ϕ

$$A_t = A_c = 2463 \text{ mm}^2.$$

Ex. 5·5. *A rectangular beam of breadth 'c' and effective depth d is reinforced with 1·2% tensile and 1·2% compressive steel. Determine moment of resistance of the section (a) by accurate method and (b) by steel beam theory neglecting concrete. Take compression reinforcement to be placed at 0·1 effective depth below top.*

Sol. $$A_c = \frac{1{\cdot}2\ bd}{100}, \quad A_t = \frac{1{\cdot}2\ bd}{100}$$

(a) Exact theory

Assume $m=18$ and critical N.A. constants as 0·4. Let n be the depth of actual neutral axis.

Taking moments of concrete and steel areas about N.A.,

$$b \times n \times \frac{n}{2} + (m-1) \times \frac{1{\cdot}2\ bd}{100} \times (n - 0{\cdot}1\ d)$$

$$= m \times \frac{1{\cdot}2\ bd}{100} \times (d-n)$$

$$\therefore \quad \frac{n^2}{2} + \frac{17 \times 1{\cdot}2\ d}{100}(n - 0{\cdot}1\ d) = \frac{18 \times 1{\cdot}2\ d}{100}(d-n)$$

$$\therefore \quad n^2 + 0{\cdot}84\ nd - 0{\cdot}4728\ d^2 = 0.$$

This gives $n = 0{\cdot}385\ d$.

Therefore, the section is under-reinforced and stress in steel reaches the maximum value. Corresponding to maximum stress t in steel, stress in concrete is given

$$c_a=\frac{t}{m}\times\frac{n}{(d-n)}=\frac{t}{18}=\frac{0{\cdot}385\,d}{(d-0{\cdot}385\,d)}$$
$$=0{\cdot}03475\,t$$

$$\text{M.R.}=b\times n\times\frac{c_a}{2}\left(d-\frac{n}{3}\right)+(m-1)\,A_s\times c_a\times\frac{n-f}{n}\times(d-f).$$

where f is the distance of centre of compression steel from top of beam.

$$\text{M.R.}=c_a\left[b-\frac{0{\cdot}385\,d}{2}\times\left(d-\frac{3{\cdot}385\,d}{3}\right)+\frac{17\times1{\cdot}2\,bd}{100}\times\frac{0{\cdot}285\,d}{0{\cdot}385\,d}\times0{\cdot}9\,d\right]$$
$$=0{\cdot}03475\,t\times3{\cdot}3038\,bd=0{\cdot}0107\,bd^2t.$$

(*b*) **Steel beam theory :**

$$\text{M.R.}=\frac{1{\cdot}2\,bd}{100}\times t\times0{\cdot}9\,d$$
$$=0{\cdot}0108\,bd^2t.$$

5·6. Shear in Doubly Reinforced Beams

Fig. 5·7 shows the distribution of shear stress in a doubly reinforced beam. Distribution of shear stress about N.A. will be as for homogeneous section and is given by $\frac{FAy}{I}$. If there is no compressive steel, shear distribution curve will be like C_1 but due to compressive steel Ay suddenly increases below level of compressive steel and hence curve will be C_2. Maximum shear intensity occurs on any plane between N.A. and tensile steel. It is given by the same formula as for a singly reinforced section, *i.e.* $q=\frac{F}{ba}$. In this case a will not be jd but will be distance between tensile force and C.G. of total compressive force.

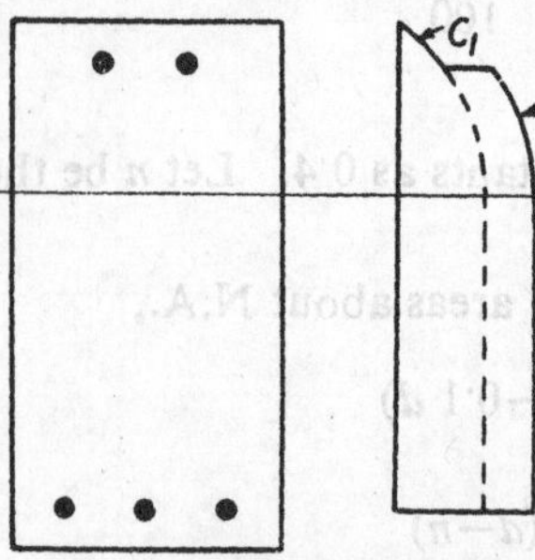

Fig. 5·7

Lever arm can be found as follows :

Total compressive force due to concrete,

$C_c = b \times n \times \frac{c_a}{2}$, where c_a is maximum compressive stress developed in concrete.

Total compressive force in steel

$$C_s = (m-1) \times A_c \times c_a \times \frac{n-f}{n}.$$

C.G. of compressive force from top of beam

$$= \frac{C_c \times \frac{n}{3} + C_s \times f}{C_c + C_s}$$

$$= \frac{\frac{bnc_a}{2} \times \frac{n}{3} + (m-1)\, A_c c_a \times \frac{n-f}{n} \times f}{\frac{bnc_a}{2} + (m-1)\, A_c c_a \times \frac{n-f}{n}}$$

$$= \frac{\frac{bn^2}{6}\ c_a + (m-1)\, A_c c_a \times \frac{n-f}{n} \times f}{\frac{bnc_a}{2} + (m-1)\, A_c c_a \times \frac{n-f}{n}}$$

$$\text{Lever arm } a = d - \frac{\frac{bn^2}{6} + (m-1) A_c \frac{n-f}{n} \times f}{\frac{bn}{2} + (m-1) A_c \frac{n-f}{n}}$$

Lever arm can also be found as follows :

If t_a is actual tensile stress in steel

$$M = t_a \times A_t \times a$$

$$\therefore \qquad a = \frac{M}{t_a \times A_t}.$$

Ex. 5·6. *In problem 5·3 if the S.F. at the section is 30,000 N, determine the intensity of shear stress and bond stress in tensile and compressive reinforcement. Also calculate the moment of inertia of the section.*

Sol. Moment of resistance $= t \times A_t \times a$

$$40{,}000 \times 1000 = 121{\cdot}5 \times 942 \times a$$

$$a = \frac{40{,}000{,}000}{121{\cdot}5 \times 952} = 349{\cdot}7 \text{ mm}$$

$$q = \frac{F}{b \times a} = \frac{30{,}000}{200 \times 349{\cdot}7}$$

$$= 0{\cdot}4289 \text{ N/mm}^2$$

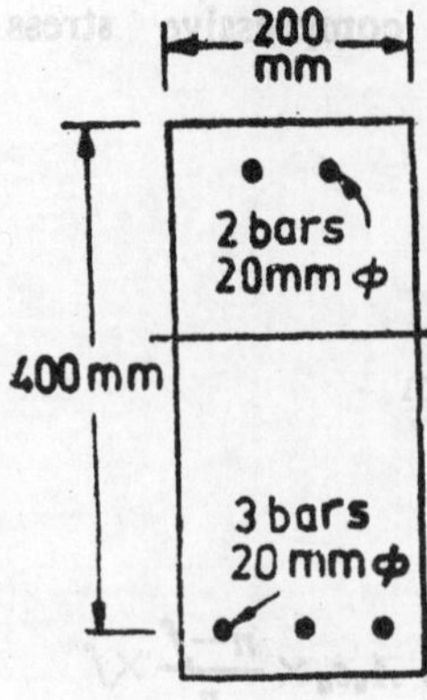

Fig. 5·8

Bond stress in tensile steel

$=\frac{30{,}000}{\Sigma O\times a}$

$=\frac{30{,}000}{3\times 62{\cdot}8\times 349{\cdot}7}$

$=0{\cdot}4553\ \text{N/mm}^2$

$I=\frac{bn^3}{3}+(m-1)\,A_c\times(n-f)^2+mA_t(d-n)^2$

$=\frac{200\times(163{\cdot}9)^3}{3}+17\times 628\times(123{\cdot}9)^2+18\times 942\times(236{\cdot}1)^2$

$=140{,}290\times 10^4\ \text{mm}^4.$

5·7. Analysis of Doubly Reinforcement Beams by Revised Method

By IS 456 revised compressive stress in steel in compression zone is to be taken as 1·5 m times the calculated compressive stress in the surrounding concrete or σ_{sc} whichever is lower.

In all cases the value of $1{\cdot}5\,mc\times\frac{n-f}{n}$ will be less than σ_{sc}.

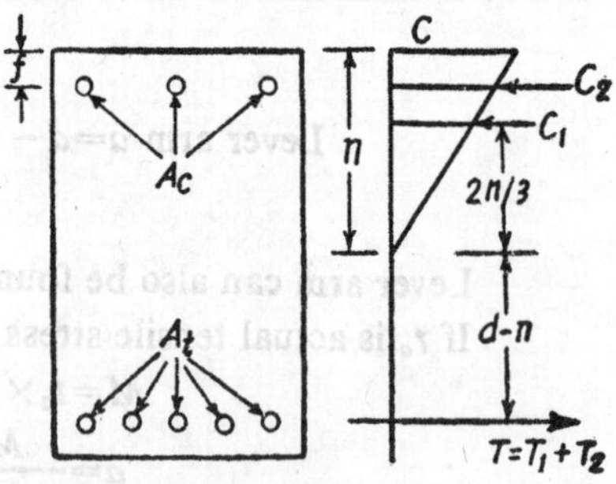

Fig. 5·9

As shown in Fig. 5·9, on compression side compressive force will be C_1 and C_2. C_1 being total force in concrete including area of compression reinforcement $=\frac{cbn}{2}$ and C_2 will be $(1{\cdot}5\,m-1)\,c\,\frac{n-f}{n}\,A_c$. On tension side tensile force will be $T_1+T_2=t\,A_t$

Now $$t=mc\left(\frac{d-n}{n}\right)$$

$$\therefore\quad \frac{cbn}{2}+(1{\cdot}5\,m-1)c\left(\frac{n-f}{n}\right)A_c=mc\left(\frac{d-n}{n}\right)A_t$$

$$\therefore\quad \frac{bn^2}{3}+(1{\cdot}5\,m-1)\,A_c\,(n-f)=mA_c\,(d-n)$$

Left side is moment of total equivalent compression area in concrete about N.A. and right side is equivalent tension area in concrete about N.A. Thus N.A. is the C.G of equivalent concrete areas.

$$\text{M.R.}=C_1\left(d-\frac{n}{3}\right)+C_2\,(d-f)$$

$$=\frac{cbn}{n}\left(d-\frac{n}{3}\right)+(1\cdot 5\,m-1)\,A_c\left(\frac{n-f}{n}\right)c\,(d-f)$$

$$=Qbd^2+c\,(1\cdot 5\,m-1)\,A_c\left(\frac{n-f}{n}\right)(d-f)$$

$$=M_1+M_2.$$

Thus total moment of resistance consists of M_1 for balanced section and M_2 provided by compression steel.

$$A_t=A_{t1}+A_{t2}$$

$$A_{t1}=\frac{M_1}{t\times a}$$

$$A_{t2}=\frac{M_2}{t\,(d-f)}$$

Area of additional tensile steel A_{t2} can be obtained by taking moments of extra equivalent concrete area given by compression steel and equivalent area of concrete given by A_{t2} about N.A.

$$\therefore\quad (1\cdot 5\,m-1)\,A_c\,(n-f)=m\,A_{t2}\,(d-n)$$

$$\therefore\quad A_{t2}=\frac{(1\cdot 5\,m-1)}{m}\,A_c\left(\frac{n-f}{d-n}\right)$$

where

$$A_c=\frac{M_2}{c\,(1\cdot 5\,m-1)\left(\frac{n-f}{n}\right)(d-f)}$$

Ex. 5 7. *Work out Ex. 5·1 by revised theory.*

Sol.

$$M=60{,}000{,}000 \text{ N-mm}$$

$$M_1=46{,}300{,}000 \text{ N-mm}$$

$$M_2=13{,}970{,}000 \text{ N-mm}$$

$$f=40 \text{ mm},\quad d=460,\quad n=184 \text{ mm}$$

$$M_2=c\,(1\cdot 5\,m-1)\,A_c\left(\frac{n-f}{n}\right)(d-f)$$

$$\therefore\quad 13{,}970{,}000=5\,(27-1)\,A_c\left(\frac{184-40}{184}\right)(460-40)$$

$$\therefore\quad A_c=326\cdot 9 \text{ mm}^2$$

Provide three bars of 12 mm ϕ,

$$A_c = 339 \text{ mm}^2$$

$$A_t = 1059{\cdot}1 \text{ mm}^2$$

Provide 4 bars of 20 mm ϕ,

$$A_t = 125{\cdot}6 \text{ mm}^2.$$

Ex. 5·8. *Work out Ex. 5·3 by revised theory.*

Sol. Let n be the depth of N.A.

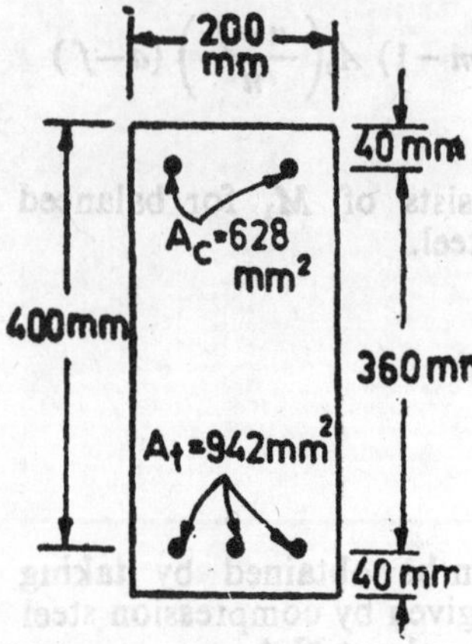

Fig. 5·10

Take moment of equivalent areas about N.A.

$$\frac{bn^2}{2} + (1{\cdot}5\, m - 1)\, A_c\, (n - f)$$

$$n = mA_t\, (d - n)$$

$$\therefore \quad \frac{200\, n^2}{2} + (27-1)628\ .\ A_c(n-40)$$

$$18 \times 942(460 - n)$$

$$= 168{\cdot}55 \text{ mm.}$$

$$\text{M.R.} = \frac{c_a\, bn}{2}\left(d - \frac{n}{3}\right)$$

$$+ c_a\, (1{\cdot}5\, m - 1)\, A_c \left(\frac{n-f}{n}\right)(d - f)$$

$$40{,}000 \times 1000 = c_a \times \frac{200 \times 168{\cdot}55}{2}\left(400 - \frac{168{\cdot}55}{3}\right)$$

$$+ c_a\, (27-1) \times 62{\cdot}8 \left(\frac{168{\cdot}55 - 40}{168{\cdot}55}\right)(400 - 40)$$

$$\therefore \quad c_a = 3{\cdot}892 \text{ N/mm}^2$$

Compressive stress in steel on compression side

$$= c_a\, (1{\cdot}5\, m - 1)\left(\frac{n-f}{n}\right)$$

$$= 3{\cdot}892\, (27-1)\left(\frac{168{\cdot}55 - 40}{168{\cdot}55}\right)$$

$$= 77{\cdot}177 \text{ N/mm}^2$$

Tensile stress in steel on tension side

$$= mc_a\left(\frac{d-n}{n}\right)$$

$$= 18 \times 3{\cdot}892\left(\frac{400 - 168{\cdot}55}{168{\cdot}55}\right)$$

$$= 96{\cdot}2 \text{ N/mm}^2.$$

Ex. 5·9. ***Work out exercise 5·1 by revised theory using high yield strength of deformed bars, c=5 N/mm², m=18.***

Sol. Bars used will be upto 20 mm ϕ

$\sigma_{st}=230$ N/mm²

$\sigma_{sc}=175$ N/mm²

$M=46{,}030{,}000$ N-mm

$M_1=Qbd^2=0\cdot652\times250\times(460)^2$

$=34{,}490{,}800$ N-mm

$M_2=46{,}030{,}000-34{,}490{,}800$

$=11{,}539{,}200$ N-mm

$A_t=A_{t1}+A_{t2}$

$=\dfrac{M_1}{tjd}+\dfrac{M_2}{t(d-f)}$

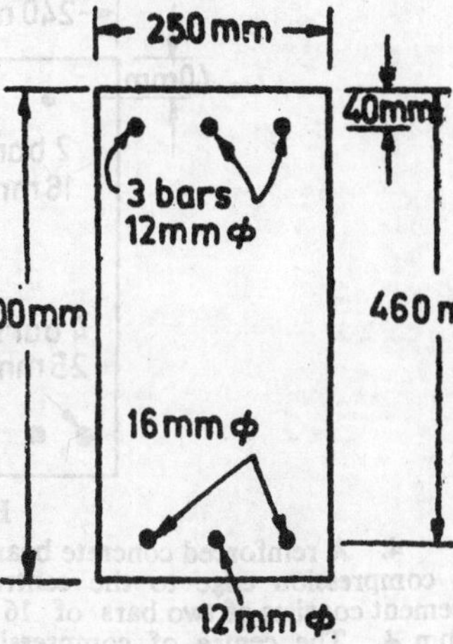

Fig. 5·11

$$=\frac{34{,}490{,}800}{230\times0\cdot904\times460}+\frac{11{,}539{,}200}{230(460-40)}$$

$$=361+119=480 \text{ mm}^2$$

Use 2 bars 16 mm ϕ, and 1 bar 12 mm ϕ

$$A_t=402+113=515 \text{ mm}^2$$

$$A_c=\frac{M_2}{c\,(1\cdot5\,m-1)\left(\dfrac{n-f}{n}\right)(d-f)}$$

$$=\frac{11{,}549{,}200}{5(27-1)\left[\dfrac{0\cdot289\times(460-40)}{0\cdot289\times460}\right](460-40)}$$

$$=302 \text{ mm}^2$$

Use 3 bars 12 mm ϕ,

$$A_c=339 \text{ mm}^2.$$

PROBLEMS

1. Design an RCC beam to resist B.M. of $1\cdot6\times10^8$ N mm. The size of the beam is to be restricted to 300 mm×920 mm, $c=5$ N/mm², $t=140$ N/mm² and $m=18$.

2. A rectangular beam 200 mm wide and 400 mm deep has to resist B.M. of 30,000 N-m at a section. Design the tensile and compressive reinforcements for the beam. Use mix M 15.

3. Find the moment of resistance of the RCC beam shown in Fig. 5·12 if the stresses in steel and concrete are not to exceed 140 N/mm² and 5 N/mm² respectively, $m=18$.

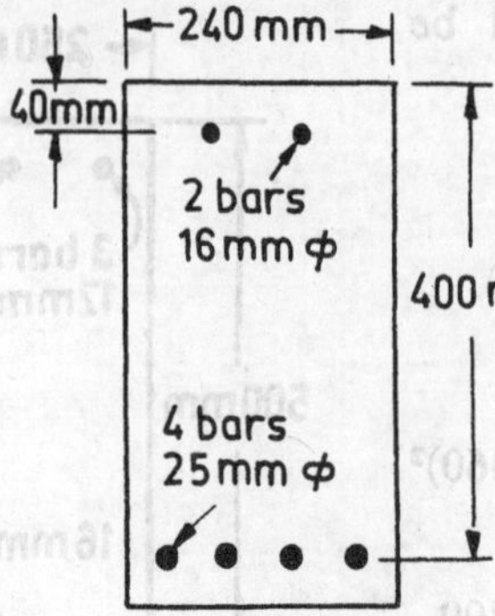

Fig. 5·12

4. A reinforced concrete beam is 200 mm wide and 400 mm deep from the compression edge to the centre of tension steel. The compression reinforcement consists of two bars of 16 mm ϕ and tensile reinforcement 3 bars of 20 mm ϕ. The centre of compressive steel is at distance of 40 mm from top edge of the beam. Find the maximum compressive stress in concrete and stresses in both compressive and tensile steels at a section, where B.M. is 35,000 N-m, $m=18$.

5. Find the moment of resistance of the RCC beam as shown in Fig. 5·13. Maximum permissible stress are $c=5$ N/mm² and $t=140$ N/mm², $m=18$.

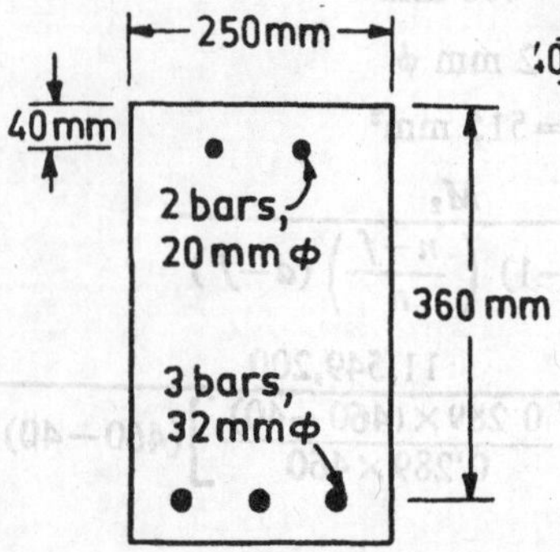

Fig. 5·13

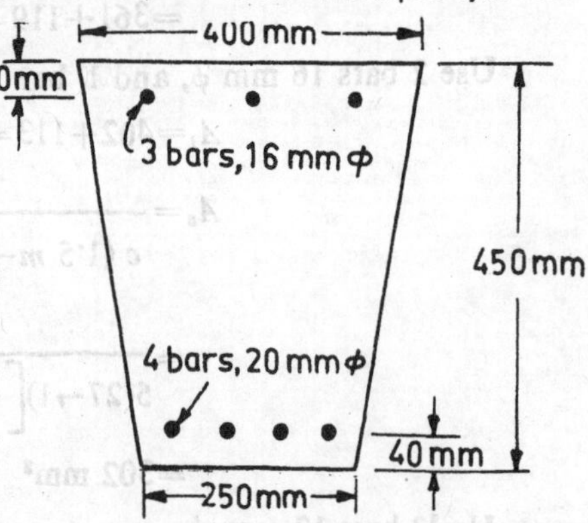

Fig. 5·14

6. In problem 1, if the S.F. at a section is 80,000 N, determine the intensity of shear stress and bond stress in tensile and compressive reinforcements.

7. Find the moment of resistance of the beam having trapezoidal section as shown in Fig. 5·14, $c=5$ N/mm², $t=140$ N/mm², $m=18$.

6

T and L Beams

6·1. In RCC construction floor slabs and beams are cast together and thus form a monolithic construction. When the beam deflects under loads, along with its some portion of the slab also deflects as shown in Fig. 6·1. Therefore, effective section of the beam is not rectangular but of T-shape, in which some portion of

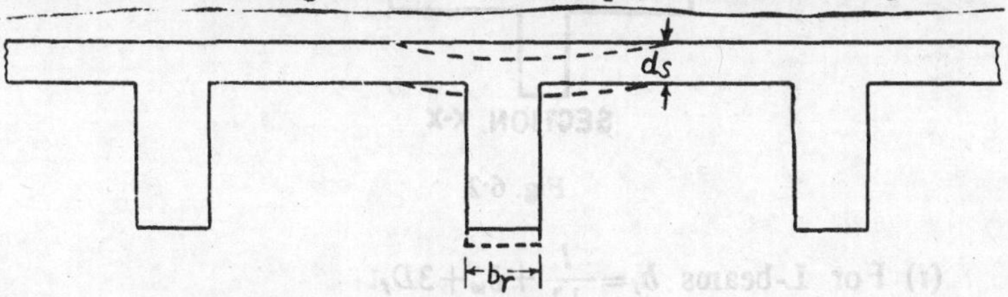

Fig. 6·1

the slap forms the flange of T and helps in resisting compressive stresses. The portion of the beam below the slab is called web or rib.

A slab which is assumed to act as a flange of a T-beam or L-beam shall satisfy the following :

(*a*) The slab shall be cast integrally witn the web, or the web and the slab shall be effectively bounded together in any other manner and

(*b*) If the main reinforcement of the slab is parallel to the beam, transverse reinforcement shall be provided as in Fig. 6 2, such reinforcement shall not be less than 60 per cent of the main reinforement at mid-span of the slab.

Effective width of flange in the absence of more accurate determination, may be taken as the following but in no case greater than breadth of the web plus half the sum of the clear distances to the adjacent beams on either side :

(*i*) For T-beams, $b_f=\frac{l_o}{6}+b_w+6D_f$

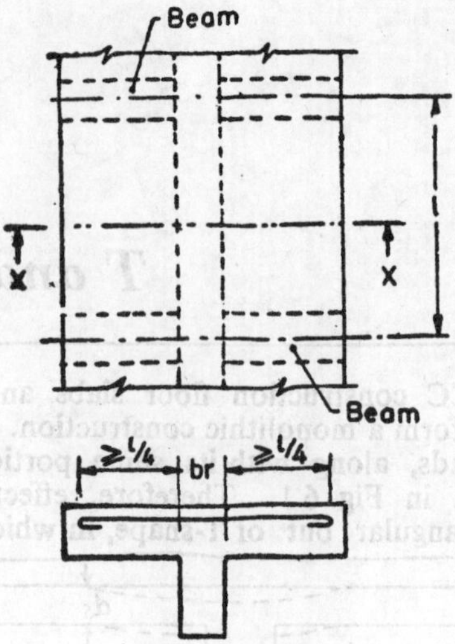

Fig. 6·2

(*i*) For L-beams $b_f=\frac{l}{12}+b_w+3D_f$.

(*ii*) For isolated beams, the effective flange width shall be obtained as below but in no case greater than actual width—

$$\text{T-beam, } b_f=\frac{l_0}{\left(\frac{l_0}{b}+4\right)}+b_w$$

$$\text{L-beam, } b_f=\frac{0{\cdot}5\, l_0}{(l_0/b)+4}+b_w$$

where, b_f= effective width of flange,

l_0=distance between point of zero moments in the beam,

b_w=breadth of the web,

D_f=thickness of flange and

b=actual width of the flange.

Note. For continuous beams and frames, l_0 may be assumed as 0·7 times the effective span. For simply supported beams l_0 will be the effective span.

6·2. Position of N.A. in T-beams

The neutral axis of a T-beam may be either in the flange or in the web depending upon the thickness of the flange, amount of steel and depth of the beam.

When N.A. lies in the flange, *i.e.* $k'd$ is less than d_s, all the formulae which have been derived for singly reinforced sections

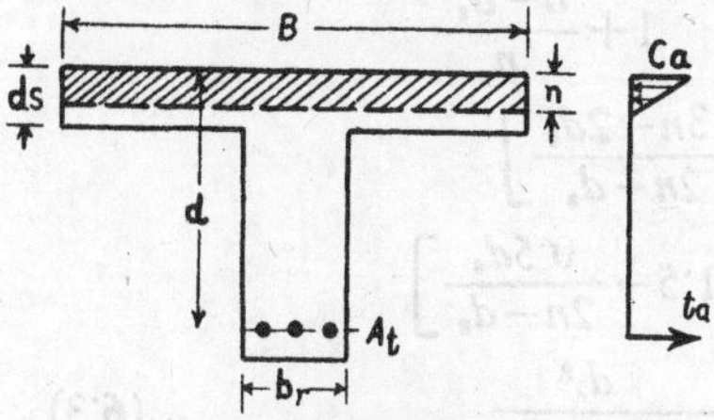

Fig. 6·3

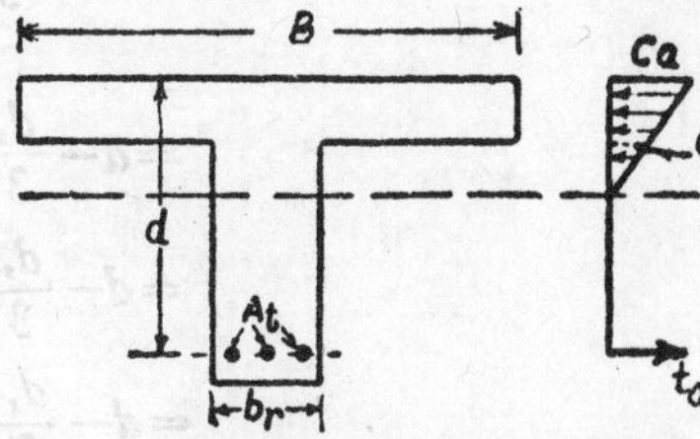

Fig. 6·4

are applicable. The width of the beam being width of the flange. If the area of the steel section is known, neutral axis will be given by,

$$\frac{Bn^2}{2}=mA_t(d-n) \qquad ...(6·1)$$

Neutral axis may lie in the web, *i.e.* $k'd$ is greater than d_s, in that case formulae derived for rectangular sections are not applicable. In such case the compression taken by portion of concrete in web is very small and generally neglected. Compression taken by flange only is taken into account. The position of neutral axis for such cases can be determined by taking moment about N.A.

$$B\times d_s\times\left(n-\frac{d_s}{2}\right)=mA_t\,(d-n) \qquad ...(6·2)$$

Lever arm for such a case can be found as follows :

Let c_a be the stress at top of the flange.

Stress at bottom of flange $c_a'=c_a\times\frac{n-d_s}{n}$, centre of gravity of compressive force will be centre of gravity of trapezium.

Let $\bar{y}$ be the distance of C.G. of compressive force from top of flange. Taking moments about top,

$$\bar{y}=\frac{c_a\times\frac{d_s}{2}\times\frac{d_s}{3}+c_a'\times\frac{d_s}{2}\times\frac{2}{3}}{c_a\times\frac{d_s}{2}\times c_a'\,\frac{d_s}{2}}$$

$$=\frac{c_a+2c_a'}{c_a+c_a'}\times\frac{d_s}{2}$$

Substituting value of c_a',

$$\bar{y}=\frac{1+2\times\frac{n\times d_s}{n}}{1+\frac{n-d_s}{n}}\times\frac{d_s}{3}$$

Lever arm $a=d-\frac{d_s}{3}\times\frac{1+2\times\frac{n-d}{n}}{1+\frac{n-d_s}{n}}$

$$=d-\frac{d_s}{3}\left[\frac{3n-2d_s}{2n-d_s}\right]$$

$$=d-\frac{d_s}{3}\left[1{\cdot}5-\frac{0{\cdot}5d_s}{2n-d_s}\right]$$

$$=d-\frac{d_s}{2}+\frac{d_s^2}{6(2n-d_s)} \quad \text{...(6·3)}$$

$$\text{Moment of resistance}=B\times d_s\times\frac{(c_a+a_a')}{2}\times a \quad \text{...(6·4)}$$

$$=t_a\times A_t\times a \quad \text{...(6·5)}$$

6·3. Depth of T-beam as a Balanced Section

Depth of a T-beam as a balanced section will be based on the assumption that neutral axis lies in the rib and that compression taken by concrete area in the rib is neglected.

For given stresses, neutral axis constant k is given by $\frac{1}{1+t/cm}$.

Depth of neutral axis $=kd$

Compressive stress at the bottom of flange

$$c'=c\times\frac{kd-d_s}{kd}.$$

Total compressive force in flange

$$=B\times d_s\times\frac{(c+c')}{2}$$

$$=\frac{B\times d_s}{2}\times\left[c+c\times\frac{kd-d_s}{kd}\right]$$

$$=\frac{Bd_sc}{2}\left[\frac{2kd-d_s}{kd}\right].$$

Moment of resistance of the beam

$$=\frac{Bd_sc}{2}\left[\frac{2kd-d_s}{kd}\right]\times \text{Lever arm } a \quad \text{...(A)}$$

Lever arm $a=d-\frac{d_s}{2}+\frac{d_s^2}{6(2n-d_s)}$

$$=d-\frac{d_s}{2}+\frac{d_s^2}{6(2kd-d_s)}$$

Moment of resistance

$$=\frac{Bd_sc}{2}\times\frac{(2kd-d_s)}{kd}\left[d-\frac{d_s}{2}+\frac{d_s^2}{6(2kd-d_s)}\right]$$

$$=\frac{Bd_sc}{2}\times\frac{(2kd-d_s)}{kd}\times\frac{[12kd^2-6d\,d_s-6kd\,d_s+3d_s^2+d_s^2]}{6(2kd+d_s)}$$

$$=\frac{Bd_sc}{12kd}\,[12kd^2-6d\,d_s(1+k)+4d_s^2]$$

$$=Bd_sc\left[d-\frac{d_s(1+k)}{2k}+\frac{d_s^2}{3kd}\right] \quad ...(6\,6)$$

This represents quadratic equation and the value of d can be found from above equation.

The value of lever arm lies between 0·88d and 0·95d. It can be assumed as 0·9d. Substituting this value of a in equation (A)

$$M=\frac{Bd_sc}{2}\left[\frac{2kd-d_s}{kd}\right]\times 0{\cdot}9d$$

$$=0{\cdot}45\,Bd_sc\,\frac{(2kd-d_s)}{kd} \quad ...(6{\cdot}7)$$

This represents a linear equation in d and value of d can be found out.

k is 0·4 for all mixes when $t=140$ N/mm². Equation (6·6) can be written as

$$\frac{M}{Bc}=d_s\left[d-\frac{d_s(1+0{\cdot}4)}{2\times 0{\cdot}4}+\frac{d_s^2}{3\times 0{\cdot}4d_s}\right]$$

$$=d_s\left[d-1{\cdot}75\,d_s+\frac{d_s^2}{1{\cdot}2d}\right].$$

In graph 12, value of M/Bc have been plotted against effective depth d for different values of d_s varying from 80 to 200 mm. Knowing the value of flange width, moment to be resisted and mix used, the value of M/Bc can be calculated and from graph, value of d can be found for particular value of d_s.

6·4. Economical Depth of T-beams

Cost of T-beam will depend on concrete provided in the rib and cost of steel reinforcement. From economy point of view, balanced section will not form suitable basis for design. This is because in the case of T-beams area of concrete in compression is large and so the depth of the beam for balanced section will be small and large area of steel will be required. As steel is much costlier than concrete, balanced section will be uneconomical. The economical depth can be found as follows :

Volume of concrete in rib per mm length

$$=b_r(d-d_s+d_c)$$

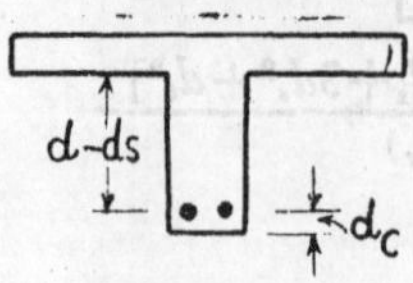

Fig. 6·5

Volume of steel per mm length

$$=A_t$$

If cost of concrete is k per cubic mm and that of steel is kr per cubic mm.

Cost of beam per mm length

$$C=kb_r(d-d_s+d_c)+kr\,A_t$$
$$=k[b_r(d-d_s+d_c)+rA_t]$$

$$A_t=\frac{M}{t\times a},\text{ where } a=0{\cdot}9d \text{ approximately,}$$

$$C=k\left[\,b_r(d-d_s+d_c)+\frac{rM}{t\times 0{\cdot}9d}\right]$$

b_r is fixed to accommodate steel and is therefore, constant, d_s and d_c are constants.

For cost to be minimum

$$\frac{dC}{dd}=0$$

$$\therefore\ k\left(b_r-\frac{Mr}{t\times 0{\cdot}9d^2}\right)=0$$

$$b_r=\frac{Mr}{t\times 0{\cdot}9d^2} \qquad \ldots(6{\cdot}8)$$

$$d=\sqrt{\frac{Mr}{0{\cdot}9b_rt}} \qquad \ldots(6{\cdot}9)$$

Taking $r=60$ for the M 15 mix and $t=140$ N/mm²

$$d=\sqrt{\frac{M\times 60}{0{\cdot}9\times 140b_r}}$$

$$=\sqrt{\frac{M}{2{\cdot}1b_r}} \qquad \ldots(6{\cdot}10)$$

Equation (6·8) can be written as

$$Mr=0{\cdot}9\times t\times b_rd^2$$

Taking $t=140$ N/mm², $Mr=126\,b_rd^2$.

In graph 13 values of Mr have been plotted against effective depth d for values of b_r varying from 200 to 600 mm.

As cost of concrete will depend on the mix, value of r will vary with the mix. Knowing the value of Mr, various value of d can be obtained for different value of b_r. Suitable values of b_r and d can be selected from considerations of shear and placing of bars.

6·5. Shear and Bond in T-beams

Fig. 6·6 shows distribution of shear stress in a T-beam. Maximum shear stress occurs between neutral axis and tensile steel. The average shear stress will be given by the same formula as for rectangular beam.

$$s=\frac{F}{b\times d}.$$

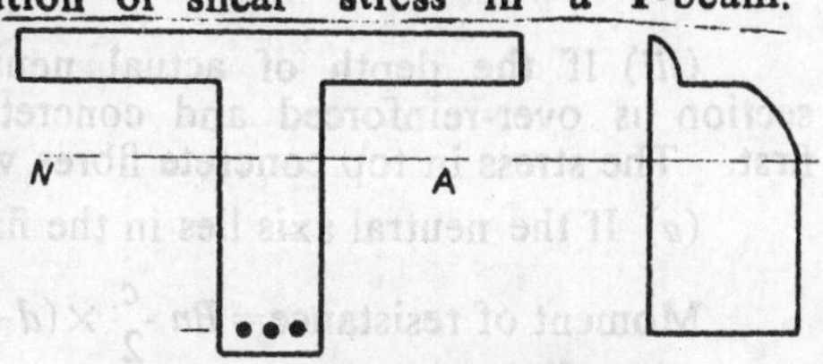

Fig. 6·6. Distribution of shear stresses in T-beam.

In this case b is breadth of the rib and d is effective depth.

6·6. Types of Problems

1. *To check the section for stresses.* Given the section and applied moment M.

Find the depth of N.A. by equating moments of equivalent areas as described in article 6·2.

Let c_a be actual stress in concrete.

If natural axis lies in flange

Moment of resistance

$$=\frac{Bnc_a}{2}\left(d-\frac{n}{3}\right)=M.$$

Value of c_a can be found from the equation.

If the neutral axis lies in the web, moment of resistance is given by

$$M=Bd_s\frac{c_a}{2}\left[1+\frac{n-d_s}{n}\right]\left[d-\frac{d_s}{2}+\frac{d_s^2}{6(2n-d_s)}\right]$$

From this value of c_a, actual stress in concrete, can be found.

Stress in steel $t_a=mc_a\dfrac{d-n}{n}$.

2. *Moment of resistance of a given T-beam.* Given the section and maximum allowable stresses.

(*i*) By taking moments of areas about neutral axis, find the value of neutral axis depth n by equation (6·1).

If n comes greater than d_s, neutral axis lies in web and use equation (6·2). This gives actual value of N.A. depth.

(*ii*) From the given permissible stresses c and t, determine the value of critical neutral axis depth constant

$$K=\frac{1}{1+t/cm}$$

Depth of critical N.A. $n_c=kd=\dfrac{1}{1+t/cm}\,d.$

(*iii*) If the depth of actual neutral axis is greater than n_c, section is over-reinforced and concrete reaches its maximum stress first. The stress in top concrete fibres will be allowable stress c.

(*a*) If the neutral axis lies in the flange.

$$\text{Moment of resistance}=Bn\,\frac{c}{2}\times(d-n/3) \qquad \text{...(6·11)}$$

as for rectangular section.

(*b*) If the neutral axis lies in the web, lever arm is given by equation (6·3).

$$\text{Moment of resistance}=B\times d_s\,\frac{(c+c')}{2}\times a$$

$$=\frac{Bd_s}{2}\left[c+\frac{n-d_s}{n}\,c\right]\left[d-\frac{d_s}{2}+\frac{d_s^2}{6(2n-d_s)}\right]$$

$$=\frac{Bd_s c}{2}\left[1+\frac{n-d_s}{n}\right]\left[d-\frac{d_s^2}{2}+\frac{d_s}{6(2n-d_s)}\right] \qquad \text{...(6·12)}$$

(*iv*) If the actual neutral axis is above the critical neutral axis, the section is under-reinforced and steel reaches maximum stress first. The stress in top fibre of concrete corresponding to maximum stress in steel will be

$$c_a=\frac{t}{m}\times\frac{n}{d-n}.$$

For this case also equations (6·11) and (6·12) can be used for computing amount of resistance but the value c_a will be used instead of c.

(*a*) If neutral axis lies in flange

$$\text{M.R.}=\frac{Bnc_a}{2}\,(d-n/3) \qquad \text{...(6·13)}$$

(*b*) If the neutral axis lies in web

$$\text{M.R.}=Bd_s\,\frac{c_a}{2}\left(1+\frac{n-d_s}{n}\right)\left[d-\frac{d_s}{2}+\frac{d_s^2}{6(2n-d_s)}\right] \qquad \text{...(6·14)}$$

3. *Design of a T-beam section.* Given the mix and thickness of slab.

(*i*) The width of rib is assumed. It should be sufficient to accommodate steel.

(*ii*) Width of flange *B* is determined.

(*iii*) Determine the effective depth of beam for balanced section from graph 12 or from equation (6·6) or (6·7).

Determine the effective depth of the beam for economical section from graph 13 or from equation (6·9).

Also determine the effective depth so that shear stress is within permissible limit.

Shear stress $q=\dfrac{F}{b_r d}$

Shear stress should not be more than 4 times the allowable value of shear stress.

$$4\sigma_{ss}=\frac{F}{b_r d} \qquad [a \text{ is assumed as } 0{\cdot}9\, d]$$

$$d=\frac{F}{4b_r\sigma_{ss}}$$

Maximum of the three values obtained is taken as effective depth. Generally depth required from point of view of economy will be maximum.

(*iv*) Calculate the steel area by taking lever arm as $d-\dfrac{d_s}{2}$.

Provide steel reinforcement.

(*v*) Check the section for stresses.

(*vi*) Check the section for shear stress. Provide shear reinforcement if necessary by bent up bars or stirrups.

(*vii*) Check-section for bond stress s_b.

Ex. 6·1. *A reinforced concrete beam shown in Fig. 6·7 is subjected to a B.M of 200,000 N-m at a section. Determine the maximum stress developed in concrete and steel m=18.*

Sol. Let *n* be depth of N.A. Assume it to lie in web. Taking moments of equivalent areas about N.A.

$$1200\times120\times(n-60)=18\times8\times314\times(600-n)$$

$$\therefore \quad n=189 \text{ mm}$$

Let *c* be the compressive stress in concrete at extreme fibres. Corresponding stress at bottom of flange

$$=c_a\times\frac{n-d_s}{n}$$

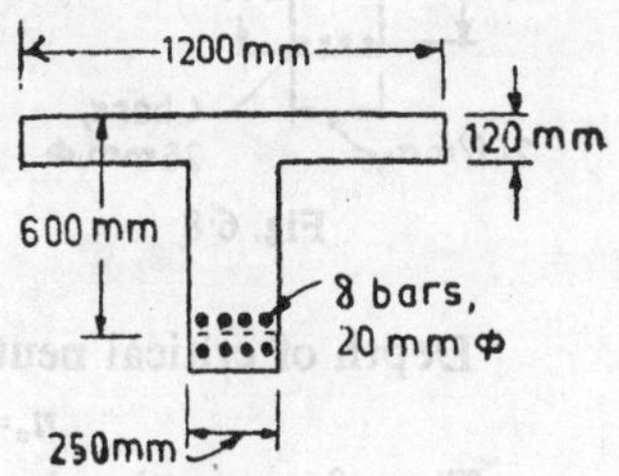

Fig. 6·7

Moment of resistance

$$=\frac{Bd_s\, c_a}{2}\left[1+\frac{n-d_s}{n}\right]\left[d-\frac{d_s}{2}+\frac{d_s^2}{6\,(2n-d_s)}\right]$$

$$200{,}000\times 1000=\frac{1200\times 120\times c_a}{2}\left[1+\frac{189-120}{189}\right]$$

$$\left[600-60+\frac{120\times 120}{6\,(2\times 189-120)}\right]$$

$$c_a=3{\cdot}705 \text{ N/mm}^2$$

Stress in steel $\quad t_a=mc_a\left(\frac{d-n}{n}\right)$

$$=18\times 3{\cdot}705\times\frac{600-189}{189}$$

$$=145 \text{ N/mm}^2$$

Alternately, $\quad M=A_t\times t_a\times a$

$$t_a=\frac{200{,}000\times 1000}{8\times 314\times 549{\cdot}3}$$

$$=145 \text{ N/mm}^2.$$

Ex. 6·2. *A reinforced cement concrete T-beam incorporating a 120 mm thick slab has a flange breadth of 1500 mm, rib breadth of 240 mm and overall depth of 440 mm. The section is reinforced with 4 bars of 25 mm ϕ in a single row. The centres of reinforcement being 40 mm from the bottom of rib. Determine the moment of resistance of the section, assuming allowable stresses as c=5 N/mm², t=140 N/mm² and m=18.*

Sol. Let n be the depth of N.A. Assume it to lie in web. Taking moments of equivalent area about N.A.

$$1500\times 120\times(n-60)=18\times 4\times 491\times(400-n)$$

$$n=117{\cdot}5<d_s$$

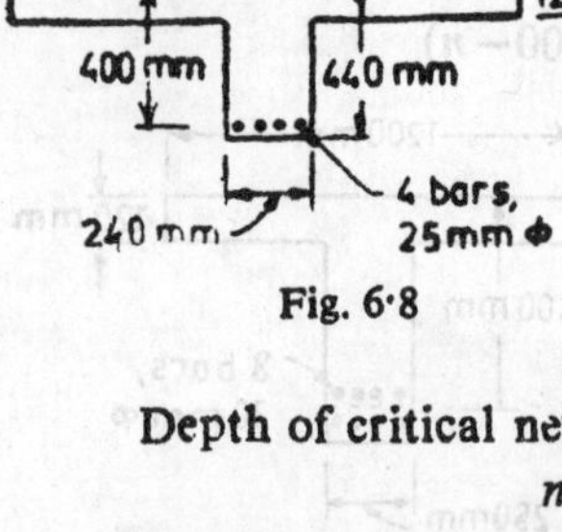

Fig. 6·8

∴ Neutral axis lies in flange.

Taking moments about N.A.

$$1500\times\frac{n^2}{2}=18\times 4\times 491\,(400-n)$$

$$n^2=18{,}880-47{\cdot}2\, n$$

$$n=115{\cdot}8 \text{ mm}$$

Depth of critical neutral axis

$$n_c=0{\cdot}4\times 400=160 \text{ mm}.$$

Therefore, section is under-reinforced. Steel reaches maximum stress.

$$\text{Lever arm} = d - n/3 = 400 - \frac{115{\cdot}8}{3} = 400 - 38{\cdot}6 = 361{\cdot}4 \text{ mm.}$$

$$\text{Moment of resistance} = t \times A_t \times a$$
$$= 140 \times 4 \times 491 \times 361{\cdot}4$$
$$= 99{,}380{,}000 \text{ N-mm.}$$

Ex. 6·3. *A road bridge of 10 m span consists of a concrete slab forming the flange of longitudinal beams with ribs at 3 m centres. If the stresses in the concrete and steel are limited to $c=5$ N/mm², $t=140$ N/mm² respectively and $m=18$, determine the thickness of the slab, depth of ribs and the main reinforcement required in main beams and slabs. The maximum B.M. in slab may be taken as $wl^2/12$ and the main beams are simply supported. Live load is 20,000 N/m².*

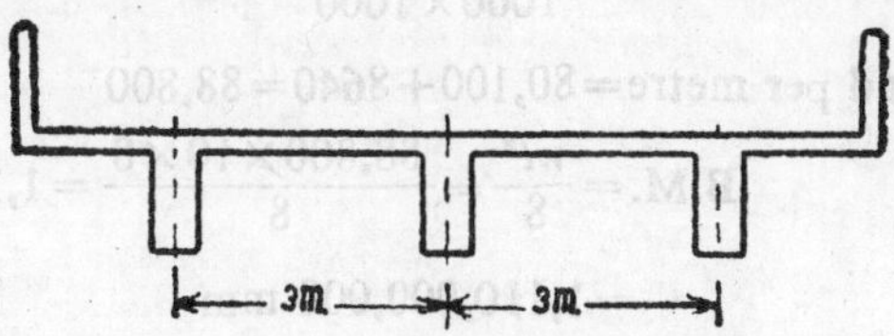

Fig. 6·9

Sol. Design of slab. Assume depth of slab as 180 mm.

Dead load of slab/sq. m.

$$= 24{,}000 \times \frac{180}{1000} = 4320 \text{ N}$$

Floor finish load (assumed) sq. m = 2400 N

Total dead load/sq. m = 6720 N

Total load per sq. m = D.L. + L.L. = 6720 + 20,000 = 26,720 N

$$\text{Maximum B.M. in slab} = \frac{wl^2}{12} = \frac{26{,}720 \times 3 \times 3}{12}$$
$$= 20{,}040 \text{ Nm}$$

$$M = Qbd^2$$
$$20{,}040 \times 1000 = 0{\cdot}87 \times 1000 \times d^2$$
$$d = \sqrt{\frac{20{,}040}{0{\cdot}87}} = 152 \text{ mm}$$

Provide overall depth of 180 mm.

Effective depth = 155 mm.

$$A_t = \frac{M}{tjd} = \frac{20{,}040 \times 1000}{140 \times 0{\cdot}87 \times 155}$$
$$= 1061 \text{ mm}^2$$

Use 12 mm ϕ bars at 100 mm c/c.

$$A_t = 1131 \text{ mm}^2$$

Distribution steel $= \frac{0{\cdot}15}{100} \times 180 \times 1000 = 270 \text{ mm}^2$

Use 8 mm ϕ bars at 180 mm c/c.

$$A_t = 279 \text{ mm}^2.$$

Design of Beam. Total load from slab per metre

$$= 26{,}720 \times 3 = 80{,}160 \text{ N.}$$

Assuming rib to be 1200 mm × 300 mm, wt. per metre

$$= \frac{1200 \times 300}{1000 \times 1000} \times 24{,}000 = 8640 \text{ N}$$

Total load per metre = 80,100 + 8640 = 88,800

$$\text{B.M.} = \frac{wl^2}{8} = \frac{88{,}800 \times 10 \times 8}{8} = 1{,}110{,}000 \text{ N-m}$$

$$= 1{,}110{,}000{,}000 \text{ mm}$$

Width of flange is least of

(i) $12\, d_s + b_r = 12 \times 180 + 300 = 2160 + 300 = 2460$ mm

(ii) $\frac{\text{Span}}{3} = \frac{10}{3} \times 1000 = 3333$ mm

(iii) Centre to centre of beams

$\therefore$ $B = 2460$ mm

By revised code,

$$\text{Flange width} = \frac{l_0}{6} + b_w + D_f$$

$$= \frac{10{,}000}{6} + 300 + 6 \times 180$$

$$= 3047 \text{ mm.}$$

Flange width of 2460 mm will be assumed.

Economical depth is given by

$$d = \sqrt{\frac{M}{2{\cdot}1\, b_r}}$$

Assuming $b_r = 300$ mm

$$d = \sqrt{\frac{1{,}110{,}000{,}000}{2{\cdot}1 \times 300}} = 1327 \text{ mm.}$$

Alternately d can be obtained from graph 16 of Appendix.

Depth for balanced section as obtained from graph 12 of Appendix is 700 mm.

Depth required for shear considerations,

$$d=\frac{F}{4b_r\,\sigma_{ss}}.$$

Maximum shear force

$$F=88{,}500\times5=442{,}500 \text{ N}.$$

$$d=\frac{442{,}500}{4\times300\times15}=737{\cdot}5 \text{ mm}.$$

$(\sigma_{ss}=0{\cdot}5)$

Provide effective depth of 1330 mm, with overall depth of 1410 mm.

Actual dead load of rib

$$=\frac{1230\times300}{1000\times1000}\times24000=8856.$$

Total load per metre$=80{,}160+8856=89{,}016$

$$\text{B.M.}=\frac{w\,l^2}{8}=\frac{89{,}016\times10^2}{8}$$

$$=1{,}112{,}700 \text{ N-m}$$

Max. shear $=89016\times5=440{,}580$ N.

Assume lever-arm $a=d-\frac{d_s}{2}=1330-\frac{180}{2}=1240$ mm.

$$A_t=\frac{M}{t\times a}=\frac{1{,}112{,}700{,}000}{140\times1240}$$

$$=6410 \text{ mm}^2$$

Provide 8 bars of 32 mm ϕ in two rows,

$A_t=6434$ mm².

Depth of rib provided is 1230 mm

Effective depth

$=1230+180-80$

$=1330$ mm.

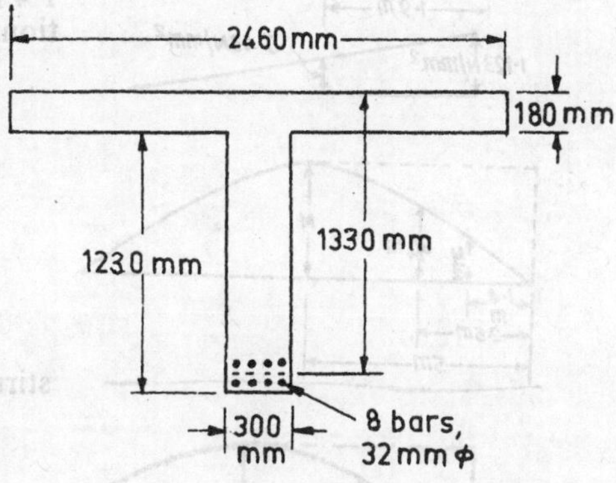

Fig. 6·10

Check for stresses.

Let n be depth of N.A. Assume it to lie in web.

Taking moments about N.A.

$$2460\times180\times(n-90)=18\times6434\times(1330-n)$$

$$n=347 \text{ mm}.$$

Depth of critical N.A.

$$n_c=0{\cdot}4\times1330=532 \text{ mm}$$

$\therefore$ **Section is under-reinforced**

$$\text{Lever arm } a = d - \frac{d_s}{2} + \frac{d_s^2}{6(2n-d_s)}$$

$$= 1330 - 90 + \frac{180 \times 180}{6(2 \times 347 - 180)}$$

$$= 1240 + 10{\cdot}5 = 1250{\cdot}5 \text{ mm.}$$

$$\text{Moment of resistance} = t \times A_t \times a$$

$$= 140 \times 6434 \times 1250{\cdot}5$$

$$= 1{,}1264{,}000{,}000 \text{ N-mm.}$$

This is less than moment of resistance.

$\therefore$ Section provided is O.K.

Check for shear.

$$\text{Average shear stress} = \frac{F}{bd} = \frac{440{,}580}{300 \times 1330} = 1{\cdot}11 \text{ N/mm}^2$$

Design of the shear reinforcement

$$\% \text{ Stress area} = \frac{6434}{300 \times 1330} \times 100 = 1{\cdot}612$$

$$\tau_c = 0{\cdot}43$$

Distance where shear stress is $\tau_c = \dfrac{0{\cdot}43}{1{\cdot}123} \times 5 = 1{\cdot}9$ m

Two bars will be bent at 2·4 m from support and 2 bars at 1·4 m from support from considerations of resisting moment.

Shear taken by two bars

$$= \sigma_s \times A_s \times \sin\theta$$

$$= 140 \times 804 \times 2 \times \frac{1}{\sqrt{2}}$$

$$= 159{,}200 \text{ N.}$$

Shear to be provided by stirrups

$$= 440{,}580 - 159{,}200$$

$$- 0{\cdot}43 \times 300 \times 1330$$

$$= 109{,}810 \text{ N}$$

Use 10 mm ϕ 2-legged stirrups.

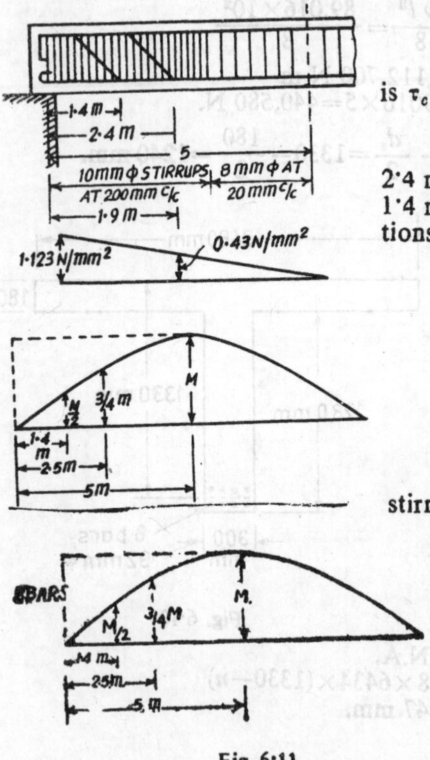

Fig. 6·11

$$\text{Spacing, } s_v = \frac{\sigma_s \times A_w \times d}{F}$$

$$= \frac{140 \times 2 \times 79 \times 1330}{109{,}810} = 268 \text{ mm}$$

Use 10 mm ϕ 2-legged stirrups at 200 mm c/c.

After 2·4 m from supports, provide nominal shear reinforcement of 8 mm ϕ,

$$s_v < \frac{2 \times 50 \times 250}{0{\cdot}4 \times 300} < 208. \text{ Say 200 mm c/c.}$$

6·7. Doubly Reinforced T-beams

When the dimensions of rib are to be restricted and the section has to resist more B.M. than it can take as balanced section, T-beam is to be designed as doubly reinforced.

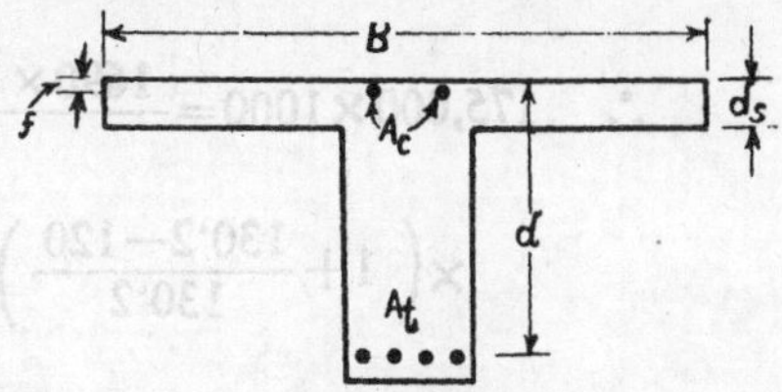

Fig. 6·12

Find the moment of resistance M_1 of T-beam as a balanced section

$$M_1 = Bd_s \frac{c}{2}\left[1 + \frac{kd - d_s}{kd}\right]\left[d - \frac{d_s}{2} + \frac{d_s^2}{6(2kd - d_s)}\right]$$

Moment of resistance to be provided by compression steel.

$$M_2 = M - M_1$$

$$M_2 = (m-1)A_c \times c \times \frac{kd - f}{kd} \times (d - f)$$

From this, value of A_s can be found

$$A_{t1} = \frac{M_1}{t \times a}$$

$$A_{t2} = \frac{M_2}{t(d - F)}$$

Total area of tensile steel A_t

$$= A_{t1} + A_{t2}.$$

Ex. 6·4. *The reinforced concrete beam shown in Fig. 6·13 is subjected to a B.M. of 175,000 N-m at a section. Determine the maximum stresses developed in concrete, compression steel and tensile steel.*

Sol. Let n be the depth of neutral axis. Assume it to lie in web.

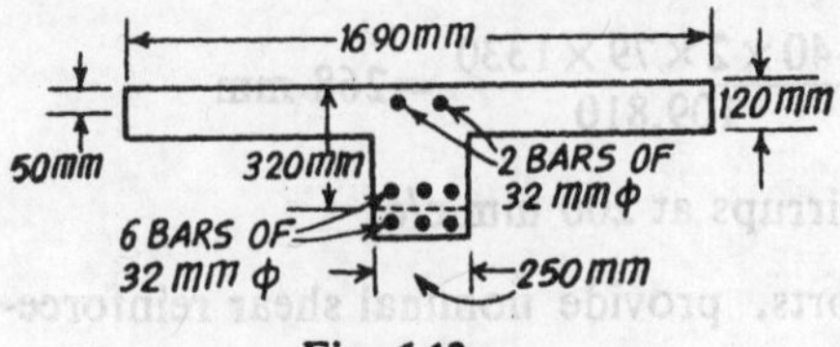

Fig. 6·13

Taking moment about N.A.

$$1690\times120\times(n-60)+17\times2\times804\times(n-50)$$

$$=18\times6\times804\times(320-n)$$

$$\therefore\quad n=130\cdot2 \text{ mm}$$

$$\text{M.R.}=\frac{B\times d_s\times c_a}{2}\left[1+\frac{n-d_s}{n}\right]\left[d-\frac{d_s}{2}+\frac{d_s^2}{6(2n-d_s)}\right]$$

$$+(m-1)A_c\times c_c\times\frac{n-f}{2}\times(d-f)$$

$$\therefore\quad 175{,}000\times1000=\frac{1690\times1200\times c_a}{2}$$

$$\times\left(1+\frac{130\cdot2-120}{130\cdot2}\right)\left[320-60+\frac{120\times120}{6(2\times130\cdot2-120)}\right]$$

$$+17\times2\times80\cdot4\times c_a\times\frac{130\cdot2-50}{130\cdot2}\times(320-50)$$

$$\therefore\quad c_a=5\cdot05 \text{ N/mm}^2.$$

Stresses in compressive steel

$$=18\times5\cdot05\times\frac{130\cdot2-50}{130\cdot2}$$

$$=18\times5\cdot05\times\frac{80\cdot2}{130\cdot2}=56 \text{ N/mm}^2.$$

Stress in tensile steel

$$=mc_a\,\frac{d-n}{n}$$

$$=18\times5\cdot05\times\frac{320-130\cdot2}{130\cdot2}$$

$$=18\times5\cdot05\times\frac{189\cdot8}{130\cdot2}=132\cdot5 \text{ N/mm}^2.$$

Ex. 6·5. *The T-beam shown in Fig. 6·14 is simply supported over a span of 10 m. Calculate the maximum uniformly distributed load inclusive of its own weight, it can carry if the allowable stresses are c=5 N/mm², t=140 N/mm², m=18.*

Sol. Let n be depth of N.A. Assume it to lie in web. Taking the moments of equivalent area about N.A.

$$1450\times100\times(n-50)+17\times2\times491\times(n-40)$$

$$18\times9\times80{\cdot}4(600-n)$$

$\therefore\quad n=241{\cdot}6$ mm

Depth of critical neutral axis

$0{\cdot}4\times600=240$ mm

$\therefore$ Section is over reinforced.

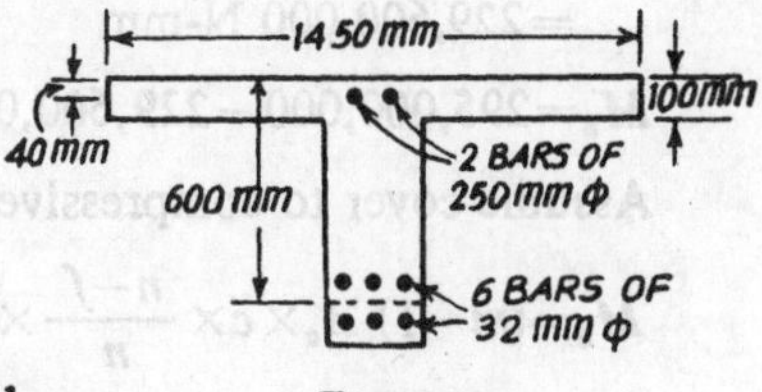

Fig. 6·14

Concrete reaches maximum stress first

$$\text{M.R.}=\frac{B\times d_s\times c}{2}\left(1+\frac{n-d_s}{n}\right)\left[d-\frac{d_s}{2}+\frac{d_s^2}{6(2n-d_s)}\right]$$

$$+(m-1)\times A_c\times c-\frac{d-f}{n}\times(d-f)$$

$$=\frac{1450\times100\times5}{2}\left(1+\frac{241{\cdot}6-100}{241{\cdot}6}\right)\left[600-50+\frac{100\times100}{6(2\times241{\cdot}6-100)}\right]$$

$$+17\times2\times491\times5\times\frac{241{\cdot}6-40}{251{\cdot}6}\times(600-40)$$

$$=357{,}700{,}000 \text{ N-mm}$$

$$\text{B.M.}=\frac{wl^2}{8}$$

$$\therefore\quad \frac{w\times10\times10}{8}\times1000=357{,}700{,}000$$

$$w=3577\times8=28{,}616 \text{ N/m.}$$

Ex. 6·6. *A reinforced concrete T-beam has a flange of 1740 mm × 120 mm and rib of 360 mm × 300 mm as shown in Fig. 6·15. What reinforcement is required at a section, where the beam is subjected to a B.M. of 295,000 N-m, c=5, N/mm², t=140 N/mm² and m=18 ?*

Sol. Assume 80 mm cover to centre of tensile steel.

$$d=360+120-80=400 \text{ mm}$$

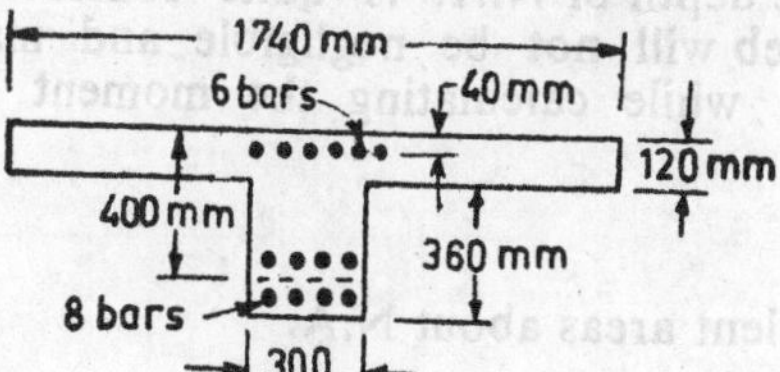

Fig. 6·15

Depth critical N.A. $=0{\cdot}4\times$)0$=160$ mm

Moment of resistance of balanced section

$$=\frac{B\times d_s\times c}{2}\left(1+\frac{n-d_s}{n}\right)\times\left[d-\frac{d_s}{2}+\frac{d_s^2}{6(2n-d_s)}\right]$$

$$M_1=\frac{1740\times120\times5}{2}\left(1+\frac{160-120}{160}\right)\left[400-60+\frac{120\times120}{6(320-120)}\right]$$

$$=229{,}600{,}000 \text{ N-mm}$$

$$M_2=295{,}000{,}000-229{,}600{,}000=65{,}400{,}000 \text{ N-mm.}$$

Assume cover to compressive steel $f=40$ mm.

$$M_2=(m-1)\,A_c\times c\times\frac{n-f}{n}\times(d-f)$$

$$65{,}400{,}000=17\times A_c\times5\times\frac{160-40}{160}\,(400-40)$$

$$=17\times A_c\times5\times\tfrac{3}{4}\times360$$

$$A_c=2850 \text{ mm}^2$$

Provide 5 bars of 25 mm ϕ, $A_c=2945$ mm^2

$$A_{t1}=\frac{M_1}{t\times a}=\frac{229{,}600{,}000}{140\times\left[d-\frac{d_s}{2}+\frac{d_s^2}{6(2n-d_s)}\right]}$$

$$=\frac{229{,}600{,}000}{140\times\left[400-60+\frac{120\times120}{6(320-120)}\right]}$$

$$=\frac{229{,}600{,}000}{140\times352}=4660 \text{ mm}^2$$

$$A_{t2}=\frac{M_2}{t(d-f)}=\frac{65{,}400{,}000}{140\times360}=1300 \text{ mm}^2.$$

Total $A_t=4660+1300$ mm^2$=5960$ mm^2

Provide 8 bars of 32 mm ϕ

$A_t=6434$ mm^2.

6·8. Moment of Resistance of T-beam taking Compression in Rib into Account

In case of beams where the depth of N.A. is quite considerable, compression taken by the web will not be negligible and may be taken into consideration while calculating the moment of resistance.

Let n be depth of N.A.

Taking moments of equivalent areas about N.A.

$$Bd_s\left(n-\frac{d_s}{2}\right)+b_r\,\frac{(n-d_s)^2}{2}=mA_t(d-n).$$

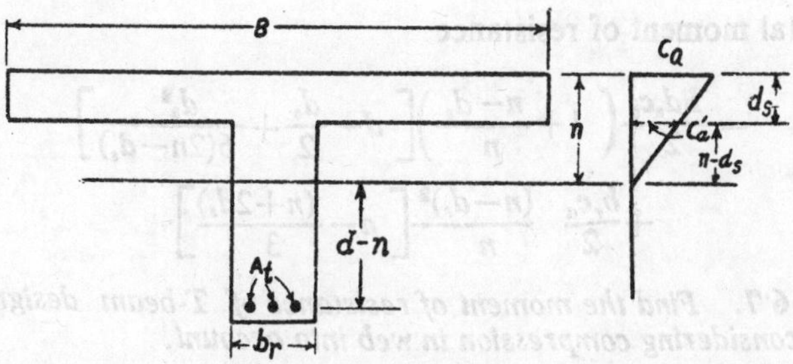

Fig. 6·16

From the above equation value of n can be determined. Knowing the position of neutral axis, the moments of resistance of the section can be found by taking moments of compressive forces about centre of tension steel. Alternately the centre of gravity of compressive forces can be found and lever arm will be distance between centre of gravity of compressive forces and tensile forces. The lever arm multiplied by total compressive force gives moment of resistance.

Let $\bar{y}$ be the distance of C.G. of compressive force from the top. Taking moments about top.

$$y=\frac{B\times d_s\times\frac{(c+c_a')}{2}\times\frac{2c_a'+c_a}{c_a+c_a}\times\frac{d_s}{3}+b_r\,(n-d_s)\times\frac{c_a'}{2}\left[d_s+\frac{(n-d_s)}{3}\right]}{B\times d_s\times\frac{(c+c_a')}{2}+b_r(n-d_c)\times\frac{c_a'}{2}}$$

where c_a' is stress at bottom of flange

$$c_a'=c_a\times\frac{n-d_s}{n}.$$

$$\text{Lever arm}=a=d-y.$$

Alternately moment of resistance can be calculated as follows :

Neutral axis is found as above. The moment of resistance will consist of that due to force in flange and that due to force in rib.

Moment of resistance due to force in flange was found to be

$$=\frac{Bd_sc_a}{2}\left(1+\frac{d+d_s}{n}\right)\left[d-\frac{d_s}{2}+\frac{d_s^2}{6(2n-d_s)}\right]$$

Moment of resistance due to force in rib

$$=b_r(n-d_s)\times\frac{c_a'}{2}\left[d-n+\frac{2}{3}\,(n-d_s)\right]$$

$$=\frac{b_r}{2}\,c_a(n-d_s)\left(\frac{n-d_s}{n}\right)\left(d-\frac{n}{3}-\frac{2}{3}\,d_s\right)$$

$$=\frac{b_rc_a}{2}\cdot\frac{(n-d_s)^2}{n}\left[d-\frac{(n+2d_s)}{3}\right]$$

Total moment of resistance

$$=\frac{Bd_s c_a}{2}\left(1+\frac{n-d_s}{n}\right)\left[d-\frac{d_s}{2}+\frac{d_s^2}{6(2n-d_s)}\right]$$

$$+\frac{b_r c_a}{2}\ \frac{(n-d_s)^2}{n}\left[d-\frac{(n+2d_s)}{3}\right].$$

Ex. 6·7. *Find the moment of resistance of T-beam designed in Ex. 6·3, considering compression in web into account.*

Sol. Let n be the depth of N.A.

Taking moment about N.A.

$$Bd_s\left(n-\frac{d_s}{2}\right)-\frac{b_r(n-d_s)^2}{2}=mA_t(d-n)$$

$$2460\times18(n-90)+\frac{30}{2}(n-180)^2=18\times6434\times(1320-n)$$

$$n=342 \text{ mm}$$

Depth of critical N.A. $=n_c=1320\times0\cdot4=528$ mm

∴ Section is under-reinforced. Steel reaches maximum stress.

∴ Stress in top fibres of concrete $=\frac{t_a}{m}\times\frac{n}{d-n}$

$$=\frac{140}{18}\times\frac{342}{1320-342}=2\cdot72 \text{ N/mm}^2$$

Moment of resistance

$$=\frac{Bd_s.c_a}{2}\left(1+\frac{n-d_s}{n}\right)\left[\frac{d-d_s}{2}+\frac{d_s^2}{6(2n-d_s)}\right]$$

$$+\frac{b_r c_a}{2}\ \frac{(n-d_s)^2}{n}\left[d-n+\frac{2}{3}(n-d_s)\right]$$

$$=\frac{2460\times180+2\cdot72}{2}\left(1+\frac{342-180}{342}\right)\left[1320-90+\frac{180\times180}{6(684-180)}\right]$$

$$+\frac{30\times2\cdot72}{2}\times\frac{(342-180)^2}{342}\left[1320-342+\frac{2}{3}(342-180)\right]$$

$$=1,132,010,000 \text{ N-mm.}$$

6·9. L-Beams

The end beams have slab on one side only and they act as L-beams. The width of the flange is taken as given in Art. 6·1.

L-beams and beams around stairs receive their load from one side only and they are subjected to considerable amount of torsion. The flange along with the rib of the L-beam will be able to resist considerable torsion. However, if the amount of torsion is larger than torsional resistance torsion reinforcement is provided in the form of longitudinal reinforcement and closely spaced stirrups. Further, as the L-beams are exposed on the outside they will be subject to temperature stresses. Therefore, on exposed face additional longitudinal bars may be provided.

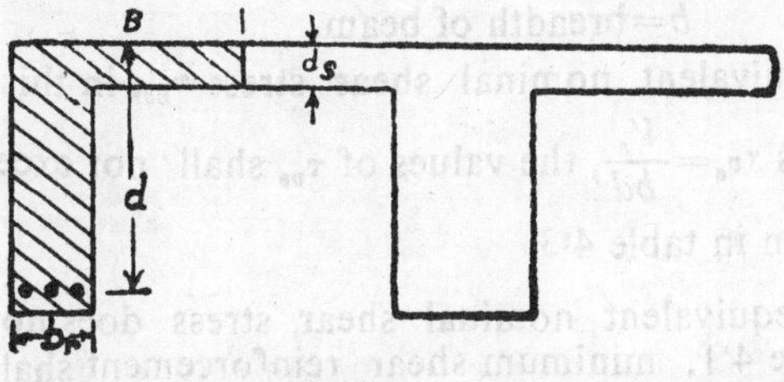

Fig. 6·17

If such beams are not restrained at the ends, *i.e.* they are simply supported, the loading will be causing unsymmetrical bending. The neutral axis will be inclined to the horizontal axis and may cut the flange of the beam. Thus one corner of flange will be stressed maximum. Allowable stresses for such beams may be reduced say by 30%.

The design principles for L-beams are the same as for T-beams.

6·10. General Notes on Design of L-beams (I.S.I.)

In general, where the torsional resistance or stiffness of members has not been taken into account in the analysis of structures, no specific calculation for torsion will be necessary, adequate control of any torsional cracking being provided by the required nominal shear reinforcement. Where the torsional resistance of stiffness of members is taken into account in the analysis, the members shall be designed for torsion.

Torsional reinforcement is not calculated separately from that required for bending and shear instead the total longitudinal reinforcement is determined for a fictitious bending moment which is function of actual bending moment and torsion ; similarly web reinforcement is determined for fictitious shear which is a function of actual shear and torsion.

The design rules apply to beams of solid rectangular cross-section. However, the rules may be applied, to flanged beams by substituting b_w for b, in which case they are generally conservative.

(*i*) **Critical section.** Sections located less than a distance d, from the face of the support may be designed for the same torsion as computed at a distance d where d, is the effective depth.

(*ii*) **Equivalent shear.** Equivalent shear, V_e, shall be calculated from the formula :

$$V_e = V + 1{\cdot}6\,\frac{T}{b}$$

where, V_e=equivalent shear
V=shear
T=torsional moment, and
b=breadth of beam.

The equivalent nominal shear stress τ_{ve}, in this case shall be calculated as $\tau_{v_e}=\frac{V_e}{bd}$, the values of τ_{ve} shall not exceed the values of $\tau_{c\ max}$ given in table 4·3.

If the equivalent nominal shear stress does not exceed τ_c as given in table 4·1, minimum shear reinforcement shall be provided as specified in Art. 4·6.

If the τ_{v_e} exceeds τ_c both longitudinal and transverse reinforcement shall be provided as follows :

(*a*) **Longitudinal Reinforcement.** The longitudinal reinfocement shall be designed to resist an equivalent bending moment M_{e1} given by $M_{ei}=M+M_t$

where, M=bending moment at the cross-section.

and $$M_t=\frac{T\left(1+\frac{D}{b}\right)}{1\cdot 7}$$

where T is the torsional moment, D is the overall depth of the beam and b is the breadth of the beam.

If the numerical value of M_t exceeds the numerical value of the moment M, longitudinal reinforcement shall be provided on flexural compression face such that the beam can also withstand an equivalent moment M_{e2} given by $M_{e2}=M_t-M$, the moment M_{e2} being taken as acting in the opposite sense to the moment M.

(*b*) **Transverse Reinforcement.** Two-legged closed hoops enclosing the corner longitudinal bars shall have an area of cross-section A_{sv}, given by

$$A_{sv}=\frac{T\times s_v}{b_1 d_1\ \sigma_{sv}}+\frac{V\times sv}{2\cdot 5 d_1\ \sigma_{sv}}$$

but the total reinforcement shall not be less than

$$\frac{(\tau_{ve}-\tau_c)\times b-s_v}{\sigma_{sv}}$$

where, T=torsional moment
V=shear force
sv=spacing of the stirrups reinforcement
b_1=centre to centre distance between corner bars in the direction of the width

d_1=centre to centre distance between corner bars in the direction of the depth

b=breadth of the member

σ_{sv}=permissible tensile stress in shear reinforcement

τ_{ve}=equivalent shear stress

τ_c=shear strength of the concrete.

Ex. 6·8. *Design the canopy and canopy beam shown in Fig. 6·18. Take live load as 750 N/m and allowable stress c=5 N/mm², t=140 N/mm² and m=18.*

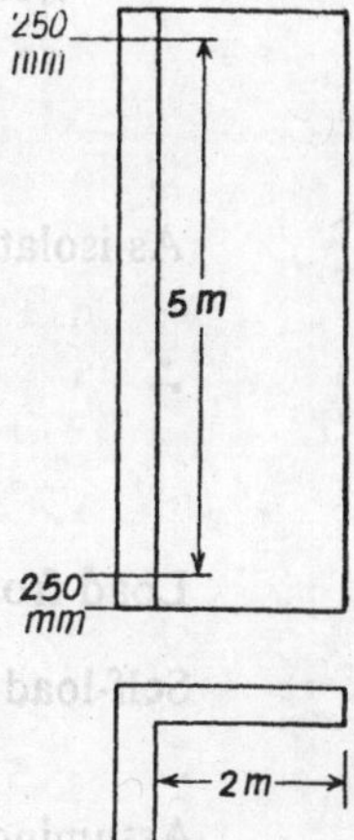

Fig. 6·18

Sol. Design of Canopy

Assume thickness of slab varying from 180 mm to 60 mm, average thickness being 120 mm.

Dead load per square meter

$$=1\times1\times\frac{120}{1000}\times25,000=3000\text{ N/m}^2$$

Finishing and plastering=500 N/m²

Live load=750 N/m²

Total=4250 N/m²

Consider 1 m width of canopy

$$\text{Max. B.M.}=\frac{wl^2}{2}=\frac{4250\times2^2}{2}$$

$$=8500\text{ N-m}$$

$$=8,500,000\text{ N-mm}$$

$$d=\sqrt{\frac{M}{Qb}}\sqrt{\frac{8.500,000}{0{\cdot}87\times1000}}$$

$$=98{\cdot}8\text{ mm}$$

Effective depth=180−15−5=160 mm

$$A_t=\frac{8,500,000}{140\times0{\cdot}87\times160}=436\text{ mm}^2$$

Provide 1 mm ϕ bars at 180 mm centre to centre.

Provide depth of 180 mm at end and depth of 60 mm at edge.

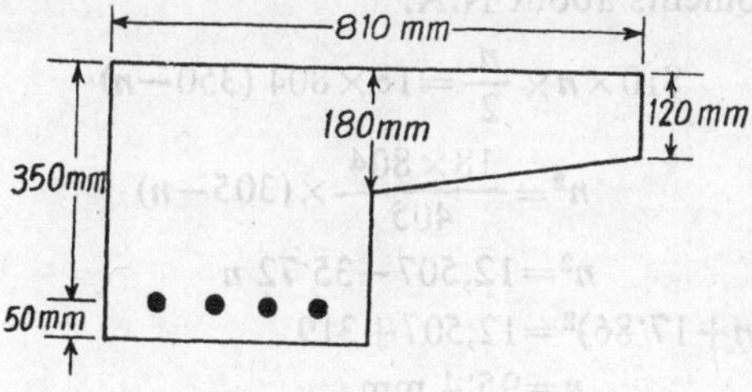

Fig. 6·19

Distribution of steel $=\frac{0{\cdot}15}{100}\times1000\times180=270\text{ mm}^2$

Provide 3 mm diameter bars at 180 mm c/c.

Design of canopy beam. The beam will be designed as L-beam. It will be subjected to torsion.

Assume the section as shown in Fig. 6·19.

Assume thickness of flange as average thickness of 120 mm.

$$\text{Width of flange}=\frac{l_0}{12}+b_w+3D_f$$

$$=\frac{5000}{12}+400+3\times120=1237\text{ mm.}$$

As isolated beam, $b_f=\frac{0{\cdot}5l_0}{l_0/b+4}+b_w$

$$\therefore\quad b_f=\frac{0{\cdot}5\times5000}{\frac{5000}{2400}+4}+400=410+400$$

$$=810\text{ mm}$$

Load from canopy $=4250\times2=8500$ N/m

Self-load of beam $=\frac{400}{1000}\times\frac{400}{1000}\times25{,}000=4000$ N/m

Total $=12{,}500$ N/m.

Assuming partial fixity at columns.

$$M_{max}=\frac{wl^2}{10}=\frac{12{,}500\times5^2}{10}=31{,}250\text{ N-m}$$

$$=31{,}250{,}000\text{ N-mm}$$

Assuming lever as $d-\frac{d_s}{2}=350-60=290$ mm

$$A_t=\frac{31{,}250{,}000}{140\times290}=770\text{ mm}^2$$

Provide 4 bars 16 mm ϕ, $A_t=804\text{ mm}^2$

Let n be the depth of N.A. Assume it to lie in flange.

Taking moments about N.A.

$$810\times n\times\frac{n}{2}=18\times804\,(350-n)$$

$$\therefore\quad n^2=\frac{18\times804}{405}\times(305-n)$$

$$\therefore\quad n^2=12{,}507-35{\cdot}72\,n$$

$$\therefore\quad (n+17{\cdot}86)^2=12{,}507+319$$

$$\therefore\quad n=95{\cdot}4\text{ mm}$$

Depth of critical N.A. $=0{\cdot}4\times350$ mm $=140$ mm.

Section is under-reinforced and steel will reach maximum stress first.

$$t_a = \frac{M}{A_t\left(d-\frac{n}{3}\right)} = \frac{31,250,000}{804\left(350-\frac{95{\cdot}4}{3}\right)}$$

$$= \frac{31,250,000}{804\times 318{\cdot}2} = 122{\cdot}2 \text{ N/mm}^2$$

$$c_a = \frac{t_a}{m}\times\frac{n}{d-n} = \frac{122{\cdot}2}{18}\times\frac{95{\cdot}4}{350-95{\cdot}4}$$

$$= 2{\cdot}54 \text{ N/mm}^2.$$

Shear stresses due to vertical loads

$$\text{Maximum shear force} = \frac{12,500\times 5}{2} = 31,250 \text{ N.}$$

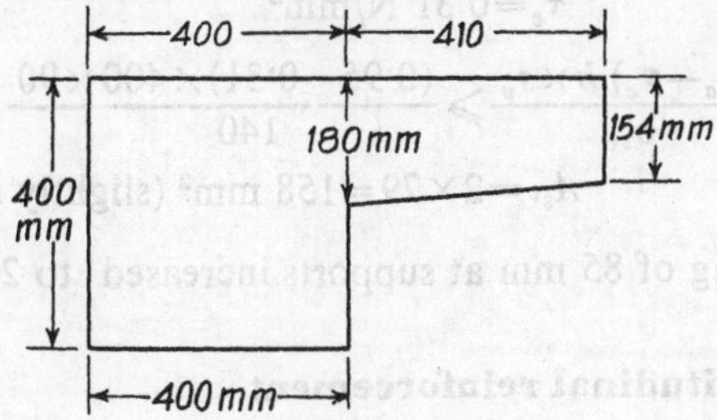

Fig. 6·20

$$q = \frac{F}{bd'} = \frac{31,250}{400\times 350} = 0{\cdot}223 \text{ N/mm}^2$$

Shear stress due to torsion

Maximum torsion $T = 4250\times 2\times(1+0{\cdot}2)\times 2{\cdot}5\times 1000$
$= 25,500,000$ N-mm.

$$V_s = V + 1{\cdot}6\,\frac{T}{b}$$

$$= 31,250 + 1{\cdot}6\times\frac{25,500,000}{400}$$

$$= 133,250 \text{ N.}$$

$$\tau_{ve} = \frac{133,250}{400\times 350} = 0{\cdot}95 \text{ N/mm}^2 < \tau_{max}$$

$$\tau_{max} = 1{\cdot}6 \text{ N/mm}^2.$$

Design of stirrups

Provide stirrups at 100 mm centres

$$A_{sv}=\frac{T\times s_v}{b_1 d_1 \sigma_{sv}}+\frac{V\times s_v}{2\cdot5 d_1\sigma_{sv}}$$

$$\therefore \qquad 2\times 79=\frac{25{,}500{,}000\times s_v}{350\times 350\times 140}+\frac{31{,}250\times s_v}{2\cdot5\times 340\times 140}$$

$$=1\cdot487 s_v+0\cdot263 s_v=1\cdot75\ s_v$$

$$\therefore \qquad s_v=\frac{2\times 79}{1\cdot75}=90\cdot3\ \text{mm}$$

$$\frac{100A_s}{bd}=\frac{100\times 804}{400\times 350}=0\cdot57$$

$$\tau_c=0\cdot31\ \text{N/mm}^2$$

$$A_{sv}\geqslant\frac{(\tau_{ac}-\tau_c)\,b\times s_v}{\sigma_{sv}}\geqslant\frac{(0\cdot95-0\cdot31)\times 400\times 90}{140}\geqslant 164\cdot6\ \text{mm}^2$$

$$A_{sv}=2\times 79=158\ \text{mm}^2\ \text{(slightly less)}$$

Use spacing of 85 mm at supports increased to 200 mm at the centre.

Design of longitudinal reinforcement

$$M_{et}=M+M_t=M+\frac{T\left(1+\frac{D}{b}\right)}{1\cdot7}$$

$$=31{,}250{,}000+\frac{25{,}500{,}000\left(1+\frac{400}{400}\right)}{1\cdot7}$$

$$=61{,}250{,}000\ \text{N-mm.}$$

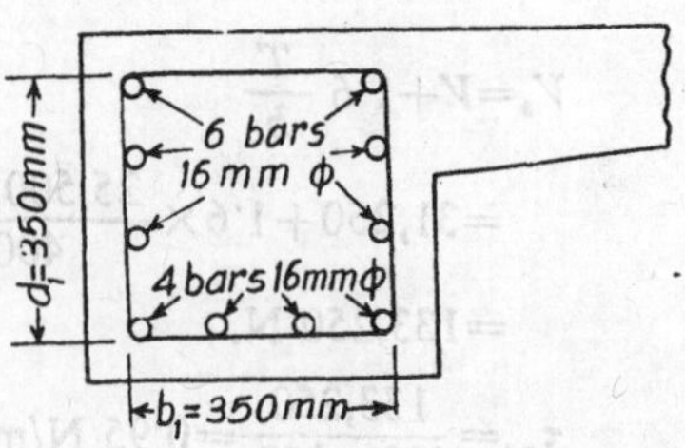

Fig. 6·21

Area of longitudinal reinforcement

$$=\frac{M_{et}}{140\times\left(d-\frac{d}{2}\right)}=\frac{61,250,000}{140\times 290}$$

$$=1509 \text{ mm}^2$$

Use 10 bars of 16 mm ϕ as shown.

Ex. 6·9. *An R.C. cantilevered canopy 3·05 m effective overhang and 6·1 m width is supported on two beams, cantilevered out from 368 mm thick walls and spaced at 3·66 m centres so that the slab overhangs each beam by 1·22 m on either side. If the ribs of beams are to be integrated with the slab so as to give a flat soffit, design the slab and the beams and sketch the details of reinforcements in beams and slabs. If the height of the wall over the beams on the anchoring length be 4 75 m, calculate the anchoring length of the beams required for a factor of safety of 1·5. Weight of water-proof concrete over slab =1470 N/m², live load 735 N/m². Permissible stresses in concrete 4·94 N/mm² and in steel 141 N/mm² and m=18.*

Sol. The slab is designed as shown in Ex. 3·5.

The slab thickness of 100 mm will be assumed for design of T-beams.

Fig. 6·22

$$\text{Flange width}=\frac{l_0}{6}+b_w+6D_f$$

$$\therefore \qquad b_f=\frac{3050}{6}+200+6\times 100$$

$$(b_w=\text{width of web}=200 \text{ mm})$$

$$\therefore \qquad b_f=1308 \text{ mm.}$$

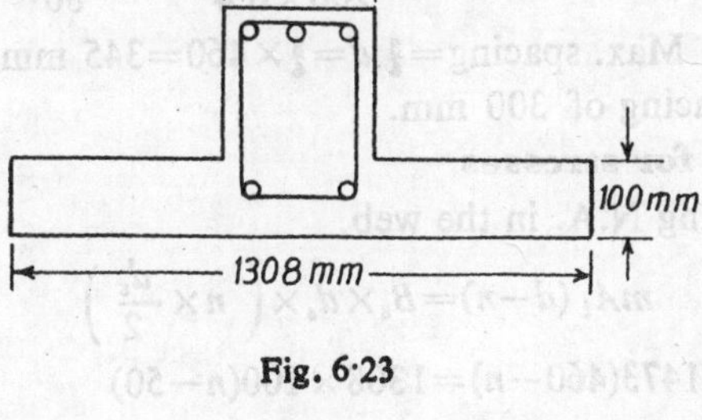

Fig. 6·23

Load of slab per square meter

$$=1\times 1\times \frac{120}{1000}\times 25{,}000+1470+735$$

$$=5205\ \text{N/m}^2$$

$$\text{Load from slab}=\frac{5205\times 6\cdot 1}{2}=15{,}875\ \text{N/m}$$

Assume load from web = 1500 N/m

$$\text{Total}=17{,}375\ \text{N/m}$$

$$\text{Max. B.M.}=17{,}375\times \frac{(3\cdot 05)^2}{2}=80{,}815\ \text{N-m.}$$

$$\text{Economic depth}=\sqrt{\frac{80{,}815\times 1000}{2\cdot 1\times 200}}=438\cdot 7\ \text{mm}$$

Overall depth of 500 mm is provided with effective depth of 460 mm.

$$\text{Lever arm}=460-50=410\ \text{mm}$$

$$A_t=\frac{M}{140\times 410}=\frac{80{,}815\times 1000}{140\times 410}=1408\ \text{mm}^2$$

Provide 3 bars of 25 mm ϕ,

$$A_t=1473\ \text{mm}^2$$

$$\text{Max. shear}=17{,}375\times \frac{3\cdot 05}{2}=26{,}497\ \text{N}$$

$$\text{Shear stress}=\frac{26{,}497}{200\times 460}=0\cdot 288\ \text{N/mm}^2$$

$$\frac{100\ A_s}{bd}=\frac{100\times 1378}{200\times 460}=1\cdot 50,\ \tau_c=0\cdot 42\ \text{N/mm}^2$$

Provide nominal shear reinforcement.

Use 8 mm ϕ stirrup

$$\frac{2\times 50}{200\times s_r}\geqslant \frac{0\cdot 4}{250}$$

$$s_v<\frac{2\times 50\times 250}{200\times 0\cdot 4}<\frac{25{,}000}{80}\nless 312\cdot 5\ \text{mm}$$

$$\text{Max. spacing}=\tfrac{3}{4}\ d=\tfrac{3}{4}\times 460=345\ \text{mm}$$

Use spacing of 300 mm.

Check for stresses

Assuming N.A. in the web,

$$mA_t\,(d-n)=B_s\times d_s\times\left(n\times\frac{d_s}{2}\right)$$

$$18\times 1473(460-n)=1308\times 100(n-50)$$

$$12,196,440-26,514n=130,800n-6,540,000$$

$$n=\frac{12,196,440+6,540,000}{26,514+130,800}=119 \text{ mm.}$$

Critical depth of neutral axis

$$=0{\cdot}4\times 460=184 \text{ mm}$$

Max. stress will be in steel.

$$\text{M.R.}=A_t\times t_a\left[d-\frac{d_s}{2}+\frac{d_s^2}{6(2n-d_s)}\right]$$

$$\therefore\quad 79,120,000=1473\ t_a\left[460-50+\frac{100^2}{6(238-100)}\right]$$

$$t_a=127{\cdot}3 \text{ N/mm}^2$$

$$\therefore\quad c_a=\frac{127{\cdot}3\times 119}{18(460-119)}=2{\cdot}47 \text{ N/mm}^2.$$

6·11. Doubly Reinforced T-beams using Revised Theory

$$M_1=Bd_s\ \frac{c}{2}\left(1+\frac{n-d_s}{n}\right)\left[d-\frac{d_s}{2}+\frac{d_s^2}{6(2n-d_s)}\right]$$

$$M_2=M-M_1$$

$$=(1{\cdot}5m-1)\ A_c\times c\left(\frac{n-f}{n}\right)(d-f)$$

$$A_t=A_{t1}+A_{t2}=\frac{M_1}{t\times a}+\frac{M_2}{t(d-f)}$$

Additional equivalent area given by compression steel

$$=1{\cdot}5(m-1)\ A_c.$$

Ex. 6·10. *Work out Ex. 6·4 by revised theory.*

Sol. Equivalent concrete area on compression side will be Bd_s and $(1{\cdot}5\ m-1)\ A_c$, while equivalent concrete area on tension side will be mA_t. This will happen when N.A. is in the web.

Take moments about N.A.

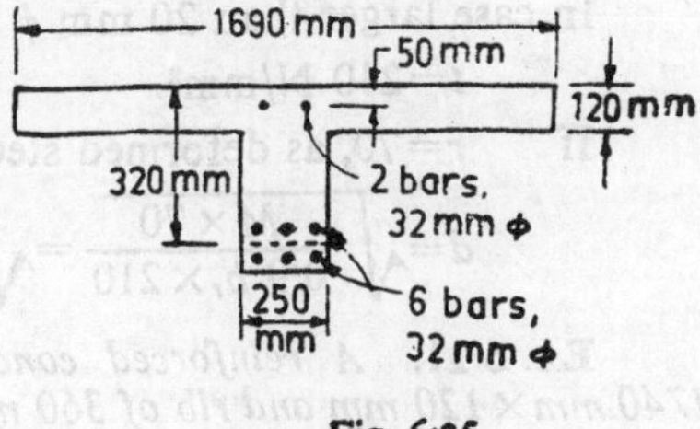

Fig. 6·25

$$1690\times 120(n-60)+(1{\cdot}5\times 18-1)\times 1\times 804\times(n-50)$$

$$=18\times 6\times 804(320-n)$$

$$\therefore\quad n=126{\cdot}9 \text{ mm}$$

$$M_1=Bd_s\ \frac{c_a}{2}\left(1+\frac{n-d_s}{n}\right)\left[d-\frac{d_s}{2}+\frac{d_s^2}{6(2n-d_s)}\right]$$

$$=1690\times 120\times \frac{c_a}{2}\left[1+\frac{126{\cdot}9-120}{126{\cdot}9}\right]\left[320-60+\frac{(120)^2}{(253{\cdot}8-120)}\right]$$

$$=29{,}715{,}000\ c_a$$

$$M_2=(1{\cdot}5m-1)A_c\times c_a\left(\frac{n-f}{n}\right)(d-f)$$

$$=(27-1)\times 2\times 804\ c_a\left(\frac{126{\cdot}9-50}{126{\cdot}9}\right)(320-50)$$

$$=6{,}841{,}000\ c_a$$

$$\therefore\quad M=M_1+M_2=29{,}715{,}000\ c_a+6{,}841{,}000\ c_a$$

$$=36{,}556{,}000\ c_a$$

$$\therefore\quad c_a=\frac{M}{36{,}556{,}000}=\frac{175{,}000{,}000}{36{,}556{,}000}=4{\cdot}787\ \text{N/mm}^2$$

Stress in compressive steel

$$=1{\cdot}5\ m\ c_a\left(\frac{n-f}{n}\right)=1{\cdot}5\times 18\times 4{\cdot}787\left(\frac{126{\cdot}9-50}{126{\cdot}9}\right)$$

$$=78{\cdot}32\ \text{N/mm}^2$$

Stress in tensile steel

$$=mc_a\left(\frac{d-n}{n}\right)=18\times 4{\cdot}787\left(\frac{320-126{\cdot}9}{126{\cdot}9}\right)$$

$$=131{\cdot}12\ \text{N/mm}^2.$$

6·12. Economical Depth of T-beams using High Yield Strergth Deformed Bars

From Art. 6·4,

$$d=\sqrt{\frac{M\times r}{0{\cdot}9\ b_r t}}$$

In case larger than 20 mm ϕ bars are used.

$$t=210\ \text{N/mm}^2.$$

If $r=70$, as deformed steel is costlier.

$$d=\sqrt{\frac{M\times 70}{0{\cdot}9\ b_r\times 210}}=\sqrt{\frac{M}{2{\cdot}7\ b_r}}.$$

Ex. 6·11. *A reinforced concrete T-beam has a flange of 1740 mm × 120 mm and rib of 360 mm × 300 mm as shown in Fig. 6 26. What high yield strength deformed bar reinforcement is required at a section, where the beam is subjected to a B.M. of 295,000 N-m. Take $c=5$ N/mm² and $m=18$.*

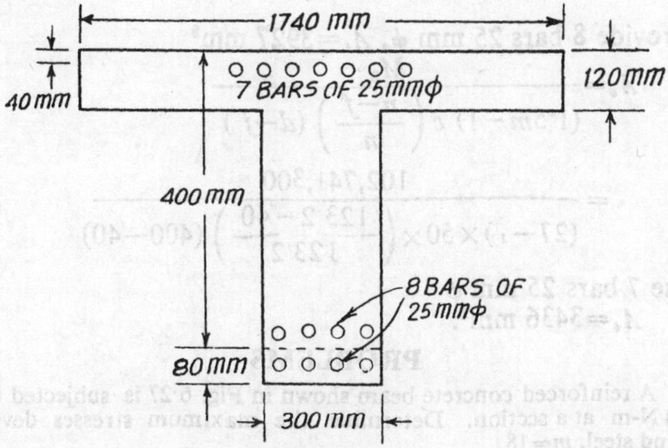

Fig. 6·26

Sol. Assume 80 mm cover to centre of tensile steel

$$d=360+120-80=400 \text{ mm}$$

Bars of greater than 20 mm diameter will be used,

$$\sigma_{st}=210 \text{ N/mm}^2$$

Depth of critical N.A. $=0{\cdot}308\ d=123{\cdot}2$ mm

Moment of resistance of balanced section

$$M_1=\frac{Bd_s c}{2}\left(1+\frac{n-d_s}{n}\right)\left[d-\frac{d_s}{2}+\frac{d_s^2}{6(2n-d_s)}\right]$$

$$=\frac{1740\times120\times5{\cdot}0}{2}\left(1+\frac{123{\cdot}2-120}{123{\cdot}2}\right)\times\left[400-60+\frac{(120)^2}{6(246{\cdot}4-120)}\right]$$

$$=192{,}258{,}700 \text{ N-mm}$$

$$M_2=M-M_1=295{,}000{,}000-192{,}258{,}700$$

$$=102{,}741{,}300 \text{ N-mm.}$$

Assume cover to compressive steel $f=40$ mm

$$A_t=A_{t1}+A_{t2}$$

$$=\frac{M_1}{t\times a}+\frac{M_2}{t(d-f)}$$

$$=\frac{192{,}258{,}700}{210\times\left[400-60+\frac{(120)^2}{6(246{\cdot}4-120)}\right]}+\frac{102{,}741{,}300}{210(400-40)}$$

$$=2550+1359$$

$$=3909 \text{ mm}^2.$$

Provide 8 bars 25 mm ϕ, $A_t = 3927 \text{ mm}^2$

$$A_c = \frac{M_2}{(1{\cdot}5m-1)\, c\left(\dfrac{n-f}{n}\right)(d-f)}$$

$$= \frac{102{,}741{,}300}{(27-1)\times 50\times\left(\dfrac{123{\cdot}2-40}{123{\cdot}2}\right)(400-40)}$$

Use 7 bars 25 mm ϕ

$A_c = 3436 \text{ mm}^2$.

PROBLEMS

1. A reinforced concrete beam shown in Fig. 6·27 is subjected to a B.M. of 100,000 N-m at a section. Determine the maximum stresses developed in concrete and steel, $m=18$.

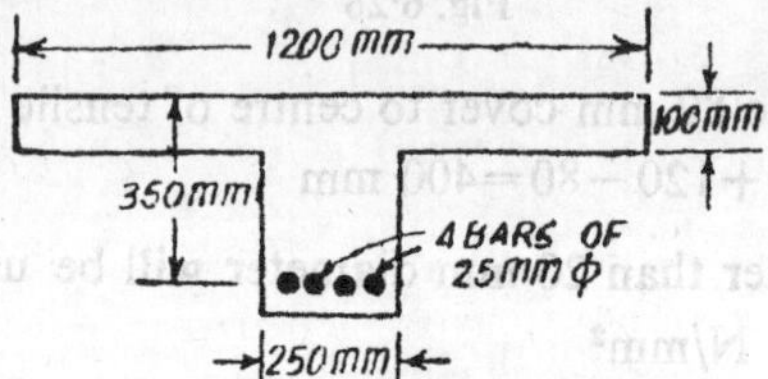

Fig. 6·27

2. A reinforced concrete T-beam, incorporating a 100 mm, thick slab has a flange width of 1540 mm, rib breadth of 24 cm and overall depth of 360 mm. The section is reinforced with 4 bars 20 mm ϕ in a single row. The centre of reinforcement being 40 mm from the bottom of the rib. Determine the moment of resistance of the section, assuming allowable stresses as $c=5 \text{ N/mm}^2$, $t=140 \text{ N/mm}^2$, $m=18$.

3. A reinforced concrete floor consists of 120 cm thick slab with beams spaced at 3 m apart. The span of the beam is 10 m. Live load on floor may be taken as 2000 N/mm^2. Design a suitable section for the beams.

4. A reinforced concrete beam shown in Fig. 6·28 is subjected to a B.M. of 100,000,000 N-mm at a section. Determine the maximum stresses developed in concrete, compressive steel and tensile steel, $m=18$.

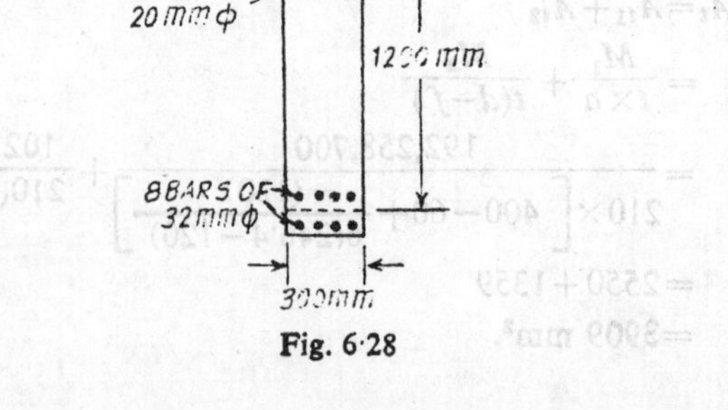

Fig. 6·28

5. The T-beam shown in Fig. 6·29 is simply supported over a span of 8 m. Calculate the maximum uniformly distributed load, inclusive of its own weight, it can carry if the allowable stresses are $c=5/\text{mm}^2$, $t=140$ N/mm², $m=18$.

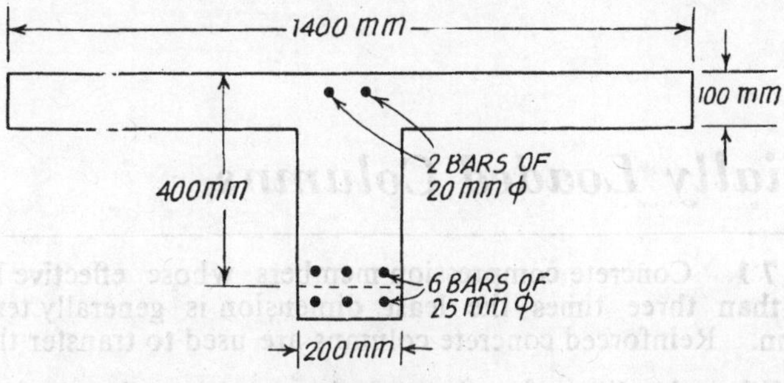

Fig. 6·29

6. Determine the moment of resistance of beam in problem 2, considering compression in web into account.

7. Design the canopy and canopy beam shown in Fig. 6·30.

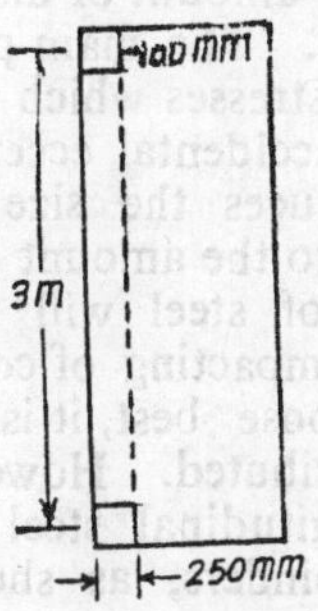

Fig. 6·30

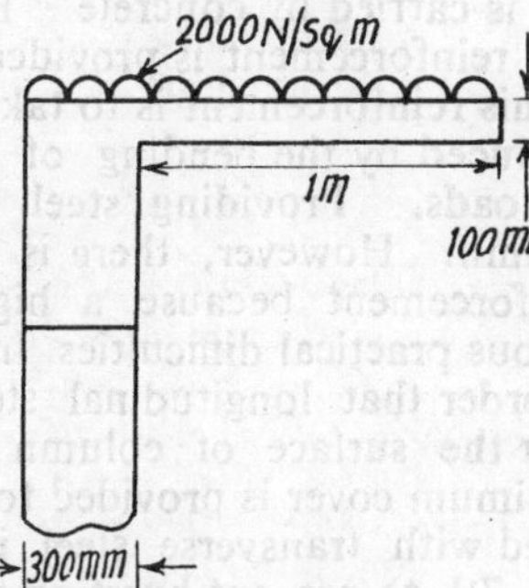

Fig. 6·31

8. Design the beam section shown in Fig. 6·31. The span of the beam is 4 m.

7

Axially Loaded Columns

7·1. Concrete compression members whose effective length is more than three times the least dimension is generally termed as column. Reinforced concrete columns are used to transfer the load.

When the line of action of the resultant thrust of the loads supported by a column coincides with the centre of gravity of the column section, the column is termed as axially loaded column.

As already discussed, concrete is strong in compression and hence it is advantageous to design a section such that all compression is carried by concrete. However, a certain amount of minimum steel reinforcement is provided in RCC columns. The main purpose of this reinforcement is to take up any tensile stresses which may be produced by the bending of the column or accidental eccentricity of loads. Providing steel reinforcement reduces the size of the column. However, there is an upper limit to the amount of steel reinforcement because a higher percentage of steel will involve serious practical difficulties in placing and compacting of concrete. In order that longitudinal steel serves its purpose best, it is placed near the surface of column and evenly distributed. However, a minimum cover is provided to steel. The longitudinal steel is provided with transverse steel in the form of binders, as shown in Fig. 7·1 to prevent bursting tendency in columns.

7·2. Short Column and Long Column

A column whose effective length does not exceed 12 times the least lateral dimension is termed as short column. For more exact calculations a column whose effective length does not exceed 50 times least radius of gyration is termed as long column. The allowable working stresses are given for short columns. For long columns reduced stresses are to be taken. The following formulas give the coefficients by which the maximum allowable values of stresses for steel and concrete are to be multiplied.

$$C_r = 1{\cdot}25 - \frac{l_{ef}}{48b}$$

where C_r = reduction coefficient,

l_{ef} = effective length of the column, and

b=least lateral dimension of column.

or $\quad C_r=1{\cdot}25-\dfrac{l_{ef}}{160\ i_{min}}$

where $\quad i_{min}$=least radius of gyration.

When in a column having helical reinforcement, the permissible load is based on the core area, the least dimension should be taken as diameter of core.

7·3. Type of Columns

Following are main types of columns :

(1) Column with longitudinal steel and lateral ties.

(2) Column with longitudinal steel and spirals.

(3) Composite column in which structural steel section is encased in concrete.

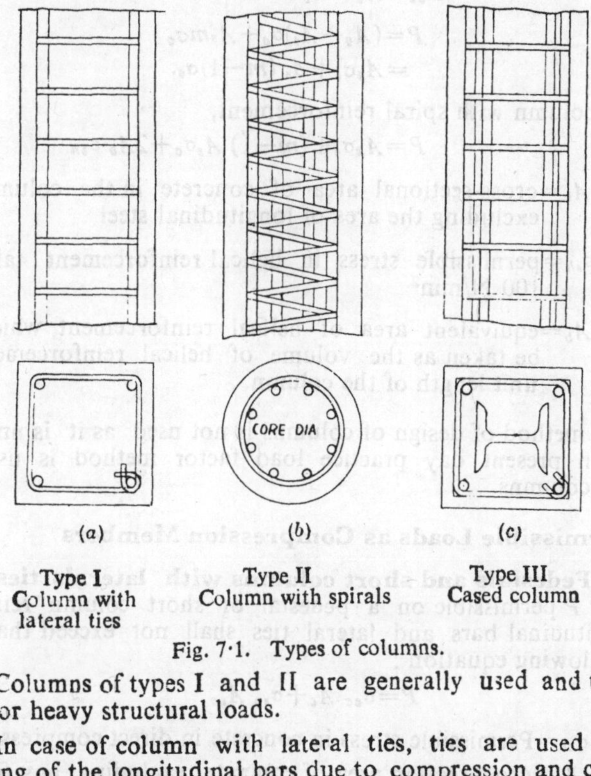

(a) Type I Column with lateral ties

(b) Type II Column with spirals

(c) Type III Cased column

Fig. 7·1. Types of columns.

Columns of types I and II are generally used and type III is used for heavy structural loads.

In case of column with lateral ties, ties are used to prevent buckling of the longitudinal bars due to compression and consequent spalling of the concrete is avoided. In case of concrete with spiral reinforcement, spirals serve the above purpose and in addition they form a lateral support to the concrete inside and having confining effect on it.

7·4. Elastic Theory for Design of Columns

Until recently all reinforced concrete columns were designed on the basis of elastic theory. In this theory, it is assumed that concrete is a perfectly elastic material within the range of stress used in design and there is no slippage between concrete and steel. Thus the deformation of steel and concrete in an axially loaded column will be same.

If σ_c is the permissible stress in concrete in direct compression stress in steel will be $m\sigma_c$.

Load carried by column $P = A_{cc} \times \sigma_c + A_s m \sigma_c$

where A_{cc} is the area of concrete and A_s is the area of steel.

If A_g is the gross cross-sectional area of section,

$$A_{cc} = A_g - A_s$$

$$\therefore \quad P = (A_g - A_s)\sigma_c + A_s m \sigma_c$$
$$= A_g \sigma_c + A_s (m-1)\sigma_c.$$

For column with spiral reinforcement,

$$P = A_k \sigma_c + (m-1) A_s \sigma_c + 2A_b \sigma_{sh}$$

where A_k = cross-sectional area of concrete in the column core excluding the area of longitudinal steel

σ_{sh} = permissible stress in helical reinforcement taken as 100 N/mm^2

A_b = equivalent area of helical reinforcement which may be taken as the volume of helical reinforcement per unit length of the column.

This method of design of columns is not used as it is uneconomical. In present day practice load factor method is used for design of columns.

7·5. Permissible Loads as Compression Members

(1) **Pedestals and short columns with laterals ties.** The axial load P permissible on a pedestal or short column reinforced with longitudinal bars and lateral ties shall not exceed that given by the following equation :

$$P = \sigma_{cc} A_c + \sigma_{sc} A_{sc}$$

where σ_{cc} = Permissible stress in concrete in direct compression,

A_c = cross-section area of concrete excluding any finishing material and reinforcing steel.

σ_{sc} = permissible compressive stress for column bars and

A_{sc} = cross-sectional area of the longitudinal steel.

(The minimum eccentricity may be deemed to be incorporated in the above equation.)

(2) **Short column with helical reinforcement.** The permissible load for columns with helical reinforcement satisfying provisions, that the ratio of volume of helical reinforcement to the volume of the core shall not be less than $0{\cdot}36\left[\frac{A_g}{A_c}-1\right]\frac{f_{ck}}{f_y}$ where, A_g=gross area of section, A_c=area of the core of the helically reinforced column measured to the outside diameter of the helix, f_{ck}= characteristic compressive strength of the concrete and f_y=characteristic strength of the helical reinforcement but not exceeding 415 N/mm², shall be 1·05 times the permissible load for similar member with lateral ties or rings.

(3) **Long columns.** The maximum permissible stress in a reinforced concrete column or part thereof having a ratio of effective column length to least lateral dimension above 12 shall not exceed that which results from the multiplication of the appropriate maximum permissible stress by the following formula :

$$c_r = 1{\cdot}25 - \frac{l_{ef}}{48b}$$

where c_r=reduction coefficient

l_{ef}=effective length of column and

b=least lateral dimension of column, for column with helical reinforcement b is the diameter of the core.

For, more exact calculations, the maximum permissible stresses in a reinforced concrete column or part thereof having a ratio of effective column length to least lateral radius of gyration above 40 shall not exceed those which result from the multiplication of the appropriate maximum permissible stresses by the coefficient c_r given by the following formula,

$$c_r = 1{\cdot}25 - \frac{l_{ef}}{160\ i_{min}},$$

where, i_{min} is the least radius of gyration.

(4) **Composite Columns.**

(*i*) **Allowable load.** The alloable axial load P on a composite column consisting of structural steel or cast-iron column thoroughly encased in concrete reinforced with both longitudinal and spiral reinforcement, shall not exceed that given by the following formula :

$$P = \sigma_{cc}\ A_c + \sigma_s t\ A_{sc} + \sigma_{mc}\ A_m$$

where, σ_{cc}=permissible stress in concrete in direct compression.

A_c=net area of concrete section, which is equal to the gross area of the concrete section $-A_{sc}-A_m$.

σ_{sc}=permissible compressive stress for column bars.

A_{sc}=cross sectional area of longitudinal bar reinforcement
σ_{mc}=allowable unit stress in metal core, not to exceed 125 N/mm² for a steel core, or 70 N/mm² for a cast iron and,
A_m=the cross-sectional area of the steel or cast iron core.

(*ii*) **Metal core and Reinforcement.** The cross sectional area of the metal core shall not exceed 20 per cent of the gross area of the column. If a hollow metal co.e is used, it shall be filled with concrete. The amount of longitudinal and spiral reinforcement and the requirements as to spacing of bars, details of splices and thickness of protective shell outside the spiral, shall conform to as for compression members A clearance of at least 75 mm shall be maintained between the spiral and metal core at all points, except that when the core consists of a structural H-column, the minimum clearance may be reduced to 50 mm.

(*iii*) **Splices and connections of metal cores** Metal cores in composite columns shall be accurately milled at splices and positive provisions shall be made for alignment of one core above another. At the column base, provisions shall be made to transfer the load to the footing at safe unit stresses. The base of the metal section shall be designed to transfer the load from the entire composite columns to the footing, or it may be designed to transfer the load from the metal section only, provided it is placed in the pier or pedestal as to leave ample section of concrete above the base for the transfer of loads from the reinforced concrete section of the column by means of bond on the vertical reinforcement and by direct compression on the concrete. Transfer of loads to the metal core shall be provided for by the use of bearing members, such as billets, brackets or other positive connections, these shall be provided at the top of the metal core and at intermediate floor levels where required. The column as a whole shall satisfy the requirements of formula given under (*i*) at any point, in addition to this reinforced concrete portion shall be designed to carry, according to (1) and (2) of Art. 7·5, as the case may be, all floor loads brought into the column at levels between the metal brackets or connections. In applying the formulas the gross area of column shall be taken to be area of the concrete section outside the metal core, and the allowable load on the reinforced concrete shall be further limited to 0·28 f_{ck} times gross sectional area of the column.

(*iv*) **Allowable load on metal core only.** The metal core of composite column shall be designed to carry safely any construction or other loads to be placed upon them prior to the encasement in concrete.

7·6. General Notes on Design of Columns (I.S.I.)

(I) **Effective Length of Compression Members.** In the absence of more exact analysis, the effective length l_{ef} of columns may be obtained as follows :

The effective length may be obtained from the ratio of l_{ef}/l, where l is the unsupported length from Fig. 7·2 and Fig. 7·3. Use

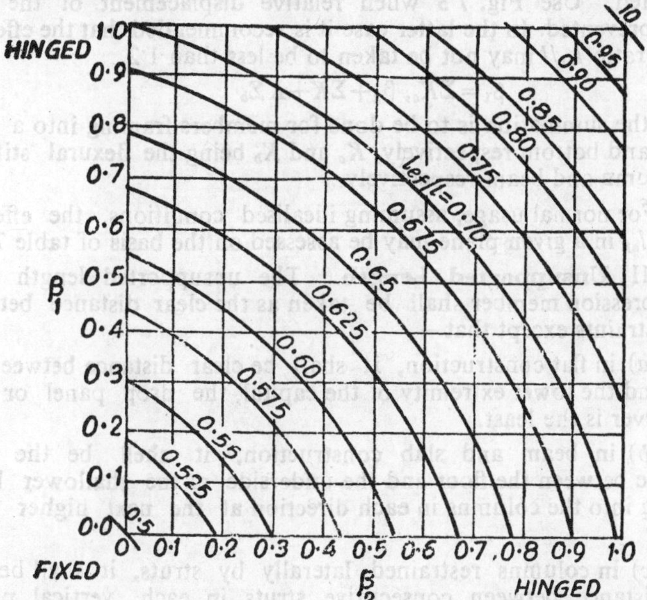

Fig. 7·2. Effective length ratio for a column in a frame with no sway.

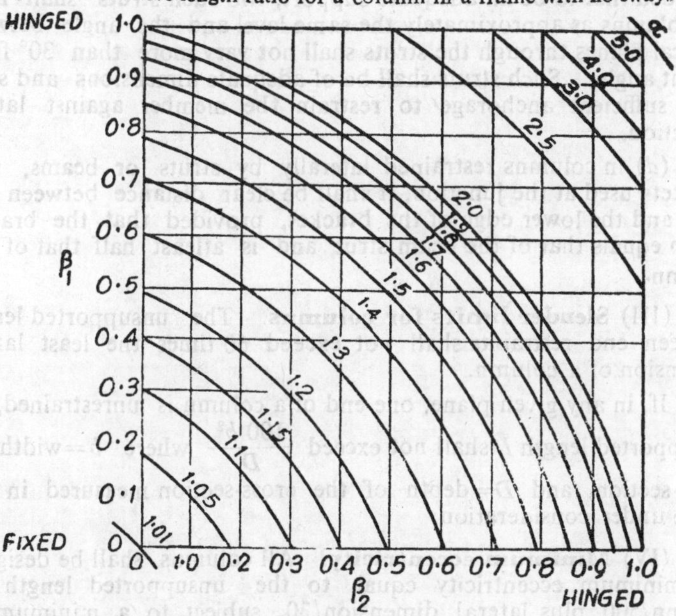

Fig. 7·3. Effective length ratio for a column in a frame without restraint against sway.

Fig. 7·2 when relative displacement of the ends of the column is prevented. Use Fig. 7·3 when relative displacement of the ends is not prevented. In the latter case it is recommended that the effective length ratio l_{ef}/l may not be taken to be less than 1·2.

$$\beta_1 = \Sigma K_c, \beta_2 + \Sigma K + K\Sigma_b$$

where the summation is to be done for members framing into a joint at top and bottom respectively, K_c and K_b being the flexural stiffness for column and beam respectively.

For normal usage, assuming idealised conditions, the effective length l_{ef} in a given plane may be assessed on the basis of table 7·1.

(II) **Unsupported Length.** The unsupported length l, of a compression member shall be taken as the clear distance between end restraints except that—

(*a*) in flat construction, it shall be clear distance between the floor and the lower extremity of the capital, the drop panel or slab whichever is the least.

(*b*) in beam and slab construction, it shell be the clear distance between the floor and the underside of the shallower beam framing into the columns in each direction at the next higher floor level.

(*c*) in columns restrained laterally by struts, it shall be the clear distance between consecutive struts in each vertical plane, provided that to be an adequate support, two such struts shall meet the columns as approximately the same level and the angle between vertical planes through the struts shall not vary more than 30° from a right angle. Such struts shall be of adequate dimensions and shall have sufficient anchorage to restrain the member against lateral deflection.

(*d*) in columns restrained laterally by struts or beams, with brackets used at the junction, it shall be clear distance between the floor and the lower edge of the bracket, provided that the bracket width equals that of the beam strut and is atleast half that of the column.

(III) **Slender limits for columns.** The unsupported length between end restraints shall not exceed 60 times the least lateral dimension of a column.

If, in any given plane, one end of a column is unrestrained, its unsupported length l, shall not exceed $\dfrac{100\, b^2}{D}$ where b=width of cross-section, and D=depth of the cross-section measured in the plane under consideration.

(IV) **Minimum eccentricity.** All columns shall be designed for minimum eccentricity equal to the unsupported length of column/500 plus lateral dimension/30, subject to a minimum of 20 mm.

Table 7·1
Effective Length of Compression Members

Degree of end restraint of Compression Member	*Symbol*	*Theoretical value of effective length*	*Recommended value of effective length*
1. Effectively held in position and restrained against rotation at both ends.		0·5 *l*	0·65 *l*
2. Effectively held in position at both ends, restrained against rotation at one end.		0·7 *l*	0·80 *l*
3. Effectively held in position at both ends, but not restrained against rotation.		0·7 *l*	1·20 *l*
4. Effectively held in position and restrained against rotation at one end, and at the other restrained against rotation but not held in position.	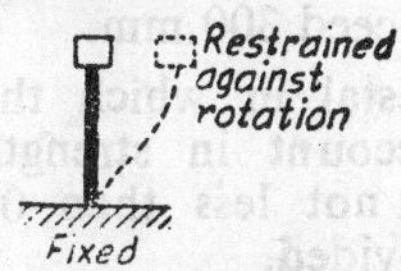	1·00 *l*	1·20 *l*
5. Effectively held in position and restrained against rotation at one end, and at the other partially restrained against rotation but not held in position.	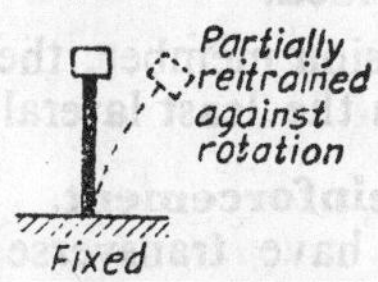	—	1·50 *l*
6. Effectively held in position at one end but not restrained against rotation, and at the other end restrained against rotation but not held in position.	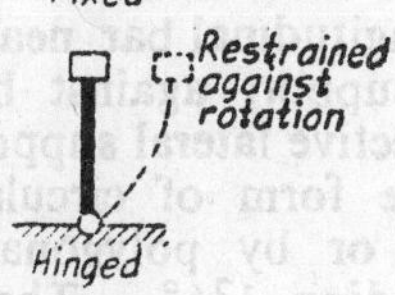	2·00 *l*	2·00 *l*
7. Effectively held in position and restrained against position at one end but not held in position nor restrained against rotation at the other end.		2·00 *l*	2·00 *l*

(V) **Longitudinal Reinforcement**

(*a*) The cross-sectional area of longitudinal reinforcement, shall be not less than 0·8 per cent nor more than 6 per cent of the gross-sectional area of the column.

The use of 6 per cent reinforcement may involve practical difficulties in placing and compacting of concrete, hence lower percentage is recommended. Where bars from the columns below have to be lapped with those in the column under consideration, the percentage steel shall usually not exceed 4 per cent.

(*b*) In any column that has a large cross-sectional area than the required to support the load, the minimum percentage of steel shall be based upon the area of concrete required to resist the direct stress and not upon the actual area.

(*c*) The minimum number of longitudinal bars provided in a column shall be four in rectangular columns and six in circular section.

(*d*) The bars shall not be less than 12 mm in diameter.

(*e*) A reinforced concrete column having helical reinforcement shall have at least six bars of longitudinal reinforcement within the helical reinforcement.

(*f*) In a helically reinforced column, the longitudinal bars shall be in contact with the helical reinforcement and equidistant around its inner circumference.

(*g*) Spacing of longitudinal bars measured along the periphery of the column shall not exceed 300 mm.

(*h*) In case of pedestal in which the longitudinal reinforcement is not taken into account in strength calculations, nominal longitudinal reinforcement not less than 0·15 per cent of the cross-sectional area shall be provided.

Pedestal is a compression member, the effective length of which does not exceed three times the least lateral dimension.

(VI) **Transverse reinforcement.** A reinforced concrete compression members shall have transverse or helical reinforcement so disposed that every longitudinal bar nearest to the compression face has effective level support against buckling subject to provisions as follow. The effective lateral suppo t is given by transverse reinforcement either in the form of circular rings capable of taking up circumferential tension or by polygonal links (lateral ties) with internal angles not exceeding 135°. The ends of the transverse reinforcement shall be properly anchored.

(*a*) **Arrangement of Transverse Reinforcement**

(*i*) If the longitudinal bars are not spaced more than 75 mm on either side, transverse reinforcement need only to go round corner and alternate bars for the purpose of providing effective lateral support (see Fig. 7·4).

(*ii*) If the longitudinal bars spaced at a distance of not exceeding 48 times the diameter of the tie are effectively tied in two directions, additional bars in between these bars need to be tied in one direction by open ties (see Fig. 7·5).

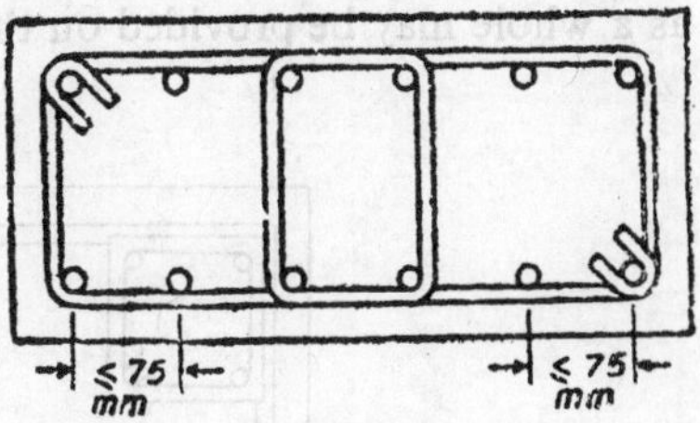

Fig. 7·4

(*iii*) Where the longitudinal reinforcing bars in a compression member are placed in more than one row, effective lateral support to the longitudinal bars in the inner rows may be

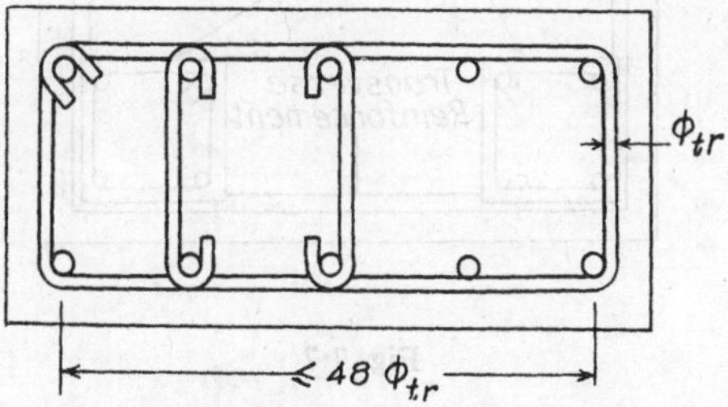

Fig. 7·5

assumed to have been provided if—

(1) transverse reinforcement is provided for the outermost row in accordance with (*i*) and (*ii*) and

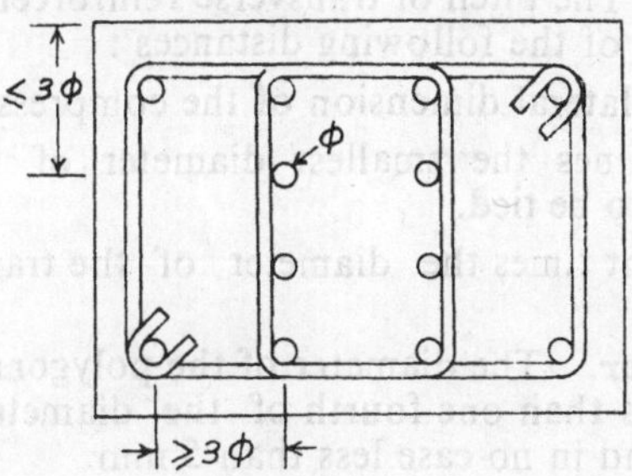

Fig. 7·6

(2) no bar of the inner row is closer to the nearest compression face than three times the diameter of the largest bar in the inner row (see Fig. 7·6).

(*iv*) Where the longitudinal bars in a compression member are grouped (not in contact) and such group adequately tied with transverse reinforcement in accordance with the above requirements, the transverse reinforcement for the compression member as a whole may be provided on the assumption that each group is

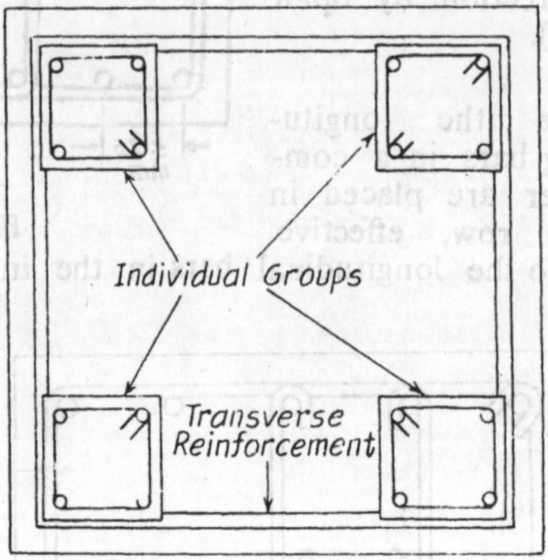

Fig. 7·7

a single longitudinal bar for purpose of determining the pitch and the diameter of the transverse reinforcement. The diameter of such transverse reinforcement need not, however, exceed 20 mm (see Fig. 7·7).

(*b*) **Pitch and diameter of lateral ties**

(*i*) **Pitch.** The pitch of transverse reinforcement shall not be more than the least of the following distances :

(1) The least lateral dimension of the compression member.

(2) Sixteen times the smallest diameter of the longitudinal reinforcement bar to be tied.

(3) Forty-eight times the diameter of the transverse reinforcement.

(*iii*) **Diameter.** The diameter of the polygonal links or lateral ties shall be not less than one-fourth of the diameter of the largest longitudinal bar, and in no case less than 5 mm.

(*c*) **Helical Reinforcement**

(*i*) **Pitch.** Helical reinforcement shall be of regular formation with the turns of the helix spaced evenly and its ends shall be anchored properly by providing one and a half extra turns of the spiral bar. Where an increased load on the column on the strength of helical reinforcement is allowed for, the pitch of helical turns shall be not more than 75 mm, nor less than 25 mm, nor less than three times the diameter of the steel bar forming the helix. In other cases the requirements of *b* shall be complied with.

(*ii*) **Diameter.** The diameter of the helical reinforcement.

(*d*) In columns where longitudinal bars are offset at a splice, the slope of the inclined portion of the bar with the axis of the column shall not exceed 1 in 6, and the portions of the bar above and below the offset shall be parallel to the axis of the column. Adequate horizontal support at the offset bends shall be treated as a matter of design, and shall be provided by metal ties, spirals, or parts of the floor construction. Metal ties or spirals so designed shall be placed near (not more than eight-bar diameter from) the point of bend. The horizontal thrust to be resisted shall be assumed as $1\frac{1}{2}$ times the horizontal components of the nominal stress in the inclined portion of the bar. Offset bars shall be bent before they are placed in the forms. Where columns faces are offset 75 mm. or more, splices of vertical bars adjacent to the offset face shall be made by separate dowels overlapped, as given in lap splices below :

(*i*) Lap splices shall not be used for bars larger then 30 mm. For larger diameters, bars may be welded, in cases where welding is not practicable, lapping of bars larger than 36 mm may be permitted, in which case additional spirals should be provided around the lapped bars.

(*ii*) Lap splices shall be considered as staggered if the centre to centre distance of the splices is not less than 1·3 times the lap length.

(*iii*) Lap length in compression shall be equal to the development length in compression, but not less than 24 ϕ.

(*iv*) When bars of two different diameters are to be spliced, the lap length shall be calculated on the basis of diameters of the smaller bar.

(*v*) In case of bundled bars, lapped splices of the bundled bars shall be made by splicing one bar at a time, such individual splices within a bundle shall be staggered.

(*e*) **Cover.** The longitudinal reinforcing bar in a column shall have concrete cover not less than 40 mm, nor less than the diameter of such rod or bar. In the case of columns of minimum dimension of 200 mm or under, whose reinforcing bar, do not exceed 12 mm, the cover of 25 mm may be used.

Ex. 7·1. *A short R.C. column is required to carry an axial load of 1,000,000 N. If the permissible compressive stresses in the concrete and steel are* $\sigma_c = 4$ *N/mm*^{2} *and* $\sigma_{sc} = 130$ *N/mm*^{2} *respectively, determine the size of the column, longitudinal reinforcement and lateral ties.*

Sol. Provide square column with 0·8% steel.

$$A_{cc}=0{\cdot}992\ A_g,\ A_s=0{\cdot}008\ A_g$$

$$1{,}000{,}000=0{\cdot}992\ A_g\times 4+0{\cdot}008\ A_g\times 130$$

$$\therefore\ A_g=\frac{1{,}000{,}000}{3{\cdot}968+1{\cdot}04}+\frac{1{,}000{,}000}{5{\cdot}008}$$

$$=199{,}680\ \text{mm}^2.$$

Using square column, side of square=446·9 mm.

Use column of 440 mm × 440 mm

$$P=1{,}000{,}000=4(440\times 400-A_s)+130\ A_s$$

$$\therefore\ 1{,}000{,}000=774{,}400+126\ A_s$$

$$\therefore\ A_s=\frac{225{,}600}{126}=1790{\cdot}5\ \text{mm}^2.$$

Provide 4 bars of 25 mm ϕ, A_s=1963 mm².

Lateral ties. $\frac{1}{4}$ th diameter of main reinforcement $=\frac{25}{4}=6{\cdot}25$ mm. Use 8 mm ϕ lateral ties.

Pitch

(1) Least dimension of the compression member=440 mm.

(2) Sixteen times the diameter of longitudinal bar=16×25 =400 mm.

(3) Forty eight times the diameter of the transverse reinforcement =48×8=384.

Use pitch of 350 mm. Provide 8 mm ϕ ties at 350 mm centres.

Ex. 7·2. *Calculate the maximum load that can be carried by the 500 mm diameter column reinforced with 6 bars of 20 mm ϕ as shown in Fig. 7·8, when it is used as a short column. Allowable stresses are $\sigma_c=4$ N/mm², $\sigma_{sc}=130$ N/mm, $\sigma_{sh}=100$ N/mm².*

Sol. Length of each spiral $=\sqrt{(\pi d)^2+p^2}=\sqrt{\pi\times(428)^2+(60)^2}$

$$=1321\ \text{mm}.$$

$$\text{Volume of spiral per 1 mm height}=\frac{\frac{\pi}{4}\times 8^2\times 1321}{60}$$

$$=1885\ \text{mm}^3$$

Volume of core per 1 mm height $\frac{\pi}{4}\times(420)^2=138{,}544$ mm

$$\frac{\text{Volume of helical reinforcement}}{\text{Volume of core}}=\frac{1885}{138{,}544}=0{\cdot}0136$$

Minimum volume=

$$0{\cdot}36\left(\frac{A_g}{A_c}-1\right)\frac{f_{ck}}{f_y}$$

$$=0{\cdot}36\left[\frac{\frac{\pi}{4}\times(500)^2}{\frac{\pi}{4}\times(424)^2}-1\right]\frac{4}{100}$$

$$=0{\cdot}00193<0{\cdot}0136$$

$$P=1{\cdot}05[\sigma_{cc}A_c+\sigma_{sc}A_{sc}]$$

$$A_{sc}=6\times\frac{\pi}{4}\times(20)^2=1885 \text{ mm}^2$$

$$A=\frac{\pi}{4}\times(500)^2-1885$$

$$=196{,}350-1885$$

$$=194{,}465 \text{ mm}^2$$

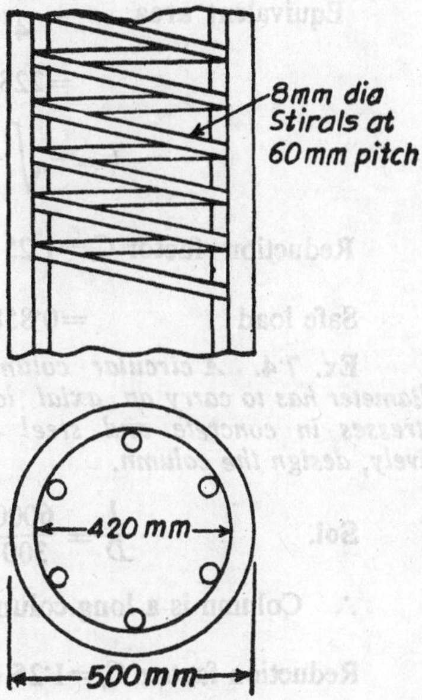

Fig. 7·8

$$\therefore \quad P=1{\cdot}05[4\times194{,}465+130\times1885]=1{,}074{,}056 \text{ N.}$$

Ex. 7·3. *What will be the load carried by the column of Ex. 7·2 if the effective height is 8 m.*

Sol. $$\frac{l}{D}=\frac{8000}{420}=19$$

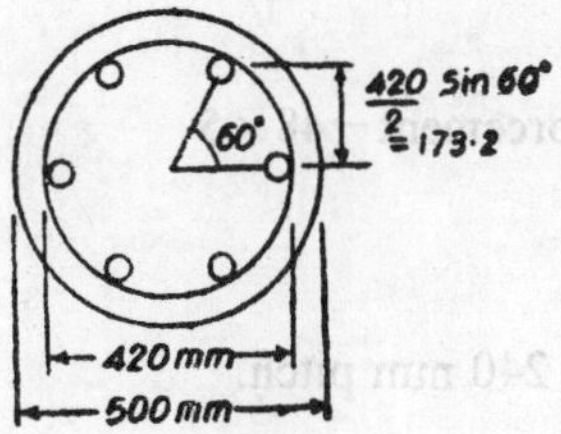

Fig. 7·9

∴ The column is a long column

$$\text{Reduction factor}=1{\cdot}25-\frac{8000}{48\times420}$$

$$=0{\cdot}8523$$

$$\text{Safe load}=1{,}074{,}056\times0{\cdot}8532$$

$$=916{,}357 \text{ N.}$$

Safe load by exact formula—

Equivalent moment of inertia

$$=\frac{\pi}{64}\times(500)^4+17\times4\times314\times(173{\cdot}2)^2$$

$$=370{,}709$$

Equivalent area $=\frac{\pi}{4}\times(500)^2+17\times6\times314$

$=2283{\cdot}5\times10^5$ mm²

$$i_{min}=\sqrt{\frac{I}{A}}=\sqrt{\frac{370{,}709\times10^4}{2283{\cdot}5\times10^2}}=127{\cdot}6 \text{ mm}$$

Reduction factor $C_r=1{\cdot}25-\frac{8000}{160\times127{\cdot}6}=0{\cdot}8582$

Safe load $=0{\cdot}8582\times1{,}074{,}056=921{,}755$ N.

Ex. 7·4. *A circular column 6 m effective height and 300 mm diameter has to carry an axial load of 420,000 N. If the allowable stresses in concrete and steel are 4 N/mm² and 130 N/mm² respectively, design the column.*

Sol. $\frac{l}{D}=\frac{6000}{300}=20$

∴ **Column is a long column.**

Reduction factor $C_r=1{\cdot}25-\frac{l_{ef}}{48\,b}=1{\cdot}25-\frac{6000}{48\times300}=0{\cdot}883$

Safe load on column $=0{\cdot}833\left[\left\{\frac{\pi}{4}\times(300)^2-A_s\right\}\times4+A_s\times1\,30\right]$

$=420{,}000$

$282{,}743-4A_s+130A_s=504{,}202$

∴ $A_s=\frac{504{,}202-282{,}743}{126}=1758 \text{ mm}^2$

Provide 6 bars of mm 20 ϕ, $A_s=6\times314=1884$ mm²

Provide 40 mm cover to main reinforcement.

1/4th of diameter of main reinforcement = 5 mm.

Pitch of circular hoops—

(1) 16 times diameter of main reinforcement $=16\times20$

$=320$ mm

(2) 48 times diameter of transverse reinforcement $=48\times5$

$=240$ mm

(3) Least lateral dimension = 300 mm

Provide 5 mm diameter circular hoops of 240 mm pitch.

Ex. 7·5. *Design a square section to carry a load of 500,000 N. Effective length of column is 6 metres. Allowable stresses are* $\sigma_c=4$ *N/mm²*, $\sigma_{sc}=130$ *N/mm²*.

Sol. Let D be the side of square column. **1% steel will be used** in column section.

$$\text{Reduction factor } C_r = 1.25 - \frac{6000}{48 \times D} = 1.25 - \frac{125}{D}$$

$$\text{Safe load on column} = \left(1.25 - \frac{125}{D}\right) [A_{cc} \times 4 + A_s \times 130]$$

$$\therefore \quad 500{,}000 = \left(1.25 - \frac{125}{D}\right)$$

$$[(D^2 - 0.01 \; D^2) \times 4 + 0.01 \; D^2 \times 130]$$

$$= \left(1.25 - \frac{125}{D}\right) [3.96 \; D^2 + 1.3D^2]$$

$$= \left(1.25 - \frac{125}{D}\right)(5.26D^2)$$

$$\therefore \quad 1.25 \; D^2 - 125 \; D = 500{,}000/5.26$$

$$\therefore \quad D^2 - 100 \; D = \frac{500{,}000}{1.25 \times 5.26} = 76{,}047$$

$$\therefore \quad (D - 50)^2 = 76{,}047 + 2500 = 78{,}547$$

$$\therefore \quad D = 330.3 \text{ mm}$$

Provide 320 mm × 320 mm column.

$$\text{Reduction factor} \quad = 1.25 - \frac{6000}{48 \times 320} = 0.8594$$

$$500{,}000 = 0.8594 \; [(320 \times 320 - A_s) \times 4 + 130 \; A_s]$$

$$\frac{500{,}000}{0.8594} = 409{,}600 - 4A_s + 130A_s$$

$$\therefore \quad A_s = 1367 \text{ mm}^2.$$

Provide 8 bars of 16 mm φ, $A_s = 1608$ mm².

Lateral ties. Provide 5 mm ϕ lateral ties.

Pitch. (*i*) 48 times the diameter of lateral ties

$$= 5 \times 48 = 240 \text{ mm}$$

(*ii*) 16 times the diameter of main reinforcement

$$= 16 \times 16 = 256 \text{ mm}$$

(*iii*) Least lateral dimension = 320 mm.

Provide 5 mm ϕ lateral ties at 240 mm centres.

Check by exact formula

Equivalent area $=320\times 320+17\times 1608$

$=102{,}400+27{,}336$

$=129{,}736\ \text{mm}^2$

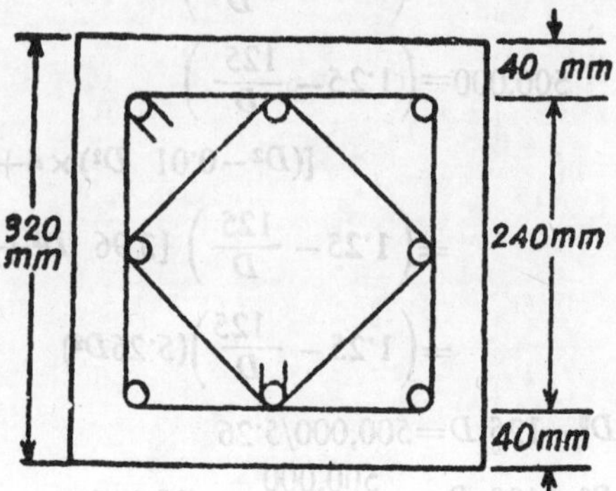

Fig. 7·10

Equivalent moment of intertia

$$=\frac{1}{12}\times 320\times(320)^3+17\times 6\times 201\times(112)^2$$

$$=113{,}060\times 10^4\ \text{mm}^2$$

$$i_{min}=\sqrt{\frac{I}{A}}=\sqrt{\frac{4{,}130{,}600\times 10^4}{129{,}736}}=93{\cdot}4\ \text{mm}$$

Reduction factor $=1{\cdot}25-\dfrac{6000}{160\times 93{\cdot}4}=0{\cdot}8485$

Safe load $=0{\cdot}8485\,[(320\times 320-1608)\times 4+1608\times 130]$

$=0{\cdot}8485\times 612{,}208=519{,}458$ N.

Ex. 7·6. *A short composite column has section as shown in Fig. 7·11. Find the safe load on the column if permissible stresses are* $\sigma_c=4/mm^2$, $\sigma_{sc}=130\ N/mm^2$ *and* $\sigma_m=125\ N/mm^2$.

Sol. $P=A_{cc}\times\sigma_c+A_s\times\sigma_{sc}+A_m\,\sigma_m$

$=(400\times 400-8\times 201-3448)\times 4$

$+8\times 201\times 130+3448\times 125$

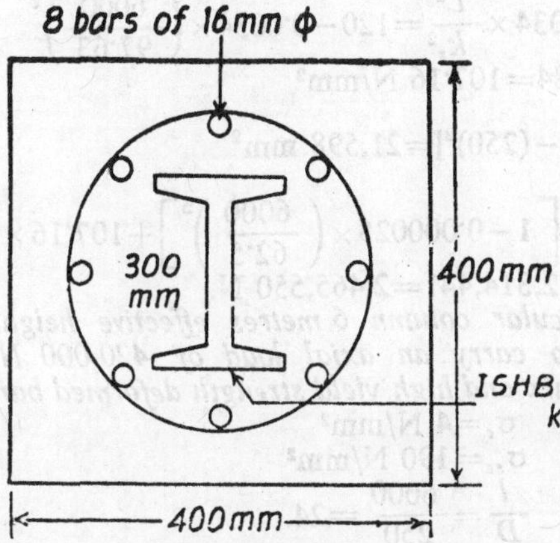

Fig. 7·11

$$\therefore \quad P=[(160{,}000-1608-3448)\times 4+1608 \times 130+3448\times 125]$$
$$=619{,}776+209{,}040+431{,}000$$
$$=1{,}259{,}816 \text{ N.}$$

Ex. 7·7. *A concrete filled pipe column consists of steel pipe with external diameter of 300 mm and internal diameter of 250 mm and is 6 m long. Find the safe load on the column,* $\sigma_c=4$ *N/mm*^{2}.

Sol. For concrete filled pipes,

$$P=A_c\sigma_c\left(1-0{\cdot}000025\frac{L^2}{k_c^2}\right)+\sigma_{sc}A_{sc}$$

L = unsupported length = 6000 mm.

$$A_c=\frac{\pi}{4}\times(250)^2=49{,}087 \text{ mm}^2$$

$$\sigma_c=4 \text{ N/mm}^2$$

$$\therefore \quad k_c=\sqrt{\frac{I_c}{A_c}}=\sqrt{\frac{\frac{\pi}{64}(250)^4}{\frac{\pi}{4}(250)^2}}=\frac{250}{4}=62{\cdot}5 \text{ mm}$$

$$k_s=\sqrt{\frac{\frac{\pi}{64}[(300)^4-(250)^4]}{\frac{\pi}{4}[(300)^2-(250)^2]}}=\sqrt{\frac{1}{16}[(300)^2+(250)^2]}$$

$$=97{\cdot}63 \text{ mm}$$

$$\sigma_{sc} = 120 - 0{\cdot}0034 \times \frac{L^2}{k_s^2} = 120 - 0{\cdot}0034 \times \left(\frac{6000}{97{\cdot}63}\right)^2$$
$$= 120 - 12{\cdot}84 = 107{\cdot}16 \text{ N/mm}^2$$

$$A_s = \frac{\pi}{4}[(300)^2 - (250)^2] = 21{,}598 \text{ mm}^2$$

$$P = 49{,}087 \times 4\left[1 - 0{\cdot}000025 \times \left(\frac{6000}{62{\cdot}5}\right)^2\right] + 107{\cdot}16 \times 21{,}598$$
$$= 151{,}109 + 2{,}314{,}441 = 2{,}465{,}550 \text{ N.}$$

Ex. 7·9. *A circular column 6 metres effective height and 250 mm diameter has to carry an axial load of 420,000 N. Design column using M 15 mix and high yield strength deformed bars.*

Sol.
$$\sigma_c = 4 \text{ N/mm}^2$$
$$\sigma_{sc} = 190 \text{ N/mm}^2$$
$$\frac{l}{D} = \frac{6000}{250} = 24.$$

Column is a long column.

$$\text{Reduction factor} \quad = 1{\cdot}25 - \frac{l}{48D} = 1{\cdot}25 - \frac{6000}{48 \times 250} = 0{\cdot}75$$

$$\text{Safe load on column} = 0{\cdot}75\left[\left\{\frac{\pi}{4} \times (250)^2 - A_s\right\} \times 4 + 190A_s\right]$$
$$420{,}000 = 0{\cdot}75[196{,}349 - 4A_s + 190A_s]$$
$$\therefore \quad A_s = \frac{363{,}651}{186} = 1955 \text{ mm}^2$$

Use 8 bars 20 mm ϕ, $A = 2514$ mm².

Ex. 7·10. *Calculate the maximum load that can be carried by the 500 mm diameter column reinforced with 6 bars of 20 mm ϕ as shown in Fig. 7·12. It is used as a short column. High yield strength deformed bars are used. Concrete mix M 15 is used.*

Sol.
$$\sigma_{sc} = 190 \text{ N/mm}^2, \; \sigma_c = 4 \text{ N/mm}^2$$
$$\sigma_{sh} = 160 \text{ N/mm}^2.$$

$$\text{Length of each spiral} = \sqrt{(\pi D^2 + p^2)}$$
$$= \sqrt{(\pi \times 428)^2 + 60^2} = 1321 \text{ mm.}$$

Volume of spiral per 1 mm height

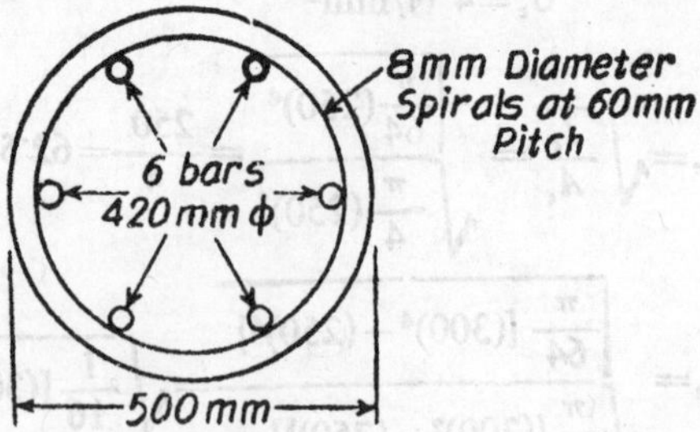

Fig. 7·12

$$A_b = \frac{\frac{\pi}{4} \times 8^2 \times 1321}{60} = 1100 \text{ mm}^3$$

Volume of concrete core per mm

$$= \frac{\pi}{4} \times (420)^2 = 138{,}544 \text{ mm}^3$$

$$\therefore \quad \frac{1100}{138{,}544} = 0{\cdot}00794$$

Minimum spiral reinforcement=

$$0{\cdot}36\left(\frac{A_g}{A_c} - 1\right)\frac{f_{ck}}{f_y} = 0{\cdot}36\left[\frac{\frac{\pi}{4} \times (500)^2}{\frac{\pi}{4} \times (420)^2} - 1\right] \times \frac{4}{160}$$

$$= 0{\cdot}0038 < 0{\cdot}00794$$

$$\text{Safe load } P = 1{\cdot}05\left[\left\{\frac{\pi}{4} \times (500)^2 - 6 \times 314\right\} \times 4 + 6 \times 314 \times 190\right]$$

$$= 1{\cdot}05[777{,}862 + 357{,}960]$$

$$= 1{\cdot}05[1{,}135{,}822] = 1{,}193{,}813 \text{ N.}$$

PROBLEMS

1. A short column 300 mm square is reinforced with 4 bars of 20 mm diameter. Calculate the permissible load on the column. Allowable stresses are $\sigma_c = 4 \text{N/mm}^2$, $\sigma_{sc} = 130 \text{ N/mm}^2$.

Design the transverse reinforcement for the column.

2. A 300 mm × 200 mm column 4 m effective height has to support an axial load of 300,000 N. Design the column reinforcement. Check the section by exact formula.

3. A short column 400 mm in diameter has core diameter of 320 mm and is reinforced with 6 bars of 20 mm diameter. The spiral ties are 3 mm ϕ at 50 mm pitch. Determine the safe load on the column $\sigma_c = 4 \text{ N/mm}^2$, $\sigma_{sc} = 130 \text{ N/mm}^2$, $\sigma_{sh} = 100 \text{ m/mm}^2$.

4. What is the safe load carried by column of Example 3, if the effective height column is 7 m ?

5. Design an RCC column 6 m long to support load of 400,000 N inclusive of self-weight.

6. Design a circular column to carry load of 1,200,000 N. The effective height of column is 8 m.

7. A short composite column has to support a load of 2,000,000 N. The size of the column is to be 400 mm × 400 mm. Design the column section.

8. **Find the allowable load on the column section shown in Fig. 7·13.**

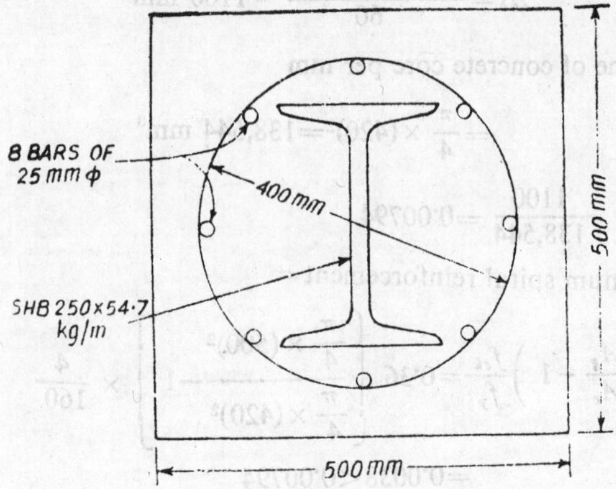

Fig. 7·13

9. **A concrete filled column consists of steel pipe with external diameter 200 mm and internal diameter 150 mm. The column is 5 m long. Find the safe load on the column. Allowable stress in concrete is 4 N/mm².**

8

Two-way Reinforced Slabs

8·1. When the slabs are supported on all four edges the load is transferred to all the four supports and, therefore, the bending moments and deflections in such slabs are considerably less than those in slabs supported on two opposite edges only. In case the slab is simply supported on all the four supports the corners get lifted up when such a slab is loaded. In case the corners are held down, the deflections in the central portion are further reduced and thus the bending moments are reduced in such slabs. Where the corners are held down due to monolithic construction of the slab with the walls or with edge beams, tnere will be torsion at the corners and. therefore, the torsion reinforcement is to be provided.

A typical beam and slab arrangement is shown in Fig. 8·1 (*a*). In this type of construction slab is spanning in two directions over

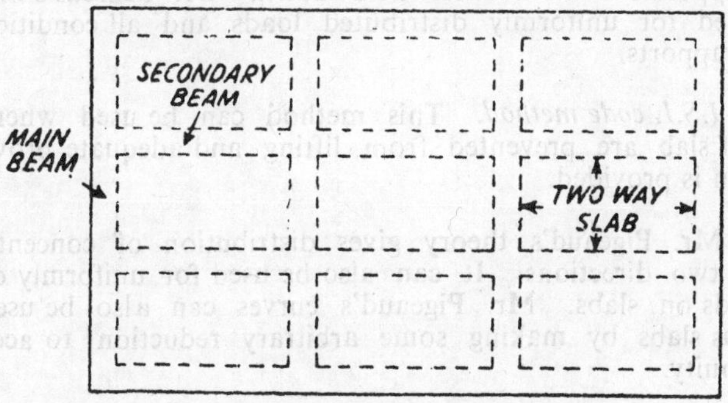

(*a*) Beam and slab arrangement

Fig. 8·1

main beams and secondary beams. Because of monolithic construction of slab with beams the corners of slab are held down. As slab is spanning in both directions steel in each direction is considered as main steel. The secondary beams are in turn supported on main beams and are designed as continuous beams. The main beams are supported on columns or walls. The load transferred to main beams from secondary beams is the reaction coming from continuous beams.

(*b*) Continuous slab

Fig. 8·1

The load carried to each support and the bending moments in slabs mentioned above are not easy to calculate if all the conditions are to be satisfied. Therefore, empirical formulae and approximate theories have been formulated which give bending moments in slabs supported on all four edges. The method applicable in particular case depends on the support condition and type of load. Bending moments can be obtained by one of the following methods :

1. The theory of Grashoff and Rankine is used for distributed loads when the slab is simply supported on all four edges and the corners are not held down.

2. The method of Dr. Marcus can be used if the slab is simply supported with corners held down. Dr. Marcus's method can be used for uniformly distributed loads and all conditions of fixity at supports.

3. *I.S.I. code method.* This method can be used where the corners of slab are prevented from lifting and adequate provision for torsion is provided.

4. Mr. Pigeaud's theory gives distribution of concentrated loads in two directions. It can also be used for uniformly distributed loads on slabs. Mr. Pigeaud's curves can also be used for continuous slabs by making some arbitrary reduction to account for continuity.

8·2. Grashoff and Rankine method for slab simply supported on four edges with corners not held down and carrying uniformly distributed loads

The slabs of single storeyed buildings where corners are not held down come in this category.

Notations

L = Length of the slab.

B = Breadth of slab.

w = Uniformly distributed load on slab in N/m².

w_L = Load carried by slab in L direction.

w_B = Load carried by slab in B direction.

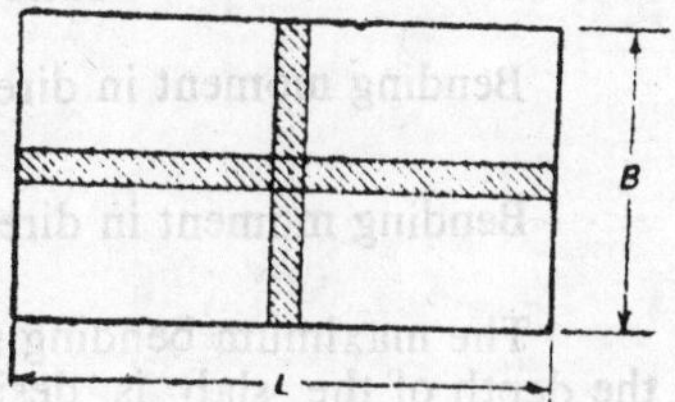

Fig. 8·2

$$w = w_L + w_B \qquad \ldots(a)$$

Consider two strips of unit width in the directions L and B respectively at the centre of slab. The deflection of each strip at centre will be same.

Central deflection of strip in direction $L = \dfrac{5 w_L L^4}{384\ EI}$

Central deflection of strip in direction $B = \dfrac{5 w_B B^4}{384\ EI}$

$$\frac{5 w_L L^4}{314\ EI} = \frac{5 w_B B^4}{384\ EI}$$

$$w_L = w_B \times \frac{B^4}{L^4} \qquad \ldots(b)$$

Substituting in (a)

$$w = w_B \times \frac{B^4}{L^4} + w_B$$

$$w_B = \frac{wL^4}{L^4 + B^4} = w \times \frac{1}{1 + \left(\frac{B}{L}\right)^4}$$

$$w_L = \frac{w \times B^4}{L^4 + B^4} = w \times \frac{1}{1 + \left(\frac{L}{B}\right)^4}$$

Let $\dfrac{L}{B} = k.$

$$w_L = w\ \frac{1}{1 + k^4} = w k_L$$

$$w_B = w\ \frac{k^4}{1 + k^4} = w k_B$$

The values of k_L and k_B depend on ratio L/B. Values of k_L and k_B for different L/B ratio are tabulated in Table 8·1.

Bending moment in direction of B, $M_B=k_B\ \dfrac{w\times B^2}{8}$

Bending moment in direction of L, $M_L=k_L\ \dfrac{w\times L^2}{8}=\dfrac{M_B}{k^2}$

The maximum bending moment occurs along shorter span and the depth of the slab is designed for this bending moment. The reinforcement for such a slab is provided in two directions, placed one above the other. The effective depth of slab in two directions will be different. The greater effective depth is provided in shorter direction in which bending moment is larger.

It is seen from Table 8·1 that for ratios of L/B greater than 2, the load carried in long direction is very small. The slabs are designed as two-way reinforced slabs for ratio of L/B upto 2. If the ratio of L/B exceeds 2, the slab is considered to span across shorter span only and the B.M. in the slab is $\dfrac{wB^2}{8}$.

The ratio of the load carried by slab in two directions as given by Table 8·1 is not same at all the points of slab. It is true only at the centre of the slab where maximum B.M. occurs.

8·3. Slabs simply supported on four edges with corners held down, *i.e.* prevented from lifting and carrying uniformly distributed loads

Slabs built into thin walls or monolithic with thin beams come into this category. The bending moment in such slabs can be found by either of the two methods, given below :

(*a*) The bending moments in such slabs in directions of B and L are $M_B=k_B'\ \dfrac{wB^2}{8}$ and $M_L=k_L'\ \dfrac{wL^2}{8}$.

The co-efficient of k_B' and k_L' in this case will be less than those for slabs with corners not prevented from lifting. The co-efficients k_B' and k_L' for different ratios of L/B are given in Table 8·2.

(*b*) *Dr. Marcus's method.* Dr. Marcus has given an approximate method for determining bending moment in slabs supported on four edges with corners prevented from lifting. The bending moments at mid span of such a slab are the mid-span bending moments as calculated by Grashoff and Rankine method multiplied by a factor C, which is less than unity as the positive bending moments in the slab are reduced due to corners restraint and torsion. This factor C is given by—

$$C=1-\frac{5}{6}\cdot\frac{k^2}{1+k^4},\ \text{where } k=\frac{L}{B}$$

Table 8·1

Ratio L/B	1·0	1·05	1·1	1·15	1·2	1·25	1·3	1·4	1·5	1·6	1·75	2·0	2·5	3·0
K_B	0·05	0·55	0·59	0·64	0·68	0·71	0·74	0·79	0·84	0·87	0·90	0·94	0·97	0·98
K_L	0·50	0·45	0·41	0·36	0·32	0·29	0·26	0·21	0·16	0·13	0·10	0·06	00 3	0·02

Table 8·2

Ratio L/B	1·0	1·05	1·1	1·15	1·2	1·25	1·3	1·4	1·5	1·6	1·75	2·0	2·5	3·0
k_B'	0·30	0·33	0·36	0·39	0·42	0·45	0·48	0·53	0·58	0·63	0·68	0·76	0·87	0·94
k_L'	0·30	0·27	0·24	0·22	0·19	0·17	0·15	0·13	0·11	0·09	0·07	0·05	0·03	0·02

Bending moment in direction of $B = M_B = Ck_B \dfrac{wB^2}{8}$.

Bending moment in direction of $L = M_L = Ck_L \dfrac{wL^2}{8}$.

The values C for different ratios of L/B can be obtained from Table 8·3.

The bending moments as obtained by Dr. Marcus's method and exact theory assuming Poisson's ratio equal to zero are almost identical.

8·4. Slabs fixed on all four edges carrying uniformly distributed loads

For slabs fixed on all the four edges the bending moments may be calculated by Dr. Marcus's method.

Short span—Support moment $M_B' = -k_B \dfrac{wB^2}{12}$

Mid span moment $M_B = +C'k_B \dfrac{wB^2}{24}$

Long span—Support moment $M_L' = -k_L \dfrac{wL^2}{12}$

Mid span moment of $M_L = +C'k_L \dfrac{wL^2}{24}$

where
$$C' = 1 - \frac{5}{18} \times \frac{k^2}{1+k^4}$$

The values of C' for different values as L/B are given in Table 8·3.

Table 8·3

L/B	1·0	1·05	1·10	1·15	1·2	1·25	1·3
C	0·583	0·586	0·591	0·598	0·610	0·622	0·635
C'	0·861	0·862	0·864	0·866	0·870	0·874	0·879
L/B	1·4	1·5	1·6	1·75	2·0	2·5	3·0
C	0·663	0·691	0·718	0·755	0·704	0·87	0·909
C'	0·888	0·897	0·907	0·919	0·935	0·957	0·970

8·5. Slabs with continuity or fixity on one, two or three edges and uniformly loaded

The bending moments in such slab can be found either by Grashof-Rankine rule with Dr. Marcus correction or by ISI code method.

(*i*) *Grash of-Rankine method with Marcus correction.* The bending moment formulae for various end conditions are tabulated in Table 8·4.

8·6. I.S.I. Code Method

(*a*) **Restrained Slabs.** When the corners of slab are prevented from lifting, the slab may be designed as follows :

The maximum bending moments per unit width in a slab are given by the following equations :

$$M_x = \alpha_x w l_x^2$$
$$M_y = \alpha_y w l_y^2$$

where α_x and α_y are coefficients given in Table 8·5.

w = total load per unit area

M_x, M_y = moments on strips of unit width spanning l_x and l_y respectively.

l_x, l_y = lengths of the shorter span and longer span respectively.

(*i*) Slabs are considered as divided in each direction into middle strips and edge strips as shown in Fig. 8·3, the middle strip being three-quarters of the width and each edge strip one-eighth of the width.

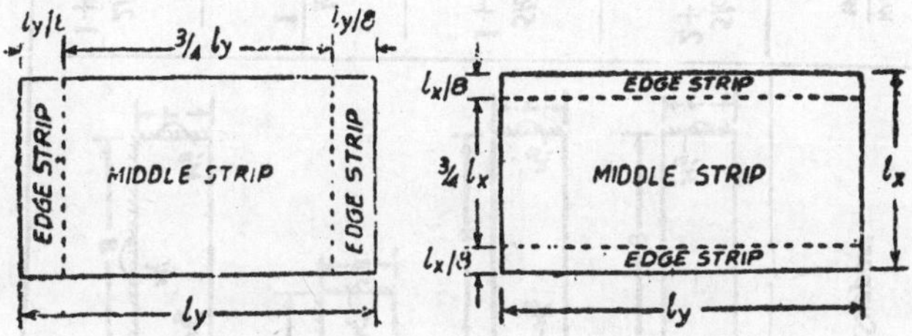

Fig. 8·3. Division of slab into middle and edge strip.

(*ii*) The maximum moments calculated apply only to the middle strips and no redistribution shall be made.

(*iii*) Tension reinforcement provided at mid-span in the strip shall extend in the lower part of the slab to within 0·25 l of a continuous edge, or 0·15 l of a discontinuous edge.

Table 8·4
Slab with Continuity of Fixity

Condition at Support	$\frac{w_B}{w}$	$\frac{w_L}{w}$	Span Moment M_B	Span Moment M_L	Multiplying factor for span: For M_B	Multiplying factor for span: For M_L	Support moment M'_B	Support moment M'_L
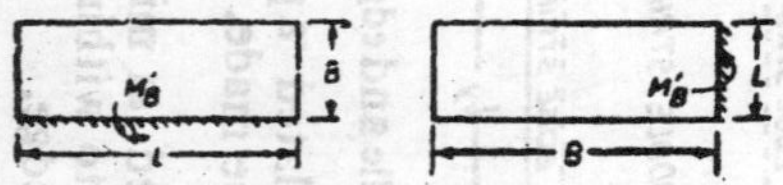	$\frac{5k^4}{2+5k^4}$	$\frac{2}{1+5k^4}$	$\frac{9}{128} w_B \times B^2$	$\frac{1}{8} w_L \times L^2$	$1-\frac{75}{32}\times\frac{k^2}{2+5k^4}$	$1-\frac{5}{3}\times\frac{k^2}{2+5k^4}$	$\frac{w_B \times B^2}{8}$	—
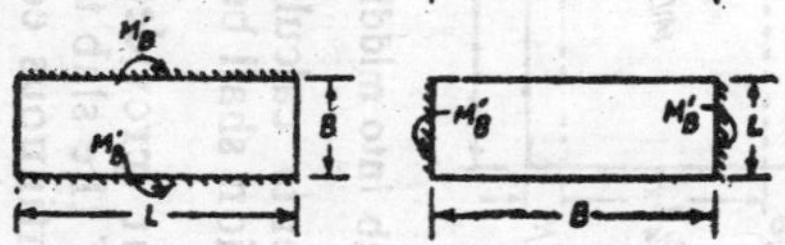	$\frac{5k^4}{1+5k^4}$	$\frac{1}{1+5k^4}$	$\frac{1}{24} w_B \times B^2$	$\frac{1}{8} w_L \times L^2$	$1-\frac{25}{18}\times\frac{k^2}{2+5k^4}$	$1-\frac{5}{6}\times\frac{k^2}{1+5k^4}$	$\frac{w_B \times B^2}{12}$	—
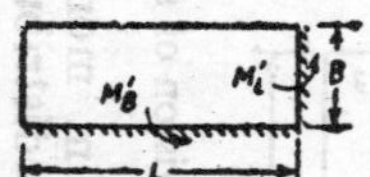	$\frac{k^4}{1+k^4}$	$\frac{1}{1+k^4}$	$\frac{9}{128} w_B \times B^2$	$\frac{9}{128} w_L \times L^2$	$1-\frac{15}{35}\times\frac{k^2}{1+k^4}$	$1-\frac{15}{32}\times\frac{k^2}{1+k^4}$	$\frac{w_B \times B^2}{8}$	$\frac{w_L \times L^2}{8}$
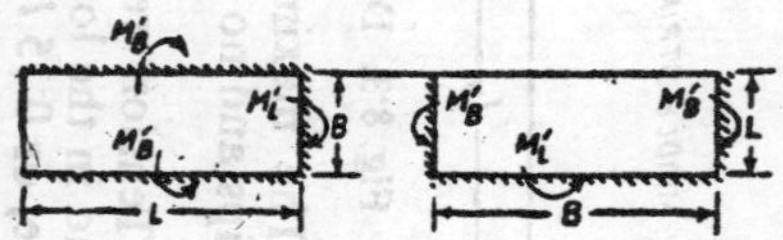	$\frac{2k^4}{1+2k^4}$	$\frac{1}{1+2k^4}$	$\frac{1}{24} w_B \times B^2$	$\frac{9}{128} w_L \times L^2$	$1-\frac{5}{9}\times\frac{k^2}{1+2k^4}$	$1-\frac{15}{32}\times\frac{k^2}{1+2k^4}$	$\frac{w_B \times B^2}{12}$	$\frac{w_L \times L^2}{8}$

M_L' = Maximum B.M. at support span L.
M_B' = Maximum B.M. at support span B.

(*iv*) Over the continuous edges of a middle strip, the tension reinforcement shall extend in the upper part of the slab a distance of 0·15 *l* from the support, and at least 50 per cent shall extend a distance of 0·3 *l*.

(*v*) At a discontinuous edge, negative moments may arise. They depend on the degree of fixity at the edge of the slab but, in general tension reinforcement equal to 50 per cent of that provided at mid-span extending 0·1 *l* into the span will be sufficient.

(*vi*) Reinforcement in edge strip, parallel to that edge, shall comply with the minimum reinforcement for slabs.

(*vii*) Torsion reinforcement shall be provided at any corner wher the slab is simply supported on both edges meeting at that corner. It shall consist of top and bottom reinforcement, each with layers of bars placed parallel to the sides of the slabs and extending from the edges of a minimum distance of one-fifth of the shorter span. The area of reinforcement in each of these four layers shall be three-quarters of the area required for the maximum mid-span moment in the slab.

(*viii*) Torsion reinforcement equal to half that described in (*vii*) above shall be provided at a corner contained by edges over only one of which the slab is continuous.

(*ix*) Torsion reinforcement need not be provided at any corner contained by edges over both of which the slab is continuous.

(*x*) Where l_y/l_x is greater than 2, the slabs shall be designed as spanning one way.

(*b*) **Simply supported slabs.** When simply supported slabs do not have adequate provision to resist torsion at corners and to prevent the corners from lifting, the maximum moments per unit width are given by the following equation :

$$M_x = \alpha_x w l_x^2$$

$$M_y = \alpha_y w l_y^2.$$

where, M_x, M_y, w, l_x, l_y are same as for restrained edges.

α_x and α_y are moment coefficients given in Table 8·6. At least 50 per cent of the tension reinforcement provided at mid-span should extend to the supports the remaining 50 per cent should extend to within 0·1 l_y or 0·1 l_x of the support.

Table 8·5
Bending Moment Coefficients for Rectangular Panels Supported on Four Sides with Provision for Torsion at Corners

Case No.	Types of panel and moments	Short span coefficients α_x								Long span coefficients α_y for all values of l_y/l_x
		1·0	1·1	1·2	1·3	1·4	1·5	1·75	2·0	
1.	**Interior Panels**									
	Negative moment at continuous edge	0·032	0·037	0·043	0·047	0·051	0·053	0·060	0·065	0·032
	Positive moment at mid-span	0·024	0·028	0·032	0·036	0·039	0·041	0·045	0·049	0·024
2.	**One Short Edge Discontinuous**									
	Negative moment at continuous edge	0·037	0·043	0·048	0·051	0·055	0·057	0·064	0·068	0·037
	Positive moment at mid-span	0·028	0·032	0·036	0·039	0·041	0·044	0·048	0·052	0·028
3.	**One Long Edge Discontinuous**									
	Negative moment of continuous edge	0·037	0·044	0·052	0·057	0·063	0·067	0·077	0·085	0·037
	Positive moment at mid-span	0·028	0·033	0·039	0·044	0·047	0·051	0·059	0·065	0·028

4.	**Two Adjacent Edges Discontinuous**									
	Negative moment at continuous edge	0·047	0·053	0·060	0·065	0·071	0·075	0·084	0·091	0·047
	Positive moment at mid-span	0·035	0·040	0·045	0·049	0·053	0·056	0·063	0·069	0·035
5.	**Two Short Edges Discontinuous**									
	Negative moment at continuous edge	0·045	0·049	0·052	0·056	0·059	0·060	0·065	0·069	—
	Positive moment at mid-span	0·035	0·037	0·042	0·043	0·044	0·045	0·049	0·521	0·035
6.	**Two Long Edges Discontinuous**									
	Negative moment at continuous edge	—	—	—	—	—	—	—	—	0·045
	Positive moment at mid-span	0·35	0·043	0·051	0·057	0·063	0·068	0·080	0·088	0·035
7.	**Two Edges Discontinuous (One Loug Edge Continuous)**									
	Negative moment at continuous edge	0·057	0·064	0·071	0·076	0·080	0·084	0·091	0·097	—
	Positive moment at mid-span	0·043	0·048	0·053	0·057	0·066	0·064	0·069	0·073	0·043

Case No.	Types of panel and moments	Short span coefficients α_x value of l_y/l_x								Loan span coefficient α_y for all values of l_y/l_x
		1·0	1·1	1·2	1·3	1·4	1·5	1·75	2·0	
8.	Three Edges Discontinuous (One Short Edge Discontinuous)									
	Negative moment at continuous edge	—	—	—	—	—	—	—	—	0·057
	Positive moment at mid-span	0·043	0·051	0·059	0·065	0·071	0·076	0·087	0·096	0·043
9.	For Edges Discontinuous									
	Positive moment at mid-span	0·056	0·064	0·072	0·079	0·085	0 089	0·100	0·107	0·056

Table 8·6
Bending Moment Coefficients for Slabs Spanning in two Directions at Right Angles, Simply Supported on four Sides

l_y/l_x	1·0	1·1	1·2	1·3	1·4	1·5	1·75	2·0	2·5	3·0
α_x	0·062	0·074	0·084	0·093	0·099	0·104	0·113	0·118	0·122	0·124
α_y	0·062	0·061	0·059	0·055	0·051	0·046	0·037	0·029	0·020	0·014

8·7 Rectangular Panel with Concentrated Loads M. Pigeaud's Theory

The curves shown in Fig. 8·5 are based on M. Pigeaud's theory and can be used for calculating bending moments on a panel freely supported along four edges with restrained corners and carrying symmetrically placed load and distributed over some defined area.

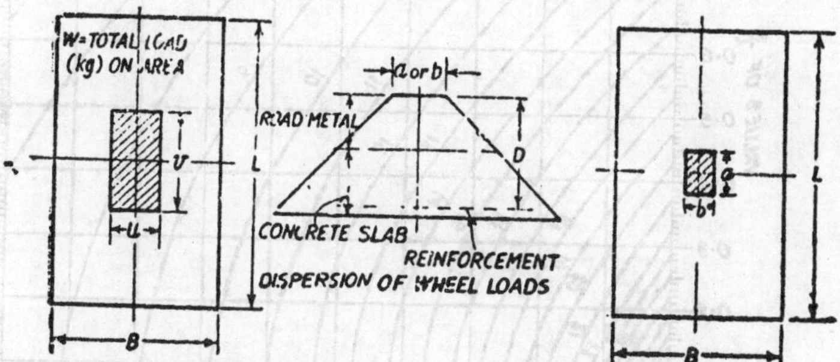

$u = b+2D$
$v = a+2D$.

a = assumed contact length
b = tyre width

Short span B.M. = $W(m_1+0{\cdot}15\ m_2)$
Long span B.M. = $W(0{\cdot}15\ m_1+m_2)$
(B.M.S. in kg. m. per metre width)

Fig. 8·4

Wheel loads and concentrated loads are assumed to be dispersed through road finish to the centre of reinforcement as shown in Fig. 8·4 and values of u and v are obtained. $u=b+2D$, $v=a+2D$. From these values of u and v raitos $\frac{u}{B}$ and $\frac{v}{L}$ are calculated, for which the bending moments m_1 and m_2 for unit load are found from curves of Fig 8·5 for the appropriate value of the ratio of spans $k=\frac{L}{B}$. For a total load of W on the area u by v the B.M. in shorter span is $W(m_1+0{\cdot}15m_2)$ and that in longer span is $W(0{\cdot}15m_1+m_2)$ in N-m per metre width of slab. In calculating above B.M. the value of Poisson's ratio is assumed as 0·15.

The curves shown in Fig. 8·5 are plotted for k=1·0, 1·25, 1·414 1 67, 2·0, 2·5 and infinity. For intermediate values of k, the values m_1 and m_2 may be interpolated. The curves of k=1·0 are applicable to a square panel, $k=\infty$ apply to panel of large length L as compared to breadth B and can be used for determining the transverse and longitudinal bending moments on a long narrow panel supported on two-edges only.

When there are two concentrated loads symmetrically disposed or an eccentric load, the resulting bending moments can be calculated from the rules given in Table 8·7.

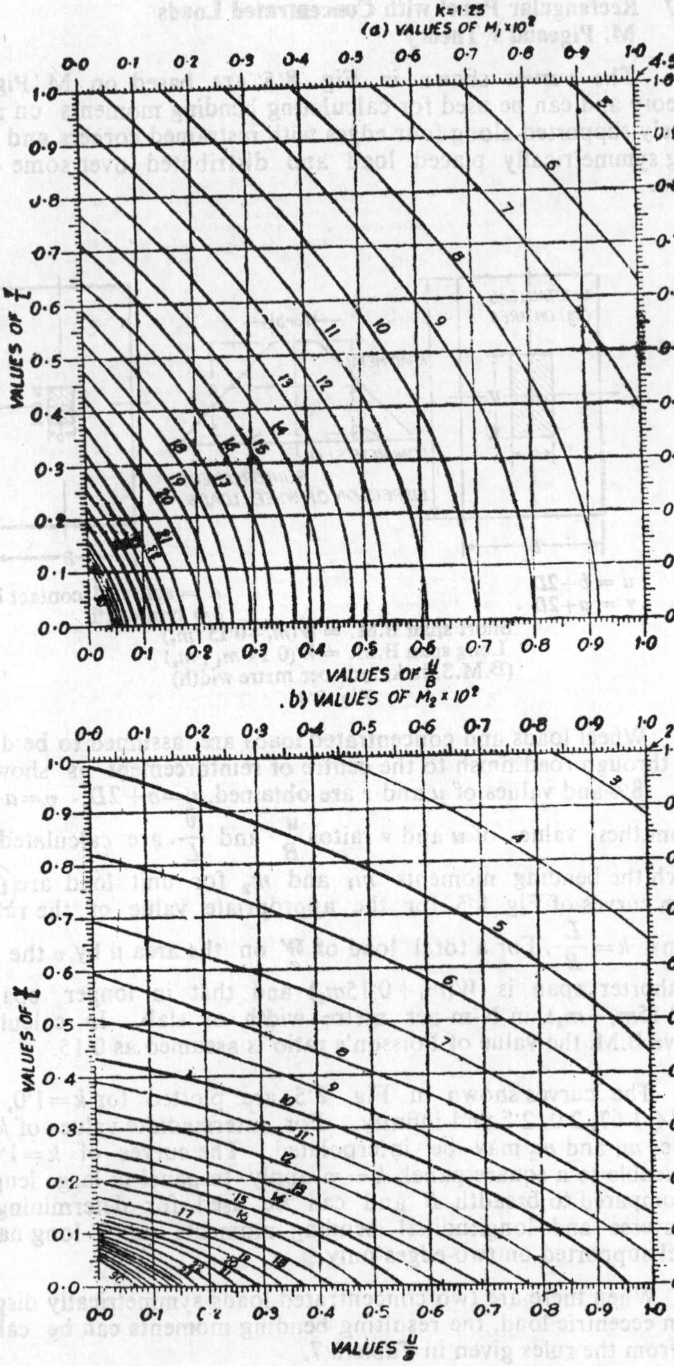
K=1·25
(a) VALUES OF M1×10²
4·52
VALUES OF v/L
VALUES OF u/B
(b) VALUES OF M2×10²
2·67
VALUES OF v/L
VALUES u/B

Fig. 8·5 (b)

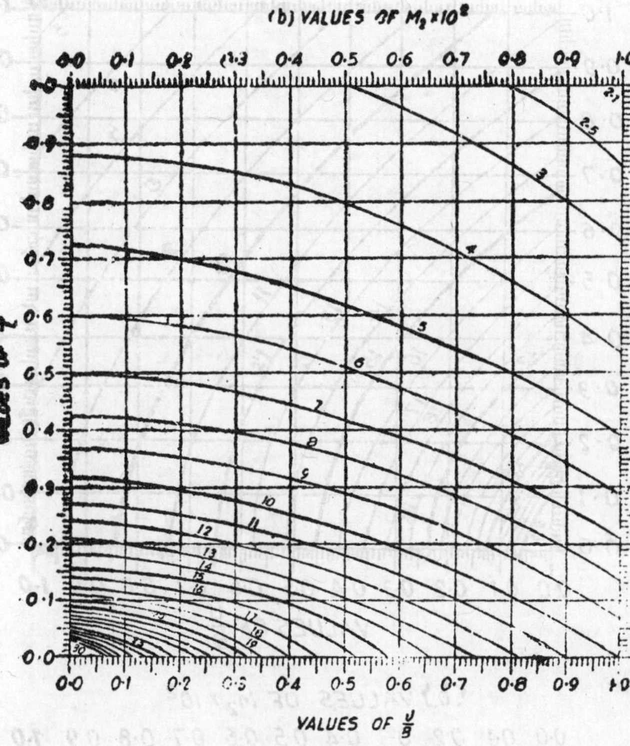

Fig. 8·5 (*b*)

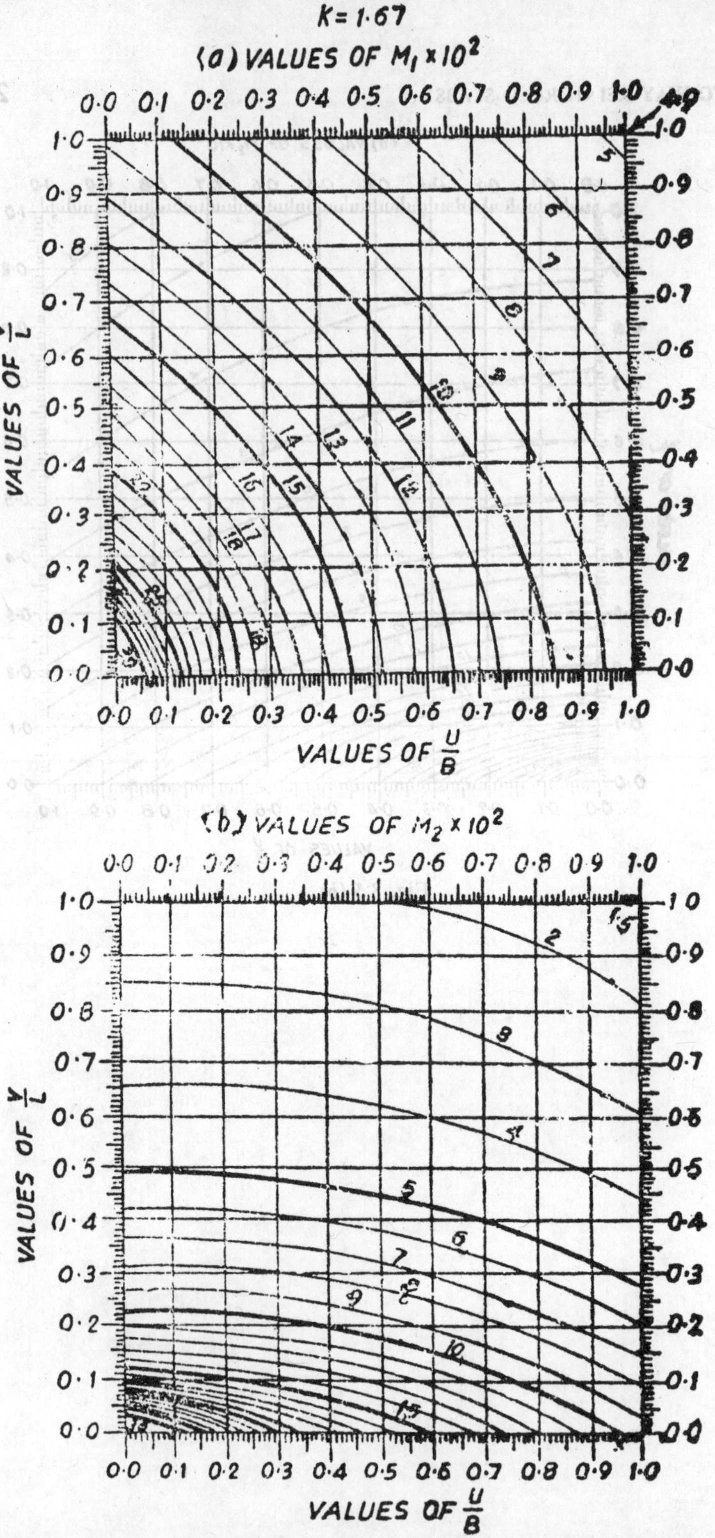
K = 1.67
(a) VALUES OF $M_1 \times 10^2$
VALUES OF $\frac{V}{L}$
VALUES OF $\frac{U}{B}$
(b) VALUES OF $M_2 \times 10^2$
VALUES OF $\frac{V}{L}$
VALUES OF $\frac{U}{B}$

Fig. 8.5 (c)

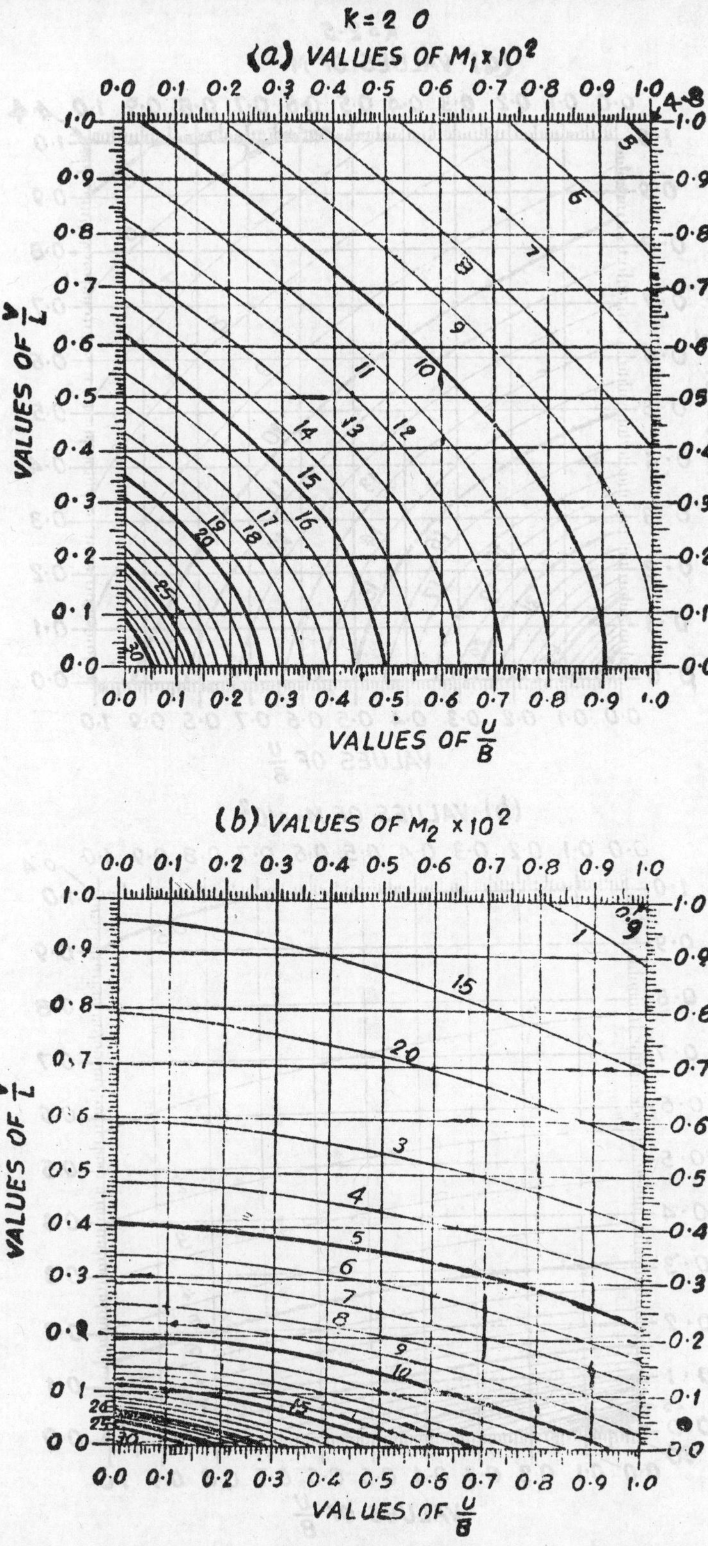

K=2 0
(a) VALUES OF $M_1 \times 10^2$
VALUES OF $\frac{V}{L}$
VALUES OF $\frac{U}{B}$
(b) VALUES OF $M_2 \times 10^2$
VALUES OF $\frac{V}{L}$
VALUES OF $\frac{U}{B}$

Fig. 8·5 (*d*)

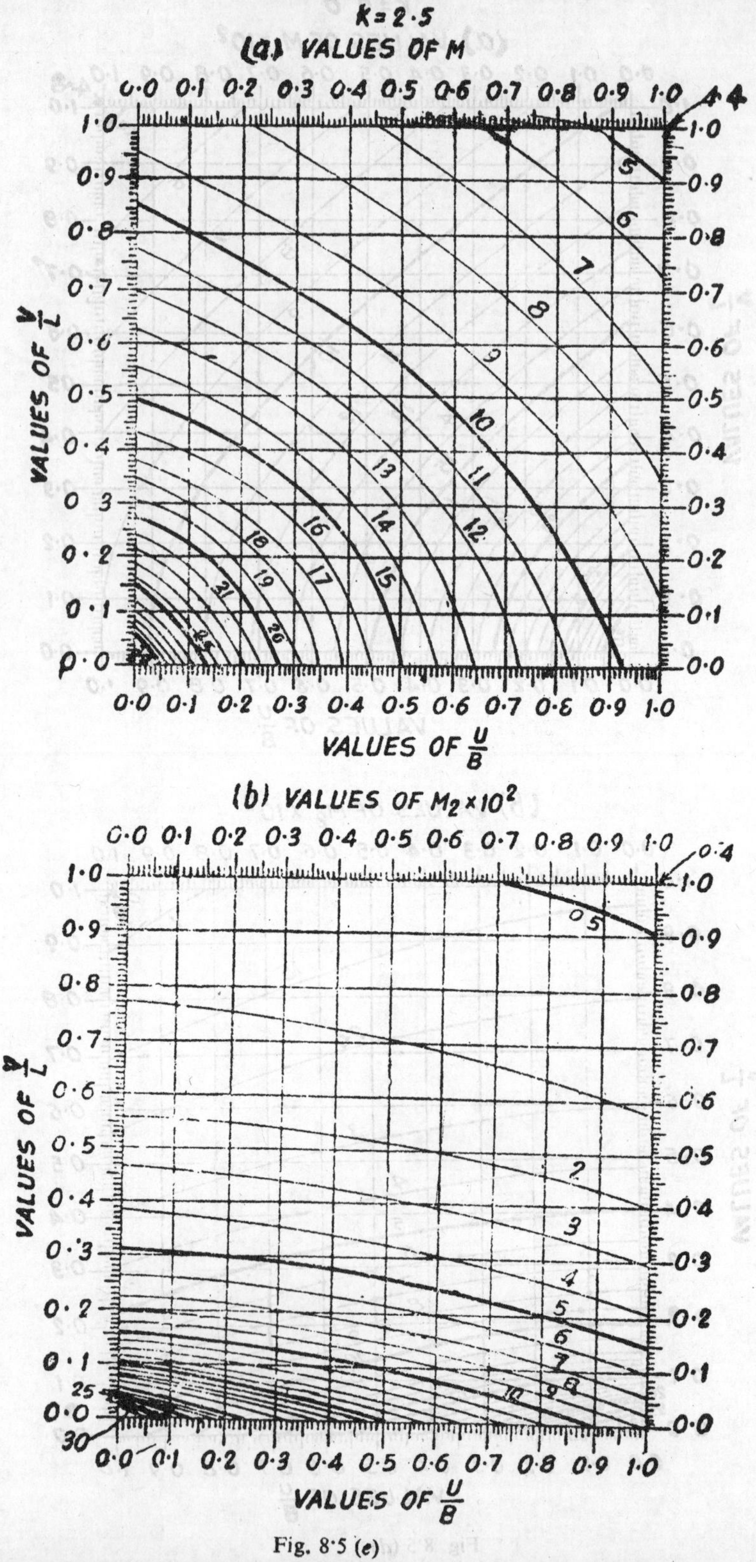
K=2·5
(a) VALUES OF M
VALUES OF V/L
VALUES OF U/B
(b) VALUES OF M2×10²
VALUES OF V/L
VALUES OF U/B

Fig. 8·5 (e)

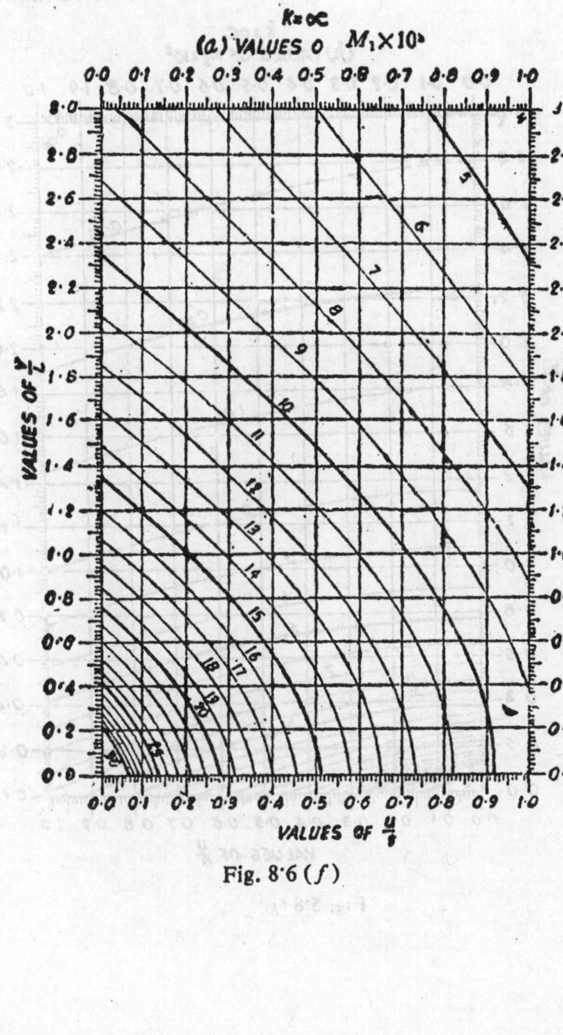
K=∞
(a) VALUES OF $M_1 \times 10^3$
VALUES OF $\frac{v}{L}$
VALUES OF $\frac{u}{B}$
0·0 0·1 0·2 0·3 0·4 0·5 0·6 0·7 0·8 0·9 1·0
3·0 2·8 2·6 2·4 2·2 2·0 1·8 1·6 1·4 1·2 1·0 0·8 0·6 0·4 0·2 0·0
4 5 6 7 8 9 10 11 12 13 14 15 16 17 18 19 20 23

Fig. 8·6 (f)

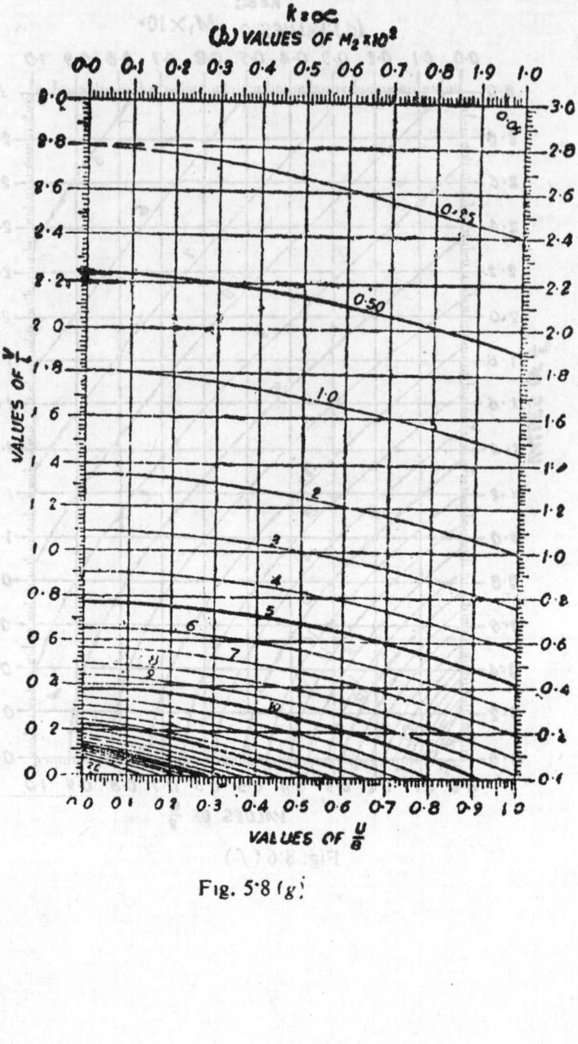

Fig. 5·8 (g)

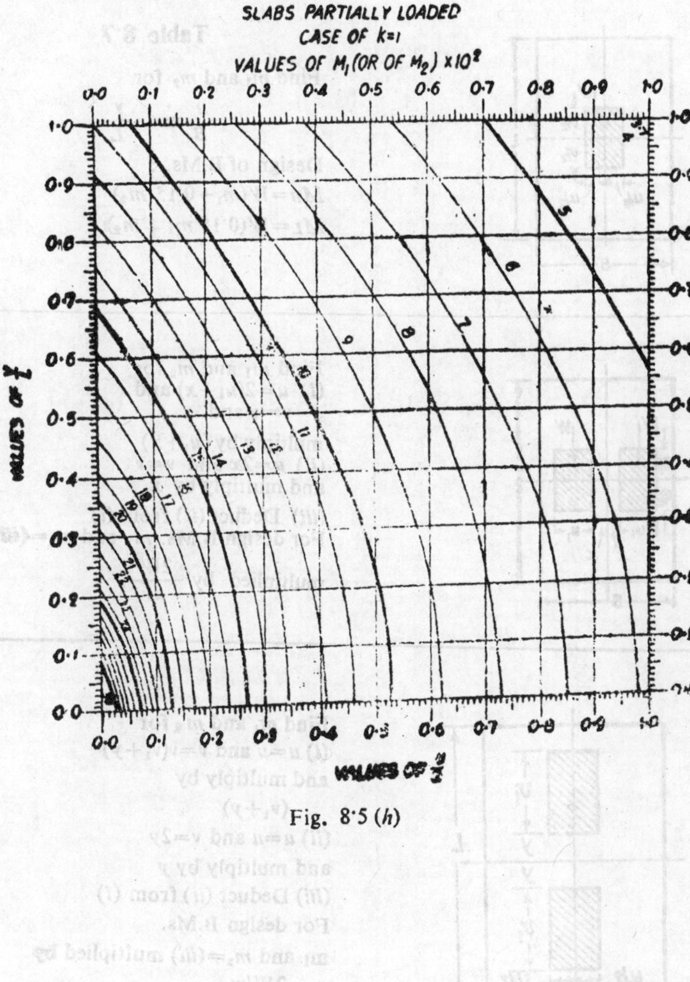
SLABS PARTIALLY LOADED
CASE OF k=1
VALUES OF M_1 (OR OF M_2) × 10²
VALUES OF v/l
VALUES OF u/l

Fig. 8·5 (*h*)

Table 8·7

Ecentric and Multiple Concentric Load

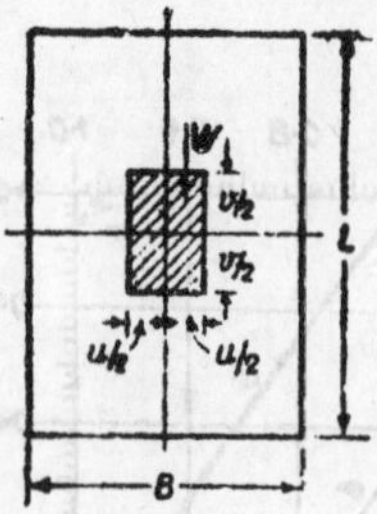

Find m_1 and m_2 for $\frac{u}{B}$ and $\frac{v}{L}$

Design of B.Ms. :

$M_B = W(m_1 + 0{\cdot}15\ m_2)$

$M_L = W(0{\cdot}15\ m_1 + m_2)$

I

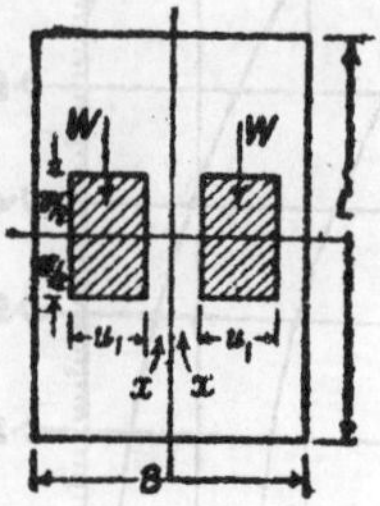

Find m_1 and m_2 for
(*i*) $u = 2(u_1 + x)$ and $v = v$ and
multiply by $(u_1 + x)$
(*ii*) $u = 2x$ and $v = v$ and multiply by x
(*iii*) Deduct (*ii*) from (*i*)
For design B.Ms. m_1 and m_2 = (*iii*) multiplied by $\frac{2W}{u_1}$.

III

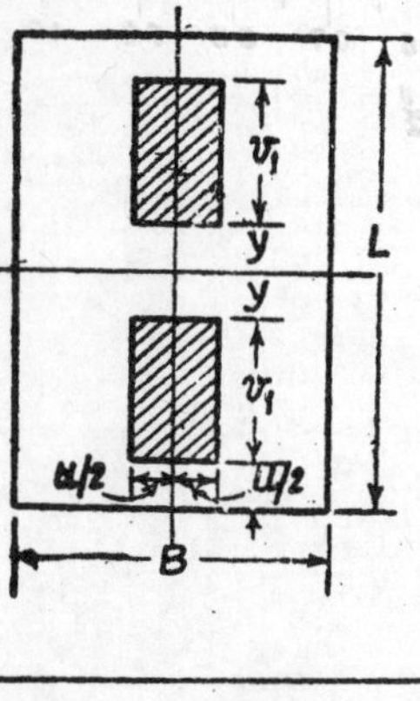

Find m_1 and m_2 for
(*i*) $u = u$ and $v = v(v_1 + y)$ and multiply by $(v_1 + y)$
(*ii*) $u = u$ and $v = 2y$ and multiply by y
(*iii*) Deduct (*ii*) from (*i*)
For design B.Ms.
m_1 and m_2 = (*iii*) multiplied by $2W/v_1$.

V

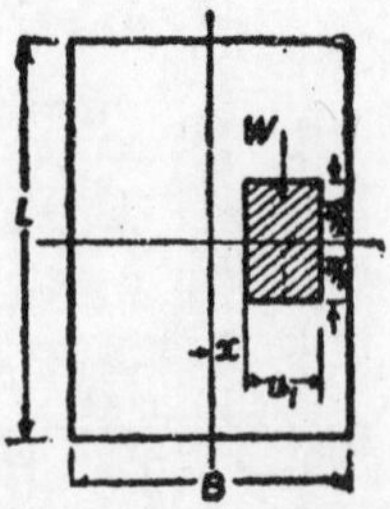

Find m_1 and m_2 for
(*i*), (*ii*) and (*iii*) as case II
For design B.Ms.
m_1 and m_2 = (*iii*) multiplied by W/u_1

<table>
<tr>
<td>Eccentric and multiple concentric loads
V</td>
<td>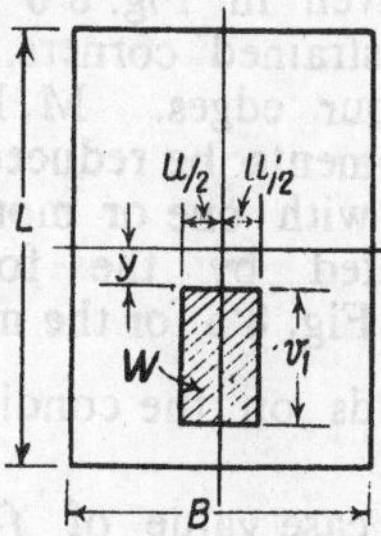
</td>
<td>Find m_1 and m_2 for

(i), (ii) and (iii) as in case III.

For design B.Ms.

m_1 and m_2=(iii) multiplied by W/v_1</td>
</tr>
<tr>
<td>VI</td>
<td>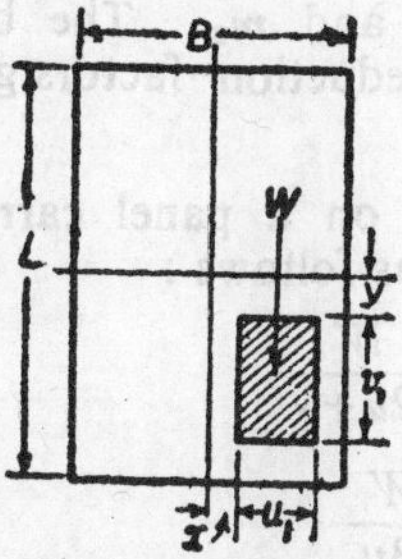
</td>
<td>Find m_1 and m_2 for
(i) $u=2(u_1+x)$ and
$v=2(v_1+y)$ and
multiply by
$(u_1+x)(v_1+y)$
(ii) $u=2x$ and $v=2y$ and
multiply by xy
(iii) $u=2(u_1+x)$ and $v=2y$
and multiply by $y(u_1+x)$
(iv) $u=2x$ and $v=2(v_1+y)$
For design B.Ms. subtract (iii)+(iv) from (i)+(ii) and multiply by W/u_1v_1.</td>
</tr>
<tr>
<td>Span ratio adjustment</td>
<td colspan="2">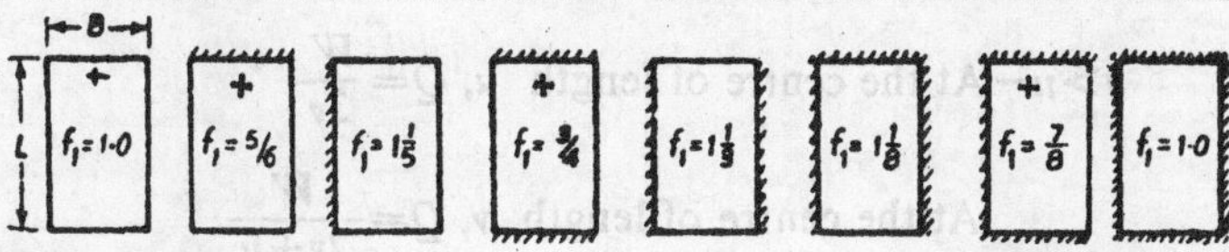

Non-continuous but monolithic with support (—)   Continuous over support (m)
$k=f_1 \frac{L}{B}$; if in cases marked thus +, $f_1 \frac{L}{B}<1·0$
Transpose B and L (u and v)</td>
</tr>
<tr>
<td>B.M. reduction factor for continuity</td>
<td colspan="2">Calculate B.Ms. in each direction as if freely supported but with $k=f_1 \frac{L}{B}\left(\text{or } k=f_1 \frac{L}{B} \text{ if } <1·0\right)$ multiplied by the following :

Co-efficients to give the design bending moments at the specified sections.

Mid-span : Interior span=0·7 ; End span=0·85

Supports ; End support=0·25 ; Penultimate=0·95
Interior (except penultimate 0·90)</td>
</tr>
</table>

The bending moment co efficients given in Fig. 8·6 are for mid-span of panels freely supported with restrained corners. If the panel is fixed or continuous along all four edges. M. Pigeaud recommends that the mid-span bending moments be reduced by 20 per cent. The bending moments for panels with one or more edges continuous over supports can be calculated by the following method : Find m_1 and m_2 from curves of Fig. 8·5 for the modified value of k, *i.e.* $k=f_1 \frac{L}{B}$, where f_1 depends on the conditions at supports and is obtained from Table 8·7. In case value of $f_1 \frac{L}{B}$ is less than unity. L and B (and consequently u and v) should be transposed throughout the calculation of m_1 and m_2. The bending moments thus obtained are multiplied by reduction factors given in Table 8 7 to give design bending moments.

The maximum S.F. per unit length on a panel carrying a concentrated load are given by M. Pigeaud as follows :

$u > v$ — At the centre of length u, $Q=\frac{W}{2u+v}$

At the centre of length v, $Q=\frac{W}{3u}$

$v > u$ — At the centre of length u, $Q=\frac{W}{3v}$

At the centre of length v, $Q=\frac{W}{2v+u}$.

8·8. General Notes on Design of Two-way Slabs (I.S.I.)

1. *Limiting stiffness of members.* Reinforced concrete should possess adequate stiffness to prevent such deformation as might impair the strength, appearance or the efficiency of the structure or produce cracks in finishes and partitions. For normal cases it may be assumed that this condition is satisfied for slabs if the ratio of span to overall depth does not exceed the following values :

Slabs	*Maximum value of span/Depth ratio*
Simply supported slabs spanning in two directions	35
Continuous slabs spanning in two directions	40

Note. In slabs spanning in two directions, the shorter of the two spans should be used to calculate the span/depth ratio.

2. *Loads on supporting beams.* Fig. 8·6 (*a*) shows a slab supported on four edges. The bending in central portion of such slab takes place in two directions shown by strips *a-a* and *b-b*. Near the support bending takes place along strips *d-d* and *e-e*. The loads on supporting beams for such slabs may be assumed in accordance with Fig. 8·6 (*b*).

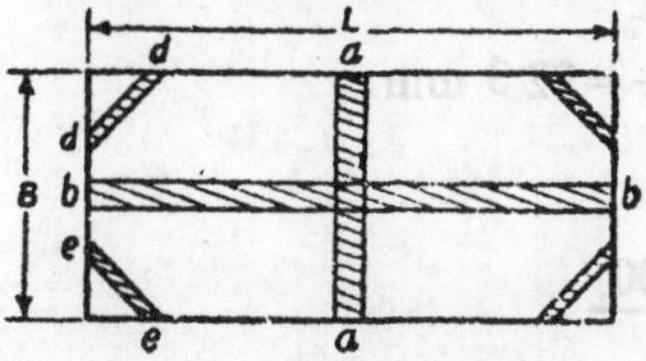

(*a*) Load in this shaded area to be carried by beam A.

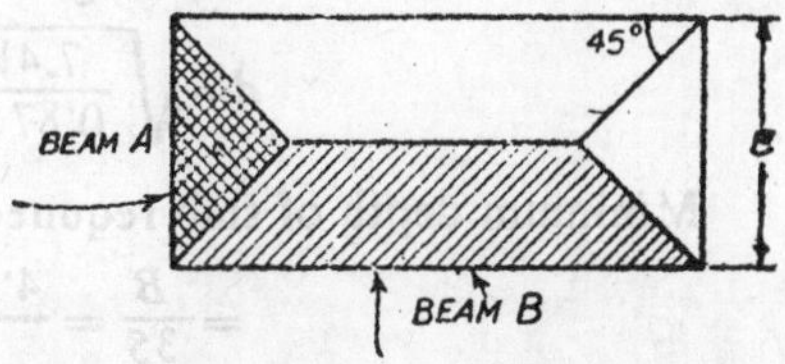

(*b*) Load included in this shaded area to be carried by beam B.

Fig. 8·6

3. *Reinforcement for torsion.* Torsion reinforcement should be provided at the corners of a slab, where corner restraint is provided except at corners contained by edges over both of which the slab is continuous. The amount of reinforcement to be provided should be in accordance with article 8·5 (*e*).

Ex. 8·1. *Design a two-way reinforced slab for a room having clear dimensions of 5 m×4 m. The superimposed load is 200 N/m². Use M 15 max.*

Sol. (*i*) *Assuming corners of slab not held down.*

Minimum depth of slab$=\frac{4000}{35}=114{\cdot}3$ mm.

Assume overall depth of slab as 120 mm and effective depth as 100 mm.

Effective span in two directions will be 5·1 m and 4·1 m.

Load per sq. metre, $w=2000+25{,}000\times1\times\frac{1\times120}{1000}$ (self-weight)

$$=2000+3000=5000 \text{ N}$$

$$w_B=\frac{wk^4}{2+k^4}, \text{ where } k=\frac{5{\cdot}1}{4{\cdot}1}=1{\cdot}244$$

$$=5000\times\frac{(1{\cdot}244)^4}{1+(1{\cdot}244)^4}$$

$$=5000\times\frac{2{\cdot}394}{3{\cdot}394}=3527 \text{ N/m}^2$$

$$w_L=1473 \text{ N/m}^2$$

$$M_B=\frac{w_B\times B^2}{8}=\frac{3527\times(4{\cdot}1)^2}{8}$$

$$=7411 \text{ N-m}=7{,}411{,}000 \text{ N-mm}$$

$$M_L = \frac{w_L \times L^2}{8} = \frac{1473 \times (5 \cdot 1)^2}{8}$$

$$= 4789 \text{ N-m} = 4{,}789{,}000 \text{ N-mm}$$

Effective depth $d = \sqrt{\frac{M}{Qb}}$

$$d = \sqrt{\frac{7{,}411{,}000}{0 \cdot 87 \times 1000}} = 92 \cdot 3 \text{ mm}$$

Minimum depth of slab required

$$= \frac{B}{35} = \frac{4 \cdot 1 \times 1000}{35}$$

$$= 117 \text{ mm}$$

Provide overall depth of 120 mm

Using 10 mm ϕ bars.

Effective depth $= 120 - 5 - 15 = 100$ mm.

Reinforcement along short span

$$A_t = \frac{7{,}411{,}000}{140 \times 0 \cdot 87 \times 100} = 608 \cdot 5 \text{ mm}^2$$

Provide 10 mm ϕ bars at 120 mm c/c, $A_t = 654$ mm²

Reinforcement along long span

Effective depth $= 100 - 10 = 90$ mm

$$A_t = \frac{4{,}789{,}000}{140 \times 0 \cdot 87 \times 90} = 437 \text{ mm}^2$$

Provide 10 ϕ bars at 170 mm centre to centre, $A_t = 462$ mm².

(*ii*) *Assuming corners held down and torsion reinforcement provided at the corners.*

(*a*) *Marcus method* :

Assume load at 5000 N/m² as in case (*i*)

$$M_B = \frac{w_B B^2}{8}\left[8 - \frac{5}{6} \frac{k^2}{1+k^4}\right]$$

$$= 7411 \left[1 - \frac{5}{6} \times \frac{(1 \cdot 244)^2}{1 + (1 \cdot 244)^4}\right]$$

$$= 7411 \times 0 \cdot 621 = 4602 \text{ N-m}$$

$$= 4{,}602{,}000 \text{ N-mm.}$$

$$M_L = 4789 \times 0 \cdot 621$$

$$= 2974 \text{ N-m} = 2{,}974{,}000 \text{ N-mm}$$

$$d=\sqrt{\frac{M}{Qb}}=\sqrt{\frac{4,602,000}{0{\cdot}87\times 1000}}$$

$$=72{\cdot}7 \text{ mm.}$$

Provide overall depth of 120 mm. Using 8 mm ϕ bars, effective depth$=120-15-4=101$ mm.

Reinforcement along long span

$$A_t=\frac{4,602,000}{140\times 0{\cdot}87\times 101}=374{\cdot}1 \text{ mm}^2$$

Use 8 mm ϕ bars at 130 mm centre to centre, $A_t=387$ mm²

Reinforcement along short span

Effective depth$=101-8=93$ mm

$$A_t=\frac{2,974,000}{140\times 0{\cdot}87\times 93}=262{\cdot}5 \text{ mm}^2$$

Use 8 mm ϕ bars at 180 mm centre to centre, $A_t=279$ mm².

Torsion reinforcement

Torsional reinforcement of $\frac{3}{4}$th of area of main reinforcement is provided in both directions parallel to sides at top and bottom upto distance of $\frac{1}{5}$th shorter span. $A_t=\frac{3}{4}\times 374{\cdot}1=280{\cdot}6$ mm².

Provided 8 mm ϕ bars at 180 mm c/c. along both directions at top and bottom for distance of$=\frac{4{\cdot}1}{5}=0{\cdot}82$ m. Provide six bars as shown in Fig. 8·7.

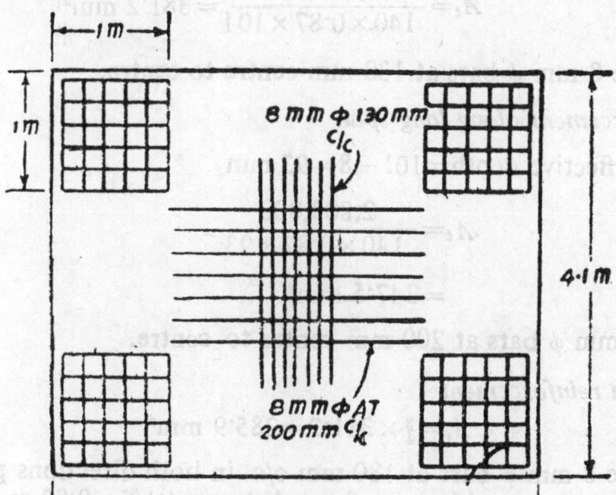

Fig. 8·7

(*b*) Empirical formula

Assume $w = 5000 \text{ N/m}^2$

$$k = \frac{5.1}{1.4} = 1.244$$

From table 8·2, $k_B' = 0.4464$, $k_L' = 0.1724$ (By interpolation)

$$M_B = \frac{k_B' \times w \times B^2}{8}$$

$$= \frac{0.4464 \times 5000 \times (4.1)^2}{8}$$

$$= 4690 \text{ N-m}$$

$$= 4{,}690{,}000 \text{ N-mm}$$

$$M_L = \frac{k_L' \times w \times L^2}{8} = \frac{0.1724 \times 5000 \times (5.1)^2}{8}$$

$$= 2803 \text{ N-m} = 2{,}803{,}000 \text{ N-mm}$$

$$d = \sqrt{\frac{M}{Qb}} = \sqrt{\frac{4690 \times 1000}{0.87 \times 1000}} = 73.4 \text{ mm}$$

Provide overall depth of 120 mm.

Using 8 mm ϕ bars, effective depth $= 120 - 15 - 4 = 101$ mm.

Reinforcement along short span

$$A_t = \frac{4{,}690{,}000}{140 \times 0.87 \times 101} = 381.2 \text{ mm}^2$$

Provide 8 mm ϕ bars at 130 mm centre to centre.

Reinforcement along long span

Effective depth $= 101 - 8 = 93$ mm

$$A_t = \frac{2{,}803{,}000}{140 \times 0.87 \times 93}$$

$$= 247.5 \text{ mm}^2$$

Use 8 mm ϕ bars at 200 mm centre to centre.

Torsion reinforcement

$$A_t = \tfrac{3}{4} \times 381.2 = 285.9 \text{ mm}^2$$

Provide 8 mm ϕ bars at 180 mm c/c in both directions parallel to sides, both at top and bottom for a distance $4.1/5 = 0.82$ m.

Provide 6 bars as shown in Fig. 8·7.

(c) **ISI Code formula**

Assume $w=4880\ \text{N/m}^2$, $k=\frac{5\cdot1}{4\cdot1}=1\cdot244=\frac{l_y}{l_x}$

$\alpha_x=0\cdot0751$; $\alpha_y=0\cdot056$ (By interpolation)

$M_B=\alpha_x\times wl_x^2$

$=0\cdot0751\times4880\times(4\cdot1)^2$

$=6161$ N-m $=6,161,000$ N-mm

$M_L=\alpha_y\times wl_y^2$

$=0\cdot056\times4880\times(4\cdot1)^2$

$=4594$ N-m $=4,594,000$ N-mm

$$d=\sqrt{\frac{M}{Qb}}=\sqrt{\frac{6,161,000}{0\cdot87\times1000}}=84\ \text{mm}$$

Provide overall depth of 120 mm.

Using 8 mm ϕ bars, effective depth

$=120-13-4=103$ mm.

Reinforcement along short span in middle strip.

$$A_t=\frac{6,161,000}{140\times0\cdot87\times103}=491\ \text{mm}^2$$

Provide 8 mm ϕ bars at 100 mm centre to centre.

Bend half the bars at $\frac{1}{5}$th span from the ends.

Reinforcement along long span in middle strip.

Effective depth $=103-8=95$ mm

$$A_t=\frac{4,594,000}{146\times0\cdot87\times95}=397\ \text{mm}^2$$

Provide 8 mm ϕ bars at 140 mm centre to centre.

Bend half bars at $\frac{1}{5}$th span from the ends.

Reinforcement in edge strip

Area of steel$=\frac{0\cdot15}{100}\times1000\times120=180\ \text{mm}^2$

Provide 8 mm ϕ bars at 250 mm c/c.

Torsion reinforcement

$A_t=\frac{3}{4}\times491=368\ \text{mm}^2$

Provide 8 mm ϕ bars at 130 mm centre to centre in both directions at top and bottom for a distance of $4\cdot1/5=0\cdot82$ m.

Provide 8 bars as shown in Fig. 8·8.

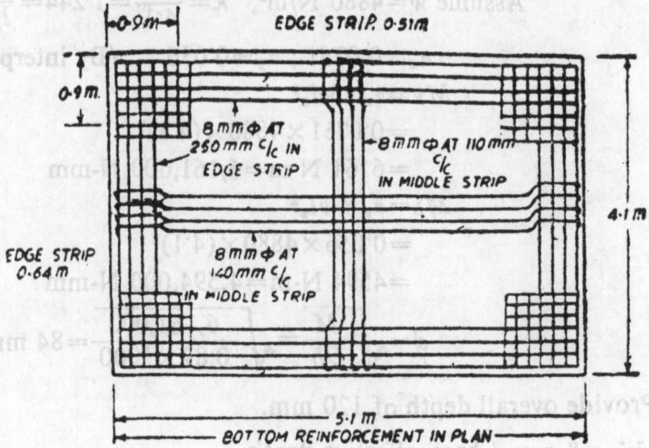

Fig. 8·8

Ex. 8·2. *Part plan of multi-storeyed office building is shown in Fig. 8·9. Design slabs S_1, S_2 and S_3 given the slab carries superimposed load of 4000 N/m². Design B_1. Use mix M 15.*

Sol. Slabs will be designed by Grashof-Rankine method with correction.

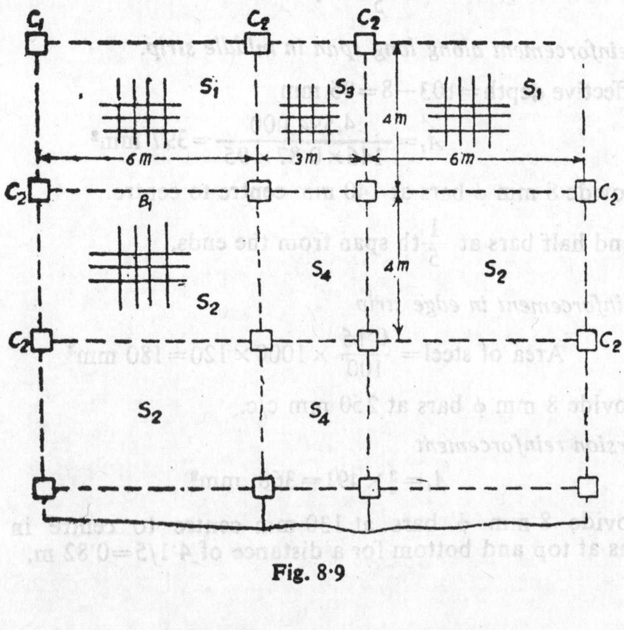

Fig. 8·9

Design of S_1 $\quad k=\dfrac{L}{B}=\dfrac{6}{4}=1{\cdot}5$

Multiplying factor for span moments

$$=1-\frac{15}{32}\times\frac{k^2}{1+k^4}$$

$$=1-\frac{15}{32}\times\frac{(1{\cdot}5)^2}{1+(1{\cdot}5)^4}$$

$$=0{\cdot}826.$$

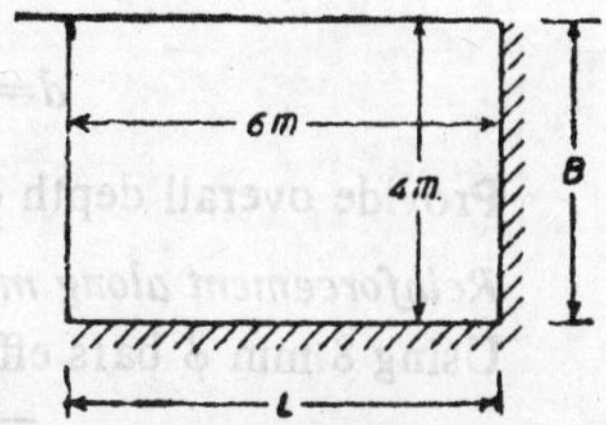

Fig. 8·10

Assume slab thickness of 160 mm.

Total load per square metre

$$=4000+25{,}000\times1\times1\times\frac{160}{1000}$$

$$=4000+4000=8000\ \text{N/m}^2$$

$$w_B=\frac{k^4}{1+k^4}\times8000=\frac{(1{\cdot}5)^4}{1+(1{\cdot}5)^4}\times8000$$

$$=0{\cdot}84\times8000=6720\ \text{N/m}^2$$

$$w_L=8000-6720=1280\ \text{N/m}^2.$$

Span moments

$$M_B=\frac{9}{128}w_B\times B^2\times\text{Multiplying factor}$$

$$=\frac{9}{128}\times6720\times4^2\times0{\cdot}826$$

$$=6244{\cdot}6\ \text{N-m}=6{,}244{,}600\ \text{N-mm}$$

$$M_L=\frac{9}{128}\times w_L\times L^2\times\text{Multiplying factor}$$

$$=\frac{9}{128}\times1280\times6^2\times0{\cdot}826$$

$$=2676\ \text{N-m}=2{,}676{,}000\ \text{N-mm}.$$

Support moments

$$M_B'=\frac{w_B\times B^2}{8}=\frac{6720\times4^2}{8}$$

$$=13{,}440\ \text{N-m}=13{,}440{,}000\ \text{N-mm}$$

$$M_L' = \frac{w_L \times L^2}{8} = \frac{1280 \times 6^2}{8}$$

$$= 5760 \text{ N-m}$$
$$= 5,760,000 \text{ N-m}$$

$$d = \sqrt{\frac{M}{Qb}} = \sqrt{\frac{13,440,000}{0{\cdot}87 \times 1000}} = 124{\cdot}3 \text{ mm}$$

Provide overall depth of 150 mm c/c.

Reinforcement along middle of short span

Using 8 mm ϕ bars effective depth

$$= 150 - 15 - 4 = 131 \text{ mm}$$

$$A_t = \frac{6.244,600}{140 \times 0{\cdot}87 \times 131} = 391{\cdot}4 \text{ mm}^2$$

Provide 8 mm ϕ bars at 120 mm c/c, $A_t = 418 \text{ mm}^2$

Reinforcement along middle of long span

Effective depth $= 131 - 8 = 123$ mm

$$A_t = \frac{2,676,000}{140 \times 0{\cdot}87 \times 123} = 178{\cdot}6 \text{ mm}^2$$

Provide minimum reinforcement 8 mm ϕ bars at 200 mm centres.

Reinforcement for negative B.M.

Reinforcement for support B

$$A_t = \frac{5,760,000}{140 \times 0{\cdot}87 \times 131} = 361 \text{ mm}^2$$

Provide 8 mm ϕ bars at 130 mm centres, $A_t = 387 \text{ mm}^2$

Reinforcement at support L

Effective depth $= 131 - 8 = 123$ mm

$$A_t = \frac{13.440,000}{140 \times 0{\cdot}87 \times 123} = 897 \text{ mm}^2$$

Provide 8 mm ϕ bars at 55 mm centres, $A_t = 922 \text{ mm}^2$

Torsion reinforcement at C_1

$$A_t = \frac{3}{4} \times 391{\cdot}4 = 293{\cdot}6 \text{ mm}^2$$

Provide 8 mm ϕ bars at 170 mm c/c both ways at top and bottom upto distance of 0·8 metre from support.

Torsion reinforcement at C_2

Half the reinforcement as for corner C_1 is to be provided.

Provide 8 mm ϕ bars at 300 mm centres.

Design of Slab S_2

$$k=\frac{L}{B}=\frac{6}{4}=1{\cdot}5.$$

Multiplying factor for span moments

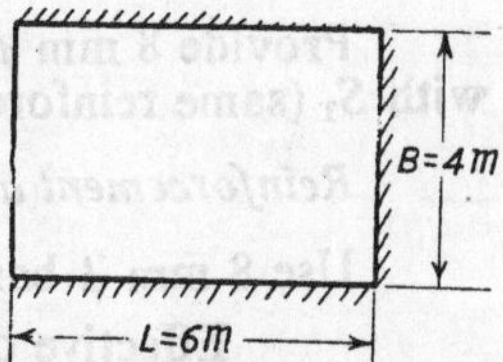

Fig. 8·11

For M_B factor $=1-\frac{5}{9}\times\frac{k^2}{1+2k^4}$

$$=1-\frac{5}{9}\times\frac{(1{\cdot}5)^2}{1+2\times(1{\cdot}5)^4}$$

$$=0{\cdot}8876.$$

For M_L factor $=1-\frac{15}{32}\times\frac{k^2}{1+2k^4}$

$$=1-\frac{15}{32}\times\frac{(1{\cdot}5)^2}{1+2\times(1{\cdot}5)^4}=0{\cdot}9053$$

Assume 150 mm overall depth,

$$w=4000+25{,}000\times1\times1\times\frac{250}{1000}$$

$$=4000+3750=7750\ \text{N/m}^2$$

$$w_B=76{,}000\times\frac{2k^4}{1+2k^4}=\frac{7750\times2\times(1{\cdot}5)^4}{1+2\times(1{\cdot}5)^4}\times7750$$

$$w_B=7053\ \text{N/m}^2$$

$$w_L=697\ \text{N/m}^2.$$

Span moments

$$M_L=\frac{w_B\times B^2}{24}\times0{\cdot}8876$$

$$=\frac{7053\times4^2}{24}\times0{\cdot}8876=4173{\cdot}5\ \text{N-m}$$

$$=4{,}173{,}500\ \text{N-mm}$$

$$M_L=\frac{9}{128}\times w_L\times L^2\times0{\cdot}9053$$

$$=\frac{9}{128}\times697\times6^2\times0{\cdot}9053$$

$$=1597\ \text{N-m}=1{,}597{,}000\ \text{N-mm}.$$

Support moments

$$M_B'=\frac{w_B\times B^2}{12}=\frac{7053\times4^2}{12}$$

$$=9404\ \text{N-m}=9{,}404{,}000\ \text{N-mm}.$$

$$M_L = \frac{w_L \times L^2}{8} = \frac{697 \times 6^2}{8}$$

$$= 3137 \text{ N-m} = 3{,}137{,}000 \text{ N-mm}$$

Provide overall depth of 150 mm as for S_1.

Provide 8 mm ϕ bars at 55 mm centres at support common with S_1 (same reinforcement).

Reinforcement at support L on either side

Use 8 mm ϕ bars.

Effective depth $= 150 - 4 - 8 - 15 = 123$ mm

$$A_t = \frac{9{,}404{,}000}{140 \times 0{\cdot}87 \times 123} = 627{\cdot}7 \text{ mm}$$

Provide 8 mm ϕ bars at 80 mm centres.

Reinforcement for +ve B.M. along shorter span

Effective depth $= 150 - 4 - 15 = 131$ mm.

$$A_t = \frac{4{,}173{,}500}{140 \times 0{\cdot}87 \times 131} = 261{\cdot}6 \text{ mm}^2$$

Provide 8 mm ϕ bars at 180 mm centre, $A_t = 279$ mm^2.

Reinforcement at support B common with S_4

Effective depth $= 131 - 8 = 123$

$$A_t = \frac{3{,}137{,}000}{140 \times 0{\cdot}87 \times 123} = 209{\cdot}4 \text{ mm}^2$$

Minimum reinforcement

$$= \frac{0{\cdot}15 \times 150 \times 1000}{100} = 225 \text{ mm}^2$$

Provide 8 mm ϕ bars at 200 mm centres, $A_t = 252$ mm^2.

Reinforcement for +ve B M. along longer span

$$A_t = \frac{1{,}597{,}000}{140 \times 0{\cdot}87 \times 123} = 106{\cdot}6 \text{ mm}^2$$

Provide minimum reinforcement 8 mm ϕ at 200 mm centres.

Torsion reinforcement at C_2

Half of the reinforcement provided at C_1 is provided at C_2.

Provide 8 mm ϕ bars at 340 mm centres both ways at top and bottom.

Design of slab S_3

$$k = \frac{L}{B} = \frac{4}{3} = 1{\cdot}33.$$

Multiplying factors for span moments

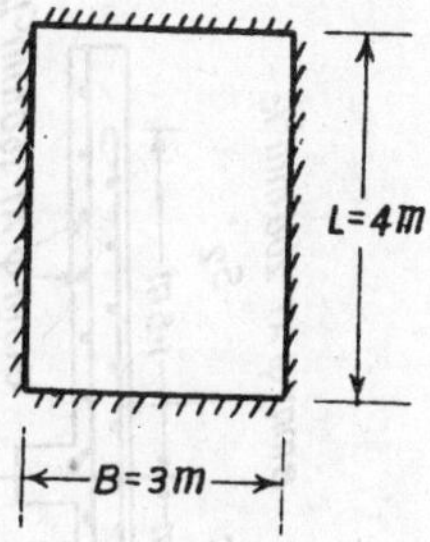

Fig. 8·11

For $M_B = 1 - \frac{5}{9} \times \frac{k^2}{1+2k^4}$

$= 1 - \frac{5}{9} \times \frac{(1 \cdot 33)^2}{1+3\times(1 \cdot 33)^4} = 0 \cdot 865$

For $M_L = 1 - \frac{15}{32} \times \frac{k^2}{1+2k^4}$

$= 1 - \frac{15}{32} \times \frac{(1 \cdot 33)^2}{1+2\times(1 \cdot 33)^4}$

$= 0 \cdot 886.$

Assume overall depth of 150 mm for slab

$$w = 7750 \text{ N/m}^2$$

$$w_B = 7750 \times \frac{2k^4}{1+2k^4} = 7750 \times 2 \times \frac{(1 \cdot 33)^4}{1+2\times(1 \cdot 33)^4}$$

$$= 6682 \text{ N/m}^2$$

$$w_L = 1068 \text{ N/m}^2.$$

Span moments

$$M_B = \frac{w_B \times B^2}{24} \times 0 \cdot 865$$

$$= \frac{6682 \times 3^2}{24} \times 0 \cdot 865 = 2167 \cdot 5 \text{ N-m}$$

$$= 2{,}167{,}500 \text{ N-mm}$$

$$M_L = \frac{9}{128} \times w_L \times L^2 \times 0 \cdot 886$$

$$= \frac{9}{128} \times 1068 \times 4^2 \times 0 \cdot 886$$

$$= 1064 \cdot 5 \text{ N-m}$$

$$= 1{,}064{,}500 \text{ N-mm.}$$

Support moments

$$M_B' = \frac{w_B \times B^2}{12} = \frac{6682 \times 3^2}{12} = 5011 \cdot 5 \text{ N-m}$$

$$= 5{,}011{,}500 \text{ N-mm}$$

$$M_L' = \frac{w_L \times L^2}{8} = \frac{1068 \times 4^2}{8} = 2136 \text{ N-m}$$

$$= 2{,}136{,}000 \text{ N-mm.}$$

Provide overall depth of 150 mm

At support common with S_1, provide 8 mm ϕ bars at 130 mm centres.

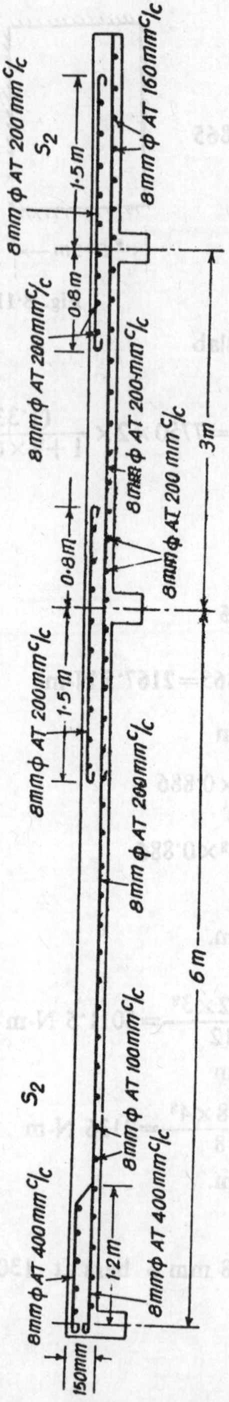

Fig. 8·13. Section $S_2 S_2$.

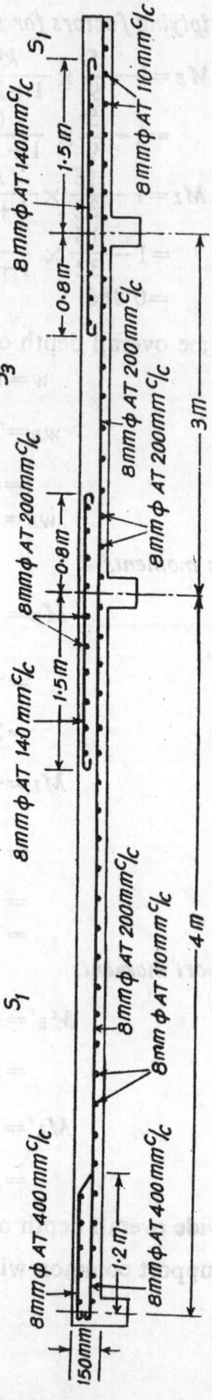

Fig. 8·14. Section through S_1, S_3, S_1.

Reinforcement at support B common with S_4

Effective depth $=150-4-13=133$ mm

$$A_t=\frac{2,136,000}{140\times0{\cdot}87\times133}=132\ \text{mm}^2$$

Minimum reinforcement $=\frac{0{\cdot}15}{100}\times150\times1000=225\ \text{mm}^2$

Provide 8 mm ϕ bars at 200 mm centres.

Reinforcement for +ve B.M. along shorter span

$$A_t=\frac{2,167,500}{140\times0{\cdot}87\times125}=142{\cdot}4\ \text{mm}^2$$

Minimum reinforcement $=\frac{0{\cdot}15}{100}\times150\times1000=225\ \text{mm}^2$

Provide 8 mm ϕ bars at 200 mm centres.

Reinforcement for +ve B.M. along longer span

Effective depth $=133-8=125$ mm

$$A_t=\frac{1,064,500}{140\times0{\cdot}87\times125}=69{\cdot}9\ \text{mm}^2$$

Minimum reinforcement $=\frac{0{\cdot}15}{100}\times150\times1000=225\ \text{mm}^2$

Provide 8 mm ϕ bars at 200 mm centres.

Design of Slab S_4

Multiplying factor for span moments

$$k=\frac{L}{B}=\frac{4}{3}.$$

Multiplying factor $=0{\cdot}882$

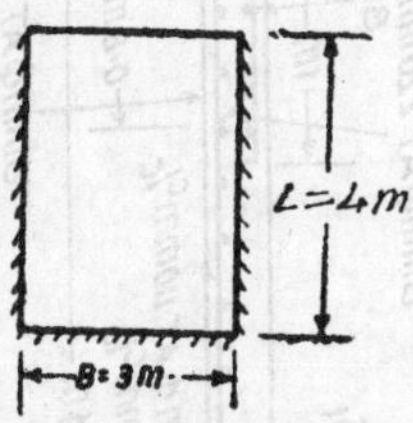

Fig. 8·15

Assume overall depth of 150 mm

$w=7600\ \text{N/m}^2$

$w_B=7600\times0{\cdot}76=5776\ \text{N/m}^2$

$w_L=1824\ \text{N/m}^2$

Span moments

$$M_B=\frac{w_B\times B^2}{24}\times0{\cdot}882=\frac{5776\times3^2}{24}\times0{\cdot}882$$

$$=1911\ \text{N-m}=1,911,000\ \text{N-mm}$$

$$M_L=\frac{w_L\times L^2}{24}\times0{\cdot}882$$

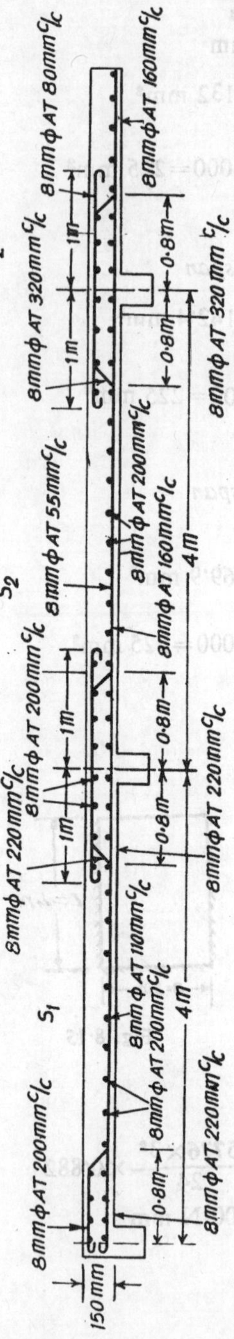

Fig. 8·16. Section through S_2, S_4, S_4.

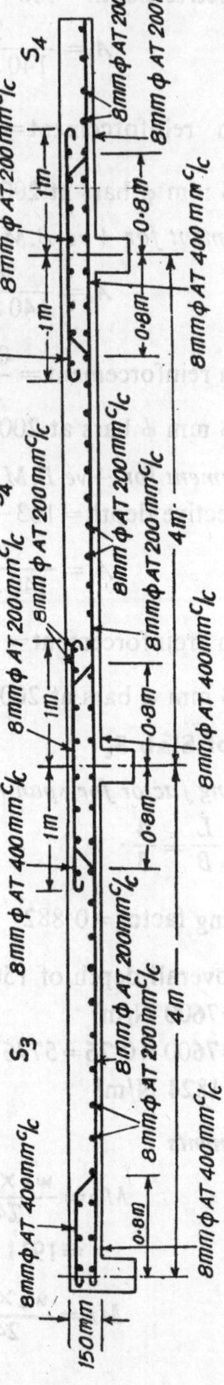

Fig. 8·17. Section through S_1, S_3, S_3.

$$=\frac{1824\times4^2}{24}\times0{\cdot}882=1073\text{ N-m}=1,073,000$$

Support moments $M_B'=\frac{w_B\times B^2}{12}=\frac{5776\times3^2}{12}=4332\text{ N-m}$

$$=4,332,000\text{ N-m}$$

$$M_L'=\frac{w_L\times L^2}{12}=\frac{1824\times4^2}{12}$$

$$=2432\text{ N-m}=2,432,000\text{ N-mm.}$$

Provide 8 mm ϕ bars at 200 mm centres at supports common with S_2 and 8 mm ϕ bars 200 mm at support common with S_3.

Reinforcement at support common with S_4

$$A_t=\frac{1,911,000}{140\times0{\cdot}87\times133}=150\text{ mm}^2$$

Provide minimum reinforcement of 8 mm ϕ bars at 200 mm centres.

Reinforcement for +ve B.M. in longer span

$$A_t=\frac{2,432,000}{140\times0{\cdot}87\times125}=215\text{ mm}^2.$$

Provide minimum reinforcement of 8 mm ϕ bars at 200 mm centres.

Reinforcement for +ve B.M. in longer span

Provide minimum reinforcement of 8 mm ϕ bars at 200 mm centres.

Design of Beam B.

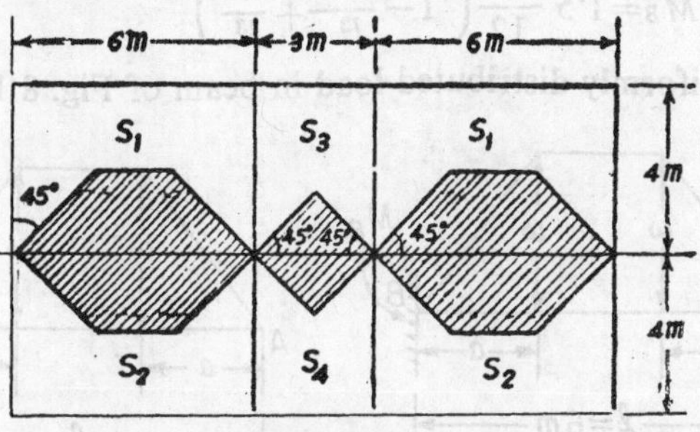

Fig. 8·18 (*a*)

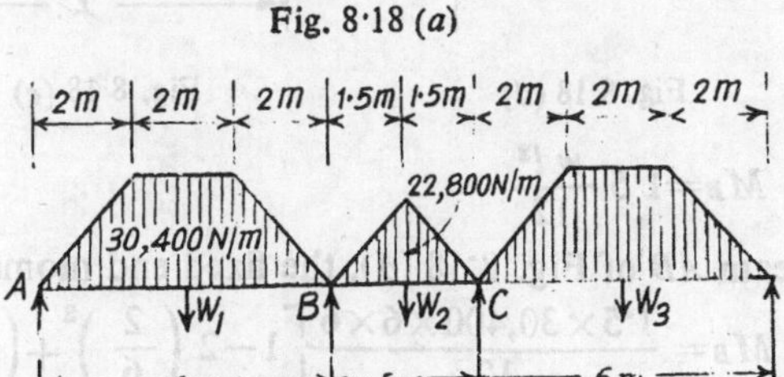

Fig. 8·18 (*b*)

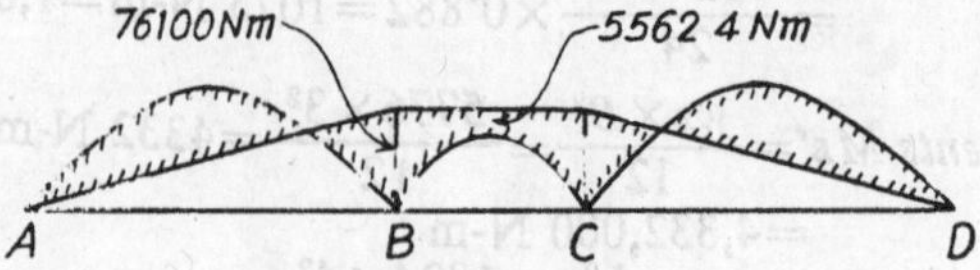

Fig. 8·18 (c)

$$W_1=W_3=2\times7600\times2\left(\frac{2+6}{2}\right)=121,600 \text{ N}$$

$$W_2=2\times7600\times1\text{·}5\times\frac{3}{2}=34,200 \text{ N}$$

Assume dead load of rib as 3000 N/m. Same depth of rib is provided for spans *AB*, *BC* and *CD*.

The continuous beam is symmetrical about centre line so distribution of moments is done by short-cut method taking *K* for span *BC* at ½ of the actual value. The fixed end moments and distribution of moments are shown in Table 8·7.

For the type of loading shown in Fig. 8·18 (*e*), fixed end moments are given by

$$M_A=M_B=\frac{wl^2}{12}\left(1-\frac{2a^2}{l^2}+\frac{a^3}{l^3}\right)$$

For end *A* simply supported and end *B* fixed [Fig. 8·18 (*d*)]

$$M_B=1\text{·}5\,\frac{wl^2}{12}\left(1-\frac{2a^2}{l^2}+\frac{a^3}{l^3}\right)$$

For uniformly distributed load in beam of Fig. 8·18 (*e*)

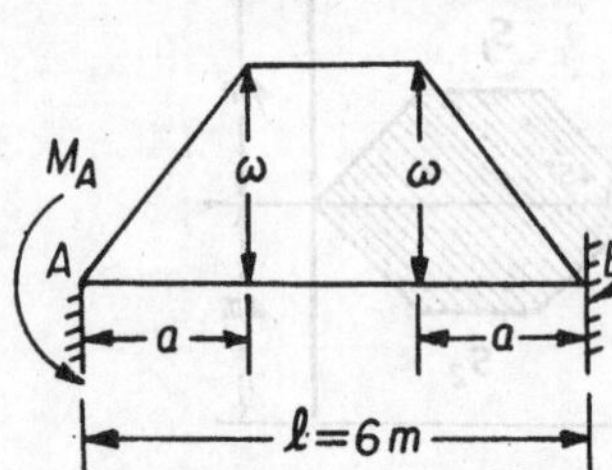

Fig. 8·18 (*d*)

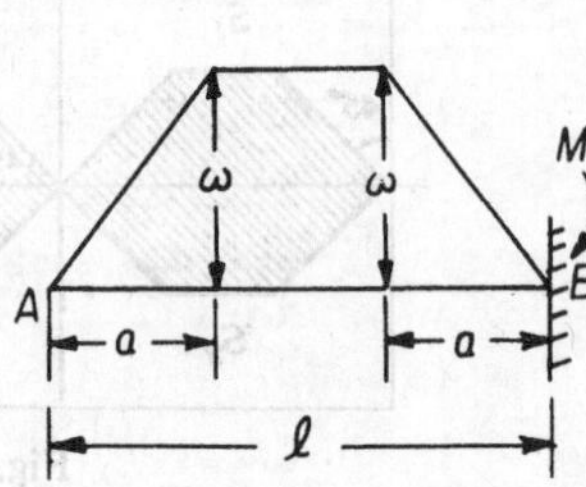

Fig. 8·18 (*e*)

$$M_B=1\text{·}5\,\frac{w\,l^2}{12}$$

For beam *AB* of Fig. 8·18 (*d*), the fixed end moment M_B is

$$M_B=\frac{1\text{·}5\times30,400\times6\times6}{12}\left[1-2\left(\frac{2}{6}\right)^2+\left(\frac{2}{6}\right)^3\right]$$

$$+\frac{1\text{·}5\times3000\times6\times6}{12}$$

$$=\frac{1{\cdot}5\times 30{,}400\times 6\times 6}{12}\left[1-\frac{2}{9}+\frac{1}{27}\right]+13{,}500$$

$$=111{,}467+13{,}500=124{,}967$$

For triangular loading on span BC in Fig. 8·18 (*b*), the fixed end moments are

$$M_B=M_C=\frac{5\,wl^2}{96}$$

and for uniform load fixed end moments are $\dfrac{w'\,l^2}{12}$.

Total fixed end moment

$$=\frac{5\,wl^2}{96}+\frac{w'\,l^2}{12}$$

$$=\frac{5\times 22{,}800\times 3\times 3}{96}+\frac{3000\times 3\times 3}{12}$$

$$=10{,}688+2250=12{,}938 \text{ N-m.}$$

Table 8·7

	A	3/7 B 4/7	C
F.E.M.	0	+124,967	−12,938
Balancing		−48012	−64017
Final moments	0	+76,955	−76955

Taking moments about B,

$$R_A\times 6+76{,}955=121{,}600\times 3+3000\times 6\times 3$$

$$\therefore\quad R_A=60{,}800+9{,}000-12{,}826=56{,}974 \text{ N}$$

Minimum +ve B.M. in AB will occur, where S.F. is zero.

Let x be the distance from A, where S.F. is zero.

$$56{,}974-30{,}400-30{,}400(x-2)-3{,}000\times x=0$$

$$x=\frac{87{,}774}{33{,}400}=2{\cdot}62 \text{ m.}$$

Maximum +ve B.M. in AB

$$=56{,}974\times 2{\cdot}62-30{,}400\times(2{\cdot}62-1{\cdot}33)$$

$$-\frac{30{,}400\times(2{\cdot}62-2)^2}{2}-3000\times\frac{(2{\cdot}62)^2}{2}$$

$$=149{,}272-39{,}200-5{,}860-10{,}300$$

$$=93{,}912 \text{ N-m.}$$

B.M. at centre BG $=22{,}800\times\dfrac{3}{2}\times\dfrac{3}{6}+\dfrac{3000\times 3^2}{8}-76{,}955$

$$=17{,}100+3{,}375-76{,}100=55{,}625 \text{ N-m.}$$

Design of AB

Assume rib width of 250 mm

Economical depth

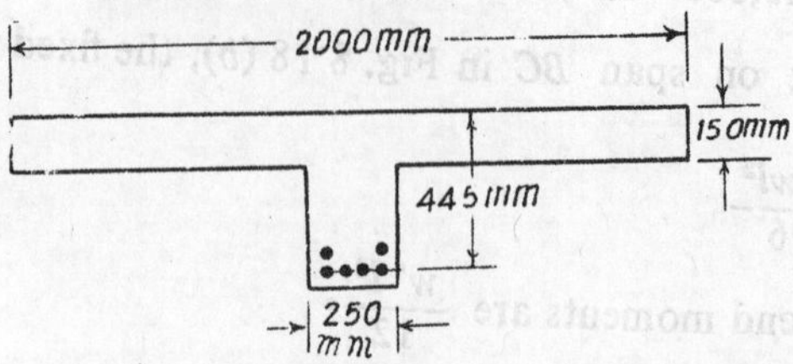

Fig. 8·19

$$d=\sqrt{\frac{M}{2{\cdot}1\, b_r}}=\sqrt{\frac{93{,}912\times1000}{2{\cdot}1\times250}}$$

$=423$ mm

Assume 500 mm overall depth.

Effective depth $=500-55=445$ mm

$$\text{Flange width } b_f=\frac{l_0}{6}+b_w+6\,D_f=\frac{6000}{6}+250+6\times150$$

$=2150$, say 2000 mm

∴ Flange width $=2000$ mm

Assume lever arm $=d-\frac{d_s}{2}=445-75=370$ mm

$$A_s=\frac{M}{t\times a}=\frac{93{,}912\times1000}{140\times370}=1813\text{ mm}^2$$

Provide 6 bars 20 mm ϕ

$$A_t=1885\text{ mm}^2$$

Let n be the depth of N.A. assumed to lie in flange. Taking moments about N.A.

$$2000\,n\times\frac{n}{2}=18\times1885(445-n)$$

$$n^2=15099-33{\cdot}93\,n$$

$$(n+17)^2=150{,}99+289$$

$$n=107\text{ mm}$$

$$\text{M.R.}=B\times n\times\frac{c}{2}\times\left(d-\frac{n}{3}\right)$$

$$93{,}912{,}000=2000\times107\times\frac{c}{2}\left(445-\frac{107}{3}\right)$$

$$c=2{\cdot}14\text{ N/mm}^2.$$

At supports

The beam will act as rectangular beam.

Effective depth for balanced section

$$=\sqrt{\frac{M}{0{\cdot}87\times250}}=\sqrt{\frac{76{,}955\times1000}{0{\cdot}87\times250}}$$

$=595$ mm.

Effective depth provided

$$=500-35=465 \text{ mm.}$$

The beam will be designed as doubly reinforced beam

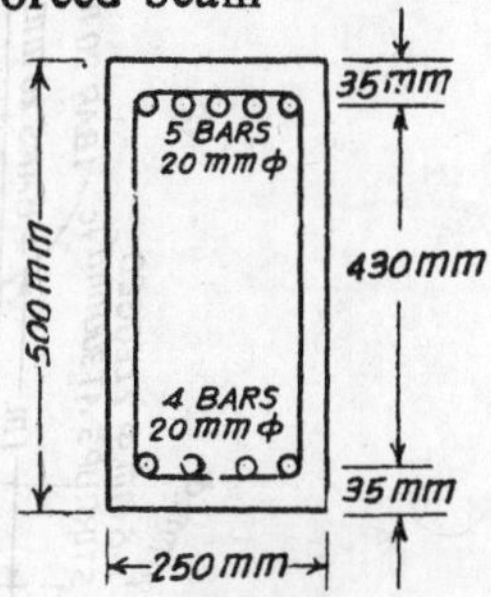

Fig. 8·20

$$M_1=0{\cdot}870\times bd^2$$

$$=0{\cdot}87\times 250\times(465)^2$$

$$=47{,}030{,}000 \text{ N-mm}$$

$$M_2=76{,}955{,}000-47{,}030{,}000$$

$$=29{,}925{,}000 \text{ N-mm}$$

$$n=0{\cdot}4\times 465=186 \text{ mm}$$

$$A_t=\frac{47{,}030{,}000}{140\times 0{\cdot}87\times 465}+\frac{29{,}925{,}000}{140\times 430}$$

$$=830+497=1327 \text{ mm}^2$$

Provide 5 bars 20 mm ϕ, area provided

$$=1571 \text{ mm}^2$$

$$A_c=\frac{29{,}925{,}000}{17\times 5\times \frac{186-35}{186}\times 430}$$

$$=\frac{29{,}925{,}000\times 186}{17\times 5\times 151\times 430}=1008 \text{ mm}^2$$

Provide 4 bars 20 mm ϕ.

Shear reinforcement

S.F. at $A=56{,}974$ N

S.F. at $B=121{,}600+18{,}000-56{,}974=82{,}626$ N

Shear stress $q=\frac{F}{bd}$

$$\therefore \qquad q=\frac{82{,}626}{465\times 250}=0{\cdot}71 \text{ N/mm}^2$$

Shear reinforcement is necessary.

$$\% \text{ steel}=\frac{1313\times 100}{465\times 250}=1{\cdot}13$$

$$\tau_c=0{\cdot}38 \text{ N/mm}^2$$

Shear force $\quad =0{\cdot}38\times 465\times 250=44{,}175$ N.

Let x metres be distance from support B, where S.F. is 44,175 N.

$$\therefore \qquad 82{,}626-\frac{30{,}400}{2}\times\frac{x^2}{2}-3000\,x=44{,}175$$

$$7600\,x^2+3000\,x=38{,}451$$

$$\therefore \qquad x^2+0{\cdot}4\,x=5{\cdot}06$$

$$(x+0{\cdot}2)^2=5{\cdot}06+0{\cdot}04=5{\cdot}10$$

$$\therefore \qquad x=2{\cdot}06 \text{ m}$$

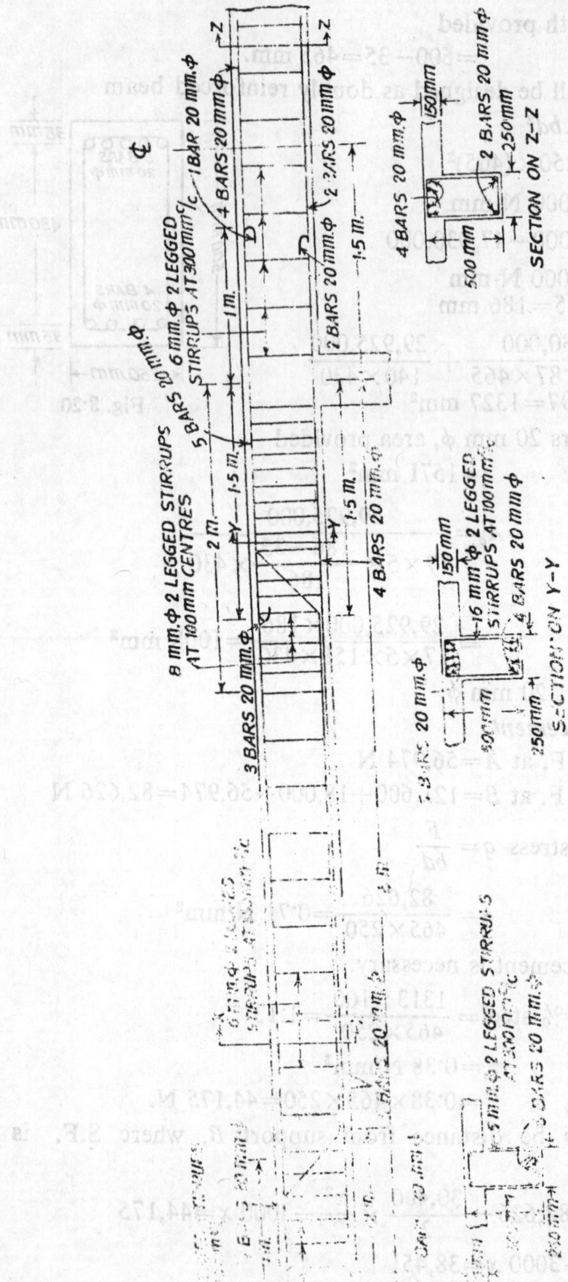
8 mm.φ 2 LEGGED STIRRUPS AT 100mm CENTRES
5 BARS 20 mm.φ
6 mm.φ 2 LEGGED STIRRUPS AT 300mm c/c
1 BAR 20 mm.φ
4 BARS 20 mm.φ
2 BARS 20 mm.φ
3 BARS 20 mm.φ
4 BARS 20 mm.φ
2 m.
1.5 m.
1 m.
1.5 m.
150mm
500 mm
250mm
SECTION ON Z-Z
6 mm φ 2 LEGGED STIRRUPS AT 100mm c/c
16 mm φ
SECTION ON Y-Y
SECTION ON X-X

Use 8 mm ϕ stirrups

$$\text{Spacing}=\frac{A_t \times t \times d}{F}=\frac{2\times 50\times 140\times 465}{(82{,}626-44{,}175)}=169 \text{ mm.}$$

Provide 8 mm ϕ 2-legged stirrups at 100 mm centres upto 2·0 metres from *B*.

Shear reinforcement near support A

Use 8 mm ϕ stirrups,

$$\text{Spacing}=\frac{2\times 50\times 140\times 465}{56{,}974-44{,}175}=508 \text{ mm.}$$

Provide 8 mm ϕ stirrups at 150 mm centres upto one metre from end *A*. In the remaining space provide nominal stirrups 6 mm ϕ at 300 mm.

Sufficient development length is provided including standard hooks.

Design of Beam BC

Provide overall depth of 500 mm.

Effective depth=465

M_1=47,030,000 N-mm

M_2=55,625,000−47,030,000=8,595,000 N-mm

$$A_t=\frac{47{,}030{,}000}{140\times 0{\cdot}87\times 465}+\frac{8{,}595{,}000}{140\times 430}$$

$$=827+143=970 \text{ mm}^2$$

Provide 4 bars 20 mm ϕ,

$$A_c=\frac{859{,}500}{17\times 5\times \frac{186-35}{186}\times 430}$$

$$=\frac{8{,}595{,}000\times 186}{17\times 5\times 151\times 430}=290 \text{ mm}^2$$

Provide 2 bars 20 mm ϕ.

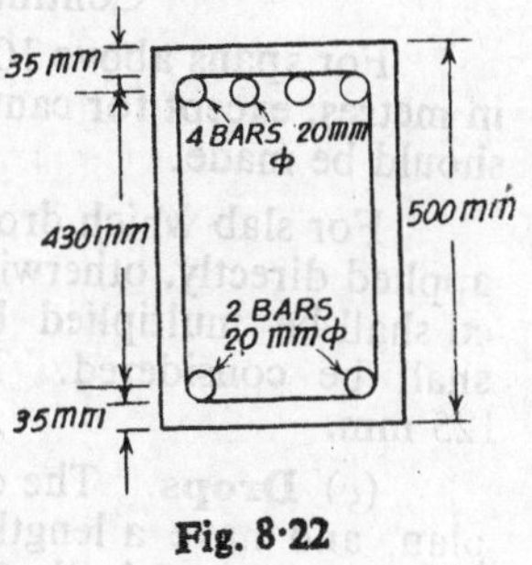

Fig. 8·22

$$\text{Shear force}=\frac{22{,}800\times 3}{2\times 2}+3000\times \frac{3}{2}=21{,}600 \text{ N.}$$

$$\text{Shear stress}=\frac{21{,}600}{250\times 465}=0{\cdot}21 \text{ N/mm}^2$$

No shear reinforcement is required.

Provide nominal shear reinforcement 6 mm ϕ, 2-legged stirrups at 300 mm centres. The section is shown in Fig. 8·22.

8·9. Flat Slab Floors

The term flat slab means a reinforced concrete slab with or without drops, supported generally without beams, by columns with or without flared column heads (see Fig. 8·23). A flat slab may be solid slab or may have recesses formed on the soffit so that the soffit comprises a series of ribs in two directions. The recesses may be formed by removable or permanent filler blocks.

(1) For the purposes of this clause, the following definitions shall apply :

(*a*) **Column strip.** Column strip means a design strip having a width of 0·25 l_2, but not greater than 0·25 l_1 on each side of the column centre-line, where l_1 is the span in the direction moments are being determined, measured centre to centre of supports and l_2 is the span transverse to l_1, measured centre to centre of supports.

(*b*) **Middle strip.** Middle strip means a design strip bounded on each of its opposite sides by the column strip.

(*c*) **Panel.** Panel means that part of a slab bounded on each of its four sides by the centre line of a column or centre-lines of adjacent spans.

(*d*) **Thickness of flat slab.** The thickness of flat slab shall be generally controlled by considerations of span to effective depth ratios as follows :

For span upto 10 m, values of span to effective depth.

Cantilever	7
Simply supported	20
Continuous	26

For spans above 10 m, the values may be multiplied by 10/span in metres, except for cantilevers in which case deflection calculations should be made.

For slab which drops, span to leffective depth ratios shall be applied directly, otherwise the span to effective depth ratios obtained shall be multiplied by 0·9. For this purpose the longer span shall be considered. The minimum thickness of slab shall be 125 mm.

(*e*) **Drops.** The drops when provided shall be rectangular in plan, and have a length in each direction not less than one-third of the panel length in that direction. For exterior panels, the width of drops at right angles to the non-continuous edge and measured from the centre line of the columns shall be equal to one-half the width of drop for interior panels.

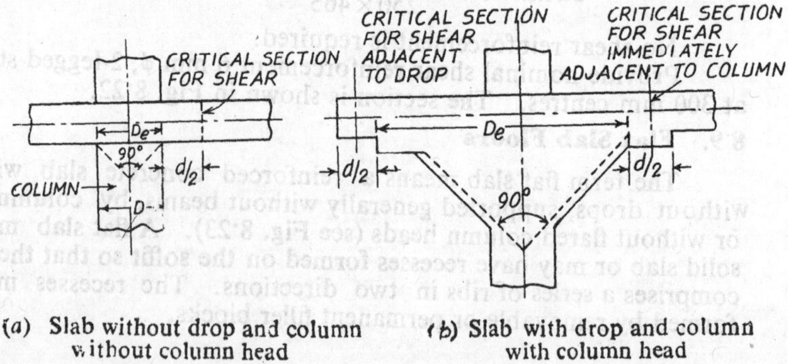

(*a*) Slab without drop and column without column head (*b*) Slab with drop and column with column head (*f*) **Column heads.** Where column heads are provided, that portion of a column head which lies within the largest circular cone or pyramid that has a vertex angle of 90° and can be included entirely within the outlines of the column head, shall be considered for design purposes (see Fig. 8·23).

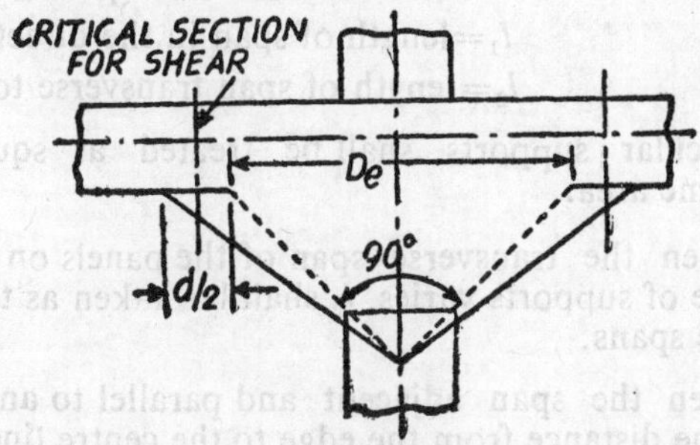

(*c*) Slab without drop and column with column head

Fig. 8·23

(2) **Determination of Bending Moment.** It shall be permissible to design the slab system by one of the following methods :

(A) The direct design method.

(B) The equivalent frame method.

8·10. Direct Design Method

(*a*) Slab systems designed by the direct method shall fulfil the following conditions :

(*i*) There shall be minimum of three continuous spans in each direction.

(*ii*) The panels shall be rectangular, and the ratio of the longer span to the shorter span within a panel shall not be greater than 2·0.

(*iii*) It shall be permissible to offset columns to a maximum of 10 per cent of the span in the direction of the offset notwithstanding the provision in (*i*).

(*iv*) The successive span lengths in each direction shall not differ by more than one third of the longer span. The end spans may be shorter than the interior spans.

(*v*) The design live load shall not exceed three times the design dead load.

(*b*) **Total Design Moment for a Span.** In the direct design method, the total design moment for a span shall be determined for a strip bounded laterally by the centre line of the panel on each side of the centre line of the supports.

(*i*) The absolute sum of the positive and average negative bending moments in each direction shall be taken as :

$$M_0 = \frac{W\, l_n}{8}$$

where, M_0=Total moment

W=design load on an area $l_2 \times l_n$

l_n=clear span extending from face to face of columns capitals, brackets or walls, but not less than 0·65 l_1.

l_1=length of span in the direction M_0 and

l_2=length of span transverse to l_1.

(*ii*) Circular supports shall be treated as square supports having the same area.

(*iii*) When the transverse span of the panels on either side of the centre-line of supports varies, l_2 shall be taken as the average of the transverse spans.

(*iv*) When the span adjacent and parallel to an edge is being considered, the distance from the edge to the centre line of the panel shall be substituted for l_2 in (*i*) above.

(c) Negative and Positive Design Moments

(*i*) The negative design moment shall be located at the face of rectangular supports, circular supports being treated as square supports having the same area.

(*ii*) In an interior span, the total design moment M_0 shall be distributed in the following proportions :

Negative design moment=0·65

Positive design moment=0·35

(*iii*) In the end span, the total design moment M_0 shall be distributed in the following proportions :

$$\text{Interior negative design moment} = 0{\cdot}75 - \frac{0{\cdot}10}{1+1/\alpha_c}$$

$$\text{Positive design moment} = 0{\cdot}63 - \frac{0{\cdot}28}{1+1/\alpha_c}$$

$$\text{Exterior negative design moment} = \frac{0{\cdot}65}{1+1/\alpha_c}$$

α_c is the ratio of flexural stiffness of the exterior columns to the flexural stiffness of the slab at a joint taken in the direction moments are being determined and is given by $\alpha_c = \frac{\Sigma k_c}{\Sigma k_s}$

where, Σk_c=sum of the flexural stiffnesses of the columns meeting at the joint and

k_s=flexural stiffness of the slab, expressed as moment per unit rotation (*iv*) It shall be permissible to modify these design moments by upto 10 per cent, so long as the total design moment M_0 for the panel in the direction considered is not less than that required by (*i*) in (*b*) above.

(*v*) The negative moment section shall be designed to resist the larger of the two interior negative design moments determined for the spans framing into a common support unless and analysis is made to distribute the unbalanced moment in accordance with the stiffness of the adjoining parts.

(*d*) **Distribution of Bending Moments Across the Panel Width.** Bending moments at critical cross-section shall be distributed to the column strips and middle strips as specified (*B*) at (*e*) as applicable.

(*e*) **Moments in Columns**

(*i*) Columns built integrally with the slab system shall be designed to resist moments arising from loads on the slab system.

(*ii*) At an interior support, the supporting members above and below the slab shall be designed to resist the moment M given by the following equation, in direct proportion to their stiffness unless a general analysis is made :

$$M=0{\cdot}8\frac{(w_d+0{\cdot}5\ w_1)\ l_2\ l_n^2-w_d'\ l_2'\ l_n'^2}{1+1/\alpha_c}$$

where, w_d, w_1=design dead and live loads respectively per unit area.

l_2=length of span transverse to the direction of M,

l_n=length of the clear span in the direction of M, measured face to face of supports.

$\alpha_c=\frac{\Sigma k_c}{\Sigma k_s}$ where k_c and k_s are as defined in (*c*), (*iii*) and w_d', l_2 and l_n' refer to the shorter span.

(*f*) **Effects of Pattern Loading.** In the direct design method, when the ratio of live load to dead load exceeds 0·5.

(*i*) The sum of the flexural stiffnesses of the columns above and below the slab. Σk_c shall be such that α_c is not less than the appropriate minimum value $\alpha_{c\ min}$ specified in Table 8·8 or

(*ii*) If the sum of the flexural stiffnesses of the columns does not satisfy (*i*), the positive design moments for the panel shall be multiplied by the coefficient β_s given by the following equation .

$$\beta_s=1+\left(\frac{2-\frac{w_d}{w_1}}{4+\frac{w_d}{w_1}}\right)\left(1-\frac{\alpha_c}{\alpha_{c\ min}}\right)$$

where α_c is the ratio of flexural stiffness of the columns above and below the slab to the flexural stiffness of the slabs at a joint taken in the direction moments are being determined and is given by,

$$\alpha_c = \frac{\Sigma k_c}{\Sigma k_s}$$

where k_c and k_s are flexural stiffness of column and slab respectively.

Table 8·8

Minimum Permissible Values of α_c

$\dfrac{\textit{Live Load}}{\textit{Dead Load}}$	Ratio $\dfrac{l_2}{l_1}$	Value of $\alpha_{c\ min}$
0·5	0·5 to 0·20	0
1·0	0·5	0·6
1·0	0·8	0·7
1·0	1·0	0·7
1·0	1·25	0·8
1·0	2·0	1·2
2·0	0·5	1·3
2·0	0·8	1·5
2·0	1·0	1·6
2·0	1·25	1·9
2·0	2·0	4·9
3·0	0·5	1·8
3·0	0·8	2·0
3·0	1·0	2·3
3·0	1·25	2·8
3·0	2·0	13·0

8·11. Equivalent Frame Method

(*a*) **Assumptions.** The bending moments and shear forces may be determined by an analysis of the structure as a continuous frame and the following assumptions may be made :

(*i*) The structure shall be considered to be made in to equivalent frame on column lines taken longitudinally and transversely through the building. Each frame consists of a row of equivalent columns or supports, bounded laterally by the centre line of the panel on each side of the centre line of the columns or supports. Frames adjacent and parallel to an edge shall be bounded by the edge and the centre line of the adjacent panel.

(*ii*) Each such frame may be analysed in its entirely, or, for vertical loading, each floor thereof and the roof may be analysed separately with its columns being assumed fixed at their remote ends. Where slabs are thus analysed separately, it may be assumed in determining the bending moment at a given support that the slab is fixed at any support two panels distant therefrom provided the slab is continuous beyond that point.

(*iii*) For the purpose of determining relative stiffness of members the moment of inertia of any slab or column may be assumed to be that of the gross section of the concrete alone.

(*iv*) Variations of moment of inertia along the axis of the slab on account of provision of drops shall be taken into account. In the case of recessed or coffered slab which is made solid in the region of the columns, the stiffening effect may be ignored provided the solid part of the slab does not extend more than 0·15 l_{ef}, into the span measured from the centre line of the column. The stiffness, effect of flared column ends may be ignored.

(*b*) **Loading Pattern**

(*i*) When the loading concern is known, the structure shall be analysed for the load concerned.

(*ii*) When the live load is variable but does not exceed three-quarters of the dead load, or the nature of the live load is such that all panels will be loaded simultaneously, the maximum moments may be assumed to occur at all sections when full design live load is on the entire slab system.

(*iii*) For other conditions of live load/dead load ratio and when all panels are not loaded simultaneously :

(I) Maximum positive moment near mid span of a panel may be assumed to occur when three-quarter of the full live load is on the panel and on alternate panels.

(II) Maximum negative moment in the slab at a support may be assumed to occur when the three-quarters of the full live load is on the adjacent panels only.

(*iv*) In no case shall design moments be taken to be less than those occurring with full design live load on all panels.

(*c*) **Negative Design Moment**

(*i*) At interior supports, the critical section for negative moment, in both the column strip and middle strip, shall be taken at the face of rectilinear support, but in no case at a distance greater than 0·175 l_1 from the centre of the column, where l_1 is the length of the span in the direction moments are being determined measured centre-to-centre of supports.

(*ii*) At exterior supports provided with brackets or capitals the critical section for negative moment in the direction perpendicular to the edge shall be taken at a distance from the face of the supporting element not greater than one-half the projection of the bracket or capital beyond the face to the supporting element.

(*iii*) Circular or regular polygon shaped supports shall be treated as square supports having the same area.

(*d*) **Modification of Maximum Moment.** Moments determined by means of the equivalent frame method, may be reduced in such proportion that the numerical sum of the positive and average negative moments is not less than the value of total design moment M_0.

(*e*) **Distribution of Bending Moment Across the Panel Width.**

(*i*) *Column strip.* Negative Moment at an Interior Support—At an interior support, the column strip shall be designed to resist 75 per cent of the total negative moment in the panel at that support.

(*ii*) *Column strip.* Negative Moment at an Exterior Support—

(*a*) At an exterior support, the column strip shall be designed to resist the total negative moment in the panel at that support.

(*b*) Where the exterior support consists of a column or a wall extending for a distance equal to or greater than three quarters of the value of l_2, the length of span transverse to the direction, moments are being determined, the exterior negative moment shall be considered to be uniformly distributed across the length l_2.

(*iii*) *Column strip.* Positive moment for each span—For each span, the column strip shall be designed to resist 60 per cent of the total positive moment in the panel.

(*iv*) *Moment in the Middle Strip.* The middle strip shall be designed on the following basis.

(I) The portion of the design moment not resisted by the column strip shall be assigned to the adjacent middle strips.

(II) Each middle strip shall be proportioned to resist the sum of the moments assigned to its two half middle strips.

(III) The middle strip adjacent and parallel to an edge supported by a wall shall be proportioned to resist twice the moment assigned to half the middle strip corresponding to the first row of interior columns.

8·12. Shear in Flat Slab

(*a*) The critical section for shear shall be at a distance $d/2$ from the periphery of the column called drop panel (see Fig. 8·23).

The shape in plan is geometrically similar to the support immediately below the slab see Fig. [8·24 (*a*), (*b*)]. For column sections with re-entrant angles, the critical section shall be taken as indicated in Fig. 8·24 (*c*), (*d*).

(*i*) In the case of column near the free edge of a slab, the critical section shall be taken as shown in Fig. 8·25.

(*ii*) When openings in flat slabs are located at a distance less than ten times the thickness of the slab from a concentrated reaction or when the openings are located within the column strip, the

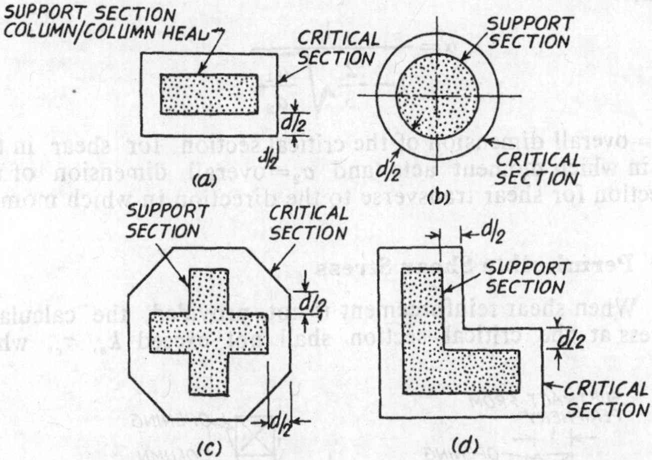

Fig. 8·24. Critical sections in plan for shear in flat slabs.

critical sections specified shall be modified so that the part of the periphery of the critical section which is enclosed by radial projections of the openings to the centroid of the section area shall be considered ineffective (see Fig. 8·26), and openings shall not encroach upon column head.

(*b*) **Calculation of shear stress.** The shear stress τ_0 shall be the sum of the values calculated according to (*i*) and (*ii*) below :

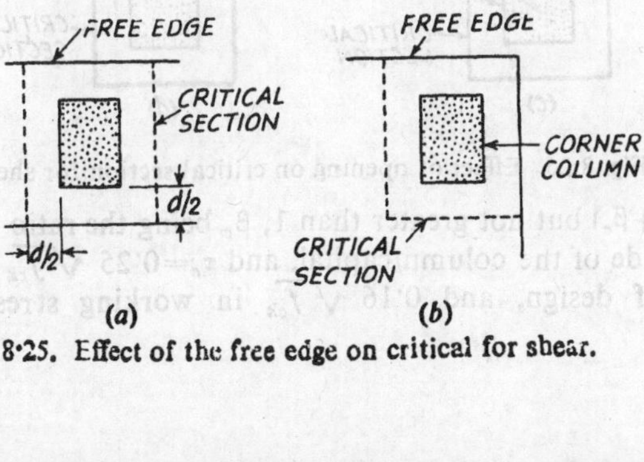

Fig. 8·25. Effect of the free edge on critical for shear.

(*i*) The nominal shear stress in flat slabs shall be taken as V/b_0d, where V is the shear force due to design load, b_0 is the periphery of the critical section and d is the effective depth.

(*ii*) When unbalanced gravity load, wind, earthquake or other forces cause transfer of bending moment between slab and column, a fraction $(1-\alpha)$ of the moment shall be considered transferred by eccentricity of the shear about the centroid of the critical section. Shear stresses shall be taken as varying linearly about the centroid of the critical section. The value of α shall be obtained from the equation,

$$\alpha = \frac{1}{1-\frac{2}{3}\sqrt{\frac{a_1}{a_2}}}$$

where, a_1=overall dimension of the critical section for shear in the direction in which moment acts and a_2=overall dimension of the critical section for shear transverse to the direction in which moment acts.

(*c*) Permissible Shear Stress

(*i*) When shear reinforcement is not provided, the calculated shear stress at the critical section shall not exceed k_s, τ_c, where

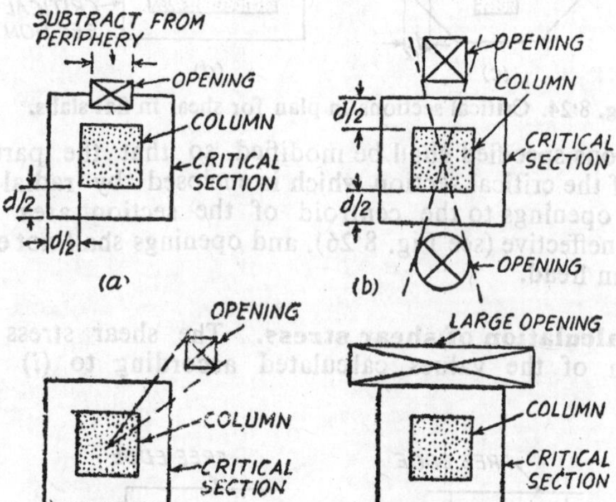

Fig. 8·26. Effect of opening on critical section for shear.

$k_s=(0{\cdot}5+\beta_c)$ but not greater than 1, β_c being the ratio of short side to long side of the column/capital, and $\tau_c=0{\cdot}25\sqrt{f_{ck}}$ in limit state method of design, and $0{\cdot}16\sqrt{f_{ck}}$ in working stress method of design.

(*ii*) When the shear stress at the critical section exceeds the value given above, but less than $1.5\tau_c$ shear reinforcement shall be provided. If the shear stress exceeds $1.5\ \tau_c$, the flat slab shall be redesigned. Shear stresses shall be investigated at successive sections more distant from the support and shear reinforcement shall be provided upto a section where the shear stress does not exceed $0.5\ \tau_c$ while designing the shear reinforcement, the shear stress carried by the concrete shall be assumed to be $0.5\ \tau_c$ and reinforcement shall carry the remaining shear.

8·13. Slab Reinforcement

(1) **Spacing.** The spacing of bars in a flat slab, shall not exceed 2 times the thickness, except where a slab is of cellular or ribbed construction.

(2) **Area of Reinforcement.** When drop panels are used : the thickness of drop panel for determination of area of reinforcement shall be the lesser of the following :

(*i*) Thickness of drop and

(*ii*) Thickness of slab plus one quarter the distance between edge of drop and edge of capital.

(3) **Minimum Length for Reinforcement**

(*i*) Reinforcement in flat slabs shall have the minimum lengths specified in Fig. 8·27. Larger lengths of reinforcement shall be provided when required by analysis.

(*ii*) When adjacent spans are unequal, the extension of negative reinforcement beyond each face of the common column shall be based on the longer span.

(4) **Anchoring Reinforcement**

(*i*) All slab reinforcement perpendicular to a discontinuous edge shall have an anchorage (straight, bent or otherwise anchored part the internal face of the spandrel beam, wall or column, an amount—

(*a*) **For positive reinforcement.** Not less than 15 cm except that with fabric reinforcement having a fully welded transverse wire directly over the support, it shall be permissible to reduce this length to one half of the width of the support or 5 cm, whichever is greater.

(*b*) **For negative reinforcement.** Such that the design stress is developed at the internal face.

(*ii*) Where the slab is not supported by a spandrel beam or wall, or where the slab cantilevers beyond the support, the anchorage shall be obtained within the slab.

8·14. Opening in Flat Slabs

Opening of any size may be provided in the flat slab if it is shown by analysis that the requirements of strength and serviceability are met. However, for openings conforming to the following, no special analysis is required :

(I) Openings of any size may be placed within the middle half of the span in each directions provided the total amount of reinforcement required for the panel without the opening is maintained.

(II) In the area common to two column strips, not more than one-eighth of the width of strip in either span shall be interrupted by the openings. The equivalent of reinforcement interrupted shall be added on all sides of the openings.

(III) In the area common to one column strip and one middle strip, not more than one-quarter of the reinforcement in either strip shall be interrupted by the openings. The equivalent of reinforcement interrupted shall be added on all sides of the openings.

(IV) The shear requirements shall be satisfied.

8·15. Bending Moments in Panel with Marginal Beams or Walls

Where the slab is supported by a marginal beam with a depth greater than 1·5 times the thickness of the slab or by a wall, then—

(I) the total load to be carried by the beam or wall shall comprise those loads directly on the wall or beam plus a uniformly distributed load equal to one-quarter of the total load on the slab, and

(II) the bending moments on the half-column strip adjacent to the beam or wall shall be one-quarter of the bending moments for the first interior column strip.

8·16. Transfer of Bending Moments to Columns

When unbalanced gravity load, wind, earthquake, or other lateral loads cause transfer of bending moment between slab and column, the flexural stresses shall be investigated using a fraction, α of the moment given by

$$\alpha = \frac{1}{1+\frac{2}{3}\sqrt{a_1/a_2}}$$

where a_1=overall dimension of the critical section for shear in the direction in which moment acts and a_2=overall dimension of the critical section for shear transverse to the direction in which moment acts.

A slab width between lines that are one and one-half slab or drops panel thickness 1·5 *D*, on each side of the column or capital may be considered effective, *D* being the size of the column.

Concentration of reinforcement over column head by closer spacing or additional reinforcement may be used to resist the moment on this section.

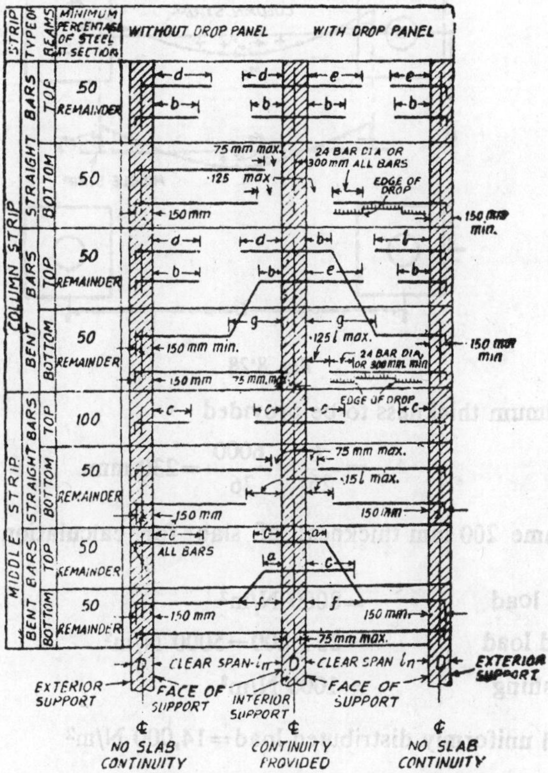

	BAR LENGTH FROM FACE OF SUPPORT						
	MINIMUM LENGTH					MAXIMUM LENGTH	
MARK	a	b	c	d	e	f	g
LENGTH	$0{\cdot}14\,l_n$	$0{\cdot}20\,l_n$	$0{\cdot}22\,l_n$	$0{\cdot}30\,l_n$	$0{\cdot}33\,l_n$	$0{\cdot}20\,l_n$	$0{\cdot}24\,l_n$

Minimum bend joint locations and extensions for reinforcement in flat slabs

Fig. 8·27

Ex. 8·3. *Design flat slab floor of a warehouse building carrying a load of 8000 N/m². The columns are spaced at 6 m centres.*

Sol., Design of Interior Panels. Flat slab with drops is provided (Fig. 8·29, 8·30 and 8·31).

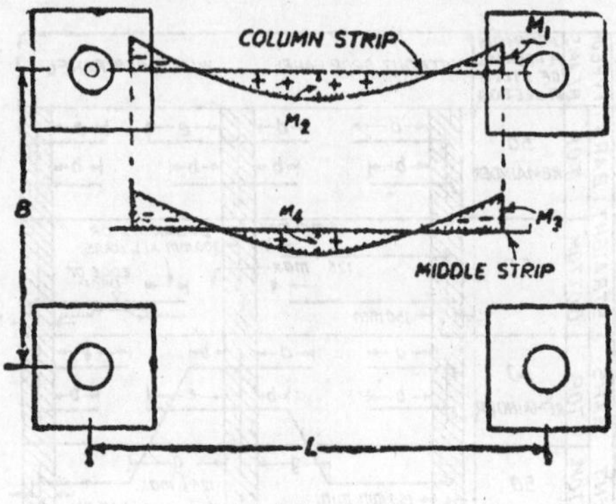

Fig. 8·28

Minimum thickness to be provided

$$=\frac{L}{26}=\frac{6000}{26}=230 \text{ mm.}$$

Assume 200 mm thickness of slab for calculations of dead load.

Live load $=8000$ N/m²

Dead load $=25\times200=5000$ N/m²

Finishing $=1000$ N/m²

Total uniformly distributed load $=14{,}000$ N/m²

Total load on the panel

$$=14{,}000\times6\times6=504{,}000 \text{ N}$$

Maximum value of $D=0{\cdot}25\,L=0{\cdot}25\,L\times\ 0=1500$ mm

Provide $D=1400$ mm.

$$M_0=\frac{W\,l_n}{8}=\frac{(14{,}000\,l_2\,l_n)\,(l_n)}{8}$$

$$=\frac{14{,}000\times6\times(6-1{\cdot}4)(6-1{\cdot}4)}{8}$$

$$=222{,}180 \text{ N-m}$$

$$=222{,}180{,}000 \text{ N-mm.}$$

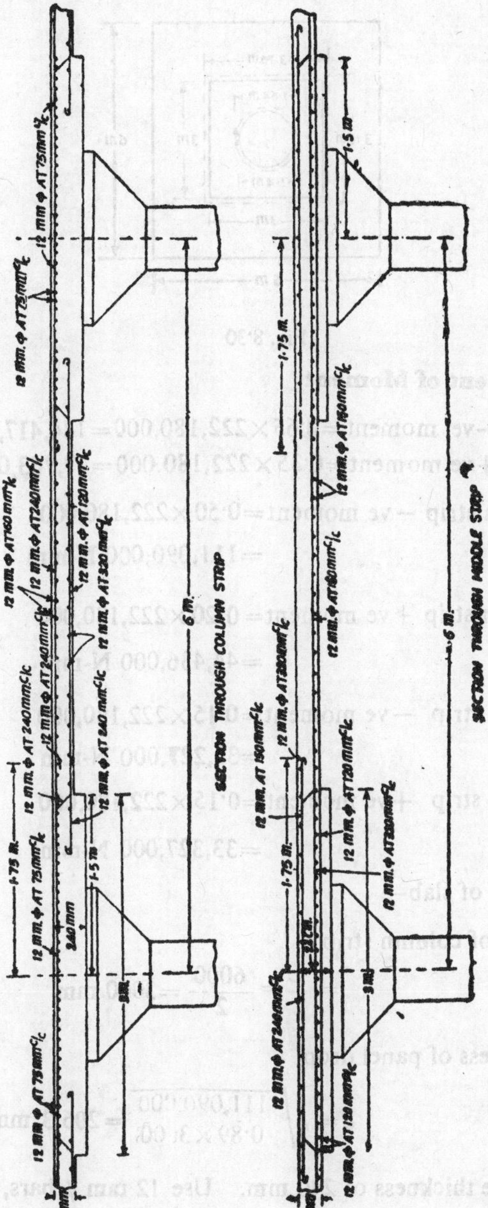
SECTION THROUGH COLUMN STRIP
SECTION THROUGH MIDDLE STRIP

Fig. 8·29

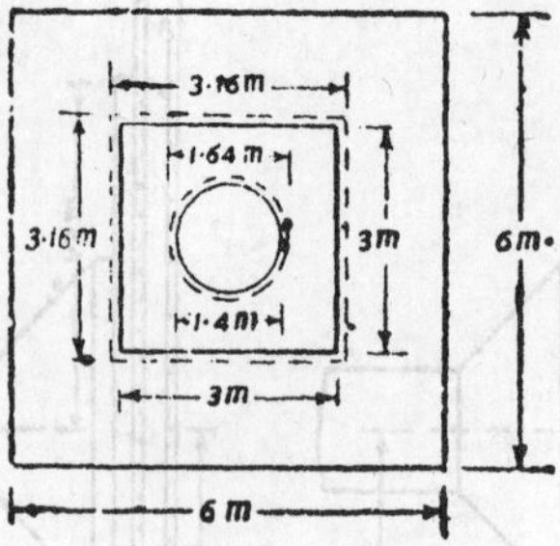

Fig. 8·30

Apportionment of Moment

$$-\text{ve moment} = 0{\cdot}65 \times 222{,}180{,}000 = 144{,}417{,}000 \text{ N-mm}$$
$$+\text{ve moment} = 0{\cdot}35 \times 222{,}180{,}000 = 77{,}763{,}000 \text{ N-mm}$$

Column strip −ve moment $= 0{\cdot}50 \times 222{,}180{,}000$
$= 111{,}090{,}000$ **N-mm**

Column strip +ve moment $= 0{\cdot}20 \times 222{,}180{,}000$
$= 44{,}436{,}000$ **N-mm**

Middle strip −ve moment $= 0{\cdot}15 \times 222{,}180{,}000$
$= 33{,}327{,}000$ **N-mm**

Middle strip +ve moment $= 0{\cdot}15 \times 222{,}180{,}000$
$= 33{,}327{,}000$ **N-mm**

Design of slab—

Width of column strip

$$= \frac{L}{2} = \frac{6000}{2} = 3000 \text{ mm}$$

Thickness of panel drop,

$$d = \sqrt{\frac{111{,}090{,}000}{0{\cdot}89 \times 3000}} = 206{\cdot}3 \text{ mm}$$

Provide thickness of 240 mm. Use 12 mm ϕ bars,

Effective depth $= 240 - 12 - 6 - 13 = 209$ **mm.**

$$A_t = \frac{111{,}090{,}000}{140 \times 0{\cdot}87 \times 206{\cdot}3} = 4364 \text{ mm}^2$$

For 3 metre width provide 12 mm ϕ bars at 75 mm centres.

Middle of column strip

$$\text{B.M.}=44{,}436{,}000 \text{ N-mm.}$$

$$d=\sqrt{\frac{44{,}436{,}000}{0.87\times3000}}=130.5 \text{ mm}$$

Provide overall depth of 160 mm,

Effective depth $=160-6-15=139$ mm.

$$A_t=\frac{44{,}436{,}000}{140\times0.87\times139}=2625 \text{ mm}^2$$

Provide 12 mm ϕ bars at 120 mm centres.

Middle strip

Width of middle strip

$=6000-3000=3000$ mm.

Effective depth $=160-6-12-15=127$ mm.

$$A_t=\frac{33{,}327{,}000}{140\times0.87\times127}=2154.5 \text{ mm}^2.$$

Provide 12 mm ϕ bars at 160 mm centres for both +ve and −ve B.M.

Check for shear

1. At a section $\frac{d_s}{2}$ distance from panel drop

Length of critical section for shear

$=4(3+0.16)=12.64$ m

$=12{,}640$ mm.

Loaded area for shear

$=(6)^2-(3.16)^2=26.01 \text{ m}^2$

Total load per sq. metre

$=8000+25\times160+1000=13{,}000 \text{ N/m}^2$

$$\text{S.F.}=13{,}000\times26.01=338{,}130 \text{ N.}$$

$$\text{Average shear stress}=\frac{F}{bd}=\frac{338{,}130}{12{,}640\times139}=0.192 \text{ N/mm}^2$$

$\tau_c=0.16\sqrt{f_{ck}}$

$=0.16\sqrt{15}=0.62 \text{ N/mm}^2$ $\therefore$ Safe

2. At section $\frac{d}{2}$ from column head.

Length of critical section for shear

$=\pi(1{\cdot}4+0{\cdot}24)=5{\cdot}152$ m

$=5152$ mm.

Load per square metre in drop panel

$=8000+25\times240+1000=15{,}000$ N/m²

Shear force $=13{,}000(6^2-3^2)+15{,}000\left[3^2-\frac{\pi}{4}(1{\cdot}64^2\right]$

$=454{,}305$ N

Shear stress $=\frac{F}{bd}=\frac{454{,}305}{5152\times207}$

$=0{\cdot}43$ N/mm² $<0{\cdot}62$ N/mm². Safe.

Sufficient anchorage length is provided for bond.

Ex. 8·4. *Design end panel and 3 span continuous edge beam ABCD of the flat slab floor shown in Fig. 8·31. Live load on the slab is 6000 N/m².*

Sol. Load per square metre will be 14,000 N/m² as in problem 8·3.

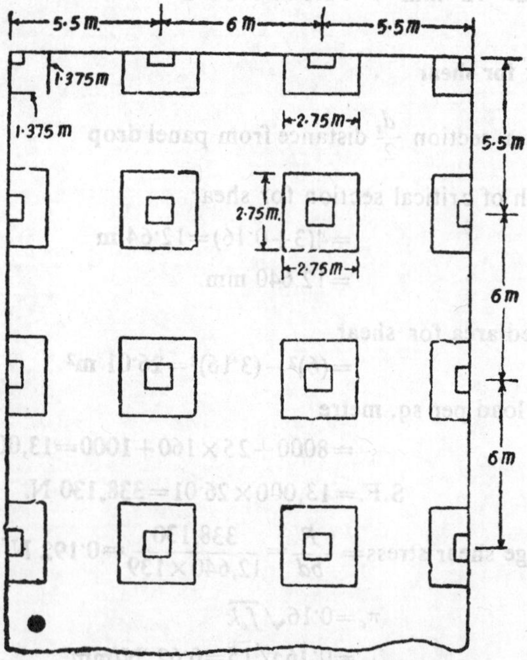

Fig. 8·31

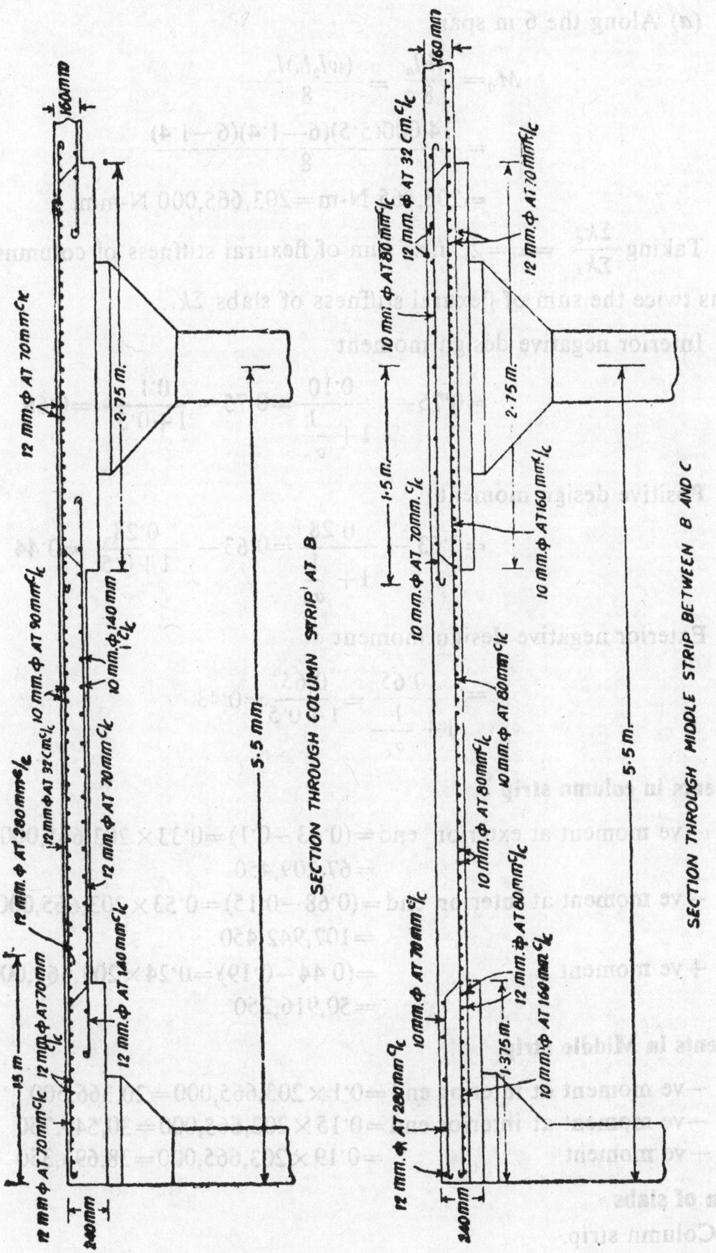
SECTION THROUGH COLUMN STRIP AT B
SECTION THROUGH MIDDLE STRIP BETWEEN B AND C
160 mm
240 mm
5·5 m
2·75 m
1·5 m
1·375 m

Fig, 8·32

(*a*) Along the 6 m span

$$M_0 = \frac{Wl_n}{8} = \frac{(wl_2l_n)l_n}{8}$$

$$= \frac{14{,}000(5{\cdot}5)(6-1{\cdot}4)(6-1{\cdot}4)}{8}$$

$$= 203{,}665 \text{ N-m} = 203{,}665{,}000 \text{ N-mm.}$$

Taking $\frac{\Sigma k_c}{\Sigma k_s} = \alpha_c = 2$, *i.e.* sum of flexural stiffness of columns Σk_c as twice the sum of flexural stiffness of slabs Σk_s.

Interior negative design moment

$$= 0{\cdot}75 - \frac{0{\cdot}10}{1+\frac{1}{\alpha_c}} = 0{\cdot}75 - \frac{0{\cdot}1}{1+0{\cdot}5} = 0{\cdot}68$$

Positive design moment

$$= 0{\cdot}63 - \frac{0{\cdot}28}{1+\frac{1}{\alpha_a}} = 0{\cdot}63 - \frac{0{\cdot}28}{1+0{\cdot}5} = 0{\cdot}44$$

Exterior negative design moment

$$= \frac{0{\cdot}65}{1+\frac{1}{\alpha_c}} = \frac{0{\cdot}65}{1+0{\cdot}5} = 0{\cdot}43$$

Moments in column strip

−ve moment at exterior end $= (0{\cdot}43 - 0{\cdot}1) = 0{\cdot}33 \times 203{,}665{,}000$
$= 67{,}209{,}450$

−ve moment at interior end $= (0{\cdot}68 - 0{\cdot}15) = 0{\cdot}53 \times 203{,}665{,}000$
$= 107{,}942{,}450$

+ve moment $= (0{\cdot}44 - 0{\cdot}19) = 0{\cdot}24 \times 203{,}665{,}000$
$= 50{,}916{,}250$

Moments in Middle strip

−ve moment ot interior end $= 0{\cdot}1 \times 203{,}665{,}000 = 20{,}366{,}500$
−ve moment at interior end $= 0{\cdot}15 \times 203{,}665{,}000 = 30{,}549{,}750$
+ve moment $= 0{\cdot}19 \times 203{,}665{,}000 = 38{,}696{,}350$

Design of slabs

Column strip,

Max. B.M. $= 107{,}942{,}450$ N-mm.

Effective depth $= \sqrt{\frac{107{,}942{,}450}{0{\cdot}87 \times 2750}} = 212{\cdot}4$ mm.

Provide overall depth of 24'mm.

Using 12 mm ϕ bars, effective depth

$$=240-12-6-15=207 \text{ mm}$$

Reinforcement at interior end,

$$A_t=\frac{107{,}942{,}450}{0{\cdot}87\times207\times140}=4281{\cdot}3 \text{ mm}^2$$

Provide 12 mm ϕ bars at 70 mm centres

Reinforcement at exterior end

$$A_t=\frac{67{,}209{,}450}{0{\cdot}87\times207\times140}=2665{\cdot}7 \text{ mm}^2$$

Provide 12 mm ϕ at 70 mm centres.

Middle of column strip

$$\text{Effective depth} = \sqrt{\frac{50{,}916{,}250}{0{\cdot}87\times2750}}=146 \text{ mm}$$

Provide overall depth of 160 mm.

Using 12 ϕ bars, effective depth

$$=160-5-15=139 \text{ mm.}$$

$$A_t=\frac{50{,}916{,}250}{0{\cdot}87\times140\times139}=3007{\cdot}4 \text{ mm}^2$$

Provide 12 mm ϕ at 70 mm centres.

Middle strip

$$\text{Effective depth} =160-6-10-15=130 \text{ mm}$$

For +ve B.M. $A_t=\dfrac{38{,}696{,}350}{140\times0{\cdot}87\times130}=2444 \text{ mm}^2$

Provide 10 mm ϕ bars at 80 mm centres

For −ve B.M. at interior end,

$$A_t=\frac{20{,}366{,}500}{140\times0{\cdot}87\times130}=1286\ 3 \text{ mm}^2$$

Provide 10 mm ϕ bars at 160 mm centres,

For −ve B.M. at interior end

$$A_t=\frac{30{,}549{,}750}{140\times0{\cdot}87\times130}=1929{\cdot}4 \text{ mm}^2$$

Provide 10 mm ϕ 80 mm centres.

The same reinforcement is provided for moment in the strip parallel to 5·5 m.

The section will be safe against shear (see Ex. 8·3).

Design of Marginal Beam. The marginal beam is designed as L-beam. The depth of marginal beam is kept greater than $1\frac{1}{2}$ times the thickness of the flat slab. It carries load directly supported by it such as walls etc. and the live and dead load $\frac{1}{4}$th of the panel taken as uniformly distributed. If the beam is made 300 mm or more wide shearing stresses due to torsion may be neglected.

Design of End Beam ABCD

Assume wall load and self load of beam as 10,000 N/m.

Load on beam from slab

$$=\frac{14{,}000\times 5{\cdot}5}{4}=19{,}250 \text{ N/m.}$$

Total load per metre run on beam

$$=19{,}250+10{,}000=29{,}250 \text{ N.}$$

The continuous beam is symmetrical about centre line so distribution of moments is done by short cut method, taking K

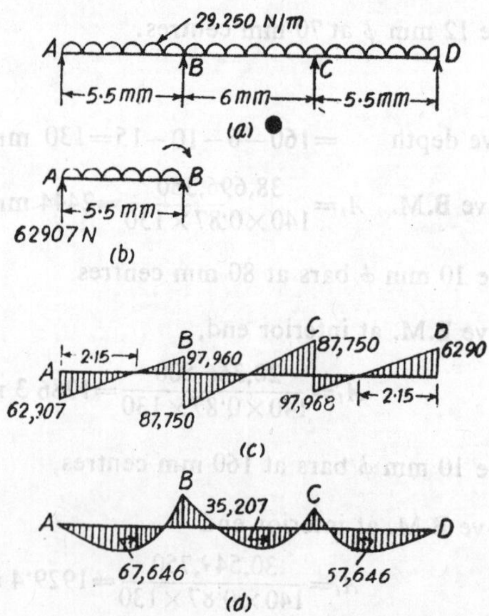

Fig. 8·33

for span *BC* as $\frac{1}{2}$ of the actual value. The fixed end moments and distributions of moments is shown in table 8·10.

Table 8·10

A		*B*	
	$\frac{18}{29}$		$\frac{11}{29}$
	+110,602		−87,750
	−14,184		−8,668
	+96,418		−96,418

$$R_A \times 5{\cdot}5 + 96{,}418 - 29{,}250 \times 5{\cdot}5 \times \frac{5{\cdot}5}{2} = 0$$

$$R_A = 62.907 \text{ N}$$

$$R_B = \frac{29{,}250 \times 17 - 2 \times 62{,}907}{2} = 186{,}718 \text{ N.}$$

The B.M. and S.F. diagrams are shown in Fig. 8·34. Keep width of beam 300 mm.

$$\text{Effective depth} = \sqrt{\frac{96{,}418 \times 1000}{0{\cdot}87 \times 300}} = 607{\cdot}8 \text{ mm}$$

Keep overall depth of 650 mm.

Use 20 mm ϕ bars.

$$\text{Effective depth} = 650 - 25 - 10 = 615 \text{ mm}$$

$$A_t = \frac{96{,}411 \times 1000}{0{\cdot}87 \times 615 \times 140} = 1287 \text{ mm}^2$$

Provide 5 bars of 20 mm ϕ.

Reinforcement for +ve B.M.

$$\text{Assume lever arm } a = d - \frac{d_s}{2} = 615 - 80 = 535 \text{ mm.}$$

$$A_t = \frac{67{,}646 \times 1000}{0{\cdot}87 \times 535 \times 100} = 1038 \text{ mm}^2$$

Provide 4 bars of 20 mm ϕ.

Reinforcement at centre of BC

$$A_t = \frac{35{,}207 \times 1000}{140 \times 535} = \mathbf{470\ mm^2}$$

Use 2 bars 20 mm ϕ.

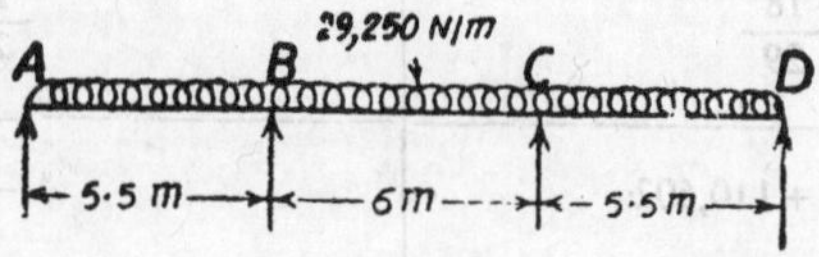

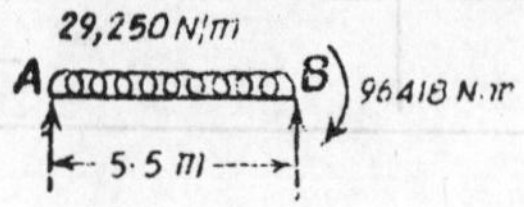

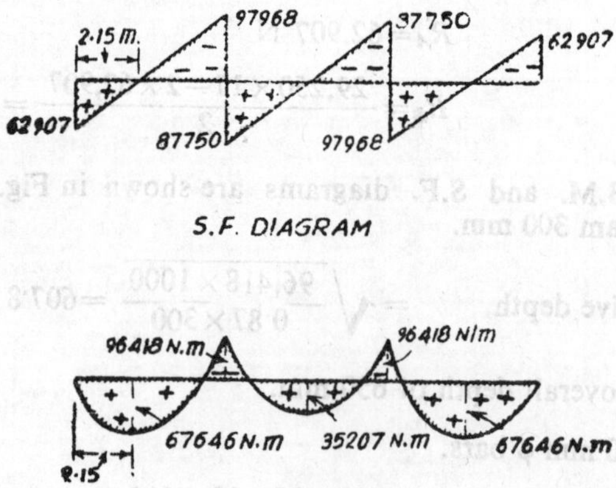

S.F. DIAGRAM

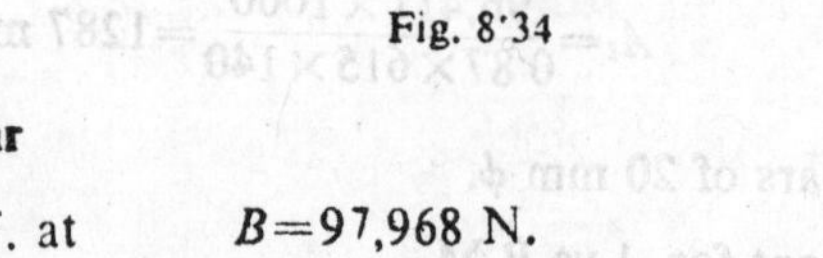

B. M. DIAGRAM

Fig. 8'34

Check for Shear

Max. S.F. at $\quad B = 97{,}968$ N.

Shear stress $\quad = \frac{V}{bd} = \frac{97{,}968}{300 \times 615} = \mathbf{0{\cdot}53\ N/mm^2}$.

% of steel $\quad = \frac{1287 \times 100}{300 \times 615} = 0{\cdot}7\%$

$\tau_a = 0{\cdot}52$

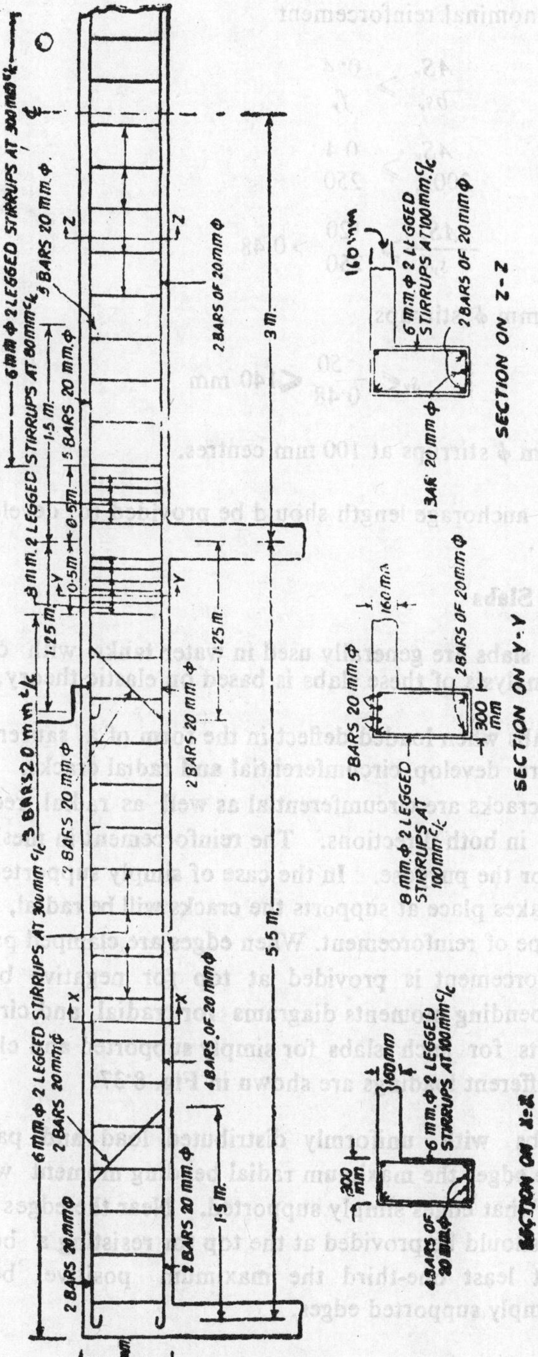
6mm Φ 2 LEGGED STIRRUPS AT 300mm c/c
3 BARS 20 mm Φ
2 LEGGED STIRRUPS AT 80mm c/c
1·5 m
5 BARS 20 mm Φ
5 BARS 20 mm Φ
8 mm. 2 LEGGED
0·5 m
0·5 m
1·25 m
3 BARS 20 mm
STIRRUPS AT 300 mm c/c
2 BARS 20 mm Φ
6mm Φ 2 LEGGED STIRRUPS AT 300mm c/c
2 BARS 20mm Φ
2 BARS 20mm Φ
2 BARS 20 mm Φ
2 BARS 20 mm Φ
2 BARS 20 mm Φ
4 BARS OF 20 mm Φ
2 BARS OF 20mm Φ
1·25 m
1·5 m
3 m
5·5 m
600 mm
160 mm
6 mm Φ 2 LEGGED STIRRUPS AT 100mm c/c
2 BARS OF 20 mm Φ
1 BAR 20 mm Φ
SECTION ON Z-Z
5 BARS 20 mm Φ
8 mm Φ 2 LEGGED STIRRUPS AT 100mm c/c
160 mm
300 mm
2 BARS OF 20mm Φ
SECTION ON Y-Y
6 mm Φ 2 LEGGED STIRRUPS AT 100mm c/c
160mm
300 mm
4 BARS OF 20 mm Φ
SECTION ON X-X

Fig. 8·35

Provide nominal reinforcement

$$\frac{AS_v}{bs_v} \geqslant \frac{0{\cdot}4}{f_y}$$

$$\therefore \quad \frac{AS_v}{300 s_v} \geqslant \frac{0{\cdot}4}{250}$$

$$\therefore \quad \frac{AS_v}{s_v} \geqslant \frac{120}{250} > 0{\cdot}48$$

Using 8 mm ϕ stirrups

$$s_v \leqslant \frac{50}{0{\cdot}48} \leqslant 140 \text{ mm}$$

Use 8 mm ϕ stirrups at 100 mm centres.

Sufficient anchorage length should be provided to develop the stress by bond.

8·17. Circular Slabs

Circular slabs are generally used in water tanks with circular bases. The analysis of these slabs is based on elastic theory.

These slabs when loaded deflect in the form of a saucer. The slabs at failure develop circumferential and radial cracks. At the centre of slab cracks are circumferential as well as radial requiring reinforcement in both directions. The reinforcement in mesh form will be ideal for the purpose. In the case of simply supported slabs if the failure takes place at supports the cracks will be radial, requiring circular type of reinforcement. When edges are clamped partially or fully reinforcement is provided at top for negative bending moment. The bending moments diagrams for radial and circumferential moments for such slabs for simply supported and clamped edges under different loadings are shown in Fig. 8·37.

For slabs with uniformly distributed load and partially clamped at the edges the maximum radial bending moment will be 80 per cent of that edges simply supported. Near the edges radial reinforcement should be provided at the top for resisting a bending moment of at least one-third the maximum positive bending moment for simply supported edges.

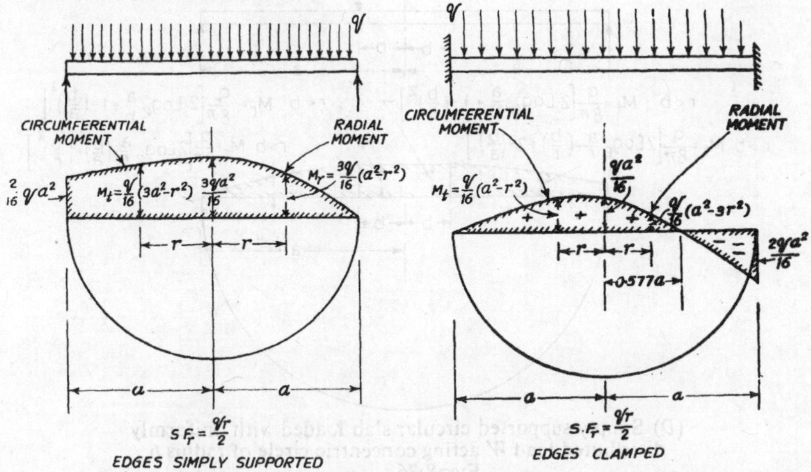

(*A*) **Uniformly distributed load *q* on whole slab.**

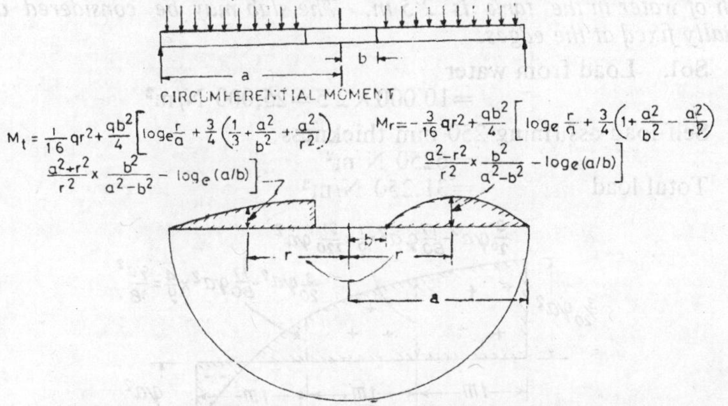

(*B*) **Simply supported circular slab with a central hole of radius '*b*' and uniformly loaded over whole area**

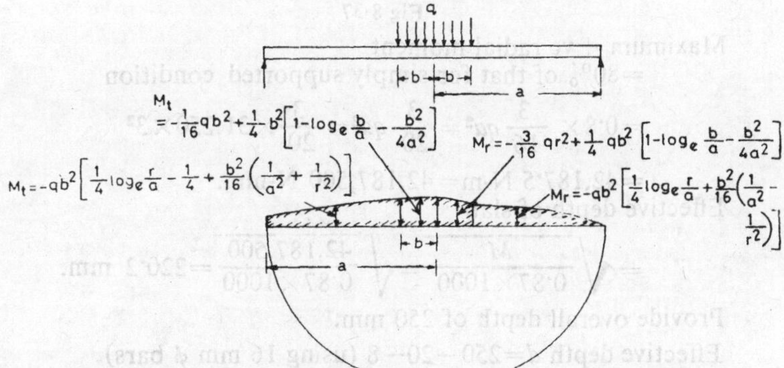

(*C*) **Simply supported circular slab with uniformly distributed.**

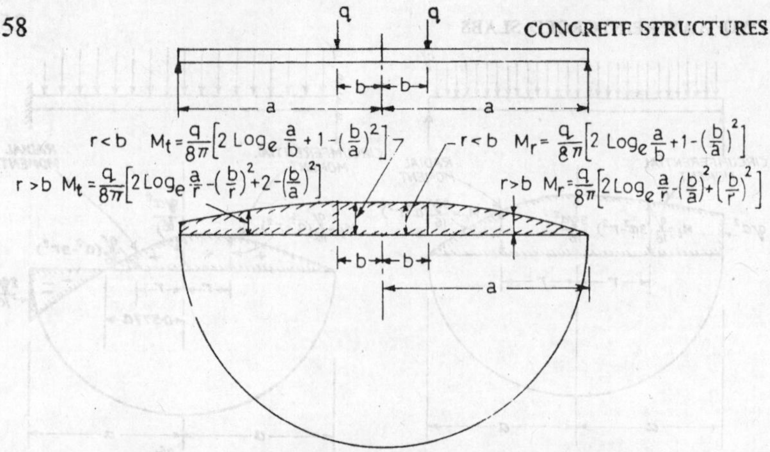

(*D*) Simply supported circular slab loaded with uniformly distributed load *W* acting concentric circle of radius *b*.

Fig. 8·36

Ex. 8·4. *Design base of a circular tank of diameter 6 m. The depth of water in the tank is 2·5 m. The slab may be considered as partially fixed at the edges.*

Sol. Load from water

$$=10{,}000\times 2{\cdot}5=25{,}000 \text{ N/m}^2$$

Self-load assuming 250 mm thickness

$$=6250 \text{ N/m}^2$$

Total load $=31{,}250$ N/m²

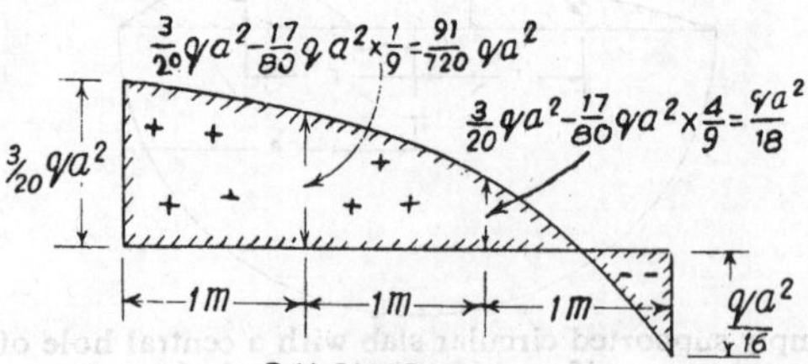

B.M. DIAGRAM

Fig 8·37

Maximum +ve radial moment

=80% of that for simply supported condition

$$=0{\cdot}8\times\frac{3}{16}qa^2=\frac{3}{20}qa^2=\frac{3}{20}\times 31{,}250\times 3^2$$

$$=42{,}187{\cdot}5 \text{ N-m}=42{,}187{,}500 \text{ N-mm.}$$

Effective depth of slab

$$=\sqrt{\frac{M}{0{\cdot}87\times 1000}}=\sqrt{\frac{42{,}187{,}500}{0{\cdot}87\times 1000}}=220{\cdot}2 \text{ mm.}$$

Provide overall depth of 250 mm.

Effective depth $d=250-20-8$ (using 16 mm ϕ bars)

$=222$ mm

$$A_t = \frac{42,187.500}{0{\cdot}87 \times 222 \times 140} = 1560 \text{ mm}^2$$

Provide 16 mm ϕ bars at 120 mm centres in both directions in the middle 2 m strip.

$-$ve B.M. at support is taken as one-third of maximum +ve B.M. *i.e.* $1/16\ qa^2$. The B.M. diagram is assumed to be parabolic and bending moments are calculated at every one metre from centre as shown in Fig. 8·37.

B.M. at one metre from centre

$$= \frac{91}{720} \times qa^2$$

$$= \frac{91}{720} \times 31,250 \times 9 = 35,547 \text{ N-m}$$

$$= 35,547,000 \text{ N-mm}$$

$$A_t = \frac{35,547,000}{0{\cdot}87 \times 222 \times 140} = 1315 \text{ mm}^2$$

Provide 16 mm ϕ bars at 150 mm centres both ways.

B.M. at 2 m from centre

$$= \frac{1}{18}\ qa^2 = \frac{1}{18} \times 31,250 \times 9$$

$$= 15,625 \text{ N-m} = 15,625,000 \text{ N-mm}$$

$$A_t = \frac{15,625,000}{0{\cdot}87 \times 222 \times 140} = 578 \text{ mm}^2$$

Provided 16 mm ϕ bars at 300 mm centre.

Radial reinforcement for $-$ve B.M.

$$-\text{ve B.M.} = \frac{1}{16}\ qa^2 = \frac{1}{16} \times 31,250 \times 9$$

$$= 17,578 \text{ N-m} = 17,578,000 \text{ N-m.}$$

$$A_t = \frac{17,578,000}{0{\cdot}87 \times 222 \times 140} = 650 \text{ mm}^2$$

Provide 16 mm ϕ bars at 300 mm centres upto distance of 60 cm from edge.

Circumferential reinforcement at the edges

Circumferential moment at the edge is assumed as 80% of that for simply supported condition.

$$\text{B.M.} = 0{\cdot}8 \times \frac{2}{16}\ qa^2 = \frac{qa^2}{10} = \frac{1}{10} \times 31,250 \times 9$$

$$= 28,125 \text{ N-m} = 28,125,000 \text{ N-mm}$$

$$A_t = \frac{28,125,000}{0{\cdot}87 \times 222 \times 140} = 1040 \text{ mm}^2$$

Provide 16 mm ϕ bars, at 300 mm centres.
The reinforcement is shown in Fig. 8·38.

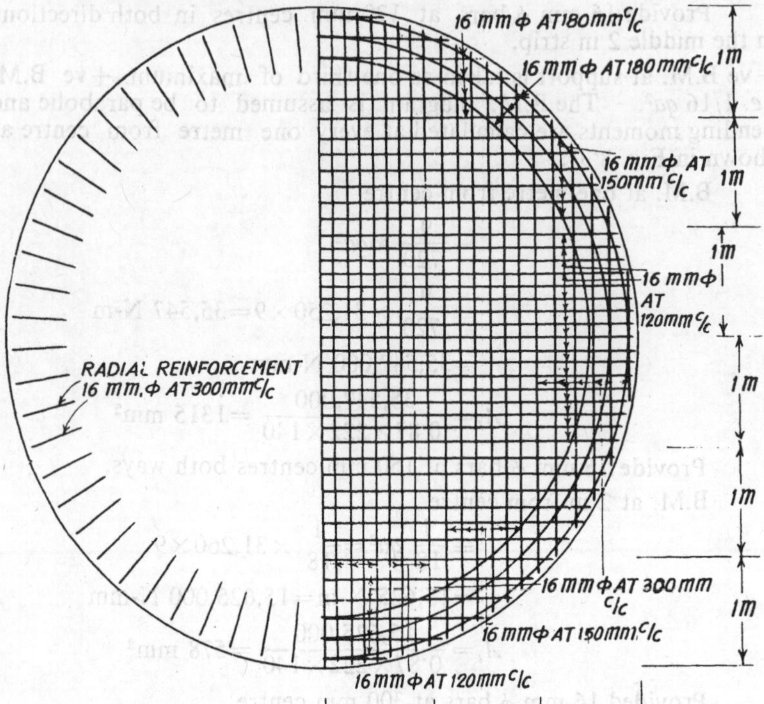

Fig. 8·38

8·18. Non-rectangular Slabs

In case a slab panel is non-rectangular, the bending moments can be obtained by taking slab as rectangular two way slab or circular slab depending on the shape of the panel.

In case of trapezoidal panel approximately symmetrical about axis at right angles to parallel sides, the slab is considered having

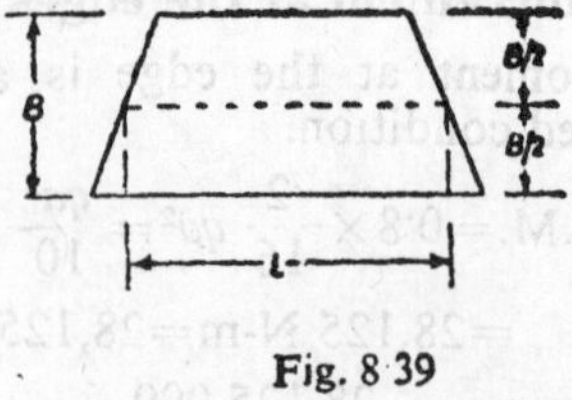

Fig. 8·39

spans as the centre to centre distance between non-parallel sides and the distance between the parallel sides as shown in Fig. 8·39.

In case of triangular panels, the design is based on circular slab of diameter equal to that of inscribed circle as shown in Fig. 8·40.

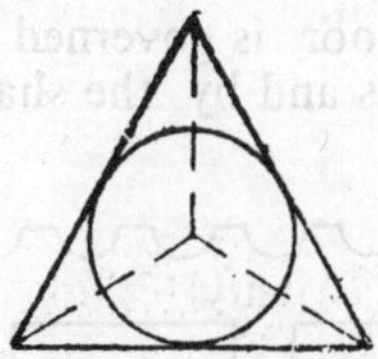

Fig. 8·40

Fig. 8·41

In case of regular polygons, the design is based on circular slab of diameter equal to average of the diameters of the inscribed and circumscribed circles (See Fig. 8·41).

For hexagon

$$D_c = 1{\cdot}08\ D$$

where D_c = Diameter of circular slab.

D = Distance between two flat sides.

For octagon

$$D_c = 1{\cdot}04\ D.$$

8·19. Hollow Floors or Ribbed Floors

Hollow floors are most economical for buildings such as hospitals, schools, hotels and blocks of flats where superimposed loads are small and the spans relatively large. This is because a solid concrete floor slab is heavy and forms a large percentage of the load to be carried by the floor itself. As discussed earlier concrete in the tensile zone is ineffective and, therefore, this area can be left out and filled with light weight material to reduce the self-weight of the slab.

There are two main types of hollow floors (*a*) hollow block floor and (*b*) moulded floors as shown in Fig. 8·42. Each type of

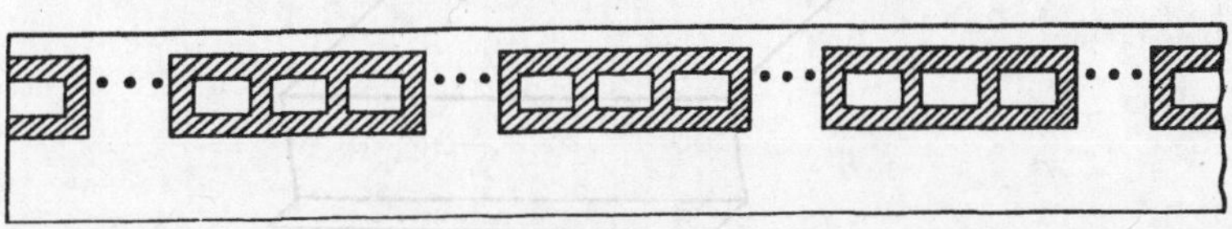

Tile floor

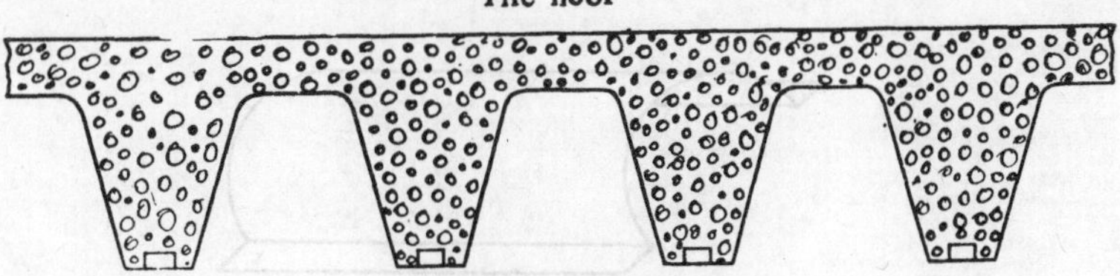

Moulded floor

Fig. 8·42. Hollow floor.

floor consists of thin reinforced concrete ribs spanning either one or two directions supporting a thin concrete floor slab above. The space between the rits is either filled with clay, asbestos or steel tiles, or left free.

The choice of type of hollow floor is governed by the spacing of the supporting beams and columus and by the shape of the floor.

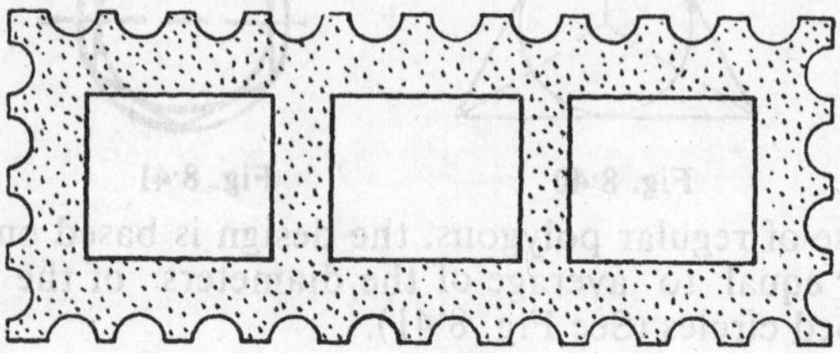

Fig. 8·43. Hollow clay blocks.

The hollow tile floor is most suitable for small irregularly shaped floors as the tiles can be easily fitted into the irregular shapes, on the other hand for large symmetrically supported floors the moulded type of floors is most suitable and economical.

Hollow clay blocks are generally 30 cm square in plan and vary in thickness from 8 to 25 cm. The surfaces are keyed for good bonding with the concrete above and the plaster below as shown in Fig. 8·43.

Steel moulds are shown in Fig. 8·44. They can be used for either temporary or permanent form work, depending on the gauge of the metal. Asbestos and fibre glass moulds are also manufactured in the same shape as steel moulds.

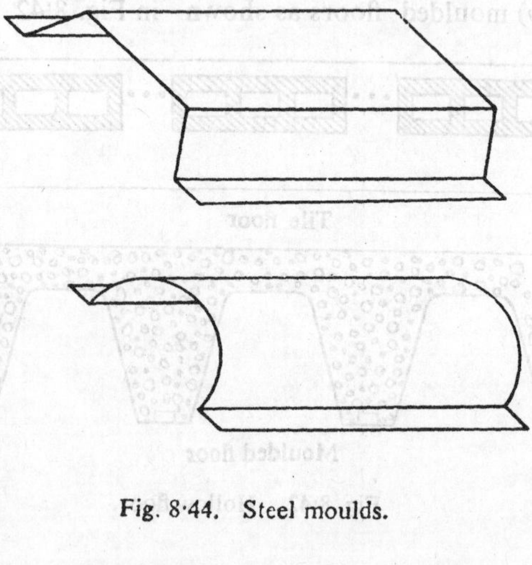

Fig. 8·44. Steel moulds.

Hollow slabs may be spanning in one direction or two directions. The slabs are designed exactly in the same way as ordinary slabs spanning between ribs. Centre to centre distance of ribs is generally not more than one metre. Ribs should have a width of not less than 6 cm.

8·20. I.S.I. Reinforcement for Ribbed, Hollow, Block or Voided Slab

(*a*) **General**. This covers the slabs constructed in one of the ways described below :

(*i*) As a series of concrete ribs with topping cast on forms which may be removed after the concrete has set.

(*ii*) As a series of concrete ribs between precast blocks which may be connected by a topping of concrete of the same strength as that used in the ribs.

(*iii*) With a continuous top and bottom face but containing voids of rectangular, oval or other shapes.

(*b*) **Analysis of Structure**. The moments and forces due to design loads on continuous slabs may be obtained by the methods given for solid slabs. Alternately, the slabs may be designed as a series of simply supported spans provided they are not exposed to weather or corrosive conditions, wide cracks may develop at the supports and the engineer shall satisfy himself that these will not impair finishes or lead to corrosion of the reinforcement.

(*c*) **Shear.** Where hollow blocks are used, for the purpose of calculating shear stress, the rib width may be increased to take account of the wall thickness of the block on one side of the rib, with narrow precast units, the width of the jointing mortar or concrete may be included.

(*d*) **Deflection.** The recommendations for deflection in respect of solids slabs may be applied to ribbed, hollow block or voided construction. The spans to effective depth ratio given for a flanged beam are applicable but when calculating the final reduction factor for web width, the rib width for hollow block may be assumed to include the walls of the blocks on both sides of the rib. For voided slabs and slabs constructed of box or I-section units, an effective rib width shall be calculated assuming all material below the upper flange of the unit to be concentrated in a rectangular rib having the same cross-sectional area and depth.

(*e*) **Size and Position of Ribs.** In-situ ribs shall be not less than 65 mm wide. They shall be spaced at centres not greater than 1·5 m apart and their depth, excluding any topping shall be not more than four times their width.

Generally ribs shall be formed along each edge parallel to the span of one way slabs. When the edge is built into a wall or rests on a beam, rib at least as wide as the bearing shall be formed along the edge.

(*f*) **Hollow Blocks and Formers.** Blocks and formers may be of any suitable material. Hollow clay tiles for the filler type shall conform to I.S.I. code—I.S.I. 3951. When required to contribute to the structural strength of a slab they shall—

(*i*) be made of concrete or burnt clay and

(*ii*) have a crushing strength of atleast 14 N/mm² measured on the net section when axially loaded in the direction of compressive stress in the slab.

(*b*) **Arrangement of Reinforcement.** The recommendations regarding maximum distance between bars apply to areas of solid concrete in this form of construction. The curtailment, anchorage and cover to reinforcement shall be as described below.

(*i*) At least 50 per cent of the total main reinforcement shall be carried through at the bottom on to the bearing and anchored.

(*ii*) Where a slab, which is continuous over supports, has been designed as simply supported, reinforcement shall be provided over the support to control cracking. The reinforcement shall have a cross-sectional area of not less than one-quarter that is required in the middle of the adjoining spans and shall extend at least one-tenth of the clearance space with adjoining spans.

(*iii*) In slabs with permanent blocks, the side cover to the reinforcement shall not be less than 10 mm. In all other cases cover shall be given as for slabs.

The construction with precast joints and hollow concrete filler block shall conform to I.S.I. 6061 (Part I) and precast joists and hollow clay filler blocks shall conform to I.S.I. 6061 (Part II).

Ex. 8·5. *Design a rib and tile floor slab for an effective span of 6 m. The load to be carried exclusive of its own weight is 4000 N/m².*

Sol. Assuming self-weight on the basis of 250 mm thickness and 20,000 N/m³

$$= \frac{250}{1000} \times 1 \times 20{,}000 = 5000 \text{ N/m}^2$$

Total load $= 5000 + 4000 = 9000$ N/m²

$$\text{Maximum B.M.} = \frac{9000 \times 6 \times 6}{8}$$

$$= 40{,}500 \text{ N-m} = 40{,}500{,}000 \text{ N-mm}$$

Using M 150 mix,

$$d = \sqrt{\frac{M}{Qb}} = \sqrt{\frac{40{,}500{,}000}{0{\cdot}87 \times 1000}} = 215 \text{ mm.}$$

$$A_t = \frac{M}{tdj} = \frac{40{,}500{,}000}{140 \times 0{\cdot}87 \times 215} = 1538 \text{ mm}^2.$$

Provide overall depth of 250 mm.

Diameter of the bar is governed by the size of the rib. Using

100 mm rib and clear distance between ribs at 300 mm centre to centre of ribs will be 400 mm.

Area of steel per rib

$$=1538\times\frac{400}{1000}=615{\cdot}2\ \text{mm}^2$$

Provide one bar of 32 mm diameter.

Thickness of slab above tile should be greater than the neutral axis depth of the slab.

Depth of N.A. $=0{\cdot}4\times215=86$ mm.

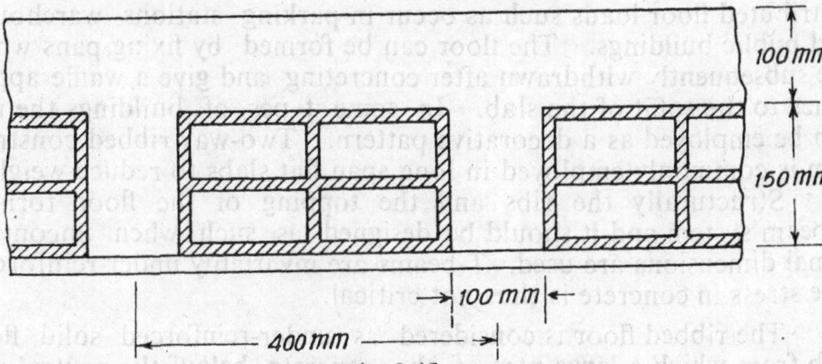

Fig. 8·45

Provide thickness of 100 mm. The floor arrangement is shown in Fig. 8·45.

Check the section as a T-beam.

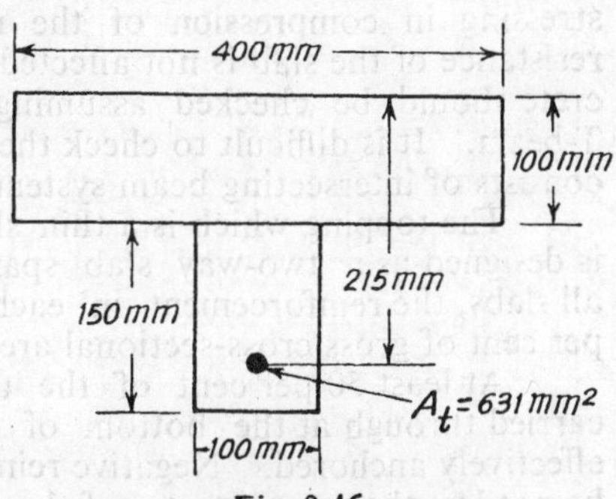

Fig. 8·46

Assuming the N.A. to be in the flange.

$$400\times n\times\frac{n}{2}=18\times631(215-n)$$

$$n=85{\cdot}66\ \text{mm}$$

Depth of critical N.A.

$$=0{\cdot}4\times215$$

$$=86\ \text{mm}$$

Section is under-reinforced. Steel reaches maximum stress

Lever arm $=d-n/3=215-\dfrac{85{\cdot}66}{3}=186{\cdot}45$ mm

Moment of resistance of the section

$$=140\times631\times186{\cdot}45$$

$$=16{,}460{,}000\ \text{N-mm}$$

This is moment of resistance for one metre.

Actual B.M. per rib (400 mm width)

$$= \frac{40{,}500{,}000 \times 400}{1000} = 16{,}200{,}000 \text{ N-m}$$

Section is Safe.

8·21. Two-way Ribbed Slabs

Ribbed slabs consist of closely spaced ribs not more than 75 cm centre to centre. Ribs span in two directions and give a waffle effect which can form an interesting pattern for a ceiling. Ribbed floors are commonly made with tapering ribs to allow easy withdrawal and re-use of the forms. Two way ribbed slabs are much stiffer than one way ribbed floors. They are used for heavy distributed floor loads such as occur in parking stations warehouses and public buildings. The floor can be formed by fixing pans which are subsequently withdrawn after concreting and give a waffle appearance to the soffit of the slab. In some types of buildings the ribs can be employed as a decorative pattern. Two-way ribbed construction is commonly employed in long span flat slabs to reduce weight.

Structurally the ribs and the topping of the floor form a T-beam system and it should be designed as such when unconventional dimensions are used. T-beams are invariably under-reinforced. The stress in concrete is thus not critical.

The ribbed floor is considered as under-reinforced solid floor slab from which a large part of the concrete below the neutral axis and some part of the concrete above the neutral axis has been removed. The slab reinforcement is concentrated in the ribs. In case the amount of concrete removed is not much as to cause over-stressing in compression of the remaining concrete, the bending resistance of the slab is not affected. The stress in the topping concrete should be checked assuming ribs and topping to act as a T-beam. It is difficult to check the compressive stress as the system consists of intersecting beam system of complexity.

The topping which is a thin slab of concrete 4 to 7 cm thick is designed as a two-way slab spanning between the ribs. As for all slabs, the reinforcement in each direction must be atleast 0·15 per cent of gross cross-sectional area.

At least 50 per cent of the total main reinforcement shall be carried through at the bottom of the rib on to the bearing and be effectively anchored. Negative reinforcement over the support shall be not less than one quarter of that in the middle of the adjoining spans and extend at least one-tenth of the clear span in adjoining spans.

In the design of ribbed floor shear is frequently critical. Due to reduction in the cross-section area of concrete, shear stresses may be high. Shear reinforcement is not practical as the ribs are very narrow. As the shear stresses are high at the supports, the shear strength can be increased by widening the ribs or by providing solid slab near the supports. In case of flat slabs, the area surrounding the column is the region of high shear stress, therefore, solid slab is provided in the portion surrounding the column.

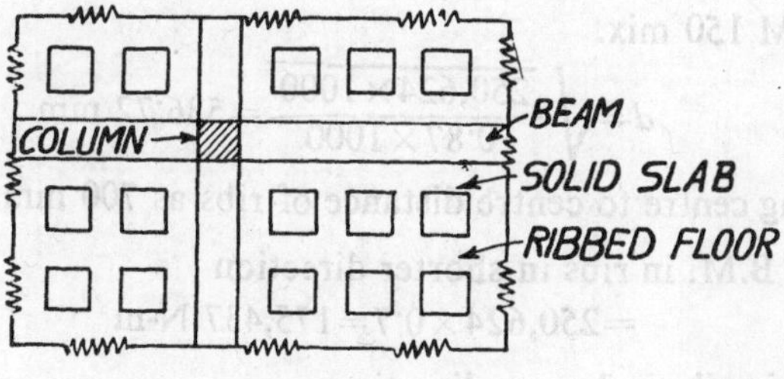

Fig. 8·47

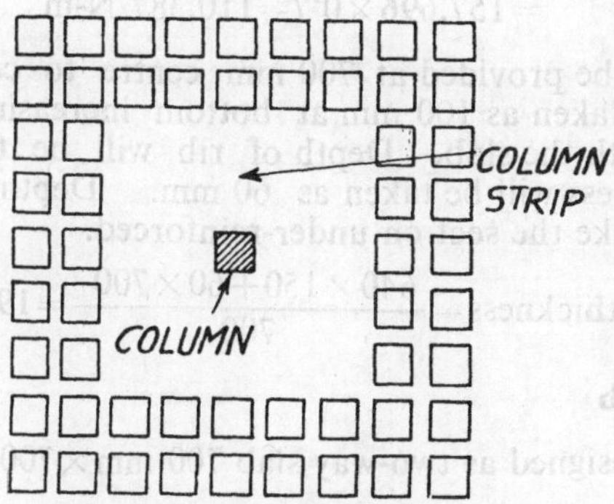

Flat floor formed with two-way rib

Fig. 8·48

Ex. 8·6. *Design roof of a hall 24 metres by 16 metres as two way ribbed floor supported on bearing walls around. It carries a superimposed live load of 5000 N/m². The finishing load may be taken as 1000 N/m².*

Sol. (Assuming panels supported on four sides with provision for torsion).

Taking average thickness as 200 mm dead load of floor

$$=\frac{200}{1000}\times 1\times 1\times 25{,}000$$

$$=5000 \text{ N/m}$$

Load per sq. metre $=5000+1000+5000=11{,}000 \text{ N/m}^2$

Designing slab by I.S.I. code

$$\frac{L}{B}=1{\cdot}5$$

Positive moment along shorter span

$$=11{,}000\times 0{\cdot}089\times 16^2$$

$$=250{,}624 \text{ N-m/metre}$$

Positive moment along longer span

$$=11{,}000\times 0{\cdot}056\times 16^2$$

$$=157{,}696 \text{ N-m/metre}$$

Use M 150 mix.

$$d=\sqrt{\frac{250{,}624\times1000}{0{\cdot}87\times1000}}=536{\cdot}72 \text{ mm}$$

Taking centre to centre distance of ribs as 700 mm

Max. B.M. in ribs in shorter direction

$$=250{,}624\times0{\cdot}7=175{,}437 \text{ N-m}$$

B.M. in ribs in longer direction

$$=157{,}696\times0{\cdot}7=110{,}387 \text{ N-m}$$

Ribs will be provided at 700 mm centre to centre. Width of the rib will be taken as 100 mm at bottom increasing to 200 mm at the junction with the slab. Depth of rib will be taken as 640 mm and slab thickness will be taken as 60 mm. Depth of 700 mm is provided to make the section under-reinforced.

$$\text{Average thickness}=\frac{640\times150+60\times700}{700}=197 \text{ mm}$$

Design of Slab

Slab is designed as two-way slab 700 mm × 700 mm continuous on all edges.

$$\text{Load}=5000+1000+60\times25=7500 \text{ N/m}^2$$

$$-\text{ve moment}=0{\cdot}032\times7500\times(0{\cdot}7)^2$$

$$=117{\cdot}6 \text{ N-m}$$

$$+\text{ve moment}=0{\cdot}024\times7500(0{\cdot}7)^2=88{\cdot}2 \text{ N-m}$$

Placing the reinforcement at centre,

Effective depth = 60 − 30 = 30 mm

$$A_t=\frac{117{\cdot}6\times1000}{0{\cdot}87\times30\times140}=32{\cdot}2 \text{ mm}^2$$

Minimum reinforcement

$$=\frac{0{\cdot}15\times1000\times60}{1000}=90 \text{ mm}^2$$

Provide 4 mm ϕ bars mesh 200 mm centres at the centre of the slab, otherwise welded prefabricated mat of appropriate strength may be used. Welded mats come in rolls.

$$A_t=\frac{20\times100}{200}=100 \text{ mm}^2$$

Design of Ribs

Max. B.M. for shorter span

$$=175{,}437 \text{ N-m}$$

Effective depth provided

$$=700-80=620 \text{ mm}$$

$$A_t=\frac{175,437\times 1000}{140(620-30)}=2124 \text{ mm}^2$$

Provide 6 bars 25 mm ϕ in three rows 2+2+2.

Max. B.M. for longer span

$$=110,387$$

$$A_t=\frac{110,387\times 1000}{140\times 590}$$

$$=1336 \text{ mm}^2$$

Provide 4 bars 25 mm ϕ in two rows 2+2.

Near the walls solid slab of width app. 200 mm is provided.

Load in shorter direction

$$=11,000\times 0{\cdot}84=9240 \text{ N/m}^2$$

$$\text{Max. shear}=9240\times\frac{16}{2}=73,920 \text{ N}$$

$$\text{Max. shear stress}=\frac{73,920}{620\times 1000}=0{\cdot}12 \text{ N/mm}^2. \quad \text{Safe.}$$

Provide 5 mm ϕ stirrups at 500 mm centre to centre.

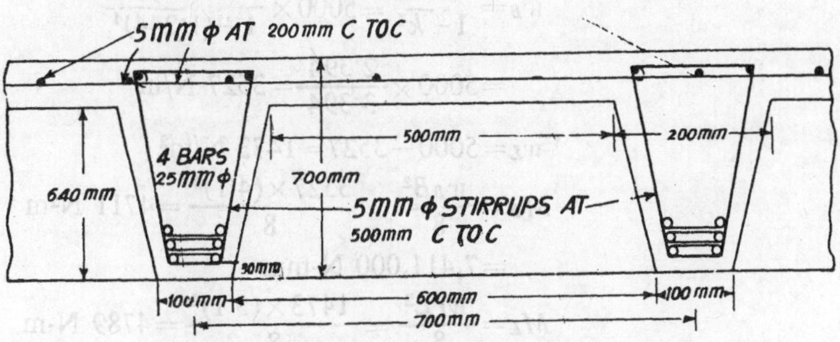

Section through rib in shorter direction

Fig. 8·49

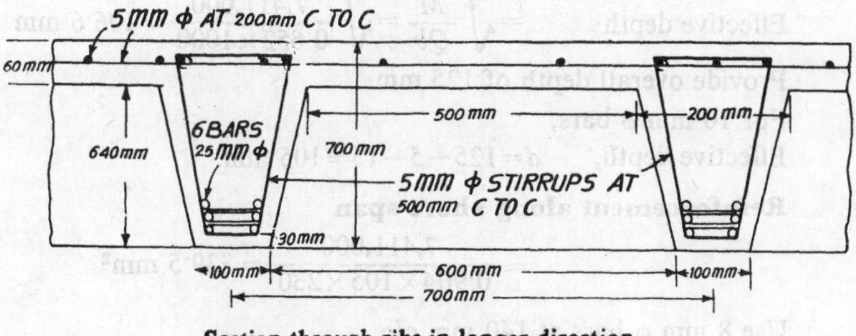

Section through ribs in longer direction

Fig. 8·50

Ex. 8·7. *Design a two-way reinforced slab for a room having clear dimensions of 5 m×4 m. The superimposed load is 2000 N/m². Use M 15 mix. Use high yield strength deformed bars.*

Sol. (*i*) Assuming corners of slab not held down.

Assume overall depth of slab as 120 mm and effective depth as 100 mm.

Effective spans in two directions will be 5·1 m×4·1 m.

Minimum depth of slab

$$=\frac{B}{35}=\frac{4\cdot1\times1000}{35}=117 \text{ mm}$$

Overall depth of 120 mm will be adopted

Self-load $$=\frac{120}{1000}\times1\times1\times25{,}000=3000 \text{ N/m}^2$$

Total load per sq. metre

$$=2000+3000=5000 \text{ N/m}^2=w$$

$$k=\frac{5\cdot1}{4\cdot1}=1\cdot244$$

$$w_B=\frac{wk^4}{1+k^4}=5000\times\frac{(1\cdot244)^4}{1+(1\cdot244)^4}$$

$$=5000\times\frac{2\cdot394}{3\cdot394}=3527 \text{ N/m}^2$$

$$w_L=5000-3527=1473 \text{ N/m}^2$$

$$M_B=\frac{w_BB^2}{8}=\frac{3527\times(4\cdot1)^2}{8}=4711 \text{ N-m}$$

$$=7{,}411{,}000 \text{ N-mm}$$

$$M_L=\frac{w_LL^2}{8}=\frac{1473\times(5\cdot1)^2}{8}=4789 \text{ N-m}$$

$$=4{,}789{,}000 \text{ N-mm}$$

Effective depth $$=\sqrt{\frac{M}{Qb}}=\sqrt{\frac{7{,}411{,}000}{0\cdot652\times1000}}=106\cdot6 \text{ mm}$$

Provide overall depth of 125 mm.

For 10 mm ϕ bars,

Effective depth, $d=125-5-15=105$ mm.

Reinforcement along short span

$$=\frac{7{,}411{,}000}{0\cdot904\times105\times230}=339\cdot5 \text{ mm}^2$$

Use 8 mm φ bars at 140 mm c/c,

$$A_t=359 \text{ mm}^2$$

Reinforcement along long span

$$d=125-8-4-15=98 \text{ mm}$$

$$A_t=\frac{4,789,000}{0\cdot904\times28\times230}=235\cdot1 \text{ mm}^2$$

Use 8 mm ϕ at 200 mm c/c,

$$A_t=250 \text{ mm}^2$$

(*ii*) Assuming corners held down and torsion reinforcement provided at the corners.

(*a*) Marcus method

Assume total load $=5000$ N/m²

$$M_B=\frac{wB^2}{8}\left[1-\frac{5k^2}{6(1+k^4)}\right]$$

$$=7411\left[1-\frac{5\times(1\cdot244)^2}{6[1+(1\cdot244)^4]}\right]$$

$$=7411\times0\cdot621=4602 \text{ N-m}$$

$$=4,602,000 \text{ N-mm}$$

$$M_L=4789\times0\cdot621=2974 \text{ N-m}$$

$$=2,974,000 \text{ N-mm}$$

$$d=\sqrt{\frac{M}{Qb}}=\sqrt{\frac{4,602,000}{0\cdot652\times1000}}=84 \text{ mm}$$

Provide overall depth of 120 mm, use 8 mm ϕ bars,

$$d=120-4-15=101 \text{ mm}$$

Reinforcement along short span

$$A_t=\frac{4,602.000}{0\cdot904\times101\times230}=219\cdot1 \text{ mm}^2$$

Use 8 mm ϕ bars at 220 mm c/c,

$$A_t=229 \text{ mm}^2$$

Reinforcement along long span

$$d=101-8=93 \text{ mm}$$

$$A_t=\frac{2,974,000}{0\cdot904\times93\times230}=153\cdot8 \text{ mm}^2$$

Use 8 mm ϕ at 280 mm c/c,

$$A_t=180 \text{ mm}^2 \text{ (minimum reinforcement)}$$

Torsion Reinforcement

$$A=\tfrac{3}{4}\times219\cdot1=164\cdot33 \text{ mm}^2$$

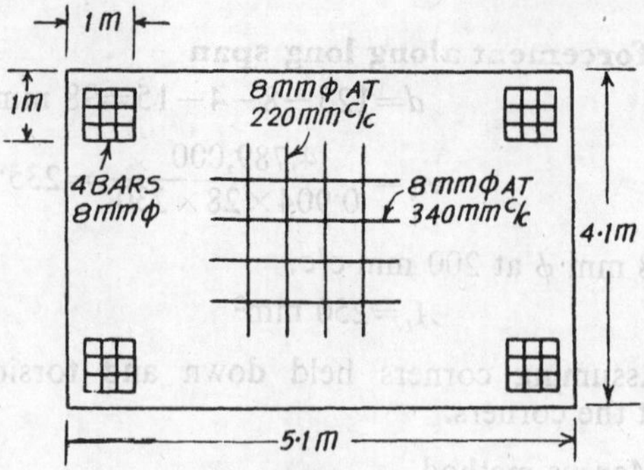

Fig. 8·51

Provide 8 mm ϕ at 320 mm c/c along both directions at top and bottom for distance of $\frac{4 \cdot 1}{5} = 0 \cdot 82$ m. Provide 4 bars. (See Fig. 8·51).

(*b*) Empirical formula,

$$w = 5000 \text{ N/m}^2$$

From Table 8·2,

$$k_B' = 0 \cdot 4664,$$

$$k_L' = 0 \cdot 1724$$

$$M_B = \frac{k_B' w B^2}{8} = \frac{0 \cdot 4664 \times 5000 \times (4 \cdot 1)^2}{8}$$

$$= 4{,}900 \text{ N-m} = 4{,}900{,}000 \text{ N-mm}$$

$$M_L = \frac{k_L' w L^2}{8} = \frac{0 \cdot 1724 \times 5000 \times (5 \cdot 1)^2}{8}$$

$$= 2803 \text{ N-m} = 2{,}803{,}000 \text{ N-mm}$$

$$d = \sqrt{\frac{M}{Qb}} = \sqrt{\frac{4{,}900{,}000}{0 \cdot 652 \times 1000}} = 87 \text{ mm}$$

Provide overall depth of 120 mm, use 8 mm ϕ reinforcement.

Effective depth $\quad d = 120 - 15 - 4 = 101$ mm

Reinforcement along short Span

$$A_t = \frac{4{,}900{,}000}{0 \cdot 904 \times 101 \times 230} = 233 \cdot 3 \text{ mm}^2$$

Use 8 mm ϕ bars at 200 mm c/c,

$$A_t = 252 \text{ mm}^2$$

Reinforcement along long span

$$d = 101 - 8 = 93 \text{ mm}$$

$$A_t = \frac{2{,}803{,}000}{0 \cdot 904 \times 93 \times 230} = 145 \text{ mm}^2$$

Use 8 mm ϕ bars at 280 mm c/c,

$$A_t = 180 \text{ mm}^2 \text{ (minimum reinforcement)}$$

Provide torsion reinforcement as in (*a*).

(*c*) I.S.I. Code Formula—

$$w = 5000 \text{ N/m}^2$$

$$k = \frac{5 \cdot 1}{4 \cdot 1} = 1 \cdot 244$$

$$\alpha_x = 0 \cdot 0751, \; \alpha_y = 0 \cdot 056$$

$$M_B = \alpha_x w l_x^2 = 0 \cdot 0751 \times 5000 \times (4 \cdot 1)^2$$

$$= 6312 \text{ N·m} = 6{,}312{,}000 \text{ N-mm}$$

$$M_L = \alpha_y w l_x^2 = 0 \cdot 056 \times 5000 \times (4 \cdot 1)^2$$

$$= 4707 \text{ N-mm} = 4{,}707{,}000 \text{ N-mm}$$

$$d = \sqrt{\frac{M}{Qb}} = \sqrt{\frac{6{,}312{,}000}{0 \cdot 652 \times 1000}} = 98 \cdot 4 \text{ mm}$$

Provide overall depth of 120 mm. Use 8 mm ϕ reinforcement

$$d = 120 - 15 - 4 = 101 \text{ mm}$$

Reinforcement along short span in middle strip

$$A_t = \frac{6{,}312{,}000}{0 \cdot 904 \times 101 \times 230} = 300 \cdot 6 \text{ mm}^2$$

Use 8 mm ϕ bars at 160 mm c/c,

$$A_t = 314 \text{ mm}^2$$

Reinforcement along long span in middle strip

$$d = 101 - 8 = 93 \text{ mm}$$

$$A_t = \frac{4{,}707{,}000}{0 \cdot 904 \times 93 \times 230} = 243 \cdot 4 \text{ mm}^2$$

Use 8 mm ϕ bars at 200 mm c/c,

$$A_t = 252 \text{ mm}^2$$

Reinforcement in edge strip

$$A_t = \frac{0 \cdot 15}{100} \times 1000 \times 120 = 180 \text{ mm}^2$$

Provide 8 mm ϕ at 280 mm c/c,

$$A_t = 180 \text{ mm}^2$$

Torsion Reinforcement

$$A_t = \frac{3}{4} \times 314 = 235 \cdot 5 \text{ mm}^2$$

Provide 8 mm ϕ at 200 mm c/c,

$$A = 252 \text{ mm}^2$$

in both directions at top and bottom for a distance of

$$\frac{4.5}{5}=0{\cdot}82 \text{ m}$$

Provide 4 bars as shown in Fig. 8·52.

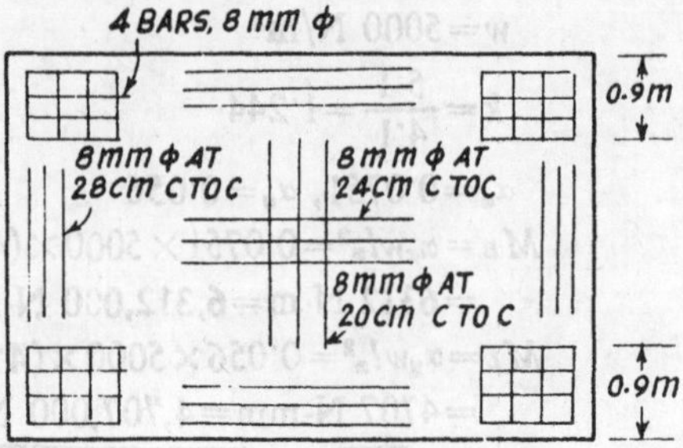

Fig. 8·52

PROBLEMS

1. The internal dimensions of a hall with 40 cm thick walls on all four sides are 4 m × 5 m. Design suitable roof to span in two directions. Live load on slab is 4000 N/m².

2. Design 7 m square interior panel of a flat slab floor with dropped panels over column capitals. Live load on floor is 5000 N/m².

Sketch a cross-section through slab and dropped panel giving details of reinforcement.

3. Design a circular slab for a room having diameter of 5 m. Assume slab to be simply supported at supports.

4. Fig. 8·53 shows plan of roof of a building. Live load on roof is 2000 N/m². Design panels S_1, S_2, S_3 and beam B_1 and B_2.

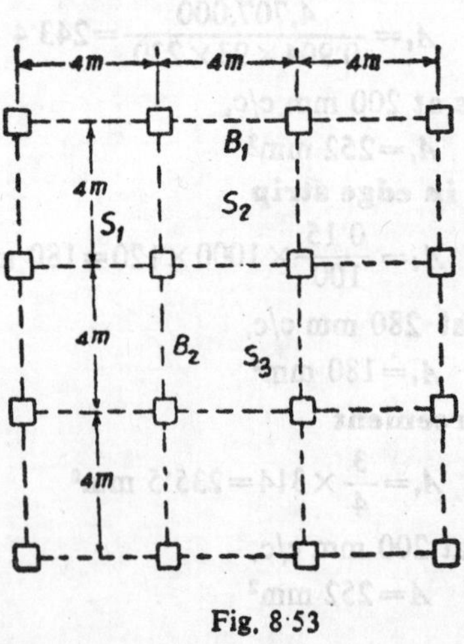

Fig. 8·53

5. A reinforced concrete floor is to be constructed for the building shown in Fig. 8·54 for the following data:

(i) Live load on floor 2000 N/m².

(ii) Weight on roofing 1000 N/m².

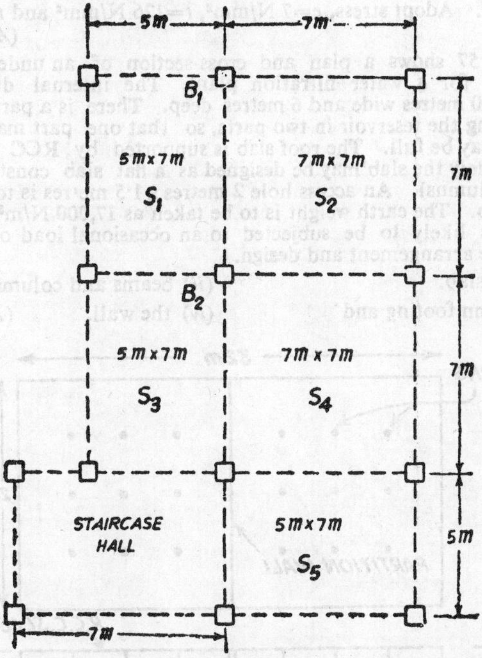

Fig. 8·54

Design slabs S_1, S_2, S_3, S_4, and S_5 and beams B_1 and B_2.

6. Design the RCC roof slab for a room 4 m × 5 m in size to support a superimposed load of 5000 N/m², if the edges are simply supported with corners held down.

7. Design 5 m × 6 m end panel of flat slab roof carrying a live load of 6000 N/m².

8. Design the roof slab shown in Fig. 8·55 for a live load of 2000 N/m².

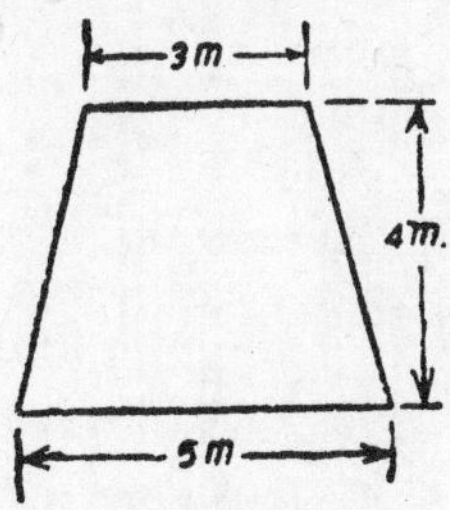

Fig. 8·55

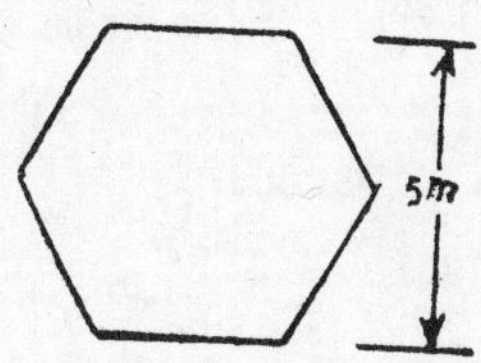

Fig. 8·56

9. Design the roof slab for hexagonal entrance hall of a building (Fig. 8·56). The live load on roof is 4000 N per square metre.

10. An R.B. floor is supported on RCC columns 0·23 m square at the corners of a rectangle 6·1 m+7·6 m with RCC beams connecting the columns along the perimeter of the rectangle. Design the floor as a two-way reinforced slab. Also design one of the 6·1 m span beams as an L-beam. Live lo ad on floor=9750 N/m^2. Adopt stress, c=7 N/mm^2, t=126 N/mm^2 and m=15.

(A.M.I.E. 1967)

11. Fig. 8·57 shows a plan and cross-section of an underground RCC reservoir required for a water filtration plant. The internal dimensions are 32 metres long, 20 metres wide and 6 metres deep. There is a partition wall at the centre dividing the reservoir in two parts, so that one part may be cleaned while the other may be full. The roof slab is supported by RCC columns and beams. (Alternately the slab may be designed as a flat slab construction, using slab resting on columns). An access hole 2 metres × 1·5 metres is to be provided in the roof of slab. The earth weight is to be taken as 17,000 N/m^3 and α=30°. The roof slab is likely to be subjected to an occasional load of 1350 N/m^2. Suggest a suitable arrangement and design.

(*i*) roof slab. (*ii*) beams and columns.

(*iii*) column footing and (*iv*) the wall. *(A.M.I.E 1969)*

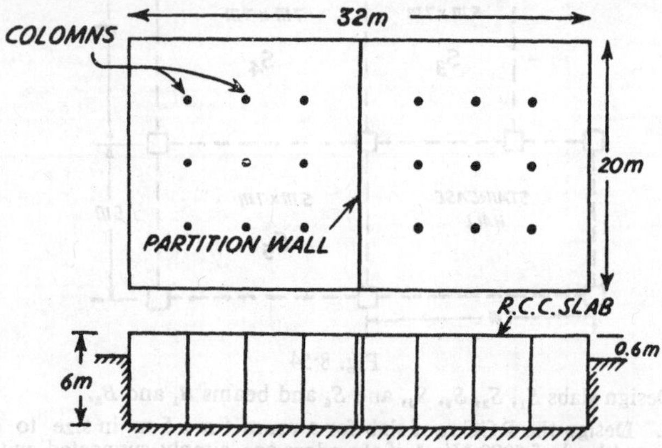

Fig. 8·57

12. An underground roof of dimension 10 m × 13 m between the existing wall 45 cm thick is to be provided with RCC roof using one column at the centre of the roof and four beams, so that the slab is divided in four panels of 5 m × 6·5 m. The roof is to carry a load of 11,500 N/m^2.

Design the slab and beams. (The slab is to be designed as a two-way reinforced slab).

(A.M.I.E. 1969)

9

Combined Bending and Direct Stresses

9·1. In concrete structures there are many cases in which members are subjected to both bending moment and direct load.

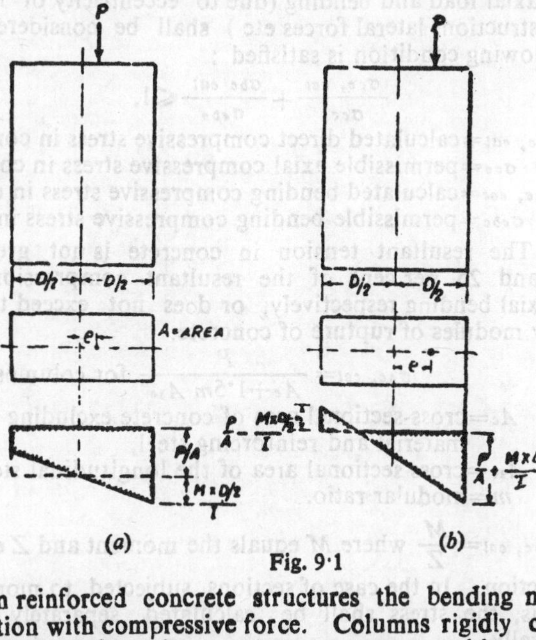

Fig. 9·1

Generally, in reinforced concrete structures the bending moment is in combination with compressive force. Columns rigidly connected to beams, columns of building and water towers subjected to wind forces and arches are some structures which are subjected to both bending and direct stresses.

On a homogeneous columns a load applied eccentrically about one of the axes of symmetry at a distance *e* from the other axis, is equivalent to the same load applied axially and a moment *P*×*e*. The stress distribution in such column is given by

$$p=\frac{P}{A}\pm\frac{My}{I}=\frac{P}{A}\pm\frac{Pey}{I}.$$

The formula can also be applied to transformed concrete section provided there is compression over the whole section or slight

tension which is permissible in concrete. However, if the eccentricity is such that it causes high tensile stresses not permissible in concrete the modification of this method has to be applied. The problem involving combined bending and direct stresses are divided into two categories (*i*) those members which have compression over the entire section Fig 9·1 (*a*), (*ii*) those members which have compression upon part of the section and tension over the remainder Fig. 9·1 (*b*).

Design of sections subjected to combined direct and bending stress is based on elastic theory.

9 2. General Notes on Design of Columns Subjected to Combined Bending and Direct Stresses (I.S.I.)

1. Design based on uncracked section. A member subjected to axial load and bending (due to eccentricity of load, monolithic construction, lateral forces etc) shall be considered safe provided following condition is satisfied :

(*a*) $$\frac{\sigma_{cc,\ cal}}{\sigma_{cc}}+\frac{\sigma_{bc\ cal}}{\sigma_{cbc}}\leqslant 1,$$

where $\sigma_{cc,\ cal}$=calculated direct compressive stress in concrete,

σ_{cc}=permissible axial compressive stress in concrete,

$\sigma_{bc,\ cal}$=calculated bending compressive stress in concrete, and

σ_{cbc}=permissible bending compressive stress in concrete

(*b*) The resultant tension in concrete is not greater than 35 per cent and 25 per cent of the resultant compression for biaxial and uniaxial bending respectively, or does not exceed three-fourths, the 7 day modules of rupture of concrete.

$$\sigma_{cc,\ cal}=\frac{P}{A_c+1{\cdot}5m\ A_{sc}}\quad \text{for columns with ties}$$

where, A_c=cross-sectional area of concrete excluding any finishing material and reinforcing steel,

A_{sc}=cross-sectional area of the longitudinal steel,

m=modular ratio.

$\sigma_{cbc,\ cal}=\frac{M}{Z}$ where M equals the moment and Z equals modulus of section. In the case of sections subjected to moments in two directions, the stress shall be calculated separately and added algebraically.

2. Design based on cracked section. If the requirements specified in (1) above are not satisfied, the stresses in concrete and steel shall be calculated by the theory of cracked section in which the tensile resistance of concrete is ignored. If the calculated stresses are within permissible stresses specified for concrete and steel the section may be assumed to be safe.

The maximum stress in concrete and steel may be found from tables and charts based on the cracked section theory or directly by determining the no-stress line which should satisfy the following requirements :

(*a*) The direct load should be equal to the algebraic sum of forces on concrete and steel.

(*b*) The moment of the external loads about any reference lines should be equal to the algebraic sum of the moment of the forces in concrete (ignoring the tensile force in concrete) and steel about the same line, and

(*c*) The moment of external loads about any other reference lines should be equal to the algebraic sum of the moment of the forces in concrete (ignoring the tensile force in concrete) and steel about the same line.

3. Members subjected to combined direct load and flexure. Member subjected to combined direct load and flexure designed by the methods based on elastic theory should be further checked for their strength under ultimate load conditions to ensure the derived margin of safety, this check is specially necessary when the bending moment is due to horizontal loads.

Table 9 1
Moments in Columns

	Moments for frames of one bay	*Moments for frames of two or more bays*
External (and similarly (loaded) columns :		
Moment at foot of upper column	$M_e \dfrac{K_u}{K_l+K_u+0{\cdot}5\,K_b}$	$M_e \dfrac{K_u}{K_l+K_u+K_b}$
Moment at head of lower column	$M_e \dfrac{K_l}{K_l+K_u+0{\cdot}5\,K_b}$	$M_e \dfrac{K_l}{K_l+K_u+K_b}$
Internal columns :		
Moment at foot of upper column	—	$M_{es} \dfrac{K_u}{K_l+K_u+K_{b1}+K_{b2}}$
Moment at head of lower column	—	$M_{es} \dfrac{K_p}{K_l+K_{b1}+K_{b1}+K_{b2}}$

Note 1

M_e=bending moments at the end of the beam framing into the column assuming fixity at the connection.

M_{es}=maximum difference between the moments at the ends of the two beams forming into opposite sides of the column each calculated on the assumption that the ends of the beams are fixed and assuming one of the beams unloaded.

K_u=stiffness of the upper column.

K_l=stiffness of the lower column.

K_b=stiffness of the beam.

K_{b1}=stiffness of beam on the one side of the column.

K_{b2}=stiffness of the beam on the other side of the column.

Note 2. For the purpose of this table, 'stiffness' of member may be obtained by dividing the moment of inertia of cross-section by the length of the member provided that the member is of constant cross-section throughout its length.

Note 3. The equation for the moment at the head of the lower columns may be used for columns in a topmost storey by taking K_u as zero.

Where the bending moment is calculated in the internal columns, it is permissible to take into account the reduction load resulting from the beam on one side of the column being fully loaded and the beam on the other side being loaded with dead load only.

9·3. Columns carrying Loads with Small Eccentricity so that Tension Develops in the Section

Consider a section *ABCD* reinforced with steel areas A_1, A_2 ......at distances y_1, y_2......from top. Let e be the eccentricity of the load from the middle section.

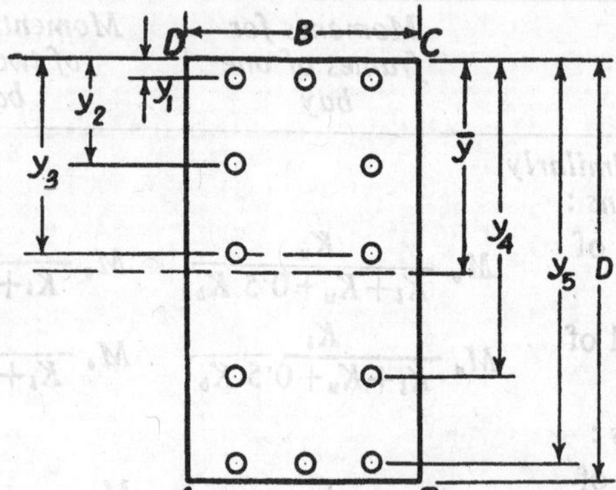

Fig. 9·2

Equivalent area of section

$$A_e = BD + \Sigma(m-1)A$$

Distance of centre of gravity of section from face *CD* is given by

$$\bar{y} = \frac{(m-1)\,A_1 y_1 + (m-1)\,A_2 y_2 + \ldots\ldots + B \times D \times D/2}{(m-1)\,A_1 + (m-1)\,A_2 + \ldots\ldots + BD}$$

$$= \frac{\Sigma(m-1)\,A \times y + \dfrac{BD^2}{2}}{\Sigma(m-1)\,A + BD}$$

Equivalent moment of inertia about centre of gravity

$$I_e=(m-1)\ A_1k_1^2+(m-1)\ A_2k_2^2+\ldots\ldots$$

$$+\frac{BD^3}{12}+BD\left(y-\frac{D}{2}\right)^2$$

$$=\Sigma(m-1)\ Ak^2+\frac{BD^3}{12}+BD\left(y-\frac{D}{2}\right)^2$$

where k_1, k_2......are the distance of centres of A_1, A_2......from C.G. of the section.

Maximum compressive stress

$$=\frac{P}{A_e}+\frac{P\left[\frac{D}{2}-y+e\right](D-y)}{I_e}$$

Minimum stress

$$=\frac{P}{A_e}-\frac{P\left[\frac{D}{2}-y+e\right]y}{I_e}$$

where e is the eccentricity of the load P.

If the section is symmetrically reinforced

$$y=\frac{D}{2}$$

$$A_e=BD+(m-1)\ \Sigma A$$

$$I_e=\frac{BD^3}{12}+(m-1)\ \Sigma Ak^2$$

Maximum stress $$=\frac{P}{A_e}+\frac{P\times e\times\frac{D}{2}}{I_e}$$

Minimum stress $$=\frac{P}{A_e}-\frac{P\times e\times\frac{D}{2}}{I_e}$$

If no tension is to develop,

$$\frac{P}{A_e}-\frac{P_eD}{2I_e}=0$$

or $$e=\frac{2I_e}{A_eD} \qquad \ldots(9\cdot1)$$

Ex. 9·1. *A column 300 mm square is reinforced with 4 bars of 2 cm diameter shown in Fig. 9·3. It is subjected to a load of 300,000 N at an eccentricity of 40 mm about one of the axis. Determine the* *maximum and minimum stresses developed in the section. Check whether the section is safe, if the permissible are*

$$\sigma_{co}=4\ N/mm^2,\ \sigma_{bc}=5\ N/mm^2,\ m=18.$$

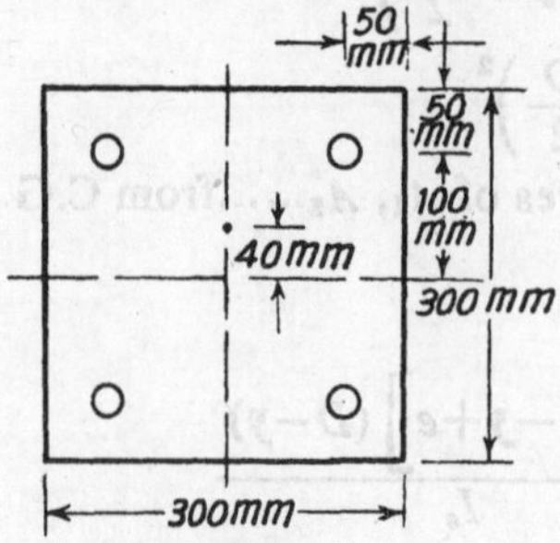

Fig. 9·3

Sol. Equivalent area A_e

$$=300\times300+(m-1)A_e$$
$$=90{,}000+17\times4\times314$$
$$=111{,}350\ mm^2$$

Equivalent moment of inertia I_e

$$=\frac{1}{12}\times300\times(300)^3+(m-1)\times4\times314\times(100)^2$$
$$=675{,}000{,}000+17\times12{,}560{,}000$$
$$=888{,}520{,}000\ mm^4$$

Stress due to axial load

$$=\sigma_{cc\,cal}=\frac{P}{A_e}=\frac{300{,}000}{111{,}350}=2{\cdot}7\ N/mm^2$$

Stress due to eccentricity of load

$$\sigma_{bc\,cal}=\frac{P\times e\times y}{I_e}=\frac{300{,}000\times40\times150}{888{,}520{,}000}$$
$$=2{\cdot}03\ N/mm^2$$

Maximum stress $=2{\cdot}7+2{\cdot}03=4{\cdot}73\ N/mm^2$

Minimum stress $=2{\cdot}7-2{\cdot}03=0{\cdot}67\ N/mm^2$

$$\frac{\sigma_{cc\,cal}}{\sigma_{cc}}+\frac{\sigma_{bc\,cal}}{\sigma_{cbc}}=\frac{2{\cdot}7}{4}+\frac{2{\cdot}03}{5}=0{\cdot}675+0{\cdot}406$$
$$=1{\cdot}081>1$$

∴ Section is not safe.

Ex. 9·2. *In Ex. 9·1, what should be the maximum eccentricity that section is safe as uncracked section ?*

Sol. Let c' be the compressive stress due to eccentricity of load.

$$\frac{\sigma_{cc\,cal}}{\sigma_{cc}}+\frac{\sigma_{bc\,cal}}{\sigma_{cbc}}=1$$

$$\frac{2{\cdot}7}{4}+\frac{\sigma_{bc\,cal}}{5}=1$$

$$\frac{\sigma_{bc\,cal}}{5}=1-0{\cdot}675=0{\cdot}325$$

$$\sigma_{bc\ cal} = 1{\cdot}625\ \text{N/mm}^2$$

$$\sigma_{bc\ cal} = \frac{P\times e\times y}{I_e} = 1{\cdot}625$$

$$\frac{300{,}000\times e\times 150}{888{,}520{,}000} = 1{\cdot}625$$

$$e = \frac{1{\cdot}625\times 888{,}520{,}000}{45{,}000{,}000} = 32\ \text{mm}.$$

Ex. 9·3. *A rectangular section 20 mm wide and 400 mm deep is reinforced with 8 bars 20 mm ϕ as shown in Fig. 9·4.*

Find the maximum eccentricity about x-axis at which a load can be applied so that no tension is developed in the section. What is the maximum value of load if the permissible stress are $\sigma_{cc}=4$ N/mm², $\sigma_{bc}=5$ N/mm², $m=18$.

Sol. Equivalent area of section

$$A_e = 180\times 400 + (m-1)\times 8\times 314$$
$$= 72{,}000 + 8\times 314\times 17$$
$$= 72{,}000 + 42{,}700 = 114{,}700\ \text{mm}^2$$

Equivalent moment of inertia

$$I_e = \frac{1}{12}\times 180\times (400)^3 + 17\times 6 \times 314\times (160)^2$$
$$= 960{,}000{,}000 + 820{,}400{,}000$$
$$= 1{,}780{,}400{,}000\ \text{mm}^4$$

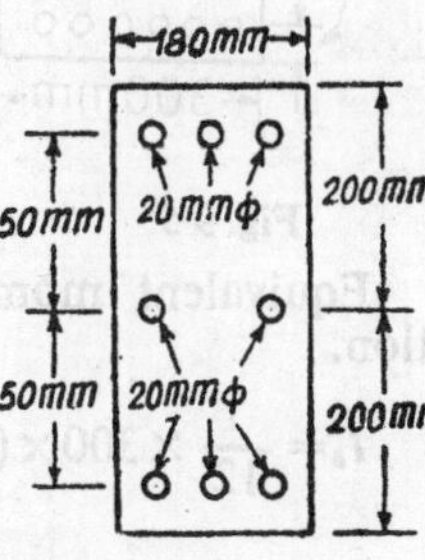

Fig. 9·4

For no tension to develop.

$$\frac{P}{A_e} - \frac{P\times e\times D/2}{I_e} = 0$$

$$e = \frac{2I_e}{A_e D} = \frac{2\times 1{,}780{,}400.000}{114{,}700\times 400} = 77{\cdot}6\ \text{mm}$$

Let P be the maximum load

$$\sigma_{cc\ cal} = \frac{P}{A_c} = \frac{P}{1147}$$

$$\sigma_{bc\ cal} = \frac{P\times e\times D/2}{Ie} = \frac{P\times 77{\cdot}6\times 200}{1{,}780{,}400{,}000}$$

$$\frac{\sigma_{bc\ cal}}{\sigma_{cc}} + \frac{\sigma_{bc\ cal}}{\sigma_{bc}} \leqslant 1$$

Maximum value of load is given by

$$\frac{\left(\dfrac{P}{114{,}700}\right)}{4} + \frac{\left(\dfrac{P\times 77{\cdot}6\times 200}{1{,}780{,}400{,}000}\right)}{5} = 1$$

$$P = \frac{1}{0{\cdot}00003927} = 255{,}000\ \text{N}$$

Ex. 9·4. *The cross-section of an arch rib is reinforced as shown in Fig. 9·5. Determine the normal thrust and B.M. on the section so that no tension develops in the section and the section is safe as uncracked section $\sigma_{co}=4\ N/mm^2$, $\sigma_{bo}=5\ N/mm^2$, m=18.*

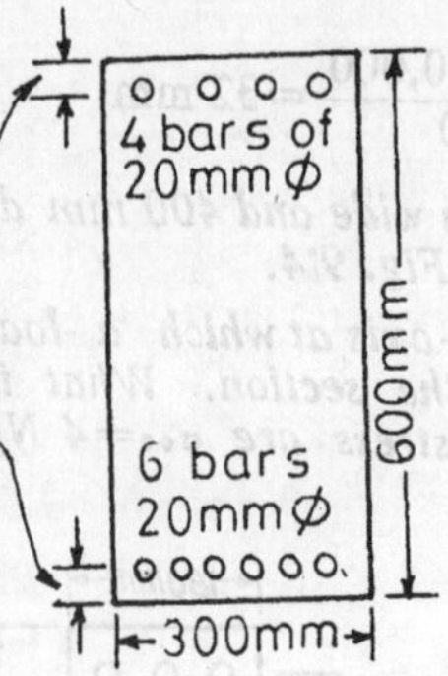

Fig. 9·5.

Sol. Equivalent area of section

$$A_e = 600\times300+17\times10\times314$$

$$=18{,}000+53{,}380=233{,}380\ mm^2$$

To determine the position of centre of gravity of section, take moments about top edge.

$$\bar{y}=\frac{300\times600\times300+4\times17\times314\times40+17\times6\times314\times560}{600\times300+10\times17\times314}$$

$$=312\ mm$$

Equivalent moment of inertia about centre of gravity of section.

$$I_e=\frac{1}{12}\times300\times(600)^3+300\times600\times(12)^2+17\times314\times4\times(272)^2+17\times314\times6\times(248)^2$$

$$=8{,}979{,}920{,}000\ mm^4$$

For no tension to develop,

$$\frac{P}{A_e}-\frac{Pe\times(D_e-\bar{y})}{I_e}=0$$

$$\therefore \quad e=\frac{I_e}{A_e\,(D-\bar{y})}=\frac{8{,}979{,}920{,}000}{233{,}380\times288}=133\ mm$$

$$\frac{\sigma_e'}{\sigma_e}+\frac{c'}{c}\ntriangleleft 1.$$

For minimum value of load

$$\frac{\left(\dfrac{P}{233{,}380}\right)}{4}+\frac{\left(\dfrac{P\times133\times312}{8{,}979{,}920{,}000}\right)}{5}=1$$

$$P=\frac{1}{0\cdot0009955}=500{,}900\ N$$

$$B.M.=P\times e=500{,}900\times133$$

$$=66{,}640{,}000\ N\text{-}mm.$$

Ex. 9·5. *Design a column carrying an axial load of 250,000 N and bending moment of 450,000 N-mm. The allowable stress in concrete area $\sigma_{co}=4\ N/mm^2$, $\sigma_{bo}=5\ N/mm^2$, m=18.*

Sol. Assume 250 mm × 250 mm section reinforced with 4 bars of 20 mm ϕ.

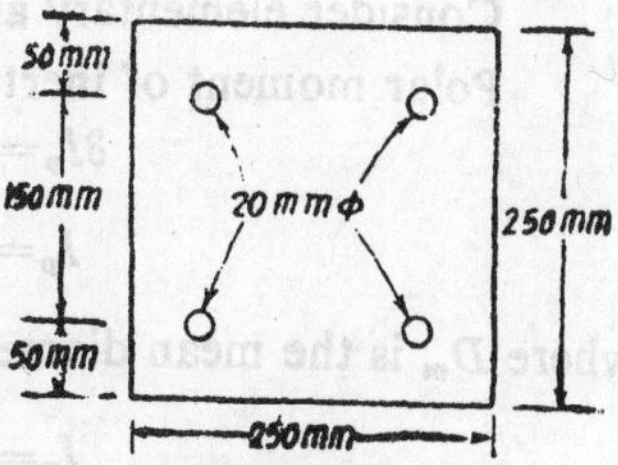

Fig. 9·6

Equivalent area

$$A_e = 250 \times 250 + (m-1) \times 4 \times 314$$
$$= 62,500 + 17 \times 4 \times 314$$
$$= 62,500 + 21,350 = 83,850 \text{ mm}^2$$

Equivalent moment of inertia

$$I_e = \frac{(250)^4}{12} + (m-1) \times 4 \times 314 \times (75)^2$$
$$= 325,520,000 + 17 \times 1,256 \times (75)^2$$
$$= 325,520,000 + 12,000,000$$
$$= 445,520,000 \text{ mm}^4$$

$$\sigma_{bc\ cal} = \frac{250,000}{83,850} = 2{\cdot}98 \text{ N/mm}^2$$

$$\sigma_{cc\ cal} = \frac{4,500,000 \times 125}{445,520,000} = 1{\cdot}26 \text{ N/mm}^2$$

$$\frac{\sigma_{cc\ cal}}{\sigma_{cc}} + \frac{\sigma_{bc\ cal}}{\sigma_{bc}} = \frac{2{\cdot}98}{4} + \frac{1{\cdot}26}{5}$$
$$= 0{\cdot}745 + 0{\cdot}252 = 0{\cdot}997 < 1.$$

∴ Section is safe.

Ex. 9·6. *Reinforced concrete circular column shown in Fig. 9·7 is subjected to a load 600,000 N and a bending moment of 30,000,000 N mm acting about one of the axis. Check whether the section is safe if the maximum allowable stresses are $\sigma_{cc} = 4$ N/mm², $\sigma_{bc} = 5$ N/mm², $m = 18$.*

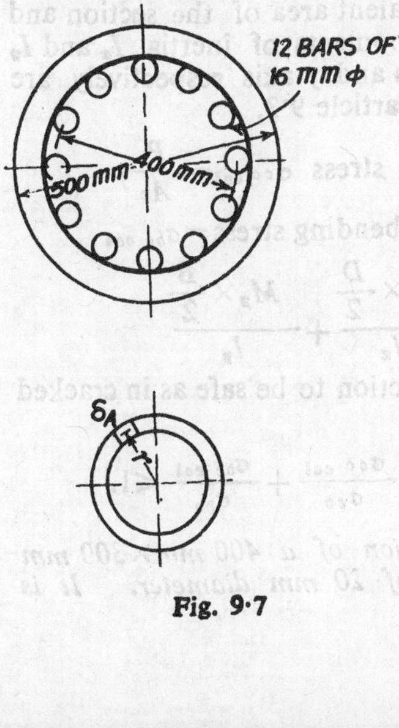

Fig. 9·7

Sol. Equivalent area of section.

$$A_e = \frac{\pi}{4} \times 500 \times 500 + (m-1) \times 12 \times 201$$
$$= 196,420 + 17 \times 12 \times 201$$
$$= 196,420 + 41,004$$
$$= 237,424 \text{ mm}^2.$$

In order to avoid undue labour in calculations for calculating equivalent moment of inertia, it is quite satisfactory to replace the longitudinal rods with angular ring having the same area as entire group of rods and having a mean diameter equal to that of the circle in which the rods are arranged.

Consider elementary area of δA of ring [Fig. 9·7 (*a*)].
Polar moment of inertia δI_p due to this area is

$$\delta I_p = \delta A \times r^2$$

$$I_p = \Sigma \delta A \times r^2 = \frac{A \times D_m^2}{4}$$

where D_m is the mean diameter

$$I_x = I_y = \frac{I_p}{2} = \frac{A \times D_m^2}{8}$$

Equivalent moment of inertia of section

$$I_e = \frac{\pi}{64} \times (500)^4 + 17 \times \frac{12 \times 201 \times (400)^2}{8}$$

$$= 3{,}067{,}961{,}576 + 820{,}080{,}000$$

$$= 3{,}888{,}041{,}576 \text{ mm}^4$$

$$\sigma_{cc\ cal} = \frac{600{,}000}{237{,}442} = 2{\cdot}53 \text{ N/mm}^2$$

$$\sigma_{cc} = \frac{30{,}000{,}000 \times 250}{3{,}888{,}041{,}576} = 1{\cdot}93 \text{ N/mm}^2$$

$$\frac{\sigma_{cc\ cal}}{\sigma_{cc}} + \frac{\sigma_{bc\ cal}}{\sigma_{bc}} = \frac{2{\cdot}53}{4} + \frac{1{\cdot}93}{5} = 0{\cdot}63 + 0{\cdot}386$$

$$= 1{\cdot}016 > 1 \qquad \therefore \text{ Section is not safe.}$$

9·4. Columns subjected to Bending about Two Axes so that no Tension develops in the Section

Corner calumns in framed structures are subjected to bending about both the axes.

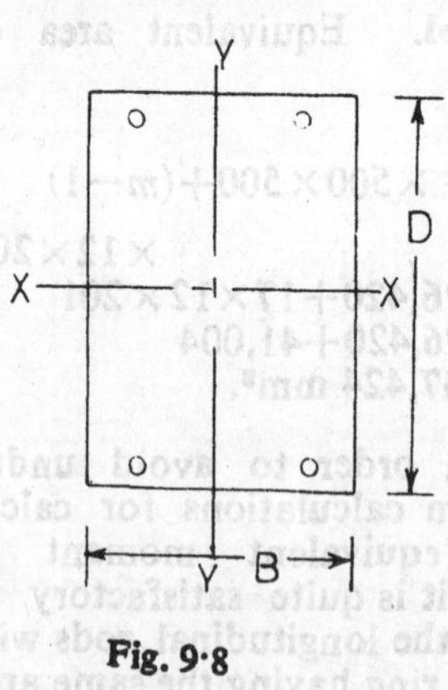

Fig. 9·8

Equivalent area of the section and equivalent moments of inertia I_x and I_y about *x*-axis and *y*-axis respectively are found as in article 9·3.

Direct stress $\sigma_{cc\ cal} = \frac{P}{A_c}$.

Total bending stress $= \sigma_{bc\ cal}$

$$= \frac{M_x \times \frac{D}{2}}{I_x} + \frac{M_y \times \frac{B}{2}}{I_y}$$

For section to be safe as in cracked section.

$$\frac{\sigma_{cc\ cal}}{\sigma_{cc}} + \frac{\sigma_{bc\ cal}}{\sigma_{bc}} < 1.$$

Ex. 9·9. *Fig. 9·2 shows the section of a 400 mm × 300 mm corner column reinforced with 4 bars of 20 mm diameter. It is* *subjected to an axial load of 350,000 N and moments of 8,000,000 N mm and about x-axis and y-axis respectively. Check the section, m=18.*

Sol. Equivalent area

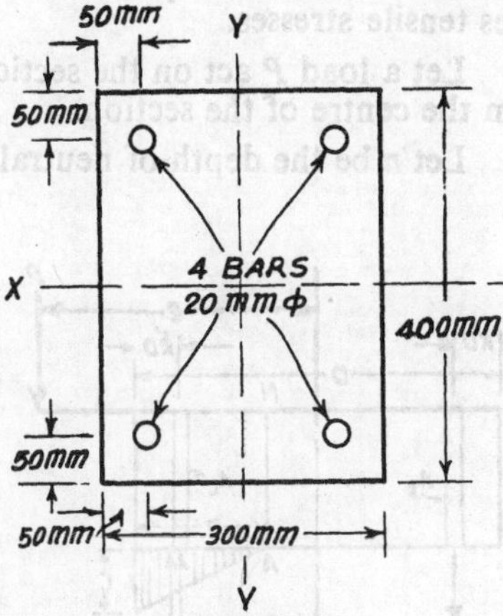

Fig. 9·9

$$A_e = 400\times300 + 17\times4\times314$$
$$= 120{,}000 + 21{,}360$$
$$= 141{,}360 \text{ mm}^2$$

$$I_x = \frac{1}{12}\times300\times400 + 17\times4 \times 314\times(150)^2$$
$$= (160{,}000 + 48{,}060)\times10^4$$
$$= 208{,}060\times10^4 \text{ mm}^4$$

$$I_y = \frac{1}{12}\times400\times(300)^3 + 17\times4 \times314\times(100)^2$$
$$= (90{,}000 + 21{,}360)\times10^4$$
$$= 111{,}360\times10^4 \text{ mm}^2$$

$$\sigma_{cc\ cal} = \frac{350{,}000}{141{,}360} = 2\cdot48 \text{ N/mm}^2$$

$$\sigma^x_{bc\ cal} = \frac{80{,}000{,}000\times200}{(208{,}060)\times10^4} = 0\cdot768 \text{ N/mm}^2$$

$$\sigma^y_{cb\ cal} = \frac{6{,}000{,}000\times150}{(111{,}360)\times10^4} = 0\cdot808 \text{ N/mm}^2$$

$$\frac{\sigma_{bc\ cal}}{\sigma_{cc}} + \frac{\sigma_{bc\ cal}}{\sigma_{bc}} = \frac{2\cdot48}{4} + \frac{0\cdot768+0\cdot808}{5}$$
$$= 0\cdot62 + 0\cdot3152$$
$$= 0\cdot9352 < 1.$$

∴ Section is safe.

9·5. Sections subjected to Large Eccentricity so that Tension is Developed

When a section is subjected to loads with large eccentricity, large tensile stresses are developed resulting in the cracking of

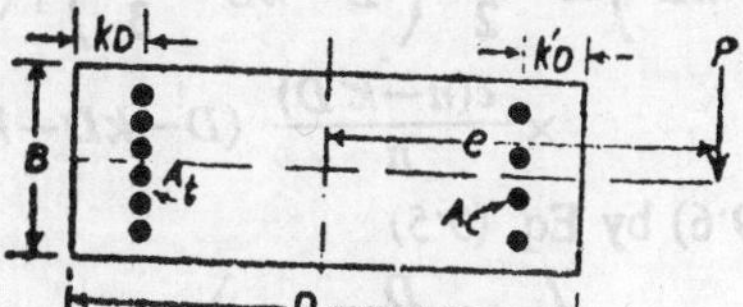

Fig. 9·10 (*a*)

concrete below neutral axis. The stress in concrete in tensile zone is neglected and only that part of concrete which is in compression is assumed to resist stresses. The steel in the compression zone of the section takes compressive stresses and that in the tensile zone takes tensile stresses.

Let a load P act on the section at an eccentricity of e measured from the centre of the section.

Let n be the depth of neutral axis.

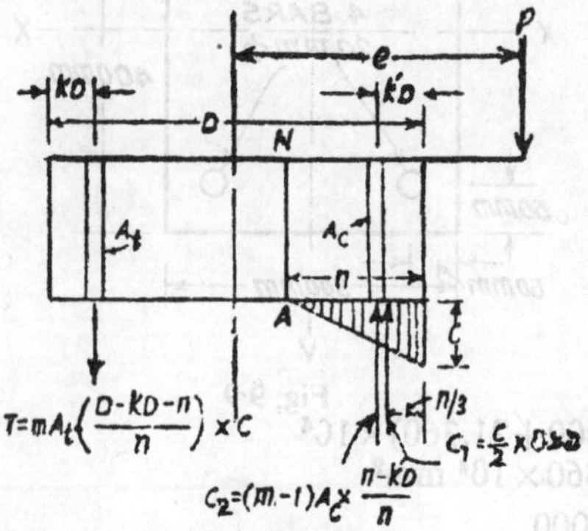

Fig. 9·10 (*b*)

σ_c be the maximum stress in concrete.

t_a be the stress in tensile steel.

t_c be the stress in compressive steel.

$k'D$ be the cover to the centre of compressive steel.

kD be the cover to the centre of tensile steel.

A_c be the area of compressive steel.

A_t be the area of tensile steel.

From the stress diagram,

$$t_a = \frac{mc\,(D-kD-n)}{n} \qquad \text{...(9·3)}$$

$$t_c = \frac{mc(c-k'D)}{n} \qquad \text{...(9·4)}$$

Total internal force $=$ Total external force

$$Bn\,\frac{\sigma_c}{2} + (m-1)\,A_c \times \frac{\sigma_c(n-k'\,D)}{n} - \frac{mA_t\sigma_c(D-k\,D-n)}{n} = P \quad \text{...(9·5)}$$

Taking moments of forces about tensile steel

$$P\left(e+\frac{D}{2}-kD\right) = \frac{Bn\sigma_c}{2}\left(D-kD-\frac{n}{3}\right) + (m-1)A_c$$

$$\times\frac{c(n-k'D)}{n}\;(D-kD-k'D) \qquad \text{...(9·6)}$$

Dividing Eq. (9·6) by Eq. (9·5)

$$\left(e+\frac{D}{2}-kD\right)$$

$$=\frac{\frac{Bn}{2}\left(D-kD-\frac{n}{3}\right)+(m-1)A_c\frac{(n-k'D)}{n}(D-kD-k'D)}{\frac{Bn}{2}+(m-1)A_c\frac{(n-k'D)}{n}-mA_t\frac{(D-kD-n)}{n}} \quad ...(9\cdot7)$$

This gives cubic equation in n which can be solved by trial and error method or graphically.

Knowing the value of n, value of σ_c can be found from equation (9·5).

Knowing the value of σ_c, stress in tensile steel and compressive steel can be calculated from equations (9·3) and (9·4).

Alternative method for finding depth of neutral axis

Consider a strip of thickness dy at a distance y from neutral axis. Stress at any point is proportional to distance from N.A. Stress in this strip $=k_1y$, where k_1 is a constant.

Total external force = Total internal force.

$$P=\int_0^n B\,dy\times k_1y+(m-1)A_c$$

$$\times k_1(n-ktD)-mA_tk'(D-k'D-n)$$

$$P=B\frac{k_1n^2}{3}+(m-1)A_ck_1$$

$$(n-k'D)-mA_ck_t(D-kD-n) \quad ...(9\cdot8)$$

$$P=k_1\left[\frac{Bn^2}{2}+(m-1)A_c'\right.$$

$$\left.(n-k'D)-mA_t(D-kD)-n)\right]$$

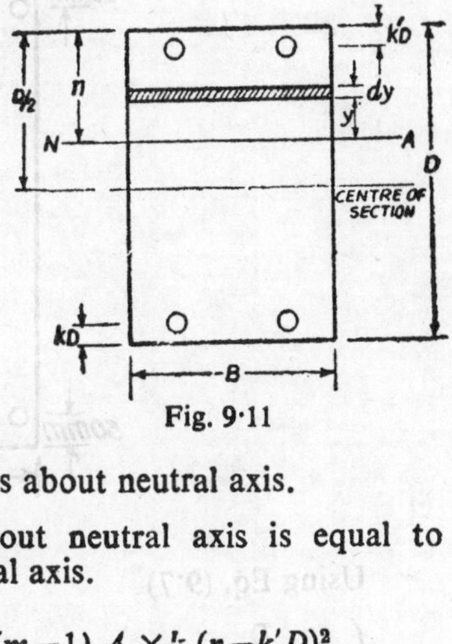

Fig. 9·11

$=k_1\times$ Moment of equivalent areas about neutral axis.

Moment of internal forces about neutral axis is equal to moment of external forces about neutral axis.

$$P\left(e-\frac{D}{2}+n\right)\int_0^n Bdy\,k_1y^2+(m-1)A_c\times k_1(n-k'D)^2$$

$$+mA_tk_1(D-kD-n)^2$$

$$\therefore \frac{P\left(e-\frac{D}{2}+n\right)}{P}=\frac{k_1\left[\frac{Bn^3}{3}+(m-1)A_c(n-k'D)^2+mA_t(D-kD-n)^2\right]}{k_1\left[\frac{Bn^2}{2}+(m-1)A_c(n-k'D)-mA_t(D-kD)-n)\right]}$$

$$e-\frac{D}{2}+n=\frac{\frac{Bn^3}{3}+(m-1)A_c(n-k'D)^2+mA_t(D-kD-n)^2}{\frac{Bn^2}{2}+(m-1)A_c(n-k'D)-mA_t(D-kD-n)} \quad ...(9\cdot9)$$

The numerator on right side represents the moment of inertia of effective section about neutral axis and denominator represents the moment of effective area about neutral axis and term on left hand side is distance of neutral axis from the point of application of load.

Ex. 9·8. *The section of an arch rib shown in Fig. 9·12 is subjected to a thrust of 400,000 N and a bending moment of 200,000,000 N-mm. Find the maximum stresses developed in steel and concrete, m=18.*

Sol. Eccentricity $e=\dfrac{200{,}000{,}000}{400{,}000}=500$ mm.

As the eccentricity is large, part of the section will be in tension. Let n be the depth of neutral axis.

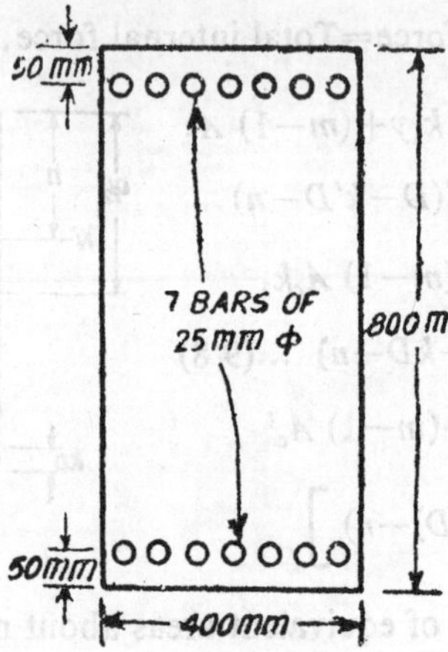

Fig. 9·12

Using Eq. (9·7)

$$\left(e+\frac{D}{2}-kD\right)$$

$$=\frac{\dfrac{Bn}{2}\left(D-kD-\dfrac{n}{2}\right)+(m-1)A_c\,\dfrac{n-k'D}{n}\,(D-kD-k'D)}{\dfrac{Bn}{2}+(m-1)A_c\,\dfrac{(n-k'D)}{n}+mA_t\,\dfrac{(D-kD-n)}{n}}$$

$$(500+400-50)$$

$$=\frac{\dfrac{400\times n}{2}\left(800-50-\dfrac{n}{3}\right)+17\times 3{,}436\times\dfrac{n-50}{n}\,(800-50-50)}{\dfrac{400\times n}{2}+17\times 3436\,\dfrac{(n-50)}{n}-18\times 3436\,\dfrac{(800-50-n)}{n}}$$

$$850=\frac{200n\left(750-\frac{n}{3}\right)+17\times 3436\times\frac{n-50}{n}\times 700}{200n+17\times 3{,}436\,\frac{(n-50)}{n}-18\times 3{,}436\,\frac{(750-n)}{n}}$$

$$850=\frac{200n^2\left(750-\frac{n}{3}\right)+17\times 3{,}436\times 700+(n-50)}{200n^2-17\times 3{,}436(n-50)-18\times 3{,}436(750-n)}$$

$$\therefore\ 170{,}000n^2-4{,}965{,}000(n-50)-5{,}288{,}000(750-n)$$

$$=150{,}000n^2-\frac{200n^3}{3}+40{,}890{,}000(n-5)$$

$$\therefore\ \frac{200n^3}{3}+20{,}000n^2+8{,}760{,}000(n-50)-52{,}880{,}000(750-n)=0$$

$$\therefore\ \frac{20n^3}{3}+20{,}000n^2+8{,}760{,}000n-438{,}000{,}000+528{,}800{,}000n$$
$$-3{,}966{,}000$$

$$\therefore\ \frac{20}{3}n^3+200n^2+61{,}640n-4{,}009{,}800=0$$

$$\therefore\qquad n^3+30n^3+9{,}246n-601{,}470=0$$

Solving by trial and error $n=470$ mm

$$\text{Total force}=400{,}000=\frac{Bn\sigma_c}{2}+(m-1)\,A_c\times\sigma_c\times\frac{n-k''D}{n}$$
$$-mA_t\,\frac{D-kD-n}{n}\times\sigma_c$$

$$400{,}000=\frac{400\times 470}{2}\times\sigma_c+17+3436\times\frac{470-50}{470}\times\sigma_c$$
$$-18\times 3436\times\frac{750-470}{470}\times\sigma$$

$$=94{,}000\sigma_c+55{,}903\sigma_c-36{,}707\sigma_c$$

$$400{,}000=113{,}106\sigma_c$$

$$\sigma_c=\frac{400{,}000}{113{,}160}=3{\cdot}535\ \text{N/mm}^2$$

$$\text{Stress in tensile steel}=m\sigma_c\ \frac{(D-kD-n)}{n}$$

$$=18\times 3{,}535\times\frac{800-50-470}{470}=37{\cdot}8\ \text{N/mm}^2$$

$$\text{Stress in compression steel}=m\sigma_c\times\frac{n-k'D}{n}$$

$$=18\times 3{,}535\times\frac{470-50}{470}=56{\cdot}9\ \text{N/mm}^2$$

9·6. Design of Eccentrically Loaded Columns Based on Average Stress

Consider a column section reinforced with p percentage of steel on each face. Let kD be cover to centre of steel.

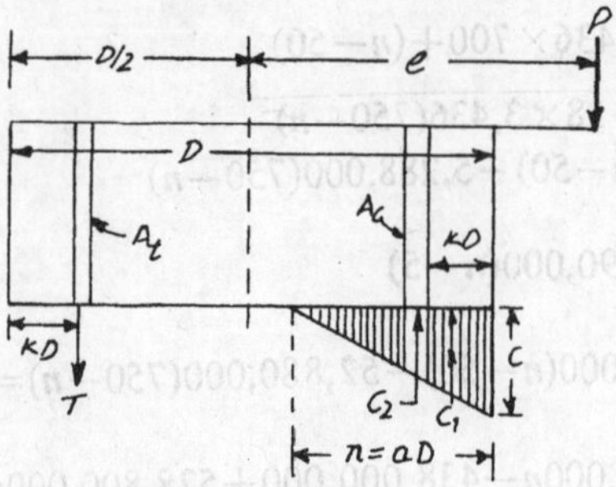

$A_t = A_c = \frac{pBD}{100}$ where p is percentage of steel used on each force.

Fig. 9·13

Let σ_c be the maximum allowable stress in concrete for the mix used, $n = aD$.

Total external load P is

$$P = Bn \times \frac{\sigma_c}{2}$$

$$= \frac{(m-1)pBD}{100} \times \sigma_c \times \frac{n-kD}{n}$$

$$+ m\frac{pBD}{100} \times \sigma_c \times \frac{D+kD-n}{n}$$

$$= \sigma_c \left[\frac{BaD}{2} + \frac{(m-1)pBD}{100} \times \frac{a-k}{a} - m\frac{pBD}{100} \times \frac{1-k-a}{a} \right]$$

For mix M 15, $m = 18$,

$$\therefore \quad \frac{P}{BD} = \frac{\sigma_c}{100} \left[50a + p\left(17 - \frac{17k}{a} - \frac{18}{a} + \frac{18k}{a} + 18 \right) \right]$$

$$= \frac{\sigma_c}{100} \left[500a + p\left(35 + \frac{k}{a} - \frac{18}{a} \right) \right] \qquad \ldots(1)$$

$$\text{Average stress on section} = \frac{P}{BD} = \frac{\sigma_c}{100} \left[50a + p\left(35 + \frac{k}{a} - \frac{18}{a} \right) \right]$$

Taking moments about centre of tensile steel

$$P\left(e + \frac{D}{2} - kD \right) = \frac{Bn\sigma_c}{2} \left(D - kD - \frac{n}{3} \right) + (m-1)A_c \times \sigma_c$$

$$\times \frac{n+kD}{n} \times (D - 2kD)$$

$$= B \times aD \times \frac{\sigma_c}{2} \times D\left(1 - k - \frac{a}{3} \right) + 17 \times \frac{pBD}{100}$$

$$\times c \times \frac{a-k}{a} \times D(1-2k)$$

$$= BD^2\sigma_c \left[\frac{a}{2}\left(1 - k - \frac{a}{3} \right) + \frac{17}{100}\, p\, \frac{(a-k)(1-2k)}{a} \right]$$

$$\frac{P}{BD^2}\left(e+\frac{D}{2}-kD\right)=\frac{\sigma_c}{100}\left[50a\left(1-k-\frac{a}{3}\right)+\frac{17p(a-k)(1-2k}{a}\right]$$

$$\frac{P}{BD}\left(\frac{e}{D}+0{\cdot}5-k\right)=\frac{\sigma_c}{100}\left[50a\left(1-k-\frac{a}{3}\right)+17p\;\frac{(a-k)(1-2k)}{a}\right].$$

Substituting P/BD from (1)

$$\frac{e}{D}+0{\cdot}5-k=\frac{50a\left(1-k-\frac{a}{3}\right)+17p\,\frac{(a-k)(1-2k)}{a}}{50a+p\left(35+\frac{k}{a}-\frac{18}{a}\right)}$$

$$\therefore\quad \frac{e}{D}=\frac{50a\left(1-k-\frac{a}{3}\right)+17p\,\frac{(a-k)(1-2k)}{a}}{50a+p\left(35+\frac{k}{a}-\frac{18}{a}\right)}-0{\cdot}5+k.$$

The graphs between e/D and average stress for value of percentage of steel p varying from 0·5 to 4 for different values of cover $k=\frac{1}{6}$, $\frac{1}{8}$ and $\frac{1}{10}$ have been drawn and shown in Appendix in graphs 14, 15, 16.

To desing a section subjected to load P and moment M, assume percentage of steel to be provided. Decide the cover to be provided to centre of reinforcement. Assume some depth D for the column and calculate e/D. For this value of e/D and cover k, find average stress value for the chosen percentage of steel. Calculate the breadth of section which will be given by $\frac{P}{D\times\text{Average stress}}$. If this value is not suitable assume different values of D and calculate B till suitable section is arrived at.

To check the design

Work out e/D and percentage of steel p. From the graphs find out average percentage stress. Maximum value of compressive stress will be $\frac{P\times100}{BD\times\text{Average percentage stress}}$.

9·7. Sections Subjected to Loads with Very Large Eccentricities

The eccentricity loaded columns may be designed by average stress method upto e/D ratio of 1·6. In case e/D ratio is more than 1·6, the section may be designed as a beam reinforced on one side for bending moment M and extra reinforcement may be provided to take up direct force P.

For mix M 15, $M=0{\cdot}87\;Bd^2$.

Assuming certain value of B, effective depth can be found out and thus overall depth D can be fixed.

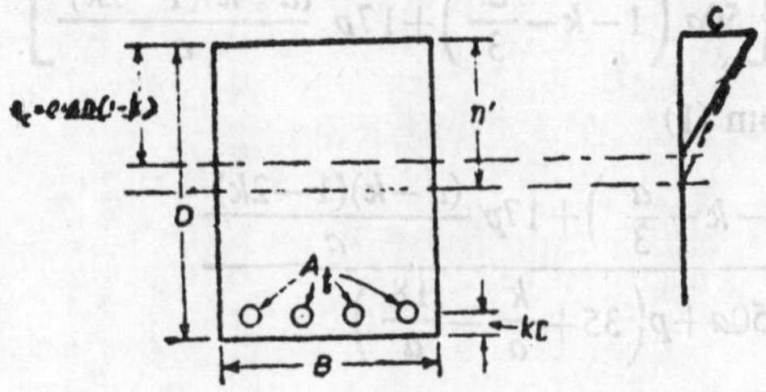

Fig. 9·14

If $n'=aD$ is the new position of neutral axis, taking into consideration effect of axial force P, then

$$P=[n'-0{\cdot}4(D-kD)]\frac{B\sigma_c}{2}$$

$$P=[a-0{\cdot}4\ (1-k)]\ \frac{DB\sigma_c}{2}$$

$$\therefore \quad \frac{P}{BD}=[0{\cdot}5a-0{\cdot}2(1-k\ ]\sigma_c$$

$$\text{Total tensile force}=0{\cdot}4\,(D-kD)\,\frac{B\sigma_c}{2}$$

Stress in tensile steel for new position of neutral axis

$$=m\sigma_c\,\frac{(D-kD-aD)}{aD}$$

$\therefore$ Area of steel required

$$=\frac{\text{Tensile force}}{\text{Stress in steel}}=\frac{0{\cdot}4\,(D-kD)\,\frac{B\sigma_c}{2}}{\frac{m\sigma_c(D-kD-aD)}{aD}}$$

$$A_t=\frac{0{\cdot}2B(D-kD)aD}{m(D-kD-aD)}=\frac{0{\cdot}2BD(1-k)a}{m(1-k-a)}$$

$$\frac{A_t}{BD}\times 100=\frac{0{\cdot}2(1-k)a}{m(1-k-a)}\times 100$$

$$\therefore \quad p=\frac{20a(1-k)}{m(1-k-a)}$$

For $\quad \mathrm{M}=15,\ m=18$

$$\therefore \quad p=\frac{20a(1-k)}{18(1-k-a)}$$

$$9p(1-k-a)=10a(1-k)$$

$$9p(1-k)-9pa=10a(1-k)$$

$$9p(1-k)=a[10-10k-9p)$$

$$\therefore \quad a=\frac{9p(1-k)}{[10-10k+9p]}$$

$$\frac{P}{BD}=[0{\cdot}5a-0{\cdot}2(1-k)]\sigma_c$$

$$M=0{\cdot}87\ Bd^2=0{\cdot}87\ BD^2(1-k)^2$$

$$\frac{MB}{P^2}=\frac{0{\cdot}87\ BD^2\ (1-k)^2\times B}{P^2}=\frac{0{\cdot}87(1-k)^2}{\left(\frac{P}{BD}\right)^2}$$

The graph between $\frac{MB}{P^2}$ and $\frac{P}{BD}$ for values of $k=\frac{1}{6}$, $\frac{1}{8}$, $\frac{1}{10}$ are shown in graph 17 of Appendix.

To design a section to resist given bending moment M and load P, assume breadth of the section B, work out $\frac{MB}{P^2}$ from the graph get the value of p percentage of steel and average stress $\frac{P}{BD}$. Knowing P and B, D is worked out and area of steel $A_t=\frac{pBD}{100}$ is computed.

Ex. 9·9. *Design a column to take load of 200,000 N and moment of 40,000,000 N-mm. Provide 1·5 per cent steel and use M 15 concrete mix.*

Sol. Assume cover to the centre of steel as $D/6$.

$$e=\frac{40{,}000{,}000}{200{,}000}=200\text{ mm}$$

Assume $D=360$ mm

$$\frac{e}{D}=\frac{200}{360}=0{\cdot}56$$

Referring to graph 14 for cover $D/6$ in Appendix, average stress for 1·5% steel and $e/D=0{\cdot}56$ is 34·8%.

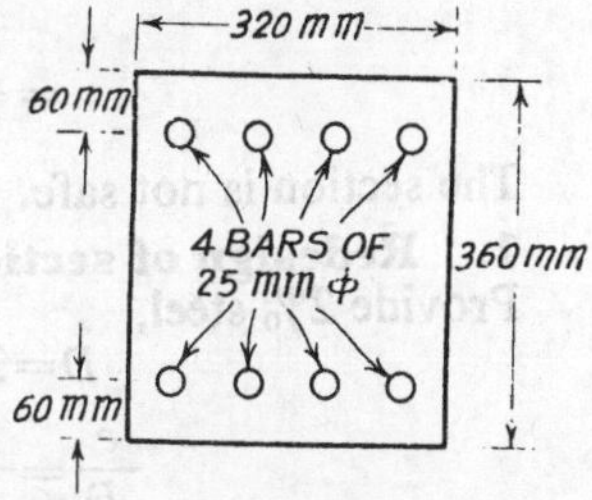

Fig. 9·15

$$\therefore\quad \frac{P}{BD}=5\times0{\cdot}348$$

$$B=\frac{400{,}000}{360\times5\times0{\cdot}384}=319\text{ mm}$$

Use 360 mm × 320 mm column.

Area of steel on each face

$$=\frac{1{\cdot}5\times319\times360}{100}=1730\text{ mm}^2$$

Provide 4 bars of 25 mm ϕ on each face.
The section is shown in Fig. 9·15.

Ex. 9·10. *A column carrying load of 200,000 N and subjected to moment of 45,000,000 N-mm is 360×270 mm in section with 4 bars of 25 mm ϕ on each face at 60 mm from the edge. If the mix used is M 15, check whether the section is safe. If the section is not safe redesign the section.*

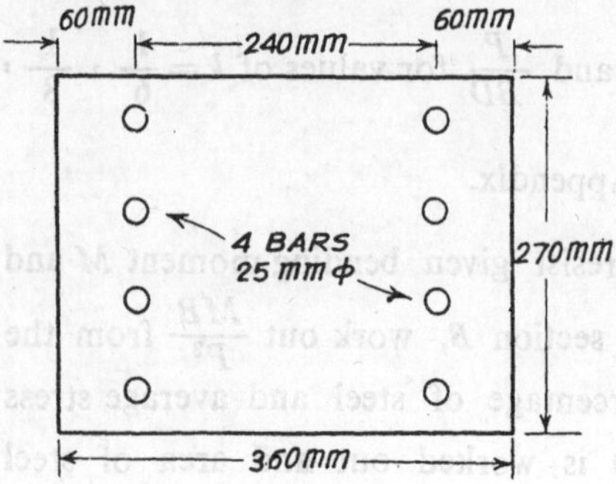

Fig. 9·16

Sol. $e=\dfrac{45,000,000}{200,000}$

$=225$ mm

$\dfrac{e}{D}=\dfrac{225}{360}=0·625$

$k=\dfrac{6}{36}=\dfrac{1}{6}$.

Percentage of steel

$p=\dfrac{A_t}{B\times D}\times 100$

$=\dfrac{4\times 491}{27\times 360}\times 200$

$=2·03\%$

Referring to graph 14 in Appendix for cover $\dfrac{D}{6}$, average stress corresponding to 2·03% steel and $\dfrac{e}{D}=0·625$ is 37%.

$$\text{Maximum stress in section}=\frac{P\times 100}{BD\times \text{Average stress}}$$

$$=\frac{200,000\times 100}{270\times 360\times 37}=5·55 \text{ N/mm}^2.$$

The section is not safe.

1. Redesign of section keeping the same depth

Provide 2% steel,

$$D=360 \text{ mm.}$$

$$\frac{e}{D}=\frac{225}{360}=0·625.$$

From graph 14 in Appendix average stress corresponding to 2% steel and $\dfrac{e}{D}=0·625$ is 37%.

$$\therefore \qquad B=\frac{200,200}{360\times 5\times 0·37}=300 \text{ mm}$$

$$\text{Steel area}=\frac{2\times 360\times 300}{100}=2160 \text{ mm}^2$$

Provide 4 bars of 25 mm ϕ on each face.

2. Redesign of section keeping the dimensions same

$$D=360 \text{ mm}, B=270 \text{ mm}$$

$$\frac{e}{D}=0{\cdot}625$$

Average stress on section

$$=\frac{200{,}000}{360\times 270}=2{\cdot}058 \text{ N/mm}^2$$

Maximum allowable stress

$$=5 \text{ N/mm}^2$$

Average stress as percentage of maximum.

Allowable stress $=\frac{2{\cdot}058}{5}\times 100=41{\cdot}16\%$

Referring to graph 14 for cover $\frac{D}{6}$ in Appendix, percentage of steel corresponding to $\frac{e}{D}=0{\cdot}625$ and average stress 41·16% is 2·5.

Area of steel $=\frac{2{\cdot}5\times 270\times 360}{100}=2430 \text{ mm}^2$

Provide 4 bars of 28 mm ϕ on each face.

Ex. 9·11. *Design a column to carry load of 100,000 N and moment of 35,000,000 N-mm. The width of the column is to be restricted to 240 mm. Use mix M 15.*

Sol. Assume cover to centre of reinforcement as $\frac{D}{6}$.

Eccentricity $e=\frac{35{,}000{,}000}{100{,}000}=350 \text{ mm}$

Assume depth of 360 mm

$$\frac{e}{D}=\frac{350}{360}=0{\cdot}973$$

Average stress on section

$$=\frac{100{,}000}{360\times 240}=1{\cdot}158 \text{ N/mm}^2$$

Maximum allowable stress

$$=5 \text{ N/mm}^2$$

Average stress as percentage of maximum allowable stress

$$=\frac{11{\cdot}58}{5}\times 100=23{\cdot}16$$

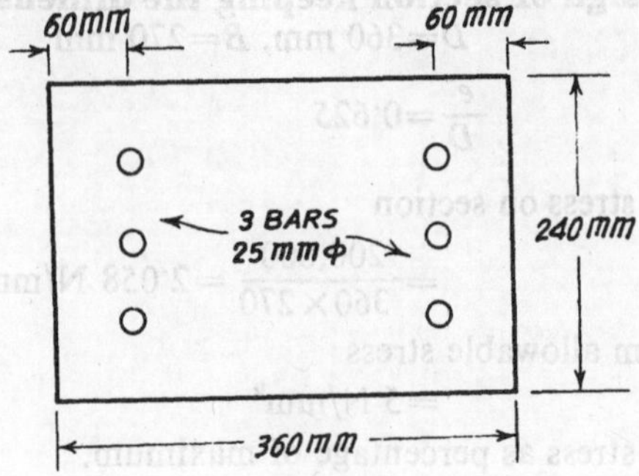

Fig. 9·17

Referring to graph 14 for cover $D/6$ in Appendix, percentage of steel corresponding to $e/D=0{\cdot}973$ and average stress 23·11% by interpolation is 1·73%.

Area of steel on each face

$$=\frac{1{\cdot}73\times360\times240}{100}=1470 \text{ mm}^2$$

Provide 3 bars of 25 mm ϕ on each face.

$$A=1473 \text{ mm}^2.$$

Ex. 9·12. *An R.C.C. section is subjected to a direct load of 500,000 N and bending moment of 50,000,000 N-mm. Design the section and check the stresses by accurate method. Use mix M15.*

Sol. Eccentricity $e=\dfrac{50{,}000{,}000}{50{,}000}=1000$ mm.

As the eccentricity is very large, the section will be reinforced only on tension side. The section will be designed as a beam and extra reinforcement provided on tension side.

$$M.R.=QBd^2$$

Keeping $B=300$ mm

$$50{,}000{,}000=0{\cdot}87\times300d^2$$

$$d=\sqrt{\frac{50{,}000{,}000}{0{\cdot}87\times300}}=438 \text{ mm.}$$

Provide overall depth of 500 mm and effective depth of 440 mm.

Depth of critical neutral axis $n_c=0{\cdot}4\times440=176$ mm.

$$P=[n'-176]\times\frac{B\sigma_c}{2}$$

$$50{,}000=[n'-176]\ \frac{300\times 50}{2}$$

$$n'=66{\cdot}7+176=242{\cdot}7\text{ mm}$$

$$A_t=\frac{0{\cdot}2\ B(D-kD)n'}{m(D-kD-n')}$$

$$=\frac{0{\cdot}2\times 300\times 440\times 242{\cdot}7}{18(440-242{\cdot}7)}$$

$$\frac{26{,}400\times 242{\cdot}7}{18\times 1973}=1800\text{ mm}^2$$

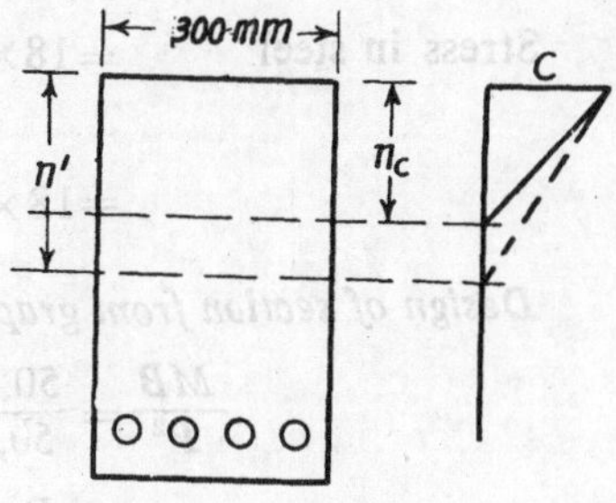

Fig. 9·18

Provide 4 bars of 25 mm ϕ, $A_t=1963$ mm^2.

Check the section

$$\left(e-\frac{D}{2}+n\right)=\frac{\dfrac{Bn^3}{2}+(m-1)A_c(n-k'D)^2+mA(D-k'D-n)^2}{\dfrac{Bn^2}{2}+(m-1)A_t(n-k'D)^2+mA_t(D-k'D-n)}$$

$$A_c=0 \qquad \ldots(9{\cdot}9)$$

$$\therefore\quad (1000-250+n)=\frac{\dfrac{300\times n^3}{3}+0+18\times 1963(440-n)^2}{\dfrac{300\times n^2}{2}+0-18\times 1963(440-n)}$$

$$(750+n)=\frac{100n^3+35{,}334(440-n)^2}{150n^2-35{,}334(440-n)}$$

Solving by trial and error, $n=251$ mm.

$$P=\frac{Bn\sigma_c}{2}=mA_t\times\sigma_c\,\frac{(D-n-kD)}{n}$$

$$\therefore\qquad \sigma_c=\frac{Pn}{\dfrac{Bn^2}{2}-mA_t(D-n-kD)}$$

$$\sigma_c=\frac{50{,}000\times 251}{300\times\dfrac{(251)^2}{2}-18\times 1963\times(440\times 251)}$$

$$=\frac{12{,}552{,}000}{9{,}470{,}000\times 6{,}675{,}000}=4{\cdot}49\text{ N/mm}^2.$$

Stress in steel $=18\times4{\cdot}49\times\dfrac{440-251}{251}$

$$=18\times4{\cdot}49\times\frac{189}{251}=64\text{ N/mm}^2.$$

Design of section from graph

$$\frac{MB}{P^2}=\frac{50{,}000{,}000\times300}{50{,}000\times50{,}000}=6$$

Assume $k=\dfrac{D}{6}$

$$B=300\text{ mm}.$$

Referring to graph 14 in Appendix for cover $\dfrac{D}{6}$, for this value of $\dfrac{MB}{P^2}=6$, average stress $\dfrac{P}{BD}$ is $0{\cdot}333$ and percentage of steel is $1{\cdot}2\%$.

$$\therefore\quad \frac{P}{BD}=0{\cdot}333$$

$$\frac{50{,}000}{300\times D}=0{\cdot}333$$

$$D\frac{50{,}000}{300\times0{\cdot}333}=500\text{ mm}.$$

$$A_t=\frac{pBD}{100}=\frac{1{\cdot}2\times300\times500}{100}=1800\text{ mm}^2.$$

Ex. 9 13. *A tonga shed projects 4 m and the frames are spaced at 4 m centres as shown in Fig. 9·19. Design the column of the frame, given that the load from the slab inclusive of self-load and weight of beam etc. is 3000 N/m². The height of the column may be taken as 5 m. Concrete mix is M15. Assume m=18.*

Sol. For the purpose of calculating the B.M. on the column the depth of the column will be assumed as 600 mm.

Maximum B.M. to which column is subjected

$$=3000\times4\times4(2-0{\cdot}3)$$
$$=81{,}600\text{ Nm}$$
$$=81{,}600{,}000\text{ N-mm}$$

Load on the column from roof

$$=3000\times4\times4=48{,}000\text{ N}$$

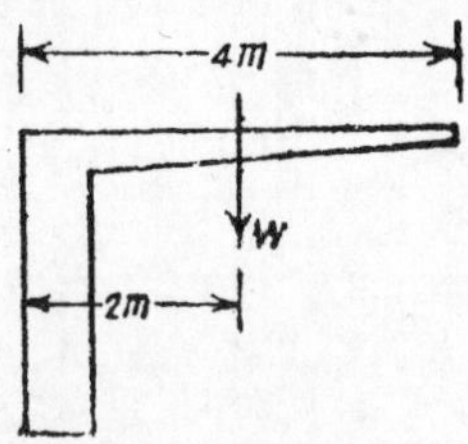

Fig. 9·19 (a)

As the section of column is subjected to large B.M. and small direct load, the column section will be designed as beam for B.M. and reinforcement provided on one face.

Assume 350 mm width of the column

Effective depth $d=\sqrt{\dfrac{M}{0{\cdot}87\times350}}$

$$=\sqrt{\frac{81{,}600{,}000}{0{\cdot}87\times350}}=517{\cdot}5 \text{ mm.}$$

Provide overall depth of 600 mm with effective depth of 540 mm.

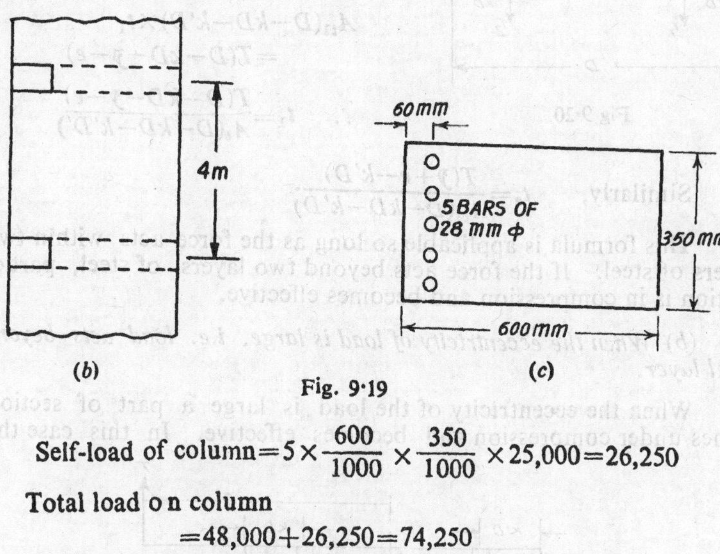

Fig. 9·19

$$\text{Self-load of column}=5\times\frac{600}{1000}\times\frac{350}{1000}\times25{,}000=26{,}250$$

Total load o n column

$$=48{,}000+26{,}250=74{,}250$$

Depth of neutral axis for balanced section

$$=0{\cdot}4\times d=2{\cdot}4\times540=216 \text{ mm.}$$

Let n' be the depth of neutral axis taking into account axial force

$$74{,}250=[n'-216]\times350\times\frac{5}{2}$$

$$n'=\frac{74{,}250\times2}{350\times5}+216=84{\cdot}9+216=300{\cdot}9 \text{ mm}$$

$$A_t=\frac{0{\cdot}2B(D-kD)n'}{m\times(D-kD-n')}=\frac{0{\cdot}2\times350\times540\times300{\cdot}9}{18\times(540-300{\cdot}9)}$$

$$=2643 \text{ mm}^2.$$

Provide 5 bars of 28 mm ϕ.

9·8. Section Subjected to Eccentric Tensile Loads

(a) *When th eccentricity is small.*

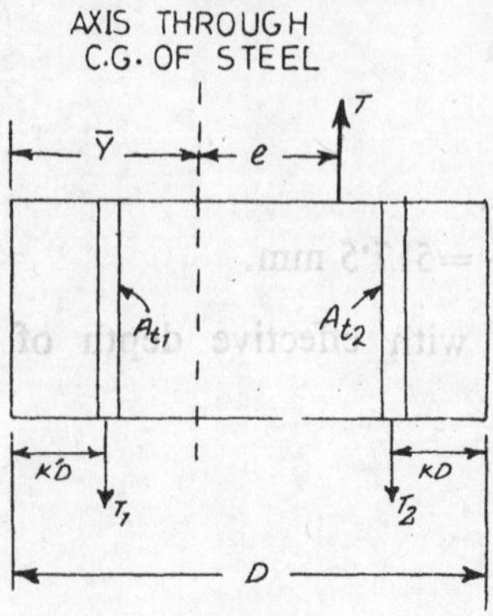

Fig 9·20.

of the load is small the entire section is in tension and entire concrete area is to be neglected. Consider a tensile force T acting at an eccentricity e (measured from the C.G. of the section) on the section shown in Fig 9·20.

Let A_{t1} and A_{t2} be areas of steel on two faces.

Taking moments about tensile steel A_{t2}.

$$A_{t1}(D-kD-k'D)\times t_1 = T(D-kD-\bar{y}-e)$$

$$\therefore \quad t_1=\frac{T(D-kD-\bar{y}-e)}{A_t(D-kD-k'D')}$$

Similarly, $$t_2=\frac{T(\bar{y}+e-k'D)}{A_{t1}(D-kD-k'D)}.$$

This formula is applicable so long as the force acts within two layers of steel. If the force acts beyond two layers of steel, part of section is in compression and becomes effective.

(b) *When the eccentricity of load is large, i.e. load acts beyond steel layer.*

When the eccentricity of the load is large a part of section comes under compression and becomes effective. In this case the

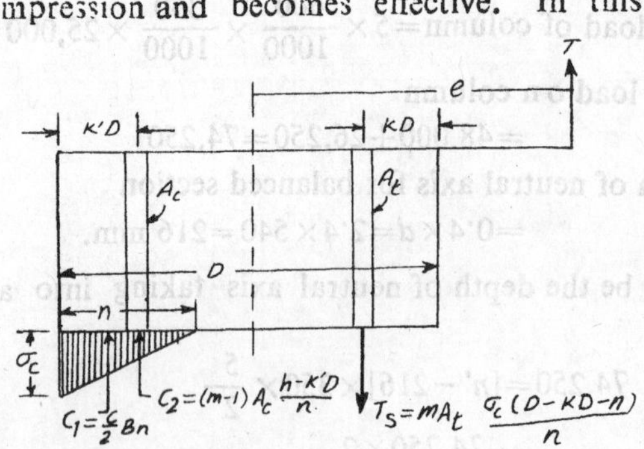

Fig. 9·21

analysis is exactly similar to the case of compressive load with large eccentricity. Equating internal forces and external forces

$$-B\frac{n\sigma_c}{2}=A_c(m-1)\frac{\sigma_c(n-k'D)}{n}+mA_t\frac{\sigma_c(D-kD-n)}{n}=+T \quad \text{...(9·10}$$

Equating moments of forces about centre of tensile steel

$$B\frac{n\sigma_c}{2}\left(D-kD-\frac{n}{3}\right)+(m-1)A_c\frac{\sigma_c(n-k'D)}{n}(D-kD'-kD)$$
$$=T\left(e-\frac{D}{2}+kD\right) \quad ...(9·11)$$

Dividing Eq. (9·11) by Eq. (9·10)

$$\left(e-\frac{D}{2}+kD\right)$$
$$=\frac{B\frac{n\sigma_c}{2}\left(D-kD-\frac{n}{3}\right)+(m-1)A_c\frac{(n-k'D)}{n}(D-kD-k'D)}{-B\frac{n}{2}-A_c(m-1)\frac{(n-k'D)}{n}+mA_t\frac{(D-kD-n)}{n}} \quad ...(9·12)$$

This represents a cubic equation similar to Eq. (9·9). This can be solved by trial and error for value of n. The values of stress in concrete and stress in compressive and tensile steel can be calculated.

Ex. 9·14. *Fig. 9·22 shows section of a bow string girder suspender subjected to a direct pull of 90,000 N and a bending moment of 7,200,000 N-mm. Find the stress in steel.*

Sol. $A_{t1}=A_{t2}=603$ mm²

Eccentricity

$$=\frac{7,200,000}{90,000}$$
$=80$ mm

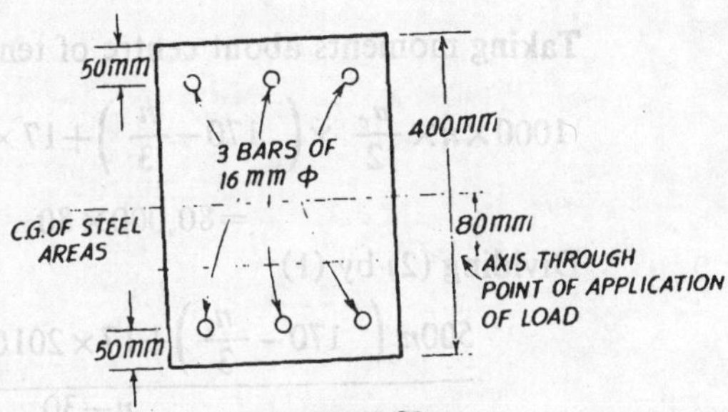

Fig. 9·22.

As the point of application of load lies between two steel areas, there are no compressive stresses induced in the section and entire concrete area is to be neglected.

Taking moments about top steel

$$At_2\times t_2\times 300=90,000\times 230$$
$$t_2=\frac{90,000\times 230}{300\times 603}=114·4 \text{ N/mm}^2$$

Taking moments about bottom steel

$$At_1\times t_1\times 300=90,000\times 70$$
$$t_1=\frac{90,000\times 70}{300\times 603}=34·8 \text{ N/mm}^2.$$

Ex. 9·15. *The slab of a tank reinforced as shown in Fig. 9·23 is subjected to a pull of 80,000 N and a bending moment of 12,000,000 N-mm. Find the maximum stresses in concrete and steel.*

Sol. Eccentricity$=\frac{12,000,000}{80,000}=150$ mm.

The point of application of loads lies outside the layer of steel hence compression is developed in a portion of the section. Let n' be the depth of neutral axis.

Area of steel on each face in 1000 mm width$=2010$ mm².

Total internal force$=$Total external force

$$-1000\times n-\frac{\sigma_c}{2}-17\times 2001\times\sigma_c\frac{(n-30)}{n}$$

$$+18\times 2001\times\frac{\sigma_c\,(170-n)}{n}=+80,000 \quad ...(1)$$

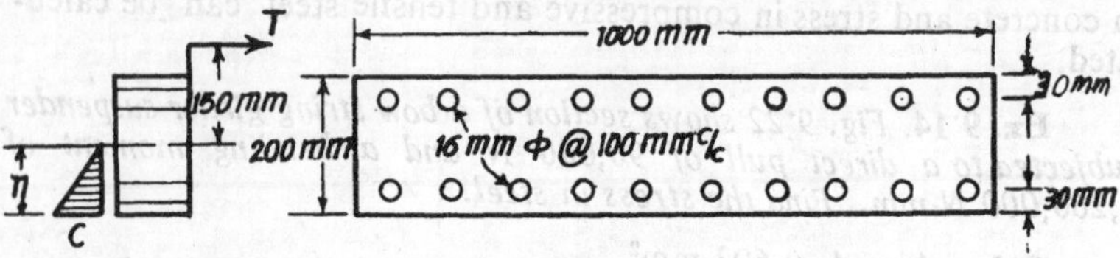

Fig. 9·23

Taking moments about centre of tensile steel

$$1000\times n\times\frac{\sigma_c}{2}\times\left(170-\frac{n}{3}\right)+17\times 2010\times\frac{\sigma_c(n-30)}{n}\times 140$$

$$=80,000\times 80 \quad ...(2)$$

Dividing (2) by (1)

$$\frac{500n\left(170-\frac{n}{3}\right)+17\times 2010\times\frac{(n-30)}{n}\times 140}{-500n-17\times 201\times\frac{n-30}{n}+18\times 201\times\frac{170-n}{n}}=+80$$

Solving by trial and error $n=45·2$ mm.

Substituting in (1)

$$-1000\times n\times\frac{\sigma_c}{2}-17\times 2010\times\sigma_c\times\frac{n+30}{n}$$

$$+18\times 2001\times\frac{170-n}{n}\times\sigma_c=80,000$$

$$\sigma_c=\frac{80,000}{65,820}=1·215 \text{ N/mm}^2.$$

Stress in compression steel

$$=18\times1.215\times\frac{4.52-30}{45.2}=7\,37\ \text{N/mm}^2$$

Stress in tensile steel $=18\times1.215\times\dfrac{170-45.2}{45.2}$

$$=60.3\ \text{N/mm}^2.$$

9.9. Circular Sections Subjected to Eccentric Loads

Consider a circular column of radius R, reinforced with steel area A_t. The centre of steel reinforcement lies at distance r from the centre of column. Let the load on the column be P at an eccentricity e. The eccentricity being such that tension is induced in a part of the section.

The longitudinal rods are replaced by an annular ring having a mean diameter equal to that of the circle in which the rods are set.

Let e be the eccentricity of the load about $y\,y$ axis. Let n be the depth of the neutral axis.

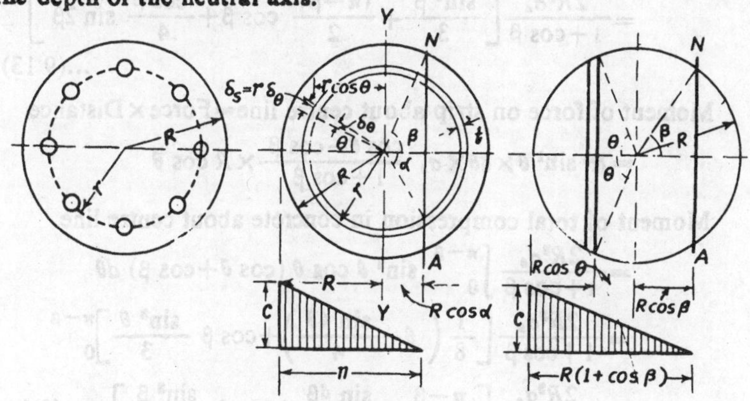

Fig. 9.24

If t is the thickness of equivalent steel ring,

$$2\pi rt=A_t.$$

Consider a thin strip subtending angle 2θ at centre and parallel to neutral axis.

Length of strip $=2R\sin\theta$

Distance of strip from centre $=R\cos\theta$

Width of strip $=\delta\,L\cos\theta)=R\sin^2\theta\,\delta\theta$

Area of the strip $=2R^2\sin^2\theta\,\delta\theta$

Intensity of stress $=\dfrac{\sigma_c\times(R\cos\theta+R\cos\beta)}{R(1+\cos\beta)}$

Total force on the strip

$$2R^2 \sin^2\theta \times \delta\theta \times \sigma_c \frac{\cos\theta+\cos\beta}{1+\cos\beta}$$

Total compression in concrete

$$=\int_0^{\pi-\beta} \frac{2R\sigma_c}{1+\cos\beta} \sin^2\theta\,(\cos\theta+\sin^2\theta\cos\beta)\,d\theta$$

$$=\frac{2R^2\sigma_c}{1+\cos\beta}\int_0^{\pi-\beta}(\sin^2\theta\cos\theta+\sin^2\theta\cos\beta)\,d\theta$$

Total compression

$$=\frac{2R^2\sigma_c}{1+\cos\beta}\left[\frac{\sin^3\theta}{3}+\frac{\cos\beta}{2}\left(\theta-\frac{\sin 2\theta}{2}\right)\right]_0^{\pi-\beta}$$

$$=\frac{2R^2\sigma_c}{1+\cos\beta}\left[\frac{\sin^3(\pi-\beta)}{3}+\frac{\cos\beta}{2}(\pi-\beta)-\frac{\cos\beta}{4}\sin 2\pi\,(\pi-\beta)\right]$$

$$=\frac{2R^2\sigma_c}{1+\cos\beta}\left[\frac{\sin^3\beta}{3}+\frac{(\pi-\beta)}{2}\cos\beta+\frac{\cos\beta}{4}\sin 2\beta\right] \quad ...(9\cdot13)$$

Moment of force on strip about centre line = Force × Distance

$$=R^2\sin^2\theta\times\delta\theta\times\sigma_c\,\frac{\cos\theta-\cos\beta}{1+\cos\beta}\times R\cos\theta$$

Moment of total compression in concrete about centre line

$$=\frac{2R^3\sigma_c}{1+\cos\beta}\int_0^{\pi-\beta}\sin^2\theta\cos\theta\,(\cos\theta+\cos\beta)\,d\theta$$

$$=\frac{2R^3\sigma_c}{1+\cos\beta}\left[\frac{1}{8}\left(\theta-\frac{\sin 4\theta}{4}\right)+\cos\beta\,\frac{\sin^3\theta}{3}\right]_0^{\pi-\beta}$$

$$=\frac{2R^3\sigma_c}{1+\cos\beta}\left[\frac{\pi-\beta}{8}+\frac{\sin 4\beta}{8}+\cos\beta\,\frac{\sin^3\beta}{3}\right] \quad ...(9\cdot14)$$

Consider small steel area $= r\delta\theta\times t$ at an angle θ

Intensity of stress on element

$$=\sigma_c\,\frac{r\cos\theta+r\cos\alpha}{R+r\cos\alpha}\,(m-1)\ \text{[effective stress]}$$

Total compression in steel above neutral axis

$$=2\int_0^{\pi-\alpha} r\,d\theta\times t\times\sigma_c\times\frac{r\cos\theta+r\cos\alpha}{R+r\cos\alpha}\,(m-1)$$

$$=\frac{2r^2\,t(m-1)\sigma_c}{R+r\cos\alpha}\int_0^{\pi-\alpha}(\cos\theta+\cos\alpha)\,d\theta$$

$$=\frac{2r^2t(m-1)\sigma_e}{R+r\cos\alpha}\Big[\sin\theta+\theta\cos\alpha\Big]_0^{\pi-\alpha}$$

$$=\frac{2r^2t(m-1)\sigma_e}{R+r\cos\alpha}\,[\sin\alpha+(\pi-\alpha)\cos\alpha] \qquad \ldots(9\cdot15)$$

Moment of compression in elementary area of steel about centre line$=\dfrac{r\cos\theta+r\cos\alpha}{R+r\cos a}\,(m-1)\times r\times d\theta\times t\times r\cos\theta.$

Moment of total compression in steel about centre line

$$=2\int_0^{\pi-\alpha}\sigma_e\,\frac{r\cos\theta+r\cos\alpha}{R+r\cos\alpha}\,(m-1)\times rd\theta\times t\times\cos\theta$$

$$=\frac{2r^3(m-1)\times t}{R+r\cos\alpha}\int_0^{\pi-\alpha}(\cos^2\theta+\cos\alpha\cos\theta)d\theta$$

$$=\frac{2r^3t(m-l)\,\sigma_e}{R+r\cos\alpha}\left[\frac{\theta}{2}+\frac{\sin 2\theta}{4}+\cos\alpha\sin\theta\right]_0^{\pi-\alpha}$$

$$=\frac{2r^3t(m-1)\sigma_e}{R+r\cos\alpha}\left[\frac{\pi-\alpha}{2}-\frac{\sin 2\alpha}{4}+\cos\alpha\sin\alpha\right]$$

$$=\frac{2r^3t(m-1)\sigma_e}{R+r\cos\alpha}\left[\frac{\pi-\alpha}{2}+\frac{\sin 2\alpha}{4}\right] \qquad (9\cdot16)$$

Similarly total tension in steel can be calculated.

Total tension in steel

$$=2\int_0^{\alpha} rd\theta\times t\times m\times\sigma_e\,\frac{r\cos\theta-r\cos\alpha}{R+r\cos\alpha}$$

$$=\frac{2r^2tm\,\sigma_e}{R+r\cos\alpha}\int_0^{\alpha}(\cos\theta-\cos\alpha)\,d\theta$$

$$=\frac{2r^2tm\sigma_e}{R+r\cos\alpha}\,(\sin\alpha-\alpha\cos\alpha) \qquad \ldots(9\cdot17)$$

Moment of total tension in steel about centre line

$$=2\int_0^{\alpha} rd\theta\times t\times m\times\sigma_e\,\frac{r\cos\theta-r\cos\theta}{R+r\cos\alpha}\times r\cos\theta$$

$$=\frac{2r^3tm\,\sigma_e}{R+r\cos\alpha}\int_0^{\alpha}(\cos^2\theta-\cos\alpha\cos\theta)\,d\theta$$

$$=\frac{2r^3tm\,\sigma_e}{R+r\cos\alpha}\left(\frac{\alpha}{2}-\frac{\sin 2\alpha}{4}\right) \qquad \ldots(9\cdot18)$$

Sum of internal forces=External force

$$\frac{2R^2\sigma_e}{1+\cos\beta}\left[\frac{\sin^3\beta}{3}+\frac{(\pi-\beta)}{2}\cos\beta+\frac{\cos\beta\sin 2\beta}{4}\right]$$

$$+\frac{2r^3\,t(m-1)\sigma_e}{R+r\cos\alpha}\,[\sin\alpha+(\pi-x)\cos\alpha]$$

$$-\frac{2r^3\,tm\sigma_e}{R+r\cos\alpha}\,[\sin\alpha-\alpha\cos\alpha]=P \qquad \ldots(9\cdot19)$$

Total moment of internal force about centre line

=moment of external force about centre line

$$\frac{2R^3\sigma_e}{1+\cos\beta}\left[\frac{\pi-\beta}{8}+\frac{\sin 4\beta}{32}+\frac{\cos\beta\sin^3\beta}{3}\right]$$
$$+\frac{2r^3t(m-1)}{R+r\cos\alpha}\left[\frac{\pi-\alpha}{2}+\frac{\sin 2\alpha}{4}\right]$$
$$+\frac{2r^3\,tm\sigma_e}{R+r\cos\alpha}\left(\frac{\alpha}{2}-\frac{\sin 2\alpha}{4}\right)$$
$$=P\times e \qquad \ldots(9{\cdot}20)$$

From the figure it is evident that

$$R\cos\beta=r\cos\alpha \qquad \ldots(9{\cdot}21)$$

The equations (9·19) and (9·20) are similar to equations (9·6) and (9·7). The relation between α and β is given by Eq. (9·21).

Dividing Eq. (9·20) by Eq. (9·19) gives and equation is terms of α and β. The equation so obtained can be solved by trial and error.

Alternately assume certain value of β and obtain corresponding value of α from Eq. (9·21), substitute these values of α and β in Eqs. (9·19) and (9·20), if both the equations are balanced the assumed value of β is correct if not then some other value of β is tried till both the equations are satisfied.

Eq. (9·19) can be written as

$$\frac{2R^2\sigma_e}{1+\cos\beta}\left(\frac{\sin^3\beta}{3}+\frac{(\pi-\beta)}{2}\cos\beta+\frac{\cos\beta\sin 2\theta}{4}\right)$$
$$+\frac{2r^2t\sigma_e}{R+r\cos\alpha}\left[(m-1)\pi\cos\alpha-\sin\alpha+\alpha\cos\alpha\right]=P$$
$$\ldots(9{\cdot}22)$$

Eq. (9·20) can be written as

$$\frac{2R^3\sigma_e}{1+\cos\beta}\left[\frac{\pi-\beta}{8}+\frac{\sin 4\beta}{32}+\cos\beta\,\frac{\sin^3\beta}{3}\right]$$
$$+\frac{r^3R\sigma_e}{R+r\cos\alpha}\left[(m-1)\pi+\alpha-\frac{\sin 2\alpha}{2}\right]=P\times e \qquad \ldots(9{\cdot}23)$$

Dividing Eq. (9·23) by Eq. (9·22)

$$\frac{2R^3}{1+\cos\beta}\left[\frac{\pi-\beta}{8}+\frac{\sin 4\beta}{32}+\frac{\cos\beta\sin^3\beta}{3}\right]+\frac{r^3t}{R+r\cos\alpha}$$

$$e=\frac{\left[(m-1)\pi+\alpha-\frac{\sin 2\alpha}{2}\right]}{\frac{2R^2}{1+\cos\beta}\left(\frac{\sin^3\beta}{3}+\frac{(\pi-\beta)}{2}\cos\beta+\frac{\cos\beta\sin 2\beta}{4}\right)+\frac{2r^2t}{R+r\cos\alpha}[(m-1)\pi\cos\alpha-\sin\alpha+\alpha\cos\alpha]}.$$

Ex. 9·15. *Design internal column of flat slab floor of Ex. 8·3.*

Solution. Area from which load comes on column $=\frac{6\times6}{4}$ $=9$ sq. metres. Load on column for B.M. $=w_d\times9+w_l\times\frac{9}{2}$, where w_d is dead load per sq. metre and w_l is live load per sq. metre. Live load is taken from half panel.

$$P=(4800+1000)\times9+8000\times\frac{9}{2}$$

$$=52,200+36,000=88,200 \text{ N}$$

B.M. in column $=0·5\times106,830,000=53,400,000$ N-mm.

Eccentricity $e=\frac{53,400,000}{88,200}=605·3$ mm.

Use column of 500 mm diameter and mean diameter of reinforcement as 400 mm.

$$R=250 \text{ mm}, r=200 \text{ mm}.$$

Using 2% steel, $A_t=\frac{\pi}{4}\times(500)^2\times\frac{2}{100}=3926 \text{ mm}^2$.

Use 8 bars of 25 mm ϕ,

$$A_t=3927 \text{ mm}^2$$

Thickness of steel shell

$$t=\frac{3927}{2\pi\times200}=3·125 \text{ mm}.$$

Use M15 concrete mix, $m=18$

$$r\cos\alpha=R\cos\beta$$

$$\therefore \quad \cos\alpha=\frac{500}{400}\cos\beta$$

$$=1·25\cos\beta$$

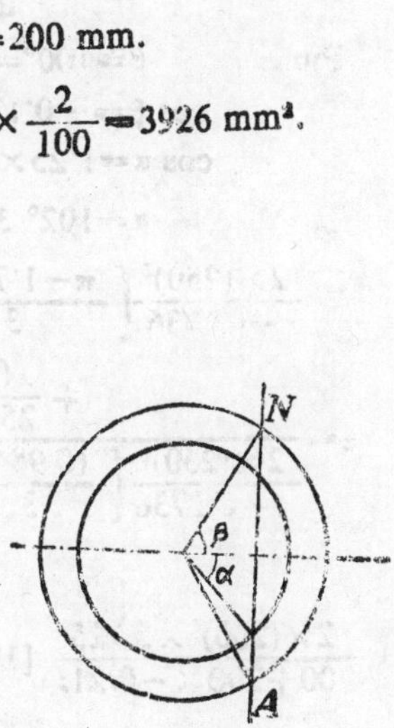

Fig. 9·25

$$e=\frac{\dfrac{2R^3}{1+\cos\beta}\left[\dfrac{\pi-\beta}{8}+\dfrac{\sin\beta}{32}+\dfrac{\cos\beta\sin^2\beta}{3}\right]+\dfrac{r^3t}{R+r\cos\alpha}\times\left[(m-1)\pi+\alpha-\dfrac{\sin 2\alpha}{2}\right]}{\dfrac{2R^3}{1+\cos\beta}\left[\dfrac{\sin^3\beta}{3}+\dfrac{(\pi-\beta)\cos\beta}{2}+\dfrac{\cos\beta\sin 2\beta}{4}\right]+\dfrac{2r^2t}{R+r\cos\alpha}\times\left[(m-1)\pi\cos\alpha-\sin\alpha+\alpha\cos\alpha\right]}$$

Put $\beta=90°$, $\quad 4\beta=850°$, $\quad \cos\beta=0$, $\quad \sin 4\beta=0$

$$\cos\alpha=1\cdot 25\times\cos\beta=0$$

$$\therefore \qquad \alpha=90°$$

$$e=\frac{\dfrac{2\times(250)^2}{1+0}\left[\dfrac{\pi-\dfrac{\pi}{2}}{8}+0+\dfrac{0\times 1^2}{6}\right]+\dfrac{(200)^3\times 3\cdot 125}{250+200\times 0}\times\left[17\pi+\dfrac{\pi}{2}-\dfrac{0}{2}\right]}{\dfrac{2\times(250)^2}{1+0}\left[\dfrac{1^3}{3}+\dfrac{\left(\pi-\dfrac{\pi}{2}\right)}{2}\times 0+\dfrac{0\times 0}{2}\right]+\dfrac{2\times(200)^2\times 3\cdot 125}{250+200\times 0}\times\left[17\pi\times 0-1+\dfrac{\pi}{2}\times 0\right]}$$

$$=\frac{798{,}130+549{,}900}{4173-50}=327 \text{ mm.}$$

Put $\qquad \beta=100°=1\cdot 7454$ radians, $\qquad 4\beta=400$

$$\cos\beta=-0\cdot 1735 \quad \sin 4\beta=\sin 40°=0\cdot 6428$$

$$\cos\alpha=1\cdot 25\times -0\cdot 1736=-0\cdot 2170$$

$$\therefore \qquad \alpha=102°\ 32'=1\cdot 7896 \text{ radians}$$

$$e=\frac{\dfrac{2\times(250)^2}{1-0\cdot 1736}\left[\dfrac{\pi-1\cdot 7454}{3}+\dfrac{0\cdot 6428}{32}+\dfrac{-0\cdot 1736\times(0\cdot 9848)^2}{3}\right]+\dfrac{(200)^3\times 3\cdot 125}{25+10\times -0\cdot 217}\left[17\pi+1\cdot 7896+\dfrac{0\cdot 4237}{2}\right]}{\dfrac{2\times(250)^2}{1-0\cdot 1736}\left[\dfrac{(0\cdot 9848)^3}{3}+\left(\dfrac{\pi-1\cdot 17454}{2}\right)\times -0\cdot 1736+\dfrac{-0\cdot 1736\times -0\cdot 3420}{4}\right]+\dfrac{2\times(200)^2\times 3\cdot 125}{200+200\times -0\cdot 217}\left[17\pi\times -0\cdot 217-0\cdot 9762+1\cdot 7896\times -0\cdot 217\right]}$$

$$=\frac{5{,}270{,}700+6{,}704{,}000}{32{,}560-15{,}670}$$

$$=\frac{11,974,700}{16,390}=730{\cdot}7 \text{ mm}$$

Let $\beta=98°=1{\cdot}7105$ radians, $4\beta=392°$

$\cos\beta=-0{\cdot}1392$, $\sin 4\beta=\sin 32°=0{\cdot}5299$

$\sin\beta=0{\cdot}9903$,

$\sin 2\beta=-0{\cdot}2756$

$\cos\alpha=1{\cdot}25\cos\beta=-0{\cdot}1740$,

$\therefore \quad \alpha=100°\ 11'=1{\cdot}7486$ radians

$\sin\alpha=0{\cdot}9843$,

$\sin 2\alpha=-0{\cdot}3480$

$$e=\frac{\dfrac{2\times(250)^3}{1-0{\cdot}1392}\left[\dfrac{\pi-1{\cdot}7105}{8}+\dfrac{0{\cdot}5299}{32}+\dfrac{-0{\cdot}1392\times(0{\cdot}9903)^3}{3}\right]+\dfrac{(200)^3\times 3{\cdot}125}{25+20\times-0{\cdot}174}\left[17\pi+1{\cdot}7486+\dfrac{0{\cdot}3480}{2}\right]}{\dfrac{2\times(250)^2}{1-0{\cdot}1392}\left[\dfrac{(9903)^3}{3}+\dfrac{\pi-1{\cdot}7105}{2}\times-0{\cdot}1392+\dfrac{-0{\cdot}1392\times-0{\cdot}2756}{4}\right]+\dfrac{2\times(200)^2\times 3{\cdot}125}{250+200\times-0{\cdot}174}\left[17\pi\times-0{\cdot}174-0{\cdot}9843+1{\cdot}7486\times-0{\cdot}174\right]}$$

$$=\frac{5,463,000+6,428,000}{33,900-12,290}=\frac{11,891,000}{21,610}=550{\cdot}5 \text{ mm}$$

Take value of $\beta=98°\ 45'$

$\beta=98°\ 45'=1{\cdot}7235$ radians, $4\beta=395°$

$\cos\beta=0{\cdot}1521$, $\sin 4\beta=\sin 35°=0{\cdot}5736$

$\sin\beta=0{\cdot}9883$

$\sin 2\beta=-0{\cdot}3007$

$\cos\alpha=1{\cdot}25\cos\beta=-0{\cdot}1901$

$\alpha=100°\ 58'=1{\cdot}7623$ radians

$\sin\alpha=0{\cdot}9817$

$\sin 2\alpha=-0{\cdot}3735$

$$e=\frac{\dfrac{2\times(250)^3}{1-0{\cdot}152}\left[\dfrac{\pi\times 1{\cdot}7235}{8}+\dfrac{0{\cdot}5736}{32}+\dfrac{(-0{\cdot}152)(0{\cdot}05883)^3}{3}\right]+\dfrac{(200)^3\times 3{\cdot}125}{25+20(-1901)}\left[17\pi+1{\cdot}7623-\left(\dfrac{-0{\cdot}3735}{2}\right)\right]}{\dfrac{2\times(250)^2}{1-{\cdot}1521}\left[\dfrac{(0{\cdot}9893)^3}{3}+\dfrac{\pi-(1{\cdot}7235)}{2}(-0{\cdot}1521)+\dfrac{(0{\cdot}1521)\ (-3007)}{4}\right]+\dfrac{2\times(200)^3\times 3{\cdot}125}{250+200(-0{\cdot}1901)}\left[17\pi(-0{\cdot}1901)(-0{\cdot}981)+1{\cdot}7623(-1901)\right]}$$

$$=\frac{5.395,000+6,560,000}{33,230-13,530}=\frac{11,955,000}{19,700}$$

$$=607{\cdot}4 \text{ mm O.K.}$$

Now $\quad P=1970\ \sigma_e$

$$\therefore \quad \sigma_e=\frac{80,200}{19,700}=4{\cdot}477 \text{ N/mm}^2. \text{ Safe.}$$

9 10. High Yield Strength Deformed Bars

The section becomes uneconomical as stresses in tensile and compressive steels are very low, and cost of deformed bars is much higher than that for plain steel.

9·11. Column with Biaxial Bending

This design method given below is approximate. Use of graphs 14, 15, 16 facilitate working out of problems.

Let section be subjected to P, M_1, M_2, where P is axial load, M_1 and M_2 are bending moments about two axes.

$$e_1=\frac{M_1}{P}\ ;\ e_2=\frac{M_2}{P}$$

choose dimension of column D_1 and D_2 in directions of e_1 and e_2 choose percentage of steel say p.

For value of p and $\frac{e_1}{D_1}$ and $\frac{e_2}{D_2}$ find the value of average stress, from these find the axial forces P_1 and P_2

$$\overline{M_2}=P_2\, e_2$$

$$\overline{M_1}=P_1\, e_1$$

If, $\quad \frac{M_1}{\overline{M_1}}+\frac{M_2}{\overline{M_2}}\leqslant 1$

The column will be safe. If $\frac{M_1}{\overline{M_1}}+\frac{M_2}{\overline{M_2}}$ is much below 1, the section may be revised by changing D_1, D_2 or p.

Ex. 9·16. *Design a column subjected to biaxial bending*

$$P=200 \text{ kN}.$$

$$M_1=25,000 \text{ kN-mm}.$$

$$M_2=15,000 \text{ kN-mm}$$

Use M15 mix.

Solution. Use 400 mm × 300 mm column, with 8 bars 25 mm c/c kept at 50 mm from faces as shown in Fig. 9·26.

$$e_1 = \frac{25{,}000}{200} = 125 \text{ mm}$$

$$e_2 = \frac{15{,}000}{200} = 75 \text{ mm}$$

$$\frac{e_1}{D_1} = \frac{125}{400} = 0{\cdot}3125$$

$$\frac{e_2}{D_2} = \frac{75}{300} = 0{\cdot}25$$

$$p = \frac{4 \times 491 \times 100}{400 \times 300} = 1{\cdot}64\%$$

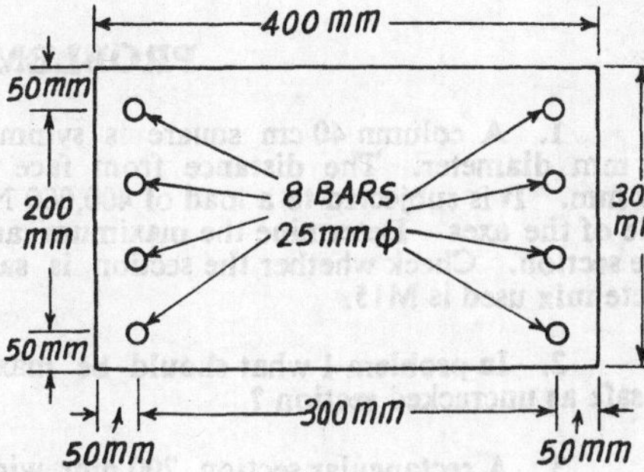

Fig. 9·26

From graph 15,

$$K = \frac{50}{400} = \frac{1}{8}$$

$$\overline{M_1} = e_1 \times \frac{\text{Percentage average stress}}{100} \times D_1 D_2 \times C$$

$$= 125 \times \frac{60}{100} \times 400 \times 300 \times 5$$

$$= 45{,}000{,}000 \text{ N-mm}$$

$$= 45{,}000 \text{ kN-m.}$$

From graph 14,

$$k = \frac{50}{300} = \frac{1}{6}$$

$$\overline{M_2} = e_2 \times \frac{\text{Percentage average stress}}{100} \times D_1 D_2 \times \sigma_c$$

$$= 75 \times \frac{77}{100} \times 400 \times 300 \times 5$$

$$= 34{,}650{,}000 \text{ N-mm}$$

$$= 34{,}650 \text{ kN-mm.}$$

$$\frac{M_1}{\overline{M_1}} + \frac{M_2}{\overline{M_2}} = \frac{25{,}000}{45{,}000} + \frac{15{,}000}{34{,}650}$$

$$= 0{\cdot}556 + 0{\cdot}433$$

$$= 0{\cdot}988 < 1. \quad \text{Safe.}$$

PROBLEMS

1. A column 40 cm square is symmetrically reinforced with 4 bars of 25 mm diameter. The distance from face to centre of reinforcement being 60 mm. It is subjected to a load of 400,000 N at an eccentricity of 50 mm about one of the axes Determine the maximum and minimum stresses developed in the section. Check whether the section is safe as uncracked section if the concrete mix used is M15.

2. In problem 1 what should be maximum eccentricity so that section is safe as uncracked section ?

3. A rectangular section 200 mm wide and 500 mm deep is reinforced with 12 bars of 20 mm ϕ as shown in Fig. 9·27.

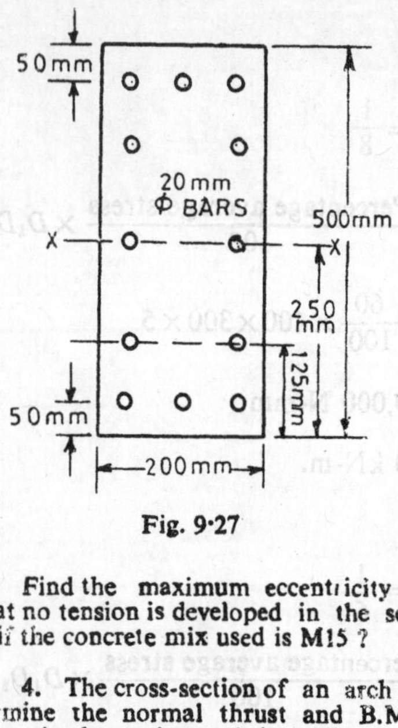

Fig. 9·27

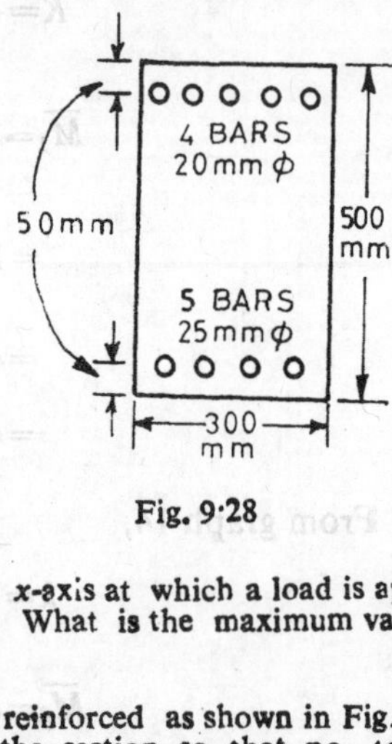

Fig. 9·28

Find the maximum eccentricity about x-axis at which a load is applied so that no tension is developed in the section. What is the maximum value of load if the concrete mix used is M15 ?

4. The cross-section of an arch rib is reinforced as shown in Fig. 9·28. Determine the normal thrust and B.M. on the section so that no tension develops in the section and the section is safe as uncracked section. Mix used is M15.

5. Design a short column carry a load of 300,000 N and bending moment of 500,000 N-mm. The mix to be used is M15.

6. A reinforced concrete circular column 400 mm in diameter is symmetrically reinforced with 8 bars of 16 mm diameter. The bars being set on a diameter of 300 mm. The column is subjected to a load of 400,000 N and a bending moment of 20,000,000 N-mm acting about one-axis. Check whether the section is safe if the concrete mix used is M15.

7. A reinforced concrete column 500 mm × 400 mm is reinforced with 4 bars of 20 mm ϕ. The centres of reinforcement being 60 mm from the face. The column is subjected to an axial load of 400,000 N and moments of 10,000,000 N-mm and 8,000,000 N-mm about axes through larger side and smaller side respectively. Check the section if m sed is M15.

8. The section of an arch rib 300 mm × 600 mm is reinforced as shown in Fig. 9·29. The section is subjected to a thrust of 30,000 N and a bending moment of 150,000,000 N-mm. Find the maximum stresses developed in steel and concrete.

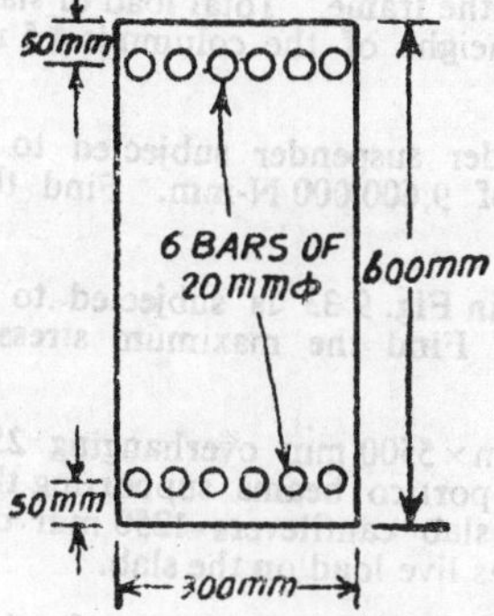

Fig. 9·29

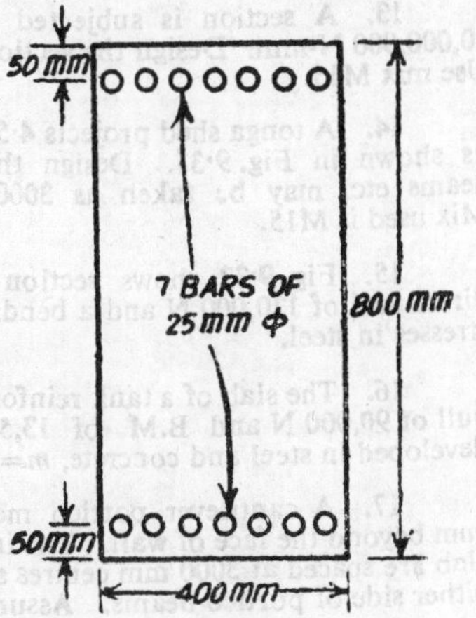

Fig. 9·30

9. The section of an arch rib shown in Fig. 9·30 is subjected to thrust of 5000,000 N and a bending moment of 250,000,000 N-mm. Find the maximum stresses developed in concrete and steel, $m=18$.

10. Design a column to take a load of 300,000 N and B.M. of 50,000,000 N-mm. Provide 2 per cent steel and use mix M15.

11. A column carrying the load of 300,000 N and subjected to the moment of 50,000,000 N-mm is 400 mm × 300 mm in section with 4 bars of 25 mm on shorter faces at 5 cm from the face. Check whether the section is safe if mix used is M15. If the section is not safe redesign the section.

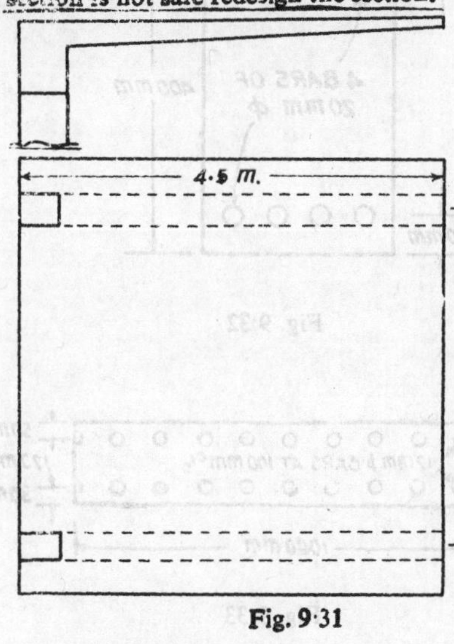

Fig. 9·31

12. Design a column to carry a load of 150,000 N and B.M. of 40,000,000 N-mm. The width of the column is to be restricted to 300 mm. Use mix M15.

13. A section is subjected to a direct load of 40,000 N and B.M. of 50,000,000 N-mm. Design the section and check the stresses by actual formula. Use mix M15.

14. A tonga shed projects 4·5 m. The frames are spaced at 3·5 m centres as shown in Fig. 9·31. Design the column of the frame. Total load of slab, beams etc. may be taken as 3000 N/m. The height of the column is 4·5 m. Mix used is M15.

15. Fig. 9·32 shows section of string girder suspender subjected to a direct pull of 120,000 N and a bending moment of 9,600,000 N-mm. Find the stresses in steel.

16. The slab of a tank reinforced as shown in Fig. 9·33 is subjected to a pull of 90,000 N and B.M. of 13,500,000 N-mm. Find the maximum stresses developed in steel and concrete, $m=18$.

17. A cantilever portico measures 2500 mm × 5500 mm overhanging 250 mm beyond the face of wall. The inverted ribs of portico beams supporting the slab are spaced at 3000 mm centres such that the slab cantilevers 1250 mm on either side of portico beams. Assume 1950 N/m² as live load on the slab.

If the beams are cast monolithic with self-supporting columns of reinforced concrete, design the column section and reinforcement. Assume columns fixed at the base, are 2750 mm above ground level and 1250 mm below the G.L.

(A.M.I.E., 1964)

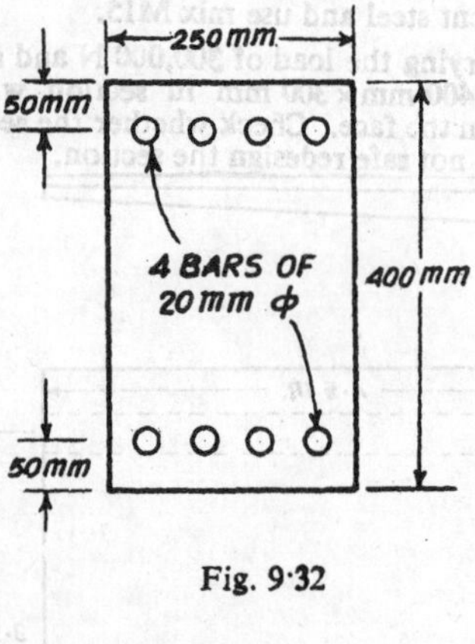

Fig. 9·32

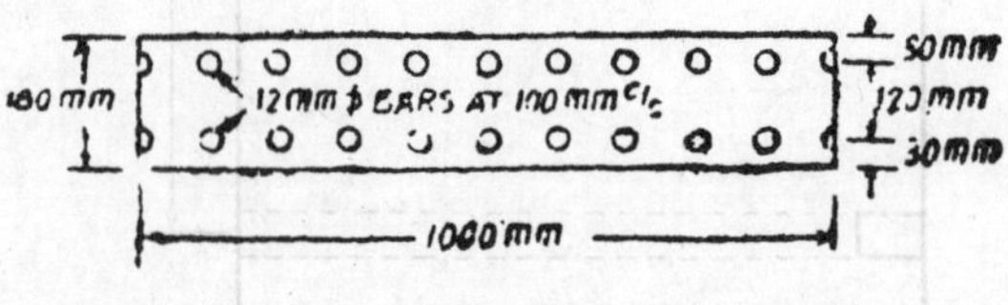

Fig. 9·33

18. The cross-section of an RCC arch rib is rectangular of overall depth 600 mm and width 300 mm. It is reinforced on both faces with equal area of steel 2580 mm^2 each the centre of steel being placed 50 mm from the respective faces The normal thrust on the section is 725,000 N acting with an eccentricity of 200 mm above the horizontal centre line. Determine the maximum stresses in the concrete and steel. Assume modular ratio, $m=15$ and position of N.A. within the section.

19. An RCC column 400 mm $\times$ 400 mm in section is reinforced with 4, 28 mm diameter bars placed at the corners. The bars have a cover of 50 mm to their centres. The column is required to carry an axial load of 350 kN in addition to bending moment about one of the principal axes of its section. If the stresses in steel and concrete are limited to 130 N/mm^2 and 7 N/mm^2 respectively, determine the maximum value of the bending moment. Assume modular ratio $m=15$ and that the neutral axis lies within the section.

(A.M.I.E. 1970)

10

RCC Footings And Foundations

10·1. RCC footings are provided to transmit the load of the structure supported by columns or walls to the soil.

Tho design is based on the assumption that the foundation is rigid so that the variation of pressure under the foundation will be linear. The distribution of pressure will be uniform if the centroid of the footing coincides with the resultant of the applied loads. This conventional method holds good only in case of rigid foundation. Such an assumption becomes less and less accurate as the footing becomes relatively more flexible.

The requirements in design of foundations are

(*i*) The pressure on the soil should not exceed the bearing capacity of the soil.

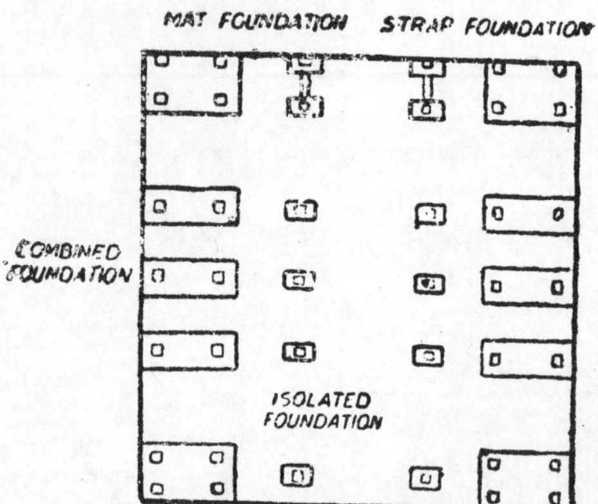

Fig. 10·1 (*a*)

(*ii*) The settlement of the structure should be within the permissible limits. Further there should be no differential settlement. Whenever possibility of differential settlement occurs, the different portions of tbe structure should be made such as to settle independently of each other.

Fig. 10·1 (*b*) and (*c*) shows two ways in which a footing may fail. Fig. 10·1 (*b*) shows how the column could push its way through the footing producing what is known as punching shear. Fig. 10·1 (*c*) shows how footing may fall due to tension crack.

10·2. Types of Foundaations

(*i*) **Continuous footings for walls.** Where the load from the wall is considerable and bearing

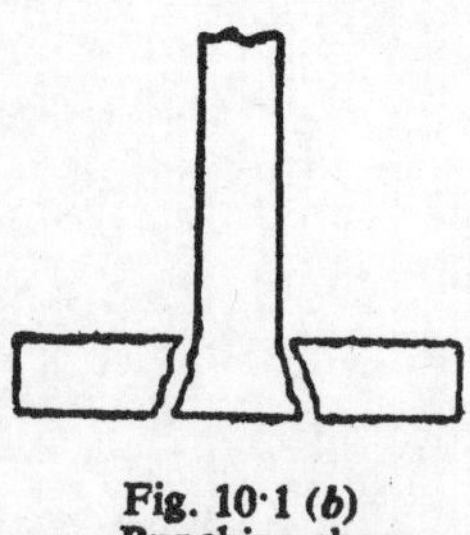
Fig. 10·1 (*b*)
Punching shear

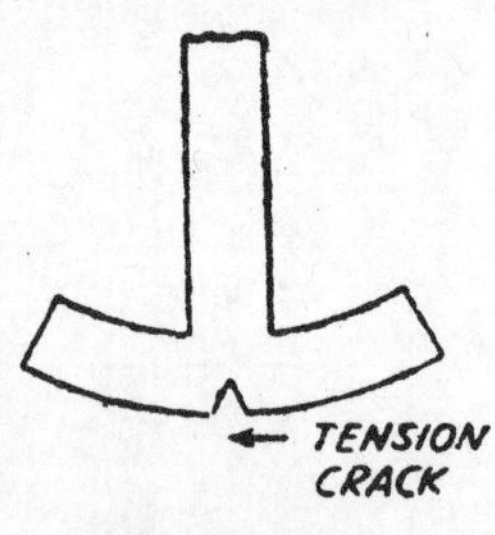

Fig. 10·1 (*c*)
Failure of footing

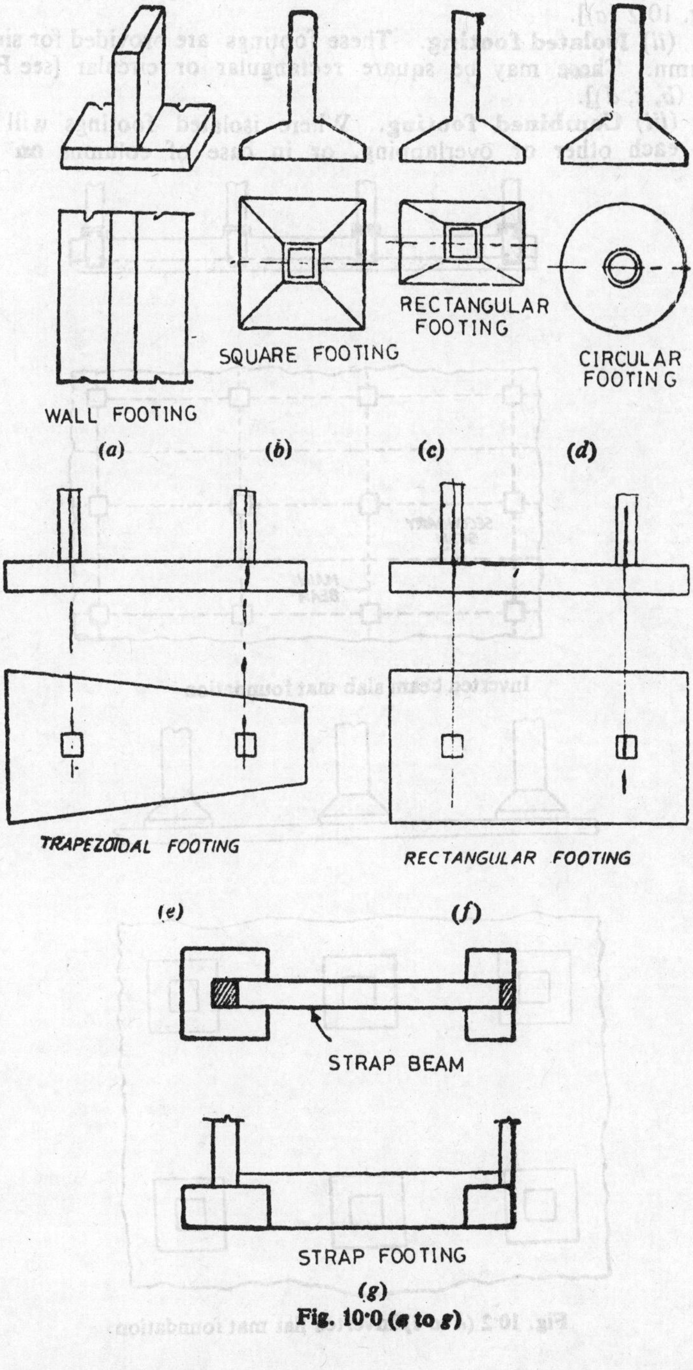
WALL FOOTING
SQUARE FOOTING
RECTANGULAR FOOTING
CIRCULAR FOOTING
(a)
(b)
(c)
(d)
TRAPEZOIDAL FOOTING
RECTANGULAR FOOTING
(e)
(f)
STRAP BEAM
STRAP FOOTING
(g)

Fig. 10·0 (*a* to *g*)

capacity of soil is low, RCC continuous footings are provided [see Fig. 10·2 (*a*)].

(*ii*) **Isolated footing**. These footings are provided for single column. These may be square rectangular or circular [see Figs. 10·2 (*b*, *c*, *d*)].

(*iii*) **Combined footing**. Where isolated footings will be near each other or overlapping, or in case of columns on the

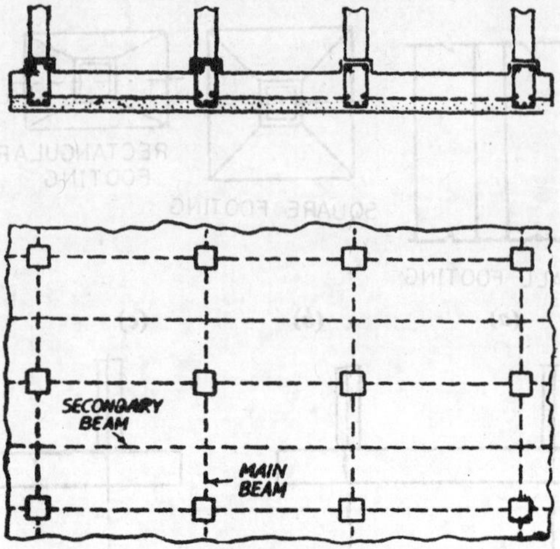

Inverted beam slab mat foundation

Fig. 10·2 (*h* to *i*) Inverted flat mat foundation

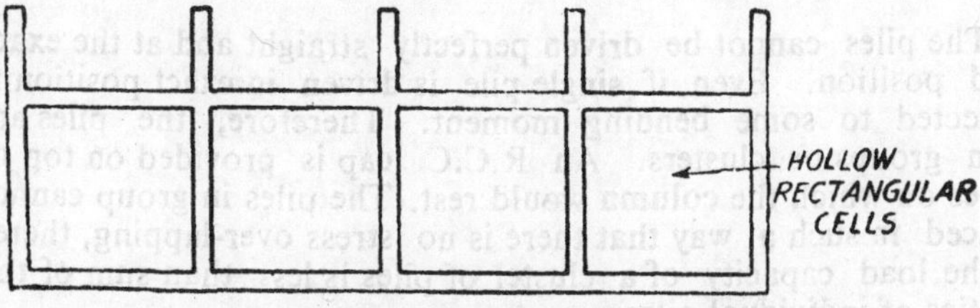

Fig. 10·2 (*j*) Cellular Raft Foundation.

property boundary, combined footings are provided for two or more columns. The footings may be rectangular or trapezoidal, [see Fig. 10·2 (*e*, *f*)].

(*iv*) **Strap footings.** For boundary columns it is necessary that the footing should not extend in the adjoing property. In such cases strap footings are used. The boundary columns and the interior columns are given spread footings which are connected by a strap beam as shown in Fig. 10·2 (*g*).

(*v*) **Mat or raft foundations.** When a number of columns in more than one row are provided with a combined footing, the foundation is called a mat foundation. The foundation may be inverted beam slab construction or flat slab construction. This type of foundation is employed where loads are heavy and bearing capacity of soil is low or where there is danger of differential settlement, [see Figs. 10·2 (*h*, *i*, *j*)].

(*vi*) **Pile foundations.** Piles are used where excessive settlement is to be eliminated and where the load is to be transferred through water or soft soil to a stratum where the bearing capacity is sufficient. Such piles transfer the load by end bearing and are known as bearing piles. The piles used to transfer the load to a depth of weak soil by skin friction along the length of the piles are known as friction piles, [see Fig. 10·2 (*k*)].

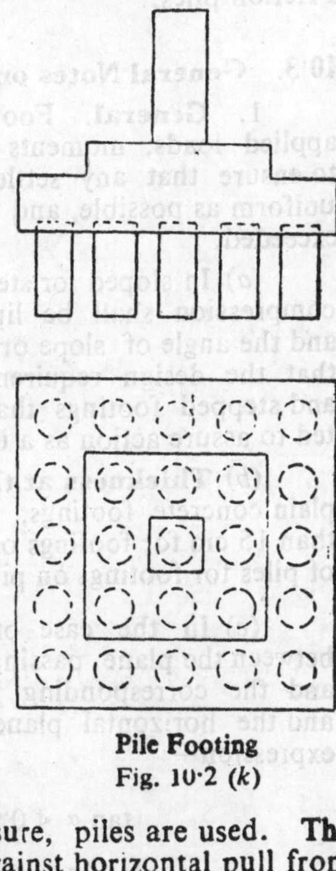

Pile Footing

Fig. 10·2 (*k*)

Piles are used to carry the foundation below the scour depth so that it should not fail even if the erosion to soil takes place in top portion. Where structures are to be anchored down against overturning moments' or hydrostatic pressure, piles are used. The piles are also used to provide anchorage against horizontal pull from sheet-filling walls and for jetties and quay walls to protect front structures against impact of ships or other floating objects.

To resist large horizontal forces batter piles are used.

The piles cannot be driven perfectly straight and at the exact desired position. Even if single pile is driven in exact position it is subjected to some bending moment. Therefore, the piles are used in groups or clusters. An R.C.C. cap is provided on top of the piles on which the column would rest. The piles in group cannot be spaced in such a way that there is no stress over-lapping, therefore, the load capacity of a cluster of piles is less than sum of the capacities of individual piles.

The following pile spacings may be used to keep the cost of the pile cap a minimum :

Type of pile	*Minimum spacing centre-to-centre of piles*
Bearing piles in hard stratum.	2 to 2½ times diameter of pile or minimum 0·8 m
Bearing piles in hard bed rock.	2 times diameter or minimum 0·6 m
Friction piles.	3 to 5 times diameter or minimum 1·0 m.

10·3. General Notes on Design of Foundation (I.S.I.)

1. **General.** Footings shall be designed to sustain the applied loads, moments and forces and the induced reactions and to ensure that any settlement which may occur shall be as nearly uniform as possible, and the safe bearing capacity of the soil is not exceeded.

(*a*) In sloped or stepped footings the effective cross-section in compression shall be limited by the area above the neutral plane, and the angle of slope or depth and location of steps shall be such that the design requirements are satisfied at every section sloped and stepped footings that are designed as a unit shall be constructed to assure action as a unit.

(*b*) **Thickness at the edge of footing.** In reinforced and plain concrete footings, the thickness at the edge shall be not less than 15 cm for footings on soils, nor less than 30 cm above the tops of piles for footings on piles.

(*c*) In the case of plain concrete pedestals, the angle *a* between the plane passing through the bottom edge of the pedestal and the corresponding junction edge of the column with pedestal and the horizontal plane (see Fig. 10·3) will be governed by the expression—

$$\tan \alpha \not< 0{\cdot}9\sqrt{\frac{100 q_0}{f_{ck}}+1}$$

where q_0 = calculated maximum bearing pressure at the base of the pedestal in N/mm^2 and

f_{ck} = characteristic strength of concrete at 28 days in N/mm^2.

2. **Moments and Forces.** (*a*) In the case of footings on piles, computation for moments and shears may be based on the assumption that the reaction from any pile is concentrated at the centre of the pile.

(*b*) For the purpose of computing stresses forces which support a round or octagonal concrete column or pedestal, the face of the column or pedestal shall be taken as the side of a square inscribed within the perimeter of the round or octagonal column or pedestal.

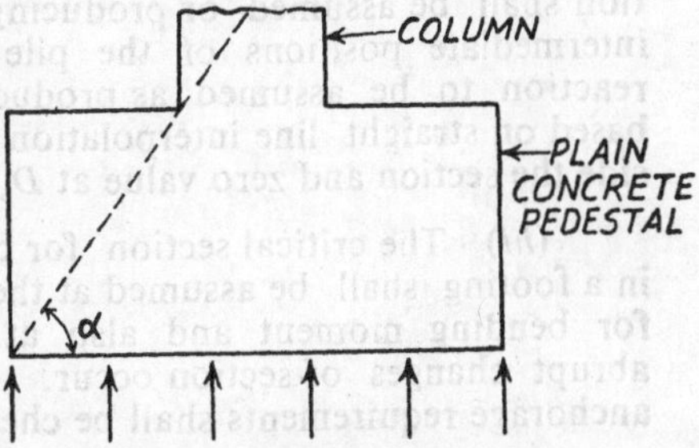

Fig. 10·3

(*c*) **Bending Moment.** (*i*) The bending moment at any section shall be determined by passing through the section a vertical plane which extends completely across the footing, and computing the moment of the forces acting over the entire area of the footing on one side of the said plane.

(*ii*) The greatest bending moment to be used in the design of an isolated concrete footing which supports a column, pedestal or wall, shall be the moment computed in the manner prescribed in (*i*) above at sections located as follows :

(1) At the face of the column, pedestal or wall, for footings supporting a concrete column, pedestal or wall.

(2) Half way between the centre line ond the edge of the wall, for footings under masonry walls.

(3) Half way between the face of the column or pedestal and the edge of the gussetted base, for footings under gussetted bases.

(*d*) **Shear and Bond.** (*i*) The shear strength of footings is governed by the more severe of the following two conditions :

(1) The footing acting essentially as a wide beam, with a potential diagonal crack extending in a plane across the entire width, the critical section for this condition shall be assumed as a vertical section located from the face of the column, pedestal or wall at a distance equal to the effective depth of the footing in case of footings on soil, and a distance equal to half the effective depth of footing for footings on piles.

(2) Two-way action of the footing, with potential diagonal cracking along the surface of truncated cone or pyramid around the concentrated load ; in this case, the footing shall be designed for shear in accordance with appropriate provisions as given in flat slabs.

(*ii*) In computing the external shear on any section of a footing supported on piles, the entire reaction from any pile of diameter D_p whose centre is located $D_p/2$ or more outside the section shall be assumed as producing shear on the section ; the reaction from any pile whose centre is located $D_p/2$ or more inside the section shall be assumed or producing no shear on the section. For intermediate positions of the pile centre, the portion of the pile reaction to be assumed as producing shear on the section shall be based on straight line interpolation between full value at $D_p/2$ outside the section and zero value at $D_h/2$ inside the section.

(*iii*) The critical section for checking the development length in a footing shall be assumed at the same planes as those described for bending moment and also at all other vertical planes where abrupt changes of section occur. If reinforcement is curtailed, the anchorage requirements shall be checked.

3. **Tensile Reinforcement.** The total tensile reinforcement at any section shall provide a moment of resistance at least equal to the bending moment on the section.

(*a*) Total tensile reinforcement shall be distributed across the corresponding resisting section as given below :

(*i*) In one-way reinforced footing, the reinforcement shall be distributed uniformly across the full width of the footing.

(*ii*) In two-way reinforced square footing, the reinforcement extending in each direction shall be distributed uniformly across the full width of the footing.

(*iii*) In two-way reinforced rectangular footing, the reinforcement in the long direction shall be distributed uniformly across the full width of the footing. For the reinforcement in the short direction, a central band equal to the width of the footing shall be marked along the length of the footing and portion of the reinforcement determined in accordance with the equation given below shall be uniformly distributed across the central band.

$$\frac{\text{Reinforcement in central band width}}{\text{Total reinforcement in short direction}}=\frac{2}{\beta+1}$$

where, β is the ratio of the long side to the short side of the footing. The remainder of the reinforcement shall be uniformly distributed in the outer portions of the footing.

4. **Transfer of Load at the Base of Columns.** The compressive stress in concrete at the base of column or pedestal shall be considered as being transferred by bearing to the top of the supporting pedestal or footing. The bearing pressure on the loaded area shall not exceed the permissible bearing stress in direct compression multiplied by a value equal to $\sqrt{A_1/A_2}$ but not greater than 2, where,

A_1=Supporting area for bearing of footing, which in sloped or stepped footing may be taken as the area of the lower base of the largest frustrum of a pyramid or cone contained wholly within the footing and having for its upper base, the area actually loaded and having side slope of one vertical to two horizontal, and

A_2=loaded area at the column base.

For working stress method of design the permissible bearing stress on full area of concrete shall be taken as $0{\cdot}25\ f_{ck}$, for limit state method of design the permissible bearing stress shall be $0{\cdot}45\ f_{ck}$.

(*a*) Where the permissible bearing stress on the concrete in the supporting or supported member would be exceeded, reinforcement shall be provided for developing the excess force, either by extending the longitudinal bars into the supporting member, or by dowels as given in (*c*) below :

(*b*) Where transfer of force is accomplished by reinforcement, the development length of the reinforcement shall be sufficient to transfer the compression or tension to the supporting member.

(*c*) Extended longitudinal reinforcement or dowels of at least 0·5 per cent of the cross-sectional area of the supported column or pedestal and a minimum of four bars shall be provided. Where dowels are used, their diameter shall not exceed the diameter of the column bars by more than 3 mm.

(*d*) Column bars of diameters longer than 36 mm in compression which can be dowelled at the footings with bars of smaller size of the necessary area. The dowel shall extend into a column, a distance equal to the development length of the column bar and into the footing, a distance equal to the development length of the dowel.

10·4. Footing for Two or More Columns

Footing for two or more columns shall be so shaped as to have the line of action of the resultant load collinear with the line of action of the resultant reaction. After completely determining the intensities of pressures under the footing at different points, the combined footing shall be designed as an inverted slab with or without beams supporting the intensities of soil pressures resulting from the applied direct forces and bending moments.

10·5. Severe Transverse Forces

Footing subjected to severe transverse forces shall be checked against sliding, the resistance to sliding being provided through friction or otherwise.

10·6. Continuous Footings

1. RCC Walls

Let W N/m be the load from the RCC wall and W_1 Newtons be the weight of the footing per metre length. If p_0 is the bearing capacity of the soil in N/m² breadth of footing $B=\dfrac{W+W_1}{p_0}$.

Net upward pressure on footing p

$$=\frac{W}{B\times 1}\ \text{N/m}^2$$

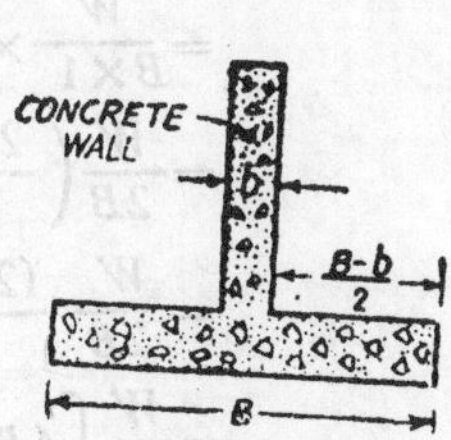

Fig. 10·4

Let b metres be the width of the wall.

Maximum B.M. at face of wall per metre length

$$=p\times 1\times\left(\frac{B-b}{2}\right)\times\left(\frac{B-b}{4}\right).$$

$$\therefore \quad M=\frac{p\times(B-b)^2}{8}\times 1100 \text{ N-mm.}$$

$$M=0{\cdot}87\times 1000\ d^2$$

$$\frac{p\times(B-b)^2}{8}\times 1000=0{\cdot}87\times 1000\times d^2$$

$$d=\sqrt{\frac{p(B-b)^2}{8\times 0{\cdot}87}} \text{ mm.}$$

From this effective depth d is calculated and overall depth fixed.

Area of steel $\quad A_t=\dfrac{M}{0{\cdot}87\times d\times 140}$ mm².

2. Masonry Walls

Let W N/m be the load from the wall and W_1 N/m be the weight of the footing. If p_0 is the bearing capacity of the soil in N/m², breadth of footing $B=\dfrac{W+W_1}{p_0}$.

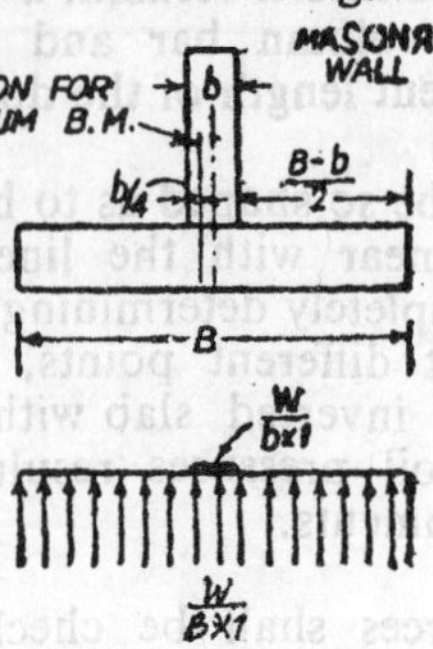

Fig. 10·5

Net upward pressure on footing

$$p=\frac{W}{B\times 1} \text{ N/m}^2$$

For masonry wall footings, the B.M. is to be calculated half-way between the middle and edge of the wall.

Maximum B.M. per metre width

$$=\frac{W}{B\times 1}\times\frac{\left(\dfrac{B-b}{2}+\dfrac{b}{4}\right)^2}{2}-\frac{W}{b\times 1}\times\frac{b}{4}\times\frac{b}{8}$$

$$=\frac{W}{2B}\left(\frac{2B-2b+b}{4}\right)^2-\frac{Wb}{32}$$

$$=\frac{W}{2B}\ \frac{(2B-b)^2}{16}-\frac{Wb}{32}$$

$$=\frac{W}{32}\left[4B-4b+\frac{b^2}{B}-b\right]$$

$$=\frac{W}{32}\left[4B-5b+\frac{b^2}{B}\right]$$

$$=\frac{W}{32B}[4B^2-5Bb+b^2]$$

$$=\frac{W}{32B}(4B-b)(B-b)$$

$$=\frac{p}{8}\left(B-\frac{b}{4}\right)\left(B-b\right)$$

$$\therefore \quad M=\frac{p}{8}\left(B-\frac{b}{4}\right)\left(B-b\right)\times 1000 \text{ N-mm.}$$

Effective depth d

$$=\sqrt{\frac{M}{0{\cdot}87\times 1000}} \text{ mm}$$

$$A_t=\frac{M}{0{\cdot}87d\times 140} \text{ mm}^2.$$

Section should be checked for diagonal shear and bond.

Ex. 10·1. *Design a footing for an RCC wall 200 mm thick carrying a load of 600,000 N/m length. The bearing capacity of the soil is 200 kN/m².*

Sol. Assume self-weight of footings as 10% of load.

Self-weight of footing $=60{,}000$ N

Total load $=660{,}000$ N

$$\text{Breadth of footing}=\frac{660{,}000}{200{,}000}=3{\cdot}3 \text{ m}$$

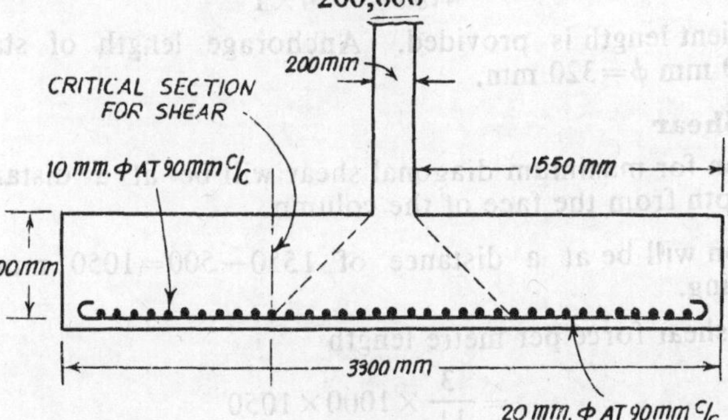

Fig. 10·6

Net upward pressure on footing

$$=\frac{600{,}000}{3300\times 1000}=\frac{2}{11} \text{ N/mm}^2$$

Maximum B.M. per metre width

$$=\frac{2}{11}\times 1000\times\frac{(1550)^2}{2}$$

$$=218{,}400{,}000 \text{ N-mm.}$$

$$d=\sqrt{\frac{M}{Qb}}=\sqrt{\frac{218,400,000}{0{\cdot}87\times1000}}=500 \text{ mm.}$$

Provide overall depth of 560 mm and effective depth of 500 mm

$$A_t=\frac{218,400,000}{0{\cdot}87\times500\times140}=3585 \text{ mm}^2$$

Using 20 mm ϕ bars,

$$\text{Spacing}=\frac{1000\times314}{3585}=87{\cdot}7 \text{ mm}$$

Provide 90 mm ϕ bars at 80 mm centres.

Longitudinal Reinforcement

Longitudinal reinforcement $=0{\cdot}15\%$ of sectional area

$$=\frac{0{\cdot}15}{100}\times1000\times560=840 \text{ mm}^2$$

Using 10 mm ϕ bars,

$$\text{Spacing}=\frac{1000\times79}{840}=94 \text{ mm}$$

Provide 10 mm ϕ bars at 90 mm centres.

Check for bond

$$\text{Development length}=\frac{\phi\sigma_s}{4\tau bd}=\frac{20\times140}{4\times1}=700 \text{ mm}$$

Sufficient length is provided. Anchorage length of standard hook for 20 mm $\phi=320$ mm.

Check of Shear

Section for maximum diagonal shear will be at a distance of effective depth from the face of the column.

Section will be at a distance of $1550-500=1050$ mm from end of footing.

Max. shear force per metre length

$$=\frac{3}{11}\times1000\times1050$$

$$=190,000 \text{ N}$$

$$\text{Shear stress}=\frac{190,000}{500\times1000}=0{\cdot}382 \text{ N/mm}^2$$

$$\text{Percentage of steel}=\frac{3930\times100}{500\times1000}=0{\cdot}79$$

$$\tau=0{\cdot}35 \text{ N/mm}^2 \text{ (from Table 4·1)}$$

The section need not be revised.

Ex. 10·2. *Design an RCC footing for a masonry wall 300 mm thick carrying load of 150,000 N per metre length. The bearing capacity of soil is 100 kN/m².*

Sol. Assume self-weight of footing as 10% of super-imposed load.

∴ Weight of footing=15,000 N/m

Total load=150,000+15,000=165,000 N/m

$$\text{Breadth of footing}=\frac{165,000}{100,000}=1\cdot65 \text{ m}$$

Provide breadth of 1·7 m.

$$\text{Net upward pressure}=\frac{150,000}{1700\times1000}=\frac{15}{170}\ \text{N/mm}^2$$

Maximum B.M. per metre width

$$=\frac{15}{170}\times1000\times\frac{(775)^2}{2}-\frac{150,000}{300}\times\frac{(75)^2}{2}$$

$$=25,090,000 \text{ N-mm.}$$

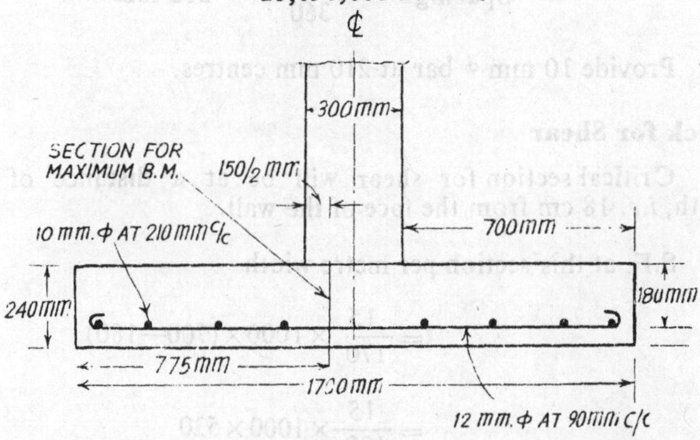

Fig. 10·7

Effective depth of footing

$$=\sqrt{\frac{25,090,00}{0\cdot87\times1000}}=169\cdot8 \text{ mm}$$

Provide overall depth of 240 mm with effective depth of 180 mm.

$$A_t=\frac{25,090,000}{0\cdot87\times180\times140}=1144 \text{ mm}^2$$

Using 12 mm, ϕ bars,

$$\text{Spacing}=\frac{1000\times 113}{1144}=98 \text{ mm.}$$

Provide 12 mm ϕ bars at 90 mm centres.

Check for Bond

Sufficient anchorage length is provided.

Longitudinal Reinforcement

Minimum reinforcement in longitudinal direction

$$=0{\cdot}15\% \text{ of cross-sectional area}$$

$$=\frac{0{\cdot}15}{100}\times 1000\times 240=360 \text{ mm}^2$$

Providing 10 mm ϕ bars,

$$\text{Spacing}=\frac{1000\times 78{\cdot}5}{360}=218 \text{ mm}$$

Provide 10 mm ϕ bar at 210 mm centres.

Check for Shear

Critical section for shear will be at a distance of effective depth, *i.e.* 18 cm from the face of the wall.

S.F. at this section per metre width

$$=\frac{15}{170}\times 1000\times (700-180)$$

$$=\frac{15}{170}\times 1000\times 520$$

$$=45{,}900 \text{ N}$$

$$\text{Shear stress}=\frac{F}{b\times d}=\frac{45{,}900}{1000\times 180}$$

$$=0{\cdot}293 \text{ N/mm}^2. \quad \text{Safe.}$$

10·7 . Isolated Footings

Let W be the load coming from the column and W_1 be the weight of the footing.

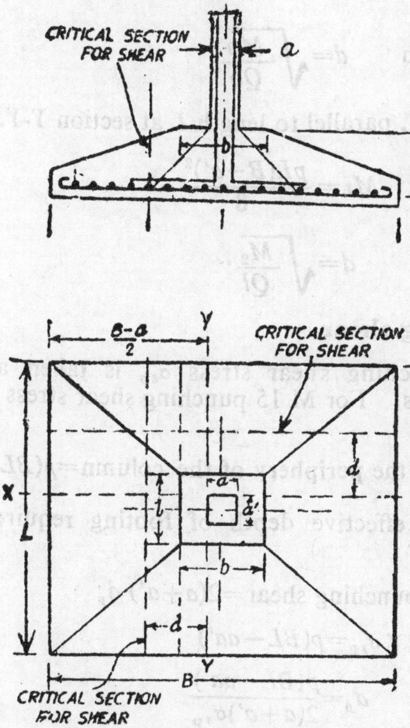

Fig. 10·8

If p_0 is the bearing capacity of soil, area of footing required

$$=\frac{W+W_1}{p_0}$$

Net upward pressure on footing due to load W will be

$$p=\frac{W}{\text{Area of footing}}$$

(i) Rectangular Footings

Let the size of the column be $a\times a'$.

Let size of the footing at top be $l\times b$ and at bottom be $L\times B$.

$$L\times B=\frac{W+W_1}{p_0}$$

Net upward pressure on footing

$$p=\frac{W}{L\times B}$$

Maximum B.M. parallel to breadth B at section X-X.

$$M_B=p\,\frac{(L-a')}{2}\times B\times\frac{(L-a')}{4}$$

$$=\frac{pB(L-a')^2}{8}.$$

Effective depth $d = \sqrt{\dfrac{M_B}{Qb}}$

Maximum B.M. parallel to length L at section Y-Y.

$$M_L = \frac{pL(B-a')^2}{8}$$

Effective depth $d = \sqrt{\dfrac{M_2}{Ql}}$.

Depth for punching shear

Allowable punching shear stress σ_{sp} is taken as twice the allowable shear stress. For M 15 punching shear stress is taken as 1 N/mm^2.

Total shear on the periphery of the column $= p(BL - aa')$.

Let d_p be the effective depth of footing required to resist punching.

Area resisting punching shear $= 2(a+a')\, d_p$

$\therefore \quad 2(a+a')d_p \times \sigma_{sp} = p(BL - aa')$

$$\therefore \quad d_p = \frac{p(BL - aa')}{2(a+a')\sigma_{sp}}$$

The maximum of effective depths as given for B.Ms. and punching is adopted.

The area of reinforcement in each direction is calculated.

The critical section for shear is taken at the distance equal to effective depth of footing from face of column and shear stress is checked at this section Critical section for bond is at face of column and bond stress is checked at this section.

***(ii)* Square Footing**

For square footing $a = a'$, $l = b$ and $L = B$

Maximum B.M. in two direction will be same, *i.e.*

$$M_L = M_B = \frac{pB(B-a)^2}{8}$$

Effective depth $d = \sqrt{\dfrac{M}{Qb}}$

Effective depth required for punching shear will be

$$d_p = \frac{p(B^2 - a^2)}{4a\, \sigma_{sp}}$$

Greater of the effective depths as given by B.M. and punching is adopted.

$$A_t = \frac{M}{jd \times 1400}$$

The area of reinforcement in each direction will be same. The critical section for shear will be at a distance of effective depth from the face of the column and shear stress is checked at this section. Critical section for bond is at the face of the column. Bond stress is checked at this section.

(*iii*) **Circular Footing**

Circular columns are provided with square footing or circular footing. The design of square footing for a circular column is similar to design of square footing for a square column.

Let R_1 be the radius of column and W be the load from the column. Let the weight of the footing be W_1 and R_2 be radius of footing.

Total load $= W + W_1$

If p_0 is the bearing capacity of soil, area of footing required

$$= \frac{W+W_1}{p_0} = \pi \times R_2^2$$

$$\therefore \quad R_2 = \sqrt{\frac{W+W_1}{\pi p_0}}$$

Net upward pressure on footing $p = \dfrac{W}{\pi R_2^2}$.

The B.M. in the footing is calculated at the edge of the column for one quadrant.

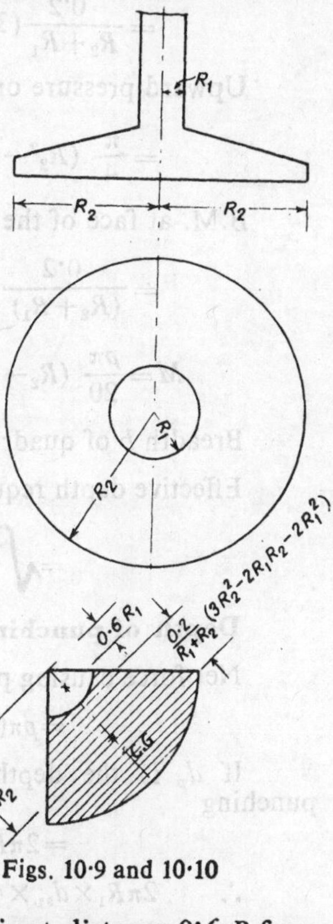

Figs. 10·9 and 10·10

The distance of C.G. of quadrant is at distance 0·6 R from the centre. Distance of the C.G. for shaded area from the centre

$$\bar{x} = \frac{\frac{\pi}{4} R_2^2 \times 0{\cdot}6 R_2 - \frac{\pi}{4} R_1^2 \times 0{\cdot}6 R_1}{\frac{\pi}{4} R_2^2 - \frac{\pi}{4} R_1^2}$$

$$=\frac{0{\cdot}6(R_2^3-R_1^3)}{R_2^2-R_1^2}=\frac{0{\cdot}6(R_2^2+R_1R_2+R_1^2)}{R_2+R_1}$$

Distance of C.G. from the face of the column

$$=0{\cdot}6\frac{(R_2^2+R_1R_2+R_1^2)}{R_2+R_1}-R_1$$

$$=\frac{0{\cdot}6(R_2^2+R_1R_2+R_1^2)-R_1R_2-R_1^2}{(R_2+R_1)}$$

$$=\frac{0{\cdot}6R_2^2-0{\cdot}4R_1R_2-0{\cdot}4R_1^2}{(R_2+R_1)}$$

$$=\frac{0{\cdot}2}{R_2+R_1}(3R_2^2-2R_1R_2-2R_1^2)$$

Upward pressure on shaded area

$$=\frac{\pi}{4}(R_2^2-R_1^2)\times p$$

B.M. at face of the column

$$=\frac{0{\cdot}2}{(R_2+R_1)}(3R_2^2-2R_1R_2-2R_1^2)\times\frac{\pi}{4}(R_2^2-R_1^2)\times p$$

$$M=\frac{p\pi}{20}(R_2-R_1)(3R_2^2-2R_1R_2-2R_1^2)$$

Breadth b of quadrant $=\pi R_1/2$

Effective depth required for B.M.

$$=\sqrt{\frac{M}{Q\pi R_1/2}}$$

Depth of punching shear

Net force causing punching

$$=p\pi(R_2^2-R_1^2)$$

If d_p is the depth required for punching, area resisting punching

$$=2\pi R_1\times d_p$$

$$\therefore\quad 2\pi R_1\times d_{sp}\times\sigma_{sp}=p\pi(R_2^2-R_1^2)$$

$$\therefore\quad d_p=\frac{p(R_2^2-R_1^2)}{2R_1\,\sigma_{sp}}$$

Greater of the depths required for bending and punching is adopted.

Area of steel for one quadrant $=M/j\sigma_s$

Same reinforcement is provided in two directions at right angles. The reinforcement is placed in the inscribed square.

The section is checked for bond stress at the face of the column. The critical section for shear will be at a distance equal to effective depth from the face of the column. The section is checked for shear stress at critical section.

Ex. 10·3. *Design a footing for a square column 400 mm × 400 mm carrying load of 1,000,000 N. The bearing capacity of the soil is 200 kN/m². Take c=5 N/mm², t=140 N/mm², m=18.*

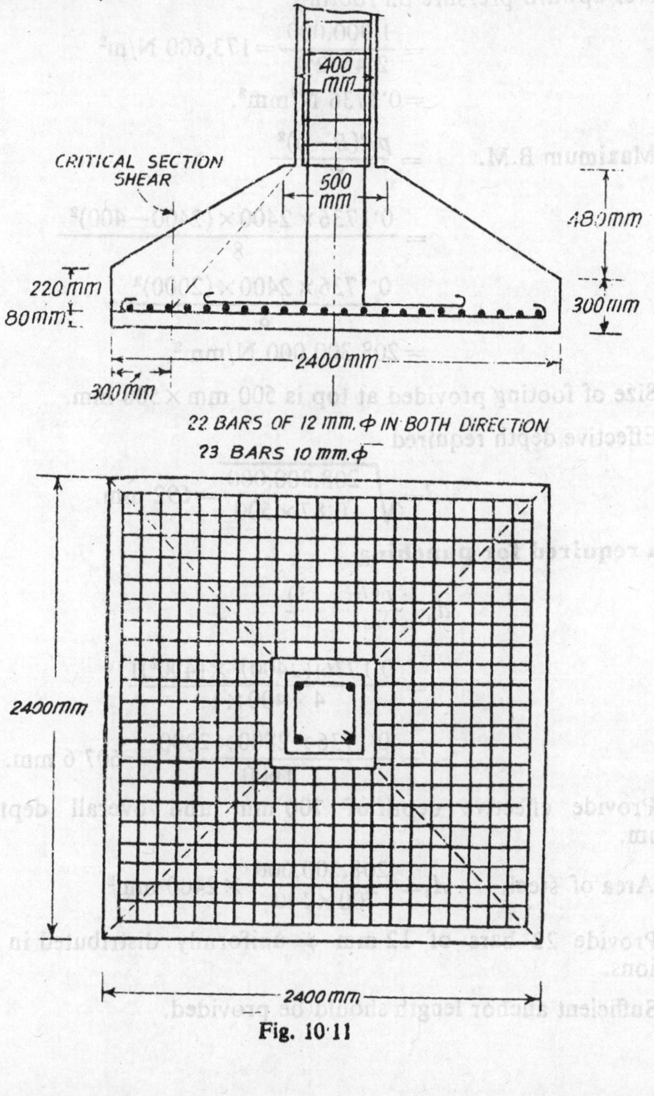

Fig. 10·11

Sol. Assume self-weight of footing as 10% of column load

Weight of footing $=1{,}000{,}000$ N

Total load $=1{,}000{,}000+100{,}000=1{,}100{,}000$ N.

Area of footing required

$$=\frac{1{,}100{,}000}{200\times 1000}=5\cdot 5\ \text{m}^2.$$

Provide 2·4 m × 2·4 m square footing.

Net upward pressure on footing

$$=\frac{1{,}000{,}000}{2\cdot 4\times 2\cdot 4}=173{,}600\ \text{N/m}^2$$

$$=0\cdot 1736\ \text{N/mm}^2.$$

$$\text{Maximum B.M.}=\frac{pB(B-a)^2}{8}$$

$$=\frac{0\cdot 1736\times 2400\times(2400-400)^2}{8}$$

$$=\frac{0\cdot 1736\times 2400\times(2000)^2}{8}$$

$$=208{,}300{,}000\ \text{N/mm}^2.$$

Size of footing provided at top is 500 mm × 500 mm.

Effective depth required

$$d=\sqrt{\frac{208{,}300{,}000}{0\cdot 87\times 500}}=692\ \text{mm}.$$

Depth required for punching

$$d_p=\frac{p(B^2-a^2)}{4a}$$

$$=\frac{0\cdot 1736[(2400)^2\times(400^2)]}{4\times 400\times 1}$$

$$=\frac{0\cdot 1736\times 2800\times 2000}{1600}=607\cdot 6\ \text{mm}.$$

Provide effective depth of 700 mm and overall depth of 780 mm.

$$\text{Area of steel,}\quad A_t=\frac{208{,}300{,}000}{700\times 140}\times 2400\ \text{mm}^2.$$

Provide 22 bars of 12 mm ϕ uniformly distributed in both directions.

Sufficient anchor length should be provided.

Check for Shear

Critical section for shear will be at a distance of effective depth, *i.e.* 700 mm from the face of the column.

Maximum shear at 700 mm from face of the column

$$=0{\cdot}1736\times300\times2400=125{,}000 \text{ N.}$$

Effective depth at the section

$$=\frac{300}{950}(780-300)+220$$

$$=141+220=361 \text{ mm}$$

Breadth of footing at top of this section

$$=500+2\times650=1800 \text{ mm}$$

Shear stress $=\dfrac{125{,}000}{361\times1800}=0{\cdot}192 \text{ N/mm}^2$. Safe.

Ex, 10·4. *Design a footing for a rectangular column 500 mm×300 mm to carry a load of 800,000 N. The bearing capacity of soil is 200 kN/m². Use mix. M 15.*

Sol. Assume self-weight of footing as 10% of load coming on footing.

Weight of footing $=80{,}000$ N

Total load $=800{,}000+80{,}000=880{,}000$ N

Area of footing required

$$=\frac{880{,}000}{200{,}000}=4{\cdot}4 \text{ m}^2$$

Adopt 2·2 m×2 m footing.

Net upward pressure on footing

$$=\frac{800{,}000}{2200\times2000}=0{\cdot}182 \text{ N/mm}^2$$

B.M. along Y-Y

$$M_y=\frac{0{\cdot}182\times850\times850}{2}\times2000$$

$$=131{,}400{,}000 \text{ N-mm.}$$

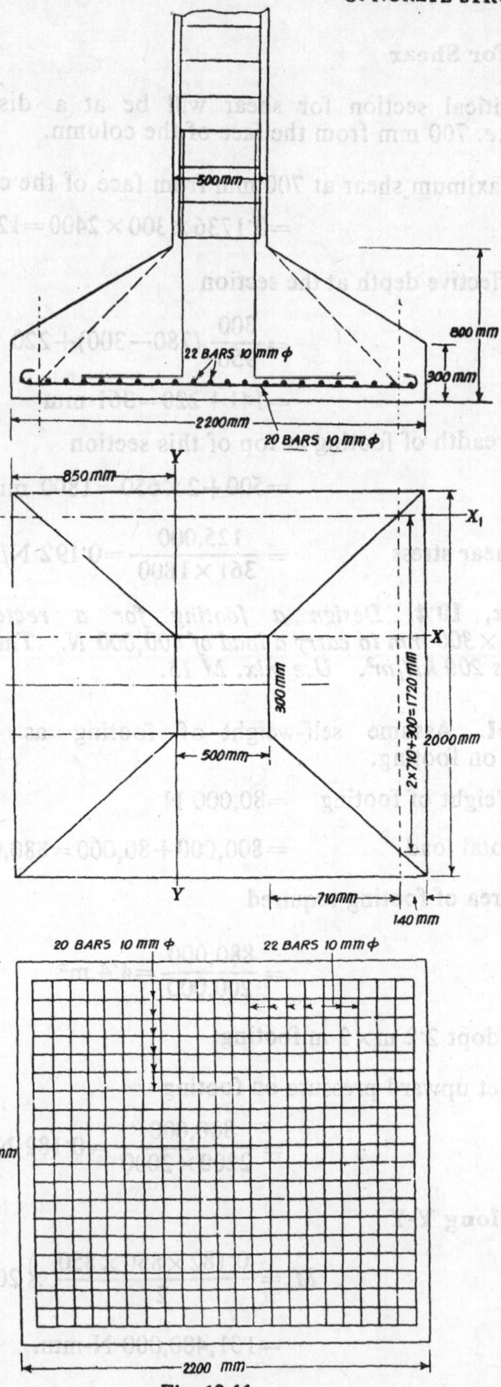
500mm
22 BARS 10 mm φ
800 mm
300 mm
2200mm
20 BARS 10 mm φ
Y
850mm
X_1
X_1
X
X
300mm
2x710+300=1720 mm
2000mm
500mm
Y
710mm
140 mm
20 BARS 10 mm φ
22 BARS 10 mm φ
2000mm
2200 mm

Fig. 10·11

Effective depth of footing required

$$d=\sqrt{\frac{M}{Qb}}=\sqrt{\frac{131,400,000}{0{\cdot}87\times330}}=709{\cdot}8 \text{ mm.}$$

B.M. along X-X

$$M_x=\frac{0{\cdot}182\times850\times850}{2}\times2200$$

$$=144,540,000 \text{ N-mm.}$$

Effective depth of footing required

$$=\sqrt{\frac{144,540,000}{0{\cdot}87\times500}}=576 \text{ mm.}$$

Depth required for punching d_p

$$=\frac{0{\cdot}182(2200\times2000-300\times500)}{1\times2(300+500)}$$

$$=483 \text{ mm.}$$

Adopt overall depth of 800 mm and clear cover of 75 mm.

Reinforcement required parallel to long direction for M_y

$$=\frac{131,400,000}{140\times0{\cdot}87\times710}$$

$$=1520 \text{ mm}^2$$

Provide 20 bars of 10 mm ϕ uniformly spaced.

Reinforcement required parallel to short direction for M_x

$$=\frac{144,540,000}{140\times0{\cdot}87\times720}$$

$$=1648 \text{ mm}^2$$

Reinforcement required in central band of 2000 mm

$$=\frac{2}{\frac{2200}{2000}+1}\times1648=\frac{2}{2{\cdot}1}\times1648$$

$$=1570 \text{ mm}^2.$$

Provide 20 bars of 10 mm ϕ in the central band and one bar of 10 mm ϕ in each of the end bands.

Check for Bond

Sufficient anchor length should be provided.

Check for Shear

Critical section for shear will be as a distance of effective depth from the face of the column.

Shear force $= 0{\cdot}182 \times 140 \times 2000 = 50{,}960$ N

Breadth of footing at this section

$$= 2 \times 710 + 300 = 1420 + 300 = 1720 \text{ mm.}$$

Depth of footing at this section

$$= 300 + \frac{140}{850} \times 500$$

$$= 300 + 82 = 382 \text{ mm.}$$

Effective depth $= 382 - 90 = 292$ mm

Shear stress $= \dfrac{50{,}960}{1720 \times 292} = 0{\cdot}101$ N/mm². Safe.

Ex. 10·5. *A circular column 500 mm diameter carries a load of 1,000,000 N. Design a circular footing for the column. Bearing capacity of soil is 200 kN/m. Use concrete mix M 15.*

Sol. Assume self-weight of footing as 10% of load on it.

Weight of footing $= 100{,}000$ N

Total load of soil $= 1{,}100{,}000$ N

Area of footing required

$$= \frac{1{,}100{,}000}{200{,}000} = 5{\cdot}5 \text{ m}^2.$$

$$\text{Diameter of footing} = \sqrt{\frac{5{\cdot}5}{\pi/4}} = 2{\cdot}64 \text{ m.}$$

Provide 2·8 m diameter footing.

Net upward pressure on footing

$$= \frac{1{,}000{,}000}{\frac{\pi}{4} \times (2800)^2} = 0{\cdot}162 \text{ N/mm}^2.$$

Maximum B.M. at the face of the column for one quadrant

$$= \frac{\pi p}{20}(R_2 - R_1)(3R_2^2 - 2R_1R_2 - 2R_1^2)$$

$$=\frac{\pi}{20}\times 0{\cdot}162(1400-250)[3\times(1400)^2-2\times 1400 \times 250-2\times(250)^2]$$

$$=\frac{\pi}{20}\times 0{\cdot}162\times 1150(5{,}880{,}000-700{,}000-125{,}000)$$

$$=\frac{\pi}{20}\times 0{\cdot}162\times 1150\times 5{,}055{,}000$$

$$=147{,}900{,}000 \text{ N-mm.}$$

Width of footing

$$=\frac{\pi}{2}\ R_1=\frac{\pi}{2}\times 250=125\pi$$

Effective depth of footing required

$$=\sqrt{\frac{147{,}900{,}000}{0{\cdot}87\times 125\pi}}=658 \text{ mm.}$$

Depth required for punching

Shear Force $=p\pi(R_2^2-R_1^2)$

$=0{\cdot}162\pi[(1400)^2-(250)^2]$

$=0{\cdot}162\pi\times 1650\times 1150$

If d_p is depth required for punching, area resisting punching

$$=2\pi\times 250\times d_p$$

$$\therefore\quad 2\pi\times 250\times d_p\times 1=0{\cdot}162\times\pi\times 1650\times 1150$$

$$d_p=\frac{0{\cdot}162\pi\times 1650\times 1150}{2\pi\times 1\times 250}=614 \text{ mm.}$$

Provide overall depth of 750 mm with effective depth of 670 mm.

$$A_t=\frac{147{,}900{,}000}{0{\cdot}87\times 670\times 140}=1810 \text{ mm}^2.$$

Provide 23 bars of 10 mm diameter in two directions at right angles to each other. These bars are placed in the inscribed square as shown in Fig. 10·12. Two bars of 10 mm ϕ are provided beyond this square on each side.

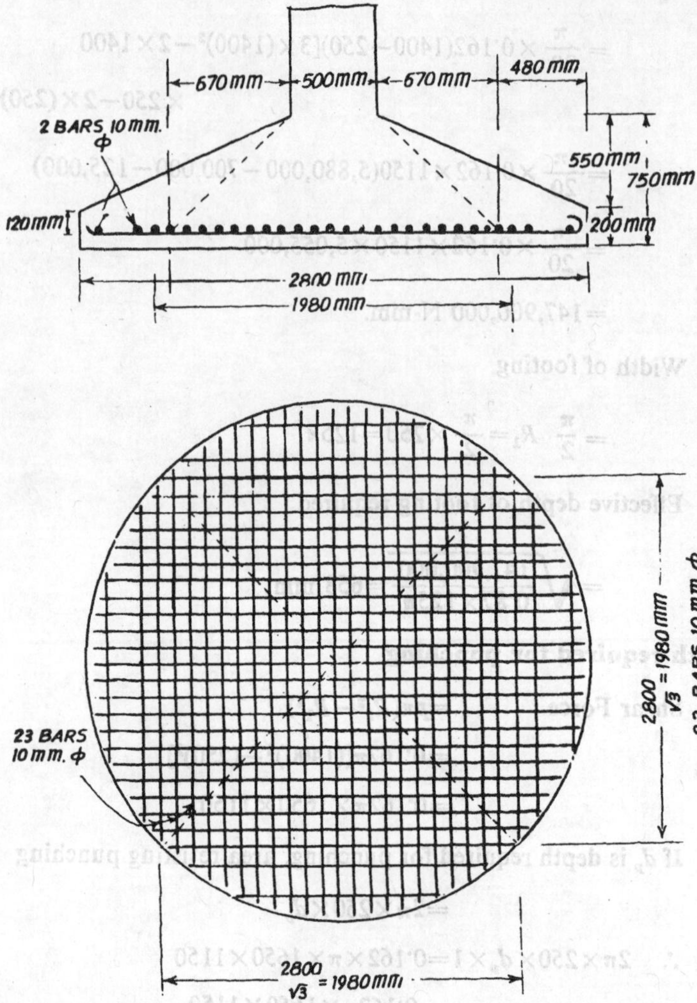

Fig. 10·12

Check for Bond

Provide sufficient anchor length.

Check for Shear

Critical section will be at a distance of effective depth, 670 mm from the face of the column.

Shear force $= p\pi[(1400)^2-(920)^2]$

$= 0{\cdot}162\pi \times 2320 \times 480 = 566{,}700$ N.

Effective depth at this section

$$=200+\frac{550}{1150}480-80$$

$$=350 \text{ mm.}$$

Shear stress $=\dfrac{566,700}{350\pi\times 1840}$

$$=0{\cdot}322 \text{ N/mm}^2. \quad \text{Safe.}$$

Ex. 10·6. *Design a square footing for the circular column of Problem 10·5.*

Sol. Area of footing required

$$=5{\cdot}5 \text{ m}^2$$

Adopt 2·4 m×2·4 m footing

Side of the inscribed square in the column

$$=\frac{500}{\sqrt{2}}=354 \text{ mm.}$$

Net upward pressure on footing

$$p=\frac{1,000,000}{2400\times 2400}=0{\cdot}1735 \text{ N/mm}^2.$$

Maximum B.M. at the side of inscribed square

$$=\frac{pB(B-a)^2}{8}$$

$$=\frac{0{\cdot}1735\times 2400(2400-354)^2}{8}$$

$$=0{\cdot}1735\times 300\times(2046)^2$$

$$=217,900,000 \text{ N-mm.}$$

Breadth of footings provided at top is 600 mm.

Effective depth required for B.M.

$$d=\sqrt{\frac{217,900,000}{0{\cdot}87\times 600}}=646 \text{ mm.}$$

Provide overall depth of 720 mm with effective depth of 650 mm.

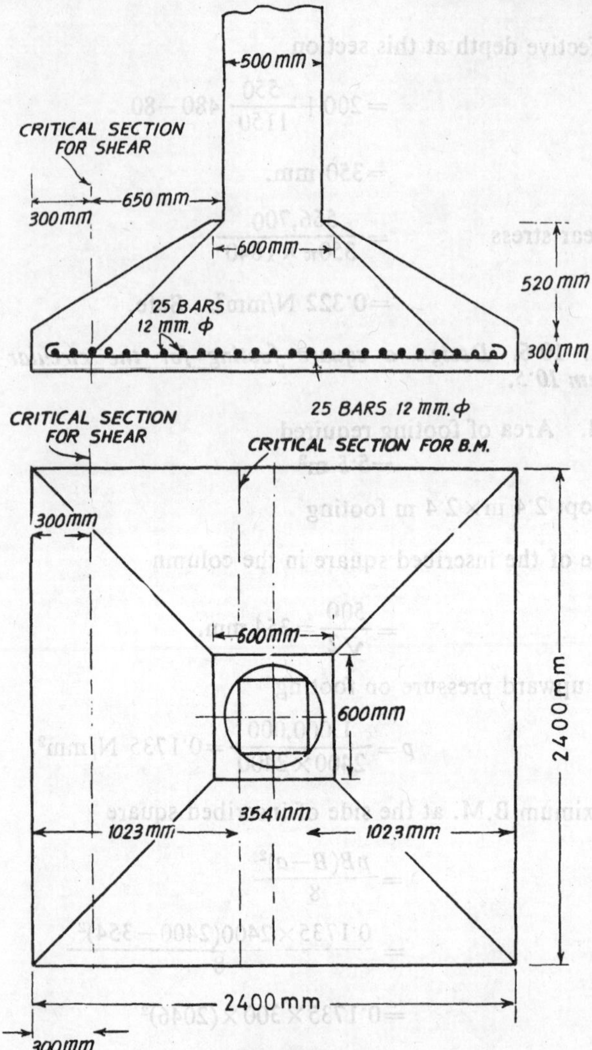

Fig. 10·13

Depth required for punching

$$\text{Punching shear} = p\left[2400\times2400-\frac{\pi}{4}\times(500)^2\right]$$

$$=0{\cdot}1735(5{,}760{,}000-196{,}400)$$

$$=965{,}400 \text{ N.}$$

If d_p is depth required for punching, area resisting punching

$$=\pi\times500\times d_p$$

$$\therefore \quad \pi\times500\times d_p\times1=965{,}400$$

$$d_p = \frac{965{,}400}{500\pi} = 615 \text{ mm.}$$

The depth provided is adequate

$$A_t = \frac{217{,}900{,}000}{0 \cdot 87 \times 650 \times 140} = 2750 \text{ mm}^2$$

Provide 25 bars of 12 mm ϕ in both directions.

Check for Bond

Provide sufficient anchor length including standard hooks.

Check for Shear

Critical section for shear will be at a distance of effective depth from face of column, *i.e.* 650 mm from face of column.

Shear force $= p \times 300 \times 2400$

$= 0 \cdot 1735 \times 300 \times 2400$

$= 124{,}900$ N.

Effective depth at this section

$$= \frac{300}{900} \times 520 + 200 - 80$$

$= 173 \cdot 3 + 120 = 293 \cdot 3$ mm.

Breadth of footing at top of this section

$= 2 \times 600 + 600 = 1800$ mm

Shear stress $= \dfrac{124{,}900}{293 \cdot 3 \times 1800} = 0 \cdot 237 \text{ N/mm}^2$. Safe.

Ex. 10·7. *Design a footing for a 400 mm × 400 mm column carrying load of 500,000 N inclusive of self-weight, and subjected to B.M of 25,000,000 N-mm. Bearing capacity of soil is 200 kN/m², use concrete mix M 15.*

Sol. Assume self-weight of footing 10% of load coming on it.

Weight $= 50{,}000$ N

Total loads on footing $= 550{,}000$ N

Area of footing required

$$= \frac{550{,}000}{200{,}000} = 2 \cdot 75 \text{ m}^2.$$

Provide 1700 mm × 1700 mm footing

Eccentricity of load $= \dfrac{25{,}000{,}000}{500{,}000} = 50$ mm.

The footing will be provided symmetrical about x-axis. About y-axis it will be so provided so that C.G. of the load coincides with C.G. of footing so that the pressure distribution at base is uniform. The footing will project 700 mm from one face of column on one side and 600 mm on other side.

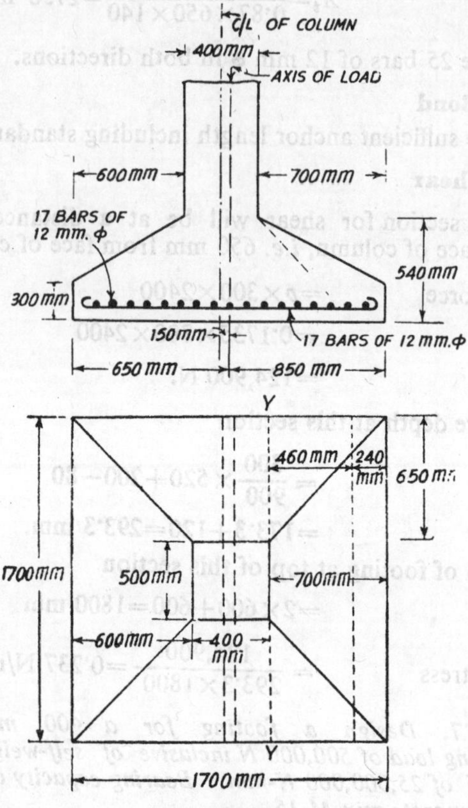

Fig. 10·14

Net upward pressure on footing

$$=\frac{500,000}{1700\times 1700}=0{\cdot}173\ \text{N/mm}^2$$

Maximum B.M. will be at section *Y-Y*

$$M=0{\cdot}173\times 1700\times 700\times \frac{700}{2}$$

$$=72,040,000\ \text{N-mm}$$

Effective depth required

$$=d=\sqrt{\frac{72,040,000}{0{\cdot}87\times 400}}=455\ \text{mm.}$$

Depth required for punching

Punching shear $=0{\cdot}173[1700\times1700-400\times400]$

$=0{\cdot}173\times(2,890,000-160,000)$

$=0{\cdot}173\times2,730,000=472,300$ N

Area resisting punching

$=4\times400\times d_p$

$\therefore\quad 4\times400\times d_p\times10=472,300$

$$d_p=\frac{472,300}{1600}=295{\cdot}2 \text{ mm.}$$

Provide overall depth of 540 mm with effective depth of 460 mm.

$$A_t=\frac{72,040,000}{0{\cdot}87\times460\times140}=1286 \text{ mm}^2.$$

Provide 17 bars of 10 mm ϕ.

Though the B.M. in other direction is slightly less, same reinforcement will be provided.

Check for Bond

Sufficient anchor length including standard hooks are provided.

Check for Shear

Critical section for shear will be at a distance of effective depth, *i.e.* 460 mm from face of column.

Maximum shear $=0{\cdot}173\times1,700\times240$

$=70,650$ N.

Effective depth of footing at this section

$$=\frac{240}{700}\times340+200-80$$

$=116{\cdot}8+120=236{\cdot}8$ mm.

Breadth of footing at this section

$$=\frac{650}{700}\times460\times2+400$$

$=854+400=1254$ mm

$$\text{Shear stress}\quad=\frac{70,650}{236{\cdot}8\times1254}$$

$=0{\cdot}238$ N/mm². Safe.

Shear stress will be safe at other sections.

Ex. 10·8. *Design a symmetrical square footing for the column of Ex. 10·7.*

Sol. Assume self-weight of footing as 50,000 N

Total load $=550,000$ N.

Let B mm be the side of square footing at base.

Maximum stress at base

$$=\frac{W}{B^2}+\frac{M}{B^3/6}=0.2 \text{ } (0.2 \text{ N/mm}^2)$$

$$\therefore \quad \frac{550{,}000}{B^2}+\frac{25{,}00{,}000\times 6}{B^3}=0.2$$

$$2{,}750{,}000+\frac{750{,}000{,}000}{B}=B^2$$

This can be solved by trial and error.

Provide 1800 mm×1800 mm footing.

Net upward pressure distribution variation will be given by

$$p=\frac{500{,}000}{1800\times 1800}+\frac{25{,}000{,}000}{(1800)^3/6}$$

$$=0.157\pm 0.026$$

$$p_{max.}=0.183 \text{ N/mm}^2$$

$$p_{min.}=0.131 \text{ N/mm}^2.$$

Maximum B.M. will be at section *Y-Y*

$$M_y=0.163\times 700\times 1800\times \frac{700}{2}$$

$$+\frac{0.02\times 700\times 1800}{2}\times 700\times \frac{2}{3}$$

$$=71{,}880{,}000+5{,}880{,}000$$

$$=77{,}760{,}000 \text{ N-mm.}$$

Effective depth required

$$=\sqrt{\frac{77{,}760{,}000}{0.87\times 400}}=472.6 \text{ mm.}$$

Provide overall depth of 560 mm with effective depth of 480 mm.

Depth required for punching

Punching shear $=500{,}000-400\times 400\times 0.157$

$=500{,}000-25{,}120=474{,}880$ N

$$d_p=\frac{474{,}880}{4\times 400\times 1}=284.3 \text{ mm}$$

Depth provided is adequate

$$A_t=\frac{77{,}760{,}000}{0.87\times 480\times 140}=1330 \text{ mm}^2$$

Provide 17 bars of 10 mm ϕ in both directions.

Check for Bond

Sufficient anchor length including standard hooks are provided.

Check for Shear

Critical section for shear will be at a distance of 480 mm from the face of the column.

$$S.F. = 0{\cdot}178 \times 1800 \times 220 + \frac{0{\cdot}005}{2} \times 1800 \times 220$$

$$= 70{,}490 + 990 = 71{,}480 \text{ N}$$

Effective depth of footing at this section

$$= 200 + \frac{360}{700} \times 220 - 80$$

$$= 120 + 113{\cdot}2 = 233{\cdot}2 \text{ mm}$$

Breadth of footing at top at this section

$$= 400 + 2 \times 480 = 1360 \text{ mm}$$

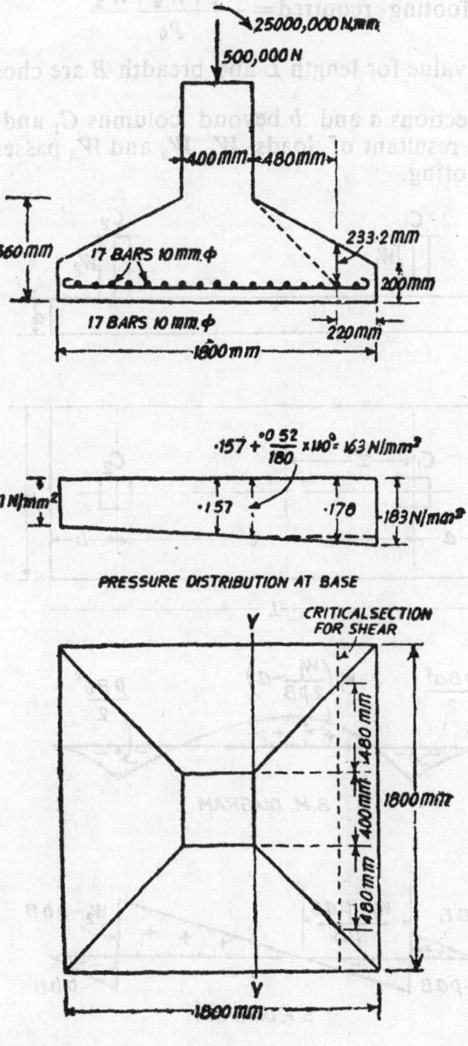

Fig. 10·15

$$\text{Shear stress} = \frac{71{,}480}{1360 \times 233{\cdot}2} = 0{\cdot}225 \text{ N/mm}^2. \quad \text{Safe.}$$

10·8. Design of Rectangular Combined Footings

Let W_1 and W_2 be the loads from columns C_1 and C_2 and l be the distance centre to centre of columns.

Let W_3 be the weight of the footing.

$$\text{Area of footing required} = \frac{W_1 + W_2 + W_3}{p_0}$$

Suitable value for length L and breadth B are chosen.

The projections a and b beyond columns C_1 and C_2 are selected so that the resultant of loads W_1, W_2 and W_3 passes through the C.G. of the footing.

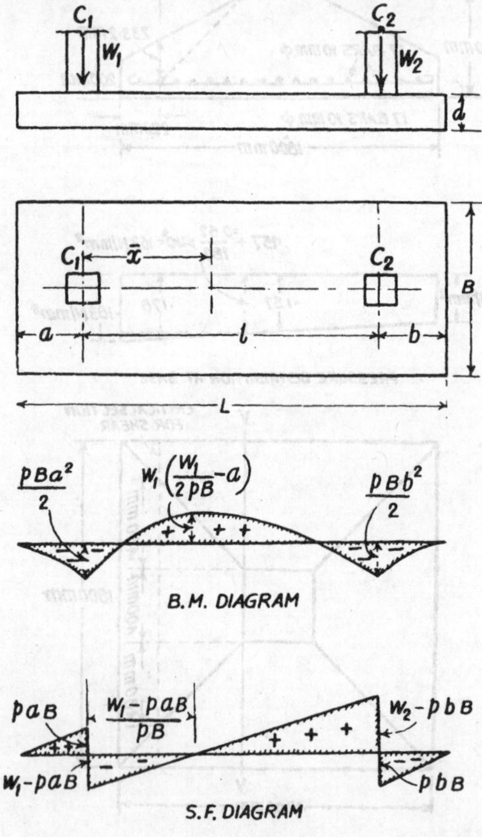

Fig. 10·16

Let $\bar{x}$ be the distance of C.G. of column loads W_1 and W_2 from C_1. Taking moments about column 1.

$$\bar{x}=\frac{W_2 l}{W_1+W_2}$$

a and b are selected such that

$$a+\bar{x}=\frac{L}{2}=b+l-\bar{x}.$$

Net upward pressure on footing$=\dfrac{W_1+W_2}{LB}$.

The B.M. and S.F. diagrams are shown in Fig. 10·16. Maximum B.M. occurs where S.F. is zero, *i.e.* at distance of

$$\frac{W_1-paB}{pB}=\frac{W_1}{pB}-a \text{ from } C_1.$$

The depth is designed for the maximum B.M. and punching shear. The reinforcement is provided to resist the required B.M. The section is checked for shear and bond and if required shear reinforcement is provided.

Transverse B.M.

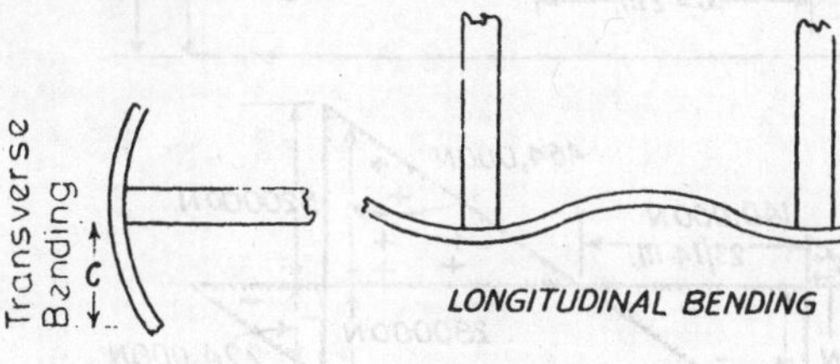

Fig. 10·17

In the portion of the footing near column, bending takes place in two directions. The footing bends in the longitudinal as well as transverse direction in the form of a saucer. In the widths of the footing equal to width of the column plus twice the effective depth of the footing transverse reinforcement is provided to take transverse bending moment. If c is the projection beyond the column face, transverse B.M. per unit width$=\dfrac{pc^2}{2}$. Reinforcement is provided to resist this B.M.

Ex. 10·9. *Design a combined footing for two columns 320 mm × 320 mm carrying load of 600,000 N and 400 mm × 400 mm carrying load of 800,000 N. The columns are spaced at 3·5 m centres. Bearing capacity of soil is 150 kN/m².*

Sol. Assume self-weight of footing as 10% of load coming on it. Load C_1=600,000 N, Load C_2=800,000 N.

$$\text{Weight of footing} = \frac{600{,}000+800{,}000}{10} = 140{,}000 \text{ N}$$

Total load=600,000+800,000+140,000 N=1,540,000 N

Area of footing required

$$= \frac{1{,}540{,}000}{150{,}000} = 10{\cdot}27 \text{ m}^2$$

Provide 2·1 m×5 m footing.

If $\bar{x}$ is the distance of C.G. of loads from column C_1.

$$\bar{x} = \frac{800{,}000\times 3{\cdot}5}{600{,}000+800{,}000} = \frac{2{,}800{,}000}{1{,}400{,}000} = 2{\cdot}0 \text{ m.}$$

Projections of 0·5 m and 1·0 m are provided beyond columns, C_1 and C_2 respectively so that C.G. of footing coincides with C.G. of loads shown in Fig. 10·18.

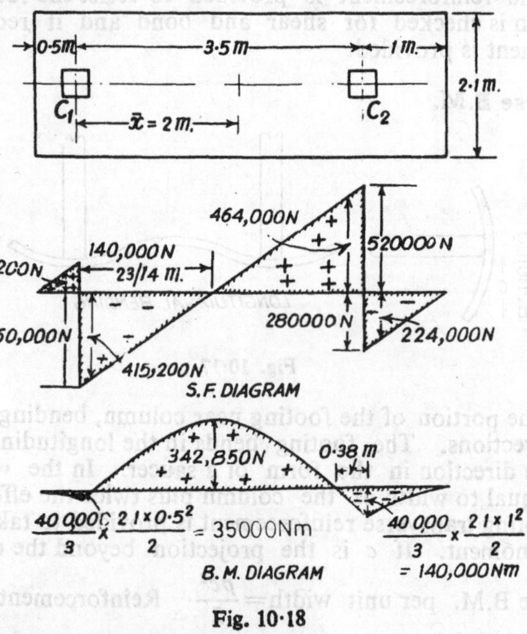

Fig. 10·18

Net upward pressure on footing

$$= \frac{600{,}000+800{,}000}{5\times 2{\cdot}1}$$

$$= \frac{140{,}000}{10{\cdot}5} = \frac{400{,}000}{3} \text{ N/m}^2$$

S.F. diagram is shown in Fig. 10·18 (*b*). Distance from centre of column, where S.F. is zero will be

$$\frac{600{,}000}{2{\cdot}1\times\dfrac{400{,}000}{3}}-0{\cdot}5=\frac{23}{14}\ \text{m}$$

Maximum B.M. will be at this point.

$$\text{Maximum B.M.}=600{,}000\times\frac{23}{14}-\frac{400{,}000}{4}\times2{\cdot}1\times\frac{\left(\frac{23}{14}+0{\cdot}5\right)^2}{2}$$

$$=\frac{1}{14}\ 13{,}800{,}000-9{,}000{,}000)$$

$$=342{,}850\ \text{N-m}$$

Effective depth required

$$=\sqrt{\frac{342{,}850\times1000}{0{\cdot}87\times2100}}=433{\cdot}1\ \text{mm.}$$

Depth required for punching

Punching shear at column 2

$$=\left(800{,}000-\frac{400{,}000}{3}\times\frac{400}{1000}\times\frac{400}{1000}\right)$$

$$=\frac{2{,}336{,}000}{3}\ \text{N}$$

If d_p is depth required for punching

$$d_p=\frac{2{,}336{,}000/3}{0{\cdot}01\times4\times400}=486\ \text{mm}$$

Provide overall depth of 580 mm with effective depth of 500 mm.

$$A_t=\frac{342{,}850\times1000}{0{\cdot}87\times500\times140}=5628\ \text{mm}^2$$

Provide 29 bars of 16 mm ϕ.

Check for Bond

Sufficient anchor length is provided.

Transverse Reinforcement near column C_1

Projection beyond column face

$$=\frac{2100-320}{2}=\frac{1780}{2}=890\ \text{mm}$$

Minimum B.M. per mm width

$$=\frac{400{,}000}{3\times1000\times1000}\times\frac{(890)^2}{2}$$

$$=52{,}810\ \text{N mm}$$

$$A_t = \frac{52{,}810}{0{\cdot}87(500-14)\times 140}$$

$$=0{\cdot}894 \text{ mm}^2$$

Using 12 mm ϕ bars, spacing

$$=\frac{113}{0{\cdot}894}=126 \text{ mm}$$

Provide 12 mm ϕ bars at 120 mm centres.

Distance in which transverse reinforcement is provided

$$=320+2\times 500=1320 \text{ mm}.$$

Provide reinforcement in width of 1400 mm.

Same reinforcement as for column one is provided in transverse direction under column 2 for width of 1400 mm.

Reinforcement in cantilever portions

Near column C_1

$$\text{B.M.}=35{,}000 \text{ N-m}$$

$$A_t=\frac{35{,}000\times 1000}{0{\cdot}87\times 500\times 140}=572 \text{ mm}^2$$

Minimum reinforcement

$$=0{\cdot}15\% \text{ of cross-sectional area}$$

$$=\frac{1{\cdot}15\times 580\times 2100}{100}=1827 \text{ mm}^2$$

Near column C_2

$$\text{B.M.}=140{,}000 \text{ N-m}$$

$$A_t=\frac{140{,}000\times 1000}{0{\cdot}87\times 500\times 140}=2299 \text{ mm}^2$$

Check for Shear

Maximum S.F. at face of column C_2

$$=520{,}000-0{\cdot}2\times \frac{400{,}000}{3}\times 2{\cdot}1$$

$$=464{,}000 \text{ N}$$

Shear stress $$=\frac{464{,}000}{540\times 2100}=0{\cdot}409 \text{ N/mm}^2$$

Taking $\tau_c=0{\cdot}34$ N/mm. Shear stress for which stirrups are to be provided $=0{\cdot}409-0{\cdot}34=0{\cdot}069$ N/mm^2.

Minimum shear reinforcement is required. Provide 8 mm ϕ 8-legged stirrups at 400 mm.

The details of reinforcement are shown in Fig. 10·19.

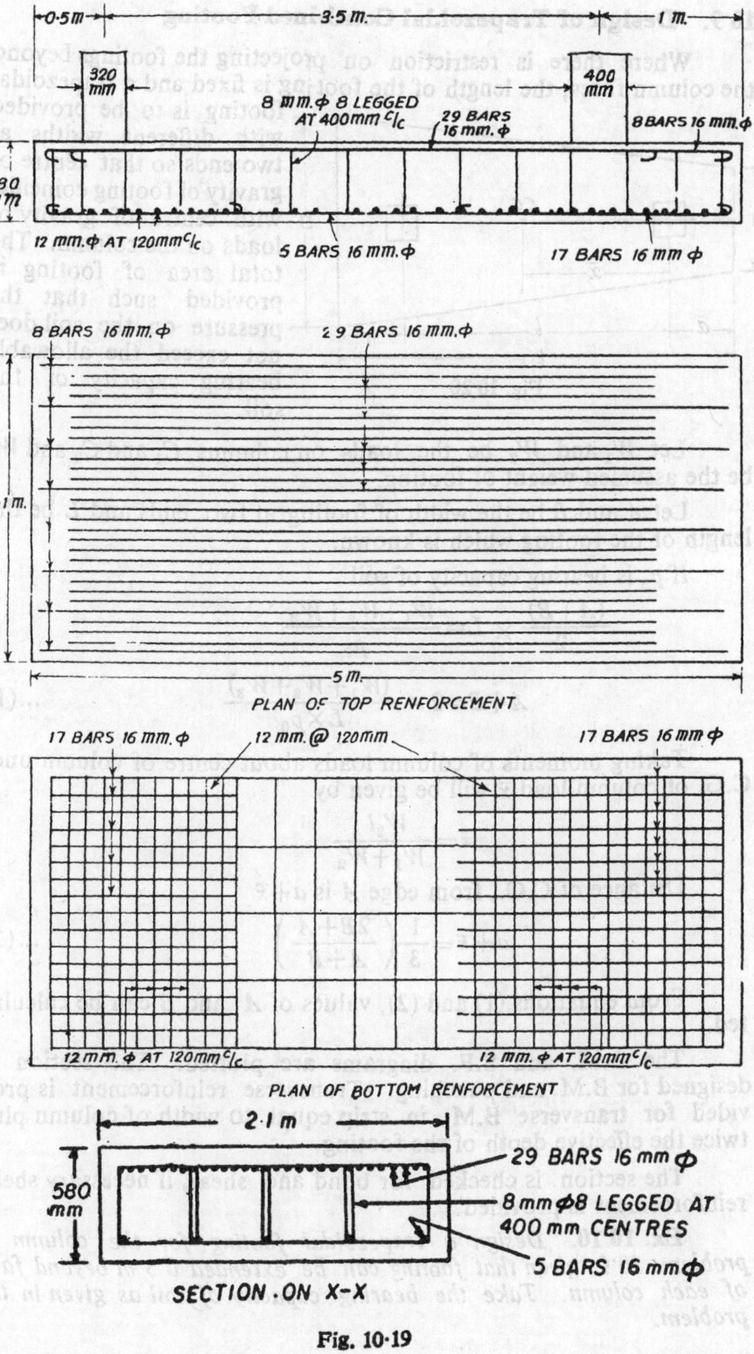
0·5 m
3·5 m.
1 m.
320 mm
400 mm
8 mm.φ 8 LEGGED AT 400mm c/c
29 BARS 16 mm.φ
8 BARS 16 mm.φ
580 mm
12 mm.φ AT 120mm c/c
5 BARS 16 mm.φ
17 BARS 16 mm φ
8 BARS 16 mm.φ
29 BARS 16 mm.φ
2·1 m.
5 m.
PLAN OF TOP RENFORCEMENT
17 BARS 16 mm.φ
12 mm. @ 120mm
17 BARS 16 mm φ
17
12 mm.φ AT 120mm c/c
12 mm. φ AT 120mm c/c
PLAN OF BOTTOM RENFORCEMENT
2·1 m
580 mm
29 BARS 16 mm φ
8 mm φ8 LEGGED AT 400 mm CENTRES
5 BARS 16 mmφ
SECTION ON X–X

Fig. 10·19

10·9. Design of Trapezoidal Combined Footing

Where there is restriction on projecting the footings beyond the column faces, the length of the footing is fixed and a trapezoidal footing is to be provided with different widths at two ends so that centre of gravity of footing coincides with centre of gravity of loads on the column. The total area of footing is provided such that the pressure on the soil does not exceed the allowable bearing capacity of the soil.

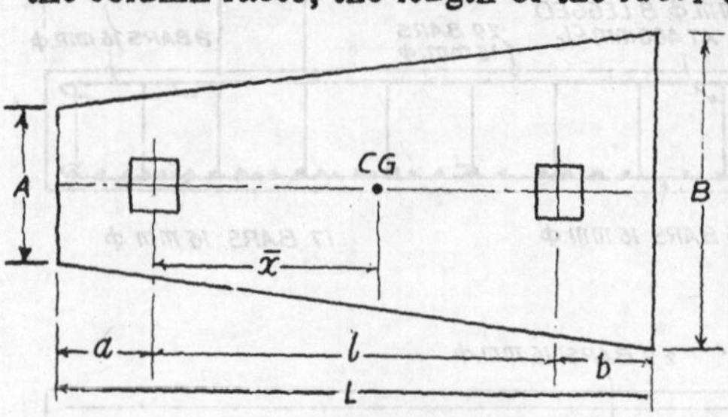

Fig. 10·20

Let W_1 and W_2 be the loads on columns C_1 and C_2 and W_3 be the assumed weight of footing.

Let A and B be the width of footing at two ends and L be the length of the footing which is known.

If p_0 is bearing capacity of soil

$$\frac{(A+B)}{2}\times L=\frac{W_1+W_2+W_3}{p_0}$$

$$A+B=2\times\frac{(W_1+W_2+W_3)}{L\times p_0} \quad ...(1)$$

Taking moments of column loads about centre of column one, C.G. of column load $\bar{x}$ will be given by

$$\bar{x}=\frac{W_2 l}{W_1+W_2}$$

Distance of C.G., from edge A is $a+\bar{x}$

$$a+\bar{x}=\frac{1}{3}\left(\frac{2B+A}{A+B}\right) \quad ...(2)$$

From equations (1) and (2), values of A and B can be calculated.

The B.M. and S.F. diagrams are plotted. The section is designed for B.M. and punching. Transverse reinforcement is provided for transverse B.M. in strip equal to width of column plus twice the effective depth of the footing.

The section is checked for bond and shear, if necessary shear reinforcement is provided.

Ex. 10·10. *Design a trapezoidal footing for the column of problem 10·9, given that footing can be extended 0·5 m beyond face of each column. Take the bearing capacity of soil as given in the problem.*

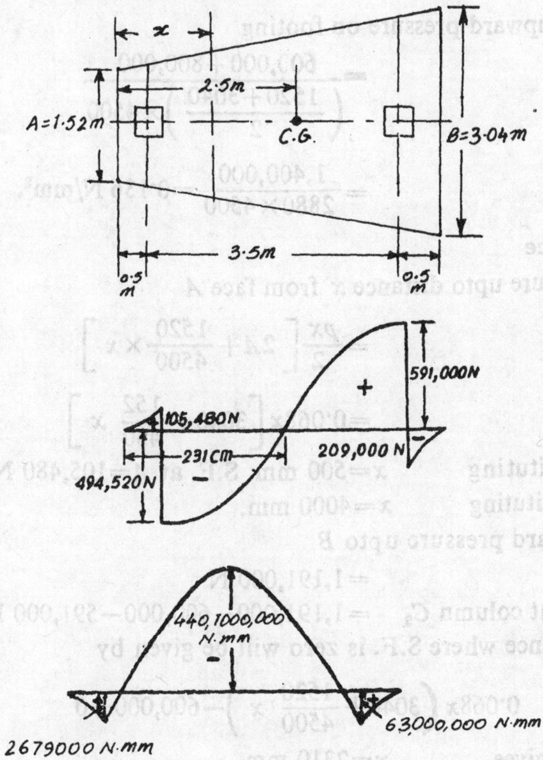

Fig. 10·21

Sol. Area of footing required$=10{\cdot}27$ m^2

Length of footing $=4{\cdot}5$ m.

Let A and B be widths of footing at two ends.

$$\left(\frac{A+B}{2}\right)\times 4{\cdot}5=10{\cdot}27$$

$$A+B=\frac{10{\cdot}27}{2{\cdot}25}=4{\cdot}56 \text{ m.} \qquad ...(1)$$

The C.G. of loads is at 2·5 m for the face A of footing

$$2{\cdot}5=\frac{4{\cdot}5}{3}\left(\frac{2B+A}{A+B}\right)=1{\cdot}5\ \frac{(2B+A)}{4{\cdot}56}$$

$$2B+A=7{\cdot}60$$

Eq. (2)$-$Eq. (1), gives $B=3{\cdot}04$ m

$$A=1{\cdot}52 \text{ m.}$$

Net upward pressure on footing

$$=\frac{600,000+800,000}{\left(\frac{1520+3040}{2}\right)\times 4500}$$

$$=\frac{1,400,000}{2880\times 4500}=0{\cdot}136 \text{ N/mm}^2.$$

Shear force

Pressure upto distance x from face A

$$=\frac{px}{2}\left[2A+\frac{1520}{4500}\times x\right]$$

$$=0{\cdot}068x\left[3040+\frac{152}{450}x\right]$$

Substituting $x=500$ mm. S.F. at $A=105,480$ N

Substituting $x=4000$ mm.

Upward pressure upto B

$=1,191,000$ N

S.F. at column C_2 $=1,191,000-600,000-591,000$ N

Distance where S.F. is zero will be given by

$$0{\cdot}068x\left(3040+\frac{1520}{4500}x\right)-600,000=0$$

This gives $x=2310$ mm.

Bending moment

Moment of total pressure at distance x

$$=p\times A\times x\times\frac{x}{2}+\frac{1520}{4500}\times x\times\frac{x}{2}\times p\times\frac{x}{3}$$

$$=0{\cdot}136\times 1520\times\frac{x^2}{3}+\frac{x^3}{3}\times 0{\cdot}136\times\frac{1520}{9000}$$

$$=0{\cdot}136x^2\left[760+\frac{152}{2700}x\right]$$

Substituting $x=500$ mm B.M. in footing at centre of column $C_1=26,790,000$ N-mm.

B.M. beyond this will be

$$=600,000(x-500)-0{\cdot}136\,x^2\left(760+\frac{152}{2700}x\right)$$

Putting $x=2310$ mm, maximum B.M.$=440,100,000$ N-mm

Putting $x=4000$, B.M.$=63,000,000$ N-m

S.F. and B.M. diagram are shown in Fig. 10·21.

Width of footing at point of maximum B.M.

$$=1520+\frac{1520}{4500}\times 2310$$

$$=1520+778=2298 \text{ mm}$$

Effective depth required

$$=\sqrt{\frac{440,100,000}{0{\cdot}87\times 2298}}=469{\cdot}2 \text{ mm.}$$

Depth required for punching shear

Punching shear for column C_2

$$=800,000-0{\cdot}136\times 400\times 400$$

$$=800,000-21,760=778,240 \text{ N}$$

If d_p is depth required for punching

$$1\times 4\times 400\times d_p=778,240$$

$$d_p=\frac{778,240}{1600}=486{\cdot}4 \text{ mm}$$

Provide overall depth of 580 mm, with effective depth of 500 mm.

$$A_t=\frac{440,100,000}{0{\cdot}87\times 500\times 140}=7220 \text{ mm}^2$$

Provide 36 bars of 16 mm diameter.

Sufficient anchor length is provided including standard hook.

Reinforcement in cantilever portions

Beyond column C_2, $A_t=\frac{63,000,000}{0{\cdot}87\times 500\times 140}=1035 \text{ mm}^2$

28 bars of 26 mm ϕ are provided as required for bond.

Beyond column C_1, $A_t=\frac{26,790,000}{0{\cdot}87\times 500\times 140}=440 \text{ mm}^2$

24 bars of 16 mm ϕ are provided as required for bond.

Transverse Reinforcement

Near column C_2

Width of footing at centre line of column C_2

$$=1520+\frac{1520}{4500}\times 4000$$

$$=1520+1350=2870 \text{ mm}$$

Projection of footing beyond face of column

$$=\frac{2870-400}{2}=\frac{2470}{2}+1235 \text{ mm.}$$

B.M. per one mm width

$$=0{\cdot}136\times\frac{(1235)^2}{2}=103,700 \text{ N}\cdot\text{mm}$$

$$A_t=\frac{103,700}{0{\cdot}87(500-16)\ 140}=1{\cdot}46 \text{ mm}^2$$

Using 10 mm ϕ bars,

$$\text{Spacing} = \frac{201}{1{\cdot}46} = 138 \text{ mm.}$$

Provide 16 mm ϕ bars at 130 mm centres.

Width of strip in which reinforcement is provided

$$= 400 + 2 \times 500 = 1400 \text{ mm.}$$

Near column C_1

Width of footing at centre line of column C_1

$$= 1520 + \frac{1520}{4500} \times 500$$

$$= 1520 + 170 = 1690 \text{ mm.}$$

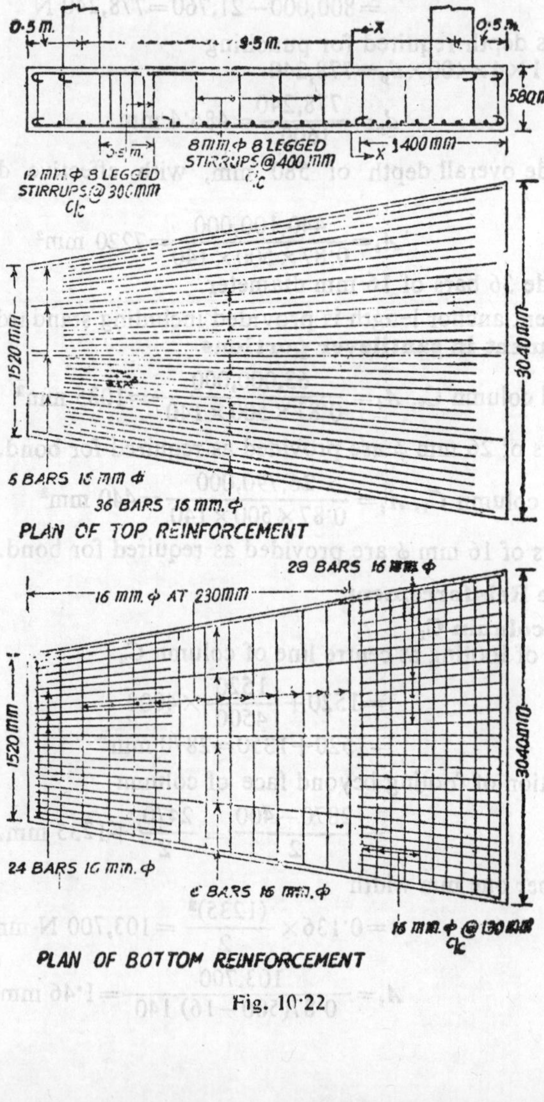

Fig. 10·22

$$\text{Projection} = \frac{1690-320}{2} = \frac{1370}{2} = 685 \text{ mm.}$$

B.M. per mm width of footing

$$= \frac{0{\cdot}136\times(685)^2}{2} = 31{,}990 \text{ N-mm.}$$

$$A_t = \frac{31{,}990}{0{\cdot}87\times(500-16)\times140} = 0{\cdot}504 \text{ mm}^2$$

Minimum reinforcement

$$= 0{\cdot}15\% \text{ of cross-sectional area}$$

$$= \frac{1{\cdot}15}{100}\times580\times1 = 0{\cdot}87 \text{ mm}^2$$

Using 16 mm ϕ bars,

$$\text{Spacing} = \frac{201}{0{\cdot}87} = 231 \text{ mm.}$$

Provide 16 mm ϕ bars at 230 centres.

Width of strip in which reinforcement is to be provided

$$= 320+2\times500 = 1320 \text{ mm.}$$

Check for Shear

Minimum shear under column $C_2 = 591{,}000$ N

$$\text{Shear stress} = \frac{591{,}000}{500\times2870}$$

$$= 0{\cdot}412 \text{ N/mm}^2$$

Maximum shear force under column $C_1 = 494{,}420$ N

$$\text{Shear stress} = \frac{494{,}420}{500\times1690} = 0{\cdot}585 \text{ N/mm}^2$$

Taking $\tau_c = 0{\cdot}34$ N/mm^2.

Net share $= (0{\cdot}585-0{\cdot}34)\times500\times1690$

$= 207{,}025$ N

Providing 8 legged 12 mm ϕ stirrups.

$$\text{Spacing} = \frac{8\times113\times140\times500}{207{,}025} = 306 \text{ mm.}$$

Provide 8-legged 12 mm ϕ stirrups at 300 mm centres. These stirrups are provided upto 0·5 m from the face of the column.

Check for shear stress at 1 m from the edge of footing

$$\text{Width of footing} = 1520+\frac{1520}{4500}\times100$$

$$= 1520+360 = 1880 \text{ mm.}$$

$$\text{S.F.} = 600{,}000 - \frac{1520+1880}{2} \times 100 \times 0{\cdot}0136$$

$$= 368{,}800 \text{ N}$$

$$\text{Shear stress} = \frac{368{,}800}{500 \times 1880} = 0{\cdot}392 \text{ N/mm}^2.$$

In the remaining portion provide 8 mm ϕ 8-legged stirrups as 400 mm centres.

Ex. 10·11. *Design a combined footing for four columns 400 mm × 400 mm spaced equally at 4 m centres and carrying loads as shown in Fig. 10·23. The footings may be extended upto 1·5 m beyond centre of boundary columns. The bearing capacity of soil is 200 kN per m².*

Sol. The footing will be projected 0·5 m beyond centre of column C_1 and 1·5 m beyond centre of column C_4. Total length of footing will be 14 m. Rectangular footing will be provided.

To find the distance of C.G. of load, take moments about centre of column C_1.

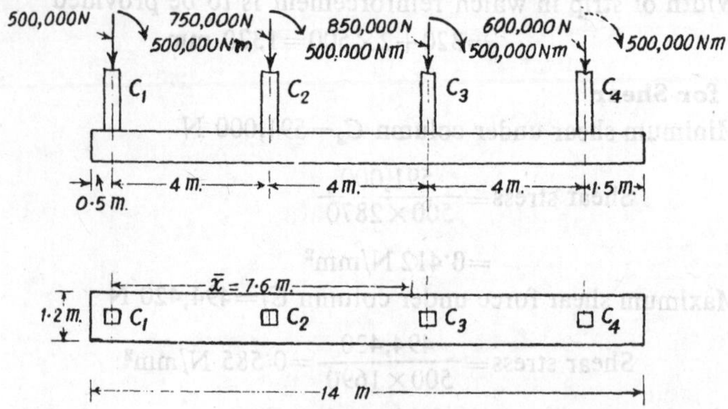

Fig. 10·23

$$\bar{x} = \frac{500{,}000 \times 4 + 750{,}000 \times 4 + 850{,}000 \times 8 + 600{,}000 \times 12}{300{,}000 + 750{,}000 + 850{,}000 + 600{,}000}$$

$$= \frac{2{,}000{,}000 + 3{,}000{,}000 + 6{,}800{,}000 + 7{,}200{,}000}{2{,}500{,}000}$$

$$= \frac{19{,}000{,}000}{2{,}500{,}000} = 7{\cdot}6 \text{ m.}$$

$$\text{Eccentricity of load} = 0{\cdot}5 + 7{\cdot}6 - 7 = 1{\cdot}1 \text{ m.}$$

Because of eccentricity of the load, there will be variation of pressure distribution at the base and there will be large bending moments produced resulting in heavier section of footing. Self-weight of footing will be assumed as 20% of the superimposed load.

Self-weight of footing $= \frac{1}{5} \times (300{,}000 + 750{,}000 + 850{,}000 + 600{,}000)$

$= 500{,}000$ N.

As the loading shown on columns is due to wind pressure the permissible stress will be increased by 33⅓ per cent and bearings capacity will be increased by 25 per cent.

Fixing the width of the footing

Let the width of the footing be B metres.

$$\text{Maximum pressure} = \frac{2{,}500{,}000 + 500{,}000}{14 \times B} + \frac{2{,}500{,}000 \times 1{\cdot}1}{\frac{B \times 14^2}{6}}$$

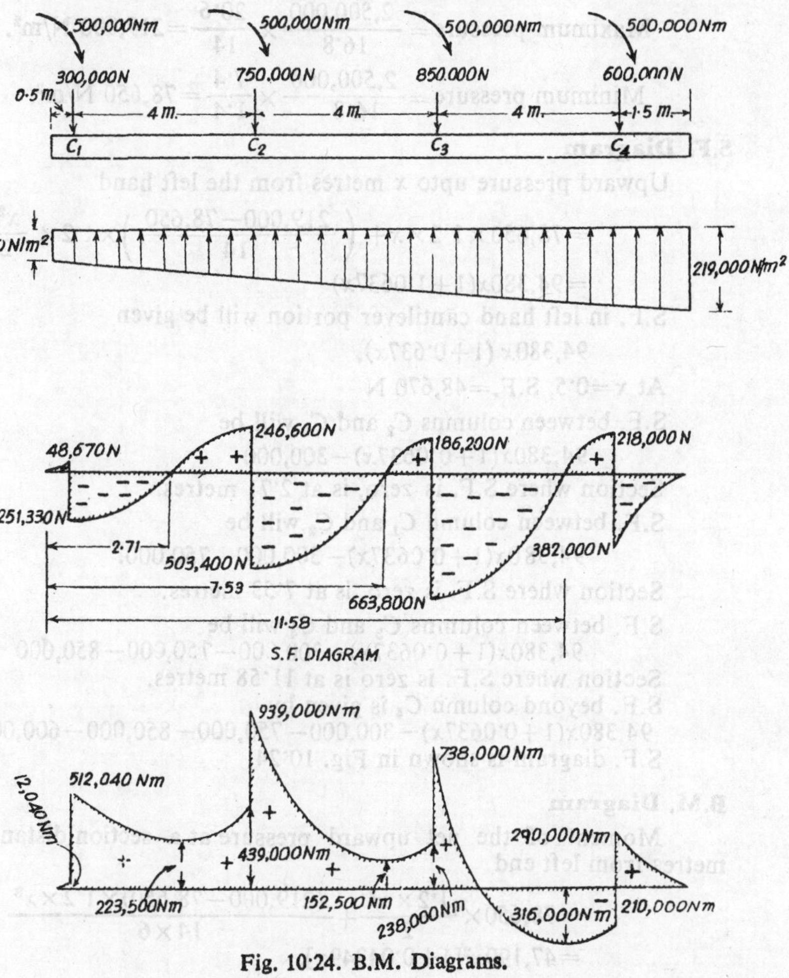

Fig. 10·24. B.M. Diagrams.

$$\frac{3,000,000}{14B}+\frac{1,179,000}{14B}=\frac{5}{4}\times 20,000=\frac{5}{4}\times 200,000$$

$$B=\frac{4,179,000}{14\times 250,000}=1{\cdot}195$$

Provide width of 1·2 m.

Net upward pressure on footing

$$=\frac{2,500,000}{14\times 1{\cdot}2}\pm\frac{2,500,000\times 1{\cdot}1}{\frac{10\times 14^2}{6}}$$

$$=\frac{2,500,000}{14\times 1{\cdot}2}\left[1\pm\frac{1{\cdot}1\times 6}{14}\right]$$

$$\text{Maximum pressure}=\frac{2,500,000}{16{\cdot}8}\times\frac{20{\cdot}6}{14}=219,000\ \text{N/m}^2.$$

$$\text{Minimum pressure}=\frac{2,500,000}{16{\cdot}8}\times\frac{7{\cdot}4}{1{\cdot}4}=78,650\ \text{N/m}^2.$$

S.F. Diagram

Upward pressure upto x metres from the left hand

$$=78,650\times 1{\cdot}2\times x+\left(\frac{219,000-78,650}{14}\right)\times 1{\cdot}2\times\frac{x^2}{2}$$

$$=94,380x(1+1{\cdot}0637x)$$

S.F. in left hand cantilever portion will be given

$94,380x\ (1+0{\cdot}637x)$.

At $x=0{\cdot}5$, S.F.$=48,670$ N

S.F. between columns C_2 and C_3 will be

$94,380x(1+0{\cdot}0637x)-300,000$

Section where S.F. is zero, is at 2·71 metres.

S.F. between column C_1 and C_2 will be

$94,380x(1+0{\cdot}0637x)-300,000-750,000.$

Section where S.F. is zero, is at 7·53 metres.

S.F. between columns C_3 and C_4 will be

$94,380x(1+0{\cdot}0637x)-300,000-750,000-850,000$

Section where S.F. is zero is at 11·58 metres.

S.F. beyond column C_4 is given by

$94,380x(1+0{\cdot}0637x)-300,000-750,000-850,000-600,000$

S.F. diagram is shown in Fig. 10·24.

B.M. Diagram

Moment of the net upward pressure at a section distance x metres from left end

$$=78,650\times\frac{1{\cdot}2\times x^2}{2}+\frac{(219,000-78,650)\times 1{\cdot}2\times x^3}{14\times 6}$$

$$=47,190x^2[1+0{\cdot}04249x].$$

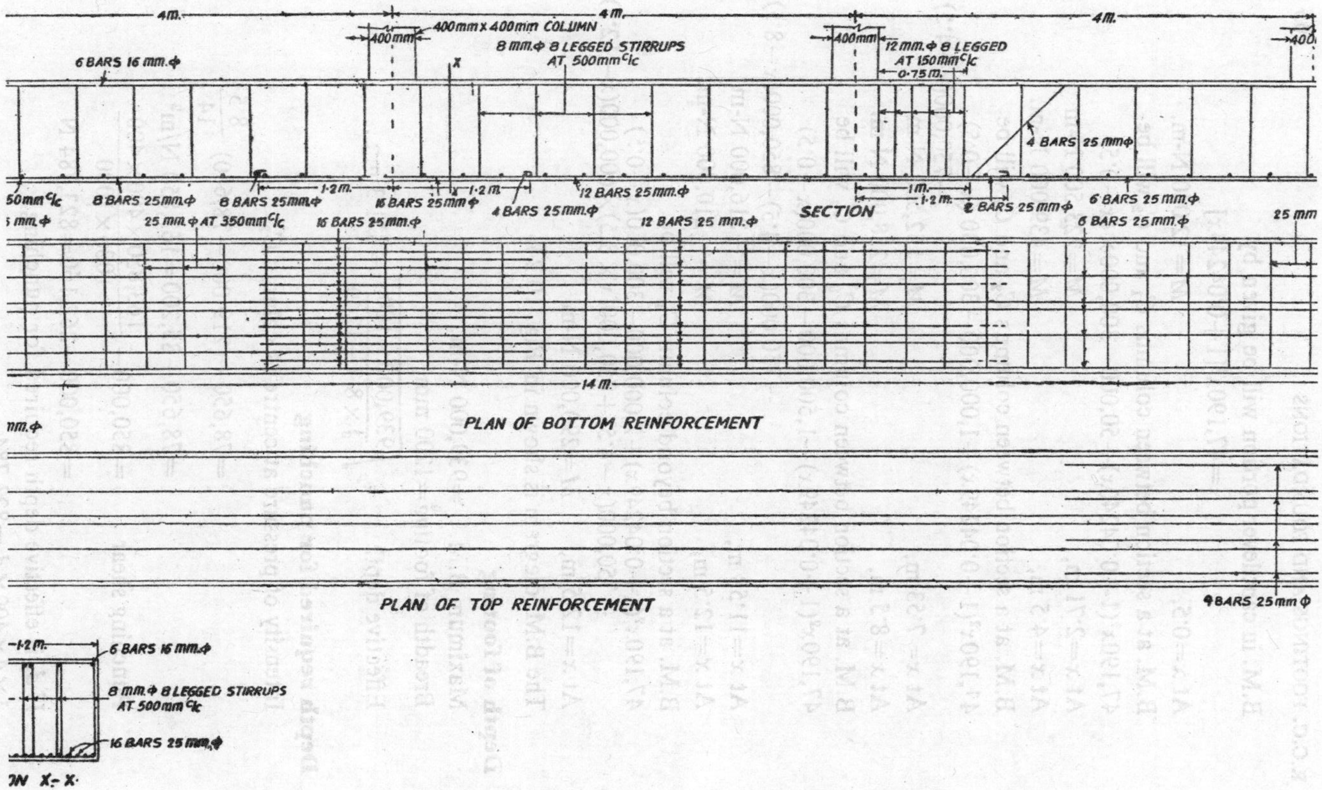

4 m.
4 m.
4 m.
400mm
400mm x 400mm COLUMN
8 mm.Φ 8 LEGGED STIRRUPS AT 500mm c/c
400mm
12 mm.Φ 8 LEGGED AT 150mm c/c
0·75 m.
6 BARS 16 mm.Φ
X
4 BARS 25 mmΦ
1·2 m.
1·2 m.
1 m.
1·2 m.
50mm c/c
8 BARS 25mm.Φ
8 BARS 25mm.Φ
16 BARS 25mm Φ
4 BARS 25mm.Φ
12 BARS 25mm.Φ
8 BARS 25mmΦ
6 BARS 25 mm.Φ
SECTION
25 mm.Φ AT 350mm c/c
16 BARS 25 mm.Φ
12 BARS 25 mm.Φ
6 BARS 25 mm.Φ
25 mm
14 m.
PLAN OF BOTTOM REINFORCEMENT
PLAN OF TOP REINFORCEMENT
4 BARS 25mm Φ
1·2 m.
6 BARS 16 mm.Φ
8 mm.Φ 8 LEGGED STIRRUPS AT 500mm c/c
16 BARS 25 mm.Φ
X - X

B.M. in cantilever portion will be given by

$$=47{,}190x^2[1+0{\cdot}04249x]$$

At $x=0{\cdot}5$, $M=12{,}040$ N-m.

B.M. at a section between columns C_1 and C_2 will be

$47{,}190x^2(1+0{\cdot}04240x)+50{,}000-300{,}000\times(x-0{\cdot}5)$

At $x=2{\cdot}71$ m, $M=223{,}500$ N-m

At $x=4{\cdot}5$ m, $M=439{,}000$ N-m.

B.M. at a section between columns C_2 and C_3 will be

$47{,}190x^2(1+0{\cdot}04249x)+1{,}000{,}000-300{,}000\ (x-0{\cdot}5)$
$-750{,}000(x-4{\cdot}5)$

At $x=7{\cdot}53$ m, $M=152{,}500$ N-m

At $x=8{\cdot}5$ m, $M=238{,}000$ N-m.

B.M. at a section between columns C_3 and C_4 will be

$47{,}190x^2(1+0{\cdot}04249x)+1{,}500{,}000-300{,}000(x-0{\cdot}5)$
$-750{,}000(x-4{\cdot}5)-850{,}000(x-8{\cdot}5)$

At $x=11{\cdot}58$ m, $M=-316{,}000$ N-m

At $x=12{\cdot}5$ m, $M=-210{,}000$ N-m.

B.M. at a section beyond column C_4 will be

$47{,}190x^2(1+0{\cdot}04249x)+2{,}000{,}000-300{,}000(x-0{\cdot}5)$
$-750{,}000(x-4{\cdot}5)+850{,}000(x\times 8{\cdot}5)\times 600{,}000(x-12{\cdot}5)$

At $x=12{\cdot}5$ m, $M=290{,}000$ N-m

The B.M. diagram is shown in Fig. 10·24.

Depth of footing

Maximum B.M. $=939{,}000$ N-m.

Breadth of footing $=1200$ mm.

Effective depth $=\sqrt{\dfrac{939{,}000\times 1000}{\frac{1}{3}\times 8{\cdot}7-1200}}=821{\cdot}3$ mm.

Depth required for punching

Intensity of pressure at centre of column C_3

$$=78{,}650+(219{,}000-78{,}650)\ \frac{8{\cdot}5}{14}$$

$$=78{,}650+85{,}200=163{,}850\ \text{N/m}^2$$

Punching shear $=850{,}000-\dfrac{163{,}850\times 400\times 400}{1000\times 1000}$

$$=850{,}000-262{,}160=823{,}784\ \text{N}$$

If d_p is effective depth required for punching,

$1\times 4\times 400\times d_p=823{,}784$

$$d_p=\frac{823{,}784}{1600}=514{\cdot}9\ \text{mm}.$$

Provide overall depth of 920 mm with effective depth of 830 mm.

Reinforcement

B.M. at column C_2=939,000 N-m.

Steel required at column C_2

$$=\frac{939{,}000\times1000}{0{\cdot}87\times830\times140\times\frac{4}{3}}=6960\ \text{mm}^2.$$

Provide 16 bars of 25 mm ϕ.

Sufficient anchor length is provided including standard hooks.

Transverse Reinforcement

Near column C_4

Net upward pressure at centre of column

$$=78{,}650+\frac{140{,}350}{14}\times12{\cdot}5$$

$$=78{,}650+125{,}300=203{,}950\ \text{N/m}^2.$$

Projection of footing beyond face of column

$$=\frac{1200-400}{2}=400\ \text{mm}.$$

B.M. per 1 mm width

$$=\frac{203{,}950}{1000\times1000}\times\frac{(400)^2}{2}\ \text{N-mm}$$

$$=16{,}316\ \text{N-mm}$$

$$A_t=\frac{16{,}316}{0{\cdot}87\times(830-25)\times140\times4/3}$$

$$=0{\cdot}1238\ \text{mm}^2.$$

Minimum reinforcement

$$=0{\cdot}15\%\ \text{of cross-sectional areas}$$

$$=\frac{0{\cdot}15}{100}\times920\times1=1{\cdot}38\ \text{mm}^2$$

Providing 35 mm ϕ bars,

$$\text{Spacing}\qquad=\frac{491}{1{\cdot}38}=356\ \text{mm}$$

Provide 25 mm ϕ bars at 350 mm centres.

Strip width in which transverse reinforcement is to be provided

$$=400+2\times830=2060\ \text{mm}$$

Same transverse reinforcement is provided throughout the length of the footings.

Check for Shear

Maximum S.F. = 663,800 N

$$\text{Shear stress} = \frac{663,800}{830 \times 1200} = 0{\cdot}6665 \text{ N/mm}^2$$

Allowable $\tau_c = 0{\cdot}34 \text{ N/mm}^2$

Near shear for which stirrups are to be provided

$$= (0{\cdot}6665 - 0{\cdot}34) \times 830 \times 1200$$

$$= 3,251,194 \text{ N}$$

Shear reinforcement is required only near column C_3.

Using 8-legged 12 mm ϕ stirrups,

$$\text{Spacing} = \frac{830 \times 8 \times 113 \times 140 \times \frac{4}{3}}{325,194}$$

$$= 430 \text{ mm.}$$

Provide stirrups @ 400 mm centres in length of m on right side of column.

Provide 8-legged 8 mm ϕ stirrups at 500 mm centre to centre remaining portion.

Details of reinforcement are shown in Fig. 10·25.

Ex. 10·12. *Design of footing for the columns of Ex. 10·11 with central rib beam.*

Sol. As the data are same as for Ex. 10·11, footing area and pressure distribution at the base will be same as Ex. 10·11. Thus the B.M. and S.F. diagrams for the rib beam will be same as for the footing of Ex. 10·11.

Design of Footing Slab

Consider 1 mm width of footing at right end.

Projection beyond beam

$$= \frac{1200 - 400}{2} = 400 \text{ mm}$$

$$\text{Maximum B.M.} = \frac{219,000}{1000 \times 1000} \times 400 \times 1 \times 200$$

$$= 17,520 \text{ N-mm}$$

Effective depth required

$$= \sqrt{\frac{17,520}{\frac{4}{3} \times 0{\cdot}87 \times 1}} = 123 \text{ mm}$$

Provide overall depth of 200 mm with effective depth of 130 mm.

$$A_t = \frac{175{\cdot}20}{0{\cdot}87 \times 130 \times 140 \times \frac{4}{3}} = 0{\cdot}83 \text{ mm}^2$$

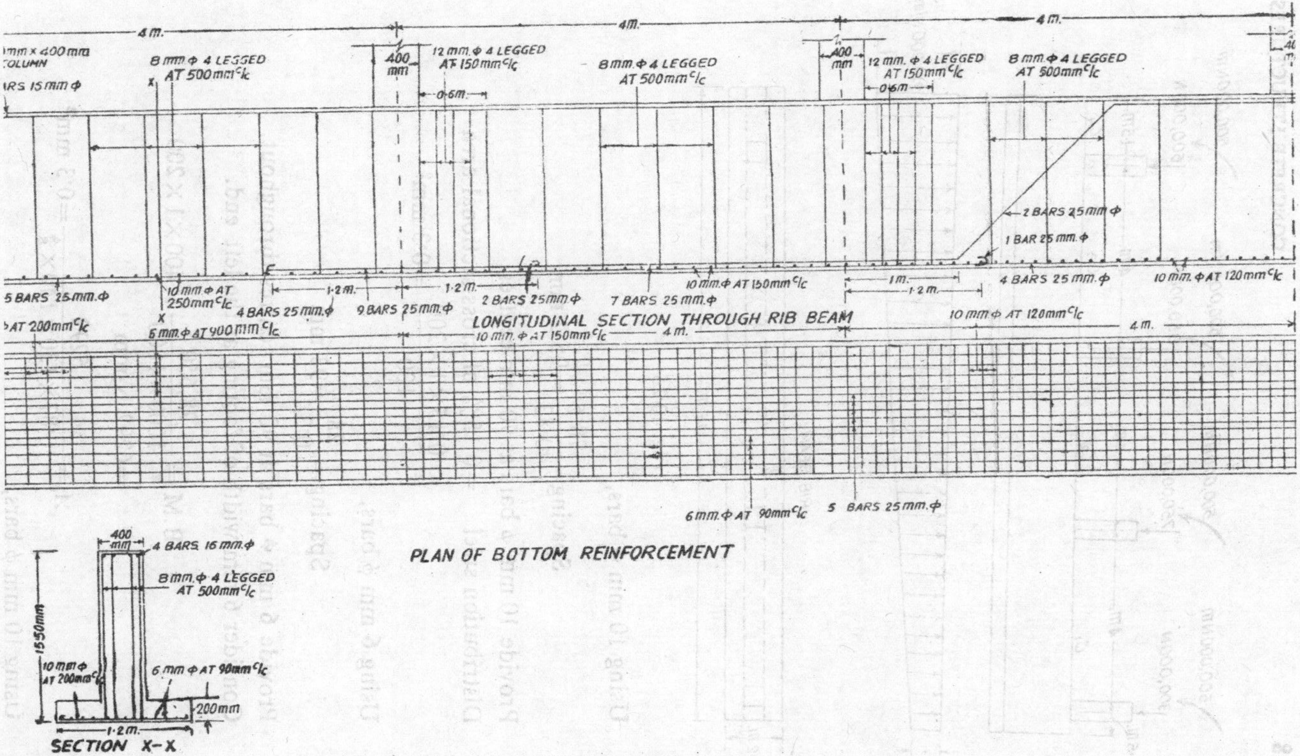
4 m.
4 m.
4 m.
mm x 400 mm COLUMN
RS 15 mm φ
8 mm φ 4 LEGGED AT 500 mm c/c
400 mm
12 mm φ 4 LEGGED AT 150 mm c/c
0·6 m.
8 mm φ 4 LEGGED AT 500 mm c/c
400 mm
12 mm φ 4 LEGGED AT 150 mm c/c
0·6 m
8 mm φ 4 LEGGED AT 500 mm c/c
2 BARS 25 mm φ
1 BAR 25 mm φ
5 BARS 25 mm φ
10 mm φ AT 250 mm c/c
4 BARS 25 mm φ
1·2 m.
9 BARS 25 mm φ
1·2 m.
2 BARS 25 mm φ
7 BARS 25 mm φ
10 mm φ AT 150 mm c/c
1 m.
1·2 m.
4 BARS 25 mm φ
10 mm φ AT 120 mm c/c
AT 200 mm c/c
6 mm φ AT 900 mm c/c
LONGITUDINAL SECTION THROUGH RIB BEAM
10 mm φ AT 150 mm c/c
4 m.
10 mm φ AT 120 mm c/c
4 m.
6 mm φ AT 90 mm c/c
5 BARS 25 mm φ
PLAN OF BOTTOM REINFORCEMENT
400 mm
4 BARS 16 mm φ
8 mm φ 4 LEGGED AT 500 mm c/c
1550 mm
10 mm φ AT 200 mm c/c
6 mm φ AT 90 mm c/c
200 mm
1·2 m.
SECTION X–X

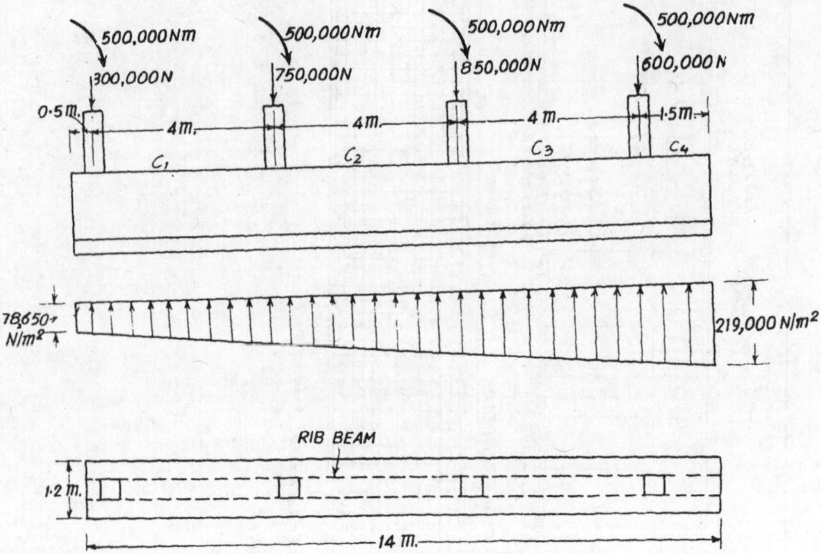

Fig. 10·25

Using 10 mm ϕ bars,

$$\text{Spacing}=\frac{78\cdot5}{0\cdot83}=95\cdot2 \text{ mm}$$

Provide 10 mm ϕ bars at 90 mm centres.

Distribution steel $=0\cdot15\%$ of cross-sectional area

$$=\frac{0\cdot15\times1\times200}{100}=0\cdot3 \text{ mm}^2$$

Using 6 mm ϕ bars,

$$\text{Spacing}=\frac{28}{0\cdot3}=93 \text{ mm}$$

Provide 6 mm ϕ bars at 90 mm centres throughout.

Consider 6 mm width of footing at the left end.

$$\text{B.M.}=\frac{78{,}650}{1000\times1000}\times400\times1\times200$$

$$=6292 \text{ N-mm}$$

$$A_t=\frac{6292}{0\cdot87\times130\times140\times\frac{4}{3}}=0\cdot3 \text{ mm}^2$$

Using 10 mm ϕ bars,

$$\text{Spacing}=\frac{79}{0\cdot3}=263\cdot3 \text{ mm.}$$

Provide 100 mm ϕ bars at 250 mm centres.

Cantilever reinforcement will be varied gradually from 10 mm ϕ bars at 90 mm centres at right end to 10 mm ϕ at 250 mm centres at left end.

Design of Rib Beam

Maximum B.M.=939,000 N-mm at column C_2

Width of beam =400 mm

Effective depth required

$$=\sqrt{\frac{939{,}000\times 1000}{\frac{4}{3}\times 0{\cdot}87\times 400}}=1439 \text{ mm}$$

Provide overall depth of 1550 mm with effective depth of 1440 mm

$$A_t=\frac{939{,}000\times 1000}{0{\cdot}87\times 1440\times 140\times \frac{4}{3}}=4020 \text{ mm}^2$$

Provide 9 bars of 25 mm ϕ.

Check for bond

Sufficient anchor length is provided including standard hooks.

Reinforcement of column C$_1$

$$A_t=\frac{512{,}040\times 1000}{0{\cdot}87\times 1440\times 140\times \frac{4}{3}}=2200 \text{ mm}^2$$

Provide 7 bars of 25 mm ϕ.

Reinforcement near column C$_2$

$$A_t=\frac{738{,}000\times 1000}{0{\cdot}87\times 1440\times 140\times \frac{4}{3}}=3170 \text{ mm}^2$$

Provide 2 bars of 25 mm ϕ.

Reinforcement between columns C$_3$ and C$_4$

$$A_t=\frac{316{,}000\times 1000}{0{\cdot}87\times 1440\times 140\times \frac{4}{3}}=1355 \text{ mm}^2$$

Provide 2 bars of 25 mm ϕ and 4 bars of 16 mm ϕ.

Reinforcement at column C$_3$

$$A_t=\frac{290{,}000\times 1000}{0{\cdot}87\times 1440\times 140\times \frac{4}{3}}=1245 \text{ mm}^2$$

Provide 3 bars of 25 mm ϕ.

Check for shear

Maximum shear stress at column C_2

$$=\frac{663{,}800}{1440\times 400}=1{\cdot}152 \text{ N/mm}^2$$

Allowable $\tau_c=0{\cdot}34$ N/mm^2

Net shear for stirrups=$(1{\cdot}152-0{\cdot}34)\times 1440\times 400=467{,}712$ N

Providing 4-legged 12 mm ϕ stirrups,

$$\text{Spacing} = \frac{1440\times4\times113\times140\times{}^{3}}{467{,}712} = 260 \text{ mm}$$

Provide spacing of 150 mm.

5 stirrups of 12 mm ϕ at 150 mm c/c are provided to the right of columns C_2, C_3 and C_4. In the remaining portion 8 mm ϕ, 4-legged stirrups at 500 mm c/c are provided.

Details of reinforcement are shown in Fig. 10·27.

10·11. Strap Footings

Strap footings are provided when an independent footing cannot be extended in the adjoining boundary. Strap beams are provided to connect two isolated footings provided under two columns or to connect a footing and a wall as shown in Fig. 10·28.

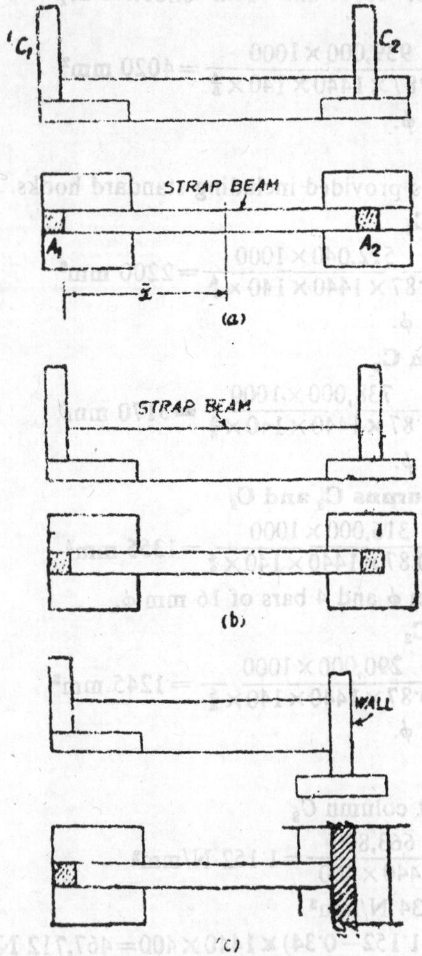

Fig. 10·28

It is assumed that the load is transferred to the soil through the footings only and the strap beam does not transfer any pressure to soil. The areas of two footings are so chosen that the pressure on the footings is uniform and the resultant pressure is in the same line as the resultant of loads.

If the loads from columns C_1 and C_2 are W_1 and W_2 respectively and total weight of the footing is W_3

$$A_1 + A_2 = \frac{W_1 + W_2 + W_3}{p_0}$$

where A_1 and A_2 are areas of two footings and p_0 is bearing capacity of soil.

The centroid of areas A_1 and A_2 will be where the resultant of loads act say $\bar{x}$ from

C_1 such that $\bar{x} = \dfrac{W_2 l}{W_1 + W_2}$.

The footings are designed as isolated footings for B.M. and punching. In case the strap beam is extended beyond the column rib beam, the footing is designed as a cantilever extended from the rib.

In the design of strap beam it is assumed that it does not take any pressure from soil and thus it is so constructed that it does not bear directly on the soil. Further it is assumed that the strap beam is rigid and transfers reaction to the footing. Therefore, the strap beam is to be designed for S.F. and B.M.

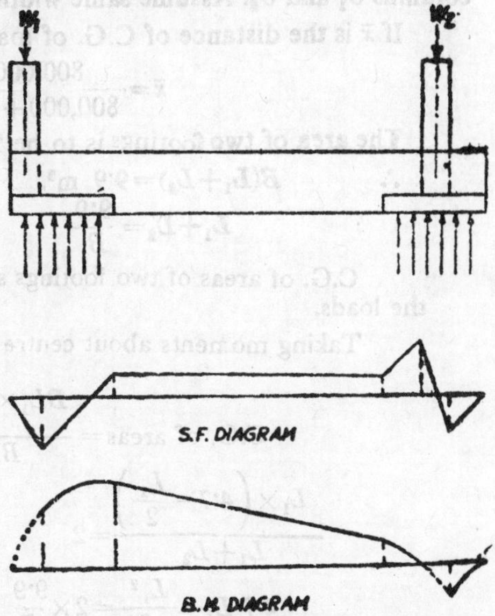

Fig. 10 29

This type of footing is economical than combined footing when the distance between the columns is large.

Example 10·23. *Design a strap footing for two columns C_1 and C_2 400 mm × 400 mm carrying loads of 800,000 N and 1,000,000 N respectively. Distance between centres of column is 4·5 mm. The column C_1 is on the property line. Bearing capacity of soil is 200 kN/m².*

Sol. Assume weight of footing as 10% of load on footing.

$$\text{Self-weight of footing} = \frac{1}{10}(800{,}000 + 1{,}000{,}000) = 180{,}000 \text{ N}$$

$$\text{Total load} = 800{,}000 + 1{,}000{,}000 + 18{,}000 = 1{,}980{,}000 \text{ N}$$

$$\text{Area of footing required} = \frac{1{,}980{,}000}{200{,}000} = 9{\cdot}9 \text{ m}^2.$$

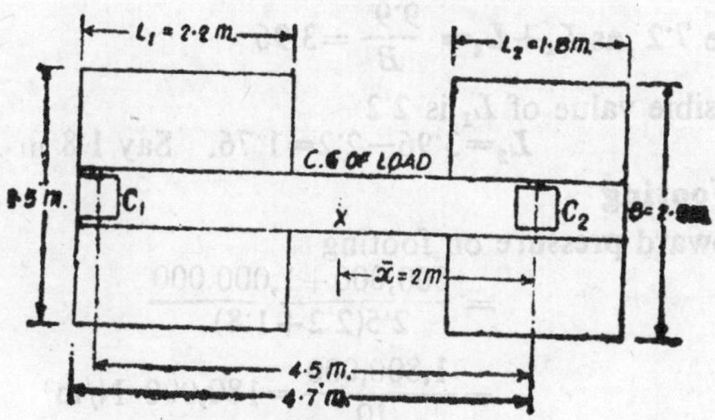

Fig. 10·30

Fixing size of Footin

Let L_1 and L_2 be the lengths of footings provided under columns C_1 and C_2. Assume same width for two footings equal to B.

If $\bar{x}$ is the distance of C.G. of loads centre of column C_2,

$$\bar{x}=\frac{800{,}000\times 4\cdot5}{800{,}000+1{,}000{,}000}=2 \text{ m.}$$

The area of two footings is to be 9·9 m².

$$\therefore \quad B(L_1+L_2)=9\cdot9 \text{ m}^2.$$

$$L_1+L_2=\frac{9\cdot9}{B}$$

C.G. of areas of two footings should coincide with the C.G. of the loads.

Taking moments about centre of column C_2.

$$\text{C.G. of areas}=\frac{BL_1\times\left(4\cdot7-\frac{L_1}{2}\right)}{BL_1+BL_2}=2 \text{ m.}$$

$$\frac{L_1\times\left(4\cdot7-\frac{L_1}{2}\right)}{L_1+L_2}=2$$

$$4\cdot7\,L_1-\frac{L_1^2}{2}=2\times\frac{9\cdot9}{B}$$

$$L_1^2-9\cdot4L_1=-\frac{39\cdot6}{B}$$

$$(L_1-4\cdot7)^2=-\frac{39\cdot6}{B}+22\cdot09$$

$$\therefore \quad L_1=\pm\sqrt{-\frac{39\cdot6}{B}+22\cdot09}+4\cdot7$$

Taking $B=2\cdot5$ m.

$$L_1=\pm\sqrt{15\cdot84+22\cdot09}+4\cdot7$$

$$=\pm2\cdot5+4\cdot7$$

L_1 cannot be 7·2 as $L_1+L_2=\frac{9\cdot9}{B}=3\cdot96$

Admissible value of L_1 is 2·2

$$L_2=3\cdot96-2\cdot2=1\cdot76. \quad \text{Say } 1\cdot8 \text{ m}$$

Design of Footing

Net upward pressure on footing

$$=\frac{800{,}000+1{,}000{,}000}{2\cdot5(2\cdot2+1\cdot8)}$$

$$=\frac{1{,}800{,}000}{10}=180{,}000 \text{ N/m}^2$$

The width of strap beam provided is 500 mm.

Projection of footing beyond strap beam

$$=\frac{2{\cdot}5-0{\cdot}5}{2}$$

$$=1{\cdot}00\text{ m}=1000\text{ mm}$$

Consider 1 mm width of footing.

$$\text{Maximum B.M.}\quad =\frac{180{,}000}{1000\times 1000}\times\frac{1000\times 1000}{2}$$

$$=90{,}000\text{ N-mm}$$

Effective depth required

$$=\sqrt{\frac{90{,}000}{0{\cdot}87\times 1}}=321{\cdot}7\text{ mm}$$

Provide overall depth of 420 mm with effective depth of 340 mm.

$$\text{Area of steel } A_t=\frac{90{,}000}{0{\cdot}87\times 340\times 140}=2{\cdot}143\text{ mm}^2$$

Provide 16 mm ϕ bars,

$$\text{Spacing}=\frac{201}{2{\cdot}143}=92{\cdot}4\text{ mm.}$$

Provide 16 mm ϕ bars at 90 mm centres.

Distribution steel $=0{\cdot}5\%$ of cross-sectional area

$$=\frac{0{\cdot}15\times 420\times 1}{100}=0{\cdot}63\text{ m}^2.$$

Providing 10 mm bars,

$$\text{Spacing}=\frac{79}{0{\cdot}63}=125{\cdot}4\text{ mm.}$$

Provide 10 mm ϕ bars at 120 mm centres.

Design of Strap Beam

Upward pressure on the beam per metre length

$$=180{,}000\times 2{\cdot}5\times 1$$

$$=450{,}000\text{ N.}$$

The loading on the beam and the S.F. and B.M. diagrams are shown in Fig. 10·31.

Maximum B.M. $=525{,}110$ Nm

$=525{,}110{,}000$ N-mm

Width of beam provided is 500 mm

Effective depth of beam

$$=\sqrt{\frac{525{,}110{,}000}{0{\cdot}87\times 500}}=1098\text{ mm}$$

Provide overall depth of 1200 mm with effective depth of 1100 mm

$$A_s = \frac{5{,}251{,}000}{0{\cdot}87 \times 1100 \times 140} = 3920 \text{ mm}^2$$

Provide 14 bars of 20 mm ϕ in two layers.

Check for Bond

Maximum shear force = 620,000 N

Sufficient anchorage length is provided.

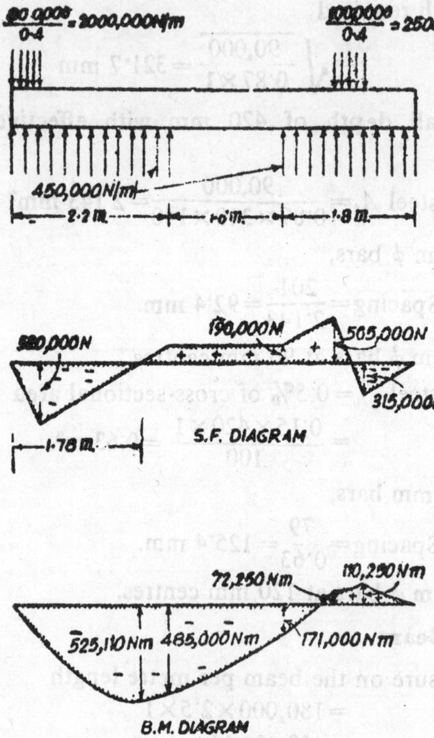

Fig. 10·31

14 bars of 20 mm ϕ are provided.

Area of steel required for negative B.M.

$$= \frac{110{,}250 \times 1000}{0{\cdot}87 \times 1100 \times 140} = 823 \text{ mm}^2$$

9 bars of 20 mm ϕ are provided at bottom near column C_2.

Check for Shear

$$\text{Shear stress} = \frac{620{,}000}{1100 \times 500} = 1{\cdot}127 \text{ N/mm}^2$$

Allowable shear stress assumed $=0{\cdot}34$ N/mm^2

Net shear for stirrups $=(1{\cdot}127-0{\cdot}34)\times 1100\times 500=432{,}850$ N

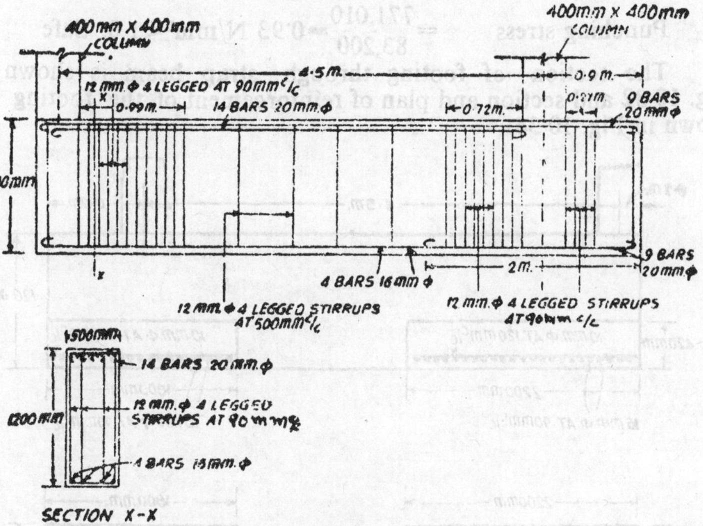

Fig. 10·32

Providing 4-legged, 12 mm ϕ stirrups,

$$\text{Spacing}=\frac{0{\cdot}87\times 1100\times 4\times 113\times 140}{432{,}850}=139{\cdot}9\text{ mm}$$

Provide stirrups at 90 mm centres.

Provide stirrups upto length of 1 m beyond face of column C_1 and upto 0·7 m beyond face of column C_2 towards left and upto 0·25 m beyond face of column C_2 towards right.

In the remaining portion provide 12 mm ϕ, 4-legged stirrups at 500 mm centres.

Check for punching under column C$_1$

Punching force

$$=800{,}000-400\times 400\times\frac{180{,}000}{1000\times 1000}$$

$$=8{,}000{,}000-28{,}800$$

$$=771{,}200\text{ N}$$

The length 500 mm will have to be punched through 1120 mm and 400 mm widths have to be punched through 340 mm.

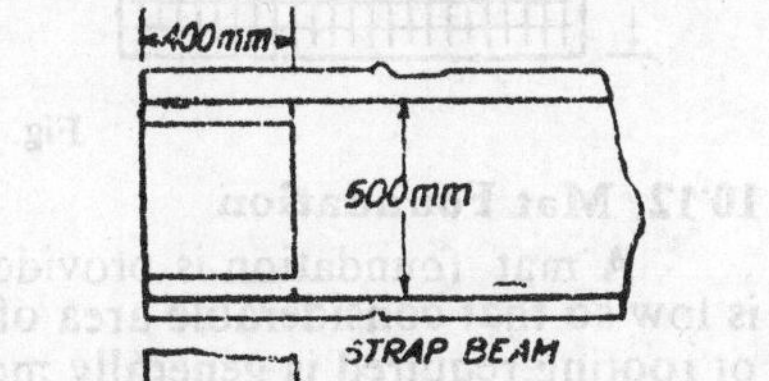

Fig. 10·33

Area resisting punching

$$= 500 \times 1120 + 2 \times 400 \times 340 = 56{,}000 + 27{,}200 = 83{,}200$$

Punching stress $= \dfrac{771{,}010}{83{,}200} = 0{\cdot}93 \text{ N/mm}^2 < 1{\cdot}0$. Safe

The section of footing through strap beam is shown in Fig. 10·32 and section and plan of reinforcement of the footing are shown in Fig. 10·34.

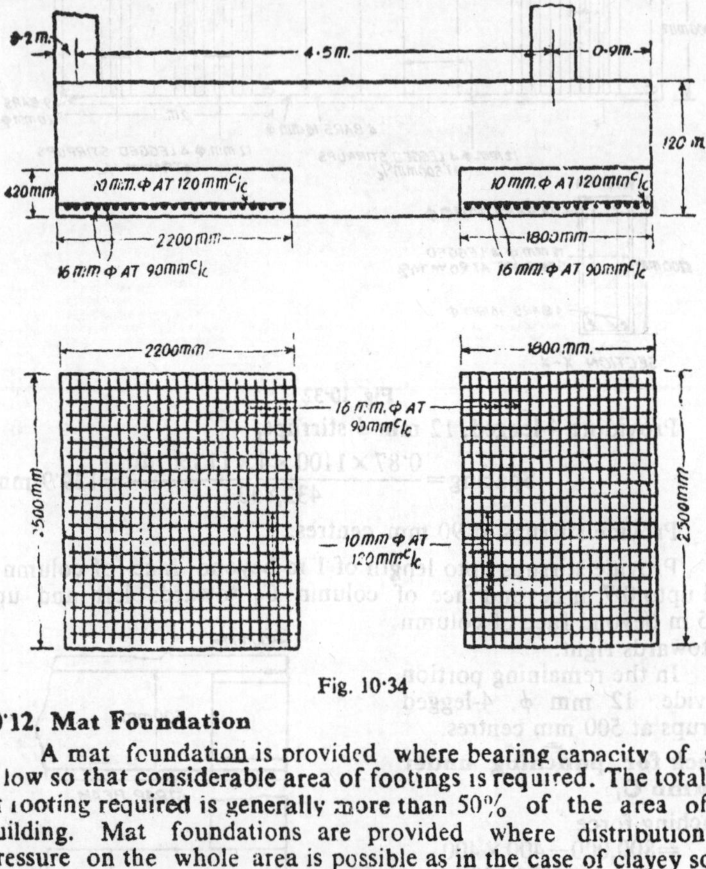

Fig. 10·34

10·12. Mat Foundation

A mat foundation is provided where bearing capacity of soil is low so that considerable area of footings is required. The total area of footing required is generally more than 50% of the area of the building. Mat foundations are provided where distribution of pressure on the whole area is possible as in the case of clayey soils. In case of stiff soils the load will be distributed on smaller area round about the column and the mat will not be effective.

In case columns are spaced uniformly and carry equal loads, the mat foundation can be looked as series of continuous combined footings in two directions. As the load is carried in two directions, a reduction factor for bending moments and shear forces may be used. The reduction factor will depend on the layout of the columns and will vary from 15 to 30%.

Beam and slab system of mat foundation is preferred when the area is extensive and the loads coming on the columns may be different or spacing of the columns is different and there is possibility of differential settlement.

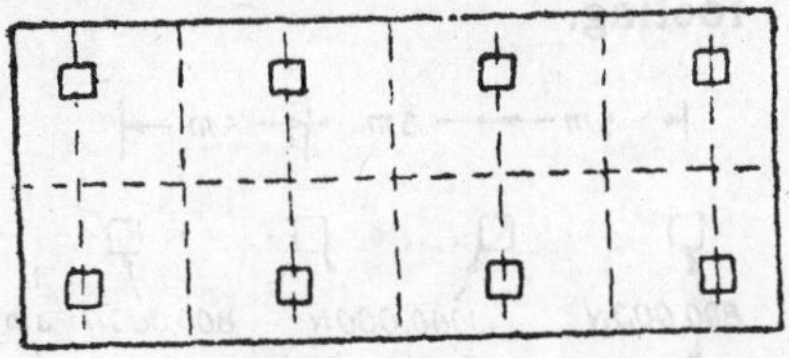

MAT FOUNDATION

To avoid cracks due to shrinkage, small areas 10 m × 10 m are constructed at a time. Construction joints are located at places of low shear stress, generally along the centre lines between the columns. The adjacent strips are filled after 24 hours. The reinforcing bars should be continuous across the joints. The concrete should be strong at the joints to transfer the shear. The section may be thickened at the joints, if necessary.

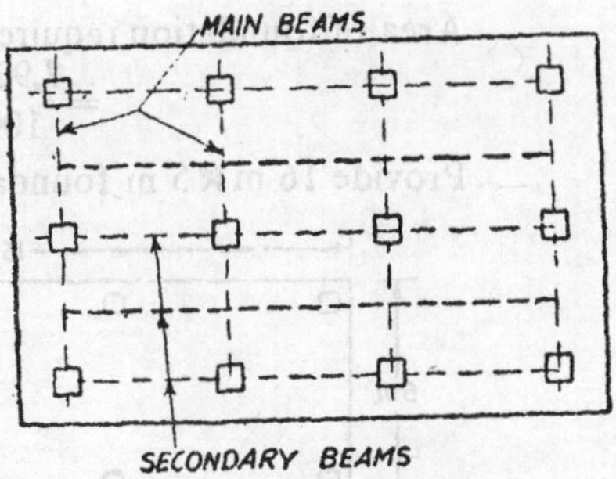

MAT FOUNDATION

Fig. 10·35.

Design of beam slab system of mat foundation

Centre of gravity of loads is worked out and eccentricities e_x and e_y along x-axis and y-axis are calculated.

The pressure of soil at any point (x, y) from C.G. will be given by $\frac{\Sigma W}{A}+\frac{\Sigma W\times e_y}{I_x}\times y+\frac{\Sigma W+e_x}{I_y}\times x$, where A is the area of the mat and I_x and I_y are moments of inertia about x-axis and y-axis respectively. The pressure on the soil will vary linearly.

The slab is designed as continuous one taking B.M. as $Wl/12$, where W is the total load on the slab per unit width of slab. As the whole structure including columns is monolithic, the secondary beams are designed for bending moment of $Wl/10$, where W is the total load on the beam. The main beams are designed to take the load from some portion of the slab, from secondary beams and any difference between upward soil pressure and column loads.

Example 10·14. *Design a mat foundation for the system of columns shown in Fig. 10·36. All columns are 400 mm × 400 mm and they carry the loads as indicated in the figure. The bearing capacity of soil is 100 kN/m².*

Sol. Assume self-weight of the footing at 10% of load on footing.

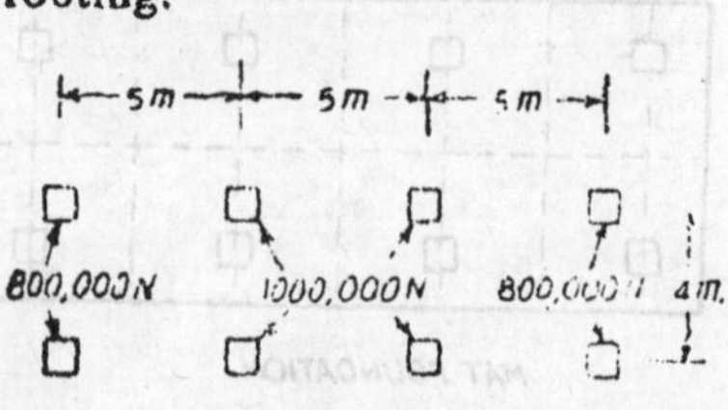

Fig. 10·36.

Self-weight of footing

$$=\frac{4\times800{,}000+4\times1{,}000{,}000}{10}$$

$=720{,}000$ N.

Total load on the foundation

$=72{,}000+4\times800{,}000+4\times1{,}000{,}000$

$=7{,}920{,}000$ N.

Area of foundation required

$$=\frac{7{,}920{,}000}{100{,}000}=79{\cdot}2\ \text{m}^2.$$

Provide 16 m × 5 m foundation as shown in Fig. 10·37.

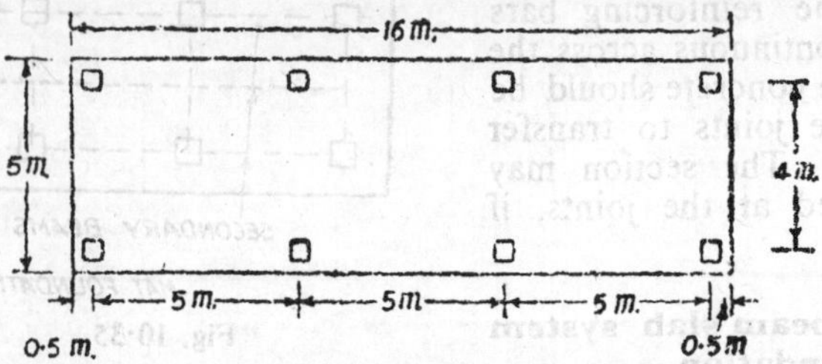

Fig. 10·37

Net upward pressure

$$=\frac{4\times800{,}000+4\times1{,}000{,}000}{16\times5}=\frac{7{,}200{,}000}{80}$$

$=90{,}000$ N-m².

The mat will be designed as a continuous footing in two directions and a reduction factor of 20% will be used for shear force and bending moment. Consider half of the mat in longitudinal direction. Upward pressure per metre length

$=90{,}000\times2{\cdot}5=225{,}000$ N.

The shear force and bending moment diagrams are shown in Fig. 10·38.

Design of footing

Maximum B.M. $=0{\cdot}8\times1{,}300{,}000=1{,}040{,}000$ N-m

$=1{,}040{,}000{,}000$ N-mm

Effective depth of footing required

$$=\sqrt{\frac{1{,}040{,}000{,}000}{0{\cdot}87\times2500}}$$

$=691{\cdot}5$ mm.

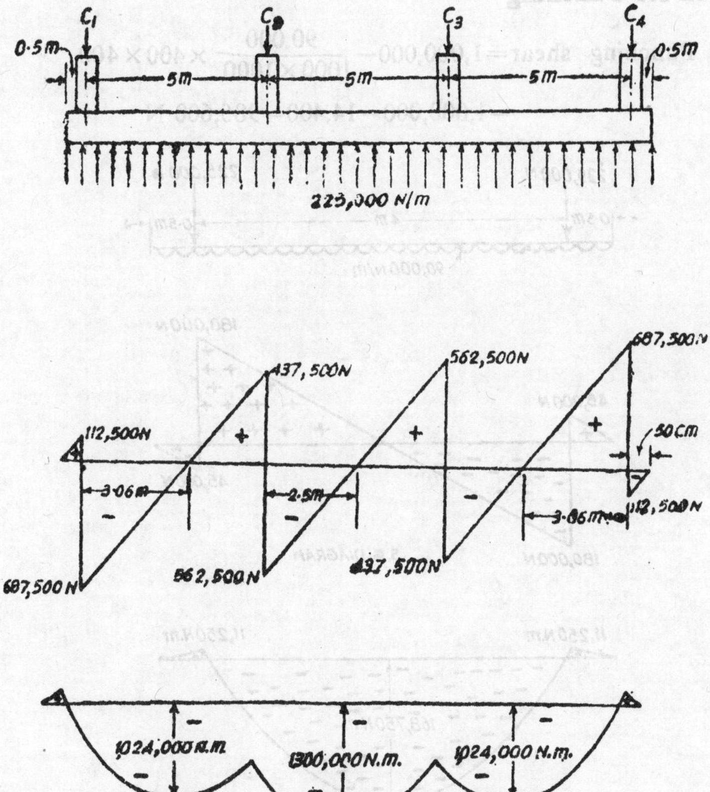

Fig. 10·38.

Provide overall depth of 800 mm with effective depth of 700 mm.

$$A_t = \frac{1,040,000,000}{0·87 \times 700 \times 140} = 12,200 \text{ mm}^2$$

Provide 25 bars of 25 mm ϕ.

Reinforcement between columns C_1 and C_2

$$A_t = \frac{1,024,000 \times 0·8 \times 1000}{0·87 \times 700 \times 140} = 9640 \text{ mm}^2$$

Provide 20 bars of 25 mm ϕ.

Check for bond

Maximum shear under column 2 = 0·8 × 562,500 N = 450,000 N

Check for Punching

$$\text{Punching shear} = 1{,}000{,}000 - \frac{90{,}000}{1000 \times 1000} \times 400 \times 400$$

$$= 1{,}000{,}000 - 14{,}400 = 985{,}600 \text{ N}$$

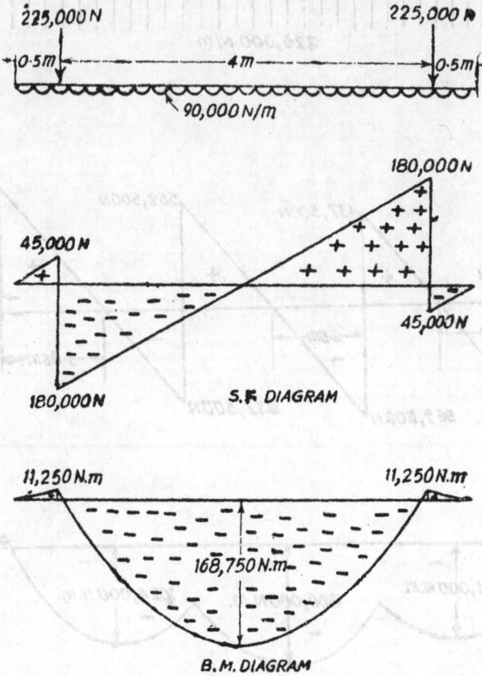

Fig. 10·39.

$$\text{Punching stress} = \frac{985{,}600}{4 \times 400 \times 700}$$

$$= 0{\cdot}88 \text{ N/mm}^2 < 1{\cdot}0. \quad \text{Safe.}$$

Check for Shear

Maximum shear force

$$= 0{\cdot}8 \times 687{,}500 \text{ kg.}$$

$$\text{Shear stress} = \frac{0{\cdot}8 \times 687{,}500}{700 \times 2500} = 0{\cdot}314. \quad \text{Safe.}$$

Design in Transverse Direction

Consider whole length of the footing.

The loading and S.F. and B.M. diagrams are shown in Fig. 10·39. Maximum B.M. $= 0{\cdot}8 \times 2{,}700{,}000$ Nm.

Effective depth required

$$= \sqrt{\frac{0{\cdot}8 \times 2{,}700{,}000 \times 1000}{0{\cdot}87 \times 16{,}000}} = 440{\cdot}4 \text{ mm}$$

Effective depth provided is 700 mm

$$A_t = \frac{0{\cdot}8 \times 2{,}700{,}000 \times 1000}{0{\cdot}87 \times 700 \times 140} = 29{,}300 \text{ mm}^2$$

Provide 60 bars of 25 mm ϕ at top.

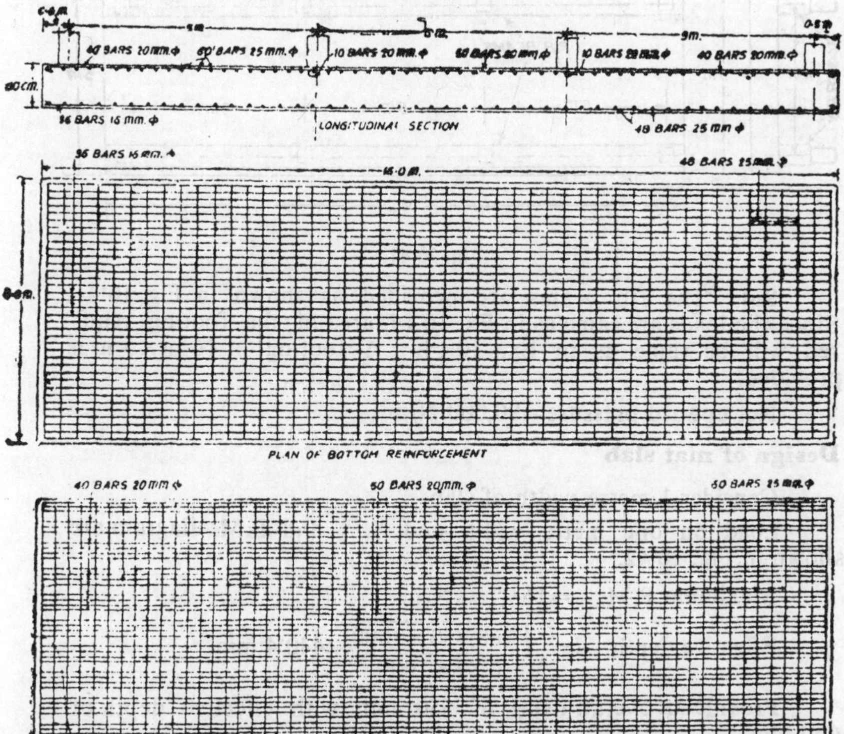

Fig. 10·40.

Reinforcement in cantilever portion

Minimum reinforcement

$$= 0{\cdot}15\% \text{ of cross-sectional area}$$

$$= \frac{0{\cdot}15 \times 16{,}000 \times 800}{100} = 19{,}200 \text{ mm}^2$$

40 bars, 20 mm ϕ have been provided.

Check for Shear

Maximum shear force$=0{\cdot}8\times 2{,}280{,}000$ N

Shear stress$=\dfrac{0{\cdot}8\times 2{,}280{,}000}{700\times 16{,}000}=0{\cdot}206$ N/mm². Safe.

No shear reinforcement is required. The arrangement for reinforcement in the footing is shown in Fig. 10·40.

Example 10·16. *Design a mat foundation for the columns of Ex. 10·4 provide two rib beams.*

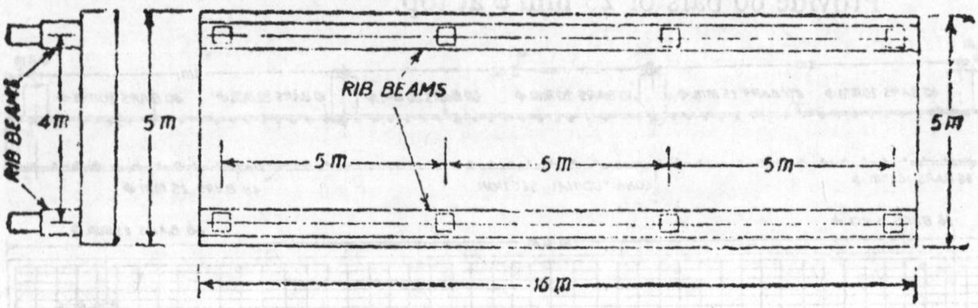

Fig. 10·41.

Sol. The rib beams are provided as shown in Fig. 10·41. Since the loads are same as in Ex. 10·14, area of the footing required will be same as in Ex. 10·14. A footing of 16 m×5 m is provided.

Net upward pressure=90,000 N/m².

Design of mat slab

Consider 1 metre width of slab.

Load on one metre width and S.F. and B.M. diagram are shown in Fig. 10·42.

Maximum B.M. $=168{,}750$ N-m

Effective depth $d=\sqrt{\dfrac{168{,}750\times 1000}{0{\cdot}87\times 1000}}=440{\cdot}4$ mm.

Provide an overall depth of 550 mm with effective depth of 450 mm

$$A_t=\frac{168{,}750\times 1000}{0{\cdot}87\times 450\times 140}=3080 \text{ mm}^2$$

Providing 20 mm ϕ bars,

$$\text{spacing}=\frac{1000\times 314}{3080}=102 \text{ mm}$$

Provide 20 mm ϕ bars, at 100 mm centres,

Distribution steel$=\dfrac{0{\cdot}15\times 550\times 1000}{100}=815$ mm²

Providing 12 mm ϕ bars,

$$\text{Spacing}=\frac{1000\times 113}{825}=137 \text{ mm.}$$

Provide 12 mm ϕ bars at 130 mm centres.

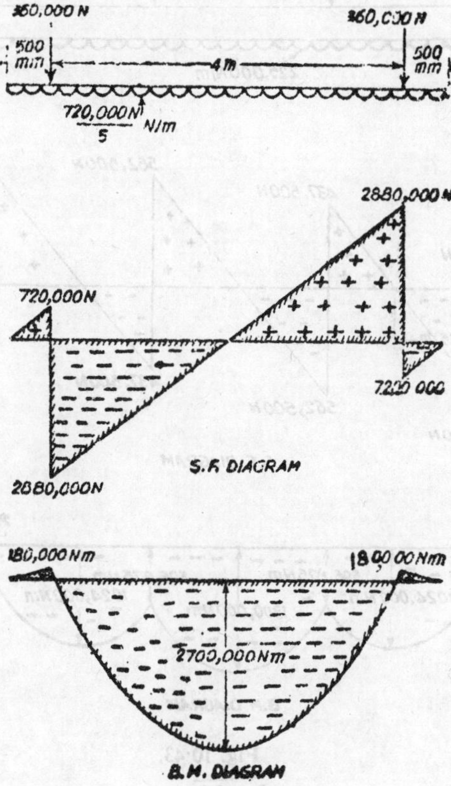

Fig. 10 42.

Reinforcement in the cantilever portion

$$A_t=\frac{11,250\times 1000}{8.87\times 450\times 140}=210 \text{ mm}^2$$

Provide 12 mm ϕ bars at 80 mm centres.

Design of beam

The loading on the rib beam is shown in Fig. 10·43 (*a*)

Net upward pressure per metre run of the beam

$$=90,000\times 2.5=225,000 \text{ N}$$

The S.F. and B.M. diagram for the beam are shown in Fig. 10·43.

Maximum B.M. $=1,300,000$ Nm.

The beam will act as T-beam. Rib width of 500 mm is provided.

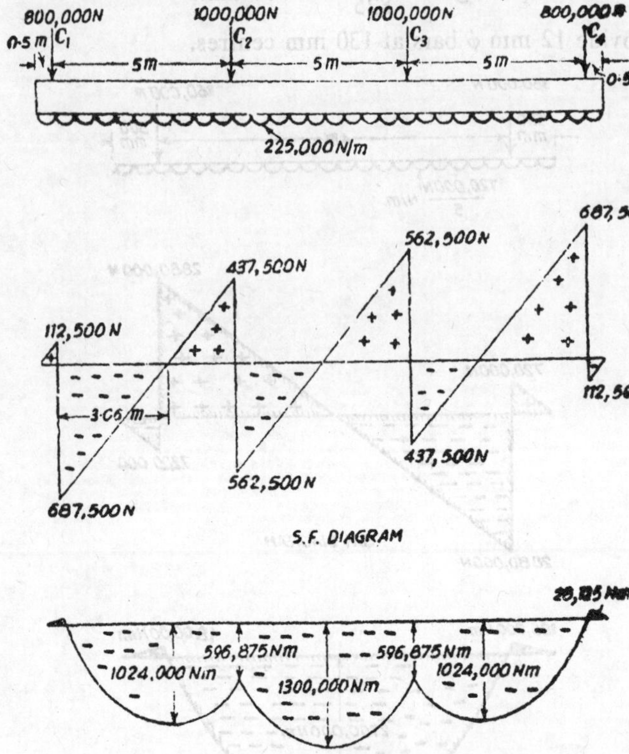

Fig. 10·43.

Economical depth $= \sqrt{\dfrac{M}{2{\cdot}1\ b_r}}$

$$d=\sqrt{\frac{1{,}300{,}000\times 1000}{2{\cdot}1\times 500}}=1113 \text{ mm}$$

Provide an overall depth of 1250 mm with effective depth of 1120 mm.

Flange width $=\dfrac{l_o}{6}+b_w+6D_f$

∴ Flange width $=167$ cm

As there is only 250 mm width on one side,

Flange width $=250+500+\dfrac{\left(\dfrac{l_o}{6}+6D_f\right)}{2}$

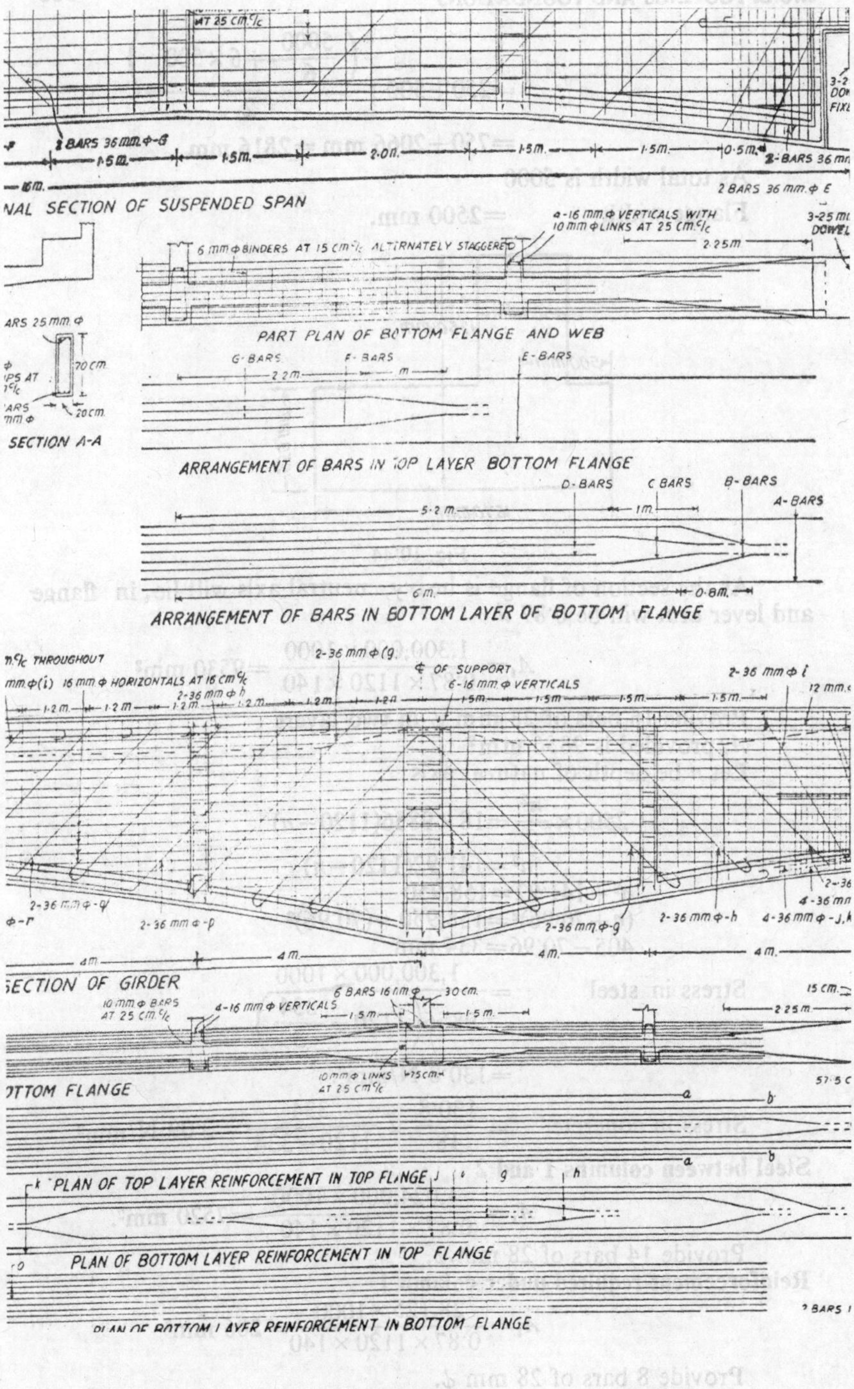

AT 25 CM.C/C
2 BARS 36 MM.Φ-G
1·5 M.
1·5 M.
2·0 M.
1·5 M.
1·5 M.
0·5 M.
2-BARS 36 MM
2 BARS 36 MM.Φ E
NAL SECTION OF SUSPENDED SPAN
4-16 MM.Φ VERTICALS WITH 10 MM Φ LINKS AT 25 CM.C/C
3-25 MM DOWEL
6 MM Φ BINDERS AT 15 CM.C/C ALTERNATELY STAGGERED
2·25 M.
ARS 25 MM.Φ
70 CM.
20 CM.
SECTION A-A
PART PLAN OF BOTTOM FLANGE AND WEB
G-BARS
F-BARS
E-BARS
2·2 M.
ARRANGEMENT OF BARS IN TOP LAYER BOTTOM FLANGE
D-BARS
C BARS
B-BARS
A-BARS
5·2 M.
1 M.
6 M.
0·8 M.
ARRANGEMENT OF BARS IN BOTTOM LAYER OF BOTTOM FLANGE
2-36 MM Φ (G)
Φ OF SUPPORT
6-16 MM.Φ VERTICALS
2-36 MM Φ I
16 MM Φ HORIZONTALS AT 16 CM C/C
2-36 MM Φ H
1·2 M.
1·5 M.
12 MM.
2-36 MM Φ-Q
2-36 MM Φ-P
2-36 MM.Φ-G
2-36 MM Φ-H
4-36 MM Φ-J,K
4 M.
4 M.
4 M.
4 M.
SECTION OF GIRDER
10 MM Φ BARS AT 25 CM.C/C
4-16 MM Φ VERTICALS
6 BARS 16 MM Φ
30 CM.
1·5 M.
1·5 M.
15 CM.
2·25 M.
10 MM Φ LINKS AT 25 CM C/C
75 CM
OTTOM FLANGE
57·5 C
PLAN OF TOP LAYER REINFORCEMENT IN TOP FLANGE
PLAN OF BOTTOM LAYER REINFORCEMENT IN TOP FLANGE
PLAN OF BOTTOM LAYER REINFORCEMENT IN BOTTOM FLANGE

$$=250+500+\frac{\left(\frac{5000}{6}+6\times 500\right)}{2}$$

$$=750+2066 \text{ mm} = 2816 \text{ mm}$$

As total width is 5000

Flange width $=2500$ mm.

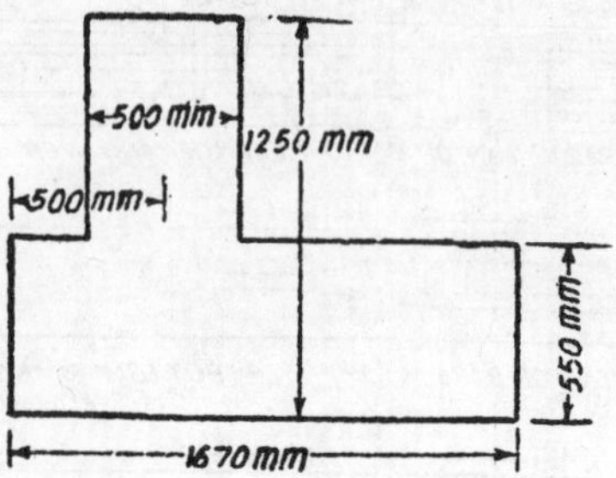

Fig. 10·44

As the section of flange is heavy, neutral axis will lie in flange and lever arm will be 0·87 d

$$A_t=\frac{1{,}300{,}000\times 1000}{0{\cdot}87\times 1120\times 140}=9530 \text{ mm}^2$$

Provine 16 bars of 28 mm ϕ in two layers
A_t provided is 9856 mm²
Let n be depth of natural axis

$$2500\times\frac{n^2}{2}=18\times 9856(1120-n)$$

$$n^2=141{\cdot}92(1120-n)$$

$$n^2+141{\cdot}92=158{,}950$$

$$(n+70{\cdot}96)^2=158{,}950+(70{\cdot}96)^2$$

$$405-70{\cdot}96=334 \text{ mm}$$

Stress in steel $=\dfrac{1{,}300{,}000\times 1000}{9856\left(1120-\dfrac{334}{3}\right)}$

$=130{\cdot}8$ N/mm².

Stress in concrete $=\dfrac{130{\cdot}8}{18}\times\dfrac{334}{1120-334}=3{\cdot}09$ N/mm².

Steel between columns 1 and 2

$$A_t=\frac{1{,}024{,}000\times 1000}{0{\cdot}87\times 1120\times 140}=7520 \text{ mm}^2.$$

Provide 14 bars of 28 mm ϕ.

Reinforcement required under column 1

$$A_t=\frac{28{,}125\times 1000}{0{\cdot}87\times 1120\times 140}=206 \text{ mm}^2$$

Provide 8 bars of 28 mm ϕ.

Check for Shear

Maximum S.F. $=687{,}500$ N

Shear stress $=\dfrac{687{,}500}{1120\times500}=1{\cdot}228$ N/mm^2

Allowable shear stress$=0{\cdot}34$ N/mm^2

Shear force for which stirrups will be provided

$=(1{\cdot}228-0{\cdot}34)\times1120\times500=497{,}280$ N

Using 6-legged 10 mm φ stirrups,

$$\text{Spacing}=\frac{1120\times6\times79\times140}{497{,}280}=149 \text{ mm.}$$

Provide 10 mm ϕ 6-legged stirrups at 90 mm centres. Provide these stirrups for a length of 2 m beyond centres of columns C_1 and C_4. The same shear reinforcement is provided between columns C_2 and C_3 for a distance of 1·5 m beyond columns C_2 and C_3.

Towards the left of columns C_3 and right of columns C_3 same shear reinforcement is provided for a distance of 1 m. In the remaining portion provide 8 mm ϕ 6-legged stirrups at 500 mm c/c.

Check for Punching

Punching force $=1{,}000{,}000-\dfrac{90{,}000}{1000\times1000}\times400\times400$

$=1{,}000{,}000-14{,}400=985{,}600$ N

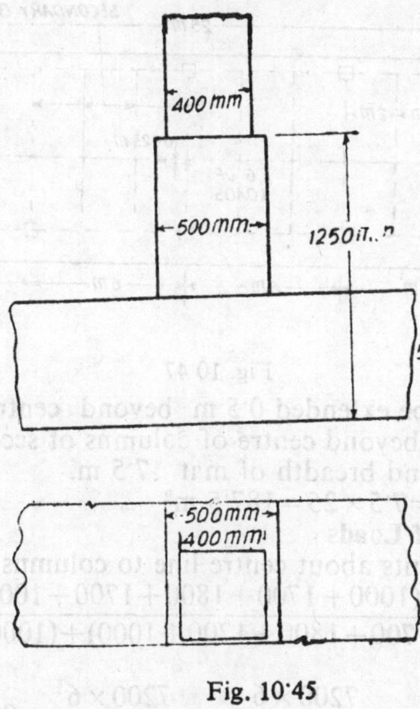

Fig. 10·45

Area resisting punching

$=2\times1120\times500+2\times400\times450$

$=1{,}120{,}000+360{,}000=1{,}480{,}000$ mm^2

Punching stress $= \dfrac{985{,}600}{1{,}480{,}000} = 0{\cdot}667$ N/mm². **Safe.**

Details of reinforcement are shown in Fig. 10·45.

Example 10·16. *Design a suitable foundation for the arrangement of columns shown in Fig. 10·46. The load carried by each column is shown in the figure. The columns in the first and second row are 500 mm × 500 mm, and in third row are 400 mm × 400 mm. The bearing capacity of the soil is 100 kN/m².*

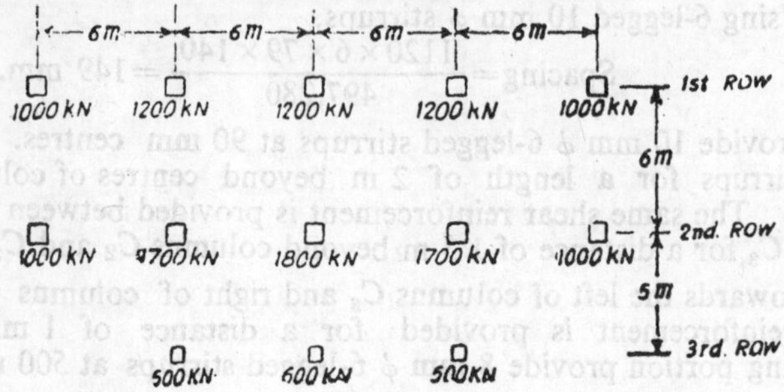

Fig. 10·46

Sol. A mat foundation will be designed for the columns of first and second row and separate combined footing will be provided for the third row.

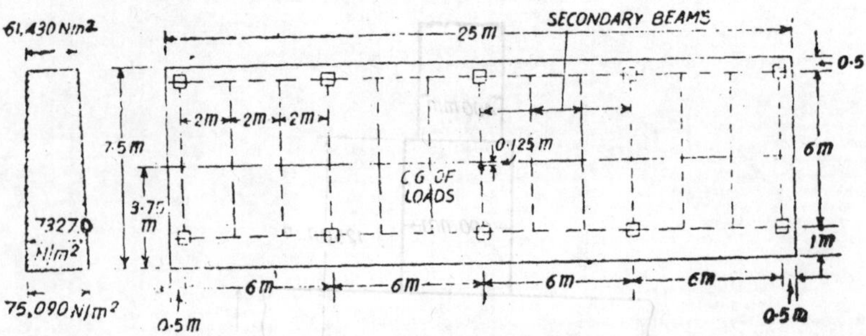

Fig. 10·47

The mat will be extended 0·5 m beyond centre of columns of first row and 1·0 m beyond centre of columns of second row. Length of the mat = 25 m and breadth of mat = 7·5 m.

Area of mat $= 7{\cdot}5 \times 25 = 187{\cdot}5$ m²

Centre of Gravity of Loads

Taking moments about centre line to columns of first row

$$\bar{y} = \frac{(1000+1700+1800+1700+1000)\times 6}{(1000+1700+1800+1700+1000)+(1000+1200+1200+1200+1000)}$$

$$= \frac{7200\times 6}{7200+5600} = \frac{7200\times 6}{12{,}800} = 0{\cdot}375 \text{ m}$$

Eccentricity of load $= 0{\cdot}125$ m.

Assume self-weight of footing as 10% of superimposed load.

Total load $=12,800+1280=14,080$ kN.

Maximum and minimum pressure at base is given by

$$p=\frac{14,080}{187{\cdot}5}+\frac{12,800\times 0{\cdot}15}{2{\cdot}5\times\frac{(7{\cdot}5)^2}{6}}$$

$$=75{\cdot}08+6{\cdot}83$$

$$p_{max}=81{\cdot}91\text{ kN/m}^2$$

$$p_{min}=68{\cdot}5\text{ kN/m}^2.$$

Pressure distribution at base due to superimposed loads only is

$$p=\frac{12,800}{187{\cdot}5}\pm\frac{12,800\times 0{\cdot}125}{25\times\frac{(7{\cdot}5)^2}{6}}$$

$$=68{\cdot}26\pm 6{\cdot}83$$

$$p_{max}=75{\cdot}09\text{ kN/m}^2=75,090\text{ N/m}^2.$$

$$p_{min}=61{\cdot}43\text{ kN/m}^2=61,430\text{ N/m}^2.$$

Design of Footing Slab

Pressure midway between rows of columns

$$=61,430+\frac{(75,090-6143)\times 3{\cdot}5}{7{\cdot}5}$$

$$=61,430+6380=67,810\text{ N/m}^2$$

Span $=2$ m

Maximum B.M. $=\frac{wl^2}{12}=\frac{67,810\times 2^2}{12}$

(Taking average pressure)

$=22,603$ N-m $=22,603,000$ N-mm

Effective depth $d=\sqrt{\frac{22,603,000}{0{\cdot}87\times 1000}}=161{\cdot}2$ mm

Provide overall depth of 240 mm with effective depth of 165 mm.

$$A_s=\frac{22,603,000}{0{\cdot}87\times 165\times 40}=1110\text{ mm}^2$$

Provide 12 mm ϕ bars,

$$\text{Spacing}=\frac{1000\times 113}{1110}=102\text{ mm}$$

Provide 12 mm ϕ bars at 100 mm centres. Some reinforcement is provided at supports

Distribution steel $=\frac{0{\cdot}15\times 240\times 1000}{100}=360$ mm^2

Provide 8 mm ϕ bars,

$$\text{Spacing}=\frac{50\times 1000}{360}=139\text{ mm}$$

Provide 8 mm ϕ bars at 130 mm centres.

Design of Cantilever Portion

Projection of slab beyond face of second row of column

$=0{\cdot}75$ m

Maximum B.M. $=\dfrac{75{,}090\times(0{\cdot}75)^2}{2}$

(Taking maximum pressure)

$=21{,}119$ Nm$=21{,}119{,}000$ N-mm

$A_t=\dfrac{21{,}119{,}000}{0{\cdot}87\times165\times140}$

$=1051$ mm².

Provide 12 mm ϕ bars at 100 mm centres and 8 mm ϕ at 130 mm centres as distribution steel.

In the projection of slab beyond face of first row of columns provide minimum reinforcement of 8 mm ϕ bars at 130 mm centres in both directions.

Design of Secondary Beam

Total load on the beam$=2\times67{,}810\times6=813{,}720$ N

Maximum B.M. will be assumed as $\dfrac{Wl}{10}$

Maximum B.M. $=\dfrac{8{,}13{,}720\times6}{10}=488{,}232$ Nm

$=488{,}232{,}000$ N-mm

The beam is designed as T-beam

Provide rib width of 400 mm.

Economical depth $d=\sqrt{\dfrac{M}{2{\cdot}1\,b_r}}=\sqrt{\dfrac{488{,}232{,}000}{2{\cdot}1\times400}}$

$=762{\cdot}5$ mm.

Provide overall depth of 880 mm with effective depth of 780 mm.

Flange width $=\dfrac{l_0}{6}=b_w+6D_f$

$=\dfrac{6000}{6}+400+6\times240=2840$ mm.

Centre-to-centre of beams

$=2000$ mm

Use 2000 mm width

Assume lever arm $=d-\dfrac{d_s}{2}=780-\dfrac{2400}{2}=660$ mm.

$A_t=\dfrac{488{,}232{,}000}{660\times140}=5280$ mm²

Provide 11 bars of 25 mm ϕ.

Check for Stresses

As the section of flange has large area the neutral axis will lie in flange.

Let n be depth of N.A.

$$\therefore \qquad 2000\times\frac{n^2}{2}=18\times 11\times 491(780-n)$$

$$n^2=75{,}830-97\text{·}2\ n$$

$$(n+48\text{·}6)^2=75{,}830+(48)^2$$

$$280-48=232\ \text{mm.}$$

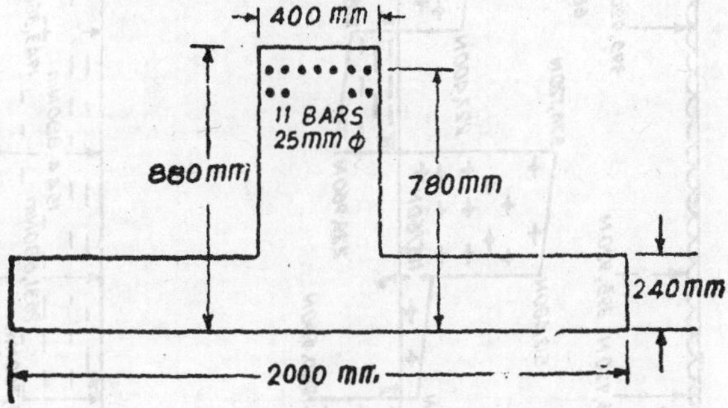

Fig. 10·48

$$\text{Stress in steel} = \frac{488{,}232{,}000}{11\times 491\times\left(780-\dfrac{232}{3}\right)}$$

$$=128\text{·}6\ \text{N/mm}^2.$$

$$\text{Stress in concrete} = \frac{128\text{·}6\times 232}{18\times(780-323)}=3\text{·}02\ \text{N/mm}^2.$$

Check for Shear

Maximum shear force at the end=417,800 N

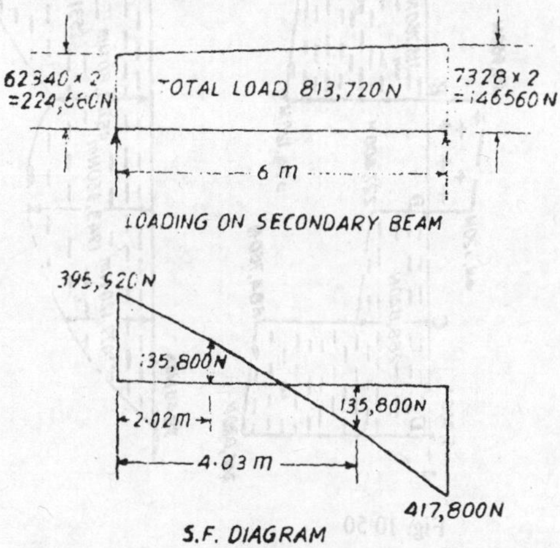

Fig. 10·49

Shear stress

$$=\frac{417{,}800}{780\times 400}$$

$$=1\text{·}339\ \text{N/mm}^2$$

Allowable

$$\tau_c=0\text{·}34\ \text{N/mm}^2$$

$$\text{Shear}=(11\text{·}339-0\text{·}34)\times 780\times 400$$

$$=311{,}688\ \text{N}$$

Two bars are bent down near the support for shear.

Shear taken by 2 bent bars

$$=t\times A_t\times\sin\theta$$

$$=140\times 2\times 419\times\frac{1}{\sqrt{2}}$$

$$=97{,}180\ \text{N.}$$

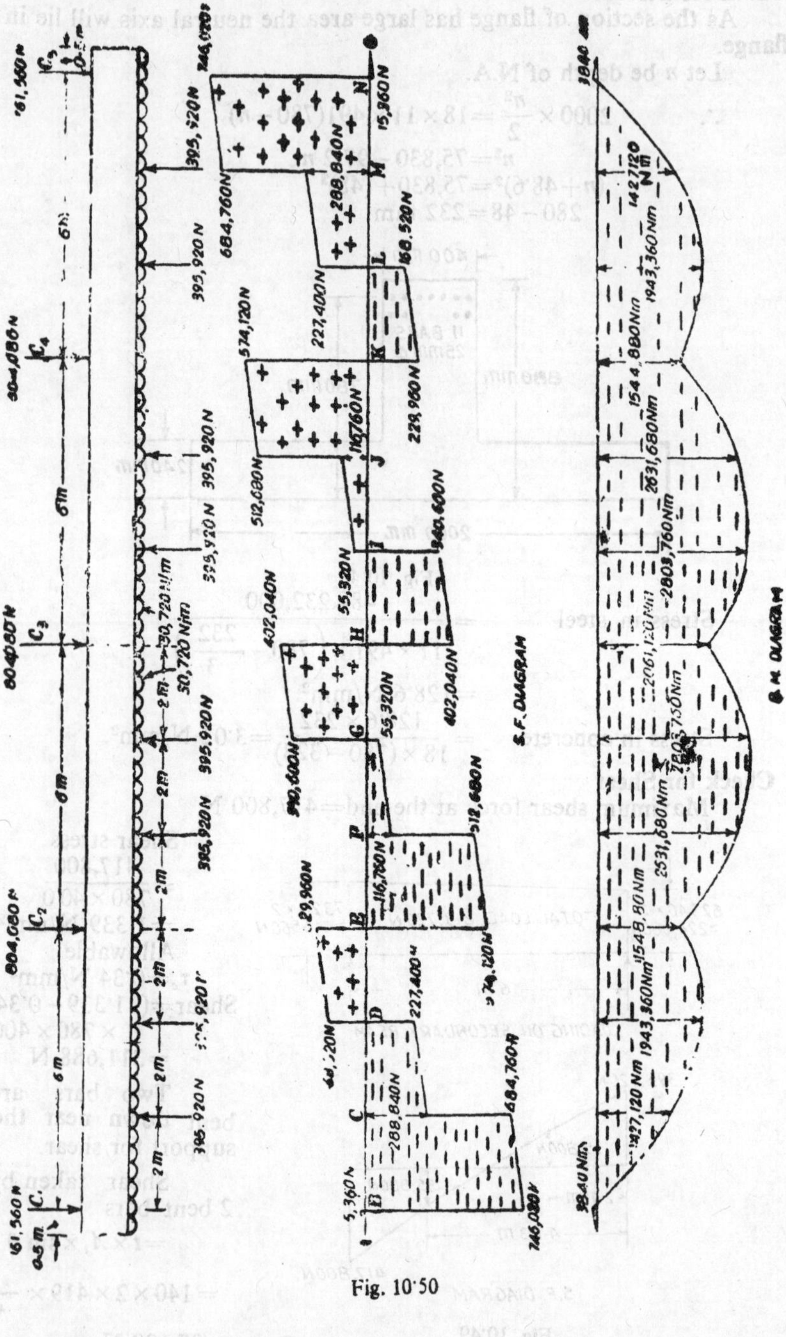

Fig. 10·50

Net S.F. for which stirrups are to be provided

$$=311,688-97,180=214,508$$

Providing 12 mm ϕ 4-legged stirrups,

$$\text{Spacing}=\frac{4\times113\times140\times780}{214,508}=230 \text{ mm}$$

Provide stirrups at 200 mm c/c upto 2·25 m from centre of beam in the remaining portion provide 8 mm ϕ 4-legged stirrups at 500 mm centres.

Check for Bond

Sufficient anchor length has been provided.

Design of main beams

Main beam 1

Upward load from secondary beams

$$=395,920 \text{ N}$$

Upward load from cantilever

$$=0·5\times61,430=39,720 \text{ N/m}$$

Net downward load from columns C_2, C_3 and C_4

$$=1,200,000-395,920$$
$$=804,080 \text{ N}$$

Net downward load from column C_1, C_5

$$=\frac{25\times30,720+8\times395,920-3\times804,080}{2}$$

$$=761,560 \text{ N}$$

The S.F. and B.M. diagrams for the beam are shown in Fig. 10·50.

Maximum B.M. $=2,803,760$ N-m

The beam is designed as a T-beam. Rib width of 500 mm is provided.

$$\text{Economical depth }=\sqrt{\frac{2,803,760\times100}{2·1\times500}}=1634 \text{ mm}$$

Provide overall depth of 1750 mm with effective depth of 1640 mm.

$$\text{Assume lever arm}=d-\frac{d_s}{2}=1640-120=1520 \text{ mm}$$

$$A_t=\frac{2,803,760\times1000}{1520\times140}=13,180 \text{ mm}^2$$

Provide 22 bars of 28 mm ϕ in three rows of 8, 8 and 6 bars.

$$A_t=13,552 \text{ mm}^2$$

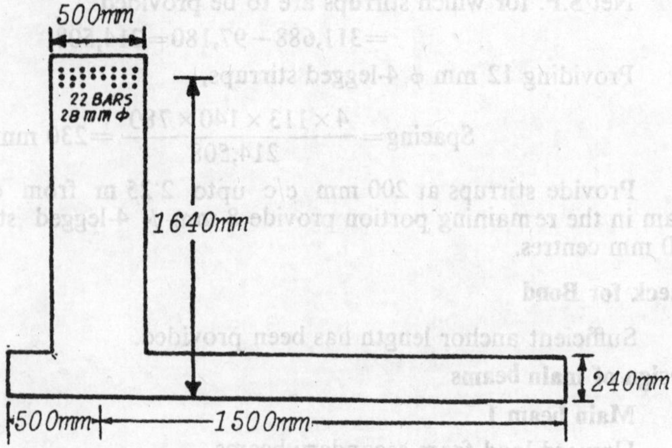

Fig. 10·51

$$\text{Width of flange} = 250+500+\frac{\left[\frac{l_o}{6}+6D_f\right]}{2}$$

$$=250+500+\frac{\left(\frac{6000}{6}+6\times240\right)}{4}$$

$$=1970 \text{ mm}$$

$$\therefore \quad B=1970 \text{ mm}$$

Say 2000 mm.

Neutral axis will lie in web

$$2000\times240(n-120)=18\times13{,}552\times(1640-n)$$

$$\therefore \quad n=\frac{2000\times240\times12+18\times13{,}552\times1640}{2000\times240+18\times13{,}552}$$

$$=\frac{57{,}600{,}000+399{,}900{,}000}{480{,}000+243{,}936}$$

$$=\frac{457{,}500{,}000}{723{,}936}=632 \text{ mm}$$

$$\text{Lever arm} = d-\frac{d_s}{2}+\frac{d_s^2}{6(2n-d_s)}$$

$$=1640-120+\frac{240\times240}{6(1264-240)}$$

$$=1520+0·76$$

$$=1527·6 \text{ mm}$$

$$\text{Stress in steel} = \frac{2{,}803{,}760\times1000}{1527·6\times13{,}552}=135·5 \text{ N/mm}^2$$

Stress in concrete $=\dfrac{1355}{18}\times\dfrac{632}{(1640-632)}$

$=\dfrac{1355}{18}\times\dfrac{632}{1008}$

$=4{\cdot}72\ \text{N/mm}^2$.

Check for Bond

As the point of contraflexure is very near the column C_1 maximum shear for checking bond will be taken as 746,020 N.

16 bars of 28 mm ϕ are provided.
In the cantilever portion provide 6 bars of 28 mm ϕ.

Check for Shear

Maximum shear force

$=746{,}200$ N

Shear stress $=\dfrac{746{,}200}{1640\times 500}=0{\cdot}91\ \text{N/mm}^2$

Allowable shear stress

$\tau_c=0{\cdot}34\ \text{N/mm}^2$

Shear for stirrups $=(0{\cdot}90-0{\cdot}34)\times 1640\times 500$

$=467{,}400$ N

Shear taken by 2 bent bars

$=140\times 616\times 0{\cdot}707\times 2=121{,}900$ N

Net shear force $=467{,}400-121{,}900=345{,}500$ N.

Provide 12 mm ϕ 4-legged stirrup,

$$\text{Spacing}=\frac{4\times 113\times 140\times 1640}{345{,}500}=400\ \text{mm}$$

Provide 12 mm ϕ 4-legged stirrups at 140 mm centres in portions *BC* and *MN*.

Shear reinforcement in portions EF and JK.

Providing 12 mm ϕ 4-legged stirrups,

$$\text{Spacing}=\frac{4\times 113\times 140\times 1640}{(574{,}120-0{\cdot}34\times 1640\times 500)}$$

$=351$ mm

Provide stirrups at 200 mm c/c.

Shear reinforcement in portion GH and HI

Providing 12 mm ϕ 4-legged stirrups,

$$\text{Spacing}=\frac{4\times 113\times 140\times 1640}{(402{,}040-0{\cdot}34\times 1640\times 500)}$$

$=842$ **mm**

Providing spacing of 500 mm centres.

In the remaining portions provide 12 mm ϕ 4-legged stirrups 500 mm centres.

Main Beam 2

Upward load from secondary beams=417,800 N

Upward pressure from cantilever per metre run

$$=\left(\frac{73{,}270+75{,}090}{2}\right)\times 1=74{,}180 \text{ N}$$

Net downward load from column

$$C_3=1{,}800{,}000-417{,}800$$
$$=1{,}382{,}200 \text{ N}$$

Net downward load from column C_2 or C_4

$$=170{,}000-417{,}800=1{,}282{,}200 \text{ N}$$

Net downward load from column C_1 or C_5

$$=\frac{25\times 74{,}180+8\times 417{,}800-2\times 1{,}282{,}200-1\times 3{,}382{,}200}{2}$$

$$=\frac{1{,}854{,}500+3{,}342{,}400-2{,}564{,}400-1{,}382{,}200}{2}$$

$$=\frac{1{,}250{,}300}{2}=625{,}150 \text{ N.}$$

S.F. and B.M. diagrams are shown in Fig. 10·52.

Maximum B.M.=1,021,710 N-m.

The beam will be designed as a T-beam.

Provide rib width of 500 mm.

$$\text{Economical depth} \quad =\sqrt{\frac{1.021{,}710\times 100}{2{\cdot}1\times 500}}=986{\cdot}4 \text{ mm}$$

Provide overall depth of 1150 mm with effective depth of 1040 mm.

$$\text{Assume lever arm}=d-\frac{d_s}{2}=1040-120=920 \text{ mm}$$

$$A_t=\frac{102{,}171\times 1000}{920\times 140}=7932 \text{ mm}^2$$

Provide 4 bars of 28 mm ϕ in two rows. A_t=8624 mm².

Check for stresses

$$\text{Flange width} \quad =250+500+\frac{\left(\frac{l_o}{6}+6D_f\right)}{2}$$

$$=250+500+\frac{\left(\frac{600}{6}+6\times 240\right)}{2}$$

$$=1970 \text{ mm, say 2000 mm.}$$

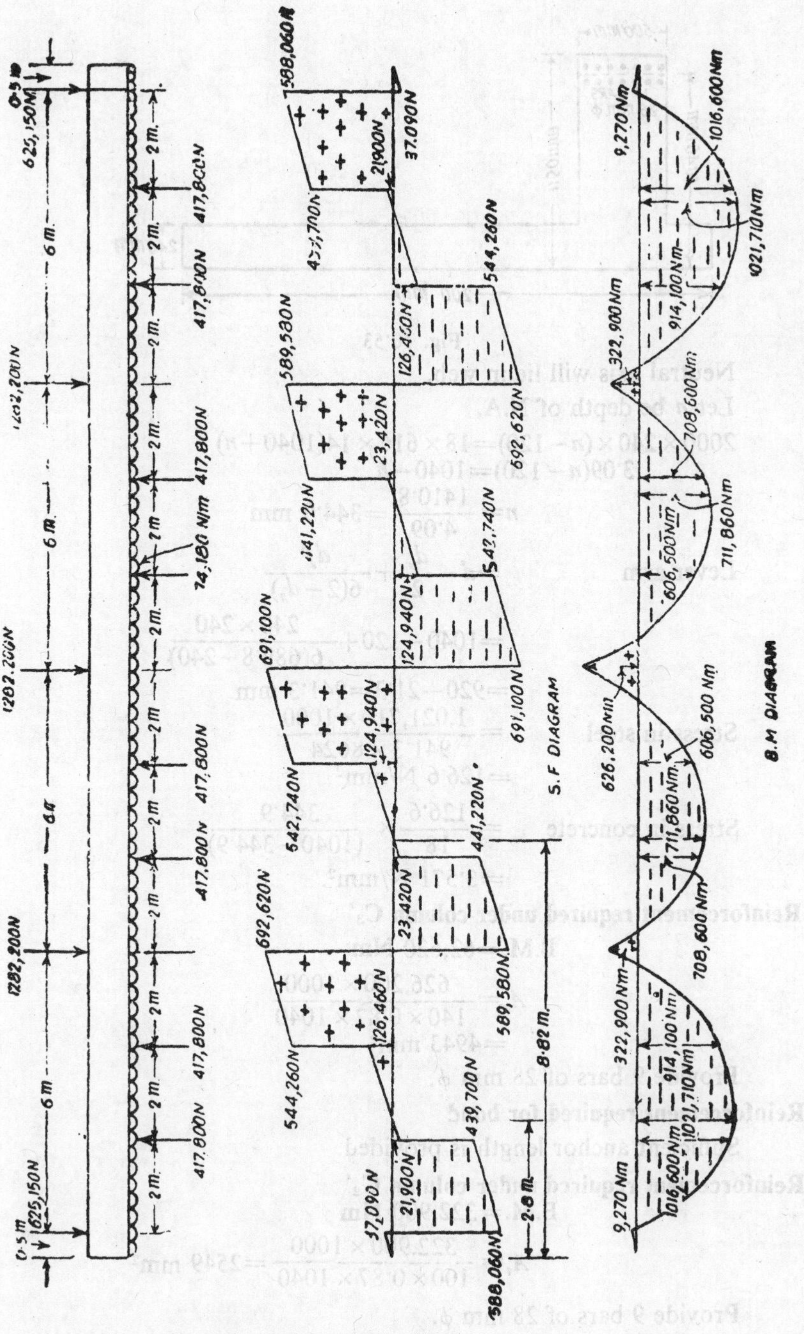

Fig. 10·52

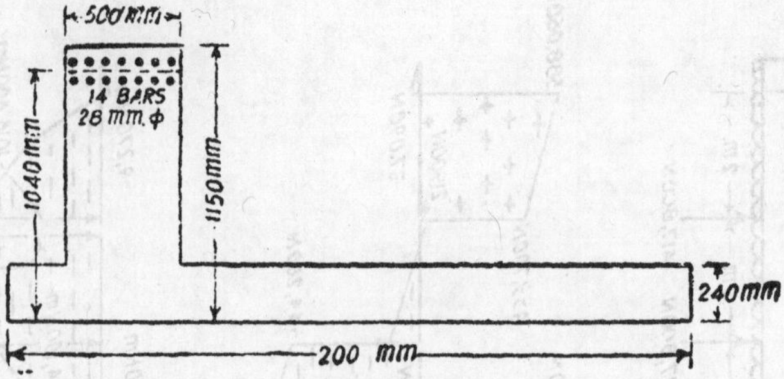

Fig. 10·53

Neutral axis will lie in web.

Let n be depth of N.A.

$$2000 \times 240 \times (n-120) = 18 \times 616 \times 14(1040+n)$$

$$3{\cdot}09(n-120) = 1040 - n$$

$$n = \frac{1410{\cdot}8}{4{\cdot}09} = 344{\cdot}9 \text{ mm}$$

Lever arm $= d - \dfrac{d_s}{2} + \dfrac{d_s^2}{6(2-d_s)}$

$$= 1040 - 120 + \frac{240 \times 240}{6(689{\cdot}8 - 240)}$$

$$= 920 + 21{\cdot}3 = 941{\cdot}3 \text{ mm}$$

Stress in steel $= \dfrac{1{,}021{,}710 \times 1000}{941{\cdot}3 \times 8624}$

$$= 126{\cdot}6 \text{ N/mm}^2$$

Stress in concrete $= \dfrac{126{\cdot}6}{18} \times \dfrac{344{\cdot}9}{(1040 - 344{\cdot}9)}$

$$= 3{\cdot}571 \text{ N/mm}^2.$$

Reinforcement required under column C_3'

$$\text{B.M.} = 62{,}620 \text{ Nm}$$

$$A_t = \frac{626{,}200 \times 1000}{140 \times 0{\cdot}87 \times 1040}$$

$$= 4943 \text{ mm}^2$$

Provide 9 bars of 28 mm ϕ.

Reinforcement required for bond

Sufficient anchor length is provided.

Reinforcement required under column C_2'

$$\text{B.M.} = 322{,}900 \text{ Nm}$$

$$A_t = \frac{322{,}900 \times 1000}{100 \times 0{\cdot}87 \times 1040} = 2549 \text{ mm}^2$$

Provide 9 bars of 28 mm ϕ.

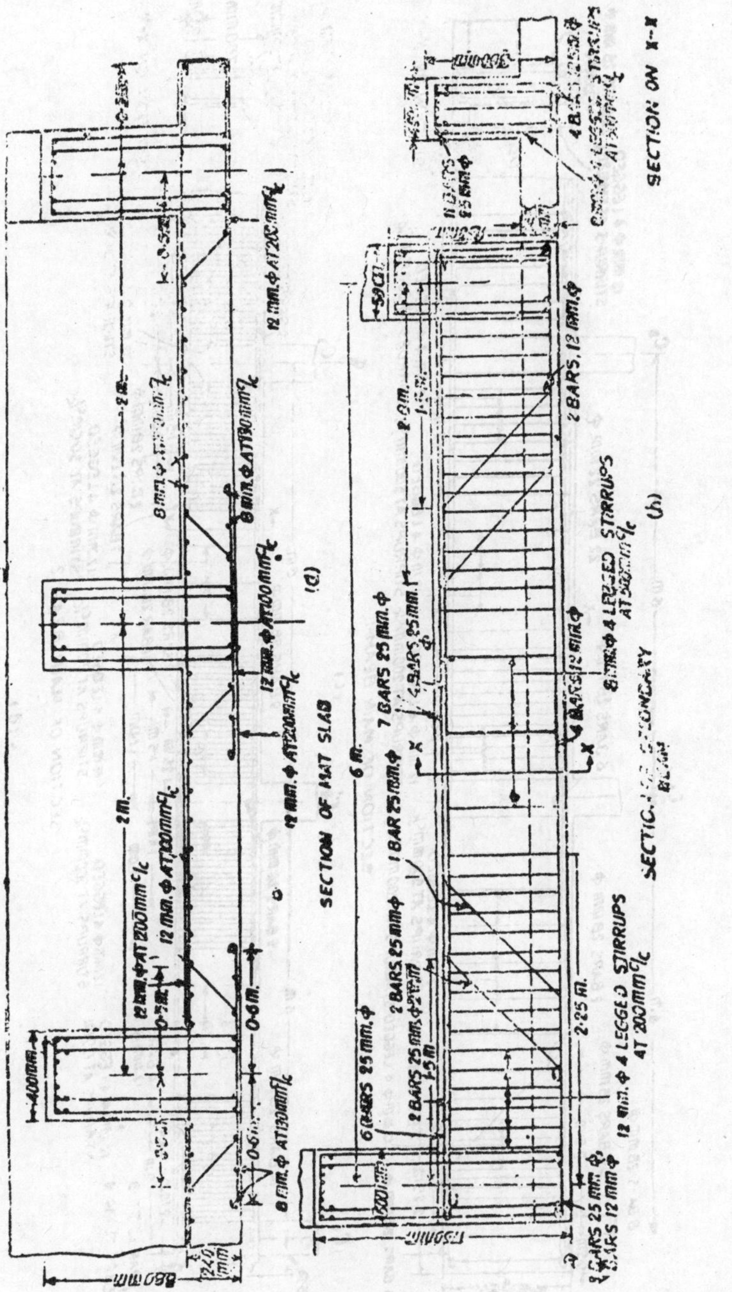

Fig. 10·54

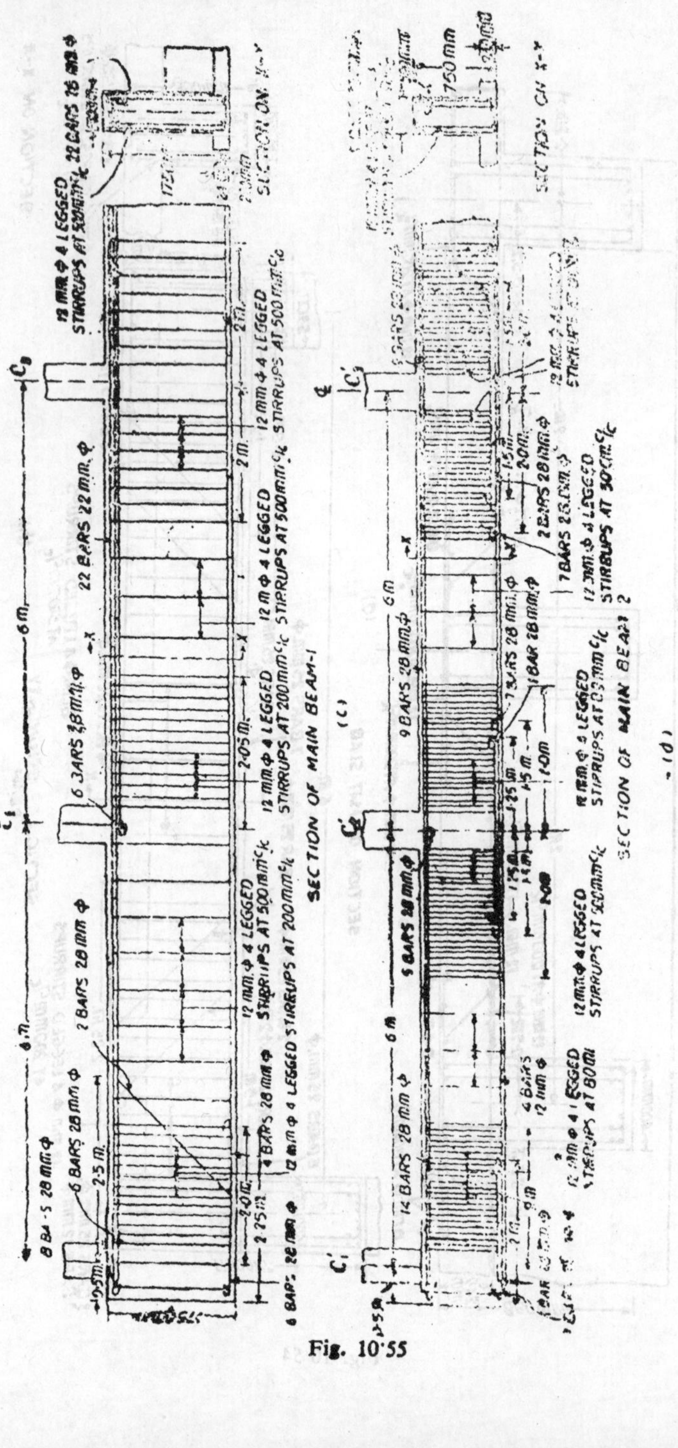

Fig. 10·55

Reinforcement required under column C_1'

Provide 8 bars of 28 mm ϕ.

Reinforcement required between columns C_2' and C_3'

$$A_t = \frac{711,860 \times 1000}{140 \times 920} = 5527 \text{ mm}^2.$$

Provide 9 bars of 28 mm ϕ.

Check for Shear

Maximum S.F. $= 691,100$ N

Shear stress $= \dfrac{691,100}{1040 \times 500}$

$= 1\cdot329 \text{ N/mm}^2.$

Allowable shear stress

$\tau_c = 0\cdot34 \text{ N/mm}^2$

Shear net $= (1\cdot329 - 0\cdot34) \times 1040 \times 500$

$= 514,280$ N.

Providing 12 mm ϕ 4-legged stirrups,

$$\text{Spacing} = \frac{4 \times 113 \times 140 \times 1040}{514,280}$$

$= 127\cdot97$ mm.

Provide stirrups at 80 mm centres on both sides of each column for a distance of 2 m. In the remaining portion provide same srirrups at 500 mm centres.

The details of reinforcement in secondary beam are shown in Fig. 10·54 and reinforcement **in main beam are shown in Fig. 10·55.**

10·13. Pile Foundations

Piles are long slender structural units used to transmit loads through soil of weaker capacity to soil having high bearing capacity. Piles are also used to resist uplift or horizontal forces. Pile foundations are generally used for bridge foundations, jetties and other structures where deep foundations are required.

Piles are classified according to the manner in which they transfer the load to sub-soil. If the stratum on which the pile **rest** is hard like rock, dense sand or gravel, the capacity of the pile is derived from the stratum upon which it rests and such piles are called end bearing or point bearing piles. If the pile does not reach impenetrable bearing stratum and is driven through penetrable soil the bearing capacity of piles is derived partly through end bearing and partly through skin friction and such piles are called friction piles. The following types of piles are generally used :

(*i*) *Timber Piles.* These piles are generally circular in section having diameter varying from 300 to 400 mm at the butt end. The line joining the centres of the butt and the tip should be within the pile. Sawn timber piles of square sections 300 mm × 300 mm, 350 mm × 350 mm and 400 mm × 400 mm and varying in lengths from 6 to 12 m are also used.

The maximum permissible stress in timber piles is taken as 4 N/mm^2. The timber piles cannot be subjected to heavy blows, therefore, carrying capacity of these piles is restricted to 250 kN.

Piles entirely below ground watertable are considered permanent and do not require any preservative treatment. The piles which are partly in water should be treated otherwise they will rot. Generally piles are treated with creosote oil.

(*ii*) *Pre-cast Concrete Piles.* These piles are cast cured in a factory and transported to site. These piles are generally square or octagonal with chamfered corners. The octagonal piles are provided with spiral reinforcement and are, therefore, more efficient in resisting driving stresses. The spirals are closely spaced near the bottom as well as the top. The piles are tapered near the ends and cast iron shoes are provided at the ends when hard driving is anticipated (Fig. 10·56).

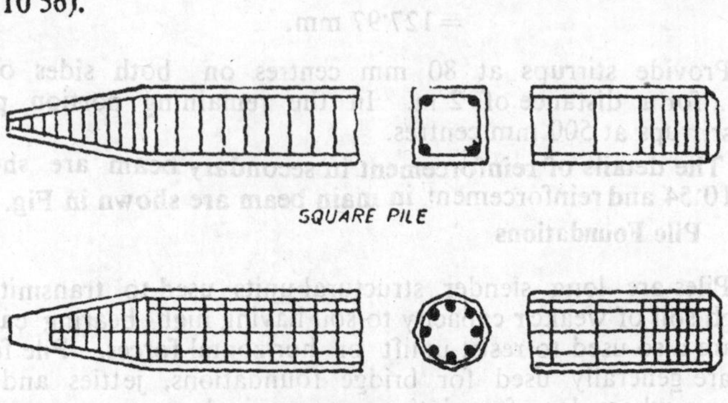

Fig. 10·56

(*iii*) *Steel Piles.* Steel piles are used as sheet piles and as bearing piles in foundations. These piles are in the form of rolled and fabricated sections. The H beams, pipe sections and box type are most commonly used. These piles have certain advantages over other types of piles *e.g.* they have greater capacity per pile, they can withstand hard driving without shattering. Because of facility of splicing on the site these piles can be used for any desired length. These piles can withstand rough handling and they have good resistance to lateral forces and buckling. Hence these piles are very suitable for marine structures where they have to resist impact forces from the berthing of ships. The biggest drawback of these piles is their susceptibility to corrosion. These piles are to be protected against sea water, where water has high salinity and where they are subjected to alternate wetting and drying conditions. The most common protection against corrosion is a coating of tar, paint or asphalt.

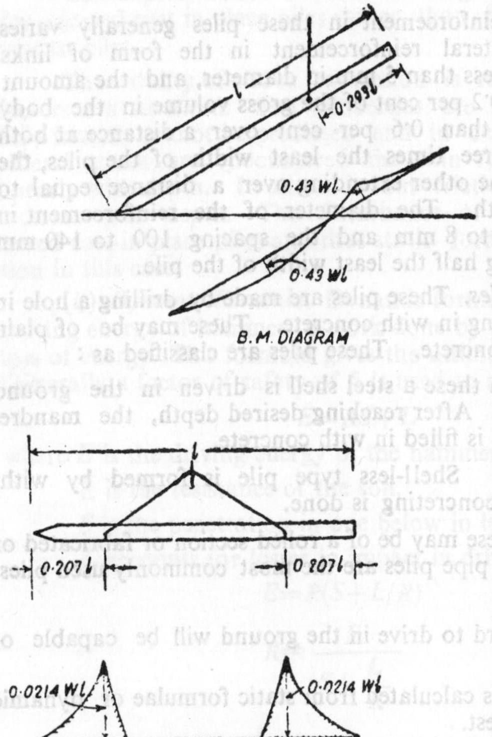

Fig. 10'58

Allowable stress in these piles is 70/mm². When these piles, bear on rock or are encased with at least 50 mm thick concrete the permissible stresses may be increased by increased by 33%.

These piles are designed for

(*a*) Handling stresses,

(*b*) Driving stresses,

(*c*) Final dead load and live load coming from the structure.

Upto 8 m length, piles are supported at one point and the other resting on the ground during handling. For bigger lengths the pile map be supported at two or more places.

For single point support the distance from butt end should be 0·293 l as shown in Fig. 10'57 for maximum B.M. to be minimum possible.

Usually the strength of pile is computed from a detailed soil study and applying well known formulae. The load test results for capacity and settlement help in verifying the computations.

Generally three test piles uniformly distributed over the site ase used for test purposes. These test piles are used in a region where soil conditions are fairly uniform. Additional test piles may be used at places where different soil conditions are encountered.

Example 10·17. *Compute the strength of a pile from the following data :*

Weight of steam hammer=40,000 N

Fall of hammer=600 mm.

Penetration in last ten blows=25 mm.

Sol. $E=W\times h=40{,}000\times 600=24{,}000{,}000$ N-mm

$$C=2{\cdot}5$$

$$S=\frac{25}{10}=2{\cdot}5$$

$$R=\frac{E}{S+C}=\frac{24{,}000{,}000}{2{\cdot}5+2{\cdot}5}=4{,}800{,}000 \text{ N}$$

Using factor of safety of 6,

$$\text{Allowable load}=\frac{4{,}800{,}000}{6}=300{,}000 \text{ N.}$$

10·15. Pile Caps

As already discussed piles are generally used in clusters. To transfer the load from a column or pier to the pile group an R.C.C.

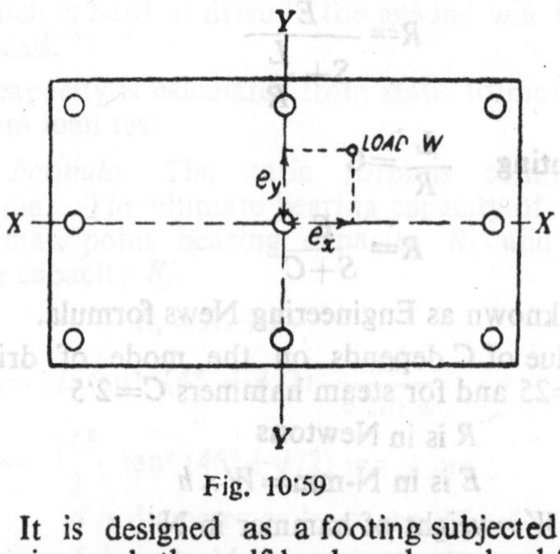

Fig. 10·59

pile cap is used. It is designed as a footing subjected to load form the column or pier and the self-load and any load from the supported soil. Under a concentric load all the piles are assumed to take equal axial loads. The piles are arranged in a geometric form so that the stresses in the cap are the minimum.

Under an eccentric loading, the pile cap is assumed to be perfectly rigid and the piles are treated as hinged in the pile cap so that no bending moment is transmitted to the pile. The force in any pile at (x, y) distance from C.G. of pile group is given by

$$P=\frac{W}{n}\pm\frac{M_x x}{\Sigma x^2}\pm\frac{M_y \times y}{\Sigma y^2}$$

where P=Load carried by pile at (x, y) distance from centroid

W=Total load transmitted

n=Number of piles

M_y=Moment of load about y-y axis

M_x=Moment of load about x-x axis.

In Fig. 10·59

$$M_y=W\times e_x$$
$$M_x=W\times e_y$$
$$n=9.$$

Design of Pile Caps

The pile caps should be extended beyond exterior piles by 100 to 150 mm. The piles are embedded about 150 mm in the cap and the pile cap reinforcement is placed about 100 mm above the head of piles, therefore, effective depth of pile is overall depth minus 250 mm.

Pile caps should be designed for bending, diagonal tension, punching shear and bond stress. For maximum diagonal tension, stresses are checked at a distance equal to the effective depth of the cap from the face of the column. In case shear stress is more, deeper cap should be used in preference to providing shear reinforcement.

The bending moments in caps carrying axially loaded columns centred directly over the centroid of the pile group are given below :

(*i*) *Cap on Three Piles.* If W is the total load on cap, load on each piles $=W/3$.

The cap is considered to consist of beams B_1 and B_2 as shown in Fig. 10·60.

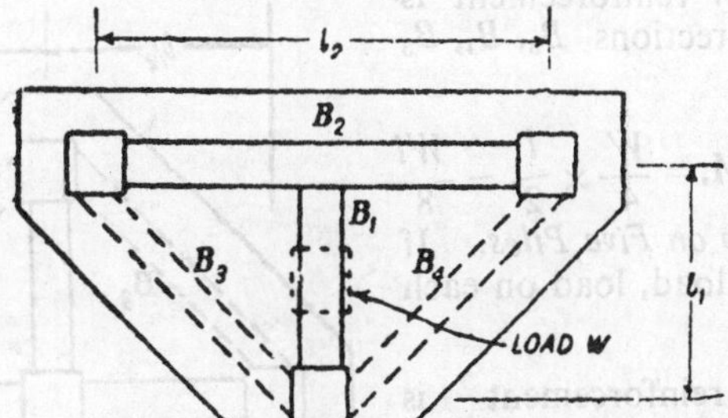

Fig. 10·60

$$\text{Maximum B.M. in } B_1=\frac{W}{3}\times\frac{2l_1}{3}=\frac{2Wl_1}{9}$$

Maximum B.M. in $B_2 = \frac{W}{3} \times \frac{l_2}{2} = \frac{Wl_2}{6}$

Secondary reinforcement is provided along directions B_3 and B_4 shown dotted, which goes round the reinforcement of the piles. The rainforcement provided is 10 mm to 12 ϕ at 15 cm centres vertically.

(*ii*) *Cap on Four Piles.* If W is the total load on cap, load on each pile $= W/4$.

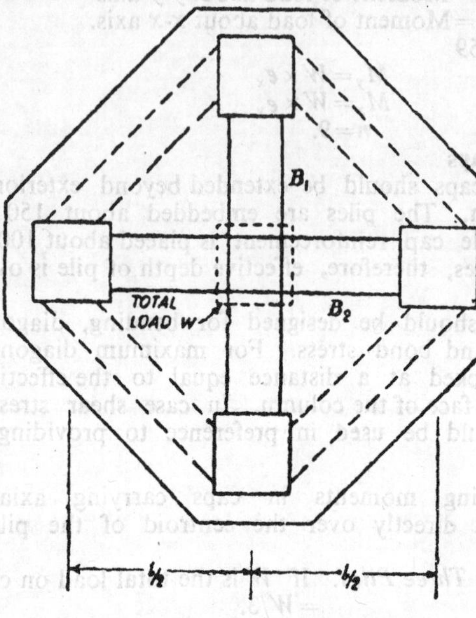

Fig. 10·61

Main reinforcement is provided in the cap in directions B_1 and B_2. Secondary reinforcement is provided in directions B_3, B_4, B_5 and B_6.

Maximum B.M. $= \frac{W}{4} \times \frac{l}{2} = \frac{Wl}{8}$

(*iii*) *Cap on Five Piles.* If W is total load, load on each pile $= W/5$.

Main reinforcement is provided in the cap in directions B_1 and B_2 and secondary reinforcement is provided in the directions B_3, B_4, B_5 and B_6.

Maximum B.M.

$$= \frac{W}{5} \times \frac{l}{2} = \frac{Wl}{10}\cdot$$

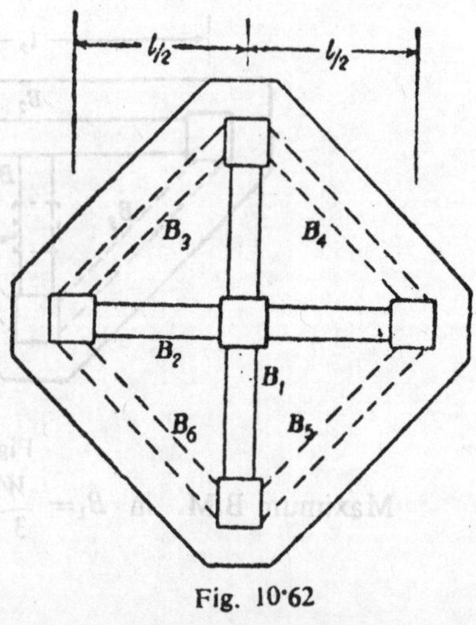

Fig. 10·62

(*iv*) *Pile cap on Six Piles.* If W is the total load, load on each pile

$$=\frac{W}{6}.$$

Main reinforcement is provided in the direction of length and secondary reinforcement is provided in direction of breadth as shown in Fig. 10·63.

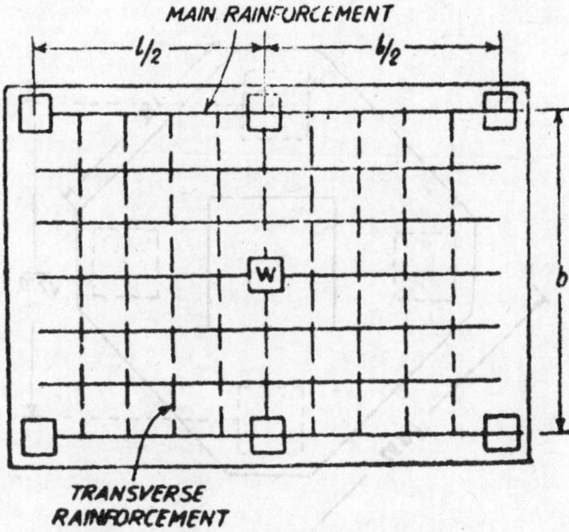

Fig. 10·63.

Maximum B.M. for width

$$b=\frac{W}{6}\times 2\times\frac{l}{2}=\frac{Wl}{6}.$$

Transverse reinforcement provided at right angles is 20 mm ϕ 200 mm centres.

The pile cap should be checked as a slab supported on four edges and carrying a concentrated load. If the bending moment in direction of breadth exceeds that in the direction of length, reinforcement should be provided for that.

Example 10·18. *Design a pile and the pile cap for a column 600 mm × 600 mm carrying a load of 2,400,000 N supported on four piles as shown. The piles are 300 mm × 300 mm and the centre-to-centre distance of pile is 1 m.*

Sol. Assume weight of pile cap as 5% of superimposed load

Weight of pile cap $=\frac{5}{100}\times 2{,}400{,}000=120{,}000$ N

Load per pile $=\frac{2{,}400{,}000+120{,}000}{4}=630{,}000$ N

Let A_s be the areas of steel required.

Load carried by pile$=[300\times 300-A_s]\times 4+130\,A_s$

$\therefore$ $630{,}000=360{,}000-4A_s+130\,A_s$

$$A_s = \frac{630{,}000-360{,}000}{126}$$

$$= \frac{270{,}000}{126} = 2143 \text{ mm}^2$$

Provide 4 bars of 28 mm ϕ.

Lateral Ties

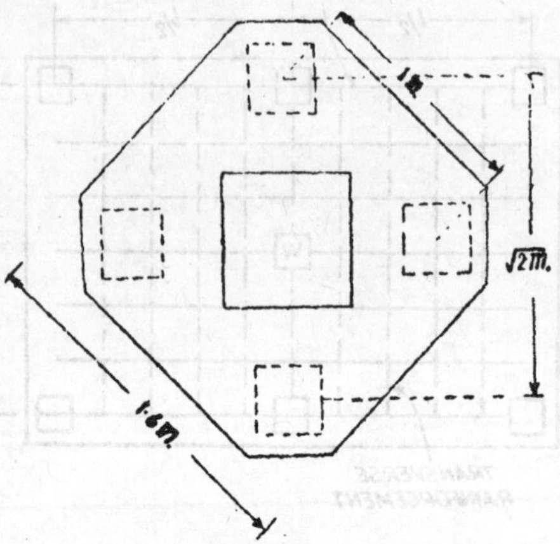

Fig. 10·64

0·2% of volume of concrete per mm height

$$= \frac{0{\cdot}2}{100} \times 300 \times 300 \times 1 = 180 \text{ mm}^2$$

Provide 8 mm ϕ ties

Volume of 1 tie $= 4 \times 220 \times 50 = 44{,}000 \text{ mm}^3$

Spacing of ties $= \frac{44{,}000}{180} = 244$ mm.

Provide 8 mm ϕ ties at 150 mm centres (spacing equal to half the load dimension of the pile).

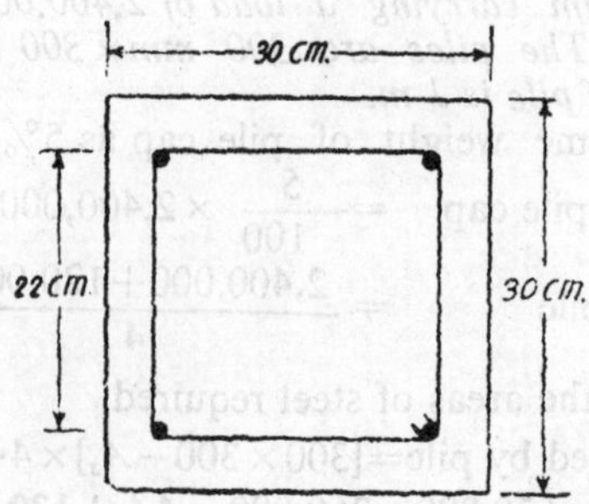

Fig. 10·65

For top and bottom 900 mm length provide 8 mm ϕ ties at 80 mm centres. Gradually increase the spacing of ties to 150 mm centre in a length of 900 mm.

Design of Pile Cap

Load per pile $=630{,}000$ N

Maximum B.M. $=630{,}000\times\frac{\sqrt{2}}{2}$

$=315{,}000\sqrt{2}$ N-m

$=445{,}400{,}000$ N-mm

Assume width of the beam

$=600$ mm (dimension of the column)

Effective depth $=\sqrt{\frac{445{,}400{,}000}{0{\cdot}87\times600}}=923{\cdot}7$ mm

Provide overall depth of 1200 mm with effective depth of

$1200-250=950$ mm

$$A_t=\frac{445{,}400{,}000}{0{\cdot}87\times950\times140}=3851\ \text{mm}^2$$

Provide 13 bars of 20 mm ϕ.

Reinforcement required for bond

Sufficient anchor length is provided including standard hooks.

Check for punching shear

Punching shear $=$Column load$=2{,}400{,}000$ N

Area resisting punching

$=4\times600\times950=2{,}280{,}000\ \text{mm}^2$

Punching stress $=\frac{2{,}400{,}000}{2{,}280{,}000}=1{\cdot}05\ \text{N/mm}>1\ \text{N/mm}^2$

Revised section

Increase the depth to 1250 mm

Effective depth $=1000$ mm

Punching stress $=\frac{2{,}400{,}000}{4\times600\times1000}=1\ \text{N/mm}^2$

Area of steel $=\frac{445{,}400{,}000}{0{\cdot}87\times1000\times140}=3657\ \text{mm}^2$

Provide 12 bars of 20 mm ϕ.

Reinforcement required for bond

Sufficient anchor length is provided.

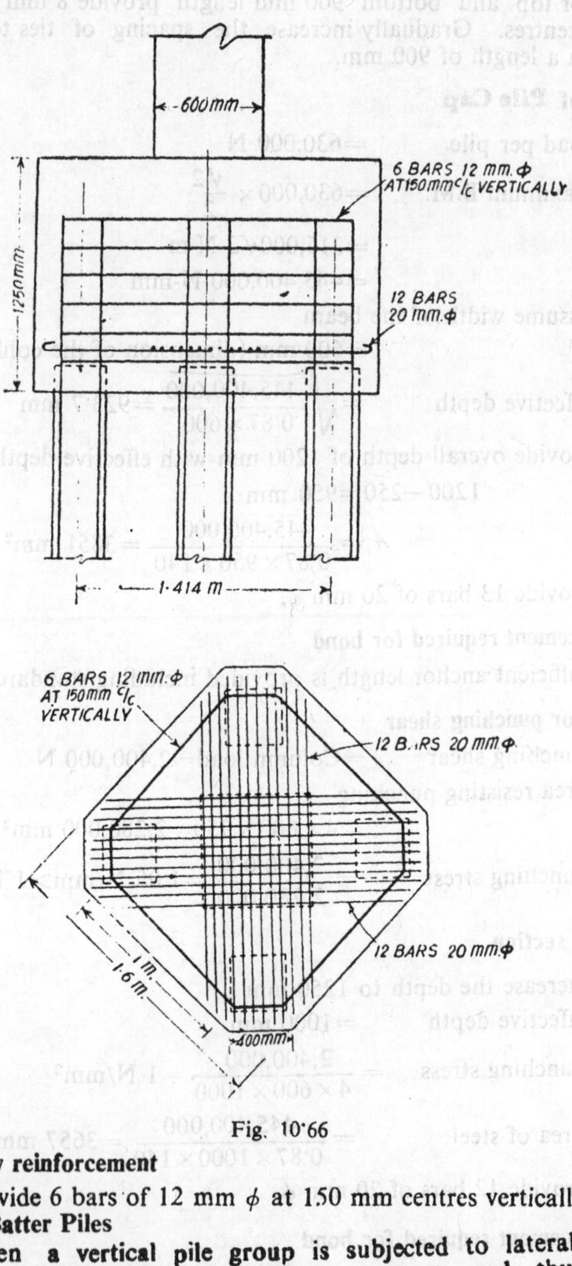

Fig. 10·66

Secondary reinforcement

Provide 6 bars of 12 mm ϕ at 150 mm centres vertically.

10·16. Batter Piles

When a vertical pile group is subjected to lateral forces, they will be subjected to bending moment and thus there will be lateral movement of the pile cap. The lateral movement is limited to 5 mm in case of buildings and 10 mm for other structures such as transmission towers. From this consideration lateral force in R C C vertical piles should be restricted to carry 5000 N otherwise suitable reinforcement should be provided to withstand flextural stresses. When the lateral load is **large**, battered piles are provided. Generally batter provided is 1 in 12. For greater batter, special type of pile driving equipment is required.

Following assumptions are made in design of pile groups having battered piles :

(*i*) A battered pile is capable of taking same axial load as a vertical pile.

(*ii*) Piles are grouped according to their batter. It is assumed that all piles in same group are subjected to same axial loads.

The analysis may be done graphically by Culmann's method or by analytical method.

Culmann's Method

(*i*) On profile of pile foundation the centre lines of different groups are marked say Group 1, Group 2 and Group 3. The resultant external force *P* is acting as shown.

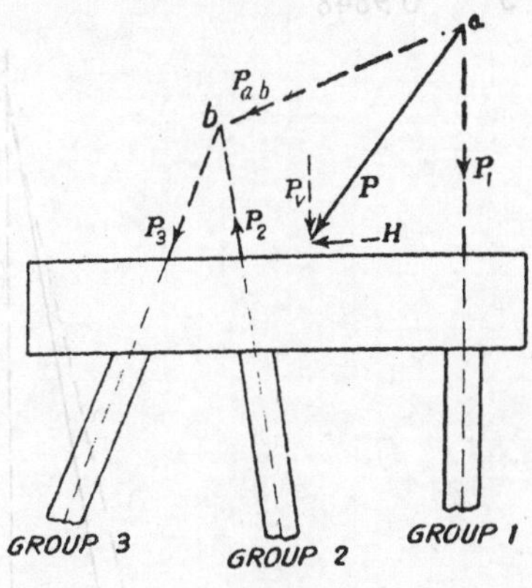

Fig. 10·67

(*ii*) Let the lines of action of any group say 1 and force *P* intersect at point *a*.

(*iii*) The lines of action of axial forces in group 2 and group 3 intersect at *b*.

(*iv*) *P* is resolved along axial direction of group 1 and along *ab*. The resolved part along group (1) say P_1 gives the axial forces in the piles of this group. If n_1 is the number of piles in group 1, the axial force in each pile in the group will be P_1/n_1.

(*v*) The component along *ab* say P_{ab} is further resolved along axial directions of piles in group 2 and group 3. Let the components be P_2 and P_3. If n_2 and n_3 are the number of piles in group 2 and group 3 respectively, the force in one pile of group 2 will be P_2/n_2 and that in one pile of group 3 will be P_3/n_3.

Example 10·19. *Fig. 10·68 shows cross-section of pile foundation with four piles in each group. The piles in group 1 are vertical and piles in group 2, 3 and 4 have same* ***batter****. Find the capacity of the piles if these piles* ***carry same*** *axial load. Find the capacity for which pile should be designed. Find the lateral displacement of pile cap if piles undergo axial deformation of 10 mm per 500,000 N load.*

Sol. Graphical construction is shown in Fig. 80·68 (*b*).

The problem can also be solved by statics.

Let θ be the slope of battered piles.

Let V_1 be the axial force in vertical piles and V_2 be the force in the battered piles.

$\Sigma V=0$ gives $4V_1+4\times 3V_2 \cos\theta = P \sin 80°$ ...(1)

$\Sigma H=0$ gives $4\times 3V_2 \sin\theta = P\times\cos 80°$ ...(2)

Taking moments about vertical piles

$4\times 3V_2 \cos\theta\times 2 = P \sin 80° \times 1{\cdot}5$ ...(3)

From 3, $12\ V_2 \cos\theta = 3/4\ P \sin 80°$

Substituting in equation (1)

$4V_1 + 3/4\ P \sin 80 = P \sin 80°$

$$V_1=\frac{P}{16}\sin 80°=\frac{3{,}000{,}000}{16}\times 0{\cdot}984=184{,}650 \text{ N}$$

Substituting value of V_1 in equation (1)

$$12V_2\cos\theta = P\sin 80° - 4\times\frac{P}{16}\sin 80° = \frac{3}{4}P\sin 80° \quad ...(4)$$

Dividing Eq. (2) by Eq. (4)

$$\frac{\cos\theta}{\sin\theta}=\frac{P\cos 80°}{2/4P\sin 80°}=\frac{4}{3}\times\frac{0{\cdot}1736}{0{\cdot}9848}=0{\cdot}2351$$

$\therefore\ \theta = 13°\ 14'$

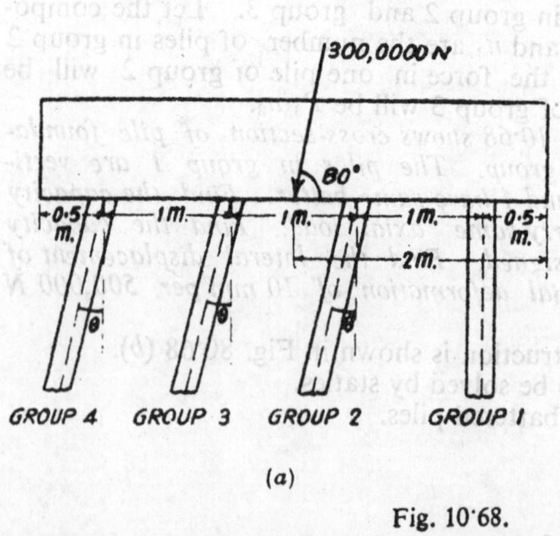

(a)

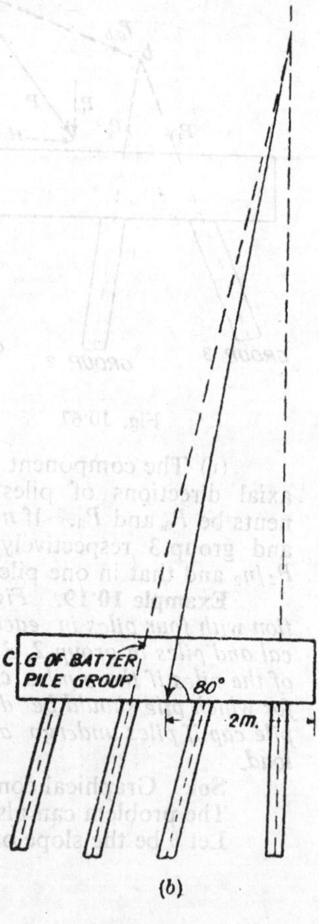

(b)

Fig. 10·68.

Fig. 10·69.

$$V_2 = \frac{3}{48} \frac{P \sin 80°}{\cos \theta}$$

$$= \frac{3 \times 3{,}000{,}000 \times 0{\cdot}9841}{48 \times 0{\cdot}9735} = 189{,}700 \text{ N}$$

Movement of vertical piles

$$= \frac{184{,}650}{500{,}000/10} = 3{\cdot}66 \text{ mm}$$

Axial movement of batter piles

$$= \frac{189{,}700}{500{,}000/10} = 3{\cdot}79 \text{ mm}$$

Lateral displacement = 1·59 mm.

10·17. Analytical Method

In this method the external force is resolved into the vertical component P_V and horizontal component P_H. The eccentricity e of vertical forces about the C.G. of pile group is found.

$$\frac{P_V}{N} + \frac{P_V \times e \times r}{\Sigma r^2}$$ where N is the number of piles, r is the distance of the pile from C.G. of pile groups and Σr^2 is sum of the squares of distances of various piles from C.G. of the pile group. Knowing the vertical component of force in a pile, axial force in a pile and horizontal component of force in a pile can be calculated. The sum of the horizontal components in all piles will balance the external horizontal force if batter is proper. Unbalanced horizontal force of 5000 N per pile is permissible.

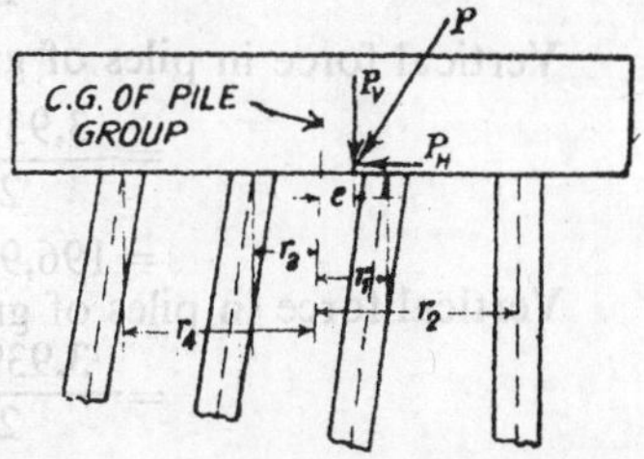

Fig. 10·70

Example 10·20. *The pile foundation shown in Fig. 10·71 has 4 piles in each group. Groups 1 and 2 have vertical piles and groups 3, 4 and 5 have battered piles having same batter. Find the batter and design capacity of piles.*

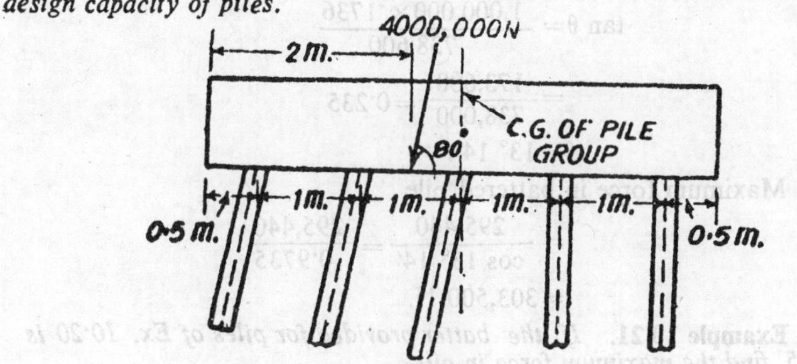

Fig. 10·71

Sol. $e=0{\cdot}5$ m

$$\Sigma r^2=4\times 2(1^2+2^2)=40$$

$$N=5\times 4=20$$

$$P_V=4{,}000{,}000 \sin 30^\circ=4{,}000{,}000\times 0{\cdot}9848=3{,}939{,}200 \text{ N}$$

Vertical force in piles of group 1

$$=\frac{3{,}939{,}200}{20}-\frac{3{,}939{,}200\times 0{\cdot}5\times 2}{40}$$

$$=196{,}960-98{,}480=98{,}480 \text{ N.}$$

Vertical force in piles of group 2

$$=\frac{3{,}939{,}200}{20}-\frac{3{,}939{,}200\times 0{\cdot}5\times 1}{40}$$

$$=196{,}960-49{,}240=147{,}720 \text{ N}$$

Vertical force in piles of group 3

$$=\frac{3{,}939{,}200}{20}-0=196{,}960 \text{ N}$$

Vertical force in piles of group 4

$$=\frac{3{,}939{,}200}{20}+\frac{3{,}939{,}200\times 0{\cdot}5\times 1}{40}$$

$$=196{,}960+49{,}240=246{,}200 \text{ N}$$

Vertical force in piles of group 5

$$=\frac{3{,}939{,}200}{20}+\frac{3{,}939{,}200\times 0{\cdot}05\times 2}{40}$$

$$=196{,}960+98{,}480=295{,}440 \text{ N}$$

Let θ be the slope of battered piles with vertical. Horizontal component of force in batter piles will be equal to vertical component $\times \tan\theta$.

$$196{,}960\times 4 \tan\theta+246{,}200\times 4\times \tan\theta+295{,}440\times 4\times \tan\theta$$

$$=4{,}000{,}000 \cos 80^\circ$$

$$738{,}600 \tan\theta=1{,}000{,}000\times 0{\cdot}1736$$

$$\tan\theta=\frac{1{,}000{,}000\times {\cdot}1736}{738{,}600}$$

$$=\frac{173{,}600}{738{,}000}=0{\cdot}235$$

$$\theta=13^\circ\ 14'.$$

Maximum force in battered pile

$$=\frac{295{,}440}{\cos 13^\circ\ 14'}=\frac{295{,}440}{0{\cdot}9735}$$

$$=303{,}500 \text{ N.}$$

Example 10·21. *If the batter provided for piles of Ex. 10·20 is 1 in 5, find the maximum force in pile.*

Sol. Horizontal component of force in piles ground above

$$=196,960\times 4 \tan\theta+246,200\times 4\tan\theta+295,440\times 4\tan\theta$$

$$=708,600\times 4\tan\theta$$

$$=738,600\times 4\times 0{\cdot}2=590,880 \text{ N}$$

Unbalanced horizontal force

$$=4,000,000\cos\theta-590,880$$

$$=40,00,000\times 0{\cdot}1736-590,880$$

$$=694,400-590,880$$

$$=103,520 \text{ N}$$

Lateral force per pile, assuming all piles to taken same force

$$=\frac{103,520}{20}=5176 \text{ N}$$

$$\tan\theta=0{\cdot}2 \qquad \therefore \cos\theta=\frac{1}{\sqrt{1{\cdot}0}}$$

Maximum force in pile

$$=\frac{295,440}{\cos\theta}$$

$$=225,440\times\sqrt{204}=331,200 \text{ N}.$$

PROBLEMS

1. Design a reinforced concrete footing for a 350 mm × 350 mm reinforced concrete column carrying an axial load of 700,000 N. Safe bearing capacity of soil is kN/m². Mix used is M15.

2. Design a rectangular footing for a 300 mm × 400 column carrying an axial load of 800,000 N. Safe bearing capacity of soil is 200 kN/m². Use mix M15.

3. Design a reinforced concrete footing for 400 mm × 400 mm column carrying axial load of 1,000,000 N and B.M. 60,000,000 N-mm. Bearing capacity of soil is 300 kN/m².

4. Design a circular footing for a 300 mm diameter circular column carrying a load of 500,000 N. The bearing capacity of soil is 100 kN/m².

5. Design a squere footing for column of problem 4.

6. Design a square pyramidal footing for a 450 mm square column reinforced with 8 bars of 25 mm diameter longitudinal bars carry a totol axial load fo 1050 kN, if the bearing capacity of the soil is 180 kN/m².

7. Design a combined rectangular footing for an exterior column 4 m × 400 mm and an interior column 500 mm × 500 mm distance 4 m apart, centre to centre carrying loads of 100,000 N and 160,000 N respectively. The space available, restricts the width of the foundation to 2 m and the length should be a minimum. Safe bearing capacity of soil is 200 kN/m².

8. A reinforced concrete foundation beam is used to transmit load from two columns to the soil. The columns are 4 m centres and the footing may project only one metre beyond the centre of the left column. The pressure distribution under the footing is to be uniform and not in excess of 180 kN/m². The weight of the footing and soil fill is equivalent to 20 kN/m².

Design the foundation beam. The left column is 300 mm × 300 mm in section and carries a load of 500 kN and the other column is 450 mm × 450 mm section, its load being 700 kN.

9. Design a footing with strap beam for two columns each 450 mm ×450 mm in section. Left hand column carries load of 1,000,000 N and the footing should not extend beyond face of this column. Right hand column carries load of 120,000 N. Centre to centre distance of columns is 4 m. Take bearing capacity of soil as 200 kN/m².

10. Design a trapezoidal footing for two columns one 400 mm×400 mm carrying 800,000 N and second 450 mm×450 mm, carrying 1,200,000 N. Centre to centre distance of columns is 3 m and the footing can extend 0·5 m beyond the face of each column. Bearing capacity of soil is 200 kN/m².

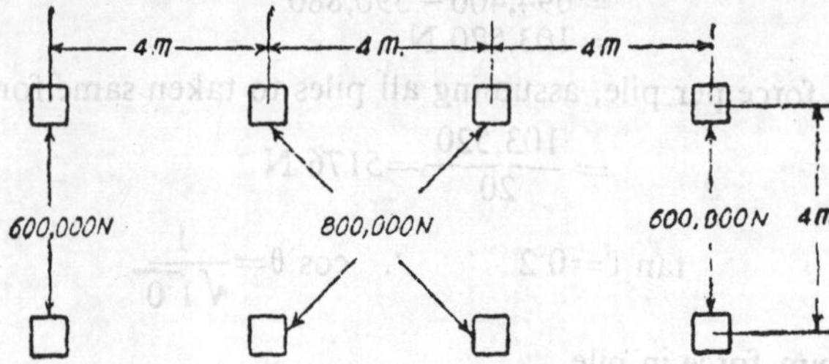

Fig. 10·72

11. Design a mat foundation for the system of columns shown in Fig. 10·72. All columns are 350 mm×350 mm and they carry loads as indicated in the figure. The bearing capacity of soil is 100 kN/m².

12. Design mat foundation for the columns of problem 11, providing two rib beams.

13. Design a suitable mat foundation for the arrangement of columns

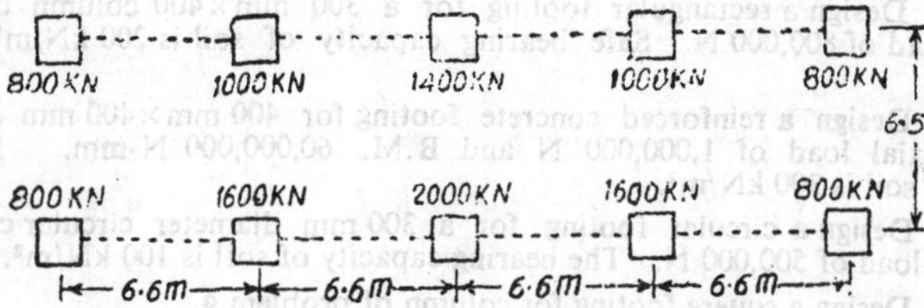

Fig. 10·73

shown in Fig. 10·73. The loads carried by the columns are shown in the figure. All columns are 500 mm×500 mm. The bearing capacity of soil is 100 kN/m².

14. Compute the allowable load on a pile from the following data :

Weight of steam hammer =3000 N

Fall of hammer =500 mm

Penetration in last ten blows =20 mm

Factor of safety =6.

15. Design a pile and pile cap for a column 500 mm×500 mm carrying a load of 1,800,000 N supported on three piles. The piles are 300 mm×300 mm are arranged at a centre to centre distance of 1 m.

16. Design a pile and pile cap for a column 600 mm×600 mm carrying a load of 3,600,000 N supported on four piles. The length of the piles is 15 m. Check the section of pile for handling stresses.

17. Design a suitable footing for a row of four column. All columns are 400 mm×400 mm and each column carries a load of 1,000,000 N. Centre to centre distance of columns is 3 m. Bearing capacity of soil is 100 kN/m².

18. Design a suitable footing for row of columns in problem 17 if each column is subjected to a clockwise moment of 5,000,000 N-mm in addition to the load of 1,000,000 N.

19. Design a suitable footing for row of columns loaded as shown in Fig. 10·74. Bearing capacity of soil is 120 kN/m². All columns are 400 mm× 400 mm.

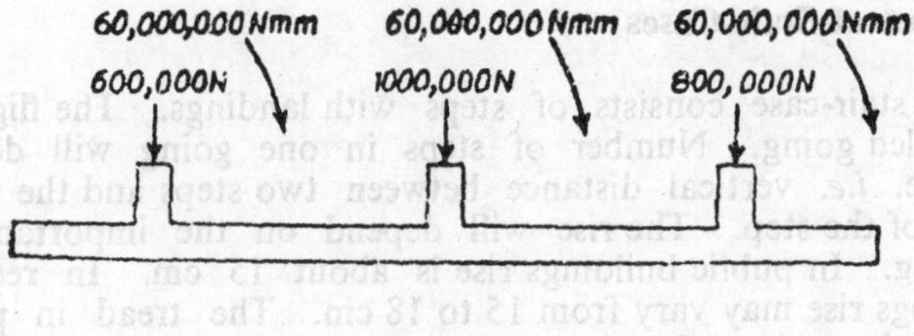

Fig. 10·74.

20. Design a combined footing for a wall column with a cross-section 400 mm×400 mm carrying a load of 1,500,000 N and an interior column 600 mm square with a load of 2,000,000 N. The distance between column centres is 4·5 and the allowable unit pressure on soil is 100 kN/m². The outside edge of the foundation coincide with one outside edge of the column.

21. A reinforced concrete column 400 mm×400 mm is required to carry an axial load of 100 kN. Calculate the longitudinal reinforcement required.

Also design a square RCC footing for the column and show by neat sketch the dimensions of the footing and the arrangement of reinforcement. Bearing capacity of soil is 200 kN/m².

22. Fig. 10·75 shows the proposal for a combined RCC foundation for two columns carrying 800 kN each. The depth of inverted T-beam and the width of the bottom slab are fixed from other considerations cannot be changed; the width of the rib of the beam and the thickness of the bottom slab.

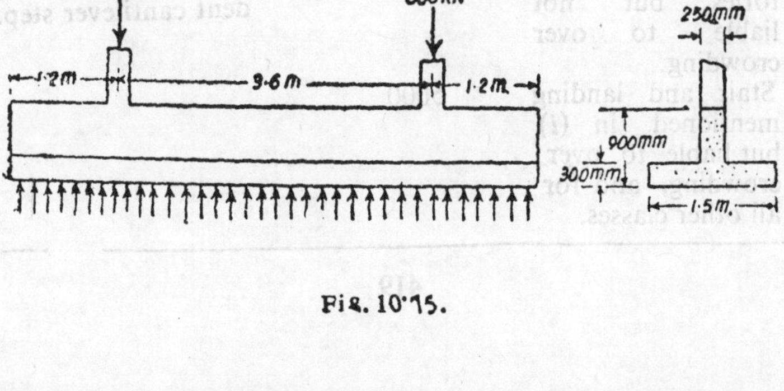

Fig. 10·75.

11

Stair Cases

11·1. Types of Stair Cases

The stair-case consists of steps with landings. The flight of steps is called going. Number of steps in one going will depend on the rise, *i.e.* vertical distance between two steps and the tread, *i.e.* width of the step. The rise will depend on the importance of the building. In public buildings rise is about 15 cm. In residential buildings rise may vary from 15 to 18 cm. The tread in public buildings is kept about 30 cm and in residential buildings the tread may vary from 20 to 30 cm. The sum of tread plus twice the rise is kept about 60 cm. The product of tread and rise is kept between 400 to 450.

Different types of stair cases used are shown in Fig. 11·1.

11·2. Live Load on Stairs (I.S.I.)

Live load on stairs shall be taken as given below.

Type of stairs and landings	*Minimum live load in N/m²*	*Alternative to minimum live load*
(*i*) Stairs in dwelling houses, tenements, hospital wards, bed rooms and private sitting rooms in hostels and dormitories, but not liable to over crowding.	3000	Subjected to minimum of 1300 N concentrated load at the unsupported end of each step for stairs constructed out of structurally independent cantilever step.
(*ii*) Stair and landing mentioned in (*i*) but liable to over crowding, and for all other classes.	5000	

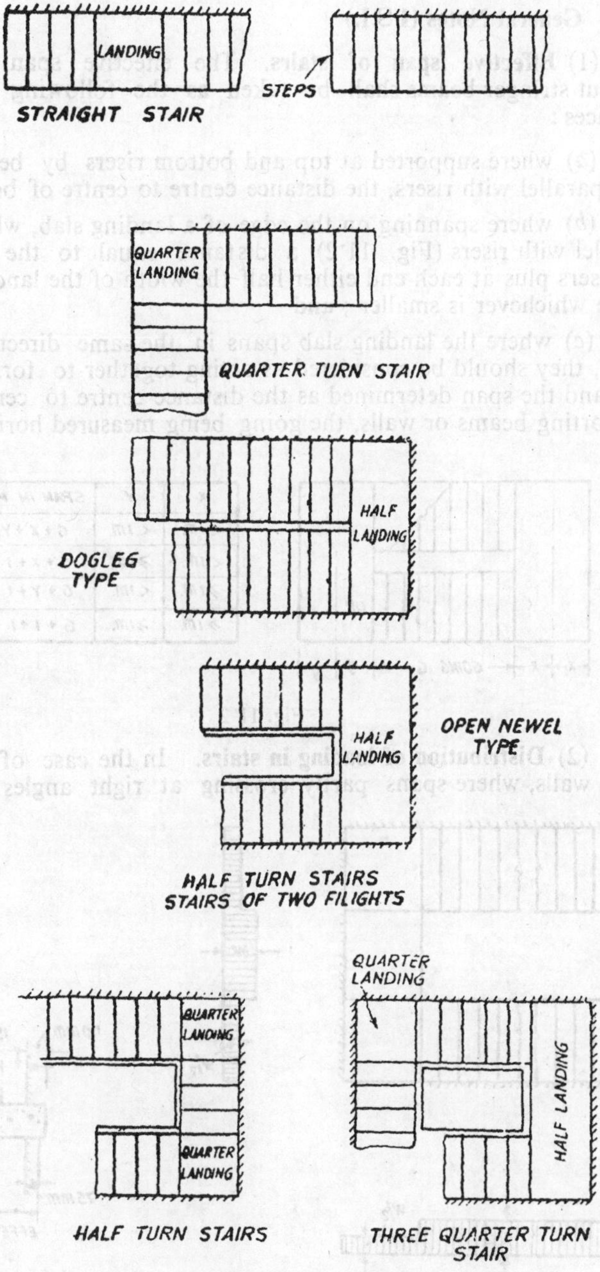
LANDING
STEPS
STRAIGHT STAIR
QUARTER LANDING
QUARTER TURN STAIR
HALF LANDING
DOGLEG TYPE
HALF LANDING
OPEN NEWEL TYPE
HALF TURN STAIRS
STAIRS OF TWO FILIGHTS
QUARTER LANDING
QUARTER LANDING
QUARTER LANDING
HALF LANDING
HALF TURN STAIRS
THREE QUARTER TURN STAIR
STAIR CASES OF THREE FLIGHT

Fig. 11·1

11·3. General Notes (I.S.I.)

(1) **Effective span of stairs.** The effective span of stairs without stringer beams shall be taken as the following horizontal distances :

(*a*) where supported at top and bottom risers by beams spanning parallel with risers, the distance centre to centre of beams ;

(*b*) where spanning on the edge of a landing slab, which spans parallel with risers (Fig. 11·2) a distance equal to the 'going' of the risers plus at each end either half the width of the landing or one metre whichever is smaller ; and

(*c*) where the landing slab spans in the same direction as the stairs, they should be considered as acting together to form a single slab and the span determined as the distance centre to centre of the supporting beams or walls, the going being measured horizontally.

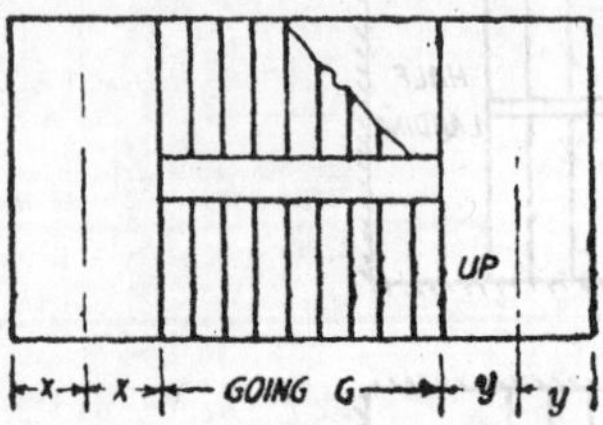

x	Y	SPAN IN METRES
< 1m.	< 1m.	G + X + Y
< 1m.	≯1m.	G + X + 1
≯1m.	< 1m.	G + Y + 1
≯1m.	≯1m.	G + 1 + 1

Fig. 11·2

(2) **Distribution of loading in stairs.** In the case of stair with open walls, where spans partly crossing at right angles occurs, the

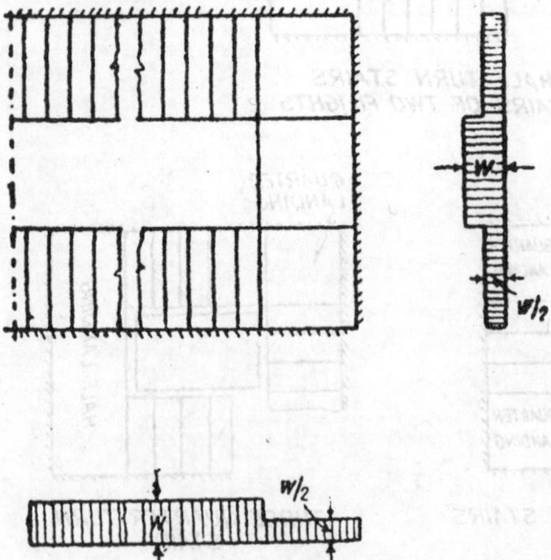

Fig. 11·3. Load on areas common to the systems to be taken as one-half in each direction loading on stairs with open wells.

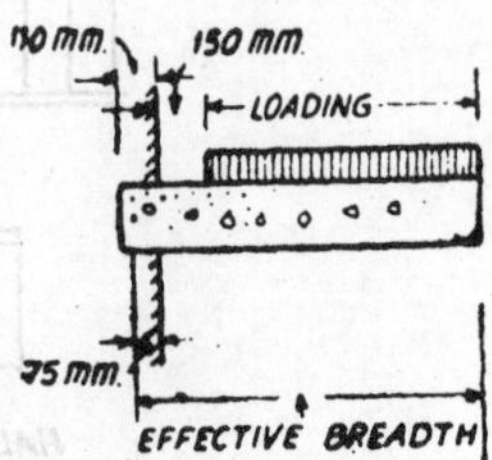

Fig. 11·4. Loading on stairs built into walls.

load on areas common to any two such spans may be taken as one-half in each direction as shown in Fig. 11·3. Where flights or landings are built into walls at a distance of not less than 110 mm and are designed to span in the direction of the flight, 150 mm strip may be deducted from the loaded area and the effective breadth of the section increased by 75 mm for purposes of design breadth (Fig. 11·4).

Example 11·1. *Design the arrangement of stair-case shown in Fig. 11·5. Live load to be taken is 3000 N/m². Rise of steps is 160 mm and tread is 250 mm.*

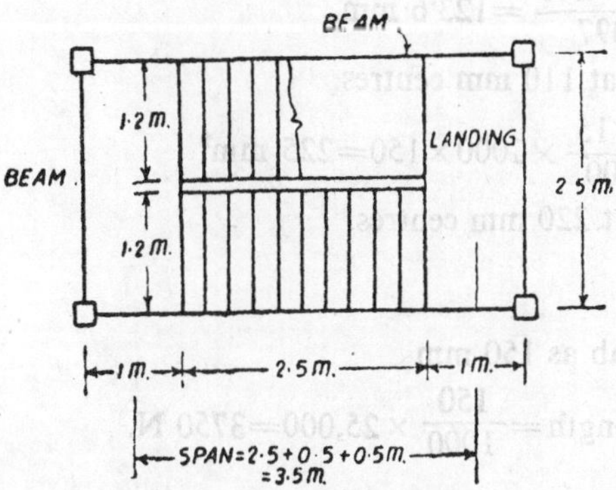

Fig. 11·5

Sol. Span centre to centre to landings

$=2\cdot5+0\cdot5+0\cdot5$

$=3\cdot5$ m.

Weight of each step

$$=\frac{250\times160}{2\times1000\times1000}\times1\cdot2\times25{,}000$$

$=600$ N.

Weight of steps per horizontal metre run

$$=600\times\frac{1000}{250}$$

$=2400$ N

Assume thickness of slab (waist) as 150 mm.

Weight of slab per horizontal metre run

$$=\frac{150}{1000}\times\frac{1\cdot2\times25{,}000}{\cos\theta}$$

$$=\frac{150}{1000}\times\frac{1\cdot2\times25{,}000}{\dfrac{250}{\sqrt{(250)^2+(160)^2}}}$$

$=5343$ N

Dead load per horizontal metre run

$=2400+5343$

$=7743$ N

Live load per metre run

$=3000\times1\cdot2=3600$ N

Total load per metre

$=7743+3600$

$=11{,}343$ N-m.

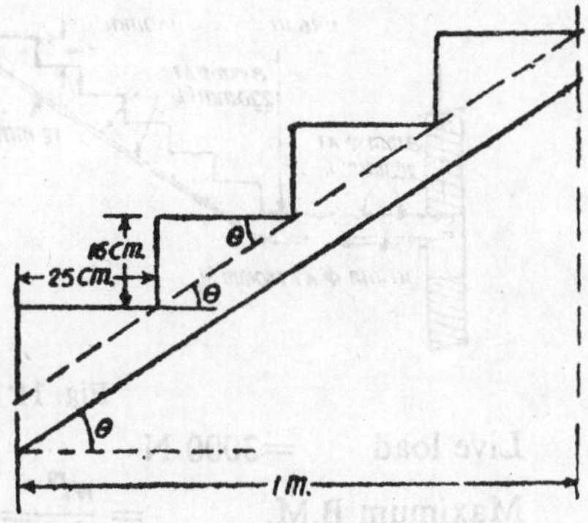

Fig. 11·6

Assume same load throughout the span.

$$\text{Maximum B.M.}=\frac{wl^2}{8}=\frac{11{,}343\times(3\cdot5)^2}{8}\text{ N-m.}$$

$=17{,}369$ N-m$=17{,}369{,}000$ mm.

Width of slab $=1{\cdot}2$ m.

Effective depth required

$$=\sqrt{\frac{17{,}369{,}000}{0{\cdot}87\times1200}}=129 \text{ mm}$$

Adopt overall depth of 150 mm with effective depth of 130 mm

$$A_t=\frac{17{,}369{,}000}{0{\cdot}87\times130\times140}=1097 \text{ mm}^2$$

Providing 12 mm ϕ bars,

$$\text{Spacing}=\frac{1200\times113}{1097}=123{\cdot}6 \text{ mm.}$$

Provide 12 mm ϕ bars at 110 mm centres.

$$\text{Distribution steel}=\frac{0{\cdot}15}{100}\times1000\times150=225 \text{ mm}^2$$

Provide 8 mm ϕ bars at 220 mm centres.

Design of landing slab

Consider 1 m width

Assume thickness of slab as 150 mm

$$\text{Dead load per metre length}=\frac{150}{1000}\times25{,}000=3750 \text{ N.}$$

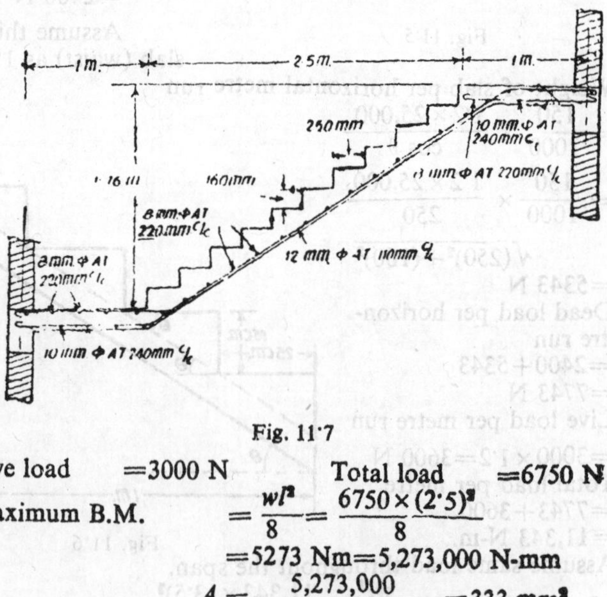

Fig. 11·7

Live load $=3000$ N Total load $=6750$ N

$$\text{Maximum B.M.}=\frac{wl^2}{8}=\frac{6750\times(2{\cdot}5)^2}{8}$$

$$=5273 \text{ Nm}=5{,}273{,}000 \text{ N-mm}$$

$$A_t=\frac{5{,}273{,}000}{0{\cdot}87\times130\times140}=333 \text{ mm}^2$$

Provide 10 mm ϕ bars at 240 mm centres.

Example 11·2. *Fig. 11·8 shows arrangement of stair-case for a building. The stairs are built in the wall along the flights. The tread is 250 mm and rise is 170 mm. Design the stair-case assuming live load as 3000 N/m².*

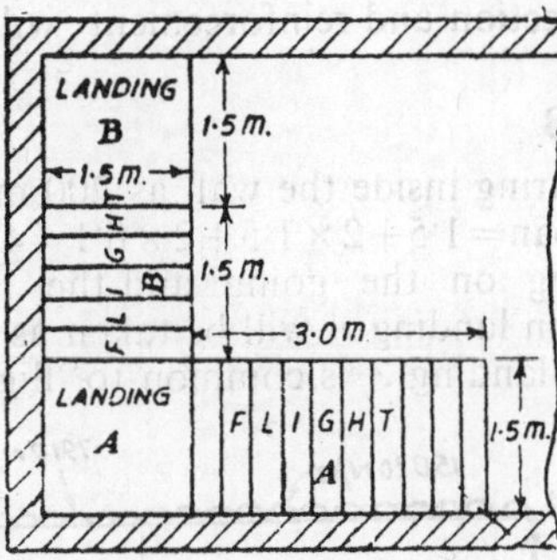

Fig. 11·8

Sol. Load of each step

$$=1{\cdot}5\times\frac{250\times170}{2\times1000\times1000}\times25{,}000$$

$$=797\text{ N.}$$

Load of steps per horizontal metre $=797\times4=3188$ N.

Assume thickness of slab (waist) as 200 mm.

Weight of slab horizontal metre run

$$=\frac{200}{1000}\times\frac{1\times1{\cdot}5\times25{,}000}{\cos\theta}$$

$$=\frac{200}{1000}\times\frac{1\times1{\cdot}5\times25{,}000}{\dfrac{250}{\sqrt{(250)^2+(170)^2}}}$$

$$=\frac{1}{5}\times1{\cdot}5\times25{,}000\times\frac{\sqrt{(250)^2+(170)^2}}{250}$$

$$=9070\text{ N.}$$

Total dead load for horizontal metre run $=3188+9070=12{,}258$ N.

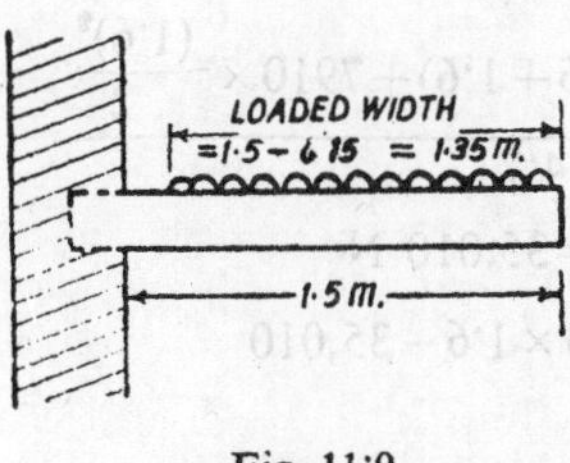

Fig. 11·9

For live load, loaded width will be equal to width of step minus 150 mm.

Loaded width

$$=1{\cdot}5-0{\cdot}15=1{\cdot}35\text{ m.}$$

Live load per metre length

$$=3000\times1{\cdot}15=4050\text{ N.}$$

Total load per horizontal metre $=12{,}258+4{,}050=16{,}308$ N.

In the design of stair case it will be assumed that the load per horizontal metre run on the landing is same as on the going though in reality the load per horizontal metre of going is more than on the landing because of weight of steps.

As the effective spans for flights A and B are almost equal, therefore, same section and reinforcement will be provided for flights A and B.

Design of Flight B

Assume bearing inside the wall as 200 mm.

Effective span$=1{\cdot}5+2\times1{\cdot}5+2\times0{\cdot}1=$**4·7 m.**

The loading on the going and the landing B will be 15,820 N/m. Loading on landing A will be taken as half of 15,820 N, *i.e.* 7910 N/m as the landing A is common to flights A and B.

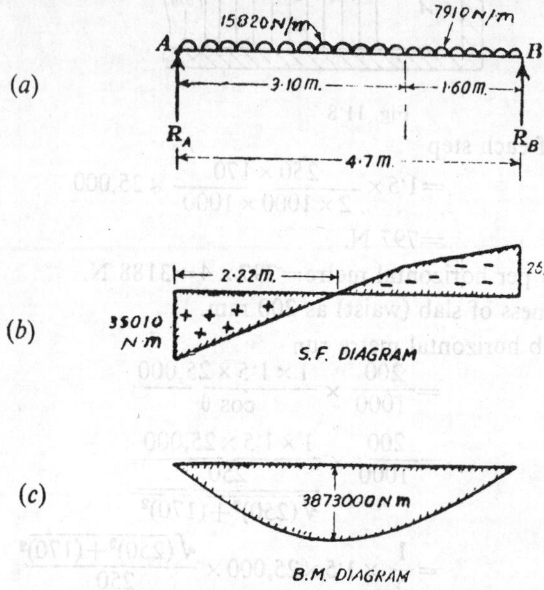

Fig. 11·10

Loading on the landing shown in Fig. 11·10 (*a*). Taking moments about B,

$$R_A=\frac{15{,}820\times3{\cdot}1\times(15{\cdot}5+1{\cdot}6)+7910\times\dfrac{(1{\cdot}6)^2}{2}}{4{\cdot}7}$$

$$=\frac{145{,}900+9561}{4{\cdot}7}=35{,}010\text{ N}$$

$$R_B=15{,}820\times3{\cdot}1+7910\times1{\cdot}6-35{,}010$$
$$=26{,}688\text{ N.}$$

Distance from A, where S.F. is zero

$$=\frac{35{,}010}{15{,}820}=2{\cdot}22\text{ m.}$$

$$\text{Maximum B.M.} \quad = 35{,}010 \times 2{\cdot}22 - \frac{15{,}820 \times (2{\cdot}22)^2}{2}$$

$$= 77{,}730 - 39{,}000 = 38{,}730 \text{ N-m.}$$

$$= 38{,}730{,}000 \text{ N-mm.}$$

Breadth of slab for design

$$= 1{\cdot}5 \text{ m} + 7{\cdot}5 \text{ mm} = 1{\cdot}575 \text{ m}$$

$$= 1575 \text{ mm.}$$

$$\text{Effective depth} \quad = \sqrt{\frac{38{,}730{,}000}{0{\cdot}87 \times 1575}} = 168{\cdot}1 \text{ mm.}$$

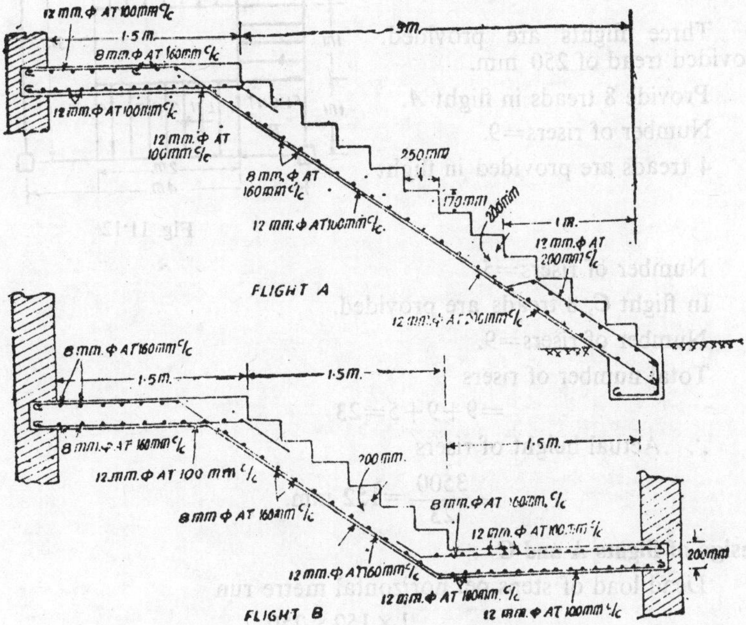

Fig. 11·11

Provide overall depth of 200 mm with effective depth of 180 mm.

$$A_t = \frac{38{,}730{,}000}{0{\cdot}87 \times 180 \times 140} = 1768 \text{ mm}^2$$

Providing 12 mm ϕ bars,

$$\text{Spacing} \quad = \frac{1575 \times 113}{1768} = 101 \text{ mm}$$

Provide 12 mm ϕ bars at 100 mm centres.

Distribution steel $= 0{\cdot}15\%$ of cross-sectional area

$$= \frac{0{\cdot}15}{100} \times 1000 \times 200 = 300 \text{ mm}^2.$$

Provide 8 mm bars at 160 mm ϕcentres.

Same section and reinforcement is provided for both the flights.

Example 11·3. *Design a suitable stair-case for a multi-storeyed building having stair-case hall 3 m × 4 m. The height between floors is 3·5 m. Live load to be assumed for design is 3500 N/m².*

Sol. An open-wall type stair case will be provided. If rise of 16 cm for each step is provided, total number of steps required will be 22.

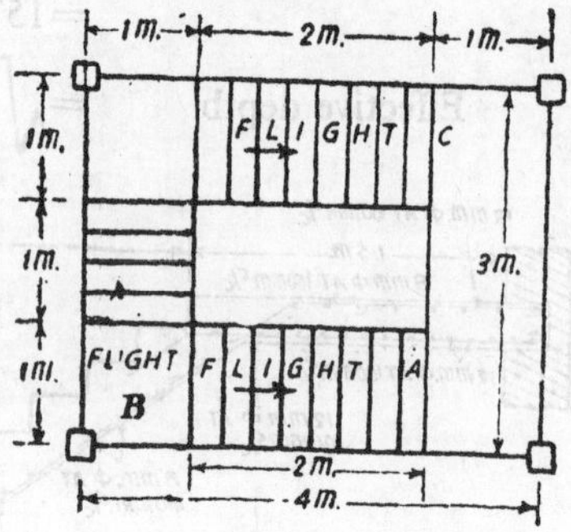

Fig. 11·12

Three flights are provided. Provided tread of 250 mm.

Provide 8 treads in flight A.

Number of risers=9.

4 treads are provided in flight B.

Number of risers=5.

In flight C, 8 treads are provided.

Number of risers=9.

Total number of risers

$$=9+9+5=23$$

∴ Actual height of risers

$$=\frac{3500}{23}=152 \text{ mm.}$$

Design of flights A and C

Dead load of steps per horizontal metre run

$$=4\times\frac{1\times152\times250}{2\times1000\times1000}\times24{,}000=1824 \text{ N}$$

Assume thickness of slab (waist) as 160 mm.

Dead load of slab per horizontal metre run

$$=1\times1\times\frac{160}{1000}\times\frac{24{,}000}{\cos\theta}=\frac{160}{1000}\times24{,}000\times\frac{\sqrt{(250)^2+(152)^2}}{250}$$

$$=4496 \text{ N}$$

Dead load per horizontal metre run

$$=1824+4496=6320 \text{ N.}$$

Live load per metre=3000×1=3000 N.

Total load per metre run=9320 N.

As the landings are spanning in both directions for flights A and B, the load on the landing will be taken as half of that on going. The loa ding is shown in Fig. 11·13.

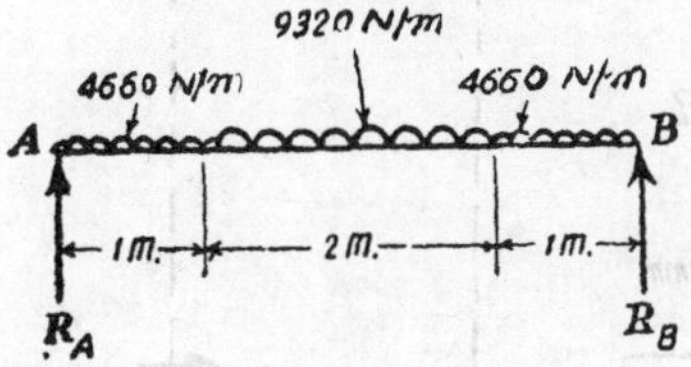

Fig. 11·13

$$R_A = R_B = 9320 + 4660$$
$$= 13{,}980 \text{ N}$$

Maximum B.M.

$$= 13{,}980 \times 2 - 4{,}660 \times 1{\cdot}5 - 9{,}320 \times 0{\cdot}5$$
$$= 27{,}960 - 6{,}990 - 4{,}660 = 16{,}310 \text{ N-m}$$
$$= 16{,}310{,}000 \text{ N-mm}$$

Effective depth $= \sqrt{\dfrac{16{,}310{,}000}{0{\cdot}87 \times 1000}} = 135$ mm

Provide overall depth of 160 mm with effective depth of 142 mm.

$$A_t = \frac{16{,}310{,}000}{0{\cdot}17 \times 142 \times 140} = 945 \text{ mm}^2$$

Provide 10 mm ϕ bars at 80 mm centres.

Distribution steel $= \dfrac{0{\cdot}15}{100} \times 1000 \times 160 = 240 \text{ mm}^2$

Provide 8 mm ϕ bars at 200 mm centres.

Flight B

Same depth of slab as for flight A is provided. Loading on the slab as shown in Fig. 11·14.

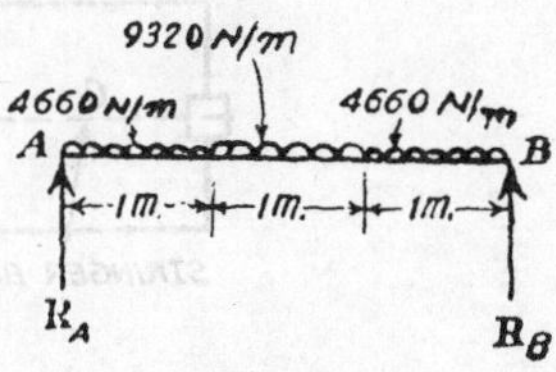

Fig. 11·14

$$R_A = R_B = 9320 \times 0{\cdot}5 + 4660$$
$$= 9320 \text{ N}$$

Maximum B.M.

$$= 9320 \times 1{\cdot}5 - 4660 \times 1 \times 1 - 9320 \times 0{\cdot}5 \times 0{\cdot}25$$
$$= 13{,}980 - 5825$$
$$= 8155 \text{ N-mm} = 8{,}155{,}000 \text{ N-m}$$

$$A_t = \frac{8{,}155{,}000}{0{\cdot}87 \times 142 \times 140} = 472{\cdot}5 \text{ mm}^2$$

Provided 10 mm ϕ bars at 160 centres.

Distribution steel

Provide 8 mm ϕ bars at 200 mm centres.

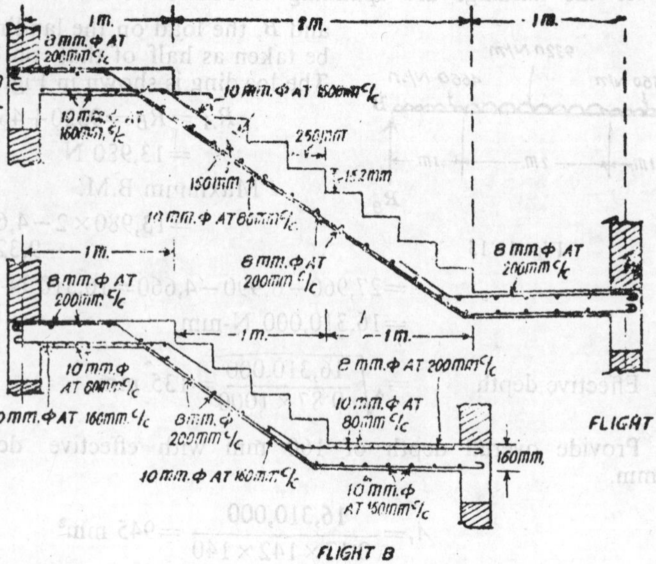

Fig. 11·15

Example 11·4. *Design a stair-case in a public building, leading from entrance hall to the upper floor. The arrangement of stair-case is shown in Fig. 11·16. The stair-case is to be designed with central stringer line. The tread is 250 mm and rise is 160 mm. The live load is to be taken as 5000 N/m².*

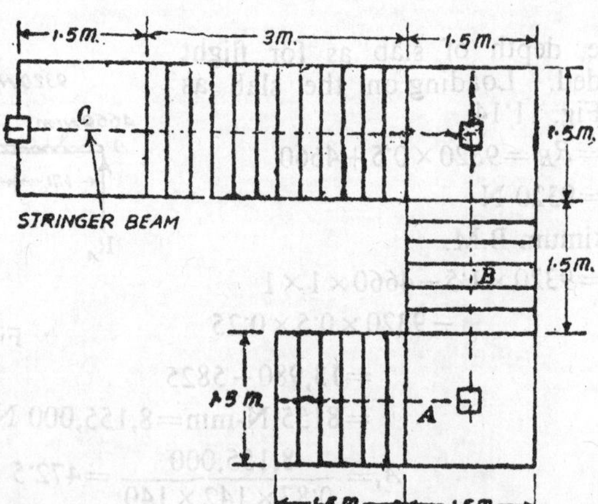

Fig. 11·16

Sol. Design of steps. Provide thickness of slab as 80 mm This slab forms flange of stringer beam.

Weight of one step and weight of slab (waist)

$$=\frac{250\times160\times1\cdot0}{2\times1000\times1000}\times25{,}000+\frac{80\times\sqrt{(250)^2+(160^2)}}{1000\times1000}\times1\cdot0\times25{,}000$$

$$=500+80\sqrt{[(250)^2+(160)^2]}\times0\cdot025=1094\text{ N.}$$

Component of this load at right angles to flight

$$=1094\times\cos\theta=\frac{250\times1094}{\sqrt{(250)^2+(160)^2}}=921\text{ N.}$$

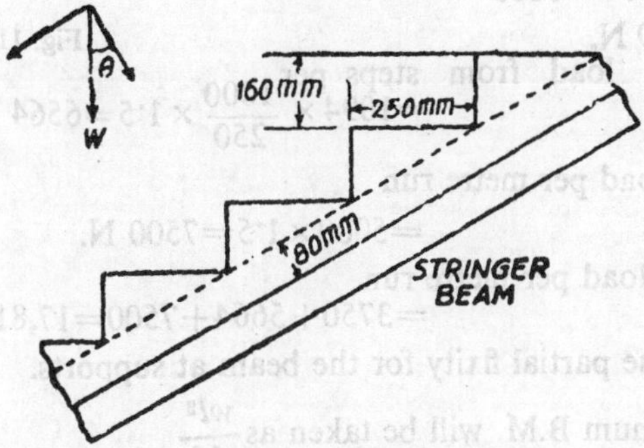

Fig. 11·17

Assume width of T-beam rib as 300 mm.

Projection of step beyond rib of stringer beam

$$=\frac{1500}{2}-150=600\text{ mm.}$$

$$\text{Dead load B.M.}=\frac{wl^2}{2}=\frac{921\times(0\cdot6)^2}{2}=165\cdot8\text{ N-m.}$$

Each step is to be designed for a concentrated live load of 1300 N.

Live load B.M. $=1300\cos\theta\times0\cdot6$ N-m.

$$=1300\times\frac{250}{\sqrt{(250)^2+(160)^2}}\times0\cdot6=657\text{ N-m}$$

Total B.M. $=165\cdot8+657=822\cdot8$ N-m $=822{,}800$ N-mm

Effective depth of step $=(160-30)\times\cos\theta+80$

$$=130\times\frac{250}{\sqrt{(250)^2+(160)^2}}+80$$

$$=109\cdot4+80=189\cdot4\text{ mm}$$

The depth provide is more than required as a balanced section.

$$A_t=\frac{822{,}800}{0\cdot87\times189\cdot4\times140}=35\cdot7\text{ mm}^2.$$

Provide one bar of 10 mm ϕ. Provide 6 mm ϕ stirrups at 150 mm centres.

Design of Stringer Beam C

Span $=1{\cdot}5+3+\frac{1{\cdot}5}{2}$

$=5{\cdot}25$ m.

Assume rib of 500 mm × 300 mm.

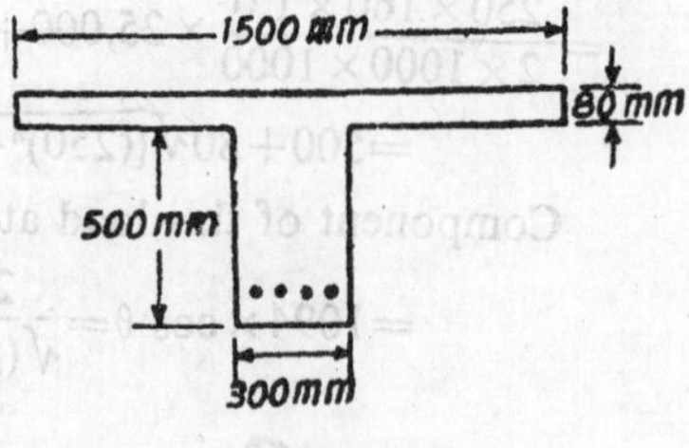

Fig. 11·18

Self-weight of rib per metre

$$\text{run}=\frac{500}{1000}\times\frac{300}{1000}\times 1\times 25{,}000$$

$=3750$ N.

Dead load from steps per metre run $=1094\times\frac{1000}{250}\times 1{\cdot}5=6564$ N.

Live load per metre run

$=5000\times 1{\cdot}5=7500$ N.

Total load per metre run

$=3750+5664+7500=17{,}814$ N.

Assume partial fixity for the beam at supports.

Maximum B.M. will be taken as $\frac{wl^2}{10}$

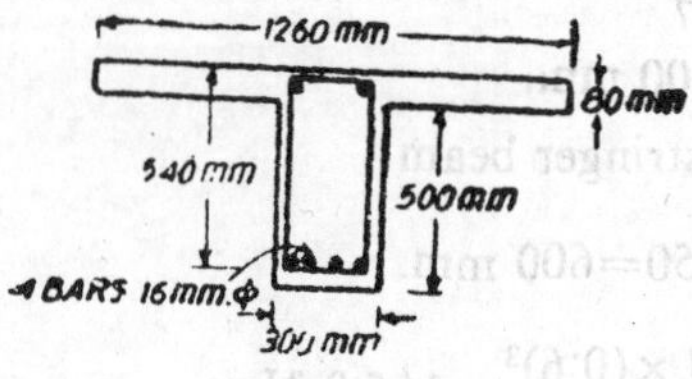

Fig. 11·19

$=17{,}814\times\frac{(5{\cdot}25)^2}{10}$

$=49{,}100$ N-m

$=49{,}100{,}000$ N-mm

Assume lever are

$=d-\frac{d_s}{2}=540-\frac{80}{2}$

$=500$ mm.

$$A_t=\frac{49{,}100{,}000}{500\times 140}=701{\cdot}4 \text{ mm}$$

Provide 4 bars of 16 mm ϕ

$A_t=804$ mm^2

$$\text{Flange width}=\frac{l_0}{\frac{l_0}{b}+4}+b_w=\frac{5250}{\frac{5250}{1500}+4}+300=1000 \text{ mm.}$$

Flange width $=1000$ mm.

Assume N.A. to lie in web. Taking moment about N.A.

$1000\times 80(n-40)=804\times 18\times(540-n)$

$5{\cdot}53(n-40)=(540-n)$

$5{\cdot}53n-221{\cdot}2=540-n$

$6{\cdot}53n=761{\cdot}2$

$n=116{\cdot}6$ mm.

Depth of critical neutral axis

$=0{\cdot}4\times 540=216$ mm.

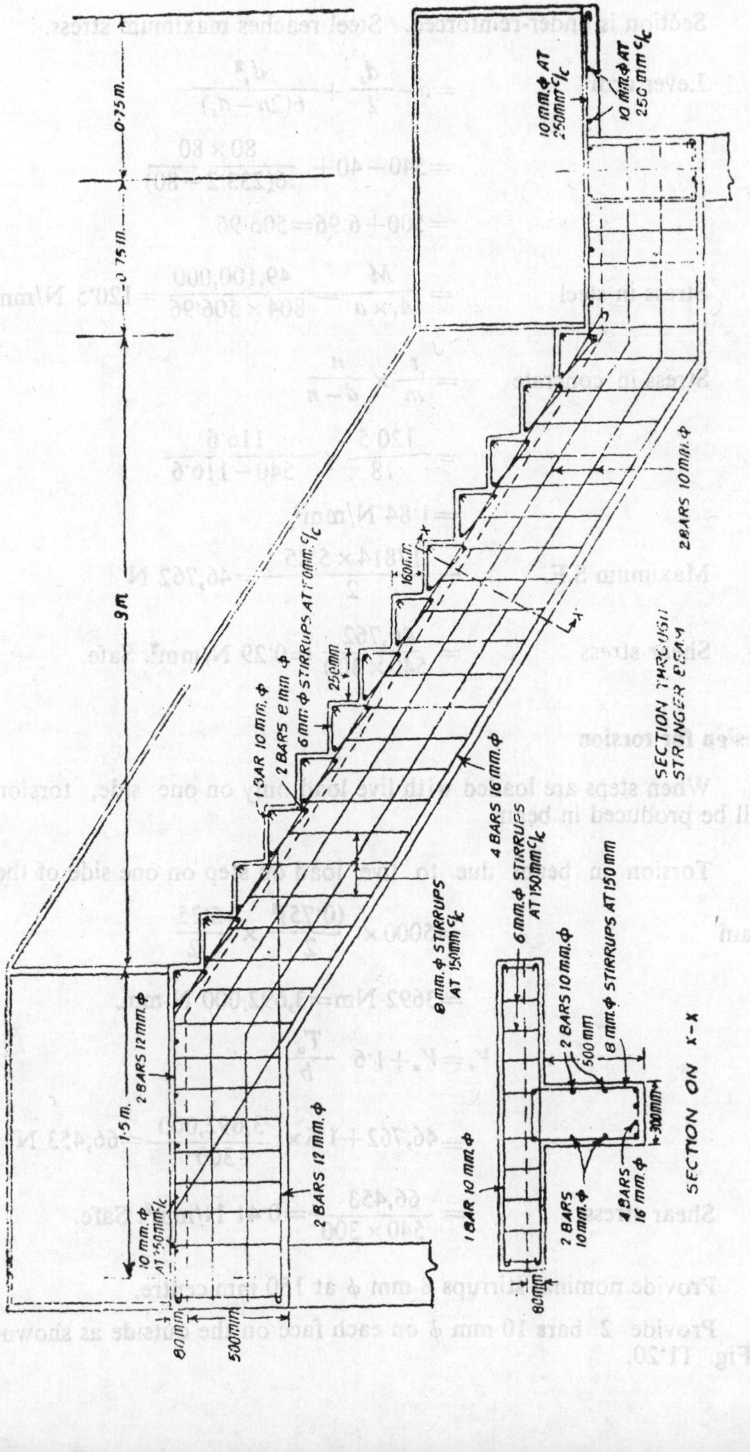
0·75 m.
0·75 m.
9 m
1·5 m.
10 mm.φ AT 250mm c/c
10 mm.φ AT 250 mm c/c
2 BARS 10 mm.φ
160 mm
250mm
1 BAR 10 mm.φ
2 BARS 8 mm φ
6 mm.φ STIRRUPS AT 150mm. c/c
4 BARS 16 mm.φ
6 mm.φ STIRRUPS AT 150 mm c/c
8 mm.φ STIRRUPS AT 150 mm c/c
2 BARS 12 mm.φ
2 BARS 12 mm.φ
80 mm
500 mm
SECTION THROUGH STRINGER BEAM
6 mm.φ STIRRUPS AT 150 mm
2 BARS 10 mm.φ
500 mm
8 mm.φ STIRRUPS AT 150 mm
300 mm
4 BARS 16 mm.φ
2 BARS 10 mm.φ
1 BAR 10 mm.φ
80 mm
SECTION ON X–X

Fig. 11·20

Section is under-reinforced. Steel reaches maximum stress.

Lever arm $= d - \frac{d_s}{2} + \frac{d_s^2}{6(2n - d_s)}$

$= 540 - 40 + \frac{80 \times 80}{6(233 \cdot 2 - 80)}$

$= 500 + 6 \cdot 96 = 506 \cdot 96.$

Stress in steel $= \frac{M}{A_t \times a} = \frac{49,100,000}{804 \times 506 \cdot 96} = 120 \cdot 5 \text{ N/mm}^2$

Stress in concrete $= \frac{t}{m} \times \frac{n}{d - n}$

$= \frac{120 \cdot 5}{18} \times \frac{116 \cdot 6}{540 - 116 \cdot 6}$

$= 1 \cdot 84 \text{ N/mm}^2$

Maximum S.F. $= \frac{17814 \times 5 \cdot 25}{2} = 46,762 \text{ N}$

Shear stress $= \frac{46,762}{540 \times 300} = 0 \cdot 29 \text{ N/mm}^2$. Safe.

Design for torsion

When steps are loaded with live load only on one side, torsion will be produced in beam.

Torsion in beam due to live load on step on one side of the beam $= 5000 \times \frac{(0 \cdot 75)^2}{2} \times \frac{5 \cdot 25}{2}$

$= 3692 \text{ Nm} = 3,692,000 \text{ N-mm}.$

$$V_e = V_u + 1 \cdot 6 \frac{T_u}{b}$$

$= 46,762 + 1 \cdot 6 \times \frac{3,692,000}{300} = 66,453 \text{ N}$

Shear stress $= \frac{66,453}{540 \times 300} = 0 \cdot 41 \text{ N/mm}^2$. Safe.

Provide nominal stirrups 8 mm ϕ at 150 mm centre.

Provide 2 bars 10 mm ϕ on each face on the outside as shown in Fig. 11·20.

PROBLEMS

1. Fig. 11·21 shows arrangement of stair-case for a building. Rise of steps is 160 mm and tread is 250 mm. Live load to be taken for design is 3000 N/m².

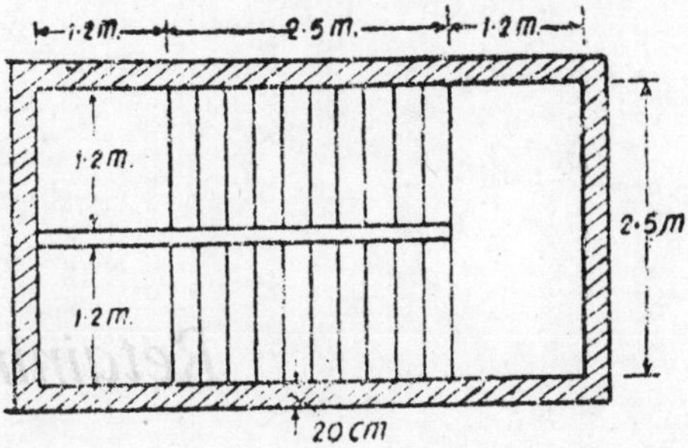

Fig. 11·21

2. Fig. 11·22 shows the arrangement of a stair-case for a warehouse. The rise of steps is 160 mm. Design the stair-case and beams *A*, *B*, *C* and *D*. Live load to be taken for design is 5000 N/m².

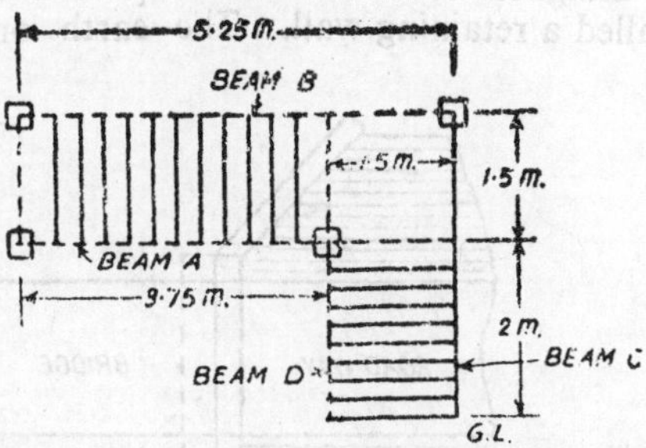

Fig. 11·22

3. Design a suitable stair-case for a multi-storeyed building having stair-case hall 2·5 m × 4·5 m. The height between floors is 3·5 m. Live load to be taken for design is 3000 N/m².

4. Fig. 11·23 shows arrangement of stair-case for a building. Rise of steps is 160 mm and tread is 250 mm. Design stair-case slab and landing beam. Live load to be taken for design is 3000 N/m².

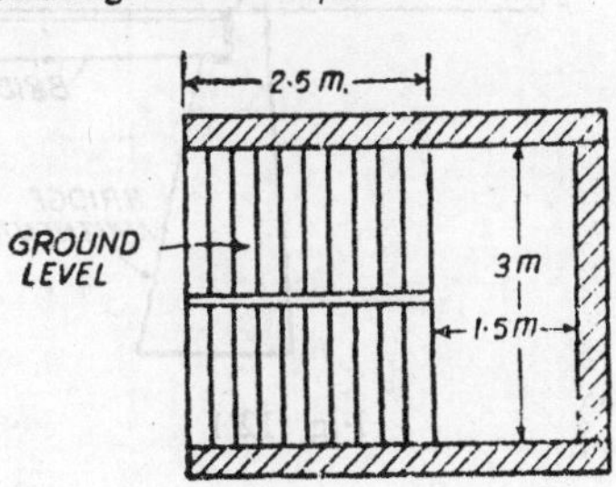

Fig. 11·23

12

Retaining Walls

12·1 General

A wall designed to maintain unequal level of ground on its two faces is called a retaining wall. The earth on the side of the

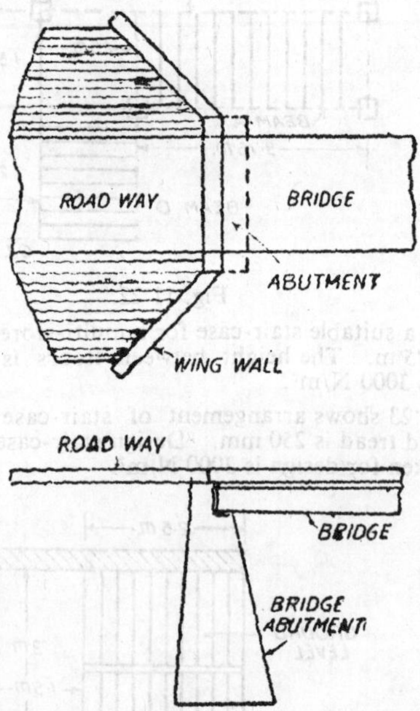

Fig. 12·1

wall where ground level is higher is called backfill and the retaining wall is to retain this earth. A retaining wall which supports the load of a bridge and also retains earth is called abutment. The walls placed at an inclination to the normal to the direction of fill and used to retain the ends of an approach fill for a bridge are known as wing walls.

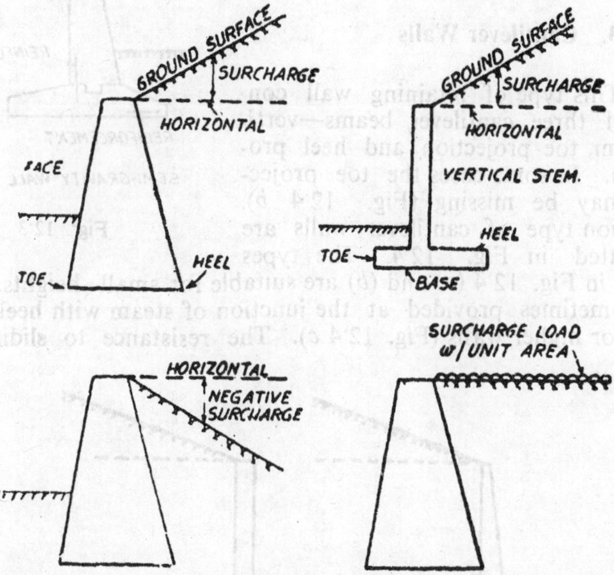

Fig. 12·2

Various portions of the retaining walls with their names are illustrated in Fig. 12·2.

12 2. Types of Retaining Walls

Following types of retaining walls are generally used.

1. Gravity Walls

This is the simplest type of retaining wall [Fig. 12·2 (*a*)]. These walls consist of mass of concrete or masonry. Main function of mass of concrete or masonry is to provide dead weight to give stability against thrust of retained earth. The unit stresses in concrete or masonry are very low and these walls are so proportioned that there are no tensile stresses developed. Because of massive construction these retaining walls are more resistant to destructive agencies than other walls.

2. Semi-gravity Walls

Gravity retaining walls require greater toe to give sufficient base width so that no tension is developed at the base. Also in gravity retaining walls fairly heavy section of stem is required. By providing some reinforcement in toe and stem as shown in Fig. 12·3, the section of retaining wall can be reduced.

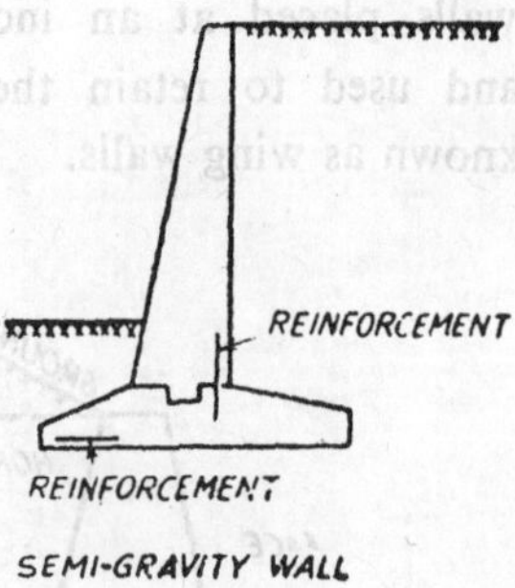

Fig. 12·3

3. Cantilever Walls

This type of retaining wall consists of three cantilever beams—vertical stem, toe projection and heel projection. In some cases the toe projection may be missing (Fig. 12·4 *b*). Common type of cantilever walls are illustrated in Fig. 12·4. The types shown in Fig. 12·4 (*a*) and (*b*) are suitable for small heights. Fillets are sometimes provided at the junction of steam with heel and toe slabs for higher walls (Fig. 12·4 *c*). The resistance to sliding of a

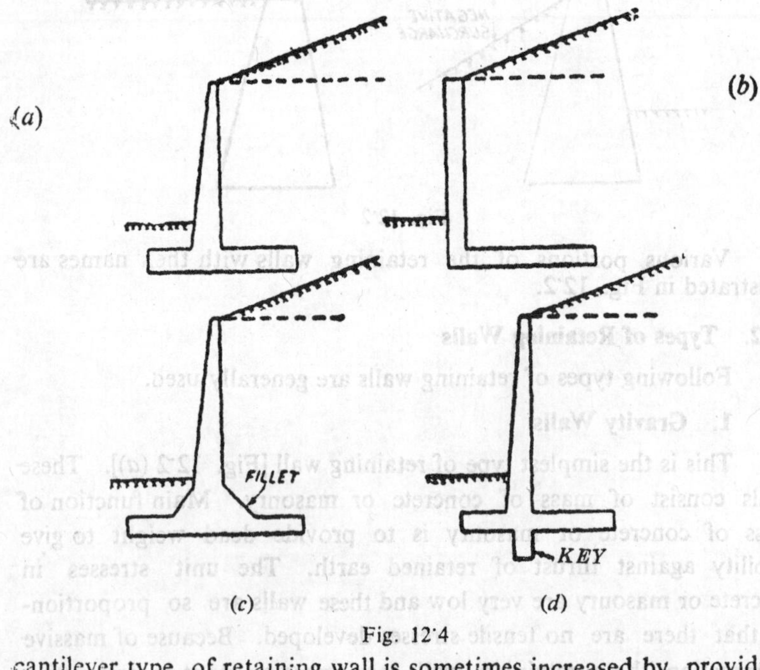

Fig. 12·4

cantilever type of retaining wall is sometimes increased by providing vertical projection known as key at the base (Fig. 12·4 *d*). The key may be provided near the toe, near heel or at the middle of base.

4. Counterfort retaining walls

The cantilever types of retaining walls are economical only up to height of about 6 to 8 m. However, for higher walls the design of cantilever slabs, stem and heel will prove uneconomical. Greater economy can be achieved by providing counterforts as shown in Fig. 12·5. The greater economy is achieved in this case as the stem and heel are not designed as cantilevers but as continuous slabs over counterforts and thus bending moments are reduced. However, in this type of retaining walls additional concrete, reinforcement and formwork is required for the counterforts.

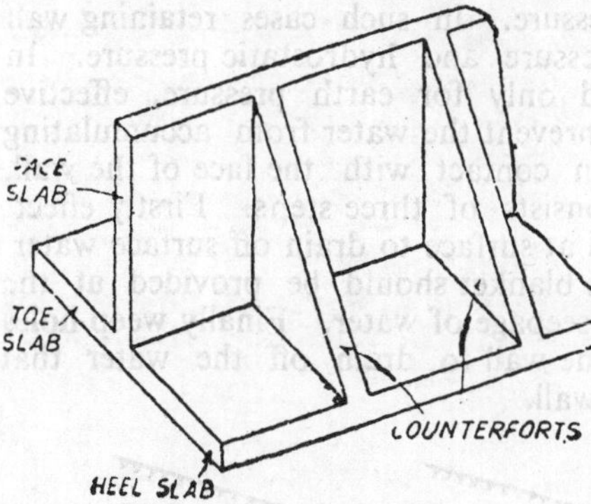

Fig. 12·5

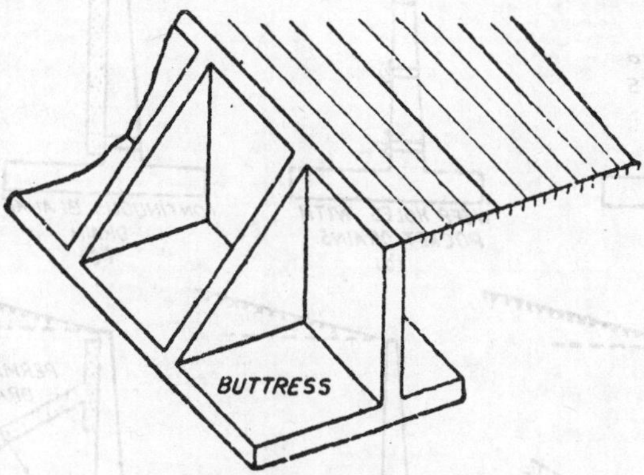

Fig. 12·6

Sometimes retaining walls with counterforts on the side of toe, in addition to counterforts on the side of the heel, are constructed but these types of walls are not commonly used.

5. Buttressed walls

A buttressed wall is similar to a counterfort type of wall but in this case counterforts are placed on the front of the wall as shown in Fig. 12·6 and are known as buttresses rather than counter forts. In this type of wall heel projection is small and hence back-fill contributes less to the stability of wall. Also buttress reduce clearance in front of the wall. These types of walls are rarely used.

12·3. Drainage of Retaining Walls

The theoretical earth pressure calculated against the face of a retaining wall takes into consideration the pressure exerted by earth and no hydrostatic pressures are considered. However, when the backfill is submerged retaining wall is to be designed to resist hydrostatic pressure in addition to pressure exerted by the submerged soil. Surface or underground water may seep into backfill and this will cause hydrostatic pressure. In such cases retaining wall should be designed for earth pressure and hydrostatic pressure. In case retaining wall is designed only for earth pressure, effective drainage should be provided to prevent the water from accumulating at the back and thus coming in contact with the face of the wall. The drainage system of walls consists of three steps. Firstly effective drainage should be provided at surface to drain off surface water Secondly, a relatively impervious blanket should be provided at the top of the backfill to reduce the seepage of water. Finally weep holes or weepers are provided inside the wall to drain off the water that accumulates at the back of the wall.

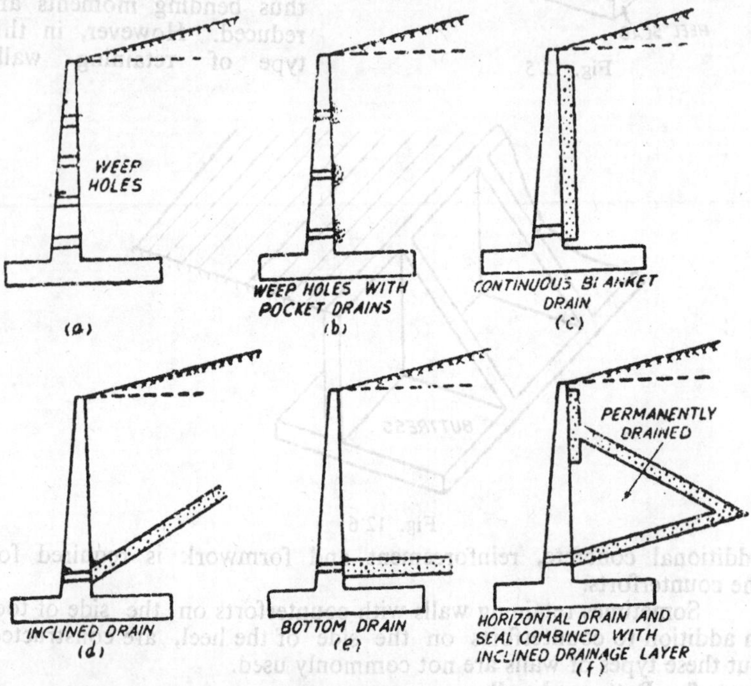

Fig. 12·7

Various types of drainage arrangements are shown in Fig. 12·7. Weep holes are generally formed by embedding 10 cm diameter pipes in the wall. The vertical distance between horizontal rows of weep holes should not exceed about 1·5 m. The horizontal spacing between weep holes depends upon the arrangement provided at the back of the wall for seepage. Cheapest method is to provide gravel or crushed stone (inverted filter) at the intake end of the weep holes. The horizontal spacing of weep holes in this case is also about 1·5 m.

12·4. State of Equilibrium in Soil

The magnitude of pressure exerted by earth depends on the state of equilibrium in the soil. Two states of equilibrium exist in soil, one is state of elastic equilibrium and other is state of plastic equilibrium. When shearing stresses in a soil mass are less than the shearing strength of soil, the soil is said to be in state of elastic equilibrium. When shearing stresses in a soil mass are equal to shearing strength of soil, the soil, is said to be in a state of plastic equilibrium. In this state every soil particle is on the verge of failure.

The earth pressure exerted is divided in three categories, *i.e.* earth pressure at rest, active earth pressure and passive earth pressure depending on the lateral deformation of the soil.

Consider a semi-infinite soil mass which has been deposited artificially and which has remained undisturbed. This soil mass is

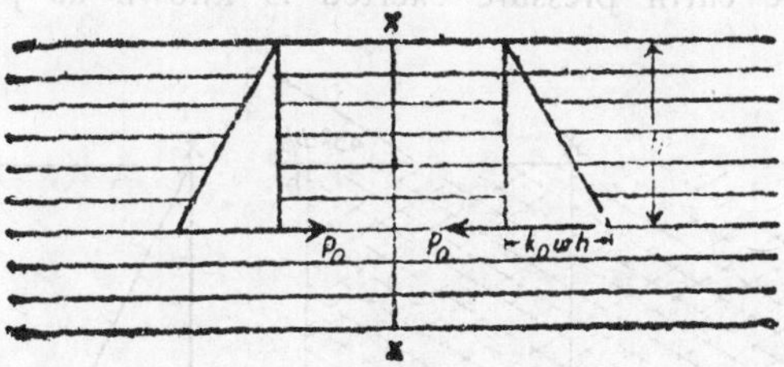

Fig. 12·8

said to be in a state of elastic equilibrium. On any plane *X-X* the pressure exerted at a depth of h is p_0 and is known as earth pressure at rest, $p_0=k_0wh$, where w is the density of the soil and k_0 is the coefficient of earth pressure at rest. The value of k_0 depends on the process in which deposit is formed. The value of k_0 generally lies between 0·4 to 0·5.

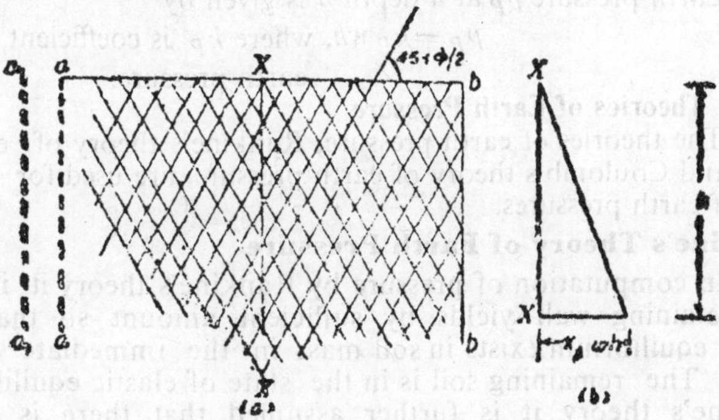

Fig. 12·9

If the soil mass discussed above is allowed to stretch laterally by sufficient amount shearing stresses are developed along several planes and these shearing stresses are equal to shearing strength of soil. The shear planes are inclined at an angle $45+\phi/2$ to the horizontal as shown in Fig. 12·9. Under these conditions the soil mass is said to be under state of plastic equilibrium and the soil is said to be in active state. The earth pressure exerted is known as active pressure. The state of pressure was investigated by Rankine and is known as Rankine active pressure.

Earth pressure p_A at depth h is given by

$p_A = k_A\ wh$, where k_A is coefficient of active earth pressure.

If the soil mass discussed above is allowed to compress instead of stretching, shearing stresses are developed along several planes and these shearing stresses are equal to the shearing strength of the soil. The shearing planes are inclined at an angle $45° - \phi/2$ to the horizontal as shown in Fig. 12·10. The whole soil mass is in state of plastic equilibrium. In this case soil is said to be in passive state and the earth pressure exerted is known as passive earth pressure.

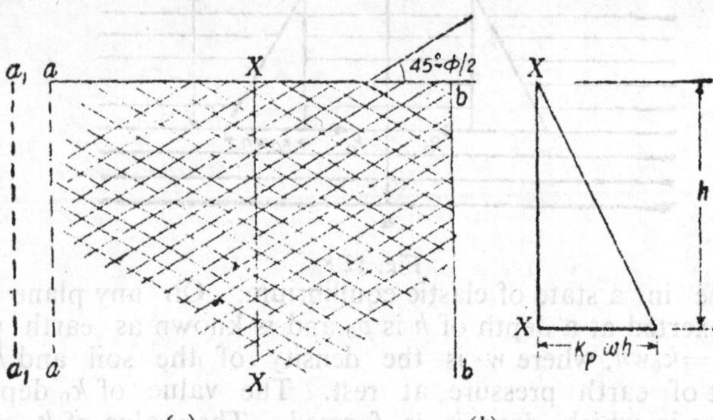

Fig. 12·10

Earth pressure p_P at a depth h is given by

$p_P = k_P\ wh$, where k_P is coefficient of passive earth pressure.

12·5. Theories of Earth Pressure

The theories of earth pressure, Rankine's theory of earth pressure, and Coulomb's theory of earth pressure, are used for computation of earth pressures.

Rankine's Theory of Earth Pressure

In computation of pressure by Rankine's theory it is assumed that retaining wall yields by sufficient amount so that state of plastic equilibrium exists in soil mass in the immediate vicinity of wall. The remaining soil is in the state of elastic equilibrium. In Rankine's theory it is further assumed that there is no friction between soil and wall.

(*a*) **Active Pressure.** When the retaining wall deforms outwards, *i.e.* soil is allowed to stretch by about 2% of the height of wall, the earth pressure exerted is active earth pressure. The active earth pressure at a depth h in a retaining wall with horizontal ground surface is given by

$$p_A = wh\ \frac{1-\sin\phi}{1+\sin\phi}\ \text{where } \phi \text{ is the angle of repose of soil.}$$

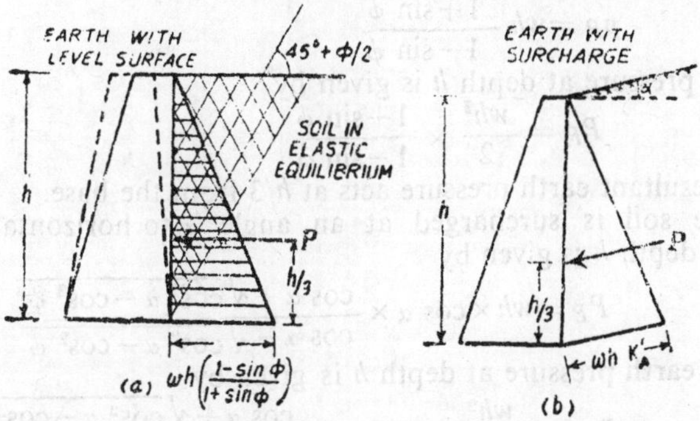

Earth with level ground surface Earth with surcharge

Fig. 12·11

Total pressure at depth h is given by

$$P_A = \tfrac{1}{2} \times wh \times h\ \frac{1-\sin\phi}{1-\sin\phi} = \frac{wh^2}{2} \times \frac{1-\sin\phi}{1+\sin\phi}$$

Resultant pressure acts at $h/3$ from base.

If the earth is surcharged at an angle α to the horizontal, pressure at depth h is given by

$$P_A = wh \times \cos\alpha \times \frac{\cos\alpha - \sqrt{\cos^2\alpha - \cos^2\phi}}{\cos\alpha + \sqrt{\cos^2\alpha - \cos^2\phi}}$$

Total pressure P_A at depth h is given by

$$P_A = \tfrac{1}{2} wh^2 \times \cos\alpha \times \frac{\cos\alpha - \sqrt{\cos^2\alpha - \cos^2\phi}}{\cos\alpha + \sqrt{\cos^2\alpha - \cos^2\phi}}$$

The resultant earth pressure acts at $h/3$ and is inclined at an angle α to the horizontal.

(*b*) **Passive Earth Pressure.** If the retaining wall deforms laterally and compresses soil, earth pressure exerted is passive earth pressure.

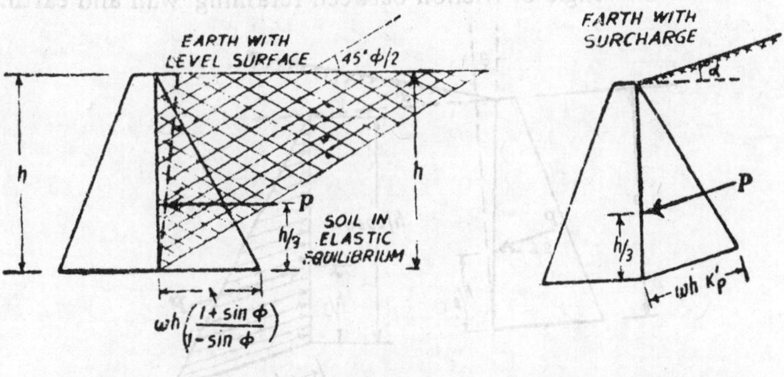

(*a*) Earth with level ground surface (*b*) Earth with surcharge

Fig. 12·12

Passive earth pressure at a depth h in a retaining wall with level ground surface is given by

$$p_P = wh \frac{1+\sin\phi}{1-\sin\phi}$$

Total pressure at depth h is given by

$$P_P = \frac{wh^2}{2} \times \frac{1+\sin\phi}{1-\sin\phi}$$

The resultant earth pressure acts at $h/3$ from the base.

If the soil is surcharged at an angle α to horizontal, earth pressure at depth h is given by

$$P_P = wh \times \cos\alpha \times \frac{\cos\alpha + \sqrt{\cos^2\alpha - \cos^2\phi}}{\cos\alpha - \sqrt{\cos^2\alpha - \cos^2\phi}}$$

Total earth pressure at depth h is given by

$$P_P = \frac{wh^2}{2} \times \cos\alpha \times \frac{\cos\alpha + \sqrt{\cos^2\alpha - \cos^2\phi}}{\cos\alpha - \sqrt{\cos^2\alpha - \cos^2\phi}}$$

The resultant pressure acts at $h/3$ and is inclined at an angle α to horizontal.

Coulomb's Theory of Earth Pressure

It is assumed in the Rankine's theory of earth pressure that the back of retaining wall is smooth and thus there is no friction between wall and earth. This assumption is however seldom justified and as the back of the retaining wall is never smooth and thus the earth pressure computed on the basis of Rankine's theory leads to erroneous results. Coulomb's theory can be applied to any boundary conditions. In the Coulomb's theory it is assumed that the retaining wall yields by sufficient amount and causes state of plastic equilibrium in soil. This is also the assumption made in Rankine's theory. In the Coulomb's theory it is further assumed that surface of sliding is plane though in reality the surface of sliding is slightly curved.

However the error due to assuming that the surface of sliding as plane is negligible for lower value of ϕ.

Active Earth Pressure

Notations : ϕ = Angle of internal friction of soil.

δ = Angle of friction between retaining wall and earth.

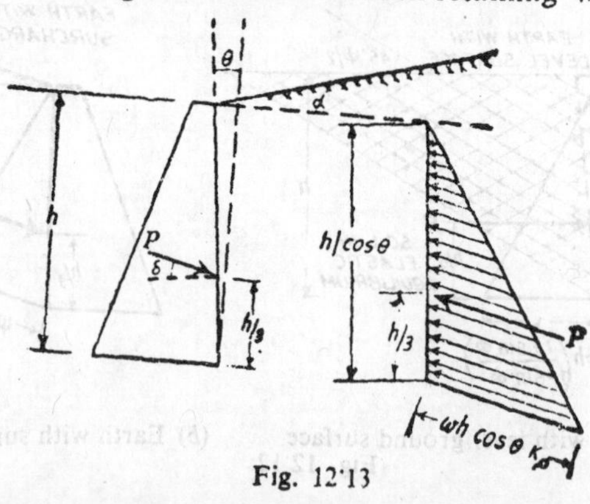

Fig. 12·13

α=Angle of surcharge.

θ=Angle made by back of wall with vertical.

Intensity of active earth pressure at a depth h is given by

$$p_A = wh \cos\theta \times k_A$$

where
$$k_A = \frac{\cos^2(\phi-\theta)}{\cos^2\theta\cos(\delta+\theta)\left[1+\sqrt{\dfrac{\sin(\delta+\phi)\sin(\phi-\alpha)}{\cos(\delta+\theta)\cos(\theta-\delta)}}\right]}$$

Total pressure upto depth h is given by

$$P_A = \tfrac{1}{2}wh^2\cos\theta\, k_A$$

The resultant earth pressure acts at lower middle third and is inclined at an angle δ to the normal to the face of the walls as shown in Fig. 12·13.

Passive Earth Pressure

Intensity of passive earth pressure at a depth h is given by

$$p_P = wh\cos\phi\, k_P$$

and total earth pressure upto depth h is given by

$$P_P = \tfrac{1}{2}wh^2\cos\theta\, k_P$$

where
$$k_P = \frac{\cos^2(\phi+\theta)}{\cos^2\theta\,(\cos(\theta-\delta)\left(1-\sqrt{\dfrac{\sin(\phi+\delta)\sin(\phi+\alpha)}{\cos(\theta-\delta)\cos(\theta-\alpha)}}\right)^2}$$

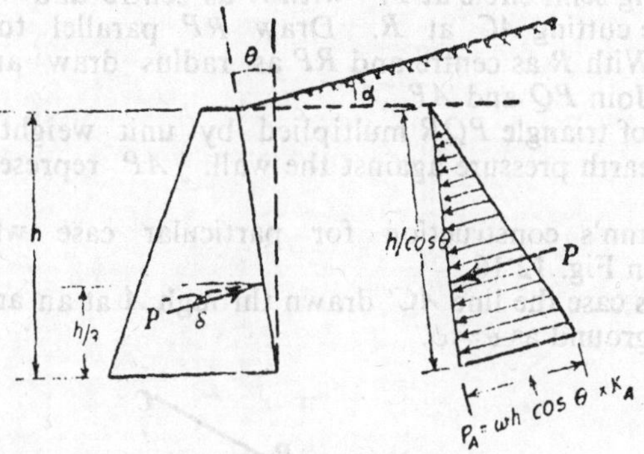

Fig. 12·14

The resultant pressure acts at lower middle third point and is inclined at angle δ to the face of the wall as shown in Fig. 12·14.

12·6. Graphical Representation of Coulomb's Theory—Rebhann's Construction

By Rebhann's construction, total active earth pressure against a retaining wall can be computed based on Coulomb's theory. The construction is illustrated in Fig. 12·15.

Let AB be the back of the wall, ϕ be angle of internal friction of soil, α be angle of surcharge and δ be angle of friction between soil and retaining wall.

Through A draw a horizontal line AD. Draw a line AC at an angle ϕ to the horizontal meeting ground surface at C. Describe a semi-circle with AC as diameter. Through B draw a line BE making

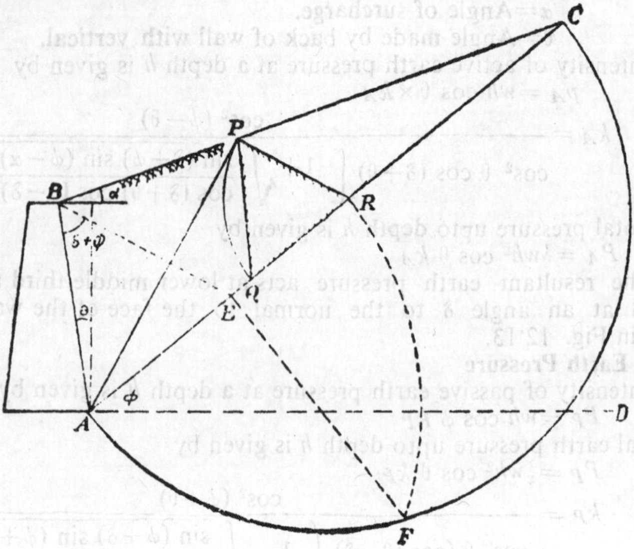

Fig. 12·15

$\delta+\phi$ with AB and meeting AC in E. At E draw a perpendicular to AC cutting semi-circle at F. with A as centre and AF as radius draw an arc cutting AC at R. Draw RP parallel to BE cutting BC at P. With R as centre and RP as radius draw an arc cutting AC in Q. Join PQ and AP.

Area of triangle PQR multiplied by unit weight of soil represents total earth pressure against the wall. AP represents plane of sliding.

Rebhann's construction for particular case when $\alpha=\phi$ is illustrated in Fig. 12·16,

In this case the line AC' drawn through A at an angle ϕ will be parallel to ground as $\alpha=\phi$.

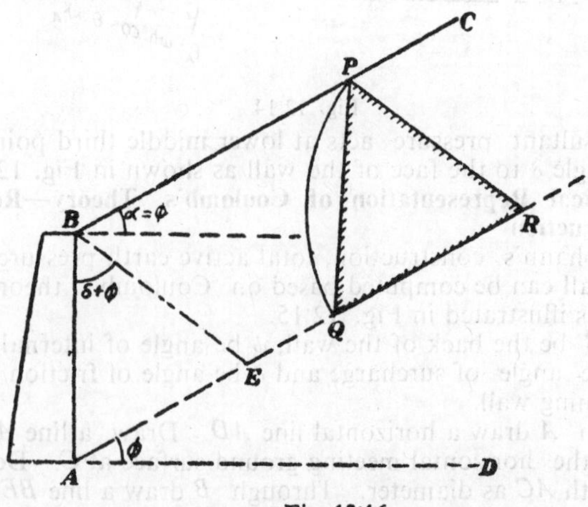

Fig. 12·16

Through B draw a line at an angle $\delta+\phi$ meeting AC' in E. Select any point P on the ground surface and through this point draw a line PR parallel to BE meeting AC' in R. With R as centre and RP as radius draw an arc meeting AC' in Q.

Join PQ. Area of triangle PQR multiplied by unit weight of soil represents total earth pressure against the wall.

12·7. Earth Pressure due to Submerged Soil

When the pores of permeable soil are filled with water. the water exerts full hydrostatic pressure. The pressure of water causes buoyant effect on the soil with the result that effective weight of soil which causes earth pressure is reduced by an amount equal to weight of water displaced by soil. The reduced weight is called submerged weight.

Let the submerged weight of soil be w_1 and unit weight of water be w_w. Pressure at depth h exerted by submerged soil will be $p_s=w_1hk$, where k is coefficient of earth pressure. Pressure exerted by water at depth h is $p_w=w_wh$.

Pressure against the face of a retaining wall for various cases of submerged soil is shown in Fig. 12·17 (*a*).

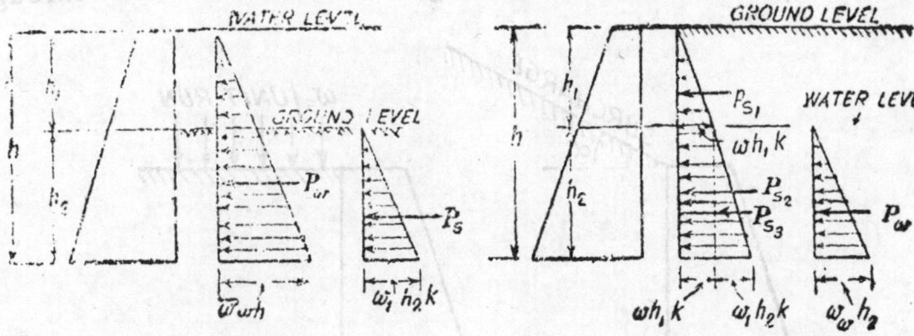

$Pw=\frac{1}{2}w_wh^2$, $P_s=\frac{1}{2}w_1h_2^2k$

$P_{S_1}=\frac{1}{2}wh_1^2k$, $P_{S_2}=wh_1h_2k$

$P_{S_3}=\frac{1}{2}w_1h_2^2$

$P_w=\frac{1}{2}w_wh_2^2$

(*i*) Water level above ground surface. (*ii*) Ground surface above water level

Fig. 12·17 (*a*)

12·8. Earth Pressures in a Braced Trench

In the previously discussed theories of earth pressures it was assumed that the retaining walls yield by sufficient amount so that

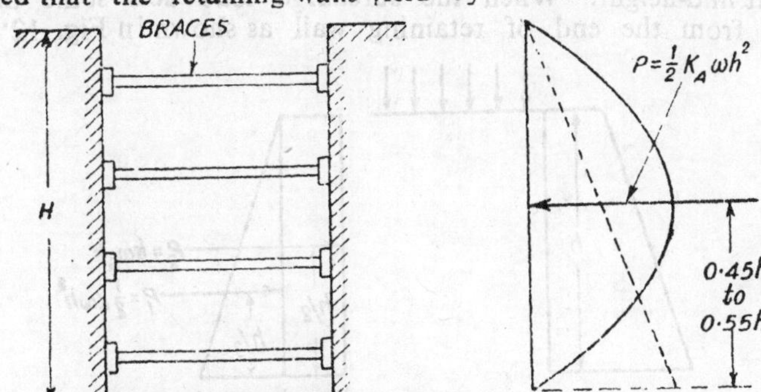

Fig. 12·17 (*b*) Earth pressures in Braced Trench

active or passive earth pressure acts. However many practical cases fail to satisfy this deflection requirement for example basement wall supported by framing at floor level shown in Fig. 12·17 (*b*) etc. In all these cases the total pressure is approximately the same as calculated by Rankine Theory of active pressure but the resultant acts at about mid-depth as shown in the figure. The distribution of earth pressure may vary but for all practical purposes it may be considered as parabolic.

12·9. Surcharge

Surcharge on a retaining wall may be due to sloping earth as shown in Fig. 12·17 (*c*) or due to the additional load acting on the earth surface. The effect of surcharge load is to give additional earth pressure which is to be taken into consideration in design. This additional pressure is exerted provided the load acts on the possible sliding plane. If the load acts outside the sliding plane there is no effect of surcharge on the retaining wall. The effect of uniform surcharge over the entire area is to give an effective equivalent to height of soil. If w_s is the weight of surcharge per unit length, the earth pressure due to surcharge is kw_s and is constant through

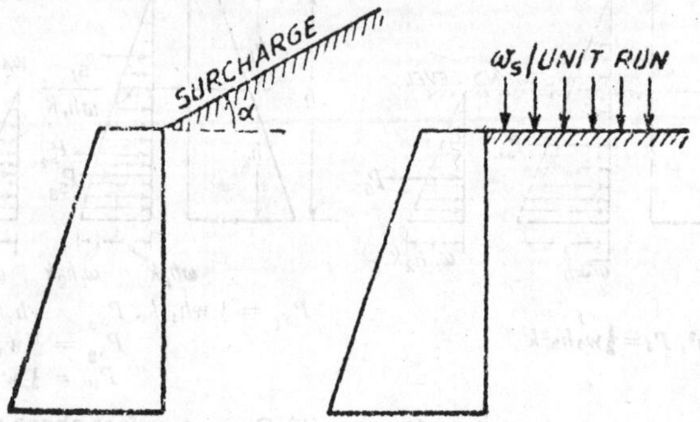

(*c*) (*d*)

Fig. 12·17

the depth. Thus the total pressure due to surcharge is kw_sh and acts at mid-height. When the surcharge load acts some distance away from the end of retaining wall as shown in Fig. 12·17 (*f*)

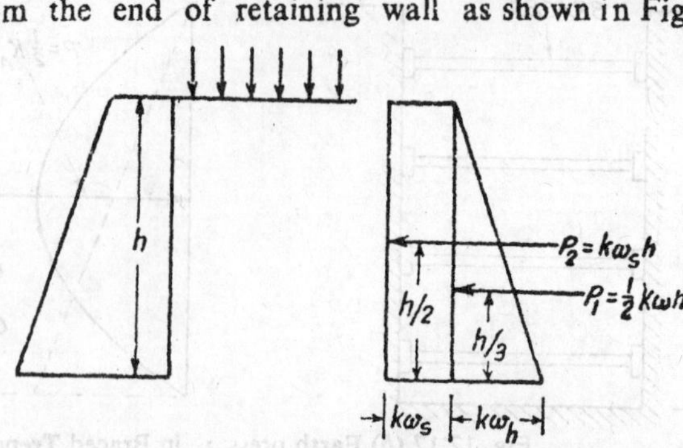

Fig. 12·17 (*e*)

its effect can be taken into consideration by assuming the pressure to be acting over reduced depth. In Fig. 12·17 (*f*), let the surcharge load start from a distance *C* from the edge of the wall. Draw a line at 45° to the horizontal cutting the face of the wall at *D*. Let the height of point *D* be h_2.

The effect of surcharge is taken as pressure kw_s, acting over height h_2. The angle 45° at which line *CD* is drawn is approximate one. Generally the angle varies between 40° and 45°. It may be noted that this is an approximate method of calculating pressure. In actual case the pressure at *D* does not change abruptly as shown in the figure.

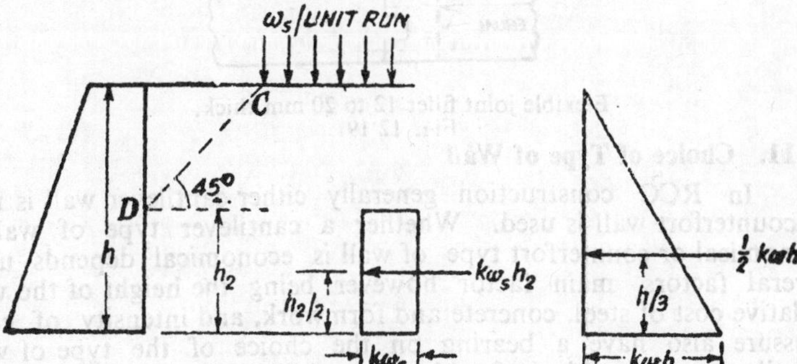

Fig. 12·17 (*f*)

12·10. General Design Requirements

To avoid failure of retaining wall following general requirements should be satisfied :

1. The factor of safety against sliding should be at least 1·5 if the passive resistance of the soil in front of the wall is neglected as it is possible that soil is removed by constructional operation, scour etc. The factor of safety against sliding should be 2 if the passive resistance of soil is taken into consideration.

2. A factor of safety of 1·5 should be maintained against overturning.

The bearing pressure at toe should be within permissible limits.

3. The thickness of stem, heel and toe should be adequate to resist the bending moments and shear forces to which they are subjected.

4. A construction joint should be provided between base slab and upper portion of wall. The construction joint should be capable of resisting the shear force caused by the lateral pressure above the joint.

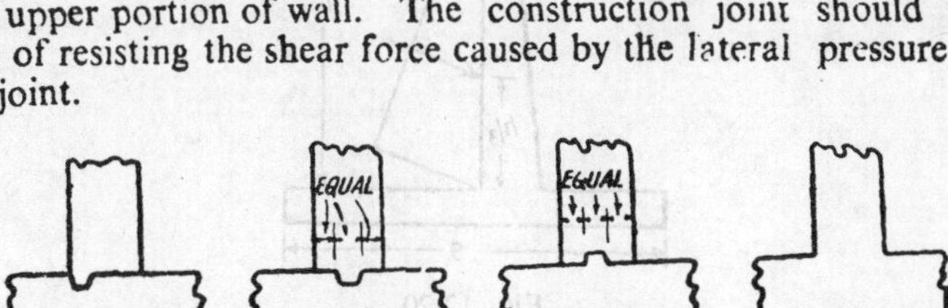

Fig. 12·18

Various types of construction joints are shown in Fig. 12·18. The horizontal shear is to be resisted by the key and dowel bars.

If the length of retaining wall exceeds 45 metres expansion joint should be provided. The spacing of these joints depends on the temperature variation, exposure to weather and the time and season of laying the concrete. However in no case should the spacing of these joints exceed 45 m.

A typical expansion joints is shown in Fig. 12·19.

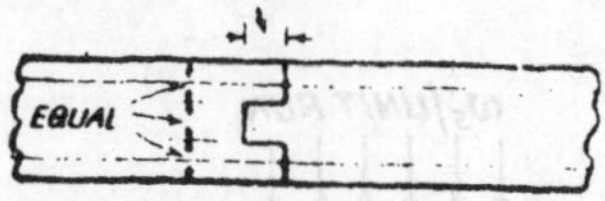

Flexible joint filler 12 to 20 mm thick.
Fig. 12·19

12·11. Choice of Type of Wall

In RCC construction generally either cantilever wall is used or counterfort wall is used. Whether a cantilever type of wall is economical or counterfort type of wall is economical depends upon several factors, main factor however being the height of the wall. Relative cost of steel, concrete and formwork, and intensity of earth pressure also have a bearing on the choice of the type of wall. Cantilever type wall is preferable where skilled labour is not available as the construction is simpler in this case and the placing of steel is much easier. In the case of counterfort retaining walls the cost of formwork is more and placing of steel is complicated.

In general cantilever type of retaining wall can be used for height upto 6 m.

12·12. Design of Cantilever Retaining Wall

In cantilever retaining wall stem acts as cantilever and is subjected to earth pressure. The maximum S.F. and B.M. occur at the junction of steam with the base.

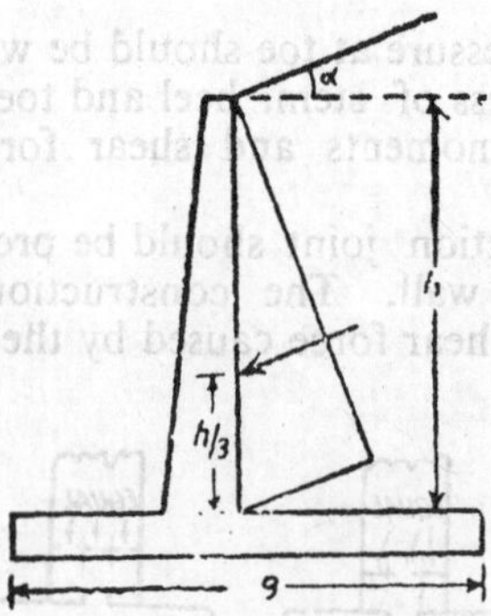

Fig. 12·20

Let h be the height of steam, α be angle of surcharge and ϕ be the angle of repose of soil.

Consider one metre length of retaining wall.

Intensity of pressure at depth h,

$$p = wh \cos\alpha \frac{\cos\alpha - \sqrt{\cos^2\alpha - \cos^2\phi}}{\cos\alpha + \sqrt{\cos^2\alpha - \cos^2\phi}}$$

Total pressure $$P = \frac{wh^2}{2}\cos\alpha \frac{\cos\alpha - \sqrt{\cos^2\alpha - \cos^2\phi}}{\cos\alpha + \sqrt{\cos^2\alpha - \cos^2\phi}}$$

The pressure is parallel to the surface of ground and acts at $h/3$ from junction of the stem with the base.

Horizontal component

$$P_H = \frac{wh^2}{2}\cos^2\alpha \left[\frac{\cos\alpha - \sqrt{\cos^2\alpha - \cos^2\phi}}{\cos\alpha + \sqrt{\cos^2\alpha + \cos^2\phi}}\right]$$

Vertical component

$$P_V = \frac{wh^2}{2}\cos\alpha\sin\alpha \left[\frac{\cos\alpha - \sqrt{\cos^2\alpha^2 - \cos^2\phi}}{\cos\alpha + \sqrt{\cos^2\alpha - \cos^2\phi}}\right]$$

Maximum B.M. $$= \frac{wh^3}{6}\cos^2\alpha \left[\frac{\cos\alpha - \sqrt{\cos^2\alpha - \cos^2\phi}}{\cos\alpha + \sqrt{\cos^2\alpha - \cos^2\phi}}\right]$$

When $\alpha = 0$, maximum B.M.

$$= \frac{wh^3}{6}\left[\frac{1 - \sin\phi}{1 + \sin\phi}\right].$$

Minimum thickness of 15 cm is generally provided at the top of stem. The thickness of stem at junction with base and reinforce ment are designed for above B.M. The reinforcement is curtailed in the top portion.

Temperature reinforcement is provided vertically and horizontally on the face of the stem.

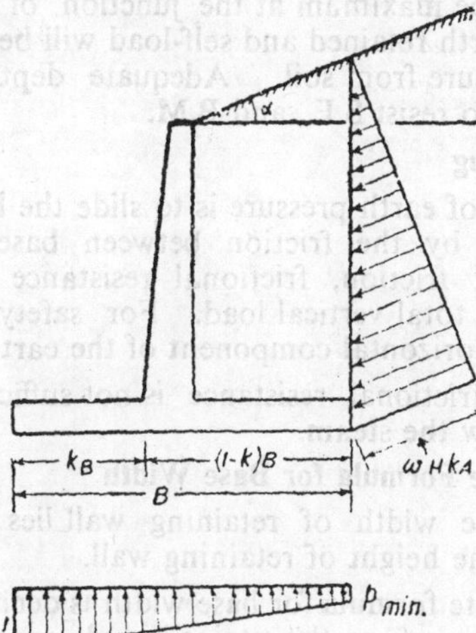

Fig. 12·21.

Pressure Distribution on the Base

The forces acting on the retaining wall are

1. Earth pressure (Fig. 12·21).
2. Weight of earth supported on heel.
3. Weight of the stem.
4. Weight of the base.

Weight of the earth on the toe portion is neglected.

The resultant of these forces should cut the base in the middle third so that no portion of the base is lifted up, *i.e.*, no tension is developed at the base.

The maximum and minimum pressures at the base are given

$$\frac{W}{B} \pm \frac{6We}{B^2}$$

where W is the sum of vertical loads and the vertical component of earth pressure and e is the eccentricity of the resultant force acting on the base.

The maximum pressure should not exceed the allowable pressure on the soil.

Design of Toe

The toe is subjected to upward pressure from the soil and self-load which acts downwards. The maximum S.F. and B.M. will occur at the face of the steam. Adequate depth and reinforcement is provided to resist S.F. and B.M.

Design of Heel

The heel is subjected to upward pressure from the soil and downward load due to self-weight and earth above the heel. The B.M. and S.F. will be maximum at the junction of heel with stem. The B.M. due to earth retained and self-load will be more than that due to upward pressure from soil. Adequate depth and reinforcement are provided to resist S.F. and B.M.

Check Against Sliding

The tendency of earth pressure is to slide the base. This effect is counterbalanced by the friction between base and the soil. If μ is co-efficient of friction, frictional resistance to sliding will be μW, where W is the total vertical load. For safety μW should be at least 1·5 times the horizontal component of the earth pressure.

In case the frictional resistance is not sufficient, key is provided preferably below the steam.

12·13. Approximate Formula for Base Width

Generally base width of retaining wall lies between 0·5 H to 0·6 H, where H is the height of retaining wall.

An approximate formula for base width is derived below assuming that the density of earth retained and concrete is same and neglecting the weight of the toe and any earth above it. The formula can be used for level ground surface.

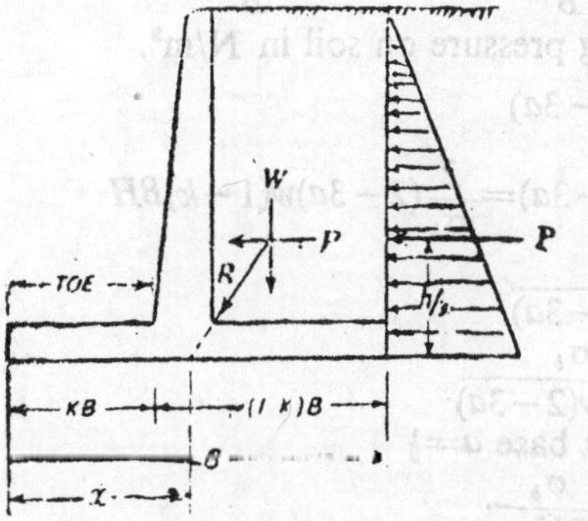

Fig. 12·22

Let H be the height of retaining wall. B be the base width, kB be the width of toe and w be the density of soil in N/m³.

Total weight will be taken as $W=w(1-k)\,BH$.

Total pressure

$$P=\frac{wH^2}{2}\left(\frac{1+\sin\phi}{1+\sin\phi}\right)$$

Let $\bar{x}$ be the distance, from toe, where the resultant strikes the base. Taking moments about this point.

$$P\times\frac{H}{3}=W\left(kB+\frac{1-k}{2}B-\bar{x}\right)$$

$$=\frac{W}{2}(B+kB-2\bar{x})$$

$$\therefore\ \frac{H^3}{6}\times w\,\frac{(1-\sin\phi)}{(1+\sin\phi)}=w\,\frac{(1-k)BH}{2}[B(1+k)-2\bar{x}]$$

$$\frac{H^2(1-\sin\phi)}{3(1-k)B(1+\sin\phi)}=B(1+k)-2\bar{x}$$

Let $\bar{x}=aB$

$$\therefore\ \frac{H^2\,(1-\sin\phi)}{3\,(1+\sin\phi)}=(1-k)B[B(1+k)-2aB]$$

$$=(1-k)B^2[1+k-2a]$$

If no tension is to develop at base, value of x should be more than $B/3$. In the limiting case $x=B/3$ or $a=1/3$. Substituting this value of a.

$$\frac{H^2}{3}\left(\frac{1-\sin\phi}{1+\sin\phi}\right)=(1-k)B^2(1+k-2/3)=\frac{B^2}{3}(1-k)\,(3k+1)$$

$$\therefore\qquad B=H\sqrt{\frac{\left(\frac{1-\sin\phi}{1-\sin\phi}\right)}{(1+k)(3k+1)}}\qquad\ldots(1)$$

Pressure distribution at the base is given by

$$p=\frac{W}{B}\pm\frac{W\left(\frac{B}{2}-\bar{x}\right)}{B^2/6}$$

$$=\frac{W}{B}\pm\frac{6W}{B^2}\left(\frac{B}{2}-x\right)$$

$$=\frac{W}{B}\pm\frac{6M}{B^2}\left(\frac{B}{2}-aB\right)$$

$$p_{max}=\frac{W}{B}+\frac{6W}{B}\left(\tfrac{1}{2}-a\right)=\frac{W}{B}\ [1+3-6a]=\frac{2W}{B}(2-3a)$$

If σ_b is the allowable bearing pressure on soil in N/m².

$$\sigma_b=\frac{2W}{B}(2-3a)$$

$$\therefore \qquad B=\frac{2W}{\sigma_b}(2-3a)=\frac{2}{\sigma_b}(2-3a)w(1-k)BH$$

$$1-k=\frac{\sigma_b}{2Hw(2-3a)}$$

$$\therefore \qquad k=1-\frac{\sigma_b}{2Hw(2-3a)}$$

If no tension is to develop at base $a=\frac{1}{3}$

$$\therefore \qquad k=1-\frac{\sigma_b}{2Hw}.$$

For pressure to be within permissible limit the above formula gives the minimum value of ratio of toe width to base width.

In Appendix graph 18 gives the relation between H and k for various values of σ_h/w. This graph is useful when bearing capacity of soil is low and ratio of σ_b/w is lower than 10. Graph 19 gives relation between B/H and ϕ for various values of k, *i.e.* $k=0$, 1/5, 1/4, 1/3 and 1/2. To fix base width for a retaining wall, determine the value of σ_b/w, from graph 18 find the value of k corresponding to height H of retaining wall. This value of k or more than this is adopted. From graph 19 obtain the ratio B/H for this value of k and value of ϕ as given for soil. Thus the value of base width B can be fixed. This is the minimum breadth which can be adopted.

Example 12·1. *Design a cantilever retaining wall to retain earth 4 m above ground level. The top of the earth is to be leveled. The density of earth 15,000 N/m³ and an angle of internal friction of soil is 30°. The safe bearing capacity of soil is 200 kN/m² and coefficient of friction between soil and wall is 0·6.*

Sol. Take the depth of base of retaining wall 1 m below ground level.

Total height of retaining wall

$$=4+1=5 \text{ m}$$

Assume $k=1/3$.

i.e. toe width $=kB=B/3$.

$$B=H\sqrt{\frac{\dfrac{1-\sin\phi}{1+\sin\phi}}{(1-k)(3k+1)}}$$

$$=5\sqrt{\frac{1/3}{(1-1/3)(3\times 1/3+1)}}$$

$$=5\sqrt{1/4}=2{\cdot}5 \text{ m}.$$

Assume width of toe $=0{\cdot}8$ m and thickness of heel slab and toe slab as 0·3 m.

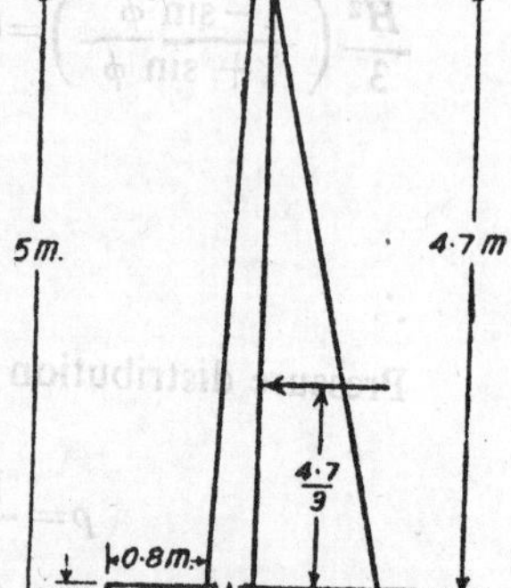

Fig. 12·23

Design of Steam

Consider 1 m length of retaining wall.

$$\text{Maximum B.M.} = \frac{wh^3}{6}\left(\frac{1-\sin\phi}{1+\sin\phi}\right)$$

$$= \frac{15{,}000\times(4\cdot7)^3}{6}\times\frac{1}{3}$$

$$= 86{,}520 \text{ Nm.} = 86{,}520{,}000 \text{ N-m.}$$

$$\text{Effective depth } d = \sqrt{\frac{86{,}520{,}000}{0\cdot87\times1000}} = 315 \text{ mm.}$$

Provide overall depth of 360 mm with effective depth of 320 mm The width of wall provided at top will be 16 cm.

$$A_t = \frac{86{,}520{,}000}{0\cdot87\times320\times140} = 2220 \text{ mm}^2.$$

Provide 12 mm ϕ at 50 mm centres, $A_t = 2262$ mm².

Curtailment of Stem Reinforcement

At depth h metres from toe, let A_t be the steel required.

$$\text{Effective depth} = \left(120+\frac{200}{4\cdot7}h\right)$$

$$\text{B.M.} = \frac{wh^3}{6}\times\frac{1-\sin\phi}{1+\sin\phi}\times1000 \text{ N-mm}$$

$$= \frac{wh^3}{6}\times1/3\times1000 = \frac{500}{9}wh^3$$

$$A_t = \frac{\frac{500}{9}wh^3}{0\cdot87\left(120+\frac{200}{4\cdot7}\right)\times140}$$

Reinforcement at depth of 3·5 m is given by

$$A_t = \frac{\frac{500}{9}\times15{,}000\times(3\cdot5)^3}{0\cdot87\left(120+\frac{200}{4\cdot7}\times3\cdot5\right)\times140}$$

$$= 1091 \text{ mm}^2.$$

Provide 12 mm ϕ bars at 100 mm centres. $A_t = 1130$ mm². Half the bars from bottom can be curtailed. The bars will be stopped at $45d$, *i.e.* $45\times12 = 540$ mm beyond theoretical curtailment point. The bars are curtailed at 2·90 m depth.

Reinforcement at depth of 2·0 m is given by

$$A_t = \frac{\frac{500}{9}\times15{,}000\times(2\cdot0)^3}{0\cdot87\left(120+\frac{200}{4\cdot7}\times2\cdot0\right)\times140}$$

$$= 270 \text{ mm}^2.$$

Half the bars from bottom are taken straight, *i.e.* 12 mm ϕ bars at 200 mm centres are provided. A_t=566 mm². The bars are stopped at 600 mm beyond this point. The bars are curtailed at 1·4 m depth.

Distribution Steel

Distribution steel based on average thickness

$$=\frac{0.15}{100}\left[\frac{160+360}{2}\right]\times 1000=390 \text{ mm}^2.$$

Provide 10 mm ϕ bars at 200 mm centres horizontally for main reinforcement. Provide 10 mm bars at 400 mm centres both ways on outer face as temperature reinforcement.

Pressure Distribution at Base

Load	*Distance of C.G. from end of toe*
$W_1=0.16\times 4.7\times 24{,}000=18{,}050$ N	1·08 m
$W_2=\frac{0.2}{2}\times 1\times 4.7\times 24{,}000=11{,}280$ N	0·92 m
$W_3=0.3\times 2.5\times 1\times 24{,}000=18{,}000$ N	1·25 m
$W_4=1.34\times 1\times 4.7\times 15{,}000=94{\cdot}470$ N	1·83 m

$\Sigma W=18{,}050+11{,}280+18{,}000+94{,}470=141{,}800$ N

$$P=\frac{wH^2}{2}\times\frac{1-\sin\phi}{1+\sin\phi}=\frac{15{,}000\times 5^2}{2}\times\frac{1}{3}$$

$=62{,}500$ N acting at 5/3 m from base.

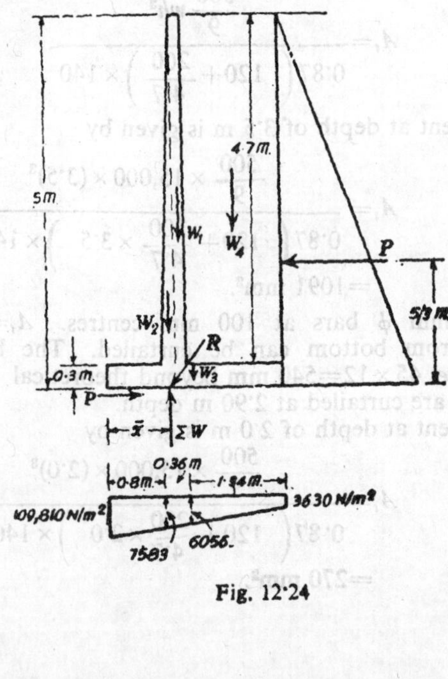

Fig. 12·24

Let $\bar{x}$ be distance where resultant cuts the base. Taking moments of external forces and reactive forces about the end of toe,

$$W_1 \times 1{\cdot}08 + W_2 \times 0{\cdot}93 + W_3 \times 1{\cdot}25 + W_4 \times 1{\cdot}83 = P \times 5/3 + \Sigma W \times \bar{x}$$

$$18{,}050 \times 1{\cdot}08 + 11{,}208 \times 0{\cdot}93 + 18{,}000 \times 1{\cdot}25 + 94{,}870 \times 1{\cdot}83$$

$$= 62{,}500 \times 5/3 + 180\,\bar{x}$$

$$\bar{x} = \frac{19{,}490 + 10{,}490 + 22{,}500 + 172{,}900 - 104{,}170}{141{,}800}$$

$$= \frac{121{,}210}{141{,}800} = 0{\cdot}86 \text{ m.}$$

Eccentricity $= \frac{2{\cdot}5}{4} - 0{\cdot}86 = 0{\cdot}39$ m.

Pressure distribution at base is given by

$$p = \frac{141{,}800}{2{\cdot}5} \pm \frac{141{,}800 \times 0{\cdot}39}{(2{\cdot}5)^2/6}$$

$$= 56{,}720 \pm 53{,}090 \qquad p_{max} = 109{,}810 \text{ N/m}^2$$

$$p_{min} = 3630 \text{ N/m}^2$$

Pressure at junction of stem with toe

$$= 109{,}810 - \frac{109{,}810 - 3630}{2{\cdot}5} \times 0{\cdot}8$$

$$= 109{,}810 - 33{,}980 = 75{,}830 \text{ N/m}^2$$

Pressure at junction of stem with heel

$$= 3630 + \frac{109{,}810 - 3630}{2{\cdot}5} \times 1{\cdot}34$$

$$= 3630 + 56{,}930 = 60{,}560 \text{ N/m}^2.$$

Design of toe

Self-load of the base per meter run

$$= 0{\cdot}3 \times 1 \times 24{,}000 = 7200 \text{ N.}$$

The forces acting on toe are pressure from soil acting upwards and self-weight acting downwards.

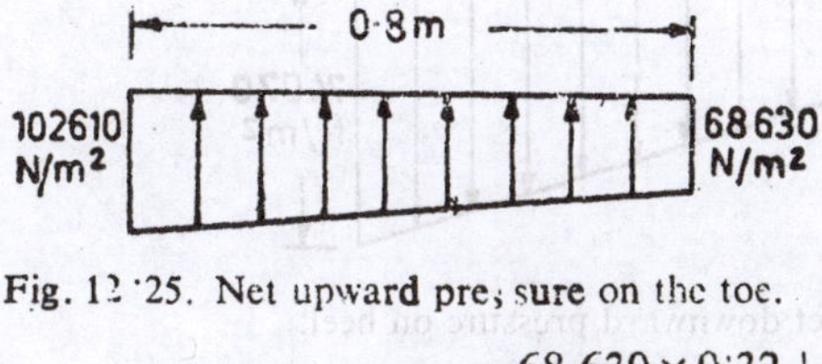

Fig. 12·25. Net upward pressure on the toe.

Maximum B.M.

$$= 68{,}630 \times \frac{(0{\cdot}8)^2}{2} + (102{,}610 - 68{,}630) \times \frac{0{\cdot}3}{2} \times \frac{2}{3} \times 0{\cdot}8$$

$$= 68{,}630 \times 0{\cdot}32 + 33{,}980 \times \frac{0{\cdot}64}{3}$$

$$= 29{,}960 + 7{,}250 = 29{,}210 \text{ N-m.}$$

$$= 29{,}210{,}000 \text{ N-mm.}$$

Effective depth, $d = \sqrt{\dfrac{29{,}210{,}000}{0{\cdot}87 \times 1000}} = 183{\cdot}2$ mm.

Provide overall depth of 300 mm with effective depth of 240 mm.

$$A_t = \frac{29,210,000}{0{\cdot}87 \times 240 \times 140} = 1000 \text{ mm}^2.$$

Provide 12 mm ϕ bars at 100 mm centres.

(Alternate bars from stem are bent and taken straight in toe)

$$\text{Distribution steel} = \frac{0{\cdot}15}{100} \times 300 \times 1000 = 450 \text{ mm}^2.$$

Provide 10 mm ϕ bars at 170 mm centres.

Check for Bond and Shear

$$\text{Maximum shear force} = \frac{(102,610+68,630)}{2} \times 0{\cdot}8 = 68,500 \text{ N}.$$

$$\text{Maximum shear stress} = \frac{68,500}{240 \times 1000}$$

$$= 0{\cdot}285 \text{ N/mm}^2. \quad \text{Safe.}$$

Sufficient anchorage length is provided including standard hooks.

Half the bars from stem are taken in the toe and remaining straight.

Downwards load due to earth and self-load per metre width

$$= 1 \times 1 \times 4{\cdot}7 \times 15,000 + 7200$$
$$= 70,500 + 7200 = 77,700 \text{ N}.$$

Maximum B.M.

$$= 17,140 \times \frac{(1{\cdot}34)}{2} + (74,070 - 171,40) \times \frac{1{\cdot}34}{2} \times \frac{2}{3} \times 1{\cdot}34$$

$$= 17,140 \times \frac{(1{\cdot}34)^2}{2} + 56,930 \times \frac{(1{\cdot}34)^2}{3}$$

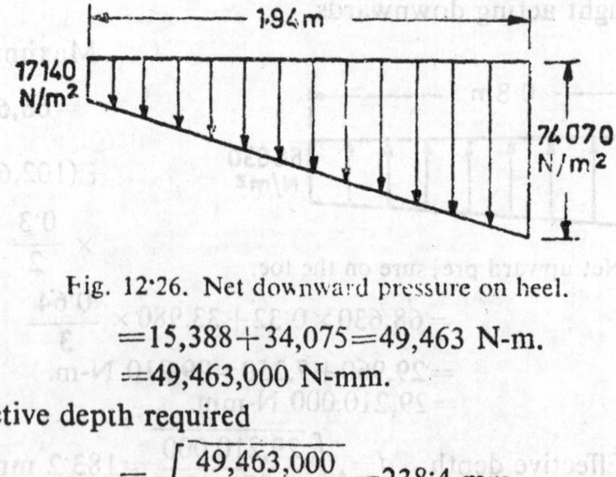

Fig. 12·26. Net downward pressure on heel.

$$= 15,388 + 34,075 = 49,463 \text{ N-m}.$$
$$= 49,463,000 \text{ N-mm}.$$

Effective depth required

$$= \sqrt{\frac{49,463,000}{0{\cdot}87 \times 1000}} = 238{\cdot}4 \text{ mm}.$$

Provide overall depth of 300 mm with effective depth of 240 mm

$$A_t = \frac{49{,}463{,}000}{0{\cdot}87 \times 240 \times 140} = 1692 \text{ mm}^2$$

Provide 16 mm ϕ bars at 110 mm centres.

Distribution Steel

Provide 10 mm ϕ bars at 170 mm centres.

Check against Sliding

Frictional resistance $= \mu W = 0{\cdot}6 \times 141{,}800 = 85{,}080$ N

Force causing sliding = Earth pressure = 62,500 N

Factor of safety $= \dfrac{85{,}080}{62{,}500} = 1{\cdot}36$. Not safe.

A key 0·5 m deep and 0·4 m wide is provided to increase resistance against sliding.

Design of Key

The section is checked at bottom of key (at $A-A$ in Fig. 12·27). The weight of key and earth shown hatched in the figure will be

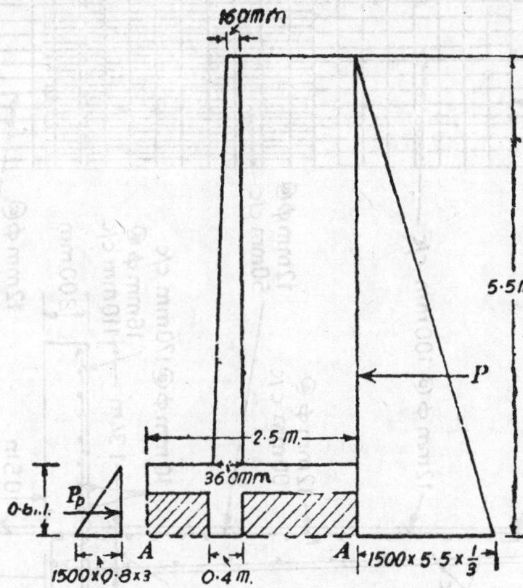

Fig. 12·27

additional weight coming at A-A. There will also be passive earth pressure in front of the wall.

Total weight $= 141{,}800 + (2{\cdot}5 - 0{\cdot}4) \times 0{\cdot}5 \times 15{,}000 + 0{\cdot}4 \times 4{\cdot}5 \times 1 \times 24{,}000$

$= 141{,}800 + 15{,}750 + 4{,}800 = 162{,}350$ N

Active earth pressure

$$P_A = \frac{wH^2}{4}\left(\frac{1-\sin\phi}{1+\sin\phi}\right)$$

$$= \frac{15{,}000 \times (5{\cdot}5)^2}{2} \times \frac{1}{3} = 75{,}625 \text{ N.}$$

$$\text{Passive pressure} \quad P_P = \frac{wH_1^2}{2}\left(\frac{1+\sin\phi}{1-\sin\phi}\right)$$

$$= \frac{15{,}000 \times (0{\cdot}8)^2}{2} \times 3 = 14{,}400 \text{ N.}$$

Net horizontal force $= 75{,}625 - 14{,}400 = 61{,}225$ N

$$\text{Factor of safety} \quad = \frac{0{\cdot}6 \times 164{,}520}{61{,}225} = 1{\cdot}61 \text{ safe.}$$

Half the bars from stem will be taken inside the key.

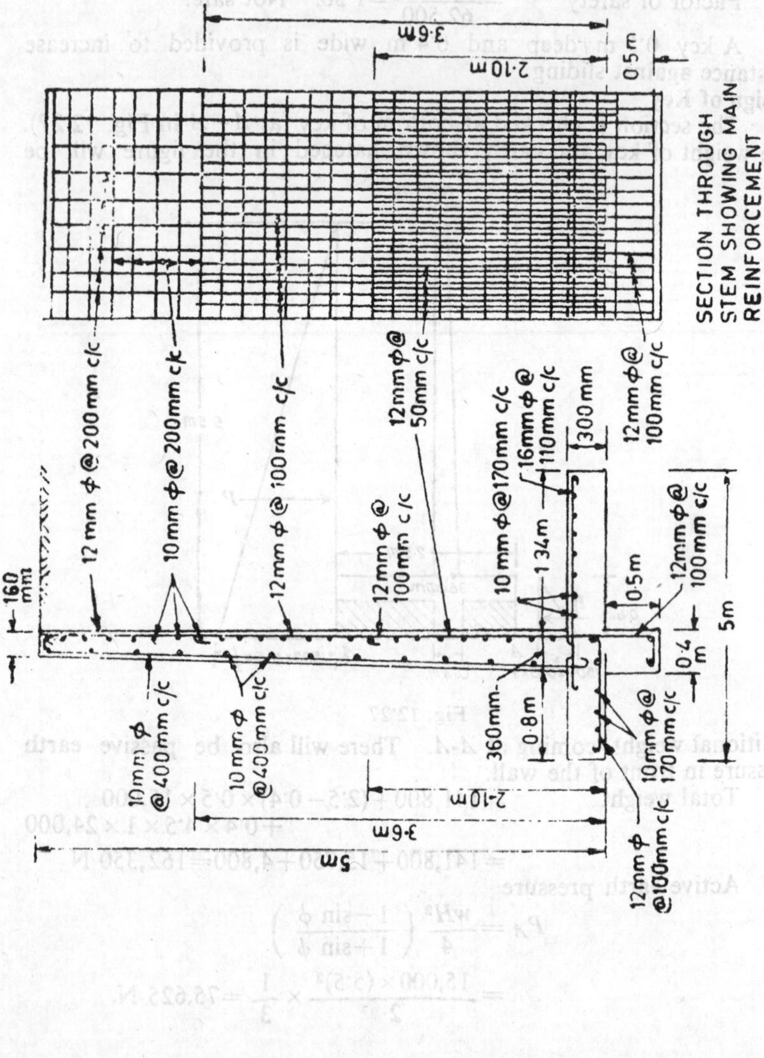

Fig. 12·28

Maximum shear at junction of key with base=Active pressure upto base level − Passive pressure upto base level.

$$\text{Maximum shear} = 62{,}500 - \frac{15{,}000 \times (0{\cdot}3)^2}{2} \times 3$$

$$= 62{,}500 - 20{,}250 = 60{,}275 \text{ N.}$$

$$\text{Maximum shear stress} = \frac{60{,}275}{360 \times 1000}$$

$$= 0{\cdot}167 \text{ N/mm}^2. \quad \text{Safe.}$$

Sufficient anchorage length is provided including standard hooks for bond.

Details of reinforcement in retaining wall are shown in Fig. 12·28.

Example 12·2. *A retaining wall is to retain level surface of earth to height of 5 m. The density of earth is 15,000 N/m³ and the angle of repose of soil is 30°. The bearing capacity of soil is 100 kN/m². Coefficient of friction = 0·6.*

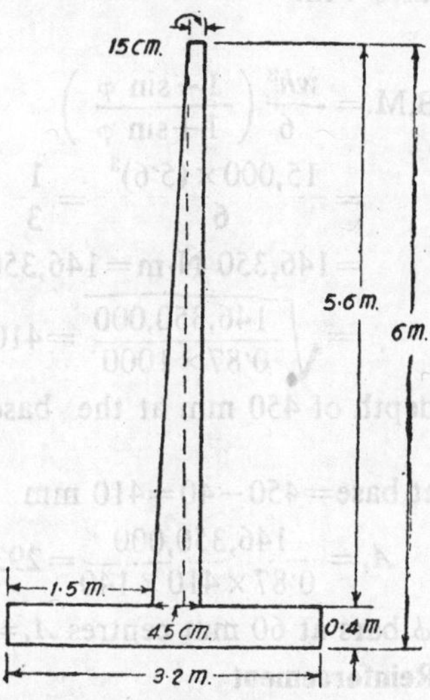

Fig. 12·29

Sol. Assume depth of bottom of base as 1 m below ground level. Total height of retaining wall = 6 m.

Design of base width of retaining wall

$$k = 1 - \frac{\sigma_b}{2H\omega(2-3a)}$$

Taking $a = \frac{1}{3}$

$$\sigma_b = 100{,}000 \text{ N/m}^2$$

$$H = 6 \text{ m}$$

$$\omega = 15{,}000 \text{ N/m}^3.$$

$$\therefore \quad k=1-\frac{100,000}{2\times 6\times 15,000\ (2-3\times 1/3)}$$

$$=1-5/9=4/9$$

$$\frac{1-\sin\varphi}{1+\sin\varphi}=\frac{1-\frac{1}{2}}{1+\frac{1}{2}}=1/3$$

$$B=H\sqrt{\frac{\frac{1-\sin\varphi}{1+\sin\varphi}}{(1-k)(3k-1)}}$$

$$=6\sqrt{\frac{1/3}{(1-4/9)(3\times 4/9+1)}}$$

$$=6\sqrt{\frac{1}{3\times 5/9\times 7/9}}=6\sqrt{\frac{9}{35}}$$

$$=3{\cdot}04 \text{ m.}$$

Provide base width of 3·2 m with toe width of 1·5 m. Assume thickness of base slab as 0·4 m.

Design of Stem

$$\text{Maximum} \quad \text{B.M.}=\frac{wh^3}{6}\left(\frac{1-\sin\varphi}{1+\sin\varphi}\right)$$

$$=\frac{15,000\times(5{\cdot}6)^3}{6}=\frac{1}{3}$$

$$=146,350 \text{ N-m}=146,350,000 \text{ N-mm}$$

$$\text{Effective depth} \quad =\sqrt{\frac{146,350,000}{0{\cdot}87\times 1000}}=410 \text{ mm}$$

Provide overall depth of 450 mm at the base and 150 mm at the top.

Effective depth at base=450−40=410 mm

$$A_t=\frac{146,350,000}{0{\cdot}87\times 410\times 140}=2931 \text{ mm}^2$$

Provide 16 mm ϕ bars at 60 mm centres A_t=3352 mm².

Curtailment of Stem Reinforcement

Let A_t be the area of steel required at depth h.

Effective depth of section at h m

$$=\left(110+\frac{300}{3{\cdot}6}h\right)$$

$$\text{B.M.}=\frac{wh^3}{6}\times\frac{1}{3}\times 1000 \text{ N-mm}$$

$$=\frac{500}{9}wh^3$$

$$\therefore \quad A_t=\frac{(500/9)\ wh^3}{0{\cdot}87\left(110+\frac{300}{5{\cdot}6}\times h\right)\times 140}$$

At $h=4$ m.

$$A_t=\frac{(500/9)\times 15{,}000\times 64}{0{\cdot}87\left(110+\frac{300}{5{\cdot}6}\times 4\right)\times 140}$$

$$=1350 \text{ mm}^2$$

Provide 16 mm ϕ bars at 120 mm centres, *i.e.* alternate bars from bottom are stopped. $A_t=16{\cdot}76$ cm².

The bars will be stopped at $45d=45\times 1{\cdot}6=720$ mm beyond this point. The bars are curtailed at 2·25 m depth.

At $h=2{\cdot}5$ m

$$A_t=\frac{(500/9)\times 15{,}000\times (2{\cdot}5)^3}{0{\cdot}87\left(110+\frac{300}{5{\cdot}6}\times 2{\cdot}5\right)\times 140}$$

$$=420 \text{ mm}^2.$$

Provide 16 mm ϕ bars at 240 mm centres, *i.e.* one-fourth of the bars from bottom are taken upto top. One-fourth of the bars from bottom are stopped at $45d=720$ mm beyound this point. The bars are curtailed at 1·75 m depth.

Distribution steel based on average thickness

$$=\frac{0{\cdot}15}{100}\times 1000\times\left(\frac{150\times 450}{2}\right)=\mathbf{450\ mm^2}.$$

Provide 10 mm ϕ bars at 150 mm centres. Provide 10 mm ϕ bars at 300 mm centres both ways on the face.

Pressure Distribution at Base

Load	*C.G. Distance from end of toe*
$W_1=0{\cdot}15\times 5{\cdot}6\times 1\times 24{,}000$ $=20{,}160$ N	1·875 m
$W_2=\frac{0{\cdot}3}{2}\times 5{\cdot}6\times 1\times 24{,}000$ $=20{,}160$ N	1·70 m
$W_3=3{\cdot}4\times 0{\cdot}4\times 1\times 24{,}000$ $=30{,}720$ N	1·60 m
$W_4=5{\cdot}6\times 1{\cdot}25\times 1\times 15{,}000$ $=105{,}000$ N	2·575 m
$P=\frac{wH^2}{2}\left(\frac{1-\sin\varphi}{1+\sin\varphi}\right)=\frac{15{,}000\times 6^2}{2}=\frac{1}{3}$ $=90{,}000$ N	2 m

$$\Sigma W=20{,}160+20{,}160+30{,}720+105{,}000$$
$$=17{,}6040 \text{ N}$$

Taking moments of all forces (external and reactions) about the end or toe,

$$W_1\times 1{\cdot}875+W_2\times 1{\cdot}70+W_3\times 1{\cdot}6+W_4\times 2{\cdot}575$$
$$=\Sigma W\times\bar{x}+90{,}000\times 2$$

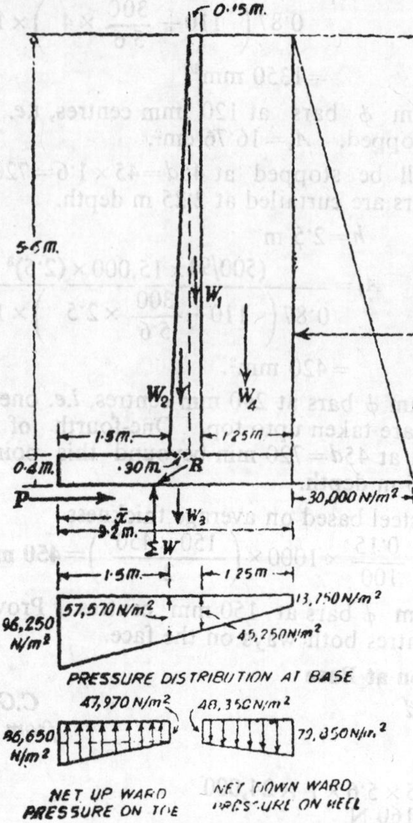

Fig. 12·30

$$20{,}160\times 1{\cdot}875+20{,}160\times 1{\cdot}70+30{,}720\times 1{\cdot}6+105{,}000\times 2{\cdot}575]$$
$$-180{,}000=176{,}040\times\bar{x}$$

$$\bar{x}=\frac{37{,}790+34{,}270+49{,}152+270{,}3750-180{,}000}{176{,}040}$$

$$=\frac{211{,}587}{176{,}040}=1{\cdot}2\text{ m.}$$

$$\text{Eccentricity}=\frac{3{\cdot}2}{2}-1{\cdot}2=0{\cdot}4\text{ m.}$$

Pressure distribution at base is given by

$$P=\frac{176{,}040}{3{\cdot}2}\pm\frac{176{,}040\times 0{\cdot}4}{(3{\cdot}2)^2/6}=55{,}000\pm 41{,}250$$

$$P_{max}=96{,}250\text{ N/m}^2$$
$$P_{min}=13{,}750\text{ N/m}^2$$

Pressure at junction of stem with toe

$$=96{,}250-\frac{(96{,}250-13{,}750)}{3{\cdot}2}\times 1{\cdot}5$$

$$=96{,}250-38{,}680=57{,}570 \text{ N/m}^2$$

Pressure at junction of stem with heel

$$=13{,}750+\frac{(96{,}250\times 13{,}750)}{3{\cdot}2}\times 1{\cdot}25$$

$$=13{,}750+31{,}500=45{,}250 \text{ N/m}^2.$$

Design of toe

Self-weight of toe per metre run

$$=0{\cdot}4\times 1\times 1\times 24{,}000=6{,}600 \text{ N/m}^2$$

Net upward pressure on toe is shown in Fig. 12·30.

$$\text{Maximum B.M.}=\frac{47{,}970\times(1{\cdot}5)^2}{2}+(86{,}650-47{,}970)\times\frac{1{\cdot}5}{2}\times 2\times\frac{1{\cdot}5}{3}$$

$$=53{,}960+29{,}10=82{,}970 \text{ N-m.}$$

$$=82{,}970{,}000 \text{ N-mm.}$$

$$\text{Effecctive depth } d=\sqrt{\frac{82{,}970{,}000}{0{\cdot}87\times 1000}}=308{\cdot}8 \text{ mm}$$

Provide overall depth of 400 mm with effective depth of 320 mm

$$A_t=\frac{82{,}970{,}000}{0{\cdot}87\times 320\times 140}=3127 \text{ mm}^2$$

Provided 16 mm ϕ bars at 60 mm centres.

Steel area provided $=3352$ mm^2

$$\text{Distribution steel}=\frac{0{\cdot}15}{100}\times 400\times 1000=600 \text{ mm}^2$$

Provide 10 mm ϕ bars at 130 mm centres.

Check for Bond and Shear

$$\text{Total shear force}=\frac{89{,}650+46{,}970}{2}\times 1{\cdot}5=100{,}960 \text{ N}$$

$$\text{Shear stress}=\frac{100{,}960}{320\times 1000}=0{\cdot}317 \text{ N/m}^2. \quad \text{Safe}$$

Sufficient anchor length is provided for bond.

Design of Heel

Load of soil per metre length of heel

$$=5{\cdot}6\times 1\times 15{,}000=84{,}000 \text{ N}$$

Self-load of heel slab $=9600$ N/m

Total downward load $=93{,}600$ N/m

Net downward pressure on heel slab is shown in Fig. 12·30.

$$\text{Maximum B.M.} = \frac{47{,}239\times(1{\cdot}25)^2}{2}+(80.015-47{,}239)\times\frac{1{\cdot}25}{2}\times\frac{2}{3}\times1{\cdot}25$$

$$=\frac{47{,}239\times(1{\cdot}25)^2}{2}+32{,}776\times\frac{(1{\cdot}25)^2}{3}$$

$$=53976 \text{ N/m.}$$

$$=53{,}976{,}000 \text{ N-mm.}$$

$$A_t=\frac{53{,}976{,}000}{0{\cdot}87\times320\times140}=1385 \text{ mm}^2$$

Provide 16 mm ϕ bars at 140 mm centres.

Distribution steel

Provide 10 mm ϕ at 130 mm centres.

Check for Sliding

Frictional Resistance $=\mu W=0{\cdot}6\times179{,}000$

Earth pressure $=90{,}000$ N

$$\text{Factor of safety} = \frac{0{\cdot}6\times179{,}000}{90{,}000}=1{\cdot}193. \text{ Not safe.}$$

Key 0·8 m deep and 0·4 m wide will be provided under the stem to increase factor of safety against sliding.

Sliding will take place along the surface at the bottom of the key, *i.e.* surface *A-A* in Fig. 12·31.

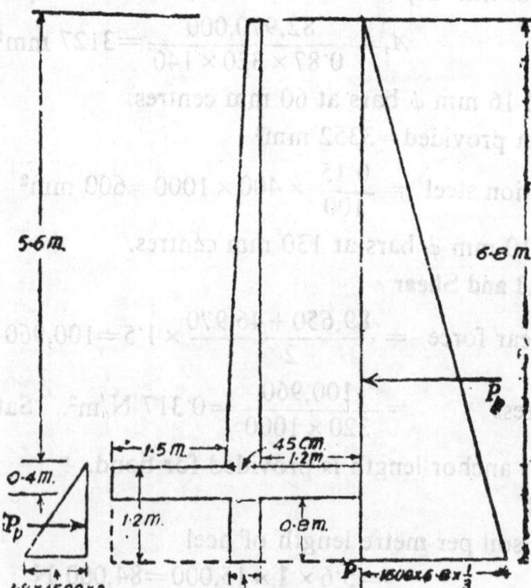

Fig. 12·31

$$\text{Total weight at } A\text{-}A = 179{,}000 + (3{\cdot}2 - 0{\cdot}4) \times 1 \times 0{\cdot}8 \times 15{,}000 + 0{\cdot}4 \times 0{\cdot}8 \times 1 \times 25{,}000$$

$$= 220{,}600 \text{ N}$$

$$\text{Total active pressure} = \frac{15{,}000 \times (6{\cdot}8)^2}{2} \times \frac{1}{3} = 115{,}600 \text{ N}$$

$$\text{Total passive pressure } P_P = \frac{15{,}000 \times (1{\cdot}2)^2}{2} \times 3 = 32{,}400 \text{ N}$$

$$\text{Net horizontal force} = 115{,}600 \times 32{,}400 = 83{,}200 \text{ N}$$

$$\text{Factor of safety} = \frac{0{\cdot}6 \times 220{,}600}{83{,}200} = 1{\cdot}59. \quad \text{Safe.}$$

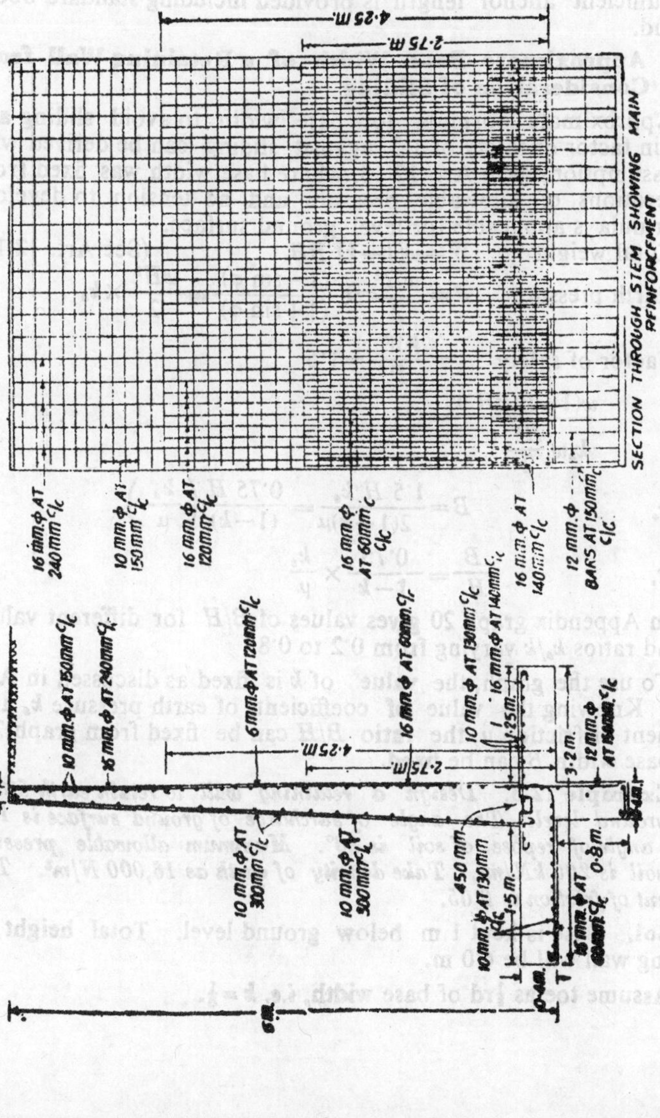

Fig. 12·28

Design of Key

Maximum shear at junction of key with base

$$= \text{Active pressure upto base level} - \text{Passive pressure upto base level}$$

$$= 90{,}000 - \frac{15{,}000 \times (0.4)^2}{2} \times 3$$

$$= 90{,}000 - 3600 = 86{,}400 \text{ N}$$

Shear stress $= \dfrac{86{,}400}{360 \times 1000} = 0.24 \text{ N/mm}^2$. Safe.

Sufficient anchor length is provided including standard hooks for bond.

12·14. Approximate Base Width of a Retaining Wall from Consideration of Sliding

Approximate formula for base width to avoid sliding and maintain factor of safety of 1·5 against sliding can be derived with same assumptions as in Art. 12·13, where base width was fixed from considerations of allowable pressure and no tension to develop. The formula is applicable for level ground surface.

Total weight $\quad W = w(1-k)\,BH$ (See Art. 12·13)

Earth pressure $\quad P = \dfrac{wH^2}{2} \times \dfrac{(1-\sin\varphi)}{(1+\sin\varphi)} = \dfrac{wH^2}{2} \times k_e$

Factor of safety $\quad = \dfrac{\mu W}{P} = 1.5$

$$\frac{w(1-k)BH\,\mu}{k_e w \dfrac{H^2}{2}} = 1.5$$

$$\therefore \quad B = \frac{1.5\,H\,k_e}{2(1-k)\mu} = \frac{0.75\,H}{(1-k)}\left(\frac{k_e}{\mu}\right)$$

$$\therefore \quad \frac{B}{H} = \frac{0.75}{1-k} \times \frac{k_e}{\mu}$$

In Appendix graph 20 gives values of B/H for different values of k and ratios k_e/k varying from 0·2 to 0·8.

To use the graph, the value of k is fixed as discussed in Art. 12·13. Knowing the value of coefficient of earth pressure k_e and coefficient of friction μ, the ratio B/H can be fixed from graph 20. Thus base width B can be fixed.

Example 12·3. *Design a retaining wall to retain earth 5·0 m above ground level. The angle of surcharge of ground surface is 15° and the angle of repose of soil is 30°. Maximum allowable pressure on the soil is 200 kN/m². Take density of earth as 16,000 N/m³. The coefficient of friction is 0·65.*

Sol. Base is kept 1 m below ground level. Total height of retaining wall will be 6·0 m.

Assume toe as ⅓rd of base width, *i.e.* $k = \frac{1}{3}$.

Fixing base width of wall from considerations of sliding

The effect of surcharge is neglected.

$$\frac{k_e}{\mu}=\frac{\frac{1}{3}}{0.65}=\frac{1}{1.95}$$

$$\frac{L}{H}=\frac{0.75}{1-k}\left(\frac{k_e}{\mu}\right)$$

$$=\frac{0.75}{1-\frac{1}{3}}\times\frac{1}{1.95}=0.578$$

$$B=0.578\times 6=3.468 \text{ m.}$$

Adopt $B=3.6$ m

Width of Toe $=1.2$ m.

Assume thickness of base slab as 400 mm.

Design of Stem

Consider 1 m length of the wall.

Pressure at base of stem

$$=wh\cos\alpha\frac{\cos\alpha-\sqrt{\cos^2\alpha-\cos^2\varphi}}{\cos\alpha+\sqrt{\cos^2\alpha-\cos^2\varphi}}$$

Maximum B.M.

$$=\frac{wh^3}{6}\cos^2\alpha\frac{\cos\alpha-\sqrt{\cos^2\alpha-\cos^2\varphi}}{\cos\alpha+\sqrt{\cos^2\alpha+\cos^2\varphi}}$$

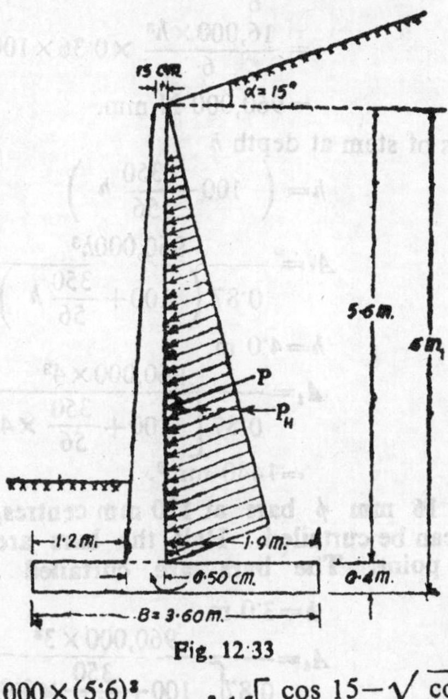

Fig. 12·33

$$=\frac{16{,}000\times(5.6)^3}{6}\cos^2 15\left[\frac{\cos 15-\sqrt{\cos^2 15-\cos^2 30}}{\cos 15+\sqrt{\cos^2 15-\cos^2 30}}\right]$$

$$=\frac{8000}{3}\times(5{\cdot}6)^3\times(0{\cdot}9659)^2\left[\frac{0{\cdot}9659-\sqrt{(0{\cdot}9659)^2-(0{\cdot}866)^2}}{0{\cdot}9659+\sqrt{(0{\cdot}9659)^2-(0{\cdot}866)^2}}\right]$$

$$=\frac{8000}{3}\times(5{\cdot}6)^3\times0{\cdot}9328\times\frac{0{\cdot}9659-0{\cdot}4227}{0{\cdot}9659+0{\cdot}4277}$$

$$=\frac{8000\times(5{\cdot}6)^3}{3}\times0{\cdot}9328\times0{\cdot}3862$$

$$=\frac{8000\times(5{\cdot}6)^3}{3}\times0{\cdot}36=168{,}600\text{ Nm}$$

$$=168{,}600{,}000\text{ N-mm}$$

$$\text{Effective depth}=\sqrt{\frac{168{,}600{,}000}{0{\cdot}87\times1000}}=440{\cdot}4\text{ mm}$$

Provide overall depth of 500 mm effective depth of 450 mm. The thickness of stem is kept 150 mm at top

$$A_t=\frac{1{,}68{,}600{,}000}{0{\cdot}87\times450\times140}=3080\text{ mm}^2$$

Provide 16 mm ϕ bars 60 mm centres, $A_t=3552\text{ mm}^2$.

Curtailment of Reinforcement

B.M. at h metres from top

$$=\frac{wh^3}{6}\times0{\cdot}36\times1000\text{ N-mm}$$

$$=\frac{16{,}000\times h^3}{6}\times0{\cdot}36\times1000$$

$$=960{,}000\text{ N-mm.}$$

Thickness of stem at depth h

$$h=\left(100+\frac{350}{56}h\right)$$

$$A_t=\frac{960{,}000h^3}{0{\cdot}87\left(100+\frac{350}{56}h\right)\times140}$$

At $h=4{\cdot}0$ m,

$$A_t=\frac{960{,}000\times4^3}{0{\cdot}87\left[100+\frac{350}{56}\times4\right]\times140}$$

$$=1440\text{ mm}^2.$$

Provide 16 mm ϕ bars at 120 mm centres, *i.e.* half the bars from bottom can be curtailed. Half the bars are stopped 750 mm beyond this point. The bars are curtailed at 3·25 m depth.

At $h=3{\cdot}0$ m

$$A_t=\frac{960{,}000\times3^3}{0{\cdot}87\left[100+\frac{350}{56}\times3{\cdot}0\right]\times140}$$

$$=740\text{ mm}^2.$$

Provide 16 mm ϕ bars at 240 mm c/c, *i.e.* ½th the bars from bottom are taken straight upto top. ½th the bars are stopped 750 mm beyond this point. The bars are curtailed at 2·25 m depth.

A_t provided is 838 mm².

Distribution Steel

Distribution steel based on average thickness

$$=\frac{0{\cdot}15}{100}\times\frac{150+500}{2}\times 1000=562{\cdot}5\ \text{mm}^2.$$

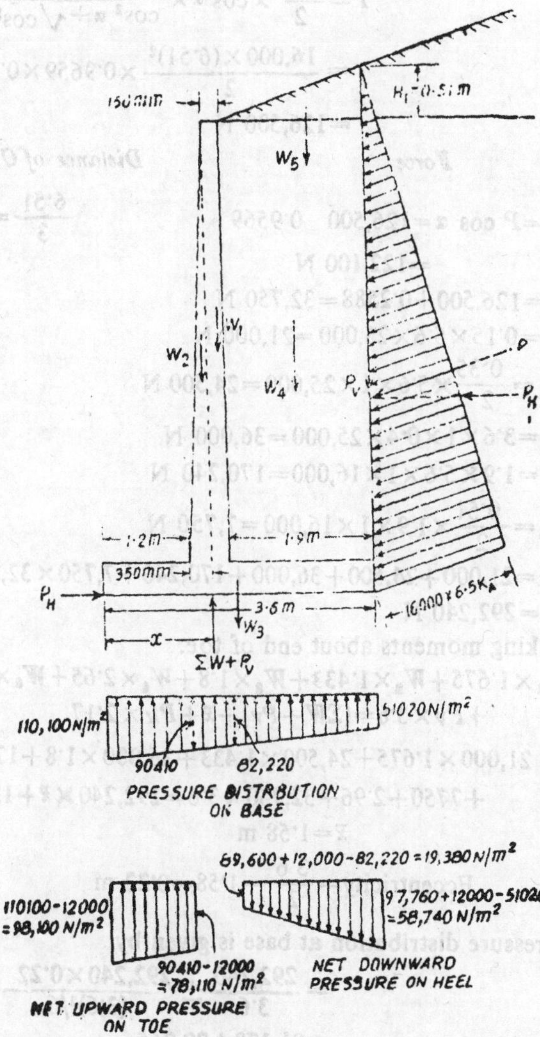

Fig. 12·34

Provide 10 mm ϕ bars at 130 mm centres as distribution steel for main reinforcement.

Provide 10 mm ϕ bars at 260 mm centres both ways in front face as temperature reinforcement.

Pressure Distribution on Base

$$H_1 = 1{\cdot}9 \tan 15° = 1{\cdot}9 \times 0{\cdot}2679 = 0{\cdot}51 \text{ m}$$

$$H + H_1 = 6 + 0{\cdot}51 \text{ m} = 6{\cdot}51 \text{ m}$$

$$P = \frac{wh^2}{2} \times \cos\alpha \times \frac{\cos\alpha - \sqrt{\cos^2\alpha - \cos^2\varphi}}{\cos^2\alpha + \sqrt{\cos^2\alpha - \cos\varphi}}$$

$$= \frac{16{,}000 \times (6{\cdot}51)^2}{2} \times 0{\cdot}9659 \times 0{\cdot}3862$$

$$= 126{,}500 \text{ N}$$

Force	*Distance of C.G. from toe*
$P_H = P\cos\alpha = 126{,}500 \quad 0{\cdot}9569 = 122{,}100$ N	$\frac{6{\cdot}51}{3} = 2{\cdot}17$ m
$P_V = 126{,}500 + 0{\cdot}2588 = 32{,}750$ N	3·6 m
$W_1 = 0{\cdot}15 \times 5{\cdot}6 \times 25{,}000 = 21{,}000$ N	1·675 m
$W_2 = \frac{0{\cdot}35}{2} \times 5{\cdot}6 \times 1 \times 25{,}000 = 24{,}500$ N	1·433 m
$W_3 = 3{\cdot}6 \times 1 \times 0{\cdot}4 \times 25{,}000 = 36{,}000$ N	1·8 m
$W_4 = 1{\cdot}9 \times 5{\cdot}6 \times 1 \times 16{,}000 = 170{,}240$ N	2·65 m
$W_5 = \frac{0{\cdot}51}{2} \times 1{\cdot}9 \times 1 \times 16{,}000 = 7{,}750$ N	2·96 m

$$\Sigma W + P_V = 21{,}000 + 24{,}500 + 36{,}000 + 170{,}240 + 7{,}750 \times 32{,}750$$
$$= 292{,}240 \text{ N}$$

Taking moments about end of toe.

$$W_1 \times 1{\cdot}675 + W_2 \times 1{\cdot}433 + W_3 \times 1{\cdot}8 + W_4 \times 2{\cdot}65 + W_5 \times 2{\cdot}96$$
$$+ P_V \times 3{\cdot}6 = (\Sigma W + P_V) - \bar{x} + P_H \times 2{\cdot}17.$$

$$\therefore 21{,}000 \times 1{\cdot}675 + 24{,}500 \times 1{\cdot}433 + 36{,}000 \times 1{\cdot}8 + 170{,}240 \times 2{\cdot}65$$
$$+ 7750 + 2{\cdot}96 + 32{,}750 \times 3{\cdot}6 = 292{,}240 \times \bar{x} + 122{,}100 \times 2{\cdot}17$$

$$\bar{x} = 1{\cdot}58 \text{ m}$$

$$\text{Eccentricity} = \frac{3{\cdot}6}{2} - 1{\cdot}58 = 0{\cdot}22 \text{ m}$$

Pressure distribution at base is given by

$$= \frac{292{,}240}{3{\cdot}6} \pm \frac{292{,}240 \times 0{\cdot}22}{(3{\cdot}6)^2/6}$$

$$= 81{,}178 \pm 29{,}765$$

$$p_{max}=110{,}943 \text{ N/m}^2$$
$$p_{min}=51{,}413 \text{ N/m}^2.$$

Pressure at junction of stem with toe

$$=110{,}943-\frac{(110{,}943-51{,}413)}{3{\cdot}6}\times 1{\cdot}2$$

$$=91{,}100 \text{ N/m}^2$$

Pressure at junction of stem with heel

$$=51{,}413+\frac{(110{,}943-51{,}413)}{3{\cdot}6}\times 1{\cdot}9$$
$$=82{,}832 \text{ N/m}^2.$$

Self-load base slab per metre $=0{\cdot}5\times 1\times 25{,}000$
$=12{,}500$ N

Load of earth on heel at the end $=6{\cdot}11\times 1\times 1\times 16{,}000$
$=97{,}760$ N/m

Load of earth on heel at junction of heel with stem
$=5{\cdot}6\times 1\times 16{,}000$
$=89{,}600$ N/m.

Net pressure diagrams on toe and heel slabs are shown in Fig. 12·34.

Design of Toe

$$\text{Maximum B.M.}=78{,}600\times\frac{(1{\cdot}2)^2}{2}+\frac{(98{,}443-78{,}600)}{2}\times 1{\cdot}2$$
$$\times\frac{2}{3}\times 1{\cdot}2=66{,}117 \text{ N-m}$$

$$=66{,}117{,}000 \text{ N-mm.}$$

$$\text{Effective depth}=\sqrt{\frac{66{,}117{,}000}{0{\cdot}87\times 1000}}=275{\cdot}7 \text{ mm}$$

Provide overall depth of 400 mm with effective depth of 330 mm.

$$A_t=\frac{66{,}117{,}000}{0{\cdot}87\times 330\times 140}=1645 \text{ mm}^2$$

Provide 16 mm ϕ bars at 110 mm centres. Half the bars from stem are taken into toe.

$$\text{Distribution steel}=\frac{0{\cdot}15}{100}\times 400\times 1000=600 \text{ mm}^2$$

Provide 10 mm ϕ bars at 130 mm centres.

Provide 10 mm ϕ bars at 260 mm centres both ways on the face of the wall as temperature reinforcement.

Check for Bond and Shear

$$\text{Maximum shear force}=\frac{98{,}443+78{,}600}{2}\times 1{\cdot}2=106{,}226 \text{ N}$$

Maximum shear stress$=\dfrac{106,226}{330\times 1000}=0.332$ N/mm². Safe.

Sufficient anchor length is provided including standard hooks for bond.

Design of Heel

$$\text{Maximum B.M.} = \frac{19,268\times(1.9)^2}{2}+\frac{(58,847-19,268)}{2}\times 1.9\times 2/3\times 1.9$$

$$=82,405 \text{ N-m}$$

$$=82,405,000 \text{ N-mm.}$$

$$\text{Effective depth} = \sqrt{\frac{82,405,000}{0.87\times 1000}}=307.7 \text{ mm}$$

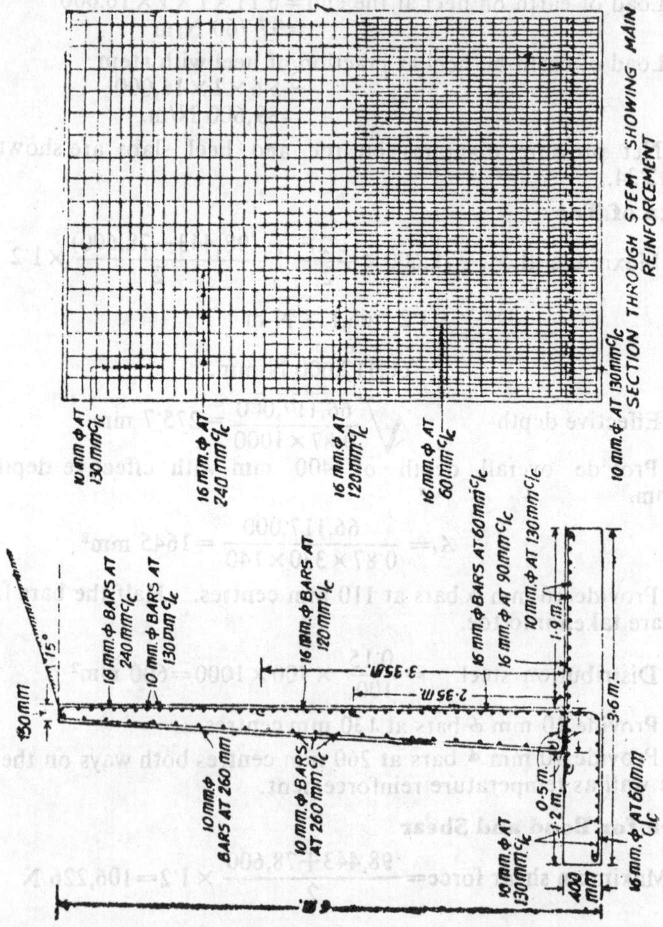

Fig. 12·35

Provided effective depth is 330 mm

$$A_t = \frac{82,405,000}{0.87 \times 330 \times 140} = 2050 \text{ mm}^2.$$

Provide 16 mm ϕ bars at 90 mm centres.

Provide 10 mm ϕ bars at 130 mm centres as distribution steel.

Check for Sliding

Frictional Resistance $= \mu W = 0.65 \times 292,240 = 189,956$ N

Earth pressure $= 122,100$ N

Factor of safety $= \frac{189,956}{122,100} = 1.56$. Safe.

12·15. Counterfort Retaining Walls

When the height of a retaining wall exceeds 6 m counterfort retaining wall generally proves economical.

For level top surface the base width is fixed as discussed in Art. 12·13 for cantilever type of retaining walls. Generally the base width varies from 0·5 H to 0·7 H, where H is the total height of retaining wall.

Spacing for Counterforts

The spacing of counterforts depends upon the height of the wall, allowable unit pressure on soil, relative cost of steel, concrete and cost of formwork. By keeping counterforts closer thickness of vertical wall and heel slab is reduced and amount of concrete and steel is reduced. However this is offset by concrete and steel required in large number of counterforts which are required, and also formwork required for counterforts. Most economical spacing of counterforts is one in which total cost is minimum. The economical spacing of counterforts varies from one-half to one-third of the height of the wall. The thickness of counterforts is governed by the space required to accommodate steel.

Design of vertical slab

Vertical slab is supported on the base and by counterforts at intervals of 2·5 to 4 m.

Vertical slab is designed as a continuous slab supported on counterforts. The lowest portion of the stem is subjected to maximum horizontal earth pressure. Taking one metre height of vertical slab and if h is the height of the vertical slab, maximum B.M. $= k_a \frac{whl}{12}$, where l is the clear distance between the counterforts, for all intermediate panels. For end panels moment is taken as $k_a \frac{whl}{10}$. The reinforcement in the stem is reduced towards the top.

Design of toe slab

The toe is designed as cantilever subjected to earth pressure acting upwards and self-load acting downwards, when no front counterforts are provided. When the front counterforts are provided the toe is designed as a continuous slab supported on counterforts.

Design of heel slab

The heel is designed as a continuous slab supported on counterforts. It is subjected to downward loads due to earth and self-load and upward force due to earth pressure.

The thickness of heel slab can be approximately fixed by the following formula :

Depth from consideration of B.M.

Neglecting the difference in unit weights of concrete and earth and assuming zero upward pressure at heel end.

$$\text{Maximum B.M.} = \frac{wH \times l^2}{12} \times 1000 \text{ N-mm.}$$

$$\text{Effective depth} = \sqrt{\frac{\frac{wHl^2}{12} \times 1000}{0{\cdot}87 \times 1000}}$$

Putting $w = 16{,}000 \text{ N/m}^3$

$$d = l\sqrt{H}\sqrt{\frac{16{,}000}{12 \times 0{\cdot}87}}$$

$$= 40\, l\sqrt{H} \text{ mm approximately.}$$

where l = Clear distance between the counterforts in metres
B = Overall height of the retaining wall.

Depth from consideration of shear

$$\text{Maximum shear force} = wH \times \frac{l}{2}$$

$$\text{Shear stress} = \frac{wH \times l/2}{d \times 1000} = 0{\cdot}40 = T_c \text{ (Assumed for M 15)}$$

$$\therefore \quad d = \frac{wHl}{800}$$

Taking $w = 16{,}000 \text{ N/m}^3$

Effective depth $= 20{\cdot}0\, Hl$ approximately in mm.

Design of counterforts

The counterfort is designed as a cantilever beam taking load from the stem portion between two counterforts. The effective depth of the counterfort is taken as distance between the junction with stem and base. The main reinforcement is provided on the inclined side. The section is checked for shear and bond.

Two legged stirrups are provided horizontally and vertically to connect the counterfort with stem and the base respectively. These stirrups are in tension.

The front counterfort is designed as a cantilever beam subjected to self-load acting downward and net earth pressure on the toe portion between two counterforts acting upwards. The section is checked for s hear and bond.

Example 12·4. *Design a counterfort retaining wall to retain earth 7 metres above the ground level. The density of earth is 15,000 N/m³ and the angle of repose of soil is 30°. The bearing capacity of soil is 150 kN/m².*

Sol. Bottom of base will be kept 1 m below ground level $H=7+1=8$ m.

Base width

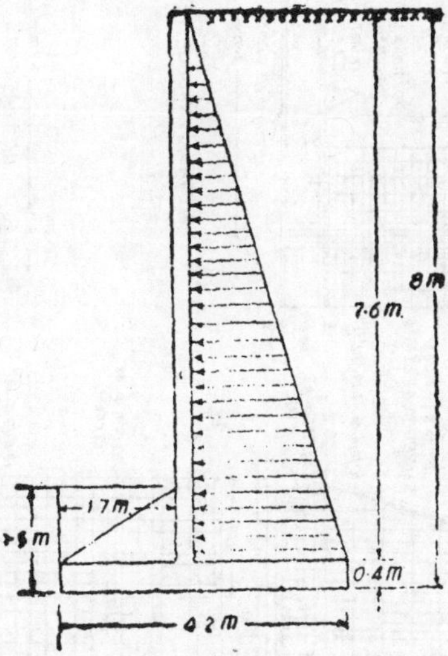

Fig. 12·36

$$\frac{\sigma_b}{w}=\frac{150,000}{15,000}=10$$

From graph 18,

$$k=0{\cdot}38$$

From graph 19,

$$B/H=0{\cdot}505$$

Adopt base width of 4·2 m with toe projection of 1·7 m.

Spacing of counterforts

The thickness of counterforts provided will be 300 mm. Counterforts will be provided at 3 m centres at the back of wall. As the toe projection is large, front counterforts are provided at 3 m centres. Clear distance between counterforts = 2·7 m.

Fixing depth of base slab

Overall depth of base slab

$$=40l\sqrt{H}+80$$
$$=40\times2{\cdot}7\times\sqrt{8}+80$$
$$=305{\cdot}5+80=385{\cdot}5 \text{ mm.}$$

Provide base thickness of 400 mm.

Design of vertical slab

Consider 1 m height of vertical slab near the base.

Earth pressure near base

$$=wh\,\frac{1-\sin\varphi}{1+\sin\varphi}$$

Maximum B.M. $\quad =wh\left(\dfrac{1-\sin\varphi}{1+\sin\varphi}\right)\times\dfrac{l^2}{12}$

$$=15,000\times7{\cdot}6\times\frac{1}{3}\times\frac{(2{\cdot}7)^2}{12}$$

$$=23,090 \text{ Nm}=23,090,000 \text{ N-mm.}$$

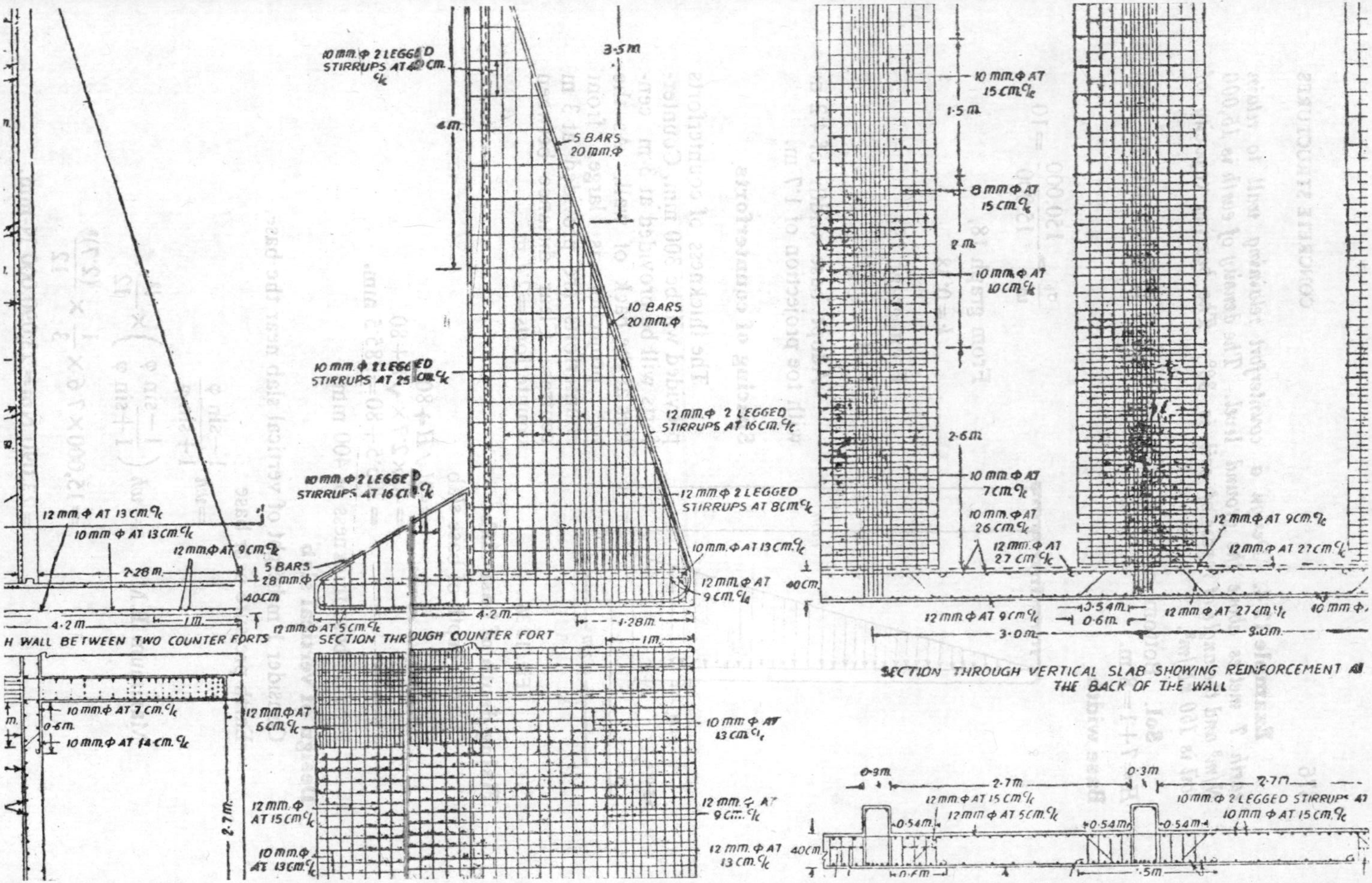

10 mm.Φ 2 LEGGED STIRRUPS AT 40 CM. c/c
3·5 m
4 m.
5 BARS 20 mm.Φ
10 BARS 20 mm.Φ
10 mm.Φ 2 LEGGED STIRRUPS AT 25 cm. c/c
12 mm.Φ 2 LEGGED STIRRUPS AT 16 CM. c/c
10 mm.Φ 2 LEGGED STIRRUPS AT 16 CM. c/c
12 mm.Φ 2 LEGGED STIRRUPS AT 8 CM. c/c
12 mm.Φ AT 13 CM. c/c
10 mm Φ AT 13 CM. c/c
12 mm.Φ AT 9 CM. c/c
5 BARS 28 mm.Φ
10 mm.Φ AT 13 CM. c/c
12 mm.Φ AT 9 CM. c/c
2·28 m.
40 CM
4·2 m.
1 m.
12 mm.Φ AT 5 CM. c/c
4·28 m.
H WALL BETWEEN TWO COUNTER FORTS
SECTION THROUGH COUNTER FORT
10 mm.Φ AT 7 CM. c/c
0·6 m.
10 mm.Φ AT 14 CM. c/c
2·7 m.
12 mm.Φ AT 6 CM. c/c
10 mm Φ AT 13 CM. c/c
12 mm.Φ AT 15 CM. c/c
12 mm.Φ AT 9 CM. c/c
10 mm.Φ AT 13 CM. c/c
12 mm.Φ AT 13 CM. c/c
10 mm.Φ AT 15 CM. c/c
1·5 m.
8 mm Φ AT 15 CM. c/c
2 m.
10 mm.Φ AT 10 CM. c/c
2·6 m.
10 mm.Φ AT 7 CM. c/c
10 mm.Φ AT 26 CM. c/c
12 mm.Φ AT 27 CM. c/c
12 mm.Φ AT 9 CM. c/c
12 mm.Φ AT 27 CM. c/c
12 mm.Φ AT 9 CM. c/c
0·54 m.
0·6 m.
12 mm Φ AT 27 CM. c/c
10 mm Φ
3·0 m.
3·0 m.
SECTION THROUGH VERTICAL SLAB SHOWING REINFORCEMENT AT THE BACK OF THE WALL
0·3 m.
2·7 m.
0·3 m
2·7 m
12 mm.Φ AT 15 CM. c/c
12 mm Φ AT 5 CM. c/c
10 mm.Φ 2 LEGGED STIRRUPS AT 10 mm Φ AT 15 CM. c/c
0·54 m.
0·54 m.
0·54 m.
40 CM.
0·6 m
1·5 m

Effective depth $=\sqrt{\dfrac{23{,}090{,}000}{0{\cdot}87\times1000}}=162{\cdot}9$ mm.

Provide overall depth of 220 mm with effective depth of 180 mm. Same thickness of slab is provided throughout the height.

$$A_t=\frac{23{,}090{,}000}{0{\cdot}87\times180\times140}=1050\text{ mm}^2.$$

Provide 10 mm ϕ bars at 70 mm centres. As the pressure decreases linearly towards the top, B.M. will decrease linearly to zero towards the top. The spacing of reinforcement will be gradually increased towards the top. Near the top 10 mm ϕ bars will be provided at 200 mm centres.

Distribution steel

Distribution steel$=\dfrac{0{\cdot}15}{100}\times220\times1000=330$ mm².

Provide 8 mm ϕ bars at 150 mm centres.

Check for Shear

Maximum shear force$=wh\left(\dfrac{1-\sin\varphi}{1+\sin\varphi}\right)\times\dfrac{l}{2}$

$$=15{,}000\times7{\cdot}6\times\frac{1}{3}\times\frac{2{\cdot}7}{2}=51{,}300\text{ N}.$$

Maximum shear stress$=\dfrac{51{,}300}{180\times1000}=0{\cdot}285$ N/mm². Safe.

Check for Bond

Sufficient anchorage length is provided for bond.

Pressure Distribution at Base

Consider one metre length of retaining wall

Force	*Distance of C.G. from end of toe*
$W_1=0{\cdot}22\times1\times7{\cdot}6\times24{,}000$ $=41{,}130$ N	1·81 m.
$W_2=0{\cdot}4\times4{\cdot}2\times1\times24{,}000$ $=40{,}320$ N	2·10 m.
$W_3=2{\cdot}28\times7{\cdot}6\times1\times15{,}000$ $=259{,}900$ N	3·06 m.

$$p=\frac{wh^2}{2}\left(\frac{1-\sin\varphi}{1+\sin\varphi}\right)=\frac{15{,}000\times8^2}{2}\times\frac{1}{3}$$

$=160{,}000$ N.

$$\Sigma W = W_1 + W_2 + W_3 = 41{,}130 + 40{,}320 + 259{,}900$$
$$= 341{,}350 \text{ N}$$

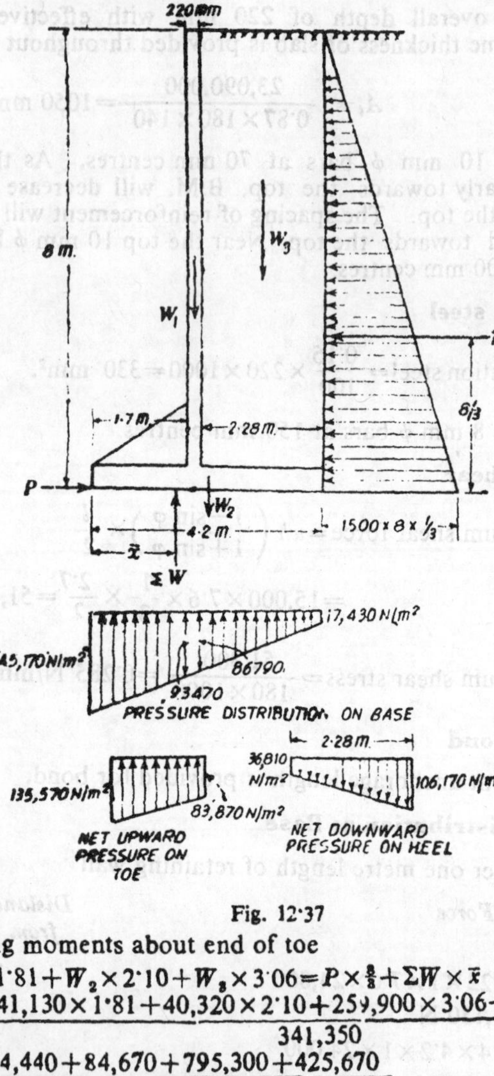

Fig. 12·37

Taking moments about end of toe

$$W_1 \times 1{\cdot}81 + W_2 \times 2{\cdot}10 + W_3 \times 3{\cdot}06 = P \times \frac{8}{3} + \Sigma W \times \bar{x}$$

$$\bar{x} = \frac{41{,}130 \times 1{\cdot}81 + 40{,}320 \times 2{\cdot}10 + 259{,}900 \times 3{\cdot}06 - 160{,}000 \times \frac{8}{3}}{341{,}350}$$

$$= \frac{74{,}440 + 84{,}670 + 795{,}300 + 425{,}670}{341{,}350}$$

$$= \frac{527{,}740}{341{,}350} = 1{\cdot}55 \text{ m.}$$

$$\text{Eccentricity} = \frac{4{\cdot}2}{2} - 1{\cdot}55$$
$$= 2{\cdot}1 - 1{\cdot}55 = 0{\cdot}55 \text{ m.}$$

Pressure distribution at base is given by

$$p=\frac{343{,}700}{4\cdot2}\pm\frac{343{,}700\times0\cdot55}{(4\cdot2)^2/6}$$

$$=81{,}833\pm64{,}298$$

$$p_{max}=146{,}131\ \text{N/m}^2$$

$$p_{min}=17{,}535\ \text{N/m}^2.$$

Pressure at junction of toe with vertical slab

$$=146{,}131-\frac{(146{,}131-17{,}535)}{4\cdot2}\times1\cdot7$$

$$=94{,}080\ \text{N/m}^2.$$

Pressure at function of heal slab with vertical slab

$$=17{,}535+\frac{(146{,}131-17{,}535)}{4\cdot2}\times2\cdot28$$

$$=87{,}344\ \text{N/m}^2.$$

Weight of base slab per metre length

$$=0\cdot40\times1\times25{,}000=10{,}000\ \text{N}.$$

Design of Toe

Net pressure on toe will vary from

$$146{,}131-10{,}000=136{,}131\ \text{N/m}^2$$

to $94{,}080-10{,}000=84{,}080\ \text{N/m}^2.$

Consider one metre width of toe at the end. Toe will act as continuous slab between counterforts.

$$\text{Maximum B.M.}\quad=\frac{1361{,}31\times(2\cdot7)^2}{12}\times1000\ \text{N-mm}$$

$$=82{,}699{,}583\ \text{N-mm}$$

$$\text{Effective depth}\quad=\sqrt{\frac{82{,}699{,}583}{0\cdot87\times1000}}=308\cdot3\ \text{mm}.$$

Provide overall depth of 400 mm with effective depth of 320 mm.

$$A_t=\frac{82{,}699{,}583}{0\cdot87\times320\times140}=2122\ \text{mm}^2$$

Provide 12 mm ϕ bars at 50 mm centres,

$$\text{Distribution steel} = \frac{0{\cdot}15}{100} \times 400 \times 1000 = 600 \text{ mm}^2.$$

Provide 100 mm ϕ bars at 130 mm centres.

Check for Shear

$$\text{Maximum shear force} = \frac{136{,}131 \times 2{\cdot}7}{2}$$

$$= 183{,}777 \text{ N}$$

$$\text{Shear stress} = \frac{183{,}777}{320 \times 1000} = 0{\cdot}574 \text{ N/mm}^2$$

$$\text{Net shear} = (0{\cdot}574 - 0{\cdot}34) \times 320 \times 1000 = 74.880 \text{ N}$$

Allowable $\tau_c = 0{\cdot}34$ N/mm²

Provide 10 mm ϕ 8-legged stirrups.

$$\text{Spacing} = \frac{8 \times 79 \times 140 \times 320}{74{,}880} = 378 \text{ mm.}$$

Use spacing of 90 mm.

These stirrups are provided for a distance of 0·5 m, and in width of one metre as shown in Fig. opp. page 460.

Check for Bond

Sufficient anchorage length is provided.

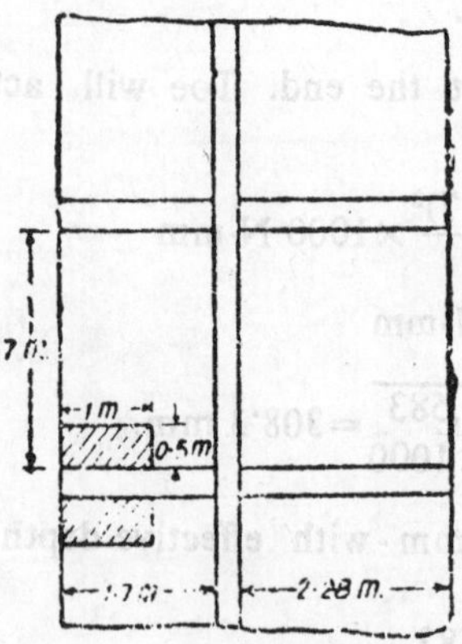

Fig. 12·38

Design of Heel Slab

Weight of earth per metre

$$= 7{\cdot}6 \times 1 \times 1 \times 15{,}000$$

$$= 114{,}000 \text{ N.}$$

Weight of heel slab

$$= 10{,}000 \text{ N/m.}$$

Pressure on heel slab will vary from

$114{,}000 + 10{,}000 - 17{,}535 = 106{,}465$ N/m²

$114{,}000 + 10{,}000 - 87{,}344 = 36{,}656$ N/m²

as shown in Fig. 12·37.

Consider one metre width of slab near end.

$$\text{Maximum B.M.} = \frac{106{,}465 \times (2{\cdot}7)^2}{12} \times 1000$$

$$= 64{,}677{,}487 \text{ N-mm}$$

$$A_t = \frac{64{,}677{,}487}{0{\cdot}87 \times 320 \times 140}$$

$$= 1659{\cdot}4 \text{ mm}^2$$

Provide 12 mm ϕ bars at 60 mm centres in one metre length.

Pressure at one metre from end of heel slab

$$= 106{,}465 - \frac{106{,}465 - 36{,}656}{2{\cdot}28} \times 1$$

$$= 75{,}847 \text{ N/m}^2$$

Area of steel required will be directly proportional to intensity of pressure

$$A_t = \frac{75{,}847}{106{,}465} \times 1659{\cdot}4$$

$$= 1182 \text{ mm}^2$$

Provide 12 mm ϕ bars at 90 mm centres in the remaining portion of heel slab.

Check for Shear

Average intensity of pressure over 1 m width

$$= \frac{106{,}465 + 75{,}847}{2} = 91{,}156 \text{ N}$$

$$\text{Max. shear force} = \frac{91{,}156 \times 2{\cdot}7}{2} = 123{,}061 \text{ N}$$

$$\text{Shear stress} = \frac{123{,}061}{320 \times 1000}$$

$$= 0{\cdot}385 \text{ N/mm}^2. \text{ Safe.}$$

Check for Bond

Sufficient anchor length is provided for bond.

Design of Counterfort at Back

Let θ be inclination of counterfort with vertical.

$$\tan \theta = \frac{2{\cdot}28}{7{\cdot}6}$$

$$\theta = 16° \ 42'.$$

The section of counterfort will be designed at 7 m below top. From counterfort is also provided at this depth.

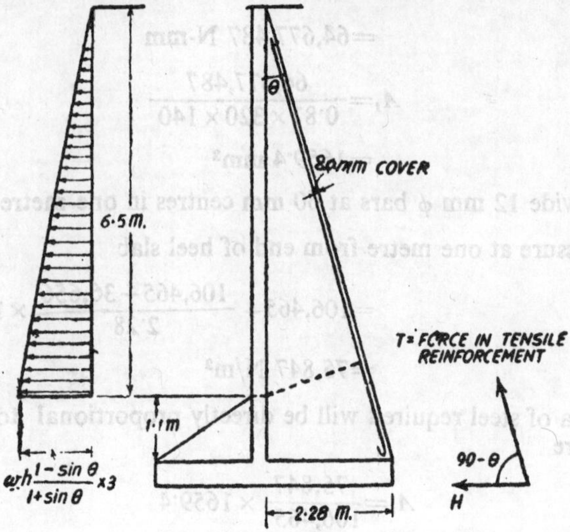

Fig. 12·39

Effective depth $=7000 \sin 16° 42' - 80$

$=2012-80=1932$ mm

Maximum B.M. $=\dfrac{wh^3}{6}\left(\dfrac{1-\sin\varphi}{1+\sin\varphi}\right)\times 3$

(centre to centre of counterforts=3 m)

$$=\frac{15{,}000\times(6\cdot5)^3}{6}\times\frac{1}{3}\times 3$$

$$=686{,}400 \text{ N-m}=686{,}400{,}000 \text{ N-mm}$$

Effective depth $=\sqrt{\dfrac{686{,}400{,}000}{0\cdot87\times300}}=1622$ mm

Effective depth provided is 1932 mm

$$A_t=\frac{686{,}400{,}000}{0\cdot87\times1932\times140}=2920 \text{ mm}^2.$$

Provide 10 bars of 20 mm φ in two rows. Half the reinforcement will be curtailed at 3·5 m from top.

Check for Shear

Shear force due to horizontal thrust at the level of front counterfort

$$=\frac{wh^2}{2}\left(\frac{1-\sin\phi}{1+\sin\varphi}\right)\times 3$$

$$=\frac{15{,}000\times(6\cdot5)^2}{2}\times\tfrac{1}{3}\times 3=316{,}900 \text{ N}$$

Net shear force $=F-\frac{M}{d'}\tan\theta$

But $d'=\frac{d}{\cos\theta}=\frac{1932}{0.9578}=2020$ mm

Net shear force $=F-\frac{M}{d}\sin\theta$

$$=319{,}600-\frac{686{,}400{,}000 \text{ N}}{1932}\times\sin 66^\circ 42'$$

$$=319{,}600-\frac{686{,}400{,}000}{1932}\times 0.2874$$

$$=319{,}600-102{,}100=217{,}500 \text{ N}$$

Shear stress $=\frac{217{,}500}{d'\times 300}$

$$=\frac{217{,}500}{2020\times 300}=359 \text{ N/mm}^2. \quad \text{Safe.}$$

Sufficient anchor length is provided for bond.

Connection of Counterfort with Stem

Force on 1 m from base $=wh\times k_A\times 2.7$

$$=15{,}000\times 6.5\times \tfrac{1}{3}\times 2.7=85{,}750 \text{ N}$$

Area of stirrups required

$$=\frac{87{,}750}{140}=627 \text{ mm}^2$$

Provide 10 mm φ 2-legged stirrups.

Spacing $=\frac{100\times 2\times 0.79}{627}=252$ mm.

Provide 2-legged stirrups 10 mm φ at 250 mm centres. As the pressure reduces towards the top gradually and is zero at top, the spacing of stirrups will be gradually increased to 400 mm.

Connection of Counterfort with Base

Consider one metre width of base slab near the end of heel.

Total downward load

$$=106{,}170\times 2.7=286{,}660 \text{ N}$$

Area of vertical stirrups required

$$=\frac{286{,}660}{140}$$

$$=255 \text{ mm}^2$$

Providing 12 mm φ 2-legged stirrups.

Spacing $=\frac{2\times 113\times 100}{2550}=88.6$ mm. Say 80 mm.

The spacing of stirrups is increased to 160 mm centres near the junction of heel slab with vertical slab.

Front Counterfort

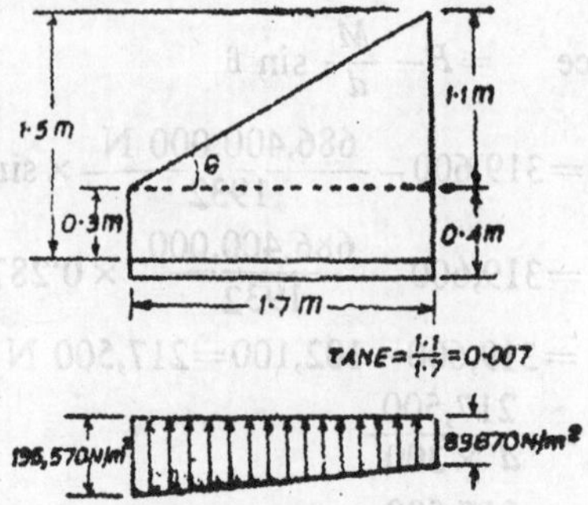

Fig. 12·40

Front counterfort is subjected to upward pressure from base slab and downward load due to self-load. Effect of self-load is neglected.

$$\text{Maximum B.M.} = 83{,}870 \times 3 \times \frac{(1{\cdot}7)^2}{2} + \frac{(135{,}570 - 83{,}870)}{2} \times 3{\cdot}0 \times 1{\cdot}7 \times 2/3 \times 1{\cdot}7$$

$$= \frac{83{,}870 \times 3 \times (1{\cdot}7)^2}{2} + 51{,}700 \times \frac{3{\cdot}0 \times (1{\cdot}7)^2}{3}$$

$$= 363{,}500 + 149{,}400$$

$$= 512{,}900 \text{ N-m}$$

$$= 512{,}900{,}000 \text{ N-mm.}$$

Effective depth required

$$= \sqrt{\frac{512{,}900{,}000}{0{\cdot}87 \times 300}} = 1400 \text{ mm.}$$

Provide overall depth of 1500 mm with effective depth of 1400 mm.

$$A_t = \frac{512{,}900{,}000}{0{\cdot}87 \times 1400 \times 140} = 3010 \text{ mm}^2$$

Provide 5 bars of 28 mm φ.

Check for Shear and Bond

$$\text{Shear force} = \frac{135{,}570 + 83{,}870}{2} \times 1{\cdot}7 \times 3$$

$$= 109{,}720 \times 5{\cdot}1 = 559{,}570 \text{ N}$$

Net shear force

$$= F - \frac{M}{d} \tan\theta$$

$$=561{,}538-\frac{514{,}912{,}000}{1410}\times\frac{1{\cdot}1}{1{\cdot}7}$$

$$=325{,}241 \text{ N}$$

Shear stress $=\frac{325{,}241}{1410\times300}=0{\cdot}769 \text{ N/mm}^2$

$\tau_c=0{\cdot}34 \text{ N/mm}^2$

Net shear $=(0{\cdot}769-0{\cdot}34)\times1410\times300$

$=181{,}467$ N

10 mm φ 4-legged stirrups are provided,

Spacing of stirrups $=\frac{4\times79\times140\times0{\cdot}87\times1410}{181{,}467}$

$=299$ mm

Provide stirrups at 160 mm centres.

Check for Bond

Sufficient anchor length is provided for bond.

Example 12·5. *Design a counterfort retaining wall to retain earth 6·5 m above ground level. The earth is surcharged at an angle of 15°. The angle of repose of the earth is 30° and its density is 16,000 N/m³. The bearing capacity of soil is 200 kN/m³. Coefficient of friction=0·6.*

Sol. Bottom of base will be provided one metre below G.L. Height of retaining wall$=6{\cdot}5+1=7{\cdot}5$ m.

Base width of retaining wall with surcharge generally lies between $0{\cdot}6\,H$ to $0{\cdot}7\,H$.

Assume base width$=5{\cdot}0$ m.

Toe width wall be assumed 1·4 m.

The counterforts are provided at 3 m centres. Width provided for each counterfort is 30 cm. Clear distance between counterforts will be 2·7 m.

Approximate thickness of base from consideration of shear

Effective depth $=20{\cdot}0\,Hl$

$=20{\cdot}0\times7{\cdot}5\times2{\cdot}7=405$ mm

Assume overall depth of 450 mm with effective depth of 370 mm.

Design of Vertical Slab

Horizontal pressure at base of vertical slab$=k_e\,wh\cos\alpha$

$$k_e=\cos\alpha\left[\frac{\cos\alpha-\sqrt{\cos^2\alpha-\cos^2\phi}}{\cos\alpha+\sqrt{\cos^2\alpha-\cos^2\phi}}\right]$$

$$=\cos15^\circ\left[\frac{\cos15-\sqrt{\cos^2 15-\cos^2 30}}{\cos15+\sqrt{\cos^2 15-\cos^2 30}}\right]$$

$$=0{\cdot}373$$

Horizontal pressure $=0{\cdot}373\times16{,}000+7{\cdot}05\times\cos 15$

$=40{,}610\ \text{N/m}^2$

Consider one metre height between counterforts

$$\text{Maximum B.M.}=\frac{40{,}610\times(2{\cdot}7)^2}{12}\times1000\ \text{N-mm}$$

$$=24{,}670{,}000\ \text{N-mm}$$

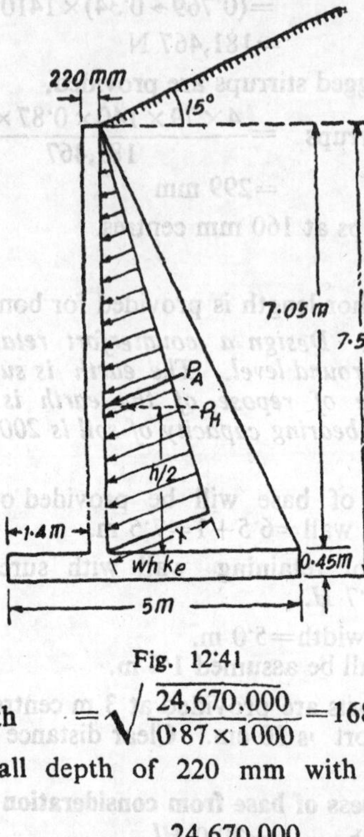

Fig. 12·41

$$\text{Effective depth}=\sqrt{\frac{24{,}670{,}000}{0{\cdot}87\times1000}}=168{\cdot}4\ \text{mm}$$

Provide overall depth of 220 mm with effective depth of 170 mm.

$$A_t=\frac{24{,}670{,}000}{0{\cdot}87\times170\times140}=1190\ \text{mm}^2.$$

Provide 12 mm φ bars at 90 mm centres. As the pressure decreases linearly to zero towards the top, the spacing of reinforcement will be gradually increased towards the top. Near the top 12 mm φ bars will be provided at 300 mm centres.

Distribution Steel

$$\text{Distribution steel}=\frac{0{\cdot}15}{100}\times220\times1000=330\ \text{mm}^2$$

Check for shear

$$\text{Maximum shear force}=40{,}610\times\frac{2{\cdot}7}{2}\ \text{N}$$

$$\text{Shear stress} = \frac{40.610 \times \frac{2{\cdot}7}{2}}{170 \times 1000}$$

$$= 0{\cdot}322 \text{ N/mm}^2. \quad \text{Safe}$$

Check for Bond

Sufficient anchor length is provided.

Pressure Distribution at Base

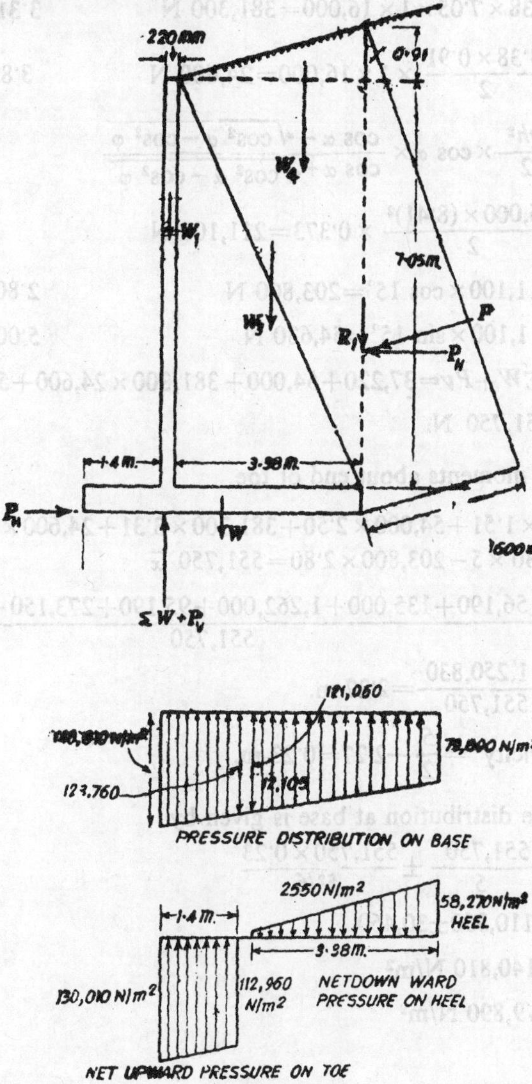

Fig. 12·42

$H_1 = 3{\cdot}38 \tan 15° = 0{\cdot}91$ m

Force	*Distance of C.G. from toe*
$W_1 = 0{\cdot}22 \times 7{\cdot}05 \times 24{,}000 = 37{,}220$ N	1·51 m
$W_2 = 5{\cdot}0 \times 0{\cdot}45 \times 1 \times 24{,}000 = 54{,}000$ N	2·50 m
$W_3 = 3{\cdot}38 \times 7{\cdot}05 \times 1 \times 16{,}000 = 381{,}300$ N	3·31 m
$W_4 = \frac{3{\cdot}38 \times 0{\cdot}91}{2} \times 1 \times 16{,}000 = 24{,}600$ N	3·87 m

$$P = \frac{wh^2}{2} \times \cos\alpha \times \frac{\cos\alpha - \sqrt{\cos^2\alpha - \cos^2\varphi}}{\cos\alpha + \sqrt{\cos^2\alpha - \cos^2\varphi}}$$

$$= \frac{16{,}000 \times (8{\cdot}41)^2}{2} \times 0{\cdot}373 = 211{,}100 \text{ N}$$

$P_H = 211{,}100 \times \cos 15° = 203{,}800$ N	2·80 m
$P_V = 211{,}100 \times \sin 15° = 54{,}630$ N	5·00 m

$$\Sigma W + P_V = 37{,}220 + 54{,}000 + 381{,}300 \times 24{,}600 + 54{,}630$$

$$= 551{,}750 \text{ N.}$$

Taking moments about end of toe

$$37{,}220 \times 1{\cdot}51 + 54{,}000 \times 2{\cdot}50 + 381{,}300 \times 3{\cdot}31 + 24{,}600 \times 3{\cdot}87 + 54{,}630 \times 5 - 203{,}800 \times 2{\cdot}80 = 551{,}750\ \bar{x}$$

$$\bar{x} = \frac{56{,}190 + 135{,}000 + 1{,}262{,}000 + 95{,}190 + 273{,}150 - 570{,}700}{551{,}750}$$

$$= \frac{1{,}250{,}830}{551{,}750} = 2{\cdot}27 \text{ m.}$$

Eccentricity $= \frac{5}{2} - 2{\cdot}27 = 0{\cdot}23$ m.

Pressure distribution at base is given by

$$p = \frac{551{,}750}{5} \pm \frac{551{,}750 \times 0{\cdot}23}{5^2/6}$$

$$= 110{,}350 \pm 30{,}460$$

$P_{max} = 140{,}810$ N/m²

$P_{min} = 79{,}890$ N/m²

Net upward pressure on toe slab

Weight of slab per metre

$$=0{\cdot}45\times 1\times 25{,}000$$
$$=11{,}250 \text{ N}$$

Net upward pressure at the end of toe

$$=141{,}778-11{,}250$$
$$=130{,}528 \text{ N/m}^2.$$

Net upward pressure at junction of vertical slab with toe slab

$$=141{,}778-\frac{141{,}778-80{,}444}{5}\times 1{\cdot}4-11{,}250$$
$$=113{,}354 \text{ N/m}^2$$

Net downward pressure on heel slab

Net downward pressure at the junction of heel slab with vertical slab

$$=7{\cdot}05\times 1\times 16{,}000+11{,}250-\left[80{,}444+\frac{141{,}778-80{,}444}{5}\times 3{\cdot}38\right]$$
$$=2144 \text{ N/m}^2$$

Net downward pressure at end of heel

$$=(7{\cdot}05+0{\cdot}91)\times 1\times 16{,}000+11{,}250-80{,}444$$
$$=58{,}166 \text{ N/m}^2$$

Net upward pressure on the toe and downward pressure on the heel is shown in Fig. 12·43.

Design of toe

Consider one metre length of retaining wall.

$$\text{Maximum B.M.} = \frac{113{,}354\times(1{\cdot}4)^2}{2}+\frac{130{,}528-113{,}354}{2}\times 1{\cdot}4\times\frac{2}{3}\times 1{\cdot}4$$
$$=122{,}307 \text{ N-m}$$
$$=122{,}307{,}000 \text{ N-mm}$$

$$\text{Effective depth required} = \sqrt{\frac{122{,}307{,}000}{0{\cdot}87\times 1000}}=375 \text{ mm}$$

Effective depth provided $=380$ mm

$$A_s=\frac{122{,}307{,}000}{0{\cdot}87\times 380\times 140}=2643 \text{ mm}^2$$

Provide 16 mm ϕ bars at 70 mm centres

Distribution Steel

$$\text{Distribution steel} = \frac{0{\cdot}15}{100} \times 450 \times 1000 = 675 \text{ mm}^2$$

Provide 10 mm φ bars at 110 mm centres.

Check for Shear

$$\text{Maximum S.F.} = \frac{130{,}528 + 113{,}354}{2} \times 1{\cdot}4 = 170{,}717 \text{ N}$$

$$\text{Shear stress} = \frac{170{,}717}{380 \times 1000} = 0{\cdot}45 \text{ N/mm}^2$$

150 mm × 150 mm fillet is provided on the side of toe. No shear reinforcement will be required.

Check for Bond

Sufficient anchor length is provided including standard hooks for bond.

Design of Heel

Heel slab acts as continuous slab between two counterforts. Consider one metre width of heel slab at the end.

$$\text{Maximum B.M.} = \frac{58{,}166 \times (2{\cdot}7)^2}{12} \times 1000$$

$$= 35{,}335{,}845 \text{ N-mm}$$

$$A_t = \frac{35{,}335{,}845}{0{\cdot}87 \times 380 \times 140} = 763{\cdot}5 \text{ mm}^2$$

Provide 10 mm φ bars at 100 mm centres throughout the width.

Check for Shear

Maximum shear force

$$= 58{,}166 \times \frac{2{\cdot}7}{2}$$

$$\text{Shear stress} = \frac{58{,}166 \times \times \frac{2{\cdot}7}{2}}{380 \times 1000}$$

$$= 0{\cdot}21 \text{ N/mm}^2. \quad \text{Safe.}$$

Check for Bond

Sufficient anchor length including standard hooks are provided.

Distribution Steel

Provide 16 mm φ bars at 240 mm centres.

Design of Counterfort

Let θ be inclination of counterforts with vertical

$$\tan\theta=\frac{3{\cdot}38}{7{\cdot}05}=0{\cdot}479$$

$$\theta=25^\circ\ 36'$$

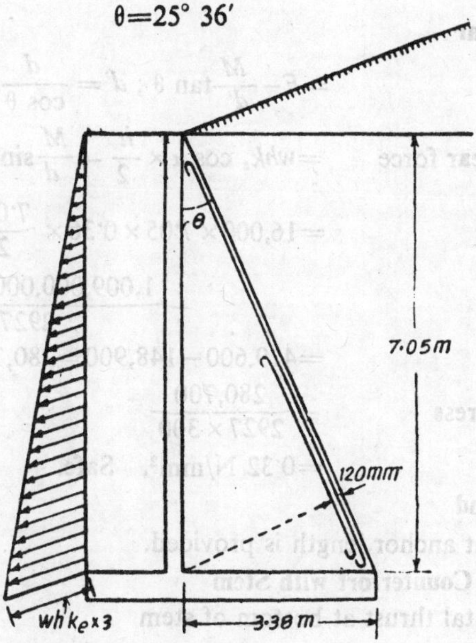

Fig. 12·43

Maximum B.M. $=whk_e\cos\theta\times\frac{h^2}{6}\times 3$

$$=wk_e\cos\alpha\times\frac{h^3}{6}\times 3$$

$$=16{,}000\times 0{\cdot}373\times\frac{\cos 15^\circ\times(7{\cdot}05)^3}{6}\times 3{\cdot}0$$

$$=1{,}009{,}000\text{ N-m}$$

$$=1{,}009{,}000{,}000\text{ N-mm}$$

Effective depth $=\sqrt{\frac{1{,}009{,}000{,}000}{0{\cdot}87\times 300}}$

$$=1966\text{ mm.}$$

Effective depth provided

$$=7050\sin\theta-12$$

$$=7050\times 0{\cdot}4321-120$$

$$=304{\cdot}70-120=2927\text{ mm}$$

$$A_t = \frac{1{,}009{,}000{,}000}{0{\cdot}87 \times 2927 \times 140} = 2840 \text{ mm}^2$$

Provide 6 bars of 25 mm φ in two rows. Four bars in one row and two bars in the second row. Two bars will be curtailed at half the height.

Check for Shear

Net S.F. $= F - \dfrac{M}{d'} \tan\theta \; ; \; d' = \dfrac{d}{\cos\theta}$

Neat shear force $= whk_e \cos\alpha \times \dfrac{h}{2} - \dfrac{M}{d} \sin\theta$

$$= 16{,}000 \times 7{\cdot}05 \times 0{\cdot}36 \times \frac{7{\cdot}05}{2}$$

$$- \frac{1{,}009{,}000{,}000 \times 0{\cdot}4321}{2927}$$

$$= 429{,}600 - 148{,}900 = 280{,}700 \text{ N}$$

Shear stress $= \dfrac{280{,}700}{2927 \times 300}$

$= 0{\cdot}32$ N/mm². Safe.

Check for Bond

Sufficient anchor length is provided.

Connection of Counterfort with Stem

Horizontal thrust at bottom of stem

$$= wh \times k_e \times \cos\alpha \times l$$

$$= 16{,}000 \times 7{\cdot}05 \times 0{\cdot}36 \times 2{\cdot}7 = 109{,}600 \text{ N}$$

Area of steel required for one metre height

$$= \frac{109{,}600}{140} = 783 \text{ mm}^2$$

Provide 2-legged 10 mm φ stirrups,

$$\text{spacing} = \frac{2 \times 0{\cdot}79 \times 1000}{783} = 200 \text{ mm}$$

The spacing of stirrups is gradually increased to 400 mm centres near top.

Connection of counterfort with heel

Consider one metre width of base slab near end of heel

Downward force $= 58{,}166 \times 2{\cdot}7$ N

Area of steel required

$$= \frac{58{,}166 \times 2{\cdot}7}{140} = 1122 \text{ mm}^2$$

Provide 8 mm φ 2-legged stirrups.

$$\text{Spacing} = \frac{2 \times 50 \times 1000}{1122} = 89{\cdot}5 \text{ mm.}$$

Provide stirrups at 90 mm centres. The spacing is gradually increased to 300 mm near the vertical slab.

Check for sliding

Frictional resistance $= \mu W = 0{\cdot}6 \times 555{,}555 = 333{,}333$ N

Horizontal force $= 203{,}800$ N

$$\text{Factor of safety} = \frac{333{,}333}{203{,}800} = 1{\cdot}64. \quad \text{Safe.}$$

12·14. Retaining Wall with Back-anchoring

When the earth to be retained is of considerable height, anchor rods are provided at the back of the wall approximately at 1/3rd the height of the retaining wall to reduce the B.M. and S.F. in the vertical slab whereby sections are much reduced. The anchor rods are embedded in concrete block of sufficient area and situated beyond plane of rupture so that the frictional force on the concrete block exceeds the tension in the rod. The area of rod is obtained by dividing tensile force in rod T by allowable stress which may be taken as 140 N/mm². The tension in the rod is found by assuming the B.M. at the foot of the stem as zero.

The rod is connected to the stem through vertical beams or horizontal beams.

In case of connection of tie rods to vertical beams, the stem is designed as a continuous slab between the vertical beams. A vertical beam is designed as cantilever. Above the rod it is subjected to earth pressure taken by the stem between two beams and below the rod it is subjected to earth pressure taken by stem between the beams and the tensile force in tie rod.

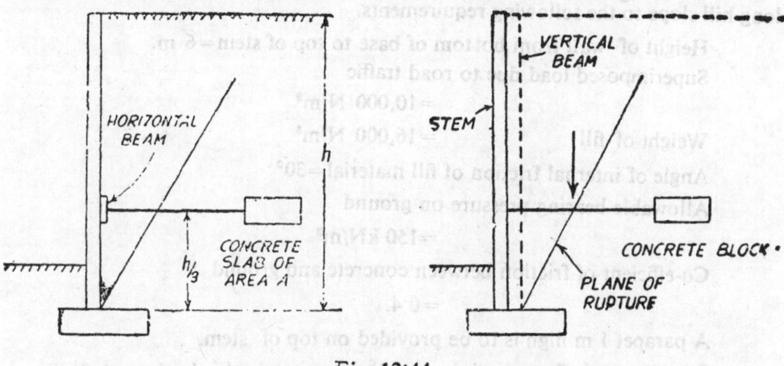

Fig. 12·44

In case of connection of rod to horizontal beams the stem is designed to act as cantilever subjected to the earth pressure above the horizontal beam. The portion below the horizontal beam is designed as cantilever subjected to earth pressure and the force in the rod. The horizontal beam is designed as a continuous beam supported by the tie rods.

The base is subjected to load of earth on the heal, self-load of the stem and base and any difference in horizontal earth pressure and the tension in the tie rod.

Let A be the area of concrete block on one face, μ be the co-efficient of friction and l be the spacing of tie rod.

In case tie rod is fixed at $h/3$.

Tension in tie rod $T=\dfrac{k_e wh^2}{2}\times l$

Area of steel rod required

$$=\frac{k_e wh^2 l}{2\times 140}$$

Area of anchor block will be given by

$$2A\mu w\frac{2h}{3}=k_e\,\frac{wh^2}{2}\times l$$

$$A=\frac{3k_e hl}{8\mu}.$$

PROBLEMS

1. Design a cantilever retaining wall to retain earth 5 m, above ground level. The earth is level with the top. Good soil suitable for foundation is available at a depth of 1·0 m. Berring capacity of soil=200 kN/m². Density of soil=1600 N/m³., ϕ=30°.

2. Design and draw in detail RCC retaining wall for a road on a side long hill slope to the following requirements.

Height of wall from bottom of base to top of stem=6 m.

Superimposed load due to road traffic

=10,000 N/m²

Weight of fill =16,000 N/m³

Angle of internal friction of fill material=30°

Allowable bearing pressure on ground

=150 kN/m²

Co-efficient of friction between concrete and ground

=0·4.

A parapet 1 m high is to be provided on top of stem.

3. A counterfort retaining wall is to retain bank of earth 8 m high. Good soil for foundation is available at a depth of 1·5 m². below ground level. Design the retaining wall. Density of soil=15,000 N/m³ Angle of internal friction of soil=30°. Bearing capacity of soil=200 kN/m².

4. Design a retaining wall to retain earth 4 m above ground level. The angle of surcharge is 20° and angle of internal friction of soil is 30°. Maximum allowable presssure on soil is 150 kN/m². Density of earth is 18,000 N/m³ μ=0·65.

5. Design a counterfort retaining wall to retain earth 7 m above the ground level. The angle of surcharge of earth is 20° and the angle of internal friction of soil is 30°. The bearing capacity of soil is 200 kN/m³. Density of earth=16,000 N/m³, μ=0·6.

6. Design a counterfort retaining wall to retain earth 6·5 m above ground level with level surface. The angle of repose of soil is 30° and its density is 15,000 N/m³. The bearing capacity of soil is 150 kN/m². Coefficient of friction=0·6.

7. Design the vertical wall of an R.C. cantilever retaining wall supporting an earth embankment 4·5 m high the top surface of which is horizontal and carries a uniform surcharge load equivalent to an extra height of 1 m of earth.

Unit weight of earth=17,000 N/m³

Angle of internal friction of earth fill=40°

$$f_c = 5{\cdot}25 \text{ N/mm}^2$$

$$f_t = 126 \text{ N/mm}^2$$

$$m = 15.$$

8. A cement concrete retaining wall is required to retain an earthen embankment for the following conditions :

Height of the embankment above the ground level=3·66 m.

The fill is level at the top and has a live load surcharge equivalent to 1·22 m, of earth.

Good foundations are available at the depth of 1·22 m below the ground level.

Weight of earth=14,000 N/m³.

Angle of internal friction=30°

Design the retaining wall including bases and show reinforcement.

9. Fig. 12·45 shown and RCC counterforted retaining wall. The counterforts are 450 mm thick and at 3·50 m centres. The top of fill is level with the top and carries a uniform surcharge load of 9800 N/m². Design the base slab *AB*.

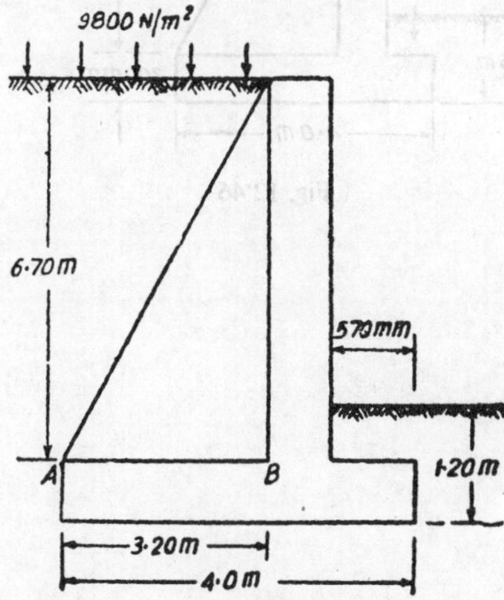

Fig. 12·45

Unit weight of earthfill=16,000 N/m³
Angle of internal friction of fill=35°

$f_c=5{\cdot}3$ N/mm²

$f_t=126$ N/mm²

$m=15$.

10. Design a reinforced concrete cantilever retaining wall to retain earth with level back fill 5 m above ground level taking the bottom of base slab 3 m, below ground level. Height of earth above may be assumed to be 3 m. Weight of earth is 18,000 N/m³ and the angle of internal friction of the earth is 30°. Following data is supplied :

Safe bearing pressure of soil ...200 kN/m²
Coefficient of friction between concrete and ground for checking sliding ...0·6
Max. allowable compression in concrete ...7·0 N/mm²
Max. allowable tension in steel reinforcements ...143 N/mm²
Modular ratio ...15

Draw (*i*) cross-sectional elevation, (*ii*) plan, and (*iii*) elevation showing detail of reinforcement.

11. A counterfort retaining wall has a cross-section as shown in Fig. 12·46. Height of wall=7·5 metres, weight of earth=20,000 N/mm³, bearing=240 kN/m², clear spacing between counterforts=2·5 metres, thickness of counterfort=400 mm³, c=5 N/mm, t=140 N/mm², m=19. ϕ for soil =30°.

Design the vertical slab and the main reinforcement in the counterfort.

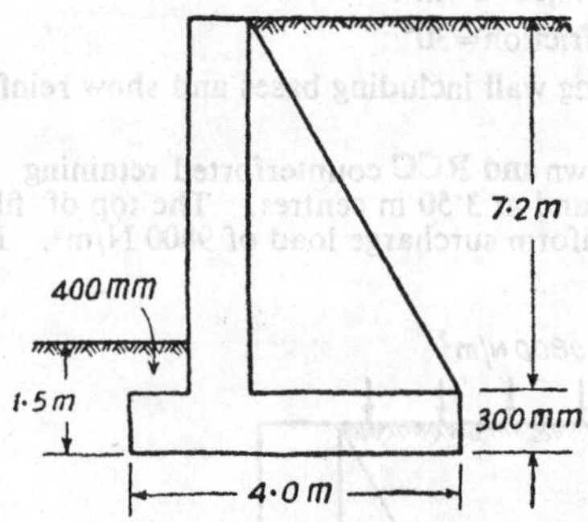

Fig. 12·46

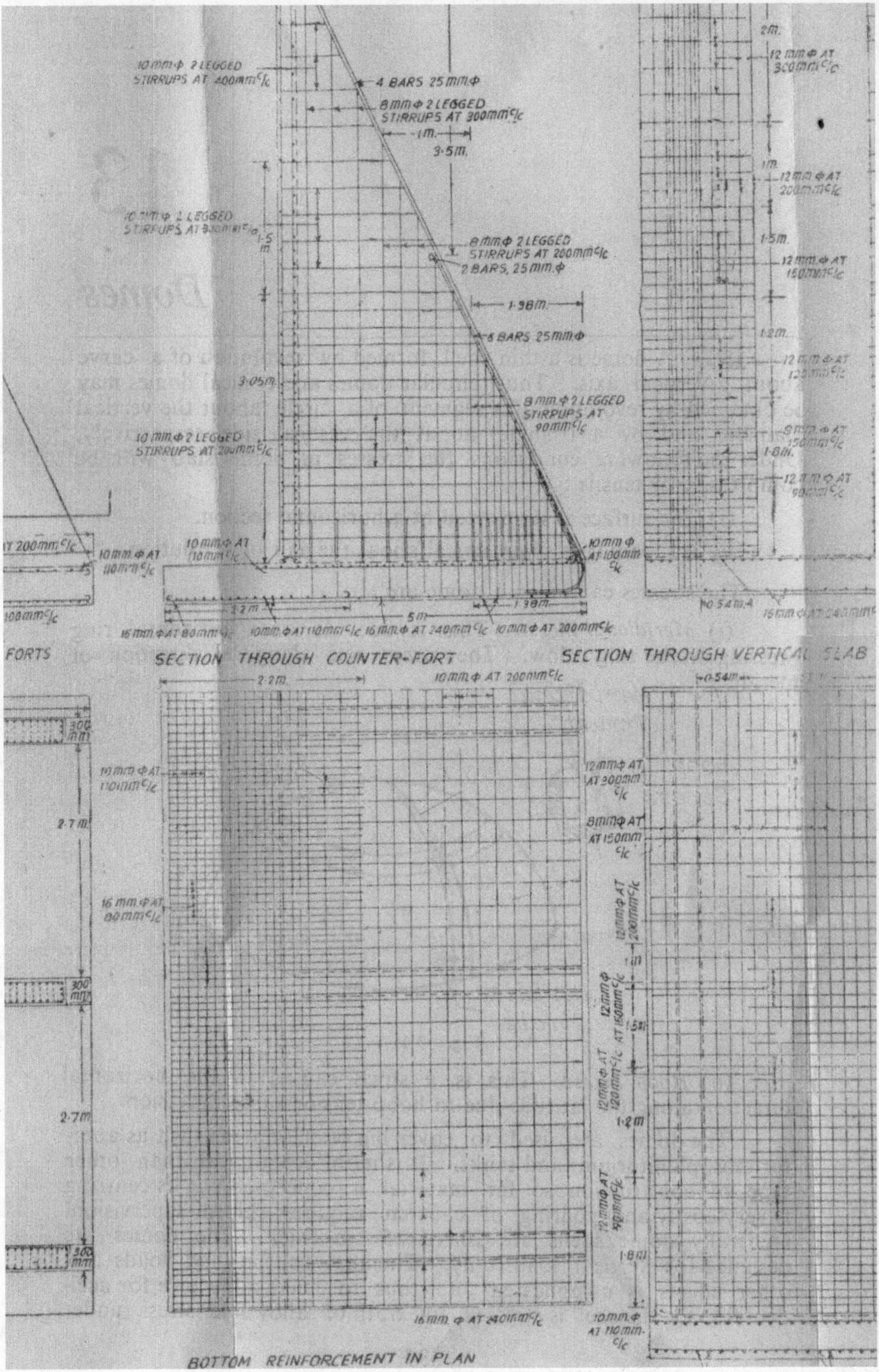

10 mm Φ 2 LEGGED STIRRUPS AT 400 mm c/c
4 BARS 25 mm Φ
8 mm Φ 2 LEGGED STIRRUPS AT 300 mm c/c
1 m.
3·5 m.
8 mm Φ 2 LEGGED STIRRUPS AT 200 mm c/c
2 BARS, 25 mm Φ
1·5 m.
1·38 m.
6 BARS 25 mm Φ
3·05 m.
8 mm Φ 2 LEGGED STIRRUPS AT 90 mm c/c
10 mm Φ 2 LEGGED STIRRUPS AT 200 mm c/c
10 mm Φ AT 110 mm c/c
10 mm Φ AT 100 mm c/c
2·2 m.
5 m.
1·38 m.
16 mm Φ AT 80 mm c/c
10 mm Φ AT 110 mm c/c
16 mm Φ AT 240 mm c/c
10 mm Φ AT 200 mm c/c
FORTS
SECTION THROUGH COUNTER-FORT
SECTION THROUGH VERTICAL SLAB
2 m.
12 mm Φ AT 300 mm c/c
1 m.
12 mm Φ AT 200 mm c/c
1·5 m.
12 mm Φ AT 150 mm c/c
1·2 m.
12 mm Φ AT 120 mm c/c
8 mm Φ AT 150 mm c/c
1·8 m.
12 mm Φ AT 90 mm c/c
0·54 m.
300 mm
2·7 m.
10 mm Φ AT 110 mm c/c
16 mm Φ AT 80 mm c/c
12 mm Φ AT 300 mm c/c
8 mm Φ AT 150 mm c/c
12 mm Φ AT 200 mm c/c
12 mm Φ AT 150 mm c/c
12 mm Φ AT 120 mm c/c
12 mm Φ AT 90 mm c/c
16 mm Φ AT 240 mm c/c
10 mm Φ AT 110 mm c/c
BOTTOM REINFORCEMENT IN PLAN

13

Domes

13·1. A dome is a thin shell formed by revolution of a curve about a vertical axis. Thus circular domes and conical domes may be obtained by revolution of a segment of a circle about the vertical diameter and by a triangle about the vertical side respectively. Under the following conditions the stresses in dome slab will be compressive or tensile :

(*i*) The surface is supported at a horizontal section.

(*ii*) All loading is symmetrical about the axis of revolution.

The stresses caused in a dome are :

(*i*) ***Meridional thrust.*** This is due to one horizontal ring pressing on the ring below. The stress acts along the direction of meridian.

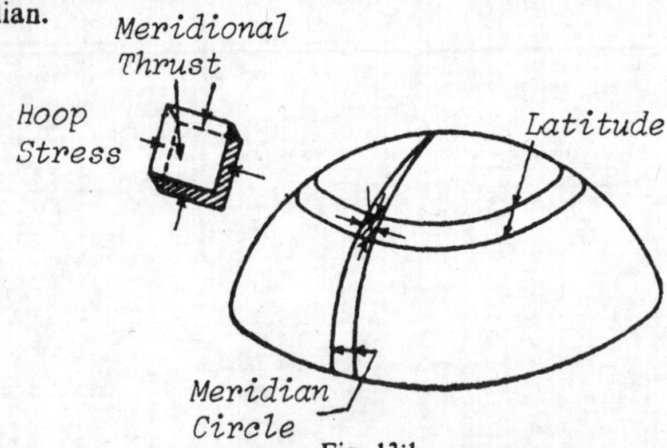

Fig. 13·1.

(*ii*) ***Hoop stress.*** This is a stress caused in the horizontal direction along the latitude due to hoop tension or compression.

The domes are used to cover big circular areas such as auditoriums, planitoriums and tanks. It is more economical than other type of construction so far material is concerned but its centring is very costly and pouring of concrete requires skilled supervision. Generally dome shells are 10 to 15 cm thick. The domes are designed for vertical loads only. The analysis for wind loads and other effects is complicated and thus to make allowance for accidental loads the shell is stressed to 2/5th of allowable stress under vertical loads. The margin is allowed for wind forces and other secondary effects.

To resist the horizontal component of the meridional thrust at the base a ring beam is provided which is subjected to hoop tension. Steel reinforcement is provided to take this hoop tension. The stress in concrete is checked which should not exceed the allowable tensile stress in concrete. Temperature and shrinkage stresses are accounted for by providing additional reinforcement of 0·2 per cent of cross-sectional area in both directions.

13·2 Analysis of Spherical Domes of Uniform Thickness

1. *Under uniformly distributed load w per unit area*

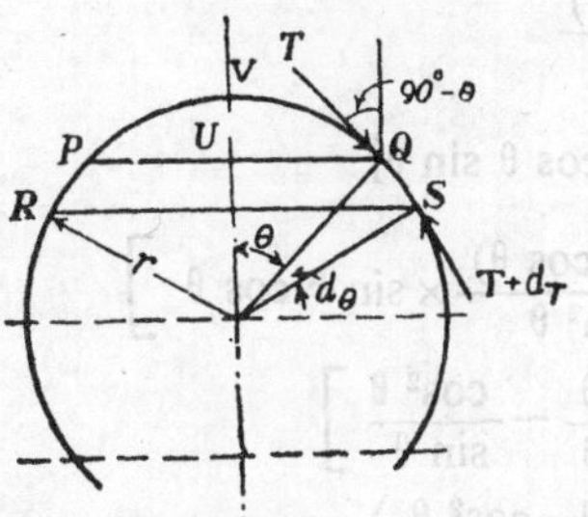

Fig. 13·2

Let r=radius of dome
t=thickness of dome shell
T=intensity of direct, *i.e.* meridional stress
H=Hoop force per unit length of surface measured on a great circle arc.

Consider a meridional circle of the dome. Consider equilibrium of a ring between two horizontal planes $P—Q$ and $R—S$. The position of the horizontal planes is defined by angle θ and $\theta+d\theta$ with the vertical as shown in Fig. 13·2. The force acting on the ring are (1) thrust T per unit length of the circle of latitude circle $P—Q$ acting tangential to surface downwards at an angle θ to horizontal, (2) thrust $T+dT$ per unit length of the circle of latitude $R-S$ acting tangentially to the surface upwards at an angle $\theta+d\theta$ to the horizontal, (3) weight of the ring acting downwards. Two thrusts acting cannot be equal as they are acting at different angles and there is also weight of the ring acting vertically downwards between them. The thrust T is caused by the weight on the domed shell PVQ whose surface area is $2\pi r \times VU$.

But $VU=r(1-\cos\theta)$

Therefore, total weight above PQ will be $2\pi r\times w\times r(1-\cos\theta)$
$=2\pi r^2 w\,(1-\cos\theta)$

Sum of the vertical components of the thrust T acting along the circumference of the circle PQ must be equal to the total weight above it.

$\therefore\ T\times 2\pi\times UQ\times\sin\theta=2\pi r^2\, w\,(1-\cos\theta)$

$T\times 2\pi\times r\sin\theta\times\sin\theta=2\pi r^2\, w\,(1-\cos\theta)$

$$T=\frac{wr\,(1-\cos\theta)}{\sin^2\theta} \qquad \ldots(13{\cdot}1)$$

The difference in thrusts T and $T+dT$ acting at angles θ and $\theta+d\theta$ respectively to the horizontal causes hoop force.

Breadth of ring $=rd\theta$.

Hoop force on the ring$=H\times rd\theta$.

Horizontal component of T will be $T \cos \theta$ and this will cause hoop tension $T \cos \theta \times$ Radius of ring at this level. Hoop tension caused bv component of $T=T \cos \theta \times r \sin \theta$. Similarly, horizontal component of the thrust $T+dT$ will cause hoop compression of $(T+dT) \cos (\theta+d\theta) \times r \sin (\theta+d\theta)$.

Difference of these two stresses causes actual hoop compression or tension on the ring considered.

Hoop force on the ring due to change in the value of T when θ increases by a small amount $d\theta$, in the limit will be

$$Hrd\theta = d[T \cos \theta \times r \sin \theta].$$

From Eq. (13·1) $T = wr\dfrac{(1-\cos \theta)}{\sin^2 \theta}$

$$\therefore \qquad H = \frac{1}{r}\,\frac{d}{d\theta}[Tr \cos \theta \sin \theta]$$

$$= wr\frac{d}{d\theta}\left[\frac{(1-\cos \theta)}{\sin^2 \theta} \times \sin \theta \cos \theta\right]$$

$$= wr\frac{d}{d\theta}\left[\frac{\cos \theta}{\sin \theta} - \frac{\cos^2 \theta}{\sin \theta}\right]$$

Differentiating, $$H = wr\left[\left(\frac{-\sin^2 \theta - \cos^2 \theta}{\sin^2 \theta}\right) - \left(\frac{-2 \sin^2 \theta \cos \theta - \cos^3 \theta}{\sin^2 \theta}\right)\right]$$

$$= wr\left[\frac{-1+2 \sin^2 \theta \cos \theta - \cos^3 \theta}{\sin^2 \theta}\right]$$

$$= wr\left[\frac{-1+2 \cos \theta - 2 \cos^3 \theta - \cos^3 \theta}{\sin^2 \theta}\right]$$

$$H = wr\left[\frac{-1+2 \cos \theta - \cos^3 \theta}{\sin^2 \theta}\right]$$

$$= wr\left[\frac{(1-\cos \theta)(\cos^2 \theta + \cos \theta - 1)}{(1-\cos^2 \theta)}\right]$$

$$= wr\frac{(\cos^2 \theta + \cos \theta - 1)}{(1+\cos \theta)} \qquad \ldots(13·2)$$

The above expression gives value of hoop stress at any section.

At crown $\quad \theta = 0$

$$\therefore \qquad H = +\frac{wr}{2}$$

Intensity of hoop stress

$$= \frac{wr}{2t} \text{ (compression)}$$

At certain value of θ hoop stress will change from compression to tension.

where $H=0, \dfrac{\cos^2\theta+\cos\theta-1}{1+\cos\theta}=0$

$$\therefore \cos^2\theta+\cos\theta-1=0$$

$$\cos\theta=0{\cdot}618$$

$$\theta=51°\ 48'.$$

Portion of dome above this angle will be in hoop compression and below this angle there will be tension. As the value of θ increases hoop stress decreases. Maximum value of hoop stress compressive that occurs at crown is $wr/2t$. For values of θ greater than 51° 48′ there is hoop tension in the dome. The value of meridional thrust at the crown is $wr/2$ and meridional stress will be $wr/2t$ at the crown and increases tó maximum at the support.

2. *Under Concentrated Load W at the Crown*

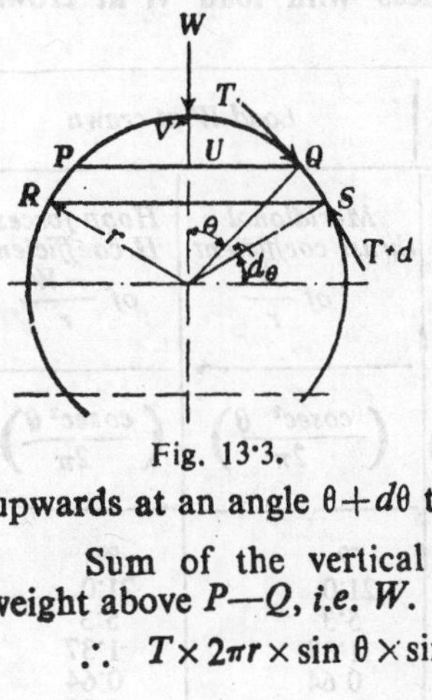

Fig. 13·3.

Consider a thin spherical shell of radius r as in first case. Consider equilibrium of elementary ring between two horizontal planes $P-Q$ and $R-S$. The positions of the planes are defined by angle θ and $\theta+d\theta$ as in the first case.

The thrust T per unit length of the circle of latitude $P-Q$, acts tangential to the surface downwards at θ to horizontal, the thrust $T+dT$ per unit length of the circle of latitude $R—S$, acts tangential to the surface upwards at an angle $\theta+d\theta$ to the horizontal.

Sum of the vertical components of thrust T is equal to the weight above $P—Q$, *i.e.* W.

$$\therefore \quad T\times 2\pi r\times\sin\theta\times\sin\theta=W$$

$$\therefore \quad T=\frac{W}{2\pi r\sin^2\theta}.$$

As in the case 1,

$$Hrd\theta=d[T\cos\theta\times r\sin\theta]$$

$$T=\frac{W}{2\pi r\sin^2\theta} \qquad \ldots(13{\cdot}3)$$

$$H=\frac{1}{r}\ \frac{d}{d\theta}\Big[T\,r\cos\theta\sin\theta\Big]$$

$$=\frac{W}{2\pi r}\ \frac{d}{d\theta}\left[\frac{\cos\theta}{\sin\theta}\right]$$

$$=\frac{W}{2\pi r}\left[\frac{-\sin^2\theta-\cos^2\theta}{\sin^2\theta}\right]$$

$$=\frac{-W}{2\pi r}\times\frac{\sin^2\theta+\cos^2\theta}{\sin^2\theta}=-\frac{W}{2\pi r}\text{ cosec}^2\theta \text{ (tension)}$$

Hoop stress $$=-\frac{W}{2\pi rt}\text{ cosec}^2\theta \qquad ...(13·4)$$

At $\theta=0$, hoop stress becomes infinite. Thus the load of latern or ornament must be spread over sufficient area and if necessary the top of dome should be thickened to spread the load.

The stresses caused by concentrated load should be superimposed on stresses due to uniformly distributed load caused by self-load or live load etc.

To ease calculations, coefficient for hoop stress and meridional stress due to uniformly distributed load and concentrated load on the crown are given in Table 13·1.

TABLE 13·1.

Thin special shell of uniform thickness with load W at crown snpported on circle of latitude.

	Uniformly distributed load		*Load W at crown*	
θ	*Meridional Thrust T co-efficients of wr*	*Hoop forces H coefficient of wr*	*Meridional thrust coefficient of* $\frac{W}{r}$	*Hoop forces H coefficient of* $\frac{-W}{r}$
	$\left(\frac{1-\cos\theta}{\sin^2\theta}\right)$	$\left(\frac{\cos^2\theta+\cos\theta-1}{1+\cos\theta}\right)$	$\left(\frac{\text{cosec}^2\theta}{2\pi}\right)$	$\left(\frac{\text{cosec}^2\theta}{2\pi}\right)$
0	0·5	+0·5	∞	∞
5	0·500	+0·496	21·0	21·0
10	0·505	+0·48	5·3	5·3
20	0·516	+0·425	1·37	1·37
30	0·537	+0·33	0·64	0·64
40	0·566	+0·20	0·38	0·38
50	0·608	+0·034	0·27	0·27
51°48	0·618	+0·0	0·26	0·26
60	0·667	−0·167	0·21	0·21
70	0·747	−0·402	0·18	0·18
80	0·838	−0·68	0·16	0·16
90	1·00	−1·0	0·16	0·16
100	1·21	−1·38	0·16	0·16
110	1·52	−1·86	0·18	0·18
120	2·00	−2·50	0·21	0·21
130	2·79	−3·45	0·27	0·27
140	4·27	−5·05	0·38	0·38
150	7·48	−8·33	0·64	0·64
160	16·60	−1·75	1·37	1·37
170	66·60	−66·60	5·3	5·3
180	∞	∞	∞	∞

13·3. Analysis of Conical Domes of Uniform Thickness

Consider a conical dome of uniform thickness t.

Let 2θ be the angle at the apex of the dome.

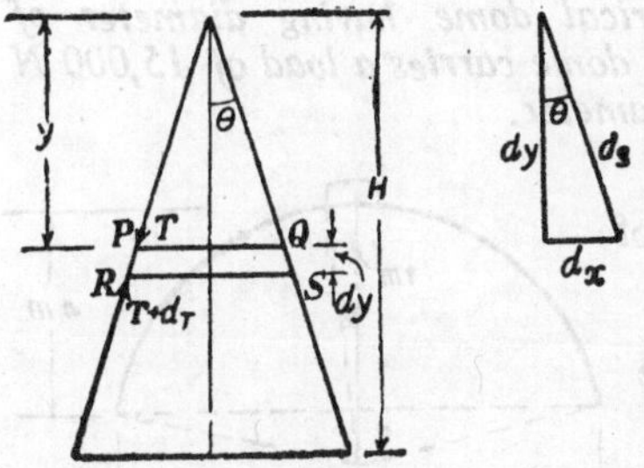

Fig. 13·4.

Let w be the intensity of uniform load on the dome. Consider an elementry ring at depth y from top and having height dy. Let T be the meridional thrust per unit length of the ring at distance y and $T+dT$ be the meridional thrust per unit length at distance $y+dy$.

The vertical component of the total meridional thrust T at level PQ is equal to the load on the dome above this level.

$$(2\pi y \tan\theta)\, T\cos\theta = \frac{2w}{2}(2\pi y \tan\theta)\frac{y}{\cos\theta}$$

$$\therefore \qquad T = \frac{w}{2} \times \frac{y}{\cos^2\theta} \qquad \ldots(13·5)$$

The horizontal component of T is $T \sin\theta$ and causes hoop compression equal to $T \sin\theta$ multiplied by the radius of ring at depth y.

Hoop tension $= T\sin\theta \times y\tan\theta = Ty\,\dfrac{\sin^2\theta}{\cos\theta}$.

At depth $y+dy$ horizontal component of thrust $T+dT$ will cause hoop compression of $(T+dT)(y+dy)\,\dfrac{\sin^2\theta}{\cos\theta}$.

The difference of these two forces will give hoop compression in the ring at all levels.

Breadth of the ring considered

$$= ds = \frac{dy}{\cos\theta}$$

If H is hoop force per unit length.

$$\frac{H\,dy}{\cos\theta} = d\left(Ty\frac{\sin^2\theta}{\cos\theta}\right)$$

$$= d\left[\frac{w}{2}\frac{y^2}{\cos^2\theta} \times \frac{\sin^2\theta}{\cos\theta}\right]$$

$$H = \frac{w}{2}\frac{\cos\theta\sin^2\theta}{\cos^3\theta} \times \frac{d}{dy}[y^2]$$

$$= \frac{w}{2} \times \frac{\sin^2\theta}{\cos^2\theta} \times 2y$$

$$= wy\tan^2\theta \qquad \ldots(13·6)$$

Intensity hoop stress$=\frac{wy\tan^2\theta}{t}$.

Dome is under hoop compression at all levels.

Example 13·1. *Design a spherical dome having diameter of 10 m at the base and rise of 4 m. The dome carries a load of 15,000 N at the crown acting on area of 1 m diameter.*

Sol. Assume thickness of dome shell as 100 mm.

Self-load of dome per square metre $=0{\cdot}1\times 25{,}000\times 1\times 1$

$=2500$ N

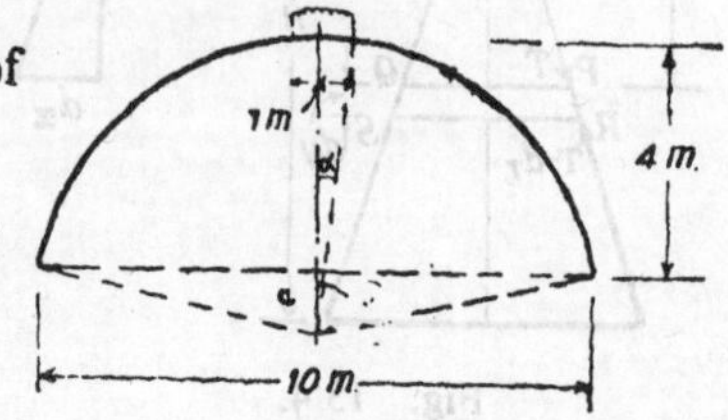

Fig. 13·5.

Radius of dome shell is given by $4\,(2R-4)=5\times 5$

$$R=\frac{1}{2}\left(\frac{25}{4}+4\right)=5{\cdot}125\text{ m}$$

$$\sin\alpha=\frac{1/2}{5{\cdot}125}=0{\cdot}0976,\ \alpha=5°\ 36'$$

$$\sin\phi=\frac{5}{5{\cdot}125}=0{\cdot}9751,\ \phi=77°\ 12'.$$

Stresses between 5° 36′ and 77° 12′

Meridional and hoop stresses between angles 5° 36′ and 77° 12′ are given by

$$\text{Meridional stress}=\frac{wR\,(1-\cos\theta)}{t\sin^2\theta}+\frac{W}{2\pi Rt\sin^2\theta}$$

$$\text{Hoop stress}=\frac{wR\,(\cos^2\theta+\cos\theta-1)}{t\,(1+\cos\theta)}-\frac{W}{2\pi Rt\sin^2\theta}$$

$w=2500\text{ N/m}^2,\ W=15{,}000\text{ N}$

$R=5{\cdot}125\text{ m},\quad t=100\text{ mm}=0{\cdot}1\text{ m}.$

$$\text{Meridional stress}=\frac{2500\times 5{\cdot}125\,(1-\cos\theta)}{0{\cdot}1\times\sin^2\theta}\times 10^{-6}$$

$$+\frac{15{,}000}{2\pi\times 5{\cdot}125\times 0{\cdot}1\times\sin^2\theta}\times 10^{-6}$$

$$=\frac{0{\cdot}128\,(1-\cos\theta)}{\sin^2\theta}+\frac{0{\cdot}0293}{2\pi\sin^2\theta}\text{ N/mm}^2$$

Hoop stress $= \dfrac{2500 \times 5{\cdot}125\ (\cos^2\theta + \cos\theta - 1)}{0{\cdot}1 \times (1+\cos\theta)} \times 10^{-6}$

$- \dfrac{W}{2\pi \times 5{\cdot}125 \times 0{\cdot}1 \times \sin^2\theta} \times 10^{-6}$ N/mm²

$= \dfrac{0{\cdot}128\ (\cos^2\theta + \cos\theta - 1)}{1+\cos\theta} - \dfrac{0{\cdot}0293}{2\pi \sin^2\theta}$ N/mm²

Table 13·2. Giving meridional and hoop stress for various values of θ

θ	*Meridional stress N/mm²*	*Hoop stress N/mm²*
10°	0·128×0·505+·0293×5·3=0·22	0·128×0·48−·0293×5·3=−·0939
20°	0·128×0·516+0·293×1·37=·106	0·128×0·425−·0293×1·37=+·0143
30°	0·128×0·537+0.293×0·64=·087	0·128×0·33−·293×0·64=+·0235
40°	0·128×0·566+0·293×0·38=·084	0·128×0 20−·0293×0·38=−·0145
50°	0·128×0·608+0·293×0·27=·086	0·128×0·034−·0293×0·27=−·0036
60°	0·128×0·667+0·293×0·21=·092	0·128×0·167−·0293×0·21=−·0275
70°	0·128×0·747+0·293×0·18=·101	−0·128×0·402−·0293×·018 =−·05467
77° 12′	0·128×0·8187+·0293×0·1673=·11	−0·128×0·596−·0293×0·1673 =−·07812

Stresses between 0° and 5° 36′

Load per unit area $= 2500 + \dfrac{15{,}000}{\frac{\pi}{4} \times l^2} = 2500 + 19{,}099$

$= 21{,}599$ N/m²

Meridional and hoop stresses are given by

Meridional stress $= \dfrac{21{\cdot}599 \times 5{\cdot}125\ (1-\cos\theta)}{0{\cdot}1 \times \sin^2\theta} \times 10^{-6}$

$= 1{\cdot}107\ \dfrac{(1 - \cos\theta)}{(\sin^2\theta)}$

Hoop stress $= \dfrac{21{,}599 \times 5{\cdot}125\ (\cos^2\theta + \cos\theta - 1)}{0{\cdot}1 \times (1+\cos\theta)} \times 10^{-6}$

$= 1{\cdot}107\ \dfrac{(\cos^2\theta + \cos\theta - 1)}{(1+\cos\theta)}.$

At 5° 36′ Meridional stress $=1{\cdot}107\times0{\cdot}5091=0{\cdot}5636$ N/mm²

Hoop stress $=1{\cdot}107\times0{\cdot}4938=0{\cdot}5466$ N/mm²

At 0°, Meridional stress $=1{\cdot}107\times0{\cdot}5=0{\cdot}5535$ N/mm²

Hoop stress $=1{\cdot}107\times0{\cdot}5=0{\cdot}5535$ N/mm²

The stresses are within permissible limits.

Steel for hoop tension

As maximum tensile stress is only 0·0939 N/mm², only nominal reinforcement of 0·2% is provided.

Steel area required in 1 mm

$$=\frac{0{\cdot}2}{100}\times100\times1=0{\cdot}2\text{ mm}^2$$

Provide 8 mm φ bars,

$$\text{Spacing}=\frac{50}{0{\cdot}2}=250\text{ mm}$$

Provide 8 mm φ bars at 250 mm centres, both along altitude and meridional circles.

Design of Ring Beam at Edges

Meridional thrust per mm length

$$=0{\cdot}11\times100=11\text{ N}$$

Horizontal component of thrust

$$=11\times\cos 77^\circ\,12'$$

Hoop tension

$$=11\times\cos 77^\circ\,12'\times\frac{10\times1000}{2}\text{ N}$$

$$=11\times0{\cdot}2147\times\frac{10{,}000}{2}=11{,}809\text{ N}$$

Area of steel required

$$=\frac{11{,}809}{140}=84{\cdot}4\text{ mm}^2$$

Provide 4 bars of 10 mm φ.

$$A_t=314\text{ mm}^2$$

Ring beam of 200 mm × 150 mm is provided.

Check for stress in concrete

Equivalent area $=200\times15+(m-1)\times A_t$

$$=3000+17\times314=35{,}338\text{ mm}^2$$

$$\text{Stress}=\frac{11{,}809}{35{,}338}=0{\cdot}334\text{ N/mm}^2.\text{ Safe.}$$

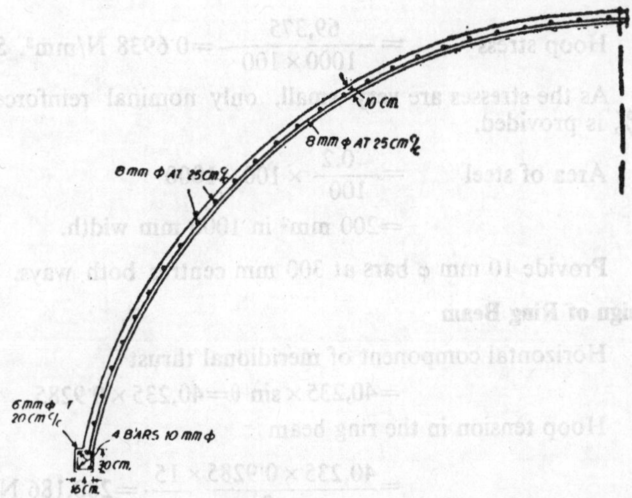

Fig. 13·6

Example 13·2 *Design a conical roof for a cylindrical tank of 15 m diameter. The rise of roof is to be 3 m. The live loads due to wind etc. may be taken as equivalent of 1200 N/m².*

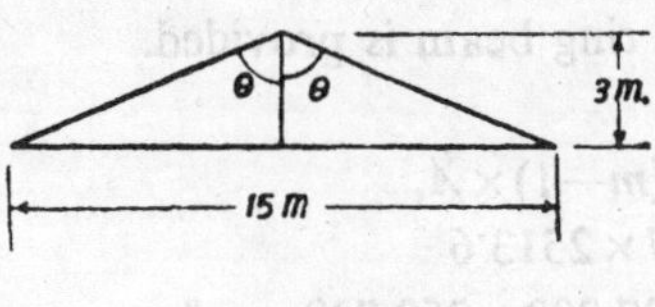

Fig. 13·7

Sol. Assume thickness of dome shell as 100 mm.

Self-load per square metre

$$0{\cdot}1 \times 1 \times 25{,}000$$
$$=2500 \text{ N}$$

Total load square metre

$$=2500+1200=3700 \text{ N}$$

$$\tan\theta=\frac{7{\cdot}5}{3}=2{\cdot}5 \therefore \theta=68° \ 12'$$

$$T=\frac{w}{2}\times\frac{y}{\cos^2\theta}$$

$$T_{max}=\frac{3700}{2}\times\frac{3}{(0{\cdot}3714)^2}=40{,}235 \text{ N/m}$$

$$H=w\tan^2\theta\times y$$

$$H_{max}=3700\times(2{\cdot}5)^2\times 3=69{,}375 \text{ N/m.}$$

Compressive Stresses in the Shell

$$\text{Meridional stress }=\frac{40{,}235}{100\times 1000}=0{\cdot}4024 \text{ N/mm}^2. \text{ Safe.}$$

Hoop stress $= \dfrac{69{,}375}{1000 \times 100} = 0{\cdot}6938$ N/mm². Safe.

As the stresses are very small, only nominal reinforcement of 0·2% is provided.

Area of steel $= \dfrac{0.2}{100} \times 100 \times 1000$

$= 200$ mm² in 1000 mm width.

Provide 10 mm φ bars at 300 mm centres both ways.

Design of Ring Beam

Horizontal component of meridional thrust

$= 40{,}235 \times \sin\theta = 40{,}235 \times 0{\cdot}9285$

Hoop tension in the ring beam

$$= \frac{40{,}235 \times 0{\cdot}9285 \times 15}{2} = 280{,}186 \text{ N}$$

Steel area required $= \dfrac{280{,}186}{140} = 2001$ mm²

Provide 8 bars of 20 mm φ.

$A_t = 2513{\cdot}6$ mm²

Section of 600 mm × 350 mm for ring beam is provided.

Equivalent concrete area

$= 600 \times 350 + (m-1) \times A_t$

$= 210{,}000 + 17 \times 2513{\cdot}6$

$= 210{,}000 = 427{,}300 = 252{,}730$ mm²

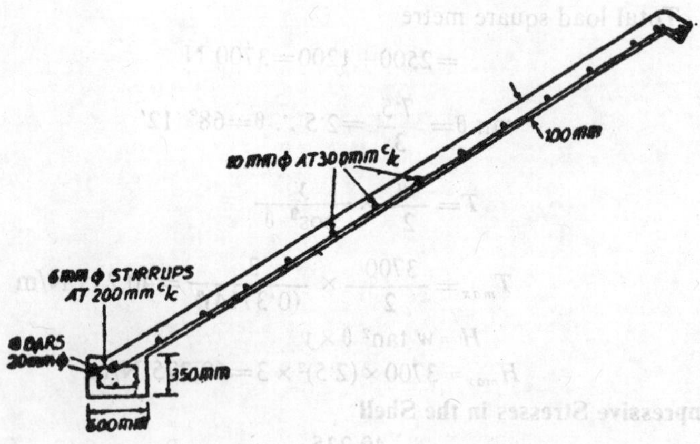

Fig. 13·8

Tensile stress in concrete

$$=\frac{280,186}{252,730}=1{\cdot}11 \text{ N/mm}^2. \text{ Safe.}$$

13·4. Cylindrical Vessels with Hemispherical Bottom

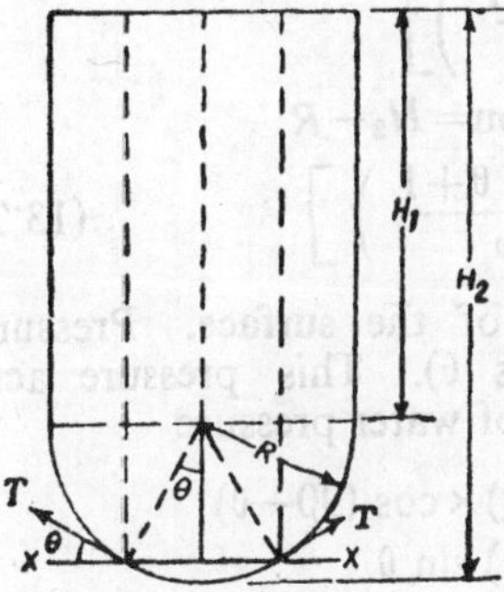

Fig. 13·9

Cylindrical tanks are sometimes provided with domed bottoms Meridional stresses and hoop stresses in such cases are found as shown below.

Consider a cylindrical tank with hemispherical bottom. Let the radius be R, height of cylindrical portion H_1 and total height be H_2. Let w be density of water. Consider a horizontal section X-X. The position of X-X is defined by angle θ made with vertical as shown in Fig. 13·9. Weight of water below section X-X.

$$=w\times\frac{2\pi R^3}{3}\left[1-\cos\theta-\frac{\cos\theta\sin^2\theta}{2}\right]$$

Weight of water in cylinder above X-X

$$=w\pi\times(R\sin\theta)^2\,(H_2-R+R\cos\theta)$$

Total weight of water

$$=\frac{2\pi w^2 R}{3}\left[1-\cos\theta-\frac{\cos\theta\sin^2\theta}{2}\right]$$
$$+\pi wR^2\sin^2\theta\,(H_2-R+R\cos\theta)$$

Let T be the meridional pull per unit length.

The vertical component of the total meridional thrust T at X-X is equal to the weight of water.

$$T\times2\pi R\sin\theta\times\sin\theta=\frac{2\pi wR^3}{3}\left(1-\cos-\frac{\cos\theta\sin^2\theta}{2}\right)$$
$$+\pi wR^2\sin^2\theta(H_2-R+R\cos\theta)$$

$$\therefore\ T=\frac{wR^2}{3\sin^2\theta}\left(1-\cos\theta-\frac{\cos\theta\sin^2\theta}{2}\right)$$
$$+\frac{wR}{2}\left(H_2-R+R\cos\theta\right)$$

$$=wR\left[\frac{R}{3\sin^2\theta}\left(1-\cos\theta-\frac{\cos\theta\sin^2\theta}{2}\right)+\frac{H_2-R}{2}+\frac{R\cos\theta}{2}\right]$$

$$wR=\left[\frac{H_2-R}{2}+\frac{R}{3}\,\frac{\left(1-\cos\theta-\frac{\cos\theta\sin^2\theta}{2}+\frac{3}{2}\cos\theta\sin^2\theta\right)}{\sin^2\theta}\right]$$

$$=wR\left[\frac{H_1-R}{2}+\frac{R}{3}\left(\frac{1-\cos\theta+\cos\theta\sin^2\theta}{\sin^2\theta}\right)\right]$$

$$=wR\left[\frac{H_2-R}{2}+\frac{R}{3}\left(\frac{1-\cos^3\theta}{\sin^2\theta}\right)\right]$$

$$=wR\left[\frac{H_2-R}{2}+\frac{R}{3}\left(\frac{\cos^2\theta+\cos\theta+1}{1+\cos\theta}\right)\right]$$

H_1=height of water in cylindrical portion$=H_2-R$

$$\therefore\ T=wR\left[\frac{H_1}{2}+\frac{R}{3}\left(\frac{\cos^2\theta+\cos\theta+1}{1+\cos\theta}\right)\right] \qquad \text{...(13·7)}$$

Let H be hoop force per unit length of the surface. Pressure of water at level of $X\text{-}X=w(H_1-R\cos\theta)$. This pressure acts normal to surface. Horizontal component of water pressure

$$=w(H_1+R\cos\theta)\times\cos(90-\theta)$$

$$=w(H_1+R\cos\theta)\sin\theta$$

Hoop tension due to water pressure

$$=w(H_1+R\cos\theta)\sin\theta\times R\sin\theta$$

$$=wR(H_1+R\cos\theta)\sin^2\theta$$

Hoop tension due meridional pull

$$=\frac{d(T\cos\theta\ R\sin\theta)}{Rd\theta}$$

Total hoop tension

$$H=\frac{d}{d\theta}\left(T\cos\theta\sin\theta\right)$$

$$+wR\,(H_1+R\cos\theta)\sin^2\theta$$

$$\frac{d}{d\theta}(T\cos\theta\sin\theta)=\frac{d}{d\theta}\left[wR\left\{\frac{H_1}{2}+\frac{R}{3}\left(\frac{\cos^2 1+\cos\theta+1}{1+\cos\theta}\right)\right\}\sin\theta\cos\theta\right]$$

$$=wR\left[H_1\frac{\cos 2\theta}{2}+\frac{R}{3}\cos 2\theta+\frac{R}{3}\cos^2\theta\times\frac{(\cos\theta-3\sin^2\theta)}{1+\cos\theta}\right]$$

$$\therefore\ H=wR\left(\frac{H_1\cos 2\theta}{2}+\frac{R}{3}\cos 2\theta+\frac{R}{3}\cos^2\theta\left(\frac{\cos\theta-3\sin^2\theta}{1+\cos\theta}\right)+\left(H_1+R\cos\theta\right)\sin^2\theta\right]$$

$$=wR\left[\frac{H_1}{2}-H_1\sin^2\theta-\frac{R}{3}\cos 2\theta+\frac{R}{3}\left\{\frac{\cos^2\theta\,(\cos\theta-\sin^3\theta)}{1+\cos\theta}+3\cos\theta\sin^2\theta\right\}+H_1\sin^2\theta\right]$$

$$=wR\left[\frac{H_1}{2}+\frac{R}{3}\left\{\left(1-2\sin^2\theta\right)+\frac{\cos^3\theta-2\cos^2\theta\sin^2\theta+3\cos\theta\sin^2\theta+3\cos^2\theta\sin^2\theta}{1-\cos\theta}\right\}\right]$$

$$=wR\left[\frac{H_1}{2}+\frac{R}{3}\left(\frac{1-2\sin^2\theta+\cos\theta-2\cos\theta\sin^2\theta+\cos^3\theta+3\cos\theta\sin^2\theta}{1+\cos\theta}\right)\right]$$

$$=wR\left[\frac{H_1}{2}+\frac{R}{3}\left(\frac{1-2\sin^2\theta+\cos\theta+\cos^3\theta+3\cos\theta\sin^2\theta}{1+\cos\theta}\right)\right]$$

$$=wR\left[\frac{H_1}{2}+\frac{R}{3}\left(\frac{1-2\sin^2\theta+2\cos\theta}{1+\cos\theta}\right)\right]$$

$$=wR\left[\frac{H_1}{2}+\frac{R}{3}\left(\frac{1-2+2\cos^2\theta+2\cos\theta}{1+\cos\theta}\right)\right]$$

$$H=wR\left[\frac{H_1}{2}+\frac{R}{3}\left(2\cos\theta-\frac{1}{1+\cos\theta}\right)\right] \quad ...(13{\cdot}8)$$

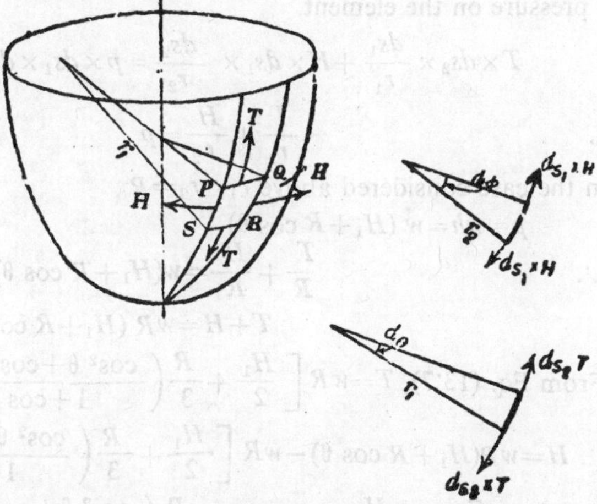

Fig. 13·10

The above formula for hoop tension can also be derived as follows :

Consider a thin walled vessel of uniform thickness subjected to uniform pressure p.

Consider an element $PQRS$ cut from the wall of the vessel by two meridional sections PS and QR and two sections PQ and SR normal to the meridians. Let,

T=tensile force per unit length in meridional direction.

H=force per unit length along parallel circle or hoop force.

ds_1=dimension of the element in meridional direction.

ds_2=dimension of the element in the direction of a parallel circle.

r_1=meridional radius of curvature.

r_2=radius of curvature of a section taken perpendicular to meridian.

The total tensile force acting on the sides of the element are $T \times ds_2$ and $H \times ds_1$.

The tensile force $T \times ds_2$ acting on the sides PQ and SR of the element have component in the direction normal to the element equal to

$$T \times ds_2 \times d\theta = T \times ds_2 \times \frac{ds_1}{r_1}$$

Similarly, the tensile forces acting on the sides PS and QR have a normal component equal to

$$H \times ds_1 \times d\varphi = H \times \frac{ds_1 \times ds_2}{r_2}$$

The sum of these normal components is in equilibrium with normal pressure on the element

$$T \times ds_2 \times \frac{ds_1}{r_1} + H \times ds_1 \times \frac{ds_2}{r_2} = p \times ds_1 \times ds_2$$

$$\therefore \quad \frac{T}{r_1} + \frac{H}{r_2} = p \qquad \ldots(13{\cdot}9)$$

In the case considered above $r_1 = r_2 = R$

$$p = wh = w\,(H_1 + R\cos\theta)$$

$$\therefore \quad \frac{T}{R} + \frac{H}{R} = w(H_1 + R\cos\theta)$$

$$T + H = wR\,(H_1 + R\cos\theta)$$

From Eq. (13·7), $T = wR\left[\frac{H_1}{2} + \frac{R}{3}\left(\frac{\cos^2\theta + \cos\theta + 1}{1 + \cos\theta}\right)\right]$

$$\therefore \; H = wR(H_1 + R\cos\theta) - wR\left[\frac{H_1}{2} + \frac{R}{3}\left(\frac{\cos^2\theta + \cos\theta + 1}{1 + \cos\theta}\right)\right]$$

$$= wR\left[H_1 - \frac{H_1}{2} + R\cos\theta - \frac{R}{3}\left(\frac{\cos^2\theta + \cos\theta + 1}{1 + \cos\theta}\right)\right]$$

$$= wR\left[\frac{H_1}{2} + \frac{R}{3}\left(3\cos\theta - \frac{\cos^2\theta + \cos\theta + 1}{1 + \cos\theta}\right)\right]$$

$$= wR\left[\frac{H_1}{2} + \frac{R}{3}\left(\frac{3\cos\theta + 3\cos^2\theta - \cos^2\theta - \cos\theta - 1}{1 + \cos\theta}\right)\right]$$

$$=wR\left[\frac{H_1}{2}+\frac{R}{3}\left(\frac{2\cos\theta+2\cos^2\theta-1}{1+\cos\theta}\right)\right]$$

$$=wR\left[\frac{H_1}{2}+\frac{R}{3}\left(2\cos\theta-\frac{1}{1+\cos\theta}\right)\right]$$

If $H_1=0$

$$T=\frac{wR^2}{3}\left(\frac{\cos^2\theta+\cos\theta+1}{1+\cos\theta}\right)\text{(Tension)}$$

$$H=\frac{wR^2}{3}\left(2\cos\theta-\frac{1}{1+\cos\theta}\right)\text{(Tension)}$$

H, will be zero when $\theta=68°\ 30'$. For values of θ greater than 68° 30′, there will be hoop compression.

13·5. Cylindrical Vessel with Conical Bottom

Consider a cylindrical vessel with conical bottom.

Let R be the radius of cylindrical portion and H_1 be its height. Let H_2 be total height of tank.

Let 2θ be the angle of cone.

Consider a horizontal section X-X, at a distance y from bottom.

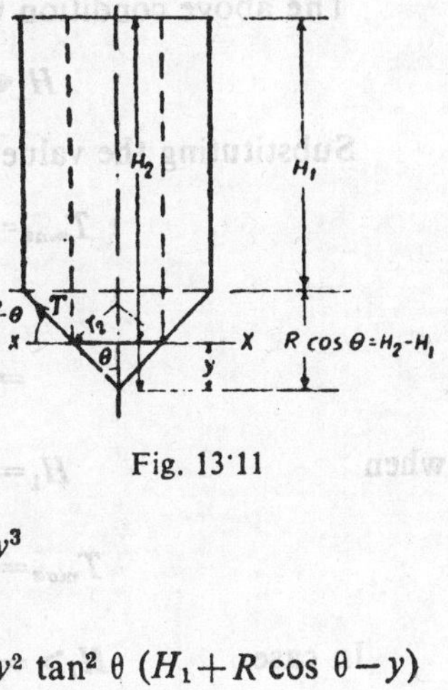

Fig. 13·11

Weight of water below X-X

$$=w\pi(y\tan\theta)^2\times\frac{y}{3}$$

$$=\frac{w\pi\tan^2\theta}{3}\times y^3$$

Weight of water above section X-X

$$=w\pi(y\tan\theta)^2\times(H_1+R\cot\theta-y)=w\pi y^2\tan^2\theta\,(H_1+R\cos\theta-y)$$

Total weight of water $=w\pi\tan^2\theta y^2\,(H_1+R\cos\theta-y)$

$$+\frac{w\pi\tan^2\theta}{3}\times y^3$$

$$=w\pi y^2\tan^2\theta\left(H_1+R\cot\theta-y+\frac{y}{3}\right)$$

$$=w\pi y^2\tan^2\theta\,[H_1+R\cot\theta-2/3y]$$

Let T be the meridional pull per unit length.

Vertical component of the total meridional thrust T at level X-X is equal to the total weight of the water.

$$T\cos\theta\times2\pi\times y\tan\theta=w\pi y^2\tan^2\theta(H_1+R\cot\theta-2/3y)$$

$$T=\frac{wy\tan\theta}{2\cos\theta}\,(H_1+R\cot\theta-2/3y)\quad\ldots(13\cdot10)$$

$$\frac{T}{r_1}+\frac{H}{r^2}=wh$$

$$\frac{1}{r_1}=0, \quad r_2=\frac{y \tan\theta}{\cos\theta}$$

$$\therefore \quad H=r\,wh=y\,\frac{\tan\theta}{\cos\theta}\,wh$$

$$=wy\,\frac{\tan\theta}{\cos\theta}\,(H_1+R\cot\theta-y) \qquad \ldots(13{\cdot}11)$$

For maximum value of T, putting

$$\frac{dT}{dy}=0 \quad \therefore \quad H_1+R\cot\theta-\frac{4}{3}y=0$$

$$y=\tfrac{3}{4}\,(H_1+R\cot\theta)$$

Maximum value of $y=R\cot\theta$

The above condition will be applicable if

$$H_1\leqslant\frac{R\cot\theta}{3}$$

Substituting the value of y in (13·10),

$$T_{max}=\frac{w\tan\theta}{2\cos\theta}\times\frac{3}{4}\,(H_1-R\cot\theta)$$

$$[H_1+R\cot\theta-\tfrac{1}{2}\,(H_1+R\cot\theta)]$$

$$=\frac{2w\tan\theta}{16\cos\theta}[H_1+R\cot\theta] \qquad \ldots(13{\cdot}12)$$

when $\quad H_1=\dfrac{R\cot\theta}{3}$

$$T_{max}=\frac{wR^2}{3\sin\theta} \qquad \ldots(13{\cdot}13)$$

In case $\quad H_1>=\dfrac{R\cot\theta}{3}$

$$T_{max}=\frac{wR}{2\cos\theta}\left[H_1+\frac{R\cot\theta}{3}\right] \qquad \ldots(13{\cdot}14)$$

For maximum value of hoop force, putting

$$\frac{dH}{dy}=0, \; H_1+R\cot\theta-2y=0$$

$$\therefore \quad y=\frac{H_1+R\cot\theta}{2}$$

This condition will be applicable if $H_1\leqslant R\cot\theta$

Substituting this value of y in (13·10)

$$H_{max}=\frac{w\tan\theta}{4\cos\theta}\,[H_1+R\cot\theta]^2 \qquad \ldots(13{\cdot}15)$$

Putting $$H_1 = R \cot \theta$$

$$H_{max} = \frac{wR^2}{\sin \theta} \quad \text{...(13·16)}$$

If $H_1 > R \cot \theta$, $$H_{max} = \frac{wRH_1}{\cos \theta} \quad \text{...(13·17)}$$

When $$H_1 = 0,$$

$$T_{max} = \frac{3\, wR^2}{16 \sin \theta} \quad \text{...(13·18)}$$

$$H_{max} = \frac{wR^2}{4 \sin \theta} \quad \text{...(13·19)}$$

EXERCISES

1. Design a spherical dome having diameter of 12 m at the base and rise of 5 m. The dome is to be designed for a live load of 1500 N/m² uniformly distributed.

2. Design a spherical dome having diameter of 15 m at the base and rise of 6 m. The dome carries lantern load of 20,000 N at crown acting on an area of 1 m diameter.

3. Design a conical roof for a cylindrical tank of 12 m diameter. The rise of the roof is to be 3 m. The live load due to wind etc. may be taken as equivalent of 1500 N/m².

4. Determine the meridional force per unit length and hoop force per unit length at section X-X of the cylindrical tank with hemispherical bottom shown in Fig. 13·12.

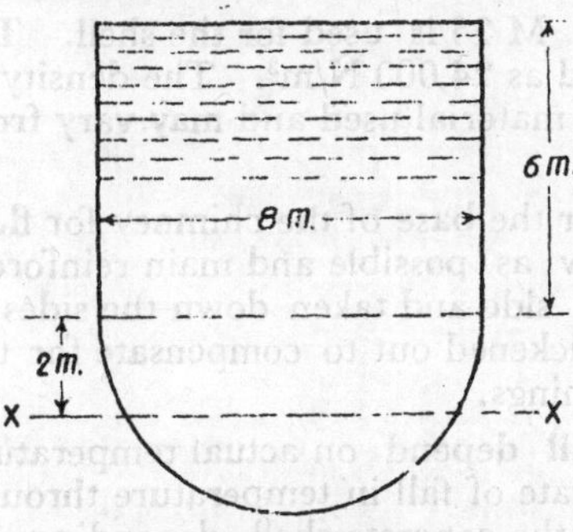

Fig. 13·12

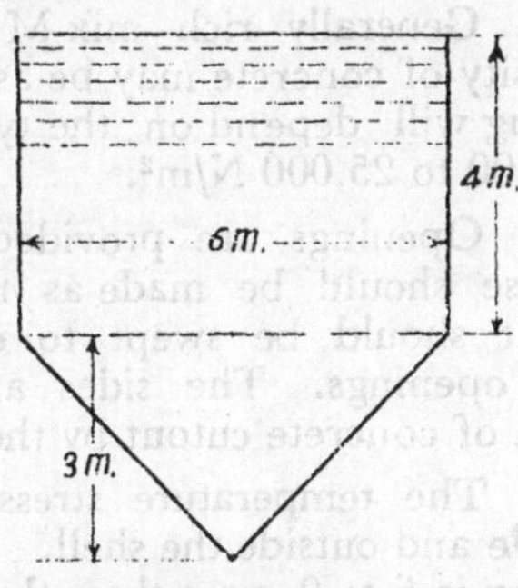

Fig. 13·13

5. Find the maximum meridional force per unit length and maximum hoop force per unit length in the cylindrical tank with conical bottom shown in Fig. 13·13.

14

R.C.C. Chimneys

14·1. Reinforced concrete chimneys are used in preference to masonry chimneys as they are light, their cost is low and they last longer. The foundation is cheaper. Generally circular slab foundation monolithic with shaft is provided.

The factors which govern the design of the shaft of chimney are self-weight, wind pressure and temperature variation.

The section is based on self-weight and wind pressure and checked for stresses due to variation in temperature inside the flue and outside.

The chimneys provided with a brick lining which may be supported by the R.C.C. shell through brackets at intervals or may be independent. There is generally air-gap provided between the R.C.C. shell and the lining.

Generally rich mix M 20 or M 25 is used for the shell. The density of concrete may be assumed as 24,000 N/m^3. The density of lining will depend on the type of material used and may vary from 18,000 to 25,000 N/m^2.

Openings are provided near the base of the chimney for flue. These should be made as narrow as possible and main reinforcement should be swept to either side and taken down the sides of the openings. The sides are thickened out to compensate for the area of concrete cutout by the openings.

The temperature stresses will depend on actual temperature inside and outside the shell. The rate of fall in temperature through lining is 6 to 9 time that through the concrete shell, depending on the kind of the lining used. The rate also depends on the air gap provided. The ratio is greater for smaller air gaps.

Temperature variation causes additional stresses in the horizontal and vertical directions. The horizontal reinforcement provided resists the horizontal stresses due to temperature and shear due to wind. The vertical reinforcement provided resists stresses due to wind, self-load and temperature.

On the windward side only steel is assumed to take tensile stresses, therefore, due to difference in temperature there is increase in tensile stresses in the reinforcement. On the leeward side steel and concrete are in compression. The effect of difference in temperature is to reduce compressive stress in steel and increase the stress in concrete. Stress in steel and on windward side and stress in concrete on the leeward side are the criterion for design.

For wind forces ISI 875 is followed.

The shape factors for chimneys are given in table 14·1.

Table 14·1

Ratio of Height to base width	*0 to 4*	*4 to 8*	*8 or over*
Type of Chimneys	*Shape Factor*		
Circular	0·7	0·7	0·7
Octagonal	0·8	0·9	1·0
Square (wind perpendicular to diagonal)	0·8	0·9	1·0
Square (wind perpendicular to face)	1·0	0·15	1·3

14·2. Stresses in Shafts due to Self-weight and Wind Pressure

Consider a hollow shaft with external diameter D and thickness of steel t_c.

Let W be the self-weight of the shaft above the section consider and P be the resultant wind force acting at h distance from the section considered. The section will be subjected to direct force W, moment $M=Ph$ and shear force P.

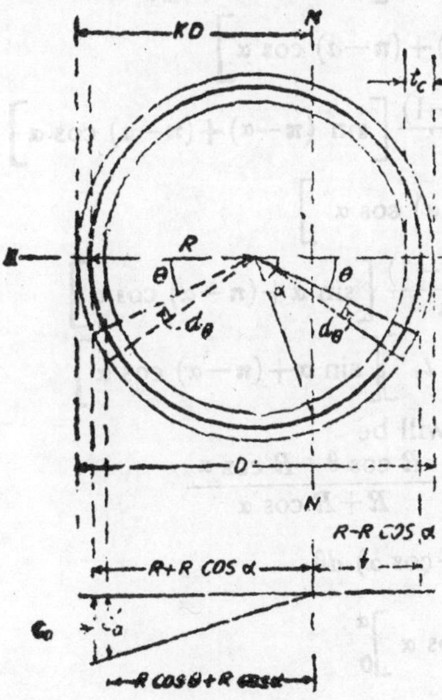

Fig. 14·1

Let A_s be the area of steel reinforcement in the shaft. The effective depth will be assumed to be diameter of the chimney as steel is situated near the end. The steel reinforcement is assumed to be in the form of ring having same area so that of steel reinforcement and situated at the centre of thickness of shell.

Let t_s be the thickness of steel ring

$$2\pi R t_s = A_s$$

$$\therefore t_s = \frac{A_s}{2\pi R},$$

where R is the radius of centre of thickness.

Let $N-A$ be the position of neutral axis at distance kD from the leeward face. The position of neutral axis is also defined by angle α as shown in Fig. 14·1.

$$kD=\frac{D}{2}+R\cos\alpha$$

Let c_a be compressive stress at centre of thickness of shell.

Consider an elementary strip $R\delta\theta$ at an angle θ from X-X. Stress in concrete at level of elementary strip

$$=c_a\times\frac{R\cos\theta+R\cos\alpha}{R+R\cos\theta}$$

Area of the strip $=R\delta\theta\times t_c$. Area of steel in the strip $=R\delta\theta\times t_s$

Total compression in concrete and steel

$$=2\int_0^{\pi-\alpha}(R\delta\theta)\,t_c\times c_a\,\frac{R\cos\theta+R\cos\alpha}{R+R\cos\alpha}$$

$$+2\int_0^{\pi-\alpha}(R\delta\theta)\,t_s\times(m-1)\,c_a\frac{R\cos\theta+R\cos\alpha}{R+R\cos\pi}$$

$$=\frac{2\,Rc_a\,t_c}{1+\cos\alpha}\int_0^{\pi-\alpha}(\cos\theta+\cos\pi)d\theta+\frac{2\,Rc_a\,t_s\,(m-1)}{1+\cos\alpha}\int_0^{\pi-\alpha}(\cos\theta+\cos\alpha)\,d\theta$$

$$=\frac{2\,Rc_a\,t_c}{1+\cos\alpha}\Big[\sin\theta+\cos\alpha\Big]_0^{\pi-\alpha}+\frac{2\,Rc_a\,t_s\,(m-1)}{1+\cos\alpha}\Big[\sin\theta+\theta\cos\alpha\Big]_0^{\pi-\alpha}$$

$$=\frac{2\,R\,c_a\,t_c}{1+\cos\alpha}\Big[\sin(\pi-\alpha)+(\pi-a)\cos\alpha\Big]+\frac{2\,Rc_a\,t_s\,(m-1)}{1+\cos\alpha}\Big[\sin(\pi-\alpha)+(\pi-\alpha)\cos\alpha\Big]$$

$$=\frac{2\,R\,c_a\,t_e}{1+\cos\alpha}\Big[\sin a\,(\pi-\alpha)\cos\alpha\Big]+\frac{2\,R\,c_a\,t_s\,(m-1)}{1+\cos\alpha}\Big[\sin\alpha+(\pi-\alpha)\cos\alpha\Big]$$

$$=\frac{2\,Rc_a}{1+\cos\alpha}\Big[t_c+(m-1)\,t_s\Big]\Big[\sin\alpha+(\pi-\alpha)\cos\alpha\Big]$$

Similarly total tension in steel will be

$$=2\int_0^{\alpha}(R\delta\theta)\,t_c\times mc_a\times\frac{R\cos\theta-R\cos\alpha}{R+R\cos\alpha}$$

$$=\frac{2\,Rc_a\,mt_s}{1+\cos\alpha}\int_0^{\alpha}(\cos\theta-\cos\alpha)\,d\theta$$

$$=\frac{2\,Rc_a\,mt_s}{1+\cos\alpha}\Big[\sin\theta-\theta\cos\alpha\Big]_0^{\alpha}$$

$$=\frac{Rc_a\, mt_s}{1+\cos\alpha}\left[\sin\alpha-\alpha\cos\alpha\right].$$

Equation sum of internal forces to external forces

$$W=\frac{2\,Rc_a}{1+\cos\alpha}\left[t_c+(m-1)\,t_s\right]\left[\sin\alpha+(\pi-\alpha)\cos\alpha\right]$$
$$-\frac{2\,Rc_a\,mt_s}{1+\cos\alpha}\left(\sin\alpha-\cos\alpha\right)$$
$$=\frac{2Rc_a}{1+\cos\alpha}\left[(t_c-t_s)\left\{\sin\alpha+(\pi-\alpha)\cos\alpha\right\}+\pi\, mt_s\cos\alpha\right] \quad \ldots(14.1)$$

Moment of compression in concrete and steel about centre line of section will be

$$=\int_0^{\pi-\alpha}(R\delta\theta)\times t_c\times c_a\times\frac{R\cos\theta+R\cos\alpha}{R+R\cos\alpha}\times R\cos\theta$$
$$+2\int_0^{\pi-\alpha}(R\delta\theta)\times(m-1)\,t_s\times c_a\times\frac{R\cos\theta+R\cos\alpha}{R+R\cos\alpha}\times R\cos\theta$$
$$=\frac{2\,R^2\,c_a\,[t_c+(m-1)\,t_s]}{1+\cos\alpha}\int_0^{\pi-\alpha}(\cos^2\theta+\cos\theta\cos\alpha)\,d\theta$$
$$=\frac{2\,R^2\,c_a\,[t_c+(m-1)\,t_s]}{1+\cos\alpha}\left[\frac{\sin 2\theta}{4}+\frac{\theta}{2}+\sin\theta\cos\alpha\right]_0^{\pi-\alpha}$$
$$=\frac{2\,R^2\,c_a[t_c+(m-1)]\,t_s}{1+\cos\alpha}\left[\frac{\sin(2\pi-2\alpha)}{4}+\frac{(\pi-\alpha)}{2}+\sin(\pi-\pi)\cos\alpha\right]$$
$$=\frac{2\,R^2\,c_a\,[t_c+(m-1)\,t_s]}{1+\cos\alpha}\left[-\frac{\sin 2\alpha}{4}+\frac{(\pi-\alpha)}{2}+\sin\alpha\cos\alpha\right]$$
$$=\frac{2\,R^2\,c_a\,[t_c+(m-1)\,t_s]}{1+\cos\alpha}\left[\frac{\sin 2\alpha}{4}+\frac{\pi-\alpha}{2}\right].$$

Moment of stress in tensile steel about the centre line of the section

$$=2\int_0^{\pi}(R\delta\theta)\times t_s\times m\times c_a\times\frac{R\cos\theta-R\cos\alpha}{R+R\cos\alpha}\times R\cos\theta$$
$$=\frac{2\,R^2\,t_s\,mc_a}{1+\cos\alpha}\int_0^{\alpha}(\cos^2\theta-\cos\theta\cos\alpha)\,d\theta$$
$$=\frac{2\,R^2\,t_s\,mc_a}{1+\cos\alpha}\left[\frac{\sin 2\theta}{3}+\frac{\theta}{2}-\sin\theta\cos\alpha\right]_0^{\alpha}$$
$$=\frac{2\,R^2\,t_s\,mc_a}{1+\cos\alpha}\left[\frac{\sin 2\alpha}{4}+\frac{\alpha}{2}-\frac{\sin 2\alpha}{2}\right]$$
$$=\frac{2\,R^2\,t_s\,mc_a}{1+\cos\alpha}\left[\frac{\alpha}{2}-\frac{\sin 2\alpha}{4}\right]$$

Some of the moments of internal forces about centre of section must be equal to moment of external force.

$$\therefore \quad M=\frac{2R^2c_a}{1+\cos\alpha}\left[t_c+(m-1)\,t_s\right]\left[\frac{\sin 2\alpha}{4}+\frac{(\pi-\alpha)}{2}\right]$$

$$+\frac{2R^2\,t_s\,mc_a}{1+\cos\alpha}\left[\frac{\alpha}{2}-\frac{\sin 2\alpha}{2}\right]$$

$$=\frac{2\,R^2\,c_a}{1+\cos\alpha}\left[(t_c-t_s)\left(\frac{\sin 2\alpha}{4}+\frac{\pi-\alpha}{2}\right)+\frac{mt_s\pi}{2}\right] \quad ...(14\cdot2)$$

Eccentricity $e=M/W$

$$=\frac{\dfrac{2R^2\,c_a}{1+\cos\alpha}\left[\left(t_c-t_s\right)\left(\dfrac{\sin 2\alpha}{4}+\dfrac{\pi-\alpha}{2}\right)+\dfrac{mt_s\pi}{4}\right]}{\dfrac{2R\,c_a}{1+\cos\alpha}\left(t_c-t_s\right)\left\{\sin\alpha+(\pi-\alpha)\cos\alpha\right\}+\pi mt_s\cos\alpha}$$

$$\therefore \quad e=R\frac{\left[\left(t_c-t_s\right)\left(\dfrac{\sin 2\alpha}{4}+\dfrac{\pi-\alpha}{2}\right)+\dfrac{mt_s\pi}{2}\right]}{(t_c-t_s)[\sin\alpha+(\pi-\alpha)\cos\alpha]+\pi mt_s\cos\alpha} \quad ...(14\cdot3)$$

The value of α which satisfies equation (14·3) is found by trial and error. Knowing value of α, position of neutral axis is known. From equation (14·1), value of c_a can be found. Knowing stress in concrete, stresses in tensile steel and compressive steel can be determined.

14·3. Stresses in Horizontal Reinforcement due to Shear Force

Let P be the shear force at the section considered.

Let D be the diameter of the chimney.

Let S mm be the pitch of horizontal reinforcement and A_t be the area of horizontal reinforcement in S mm.

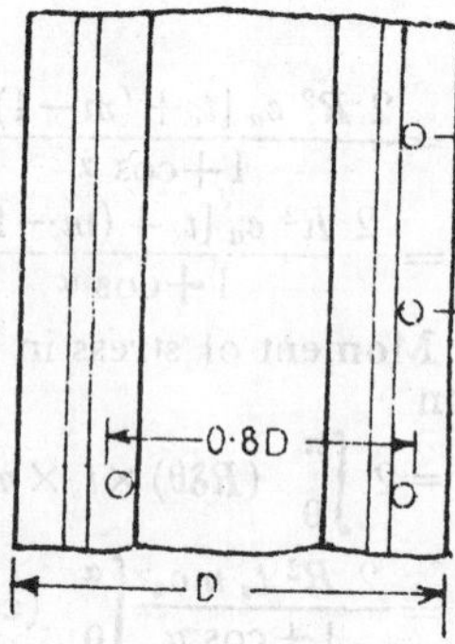

Fig. 14·2

Area of steel resisting shear in one metre height

$$=2\,A_t\times\frac{1000}{S}$$

If t is the stress in steel, force resisted

$$=2\;A_t\times\frac{1000}{S}\times t.$$

It is assumed that horizontal distance between reinforcement on both sides is 0·8 D.

Shear per one metre $=\dfrac{P}{b\times a}\times b\times 1000=\dfrac{1000\,P}{a}$

Assuming lever arm $a=0{\cdot}8\ D$

$$\text{Shear}=\frac{1000\ P}{0{\cdot}8\ D}$$

$$\therefore \quad 2A_t\times\frac{1000\times t}{S}=\frac{1000\ P}{0{\cdot}8\ D}$$

$$t=\frac{P\times S}{1{\cdot}6\ A_t\ D}$$

D is in mm S in mm and A_t is in mm².

14·4. (*A*) Stresses Due Difference of Temperature

In a chimney there is difference in temperature between inside face and outside face. If the two faces were free to expand inside face will expand more than outside face. This expansion will cause the section to curve cutside. In case of chimneys this differential expansion is prevented. The only movement possible is the increase in length in the vertical direction. Due to prevention of free expan-sion stresses are caused in the section.

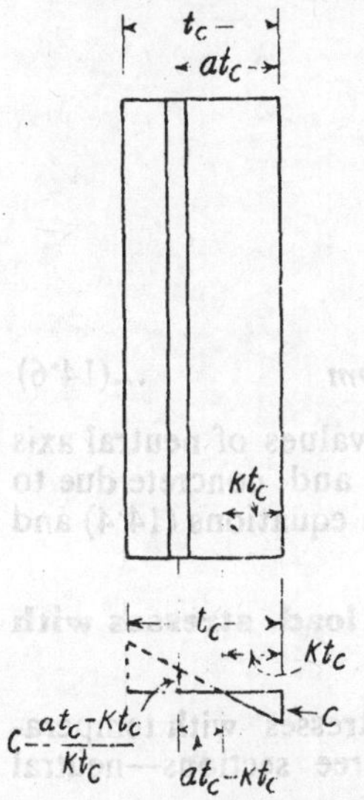

Consider a vertical section through an element of concrete shaft. Let T°C be the temperature difference between inside and outside faces. The temperature gradient is assumed to be a straight line.

Let e be the strain due to temperature difference of T°C. The coefficient of expansion for steel and concrete is the same. Let it be α.

Rise of temperature in reinforcement due to difference of temperature of T°C is $(1-a)\ T$.

Free expansion of steel$=(1-a)\ \alpha T$, where α is coefficient of expansion. The tenile stress in steel is due to the difference between that due to strain e and due temperature rise $(1-a)\ T$

$\therefore$ Stress in steel $t=E_s\ [e-(1-a)\ \alpha T]$.

Maximum stress in concrete will be on the inside.

Maximum stress in concrete

$$c=E_c\ (T\alpha-e)$$

Let neutral axis be at kt_c from inner side.

At the neutral axis, free expansion will be equal to strain e.

$$\therefore \quad (1-k)\ T\alpha=e$$

$$\therefore \quad t = E_s\,[(1-k)\,T\alpha - (1-a)\,T\alpha]$$

$$= E_s\,\alpha\,T\,(a-k) \quad \ldots(14{\cdot}4)$$

$$c = E_c\,[T\alpha - (1-k)\,T\alpha]$$

$$= E_c\,T\alpha k \quad \ldots(14{\cdot}5)$$

Total force in the section is not altered due to temperature difference.

$$\therefore \quad \tfrac{1}{2}c \times kt_c = t_s \times t$$

where t_s is area of steel reinforcement per unit width equal to $A_s/2\pi R$.

Let $\quad t_s = pt_c$

$$\therefore \quad \tfrac{1}{2}c \times kt_c = pt_c \times t$$

$$\therefore \quad t = \frac{ck}{2p}$$

But $\quad t = mc\,\dfrac{at_c - kt_c}{kt_c} = mc\,\dfrac{a-k}{k}$

$$\therefore \quad \frac{ck}{2p} = mc\,\frac{a-k}{k}$$

$$\therefore \quad k^2 = 2pm\,(a-k)$$

$$k^2 + 2pmk = 2pma$$

$$(k+pm)^2 = 2pma + p^2m^2$$

$$\therefore \quad k = +\sqrt{2pma + p^2m^2} - pm \quad \ldots(14{\cdot}6)$$

Knowing the values of p, m and a, the values of neutral axis constant k can be calculated. Stresses in steel and concrete due to temperature difference can be calculated from equations (14·4) and (14·5).

(*B*) Combined effect of wind and self-load stresses with temperature difference stresses

Combined effect of wind and self-load stresses with temperature difference stresses is to be considered at three sections—neutral axis, compression zone and tension zone.

1. *At Neutral axis*

At neutral axis there are no stresses due to external forces, only there will be stresses due to temperature difference.

Stress in steel $= E_s\alpha T(a-k)$

Stress in concrete $= E_c\alpha Tk.$

The value of k can be found from Eq. (14·6).

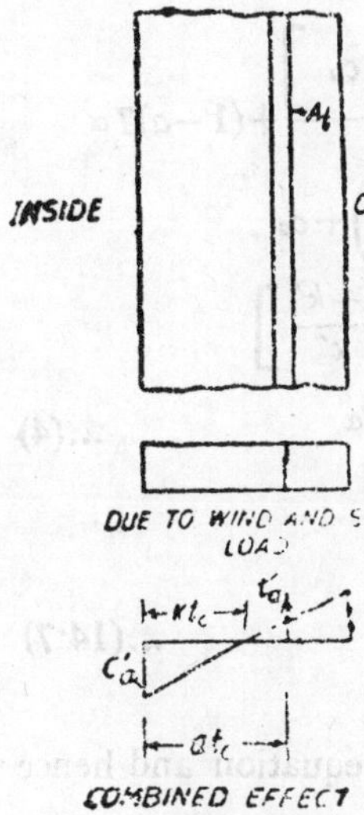

Fig. 14·4

2 *On Leeward side*

Whole thickness is in compression due to effect of self-load and wind.

Let c_a be the compressive stress in concrete, assumed uniform over the thickness.

Due to effect of temperature difference, the compressive stress in steel will decrease and compressive stress in concrete will increase.

Let $k't_c$ be the position of neutral axis due to combined effect and c_a' be compressive stress in concrete.

There is no change in total compression due to self-load and wind load

$$\therefore\ c_a t_c+(m-1)t_s \times c_a=\frac{c_a'-k't_c}{2}-t_s \times t_a'$$

Let $\qquad t_s=pt_s$

$$\therefore\ c_a t_c+(m-1)pt_c \times c_a=\frac{c_a' \times k't_c}{2}-pt_c \times m \times c_a' \times \left(\frac{a-k'}{k'}\right)$$

$$\therefore\quad c_a t_c[1+(m-1)p]=c_a' t_c\left[\frac{k'}{2}-mp\left(\frac{a-k'}{k'}\right)\right]$$

$$\therefore\quad c_a[1+(m-1)p]=c_a'\left[\frac{k'}{2}-mp\left(\frac{a-k'}{k'}\right)\right] \qquad \ldots(1)$$

$$\therefore\quad c_a'=\frac{c_a[1+m-1)p]}{\frac{k'}{2}-mp\left(\frac{a-k'}{k'}\right)}$$

Change in stress in concrete at inside face

$$=c_a'-c_a$$

$$\therefore\quad c_a'-c_a=c=E_c[T\alpha-e] \qquad \ldots(2)$$

Change in stress in steel $=t_a'+mc_a$

$$\therefore\quad t_a'+mc_a=t=E_s[e-(1-a)T\alpha] \qquad \ldots(3)$$

But $\qquad t_a'=c_a' m\left(\frac{a-k'}{k'}\right)$

$$\therefore\ \text{From (3)},\quad \frac{c_a' m\left(\frac{a-k'}{k'}\right)+mc_a}{E_s}+(1-a)\,T\alpha=0$$

$$\text{From (2)},\quad T\alpha-\frac{c_a'-c_a}{E_s}=e$$

$$\therefore \quad T\alpha - \frac{c_a' - c_a}{E_c} = \frac{m\left[c_a'\left(\dfrac{a-k'}{k'}\right)+c_a\right]}{E_s} + (1-a)T\alpha$$

$$\therefore \quad E_c[T\alpha - (1-a)T\alpha] = c_a' - c_a + c_a'\left(\frac{a-k'}{k'}\right) + c_a$$

$$E_c a T\alpha = c_a' + \left[c_a'\left(\frac{a-k'}{k'}\right)\right] = c_a'\left[1+\frac{a-k'}{k'}\right]$$

$$\therefore \quad c_a' = \frac{E_c a T\alpha}{\left[1+\dfrac{a-k'}{k'}\right]} = E_c k' T\alpha \qquad \text{...(4)}$$

Equating (1) and (4)

$$\frac{c_a[1+(m-1)p]}{\dfrac{k'}{2} - mp\left(\dfrac{a-k'}{k'}\right)} = E_c k' T\alpha \qquad \text{...(14·7)}$$

The value of k' can be found from above equation and hence stresses c_a' and t_a' can be calculated.

In case value of k' comes from more than t_c whole thickness of shell is in compression and the stress can be analysed in the same way as discussed above.

3. *On windward side*

Whole thickness is in tension due to effect of self-load and wind. Concrete does not take any tension and whole of tension is taken by steel. Let t_a be tensile stress in steel. Due to effect of temperature difference, some compressive stress will develop in concrete on inside face and tensile stress in steel will increase.

Let $k' t_c$ be the position of neutral axis due to combined effect and c_a' be compressive stress in concrete.

There is no change in total force on the section due to temperature difference.

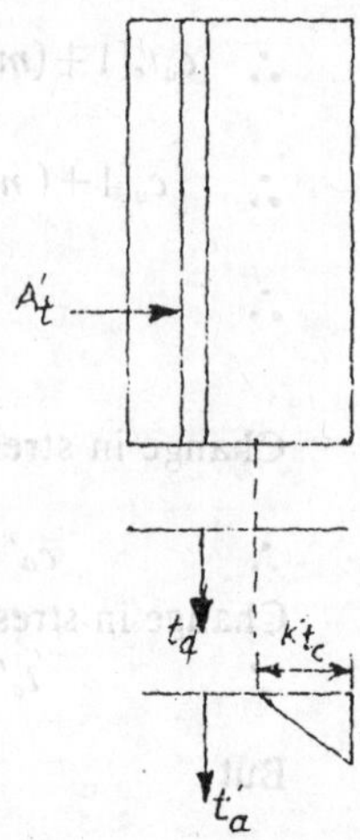

Fig. 14·5

$$t_s \times t_a = t_s \times t_a' - \frac{c_a'}{2} \times k' t_c$$

But $\qquad t_a' = mc_a'\left(\dfrac{a-k'}{k'}\right)$

Let $\qquad t_s = pt_c$

$$\therefore \quad pt_c\, t_a = pt_c \times mc'a\left(\frac{a-k'}{k'}\right) - \frac{c_a'}{2} \times k' t_c$$

$$\therefore \quad pt_a = \left[pm\left(\frac{a-k'}{2}\right) - \frac{k'}{2} \right] c_c'$$

$$\therefore \quad c_c' = \frac{pt_a}{\left[pm\left(\frac{a-k'}{k'}\right) - \frac{k'}{2} \right]} \qquad \ldots(1)$$

Change in strain in concrete at inside face

$$= \frac{t_a}{E_s} + \frac{c_a'}{E_s}$$

$$\frac{t_a}{E_S} + \frac{c_c'}{E_c} = T\alpha - e$$

$$\therefore \quad e = T\alpha - \frac{t_a}{E_s} - \frac{c_s'}{E_s} \qquad \ldots(2)$$

Change in strain in steel

$$\frac{t_o' - t_a}{E_s} = [e - (1-a)\alpha T]$$

$$\therefore \quad e = \frac{t_a' - t_a}{E_s} + (1-a)\alpha T \qquad \ldots 3)$$

Equating (2) and (3)

$$T\alpha - \frac{t_a}{E_s} - \frac{c_a'}{E_c} = \frac{t_a' - t_a}{E_s} + (1-a)\alpha T$$

$$a\alpha T - \frac{c_a'}{E_s} = \frac{t_a'}{E_s}$$

$$c_a' = \left[a\alpha T - \frac{t_a'}{E_s} \right] E_c$$

$$= a\alpha T E_c - \frac{t_a'}{m}$$

Substituting the value of t_a'

$$\therefore \quad c_a' = a\alpha T_c - c_a' \left(\frac{a-k'}{k'}\right)$$

$$\therefore \quad \frac{a}{k'} c_a' = a\alpha T E_c$$

$$\therefore \quad c_a' = \alpha T E_c k' \qquad \ldots 4)$$

Equating (1) and (4)

$$\frac{pt_c}{pm\left(\frac{a-k'}{k'}\right) - \frac{k'}{2}} = \alpha T E_c k \qquad \ldots(14\cdot 8)$$

From the above equation value of k' can be found, knowing the value of k' stresses c_a' and t_a' can be calculated.

4. Stresses in horizontal reinforcement due to temperature rise

Due to difference in temperature the inner surface being at higher temperature is prevented from natural expansion and, therefore, will get compressed and outer surface will expand more than natural expansion and will be in tension. Tensile stresses will be produced in horizontal reinforcement.

Consider 1 mm height of the shell. The compressive force in concrete on the inner side will be equal to the tensile force in the horizontal reinforcement. Let $k't_c$ be the position of natural axis.

$$\therefore \quad \frac{c_a'}{2} k't_c = A't \times t_a'$$

where $A't$ is area of horizontal reinforcement per unit height.

If A_t is the cross-sectional area of horizontal reinforcement

S is the spacing, $\quad A_t' = \frac{A_t}{S} = pt_c$

$$t_a' = mc_a'\left(\frac{a'-k'}{k'}\right) \qquad \text{...(1)}$$

$$\therefore \quad \frac{c_a'}{2} k't_c = pt_c \times mc_a'\left(\frac{a'-k'}{k'}\right)$$

$$\therefore \quad \frac{k'}{2} = mp\left(\frac{a'-k'}{k'}\right)$$

$$k'^2 + 2pmk' = 2\,pma'$$

$$(k'+pm)^2 = 2pma' + p^2m^2$$

$$\therefore \quad k' = \sqrt{2pma' + p^2m^2} - pm \qquad \text{...(2)}$$

Let e be the actual strain

Stress in concrete $\quad = c'a = (\alpha T - e)E_c \qquad$ ...(3)

Stress in steel $\quad = t_a' = [e - \alpha T(1-a')]E_s \qquad$ ...(4)

From (3) $\quad e = \alpha T - \frac{c_a'}{E_s}$

From (4) $\quad e = \frac{t'_a}{E_s} + \alpha T(1-a')$

Equating, $\alpha T - \frac{c'_a}{E_c} = \frac{t'_a}{E_s} + \alpha T(1-a)$

$$\frac{t_a'}{E_s} + \frac{c_a'}{E_c} = \alpha T a'$$

$$t_a' + mc_a' = F_s \alpha T a' \qquad \text{...(14·9)}$$

From this, values of t'_a and c'_a can be found by making use of equations (1) and (2).

Ex. 14·1. *Design a concrete chimney of 60 m height having external diameter of 4 m throughout.*

Sol. 10 cm thick lining is provided, which is supported at every 6 m. Air gap of 150 mm is provided between the lining and inside of shell.

The temperature difference between inside and outside will be assumed at 70°C.

Mix M 25 is used, $c = 8{\cdot}5\ \text{N/mm}^2$, $m = 11$.

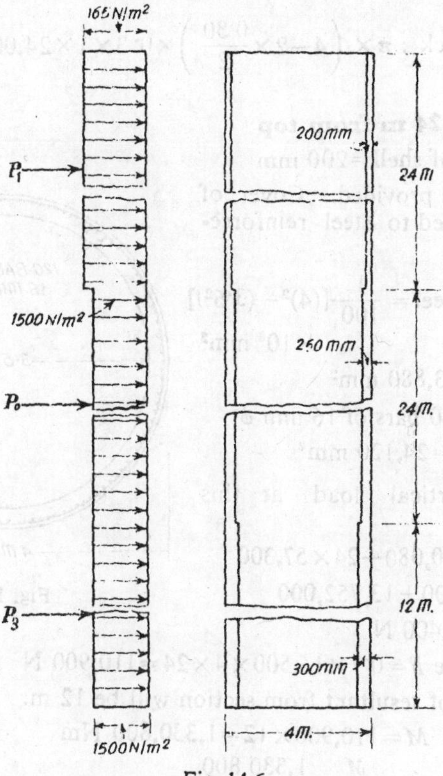

Fig. 14·6

The height of chimney is divided in three portions of 24 m, 24 m and 17 m. Thickness of shell provided is 200 mm for top 24 m 250 mm for next 24 m and 300 mm for the bottom 12 m.

Wind pressure is calculated for each of the three portions. The intensity of wind pressure is calculated at the mid height of the portions and will be assumed to be constant for the portion. The intensity of wind pressure for bottom two portions is 1500 N/m² and that for topmost portion is 1650 N/m² for Delhi area. Density of lining is assumed as 20,000 N/m³

Weight of lining for 6 m length

$$= \pi[4 - 2(0{\cdot}2 + 0{\cdot}15 + 0{\cdot}05)] \times 0{\cdot}1 \times 6 \times 20{,}000$$
$$= 120{,}800 \text{ N}$$

Weight of concrete per metre height

$$200 \text{ mm thick} : \pi \times \left(4-2\times\frac{0\cdot2}{2}\right)\times0\cdot2\times1\times24{,}000=57{,}300 \text{ N}$$

$$250 \text{ mm thick} : \pi \times \left(4\times2\times\frac{0\cdot25}{2}\right)\times0\cdot25\times1\times24{,}000$$

$$=70{,}700 \text{ N}$$

$$300 \text{ mm thick} : \pi \times \left(4-2\times\frac{0\cdot30}{2}\right)\times0\cdot3\times1\times24{,}000$$

$$=83{,}200 \text{ N}$$

Section at 24 m from top

Thickness of shell = 200 mm

1% steel is provided. Cover of 50 mm is provided to steel reinforcement.

$$\text{Area of steel} = \frac{1}{100}[(4)^2-(3\cdot6^2)]\times10^6 \text{ mm}^2$$

$$=23{,}880 \text{ mm}^2$$

Provide 120 bars of 16 mm ϕ

Steel area = 24,120 mm²

Total vertical load at this section

$$W=4\times120{,}080+24\times57{,}300$$

$$=483{,}200+13{,}752{,}000$$

$$=1{,}858{,}400 \text{ N}$$

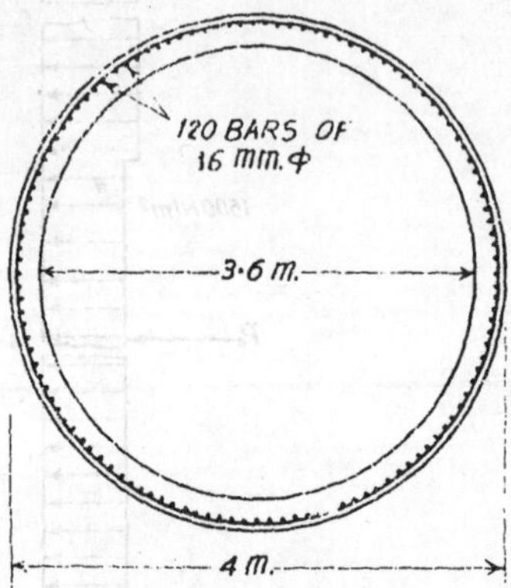

Fig. 14·7

Wind force $P=0\cdot7\times16{,}500\times4\times24=110{,}900$ N

Distance of resultant from section will be 12 m.

$$M=110{,}900\times12=1{,}330{,}800 \text{ Nm}$$

$$\text{Eccentricity } e=\frac{M}{W}=\frac{1{,}330{,}800}{1{,}858{,}400}$$

$$=0\cdot716 \text{ m}=716 \text{ mm.}$$

Check of section for stresses

$$\text{Equivalent area} =\frac{\pi}{4}[(4)^2-(3\cdot6)^2]\times10^6+(m-1)\times24{,}120$$

$$=2{,}388{,}000+10\times24{,}120=2{,}629{,}200 \text{ mm}^2$$

Equivalent thickness of steel ring

$$t_s=\frac{24{,}120}{\pi\times3900}=2 \text{ mm nearly}$$

Equivalent moment of inertia

$$=\frac{\pi}{64}(D^4-d^3)+(m-1)\times\frac{\pi\times3900}{2}\times t_s\times\left(\frac{3900}{2}\right)^2$$

$$=\frac{\pi}{64}[(4000)^4-(3600)^4]+10\times\frac{\pi\times(3900)^3}{8}$$

$$=(430,400,000+46,600,000)10^4$$

$$=477\times10^{10}\ \text{mm}^4.$$

For no tension to develop allowable eccentricity

$$=\frac{2I}{AD}$$

$$=\frac{2\times477\times10^{10}}{2,629,200\times4000}=907\ \text{mm}.$$

Actual eccentricity is less than this value, hence entire section is in compression.

Maximum compressive stress in concrete $=\frac{W}{A}+\frac{M}{I/D/2}$

$$=\frac{1,858,400}{26,292}+\frac{13,308,000\times1000\times2000}{477\times10^{10}}$$

$$=0{\cdot}7069+0{\cdot}279=0{\cdot}9859\ \text{N/mm}^2.\ \text{Safe}$$

Section at 48 m from top

Thickness of shell $=250$ mm

Provide 1% steel

Area of steel $=\frac{1}{100}\times\frac{\pi}{4}[(4000)^2-(3500)^2]$

$$=\frac{\pi}{400}\times7500\times500=29,450\ \text{mm}^2$$

Provide 120 bars of 20 mm ϕ.

Area of steel provided $=37,680\ \text{mm}^2$

Total vertical weight = Weight of lining + Weight of top 24 m of shell + Weight of next 24 m of shell

$$=8\times120,800+1,375,200+24\times70,700$$

$$=966,400+1,375,200+1,696,800$$

$$=4,038,400\ \text{N}$$

Moment $=110,900\times36+0{\cdot}7\times1500\times24\times4\times12$

$$=3,992,400+1,209,600=5,202,000\ \text{Nm}$$

Eccentricity $=\frac{M}{W}=\frac{5,202,000}{4,308,400}$

$$=1{\cdot}288\ \text{m}=1288\ \text{mm}$$

Check of section for stresses

Equivalent area A

$$=\frac{\pi}{4}[(4000)^2-3500^2]+(m-1)\times 37{,}680$$

$$=\frac{\pi}{4}\times 7500\times 500+10\times 37{,}680$$

$$=29{,}450{,}000+376{,}800$$

$$=3{,}321{,}800 \text{ mm}^2$$

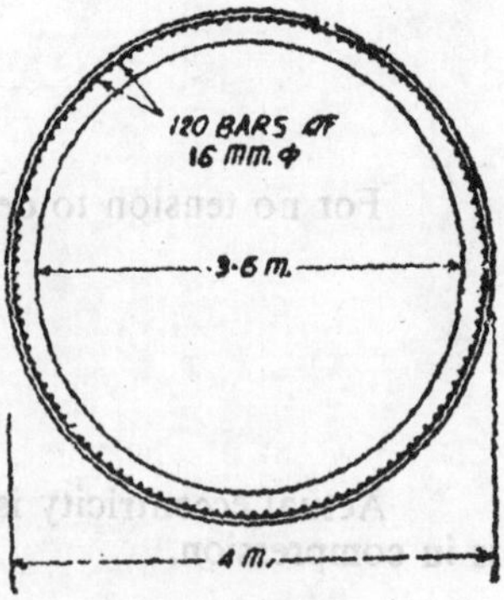

Fig. 14·8

Equivalent thickness of steel ring

$$t_s=\frac{37{,}680}{\pi\times 3900}=3{\cdot}1 \text{ mm}$$

Equivalent moment of intertia

$$=\frac{\pi}{64}\left[(4000)^4-(3500^4)\right]+\frac{10\times\pi\times(3900)^3}{8}\times 3{\cdot}1$$

$$=(520{,}000{,}000+72{,}240{,}000)\times 10^4$$

$$=592{\cdot}25\times 10^{10} \text{ mm}^4.$$

Allowable eccentricity for no tension to develop

$$=\frac{2I}{AD}=\frac{2\times 592{\cdot}24\times 10^{10}}{3{,}321{,}800\times 4000}=891{\cdot}3 \text{ mm}$$

Actual eccentricity is more than this. There will be some tension induced.

Maximum and minimum stresses are given by

$$P=\frac{4{,}166{,}104}{3{,}321{,}800}\pm\frac{5{,}202{,}000\times 1000\times 2000}{592{,}24\times 10^{10}}$$

$$=1{\cdot}25\pm 1{\cdot}756$$

$$P_{max}=2{\cdot}971 \text{ N/mm}^2$$

$$P_{min}=-0{\cdot}441 \text{ N/mm}^2.$$

The tension development is very small. Allowable tension in concrete is 0·8 N/mm². Hence section is safe.

Section at base

Provide 1% steel,

$$\text{Steel area}=\frac{1}{100}\times\frac{\pi}{4}\left[(4000)^2-(3400)^2\right]$$

$$=\frac{\pi}{400}\times 7400\times 600=348{,}100 \text{ mm}^2$$

Provide 120 bars of 20 mm ϕ,

$$A_t=37{,}680 \text{ mm}^2$$

Total vertical load $=10\times 120{,}800+24\times 59{,}690+24\times 73{,}631$
$+12\times 87{,}179$

$=5,453,852$

Moment M $=1650\times0{\cdot}7\times4\times24\times48+1500\times0{\cdot}7\times4\times24\times24+1500\times0{\cdot}7\times4\times12\times6$

$=5,322,000+2,419,200+302,400=8,403,600$ Nm

Eccentricity $e=\dfrac{8,043,600}{5,453,852}=1{\cdot}475$ m $=1475$ mm

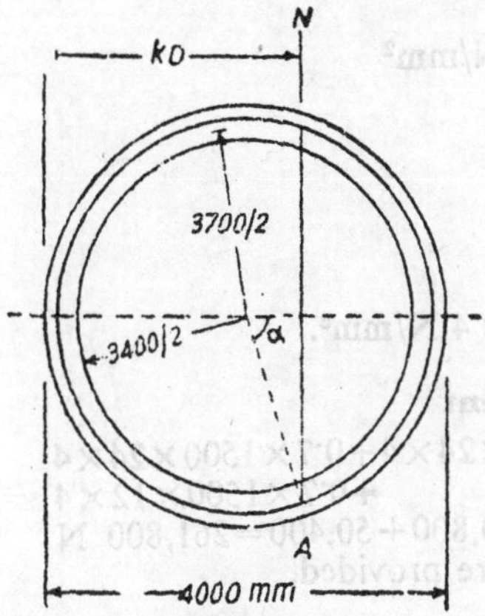

Fig. 14·9

As the eccentricity is quite considerable, part of the section will be in tension.

Check for Stresses

Steel area is assumed to be in the form of circular ring at the centre of thickness.

Let t_s be thickness of steel ring

$\pi\times3700\times t_s=37,680$

$$t_s=\frac{37,980}{3700\ \pi}=3{\cdot}2 \text{ mm}$$

$$t_c=300 \text{ mm}$$

$$m=11.$$

Let the N.A. be at a distance *kD*. The position of neutral axis is denoted by angle α as shown in Fig. 14·9.

$$e=\frac{R\left[(t_c-t_s)\left(\frac{\sin 2\alpha}{4}+\frac{\pi-\alpha}{2}\right)+\frac{m\,t_s\,\pi}{2}\right]}{(t_c-t_s)[\sin\alpha+(\pi-\alpha)\cos\alpha]+\pi m t_s\cos\alpha}$$

$$=\frac{1850\left[(300-3{\cdot}2)\left(\frac{\sin 2\alpha}{4}+\frac{\pi-\alpha}{2}\right)+11\times\frac{3{\cdot}2\times\pi}{2}\right]}{(300-3{\cdot}2)[\sin\alpha+(\pi-\alpha)\cos\alpha]+\pi\times11\times3{\cdot}2\times\cos\alpha}$$

$$=\frac{1850\left[296{\cdot}8\left(\frac{\sin 2\alpha}{2}+\frac{\pi-\alpha}{2}\right)+0{\cdot}553\right]}{296{\cdot}8\,[\sin\alpha+(\pi-\alpha)\cos\alpha]+110{\cdot}6\cos\alpha}$$

Putting $\alpha=78°$.

$$e=\frac{1850\left[296{\cdot}8\left(\frac{0{\cdot}4067}{4}+\frac{\pi-1{\cdot}36}{2}\right)+55{\cdot}3\right]}{296{\cdot}8[0{\cdot}9781+(\pi-1{\cdot}36)\times0{\cdot}2079]+110{\cdot}6\times0{\cdot}2097}$$

$=1529$ mm

This nearly agrees with actual eccentricity

$$W=\frac{2Rc_a}{1+\cos\alpha}\,[(t_c-t_s)\{\sin\alpha+(\pi-\alpha)\cos\alpha\}+\pi m t_s\cos\alpha]$$

$$5{,}453{,}852=\frac{3700\times c_a}{1+0{\cdot}2079}\left[(300-3{\cdot}2)\{0{\cdot}9781+(\pi-1{\cdot}36)0{\cdot}2079\}+\pi\times 11\times 3{\cdot}2\times 0{\cdot}2079\right]$$

$$=\frac{3700\times c_a}{1{\cdot}2079}\left[296{\cdot}8\times 1{\cdot}3486+21{\cdot}1\right]$$

$$=\frac{3700-c_a}{1{\cdot}2079}\times 423{\cdot}14$$

$$\therefore\ c_a=\frac{5{,}453{,}852\times 1{\cdot}2079}{3700\times 423{\cdot}14}=4{\cdot}208\ \text{N/mm}^2$$

$$t_a=mc_a\times\frac{R(1-\cos\alpha)}{R(1+\cos\alpha)}$$

$$=mc_a\times\frac{(1-\cos\alpha)}{(1+\cos\alpha)}$$

$$=\frac{11\times 4{\cdot}208(1-0{\cdot}2079)}{1+0{\cdot}2079}=30{\cdot}4\ \text{N/mm}^2.$$

Design of horizontal reinforcement

Total horizontal force $=0{\cdot}7\times 1650\times 24\times 4+0{\cdot}7\times 1500\times 24\times 4+0{\cdot}7\times 1500\times 12\times 4$

$=110{,}600+100{,}800+50{,}400=261{,}800$ N

12 mm ϕ hoops at 100 mm centres are provided.

Stress in steel $\quad t=\dfrac{P\times S}{1{\cdot}6\times A_t D}$

$$=\frac{261{,}800\times 100}{1{\cdot}6\times 113\times 4000}=36{\cdot}14\ \text{N/mm}^2.$$

Stresses due to temperature difference

Leeward side

$$a=\frac{5}{6}=0{\cdot}833$$

$$p=\frac{37{,}680/\pi\times 3700}{300}=0{\cdot}0108$$

$$E_c=\frac{200}{11}\ \text{kN/mm}^2$$

$$=200{,}000\ \text{N/mm}^2$$

$$T=70°\text{C}$$

$$\alpha=11\times 10^{-6}/°\text{C}$$

$$c_a=4{\cdot}208\ \text{N/mm}^2$$

$$\frac{c_a[1+(m-1)p]}{\dfrac{k'}{2}-mp\left(\dfrac{a-k'}{k}\right)}=E_e k' T\alpha$$

$$\therefore\ \frac{4{\cdot}208[1+10\times 0{\cdot}0108]}{0{\cdot}5\,k'-11\times 0{\cdot}01080\left(\dfrac{0{\cdot}833-k'}{k'}\right)}$$

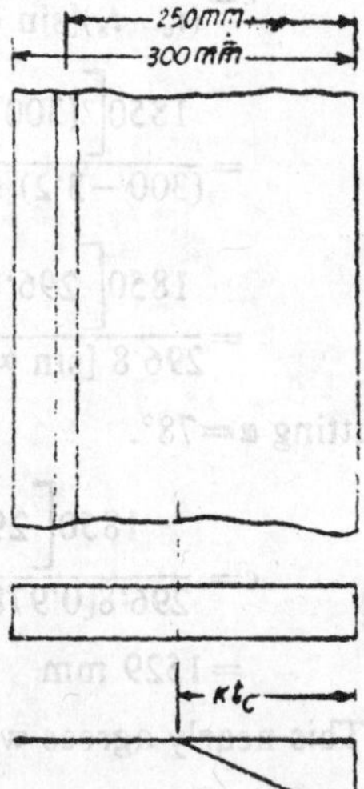

Fig. 14·10

$$=\frac{2\times10^5}{11}\times k'\times70\times11\times10^{-6}$$

$$\therefore\quad \frac{4.662}{0.5k'^2-0.1+0.1188k}=14$$

$$\therefore\quad \frac{4.662}{14}=0.5k'^2-0.1+0.1188\,k'$$

$$\therefore\quad k'^2+0.2376\ k'=0.866$$

$$\therefore\quad (k'+0.1188)^2=0.866+0.0141=0.8801$$

$$\therefore\quad k'+0.1188=0.938$$

$$\therefore\quad k'=0.8192$$

$$c_a'=E_c k' T\alpha$$

$$=\frac{2\times10^5}{11}\times0.8192\times70\times11\times10^{-6}$$

$$=11.47\ \text{N/mm}^2$$

Allowable stress $=\frac{4}{3}\times9.5$

(As wind stresses are taken into account)

$=11.33$ N/mm² (slightly less than actual)

$$t_a'=mc_a'\left(\frac{a-k'}{k'}\right)$$

$$=11\times11.47\left(\frac{0.833-0.8192}{0.8192}\right)$$

$$=2.13\ \text{N/mm}^2.$$

Windward Side

$$\frac{pt_a}{pm\left(\frac{a-k'}{k'}\right)-\frac{k'}{2}}=aTE_ck'$$

$$t_a=30.4\ \text{N/mm}^2$$

$$\frac{0.0108\times30.4}{0.0108\times11\left(\frac{0.833-k'}{k'}\right)-0.5\ k'}=11\times10^{-6}\times70\times\frac{2\times10^5}{11}\times k'$$

$$=14\ k'$$

$$0.023=0.099-0.1188\ k'-0.5\ k'^2$$

$$k'^2+0.2376\ k'=0.152$$

$$(k'^2+0.1188)^2=0.152+0.0141$$

$$k'+0.1188=0.1661$$

$$\therefore\quad k'=0.2888$$

$$c_a'=\alpha TE_ck'$$

$$=11\times10^{-6}\times70\times\frac{2\times10^5}{11}\times0.2888$$

$$=4.04\ \text{N/mm}^2$$

$$t_a'=mc_a'\left(\frac{a-k'}{k'}\right)$$

$$=11\times4{\cdot}04\left(\frac{0{\cdot}833-0{\cdot}2888}{0{\cdot}2888}\right)$$

$$=83{\cdot}74 \text{ N/mm}^2. \quad \text{Safe.}$$

At Neutral axis

$$k=\sqrt{2\,pma+p^2m^2}-pm$$

$$=\sqrt{2\times0{\cdot}0108\times11\times0{\cdot}833+(0{\cdot}0108\times11)^2}-0{\cdot}0108\times11$$

$$=\sqrt{0{\cdot}198+0{\cdot}0141}-0{\cdot}1188$$

$$=0{\cdot}4612-0{\cdot}1188=0{\cdot}3424$$

$$t_a'=E_s\alpha T\ (a-k)$$

$$=2\times10^5\times11\times10^{-6}\times70\times(0{\cdot}833-0{\cdot}3424)$$

$$=1{\cdot}4\times11\times0{\cdot}4906=75{\cdot}55 \text{ N/mm}^2$$

$$c_a'=E_cT\alpha k$$

$$=\frac{2\times10^5}{11}\times11\times10^{-6}\times70\times0{\cdot}3424$$

$$=14\times0{\cdot}3424=4{\cdot}79 \text{ N/mm}^2. \quad \text{Safe.}$$

Stresses in horizontal reinforcement due to temperature difference

Reinforcement used is 12 mm ϕ at 100 mm centres.

$$p=\frac{A_s}{S\times t_e}=\frac{113}{100\times3\times100}=0{\cdot}00377.$$

$$a'=0{\cdot}833$$

$$m=11$$

$$k'=\sqrt{2\times0{\cdot}00377\times11\times0{\cdot}833+(11\times0{\cdot}00377)^2}-0{\cdot}00377\times11$$

$$=0{\cdot}3351-0{\cdot}04150=0{\cdot}2936$$

$$t_a'+11\,c_a'=2\times10^6\times11\times10^{-6}\times70\times0{\cdot}833=128{\cdot}3$$

But $$t_a'=mc_a'\left(\frac{a-k'}{k'}\right)$$

$$=11\times c_a'\times\left(\frac{0{\cdot}838-0{\cdot}2936}{0{\cdot}2936}\right)$$

$$=\frac{11\times0{\cdot}5394}{0{\cdot}2936}\times c_a'=20{\cdot}2\,c_a'$$

$$20{\cdot}2\,c_a'+11\,c_a'=128{\cdot}3$$

$$c_a'=\frac{128{\cdot}3}{31{\cdot}2}=4{\cdot}11 \text{ N/mm}^2$$

$$t_a'=20{\cdot}2\times4{\cdot}11=83{\cdot}02 \text{ N/mm}^2$$

Total stress in steel $=83{\cdot}02+36{\cdot}14$

$=119{\cdot}16$ N/mm². Safe.

Design of Chimney Base

Total vertical load at the base

$=5{,}453{,}852$ N

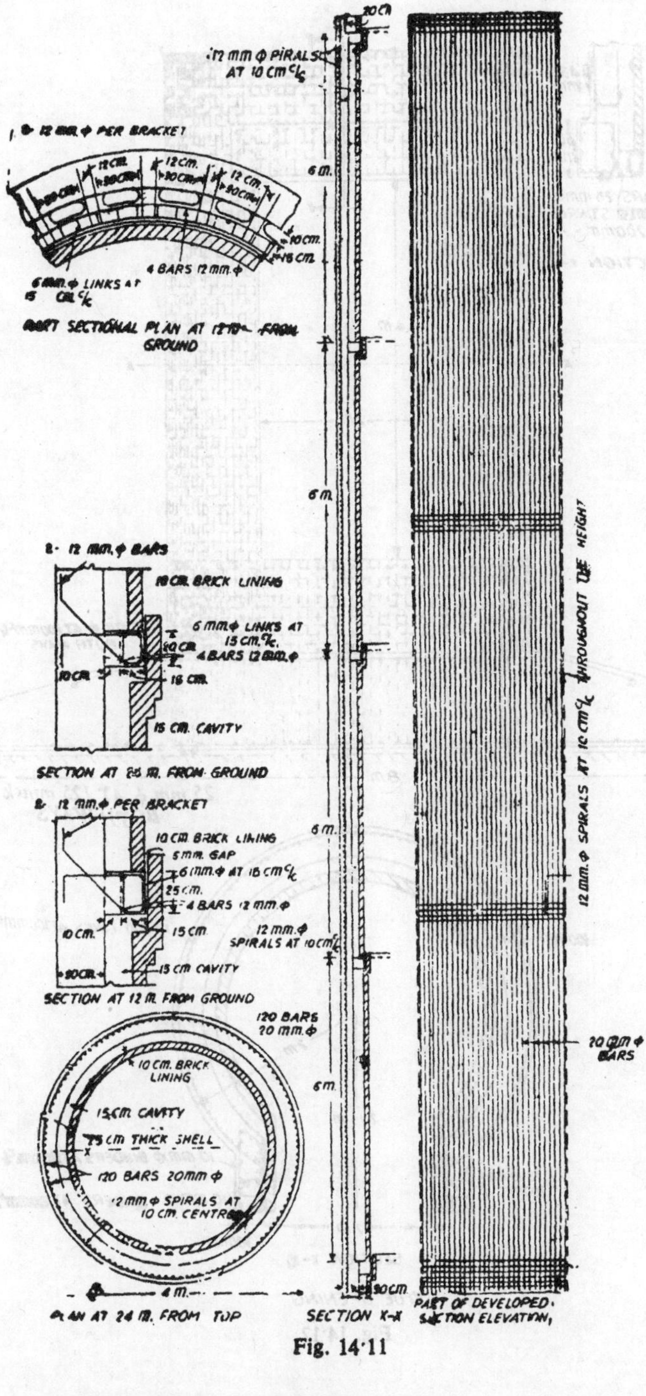
12 MM ϕ PIRALS AT 10 CM c/c
12 MM ϕ PER BRACKET
12 CM.
30 CM.
10 CM.
15 CM.
4 BARS 12 MM ϕ
6 MM ϕ LINKS AT 15 CM. c/c
PART SECTIONAL PLAN AT 12 M FROM GROUND
6 M.
2 - 12 MM ϕ BARS
10 CM. BRICK LINING
6 MM ϕ LINKS AT 15 CM. c/c
20 CM.
4 BARS 12 MM ϕ
10 CM.
15 CM.
15 CM. CAVITY
SECTION AT 25 M. FROM GROUND
2 - 12 MM ϕ PER BRACKET
10 CM. BRICK LINING
5 MM. GAP
6 MM ϕ AT 15 CM c/c
25 CM.
4 BARS 12 MM ϕ
10 CM.
15 CM.
12 MM ϕ SPIRALS AT 10 CM c/c
30 CM.
15 CM CAVITY
SECTION AT 12 M. FROM GROUND
120 BARS 20 MM ϕ
10 CM. BRICK LINING
15 CM. CAVITY
25 CM THICK SHELL
120 BARS 20MM ϕ
12 MM ϕ SPIRALS AT 10 CM. CENTRES
4 M.
PLAN AT 24 M. FROM TOP
20 CM
30 CM
SECTION X-X
12 MM ϕ SPIRALS AT 10 CM c/c THROUGHOUT THE HEIGHT
20 MM ϕ BARS
PART OF DEVELOPED SECTION ELEVATION

Fig. 14·11

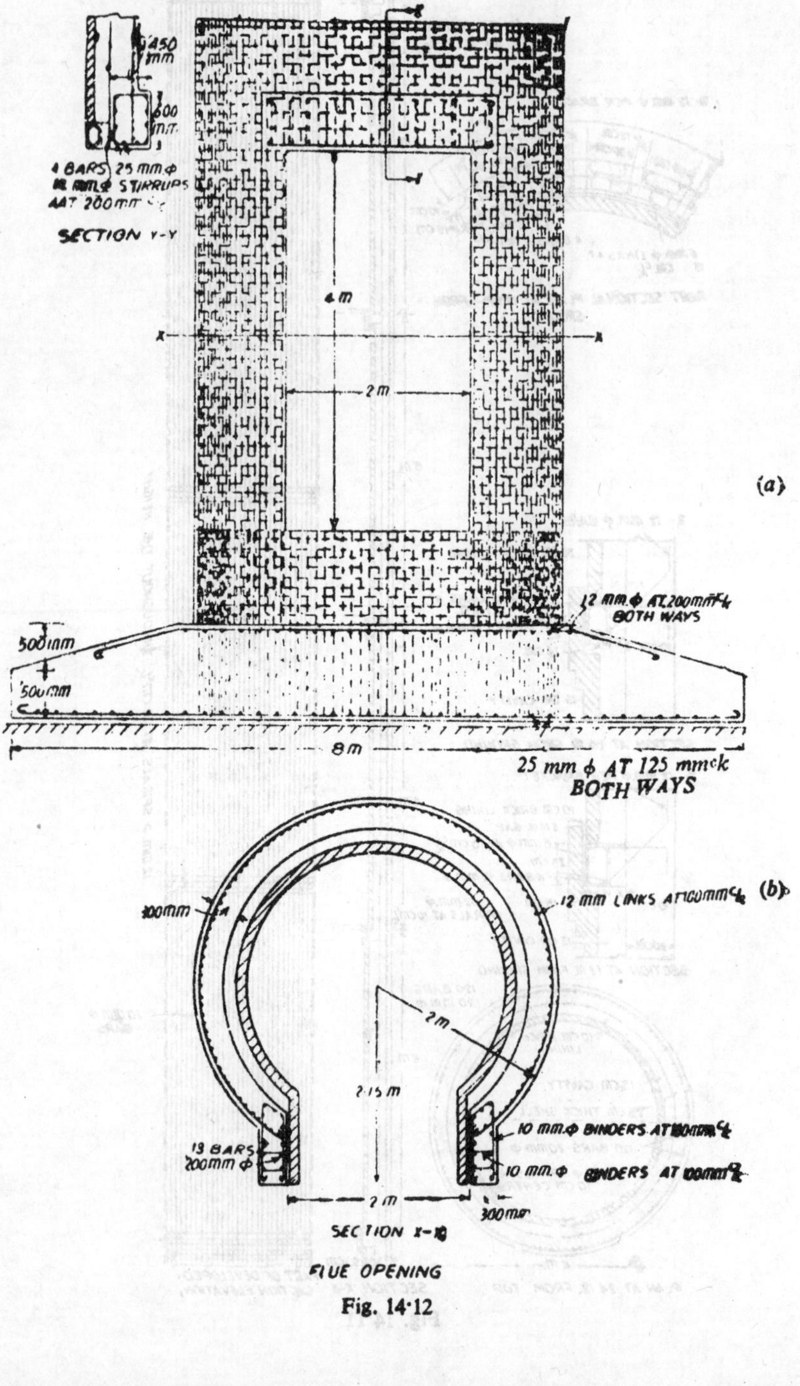
450 mm
600 mm
4 BARS 25 mm.φ
SECTION Y-Y
4 m
2 m
12 mm.φ AT 200 mm
BOTH WAYS
500 mm
500 mm
8 m
25 mm φ AT 125 mm c/c
BOTH WAYS
(a)
300 mm
12 mm LINKS AT 100 mm c/c
(b)
2 m
2.15 m
13 BARS
200 mm φ
10 mm.φ BINDERS AT 100 mm c/c
10 mm.φ BINDERS AT 100 mm c/c
2 m
300 mm
SECTION X-X
FLUE OPENING

Fig. 14·12

Total moment at the base $=8,043,600$ Nm $=M$.
Let allowable bearing pressure be 150 kN/m²
Self-weight of footing at 10% $=545,400$ N
Total load on base $=5,999,252$
Let D metres be the diameter of footing,

for no tension to occur, $\dfrac{W}{A}-\dfrac{M}{Z}=0$

$$=\frac{5,999,252}{\frac{\pi}{4}\times D^2}-\frac{8,043,600}{\frac{\pi}{32}\times D^3}$$

$$\therefore\quad 5,999,252-\frac{8,043,600\times 8}{D}=0$$

$$\therefore\quad D=\frac{8,043,600\times 8}{5,999,252}=10{\cdot}73 \text{ m.}$$

Adopt a diameter of 11·2 m.

The load on the base from chimney is taken as annular loading on the mean diameter. The load coming on the base is

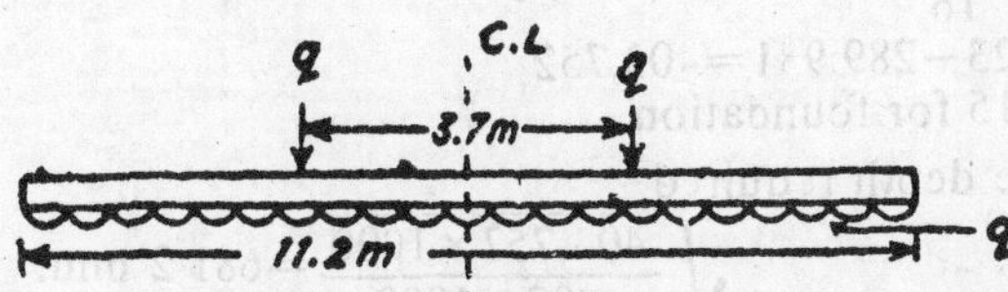

Fig 14·13

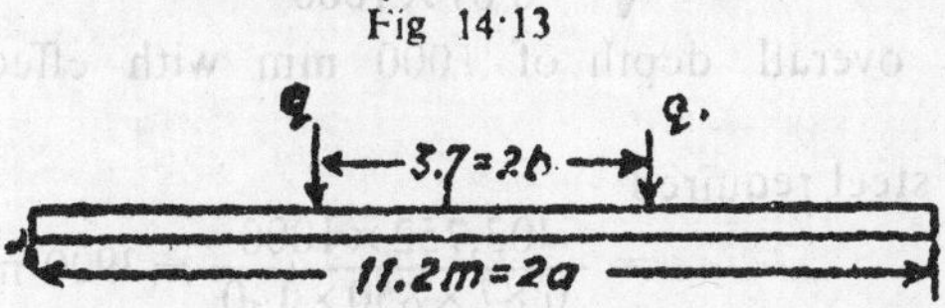

Fig. 14·14

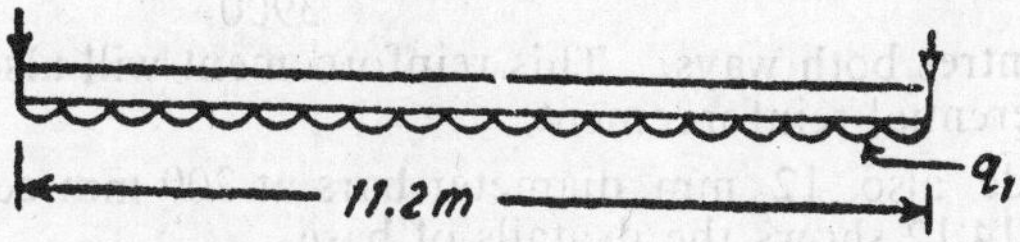

Fig. 14·15

shown in Fig. 14·13. The bending moments in the base can be found by superimposing two types of loading shown in Figs 14·14 and 14·15.

Let W be the total load

$$=5,453,852 \text{ N}$$

$$q_1=\frac{5,453,852}{\frac{\pi}{4}\times(11{\cdot}2)^2}=55,358 \text{ N/m}^2$$

Maximum bending moment in the section is governed by radial moment.

Refer to Fig. 8·36, radial moment in the centre of section is given by

$$M_r = \frac{W}{8\pi}\left[2\log_e\frac{a}{b}+1-\left(\frac{b}{a}\right)^2\right]-\frac{3}{16}q_1a^2$$

$$=\frac{5{,}453{,}852}{8\pi}\left[2\log_e\frac{5\cdot6}{1\cdot85}+1-\left(\frac{1\cdot85}{5\cdot6}\right)^2\right]-\frac{3}{16}\times 55{,}358\times(5\cdot6)^2$$

$$=\frac{5{,}453{,}852}{8\pi}\left[2\cdot216+1-0\cdot0191\right]-325{,}505$$

$$=693{,}733-325{,}505=368{,}228\ \text{N/m}^2$$

Maximum radial moment occurs at $r=1\cdot85$ m.

$$M_r=\frac{W}{8\pi}\left[2\log_e\frac{a}{b}+1-\left(\frac{b}{a}\right)^2-\frac{3q_1}{16}(a^2-b^2)\right.$$

$$=\frac{5{,}453{,}852}{8\pi}\left[2\log_e\frac{5\cdot6}{1\cdot85}+1-\left(\frac{1\cdot85}{5\cdot6}\right)^2\right]-\frac{3\times55{,}358}{16}[(5\cdot6)^2-(1\cdot85)^2]$$

$$=693{,}723-289{,}981=403{,}752$$

Use M 15 for foundation.

Effective depth required

$$\sqrt{\frac{403{,}752\times1000}{0\cdot87\times1000}}=681\cdot2\text{ mm.}$$

Provide overall depth of 1000 mm with effective depth of 850 mm.

Area of steel required

$$=\frac{403{,}752\times1000}{0\cdot87\times850\times140}=3900\text{ mm}^2$$

Provide 25 mm diameter bars at $\frac{1000\times491}{3900}=125\cdot9$ mm, say 125 mm centres both ways. This reinforcement will also take care of circumferential reinforcement.

Provide also 12 mm diameter bars at 200 mm centres on the top. Fig. 14·12 shows the details of base.

PROBLEMS

1. Design a concrete chimney of 40 m height and 3 m diameter throughout. The chimney is to be designed for wind pressure in Bombay area.

2. Design a concrete chimney of height 80 m. The external diameter of shaft is 4 m at the top and 5 m at the bottom. The chimney is to be erected in Delhi area.

3. Design the footing for a circular RCC Chimney 30·5 m high above ground level 3·66 m external diameter and 3·2 m internal diameter. There is a brick lining on the inside 11·4 cm thick for the full height, having a mean diameter of 2·9 m. The maximum wind pressure may be assumed as 980 N/m² on the projected surface, and the maximum bearing pressure on soil may be reckoned as 164 kN/m². Adopt the following values of stresses and modular ratio.

$c=7$ N/mm², $t=140$ N/mm², $m=14$ (*A.M.I.E. 1967*)

15

Beams Curved in Plan

15.1. Curved beams are used to support circular reservoirs, curved balconies and curved ramps. In curved beams the centre of gravity of loads acting on a span lies outside the line joining the supports. It causes overturning of the beams unless it is fixed at the supports or is continuous over the supports. As the reactions and the loads do not lie along the axis at any point of the beam, there will be torsional moment in addition to bending moment and shear force. The beam is to be designed for torsion as well as bending moment and shear force.

In case of circular beams supported by symmetrically placed columns the columns will give vertical reaction. Due to symmetry the torsion at the centre of curved beam between two consecutive supports will be zero. Also the torsional moment at the supports will be zero. There will be only shear force and bending moment at the supports.

15.2. Circular beams loaded uniformly and supported on symmetrically placed columns

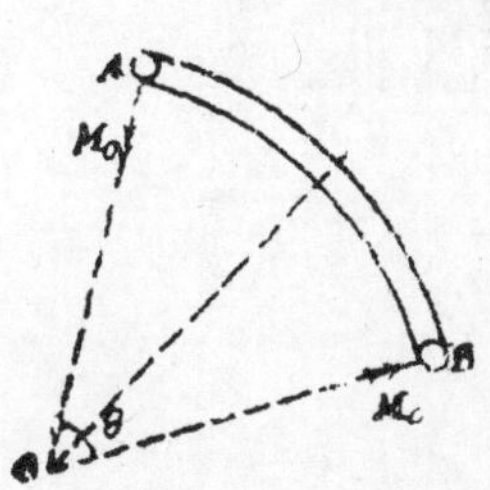

Fig. 15.1

Consider portion AB of beam between two consecutive columns.

Let the angle subtended at the centre by two consecutive columns A and B be θ.

The load on the portion AB of the beam $= wR\theta$, where w is the load per unit length.

The centre of gravity of the load will lie at distance, $\dfrac{R \sin \theta/2}{\theta/2}$ from the centre.

Let M_0 be the bending moment and F_0 be the shear force at the supports.

From symmetry $F_0 = \frac{wR\theta}{2}$

Representing the moments at suppports by vectors, the direction of moments at supports is shown in Fig 15·2.

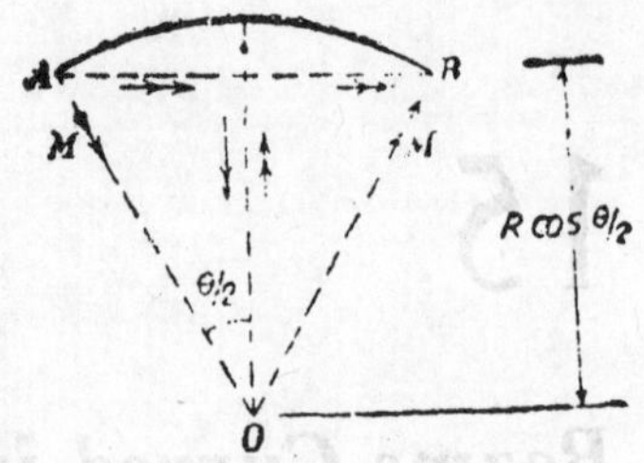

Fig. 15·2

The moment M_0 at support can be resolved along the chord AB and at right angles to it. The two components will be $M_0 \sin\theta/2$ and $M_0 \cos\theta/2$ respectively.

Taking moments of all forces about line AB

$$2M_0 \sin\theta/2 = wR\theta\left[\frac{R\sin\theta/2}{\theta/2} - R\cos\theta/2\right]$$

$$\therefore \quad M_0 = wR^2\left[\frac{\theta}{2\sin\theta/2}\times\frac{\sin\theta/2}{\theta/2} - \theta\times\frac{\cos\theta/2}{2\sin\theta/2}\right]$$

$$= wR^2\left[1 - \frac{\theta\cot\theta/2}{2}\right] \quad ...(15·2)$$

To find S.F., B.M. and torsional moment at a point P, at angle ϕ from one support.

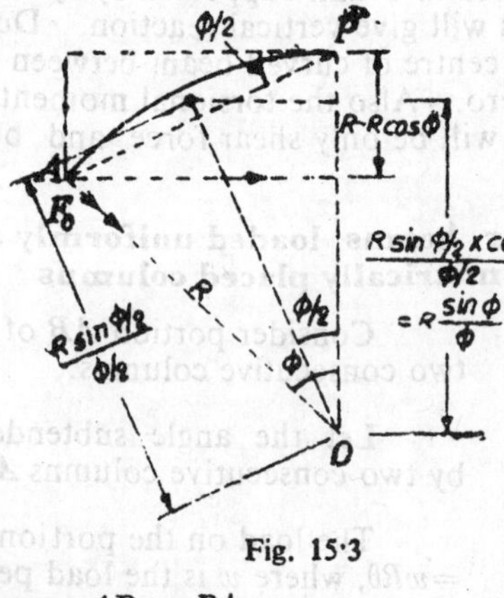

Fig. 15·3

Load between $AP = wR\phi$

$$\text{S.F. at } P = \frac{wR\theta}{2} - wR\phi = wR\,(\theta/2 - \phi) \quad ...(15·3)$$

The distance of centre of gravity of load between AP from centre $O = \frac{R\sin\phi/2}{\phi/2}$.

The direction of vector representing B.M. at P will act along PO and direction of vector representing torsion at P will act at right angles to PO.

$$M\phi = F_0 R \sin\phi - M_0 \cos\phi - wR\phi \frac{R \sin\phi/2}{\phi/2} \times \sin\phi/2$$

$$= \frac{wR^2\theta}{2} \times R \sin\phi - wR^2\left(1 - \theta\ \frac{\cot\theta/2}{2}\right)\cos\phi$$
$$-wR^2 \times 2\sin^2\phi/2$$

$$= wR^2\left[\frac{\theta}{2}\sin\phi - \cos\phi + \frac{\theta}{2}\cot\theta/2\cos\phi - 2\sin^2\phi/2\right]$$

$$= wR^2\left[-1 + \frac{\theta}{2}\sin\phi + \frac{\theta}{2}\cot\frac{\theta}{2}\cos\phi\right] \qquad ...(15\cdot4)$$

$$[\because\ -\cos\phi - 2\sin^2\phi/2 = -1]$$

$$T\phi = M_0 \sin\phi - \frac{wR\theta}{2} \times 2R\sin^2\phi/2 + wR\phi$$
$$\left(R - \frac{R\sin\phi/2}{\phi/2} \times \cos\phi/2\right)$$

$$= wR^2\left(1 - \theta\frac{\cos\theta/2}{2}\right)\sin\phi - wR^2\theta\sin^2\theta/2 + wR^2$$
$$(\phi - 2\sin\phi/2\cos\phi/2)$$

$$= wR^2\left[\sin\phi - \theta\frac{\cot\theta/2}{2}\sin\phi - \theta\sin^2\phi/2 + \phi - \sin\phi\right]$$

$$= wR^2\left[\phi - \theta\sin^2\frac{\phi}{2} - \theta\ \frac{\cot\theta/2}{2}\sin\phi\right]$$

$$= wR^2\left[\phi - \frac{\theta}{2} + \frac{\theta}{2}\cos\phi - \frac{\theta}{2}\cot\frac{\theta}{2}\sin\phi\right] \qquad ...(15\cdot5)$$

To obtain maximum value of torsional moment $dT\phi/d\phi$ is put equal to zero. This gives value of ϕ for maximum value of T. Substituting in (15·5), maximum value of T is obtained.

$$\frac{dT\phi}{d\phi} = 0 = 1 - \frac{\theta}{2}\sin\phi + \frac{\theta}{2}\cot\frac{\theta}{2}\cos\phi$$

$$M\phi = 0 = -1 + \frac{\theta}{2}\sin\phi + \frac{\theta}{2}\cot\frac{\theta}{2}\cos\phi$$

$\therefore$ Point of max. torsion will be point of contraflexure

Let, support moment $= kwR^2 \times \theta$

Mid-span moment $= k'wR^2 \times \theta$

Maximum torsion $= k''wR^2 \times \theta$

Let, ϕ' be the angle from support where maximum torsion occurs. Table 15·1 gives values of k, k', k'' and ϕ' for circular beams resisting on number of supports.

Table 15·1. Coefficients for B.M. and Torsion in Circular Beams

Number of Supports	θ	k	k′	k″	*Volume of φ′ for maximum twisting moment*
4	00	0·137	0·07	0·021	19 1/4°
5	72°	0·108	0·054	0·014	15 1/4°
6	60°	0·089	0·045	0·009	12 3/4°
7	51·4°	0·077	0·037	0·007	10 1/4°
8	45°	0·066	0 032	0·005	9 1/2°
9	40°	0 060	0·027	0·004	8 1/2°
10	36°	0·054	0·023	0 003	7 1/2°
12	30°	0·045	0 017	0·002	6 1/4°

15·3. Semicircular beam simply supported on three supports equally spaced

Let semicircular beam *ABC* be supported on three equally spaced supports *A*, *B* and *C*. Let *R* be the radius of the centre line of the beam and *w* be the load per unit length of the beam.

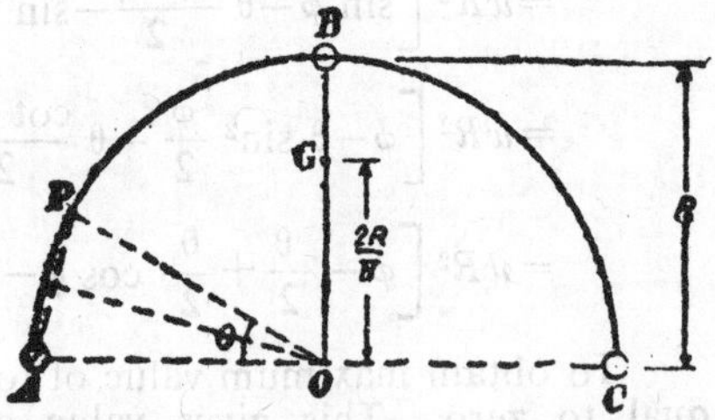

Fig. 15·4

Total load on the semi-circular beam $=wR\pi, \phi=\pi$.

Distance of C.G. of load from centre *O*

$$=\frac{R\sin\phi/2}{\phi/2}=\frac{R\sin\pi/2}{\pi/2}=2R/\pi$$

Let R_1 be the reaction at supports *A* and *C* and R_2 be the reaction at *B*.

Taking moments of reactions about line passing through *B* parallel to *AC*.

$$2R_1\times R=wR\pi\left(R-\frac{2R}{\pi}\right)$$

$$\therefore \quad R_1=\frac{wR}{2}(\pi-2)$$

$$\therefore \quad R_2=wR\pi-2\times\frac{wR}{2}(\pi-2)$$

$$=2wR.$$

B.M. at a Point P at θ°

Load on portion $AP = wR\theta$

Distance of C.G. of load from O

$$= \frac{R \sin \theta/2}{\theta/2}$$

$$M_\theta = R_1 \times R \sin \theta - wR\theta \left[\frac{R \sin \theta/2}{\theta/2} \times \sin \theta/2\right]$$

$$= \frac{wR}{2}(\pi - 2) \times R \sin \theta - 2wR^2 \sin^2 \theta/2$$

$$= wR^2\left[\frac{(\pi-2)}{2} \sin \theta - 2 \sin^2 \theta/2\right]$$

At $\theta = \frac{\pi}{2}$ $$M_B = wR^2\left[\frac{(\pi-2)}{2} - 2 \times \frac{1}{2}\right]$$

$$= wR^2 \times \frac{\pi - 4}{2} = -0.429\, wR^2$$

For maximum +ve B.M., putting $\frac{dM_\theta}{d\theta} = 0$

$$\therefore \quad \frac{dM_\theta}{d\theta} = wR^2\left[\frac{(\pi-2)}{2} \cos \theta - 2 \times 2 \sin \theta/2 \times \tfrac{1}{2} \cos \theta/2\right] = 0$$

$$\therefore \quad \left(\frac{\pi-2}{2}\right) \cos \theta = 2 \sin \frac{\theta}{2} \cos \frac{\theta}{2}$$

$$\therefore \quad \tan \theta = \frac{\pi - 2}{2}$$

$$\theta = 29° \; 44'$$

$$M_{max} = wR^2\left[\frac{\pi-2}{2} \times 0.496 - (1 - 0.8683)\right]$$

$$= wR^2[(\pi - 2) \times 0.248 - 0.1317]$$

$$= 0.1520\, wR^2$$

$$T_\theta = R_1(R - R \cos \theta) - wR\theta\left[R - \frac{R \sin \theta/2}{\theta/2} \times \cos \theta/2\right]$$

$$= \frac{wR}{2}(\pi - 2)R(1 - \cos \theta) - wR^2(\theta - 2 \sin \theta/2 \cos \theta/2)$$

$$= wR^2\left[\left(\frac{\pi-2}{2}\right) - \left(\frac{\pi-2}{2}\right) \cos \theta - \theta + \sin \theta\right]$$

For maximum value of twisting moment, put $\frac{dT_\theta}{d\theta} = 0$

$$wR^2\left[0 + \frac{\pi-2}{2} \sin \theta - 1 + \cos \theta\right] = 0$$

$$\therefore \quad \frac{\pi-2}{2} \sin \theta - 1 + \cos \theta = 0$$

$$0{\cdot}571 \sin\theta - 1 + \cos\theta = 0$$
$$0{\cdot}571 \sin\theta - 1 = -\cos\theta$$
$$(0{\cdot}571)^2 \sin^2\theta - 1{\cdot}142 \sin\theta + 1 = \cos^2\theta = 1 - \sin^2\theta.$$
$$(0{\cdot}571)^2 \sin^2\theta + \sin^2\theta = 1{\cdot}142 \sin\theta$$
$$1{\cdot}326 \sin^2\theta = 1{\cdot}142 \sin\theta$$
$$\sin\theta = \frac{1{\cdot}142}{1{\cdot}326} = 0{\cdot}8614$$
$$\therefore \quad \theta = 52°\ 49$$
$$T_{max} = wR^2[0{\cdot}571 - 0{\cdot}571 \times 0{\cdot}5078 - 1{\cdot}039 + 0{\cdot}8614]$$
$$= wR^2[0{\cdot}571 \times 0{\cdot}4922 - 0{\cdot}1776]$$
$$= 0{\cdot}103\ wR^2.$$

Example 15·1. *Design a semi-circular beam supported on three equally spaced columns. The centres of columns are on a curve of diameter 10 metres. The superimposed load on the beam per metre length is 15,000 N.*

Sol. Assume size of beam as 800 mm × 400 mm

Self-load of beam per metre length

$$= 0{\cdot}025 \times 800 \times 400 = 8000 \text{ N}$$

Total load per metre length = 15,000 + 8000 = 23,000 N

Maximum B.M. occurs at middle support.

$$\text{Maximum B.M.} = -0{\cdot}429\ wR^2$$
$$= -0{\cdot}429 \times 23{,}000 \times 5^2 \text{ N-m.}$$
$$= -246{,}675 \text{ Nm} = -246{,}675{,}000 \text{ N-mm}$$

Mix. M 15 is used.

Effective depth required

$$= \sqrt{\frac{246{,}675{,}000}{0{\cdot}87 \times 400}} = 841{\cdot}9 \text{ mm}$$

Actual depth provide is 800 − 50 = 750 mm

The beam is designed as doubly reinforced beam.

Moment of resistance as balanced section,

$$M_1 = 0{\cdot}87 \times 400 \times (750)^2 = 195{,}800{,}000 \text{ N-mm}$$
$$M_2 = 246{,}675{,}000 - 195{,}800{,}000$$
$$= 50{,}875{,}000 \text{ N-mm}$$

Area of tensile steel required

$$= \frac{195{,}800{,}000}{0{\cdot}87 \times 750 \times 140} + \frac{50{,}875{,}000}{700 \times 140}$$
$$= 2663 \text{ mm}^2$$

Provide 6 bars of 25 mm ϕ

Depth of N.A. = 0·4 d = 0·4 × 750

$$= 300 \text{ mm}$$

$$A_c = \frac{50{,}875{,}000}{17 \times 5 \times \frac{300-50}{300} \times 700}$$
$$= \frac{50{,}875{,}000}{17 \times 5 \times 250 \times 700} \times 300$$
$$= 1000 \text{ mm}^2$$

Provide 5 bars of 16 mm ϕ.

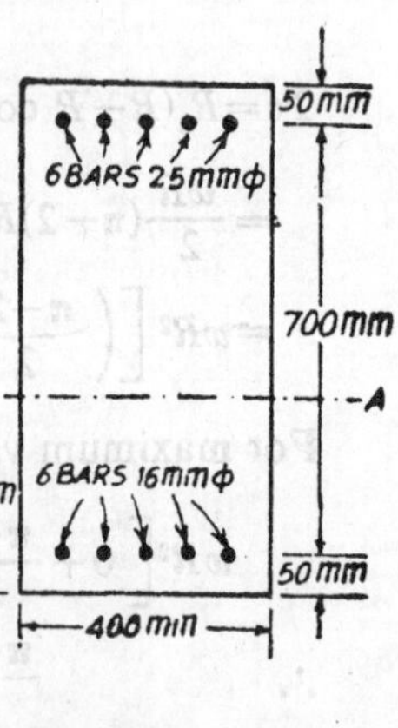

Fig 15·5

Maximum +ve B.M.

Max. +ve B.M. $=0{\cdot}152\, wR^2$

$=0{\cdot}152\times 23{,}000\times 5^2$

$=87{,}400$ Nm $=87{,}400{,}000$ N-mm

$$A_t=\frac{87{,}400{,}000}{0{\cdot}87\times 750\times 140}=957 \text{ mm}^2$$

Provide 5 bars of 16 mm ϕ.

Check for Shear and Bond

Shear at end column $=\frac{wR^2}{2}(\pi-2)$

$$=\frac{23{,}000\times 50}{2}\times 1{\cdot}142=65{,}665 \text{ N}$$

Shear stress $=\frac{65{,}665}{750\times 140}=0{\cdot}219$ N/mm². Safe.

Sufficient anchorage length is provided.

Shear at the middle column

$$=\frac{2wR}{2}=\frac{2\times 23{,}000\times 5}{2}$$

$=115{,}000$ N

Shear stress $=\frac{115{,}000}{750\times 400}=0{\cdot}383$ N/mm²

Provide 12 mm ϕ stirrups at 300 mm centres.

Sufficient anchorage length is provided.

Design for torsion

Maximum twisting moment

$=0{\cdot}103\, wR^2$

$=0{\cdot}103\times 23{,}000\times 25=59{,}225$ Nm

$=59{,}225{,}000$ N-mm

Shear at the point of maximum twisting moment

$=R_1-wR\theta, \qquad (\theta=59°\ 29')$

$=65{,}665-23{,}000\times 5\times 1{\cdot}039$

$=53{,}820$ N

Shear stress due to shear force

$$=\frac{53{,}820}{750\times 400}=0{\cdot}179 \text{ N/mm}^2. \text{ Safe.}$$

$$V_e=V_c+1{\cdot}6\frac{T}{b}=53{,}820+1{\cdot}6\times\frac{59{,}255{,}000}{400}$$

$=290{,}720$ N

$$v_{ve}=\frac{290{,}720}{750\times 400}=0{\cdot}969 \text{ N/mm}^2<\tau_{max}\ (1{\cdot}6)$$

Taking τ_c $=0{\cdot}34$

Extra shear $=(0{\cdot}969-0{\cdot}34)\times750\times400=188{,}700$ N

Design of stirrups. Use 12 mm ϕ stirrups

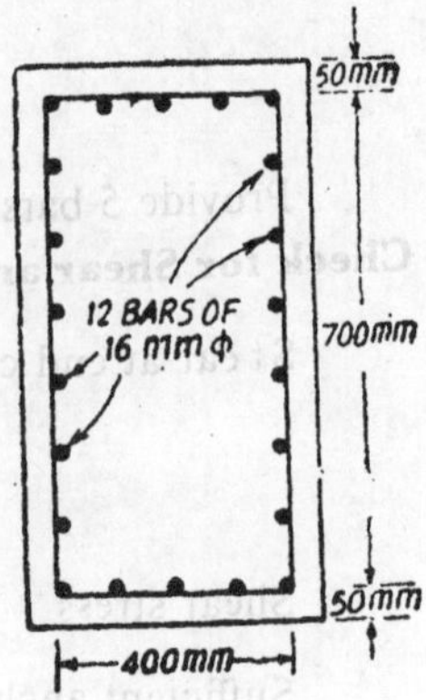

Fig. 15·6

$$\text{Spacing}=\frac{2\times113\times130\times750}{188{,}700}$$

$$=116{\cdot}8 \text{ mm}$$

Provide 12 mm ϕ stirrups at 115 mm centres to take shear due to shear force and twisting moment.

Longitudinal reinforcement for torsion

$$Me_1=M+M_t$$

$$=246{,}675{,}000+T\left(\frac{1+D/b}{1{\cdot}7}\right)$$

$$Me_1=246{,}675{,}000+59{,}225{,}000\left(\frac{1+\frac{800}{400}}{1{\cdot}7}\right)$$

$$=351{,}189{,}706 \text{ N-mm}$$

$$A_t=\frac{351{,}189{,}706}{0{\cdot}87\times750\times140}=3844 \text{ mm}^2$$

Extra reinforcement

$$=3844-2663$$

$$=1181 \text{ mm}^2$$

Use 12 bars of 16 mm ϕ as shown.

PROBLEMS

1. Design a circular beam supported on six equally spaced columns superimposed load on the beam is 20,000 N/m. The centres of columns lie on a circle of 10 m diameter.

2. Design a semi-circular beam supported on 3 columns equally spaced. The centres of columns are on a curve of diameter 8 m. The superimposed load on the beam per metre length is 2000 N.

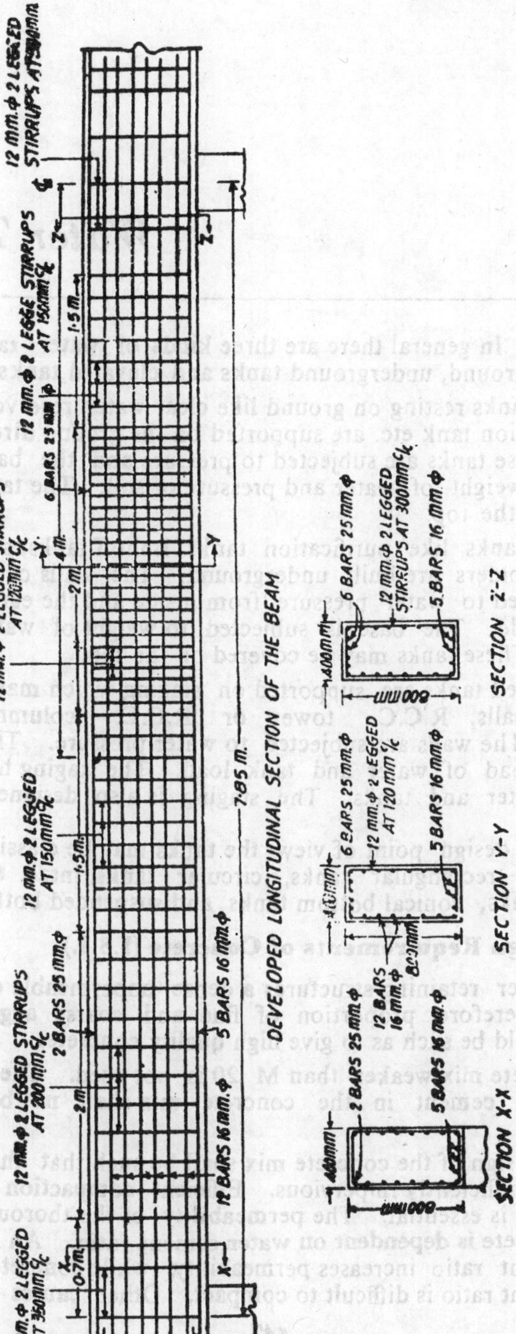
12 mm.ϕ 2 LEGGED STIRRUPS AT 350 mm c/c
12 mm.ϕ 2 LEGGED STIRRUPS AT 200 mm c/c
12 mm.ϕ 2 LEGGED AT 150 mm c/c
12 mm.ϕ 2 LEGGED STIRRUPS AT 120 mm c/c
12 mm.ϕ 2 LEGGED STIRRUPS AT 150 mm c/c
12 mm.ϕ 2 LEGGED STIRRUPS AT 300 mm
2 BARS 25 mm.ϕ
6 BARS 25 mm.ϕ
12 BARS 16 mm.ϕ
5 BARS 16 mm.ϕ
0·7 m
2 m
1·5 m
4 m
7·85 m
DEVELOPED LONGITUDINAL SECTION OF THE BEAM
400 mm
800 mm
SECTION X-X
SECTION Y-Y
SECTION Z-Z

Fig. 15·7

16

Water Tanks

16·1. In general there are three kinds of water tanks—tanks resting on ground, underground tanks and elevated tanks.

The tanks resting on ground like clear water reservoir, settling tanks, aeration tank etc. are supported on the ground directly. The walls of these tanks are subjected to pressure and the base is subjected to weight of water and pressure of soil. The tanks may be covered on the top.

The tanks like purification tanks, Imhoff tanks, septic tanks, and gas holders are built underground. The walls of these tanks are subjected to water pressure from inside and the earth pressure from outside. The base is subjected to weight of water and soil pressure. These tanks may be covered on the top.

Elevated tanks are supported on staging which may consist of masonry walls, R.C.C. tower or R.C.C. columns braced together. The walls are subjected to water pressure. The base has to carry load of water and tank load. The staging has to carry load of water and tanks. The staging is also designed for wind forces.

From design point of view the tanks may be classified as per their shape—rectangular tanks, circular tanks, intze type tanks, spherical tanks, conical bottom tanks and suspended bottom tanks.

16·2. Design Requirements of Concrete (I.S.I.)

In water retaining structures a dense unpermeable concrete is required, therefore, proportion of fine and coarse aggregates to cement should be such as to give high quality concrete.

Concrete mix weaker than M 20 is not used. The minimum quantity of cement in the concrete mix shall not be less than 300 N/m^3.

The design of the concrete mix shall be such that the resultant concrete is sufficiently impervious. Efficient compaction preferably by vibration is essential. The permeability of the thoroughly compacted concrete is dependent on water cement ratio. An increase in water cement ratio increases permeability, while concrete with low water cement ratio is difficult to compact. Other causes of leakage in concrete are defects such as segregation and honey combing. All joints should be made water-tight as these are potential sources of leakage.

Design of liquid retaining structures is different from ordinary RCC structures as it requires that concrete should not crack and hence tensile stresses in concrete should be within permissible limits.

A reinforced concrete member of liquid retaining structure is designed on the usual principles ignoring tensile resistance of concrete in bending. Additionally it should be ensured that tensile stress on the liquid retaining face of the equivalent concrete section does not exceed the permissible tensile strength of concrete as given in table 16·1. For calculation purposes the cover is also taken into concrete area.

Cracking may be caused due to restraint to shrinkage, expansion and contraction of concrete due to temperature or shrinkage and swelling due to moisture effects. Such restraint may be caused by—

(*i*) the interaction between reinforcement and concrete during shrinkage due to drying.

(*ii*) the boundary conditions.

(*iii*) the differential conditions prevailing through the large thickness of massive concrete.

Use of small size bars placed properly, lead to closer cracks but of smaller width. The risk of cracking due to temperature and shrinkage effects may be minimised by limiting the changes in moisture content and temperature to which the structure as a whole is subjected. The risk of cracking can also be minimised by reducing the restraint on the free expansion of the structure with long walls or slab founded at or below ground level, restraint can be minimised by the provision of a sliding layer. This can be provided by founding the structure on a flat layer of concrete with interposition of some material to break the bond and facilitate movement.

In case length of structure is large it should be sub-divided into suitable lengths separated by movement joints, specially where sections are changed the movement joints should be provided.

Where structures have to store hot liquids, stresses caused by difference in temperature between inside and outside of the reservoir should be taken into account.

The coefficient of expansion due to temperature change may be taken as $11\times10^{-6}/°C$ and coefficient of shrinkage may be taken as 450×10^{-6} for initial shrinkage and 200×10^{-6} for drying shinkage.

16·3. Joints in Liquid Retaining Structures

Joints are classified as given below.

(*a*) **Movement Joints.** There are three types of movement joints.

(*i*) *Contraction Joint.* It is a movement joint with deliberate discontinuity without initial gap between the concrete on either side of the joint. The purpose of this joint is to accommodate contraction of the concrete. The joint is shown in Fig. 16·1 (*a*).

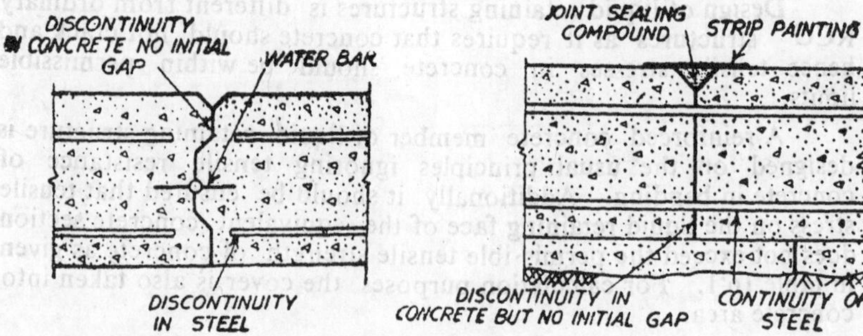

(*a*) (*b*)

Fig. 16·1

A contraction joint may be either complete contraction joint or partial contraction joint. A complete contraction joint is one in which both steel and concrete are interrupted and a partial contraction joint is one in which only the concrete is interrupted, the reinforcing steel running through as shown in Fig. 16·1 (*b*).

(*ii*) *Expansion Joint*. It is a joint with complete discontinuity in both reinforcing steel and concrete and it is to accommodate either expansion or contraction of the structure. A typical expansion joint is shown in Fig. 16·2.

This type of joint requires the provision of an initial gap between the adjoining parts of a structure which by closing or opening accommodates the expansion or contraction of the structure.

(*iii*) *Sliding Joint*. It is a joint with complete discontinuity in both reinforcement and concrete and with special provision to facilitate movement in plane of the joint. A typical joint is shown in Fig. 16·3. This type of joint is provided between wall and floor in some cylindrical tank designs.

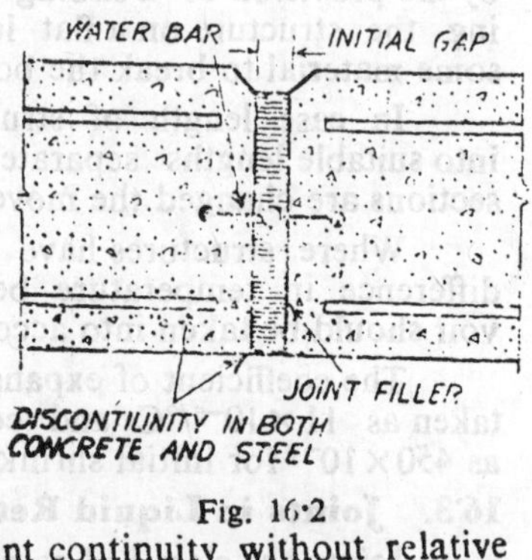

Fig. 16·2

(*b*) **Construction Joints**. This type of joint is provided for convenience in construction. Arrangement is made to achieve subsequent continuity without relative movement. One application of these joints is between successive lifts in a reservoir wall. A typical joint is shown in Fig. 16·4. The number of joints should be as small as possible and these joints should be kept from possibility of percolation of water.

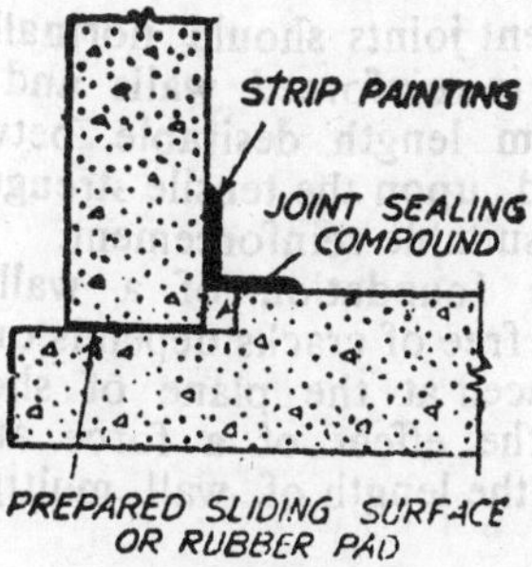

Fig. 16·3

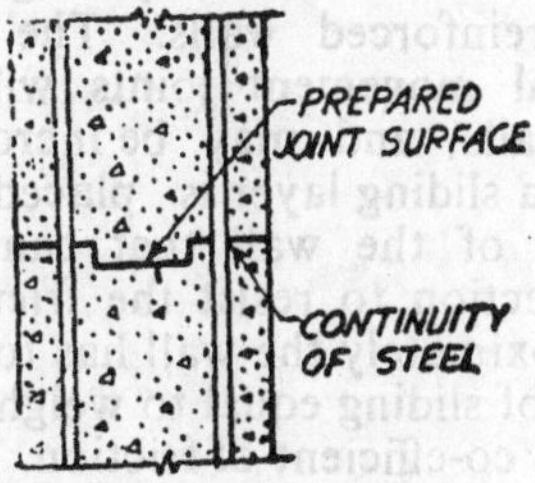

Fig. 16·4

(*c*) **Temporary Open Joints.** A gap is sometimes left temporarily between the concrete of adjoining parts of a structure which

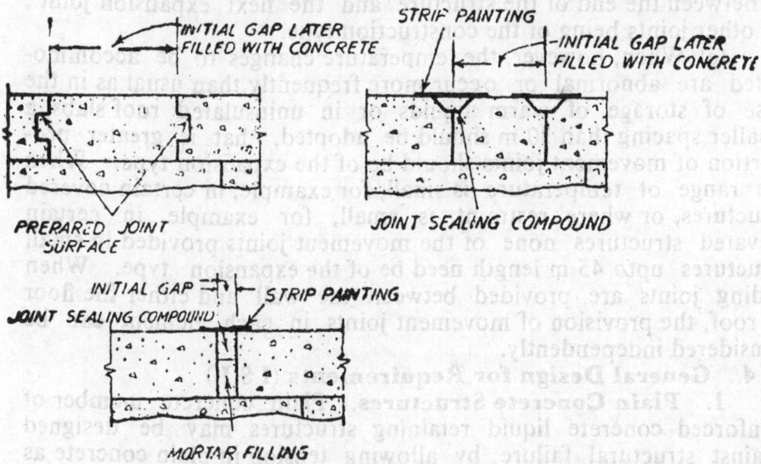

Temporary open joints
Fig. 16·5

after a suitable interval and before the structure is put to use is filled with mortar or concrete completely as in Fig. 16·5 (*a*) or as shown in Fig. 16·5 (*b*) and (*c*) with suitable jointing materials. In the first case width of the gap should be sufficient to allow the sides to be prepared before filling.

Spacing of joints. Unless alternative effective means are taken to avoid cracks by allowing for the additional stresses that may be induced by temperature or shrinkage changes or by unequal settlement, movement joints should be provided at the following spacings :

(*a*) In reinforced concrete floors, movement joints should be spaced at not more than 7·5 m apart in two directions at right angles. The wall and floor joints should be in line except where sliding joints occur at the base of the wall in which correspondence is not so important.

(*b*) For floors with only nominal percentage of reinforcement (smaller than the minimum specified) the concrete floor should be cast in panels with sides not more than 4·5 m.

(c) In concrete walls, the movement joints should normally be placed at a maximum spacing of 7·5 m in reinforced walls and 6 m in unreinforced walls. The maximum length desirable between vertical movement joints will depend upon the tensile strength of the walls, and may be increased by suitable reinforcement. Thus when a sliding layer is placed at the foundation of a wall, the length of the wall that can be kept free of cracks depends on the wall section to resist the friction induced at the plane of sliding. Approximately the wall has to stand the effect of a force at the place of sliding equal to weight of half the length of wall multiplied by the co-efficient of friction.

(d) Amongst the movement joints in floors and walls as mentioned above expansion joints should normally be provided at a spacing of not more than 30 m between successive expansion joints or between the end of the structure and the next expansion joint; all other joints being of the construction type.

(e) When, however, the temperature changes to be accommodated are abnormal or occur more frequently than usual as in the case of storage of warm liquids or in uninsulated roof slabs, a smaller spacing than 30 m should be adopted, that is greater proportion of movement joints should be of the expansion type. When the range of temperature is small, for example, in certain covered structures, or where restraint is small, for example, in certain elevated structures none of the movement joints provided in small structures upto 45 m length need be of the expansion type. When sliding joints are provided between the wall and either the floor or roof, the provision of movement joints in each element can be considered independently.

16 4. General Design for Requirements (I.S.I.)

1. Plain Concrete Structures. Plain concrete member of reinforced concrete liquid retaining structures may be designed against structural failure by allowing tension in plain concrete as per the permissible limits for tension in bending. This wall automatically take care of failure due to cracking. However, nominal reinforcement shall be provided, for plain concrete structural members.

2. Permissible Stresses in Concrete

(a) **For resistance to cracking.** For calculations relating the resistance of members to cracking, the permissible stresses in tension (direct and due to bending) and shear shall confirm to the values specified in Table 16·1. The permissible tensile stresses due to bending apply to the face of the member in contact with the liquid. In members less than 225 mm thick and in contact with liquid on one side these permissible stresses in bending apply also to the face remote from the liquid.

(b) **For strength calculations.** In strength calculations the permissible concrete stresses shall be in accordance with Table 16·1. When the calculated shear stress in concrete alone exceeds the permissible value, reinforcement acting in conjunction with diagonal compression in the concrete shall be provided to take the whole of the shear.

Table 16·1
Permissible concrete stresses in calculations relating to resistance to cracking

Grade of concrete	*Permissible stress in N/mm²* *Tension* *Direct*	*Due to Bending*	*Shear* $\left(=\frac{Q}{bd}\right)$
M 15	1·1	1·5	1·5
M 20	1·2	1·7	1·7
M 25	1·3	1·8	1·9
M 30	1·5	2·0	2·2
M 35	1·6	2·2	2·5
M 40	1·7	2·4	2·7

3. **Permissible Stresses in Steel**

(*a*) **For resistance to cracking.** When steel and concrete are assumed to act together for checking the tensile stress in concrete for avoidance of crack, the tensile stress in steel will be limited by the requirement that the permissible tensile stress in the concrete is not exceeded so the tensile stress in steel be equal to the product of modular ratio of steel and concrete, and the corresponding allowable ensile stress in concrete.

(*b*) **For strength calculations.** In strength calculations the permissible stress shall be as follows :

(*i*) Tensile stress in member in direct tension. 100 N/mm²

(*ii*) Tensile stress in member in bending on liquid retaining face of members or face away from liquid for members less than 215 mm thick. 100 N/mm²
On face away from liquid for members 225 mm or more in thickness. 125 N/mm²

(*iii*) Tensile stress in shear inforcement.
For members less than 225 mm thickness. 100 N/mm²
For members 225 mm or more in thickness. 125 N/mm²

(*iv*) Compressive stress in columns subjected to direct load. 125 N/mm²

Note 1. Stress limitations for liquid retaining faces shall also apply to :

(*a*) Other faces within 225 mm of liquid retaining face.

(*b*) Outside or external face of structures away from the liquid but placed in water logged soil upto the level of highest subsoil water level.

Note 2. The permissible stress of 100 N/mm² in (*i*), (*ii*) and (*iii*) may be increased to 112·5 N/mm² in case of deformed bars and in case of plain mild steel bars when the cross reinforcement is spot welded to the main reinforcement.

4. Stresses due to Drying Shrinkage or Temperature Change.

(*i*) Stress due to drying shrinkage or temperature change may be ignored provided that—

(*a*) the permissible stresses specified above in (*ii*) and (*iii*) are not otherwise exceeded.

(*b*) adequate precautions are taken to avoid cracking of concrete during the construction period and until the reservoir is put into use.

(*c*) recommendation regarding joints given in article 16·3 and for suitable sliding layer beneath the reservoir are completed with, or the reservoir is to be used only for the storage of water or aqueous liquids at or near ambient temperature and the circumstances are such that the concrete will never dry out.

(*ii*) Shrinkage stresses may however be required to be calculated in special cases, when a shrinkage co-efficient of 300×10^{-6} may be assumed.

(*iii*) When the shrinkage stresses are allowed, the permissible stresses, tensile stresses in concrete (direct and bending) as given in Table 16·1 may be increased by 33⅓ per cent.

5. Floors

(*i*) **Provision of movement joints.** Movement joints should be provided as discussed in article 16·3.

(*ii*) **Floors of tanks resting on ground.** If the tank is resting directly over ground, flow may be constructed of concrete with nominal percentage of reinforcement (smaller than the minimum specified in 8) provided that it is certain that the ground will carry the load without appreciable subsidence in any part and that the concrete floor is cast in panels with sides not mote than 4·5 m with contraction or expansion joints between. In such cases a screed or concrete layer not less than 75 mm thick shall first be placed on the ground and covered with a sliding layer of bitumen paper or other suitable material to destroy the bond between the screed and floor concrete.

In normal circumstances the screed layer shall be of grade not weaker than M 10, where injurious soils or aggressive water are expected, the screed layer shall be of grade not weaker than M 15 and if necessary a sulphate resisting or other special cement should be used.

(*iii*) **Floor of tanks resting on supports**

(*a*) If the tank is supported on walls or other similar supports the floor slab shall be designed as floors in buildings for bending moments due to water load and self-weight.

(*b*) When the floor is rigidly connected by the walls (as is generally the case) the bending moments at the junction between the walls and floors shall be taken into account in the design of floor together with any direct forces transferred to the floor from the walls or from the floor to the wall due to suspension of floor from the wall.

If the walls are non-monolithic with the floor slab, such as in cases, where movement joints have been provided between the floor slab and walls, the floor shall be designed only for the vertical loads on the floor.

(*c*) In continuous T-beams and L-beams with ribs on the side remote from the liquid, the tension in concrete on the liquid side at the face of the supports shall not exceed the permissible stresses for controlling cracks in concrete. The width of the slab shall be determined in usual manner for calculation of the resistance to cracking of T-beam, L-beams section at supports.

(*d*) The floor slab may be suitably tied to the walls by rods properly embedded in both the slab and the walls. In such-cases no separated beam (curved or straight) is necessary under the wall, provided the wall of the tank itself is designed to act as a beam over the supports under it.

(*e*) Sometimes it may be economical to provide the floors of circular tank, in the shape of dome. In such cases the dome shall be designed for the vertical loads of the liquid over it and the ratio of its rise to its diameter shall be so adjusted that the stresses in the dome are, as far as possible, wholly compressive. The dome shall be supported at its bottom on the ring beam which shall be designed for resultant circumferential tension in addition to vertical loads.

6. Walls

(*i*) **Provision**

(*a*) **Sliding joints at the base of the wall.** Where it is to allow the walls to expand or contract separately from the floor or to prevent moments at the base of the wall owing to fixity to the floor, sliding joints may be employed.

(*b*) The spacing of vertical movement joints should be as discussed in article 16·3 while the majority of the joints may be of the partial or complete contraction type, sufficient joints of the expansion type should be provided to satisfy the requirements given in article 16·3.

(*ii*) **Pressure on walls**

(*a*) In liquid retaining structures with fixed or floating covers the gas pressure developed above liquid surface shall be added to the liquid pressure.

(*b*) When the wall of liquid retaining structure is built in ground, or has earth embanked against it, the effect of earth pressure shall be taken into account.

(*iii*) **Walls of Tanks Rectangular or Polygonal in Plan.** While designing the walls of rectangular or polygonal concrete tanks, the following points should be borne in mind.

(*a*) In plane walls, the liquid pressure is resisted by both vertical and horizontal bending moments. An estimate should be made of the proportion of the pressure resisted by bending moments in the vertical horizontal planes. The direct horizontal tension caused by the direct pull due to water pressure on the end walls, should be added to that resulting from horizontal bending moments. On liquid retaining faces, the tensile stresses due to the combination of direct horizontal tension and bending action shall satisfy the following condition :

$$\frac{t'}{t}+\frac{\sigma_c' t}{\sigma_c'}\leqslant 1$$

where t' = calculated direct tensile stress in concrete.

t = permissible direct tensile in stress concrete (Table 16·1)

$\sigma_c' t$ = calculated tensile stress due to bending in concrete.

$\sigma_c t$ = permissible tensile stress due to bending in concrete.

(*d*) At the vertical edges where the walls of reservoir are rigidly joined, horizontal reinforcement and haunch bars should be provided to resist the horizontal bending moments even if the walls are designed to withstand the whole load as vertical beams or cantilever without lateral supports.

(*e*) In the case of rectangular or polygonal tanks, the side walls act as two-way slab, whereby the wall is continued or restrained in the horizontal direction, fixed or hinged at the bottom and hinge or free at the top. The walls thus act as thin plates subjected to triangular loading and with boundary conditions varying between full restraint and free edge. The analysis of moment and forces may be made on the basis of any recognised method.

(*iv*) **Walls of Cylindrical Tanks.** While designing walls of cylindrical tanks the following points should be borne in mind :

(*a*) Walls of cylindrical tanks are either cast monolithically with the base or are set in grooves and key-ways (movement joints). In either case deformation of wall under influence of liquid pressure is restricted at and above the base. Consequently, only part of the triangular hydrostatic load will be carried by ring tension and part of the load at bottom will be supported by cantilever action.

(*b*) It is difficult to restrict rotation or settlement of the base slab and it is advisable to provide vertical reinforcement as if the walls were fully fixed at the base, in addition to the reinforcement required to resist horizontal ring tension for hinged at base, conditions of walls, unless the appropriate amount of fixity at the base is established by analysis with due consideration to the dimensions of the base slab the type of joint between the walls and slab, and, where applicable, the type of soil supporting the base slab.

7. Roofs

(*i*) **Provision of movement Joints.** To avoid the possibility of sympathetic cracking, it is important to ensure that movement joints in the roof correspond with those in the walls, if roof and walls are monolithic. If, however, provision is made by means of a sliding joint for movement between the roof and the wall correspondence of joints is not so important.

(*ii*) **Loading.** Field covers of liquid retaining structures should be designed for gravity loads, such as the weight of roof slab, earth cover if any, live loads and mechanical equipment. They should also be designed for upward load if the liquid retaining structure is subjected to internal gas pressure.

A superficial load sufficient to ensure safety with the unequal intensity of loading which occurs during the placing of the earth cover should be allowed for in designing roofs. The engineer should specify a loading under these temporary conditions which should not be exceeded. In designing the roof, allowance should be made for the temporary condition of some spans loaded and other spans unloaded, even though in the final state the load may be small and evenly distributed.

(*iii*) **Water tightness**. In case of tanks intended for the storage of water for domestic purposes, the roof must be made water-tight. This may be achieved by limiting the stresses as for rest of the tank, or by the use of the covering of the waterproof membrane or by providing slopes to ensure adequate drainage.

(*iv*) **Protection against corrosion**. Protection measures shall be provided to the underside of the roof to prevent it from corrosion due to condensation.

8. Minimum Reinforcement

(*a*) The minimum reinforcement in walls, floors and roofs in each of two directions at right angles shall have an area of 0·3 per cent of the concrete section in that direction for sections upto 100 mm thickness. For sections of thickness greater than 100 mm and less than 450 mm the minimum reinforcement in each of the two directions shall be linearly reduced from 0·3 per cent for 100 mm thick section to 0·2 per cent for 450 mm thick sections. For sections of thickness greater than 450 mm minimum reinforcement in each of the two directions shall be kept at 0·2 per cent. In concrete sections of thickness 225 mm or greater, two layers of reinforcement steel shall be placed one near each face of the section to make up the minimum reinforcement.

(*b*) In special circumstances floor slabs may be constructed with percentage of reinforcement less than specified above. In no case the percentage of reinforcement in any member be less than 0·15% of gross sectional area of the member.

9. Minimum Cover to Reinforcement

(*a*) For liquid faces of parts of members either in contact with the liquid (such as inner faces of roof slab) the minimum cover to all reinforcement should be 25 mm or the diameter of the main bar whichever is greater. In the presence of the sea water and soils and water of corrosive characters the cover should be increased by 12 mm but this additional cover shall not be taken into account for design calculations.

(*b*) For faces away from liquid and for parts of the structure neither in contact with the liquid on any face, nor enclosing the space above the liquid, the cover shall be as for ordinary concrete member.

16·5. Tanks Resting on Ground. For small capacities rectangular tanks are generally used and for bigger capacities circular tanks are used. The walls of circular tanks may have flexible joints or rigid joints at the base.

16·6. Circular Tanks with Flexible Joint at the Base. In these tanks walls are subjected to hydrostatic pressure. The tank wall is designed as thin cylinder.

At the base, maximum pressure

$$=wH.$$

This causes hoop tension per metre

$$=\frac{wHD}{2}$$

where

w is the density of water in N/m³ taken as 10,000 N/m³.

i.e. 10 kN/m³ $\left[\begin{array}{l}\text{kN}=\text{kilo Newton}=10^3\ \text{N}\\ \text{MN}=\text{Mega Newton}=10^6\ \text{N}\end{array}\right]$

H is depth of water in metres

D is the diameter of the tank in metres.

Steel area required at the base for one metre height

$$=\frac{wHD}{2\times 100}\ \text{mm}^2$$

If *t* is the thickness of wall in mm, tensile stress in concrete

$$=\frac{wHD/2}{1000t+(m-1)A_t}\ \text{N/mm}^2$$

As the hoop tension reduces gradually to zero at top, the reinforcement is gradually reduced to minimum reinforcement at top. The main reinforcement consists of circular hoops. Vertical reinforcement equal to 0·3% of concrete area is provided and hoop reinforcement is at to this reinforcement. In smaller tanks main reinforcement is placed near the outer face. For bigger tanks the wall thickness is more and the reinforcement is placed on both faces.

Though it is assumed the base is flexible, but in reality there will always be some restraint at the base and some pressure will be resisted by cantilever action of the wall. The minimum verical reinforcement provided will be adequate to resist bending stresses caused by cantilever action.

Ex. 16·1. *Design a circular tank flexible base for capacity of 500,000 litres.*

Sol. Depth of 4 m is provided with free board of 0·2 m. If *D* is the diameter of the tank, capacity of the tank will be

$$\frac{\pi}{4}\times D^2\times 3{\cdot}8=\frac{500{,}000\times 10^6}{10^9}$$

$$D=\sqrt{\frac{500\times 4}{3{\cdot}8\pi}}=12{\cdot}94\ \text{m}.$$

Provide diameter of 13 m.

Maximum hoop tension at base for one metre height

$$=\frac{wHD}{2}=\frac{10{,}000\times 4\times 13}{2}=260{,}000\ \text{N}$$

Area of steel required

$$=\frac{260{,}000}{100}=2600\ \text{mm}^2$$

Provide 16 mm ϕ bars at 70 mm centres.

$$A_t=2873\ \text{mm}^2$$

Use mix M 20. Allowable stress in tension

$$=1{\cdot}2\ \text{N/mm}^2$$

Let t be thickness of wall in mm

$$\text{Tensile stress} = \frac{260{,}000}{1000t+(13-1)\times 2873} = 1{\cdot}2$$

$$t = \frac{260{,}000}{1{\cdot}2\times 1000} - \frac{1{\cdot}2\times 2873}{1000}$$

$$= 216{\cdot}7 - 34{\cdot}48 = 251{\cdot}18 \text{ mm}$$

Provide thickness of 260 mm.

Hoop reinforcement will be provided on both faces. 16 mm ϕ bars at 140 mm centres are provided on each face.

Vertical reinforcement

$$= \left[0{\cdot}3 - 0{\cdot}1\times\frac{(360-100)}{350}\right]\% = 0{\cdot}26\%.$$

Area of vertical reinforcement

$$= \frac{0{\cdot}26}{100}\times 260\times 1000 = 675 \text{ mm}^3$$

Area required on each face $= 338$ mm².

Provide 10 mm ϕ at 220 mm centres on each face.

Main Reinforcement at Different Depths at 3 m. from Top.

$$\text{Hoop tension} = \frac{10{,}000\times 3\times 13}{2} = 195{,}000 \text{ N}$$

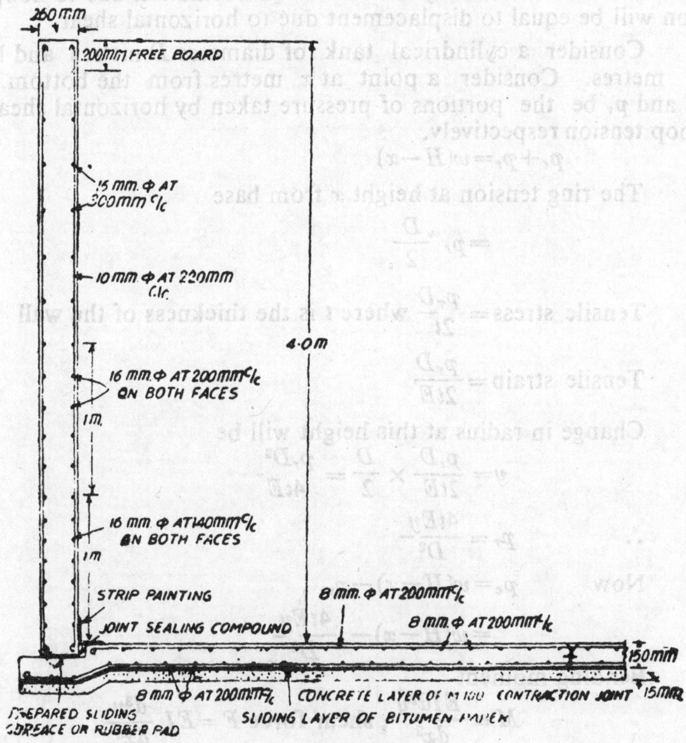

Fig. 16·6

Steel area $=\frac{195,000}{100}=1950 \text{ mm}^2$

Provide 16 mm ϕ bars at 200 mm centres on each face.

At 2 m from top

Hoop tensile reinforcement $=\frac{10,000\times2\times13}{2\times100}=1300 \text{ mm}^2$.

Provide 16 mm ϕ bars at 300 mm centres on each face.

Same reinforcement is provided for remaining portion.

Design of base. 150 mm thick base slab is provided.

Minimum reinforcement required

$$=\frac{0.3}{100}\times150\times1000=450 \text{ mm}^2.$$

Provide 8 mm ϕ bars at 200 mm centres in both directions at top and bottom.

16·7. Circular tanks with restraint at base and roof. Due to restraint at the base and roof, the walls are subjected to horizontal shear besides hoop tension. The amount of load transferred by horizontal shear and hoop tension will depend upon the end conditions. At any section the deformation due to hoop tension will be equal to displacement due to horizontal shear.

Consider a cylindrical tank of diameter D metres and height H metres. Consider a point at x metres from the bottom. Let p_c and p_r be the portions of pressure taken by horizontal shear and hoop tension respectively.

$$p_c+p_r=w(H-x)$$

The ring tension at height x from base

$$=p_r\frac{D}{2}$$

Tensile stress $=\frac{p_rD}{2t}$ where t is the thickness of the wall

Tensile strain $=\frac{p_rD}{2tE}$

Change in radius at this height will be

$$y=\frac{p_rD}{2tE}\times\frac{D}{2}=\frac{p_rD^2}{4tE}$$

$$\therefore \quad p_r=\frac{4tEy}{D^2}$$

Now $\quad p_c=w(H-x)-p_r$

$$=w(H-x)-\frac{4tEy}{D^2}$$

Bending moment

$$M=\frac{EId^2y}{dx^2}, \text{ shear force } F=EI\frac{d^3y}{dx^3}$$

Loading intensity

$$p_c = EI\frac{d^4y}{dx^4}$$

If the effect of the lateral restraint is taken into account modified flextural rigidity will be

$$= \frac{Et^3}{12(1-\mu^2)} \qquad (\mu = \text{Poisson's ratio})$$

$$\therefore \quad \frac{Et^3}{12(1-\mu^2)}\frac{d^4y}{dx^4} = w(H-x) - \frac{4tEy}{D^2}$$

$$\therefore \quad \frac{d^4y}{dx^4} + \frac{48(1-\mu^2)y}{t^2D^2} = \frac{12w(1-\mu^2)}{Et^3}\frac{(H-x)}{}$$

Putting $\quad \alpha = \sqrt[4]{\dfrac{12(1-\mu^2)}{t^2D^2}}$

$$\therefore \frac{d^4y}{dx^4} + 4\alpha^4 y = \frac{12w(H-x)}{Et^3}\cdot\frac{(1-\mu^2)}{}$$

The solution of this differential equation is

$$y = e^{\alpha x}(C_1\cos\alpha x + C_2\sin\alpha x) + e^{-\alpha x}(C_3\cos\alpha x + C_4\sin\alpha x) + \frac{w(H-x)D^2}{4Et}$$

The values of constants C_1, C_2, C_3, and C_4 will depend on the restraint provided at top and bottom.

The value of μ may be taken as 0·2.

1. Circular tanks fixed at base and free at top. At top the shear force and B.M. will be zero. At the base slope and deflection will be zero. Applying these four conditions, four equations will be obtained which can be solved and constants C_1, C_2, C_3 and C_4 evaluated.

$$\text{At } x = H,\ M = 0 \quad \therefore\ EI\frac{d^2y}{dx^2} = 0$$

$$\text{At } x = H,\ F = 0 \quad \therefore\ EI\frac{d^3y}{dx^3} = 0$$

$$\text{At } x = 0,\ \theta = 0 \quad \therefore\ EI\frac{dy}{dx} = 0$$

$$\text{At } x = 0,\ y = 0 \quad \therefore\ EIy = 0$$

Knowing four constants solution of elastic curve is known. Hence values of p_c and p_r can be found at different heights. Table 16·2 gives coefficients for ring tension and B.M. at various heights and shear at base.

Table 17·2 A. CYLINDRICAL T.

Coefficients for Tension in Cir

Triangular Load

T = Coefficient × w HR kg. pe

Positive sign indicates tension

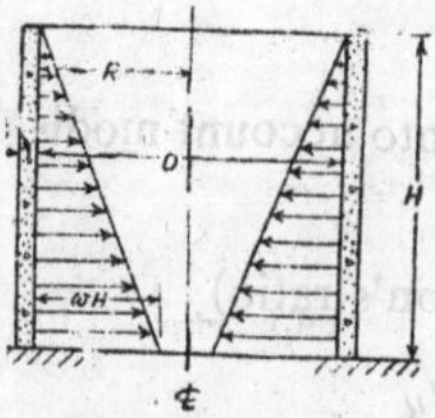

H	0·8H	0·9H	1·0H
)2	−·0529	−·0816	−·1205
38	−·0224	−·0465	−·0795
22	−·0108	−·0311	−·0602
58	−·0051	−·0232	−·0505
75	−·0021	−·0185	−·0436
77	+·0012	−·0119	−·0333
69	+·0023	−·0080	−·0268
59	+·0028	−·0058	−·0222
51	+·0029	−·0041	−·0187
38	+·0029	−·0022	−·0146
29	+·0028	−·0012	−·0122
23	+·0026	−·0005	−·0104
19	+·0023	−·0001	−·0090
13	+·0019	+·0001	−·0079

$\frac{H^2}{Dt}$	0·0H	0·1H	0·2H
0·4	+0·149	+0·134	+0·120
0·8	+0·263	+0·239	+0·215
1·2	+·0283	+0·271	+0·254
1·6	0·265	+0·268	+0·268
2·0	+0·234	0·251	+0·273
3·0	+0·134	+0·203	+0·267
4·0	+0·067	+0·164	+0·256
5·0	+0·025	+0·137	+0·245
6·0	+0·018	+0·119	+0·234
8·0	−0·011	+0·104	+0·218
10·0	−0·011	+0·098	+0·208
12·0	−0·005	+0·097	+0·202
14·0	−0·002	+0·098	+0·200
16·0	0·000	+0·099	+0·199
	Co-efficient at		
	0·75H	0·80H	0·85H
20	+0·716	+0·654	+0·520
24	+0·746	+0·702	+0·577
32	+0·782	+0·768	+0·663
40	+0·800	+0·805	+0·731
48	+0·791	+0·828	+0·785
56	+0·763	+0·838	+0·824

Table 16·2 C

CYLINDRICAL TANKS WITH FIXED BASE AND FREE TOP

Coefficients for Shear at Base

Shear V = Coefficient × wH^2 N

Positive sign indicates shear acting inwards.

$\frac{H^2}{Dt}$	0·4	0·8	1·2	1·6	2·0	3·0	4·0	5·0	6·0	8·0	10·0	12·0	14·0
Coefficient	+0·436	+0·374	+0·339	0·317	0·299	+0·262	+0·236	+0·213	+0·197	+0·174	+0·158	+0·145	+0·135

H^2/Dt	16·0	20·0	24·0	32·0	40·0	48·0	56·0
Coefficient	+0·127	+0·114	+0·102	0·089	0·080	0·072	+0·067

2. Circular tanks hinged at base and free at top. At the top shear force and bending moment will be zero. At the base deflection and B.M. will be zero

At $x=H$, $-M=EI\frac{d^2y}{dx^2}=0$

At $x=H$, $-F=EI\frac{d^3y}{dx^3}=0$

At $x=0$, $EIy=0$

At $x=0$, $-M=EI\frac{d^2y}{dx^2}=0$.

Thus four equations are obtained. These four equations can be solved for four constants. Hence values of p_c and p_r can be found.

Table 16·3 A, B and C give coefficient for ring tension and B.M. at various heights and shear at

3. Cylindrical tank with top slab and fixed base. At top deflection will be zero and slope will be as for the slope of top slab. These conditions will give two equations. At bottom, slope and deflection will be zero. Thus four equations are obtained which can be solved for constants C_1, C_2, C_3 and C_4. Hence values of p_c and p_r can be calculated. B.M. and ring tension at various heights can be calculated.

4. Cylindrical tanks with domes at top and bottom. The slopes and deflections of the wall at top and bottom will be equal to slopes and deflections for the domes at top and bottom respectively. Thus four equations can be formed which can be solved for four constants. Knowing equation of elastic curve, values of p_c and p_r can be calculated. Hence B.M. and ring tension at various heights can be calculated.

Example 16·2. *Design the tank of Problem 16·1, if the tank is fixed at the base and free at the top.*

Sol. $D=13$ m, $H=4$ m.

Assume thickness of wall=150 mm

$$\frac{H^2}{Dt}=\frac{4^2}{13\times0{\cdot}15}=8{\cdot}20$$

Coefficients for hoop tension and B.M. for various heights are found by interpolation from Table 16·2. These coefficients are given in Table 16·4.

Maximum hoop tension occurs at 0·6 H from top.

Maximum hoop tension

$$=0{\cdot}5783\times w\ \frac{HD}{2}$$

$$=0{\cdot}5783\times10{,}000\times\frac{4\times13}{2}$$

$$=147{,}360\text{ N}$$

Table 16·3 C

CYLINDRICAL TANSK WITH HINGED BASE AND FREE TOP

Coeffi.ients for Shear at Base

Shear V = Coefficient $\times wH^2$ N,

Positive sign indicates shear acting inwards.

H^2/Dt	0·4	0·8	1·2	1·6	2·0	3·0	4·0	5·0	6·0	8·0	10·0	12·0	14·0	16·0
Coefficient	+0·245	+0·234	+0·220	+0·204	+0·189	+0·158	+0·137	+0·121	+0·110	+0·096	+0·087	+0·079	+0·073	+0·068

Table 16·4

Depth	*Coefficient for hoop tension*	*Depth*	*Coefficient for B.M.*
0·0*H*	−0·011	0·1*H*	0·0000
0·1*H*	+0·1034	0·2*H*	+0·00009
0·2*H*	+0·2170	0·3*H*	+0·00019
0·3*H*	+0·3338	0·4*H*	+0·00076
0·4*H*	+0·4424	0·5*H*	+0 00151
0·5*H*	+0·5348	0·6*H*	+0 00271
0·6*H*	+0·5783	0·7*H*	+0·00371
0·7*H*	+0·5359	0·8*H*	+0·00289
0·8*H*	+0·3869	0·9*H*	−0·0021
0·9*H*	+0·1538	1*H*	−0·01436
H	—		

Steel area required $=\dfrac{147{,}360}{\cdot 100}=1473{\cdot}6\ \text{mm}^2$

Provide 12 mm ϕ bars at 140 mm centres on both faces $A_t=1616\ \text{mm}^2$.

The reinforcement is provided for 0·8*H* to 0·3 *H* from top. In the remaining portion provide 12 mm ϕ bars at 200 mm centres.

Tensile stress in concrete

$$=\frac{147{,}360}{1000\times 150+(m-1)\times 1616}$$

$$=\frac{147{,}360}{150{,}000+12\times 1616}$$

$$=\frac{147{,}360}{169{,}392}=0{\cdot}87\ \text{N/mm}^2$$

Maximum positive B.M. (tension outside) occurs at 0·7*H* from top. Maximum +ve B.M. per metre height

$$=\text{coefficient}\times wH^3$$

$$=0{\cdot}0037\times 10{,}000\times 4^3$$

$$=+2374\ \text{Nm}$$

Maximum −ve B.M. (tension inside)

$$=0{\cdot}01436\times 10{,}000\times 4^3$$

$$=9191\ \text{Nm}$$

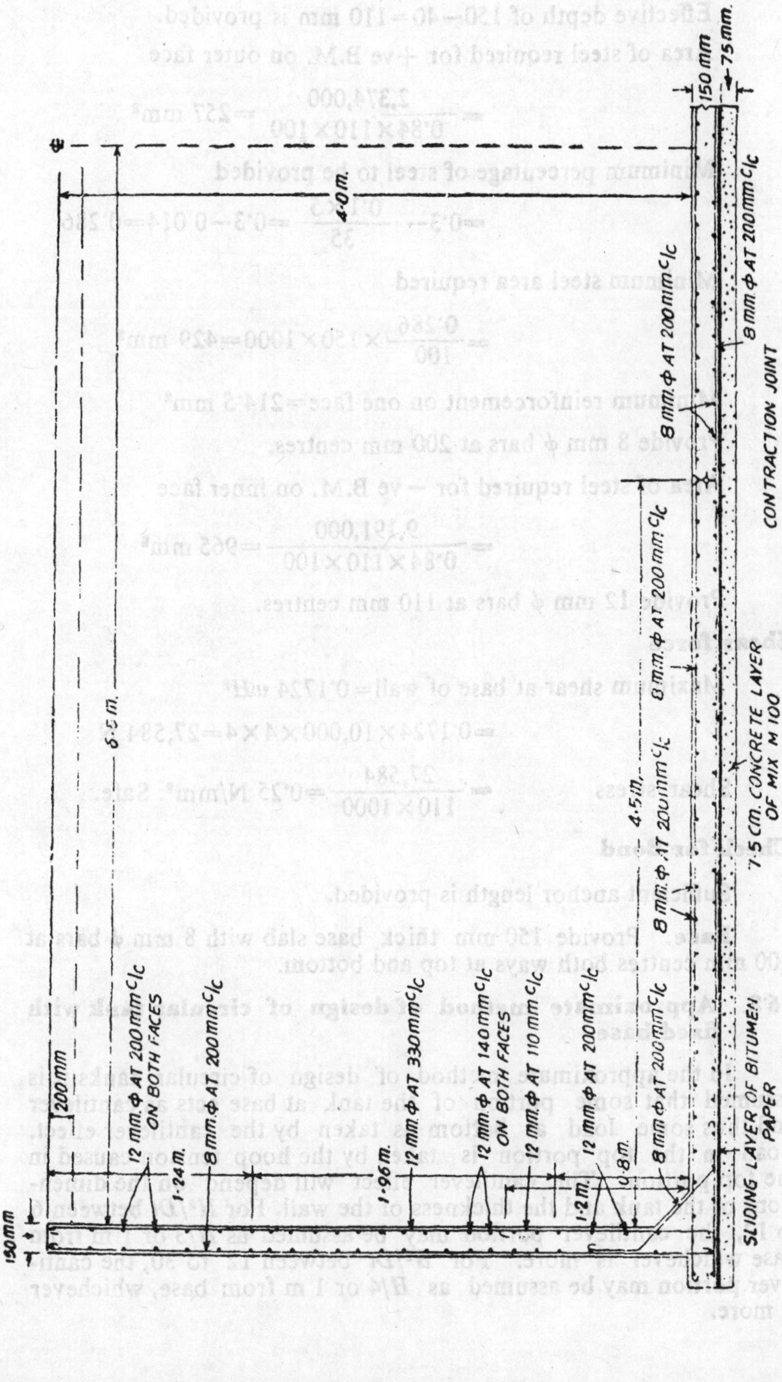
150 mm
75 mm
4·0 m.
8·5 m.
4·5 m.
150 mm
200 mm
1·24 m.
12 mm φ AT 200 mm c/c ON BOTH FACES
8 mm φ AT 200 mm c/c
1·96 m.
12 mm φ AT 330 mm c/c
12 mm φ AT 140 mm c/c ON BOTH FACES
12 mm φ AT 110 mm c/c
1·2 m.
12 mm φ AT 200 mm c/c
0·8 m.
8 mm φ AT 200 mm c/c
8 mm φ AT 200 mm c/c
8 mm φ AT 200 mm c/c
8 mm φ AT 200 mm c/c
8 mm φ AT 200 mm c/c
8 mm φ AT 200 mm c/c
SLIDING LAYER OF BITUMEN PAPER
7·5 CM. CONCRETE LAYER OF MIX M 100
CONTRACTION JOINT

Fig. 16·7

Effective depth of 150−40=110 mm is provided.

Area of steel required for +ve B.M. on outer face

$$=\frac{2,374,000}{0{\cdot}84\times110\times100}=257 \text{ mm}^2$$

Minimum percentage of steel to be provided

$$=0{\cdot}3-\frac{0{\cdot}1\times5}{35}=0{\cdot}3-0{\cdot}014=0{\cdot}286$$

Minimum steel area required

$$=\frac{0{\cdot}286}{100}\times150\times1000=429 \text{ mm}^2$$

Minimum reinforcement on one face=214·5 mm²

Provide 8 mm ϕ bars at 200 mm centres.

Area of steel required for −ve B.M. on inner face

$$=\frac{9,191,000}{0{\cdot}84\times110\times100}=965 \text{ mm}^2$$

Provide 12 mm ϕ bars at 110 mm centres.

Shear force

Maximum shear at base of wall=0·1724 wH^2

$$=0{\cdot}1724\times10,000\times4\times4=27,584 \text{ N}$$

Shear stress $=\dfrac{27,584}{110\times1000}=0{\cdot}25 \text{ N/mm}^2$. Safe.

Check for Bond

Sufficient anchor length is provided.

Base. Provide 150 mm thick base slab with 8 mm ϕ bars at 200 mm centres both ways at top and bottom.

16·8. Approximate method of design of circular tank with fixed base

In the approximate method of design of circular tanks it is assumed that some portion of the tank at base acts as cantilever and thus some load at bottom is taken by the cantilever effect. Load in the top portion is taken by the hoop tension caused in the top portion. The cantilever effect will depend on the dimensions of the tank and the thickness of the wall. For H^2/Dt between 6 to 12, the cantilever portion may be assumed as $H/3$ or 1 m from base whichever is more. For H^2/Dt between 12 to 30, the cantilever portion may be assumed as $H/4$ or 1 m from base, whichever is more.

In Fig. 16·8, AB is the height of tank and ADC pressure diagram. Portion $BD'=h$ is assumed to act as cantilever. ADB is taken as pressure causing hoop tension and DBC is taken as cantilever load. The maximum hoop tension occurs at D and maximum B.M. occurs at the base.

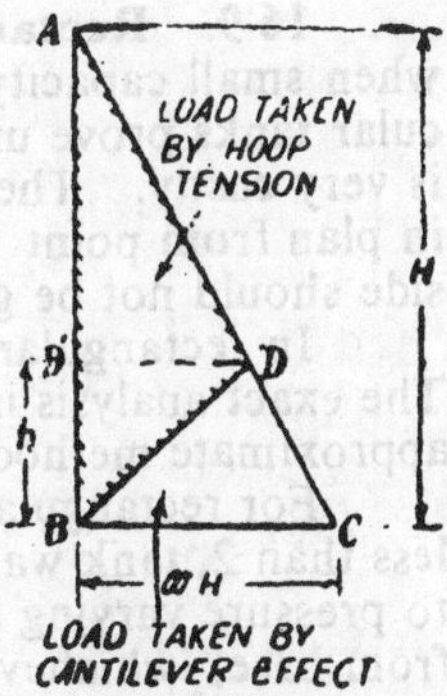

Fig. 16·8

The steel for hoop tension is provided on both faces. For the bottom portion BD reinforcement for hoop tension is provided in addition to steel required for bending.

Ex 16·3. *Design the water tank of Problem 16·1 by approximate method.*

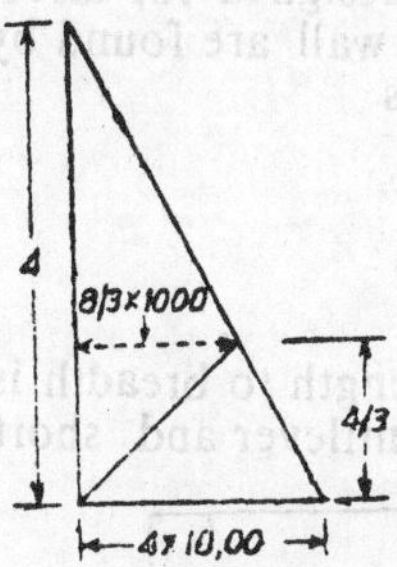

Fig. 16·9

Sol. $D=13$ m, $H=4$ m

Assume thickness of wall $t=150$ mm

$$\frac{H^2}{Dt}=\frac{4\times 4}{13\times 0{\cdot}15}=8{\cdot}276$$

It is assumed that bottom $H/3$, *i.e.* 4/3 m acts as cantilever.

$$\text{Maximum hoop tension}=\frac{pD}{2}$$

$$=\frac{80{,}000}{3}\times\frac{13}{2}$$

$$=173{,}330 \text{ N per metre height.}$$

Area of steel required

$$=\frac{173{,}330}{100}=1733 \text{ mm}^2$$

Provide 12 mm ϕ bars at 130 mm centres on both faces

$$\text{Maximum B.M.} \quad =\frac{1}{2}\times 40{,}000\times\frac{4}{3}\times\frac{1}{3}\times\frac{4}{3}$$

$$=\frac{320{,}000}{27}=11{,}850 \text{ Nm}$$

Effective depth $=150-40=110$ mm

Area of steel required

$$=\frac{11{,}850\times 1000}{0{\cdot}84\times 110\times 100}=1284 \text{ mm}^2$$

Provide 12 mm ϕ bars at 80 mm centre.

Distribution steel

$$\% \text{ reinforcement} \quad =\frac{0{\cdot}3-0{\cdot}1\times(150-100)}{(450-100)}=0{\cdot}286$$

$$\text{Steel area} \quad =\frac{0{\cdot}286}{100}\times 150\times 1000=429 \text{ mm}^2$$

Provide 8 mm ϕ bars at 220 mm centres on each face.

16·9. Rectangular tank. Rectangular tanks are provided when small capacity tanks are required. For small capacities circular tanks prove uneconomical as the formwork for circular tanks is very costly. The rectangular tanks should be preferably square in plan from point of view of economy. It is desirable that longer side should not be greater than twice the smaller side.

In rectangular tanks moments are caused in two directions. The exact analysis is rather difficult and such tanks are designed by approximate methods.

For rectangular tanks in which ratio of length to breadth is less than 2, tank walls are designed as continuous frame subjected to pressure varying from zero at the top to maximum at $H/4$ or 1 m from base, whichever is more. The bottom portion $H/4$ or 1 m whichever is more is designed as cantilever. In addition to bending walls are subjected to direct tension caused by the hydrostatic pressure on the walls. The section is to be designed for direct tension and bending. Bending moments in the wall are found by moment distribution. Direct tension in long walls

$$=\frac{w(H-h)\times B}{2}$$

and direct tension in short walls $=\dfrac{w(H-h)\times L}{2}$

For rectangular tanks of which ratio of length to breadth is greater than 2, the long walls are designed as cantilever and short

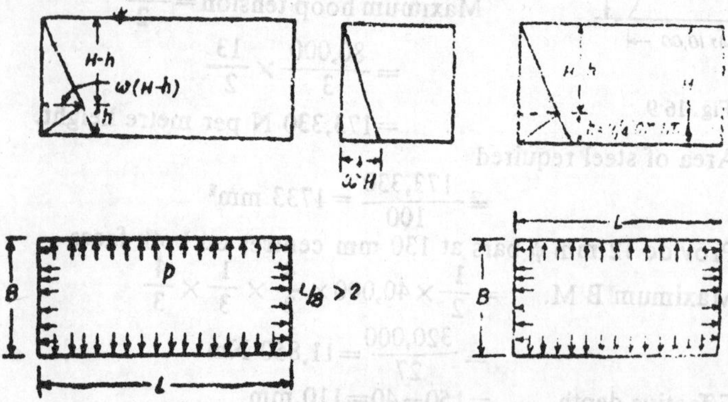

(a) Fig. 16·10 (b)

walls as slabs supported on long wall. Bottom portion of short walls $H/4$ or 1 m whichever is more, is designed as cantilever.

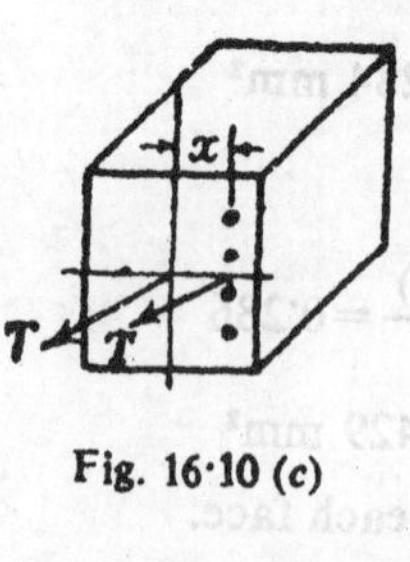

Fig. 16·10 (c)

Maximum B.M. in long walls at base $=\dfrac{1}{2}\,wH-H\times\dfrac{H}{3}=\dfrac{wH^2}{6}$. In the short walls minimum B.M. occurs at support and is given by $\dfrac{w(H-h)B}{12}$. B.M. at centre of short wall is taken as $\frac{w(H-h)B^2}{6}$. For bottom portion of short wall, which is designed as cantilever maximum B.M. is given by $\frac{wHh^2}{6}$ or $\frac{wH\times 1}{6}$ whichever is greater. In addition to B.M. short walls and long walls are subjected to direct tension. Direct tension on long walls is given by $\frac{w(H-h)\times B}{6}$. For short walls it is assumed that end one metre width of long wall contributes to direct tension on the short walls. Direct tension on short wall is $w(H-h)$.

Design of section for tension. It is assumed that entire tension is taken by steel. Let T be tension. Net B.M.$=M-T\times x$. Steel reinforcement is provided for B.M. of $M-T\times x$ and direct tension T as shown in Fig. 16·10 (c).

Ex. 16·4. *Design a rectangular tank for a capacity of 80,000 litres.*

Sol. Provide height of 3·5 m for tank with free board of 150 mm.

Effective height $=3350$ mm

Volume of tank required

$=80{,}000$ litres$=80{,}000\times 10^6$ mm^3

Area of tank to be provided

$$=\frac{80{,}000\times 10^6}{3350}$$

$=23{\cdot}88\times 10^6$ mm$^2=23{,}880{,}000$ mm^2.

Provide length of 6000 mm and breadth of 4000 mm

$$\frac{L}{B}=\frac{6000}{4000}=1{\cdot}5<2$$

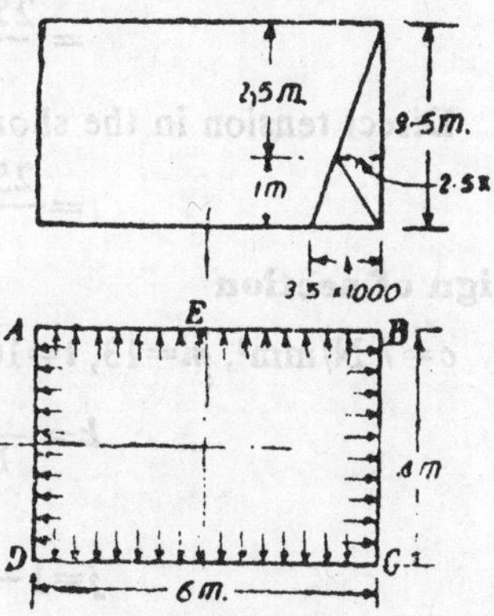

Fig. 16·11

The walls of tanks are to be designed as continuous slab.

$$\frac{H}{4}=\frac{3{\cdot}50}{4}=0{\cdot}875 \text{ m}$$

Bottom 1 m of tank will be designed as cantilever.

Pressure at depth of 2·5 m

$p=wh=2{\cdot}5\times 10{,}000$

$=25{,}000$ N/m^2.

Moments in the walls are found by moment distribution. As the frame is symmetrical about both axes moment distribution is done for one-quarter of the tank only.

Joint	A	
Member	AB	AD
Distribution Factors	0·4	0·6
Fixed Moment	$-3p$	$+\frac{4}{3}p$
Balancing	$+\frac{2}{3}p$	$+p$
Final	$+\frac{7}{3}p$	$+\frac{7}{3}p$

Moment at support $=\frac{7}{3}\ p=\frac{7}{3}\times 25{,}000=58{,}330$ Nm.

B.M. at centre of long span

$$=\frac{25{,}000\times 6^2}{8}-58{,}330=112{,}500-58{,}330$$

$$=54{,}170 \text{ Nm}$$

B.M. at centre of shorter span

$$=\frac{25{,}000\times 4^2}{8}-58{,}330$$

$$=50{,}000-58{,}330=-8{,}330 \text{ Nm}$$

Maximum B.M. $=58{,}330$ Nm

Direct tension in long wall

$$=\frac{25{,}000\times 4}{2}=50{,}000 \text{ N}$$

Direct tension in the short wall

$$=\frac{25{,}000\times 6}{2}=75{,}000 \text{ N}$$

Design of section

$c=7$ N/mm², $m=13$, $t=100$ N/mm² on water face

$$k=\frac{1}{1+t/cm}=\frac{1}{1+\frac{100}{7\times 13}}=0{\cdot}48$$

$$j=1-\frac{k}{3}=1-0{\cdot}16=0{\cdot}84$$

$$Q=\tfrac{1}{2}\ ckj=\tfrac{1}{2}\times 7\times 0{\cdot}48\times 0{\cdot}84=1{\cdot}411.$$

Considering effect of bending only, effective depth required

$$=\sqrt{\frac{58{,}330\times 1000}{1{\cdot}411\times 1000}}=203{\cdot}2 \text{ mm}$$

Provide overall depth of 250 mm with effective depth of 215 mm.

Net moment$=M-T\times x$

Area of steel

$$=\frac{M-T\times x}{0.84\times 100}+\frac{T}{1000}$$

$$=\frac{58,330\times 10000-50,000(215.0-125.0)}{0.84\times 215\times 100}+\frac{50,000}{100}$$

$=2981+500=3481$ mm².

Provide 20 mm ϕ bars at 80 mm centres.
Area of steel provided$=3928$ mm².

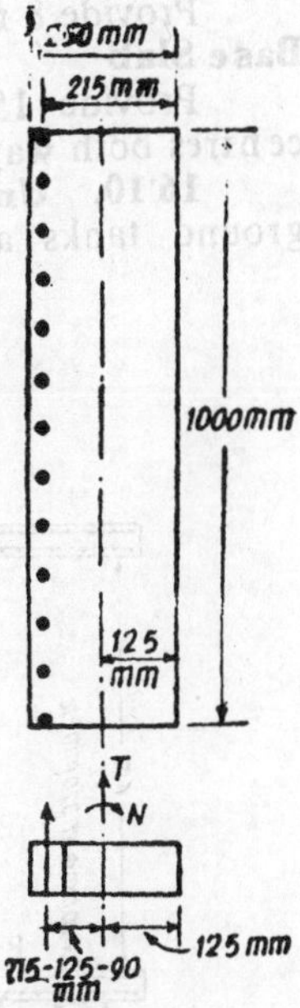

Fig. 16·12

Steel at centre of span

Tension occurs away from water face, $c=7$ N/mm², $t=125$ N/mm², $m=13$, $n=0.42\ d$, $j=0.86\ d$, $Q=1.264$.

Area of steel

$$=\frac{54,170,000-50,000(215.0-125.0)}{0.86\times 215\times 125}+\frac{50,000}{100}=2149+500=2649 \text{ mm}^2.$$

Half the bars from inner face at support are bent in outer face providing area of $\frac{3928}{2}=1964$ mm².

Remaining area to be provided$=2649-1964=685$ mm².

Additional reinforcement of 16 mm ϕ bars at 160 mm centres is provided.

At the centre of short span B.M. is of negative sign, however nominal reinforcement is provided on that face.

Cantilever Moment

Cantilever moment$=3.5\times 10,000\times\frac{1}{2}\times 1\times\frac{1}{3}=5833$ Nm.

Area of steel required$=\frac{5833\times 1000}{0.84\times 215\times 100}=323$ mm²

Provide 8 mm ϕ bars at 150 mm centres.

Ditribution steel

Distribution steel $=0.3-\frac{0.1\times(250-100)}{(450-100)}$

$=0.3-0.043=0.257\%$

$$\text{Area of steel} = \frac{0.257}{300} \times 250 \times 1000 = 642.5 \text{ mm}^2$$

Area on each face $= 321.2$ mm²

Provide 8 mm ϕ bars at 150 mm centres.

Base Slab

Provide 150 mm thick slab with 8 mm ϕ bars at 200 mm centres both ways at top and bottom.

16·10. Underground Tanks The design principles of underground tanks are same as for tanks resting on the ground. The

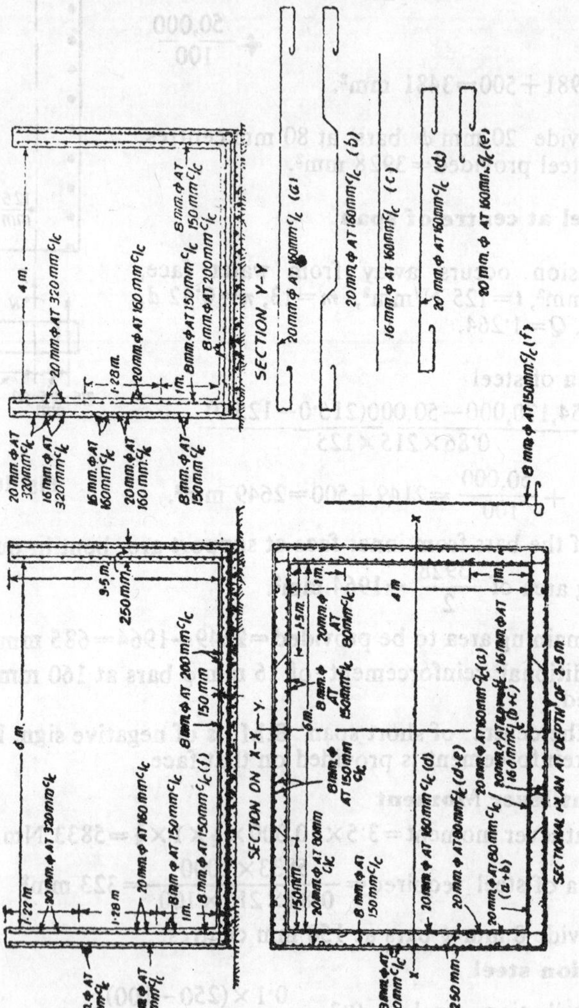

Fig. 16·13

walls of the underground tanks are subjected to internal water pressure and outside earth pressure. The section of wall is designed for water pressure and earth pressure acting separately as well as acting simultaneously.

Whenever there is possibility of water table to rise, soil becomes saturated and earth pressure exerted by saturated soil should be taken into consideration.

Ex. 16.5. *Design an underground reservoir 12 m × 5 m × 4 m deep. The angle of repose of soil is 30° and density of soil is 16,000 N/m². The soil is saturated.*

Sol. As the ratio of length of tank to its breadth is greater than 2, the long walls will be designed as cantilevers and the top portion of the short walls will be designed as slab supported by long walls. Bottom one metre of short walls will be designed as cantilever slab.

Design of long wall

1. Pressure of saturated soil acting from outside and no water pressure from inside. Earth pressure at base will be due to water pressure plus due to submerged weight of soil.

$$p = 10{,}000 \times 4 + (16{,}000 - 10{,}000) \times \tfrac{1}{3} \times 4$$
$$= 40{,}000 + 8000 = 48{,}000 \text{ N/m}^2$$

Maximum B.M. at base of long wall

$$= 48{,}000 \times \frac{4}{2} \times \frac{4}{3}$$

$$= 128{,}000 \text{ Nm} = 128{,}000{,}000 \text{ N-mm}$$

Effective depth required

$$= \sqrt{\frac{128{,}000{,}000}{1{\cdot}411 \times 1000}} = 301{\cdot}1 \text{ mm}$$

Provide overall depth of 350 mm with effective depth of 310 mm.

$$\text{Area of steel} = \frac{128{,}000{,}000}{0{\cdot}84 \times 100 \times 310} = 4828 \text{ mm}^2$$

Provide 20 mm ϕ bars at 60 mm centres on outside face. Area of steel provided = 5236 mm².

Reinforcement is curtailed in the same manner as in the second case.

Direct compression in long walls. Direct compression is caused in long walls because of earth pressure action on short walls which act as slab supported on long walls.

Direct compression at 1 m above base

$$= 36{,}000 \times \frac{4}{2} = 72{,}000 \text{ N.}$$

This will be taken by wall and the distribution steel provided.

2. Water pressure acting from inside and no earth pressure acting from outside.

Maximum water pressure at base

$$=10{,}000\times 4=40{,}000 \text{ N/m}^2$$

Maximum B.M. $\quad =40{,}000\times\dfrac{4}{2}\times\dfrac{4}{3}$

$$=\frac{320{,}000}{3}=106{,}667 \text{ Nm}$$

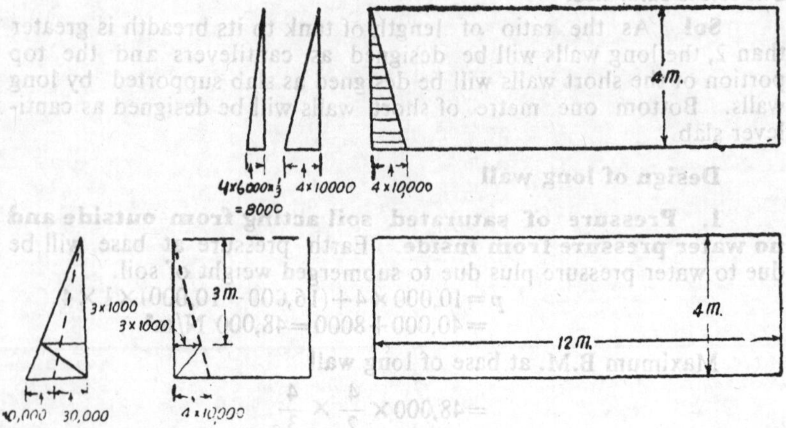

Fig. 16·14

Area of steel required

$$=\frac{106{,}667\times 1000}{0{\cdot}84\times 100\times 310}=4096 \text{ mm}^2.$$

Provide 20 mm ϕ bars at 70 mm centres on inside face. Area of steel provided $=4486$ mm^2.

Curtailment of reinforcement. Let A_{th} be the reinforcement required at depth h. The B.M. at any depth is proportional to h^3.

$$\therefore \quad \frac{A_{th}}{A_t}=\frac{h^3}{H^3}$$

$$\therefore \quad h=\sqrt[3]{\frac{A_{th}}{A_t}}\times H$$

Depth where half the bars can be curtailed, $A_{th}=\frac{1}{2}A_t$

$$\therefore \quad h=\sqrt[3]{\tfrac{1}{2}}\times 4=3{\cdot}175 \text{ m}.$$

Taking bond length as 20 d, the bars are curtailed at 1·25 m from base. 20 mm ϕ at 120 mm c/c are provided on outside and 20 mm ϕ at 140 mm c/c inside.

Depth where only $\frac{1}{4}$th reinforcement is required $A_{th}=\frac{1}{4}A_t$.

$$h=\sqrt[3]{\tfrac{1}{4}}\times 4=2{\cdot}52 \text{ m.}$$

Half reinforcement, *i.e.* 20 mm ϕ at 240 mm centres are provided on outside and 20 mm ϕ at 280 c/c inside at 2 m and above from base.

Distribution steel

% of distribution steel

$$=0{\cdot}3-\frac{0{\cdot}1\times(320-100)}{(450-100)}=0{\cdot}3-0{\cdot}06=0{\cdot}24$$

Area of steel $=0{\cdot}24\times\dfrac{320\times 100}{1000}=768 \text{ mm}^2.$

Area to be provided on each face $=384 \text{ mm}^2$

Provide 8 mm ϕ bars at 130 mm centres.

Direct tension in long walls. Direct tension is caused in long walls because of water pressure acting on short walls which act as slab supported on long walls.

Direct tension at 1 m above base

$$=3\times 10{,}000\times\frac{4}{2}=60{,}000 \text{ N}$$

Area of steel required $=\dfrac{60{,}000}{100}=600 \text{ mm}^2.$

Area of distribution steel provided is 760 mm^2.

Distribution steel will take direct tension.

Design of short wall

1. Water pressure acting from inside and no earth pressure acting from outside.

Bottom one metre acts as cantilever and remaining 3 m acts as slab supported on long walls.

Water pressure at depth of 3 m

$$=10{,}000\times 3=30{,}000 \text{ N/m}^2$$

Maximum B.M. is assumed as

$$=\frac{wl^2}{12} \text{ at supports.}$$

B.M. at support $=\dfrac{wl^2}{12}=\dfrac{30{,}000\times 4^2}{12}=40{,}000 \text{ Nm.}$

Direct tension from end one metre of long wall

$$=3000\times 1=30{,}000 \text{ N}$$

Net B.M. $=M-T\times x=40{,}000-30{,}000\times 0{\cdot}12$

$$=36{,}400 \text{ Nm}$$

Area of steel $=\dfrac{36{,}400\times 1000}{0{\cdot}84\times 310\times 100}+\dfrac{30{,}000}{100}$

(Tension inside)

$=1397+300=1697\ \text{mm}^2.$

Provide 12 mm ϕ bars at 60 mm centres in one metre length. As the bending moment is proportional to the depth of water, reinforcement will vary linearly with depth of water. At depth of 2 m, 12 mm ϕ bars at 120 mm centres are provided. At depth of 1 m, 12 mm ϕ bars at 180 mm centres are provided.

B.M. at centre of span

$=\dfrac{wl^2}{16}=\dfrac{30{,}000\times 4^2}{16}=30{,}000\ \text{Nm}.$

Net B.M. $=M-T\times x=30{,}000-30{,}000\times 0{\cdot}12$

$=26{,}400\ \text{Nm}.$

Area of steel $=\dfrac{26{,}400\times 1000}{0{\cdot}86\times 310\times 100}+\dfrac{30{,}000}{100}$

$=1014+300=1314\ \text{mm}^2.$

Actually the section is doubly reinforced as steel is to be provided on front face for the bending moment produced due to outside pressure.

12 mm ϕ bars are provided at 600 mm centres.

A_t provided $=1885\ \text{mm}^2.$

The reinforcement is reduced linearly towards top. At depth of 2 metres 12 mm ϕ bars are provided at 120 mm centres. At depth of 1 metre 12 mm ϕ bars are provided at 180 mm centres.

Design of bottom one metre

B.M. $=10{,}000\times 4\times \frac{1}{2}\times \frac{1}{3}=6670\ \text{Nm}$

Area of steel $=\dfrac{6670\times 1000}{0{\cdot}84\times 310\times 100}=256\ \text{mm}^2$

Minimum reinforcement of 8 mm ϕ at 130 mm centres is provided.

Distribution steel of 8 mm ϕ at 130 mm centres is provided on both faces.

2. Pressure of saturated soil acting from outside and no water pressure from inside

Direct compression due to cantilever action of one metre length of long wall$=36{,}000$ N.

Pressure $p=36{,}000\ \text{N/m}^2.$

B.M. at support $=\dfrac{pl^2}{12}=\dfrac{36{,}000\times 4\times 4}{12}$

$=48{,}000\ \text{Nm}$

B.M. at the centre of span

$$=\frac{36{,}000\times 4\times 4}{16}$$

$$=36{,}000 \text{ Nm}$$

Area of steel required, considering bending only,

$$A_t=\frac{48{,}000\times 1000}{0{\cdot}86\times 100\times 310}$$

$$=1843 \text{ mm}^2. \qquad \text{(Tension outside)}$$

Provide 12 mm ϕ bars at 60 mm centres

A_t provided $=1885 \text{ mm}^2$.

Actually the section is doubly reinforced and area of steel required will be less. The effect of direct compressive force is not considered as it is very small and compressive reinforcement is provided.

Area of steel at centre of span

$$=\frac{36{,}000\times 1000}{0{\cdot}84\times 100\times 310}$$

$$=1381 \text{ mm}^2.$$

Provide 12 mm ϕ at bars at 60 mm centres.

Design of bottom one metre

$$\text{B.M.}=\tfrac{1}{2}\times 48{,}000\times \tfrac{1}{3}=8000 \text{ Nm}$$

$$\text{Area of steel}=\frac{8000\times 1000}{0{\cdot}84\times 100\times 31}=307 \text{ mm}^2$$

Provide 8 mm ϕ bars at 130 mm centres.

Base slab

Check Against Uplift

Protection of 0·3 m is provided beyond the face of the walls to add to stability against uplift.

Weight of vertical walls.

Long walls $=2\times 12{\cdot}7\times 0{\cdot}35\times 4\times 24{,}000=853{,}440$ N

Short walls $=2\times 4\times 0{\cdot}35\times 24{,}000=268{,}800$ N

Base slab $=13{\cdot}3\times 5{\cdot}30\times 0{\cdot}4\times 24{,}000=676{,}740$ N

Weight of earth on projection $=2(13{\cdot}3+4{\cdot}7)\times 4\times 0{\cdot}3\times 16{,}000$
$=691{,}200$ N

Uplift pressure of water at bottom of tank

$=5{\cdot}3\times 13{\cdot}3\times 4{\cdot}3\times 1000=3{,}102{,}000$ N

Total downward weight

$=853{,}440+261{,}800+676{,}700+691{,}200$

$=2{,}490{,}140$ N

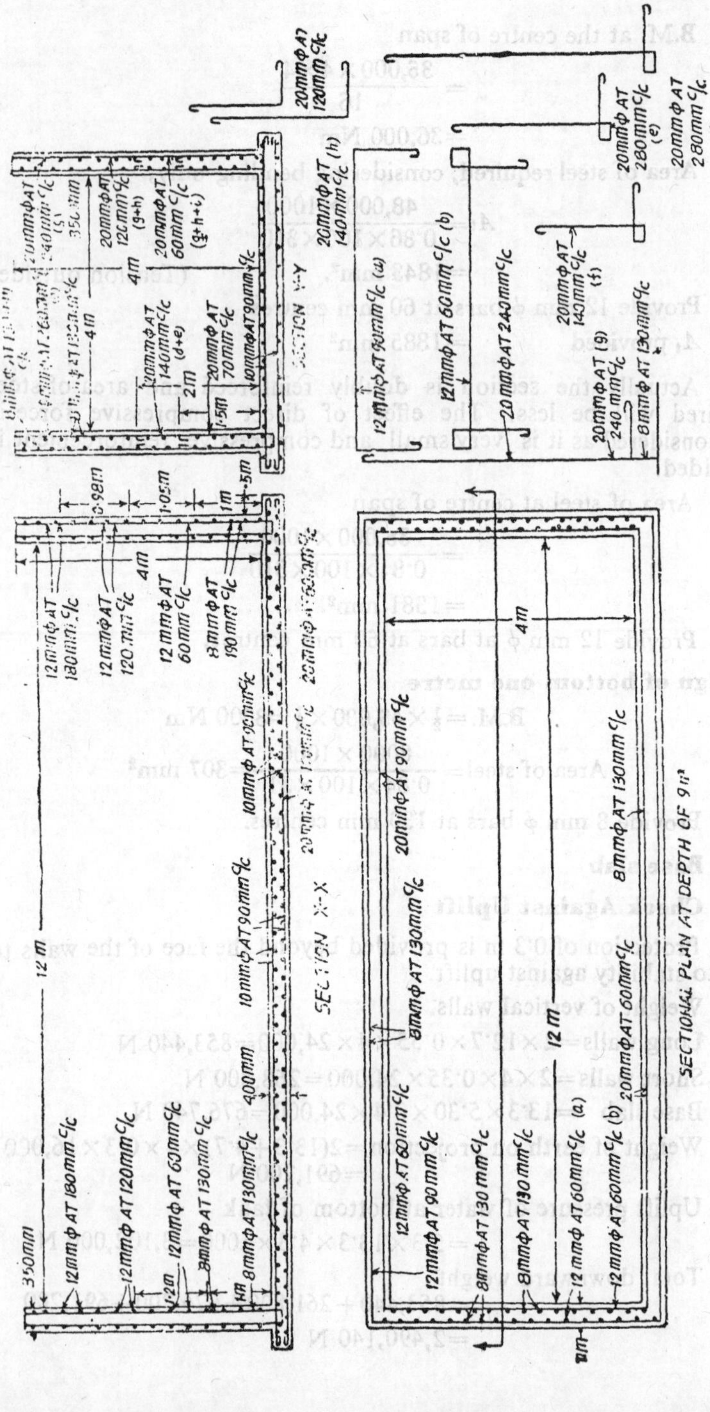
SECTION X-X
SECTION Y-Y
SECTIONAL PLAN AT DEPTH OF 9 m
12 m
4 m
350 mm
12 mmφ AT 180 mm c/c
12 mmφ AT 120 mm c/c
12 mmφ AT 60 mm c/c
8 mmφ AT 130 mm c/c
10 mmφ AT 90 mm c/c
20 mmφ AT 90 mm c/c
20 mmφ AT 130 mm c/c
8 mmφ AT 130 mm c/c
20 mmφ AT 60 mm c/c
12 mmφ AT 90 mm c/c
12 mmφ AT 60 mm c/c (a)
12 mmφ AT 60 mm c/c (b)
20 mmφ AT 240 mm c/c
12 mmφ AT 90 mm c/c (a)
12 mmφ AT 60 mm c/c (b)
20 mmφ AT 140 mm c/c (f)
20 mmφ AT 280 mm c/c (e)
20 mmφ AT 280 mm c/c (d)
20 mmφ AT 120 mm c/c
20 mmφ AT 140 mm c/c (h)
20 mmφ AT 60 mm c/c (g+h+i)
20 mmφ AT 70 mm c/c
12 mmφ AT 140 mm c/c (d+e)
10 mmφ AT 90 mm c/c
8 mmφ AT 190 mm c/c
0·98 m
1·05 m
1 m
·5 m
1·5 m
2 m
400 mm

Fig. 16·15

Frictional resistance required

$$=3,102,000-2,490,120=611,880 \text{ N}$$

Pressure of submerged earth and water at depth of 4·4 m

$$=10,000\times4{\cdot}4+\tfrac{1}{3}(16000-1000)\times4{\cdot}4$$
$$=44,000+8,800=52,800 \text{ N}$$

Total pressure per one metre length of walls

$$=\tfrac{1}{2}\times4{\cdot}4\times5280=116,160 \text{ N}$$

As the soil is saturated, the angle of friction of submerged soil will be low. Assuming coefficient of friction as 0·15, frictional resistance per metre length of wall

$$=0{\cdot}15\times116,160=17,424 \text{ N}$$

Total frictional resistance of four sides

$$=2(5{\cdot}3+13{\cdot}3)\times17424$$
$$=647973>611,800 \quad \text{Safe.}$$

Design of Base Slab

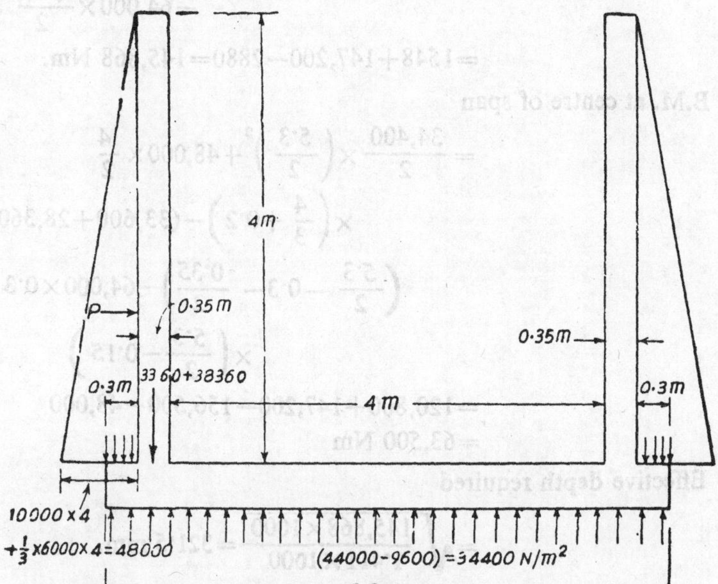

Fig. 16·16

Consider one metre length of slab.

Upward pressure of water per sq. metre

$$=4{\cdot}4\times10,000=44,000 \text{ N/m}^2.$$

Self-weight of slab $=1\times1\times0{\cdot}4\times24,000=9600 \text{ N/m}^2.$

Net upward pressure

$$=44,000-9600=34,400 \text{ N/m}^2$$

Weight of wall per metre run

$$=0{\cdot}35\times 4\times 24,000=33,600 \text{ N}$$

Weight of earth on projection

$$=16,000\times 4{\cdot}0=64,000 \text{ N/m}^2.$$

Net unbalanced force

$$=34,400\times 5{\cdot}3-2(33,600+64,000\times 0{\cdot}5)$$
$$=182,320-105,600=76,720 \text{ N}.$$

Reaction on each wall

$$=\frac{76,720}{2}=38,360 \text{ N}.$$

B.M. at edge of cantilever portion

$$=\frac{34,400\times(0{\cdot}3)^2}{2}+48,000\times\frac{4}{2}\left(\frac{4}{3}+0{\cdot}2\right)-64,000\times\frac{(0{\cdot}3)^2}{2}$$

$$=1548+147,200-2880=145,868 \text{ Nm}.$$

B.M. at centre of span

$$=\frac{34,400}{2}\times\left(\frac{5{\cdot}3}{2}\right)^2+48,000\times\frac{4}{2}\times\left(\frac{4}{3}+0{\cdot}2\right)-(33,600+28,360)\left(\frac{5{\cdot}3}{2}-0{\cdot}3-\frac{0{\cdot}35}{2}\right)-64,000\times 0{\cdot}3\times\left(\frac{5{\cdot}3}{2}-0{\cdot}15\right)$$

$$=120,800+147,200-156,500-48,000$$
$$=63,500 \text{ Nm}$$

Effective depth required

$$=\sqrt{\frac{145,868\times 1000}{1{\cdot}411\times 1000}}=321{\cdot}5 \text{ mm}.$$

Provide overall depth of 400 mm with effective depth of 350 mm.

Area of steel required at support

$$=\frac{145,868\times 1000}{0{\cdot}84\times 100\times 350}=4958 \text{ mm}^2.$$

Provide 20 mm ϕ at bars at 60 mm centres,

Reinforcement at centre of span

$$=\frac{63{,}500\times 1000}{0{\cdot}84\times 100\times 350}=2160 \text{ mm}^2.$$

Provide 20 mm ϕ bars at 120 mm centres.

Minimum reinforcement

$$=0{\cdot}3-0{\cdot}1\times\frac{300}{350}=0{\cdot}3-0{\cdot}09=0{\cdot}21$$

$$\text{Reinforcement} \quad =\frac{0{\cdot}21}{100}\times 400\times 1000=840 \text{ mm}^2.$$

Provide 10 mm ϕ bars at 90 mm c/c as distribution steel at bottom and 10 mm ϕ bars at 90 mm c/c both ways at top.

Maximum S.F. at edge

$$=33{,}600+38{,}360+64{,}000\times 0{\cdot}3-34{,}400(0{\cdot}3+0{\cdot}35)$$

$$=33{,}600+38{,}360+19{,}200-22{,}360=68{,}800 \text{ N}$$

$$\text{Shear stress}=\frac{68{,}800}{1000\times 350}=0{\cdot}197 \text{ N/mm}^2. \text{ Safe.}$$

16·11. Overhead Tanks

These tanks may be rectangular or circular. The tanks are supported on staging which consists of masonry tower or a number of columns braced together. The tank walls are designed in the same way as the walls of tanks resisting on the ground. The base slab of circular tanks is designed as circular slab supported on masonry of circular beam at the end. The slab of rectangular tanks is designed as two-way slab if length is less than twice the breadth, if length is more than twice the breadth the slab is designed as one-way slab. The base slab is subjected to bending moment at the end and to direct tension, caused by the water pressure acting on vertical walls.

For large tanks base slab is supported on series of beams supported on columns.

The staging consists of a number of columns braced together at intervals. The columns are assumed to be fixed at the braces as well as to elevated tank, therefore, effective length of column is taken as distance between bracings.

The wind force acting on the tank and staging produces tension on the windward side columns and compression on the leeward side columns. The force in any column is proportional to its distance from C.G. of the column group.

Let P be the total wind force acting at height h from the base and r_1, r_2...be the distances of the columns from the C.G. of column group, measured parallel to the direction of the wind.

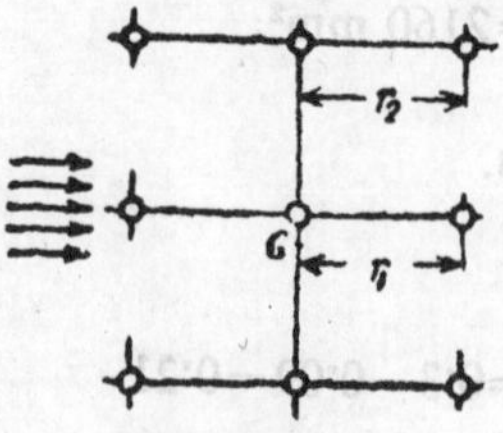

Fig. 16·17

Force F_1 in column 1 at distance r_1 from C.G. of the column group is given by

$$F_1=\frac{Phr_1}{r_1^2+r_2^2+\dots}$$

$$=\frac{Phr_1}{\Sigma r^2}$$

In the design, it is assumed that horizontal shear taken by inner columns is twice that taken by outer columns.

The bracings are designed for B.M. and shear. Same reinforcement is provided at top and bottom as the wind may blow from one side or the other.

Circular tanks are sometimes provided with inclined columns. In such cases the vertical component of the force in each column is found as given above. The horizontal shear in each column is given by deducting the sum of horizontal components of the forces in the columns from the total wind force and dividing by number of columns.

The moments in the inclined braces meeting at a column can be found as follows :

The axes of moments in column above and below the brace will be at right angles to the direction of the wind. The axes of moments in the two braces will be at right angles to their axes. By completing the triangle of moments, the moments in the braces can be found.

Let O be the column and OA and OB be the braces meeting at column O. Oa is drawn perpendicular to OA and Ob is drawn perpendicular to OB and ab is drawn at right angles to the direction of wind, ab gives moment in the column Oa and Ob will give the moment in the column Oa and Ob will give the moments in braces OA and OB respectively. For moment to be maximum in OA, the wind should blow at right angles to the brace OB. In such a case triangle of moments will be right angled triangle and side Oa will be hypotenuse.

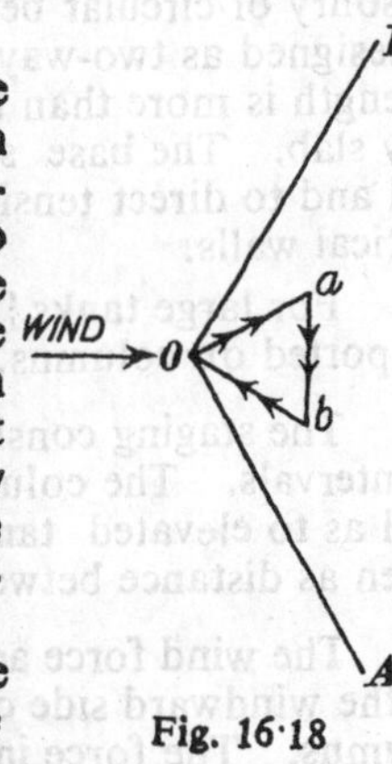

Fig. 16·18

Foundation for elevated tanks. The foundation for elevated tank columns may be combined foundation in the form of raft or independent footing may be provided for each column.

Ex. 16·6. *Design a circular tank having diameter of 6 m and height of 3 m. The tank is supported on masonry tower.*

Sol. The tank is covered with domed roof. Rise of 1 m is provided for the dome.

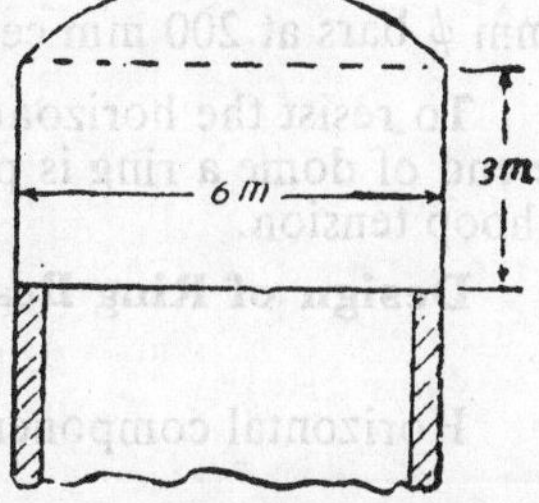

Fig. 16·19 (*a*)

Radius of dome is given by

$$3\times3=1\times(2R-1)$$

$\therefore \quad R=5$ m.

Thickness of 100 mm is provided for dome.

Assume density of concrete 2400 N/m³.

Self-load per square metre

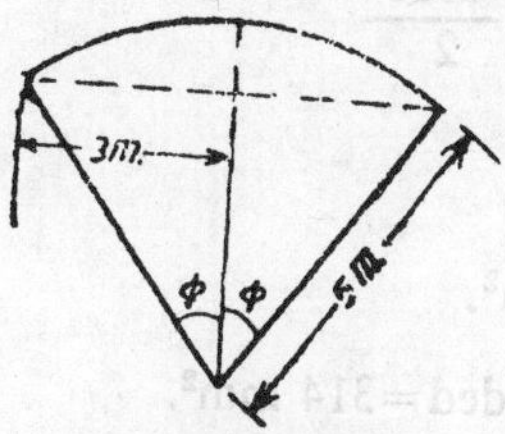

Fig. 16·19 (*b*)

$$=\frac{100}{1000}\times1\times1\times24{,}000$$

$$=2400 \text{ N.}$$

Live load per square metre

$$=1500 \text{ N.}$$

Total load $=3900$ N/m².

Hoop stress at any angle θ is given by

$$\frac{wR}{t}\times\frac{(\cos^2\theta+\cos\theta-1)}{(1+\cos\theta)}$$

Meridional stress $=\dfrac{wR(1-\cos\theta)}{t\sin^2\theta}$

$$\sin\phi=\frac{3}{5}=0{\cdot}6, \qquad \therefore \quad \phi=36^\circ\ 52'$$

$$\cos\phi=\frac{4}{5}.$$

Meridional stress $=\dfrac{3900\times5}{0{\cdot}1}\ \dfrac{(1-\cos\theta)}{\sin^2\theta}\times10^{-6}$ N/mm²

$$=0{\cdot}195\times\frac{(1-\cos\theta)}{\sin^2\theta}.$$

Hoop stress $=\dfrac{3900\times5\,(\cos^2\theta+\cos\theta-1)}{0{\cdot}1\ (1+\cos\theta)}\times10^{-6}$ N/mm²

$$=\frac{0{\cdot}195\ (\cos^2\theta+\cos\theta-1)}{1+\cos\theta}\ \text{N/mm}^2.$$

Maximum meridional stress occurs at

$$\theta=\phi=36^\circ\ 52'.$$

Meridional stress $=0{\cdot}195\times\dfrac{(1-0{\cdot}8)}{0{\cdot}6\times0{\cdot}6}$

$$=0{\cdot}108\ \text{N/mm}^2.$$

Maximum hoop stress occurs at

$$\theta=0^\circ.$$

Hoop stress $=0{\cdot}195\times\dfrac{(1+1-1)}{1+1}=0{\cdot}0975$ N/mm².

Stresses are very small. Provide nominal reinforcement of 8 mm ϕ bars at 200 mm centres both ways.

To resist the horizontal component of the meridional thrust at the end of dome a ring is provided. The ring beam will be subjected to hoop tension.

Design of Ring Beam. Meridional thrust per 1 mm length

$=0{\cdot}108\times100\times1=10{\cdot}8$ N.

Horizontal component of meridional thrust

$=10{\cdot}8\cos\phi=10{\cdot}8\times0{\cdot}8.$

Hoop tension $=\dfrac{pd}{2}=10{\cdot}8\times0{\cdot}8\times\dfrac{6000}{2}$

$=25{,}920$ N.

Area of steel required

$=\dfrac{25{,}920}{140}=185$ mm².

Provide 4 bars of 10 mm ϕ. Area provided $=314$ mm².

6 mm ϕ stirrups at 200 mm centres are provided.

200 mm × 150 mm section of ring beam is provided.

Stress in concrete. Equivalent area

$=200\times150+12\times314{\cdot}0=30{,}000+3768$

$=33{,}768$ mm².

Tensile stress $=\dfrac{25{,}920}{33{,}768}=0{\cdot}768$ N/mm². Safe.

Design of Cylindrical Walls. As the dome is designed on membrane theory, the tank wall is assumed free at top.

$H=3$ m, $D=6$ m. Assume thickness of tank wall $=150$ mm.

$$\frac{H^2}{Dt}=\frac{3^2}{6\times0{\cdot}15}=10.$$

From table 16·2 (*b*)

Maximum hoop tension

$=0{\cdot}608\,wHR$

$=0{\cdot}608\times10{,}000\times3\times3=54{,}720$ N.

Maximum hoop tension occurs at depth of

$0{\cdot}6\,H=1{\cdot}8$ m from top.

Maximum −ve B.M. $=0{\cdot}0122\,wH^3$

$=0{\cdot}0122\times10{,}000\times27=3294$ Nm.

Maximum −ve B.M. occurs at base.

Maximum +ve B.M. (from Table 16·2 A)

$$=0{\cdot}0029\times10{,}000\times27=783 \text{ Nm.}$$

Maximum +ve B.M. occurs at depth of

$$0{\cdot}7\,H=2{\cdot}1 \text{ m from top.}$$

Maximum shear at base

$$=0{\cdot}158\,wH^2$$

$$=0{\cdot}158\times10{,}000\times9=14{,}220 \text{ N.}$$

Steel required for hoop tension

$$=\frac{54{,}720}{100}=547{\cdot}2 \text{ mm}^2.$$

Provide 8 mm ϕ bars at 180 mm centres on both faces.

Minimum steel required

$$=0{\cdot}3-\frac{0{\cdot}1\times(150-100)}{(45-10)}$$

$$=0{\cdot}286\%.$$

Steel required $$=\frac{0{\cdot}286\times150\times1000}{100}=429 \text{ mm}^2.$$

Provide 8 mm ϕ bars at 180 mm centres throughout the height.

Steel required for −ve B.M.

Effective depth $=150-40=110$ mm.

$$A_t=\frac{3294\times10{,}000}{0{\cdot}84\times110\times100}=355 \text{ mm}^2.$$

Provide 8 mm ϕ bars at 140 mm centres. Every third bar will be stopped at 1 m from base, the remaining bars will be taken upto top to serve as distribution steel. On the other face 8 mm ϕ bars at 210 mm centres are provided. The reinforcement will resist +ve B.M.

Design of bottom slab. The bottom slab will be treated a s having edges clamped.

Assume 22 cm thickness of slab.

Self load of slab $=0{\cdot}22\times1\times1\times24{,}000=5280$ N/m².

Load of water $=3\times10{,}000=30{,}000$ N/m².

Total load $=30{,}000+5280-35{,}280$ N/m².

In circular slab of radius a and uniformly loaded with load of intensity q.

Circumferential moment $=\frac{q}{16}a^2.$

Radial moment $+\text{ve}=\dfrac{q}{16}a^2.$

$$-\text{ve}=\frac{2q}{16}a^2.$$

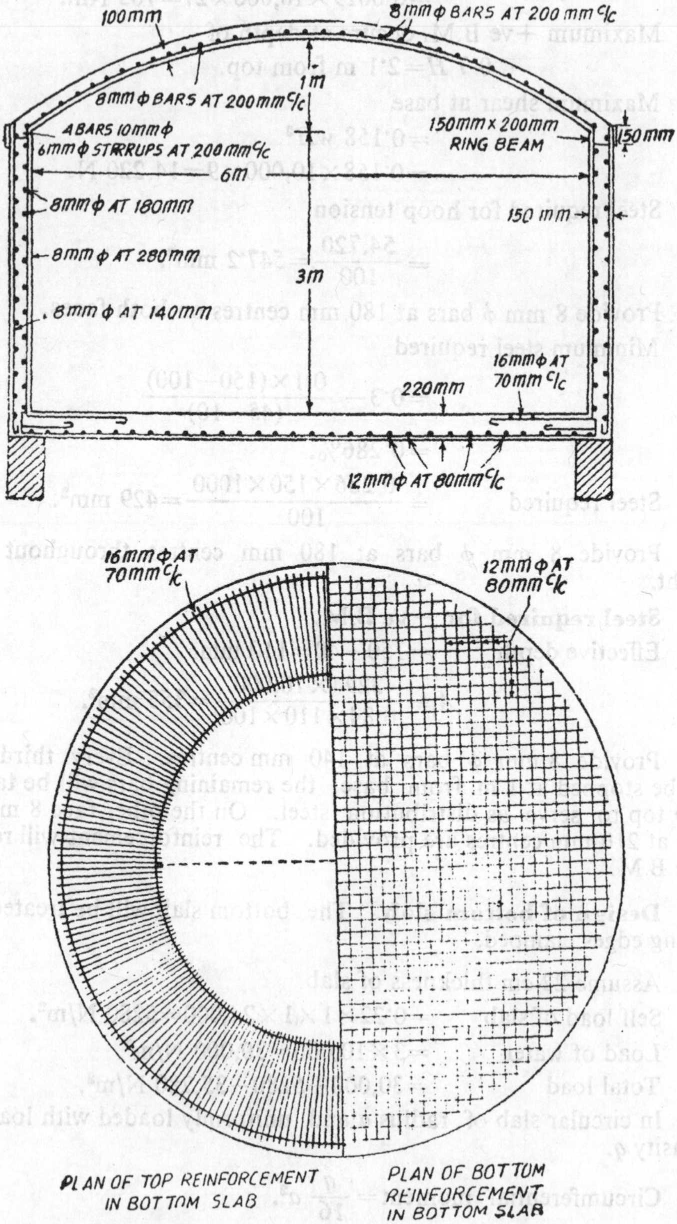

Fig. 16·19

Maximum radial negative moment

$$=\frac{2\times 35{,}280\times 3^2}{16}=39{,}690 \text{ Nm.}$$

Effective depth required

$$=\sqrt{\frac{39{,}690\times 1000}{1{\cdot}411\times 1000}}$$

(tension on water side)

$$=168 \text{ mm.}$$

Provide overall depth of 220 mm with effective depth of 180 mm.

$$A_t=\frac{39{,}690\times 1000}{0{\cdot}84\times 180\times 100}=2620 \text{ mm}^2.$$

Provide 16 mm ϕ bars at 70 mm centres 1·2 m from edge.

$$+\text{ve radial moment}=\frac{q}{16}a^2$$

$$=\frac{35{,}280\times 3^2}{16}=19{,}845 \text{ Nm.}$$

Circumferential moment

$$=19{,}845 \text{ Nm.}$$

Area of steel $=\dfrac{19{,}845\times 1000}{0{\cdot}84\times 180\times 100}$ (tension outside)

$$=1300 \text{ mm}^2.$$

Provide 12 mm ϕ bars at 80 mm centres both way throughout.

16·12. Elevated Rectangular Tanks

Design of elevated tanks at open at top is similar to the tanks resting on the ground. If the tank is covered at the top the vertical walls are considered to be supported at the top. In such cases, if the ratio of width of tank wall to height of tank is between 0·5 to 2, the coefficients for moments in two directions vertical and horizontal can be obtained from Table 16·5. The walls are designed for these moments. In case the width of the tank wall is greater than twice the height of tank it will be designed spanning vertically, simply supported at top and fixed at the base. In case the width of tank wall is less than half the height of tank, it is designed spanning horizontally and fixed at the junctions.

The roof slab is designed as two-way slab if the length does not exceed twice the breadth of the tank. In case the length of tank is more than twice the breadth of the tank, the slab is designed one-way spanning along the shorter span.

The design of bottom slab depends on the system of columns and beams provided.

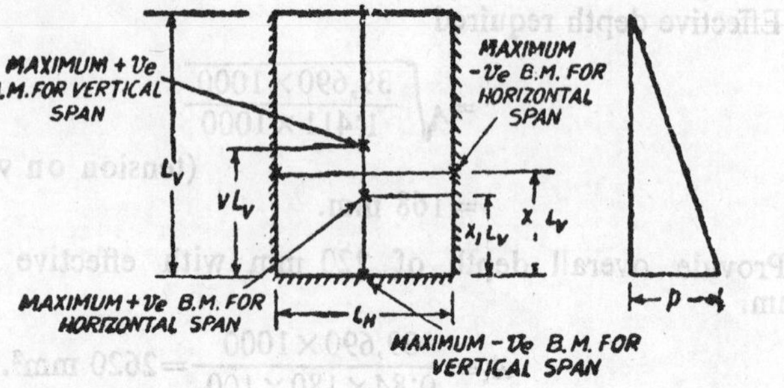

Fig. 16·20

B.M.$=kpl^2$

where l is the span,

k is coefficient given in Table 16·5

and p is pressure, $p=w \times L_V$.

For $\frac{L_H}{L_V}>2$, whole load will act along the vertical span and

maximum $-$ve B.M.$=\frac{pL_V^2}{15}$ and $+$ve B.M.$=\frac{pL_V^2}{33{\cdot}5}$.

Table 16·5

$\frac{L_H}{L_V}$	X	X_1	V	*Max. −ve B.M. for vertical span*	*Max. +ve B.M. for vertical span*	*Max. +ve B.M. for horizontal span*	*Max. +ve B.M. for horizontal span*
0·5	0·11	0·40	0·25	0·007	0·006	0·050	0·034
1·0	0·33	0·47	0·40	0·026	0·013	0·026	0·012
1·5	0·42	0·48	0·45	0·043	0·027	0·013	0·004
2·0	0·45	0·49	0·48	0·055	0·032	0·002	0·001

For $\frac{L_H}{L_V}<0{\cdot}5$, whole load will act along the horizontal span

and maximum $-$ve B.M.$=\frac{pL_H^2}{12}$ and maximum$+$ve B.M.$=\frac{pL_H^2}{16}$.

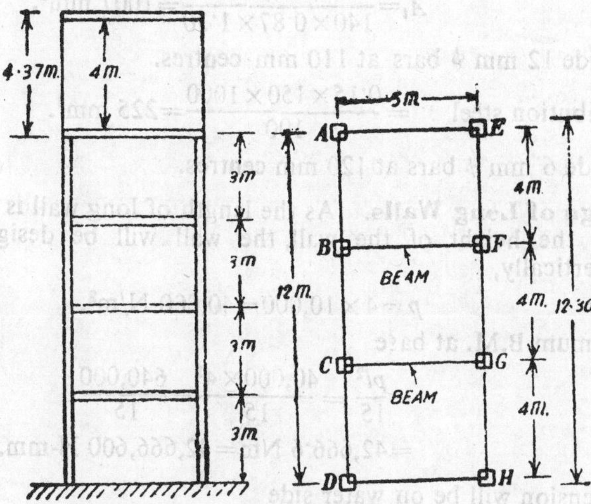

Fig. 16·21

Ex. 16·7. *Design an elevated rectangular tank 12 m×5 m×4 m high. The bottom of tank is 12 m above ground level. The tank is covered at top.*

Bearing capacity of soil is 150 kN/m².

Sol. The tank will be supported on 8 columns, spaced at 4 m centres as shown in Fig. 16·21. The columns are braced together at intervals of 3 m.

Concrete mix used is M 20.

Design of Roof Slab. Roof slab is designed for a live load of 1500 N/m².

Assume thickness of roof slab as 150 mm.

Self-load	$=0·15\times1\times1\times24{,}000=3600$ N/m²,
Live load	$=1500$ N/m².
Total load	$=5100$ N/m².

As the ratio of length to breadth of slab is greater than 2, slab will span in shorter direction.

$$\text{Maximum B.M.} \quad =\frac{wl^2}{8}=\frac{5100\times5^2}{8}$$

$$=15937·5 \text{ Nm}=15{,}937{,}500 \text{ N-mm}.$$

For mix. M 200, $c=7$ N/mm², $t=140$ N/mm², $m=13$, $Q=1·210$.

$$\text{Effective depth} \quad =\sqrt{\frac{15{,}937{,}500}{1·21\times1000}}=114·8 \text{ mm}.$$

Provide overall depth of 150 mm with effective depth of 130 mm.

$$A_t=\frac{15,937,500}{140\times0{\cdot}87\times1{\cdot}30}=1007\text{ mm}^2.$$

Provide 12 mm ϕ bars at 110 mm centres.

Distribution steel $=\frac{0{\cdot}15\times150\times1000}{100}=225\text{ mm}^2.$

Provide 6 mm ϕ bars at 120 mm centres.

Design of Long Walls. As the length of long wall is greater than twice the height of the wall, the wall will be designed as spanning vertically,

$$p=4\times10,000=40,000\text{ N/m}^2.$$

Maximum B.M. at base

$$=\frac{pl^2}{15}=\frac{40,000\times4^2}{15}=\frac{640,000}{15}$$

$$=42,666{\cdot}6\text{ Nm}=42,666,600\text{ N-mm}.$$

The tension will be on water side

$c=7,\quad t=100,\quad Q=1{\cdot}411,\quad n=0{\cdot}84.$

Effective depth $=\sqrt{\frac{42,666,600}{1{\cdot}411\times1000}}=173{\cdot}9\text{ mm}.$

Provide overall depth of 220 mm with effective depth of 180 mm.

Area of steel required

$$=\frac{42,666,600}{0{\cdot}84\times100\times180}=2805\text{ mm}^2.$$

Provide 20 mm ϕ bars at 110 mm centres.

A_t provided $=2856\text{ mm}^2.$

Maximum +ve B.M.

$$=\frac{pLv^2}{33{\cdot}5}=\frac{40,000\times4^2}{33{\cdot}5}$$

$$=19,110\text{ Nm}=19,110,000\text{ N-mm}.$$

Area of steel required

$$=\frac{19,110,000}{0{\cdot}84\times100\times180}=1260\text{ mm}^2$$

Provide 12 mm ϕ bars at 90 mm centres.

Distribution Steel

Distribution steel $=\left(0{\cdot}3-\frac{0{\cdot}1\times120}{350}\right)\%=0{\cdot}266\ \%$

Steel area $=\frac{0{\cdot}266\times120\times1000}{100}=585{\cdot}2\text{ mm}^2.$

Half the reinforcement is provided on each face. Provide 8 mm ϕ at 170 mm centres on each face. There will be direct load on the long wall due to self-weight of the roof. It is assumed that whole load from roof is transferred to long walls.

Load per metre length of wall

$$=5100\times\frac{5}{2}+5280\times4$$

$$=12{,}750+21{,}120=33{,}870 \text{ N.}$$

Direct load is very small, its effect is neglected. Actually reinforcement is provided on both faces vertically as well as horizontally.

There will be direct tension in long walls due to pressure on short walls but its effect is negligible.

Design of Short Walls

$$\frac{L_H}{L_V}=\frac{5}{4}=1{\cdot}25.$$

Coefficients for B.M. are obtained from Table 16·5 by linear interpolation.

$$p=4\times10{,}000=40{,}000 \text{ N/m}^2$$

Maximum −ve B.M. for vertical span

$$=0{\cdot}0345\times40{,}000\times4^2=22{,}080 \text{ Nm}$$

Maximum +ve B.M. for vertical span

$$=0{\cdot}02\times40{,}000\times4^2=12{,}800 \text{ Nm}$$

Maximum —ve B.M. for horizontal span

$$=0{\cdot}0195\times10{,}000\times5^2=19{,}500 \text{ Nm.}$$

Maximum +ve B.M. for horizontal span

$$=0{\cdot}008\times40{,}000\times5^2=8000 \text{ Nm.}$$

Overall thickness of 220 mm is provided.

Steel in Vertical Span

For −ve B.M. $=\dfrac{22{,}080\times1000}{0{\cdot}84\times180\times100}$ [tension on water face]

$=1460 \text{ mm}^2.$

Provided 12 mm ϕ bars at 70 mm centres

For +ve B.M. $=\dfrac{12{,}800\times1000}{0{\cdot}84\times180\times100}$ [tension outside]

$=840 \text{ mm}^2$

Provided 10 mm ϕ bars at 90 mm centres.

Steel in Horizontal Span

For $-$ve B.M. $= \dfrac{19,500\times1000}{0\cdot84\times(180-6-6)\times100}$

[tension on water side]

$=1400\ \text{mm}^2$.

Provide 12 mm ϕ bars at 80 mm centres.

For $+$ve B.M. $= \dfrac{8000\times1000}{0\cdot84\times168\times100}$ [tension outside]

$=562\ \text{mm}^2$.

Provide 12 mm ϕ bars at 170 mm centres.

There will be small amount of direct tension in short walls due to water pressure on long walls but its effect will be very small and its neglected.

Design of base slab. The slab is built monolithic with beam on all the four edges. For finding bending moment coefficients it will be considered continuous on all four edges.

Bending moment coefficients are obtained by interpolation from Table 8·5. $\dfrac{L}{B}=\dfrac{5}{4}=1\cdot25$.

Assume 22 cm thick slab.

Self-load$=0\cdot22\times1\times1\times24,000=5280\ \text{N/m}^2$.

Load of water $4\times10,000 \quad =40,000\ \text{N/m}^2$

Total load $\quad =45,280\ \text{N/m}^2$

B.M. in Short Span

$-$ve B.M. at continuous edge

$=0\cdot0475\times45,280\times4^2=39,850$ Nm.

$+$ve B.M. at mid-span

$=0\cdot036\times45,280\times4^2=26,100$ Nm.

B.M. in Long Span

$-$ve B.M. at continuous edge

$=0\cdot033\times45,280\times4^2=23,900$ Nm.

$+$ve B.M at mid-span

$=0\cdot025\times45,280\times4^2=18,120$ Nm.

Maximum B.M.$=39,850$ Nm (tension on water side)

$c+7\ \text{N/mm}^2$, $t=100\ \text{N/mm}^2$, $Q=1\cdot411$, $n=0\cdot48$ $j=0\cdot84$

Effective depth required

$$=\sqrt{\frac{39,850\times1000}{1\cdot411\times1000}}=168\cdot5\ \text{mm}.$$

Provide overall depth of 220 mm with effective depth of 180 mm.

Steel in Short Span

Steel required for −ve B.M. at continuous edge

$$=\frac{39{,}850\times1000}{0{\cdot}84\times100\times180}=2630\ \text{mm}^2.$$

Provide 16 mm ϕ bars at 70 mm centres.

A_t provided $=28{\cdot}3\ \text{mm}^2$.

$$\text{Steel for +ve B.M.}=\frac{26{,}100\times1000}{0{\cdot}84\times100\times180}=1728\ \text{mm}^2.$$

Provide 16 mm ϕ bars at 110 mm centres, steel area = 1827 mm

Steel in Long Span

Effective depth $=180-6-6=168$ mm.

$$\text{Steel for −ve B.M.}=\frac{23{,}900\times1000}{0{\cdot}84\times.00\times164}=1735\ \text{mm}^2.$$

Provide 16 mm ϕ bars at 110 mm centres

$A_t=1827\ \text{mm}^2$.

$$\text{Steel for +ve B.M.}=\frac{18{,}120\times1000}{0{\cdot}84\times100\times164}=1300\ \text{mm}^2.$$

Half the bars from support are bent to the mid-span giving 16 mm ϕ at 220 mm centres and steel area of $\frac{1827}{2}$.

$$\text{Remaining steel required}=1300-\frac{1827}{2}$$

$$=386\ \text{mm}^2.$$

10 mm ϕ bars at 220 mm centres are provided.

Design of Beams

Design of Beam B_1

Load from slab $=w=45{,}500\ \text{N/m}^2$.

Assume size of rib as 480 mm × 300 mm

Self-weight $=0{\cdot}48\times0{\cdot}3\times25{,}000$

$=3600$ N/m.

The loading on the beam is shown in Fig. 16·22 (*b*).

As the beam is built monolithic with columns, partial fixity is assumed at the ends. B.M. taken for design purpose will be 0·8 times simply supported bending moment.

$$\text{Max. B.M.}=0{\cdot}8\left[6w\times2{\cdot}5-\frac{4w}{2}\times2\left(\frac{2}{3}+0{\cdot}5\right)-4w\times0{\cdot}5\times0{\cdot}25+\frac{3600\times5^2}{8}\right]$$

$$=0{\cdot}8\left[15w-4w\times\frac{7}{6}-0{\cdot}5w+11{,}250\right]$$

$$=0{\cdot}8\ [9{\cdot}83w+11{,}250[$$

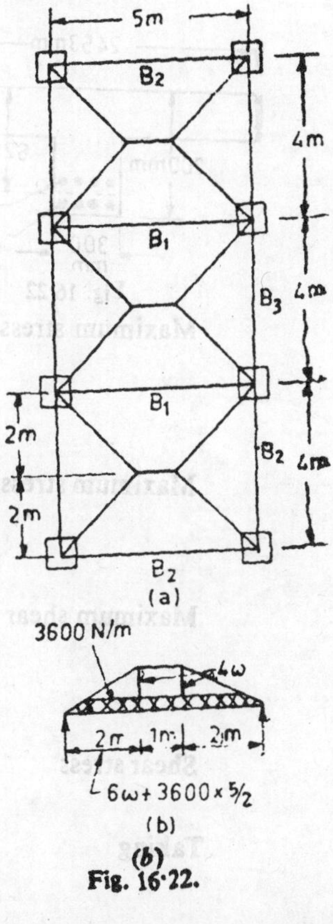

Fig. 16·22.

$=0.8\ [9.83\times 45{,}500+11{,}250]$
$=366{,}812$ N/m $=366{,}812{,}000$ N-mm

Concrete mix used is M 20.
The beam is designed as T-beam.

Effective depth $=700-80=620$ mm

Assume lever arm $=620-\dfrac{220}{2}=510$ mm.

Area of steel $=\dfrac{366{,}812{,}000}{510\times 140}=5137$ mm.

Provide 4 bars of 32 mm ϕ and 4 bars of 25 mm ϕ.
Steel area provided $=3217+1963=5180$ mm².

Check for Stress

Flange width $b_f=\dfrac{l_0}{6}+bw+6D_f$

$=\dfrac{5000}{6}+300+6\times 220=2453$ mm

Flange width $=2453$ mm

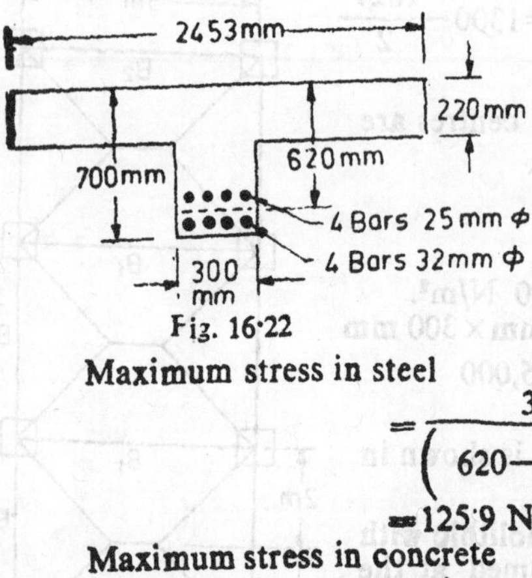

Fig. 16·22

Assume neutral axis to lie in the flange.

$2453\times\dfrac{n^2}{2}=13\times 513\ 7(620-n)$

$n^2+54.4\ n=33{,}758$

$(n+27\ 2)^2=33{,}758+739.8$

$\therefore\quad n=185.7-27.2$
$=158.5$ mm

Maximum stress in steel

$=\dfrac{366{,}812{,}000}{\left(620-\dfrac{158.5}{3}\right)\times 5137}$

$=125.9$ N/mm²

Maximum stress in concrete

$=\dfrac{t}{m}\times\dfrac{d}{d-n}=\dfrac{125.9\times 158.5}{13\times(620-158.5)}$

$=3.33$ N/mm². Safe.

Maximum shear force

$=45{,}500\times 6+3600\times\dfrac{5}{2}$

$=273{,}000+9000=282{,}000$ N

Shear stress $=\dfrac{282{,}000}{620\times 300}$

$=1.52$ N/mm²

Taking $\tau_c=0.39$ N/mm²

Net shear $=(1{\cdot}52-0{\cdot}39)\times620\times300=210{,}180$ N

Shear at 2 m from support

$$=6\times45{,}500-\tfrac{1}{2}\times4\times45{,}500\times2+\frac{3600\times5}{2}-3600\times2$$

$=92{,}800$ N

∴ Minimum shear reinforcement is required beyond this point. 4 bars of 25 mm ϕ, two at a time, and 2 bars of 32 mm ϕ are bent at intervals of 0·4 m.

Shear taken by two bars of 32 mm ϕ

$=2\times125\times804\times0{\cdot}707=142{,}000$ N

Net shear $=210{,}180-142{,}000=68{,}180$ N

Providing 12 mm ϕ 2-legged stirrups,

$$\text{Pitch}=\frac{2\times113\times125\times620}{68{,}180}=557 \text{ mm}$$

12 mm ϕ 2-legged stirrups at 80 mm centres are provided.

Shear at 0·5 m from support

$$=282{,}000-\frac{0{\cdot}5}{2}\times w-\frac{3600}{2}$$

$=282\,000-11375-1800=268{,}825$ N

At this section two bars of 25 mm ϕ bent up are effective.

Shear taken by 2 bars 25 mm ϕ

$=2\times125\times491\times0{\cdot}707=87{,}000$ N

Shear to be taken by stirrups

$=268{,}825-87{,}000-0{\cdot}39\times620\times300$

$=109{,}285$ N

Providing 12 mm ϕ 2-legged stirrups,

$$\text{Pitch}=\frac{2\times113{\cdot}0\times125\times620}{109{,}285}=160 \text{ mm}$$

Providing 12 mm ϕ 2-legged stirrups at 80 mm centres.

Shear at 1·5 m from support

$$=282{,}000-\frac{3w\times1{\cdot}5}{2}-\frac{3600\times3}{2}$$

$=174{,}225$ N

Provide 12 mm ϕ 2-legged stirrups at 80 mm centres upto 2 m from support. In the remaining portion 12 mm ϕ stirrups at 500 mm centres are provided.

Design of Beam B_2. The side wall will serve the purpose of beam. It is assumed that weight of roof is taken by long walls.

Self-weight of wall per metre run $=220\times25\times4=22{,}000$ N

Weight of water and base slab will be half of that coming on beam B_1.

Maximum B.M. at centre

$$=0{\cdot}8\left[\frac{22{,}000\times5^2}{8}+3\times45{,}500\times2{\cdot}5-2\times45{,}500\times\tfrac{1}{2}\times2\left(\frac{2}{3}+0{\cdot}5\right)+2\times45{,}500\times0{\cdot}5\times0{\cdot}25\right]$$

$$=0{\cdot}8\left[\frac{22{,}000\times5^2}{8}+3\times45{,}500\times2{\cdot}5-2\times45{,}500\times\frac{7}{6}-\frac{45{,}500}{4}\right]$$

$=233{,}967$ Nm.

Overall depth $=4220$ mm.

Assume effective depth $=4160$ mm.

Area of steel required

$$=\frac{233{,}967\times1000}{0{\cdot}86\times4120\times125}=528\text{ mm}^2$$

Provide 3 bars of 20 mm ϕ.

Shear force $=3\times45{,}500+22{,}000\times\frac{5}{2}$

$=191{,}500$ N

The section will be safe as depth provided is quite considerable.

Provide 8 mm ϕ 2-legged stirrups at 300 mm centres.

Design of beam B_3. The side walls will serve as girders.

Load of roof and wall $=5250\times2{\cdot}5+22{,}000=35{,}125$ N/m.

Load from bottom slab will be triangular as shown in Fig. 16·23 (*a*).

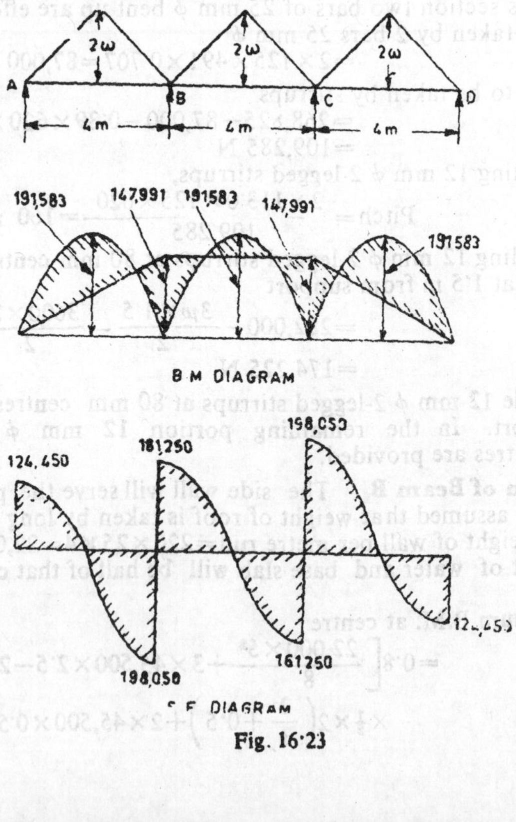

Fig. 16·23

The beam is sym- metrical and symmetrically loaded. Moments are found by moment distribution by cutting the beam at centre of span BC and taking $k=\frac{1}{2}$.

Due to triangular loading fixed end moments are

$$\frac{5wl}{48}=\frac{5}{48}\times 2w\times\frac{4}{2}\times 4=\frac{5}{3}w=\frac{5}{3}\times 45{,}500$$
$$=75{,}833 \text{ Nm.}$$

Due to uniformly distributed load fixed end moments are

$$\frac{wl^2}{12}=\frac{35{,}125\times 4^2}{12}=46{,}833 \text{ Nm.}$$

Total F.E.M. $75{,}833+46{,}833=122{,}666$

	A	3/5 B	2/5
F.E.M.	−122,666	+122,666	−122,666
Release	+122,666		
Carry over		+61,333	
		+183,999	−122,666
Balance		−36,800	−24,533
Final		+147,199	−147,199

B.M. at centre of AB

$$=\frac{8w}{3}+\frac{35{,}125\times 4^2}{8}-\frac{147{,}199}{2} \quad \ldots(w=45{,}500)$$
$$=121{,}333+70{,}250-73{,}600$$
$$=191{,}583-73{,}600=117{,}983$$

B.M. at centre of BC

$$=191{,}583-147{,}199=44{,}384 \text{ Nm.}$$

Taking moment about B,

$$R_A\times 4+147{,}199-35{,}125\times 4\times 2-\tfrac{1}{3}\times 4\times 2\times 45{,}500\times 2=0.$$
$$R_A=124{,}450 \text{ N}$$
$$R_D=124{,}450 \text{ N.}$$
$$R_B=R_C=\tfrac{1}{2}[35{,}125\times 12+3\times\tfrac{1}{2}\times 4\times 2\times 45{,}500-2\times 124{,}450]$$
$$=359{,}300 \text{ N.}$$

S.F. diagram is shown in Fig. 16·23

Maximum B.M. $=147{,}199$ Nm.

Overall depth of wall $=4220$ mm.

Assume effective depth $=4160$ mm.

Area of steel required $=\dfrac{147{,}199\times 1000}{0{\cdot}84\times 4160\times 100}=421 \text{ mm}^2.$

Provide 3 bars of 16 mm ϕ.

Area of steel required at centre of AB

$$=\frac{117{,}983\times 1000}{0.86\times 4160\times 125}=263.8 \text{ mm}^2.$$

Provide 3 bars of 12 mm ϕ.

Area of steel at centre of BC

$$=\frac{44{,}384\times\times 1000}{0.86\times 4160\times 125}=99 \text{ mm}^2.$$

2 bars of 16 mm ϕ are provided.

Shear

Maximum shear force = 198,050 N.

Shear stress $=\frac{198{,}050}{220\times 4160}=0.22 \text{ N/mm}^2$. Safe.

Provide 8 mm ϕ stirrups at 300 mm centres.

Design of Columns

Load on column A when tank full = Reaction from beams B_2 and B_3

$$=191{,}500+124{,}450=315{,}950$$

Load on column B = 359,300 + 282,000 = 641,300

Total weight of water

$$=12\times 5\times 4\times 10{,}000=2{,}400{,}000 \text{ N}.$$

(For calculation of weight of water centre line dimensions are taken because while calculating load on beams centre line dimensions were taken).

Columns B, C, F and G will take double the load of water taken by A, F, D and H.

Weight of water taken on column A

$$=\frac{2{,}400{,}000}{12}=200{,}000 \text{ N}.$$

Weight of water taken by column B

$$=400{,}000 \text{ N}.$$

Load on column A when tank is empty

$$=315{,}950-200{,}000=115{,}950$$

Load on column B when tank is empty

$$641{,}300-400{,}000=241{,}300 \text{ N}.$$

Wind Forces

Intensity of wind pressure is assumed as 1500 N/m²

Wind force acting on water tank

$$=4.37\times 12.22\times 1500$$

$$=80{,}030 \text{ N acting at } 14.185 \text{ from base.}$$

Wind Force on Staging

On columns $=4\times 0.3\times 12\times 1500=21{,}600$ N.

On bracing $=3\times 0.3\times 12\times 1500=16{,}200$ N.

Total wind force on staging = 21,600 + 16,200 = 37,800 N acting at 6 m from base.

Moment at base $= 80{,}030 \times 14{,}185 + 37{,}800 \times 6$

$= 1{,}136{,}000 + 226{,}800 = 1{,}362{,}800$ Nm.

$\Sigma r^2 = 8 \times (2{\cdot}5)^2 = 50.$

Force in column $= \dfrac{Mr}{\Sigma r^2} = \dfrac{1{,}362{,}800 \times 2{\cdot}5}{50}$

$= 68{,}140$ N.

There will be downward force of 68,140 N in each of leeward columns and upward force on 68,140 N in each of windward columns.

Self-weight of column $= 0{\cdot}3 \times 0{\cdot}4 \times 12 \times 25{,}000 = 36{,}000$ N.

Maximum force in column

$= 641{,}300 + 68{,}140 + 36{,}000$

$= 745{,}440$ N.

Minimum force in column

$= 115{,}950 - 68{,}140 + 36{,}000 = 83{,}810$ N

No upward load is produced in column when tank is empty.

Design of column

Section of column is designed for dead load and checked for wind stresses.

Maximum dead load

$= 641{,}300 + 36{,}000$

$= 677{,}300$ N

Section of 300 mm × 400 mm is provided.

$677{,}300 = (300 \times 400 - A_c) \times 5 + A_c \times 130$

$= 600{,}000 - 5A_c + 130\,A_c$

$\therefore \quad A_c = \dfrac{77{,}300}{125} = 618{\cdot}4 \text{ mm}^2$

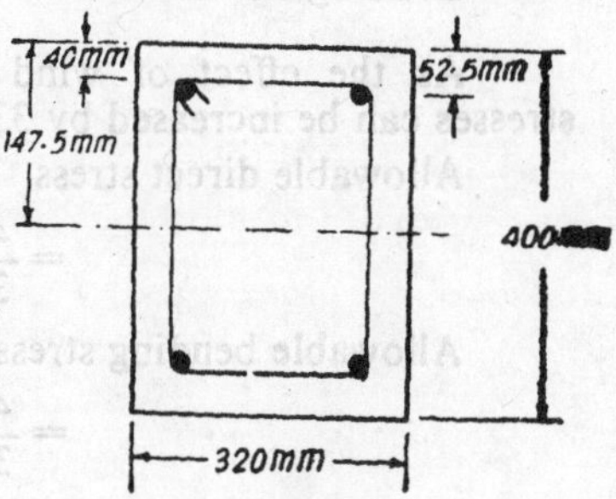

Fig. 16·24

4 bars of 20 mm ϕ are provided.

Area of steel provided

$= 1267 \text{ mm}^2$

Minimum diameter of ties

$= \dfrac{20}{4} = 5{\cdot}00$ mm

Provide 6 mm ϕ ties at 300 mm centres.

Check for stresses with wind blowing

Total horizontal force

$= 80{,}030 + 37{,}800 = 117{,}830$ N

Force taken by each column

$= \dfrac{117{,}830}{8} = 14{,}730$ N

Maximum B.M. $=14{,}730\times1\cdot5=22{,}100$ Nm
Maximum load $=745{,}440$ N

Eccentricity $=\dfrac{22{,}100\times1000}{745{,}440}=29\cdot65$ mm

As the eccentricity is very small no tension is induced.

Equivalent area of section

$=300\times400+12\times4\times314\cdot2$

$=120{,}000+15{,}084=135{,}084$ mm²

Equivalent moment of inertia

$=\dfrac{1}{12}\times300\times(400)^3=12\times4\times314\cdot2\times(147\cdot5)^2$

$=(160{,}000+32{,}790)\times10^4$

$=192{,}790\times10^4$ mm⁴

Direct stress $=\dfrac{745{,}440}{135{,}084}=5\cdot52$ N/mm²

Bending stress $=\dfrac{22{,}100\times1{,}000\times150}{192{,}790\times10^4}=1\cdot718$ N/mm²

As the effect of wind is taken into consideration allowable stresses can be increased by $33\frac{1}{3}$%.

Allowable direct stress

$=\dfrac{4}{3}\times5=6\cdot667$ N/mm².

Allowable bending stress

$=\dfrac{4}{3}\times7=9\cdot333$ N/mm²

$$\frac{5\cdot52}{6\cdot667}+\frac{1\cdot718}{9\cdot333}=0\cdot828+0\cdot1814$$

$=1\cdot0094$ slightly greater than $1\cdot00$

Design of Braces

Maximum B.M. in the brace

$=2\times22{,}100=44{,}200$ Nm

$=44{,}200{,}000$ N-mm.

300 mm × 400 mm brace is provided.

Effective depth $=400-40=360$ mm

Moment of resistance of section

$Qbd^2=1\cdot21\times300\times(360)^2$

$=47{,}040{,}000$ N-m

Actual B.M. is 44,200,000 N-mm

Area of steel required

$$=\frac{4,420,000}{0.87\times360\times140}=1018\ mm^2$$

Provide 4 bars of 20 mm ϕ both at top and bottom as the wind may blow from either side.

Maximum shear force

$$=\frac{2\times12,100}{8}=8,840$$

Shear force is very small. Shear stress is negligible.

Provide 6 mm ϕ stirrups at 300 mm centres.

Design of foundation. For foundation concrete mix of M 15 is used.

Maximum dead load = 677,300 N

Assume self-load = 68,000 N

Total load = 745,300 N

Area of footing required

$$=\frac{745,300}{150,000}=4.97\ m^2$$

Provide 2·2 m × 2·3 m footing

Net upward pressure on footing

$$=\frac{677,300}{2.2\times2.3}=133,854\ N/m^2$$

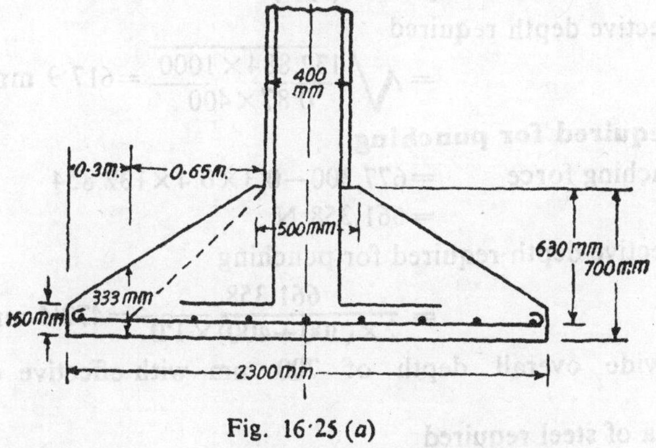

Fig. 16·25 (a)

Maximum B.M. $=2.2\times0.95\times\frac{133,854\times0.95}{2}$

$=132,884$ Nm

(a)

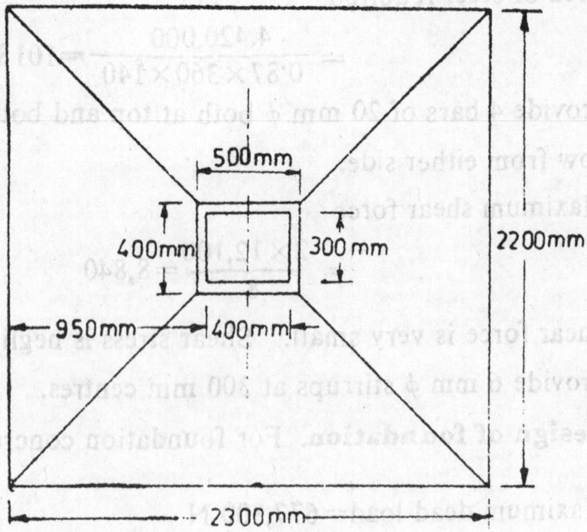

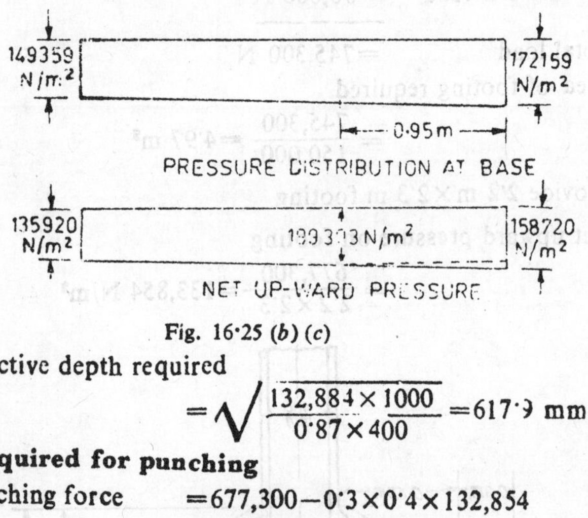

Fig. 16·25 (b) (c)

Effective depth required

$$= \sqrt{\frac{132,884 \times 1000}{0.87 \times 400}} = 617.9 \text{ mm.}$$

Depth required for punching

Punching force $= 677,300 - 0.3 \times 0.4 \times 132,854$

$= 661,358$ N

Effective depth required for punching

$$= \frac{661,358}{2 \times (300+400) \times 1.0} = 472.4 \text{ mm}$$

Provide overall depth of 700 mm with effective depth of 630 mm.

Area of steel required

$$= \frac{132,884 \times 1000}{0.87 \times 630 \times 140} = 1732 \text{ mm}^2$$

Provide 9 bars of 16 mm ϕ in both directions.

Check for diagonal shear

Maximum shear $=0{\cdot}3\times2{\cdot}2\times133{,}854=88{,}344$ N

Shear stress $=\dfrac{88{,}344}{263\times1{,}600}=0{\cdot}21$ N/mm². Safe.

Check for stresses due to wind load

Maximum vertical load

$=745{,}440$ N

Self-load $=68{,}000$ N

Total load $=813{,}440$ N

B.M. $=22{,}100$ Nm.

Pressure distribution on base is given by

$$p=\frac{813{,}440}{2{\cdot}2\times2{\cdot}3}\pm\frac{22{,}100}{2{\cdot}2\times(2{\cdot}3)^2/6}$$

$$=160{,}759\pm11{,}400$$

$$p_{max.}=172{,}159 \text{ N/m}^2.$$

$$p_{min.}=149{,}359 \text{ N/m}^2.$$

Allowable pressure on ground considering wind effects

$$=\frac{5}{4}\times150{,}000=187{,}500 \text{ N/m}^2.$$

Pressure at base due to self-weight of footing

$$=\frac{68{,}000}{2{\cdot}2\times2{\cdot}3}=13{,}439 \text{ N/m}^2$$

Net pressure on footing will vary from

158,720 N/m² (172,159−13,439)

(149,359−13,439)

B.M. at the edge of column

$$=\left[149{,}303\times\frac{0{\cdot}95\times0{\cdot}95}{2}+\frac{(158{,}720-149{,}303)}{2}\times0{\cdot}95\times\frac{2}{3}\times0{\cdot}95\right]\times2{\cdot}2$$

$$=154{,}453 \text{ Nm.}$$

As the effect of wind is taken into consideration, stresses can be increased by $33\frac{1}{3}$%.

Effective depth required

$$=\sqrt{\frac{154{,}453\times1000}{4/3\times0{\cdot}87\times400}}$$

$$=577 \text{ mm.}$$

Effective depth provided is 630 mm.

Hence O.K.

Area of steel required

$$= \frac{154,453 \times 1000}{0.87 \times 630 \times 140 \times \frac{4}{3}} = 1510 \text{ mm}^2.$$

Area of steel provided is 1700 mm². Safe.

16·13. Intze-type Tanks. Circular tanks for large capacities prove to be uneconomical if flat bottom slab is provided. In such cases it becomes economical to provide intze-type tanks. Intze-type tank consists of top dome supported on ring beam which rests on cylindrical walls. The walls are supported on conical base and also bottom dome is provided. The conical base and bottom dome are made in such a way that the horizontal thrust from conical base is balanced by that from bottom dome.

The conical base and bottom dome are supported on a circular beam which are supported on a number of columns which are generally inclined. The foundation for columns consists of continuous circular footing or raft foundation.

For very large capacities the tank is divided into two compartments by means of partition wall supported on circular beam.

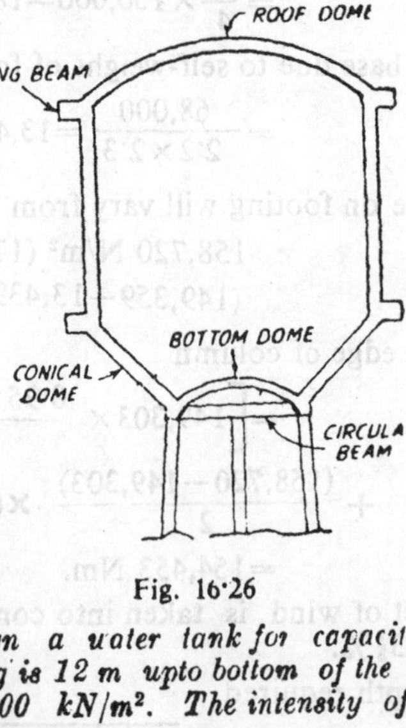

Fig. 16·26

Ex. 16·8. *Design a water tank for capacity of 600,000 litres. The height of the staging is 12 m upto bottom of the tank. The bearing capacity of soil is 200 kN/m². The intensity of wind pressure may be taken as 1500 N/m².*

Sol. 12 metres diameter is adopted for cylindrical portion. For this portion height adopted is 5 m. The tank dimensions are shown in Fig. 16·27.

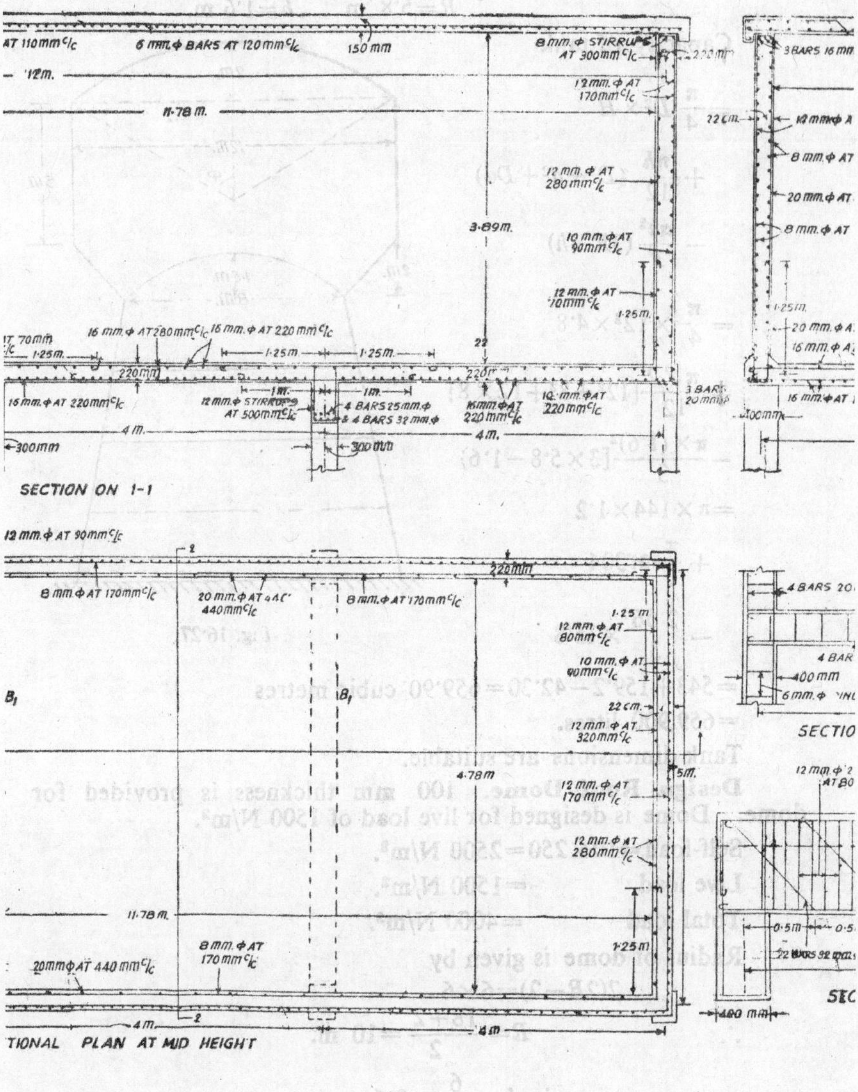

AT 110mm c/c
6 mm.φ BARS AT 120mm c/c
150 mm
8 mm.φ STIRRUPS AT 300mm c/c
12m.
11·78 m.
12 mm.φ AT 170mm c/c
12 mm.φ AT 280mm c/c
3·89m.
10 mm.φ AT 90mm c/c
12 mm.φ AT 70mm c/c
1·25m.
16 mm.φ AT 280mm c/c
16 mm.φ AT 220 mm c/c
1·25m.
220mm
15 mm.φ AT 220mm c/c
12 mm.φ STIRRUPS AT 500mm c/c
4 BARS 25 mm.φ & 4 BARS 32 mm.φ
16mm.φ AT 220 mm c/c
10 mm.φ AT 220mm c/c
3 BARS 20 mmφ
4 m.
300mm
SECTION ON 1-1
3 BARS 16 mm
22 cm.
20 mm.φ AT
8 mm.φ AT
12 mm.φ AT 90mm c/c
8 mm.φ AT 170mm c/c
20 mm.φ AT 440mm c/c
8 mm.φ AT 170mm c/c
220mm
1·25 m
12 mm.φ AT 80mm c/c
10 mm.φ AT 90mm c/c
22 cm.
12 mm.φ AT 320mm c/c
4·78m
5m.
12 mm.φ AT 170 mm c/c
12 mm.φ AT 280mm c/c
11·78 m.
8 mm.φ AT 170 mm c/c
20mmφ AT 440 mm c/c
1·25m
4 m.
TIONAL PLAN AT MID HEIGHT
4 BARS 20
4 BAR
400 mm
6 mm.φ
SECTIO
0·5m
22 BARS 32 mm.
400 mm
SEC

Concrete mix of M 20 is used.

Free board of 200 mm is provided.

Radius of Bottom Dome

$$1{\cdot}6(2R-1{\cdot}6)=4^2$$
$$2R-1{\cdot}6=20$$
$$R=5{\cdot}8 \text{ m} \qquad h=1{\cdot}6 \text{ m.}$$

Capacity of tank

$$=\frac{\pi}{4}D^2\times H$$
$$+\frac{\pi h}{12}(D^2+d^2+Dd)$$
$$-\frac{\pi h^2}{3}(3R-h)$$
$$=\frac{\pi}{4}\times 12^2\times 4{\cdot}8$$
$$+\frac{\pi\times 2}{12}[12^2+8^2+12\times 8]$$
$$-\frac{\pi\times(1{\cdot}6)^2}{3}[3\times 5{\cdot}8-1{\cdot}6)$$
$$=\pi\times 144\times 1{\cdot}2$$
$$+\frac{\pi}{6}\times 304$$
$$-\frac{2{\cdot}56}{3}\times 15{\cdot}8$$

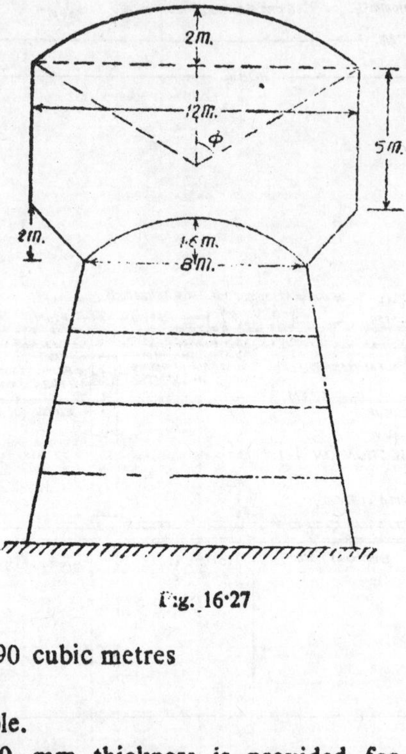

Fig. 16·27

$=543+159{\cdot}2-42{\cdot}30=659{\cdot}90$ cubic metres

$=659{,}900$ litres.

Tank dimensions are suitable.

Design Roof Dome. 100 mm thickness is provided for dome. Dome is designed for live load of 1500 N/m².

Self-load $=10\times 250=2500$ N/m².

Live load $=1500$ N/m².

Total load $=4000$ N/m².

Radius of dome is given by

$$2(2R-2)=6\times 6$$
$$\therefore \quad R=\frac{18+2}{2}=10 \text{ m.}$$
$$\sin\phi=\frac{6}{10}=0{\cdot}6$$
$$\therefore \quad \phi=36^\circ\ 52'$$

Hoop stress at any angle θ is given by

$$=\frac{wR}{t}\left[\frac{\cos^2\theta+\cos\theta-1}{1+\cos\theta}\right]$$

Maximum hoop stress occurs at

$$\theta=0$$

Maximum hoop stress

$$=\frac{4000\times10}{0{\cdot}1}\times\frac{1}{2}$$

$$=200{,}000\ \text{N/m}^2.$$

$$=0{\cdot}2\ \text{N/mm}^2.\quad \text{Safe.}$$

Nominal steel of 8 mm ϕ bars at 200 mm centres are provided both ways.

Meridional stress at any angle θ is given by

$$\frac{wR}{t}\left(\frac{1-\cos\theta}{\sin^2\theta}\right)=\frac{wR}{t(1+\cos\theta)}$$

Maximum meridional stress occurs at

$$\theta=36°\ 52'$$

Meridional stress $=\dfrac{4000\times10}{0{\cdot}1(1+0{\cdot}8)}$

$$=222{,}222\ \text{N/m}^2$$

$$=0{\cdot}212\ \text{N/mm}^2.\quad \text{Safe.}$$

Design of Ring Beam at Top

Horizontal thrust per mm length

$$=\text{Meridional thrust}\times\cos\phi\times R$$

$$=0{\cdot}212\times\frac{8}{10}\times100=16{\cdot}96\ \text{N.}$$

Hoop tension $=\dfrac{pd}{2}=16{\cdot}96\times\dfrac{12{,}000}{2}$

$$=101{,}760\ \text{N.}$$

Steel area required $=\dfrac{101{,}760}{140}=727\ \text{mm}^2.$

Provide 8 bars of 12 mm ϕ.

$$A_t=905\ \text{mm}^2.$$

Provide 300 mm × 300 mm ring beam.

Equivalent area $=300\times300+12\times905$

$$=90{,}000+10{,}860=100{,}860\ \text{mm}^2$$

Tensile stress $=\dfrac{101{,}760}{100{,}860}$

$$=1{\cdot}01\ \text{N/mm}^2. < 1{\cdot}2\ \text{N/mm}^2.\ \text{Safe.}$$

6 mm ϕ stirrups at 300 mm centres are provided.

Design of Cylindrical Walls. It is assumed that ring beam at top and bottom are free to expand and hence cylindrical wall will be assumed free at top and bottom.

Hoop tension at base $=\dfrac{pD}{2}=\dfrac{10,000\times 5\times 12}{2}$

$=300,000$ N

Area of steel required

$=\dfrac{300,000}{100}=3000 \text{ mm}^2$

Area required in each face $=1500 \text{ mm}^2$.

Provide 12 mm ϕ bars at 700 mm centres on both faces.

Steel area provided $=3232 \text{ mm}^2$.

As the pressure varies linearly with depth, hoop tension will vary linearly with depth. Area of reinforcement required will also vary linearly. Spacing of reinforcement is gradually increased towards top.

Thickness of wall. Let t be the thickness of wall at base.

Equivalent areas $=t\times 1000+12\times 3232$

Tensile stress $=\dfrac{300,000}{t\times 1000+12\times 3232}=1{\cdot}2 \text{ N/mm}^2$.

$\therefore\quad t=\dfrac{250,000-12\times 3232}{1000}$

$\therefore\quad =\dfrac{250,000-33,784}{1000}=211{\cdot}2$ mm.

Provide 250 mm thick wall at base and 150 mm thickness at top.

At depth of 4 m, steel area required

$=\dfrac{4}{5}\times 3000=2400 \text{ mm}^2$

Provide 12 mm ϕ bars at 90 mm centres.

At depth of 3 m steel area required.

$=\dfrac{3}{5}\times 3000=1800 \text{ mm}^2$

Provide 12 mm ϕ bars at 120 mm centres.

At depth of 2 m, steel area required

$=\dfrac{2}{5}\times 3000=1200 \text{ mm}^2$

Provide 12 mm ϕ bars at 180 mm centres upto top.

Distribution steel. Average thickness of wall

$=\dfrac{150+250}{2}=200$ mm

Percentage of distribution steel

$=0{\cdot}3-0{\cdot}1\times\dfrac{(200-100)}{(450-100)}$

$=0{\cdot}3-0{\cdot}03=0{\cdot}27.$

Area of distribution steel

$$=\frac{0.27\times200\times1000}{100}=540\ \text{mm}^2$$

Area to be provided on each face $=270\ \text{mm}^2$

Provide 8 mm ϕ bars at 180 mm centres on both faces.

Design of Ring Beam at Junction of Cylindrical Wall and Conical Wall. 600 mm × 500 mm ring beam is provided.

Vertical load coming at the junction of cylindrical wall with conical is transferred to the circular beam meridional thrust in the conical wall or conical dome. The horizontal component of meridional thrust will cause hoop tension at junction of conical wall and cylindrical wall. A ring beam is to be provided to take up this hoop tension.

Vertical load of dome per metre length

$$=\text{Area}\times\text{Meridional stress}\times\sin\phi$$
$$=100\times1000\times0.212\times0.6=12{,}720\ \text{N}$$

Vertical load of cylindrical wall per metre length

$$=\frac{0.15+0.25}{2}\times5\times25{,}000=25{,}000\ \text{N}$$

Weight of ring beam at top per metre length

$$=(0.3-0.15)\times0.3\times25{,}000=1125\ \text{N}.$$

Weight of ring beam cylindrical wall per metre length

$$=(0.5-0.25)\times0.6\times25{,}000=3750\ \text{N}$$

Total vertical load at junction of cylindrical wall and conical dome $=12{,}720+25{,}000+1125+3750=42{,}595$ N

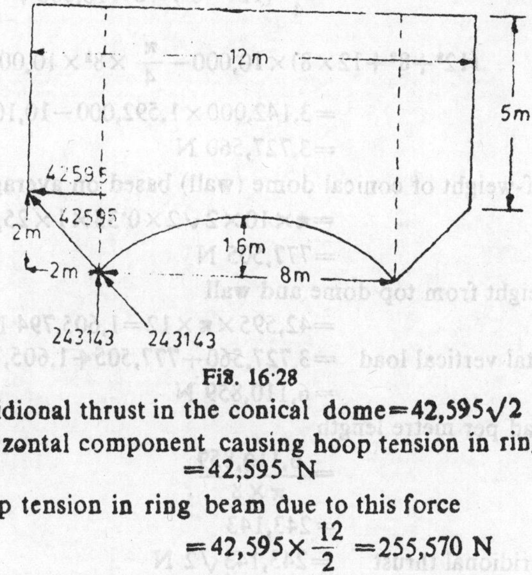

Fig. 16·28

Meridional thrust in the conical dome $=42{,}595\sqrt{2}$

Horizontal component causing hoop tension in ring beam $=42{,}595$ N

Hoop tension in ring beam due to this force

$$=42{,}595\times\frac{12}{2}=255{,}570\ \text{N}$$

Hoop tensions in ring beam due to pressure of water

$$=10,000\times5\times\frac{12}{2}\times0{\cdot}6$$

$$=180,000 \text{ N}$$

Total hoop tension $=180,000+255,570=435,570$

Area of steel required

$$=\frac{435,570}{100}=4355{\cdot}7 \text{ mm}^2$$

Provide 16 bars of 20 mm ϕ.

A_t provided $=5027 \text{ mm}^2$

Provide 8 mm ϕ stirrups at 200 mm centres.

Check for Stresses

Equivalent area $=600\times500+12\times5027$

$$=300,000+60,324=360,324 \text{ mm}^2$$

Tensile stress $=\frac{435,570}{360,324}$

$$=1{\cdot}209 \text{ N/mm}^2 > 1{\cdot}2 \text{ slightly.}$$

Design of Conical Dome. Assume thickness of 35 cm for conical dome.

Weight of water above conical dome

$$=\frac{\pi}{4}(12^2-8^2)\times5\times10,000+\frac{\pi\times2}{12}\times$$

$$(12^2+8^2+12\times8)\times10,000-\frac{\pi}{4}\times8^2\times10,000\times2$$

$$=3,142,000\times1,592,000-10,106,440$$

$$=3,727,560 \text{ N}$$

Self-weight of conical dome (wall) based on average diameter

$$=\pi\times10\times2\sqrt{2}\times0{\cdot}35\times1\times25,000$$

$$=777,505 \text{ N}$$

Weight from top dome and wall

$$=42,595\times\pi\times12=1,605,794 \text{ N}$$

Total vertical load $=3,727,560+777,505+1,605,794$

$$=6,110,859 \text{ N}$$

Load per metre length

$$=\frac{6,110,859}{\pi\times8}$$

$$=243,143$$

Meridional thrust $=243,143\sqrt{2}$ N

Meridional stress $= \dfrac{243{,}143\sqrt{2}}{350 \times 1000}$

$= 0{\cdot}982$ N/mm². Safe.

Hoop stress at depth 'x' metres from top of conical dome. Water pressure at x metres below top of conical dome

$$= p = (H+x) \times 10{,}000.$$

Weight of conical dome per square metre

$$W = t \times 25{,}000$$

where t is thickness in metres.

Total horizontal force at this section

$$= p \text{ cosec } \alpha + W \cot \alpha.$$

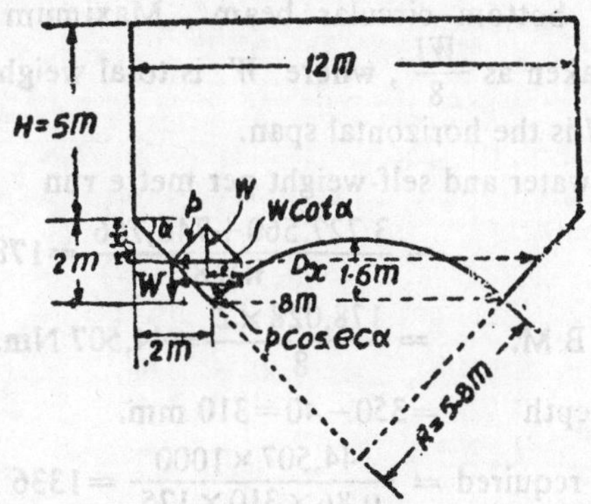

Fig. 16·29

Diameter D_x at this section

$$= D - \frac{2x}{\tan \alpha}$$

Hoop tension $= (p \text{ cosec } \alpha + W \cot \alpha) \dfrac{D_x}{2}.$

In this case $D = 12$ m, $H = 5$ m, $\alpha = 45°$, $t = 0{\cdot}35$ m.

∴ Hoop tension

$$= \left[(5+x)\ 10{,}000\sqrt{2} + 0{\cdot}35 \times 25{,}000 \times 1 \right] \left[\frac{12 - \frac{2x}{1}}{2} \right]$$

$$= \left[50\ 000\sqrt{2} + 10{,}000\sqrt{2}x + 8750 \right] \left(\frac{12-2x}{2} \right)$$

$= (79{,}461 + 14\ 14\ \text{'}x)\ (6-x)$

Hoop tension for different values of x will be

x		*Hoop tension*
0	...	476,766 N
1	...	468,015 N
2	...	430,980 N

$$\text{Steel area required} = \frac{476,766}{100} = 4,768 \text{ mm}^2$$

Provide 16 mm φ bars at 80 mm centres on both faces.
Steel area provided = 5028 mm²
Tensile stress in concrete

$$= \frac{476,766}{350 \times 1000 + 12 \times 5028}$$

$$= 1{\cdot}162 \text{ N/mm}^2 < 1{\cdot}2. \text{ Safe.}$$

The conical dome will also be designed as a slab between top ring beam and bottom circular beam. Maximum B.M. will be approximately taken as $\frac{Wl}{8}$, where W is total weight of water and self-weight and l is the horizontal span.

Weight of water and self-weight per metre run

$$= \frac{3,727,560 + 746,766}{\pi \times 8} = 178,028 \text{ N}$$

$$\text{Maximum B.M.} = \frac{178,028 \times 2}{8} = 44,507 \text{ Nm.}$$

$$\text{Effective depth} = 350 - 40 = 310 \text{ mm.}$$

$$\text{Steel area required} = \frac{44,507 \times 1000}{0{\cdot}86 \times 310 \times 125} = 1336 \text{ mm}^2.$$

Provide 12 mm ϕ bars at 80 mm centres.

Distribution steel

$$\% \text{ reinforcement} = 0{\cdot}3 - \frac{0{\cdot}1 \times (350-100)}{450-100} = 0{\cdot}229$$

$$\text{Area of steel} = \frac{0{\cdot}229 \times 350 \times 1000}{100} = 801{\cdot}5 \text{ mm}^2.$$

Area required on each face = 400·8 mm².
Provide 8 mm ϕ bars at 120 mm centres.

Design of bottom dome

Radius of dome is obtained by

$$1{\cdot}6\,(2R - 1{\cdot}6) = 4^2$$

$$R = \frac{10 + 1{\cdot}6}{2} = 5{\cdot}8 \text{ m.}$$

$$\sin \phi = \frac{4}{5{\cdot}8} = 0{\cdot}69, \; \phi = 43° \; 38'$$

200 mm thickness is provided for dome.

Self-weight of dome per square metre

$$=0.20\times 25{,}000=5000 \text{ N.}$$

Weight of water above dome

$$=10{,}000\left\{\frac{\pi}{4}\times 8^2\times 7.0-\frac{\pi\times 2^2}{3}(3\times 5.8-2)\right\}$$

$$=10{,}000\,[352-64.52]=10{,}000\times 287.48$$

$$=28{,}741{,}800 \text{ N.}$$

Surface area of dome

$$=2\pi Rh=2\pi\times 5.8\times 1.6.$$

Weight of water per square metre of surface area

$$=\frac{2{,}874{,}800}{2\pi\times 5.8\times 1.6}=49{,}320 \text{ N}$$

Total load per square metre of surface area

$$=49{,}320+5000=54{,}320 \text{ N}$$

Meridional stress is given by

$$=\frac{wR}{t\,(1+\cos\theta)}$$

Maximum meridional stress at $\theta=43°\ 38'$

$$=\frac{54{,}320\times 5.8}{0.2\,(1+0.724)}$$

$$=913{,}734 \text{ N/m}^2=0.914 \text{ N/mm}^2. \quad \text{Safe.}$$

Hoop stress $\quad =\dfrac{wR}{t}\left(\dfrac{\cos^2\theta+\cos\theta-1}{1+\cos\theta}\right)$

At $\theta=0$; hoop stress$=\dfrac{54{,}320\times 5.8}{0.2}\times\dfrac{1}{2}$

$$=787{,}640 \text{ N/m}^2.$$

$$=0.788 \text{ N/mm}^2. \quad \text{Safe.}$$

Minimum reinforcement to be provided is 0.2% of cross-sectional area

$$=\frac{0.2\times 200\times 1000}{100}=400 \text{ mm}^2.$$

Provide 8 mm ϕ bars at 120 mm centres both ways.

Design of Circular Beam. There will be hoop stress caused in the circular beam due to resultant borizontal thrust from conical dome and spherical dome at the bottom.

Horizontal thrust on circular beam

$=[0{\cdot}982\times350\times1000\times\cos 45°$
$-0{\cdot}914\times200\times1000\times\cos\phi]$;

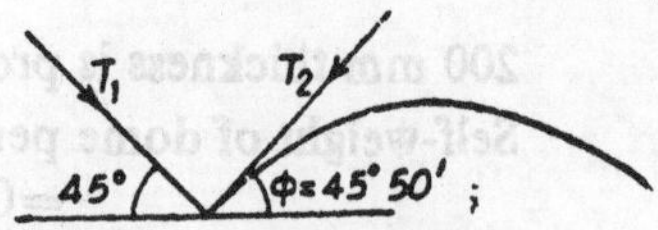

Fig. 16·30

$=[0{\cdot}982\times350\times1000\times0{\cdot}707$
$-0{\cdot}914\times200\times1000\times0{\cdot}724]$

$$=242{,}996-132{,}347=110{,}649 \text{ N/m.}$$

Vertical load on beam per metre run

$$=T_1\sin 45+T_2\sin\phi$$
$$=[0{\cdot}982\times350\times1000\times0{\cdot}707+0{\cdot}914\times200\times1000\times0{\cdot}69]$$
$$=242{,}996+126{,}132=369{,}128 \text{ N/m}$$

Section of 1000 mm × 500 mm is assumed for the beam.

Self-weight of the beam

$$=1\times0{\cdot}5\times1\times25{,}000=12{,}500 \text{ N.}$$

Total load per metre run

$$=369{,}128+12{,}500=381{,}628 \text{ N.}$$

The beam will be supported on 8 columns.

Maximum +ve B.M. $=0{\cdot}03\ wR^2\times\theta$

Maximum −ve B.M. $=0{\cdot}066\ wR^2\times\theta$

Maximum torsional moment

$$=0{\cdot}005\ wR^2\times\theta.$$

(at angle of 9½° from support)

$$\theta=45°.$$

Maximum +ve B.M.

$$=0{\cdot}03\times381{,}628\times4^2\times\frac{\pi}{4}=143{,}870 \text{ Nm.}$$

Maximum −ve B.M.

$$=0{\cdot}066\times381{,}628\times4^2\times\frac{\pi}{4}=316{,}515 \text{ Nm.}$$

Maximum torsion $=0{\cdot}005\times381{,}628\times4^2\times\frac{\pi}{4}=23{,}978$ Nm.

S.F. at support $=\dfrac{381{,}628\times4\times\pi/4}{2}=599{,}460$ N.

Shear force at point of maximum torsion

$$=599{,}460-\frac{599{,}460\times7{\cdot}5}{45/2}=399{,}640 \text{ N.}$$

Effective depth required

$$=\sqrt{\frac{316,515\times1000}{500\times1{\cdot}411}}$$

(tension near water surface)

$=670$ mm.

Provide overall depth of 1000 mm from shear considerations. Effective depth of 940 mm is provided.

Area of steel required

$$=\frac{316,515\times1000}{0{\cdot}84\times940\times100}=4009\text{ mm}^2.$$

Provide 9 bars of 25 mm ϕ $A_t=4418$ mm².

Area of steel at centre

$$=\frac{143,870\times1000}{0{\cdot}86\times940\times125}\text{ (tension outside)}$$

$=1424$ mm².

Provide 5 bars of 20 mm ϕ.

Shear stress at support

$$=\frac{599,460}{940\times500}=1{\cdot}28\text{ N/mm}^2.$$

Shear stress at point of maximum torsion

Due to S.F. shear stress

$$=\frac{399,640}{940\times500}$$

$=0{\cdot}85$ N/mm².

Equivalent shear $=V+1{\cdot}6\,\dfrac{T}{b}$

$$=399,640+1{\cdot}6\times\frac{23,978\times1000}{500}$$

$=399,640+76,730$

$=476,370$ N

Total shear stress $=\dfrac{476,370}{940\times500}=1{\cdot}01$ N/mm².

Taking $\tau_c=0{\cdot}39$ N/mm².

Net shear $=(1{\cdot}01-0{\cdot}39)\times940\times500$

$=291,400$ N.

Shear Reinforcement

At Support. Provide 12 mm ϕ 6 legged stirrups.

$$\text{Spacing}=\frac{6\times113\times100\times940}{599{,}460}=106{\cdot}3 \text{ mm.}$$

Provide 12 mm ϕ o legged stirrups at 90 mm centres.

Shear reinforcement at point of maximum torsion

Taking same pitch of 90 mm,

$$90=\frac{A_v\times100\times940}{291{,}400}$$

$$\therefore \quad A_v=\frac{90\times291{,}400}{120\times940}$$

$$=232{\cdot}5 \text{ mm}^2$$

Use 12 mm ϕ 4 legged stirrups.

Longitudinal steel for torsion and bending

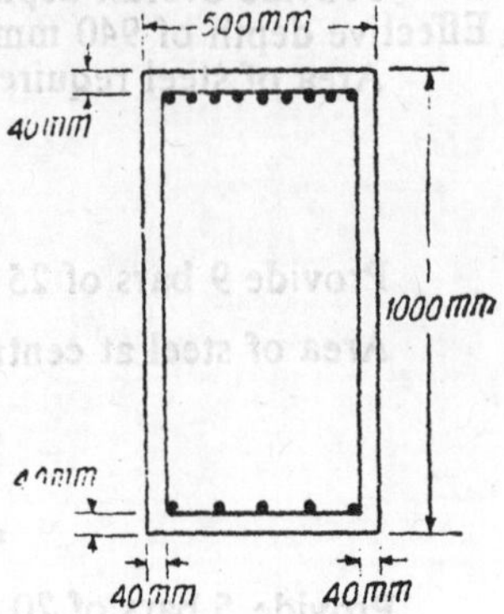

Fig. 16·31

$$M_{c1}=M+M_t$$

$$M_{et}=T\,\frac{\left(1+\frac{D}{b}\right)}{1{\cdot}7}$$

$$=\frac{23{,}978\times1000\left(1+\frac{1000}{500}\right)}{1{\cdot}7}$$

$$=42{,}314{,}118 \text{ N/mm}$$

$$M_{et}=143{,}870{,}000+42{,}314{,}118$$

$$=186{,}184{,}118$$

$$A_t=\frac{186{,}184{,}118}{0{\cdot}86\times940\times125}$$

$$=1842{\cdot}5 \text{ mm}^2$$

Provide 5 bars of 20 mm ϕ and 8 bars of 12 mm ϕ on two faces.

Design of columns. The columns will be splayed as shown in Fig. 16·32.

Vertical load on each column

$$=599{,}460\times2=1{,}198{,}920 \text{ N}$$

Self-weight of column

$$=\frac{\pi}{4}\times(0{\cdot}5)^2\times\sqrt{145}\times25{,}000=59{,}109 \text{ N}$$

Total vertical load $=1,198,920+59,109=1,258,029$ N.

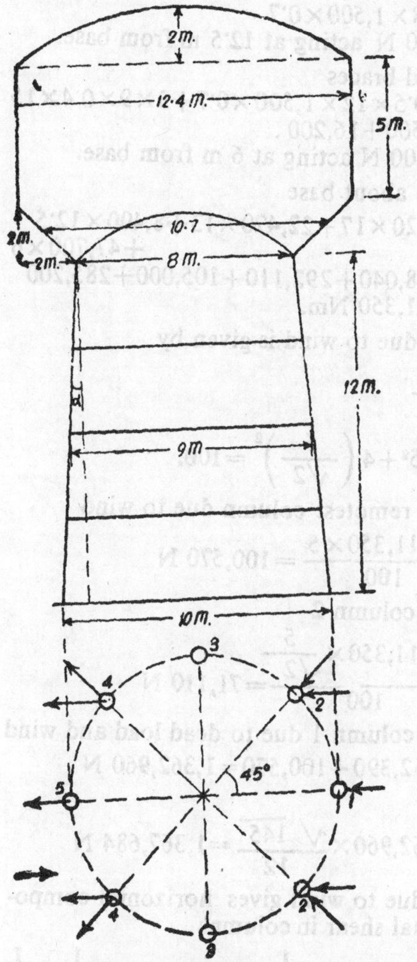

Fig. 16·32

Axial load in the column

$=1,258,029 \sec x$

$=1,258,029\times\dfrac{\sqrt{145}}{12}$

$=1,262,390$ N

Weight of water in tank

$=3,727,560+2,874,800$

$=6,602,360$ N

Weight of water per column

$=\dfrac{6,602,360}{8}$

$=825,300$ N

Vertical load on column when tank empty

$=1,262,390-825,300$

$=437,090$

Axial load in column

$=437,090\times\dfrac{\sqrt{145}}{12}$

$=438,605$ N

500 mm diameter column is provided.

Let A_s be area of steel required.

Load $P=1,262,390$

$=\left[\dfrac{\pi}{4}\times(500)^2-A_s\right]\times 5+A_s\times 130$

$=312,500\pi+125\,A_s$

$A_s=\dfrac{1,262,390-981,900}{125}$

$=2244$ mm².

Provide 8 bars of 20 mm ϕ. Area of steel provided$=2,514$ mm².
6 mm ϕ ties at 250 centres are provided.

Stresses due to wind

Intensity of wind pressure$=1500$ N/m²

Reduction factor $=0·7$

Wind force on top dome and cylindrical wall

$=(5+1)\times 12·4\times 1,500\times 0·7$

$=78,120$ N acting at 17 m from base

Wind force on slant wall

$=10·7\times 2\times 1,500\times 0·7$

$=22,470$ N acting at 18 m from base.

Wind force on circular beam

$$=1\times8\times1{,}500\times0{\cdot}7$$
$$=8{,}400 \text{ N acting at } 12{\cdot}5 \text{ m from base.}$$

Wind force on columns and braces

$$=5\times0{\cdot}5\times12\times1{,}500\times0{\cdot}7+3\times9\times0{\cdot}4\times15$$
$$=31{,}500+16{,}200$$
$$=47{,}700 \text{ N acting at 6 m from base.}$$

Total moment of pressures about base

$$=78{,}120\times17+22{,}470\times13+8{,}400\times12{\cdot}5+47{,}700\times6$$
$$=1{,}328{,}040+292{,}110+105{,}000+285{,}200$$
$$=2{,}011{,}350 \text{ Nm.}$$

Vertical load in a column due to wind is given by

$$=\frac{Mr}{\Sigma r^2}$$

$$\Sigma r^2=2\times5^2+4\left(\frac{5}{\sqrt{2}}\right)^2=100.$$

Maximum vertical load in remotest column due to wind

$$=\frac{2{,}011{,}350\times5}{100}=100{,}570 \text{ N}$$

Maximum vertical load in column 2

$$=\frac{2{,}011{,}350\times\frac{5}{\sqrt{2}}}{100}=71{,}110 \text{ N}$$

Maximum vertical load in column 1 due to dead load and wind

$$=1{,}262{,}390+100{,}570=1{,}362{,}960 \text{ N}$$

Axial load in column

$$=1{,}362{,}960\times\frac{\sqrt{145}}{12}=1{,}367{,}684 \text{ N}$$

Axial force in columns due to wind gives horizontal component which reduces the horizontal shear in columns

$$=2\times100{,}570\times\frac{1}{2}+4\times71{,}110\times\frac{1}{12}\times\frac{1}{\sqrt{2}}$$
$$=16{,}750+16{,}750=33{,}500 \text{ N}$$

Net horizontal force at base

$$=78{,}120+22{,}470+8{,}400+47{,}700-33{,}500$$
$$=123{,}190 \text{ N}$$

Horizontal shear per column

$$=\frac{123{,}190}{8}=15{,}400 \text{ N}$$

Maximum B.M. in column at junction with bottom brace

$$=15{,}400\times1{\cdot}5=23{,}100 \text{ Nm}$$

Check for stresses

Direct load = 1,367,684 N

B.M. = 23,100 Nm

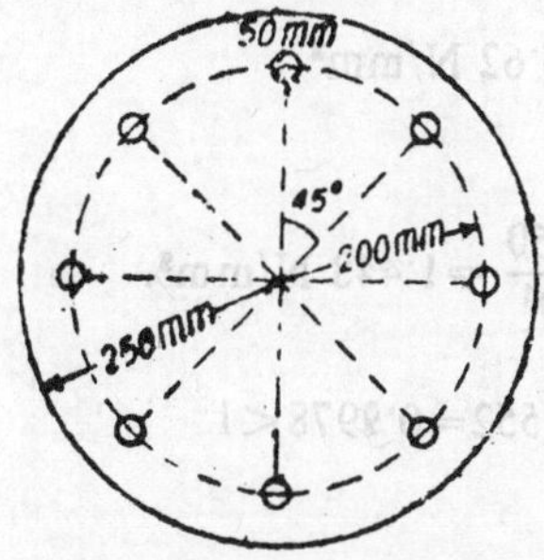

Fig. 16.33

Equivalent area of section

$= \frac{\pi}{4} \times (500)^2 + 12 \times 2514$

$= 196,300 + 30,200$

$= 226,500 \text{ mm}^2.$

Equivalent moment of inertia

$= \frac{\pi}{64} \times (50)^2 + 32 \times [314 \times 2 \times (200)^2 + 4(200 \times 0.707)^2 \times 314]$

$= (306,700 + 60,288) \times 10^4$

$= (366,988 \times 10^4) \text{ mm}^4$

Maximum stress in concrete due to direct load

$$= \frac{1,367,684}{226,500} = 6.04 \text{ N/mm}^2$$

Maximum stress in concrete due to B.M.

$$= \frac{23,100 \times 1,000 \times 250}{(366,988) \times 10^4} = 1.573 \text{ N/mm}^2$$

Allowable direct stress in concrete

$$= \frac{4}{3} \times 5 = 6.67 \text{ N/mm}^2$$

Allowable bending stress in concrete

$$= \frac{4}{3} \times 7 = 9.33 \text{ N/mm}^2$$

$$\frac{6.04}{6.67} + \frac{1.573}{9.33} = 0.9055 + 0.1686 = 1.0741 > 1.$$

∴ Section is not suitable. Steel area in the section is increased. 8 bars of 25 mm ϕ are provided. Area of steel provided $= 3927 \text{ mm}^2$

Equivalent area

$= \frac{\pi}{4} \times (500)^2 + 12 \times 3927$

$= 196,300 + 47,124$

$= 243,424 \text{ mm}^2$

Equivalent moment of inertia

$= \frac{\pi}{64} \times (500)^4 + 12 \times [1,491 \times 2 \times (197.5)^2 + 4 \times 49] \times (197.5 \times 0.707)$

$= (306,700 + 91,940) \times 10^4$

$= (398,640 \times 10^4) \text{ mm}^4$

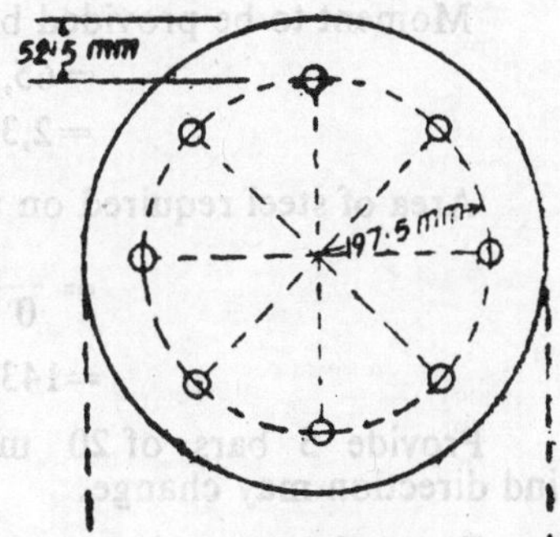

Fig. 16.34

Stress due to direct load

$$=\frac{1,367,684}{243,424}=5{\cdot}62 \text{ N/mm}^2$$

Stress due to bending

$$=\frac{23,100,000\times 250}{(398,610)\times 10^4}=1{\cdot}448 \text{ N/mm}^2.$$

$$\therefore \quad \frac{5{\cdot}62}{6{\cdot}67}+\frac{1{\cdot}448}{9{\cdot}33}=0{\cdot}8426+0{\cdot}1552=0{\cdot}9978<1$$

$\therefore$ The section is suitable.

Design of Braces. Maximum B.M. in a brace occurs when wind is blowing at right angles to adjoining brace. In Fig. 16·35 Maximum B.M. in brace *BC* occurs when wind is blowing at right angles to *AB*. *Bac* is the triangle of moments.

Moment in brace *BC*
$=2\times 2,310\times\sqrt{2}$
$=46,200\sqrt{2}$
$=65,380$ Nm
$=65,380,000$ N-mm

400 mm × 400 mm section for brace is provided with same reinforcement on both faces.

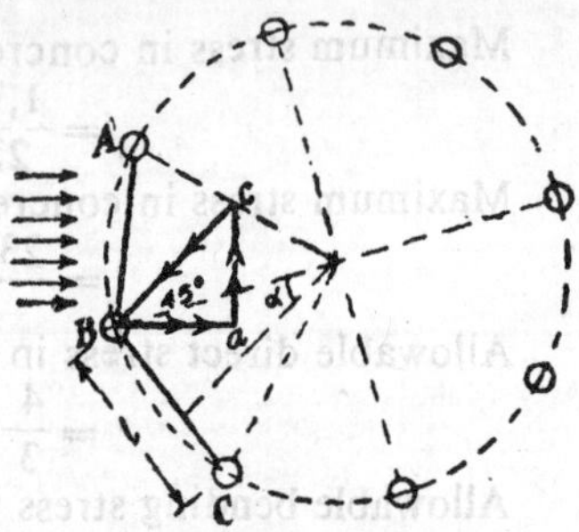

Fig. 16·35

Assume effective depth

$$=360 \text{ mm}$$

Moment of resistance of section as singly reinforced section

$$=1{\cdot}213\times 400\times(360)^2=63,000,000 \text{ N-mm}$$

Moment to be provided by compression steel

$$=65,380,000-63,000,000$$
$$=2,380,000 \text{ N-mm}$$

Area of steel required on tension side

$$=\frac{63,000,000}{0{\cdot}87\times 360\times 140}+\frac{2,380,000}{320\times 140}$$
$$=1435+62=1497 \text{ mm}^2$$

Provide 5 bars of 20 mm ϕ both at top and bottom as the wind direction may change.

Check for shear. Shear force in the brace

$$=\frac{2M}{l}, \text{ where } l \text{ is the length of brace}$$

$$=\frac{2\times 65{,}380}{2\times\frac{9\cdot5}{2}\sin\alpha}$$

$$=\frac{2\times 65{,}380}{9\cdot5\times 0\cdot3827}=35{,}900\text{ N}$$

Shear stress $=\frac{35{,}900}{380\times 400}=0\cdot249\text{ N/mm}^2$. Safe.

Provide 6 mm ϕ stirrups at 300 mm centres.

Design of Foundation

Total load on foundation when tank is full

$$=1{,}258{,}029\times 8=10{,}064{,}232\text{ N}$$

Assume self-load at 10%

$$=1{,}006{,}423\text{ N}$$

Total load $=11{,}070{,}655$ N

Area of foundation required

$$=\frac{11{,}070{,}655}{200{,}000}=55\cdot35\text{ m}^2$$

Provide 12 m outside diameter of raft and 8 m inside diameter.

$$\text{Area provided}=\frac{\pi}{4}(12^2-8^2)=62\cdot84\text{ m}^2.$$

Net upward pressure

$$=\frac{10{,}064{,}232}{62\cdot84}=160{,}156\text{ N/m}^2$$

Design of Bottom Circular Beam

Load per metre run of circular beam

$$=\frac{10{,}064{,}232}{\pi\times 10}=320{,}354\text{ N}$$

Maximum $-$ve B.M. at supports

$$=0\cdot066\,wR^2\times\theta$$

$$=0\cdot066\times 320{,}354\times 5^2\times\frac{\pi}{4}$$

$$=415{,}149\text{ Nm.}$$

Maximum $+$ve B.M. at centre

$$=0\cdot03\,wR^2\times\theta$$

$$=0\cdot03\times 320{,}354\times 5^2\times\frac{\pi}{4}$$

$$=188{,}704\text{ Nm.}$$

Maximum torsional moment

$$=0{\cdot}05\, wR^2\times\theta$$

$$=0{\cdot}005\times320{,}354\times5^2\times\frac{\pi}{4}$$

$$=31{,}451 \text{ Nm.}$$

at angle of 9·5° from support.

Maximum S.F. at supports

$$=\frac{320{,}354\times5\times\pi/4}{2}=629{,}014 \text{ N}$$

S.F. at point of maximum torsion

$$=629{,}014-\frac{629{,}014\times9{\cdot}5}{45/2}=363{,}430 \text{ N}$$

Design of Section. M 20 concrete mix is used.

Width of 500 mm is provided for the beam.

Effective depth required

$$=\sqrt{\frac{415.149\times1000}{1{\cdot}213\times500}}=827 \text{ mm}$$

Provide depth of 900 mm with effective depth of 840 mm

Area of steel required

$$=\frac{415{,}149\times1000}{0{\cdot}87\times840\times140}=4058 \text{ mm}^2$$

Provide 9 bars of 25 mm ϕ.

A_t provided $=4418 \text{ mm}^2$

Area of steel required at centre

$$=\frac{188{,}704\times1000}{0{\cdot}87\times840\times140}=1844 \text{ mm}^2$$

Provide 6 bars of 20 mm ϕ.

Shear at Support. Shear stress at support

$$=\frac{629{,}014}{840\times500}=1{\cdot}50 \text{ N/mm}^2$$

Taking $\tau_c=0{\cdot}39 \text{ N/mm}^2$

Net shear $=(1{\cdot}50-0{\cdot}39)\times840\times500=466{,}200 \text{ N}$

Provide 12 mm ϕ, 4 legged stirrups, spacing

$$=\frac{4\times113\times140\times840}{466{,}200}=114 \text{ mm.}$$

Provide 12 mm ϕ, 4 legged stirrups at 110 mm centres.

At point of maximum torsion

$$V_e = V + 1{\cdot}6 \times \frac{T}{b}$$

$$= 363{,}430 + 1{\cdot}6 \times \frac{31{,}451{,}000}{500}$$

$$= 464{,}073 \text{ N}$$

Provide 12 mm ϕ 4 legged stirrups at 110 mm centres.

Longitudinal steel for torsion. Volume of steel required in legs of stirrups for torsion per cm length

$$M_{e1} = M + M_t = 188{,}704 \times 1000 + \frac{T\left(1 + \frac{D}{b}\right)}{1{\cdot}7}$$

$$= 188{,}704{,}000 + \frac{31{,}451{,}000\left(1 + \frac{900}{500}\right)}{1{\cdot}7}$$

$$= 240{,}505{,}647 \text{ N-mm}$$

$$A_t = \frac{240{,}505{,}647}{0{\cdot}87 \times 840 \times 140} = 2351 \text{ mm}^2$$

$$\text{Longitudinal steel} = 2351 - 1844 = 507$$

Provide 8 bars of 12 mm ϕ, 4 bars on each face.

Design of raft slab

Consider one metre width of slab.

Projection of slab beyond face of beam

$$= \frac{2 - 0{\cdot}5}{2} = 0{\cdot}75 \text{ m.}$$

$$\text{Maximum B.M.} = \frac{160{,}156 \times (0{\cdot}75)^2}{2} \text{ Nm.}$$

$$= 45{,}044 \text{ Nm.}$$

Effective depth required

$$= \sqrt{\frac{45{,}044 \times 1000}{1{\cdot}213 \times 1000}} = 192{\cdot}7 \text{ mm.}$$

Provide 250 mm thick slab with effective depth of 210 mm.

Area of steel required

$$= \frac{45{,}044 \times 1000}{0{\cdot}87 \times 210 \times 140} = 1761 \text{ mm}^2.$$

Provide 16 mm ϕ bars at 110 mm centres.

Area of distribution steel

$$= \frac{0{\cdot}15 \times 250 \times 1000}{100} = 375 \text{ mm}^2.$$

Provide 8 mm ϕ bars at 130 mm centres.

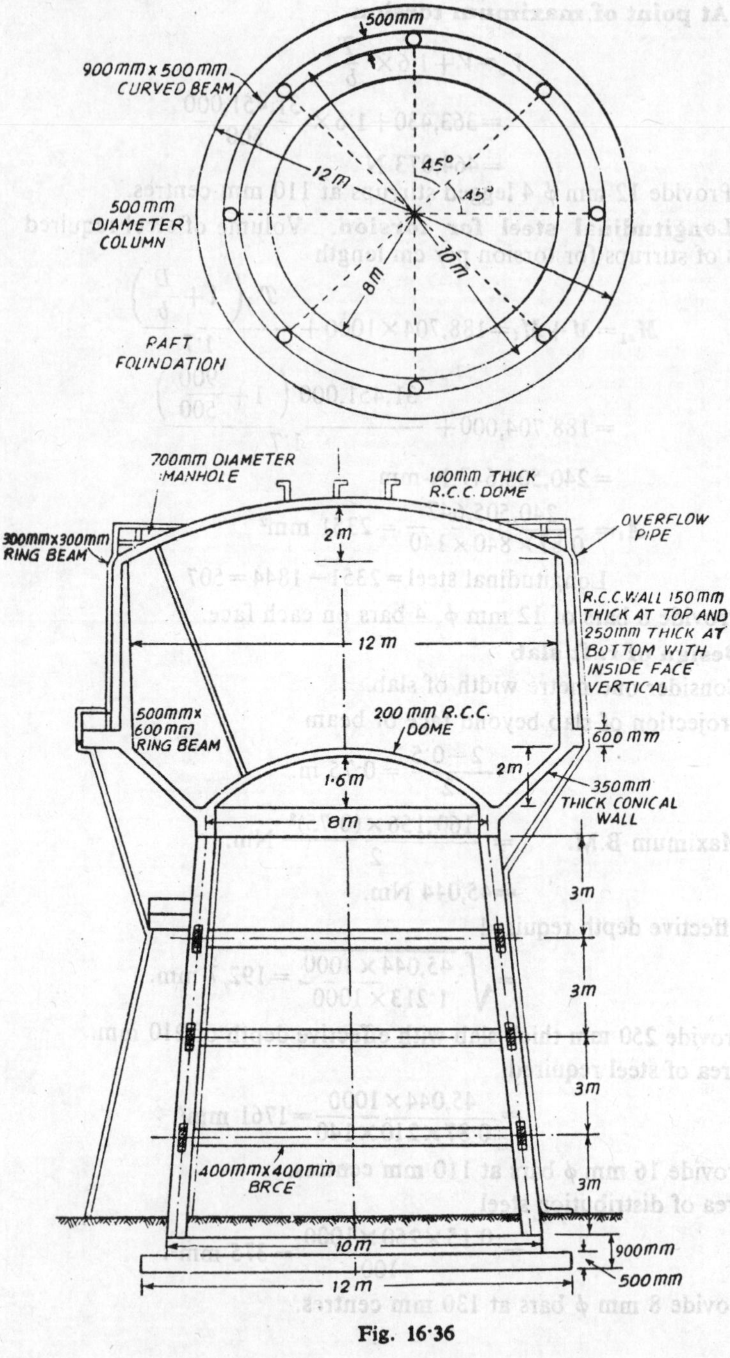
500mm
900mm x 500mm
CURVED BEAM
12 m
45°
45°
500mm
DIAMETER
COLUMN
10 m
8m
RAFT
FOUNDATION
700mm DIAMETER
MANHOLE
100mm THICK
R.C.C. DOME
300mm x 300mm
RING BEAM
2 m
OVERFLOW
PIPE
R.C.C. WALL 150mm
THICK AT TOP AND
250mm THICK AT
BOTTOM WITH
INSIDE FACE
VERTICAL
12 m
500mm x
600mm
RING BEAM
200 mm R.C.C.
DOME
600 mm
1·6 m
2m
350mm
THICK CONICAL
WALL
8m
3m
3m
3m
400mm x 400mm
BRCE
3m
10 m
900mm
12 m
500mm

Fig. 16·36

The effect of wind is not considered in design of raft as due to wind effect, stresses can be increased by $33\frac{1}{3}$%. The increase in stresses due to wind comes out to be less than $33\frac{1}{3}$%.

Check for sliding

Weight on foundation when tank empty

$$=11{,}070{,}655-6{,}602{,}360$$
$$=4{,}468{,}295 \text{ N.}$$

Net horizontal force at base

$$=123{,}190 \text{ N.}$$

Taking coefficient of friction as 0·5,

$$\text{Factor of safety} = \frac{0{\cdot}5\times 4{,}468{,}295}{123{,}190} = 18{\cdot}1. \quad \text{Safe}$$

Details of reinforcement in tank and foundation are shown in Fig. 16·36.

PROBLEMS

1. Design a rectangular tank for capacity of 100,000 litres. The depth of tank is to be 3 m and length to breadth ratio 1·5 The tank is resting on ground.

2. Design a circular tank resting on ground and having capacity of 250,000 litres.

3. Design an underground rectangular tank 3 m × 7 m and one metre deep. The tank is covered at top.

4. Design an Intze-type tank for capacity of 1,000,000 litres. The tank is to be divided in two compartments. The height of the tank is 15·16 m above ground level. The bearing capacity of soil is 150 kN/m².

5. Design the side wall, floor and beams (four in number on four columns) of small circular elevated tank to hold 25,000 litres of water. Depth is limited to 2·5 m. Use M 20 concrete mix.

6. Design an open rectangular tank in wet ground to hold 200,000 litres of water, the long side being limited to 12 m Depth of water is to be 3 m. Top of tank is at ground level, $\phi=30°$. Density of earth = 16,000 N/m³.

7. A reinforced concrete water reservoir is to be constructed in two approximately square compartments on the top of a hill. Following are the important features of the reservoir :

Capacity	Twenty million litres
Full supply level	R.L. 102 0 m.
Floor level	R.L. 97·50 m.
Free board above F.S.L.	300 mm.
Ground level on the flat plot of land on the top of hill	R.L. 98·10 m.
Foundation level for walls and columns	R.L. 96·10 m.
Maximum bearing pressure for foundation	200 kN/m².
Filling above the roof slab	1 metre earth filling
Filling all round the reservoir will have a 1·5 m top width in level with the filling on the reservoir top and sloping out at $1\frac{1}{2}$ horizontal to 1 vertical.	

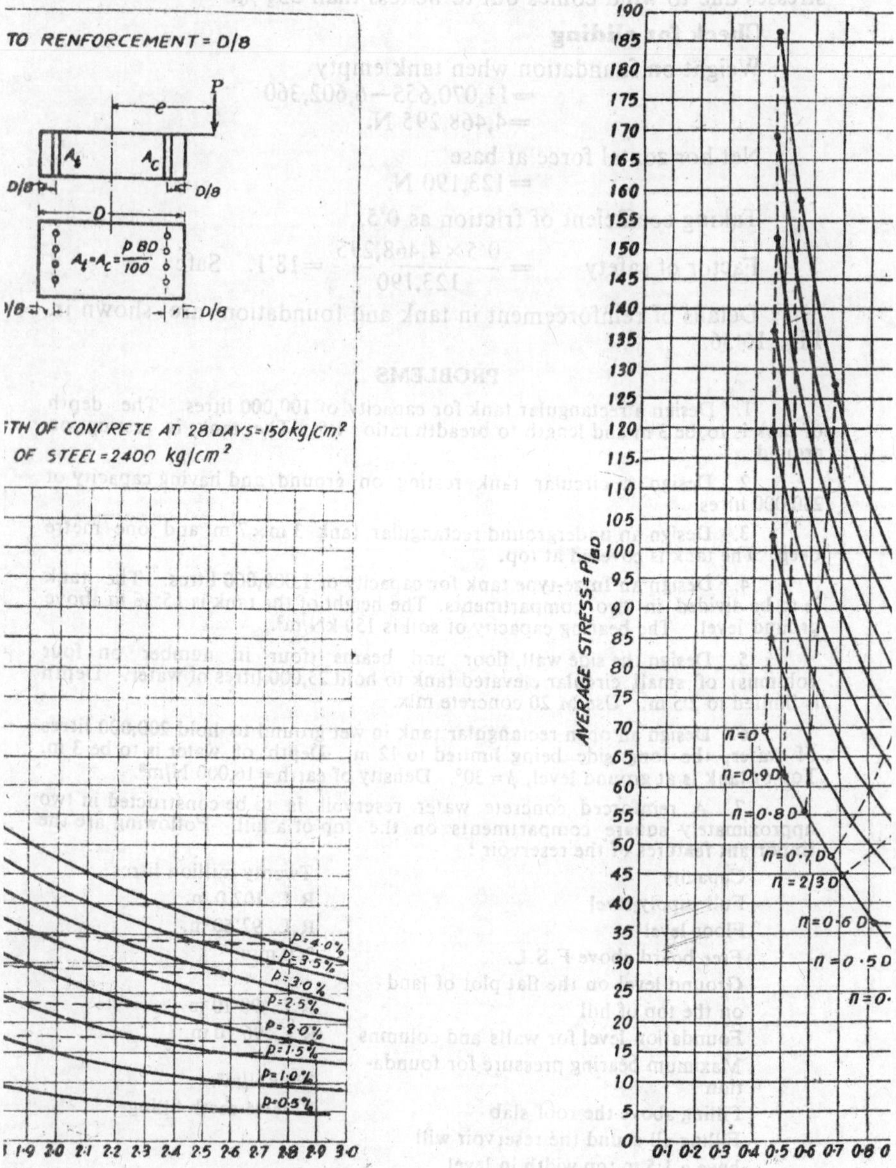

TO RENFORCEMENT = D/8
P
e
A_s
A_c
D/8
D/8
D
$A_t = A_c = \frac{p\,BD}{100}$
D/8
D/8
TH OF CONCRETE AT 28 DAYS=150 kg/cm²
OF STEEL=2400 kg/cm²
P=4.0%
P=3.5%
P=3.0%
P=2.5%
P=2.0%
P=1.5%
P=1.0%
P=0.5%
1·9 2·0 2·1 2·2 2·3 2·4 2·5 2·6 2·7 2·8 2·9 3·0
AVERAGE STRESS = P/BD
190 185 180 175 170 165 160 155 150 145 140 135 130 125 120 115 110 105 100 95 90 85 80 75 70 65 60 55 50 45 40 35 30 25 20 15 10 5 0
n=D
n=0·9D
n=0·8D
n=0·7D
n=2/3D
n=0·6D
n=0·5D
n=0·
0·1 0·2 0·3 0·4 0·5 0·6 0·7 0·8 0

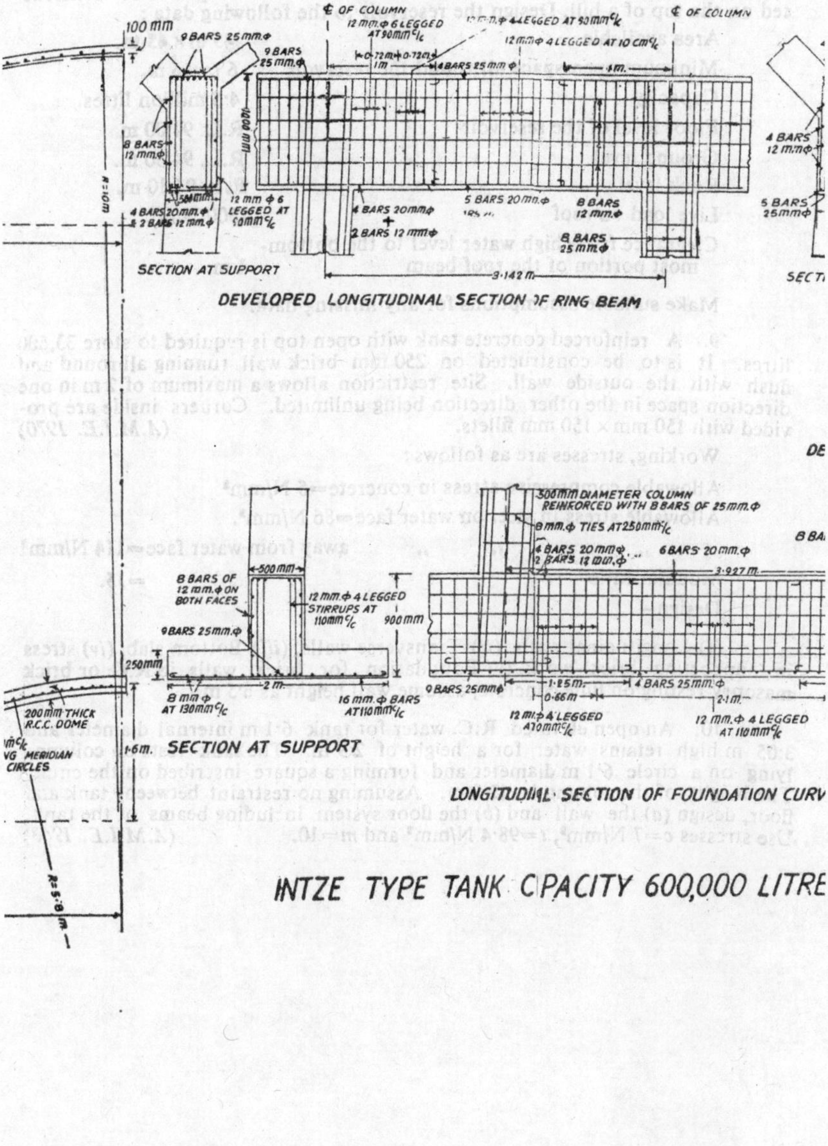

100 mm
9 BARS 25 mm.φ
8 BARS 12 mm.φ
4 BARS 20 mm.φ
& 2 BARS 12 mm.φ
12 mm φ 6 LEGGED AT 90 mm c/c
SECTION AT SUPPORT
R=10 M
℄ OF COLUMN
12 mm.φ 6 LEGGED AT 90 mm c/c
12 mm.φ 4 LEGGED AT 90 mm c/c
12 mm.φ 4 LEGGED AT 10 cm c/c
4 BARS 25 mm φ
1000 mm
5 BARS 20 mm.φ
8 BARS 12 mm.φ
8 BARS 25 mm.φ
4 BARS 20 mm φ
2 BARS 12 mm φ
3.142 m
DEVELOPED LONGITUDINAL SECTION OF RING BEAM
4 BARS 12 mm.φ
5 BARS 25 mm φ
500 mm
8 BARS OF 12 mm.φ ON BOTH FACES
12 mm.φ 4 LEGGED STIRRUPS AT 110 mm c/c
900 mm
9 BARS 25 mm.φ
250 mm
8 mm φ AT 130 mm c/c
16 mm.φ BARS AT 110 mm c/c
2 m
200 mm THICK R.C.C. DOME
MERIDIAN CIRCLES
1.6 m.
SECTION AT SUPPORT
500 mm DIAMETER COLUMN REINFORCED WITH 8 BARS OF 25 mm.φ
8 mm.φ TIES AT 250 mm c/c
4 BARS 20 mm φ
2 BARS 12 mm.φ
6 BARS 20 mm.φ
3.927 m.
1.25 m.
0.66 m
9 BARS 25 mm φ
4 BARS 25 mm.φ
2.1 m.
12 mm φ 4 LEGGED AT 70 mm c/c
12 mm.φ 4 LEGGED AT 110 mm c/c
LONGITUDINAL SECTION OF FOUNDATION CURV
R=5.8 m
INTZE TYPE TANK CIPACITY 600,000 LITRE

Design the surrounding and intermediate R.C. walls, R.C.C. columns, column footings, roof slab etc. and draw dimensioned drawings of all reinforced cement works. Surrounding walls must be designed for two conditions (*a*) Reservoir full outside earth filling removed, (*b*) Reservoir empty and outside earth filling in portion.

Assume any data missing.

8. A reinforced concrete water reservoir closed at top is to be constructed on the top of a hill. Design the reservoir to the following data :

Area available	45 m × 45 m.
Minimum open space all round the reservoir	6 m × 6 m.
Capacity	4·5 million litres
Floor level of the reservoir	R.L. 94·00 m
Ground level	R.L. 94·30 m.
Rock level	R.L. 94·30 m.
Live load on roof	1500 N/m².
Clearance from high water level to the bottom-most portion of the roof beam	1 m.

Make suitable assumptions for any missing data.

9. A reinforced concrete tank with open top is required to store 33,600 litres. It is to be constructed on 250 mm brick wall running all round and flush with the outside wall. Site restriction allows a maximum of 2 m in one direction space in the other direction being unlimited. Corners inside are provided with 150 mm × 150 mm fillets. (*A.M.I.E. 1970*)

Working, stresses are as follows:

Allowable compressive stress in concrete = 5 N/mm²

Allowable stress in steel on water face = 86 N/mm².

" " " " away from water face = 114 N/mm²

Modular ratio = 15.

Design—

(*i*) Longitudinal walls (*ii*) Transverse walls, (*iii*) Bottom slab, (*iv*) stress in supporting brick work, (*v*) Foundation for brick walls in R.C. or brick masonry resting on lime concrete, assume wall height as 6·5 m.

10. An open elevated R.C. water for tank 6·1 m internal diameter and 3·05 m high retains water for a height of 2·9 m. The tank rests on columns lying on a circle 6·1 m diameter and forming a square inscribed on the circle. The height of the column is 12·2 m. Assuming no restraint between tank and floor, design (*a*) the wall and (*b*) the floor system including beams of the tank. Use stresses $c = 7$ N/mm², $t = 98·4$ N/mm² and $m = 10$. (*A.M.I.E. 1963*)

17

Design of Aqueducts and Box Culverts

17·1. Aqueducts. The RCC aqueducts are provided where water is to be carried over rivers or canals. The vertical walls of aqueducts are designed as cantilevers subjected to water pressure and also as beams spanning between supports and carrying load of the horizontal slab. The horizontal slab has to carry weight of water and self-load. It is subjected to fixing moments and direct tension due to pressure of water on vertical walls.

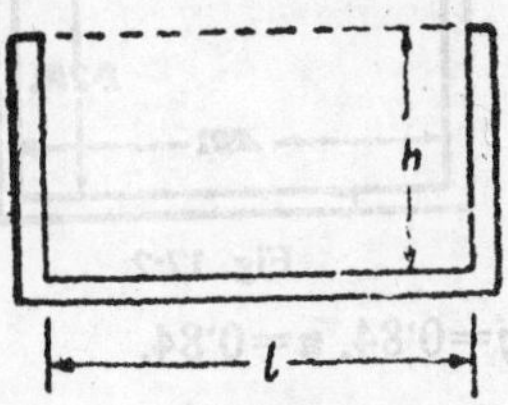

Fig. 17·1

Let l be the span of the horizontal slab and h be the depth of water in aqueduct. Moment at the end of horizontal slab will be equal to moment of vertical wall $=\frac{wh^3}{6}$

Simply supported B.M. at centre of slab will be $\frac{whl^2}{8}$.

Net B.M. at centre of slab

$$=M=\frac{whl^2}{8}-\frac{wh^3}{6}$$

For B.M. to be maximum

$$\frac{dM}{dh}=0 \therefore \frac{wl^2}{8}-\frac{wh^2}{2}=0$$

$$l=2h$$

$$M=\frac{wl^3}{15}-\frac{wl^3}{48}=\frac{wl^3}{24}$$

∴ Maximum +ve B.M. in the slab occurs when the depth of water is half the span length.

Generally the depth of water in channel is less than half the span. In such a case maximum B.M. at centre occurs when channel is full.

In case depth of water in channel is quite large, there is possibility of —ve B.M. at mid-span of horizontal slab. This is possible when fixing moment at the end is greater than simply supported moment at the centre of span.

i.e. $$\frac{wh^3}{6} \triangleright \frac{whl^2}{8} \quad \text{or} \quad h > l\sqrt{0{\cdot}075}$$

or $$h \triangleright 0{\cdot}866\, l.$$

Ex. 17·1. ***An aqueduct 4 m wide has to carry 2 m depth of water. Aqueduct has span of 9 m. Free board of 0·2 m is to be provided.***

Sol. Concrete mix. of M 20 is used.

Design of vertical wall thickness

Consider 1 m length of channel.

Fig. 17·2

$$\text{Maximum B.M.} = \frac{wh^3}{6}$$

$$= \frac{10{,}000 \times (2{\cdot}2)^3}{6}$$

$$= 17{,}740 \text{ N-m.}$$

Tension occurs on water side

$c = 7$ N/mm², $t = 100$ N/mm², $Q = 1{\cdot}411$, $j = 0{\cdot}84$, $n = 0{\cdot}84$.

Effective depth required

$$= \sqrt{\frac{17{,}740 \times 1000}{1{\cdot}411 \times 1000}} = 112 \text{ mm.}$$

As the vertical wall is designed as a girder spanning between supports, thickness of vertical wall will be taken as

$$\frac{\text{Span}}{30} = \frac{9 \times 1000}{30} = 300 \text{ mm.}$$

Effective depth $= 300 - 50 = 250$ mm.

Area of steel required

$$= \frac{17{,}740 \times 1000}{0{\cdot}84 \times 250 \times 100} = 880 \text{ mm}^2$$

Provide 10 mm ϕ bars at 800 mm centres. Half the bars are curtailed at 1 m from the base.

Distribution Steel

Percentage of distribution steel

$$= 0{\cdot}3 - \frac{200 \times 0{\cdot}1}{350} = 0{\cdot}24$$

$$A_t = \frac{0{\cdot}24 \times 300 \times 1000}{100} = 720 \text{ mm}^2$$

Provide 8 mm ϕ bars at 130 mm centres on both faces.

Design of horizontal slab

Thickness of slab is assumed as 250 mm.

Effective span of slab

$$=4+0.3=4.3 \text{ m.}$$

Consider 1 metre length of aqueduct.

Load of water per square metre of slab

$$=2.2\times10{,}000=22{,}000 \text{ N.}$$

Self-weight $=1\times1\times0.25\times25{,}000=6{,}250$ N

Total load $=28{,}250$ N/m²

Fixing moment at the end of slab = Load × Lever arm

Taken at centre of slab thickness $=\dfrac{10{,}000\times2.2\times2.2}{2}\times\left(\dfrac{2.2}{3}+0.125\right)$

$$=20{,}800 \text{ N-m}$$

Maximum B.M. at centre of span

$$=\frac{28{,}250\times(4.3)^2}{8}-20{,}800$$

$$=65{,}293-20{,}800=44{,}493 \text{ N-m}$$

Tension occurs on outside, $c=7$ N/mm², $t=125$ N/mm².
$Q=1.264$

$$j=0.86$$

Effective depth $=\sqrt{\dfrac{44{,}493\times1{,}000}{1.264\times1{,}000}}=187.6$ mm.

Provide overall depth 250 mm with effective depth of 210 mm.

Area of steel required

$$=\frac{44{,}493\times1{,}000}{0.86\times210\times125}=1971 \text{ mm}^2$$

Provide 16 mm ϕ bars at 100 mm centres.

Area of steel required near support

$$=\frac{20{,}800\times1{,}000}{0.84\times210\times100}=1179 \text{ mm}^2.$$

Half the bars from centre of span are bent up and additional 10 mm ϕ bars at 200 mm centres are provided.

Distribution steel

Percentage of distribution steel $=0.3-\dfrac{0.1\times150}{450-100}=0.257$

Area of distribution steel

$$=\frac{0.257\times250\times1{,}000}{100}=642.5 \text{ mm}^2$$

Provide 8 mm ϕ bars at 150 mm centres both faces.

Design of side wall as beams

Load from slab $=28{,}250\times\frac{4}{2}=56{,}500$ N/m

Self-load $=0{\cdot}3\times2{\cdot}45\times25{,}000=18{,}375$ N/m

Total load $=74{,}875$ N/m

Maximum B.M. $=\frac{wl^2}{8}=\frac{74{,}875\times9^2}{8}=758{,}109$ N-m

$c=7$ N/mm², $t=125$ N/mm², $Q=1{\cdot}264$, $j=0{\cdot}86$

Effective depth required

$$=\sqrt{\frac{758{,}109\times1{,}000}{1{\cdot}264\times300}}=1414 \text{ mm.}$$

Overall depth provided is 2,450 mm with effective depth of 2370 mm.

Area of steel $=\frac{758{,}109\times1{,}000}{0{\cdot}86\times2{,}970\times125}$

$=2976$ mm².

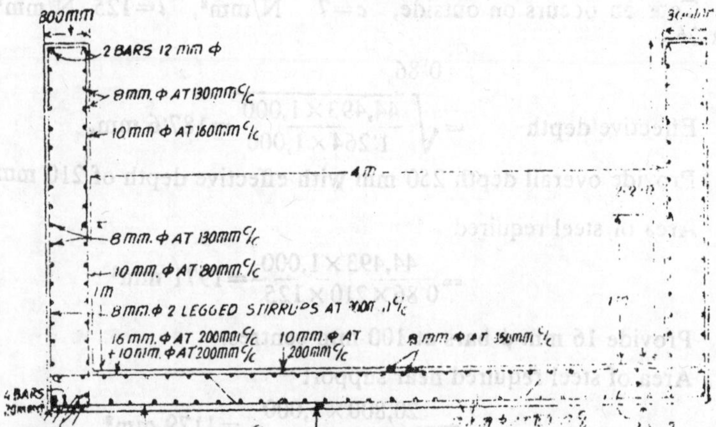

Fig. 17·3

Provide 4 bars 25 mm ϕ and 4 bars 20 mm ϕ at two [illegible]

A_t provided $=3219$ mm²

Shear stress $=\frac{74{,}875\times9/2}{2370\times300}=0{\cdot}474$ N/mm²

Provide 8 mm ϕ two-legged stirrups at 300 mm centres.

17·2. Box Culverts

Box culverts are used for conveyance of water at places where bearing capacity of soil is low.

The top slab of box culvert is subjected to earth above it and live load. The side walls are subjected to earth pressure from outside and water pressure from inside. The bottom slab is supported on soil throughout and is subjected to water load from top and pressure of soil from bottom. The bending moment in box culvert are worked out by moment distribution for the following conditions of loading :

(*i*) No water in the box. In this case side walls are subjected to earth pressure from outside.

(*ii*) Water in the box. In this case side walls are subjected to pressure of water from inside and earth pressure from outside.

Ex. 17·2. *Design a box culvert with inside dimensions 3 m × 3 m. The box culvert has to carry superimposed dead load of 10,000 N/m² and live load of 50,000 N/m². The density of earth is 16,000 N/m³ ; angle of repose of soil is 30°.*

Sol. Thickness of 300 mm is assumed for the box culvert. For the purpose of calculations of bending moments centre line dimensions are considered.

Consider one metre length of box culvert.

The bending moments in box culvert are found for the following conditions of loading :

1. Live load and dead load acting on the top and no water pressure from inside.

2. Dead load and live load acting on the top and water pressure acting inside.

3. Live load not acting on the sides of the culvert and water acting inside.

Case I. *Live load and dead load acting on the top and no water pressure from inside.*

Self-weight of top slab

$$=0{\cdot}3\times1\times1\times25{,}000=7500 \text{ N/m}^2.$$

Live load and superimposed dead load = 60,000 N/m²

Total load on top $=67{,}500$ N/m²

Weight of side wall $=3{\cdot}3\times0{\cdot}3\times25{,}000=24{,}750$ N

$$\text{Pressure at the base}=\frac{67{,}500\times3{\cdot}3+2\times24{,}750}{3{\cdot}3}=82{,}500 \text{ N/m}^2$$

Coefficient of earth pressure

$$k=\frac{1-\sin\phi}{1+\sin\phi}=\frac{1}{3}$$

Earth pressure at depth

$$h=60{,}000\times\tfrac{1}{3}+16{,}000\times h\times\tfrac{1}{3}$$

$$=20{,}000+\frac{16{,}000}{3}\,h$$

Earth pressure will vary from 20,000 N/m² at to p to 37,600 N/m² at depth of 3·3 m.

As the box culvert is symmetrical and symmetrically loaded, moment distribution is done for half of the frame by taking stiffness of the horizontal members as half of the actual stiffness.

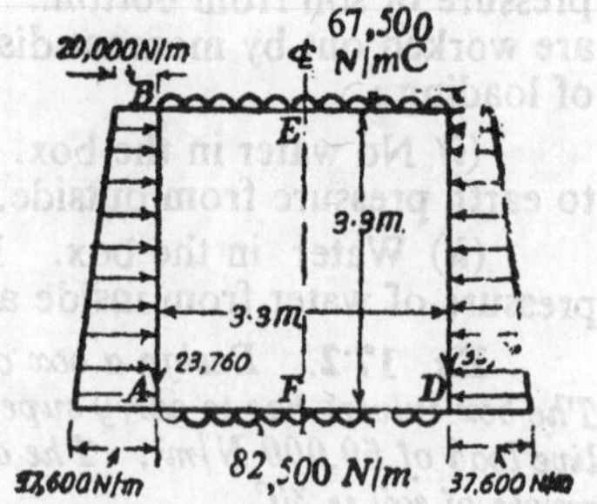

Fig. 17·4

As the thickness of the box culvert is uniform and it is of square section the stiffness of all the members will be same. Moment distribution is done for half of the box culvert, *i.e.* portion *FABE*.

k for $AB=1$ and k for BE and AF is $\frac{1}{2}$.

Distribution factors will be

$$r_{BA}=\frac{1}{1+1/2}=\frac{2}{3} \qquad r_{BE}=\frac{1/2}{1+1/2}=\frac{1}{3}$$

Similarly, $\quad r_{AB}=\frac{2}{3} \qquad r_{AF}=\frac{1}{3}$

Fixed end moments will be

$$\overline{M_{BC}}=\frac{-wl^2}{12}=-\frac{67{,}500\times(3\cdot3)^2}{12}=-61{,}256$$

$$\overline{M_{AD}}=+\frac{wl^2}{12}=+\frac{82{,}500\times(3\cdot3)^2}{12}=+74{,}869$$

$$M_{AB}=\frac{-wl^2}{12}-\frac{Wl}{10}$$

$$=\frac{-20{,}000\times(3\cdot3)^2}{12}-\tfrac{1}{2}\times(37{,}600-20{,}000)\times\frac{3\cdot3\times3\cdot3}{10}$$

$$=-18{,}150-9580=-27{,}730 \text{ N-m}$$

$$\overline{M_{AB}}=\frac{wl^2}{12}+\frac{Wl}{15}$$

$$=\frac{20{,}000\times(3\cdot3)^2}{12}+\tfrac{1}{2}\,(37{,}600-20{,}000)\times\frac{3\cdot3\times3\cdot3}{15}$$

$$=18{,}150+6390=24{,}540 \text{ N-m}$$

Moment distribution is shown in Table 17·1.

Table 17·1

Joint	A		B	
Member	*AD*	*AB*	*BA*	*BC*
Distribution factor	1/3	2/3	2/3	1/3
F.E.M. Balance	+74,869 −15,713	−27,730 −31,426	+24,540 +24,477	−61,256 +12,239
C.O Bal.	— −4080	+12,239 −8159	−15,730 +10,487	— −5243
C.O. Bal.	— −1748	+5244 −3496	−4080 +2770	— +1360
C.O. Bal.	— −453	+1360 − 907	−1748 +1165	— +583
C.O. Bal.	— −194	+583 −389	−454 +303	— +151
[illegible]	— −31	+152 −101	−195 +130	— +65
[illegible]	— −22 +52,608	+65 −43 −52,608	−51 +34 +41,598	— +17 −41,598

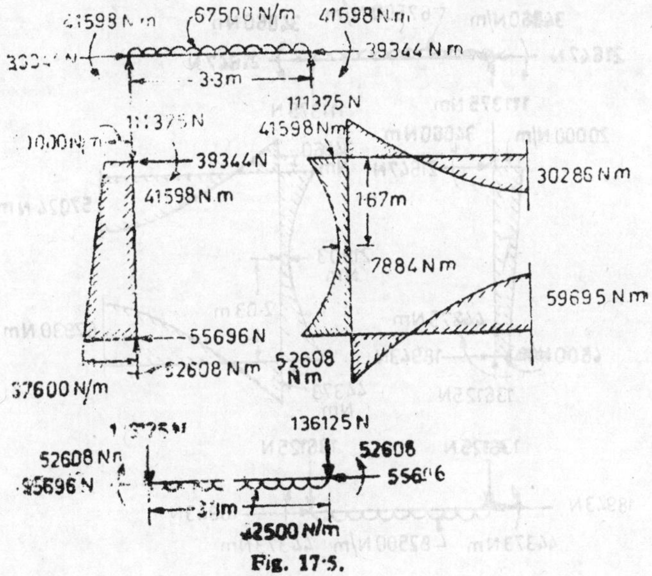

Fig. 17·5.

B.M. diagram and free body diagrams for various members are shown in Fig. 17·5.

Case II. ***Live load and dead load on top and water pressure from inside.***

Fixed end moments and distribution of moments are shown in Table 17·2. Water pressure will vary from zero at top to 10,000 × 3·3 at bottom. Net pressure will vary from 20,000 N/m at top 4600 N/m at bottom, acting from outside.

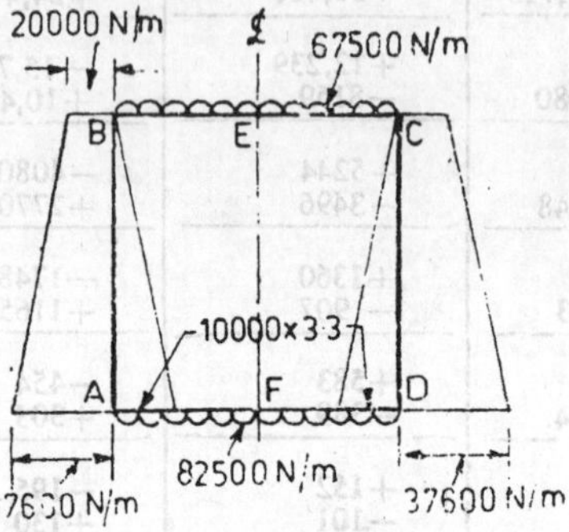

Fig. 17·6

B.M. diagram and free body diagrams for various members are shown in Fig. 17·7.

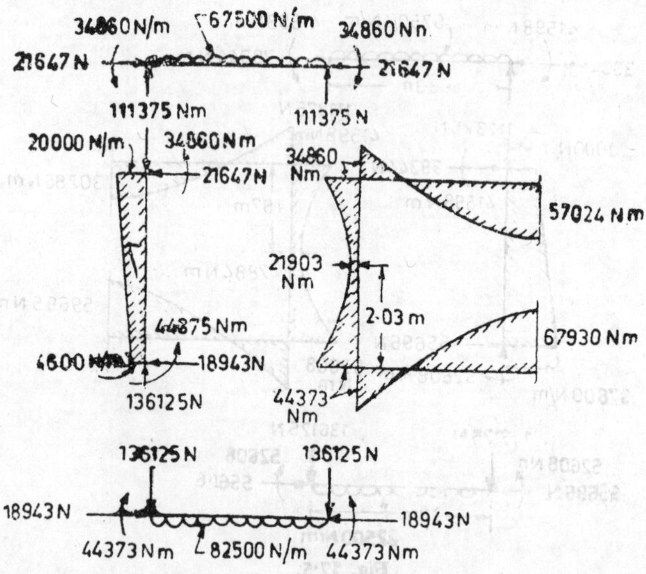

Fig. 17·7

Table 17.2

Joint	A		B	
Member	*AD*	*AB*	*BA*	*BC*
Distribution factor	1/3	2/3	2/3	1/3
F.E.M.	+74,869	−9770	+12,560	−61,256
Bal.	−21,700	−43,399	+32,464	+16,232
C.O.		+16,23 2	−21,700	
Bal.	−5411	−10,82 1	+14,467	+7233
C.O.		+7234	−5411	
Bal.	−2411	−4823	+3607	+1804
C.O.		+1804	−2412	
Bal.	−601	−1203	+1608	+804
C.O.		+804	−602	
Bal.	−268	−536	+401	+201
C.O.		+201	−268	
Bal.	−67	−134	+179	+89
C.O.		+90	−67	
Bal.	−30	−60	+44	+23
C.O.		+23	−30	
Bal.	−8	−15	+20	+10
Final Moment	+44,373	−44,37 3	+34, 860	−34,860

Case II. *Live load not acting on the side of culvert. The loading in shown in Fig. 17.8.*

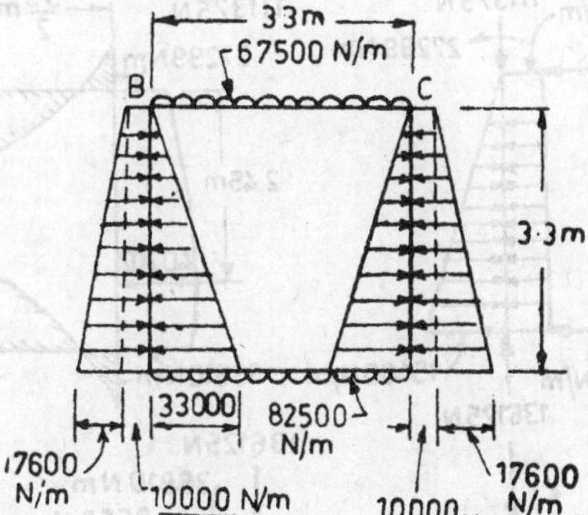

Fig. 17.8.

Fixed end moments and distribution of moments are shown in Table 17.3.

Table 17·3

Joint	A		B	
Member	AD	AB	BA	BC
Distribution factor	1/3	2/3	2/3	1/3
F.E.M. Bal.	+74,869 −26,743	+5360 −53,486	−2565 +42,547	−61,256 +21,274
C.O. Bal.	 −7091	+21,274 −14,183	−26,743 +17,829	 +8914
C.O. Bal.	 −2972	+8915 −5943	−7092 +4728	 +2364
C.O. Bal.	 −788	+2364 −1576	−2972 +1981	 +991
C.O. Bal.	 −330	+991 −661	−788 +525	 +263
C.O. Bal.	 −88	+263 −175	−331 +221	 +110
C.O. Bal.	 −37	+111 −74	−88 +59	 +29
C.O. Bal.	 −10	+30 −20	−37 +25	 +12
Final Moments	+36,810	−36,810	+27,299	−27,299

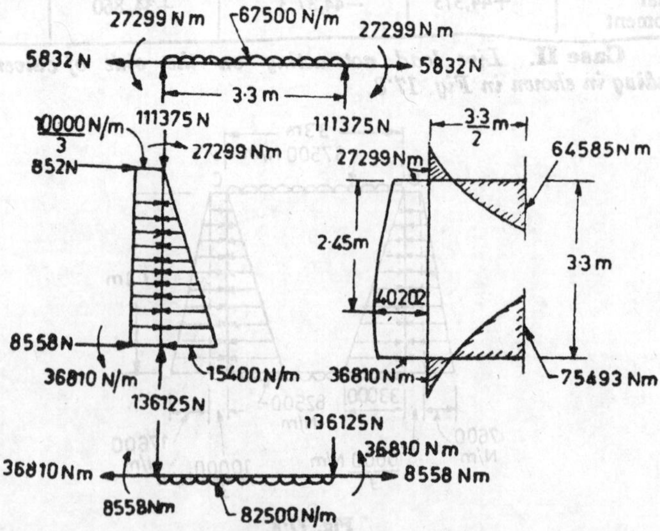

Fig. 17·9

Design of top slab

Centre of span. B.Ms. in differen t cases are

Case I. B.M. =50,286 N-m Direct compression=39,344

Case II. B.M. =57,024 N-m Direct tension=21,647 N

Case III. B.M. =64,585 N-m Direct compression=5832 N.

The section is designed for B.M. of 64,585 N-m. The effect of direct tension is neglected as it is very small.

Overall depth provided is 300 mm.

Effective depth =300−50=250 mm.

Tension occurs on water side

$c=7$ N/mm², $t=100$ N/mm², $Q=1{\cdot}411$, $n=0{\cdot}48$.

Moment of resistance of section

$=Qbd^2=1{\cdot}411\times1000\times(250)^2=88{,}180{,}000$ N-mm.

Section provided is adequate.

Area of steel required

$$=\frac{64{,}585\times1000}{0{\cdot}84\times250\times100}=3075 \text{ mm}^2.$$

Provide 20 mm ϕ bars at 100 mm centres.

A_t provided =3142 mm².

Design of section at support

Maximum B.M. =41,598 N-m.

Area of steel required

$$=\frac{41{,}598\times1000}{0{\cdot}86\times250\times125}=1548 \text{ mm}^2.$$

Half the bars from centre are bent up and additional bars 10 mm ϕ at 200 mm centres are provided.

There is direct compression in the slab in addition to B.M. but its effect is not considered as actually slab is reinforced both at top and bottom.

Distribution steel

Percentage of distribution steel

$$=0{\cdot}3-\frac{0{\cdot}1\times200}{350}=0{\cdot}243\%.$$

Area of steel $$=\frac{0{\cdot}243\times300\times1000}{100}=729 \text{ mm}^2.$$

Provide 8 mm ϕ bars at 130 mm centres both at top and bottom.

Design of bottom slab

Centre of span

Case I. B.M. =59,695 N-m Direct compression=55,695 N

Case II. B.M. =67,930 N-m Direct compression=18,943 N

Case III. B.M.=75,493 N-m Direct tension =8558 N

The section is designed for B.M. of 75,493 N-m. The effect of direct tension is neglected as it is very small.

Area of steel required

$$=\frac{75{,}493\times 1000}{0{\cdot}84\times 250\times 100}=3595\ \text{mm}^2.$$

Provide 20 mm ϕ bars at 80 mm centres.

A_t provided $=3928\ \text{mm}^2$.

Design of section at support

Maximum B.M. $=52{,}608$ N-m

Area of steel $=\dfrac{52{,}608\times 1000}{0{\cdot}86\times 250\times 125}=1958\ \text{mm}^2$.

Half the bars from centre are bent towards support and additional 10 mm ϕ bars at 160 mm centres are provided.

Distribution steel. Provide 8 mm ϕ bars at 130 mm centres at top and bottom.

Design of side walls

Maximum B.M. $=52{,}608$ N-m.

Direct compression $=136{,}125$ N.

Eccentricity $e=\dfrac{52{,}608\times 1000}{136{,}125}=386{\cdot}5$ mm.

$$\frac{e}{D}=\frac{386{\cdot}5}{300}=1{\cdot}29<1{\cdot}5.$$

The section will be equally reinforced on both sides, 20 mm ϕ bars at 150 mm centres are provided on each face.

Area of steel on each face $=2095\ \text{mm}^2$.

The depth of neutral axis is calculated from the cubic equation.

$$(e+100)=\frac{\dfrac{1000\,n}{2}\left(250-\dfrac{n}{3}\right)+12\times 2095\left(\dfrac{n-50}{n}\right)\times 200}{\dfrac{1000\times n}{2}+12\times 2095\times\dfrac{n-50}{n}-13\times 2095\times\dfrac{250-n}{n}}$$

$$(388+100)=\frac{125{,}000n+\dfrac{500\,n^2}{n}+5{,}028{,}000-\dfrac{250{,}400{,}000}{n}}{500\,n+25{,}140-\dfrac{1{,}257{,}000}{n}-\dfrac{6{,}803{,}750}{n}+27{,}235}$$

This is solved by trial and error.

$$n=105\ \text{mm app.}$$

Force $P=c\left[500\,n+25{,}140-\dfrac{1{,}257{,}000}{n}-\dfrac{6{,}808{,}750}{n}+27{,}235\right]$

$$\therefore\ 136{,}125=c\left[500\times 105+25{,}140-\frac{1{,}257{,}000}{105}-\frac{6{,}808{,}750}{105}+27{,}235\right]$$

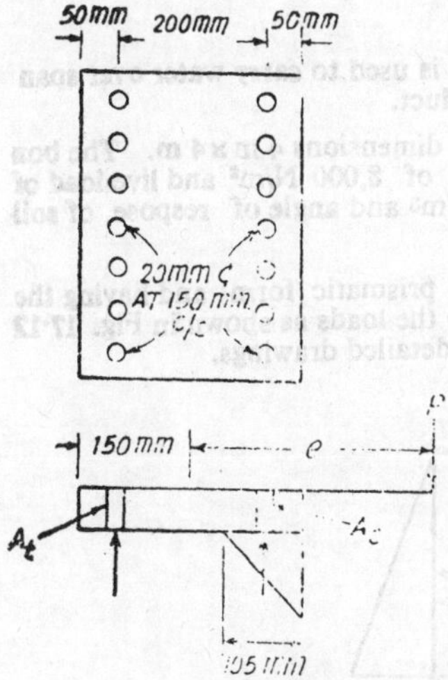

Fig. 17·10

$$c=\frac{136{,}125}{28{,}058}$$

$$=4·85 \text{ N/mm}^2$$

Stress in steel

$$=13\times 4·85\times\frac{205-150}{105}$$

$$=87·1 \text{ N/mm}^2$$

Section provided is safe.

Provide distribution steel 8 mm ϕ bars at 130 mm centres on both faces.

Details of reinforcement are shown in Fig. 17·11.

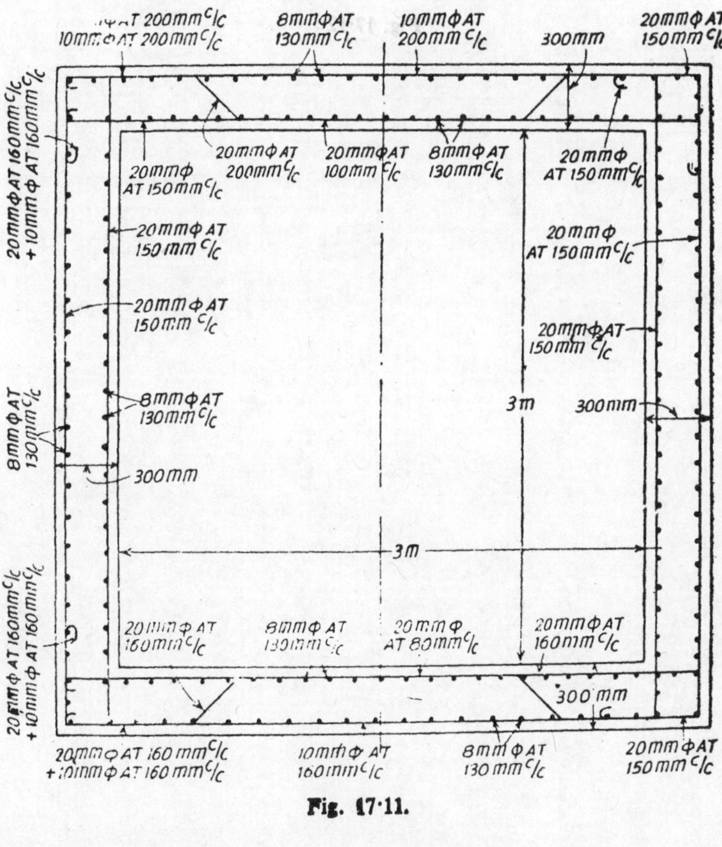

Fig. 17·11.

PROBLEMS

1. An aqueduct 3 m wide and 2 m deep is used to carry water over span of 7 m. Design a suitable section for the aqueduct.

2. Design a box culvert with inside dimensions 4 m × 4 m. The box culvert has to carry superimposed dead load of 8,000 N/m² and live load of 40,000 N/m². The density of earth is 18,000 N/m³ and angle of respose of soil is 30°.

3. A reinforced concrete box culvert of prismatic form and having the same section for all members is subjected to the loads as shown in Fig. 17·12 per metre width. Design the culvert and show detailed drawings.

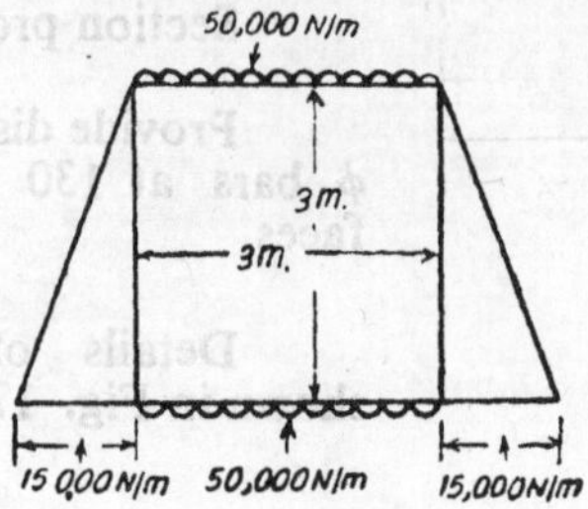

Fig. 17·12

18

Portals and Building Frames

18·1. Portals Frames. For large span roof construction concrete is not generally economical. For such roofs steel trusses prove to be economical. However, concrete construction has several advantages like fire resistance and negligible maintenance cost.

Different types of concrete roofs in general use are, simple girders, concrete trusses, rigid frames, shells and arches.

A portal frame consists of girder built monolithic with supporting columns having rigid joint at junction of girder and column. A portal frame may have horizontal girder or rigid roof. The portal frames with rigid or pitched roof are generally used for workshop sheds. The frames may be fixed or hinged at the base.

The analysis of portal frames is done by the moment distribution. The loads coming on girder and columns are computed. Bending moments in various members are found by moment distribution. From these shear forces and axial forces in various members are computed. The members are designed for B.M., S.F. and axial forces.

In case the base is fixed, moment is transmitted to the foundation. The foundation is designed for vertical load and B.M.

Ex. 18·1. *The roof of a 12 m wide hall is supported on portal frames at 4 m intervals. The height of the portal frame is 6 m. The live load on the roof is 1500 N/m². Design the roof and portal frames.*

Bearing capacity of soil = 100 kN/m².

Sol. Design of Roof Slab. Assume thickness of roof slab as 120 mm.

Dead load of slab $w_d = 1 \times 1 \times 0{\cdot}12 \times 25{,}000 = 3000\ \text{N/m}^2$

Live load $w_l = 1500\ \text{N/m}^2$.

The slab is designed as continuous slab between portal frames.

Maximum B.M. is taken as $\frac{w_d l^2}{10}+\frac{w_l l^2}{9}$ (*see* I.S. 456)

$$=\frac{3000\times4^2}{10}+\frac{1500\times4^2}{9}$$

$$=4800+2667=7467 \text{ N-m.}$$

Concrete mix used is M 15

Effective depth required

$$=\sqrt{\frac{7467\times1000}{0{\cdot}87\times1000}}=92{\cdot}6 \text{ mm}$$

Provide overall depth of 120 mm with effective depth of 100 mm.

Area of steel required

$$=\frac{7467\times1000}{0{\cdot}87\times100\times140}=613 \text{ mm}^2$$

Provide 10 mm ϕ bars at 120 mm centres.

Distribution steel $=\frac{0{\cdot}15\times120\times1000}{100}=180 \text{ mm}^2$

Provide 6 mm ϕ bars at 150 mm centres.

Design of Portal Frames

Load on portal frame from slab

$$=4\times(3000+1500)$$
$$=4\times4500$$
$$=18{,}000 \text{ N/m}$$

Depth of beam will vary from $\frac{1}{12}$th to $\frac{1}{20}$th of span depending on loading on the frame.

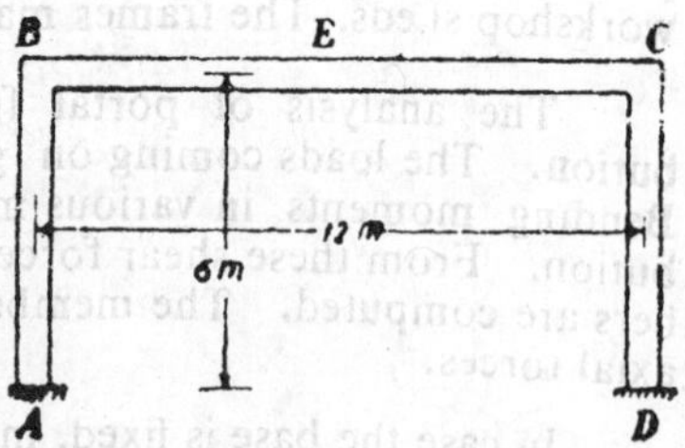

Fig. 18·1

Assume size of beam as 700 mm × 250 mm and dimensions of vertical members as 550 mm × 350 mm

Self-load of beam $=\frac{(700-120)}{1000}\times\frac{350}{1000}\times25{,}000$

$$=5075 \text{ N/m}$$

Total load $=18{,}000+5075=23{,}075$ N/m

Moment of inertia of beam

$$I_B=\frac{350\times(700)^3}{12}$$

Moment of inertia of column

$$I_C = \frac{350 \times (550)^3}{12}$$

$$\frac{I_B}{I_C} = \left(\frac{700}{550}\right)^3$$

$$I_B = 2{\cdot}061\ I_C.$$

As the frame is symmetrical and symmetrically loaded moment distribution is done for half of the frame by taking stiffness of the beam as half of the actual stiffness.

Distribution factor for beam

$$= \frac{\frac{I_B}{12} \times \frac{1}{2}}{\frac{I_B}{12} \times \frac{1}{2} + \frac{I_C}{6}} = \frac{\frac{2{\cdot}061}{24}}{\frac{2{\cdot}061}{24} + \frac{1}{6}}$$

$$= \frac{2{\cdot}061}{6{\cdot}061} = 0{\cdot}34$$

Distribution factor for column

$$= 0{\cdot}66$$

Fixed end moment $= \frac{wl^2}{12} = \frac{23{,}075 \times 12^2}{12} = 276{,}900$ Nm

Table 18·1

Joint		B	
Member	AB	BA	BC
Distribution factor	—	0·66	0·34
F.E.M. Bal.	—	— +182,754	−276,900 + 94,146
C.O.	+91,377	—	—
Final Moment	+91,377	+182,754	−182,754

The B.M. diagram for the frame is shown in Fig. 18·2.

Design of Beam

Maximum B.M. =232,596,000 N-mm

The beam is designed as a T-beam.

$$\text{Economical depth} = \sqrt{\frac{M}{2{\cdot}1\, b_r}}$$

$$= \sqrt{\frac{225{,}800{,}000}{2{\cdot}1 \times 350}} = 555 \text{ mm.}$$

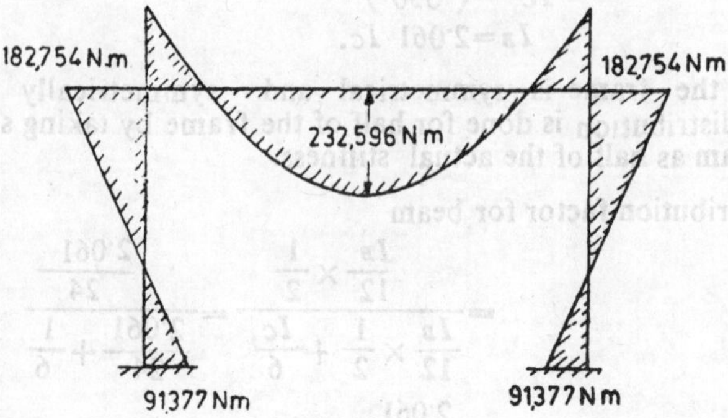

Fig. 18·2.

Overall depth of 700 mm with effective depth of 660 mm is provided. Greater depth for beam is provided as the beam has to act as rectangular beam at supports.

$$\text{Assume lever arm} = d - \frac{d_s}{2} = 660 - 60 = 600 \text{ mm.}$$

$$A_t = \frac{232{,}596{,}000}{600 \times 140} = 2769 \text{ mm}^2.$$

Provide 6 bars of 25 mm ϕ

$$A_t = 2945 \text{ mm}^2.$$

Check for stresses. Flange width

$$b_f = \frac{l_o}{6} + b_w + 6D_f$$

$$= \frac{12{,}000}{6} + 350 + 6 \times 120$$

$= 2000 + 350 + 720$

$= 3070$ mm.

$\therefore$ Flange width

$= 3070$ mm.

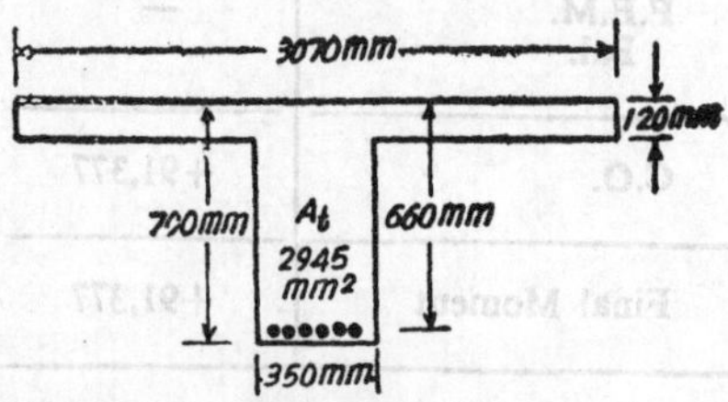

Fig. 18·3

Let n be depth of N.A.
Assume it to lie in web.

Taking moments about top,

$$n = \frac{3070 \times 120 \times 60 + 18 \times 2945 \times 660}{3070 \times 120 + 18 \times 2945}$$

$= 135{\cdot}5$ mm.

Lever arm $=d-\dfrac{d_s}{2}+\dfrac{d_s^2}{6(2n-d_s)}$

$=660-60+\dfrac{(120)^2}{6(271-120)}$

$=615{\cdot}9$ mm.

Stress in steel $=\dfrac{M}{A_t \times a}=\dfrac{232{,}596{,}000}{2945\times615{\cdot}9}$

$=128{\cdot}2$ N/mm².

Stress in concrete $=\dfrac{128{\cdot}2}{18}\times\dfrac{135{\cdot}5}{660-135{\cdot}5}$

$=1{\cdot}84$ N/mm². Safe.

Design of section at support. The beam is designed as rectangular beam.

Moment of resistance as singly reinforced balanced section

$=Qbd^2=0{\cdot}87\times350\times(660)^2$

$=132{,}600{,}000$ N-mm.

Actual B.M. $=182{,}754{,}000$ N-mm.

The beam is designed as doubly reinforced beam.

Moment of resistance to be provided by compression steel

$=182{,}754{,}000-132{,}600{,}000$

$=50{,}154{,}000$ N-mm.

Assume 3·75 cm cover to center of compression steel.

Steel on tension side

$=\dfrac{132{,}600{,}000}{0{\cdot}87\times660\times140}+\dfrac{50{,}154{,}000}{622{\cdot}5\times140}$

$=1650+575{\cdot}5=2225{\cdot}5$ mm².

Provide 6 bars of 25 mm ϕ.

Depth of N.A. $=0{\cdot}4\times660=264$ mm.

Area of steel on compression side

$$=\dfrac{50{,}154{,}000}{5\times17\times\dfrac{264-40}{264}\times622{\cdot}5}$$

$=1117$ mm².

Provide 3 bars of 25 mm ϕ.

Check for shear

S.F. at support $=23{,}675\times6=138{,}450$ N.

Shear stress $=\dfrac{138{,}450}{660\times350}=0{\cdot}6$ N/mm².

Take $\tau_c = 0{\cdot}34$ N/mm^2.

Shear reinforcement is required in end at 1·5 m.

Two bars of 25 mm ϕ are bent at 0·9 m from centre of vertical member and two at 1·8 m from centre of vertical member.

Shear taken by two bars of 25 mm ϕ

$$=2\times 491\times 140\times 0{\cdot}707=97{,}300 \text{ N.}$$

Net shear $=138{,}450-97{,}300=41{,}150$

Shear resistance $=0{\cdot}34\times 660\times 350$

$=78{,}540$ N.

Provide 8 mm ϕ 2 legged stirrups at 200 mm centres upto 1·5 m from support. In the remaining portion 6 mm ϕ 2 legged stirrups at 300 mm centres are provided.

Design of vertical members

Design of section at top

Axial force $=138{,}450$ N

B.M. $=182{,}754{,}000$ N-mm

Eccentricity $e=\dfrac{182{,}754{,}000}{138{,}450}=1320$ mm

12 bars of 25 mm ϕ are provided on each face.

$$A_t=A_c=5891 \text{ mm}^2$$

The section is checked for stresses.

Depth of neutral axis is obtained from the equation.

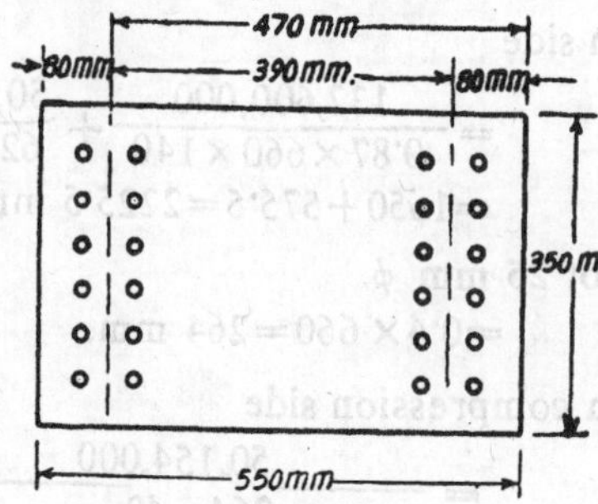

Fig. 18·4

$$\left(e+\frac{390}{2}\right)=\frac{350\times\frac{n}{2}\times\left(470-\frac{n}{3}\right)+17\times 5891\times\frac{n-80}{n}}{350\times\frac{n}{2}+17\times 5891\times\frac{n-80}{n}-18\times 5891\,\frac{470-n}{n}}$$

$$\therefore\ (1320+195)=\frac{82{,}250\,n-\frac{350}{6}\,n^2+3{,}905{,}900-\frac{33{,}247{,}000}{n}}{1750n+100{,}150-\frac{8{,}012{,}000}{n}-\frac{49{,}834{,}000}{n}+106{,}030}$$

Solving by trial and error, depth of neutral axis approximately is equal to 260 mm.

$$\text{Force } P = c\left[1750n + 100{,}150 - \frac{8{,}012{,}200}{n} - \frac{49{,}834{,}000}{n} + 106{,}030\right]$$

$$138{,}450 = c\left[175\times260 + 206{,}180 - \frac{5{,}784{,}600}{260}\right]$$

$$c = \frac{138{,}450}{45{,}500 + 206{,}180 - 222{,}500} = \frac{138{,}450}{29{,}180}$$

$$= 4{\cdot}745 \text{ N/mm}^2$$

$$\text{Stress in tensile steel } t = 18\times4{\cdot}745\times\frac{470-260}{260}$$

$$= 68{\cdot}99 \text{ N/mm}^2.$$

Design of lateral ties

Diameter of ties $= \frac{1}{4}\times$ Diameter of main reinforcement

$$= \frac{25}{4}$$

Use 8 mm ϕ ties.

Pitch of ties should be least of :

(*i*) Least lateral dimension = 350 mm

(*ii*) 48 times diameter of ties $= 48\times8 = 384$ mm

Provide 8 mm ϕ ties at 300 mm.

Design of section at base

Load at base $= 138{,}450 + 0{\cdot}55\times0{\cdot}35\times25{,}000\times6$

$= 167{,}325$ N

B.M. $= 91{,}377{,}000$ N-mm

Eccentricity $e = \dfrac{91{,}377{,}000}{167{,}325} = 546{\cdot}1$

6 bars of 25 mm ϕ are provided on each face and section is checked for stresses.

$$A_t = A_c = 2945 \text{ mm}^2$$

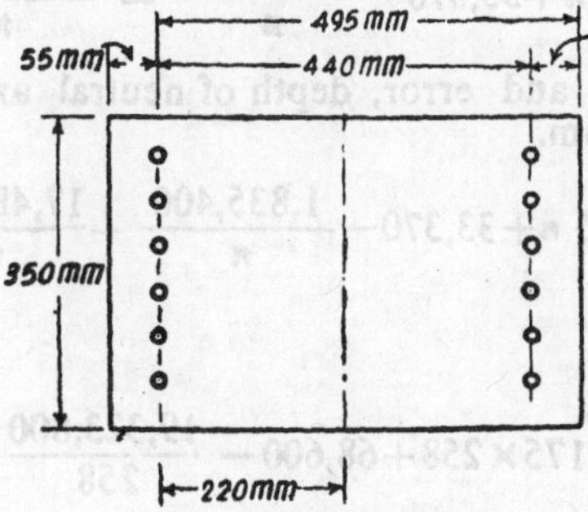

Fig. 18·5

Depth of neutral axis is found from the equation

$$(e+220)=\frac{350\times\frac{n}{2}\times\left(495-\frac{n}{3}\right)+17\times2945\times\frac{n-55}{n}\times440}{350\times\frac{n}{2}+17\times2945\times\frac{n-55}{n}-18\times2945\times\frac{495-n}{n}}$$

$$\therefore\quad 546{\cdot}1+220=\frac{86{,}625n-\frac{350}{6}n^2+22{,}030{,}800-\frac{1{,}211{,}694{,}000}{n}}{175n+50{,}070\ \frac{2{,}753{,}900}{n}-\frac{26{,}061{,}800}{n}+53{,}020}$$

Solving by trial and error depth of neutral axis approximately is equal to 280 mm.

$$\text{Force } P=c\left[175\,n+50{,}070-\frac{2{,}753{,}900}{n}-\frac{26{,}061{,}800}{n}+53{,}020\right]$$

$$\therefore\quad 167{,}325=c\left[175\times280+103{,}090-\frac{28{,}815{,}700}{280}\right]$$

$$\therefore\quad c=3{\cdot}40\ \text{N/mm}^2.$$

The stress in concrete is very low. Reinforcement on each face is reduced. Provide 4 bars of 25 mm ϕ on each face.

$$A_t=A_c=1963\ \text{mm}^2.$$

$$(e+220)=\frac{350\times\frac{n}{2}\left(495-\frac{n}{3}\right)+17\times1963\left(\frac{n-55}{n}\right)\times400}{350\times\frac{n}{2}+17\times1963\times\frac{n-55}{n}-18\times1963\times\frac{495-n}{n}}$$

$$\therefore\ (546{\cdot}1+220)=\frac{86{,}625\,n-\frac{350}{6}n^2+14{,}682{,}800-\frac{807{,}554{,}000}{n}}{175\,n+33{,}370-\frac{1{,}835{,}400}{n}-\frac{17{,}488{,}400}{n}+35{,}330}$$

Solving by trial and error, depth of neutral axis is approximatelv equal to 258 mm.

$$\text{Force } P=c\left[175\,n+33{,}370-\frac{1{,}835{,}400}{n}-\frac{17{,}488{,}400}{n}+35.330\right]$$

$$\therefore\quad 167{,}325=c\left[175\times258+68{,}600-\frac{19{,}323{,}800}{258}\right]$$

$$\therefore\quad c=4{\cdot}3\ \text{N/mm}^2.$$

Stress in steel $=18\times4{\cdot}3\times\dfrac{495-258}{258}$

$=71{\cdot}1$ N/mm².

Design of foundation

Load coming on foundation

$=167{,}325$ N.

Assume self-weight of foundation as 10%

$=16{,}733$ N.

Total load $=167{,}325+16{,}733=184{,}058$ N.

B.M. $=91{,}377{,}000$ N-mm.

Eccentricity $e=\dfrac{91{,}377{,}000}{184{,}058}=496{\cdot}5$

$=$say 500 mm.

The foundation is so provided that C.G. of loads coincides with C.G. of footing so that pressure distribution at base is uniform.

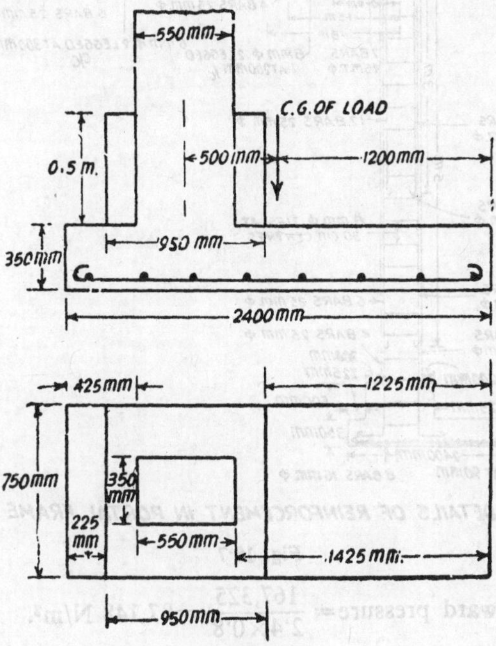

Fig. 18·6

Width of 800 mm is provided for footing.

$$\text{Length required} = \frac{184{,}058}{0{\cdot}8\times 100{,}000}$$

$$=2{\cdot}301 \text{ m.}$$

Provide length of 2·40 m.

A concrete pedestal 800 mm×950 mm×500 mm deep is provided as shown in Fig. 18·6.

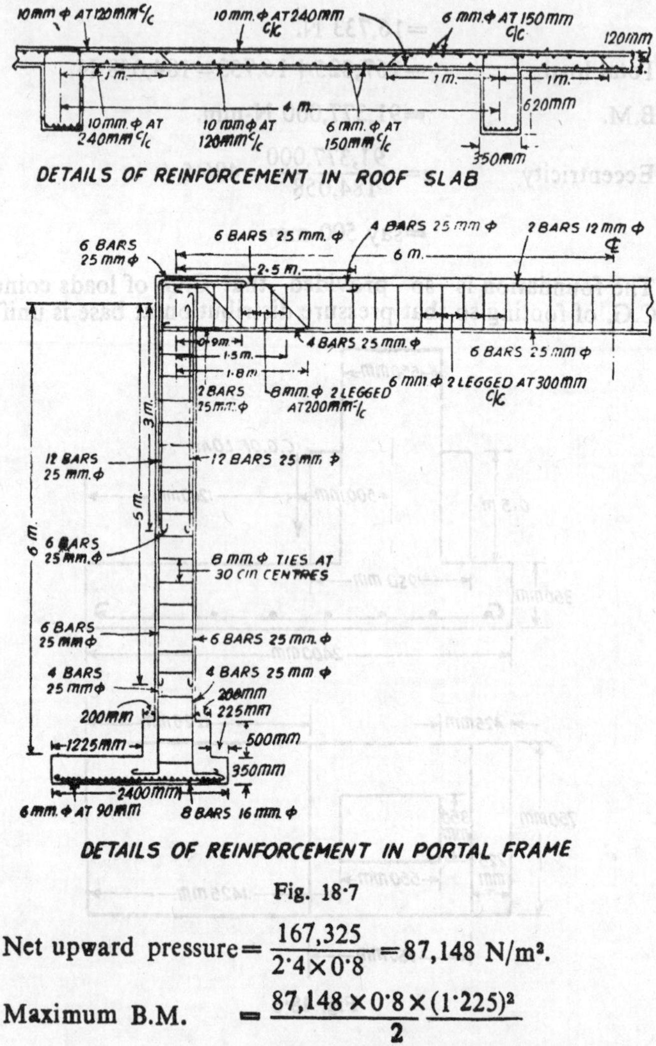

Fig. 18·7

$$\text{Net upward pressure} = \frac{167{,}325}{2{\cdot}4\times 0{\cdot}8} = 87{,}148 \text{ N/m}^2.$$

$$\text{Maximum B.M.} = \frac{87{,}148\times 0{\cdot}8\times (1{\cdot}225)^2}{2}$$

$$=52{,}311 \text{ N-m} = 52{,}311{,}000 \text{ N-mm.}$$

Effective depth required

$$=\sqrt{\frac{52{,}311{,}000}{0{\cdot}87\times800}}=274{\cdot}2\text{ mm.}$$

Depth required for punching

If d is depth required for punching

$$2\times800\times d\times1=167{,}325$$

$$\therefore \qquad d=\frac{167{,}325}{1600}=104{\cdot}6\text{ mm.}$$

Provide overall depth of 350 mm with effective depth of 280 mm.

$$\text{Area of steel}=\frac{52{,}311{,}000}{0{\cdot}87\times280\times140}$$

$$=1534\text{ mm}^2.$$

Provide 8 bars of 16 mm ϕ.

Distribution steel

Area of distribution steel

$$=\frac{0{\cdot}15\times350\times1000}{100}=525\text{ mm}^2.$$

Provide 8 mm ϕ bars at 90 mm centres.

Details of reinforcement in frame are shown in Fig. 18·7.

Ex. 18·2. *Design the hinge between portal frame and its foundation for the following:*

(*i*) *Load from column = 180,000 N.*

(*ii*) *Shear at the hinge = 60,000 N.*

(*iii*) *Distance of hinge from base = 1 m.*

(*iv*) *Size of the column is 500 mm × 300 mm.*

Also design a suitable foundation if the bearing capacity of soil is 150 kN/m².

Sol. In order the hinge is effective, the stress in hinge section should reach half the ultimate stress.

$\therefore$ Stress in hinge should reach

$$\frac{15}{2}=7{\cdot}5\text{ N/mm}^2.$$

Area of hinge required

$$=\frac{180{,}000}{7{\cdot}5}=24{,}000\text{ mm}^2.$$

Provide 300 mm × 80 mm hinge.

Shear on the hinge = 60,000 N.

Let A_t be area of bars required to take shear and θ be inclination of bars to vertical. Arrangement of bars is shown in Fig. 18·8.

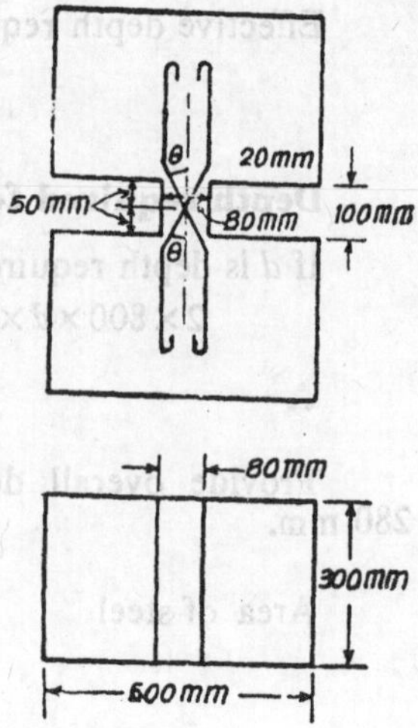

Fig. 18·8

$$A_t \times \sin\theta \times 140 = 60{,}000$$

$$\therefore \quad A_t = \frac{60{,}000}{140 \sin\theta}$$

$$\tan\theta = \frac{2}{5} = 0·4$$

$$\theta = 21° \; 48'$$

$$\sin\theta = 0·3714$$

$$\therefore \quad A_t = \frac{60{,}000}{140 \times 0·3714} = 1154 \text{ mm}^2.$$

Provide 4 bars 16 mm ϕ on each side.

Design of foundation

Assume self-weight of foundation

= 10% of load = 18,000 N.

Total direct load = 180,000 + 18,000

= 198,000 N.

B.M. at base = 60,000 × 1 = 60,000 N-m.

= 60,000,000 N-mm.

$$\text{Eccentricity} = \frac{60{,}000{,}000}{198{,}000} = 303 \text{ mm}.$$

Assume 0·7 m × 3 m foundation.

Concrete pedestal of 900 mm × 700 mm × 400 mm deep is provided as shown in Fig. 18·9.

Pressure distribution at base is given by

$$p = \frac{198{,}000}{3 \times 0·7} \pm \frac{60{,}000 \times 1}{0·7 \times \frac{3^2}{6}}$$

$$= 94{,}290 \pm 55{,}560$$

$$p_{max} = 149{,}850 \text{ N/m}^2.$$

and

$$p_{min} = 38{,}730 \text{ N/m}^2.$$

Net pressure at base is given by

$$p = \frac{180{,}000}{3 \times 0·7} \pm \frac{60{,}000}{0·7 \times \frac{3^2}{6}}$$

$$= 84{,}860 + 55{,}560$$

$$= 140{,}420$$

$\therefore$ $p_{max} = 140{,}420$ N/m².

$p_{min} = 29{,}300$ N/m².

Pressure at edge of pedestal

$$= 140{,}420 - \frac{140{,}420 - 29{,}300}{3} \times 1{\cdot}05$$

$$= 140{,}420 - 38{,}890 = 101{,}530 \text{ N/m}^2.$$

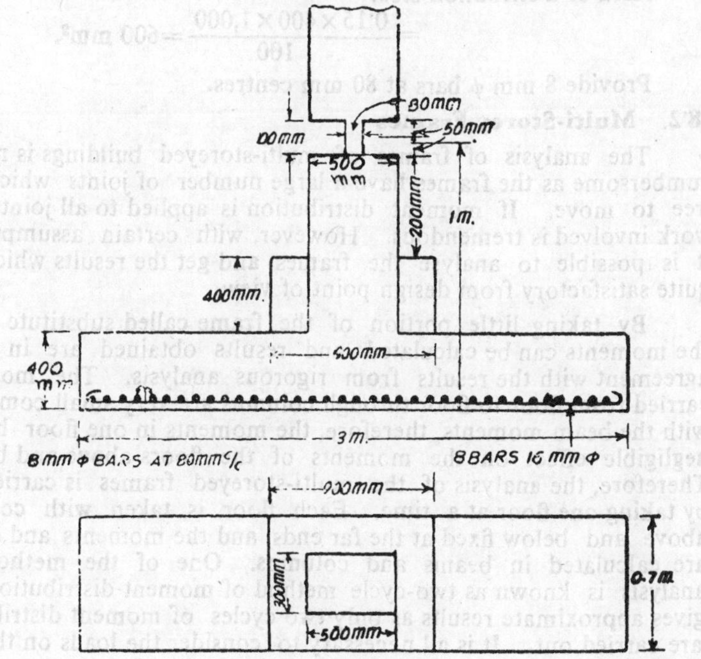

Fig. 18·9

Maximum B.M. $= 101{,}530 \times 0{\cdot}7 \times \frac{(1{\cdot}05)^2}{2} + (140{,}420$

$$- 101{,}530) \times 0{\cdot}7 \times 1{\cdot}05 \times \frac{2}{3} \times 1{\cdot}05$$

$$= 39{,}160 + 20{,}010 = 59{,}170 \text{ N-m.}$$

Effective depth required

$$= \sqrt{\frac{59{,}170 \times 1000}{0{\cdot}87 \times 700}} = 311{\cdot}8 \text{ mm.}$$

Provide overall depth of 400 mm with effective depth of 320 mm.

Area of steel $=\dfrac{59{,}170\times 1000}{0{\cdot}77\times 320\times 140}=1518\text{ mm}^2.$

Provide 8 bars of 16 mm ϕ.

Distribution steel

Area of distribution steel

$$=\frac{0{\cdot}15\times 400\times 1{,}000}{100}=600\text{ mm}^2.$$

Provide 8 mm ϕ bars at 80 mm centres.

18·2. Multi-Storey Frames

The analysis of frames of multi-storeyed buildings is rather cumbersome as the frames have a large number of joints which are free to move. If moment distribution is applied to all joints, the work involved is tremendous. However, with certain assumptions, it is possible to analyse the frames and get the results which are quite satisfactory from design point of view.

By taking little portion of the frame called substitute frame the moments can be calculated and results obtained are in good agreement with the results from rigorous analysis. The moments carried from floor to floor through columns are very small compared with the beam moments, therefore, the moments in one floor have a negligible effect on the moments of the floors above and below. Therefore, the analysis of the multi-storeyed frames is carried out by taking one floor at a time. Each floor is taken with columns above and below fixed at the far ends, and the moments and shears are calculated in beams and columns. One of the methods of analysis is known as two-cycle method of moment distribution. It gives approximate results as only two cycles of moment distribution are carried out. It is all necessary to consider the loads on the two nearest spans on each side of the joint in question.

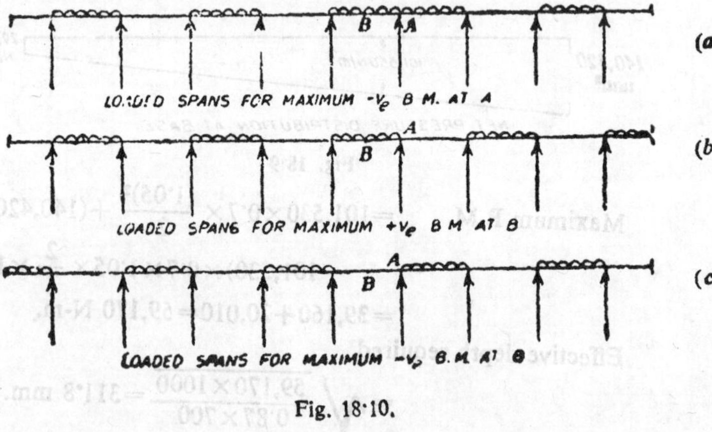

Fig. 18·10.

Analysis for Vertical Loads. The beams should be loaded as follows for maximum effect :

(*i*) For maximum negative moment at support, the load should be placed on the two spans adjacent to the support and on the alternative spans [*See* Fig. 18·10 (*a*)].

(*ii*) For maximum positive moment at the mid-point of a span, the loads should be placed on the span and on the alternate spans [*See* Fig. 18·10 (*b*)].

(*iii*) The maximum negative moment at the mid-point of a span, the load should be placed on the spans adjacent to the span under consideration and on the alternate spans [*See* Fig. 18·10 (*c*)].

The maximum moments in columns occur when alternate spans are loaded. There will be two sets of alternate loading as shown in

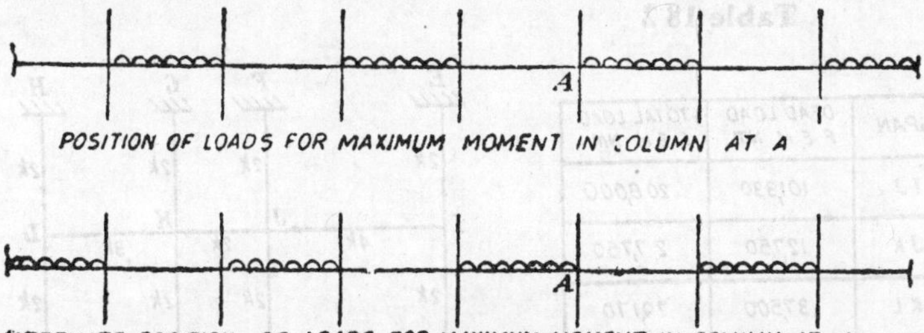

Fig. 18·11.

Fig. 18·11. Two sets of maximum moments are computed at the ends of all the columns framing into the beam. The column is designed to resist the stresses produced by every combination of the axial load and corresponding moment.

Ex. 18·3. *Analyse the multi-storeyed frame shown in Fig. 18·12. The frames are at 4 m intervals. Dead load is 3500 N/m² and live load is 5000 N/m².*

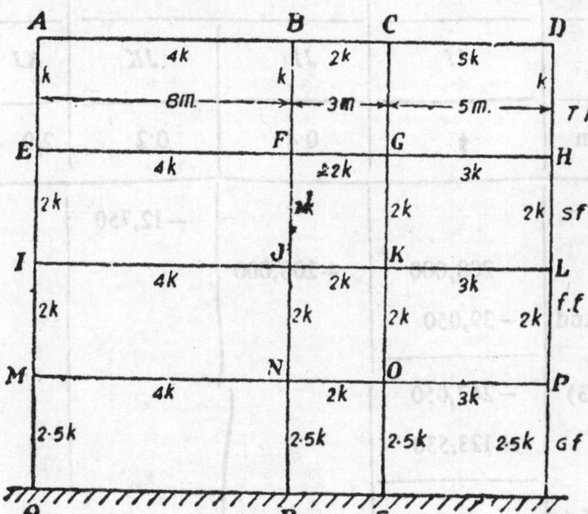

Fig. 18·12.

Sol. The second floor analysis of the frame is done here by isolating it as shown in Fig. 18·13. Similar analysis can be carried out for other floors. The stiffness of various members are assumed as shown in Fig. 18·12.

Self-weight of the beams *IJ*, *JK* and *KL* are assumed 5000 N/m, 390 N/m and 4000 N/m respectively.

Dead load from slab per metre run of girder

$$=3500\times4=14{,}000 \text{ N}$$

Live load per metre run of girder

$$=5000\times4=20{,}000 \text{ N}$$

Fixed end moments for dead load and total load for girders *IJ*, *JK* and *KL* are shown in Table 18·2.

Table 18·2

SPAN	DEAD LOAD F E M Nm	TOTAL LOAD F E M Nm
IJ	101,330	208,000
JK	12,750	27,750
KL	37,500	79,170

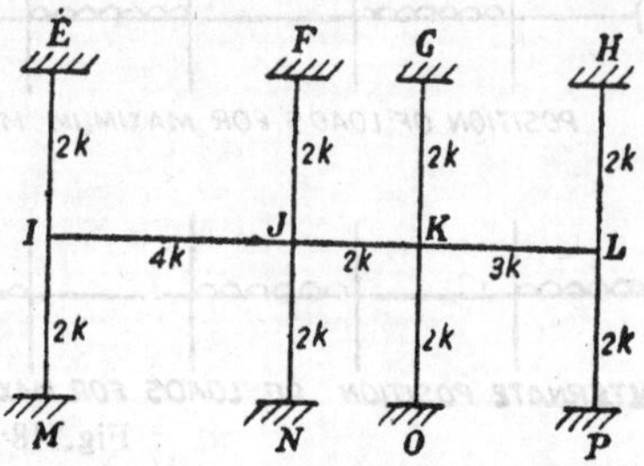

Fig. 18·13.

Moment distribution for maximum –ve B.M. at joints

Joint I

Table 18·3

Joint	I	J		K		L
Member	*IJ*	*JI*	*JK*	*KJ*	*KL*	*LK*
Distribution	½	0·4	0·2	2/9	2/3	3/7
1. D.L.F.E.M.			–12,750			
2. T.L.F.E.M.	–208,000	+208,000				
3. Distribution and carry over	–39,050					
4. Add (2) and (3)	–247,050					
5. Distribute	+123,530					
6. Total (sum of 4 and 5)	–123,520					

Joint J
Table 18·4

Joint	I	J		K		L
Member	IJ	JI	JK	KJ	KL	LK
Distribution factors	1/2	0·4	0·2	2/9	1/3	3/7
1. D.L.F.E.M.					−37,500	
2. T.L.F.E.M.	−208,000	+208,000	−27,750	+27,750		
3. Distribution carry over		+ 52,000	+ 1080			
4. Add (2) and (3)		+260,000	−26,670			
5. Distribute		−93,340	−46,670			
6. Total (sum of 4 and 5)		+166,660	−73,340			

Joint K
Table 18·5

Joint	I	J		K		L
Member	IJ	JI	JK	KJ	KL	LK
D.F.	1/2	0·4	0·1	2/9	1/3	3/7
1. D.L.F.E.M.	−101,330	+101,330				
2. T.L.F.E.M.			−27,750	− 27,750	−79,170	−79,170
3. Distribute and carry over				− 7360	−16,970	
4. Add (2) and (3)				+20,390	−96,140	
5. Distribute				+16,380	+25,250	
6. Total (sum of 4 and 5)				+36,770	−70,890	

Joint L
Table 18·6

Joint	I	J		K		L
Member	*IJ*	*JI*	*JK*	*KJ*	*KL*	*LK*
D.F	1/2	0·4	0·2	2/9	1/3	3/7
1. D.L.F.E.M.			−12,750	+12,750		
2. T.L.F.E.M.					−79,170	+79,170
3. Distribute and carry over						+11,070
4. Add (2) and (3)						−90,240
5. Distribute						−38,670
6. Total (sum of 4 and 5)						+51,570

The moment distribution is carried out as follows : The spans by side of joint are loaded with total load and next spans with dead loads. For −ve B.M. at joint *I*, span *IJ* is loaded with total load and *JK* with dead load. Effect of further spans is neglected. Effect of distribution at joint *J* is carried to point *I* in step 3. Second distribution is done at joint *I* in step 5.

Similarly for joint *J*, spans *IJ* and *JK* are loaded with total load and span *LK* with dead load. Effect of distribution at joints *I* and *K* are carried to joint *J* in step 2. Second distribution is done at joint *J* in step 5.

(ii) Moment distribution for Maximum +ve B.M. at mid-spans.

Mid-span of IJ
Table 18·7

Joint	I	J		K		L
Member	*IJ*	*JI*	*JK*	*KJ*	*KL*	*LK*
Distribution factors	1/2	0·4	0·2	2/9	1/3	3/7
1. D.L.F.L.M.			−12,750	+12,750		
2. T.L.F.E.M.	−208,000	+208,000			−79,170	
3. Distribute	+104,000	−78,100		+14,760		
4. Carry over	−39,050	+52,000	+7380			
5. Distribute	+19,530	−23,750				
Total	−123,520	+158,150				

+ve B.M. at mid-span

$$=\frac{wl^2}{8}-\frac{(123,520+158,150)}{2}=\frac{2}{3}\times 208,000-140,840$$

$$=312,000-140,840=171,160 \text{ N-m}$$

Middle of JK—Table 18·8

Joint	I	J		K		L
Member	IJ	JI	JK	KJ	KL	LK
Distribution factors	½	0·4	0·2	2/9	1/3	3/7
1. D.L.F.E.M.	−101,330	+101,330			−37,500	−37,500
2. T.L.F.E.M.			−27,750	+27,750		
3. Distribute	+5067		−14,720	−2170		−16,070
4. Carry over		+25,340	+1090	−7360	−8040	
5. Distribute			−5290	+3420		
Total moments			−46,670	+24,980		

+ve B.M. at mid-span

$$=\frac{wl^2}{8}-\left(\frac{49,670+25,980}{2}\right)=\frac{3}{2}\times 27,750-36,330$$

$$=41,630-36,330=53,000 \text{ N-m.}$$

Middle of KL—Table 18·9

Joint	I	J		K		L
Member	IJ	JI	JK	KJ	KL	LK
Distribution factors	½	0·4	0·2	2/9	1/3	3/7
1. D.L.F.E.M.			−12,750	+12,750		
2. T.L.F.E.M.		+208,000			−79,170	+79,170
3. Distribute			−39,050		+22,140	−33,930
4. Carry over				−19,530	−16,970	+11,070
5. Distribute					+12,170	−4740
Total					−61,830	+51,570

$$+\text{ve B M. at centre}=\frac{wl^2}{8}-\left(\frac{61,830+51,570}{2}\right)$$

$$=\frac{3}{2}\times 79,170-56,700$$

$$=118,760-56,700=62,060 \text{ N-m.}$$

The moment distribution is carried out as follows : For +ve B.M. at mid point of IJ, span IJ is loaded with total load, span JK with dead load and KL with total load. First distribution is done in step 3 at points, I, J and K. The moments are carried over in between IJ and from K and J in step 4. Second distribution is done in step 5. Moments are summed up to get total moments at supports. Once supports moments are known at I and J, maximum +ve moment at centre is worked out as shown. Analysis has been done for +ve B.M. at middle of spans IJ, JK and LK in Tables 18·7, 18·8 and 18·9 respectively.

Minimum moment (maximum −ve B.M.) at centre of span JK. Moment distribution is carried out in Table 18·10 for minimum moment at centre of span JK.

Table 18·10

Minimum Moment at Centre of JK

Joint	I	J		K		L
Member	IJ	JI	JK	KJ	KL	LK
Distribution factors	1/2	0·4	0·2	2/9	1/3	3/7
1. D.L.F.E M.			−12,750	+12,750		
2. T.L.F.E.M.	−208,000	+208,000			−79,170	+79,170
3 Distribute	+104,000	−78,100	−39,050	+14,760	+22,140	−33,930
4. Carry over		+52,000	+7380	−19,530	−16,970	
5. Distribute			−11,880	+ 8110		
Total			−56,300	+16,090		

−ve B.M. at centre of span

$$=\frac{w_d l^2}{8}-\left(\frac{56,300+16,090}{2}\right)$$

$$=\frac{2}{3}\times 12,750-36,200=19,130-36,200$$

$$=-17,070 \text{ N-m.}$$

The moment distribution, is carried out as follows : To find minimum moment at centre of span *JK*, the adjoining spans *IJ* and *KL* are loaded with total load and the span *JK* with dead load. First distribution is carried out in step 3. In step 4, the moments are carried to joints *J* and *K* from joints *I* and *L* as well as in between themselves. In step 5, second distribution is carried out. Total moments are found at joints. The minimum moment at centre is found as shown above. If the moment is negative reinforcement is to be provided at top at the centre of span.

Maximum moments in columns. Two possibilities of loads for maximum moments in columns were shown in Fig. 18·11. Moment distribution is done for these two cases to get maximum B M. in column.

Moments in Columns

I

Table 18 11

Joint		*J*		*K*		*L*
Distribution factors for columns						
Top	1/4	0·2		2/9		2/7
Bottom	1/4	0·2		2/9		2/7
Member (beams)	***IJ***	***JI***	***JK***	***KJ***	***KL***	***LK***
Distribution factors	¼	0·4	0·2	2/9	1/3	3/7
1. Total load F.E.M.	−208,000	+208,000	−12,750	+12,750	−79,170	+79,170
2. Distribution and carry over	−39,050	+52,000	+7380	−19,530	−16,970	+11,070
3. Add (1) and (2)	−247,050	−260,050	+5370	− 6780	−96,140	+90,240
4. Distribute Top column	+61,760	−58,070		+22,870		−25,870
Bottom column	+61,760	−58,070		+22,870		−25,870

II
Table 18·12

Joint	I	J		K		L
Distribution factors for columns Top	1/4	0·2		2/9		2/7
Bottom	1/4	0·2		2/9		2/7
Member (beams)	IJ	JI	JK	KJ	KL	LK
Distribution factors for beams	½	0·4	0·2	2/9	1/3	3/7
1. F.E.M.	−101,330	+101,330	−27,750	+27,750	−37,500	+37,500
2. Distribute and carry over	−14,720	+25,330	−1080	+7360	−8084	+1630
3. Add (1) and (2)	−116,050	+126,660	−28,830	+35,110	−45,540	+32,130
4. Distribute Top column	+29,010	−19,570		+2320		−11,180
Bottom column	+29,010	−19,570		+2320		−11,180

The moment distribution is carried as follows: Alternate spans are loaded with total load and remaining with dead load. Effect of distribution is carried to opposite joint in step 2. Total moments are worked out at each joint by addition in step 3. Second distribution is carried out in step 4. Maximum B M. in columns are found. Corresponding to above two position of load, axial force, in columns are found.

After analysing, the members are designed for moments, shears and axial force thus obtained. In case there is difference in size of the members from assumed values, the modified values of stiffness are assumed and the analysis is carried out again and members redesigned.

18·3. Analysis of building frames under wind loads and seismic forces

The effect of wind forces and horizontal forces due to seismic effect is to cause vertical forces in columns and moments in all the members. In analysing the frames for horizontal loads it is assumed that horizontal forces are transferred to joints. As the exact method of analysis for frames is cumbersome, approximate methods are used. Approximate methods in common use are (*i*) Portal method (*ii*) Cantilever method.

1. Portal Method

In this method it is assumed that points of contraflexure lie at centres of members-columns and beams. The shear taken by interior columns is double that taken by external columns. By these assumptions the frame becomes determinate and thus moments, shear forces and axial forces in all members can be worked out.

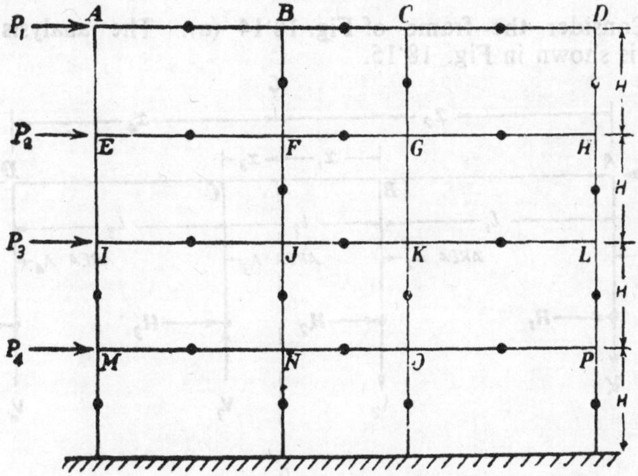

Fig. 18·14 (*a*)

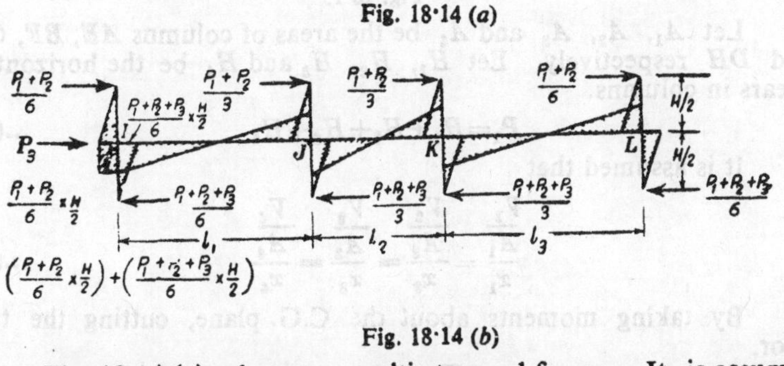

Fig. 18·14 (*b*)

Fig. 18·14 (*a*) shows a multi-storeyed frame. It is assumed that P_1, P_2, P_3 and P_4 are horizontal forces transmitted to the joints. Points of contraflexure are assumed at mid-points of different members.

In Fig. 18·14 (*b*) analysis is shown for second floor. Vertical shear at *I* in member *IJ* will be

$$\frac{\left[\left(\frac{P_1+P_2}{6} \times \frac{H}{2}\right) + \left(\frac{P_1+P_2+P_3}{6} \times \frac{H}{2}\right)\right] + \left[\left(\frac{P_1+P_2}{3} \times \frac{H}{2} + \frac{P_1+P_2+P_3}{3} \times \frac{H}{2}\right)\right]}{l_1}$$

Code permits 33⅓% increase in stresses when effect of wind or seismic load is taken into account.

II. Cantilever Method

In this method the points of contraflexure are assumed at the centres of all members, columns as well as beams. It is further assumed that direct stresses in columns are proportional to their distances from the C.G. of the frame.

Consider the frame of Fig. 18·14 (*a*). The analysis for top storey is shown in Fig. 18·15.

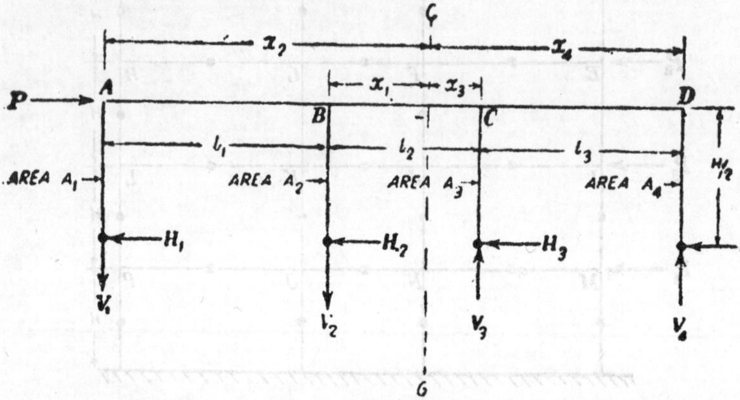

Fig. 18·15

Let A_1, A_2, A_3 and A_4 be the areas of columns *AE, BF, CG* and *DH* respectively. Let H_1, H_2, H_3 and H_4 be the horizontal shears in columns.

$$P_1=H_1+H_2+H_3+H_4 \quad \text{...(1)}$$

It is assumed that

$$\frac{\frac{V_1}{A_1}}{x_1}=\frac{\frac{V_2}{A_2}}{x_2}=\frac{\frac{V_3}{A_3}}{x_3}=\frac{\frac{V_4}{A_4}}{x_4} \quad \text{...(2)}$$

By taking moments about the C.G. plane, cutting the top floor.

$$(H_1+H_2+H_3+H_4)\times\frac{H}{2}=V_1x_1+V_2x_2+V_3x_3+V_4x_4$$

$$\therefore \quad V_1x_1+V_2x_2+V_3x_3+V_4x_4=\frac{P_1H}{2} \quad \text{...(3)}$$

Equations (2) and (3) can be solved for V_1, V_2, V_3 and V_4.

By taking moments about centre of *AB* (Fig. 18·15).

$$H_1\times\frac{H}{2}=V_1\times\frac{l_1}{2}$$

$$H_1=V_1\,\frac{l_1}{H}$$

By taking moments about centre of BC,

$$(H_1+H_2)\times \frac{H}{2}=V_1\left(l_1+\frac{l_2}{2}\right)+V_2+\frac{l_2}{2}$$

$$=V_1l_1+(V_1+V_2)\frac{l_2}{2}$$

$$\therefore \quad H_1+H_2=\frac{2}{H}\left[V_1l_1+\frac{(V_1+V_2)l_2}{2}\right]$$

Knowing H_1, H_2 can be evaluated.

H_3 and H_4 can be evaluated in a similar manner.

Knowing values of H_1, H_2, H_3 and H_4, bending moments in columns and beams can be worked out.

After finding out forces in top storey, the forces in columns in next below storey can be worked out.

The portal and cantilever methods of analysis are simple as the assumptions make the structure determinate and can be analysed by statics only. These methods do not take the elastic behaviour of the structure depending on the relative member rigidities. If the member is regular, these methods give very little error. But many bents are irregular and if relative member rigidities are to be taken in account other procedures like factor method are employed.

The portal method is satisfactory with moderate height-width ratio and structure is regular and member stiffness form a regular pattern.

The cantilever method is best applied to slender building frames having the general characteristics of a cantilever beam. If the height of the structure is of the same order of magnitude or less than its breadth, cantilever method is inferior to portal method.

III. The factor method

Consider the frame shown in Fig. 18·16. The members have been assigned relative K values.

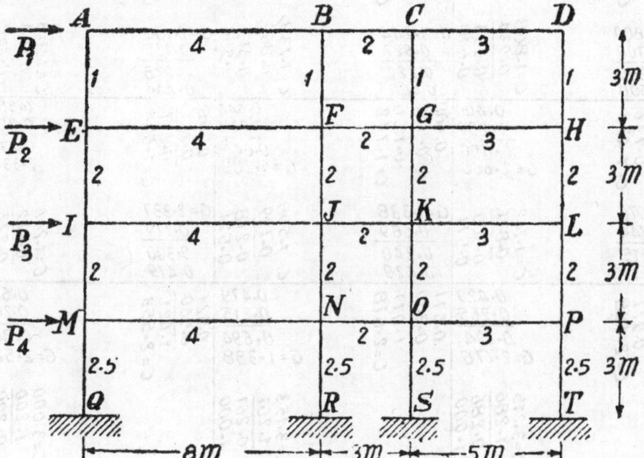

Fig. 18·16.

(i) The girder factor g for each joint of each girder is calculated. Girder factor is defined as ratio of sum of stiffnesses of columns Σk_c meeting at the joint to sum of stiffnesses of all members Σk meeting at the joint, $g = \frac{\Sigma k_c}{\Sigma k}$

Thus girder factor for girder AB at $A = \frac{1}{5} = 0{\cdot}2$.

Girder factor for QM at Q

$$= \frac{2{\cdot}5}{\alpha} = 0.$$

Stiffness of fixed end is α, therefore, $\Sigma k = \alpha$.

Girder factors for girder at J

$$= \frac{2+2}{4+2+2+2} = 0{\cdot}4.$$

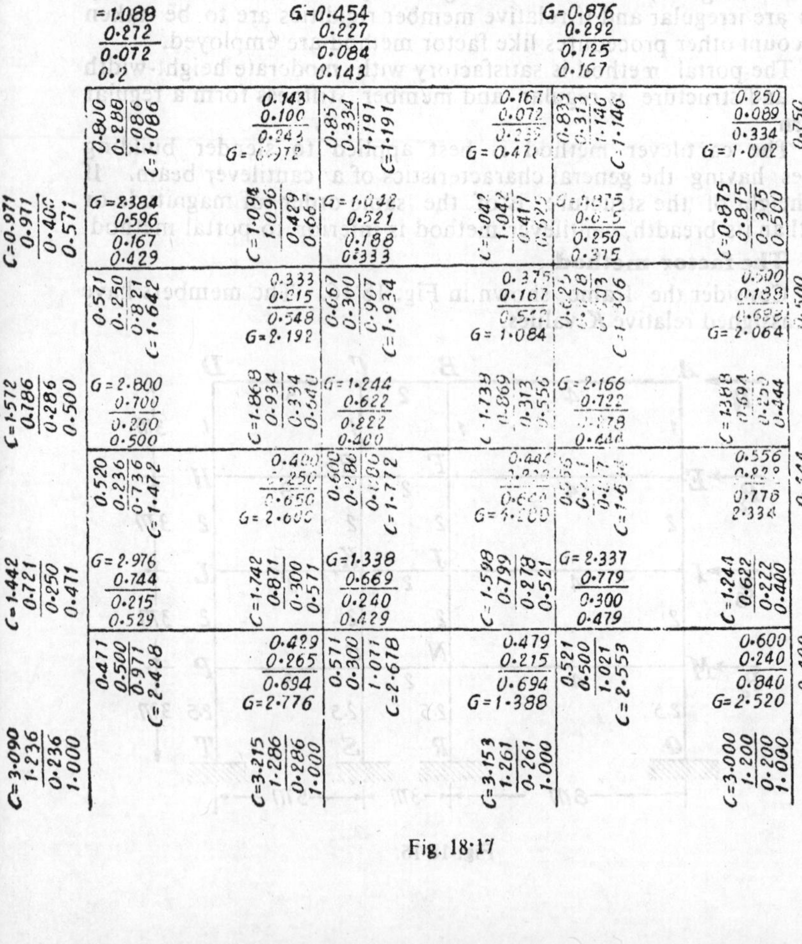

Fig. 18·17

(*ii*) Column factor c for each column for each joint is calculated $c=1-g$. The column factor at the fixed end of a column at fixed end will be 1.

These values of girder factors and column factors are shown against corresponding ends in Fig. 18·17.

(*iii*) To these factors at the ends add half value of factors at far ends.

(*iv*) Multiply the same obtained at step 3 by stiffness value of the member. Call these as girder moment factor G and column moment factor C as shown.

(*v*) The sum of the two C values at the ends of column divided by the column height is the relative column shear, and sum of all column shear is the storey shear. The column end moments will be in proportion to column moment factor.

Thus shear in column AE

$$=\frac{-P_1\times(0{\cdot}971+1{\cdot}086)}{(0{\cdot}971+1{\cdot}086)+(1{\cdot}096+1{\cdot}91)+(1{\cdot}042+1{\cdot}146)+(0{\cdot}875+1{\cdot}000)}$$

$$=\frac{-P_1\times2{\cdot}057}{2{\cdot}057+2{\cdot}287+2{\cdot}188+1{\cdot}875}$$

$$=\frac{-P_1\times2{\cdot}057}{8{\cdot}407}=-0{\cdot}2446\ P_1$$

$$M_{AE}=+0{\cdot}2446\ P_1\times3\times\frac{1{\cdot}086}{2{\cdot}057}$$

$$=--3875\ P_1$$

$$M_{EA}=-0{\cdot}2446\ P_1\times3\times\frac{0{\cdot}971}{2{\cdot}507}$$

$$=0{\cdot}3464\ P_1$$

(*vi*) The girder moments are obtained from statics and simple proportion. At joint A,

$M_{AB}=-M_{AE}$. At joint F,

$$M_{FE}+M_{FG}=-(M_{FB}+M_{FJ})\text{ and}$$

$$\frac{M_{FE}}{M_{FG}}=\frac{G_{FE}}{G_{FG}}$$

where G_{FE} and G_{FG} are corresponding girder moment factors.

$$M_{AB}=-[-0{\cdot}3875\ P_1]=0{\cdot}3875\ P_1$$

Shear in column $FB=-\dfrac{P_1(1{\cdot}096+1{\cdot}191)}{8{\cdot}407}$

$$=-\frac{P_1\times2{\cdot}287}{8{\cdot}407}$$

$$=-0{\cdot}2720\ P_1$$

$$M_{FB} = -0{\cdot}2720\ P_1 \times 3 \times \frac{1{\cdot}096}{2{\cdot}287}$$

$$= -0{\cdot}3910\ P_1$$

Shear in column FJ

$$= \frac{-(P_1+P_2)(1{\cdot}868+1{\cdot}934)}{(1{\cdot}572+1{\cdot}642)+(1{\cdot}868+1{\cdot}934)+(1{\cdot}738+1{\cdot}806)+(1{\cdot}388+1{\cdot}444)}$$

$$= \frac{-(P_1+P_2)\times 3{\cdot}802}{3{\cdot}214+3{\cdot}802+3{\cdot}544+2{\cdot}832}$$

$$= \frac{-(P_1+P_2)\times 3802}{13{\cdot}392}$$

$$= -0{\cdot}2839\ (P_1+P_2)$$

$$M_{FJ} = 0{\cdot}2839\ (P_1+P)\times 3 \times \frac{1{\cdot}934}{3{\cdot}802}$$

$$= -0{\cdot}4333\ (P_1+P_2)$$

Now $\quad M_{FE}+M_{FG} = -[-0{\cdot}3910\ P_1 - 0{\cdot}4333\ (P_1+P_2)]$

$$= 0{\cdot}8243\ P_1 + 0{\cdot}4333\ P_2.$$

$$\frac{M_{FE}}{M_{FG}} = \frac{2{\cdot}192}{1{\cdot}042}$$

$$\therefore \quad M_{FE} = \frac{2{\cdot}192}{3{\cdot}234}(M_{FE}+M_{FG})$$

$$= 0{\cdot}5588\ P_1 + 0{\cdot}2937\ P_2$$

$$M_{FG} = \frac{1{\cdot}042}{3{\cdot}234}(M_{FE}+M_{FG})$$

$$= 0{\cdot}2655\ P_1 + 0{\cdot}1396\ P_2.$$

18·4. Kani's Method for Analysis of Frames

In this method suitable stiffness for various members are assumed. Moments are worked out for the whole frame due to dead load, live load and wind loads. The sections are designed for the bending moment, shear force and axial force. If there is marked difference between assumed values of stiffnesses and actual values of stiffnesses, modified values of stiffness are assumed and the analysis is done again.

Consider member AB of frame.

Let θ_{AB} be slope at A and θ_{BA} be slope at B.

Moment at A will consist of fixed end moment M_{AB}, considering AB as fixed beam, and moment caused by rotation θ_A at A and rotation θ_B at B.

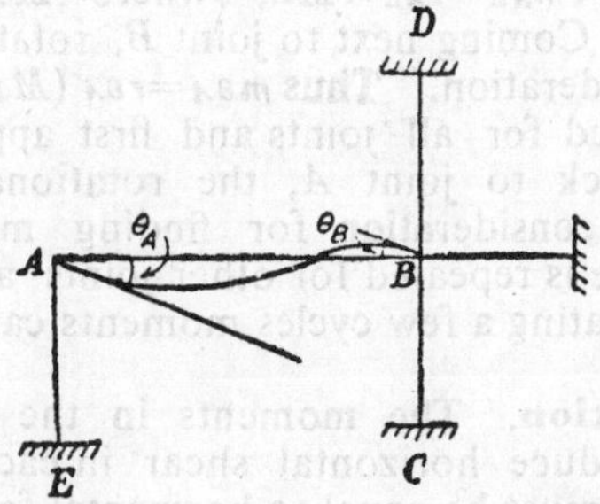

Fig. 18·18.

Moment caused by rotation θ_A will be

$$=2\ Ek\ (2\theta_A)=2\times 2\ Ek\ \theta_A=2m_{AB}$$

where k is stiffness of member AB.

$$m_{AB}=2\ Ek\ \theta_A$$

Moment at A caused by rotation θ_B at B

$$=2\ Ek\ \theta_B$$
$$=m_{BA}$$

Total moment for AB,

$$M_{AB}=\overline{M_{AB}}+2m_{AB}+m_{BA}$$

Sum of moments at a joint should be zero

$$\therefore \quad \Sigma(\overline{M_{AB}}+2m_{AB}+m_{BA})=0$$

$$\therefore \quad \Sigma m_{AB}=-\tfrac{1}{2}\times[\Sigma\overline{M_{AB}}+\Sigma m_{BA}]$$

At joint A, moments in various members will be proportional to stiffnesses.

$$\therefore \quad m_{AB}=\frac{k_{AB}}{\Sigma k}\times\Sigma M_{AB}=\frac{-k_{AB}}{2\Sigma k}(\Sigma\overline{M_{AB}}+\Sigma m_{BA})$$

where k_{AB} is stiffness of AB and Σk is sum of stiffnesses of all members meeting at A.

$-\dfrac{k_{AB}}{2\Sigma k}$ is called the rotation factor for AB at A written as r_{AB}. The rotation factors for such members at each joint are worked out. The unbalanced moment at a joint which is sum of the fixed end moment is computed and written inside the circle at each joint. Thus unbalanced moment at A, $M_A = \Sigma\overline{M_{AB}}$. Taking first joint A in Fig. 18·16 and assuming no rotation at any other joint $m_{AB} = r_{AB} \times M_A$ and $m_{AE} = r_{AE} \times M_A$, where M_A is unbalanced moment at joint A. Coming next to joint B, rotation effect of end A is taken into consideration. Thus $m_{BA} = r_{BA}(M_B + m_{AB})$. Similar procedure is adopted for all joints and first approximation completed. Coming back to joint A, the rotational effect at B and E will be taken into consideration for finding modified value of m_{AB}. The procedure is repeated for other joints and second cycle completed. By repeating a few cycles moments can be found.

Sway Correction. The moments in the top and bottom columns would produce horizontal shear in each column. Total shear in each storey must be equal to horizontal forces acting on the frame above that storey.

Analysis for vertical loads. In case there are only vertical loads acting, the total shear in each storey should be zero. If the sum of moments at top and bottom of all the columns in the same storey is not zero, for same height of columns the moment is balanced in proportion of the stiffnesses of columns, let δ be the displacement of the frame.

In a column AB, moment m_{AB}' caused by this displacement will be $\dfrac{6EI\delta}{h^2} = \dfrac{C \times k_{AB}}{h}$, where $C = 6E\delta$ and k_{AB} = stiffness $= \dfrac{1}{h}$.

If h is same for columns,

$$m_{AB}' = C_1 k_{AB}, \text{ where } C_1 = \frac{C}{h}$$

$$= m_{BA}'$$

If m_{AB}' is moment due to displacement at A, total moment at A for columns AB will be

$$M_{AB} = \overline{M_{AB}} + 2m_{AB} + m_{BA} + m_{AB}'$$

and $M_{BA} = \overline{M_{BA}} + m_{AB} + 2m_{BA} + m_{AB}'$

For vertical loads

$\overline{M_A} = \overline{M_{BA}} = 0$ for columns.

Total moment for all columns should be zero

$$\therefore \Sigma(M_{AB} + M_{BA}) = 0$$

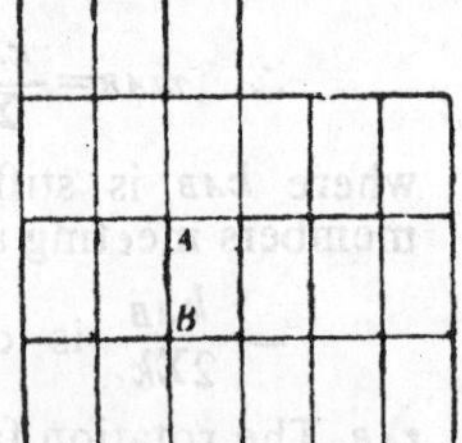

Fig. 18·19

$$\therefore \quad \Sigma(2m_{AB}+m_{BA}+m_{BA}'+2m_{BA}+m_{AB}+m_{BA}')=0$$

$$\therefore \quad 3\Sigma(m_{AB}+m_{BA})=-2m_{AB}'$$

$$\Sigma m_{AB}'=-\frac{3}{4}(\Sigma m_{AB}+m_{BA})$$

$$m_{AB}'=\frac{k_{AB}}{2\Sigma k}\times\Sigma(m_{AB}'),$$ where k_{AB} is stiffness of column AB and Σk is sum of stiffness of all columns.

$$m_{AB}'=-\frac{3}{2}\;\frac{k_{AB}}{\Sigma k}\Sigma m_{AB}+m_{BA}$$

$-\frac{3}{2}\frac{k_{AB}}{\Sigma k}$ is known as displacement factor and m_{AB}' is known as displacement contribution.

After completing first cycle of distribution, sway correction is applied and then second cycle is started. In the second cycle effect of displacement contribution is taken into consideration.

It must be remembered that in above derivation it is assumed that height is same for all columns. Above formulae will be modified if the columns have unequal heights.

Analysis for horizontal loads. Let P be the horizontal force above the storey for which horizontal shear is calculated.

$$\frac{\Sigma(M_{AB}+M_{BA})}{h}+P=0.$$ For equal heights coulmn.

$$\Sigma(2m_{AB}+m_{BA}+m_{AB}'+2m_{BA}+m_{BA}+m_{BA}')Ph=0$$

$$\Sigma(m_{AB}'+m'_{BA})=-Ph-3\,\Sigma(m_{AB}+m_{BA})$$

$$m_{BA}'=m_{AB}'$$

$$\Sigma m_{AB}'=-\frac{3}{2}\left[\frac{Ph}{3}+\Sigma(m_{AB}+m_{BA}\right]$$

$$m_{AB}'=\frac{k_{AB}}{\Sigma k}\Sigma m_{AB}'$$

$$m_{AB}'=-\frac{3}{2}\frac{k_{AB}}{\Sigma k}\left[\frac{Ph}{3}+\Sigma(m_{AB}+m_{BA})\right]$$

$\frac{Ph}{3}$ is called storey moment denoted by M_r.

In this case analysis is exactly the same as in the first case but only effect of storey moment $Ph/3$ is taken into consideration when applying sway correction.

Columns with Unequal Heights

Consider frame having columns of unequal heights as shown in Fig. 18·18. Let P be the wind force up to the bottom of storey.

If H_{AB} is the horizontal force at base of column AB,

$$H_{AB}=\frac{M_{AB}+M_{BA}}{h_{AB}}$$

Similarly

$$H_{CD}=\frac{M_{CD}+M_{DC}}{h_{CD}}$$

and so on.

Total shear in a storey must be equal to horizontal force.

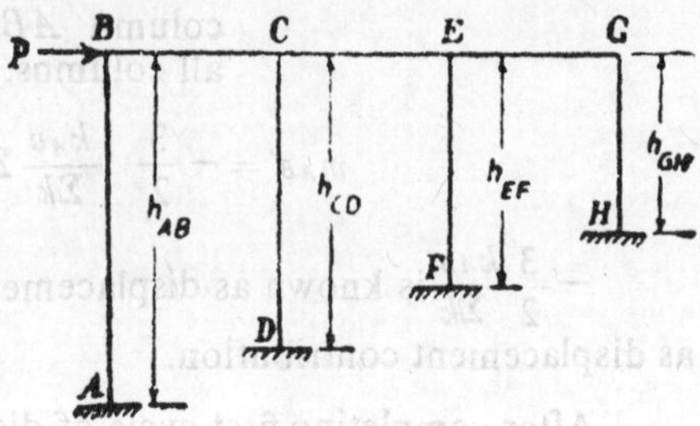

Fig. 18·20

$$\therefore \quad \Sigma H+P=0$$

where ΣH is total horizontal shear in all columns.

$$\therefore \quad \Sigma\frac{M_{AB}+M_{BA}}{h_{BA}}+P=0.$$

Now if any reference height h_R is chosen (generally maximum height of column is taken as reference height h_R)

$$H_{AB}=\frac{M_{AB}+M_{BA}}{h_R}\times\frac{h_R}{h_{AB}}=\frac{M_{AB}+M_{BA}}{h_R}\times C_{AB}$$

$$H_{CD}=\frac{M_{CD}+M_{DC}}{h_R}\times\frac{h_R}{h_{CD}}=\frac{M_{CD}+M_{DC}}{h_R}\times C_{CD}$$

and so on

where $C_{AB}=\frac{h_R}{h_{AB}}$, $C_{CD}=\frac{h_R}{h_{CD}}$ and so on

$$\Sigma H+P=0$$

$$\therefore \quad \Sigma\frac{M_{AB}+M_{BA}}{h_R}\times C_{AB}+P=0$$

$$\because \quad \frac{1}{h_R}\Sigma(m_{AB}+m_{BA}+m_{AB}'+2m_{BA}+m_{AB}+m_{BA}')C_{AB}+P=0$$

$$\Sigma(3m_{AB}+3m_{BA}+2m_{AB}')C_{AB}+Ph_R=0$$

$$\Sigma m_{AB}\ C_{AB}=-\frac{3}{2}\left[\frac{Ph_R}{3}+\Sigma(m_{AB}+m_{BA})C_{AB}\right] \quad ...(1)$$

Now due to sway, the moment produced at the end of member.

$$m_{AB}'=\frac{6E\,I_{AB}\delta}{h_{AB}^2}=\frac{6E\,\delta k_{AB}}{h_{AB}}=C\,\frac{k_{AB}}{h_{AB}}$$

where $C=6E\,\delta$

$$\frac{m_{AB}'}{h_{AB}}=\frac{6E\,\delta_{AB}}{h_{AB}^2}=C\,\frac{k_{AB}}{h_{AB}^2}$$

$$m_{AB}'\times\frac{h_R}{h_{AB}}=C\,\frac{k_{AB}\times h_R^2}{h_{AB}^2}\times\frac{1}{h_R}$$

$$m_{AB}'\,C_{AB}=\frac{C}{h_R}\,k_{AB}\,C_{AB}^2$$

$$\Sigma m_{AB}'\,C_{AB}=\frac{C}{h_R}\,\Sigma k_{AB}\,C_{AB}^2$$

$$C=\frac{\Sigma m_{AB}'\,C_{AB}}{\frac{1}{h_R}\,\Sigma k_{AB}C_{AB}^2}$$

$$m_{AB}'=C\frac{k_{AB}}{h_{AB}}$$

$$=\frac{\Sigma m_{AB}'\,C_{AB}}{\frac{1}{h_R}\,\Sigma k_{AB}\,C_{AB}^2}\times\frac{k_{AB}}{h_{AB}}$$

$$=k_{AB}\,\frac{h_R\,\Sigma m_{AB}'\,C_{AB}}{h_{AB}\,\Sigma k_{AB}\,C_{AB}^2}$$

$$=-\frac{k_{AB}\,C_{AB}}{k_{AB}\,C_{AB}}\,\Sigma_{AB}'\,C_{AB}$$

Substituting value of $\Sigma m_{AB}'\,C_{AB}$ from (1)

$$m_{AB}'=-\frac{3}{2}\,\frac{k_{AB}\,C_{AB}}{\Sigma k_{AB}\,C_{AB}^2}\left[\frac{P h_R}{3}+\Sigma(m_{AB}+m_{BA})\,C_{AB}\right]$$

Thus $-\frac{3}{2}\,\frac{k_{AB}\,C_{AB}}{\Sigma\,k_{AB}\,C_{AB}^2}=\delta_{AB}$ is displacement factor.

and $\frac{Ph_R}{3}$ is storey moment.

The analysis will be similar to that for frames with equal heights of columns, only in this case new displacement factors are to be used.

In case there are no horizontal forces acting

$$m_{AB}'=-\frac{3}{2}\,\frac{k_{AB}\,C_{AB}}{\Sigma k_{AB}\,C_{AB}^2}\,\Sigma(m_{AB}+m_{BA})\,C_{AB}$$

In case all the columns are of same height $C_{AB}=1$.

$$\therefore\quad m_{AB}'=-\frac{3}{2}\,\frac{k_{AB}}{\Sigma k_{AB}}\,\Sigma(m_{AB}+m_{BA})$$

This is same as derived earlier for columns with equal heights

Columns with unequal heights and some ends being hinged.

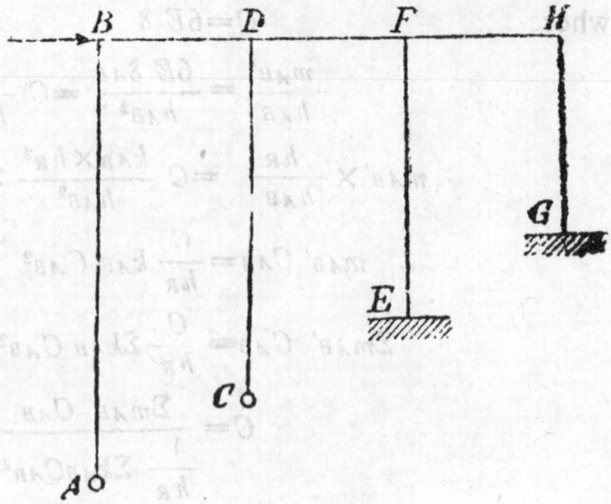

Fig. 18·19.

Consider frame having columns of unequal heights and some having hinged ends like AB and CD and some having fixed ends like EF and GH.

Let H_A, H_C, H_E and H_G be the horizontal shears at A, C, E and G.

$$H_A+H_C+H_E+H_G+P=0$$

$$\frac{M_{BA}}{H_{AB}}+\frac{M_{CD}}{H_{CD}}+\frac{M_{EF}+M_{FE}}{H_{EF}}+\frac{M_{GH}+M_{HG}}{H_{GH}}+P=0$$

Putting $\frac{h_R}{H_{AB}}=C_{AB}$, $\frac{h_R}{H_{CD}}=C_{CD}$ and so on.

$$\frac{M_{BA}\,C_{AB}}{h_R}+\frac{M_{CD}\,C_{DC}}{h_R}+\frac{(M_{EF}+M_{EE})\,C_{EF}}{h_R}+\frac{(M_{GH}+M_{GH})C_{GH}}{h_R}+P=0$$

$$\therefore \Sigma M_{BA}\,C_{AB}+\Sigma\,(M_{EF}+M_{FE})\,C_{EF}+Ph_R=0 \qquad \dots(i)$$

$$M_{BA}=2m_{BA}+m'_{BA}$$

$$M_{EF}=2m_{EF}+m_{FE}+m'_{EF}$$

$$M_{FE}=2m_{FE}+m_{EF}+m'_{FE}$$

Hence m_{BA} is the rotation moment at B. The rotation factors are obtained at B by taking stiffness of AB as $\frac{3}{4}$th otherwise, m'_{BA} is the sway moment.

$$\therefore \quad \Sigma M_{BA}\,C_{AB}=2\Sigma m_{BA}\,C_{AB}+\Sigma m'_{BA}\,C_{AB}$$

$$\Sigma(M_{EF}+M_{FE})\,C_{EF}=3\Sigma(m_{EF}+m_{FE})C_{EF}+2\Sigma m'_{EF}\,C_{EF}$$

$$\therefore \quad m'_{EF}=m'_{FE}.$$

Putting these values in equation (i),

$$2\Sigma m_{BA}\,C_{AB}+\Sigma m'_{BA}\,C_{AB}+3\Sigma(m_{EF}+m_{FE})\,C_{EF}+2\Sigma m'_{EF}\,C_{EF}+Ph_R=0$$

$$\therefore \quad \Sigma m'_{BA}\,C_{AB}+2\Sigma m'_{EF}\,C_{EF}=-[Ph_R+2\Sigma m_{BA}\,C_{AB}+3\Sigma(m_{EF}+m_{FE})\,C_{EF}] \qquad \dots(ii)$$

$$m'_{BA}=\frac{3E\Delta\ K_{AB}}{H_{AB}}=\frac{6E\Delta}{h_R}\,\frac{K_{AB}\,C_{AB}}{2}$$

$$=\frac{CK_{AB}\,C_{AB}}{2}$$

$$m'_{EF}=\frac{6E\Delta K_{EF}}{H_{EF}}=\frac{6E\Delta K_{EF}\,C_{EF}}{h_R}$$

$$=CK_{EF}\,C_{EF}$$

$\therefore$ $\Sigma m'_{AB}\,C_{AB}+2\Sigma m'_{EF}\,C_{EF}$

$$=\frac{C}{2}\,\Sigma k_{AB}\,C_{AB}^2+2C\,\Sigma k_{EF}\,C_{FE}^2$$

Putting $\quad C'_{AB}=\dfrac{C_{AB}}{2}$

$\Sigma m'_{BA}\,C_{AB}+2\Sigma m'_{EF}\,C_{EF}$

$$=2C[\Sigma k_{AB}C'_{AB}{}^2+k_{EF}\,C_{EF}{}^2] \qquad ...(iii)$$

From (*ii*) and (*iii*),

$$C=\frac{-[Ph_R+2\Sigma m_{BA}\,C_{AB}+3\Sigma(m_{EF}+m_{EF})C_{EF}]}{2[\Sigma k_{AB}\,C'_{AB}{}^2+k_{EF}\,C_{EF}{}^2]}$$

$$\therefore\quad m'_{AB}=\frac{-k_{AB}\,C'_{AB}[Ph_R+4\Sigma m_{BA}\,C'_{AB}+3\Sigma'(m_{EF}+m_{EF})C_{EF}]}{2[\Sigma k_{AB}\,C'_{AB}{}^2+k_{EF}\,C_{EF}{}^2]}$$

$$=\frac{-3k_{AB}C'_{AB}\left[\dfrac{Ph_R}{3}+\dfrac{4}{3}\Sigma m_{BA}\,C'_{AB}+\Sigma(m_{EF}+m_{EF})C_{EF}\right]}{2[\Sigma k_{AB}\,C'_{AB}+k_{EF}\,C_{EF}{}^2]}$$

$$m'_{EF}=\frac{-3k_{EF}C'_{AB}\left[\dfrac{Ph_R}{3}+\dfrac{4}{3}\Sigma m_{BA}\,C'_{AB}+\Sigma(m_{EF}+m_{EF})\,C_{EF}\right]}{2[\Sigma k_{AB}\,C'_{AB}{}^2+k_{EF}\,C_{EF}{}^2]}$$

In case all columns have ends hinged,

$$m'_{BA}=\frac{-3k_{AB}\,C'_{AB}\left[\dfrac{Ph_R}{3}+\dfrac{4}{3}\Sigma m_{BA}\,C'_{AB}\right]}{2\Sigma k_{AB}\,C'_{AB}{}^2}$$

In case all columns are of same height and have ends hinged

$$C'_{BA}=\tfrac{1}{2}$$

$$\therefore\quad m'_{BA}=\frac{-3k_{AB}\left[\dfrac{Ph_R}{3}+\dfrac{2}{3}\Sigma m_{AB}\right]}{\Sigma k_{AB}}$$

In case all columns have fixed ends,

$$m'_{EF}=\frac{-3k_{EF}\,C_{FE}\left[\dfrac{Ph_R}{3}+\Sigma(m_{EF}+m_{FE})\,C_{EF}\right]}{3\Sigma k_{EF}\,C_{EF}{}^2}$$

If all columns have fixed ends and have same heights,

$$C_{EF}=1$$

$$m'_{EF}=\frac{-3k_{EF}\left[\frac{Ph_R}{3}+\Sigma(m_{EF}+M_{FE})\right]}{2\,\Sigma k_{EF}}$$

Ex. 18·4. *Analyse the continuous beam shown in Fig. 18·20 by Kani's method.*

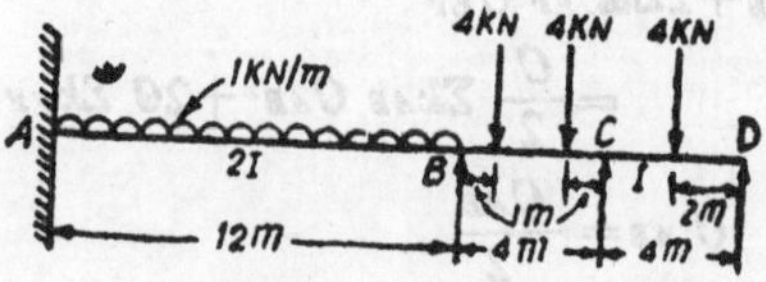

Fig. 18·20.

Sol. $\overline{M_{AB}}=-\frac{1\times12^2}{12}=-12$ kNm.

$\overline{M_{BA}}=+\frac{1\times12^2}{12}=+12$ kNm.

$\overline{M_{BC}}=-\left[\frac{4\times3^2\times1}{4\times4}+\frac{4\times3\times1^2}{4\times4}\right]$

$=-[2{\cdot}25+0{\cdot}75]=-3$ kNm.

$\overline{M_{CB}}=+\left[\frac{4\times3^2\times1}{4\times4}+\frac{4\times3\times1^2}{4\times4}\right]$

$=+[2{\cdot}25+0{\cdot}75]=+3$ kNm.

$\overline{M_{CD}}=-\frac{4\times4}{8}=-2$ kNm.

$\overline{M_{DC}}=+\frac{4\times4}{8}=+2$ kNm.

Rotation Factors

$$r=-\frac{1}{2}\frac{k}{\Sigma k}$$

Joint B

$$r_{BA}=-\frac{1}{2}\left[\frac{2I/12}{2I/12+I/4}\right]=-\frac{1}{2}\left[\frac{2}{5}\right]=-0{\cdot}2$$

$$r_{BC}=-\frac{1}{2}\left[\frac{I/4}{2I/12+I/4}\right]=-\frac{1}{2}\left[\frac{2}{5}\right]=-0{\cdot}3$$

Joint C

$$r_{CB}=-\frac{1}{2}\left[\frac{I/4}{I/4+I/4}\right]=-0{\cdot}25$$

$$r_{CD}=-\frac{1}{2}\left[\frac{I/4}{I/4+I/4}\right]=-0{\cdot}25$$

The fixed end moments and rotation factors are calculated. The process of distribution is shown in Fig. 18·21. Two concentric circles are drawn at each joint. The fixed end moments are written above the beams, outside the circles. Rotation factors are written in the space between inner circle and outer circle, towards each side of the member. Unbalanced moment at a joint is written in the inner circle. Rotation contributions are written below the beams.

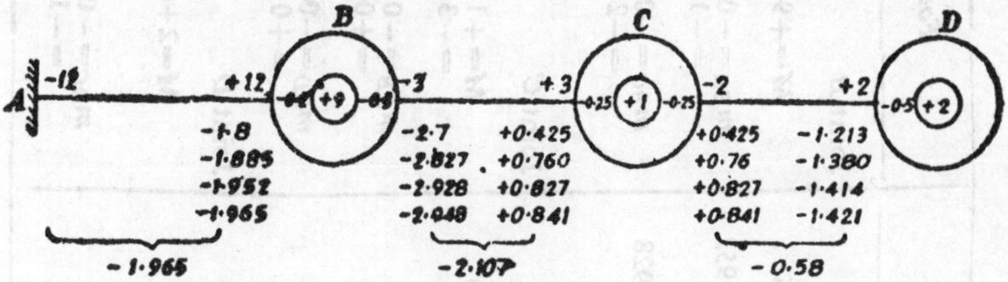

Fig. 18·21.

The process is started at joint *B*. Initially for joint *B*, rotation contribution of end *C* is assumed as zero. The rotation contribution of end *A* is zero as it is fixed end. Thus the unbalanced moment of +9 at joint *B* is distributed for *BA* and *BC* in proportion of rotation factors −0·2 and −0·3 respectviely giving

$$m_{BA} = -1{\cdot}8 \text{ and } m_{BC} = -2{\cdot}7.$$

Next distribution is carried out at joint *C*. Rotation contribution of *D* is assumed as zero and that of end *B* is now −2·7. The unbalanced moment of −2·7+1=−1·7 is distributed for *CB* and *CD* in proportion of rotation factors −0·25 and −0·25 giving

$$m_{CB} = m_{CD} = +0{\cdot}425.$$

Next at joint *D*, rotation contribution of *C* is taken as +0·425. Unbalanced moment will be +2+0·425=+2·425. Distribution will be only for *DC* and will be −0·5×2·425=−1·213. Thus first cycle is completed.

In the second cycle, at joint *B*, rotation contribution of end *A* is zero and that of *C* is now taken as +0·425. Unbalanced moment of +9+0·425=+9·425 is distributed for *BA* and *BC* in proportion of rotation factors. This process is continued till the variation in rotation contributions is negligible. The calculations for four cycles are given in Table 18·13.

Final moment at a joint is given by

$$M_{AB} = \overline{M}_{BA} 2m + {}_{AB} + m_{BA}$$

The values of m_{AB} for members are written at the centre after final cycle. So that final moment at a joint is this value plus rotation contribution of the end plus fixed end moment.

$$M_{AB} = -12 - 1{\cdot}965 = -13{\cdot}965 \text{ kNm.}$$
$$M_{BA} = +12 - 1{\cdot}965 - 1{\cdot}965 = +8{\cdot}07 \text{ kNm.}$$
$$M_{BC} = -3 - 2{\cdot}107 - 2{\cdot}948 = -8{\cdot}055$$
$$M_{CB} = +3 - 2{\cdot}107 + 0{\cdot}841 = +0{\cdot}734$$
$$M_{CD} = -2 - 0{\cdot}580 + 0{\cdot}841 = -0{\cdot}739$$
$$M_{DC} = +2 - 0{\cdot}580 - 1{\cdot}421 = -0{\cdot}001$$

Table 18·13

First cycle	*Second cycle*	*Third cycle*	*Fourth cycle*
Joint *B*	Joint *B*	Joint *B*	Joint *B*
$M=+9$	$M=+9+0{\cdot}425=+9{\cdot}425$	$M=+9+0{\cdot}76=+9{\cdot}76$	$M=+9+0{\cdot}827=9{\cdot}827$
$m_{BA}=-0{\cdot}2\times 9=-1{\cdot}8$	$m_{BA}=0{\cdot}2\times 9{\cdot}425$ $=-1{\cdot}885$	$m_{BA}=-0{\cdot}2\times 9{\cdot}76=-1{\cdot}952$	$m_{BA}=-0{\cdot}2\times 9{\cdot}827$ $=-1{\cdot}965$
$m_{BC}=-0{\cdot}3\times 9=-2{\cdot}7$	$m_{BC}=-0{\cdot}3\times 9{\cdot}425$ $=2{\cdot}827$	$m_{BC}=-0{\cdot}3\times 9{\cdot}76=-2{\cdot}928$	$m_{BC}=-0{\cdot}3\times 9{\cdot}827$ $=-2{\cdot}948$
Joint *C*	Joint *C*	Joint *C*	Joint *C*
$M=+1-2{\cdot}7=-1{\cdot}7$	$M=+1-2{\cdot}827-1{\cdot}213$ $=-3{\cdot}040$	$M=+1-2{\cdot}928-1{\cdot}380$ $=-3{\cdot}308$	$M=+1-2{\cdot}948-1{\cdot}404$ $=-3\ 362$
$m_{CB}=-0{\cdot}25\times(-1{\cdot}7)$ $=+0{\cdot}425$	$m_{CB}=-0{\cdot}25\times(-3{\cdot}04$ $=+0{\cdot}760$	$m_{CB}=-0{\cdot}25\times(-3{\cdot}308)$ $=+0{\cdot}827$	$m_{CB}=-0{\cdot}25\times(-3{\cdot}362)$ $=+0{\cdot}841$
$m_{CD}=-0{\cdot}25\times(-1{\cdot}7)$ $=+0{\cdot}425$	$m_{CD}=-0{\cdot}25\times(-3{\cdot}04)$ $=+0{\cdot}760$	$m_{CD}=-0{\cdot}25\times(-3{\cdot}308)$ $=+0{\cdot}827$	$m_{CD}=-0{\cdot}25\times(-3{\cdot}362$ $=+0{\cdot}841$
Joint *D*	Joint *D*	Joint *D*	Joint *D*
$M=+2+0{\cdot}425$ $=2{\cdot}425$	$M=+2+0{\cdot}760=+2{\cdot}760$	$M=+2+0{\cdot}827$ $=+2{\cdot}827$	$M=2+0{\cdot}841=2{\cdot}841$
$m_{DC}=-0{\cdot}5\times 2{\cdot}425$ $=-1{\cdot}213$	$m_{DC}=-0{\cdot}5\times 2{\cdot}760$ $=-1{\cdot}380$	$m_{CD}=-0{\cdot}5\times 2{\cdot}827$ $=-1{\cdot}414$	$m_{DC}=-0{\cdot}5\times 2{\cdot}841$ $=-1{\cdot}421$

Errors. Total moment at joint B

$$=M_{BA}+M_{BC}=+8{\cdot}07-8{\cdot}055=-0{\cdot}015$$

At joint C total moment

$$=M_{CB}+M_{CD}=+0{\cdot}734-0{\cdot}739=-0{\cdot}005$$

Moment at support $D=-0{\cdot}001$

Actual sum of moments at joints B, C and D should be zero. This error is extremely small and negligible.

For a member as CD, hinged at D, the stiffness of CD is taken $\frac{3}{4}$ of actual and fixed end moment is found for this condition and also rotation factors and then joint D is not considered.

$$\overline{M_{CD}}=-2-\tfrac{3}{3}=-3.$$

$$r_{CD}=-\tfrac{1}{2}\times\left[\frac{\frac{3}{4}\times\frac{I}{4}}{\frac{3}{4}\times\frac{I}{4}+\frac{I}{4}}\right]=-\frac{3}{14}$$

$$r_{CB}=-\tfrac{1}{2}\left[\frac{\frac{I}{4}}{\frac{I}{4}+\frac{3}{4}\times\frac{I}{4}}\right]=-\frac{4}{14}$$

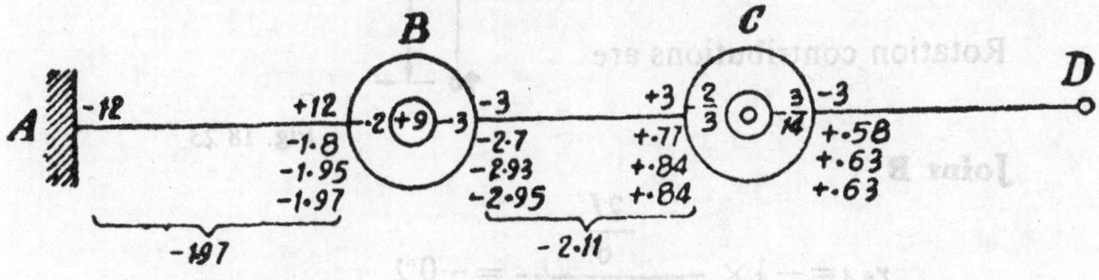

Fig. 18·22

The calculations are shown in Table 18·14.

Table 18·14

First cycle	*Second cycle*	*Third cycle*
Joint B $M=+9$	Joint B $M=+9+0{\cdot}77$ $=+9{\cdot}77$	Joint B $M=+9+0{\cdot}84$ $=+9{\cdot}84$
$m_{BA}=-0{\cdot}2\times9$ $=-1{\cdot}8$	$m_{BA}=-0{\cdot}2\times9{\cdot}77$ $=-1{\cdot}95$	$m_{BC}=0{\cdot}2\times9{\cdot}84$ $=-1{\cdot}97$
$m_{BC}=-0{\cdot}3\times9$ $=-2{\cdot}7$	$m_{BC}=-0\ 3\times2\ 77$ $=-2{\cdot}93$	$m_{BC}=-0{\cdot}3\times9{\cdot}84$ $=-2{\cdot}95$
Joint C $M=-2{\cdot}7$	Joint G $M=-2{\cdot}93$	Joint C $M=+-2{\cdot}95$
$m_{BC}=-\frac{2}{7}\times-2{\cdot}7$ $=+0{\cdot}77$	$m_{BC}=-\frac{2}{7}\times-2{\cdot}93$ $=+0{\cdot}84$	$m_{CB}=-\frac{2}{7}\times-2{\cdot}95$ $=+0{\cdot}84$
$m_{CD}=-\frac{3}{14}=-2{\cdot}7$ $=+0{\cdot}58$	$m_{CD}+-\frac{3}{14}\times-2{\cdot}93$ $=+0{\cdot}63$	$m_{CA}=-\frac{3}{14}\times-2{\cdot}95$ $=+0{\cdot}63$

Final moments

$$M_{AB} = -1{\cdot}97 - 12 = -13{\cdot}97 \text{ kNm.}$$
$$M_{BA} = -1{\cdot}97 - 1{\cdot}97 + 12 = +8{\cdot}06 \text{ kNm.}$$
$$M_{BC} = -2{\cdot}11 - 2{\cdot}95 - 3 = +8{\cdot}06 \text{ kNm.}$$
$$M_{CB} = -2{\cdot}11 - 0{\cdot}84 + 3 = -1{\cdot}73 \text{ kNm.}$$
$$M_{CD} = +0{\cdot}93 + 0{\cdot}63 - 3 = -1{\cdot}74 \text{ kNm.}$$

$M_{CB} + M_{CD} = +1{\cdot}73 - 1{\cdot}74 = -0{\cdot}01$. The error is negligible.

Ex. 18·5. *Analyse the frame shown in Fig. 18·23 by Kani's method.*

Solution. $$\overline{M_{BC}} = -\frac{1 \times 6^2}{12} = -3 \text{ kNm}$$

$$\overline{M_{CB}} = -\frac{1 \times 6^2}{12} = +3 \text{ kNm}$$

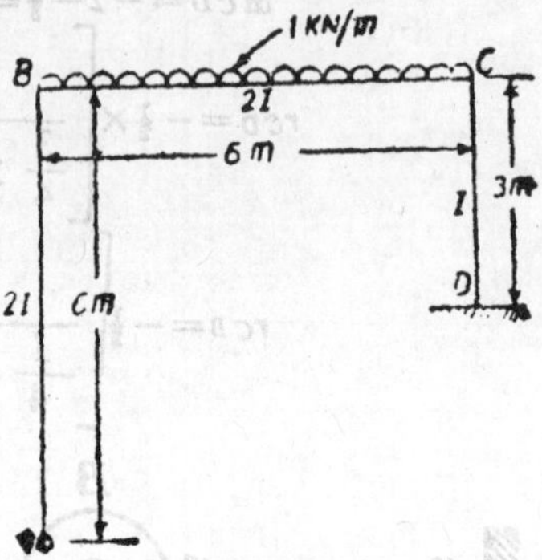

Fig. 18·23

Rotation contributions are

Joint B

$$r_{BA} = -\tfrac{1}{2} \times \frac{\frac{2I}{6}}{\frac{2I}{6} + \frac{3I}{6}} = -0{\cdot}2$$

$$r_{BC} = -\tfrac{1}{2} \times \frac{\frac{3I}{6}}{\frac{2I}{6} + \frac{3I}{6}} = -0{\cdot}3$$

Joint C

$$r_{CB} = -\tfrac{1}{2} \times \frac{3I/6}{3I/6 + I/3} = -0{\cdot}3$$

$$r_{CD} = - \times \tfrac{1}{2} \frac{I/3}{3I/6 + I/3} = -0{\cdot}2$$

Reference height is taken as 6 m.

$$C_{AB} = \frac{h_{AB}}{h_{AB}} = 1.$$

$$C_{CD} = \frac{6}{3} = 2.$$

Displacement factors

$$\delta_{AB} = \frac{-\frac{3}{2} k_{AB} C_{BA}}{\Sigma k_{AB} C^2_{AB}}$$

$$=\frac{-\frac{3}{2}\times\frac{2I}{6}\times 1}{\left(\frac{2I}{6}\times 1^2+\frac{I}{3}\times 2^2\right)}$$

$$=\frac{-\frac{3}{2}\times\frac{1}{3}}{\frac{1}{3}+\frac{4}{3}}=-\frac{3}{2}\times\frac{1}{5}$$

$$=-0{\cdot}3.$$

$$\delta_{CD}=\frac{-\frac{3}{2}\times\frac{I}{3}\times 2}{\frac{2I}{6}\times 1^2+\frac{I}{3}\times 2^2}$$

$$=\frac{\frac{3}{2}\times\frac{2}{3}}{\frac{5}{3}}=-0{\cdot}6.$$

The process of distribution is shown in Fig. 18·24. Two concentric circles are drawn at each joint. The fixed end moments are written above the beams, outside the circles. The rotation factors are written in the space between inner circles and outer circles, towards each side of the member. Unbalanced moment at a

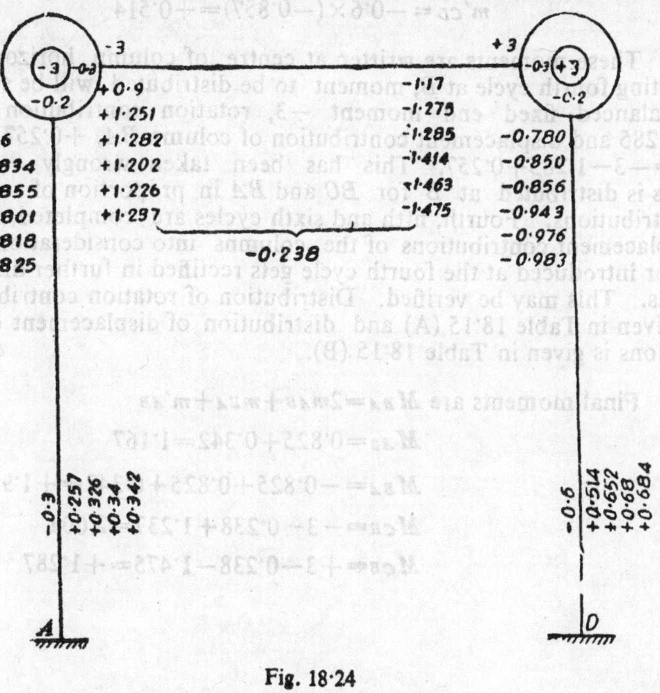

Fig. 18·24

joint is written in the inner circle. Rotation contributions are written below the line for horizontal member and for vertical members rotation contributions are written towards left side of vertical line. Displacement contributions for each column are written horizontally at centre of column. The process is started at joint B. Initially the rotation contribution of end C is assumed as zero. The rotation contribution of end A is zero as it is a fixed end. Thus the unbalanced moment of -3 is distributed for BA and BC in proportion of rotation factors $-0{\cdot}2$ and $-0{\cdot}3$ respectively giving $m_{BA}=+0{\cdot}6$ and $m_{BC}=+0{\cdot}9$. Next distribution is carried out at joint C. Rotation contribution of end D is zero as it is a fixed end and rotation contribution of end B is now taken as $+0{\cdot}9$. The unbalanced moment of $+3+0{\cdot}9=+3{\cdot}9$ is distributed for CB and CD in proportion of rotation factors $-0{\cdot}3$ and $-0{\cdot}2$ giving $m_{CB}=-1{\cdot}17$, $m_{CD}=-0{\cdot}78$. Thus first cycle is completed. In the second cycle, at joint B rotation contribution of end A is zero and that of C is now taken as $-1{\cdot}17$. Unbalanced moment of $-3-1{\cdot}17=-4{\cdot}17$ is distributed for BA and BC in proportion of rotation factors. Three cycles are completed in this way ; then sway correction is applied. The distribution is shown in Table 18·15. After third cycle $\Sigma(m_{AB}+m_{BA})\,C_{AB}=(0{\cdot}855+0)\times1+(-8{\cdot}856+0)\times2=0{\cdot}855-1{\cdot}712=-0{\cdot}857$.

$$m'_{AB}=-0{\cdot}3\times(-0{\cdot}857)=+0{\cdot}257$$

$$m'_{CD}=-0{\cdot}6\times(-0{\cdot}857)=+0{\cdot}514$$

These moments are written at centre of column horizontally. Starting fourth cycle at B, moment to be distributed will be sum of unbalanced fixed end moment -3, rotation contribution of C, $-1{\cdot}285$ and displacement contribution of column BA, $+0{\cdot}257$. Thus $M=-3-1{\cdot}285+0{\cdot}257$. This has been taken wrongly $-4{\cdot}005$. This is distributed at B for BC and BA in proportion of rotation contributions. Fourth, fifth and sixth cycles are completed, taking displacement contributions of the columns into consideration. The error introduced at the fourth cycle gets rectified in further distributions. This may be verified. Distribution of rotation contributions is given in Table 18·15 (A) and distribution of displacement contributions is given in Table 18·15 (B).

Final moments are $M_{BA}=2m_{AB}+m_{BA}+m'_{AB}$

$$M_{AB}=0{\cdot}825+0{\cdot}342=1{\cdot}167$$

$$M_{BA}=-0{\cdot}825+0{\cdot}825+0{\cdot}342=+1{\cdot}992$$

$$M_{CB}=-3-0{\cdot}238+1{\cdot}237=2{\cdot}001$$

$$M_{CB}=+3-0{\cdot}238-1{\cdot}475=+1{\cdot}287$$

Table 18·15 (A)

First cycle	*Second cycle*	*Third cycle*	*Fourth cycle*	*Fifth cycle*	*Sixth cycle*
Joint *B*	Joint *B*	Joint *B*	Joint B	Joint *B*	Joint *B*
$M=-3$	$M=-3-1{\cdot}17$ $=-4{\cdot}17$	$M=-3-1{\cdot}275$ $=4{\cdot}275$	$M=-3-1{\cdot}285$ $+0{\cdot}257$ $=-4\ 005$	$M=-3-1{\cdot}411$ $+0{\cdot}326$ $=-4{\cdot}088$	$M=-3-1{\cdot}463$ $+0{\cdot}34$ $=-4\ 123$
$m_{BA}=-0{\cdot}2\times(-3)$ $=+0{\cdot}6$	$m_{BA}=-0{\cdot}2\times(-4{\cdot}17)$ $=+0{\cdot}834$	$m_{BA}=-0{\cdot}2\times(-4{\cdot}275)$ $=+0{\cdot}855$	$m_{EA}=-0{\cdot}2\times(-4{\cdot}005)$ $=+0{\cdot}801$	$m_{BA}=-0{\cdot}2\times$ $(+0{\cdot}088)$ $=+0{\cdot}818$	$m_{AB}=-0{\cdot}2\times$ $(-4{\cdot}125)$ $=+0{\cdot}825$
$m_{BC}=-0{\cdot}3\times(-3)$ $=+0{\cdot}9$	$m_{BC}=-0{\cdot}2\times(-4{\cdot}17)$ $=+1{\cdot}251$	$m_{BC}=-0{\cdot}3\times(-4{\cdot}275)$ $=+1{\cdot}282$	$m_{BC}=-0{\cdot}3\times\times(4{\cdot}005)$ $=+1{\cdot}202$	$m_{BC}=-0{\cdot}3\times(4\ 088)$ $=+1{\cdot}226$	$m_{BC}=-0{\cdot}3\times$ $(-4\ 123)$ $=+1{\cdot}237$
Joint	Joint *C*	Joint C	Joint C	Joint C	Joint C
$M=+3+0{\cdot}9=+3{\cdot}9$	$M=+3+1{\cdot}251$ $=+4{\cdot}251$	$M=+3+1{\cdot}282$ $=+4{\cdot}282$	$M=3+1{\cdot}202$ $+0{\cdot}514$ $=+4{\cdot}716$	$M=+3+1{\cdot}226$ $+0{\cdot}652$ $=+4{\cdot}878$	$M=+3+1{\cdot}237$ $+0\ 68$ $=+4\ 917$
$m_{CD}=-0{\cdot}2\times3{\cdot}9$ $=-1{\cdot}17$	$m_{CD}=-0\ 3\times4{\cdot}251$ $=-1{\cdot}275$	$M_{CB}=-0{\cdot}3\times4{\cdot}282$ $=-1{\cdot}285$	$m_{CD}=-0{\cdot}3\times4{\cdot}716$ $=-1{\cdot}414$	$m_{CB}=-0{\cdot}3\times4{\cdot}878$ $=-1{\cdot}463$	$M_{CB}=-0{\cdot}3\times4{\cdot}917$ $=-1{\cdot}475$
$m_{CD}=-0\ 2\times3{\cdot}9$ $=-0{\cdot}78$	$m_{CD}=-0{\cdot}2\times4{\cdot}257$ $=-0{\cdot}850$	$M_{CD}=-0\ 2\times4{\cdot}282$ $=-0{\cdot}856$	$m_{CD}=-0\ 2\times4{\cdot}716$ $=-0{\cdot}943$	$m_{CD}=-0\ 2\times4{\cdot}878$ $=-0{\cdot}976$	$m_{CD}=-0{\cdot}2\times4\ 917$ $=-0{\cdot}983$

After third cycle	*After fourth cycle*	*After fifth cycle*	*After sixth cycle*
$\Sigma(m_{AB}+m_{BA})C_{AB}$ $=0{\cdot}855\times 1$ $=(-0{\cdot}856)\ \ 2$ $-0{\cdot}855-8{\cdot}712$ $=-0{\cdot}857$ $m'_{AB}=-0{\cdot}3$ $\times(-0{\cdot}857)$ $=+0{\cdot}257$ $m'_{CD}=-0{\cdot}6$ $\times-(0{\cdot}867)$ $=+0{\cdot}514$	$\Sigma(m_{AB}+m_{BA}(C_{AB}$ $=(0{\cdot}801)\times 1$ $+(-0{\cdot}943)\times 2$ $=0{\cdot}801-1{\cdot}886$ $=-1{\cdot}085$ m'_{AB} $=-0{\cdot}3(-1{\cdot}086)$ $=+0\ 327$ m'_{CD} $=-0\ 6(-1{\cdot}085)$ $=+0{\cdot}652$	$\Sigma(m_{AB}+m_{BA}(C_{AB}$ -818×1 $+(-0{\cdot}976)\times 2$ $=0{\cdot}818-1{\cdot}052$ $=-1{\cdot}034$ m'_{AB} $=-0{\cdot}3(1{\cdot}034)$ $=+0{\cdot}34$ m'_{CD} $=+0{\cdot}5(-1{\cdot}034)$ $=+0{\cdot}68$	$\Sigma(m_{BB}+m_{BA}(C_{AB}$ $=0{\cdot}825\times 1$ $+(-0{\cdot}683)\times 2$ $=0{\cdot}825-1{\cdot}966$ $=-1{\cdot}143$ m'_{AB} $=-0{\cdot}3(1{\cdot}144)$ $=+0{\cdot}342$ m'_{CD} $=-0{\cdot}6(1{\cdot}143)$ $=+0{\cdot}684$

$$M_{CD}=-0{\cdot}983-0{\cdot}983+0{\cdot}684=-1{\cdot}282$$
$$M_{DC}=-0{\cdot}983+0{\cdot}684=-0{\cdot}299$$

Error at joints

$$M_{BA}+M_{BC}=1{\cdot}992-2{\cdot}001=-0{\cdot}009$$
$$M_{CB}+M_{CD}=1{\cdot}287-1{\cdot}282=+0{\cdot}005$$

Actually sum of moments at joints should be zero. This error is negligible.

Ex. 18·6. *Analyse the frame shown in Fig. 18·25 by Kani's method.*

Solution. Rotation factors

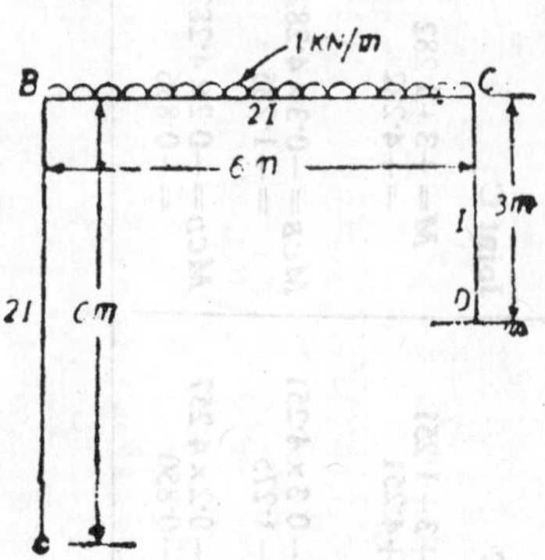

Fig. 18·25

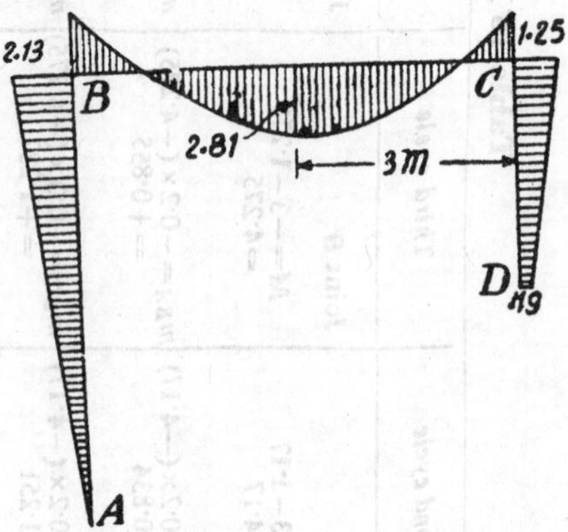

Fig. 18·26

$$r_{BA}=-\frac{1}{2}\times\frac{\frac{3}{4}\times\frac{2I}{6}}{\left(\frac{3}{4}\times\frac{2I}{6}+\frac{2I}{6}\right)}=-\frac{3}{14}$$

$$r_{BC}=-\frac{1}{2}\times\frac{\frac{2I}{6}}{\left(\frac{3}{4}\times\frac{2I}{5}+\frac{2I}{6}\right)}=-\frac{4}{14}.$$

$$r_{CB} = -\frac{1}{2} \times \frac{\frac{2I}{6}}{\left(\frac{2I}{6} + \frac{I}{3}\right)} = -\frac{1}{14}$$

$$r_{CD} = -\frac{1}{2} \times \frac{\frac{I}{3}}{\left(\frac{2I}{6} + \frac{I}{3}\right)} = -\frac{1}{3}.$$

Displacement Factors

$$\delta_{AB} = \frac{-3k_{AB}\, C'_{AB}}{2[\Sigma_{AB}\, C'_{AB}{}^2 + k_{CD}\, C_{CD}{}^2]}$$

$$C_{AB} = \frac{6}{6} = 1$$

$$\therefore \quad C'_{AB} = \tfrac{1}{2}.$$

$$C_{CD} = \frac{6}{3} = 2.$$

$$\therefore \quad \delta_{AB} = \frac{-3 \times \frac{2I}{6} \times \frac{1}{2}}{2\left[\frac{2I}{6} \times \frac{1}{4} + \frac{I}{3} \times 4\right]} = -\frac{3}{17}$$

$$\delta_{CD} = \frac{\times 3k_{CD} \quad C_D}{2[\Sigma k_{AB}\, C'_{AB}{}^2 + k_{CD}\, C_{CD}{}^2]}$$

$$= \frac{-3 \times \frac{I}{3} \times}{2\left[\frac{2I}{6} \times \frac{1}{4} + \frac{I}{3} \times 4\right]} = \frac{-12}{17}$$

$$\overline{M_{BC}} = \frac{-1 \times 6^2}{12} = -3 \text{ kNm}.$$

$$\overline{M_{CB}} = \frac{1 \times 6^2}{12} = +3 \text{ kNm}.$$

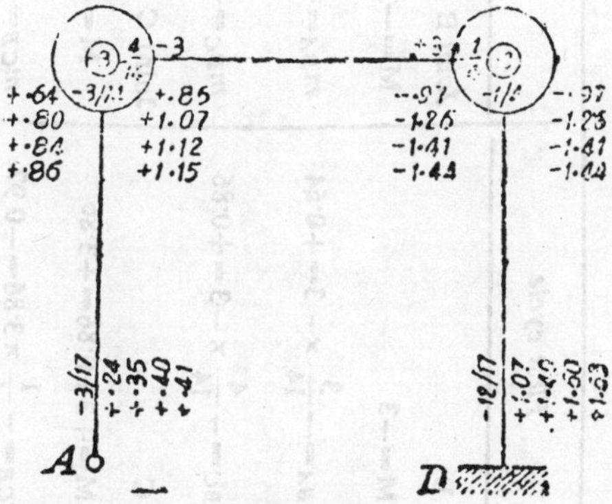

Fig. 18·27

Table 18·6 (A)

First cycle	*Second cycle*	*Third cycle*	*Fourth cycle*
Joint B	*Joint* B	Joint B	Joint B
$M=-3$	$M=-3+0{\cdot}07+0{\cdot}24=-3{\cdot}73$	$M=-3-1{\cdot}26+0{\cdot}35=-3{\cdot}91$	$M=-3-1{\cdot}41+0{\cdot}40=-4{\cdot}01$
$m_{BA}=-\frac{3}{14}\times-3=+0{\cdot}64$	$m_{BA}=-\frac{3}{14}\times-3{\cdot}73=+0{\cdot}80$	$m_{BA}=-\frac{3}{14}\times-3{\cdot}91=-0{\cdot}84$	$m_{BA}=-\frac{3}{14}\times-4.01=+0{\cdot}86$
$m_{BC}=-\frac{4}{14}\times-3=+0{\cdot}86$	$m_{BC}=-\frac{4}{14}\times-3{\cdot}73=+1{\cdot}07$	$m_{BC}=-\frac{4}{14}\times-3{\cdot}91-+1{\cdot}12$	$m_{BC}=-\frac{4}{14}\times-4{\cdot}01=+1{\cdot}15$
Joint C	Joint *C*	Joint *C*	Joint C
$M=+3+0{\cdot}86=+3{\cdot}86$	$M=+3+1{\cdot}07+1{\cdot}07=+5{\cdot}14$	$M=+3+1{\cdot}12+140=+5{\cdot}62$	$M=+3+1{\cdot}15\bigtriangledown1{\cdot}60=5{\cdot}75$
$m_{CB}=-\frac{1}{4}\times3{\cdot}86=-0\ 97$	$m_{CB}=-\frac{1}{4}\times5{\cdot}14=-126$	$m_{CB}\quad\frac{1}{4}\times5{\cdot}62=-141$	$m_{CB}=-\frac{1}{4}\times5{\cdot}75=-1{\cdot}44$
$m_{CD}=-\frac{1}{4}\times3{\cdot}86=-0{\cdot}97$	$m_{CD}=-\frac{1}{4}\times5\ 14=-1{\cdot}26$	$m_{CD}\rightarrow-\frac{1}{4}\times5{\cdot}62=-1{\cdot}41$	$m_{CD}=-\frac{1}{4}\times5{\cdot}75=-1{\cdot}44$

Table 18·16 (B) Sway Correction

First cycle	*Second cycle*	*Third cycle*	*Fourth cycle*
$M=\frac{0{\cdot}64\times2}{3}-2\times0{\cdot}97=-1{\cdot}51$	$M=\frac{0{\cdot}80\times2}{3}-2\times1{\cdot}26=+1{\cdot}99$	$M=-\frac{0{\cdot}84\times2}{3}-1{\cdot}41\times2=-2{\cdot}46$	$M=+\frac{0{\cdot}86\times2}{3}-2\times1{\cdot}4=-2{\cdot}31$
$m'_{BA}=-\frac{3}{17}=-1{\cdot}51=+0{\cdot}24$	$m'_{BA}=-\frac{3}{17}\times-1{\cdot}99=+0{\cdot}35$	$m'_{BA}=-\frac{3}{17}\times-2{\cdot}26=+0{\cdot}430$	$m'_{BA}=-\frac{3}{17}\times-2{\cdot}31=+0{\cdot}41$
$m'_{CD}=-\frac{17}{12}\times-1{\cdot}51=+1{\cdot}07$	$m'_{CD}=-\frac{17}{12}\times-1{\cdot}99=+1{\cdot}40$	$m'_{CD}=-\frac{12}{17}\times-2{\cdot}26=+1{\cdot}60$	$m'_{CD}=-\frac{12}{17}\times-2{\cdot}31=+1{\cdot}63$

Final moments

$M_{BA}=2\times0{\cdot}86+0{\cdot}41=1{\cdot}72+0{\cdot}41=+2{\cdot}13$ kNm

$M_{BC}=2\times1{\cdot}15-1{\cdot}44-3=2{\cdot}30-444=-2{\cdot}14$ kNm

$M_{CB}=2\times-1{\cdot}44+1{\cdot}15+3=-2{\cdot}88+4{\cdot}15=+1{\cdot}27$ kNm

$M_{CD}=2\times-1{\cdot}63=-2{\cdot}88=1{\cdot}63=-1{\cdot}25$ kNm

$M_{DC}=-1{\cdot}44+1{\cdot}63=+0{\cdot}19$ kNm

The B.M. diagram is shown is Fig. 18·26.

Ex. 18·7. ***Analyse the frame shown in Fig. 18·28 by Kani's method.***

Sol. As the loading is on joints only, the fixed end moments are zero.

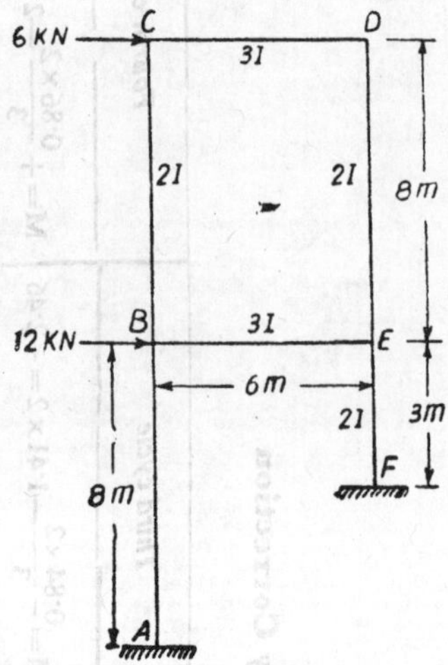

Fig. 18·28

Rotation factors will be

Joint C

$$r_{CD}=-\frac{1}{2}\times\frac{\frac{3I}{6}}{\frac{3I}{5}+\frac{2I}{8}}$$

$$=-\frac{1}{2}\times\frac{\frac{1}{2}}{\frac{1}{2}+\frac{1}{4}}=-\frac{1}{3}$$

$$r_{CB}=-\frac{1}{2}\times\frac{\frac{2I}{8}}{\frac{3I}{6}+\frac{3I}{8}}=-\frac{1}{6}$$

Similarly at joint D

$$r_{DC}=-\frac{1}{3}$$

$$r_{DE}=-\frac{1}{6}$$

Joint D

$$r_{BE}=-\frac{1}{2}\times\frac{\frac{3I}{6}}{\frac{3I}{6}+\frac{4I}{8}+\frac{2I}{8}}=-0{\cdot}2$$

$$r_{BA}=-\frac{1}{2}\times\frac{\frac{4I}{6}}{\frac{3I}{6}+\frac{4I}{8}+\frac{2I}{8}}=-0{\cdot}2$$

$$r_{BC}=-\frac{1}{2}\times\frac{\frac{2I}{8}}{\frac{3I}{6}+\frac{4I}{8}+\frac{2I}{8}}=-0{\cdot}1$$

Joint E

$$r_{EB}=-\frac{1}{2}\times\frac{\frac{3I}{6}}{\frac{3I}{6}+\frac{2I}{4}+\frac{2I}{8}}=-0{\cdot}2$$

$$r_{EF}=-\frac{1}{2}\times\frac{\frac{2I}{6}}{\frac{3I}{6}+\frac{2I}{4}+\frac{2I}{8}}=-0{\cdot}2$$

$$r_{ED}=-\frac{1}{2}\times\frac{\frac{2I}{6}}{\frac{3I}{6}+\frac{2I}{4}+\frac{2I}{8}}=-0{\cdot}1$$

Displacement factors

Upper storey

All vertical members have same height

$$\delta_{CB}=\delta_{DE}=-\frac{3}{2}\frac{k_{AB}}{\Sigma k_{AB}}$$

$$=-\frac{3}{2}\times\frac{\frac{2I}{8}}{\frac{2I}{8}+\frac{2I}{8}}=-0{\cdot}75$$

Lower Storey

Let $h_B=8$ m.

$$C_{AB}=1,\ C_{EF}=\frac{8}{4}=2$$

$$\delta_{AB}=-\frac{2}{3}\times\frac{k_{AB}\,C_{AB}}{\Sigma k_{AB}\,C^2_{AB}}$$

$$=-\frac{3}{2}\times\frac{\frac{4I}{8}\times 1}{\frac{4I}{8}\times 1+\frac{2I}{4}\times 2^2}$$

$$=-\frac{3}{2}\times\frac{\frac{1}{2}}{\frac{1}{2}+2}$$

$$=-0{\cdot}3$$

$$\delta_{EF}=-\frac{3}{2}\times\frac{\frac{2I}{4}\times 2}{\frac{4I}{8}+\frac{2I}{4}\times 2^2}=-0{\cdot}6$$

Storey moment for upper storey$=\frac{6\times 8}{3}=16$ kNm.

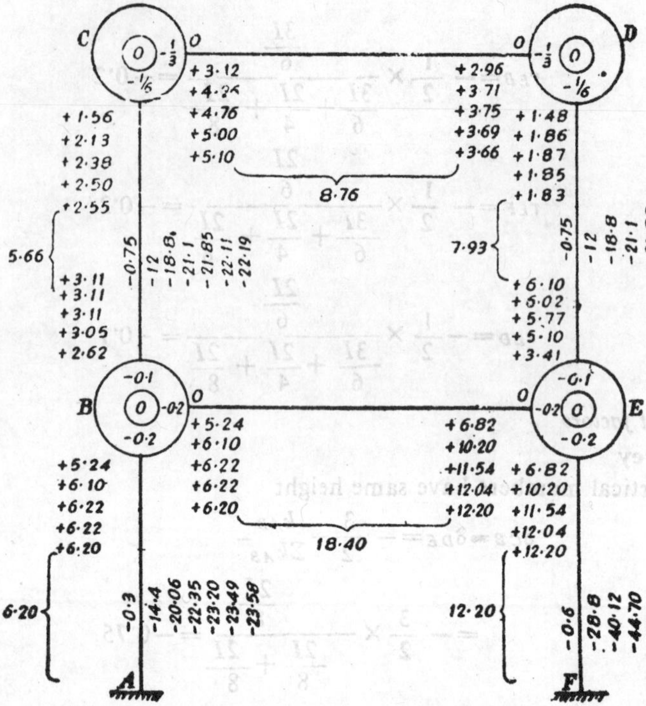

Fig. 18·29

Storey moment for lower storey$=(6+12)\times\frac{8}{3}=48$ kNm.

In this case all fixed end moments are zero and thus unbalanced moments at all joints are zero. The process is started with distribution of storey moments. Storey moment of 16 kNm for upper storey is distributed in upper storey columns in ratio of displacement factors. Similarly storey moment for lower storey is distributed in lower storey columns in ratio of displacement factors.

First cycle is started at joint *B*. Unbalanced moment at joint will be sum of displacement contributions of upper and lower storeys, *i.e.*$-12-14{\cdot}2=-26{\cdot}2$. This moment is distributed in members *BA*, *BE* and *BC* in proportion of rotation contributions. Next going to joint *C*, unbalanced moment will be displacement contribution -12 of upper storey and rotation contribution $+2{\cdot}62$ of joint *B*. Sum of these moments$=-1.+2{\cdot}62=-9{\cdot}38$ is distributed in ratio of rotation factors. This process is carried out at joints *D* and *E* and first cycle completed. After first cycle displacement contributions are calculated taking into consideration rotation

Table 18·17 (A)

First cycle	*Second cycle*	*Third cycle*	*Fourth cycle*	*Fifth cycle*
Joint *B* $M=-14·2-12$ $=-26·2$ $m_{BC}=-0\ 1\times(-16·2)$ $=+2\ 62$ $m_{BE}=-0·2\times(-26·2)$ $=+5·24$ $m_{BA}=5·24$	Joint *B* $M=-29·06-18·8$ $+1·56+6·82$ $=-30\ 48$ $m_{BC}=+3·05$ $m_{BE}=+6·1$ $m_{BA}=+6·1$	Join *B* $M=-22·35-21·10$ $+2·13+10·20$ $=-31·12$ $m_{BC}=+3·11$ $m_{BE}=+6·22$ $m_{BA}=+6·22$	Joint *B* $M=-23·20+2·38$ $=-21·85+11·54$ $=-31·13$ $m_{BC}=+3·11$ $m_{BE}=+6·22$ $m_{BA}=+6·22$	Joint *B* $M=-23·49-22·11$ $+2·5+12·04$ $=-31·06$ $m_{BC}=+3·11$ $m_{BF}=+6·20$ $m_{BA}=+6·20$
Joint *C* $M=-1+2·62=-9·28$ $m_{CD}=+3·12$ $m_{CB}=+1·56$	Joint *C* $M=-18·8+3·05+2·96$ $=-12·76$ $m_{CD}=+4·26$ $m_{CB}=+2·13$	Joint *C* $M=-21·1+3·11+3·71$ $=-14·28$ $m_{CD}=+4·76$ $m_{CB}=+2·38$	Joint *C* $M-21·85+3·11+3·75$ $=-14\ 99$ $m_{CD}=+5\ 0$ $m_{CB}=+2·5$	Joint *C* $M=-22·11+3·11+3·62$ $=-15·31$ $m_{CD}=+5·1$ $m_{CD}=+2·55$
Joint *D* $M=-12+3·12$ $=-8·88$ $m_{DC}=+2·96$ $m_{DE}=+1·48$	Joint *D* $M=-18·8+4·26+3·41$ $=-11·13$ $m_{DC}=+3\ 71$ $m_{CE}=+1·86$	Joint *D* $M=-21·1+4·76+5·10$ $=-11·24$ $m_{DC}=+3\ 75$ $m_{DE}=+1·87$	Joint *D* $M=-21·85+6·0+5·77$ $=-11·08$ $m_{CD}=+3·69$ $m_{DE}=+1·87$	Joint *D* $M=-22·11+5·10$ $+6·02$ $=-10·99$ $m_{DC}=+3·66$ $m_{DE}=+1·83$
Joint *E* $M=-22-28·8+5·24$ $+1·48$ $=-34·8$ $m_{ED}=+3·41$ $m_{EB}=+6·82$ $m_{EF}=+6·82$	Joint *E* $M=-18·8-40·12$ $+6·10+1·86$ $=-50·96$ $m_{ED}=+5·1$ $m_{EB}=+10·2$ $m_{EE}=+10·2$	Joint *E* $M=-44·7-21·1$ $+1\ 85+6·22$ $=-57·71$ $m_{ED}=+5·77$ $m_{EB}=+11·54$ $m_{EF}=+11·54$	Joint *E* $M=-46·40-21·85$ $+1\ 85+6·22$ $=-60·18$ $m_{ED}=+6·02$ $m_{EB}=+12·04$ $m_{EF}=+12·04$	Joint *E* $M=-46·98-22·11$ $+1·85-6·20$ $=-61·04$ $m_{ED}=+6·10$ $m_{EB}=+12·20$ $m_{EF}=+12·20$

Table 18·17 (B)
Sway Correction

	Before first cycle	After first cycle	After second cyele	After third cycle	After fourth cycle	After fifth cycle
Sway correction for upper storey	M=16 m'_{BE}=0·75=16 =−12 m'_{DE}=−12	M=16+1·56 +1·41 +2·62 +3·41 =25·07 $m'_{CC}=m'_{DC}$ =−0·75 ×25 07 =−18 8	M=16+2 13 +3·05 +1·86+5·10 =28·14 $m'_{CD}=m'_{CD}$ =−0·75 ×28·14 =−21·1	M=16+3·11 +2·38+5·77 +1 87 =29·13 $m'_{CD}=m'_{CD}$ =−0·75 ×29·18 =−21·85	M=16+2·50 +1·85 +3·11 +6·02 =+29·48 $m'_{CD}=m'_{CD}$ =−0·75 ×29·48 =−22·11	M=16+2·55 +1·83 +3·11 +6·10 =29·59 $m'_{CD}=m'_{DC}$ =−0·75 ×29·59 =−22·19
Sway correction for lower storey	M=48 m'_{AB}=−0·3×48 =−14·4 m'_{FE}=−0·6×48 =−28·8	M=48+5·24 +26·82 =66·88 m'_{AB}=−0·3 ×66·68 =−20·06 m'_{EF}=−0·6 ×66·88 =−40·12	M=148+6·10 +2×10·20 =74·2 m'_{AD}=−0·3 ×74·5 =−23·35 m'_{EF}=−0·6 =74·5 =−44 70	M=48+6·22 + 11·54×2 =77·3 m'_{AB}=−0·3 ×77·3 =−23·2 m'_{EF}=−0·6 ×77·3 =−46·4	M=48+6·22 +2×12·04 =78·3 m'_{AB}=−0·3 −78·3 =−23·49 m'_{EF}=−0·6 −78·3 =−46·98	M=48+6·20 +2×12·20 =+78·60 m'_{AB}=−0 3 +78 60 =−23·58 m'_{EF}=−0·6 +78·60 =−47·16

contributions on the first cycle. Next second cycle is started. This process is continued till the variation in rotation contributions and displacement contribution is negligible. Entire process is shown in Table 18·17 *A* and Fig. 18·17 *B*.

Final moments are,

For horizontal members, $M_{AB}=2m_{AB}+m_{BA}+\overline{M_{AB}}$

$M_{CD}=8{\cdot}76+5{\cdot}10=13{\cdot}86$ kN m

$M_{DC}=8{\cdot}76+3{\cdot}66=12{\cdot}42$ kN m

$M_{BE}=18{\cdot}40+6{\cdot}20=+24{\cdot}60$ kN m

$M_{EB}=18{\cdot}40+12{\cdot}20=30{\cdot}60$ kN m

For vertical members $M_{AB}=2m_{AB}+m_{BA}+m'_{AB}$

$M_{AB}=+6{\cdot}20-23{\cdot}58 \quad =-17{\cdot}38$ kN m

$M_{BA}=6{\cdot}20+6{\cdot}20-23{\cdot}58 \quad =-11{\cdot}18$ kN m

$M_{BC}=5{\cdot}66+3{\cdot}11-22{\cdot}19 \quad =-13{\cdot}42$ kN m

$M_{CB}=5{\cdot}66+2{\cdot}55-22{\cdot}19 \quad =-13{\cdot}98$ kN m

$M_{FE}=+12{\cdot}20+47{\cdot}16 \quad =-34{\cdot}96$ kNm

$M_{EF}=12{\cdot}20+12{\cdot}20-47{\cdot}16=-22{\cdot}76$ kN m

$M_{ED}=7{\cdot}93+6{\cdot}10-22{\cdot}19 \quad =-8{\cdot}16$ kN m

$M_{DF}=7{\cdot}93+1{\cdot}83-22{\cdot}19 \quad =-12{\cdot}43$ kN m

The moments are slightly unbalanced at joints. Better results can be obtained by carrying out one more cycle.

Ex. 18·8. ***Analyse the frame shown in Fig. 18·30 for vertical loads. Spacing of frames is 4 m. The loads are assumed as in Ex. 18·3.***

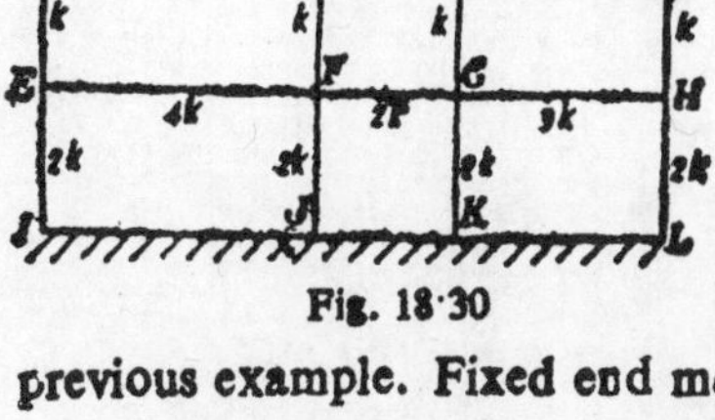

Fig. 18·30

Sol. The stiffness for various members are assumed as shown in Fig. 18·30. The fixed end moments are found for total load on span. Self-weight of beams is assumed as in Ex. 18·3. As the total load on span is same as in Ex' 18·3, fixed end moments in this case will be same as in previous example. Fixed end moments are shown in Table 18·18.

Rotation factors at joint A.

$$\text{Rotation factor} = -\frac{1}{2}\frac{k}{\Sigma k}$$

$$r_{AB}=-\frac{1}{2}\frac{4k}{(k+4k)}=-0{\cdot}4$$

$$r_{AE}=-\frac{1}{2}\frac{k}{(k+4k)}=-0{\cdot}1$$

Rotation factors at other joints are calculated in similar manner.

Displacement factors for each storey.

$$\text{Displacement factor} = \frac{-3}{2}\ \frac{k}{\Sigma k} = -\frac{3}{2}\ \frac{k}{4k}$$
$$= -\frac{3}{8}.$$

Table 18·18

Member	F.E.M.Nm
AB and *EF*	208,000
BC and *FG*	27,750
CD and *GH*	79,170

Cycles of distribution of moments are shown in Fig. 18·31. The sway correction is applied after second cycle upwards as this gives results which converage easily. Exact calculations for moments at each joint are shown in Fig. 18·19.

In Fig. 18·31 fixed end moments are written above the beams and rotation contributions below the beams. Rotation contributions for columns are written towards left sides. Displacement contributions for each column are written horizontally at the centre of column.

Table 18·19 (A)

First cycle	*Second cycle*	*Third cycle*	*Fourth cycle*
Joint A	**Joint A**	**Joint A**	**Joint A**
M=20,800	M=−20,900 −7528 ÷1337 =−20,991	M=−20,800 −8316 +1634 −216 =−27,698	M=−20,800 −8490 +1682 −2750 =−27,883
m_{AB}=8320 m_{AF}=2080	m_{AB}=+10,796 m_{AF}=+2699	m_{AB}=10,796 m_{AF}=2770	m_{AB}=+11.153 m_{AF}=+2788
Joint B	**Joint B**	**Joint B**	**Joint B**
M=+18,025 +8320 +26,345	M=+18,025 +10,793 +1484 −1194 =+29,111	M=+18,025 +11,080 +2114 −1288 −216 =+29,715	M=+18,025 +11,153 +2162 −1279 −275 =+29,768
m_{BA}=−7528 m_{AC}=−3764 m_{AF}=−1882	m_{BA}=−8316 m_{BC}=−4158 m_{BF}=−2079	m_{BA}=−8490 m_{BC}=−4245 m_{BF}=−2123	m_{BA}=−8505 m_{BC}=−4253 m_{BF}=−2126
Joint C	**Joint C**	**Joint C**	**Joint C**
M=−5142 −3764 =−8906	M=−5142 −4158 −3840 +424 =−12,680	M=−5142 −4245 −3906 +540 −216 =−12,962	M=−5142 −4253 −3846 +207 −225 =−12,949
m_{CB}=+1484 m_{CD}=+2227 m_{CG}=+742	m_{CB}=+2114 m_{CD}=+3161 m_{CG}=+1057	m_{CB}=+2162 m_{CD}=+3243 m_{CG}=+1081	m_{CB}=+2158 m_{CD}=+3237 m_{CG}=+1079
Joint D	**Joint D**	**Joint D**	**Joint D**
M=+7917 +2227 =10,144	M=+7917 +3161 +680 =+10,398	M=+7917 +3243 −686 −216 =+10,258	M=+7917 +3237 −664 −275 =+10,215
m_{DG}=−3804 m_{DH}=−1268	M_{DC}=−3906 m_{DH}=−1302	M_{DC}=−3846 m_{DH}=−1282	M_{DC}=−3831 m_{DH}=−1277
Joint E	**Joint E**	**Joint E**	**Joint E**
M=−20,800 +2080 =−18,720	M=−20,800 −4776 +2699 =−22,877	M=−20,800 −5152 +2770 −216 −150 =−23,548	M=−20,800 −5188 +2788 −265 −216 =−23,681

(Contd.)

First cycle	*Second cycle*	*Third cycle*	*Fourth cycle*
$m_{EA}=+13,370$ $m_{EF}=+53,480$ $m_{EI}=+26,470$	$m_{EA}=+16,340$ $m_{EF}=+65,340$ $m_{EI}=+32,680$	$m_{EA}=+16,820$ $m_{EF}=+67,280$ $m_{EI}=+33,640$	$m_{EA}=+16,920$ $m_{BF}=+67,690$ $m_{EI}=+33,840$
Joint F	**Joint F**	**Joint F**	**Joint F**
$M=+180,250$ $+53,480$ $-18,820$ $=+214,910$	$M=+180,250$ $+65,360$ $+8480$ $-20,790$ $=+23,300$	$m=+181,250$ $+6728+10,800$ $-2123+2160$ -1500 $=+23,440$	$M=+180,250$ $+67,690+11,350$ $+21,260-2750$ -2750 $=+233,120$
$m_{FB}=-11,940$ $m_{EF}=-47,760$ $m_{FG}=-23,880$ $m_{FJ}=-23,880$	$m_{FB}=-12,880$ $m_{FE}=-51,520$ $m_{FG}=-25,760$ $m_{FJ}=-25,760$	$m_{FC}=-12,970$ $m_{FE}=-51,880$ $m_{FG}=-25,940$ $m_{FJ}=-25,940$	$m_{FB}=-12,950$ $m_{FE}=-51,800$ $m_{FG}=-25,900$ $m_{FJ}=-25,900$
Joint G	**Joint G**	**Joint G**	**Joint G**
$M=-51,420$ $-23,880$ $+7420$ $+67,880$	$M=-51,420$ $-25,760$ $-19,800$ $+10,570$ $=-86,410$	$M=-51,420$ $-25,9$[illegible]0 $-20,580$ $-10,810$ -2[illegible]60 -1500 $=-90,790$	$M=-51,420$ $-25,900$ $-19,930$ $-10,790$ -2750 -2160 $=-91,370$
$m_{GC}=+4240$ $m_{GF}=+8480$ $m_{GH}=+12,720$ $m_{GK}=+8480$	$m_{GC}=+5400$ $m_{GF}=+10,800$ $m_{GH}=+16,200$ $m_{GK}=+10,800$	$m_{GC}=+5670$ $m_{GF}=+11,350$ $m_{GH}=+17,020$ $m_{GK}=+11,350$	$m_{GC}=+5710$ $m_{GF}=+11,420$ $m_{GH}=+17,130$ $m_{GK}=+11,420$
Joint H	**Joint H**	**Joint H**	**Joint H**
$M=+75,170$ $+12,720$ $-12,680$ $=+72,210$	$M=+75,170$ $+16,200$ $+13,$[illegible]20 $=+82,350$	$M=+79,170$ $+17,020$ $-12,820$ -2160 -1500 $=+79,710$	$M=+79,170$ $+17,130$ $-12,770$ -2750 -2160 $=+7$[illegible]20
$m_{HD}=-6600$ $m_{HG}=-19,800$ $m_{HL}=-13,200$	$m_{HD}=-6860$ $m_{HG}=-20,580$ $m_{HL}=-13,720$	$m_{HD}=-6640$ $m_{HG}=-19,930$ $m_{HL}=-13,280$	$m_{HD}=-6550$ $m_{HG}=-19,660$ $m_{HL}=-13,100$

Sway correction
Table 18·19 (b)

	After second cycle	*After third cycle*	*After Fourth cycle*
Sway correction for upper storey	$M=+26{,}990+16{,}340+10{,}570+5400-20{,}790-12{,}880-13{,}020-6860=+5750$ $m'_{AF}=m'_{AF}$ $m'_{CG}=m'_{DH}$ $=-\frac{3}{8}\times 5750$ $=-2160$	$M=27{,}770+16{,}820+10{,}810+5670-21{,}230-12{,}970-12{,}770-6640=7340$ $m''_{AE}=m'_{BF}$ $=m'_{CG}=m'_{DH}$ $=-\frac{3}{8}\times 7340$ $=-2750$	$M=27{,}880+16{,}020+10{,}790+5710-21{,}260-12{,}950=12{,}770-6550=+7770$ $m'_{AE}=m'_{BF}$ $m'_{CG}=m'_{DH}$ $=\frac{-3}{8}\times 7770$ $=-2910$
Sway correction for lower storey (Moment at fixed end will be half at the upper end of lower storey)	$=(+32{,}680+10{,}800-25{,}700-13{,}720)$ $=4000$ $m'_{EJ}=m'_{FJ}$ $=m'_{GK}=m'_{HL}$ $=-\frac{3}{8}\times 4000$ $=1500$	$M=(+33{,}750+11{,}440-25{,}880-13{,}180)$ $=5770$ $m'_{EJ}=m'_{FJ}$ $=m'_{GK}=m''_{HL}$ $=-\frac{3}{8}\times 5770$ $=2160$	$M=(33{,}840+11{,}420-25{,}900-13{,}100)$ $=6260$ $m'_{EI}=m'_{FJ}$ $m'_{GK}=m'_{HL}$ $=-\frac{3}{8}\times 6260$ $=-2348$

Final Moments are :

Moments in horizontal members

$M_{AB}=-208{,}000+111{,}530+26{,}490=-69{,}980$

$M_{BA}=+208{,}000-85{,}040+26{,}490=+149{,}450$ Nm

$M_{BC}=-27{,}750-42{,}520-20{,}940=-91{,}210$ Nm

$M_{CB}=+27{,}750+21{,}580-20{,}940=+28{,}390$ Nm

$M_{CD}=-79{,}170+32{,}370-5940=-52{,}740$ Nm

$M_{DC}=+79{,}170-38{,}310-5940=+34{,}920$ Nm

$M_{EF}=-208{,}000+67{,}690+15{,}890=-124{,}420$ Nm

$M_{FE}=+208{,}000-51{,}800+15{,}890=+172{,}090$ Nm

$M_{FG}=+27{,}750-25{,}900-14{,}480=-68{,}130$ Nm

$M_{GF}=+27{,}750+11{,}420-14{,}480=+24{,}960$ Nm

$M_{GH}=-79{,}170+17{,}130-2530=-64{,}570$ Nm

$M_{HG}=+79{,}170-19{,}660-2530=+56{,}980$ Nm

Moments in vertical members

$M_{AE}=+27{,}880+44{,}800-2910=+69{,}770$ Nm

$M_{EA}=+26{,}920+44{,}800-2910=+51{,}810$ Nm

$M_{BF}=-21{,}260-34{,}210-2910=-58{,}380$ Nm

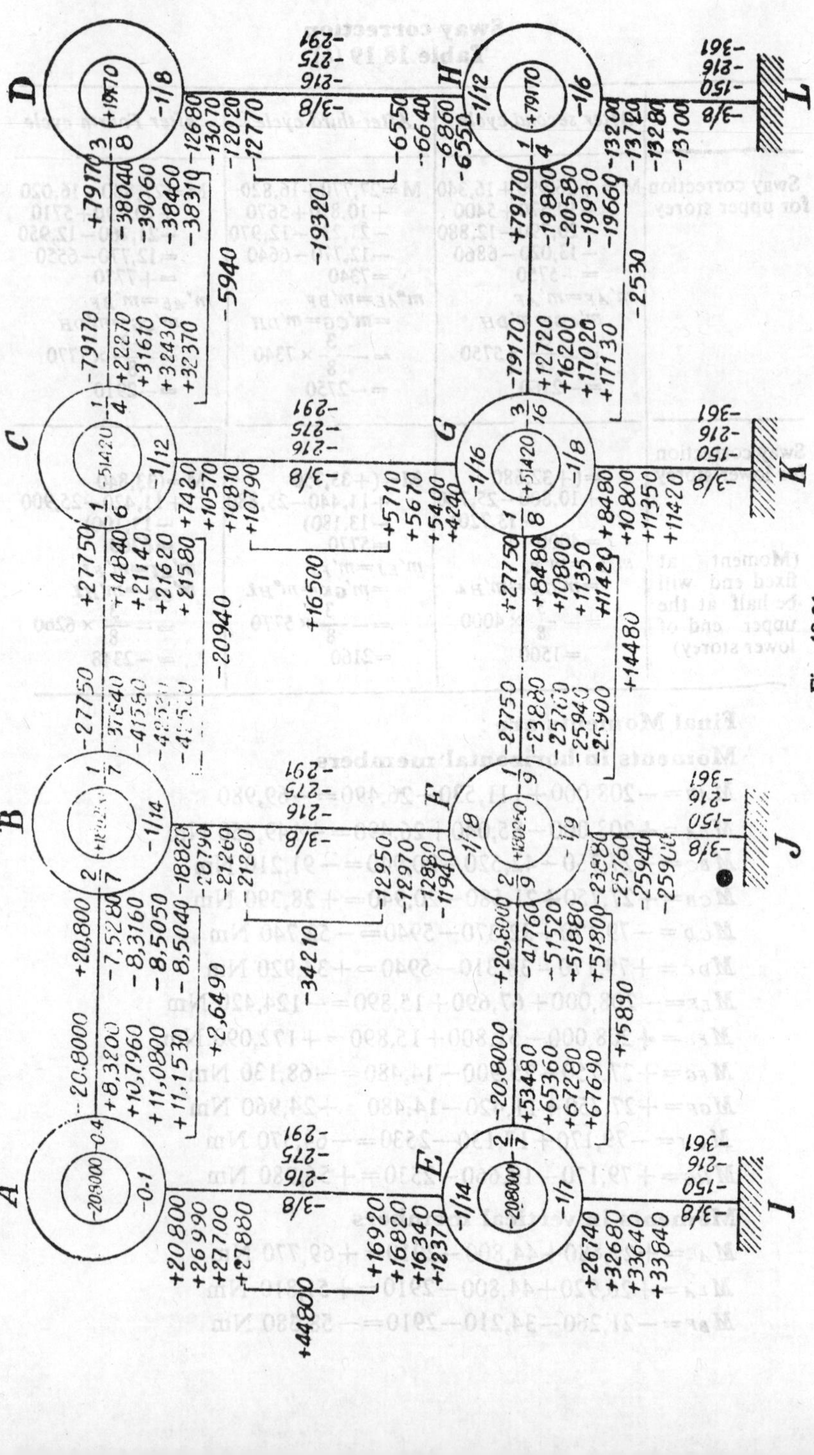

Fig. 18·31

$M_{FB}=-12{,}910-34{,}210-2910=-50{,}030$ Nm

$M_{CG}=+10{,}790+15{,}600-2910=+24{,}380$ Nm

$M_{GC}=+5710+16{,}500-2910=+19{,}300$ Nm

$M_{DH}=-12{,}770-19{,}320-2910=-35{,}000$ Nm

$M_{HD}=-6550-19{,}320-2910=-28{,}780$ Nm

$M_{EI}=+2\times33{,}840-2348=+65{,}332$ Nm

$M_{IE}=+33{,}480-2348=+31{,}492$ Nm

$M_{FJ}=-2\times25{,}900-2398=-54{,}198$ Nm

$M_{JF}=-25{,}900-2348=-28{,}248$ Nm

$M_{GK}=+2\times11{,}420-2348=+20{,}492$ Nm

$M_{KG}=+11{,}420-2348=+9072$ Nm

$M_{HL}=-2\times13{,}100-2348=-28{,}548$ Nm

$M_{LH}=-13{,}103-2348=-15{,}448$ Nm

PROBLEMS

1. In a multi-storeyed building the frames (shown in Fig. 18·32) are spaced at 4 m intervals. Dead load from slab is 3000 N/m^2 and live load is 5000 N/m^2. Design the frame by two-cycle method.

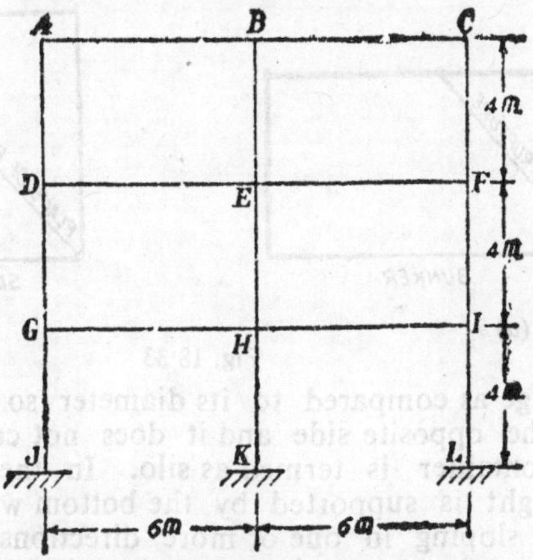

Fig. 18·32

2. In the frame shown in Fig. 18·32, wind load transferred to joints *AD* and *G* are 12,000 N and 14,000 N and 24,000 N respectively. Analyse the frame by (*i*) Portal method (*ii*) Cantilever method.

3. Design the frame of Fig 18·32 by Kani's method for dead loads of Problem 1 and wind loads of Problem 2.

19

Bunkers and Silos

19·1. Containers used for the storage of grains, cement, coal etc. are known as bunkers and silos. The essential difference between silos and bunkers lies in the ratio of their dimensions, *i.e.* ratio of height to diameter which governs the design of these structures. A shallow container whose diameter is large as compared to height is termed as bunker. In such a structure the plane of rupture between the wedge which causes maximum pressure and the remaining fill cuts the top horizontal surface and it does not cut the opposite side of the bunker. On the other hand, if the height of the container is large as compared to its diameter so that the plane of rupture cuts the opposite side and it does not cut top horizontal surface, the container is termed as silo. In the case of bunker the vertical weight is supported by the bottom which is generally hopper shaped sloping in one or more directions. In the case of silos as the plane of rupture does not pass through top horizontal surface only a small portion of vertical weight of stored material is supported by the hopper bottom and large amount of vertical weight is supported by the side walls and the stored material.

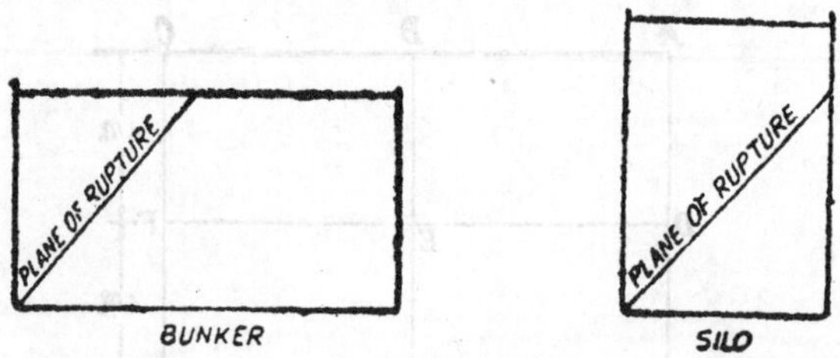

(*a*) (*b*)

Fig. 18·33

The exact analysis of these structures is not possible as there are several factors which affect the pressure exerted. The factors like impact due to filling, emptying process, jamming and location of discharging hole have considerable effect on design and hence exact analysis is rather difficult. However various empirical theories have been put forward for the design and these yield results which are in good agreement with the experimental results.

19·2. Design of Bunkers

A bunker may be of rectangular form or circular form. Rectangular bunkers are provided with hopper shaped bottoms sloping in one or more directions. Circular bunkers are generally provided with conical bottoms.

Whether a circular bunker is economical or a square bunker is economical depends on the relative cost of steel to concrete. For same capacity circular bunker requires less concrete and formwork. However the saving due to reduction in area of formwork is offset by the increased cost of circular formwork. A rectangular bunker requires large quantity of concrete but less steel as compared with circular bunker. Thus relative cost of steel and concrete will govern the choice of type of bunker.

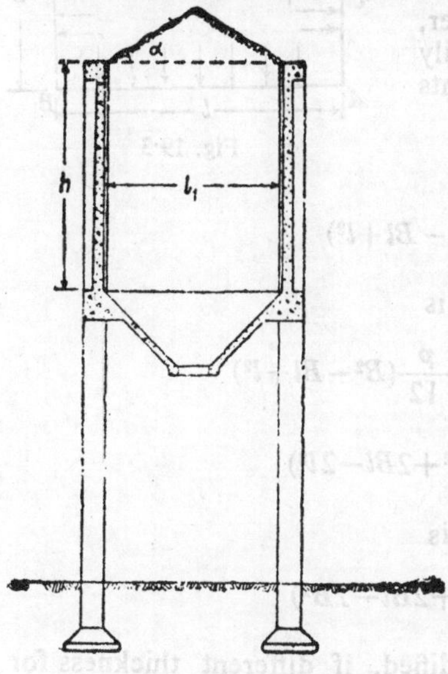

Fig. 19·2

Rectangular Bunker. Generally opening of at least 500 mm × 500 mm is provided at hopper bottom for discharging. Sufficient slope is given to the hopper so that material may be discharged effectively.

Design of vertical walls. The vertical walls are subjected to the pressure of retained material. The pressure against the vertical wall is generally calculated by Rankine's formula.

The pressure at depth h is given by

$$p = wh \cos\alpha \times \frac{\cos\alpha - \sqrt{\cos^2\alpha - \cos^2\phi}}{\cos\alpha + \sqrt{\cos^2\alpha - \cos^2\phi}}$$

where w = density of filling.

ϕ = angle of internal friction of filling.

α = angle of surcharge.

The pressure acts in direction parallel to the surcharge of retained material. Horizontal component $p_h = p \cos\alpha$.

If $\alpha = \phi$, pressure $p = wh \cos\phi$ and horizontal component $p_h = wh \cos^2\phi$.

If angle of surcharge is zero, *i.e.* top is level surface.

Pressure p is given by

$$p = wh \frac{1 - \sin\phi}{1 + \sin\phi}.$$

If the height of the wall is large as compared to its length the wall is designed as a continuous horizontal slab supported on columns. If the bunker is a square bunker maximum B.M. may be taken as $\frac{wl^2}{12}$ for design purposes.

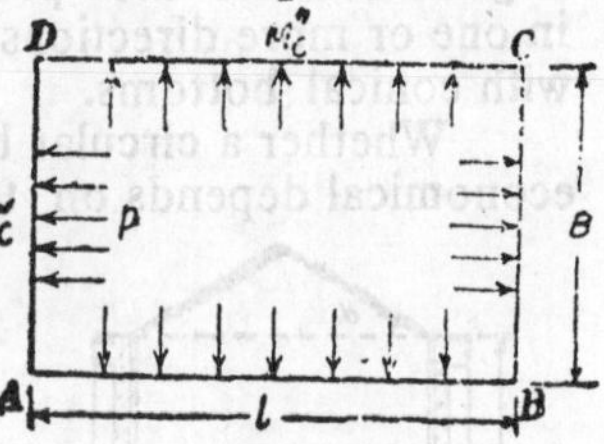

Fig. 19·3

Let l be the length of bunker, and B be its breadth. It can be easily shown by theorem of three moments that

$$M_A = M_B = M_C = M_D$$

$$= \frac{-p}{12}(B^2 - Bl + l^2)$$

+ve B.M. at centre of span B is

$$M'_C = \frac{pB^2}{8} - \frac{p}{12}(B^2 - Bl + l^2)$$

$$= \frac{p}{24}(B^2 + 2Bl - 2l^2)$$

+ve B.M. at centre of span l is

$$M'_c = \frac{p}{24}(l^2 + 2Bl - 2B^2)$$

These formulas can be modified, if different thickness for walls are used, by using different moments of inertias.

If the height h of the side walls is small as compared with the length of the wall, the side wall will span vertically. In such a case maximum −ve B.M. may be taken as $ph^2/15$ and maximum +ve B.M. may be taken as $ph^2/33{\cdot}5$. This occurs at height of 0·477 h above the bottom of the wall.

The walls are also to be designed for direct tension due to pressure of retained material. Direct tension on wall l will be $pB/2$ and that on B will be $pl/2$.

The thickness of wall is designed for the above B.M. At top minimum thickness of 300 mm is provided to allow for attachments of conveyor supports etc. The thickness is provided for depth of about 250 mm.

In addition to acting as horizontal slab, vertical walls are to be designed as beams supported between columns, the weight of the materials contained is transferred to vertical walls by hopper bottoms. The amount of vertical weight shared by each wall will depend on the ratio of length to breadth of bunker. In case of square bunkers the vertical load shared by each wall will be one-fourth. When vertical walls are designed as beams the loads coming on the beam will consist of weight of retained material, weight of hopper bottom and self-weight of the wall. Maximum B.M. in the beam is taken as $wl^2/8$. Very small quantity of steel is required as the height of the wall (depth of beam) is quite considerable. Actually the vertical wall acts as a deep beam and actual distribution of stresses in such wall is very complicated.

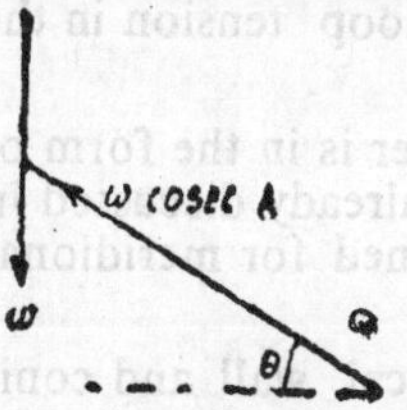

Fig. 16·4 (a)

Hopper bottom. At the bottom of side wall there will be vertical pull in downward direction due to the weight of the retained material, weight of the sloping bottom, lining and opening. This will give rise to direct tension in the sloping bottom. If w is the weight of above mentioned portions per unit length, maximum direct tension in hopper bottom per unit length will be w cosec θ, where θ is the angle made by hopper bottom with horizontal. The direct tension in the hopper bottom will gradually decrease to zero at opening.

In addition to designing hopper bottom for direct tension it is to be designed as slab spanning horizontally between the intersections of adjacent sloping faces. In such a case horizontal span of slab will increase from minimum at opening to maximum at junction with vertical wall. The pressure due to weight of retained material will decrease from maximum at opening to minimum at junction with vertical wall as the span of slab increases. The section of the slab should be designed at centre. The load coming on the slab will consist of (a) the normal component p_n due to the weight of retained material and (b) normal component of the weight of the slab and lining.

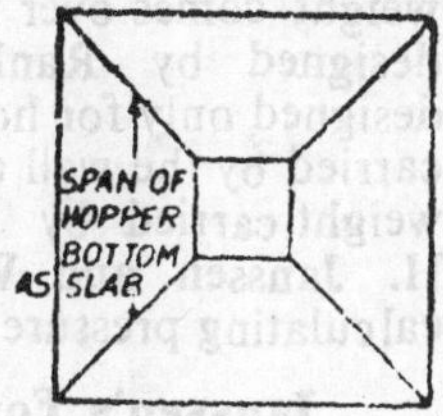

The normal pressure at depth h from retained material will be

$$p_n = wh \cos^2 \theta + p_h \sin^2 \theta$$

where p_h is horizontal pressuse at depth h

and θ = angle between the sloping side of the hopper and the horizontal.

Fig. 19·4 (b)

Normal pressure due to self-weight of slab and lining will be $w_d \times \cos \theta$, where w_d is the weight of slab and lining per unit length.

Design of columns. The columns are designed for vertical load coming on them. The section of columns is checked for effect of wind stresses. Allowable stresses can be increased by $33\frac{1}{3}\%$ when the effect of wind is taken into consideration.

Design of Circular Bunkers. The vertical walls of circular bunkers are desigend for the hoop tension caused due to pressure of retained material. If p_h is the horizontal component of pressure at depth h hoop tension at depth h will be $p_h d/2$, where d is the diameter of the bunker. The thickness of the wall of bunker is designed for the hoop tension. Minimum thickness to be provided is 120 mm. In addition to reinforcement for hoop tension in the form of hoops, vertical reinforcement is provided.

The sloping bottom of the circular bunker is in the form of conical dome. The design of conical dome is already discussed in chapter 13 on Domes. The conical dome is designed for meridional stresses and hoop stresses caused in it.

The ring beam provided at junction of vertical wall and conical dome should be designed for hoop tension.

19.3. Silos

As already mentioned, if in a container plane of rupture cuts the opposite side, the container is termed as silo. In such containers large portion of the weight of contained materials is supported by the friction between material and wall and only a small portion of weight comes over the hopper bottom. Hence the silos cannot be designed by Rankine's theory as in that case vertical wall be designed only for horizontal pressure and the effect of vertical load carried by the wall cannot be taken into consideration. The vertical weight carried by the wall causes direct compression in the wall. H. Janssen and W. Airy have put forward their theories for calculating pressure in such cases.

Janssen's Formula. The derivation of Janssen's formula is given below. (Refer Reinforced Concrete, Water towers, Bunkers, Silos and Gantries by W.S. Gray).

The following terms are used in the formula ;

w=density of the contained material

h=depth below the top of the bin where pressures are desired

A=Area of the bin, taking horizontal section

U=Perimeter of the horizontal section

$R=A/U$ the hydraulic mean depth of the section

ϕ=angle of repose of the filling

ϕ'=angle of friction of the filling on the walls of the bin

$\mu'=\tan \phi'$

p_h=intensity of horizontal pressure on the walls

p_v=intensity of vertical pressure in the walls

$$k=\frac{p_h}{p_v}=\frac{1-\sin \phi}{1+\sin \phi}$$

Consider elementary volume of thickness dh at depth h. Let p_v be the pressure above the slice and p_v+dp_v be the pressure below the slice. Let p_h be the horizontal pressure at this level and f be the frictional stress. Equating the vertical forces.

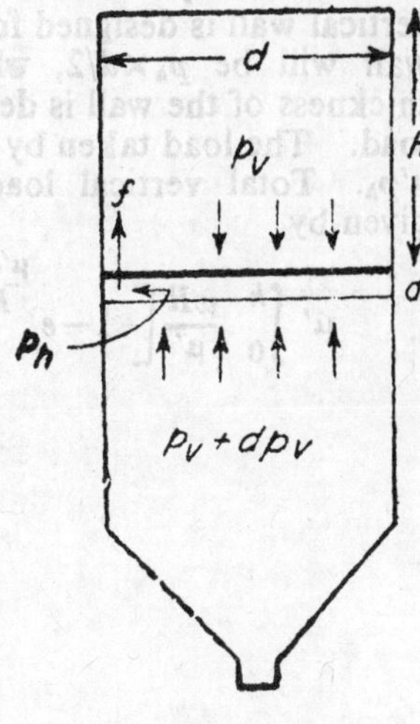

Fig. 19·5

$$p_v \times A \times wAdh$$

$$=(p_v+dp_v)A=fUdh.$$

Now $\quad f=\mu' p_h$

$\therefore \quad wAdh=Adp_v+\mu' p_h U dh$

Dividing by A,

$$wdh=dh_v+\frac{\mu' ph}{R}dh.$$

$$\therefore \quad dp_v=\left(w-\frac{\mu' p_h}{R}\right)dh.$$

Now $p_h=kp_v$.

$$dp_v=\left(w-\frac{\mu' kp_v}{R}\right)dh.$$

$$\therefore \quad \frac{dh_v}{\left(w-\dfrac{\mu' kp_v}{R}\right)}=dh.$$

$$\therefore \quad \log_e \frac{\left(w-\dfrac{\mu' kp_v}{R}\right)}{-\dfrac{\mu' k}{R}}=h+C \qquad \text{where } C=\text{constant}$$

$$\therefore \quad \log_e\left(w-\frac{\mu' kp_v}{R}\right)=\frac{-\mu' kh}{R}+C' \qquad \text{where } C'=\text{constant}$$

At $\qquad h=0, p_v=0.$

$$\therefore \quad \log w=C'$$

$$\log_e \frac{w-\dfrac{\mu' kp_v}{R}}{w}=\frac{-\mu' kh}{R}$$

$$\therefore \quad \frac{w-\dfrac{\mu' kp_v}{R}}{w}=e^{-\mu' kh/R}$$

$$\therefore \quad 1-\frac{\mu' kp_v}{wR}=e^{-\mu' kh/R}$$

$$\therefore \quad p_v=\frac{wR}{\mu' k}\left[1-e^{-\mu' kh/R}\right]$$

$$\therefore \quad p_h=\frac{wR}{\mu'}\left[1-e^{-\mu' kh/R}\right]$$

Generally the silos are circular with conical bottom. The vertical wall is designed for hoop tension. Hoop tension in vertical wall will be $p_h \times d/2$, where d is the diameter of the silo. The thickness of the wall is designed for hoop tension as well as vertical load. The load taken by the wall is due to friction and is equal to $\mu' p_h$. Total vertical load taken by the wall for any depth h is given by

$$\mu' \int_0^h \frac{wR}{\mu'}\left[1-e^{\frac{\mu' kh}{R}}\right] U\, dh$$

$$= w \times A \int_0^h \left(1-e^{\frac{-\mu' kh}{R}}\right) dh$$

$$= wA\left[h+\frac{R}{\mu' k}\, e^{\frac{-\mu' kh}{R}}\right]_0^h$$

$$= wA\left[h \times \frac{R}{\mu' k}\, e^{-\mu' kh/R} - \frac{R}{\mu' k}\right]$$

$$= wA\left[h-\frac{R}{\mu' k}\left(1-e^{\frac{-\mu' kh}{R}}\right)\right]$$

$$= wAh - Ap_v = A(wh - p_v)$$

The bottom conical dome is designed in the same manner as for bunker. The surcharge over the conical portion is taken as p_v and Rankine's formula applied for horizontal pressure.

W. Airy's Formula. By Airy's formula it is possible to calculate horizontal pressure per unit length of periphery and position of plane of rupture. Once the horizontal pressure is known vertical pressure on vertical wall can be calculated and hence vertical load taken by vertical wall can be found.

Two cases arise in practice one is when the plane of rupture cuts the top horizontal surface and it does not cut opposite side

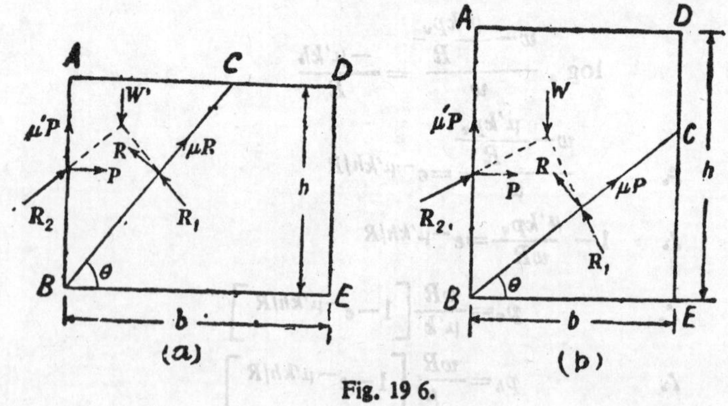

Fig. 19 6.

and second case is when plane of ruprure cuts opposite side (*see* Fig. 19·6).

The position of plane of rupture is defined by angle θ.

Case 1. *Where plane of rupture cuts the top horioznlal surface and it does not cut opposite side.*

Taking the wedge ABC of unit thickness, let W be the length of slice. Let P be the reaction from the wall, R the reaction from the material on BC. Let $\mu' P$ and μR be frictional forces from wall and material as shown.

Now $$AC = h \cot\theta$$

$$W = \tfrac{1}{2} \times h \cot\theta \times h \times w = \frac{wh^2}{2} \cot\theta$$

Where w=density of material stored and θ is the inclination of plane of rupture.

Resolving the forces acting on the wedge, in vertical and horizontal directions

$$W = \mu' P + \mu R \sin\theta + R\cos\theta$$

$$\therefore \quad R = \frac{W - \mu' P}{(\mu \sin\theta + \cos\theta)} \qquad \ldots(i)$$

$$P + \mu R \cos\theta = R \sin\theta$$

$$\therefore \quad R = \frac{P}{(\sin\theta - \mu\cos\theta)} \qquad \ldots(ii)$$

$$\therefore \quad \frac{P}{\sin\theta - \mu\cos\theta} = \frac{W - \mu' P}{\mu\sin\theta + \cos\theta}$$

$$\therefore \quad P(\mu\sin\theta + \cos\theta) + P\mu'(\sin\theta - \mu\cos\theta) = W(\sin\theta - \mu\cos\theta)$$

$$P[\sin\theta(\mu + \mu') + \cos\theta(1 - \mu\mu')] = W(\sin\theta - \mu\cos\theta)$$

$$\therefore \quad P = \frac{W(\sin\theta - \mu\cos\theta)}{\sin\theta(\mu - \mu') + \cos\theta(1 - \mu\mu')}$$

Dividing each term by $\cos\theta$,

$$P = \frac{W[\tan\theta - \mu]}{\tan\theta(\mu + \mu') + (1 - \mu\mu')} = \frac{wh^2}{2} \times \frac{\cot\theta[\tan\theta - \mu]}{\tan\theta(\mu + \mu') + (1 - \mu\mu')}$$

$$= \frac{wh^2}{2} \times \frac{(\tan\theta - \mu)}{(\mu - \mu')\tan^2\theta + (1 - \mu\mu')\tan\theta}$$

For max. value of P, $\dfrac{dP}{d\theta} = 0$

Let $$P = \frac{wh^2}{2}\left(\frac{u}{v}\right)$$

where u and v are functions of θ.

$$\therefore \quad \frac{dP}{dv}=\frac{wh^2}{2}\left[\frac{udv-vdu}{v^2}\right]=0$$

$$\therefore \quad udv=vdu$$

$$\therefore \quad \frac{u}{v}=\frac{du}{dv}$$

$$\therefore \quad =\frac{\tan\theta-\mu}{(\mu+\mu')\tan^2\theta+(1-\mu\mu')\tan\theta}$$

$$=\frac{\sec^2\theta}{2(\mu+\mu')\tan\theta\sec^2\theta+(1-\mu\mu')\sec^2\theta}$$

$$=\frac{1}{2(\mu+\mu')\tan\theta+(1-\mu\mu')}$$

$$\therefore \quad (2\mu+\mu')\tan^2\theta-2\mu(\mu+\mu')\tan\theta+(1-\mu\mu')\tan\theta-\mu(1-\mu\mu')$$
$$=(\mu+\mu')\tan^2\theta+(1-\mu\mu')\tan\theta$$

$$\therefore \quad \tan^2\theta(\mu+\mu')-2\mu(\mu+\mu')\tan\theta-\mu(1-\mu\mu')=0.$$

$$\therefore \quad \tan^2\theta-2\mu\tan\theta-\frac{\mu(1-\mu\mu')}{\mu+\mu'}=0$$

$$\therefore \quad \tan\theta=\frac{2\mu+\sqrt{4\mu^2+\dfrac{4\mu(1-\mu\mu)}{\mu+\mu'}}}{2}$$

$$=\mu+\sqrt{\frac{\mu^3+\mu^2\mu'+\mu-\mu^2\mu'}{\mu+\mu'}}$$

$$=\mu+\sqrt{\frac{\mu(1+\mu^2)}{\mu+\mu'}}$$

$$\therefore \quad P=\frac{wh^2}{2}\times\frac{\left[\mu+\sqrt{\dfrac{\mu(1+\mu^2)}{\mu+\mu'}}-\mu\right]}{\tan\theta(\mu+\mu')\tan\theta+(1-\mu\mu')}$$

$$=\frac{wh^2}{2}\times\frac{\sqrt{\dfrac{\mu(1+\mu^2)}{\mu+\mu'}}}{\left[\mu+\sqrt{\dfrac{\mu(1+\mu^2)}{\mu+\mu'}}\right]\left[(\mu+\mu')\left\{\mu+\sqrt{\dfrac{\mu(1+\mu^2)}{\mu+\mu'}}\right\}+(1-\mu\mu')\right]}$$

$$P=\frac{wh^2}{2}\times\frac{\sqrt{\dfrac{\mu(1+\mu^2)}{\mu+\mu'}}}{\left[\mu+\sqrt{\dfrac{\mu(1+\mu^2)}{\mu+\mu}}\right]\left[\mu^2+1+\sqrt{\mu(1+\mu^2)(\mu+\mu')}\right]}$$

$$\therefore \quad P=\frac{wh^2}{2}$$

$$\times\frac{1}{\left[\sqrt{\frac{\mu}{\frac{\mu(1+\mu^2)}{\mu+\mu'}}}+1\right][\mu^2+1+\sqrt{\mu(1+\mu^2)(\mu+\mu')}]}$$

$$=\frac{wh^2}{2}\times\frac{1}{\left[\sqrt{\frac{(\mu+\mu')\mu}{(1+\mu^2)}}+1\right]\left[\mu^2+1+\sqrt{\mu(1+\mu^2)(\mu+\mu')}\right]}$$

$$=\frac{wh^2}{2}\times\frac{1}{\mu^2+1+(\mu^2+1)\sqrt{\frac{(\mu+\mu')\mu}{1+\mu^2}}+\sqrt{\mu(1+\mu^2)(\mu+\mu')}+\mu(\mu+\mu')}$$

$$=\frac{wh^2}{2}\times\frac{1}{2\mu^2+1+\mu\mu'+2\sqrt{\mu(\mu+\mu')(1+\mu^2)}}$$

$$=\frac{wh^2}{2}\times\frac{1}{(\mu^2+\mu\mu')+2\sqrt{\mu(\mu+\mu')(1+\mu^2)}+(1+\mu^2)}$$

$$=\frac{wh^2}{2}\times\frac{1}{[\sqrt{\mu(\mu+\mu')}]^2-2\sqrt{\mu(\mu+\mu')(1+\mu^2)}+[\sqrt{(1+\mu^2)}]^2}$$

$$=\frac{wh^2}{2}\times\frac{1}{[\sqrt{\mu(\mu+\mu')}+\sqrt{(1+\mu^2)}]^2}$$

$$P_h=\frac{dP}{dh}=wh\times\frac{1}{[\sqrt{\mu(\mu+\mu')}+\sqrt{(1+\mu^2)}]^2}$$

Vertical load taken by walls

$$\pi bP\times\mu'=\frac{wh^2\pi b}{2}\times\frac{\mu'}{[\sqrt{\mu(\mu+\mu')}+\sqrt{(1+\mu)^2}]^2}$$

The depth upto which silo will act as shallow will be given by

$$\tan\theta=\frac{h}{b}=\mu+\sqrt{\frac{\mu(1+\mu^2)}{\mu+\mu'}}$$

$$\therefore \quad h=b\left[\mu+\sqrt{\frac{\mu(1+\mu^2)}{\mu+\mu'}}\right].$$

Case II. *Plane of rupture cuts the opposite side*

$$CD=h-a\tan\theta.$$

$$W=wbh-w\frac{b}{2}\times b\tan\theta$$

$$=wb\left[h-\frac{b\tan\theta}{2}\right]$$

As worked out previously,

$$P=\frac{W[\tan\theta-\mu]}{\tan\theta\,(\mu+\mu')+(1-\mu\mu')}$$

$$=wb\times\frac{\left[h-\dfrac{b\tan\theta}{2}\right]\left[\tan\theta-\mu\right]}{\tan\theta\,(\mu+\mu')+(1-\mu\mu')}$$

$$=wb\times\frac{wb\left[\left(h+\dfrac{b\mu}{2}\right)\tan\theta-\dfrac{2}{b}\tan^2\theta-h\mu\right]}{\tan\theta\,(\mu+\mu')+(1-\mu\mu')}$$

Now for max. value of P, $\dfrac{dP}{d\theta}=0$

$$\therefore\quad\frac{\left(h+\dfrac{b\mu}{2}\right)\tan\theta-\dfrac{b}{2}\left(\tan^2\theta\right)-h\mu}{\tan\theta(\mu+\mu')+(1-\mu\mu')}$$

$$=\frac{\left(h+\dfrac{b\mu}{2}\right)\sec^2\theta-b\tan\theta\sec^2\theta}{(\mu+\mu')\sec^2\theta}$$

$$\therefore\quad(\mu+\mu')\left[\left(h+\frac{b\mu}{2}\right)\tan\theta-\frac{b}{2}\tan^2\theta-h\mu\right]$$

$$=\left[\left(h+\frac{b\mu}{2}\right)-b\tan\theta\right]\left[\tan\theta\,(\mu+\mu')+(1-\mu\mu')\right]$$

$$\therefore\quad\left(h+\frac{b\mu}{2}\right)(\mu+\mu')\tan\theta-\frac{b}{2}(\mu+\mu')\tan^2\theta-h\mu(\mu-\mu')$$

$$=\left[\left(h+\frac{b\mu}{2}\right)(\mu+\mu')-b(1-\mu\mu')\right]\tan\theta-b\,(\mu+\mu')\tan^2\theta$$
$$+(1-\mu\mu')\left(h+\frac{\mu b}{2}\right)\Bigg]$$

$$\therefore\quad\frac{b}{2}\,\mu+\mu')\tan^2\theta+b(1-\mu\mu')\tan\theta-h\mu(\mu+\mu')$$
$$-\left(1-\mu\mu'\right)\left(h+\frac{\mu b}{2}\right)=0$$

$$\therefore\quad\tan^2\theta+\frac{2(1-\mu\mu')}{\mu+\mu'}\tan\theta-\frac{2h\mu}{b}-\frac{2(1-\mu\mu')}{b(\mu+\mu')}\left(h+\frac{b\mu}{2}\right)=0$$

$$\therefore\quad\tan^2\theta+\frac{2(1+\mu\mu')}{\mu+\mu'}\tan\theta$$
$$-\frac{2h\mu^2+2h\mu\mu'+2h+b\mu-2h\mu\mu'-b\mu^2\mu'}{b(\mu+\mu')}$$

$$\therefore\quad \tan^2 2\theta+\frac{2(1-\mu\mu')}{\mu+\mu'}\tan\theta-\frac{2h(\mu^2+1)+b\mu(1-\mu\mu')}{b(\mu+\mu')}=0.$$

$$\therefore\quad \tan\theta=\frac{\dfrac{-2(1-\mu\mu')}{\mu+\mu'}+\sqrt{\dfrac{4(1-\mu\mu')^2}{(\mu+\mu')^2}+\dfrac{4[2h(\mu^2+1)+b\mu(1-\mu\mu')]}{b(\mu+\mu')}}}{2}$$

$$=\frac{-(1-\mu\mu')}{\mu+\mu'}+\sqrt{\frac{b(1-\mu\mu')+(\mu+\mu')[2h(\mu^2+1)+b\mu(1+\mu\mu')]}{b(\mu+\mu')^2}}$$

$$=\frac{1}{\mu+\mu'}\Big[-(1-\mu\mu')+\sqrt{(1-\mu\mu')^2+(\mu+\mu')\left\{\frac{2h}{b}(\mu^2+1)+\mu(1-\mu\mu')\right\}}\Big]$$

$$=\frac{1}{\mu+\mu'}\Big[-(1-\mu\mu')+\sqrt{(1-\mu\mu')[1-\mu\mu'+\mu^2+\mu\mu'']+\frac{2h}{b}(\mu^2+1)(\mu+\mu')}\Big]$$

$$=\frac{1}{\mu+\mu'}\Big[(1-\mu\mu')+\sqrt{(1-\mu\mu')(1+\mu^2)+\frac{2h}{b}(1+\mu^2)(\mu+\mu')}\Big]$$

$$\therefore\quad P'=wb\,\frac{\left[\dfrac{2h-b\tan\theta}{2}\right][\tan\theta-\mu]}{(\mu+\mu')\tan\theta+(1-\mu\mu')}$$

$$=\frac{wb}{2}\times\frac{[2h-b\tan\theta][\tan\theta-\mu]}{(\mu+\mu')\tan\theta+(1-\mu\mu')}$$

$$=\frac{wb}{2(\mu+\mu')^2}\left[\sqrt{\frac{2h}{b}(\mu+\mu')+(1-\mu\mu')}-\sqrt{1+\mu^2}\right]^2$$

$$P_h=\frac{dP}{dh}=\frac{wb}{(\mu+\mu')}\left\{1-\frac{\sqrt{(1+\mu^2)}}{\sqrt{\dfrac{h}{b}(\mu+\mu')+(1-\mu\mu')}}\right\}$$

Vertical load taken by wall $=\pi bP\mu'$.

Surcharge for design of conical hopper.

Let P_H be the total horizontal force for H depth of cylindrical portion.

$$\text{Surcharge pressure} = \frac{\frac{\pi b^2}{4} wH - \pi b\, P_H \mu'}{\frac{\pi b^2}{4}}$$

$$= wH - \frac{4P_H \mu'}{b}.$$

Once the intensity of horizontal pressure is known the wall of the container can be designed for hoop tension and vertical load of material supported by it. The vertical load supported by wall will be $\mu' P$. The hopper bottom can be designed in the same way as for bunkers.

The value of μ and μ' for various materials are given in Table 19·1. The values of constants to be used in Janson's and Airs's formulae are given in Tables is 19·2

Table 19·1

Material	*Density*	*Coefficient of fraction*		$k = \frac{1-\sin\phi}{1+\sin\phi}$
		Filling on filling $\mu = \tan\phi$	*Filling on concrete* μ'	
Cement	1440	0·316	0·554	0·5371
Coal	800	0·700	0·700	0·2709
Anthra cite	835	0·510	0·510	0·3753
Coke	450	0·839	0·839	0·2174
Sand	1600	0·674	0·577	0·2830
Whea	805	0·466	0·444	0 4062

Ex. 19·1. *Design a bunker with hopper bottom to store coal, for capacity of 200 kN. Angle of repose of coal is 30°. The stored coal is to be surcharged at an angle of repose. Weight of coal is 800 N/m³.*

Table 19 2
Values of $(1-e^{-x})$ in which $x=\mu P/R$ in JansSen's formula

x	$1-e^{-x}$	x	$1-e^{-x}$	x	$1-e^{-x}$	x	$1-e^{-x}$	x	$1-e^{-x}$	x	$1-e^{-x}$
0·01	0·010	0·31	0·267	0·61	0·457	0·91	0·597	1·42	0·758	2·05	0·871
0·02	0·020	0·32	0·274	0·62	0·462	0·92	0·599	1·44	0·763	2·10	0·878
0·03	0·030	0·33	0·281	0·63	0·467	0·93	0·605	1·46	0·768	2·15	0·883
0·04	0·040	0·34	0·288	0·64	0·473	0·94	0·609	1·48	0·772	2·20	0·889
0·05	0·049	0·35	0·295	0·65	0·478	0·95	0·613	1·50	0.777	2·25	0·895
0·06	0·058	0·36	0·302	0·66	0·483	0·96	0·617	1·52	0·781	2·30	0·900
0·07	0·068	0·37	0·309	0·67	0·488	0·97	0·621	1·54	0·786	2·35	0·905
0·08	0·077	0·38	0·316	0·68	0·493	0·98	0·625	1·56	0·790	2·40	0·909
0·09	0·086	0·39	0·323	0·69	0·498	0·99	0·628	1·58	0·794	2·45	0·914
0·10	0·095	0·40	0·330	0·70	0·503	1·00	0·632	1·60	0·798	2·50	0·918
0·11	0·104	0·41	0·336	0·71	0·508	1·02	0·639	1·62	0·802	2·55	0·922
0·12	0·113	0·42	0·343	0·72	0·512	1·04	0·646	1·64	0·806	2·60	0·926
0·13	0·122	0·43	0·349	0·73	0·528	1·06	0·653	1·66	0·811	2·65	0·929
0·14	0·131	0·44	0·356	0·74	0·523	1·08	0·660	1·68	0·814	2·70	0·933
0·15	0·139	0·45	0·362	0·75	0·528	1·10	0·667	1·70	0·817	2·75	0·936
0·16	0·148	0·46	0·369	0·76	0·532	1·12	0·664	1·72	0·821	2·80	0·939
0·17	0·156	0·47	0·375	0·77	0·537	1·14	0·680	1·74	0·824	2·85	0·941
0·18	0·165	0·48	0·381	0·78	0·542	1·16	0·687	1·76	0·828	2·90	0·945
0·19	0·173	0·49	0·387	0·79	0·546	1·18	0·693	1·78	0·831	2·95	0·948
0·20	0·181	0·50	0·393	0·80	0·551	1·20	0·699	1·80	0·835	3·00	0·950
0·21	0·190	0·51	0·400	0·81	0·555	1·22	0·705	1·82	0·838		
0·22	0·198	0·52	0·405	0·82	0·560	1·24	0·711	1·84	0·841		
0·23	0·205	0·53	0·411	0·83	0·564	1·26	0·716	1·86	0·844		
0·24	0·213	0·54	0·417	0·84	0·568	1·28	0·722	1·88	0·847		
0·25	0·221	0·55	0·423	0·85	0·573	1·30	0·727	1.90	0·850		
0·26	0·229	0·56	0·429	0·86	0·577	1·32	0·733	1·92	0·853		
0·27	0·237	0·57	0·435	0·87	0·581	1·34	0·738	1·94	0·856		
0·28	0·244	0·58	0·440	0·88	0·585	1·36	0·743	1·96	0·859		
0·29	0·252	0·59	0·446	0·89	0·589	1·38	0·748	1·98	0·862		
0·30	0·259	0·60	0·451	0·90	0·593	1·40	0·753	2·00	0·865		

Table 19·3

Aid to Airy's formula

Material	*Cement*	*Coal*	*Anthracite*	*Coke*	*Sand*	*Wheat*
μ	0·316	0·700	0·510	0·839	0·674	0·466
μ'	0 554	0·700	0·510	0·839	0·577	0·444
$\mu+\mu'$	0·870	1·400	1·020	1 678	1·251	0 910
μ^2	0·100	0·490	0·260	0·704	0·454	0·217
$\mu\mu'$	0·175	0·490	0·260	0·704	0·389	0·207
$\sqrt{1+\mu^2}$	1·049	1·221	1·123	1·305	1·206	1·104
$\sqrt{\mu(\mu+\mu')}$	0 525	0 990	0·721	1·187	0·915	0·651
$\frac{1}{[\sqrt{\mu(\mu+\mu')}+\sqrt{1+\mu^2}]^2}$	0·404	0·258	0·294	0·161	0·222	0·325
$\sqrt{\frac{\mu(1+\mu^2)}{\mu+\mu'}}$	0·948	1·563	0·761	1·762	1·559	1·256
$1-\mu\mu'$	0·825	0·510	0·740	0·296	0·611	0·793
$(\mu+\mu')^2$	0·757	1·960	0·040	2·816	1·554	0·838

Sol. Square bunker with hopper bottom is provided. For square portion height of 3·5 m and side of 2·5 m are provided. Hopper portion has height of 1 m and opening of 500 mm × 500 mm.

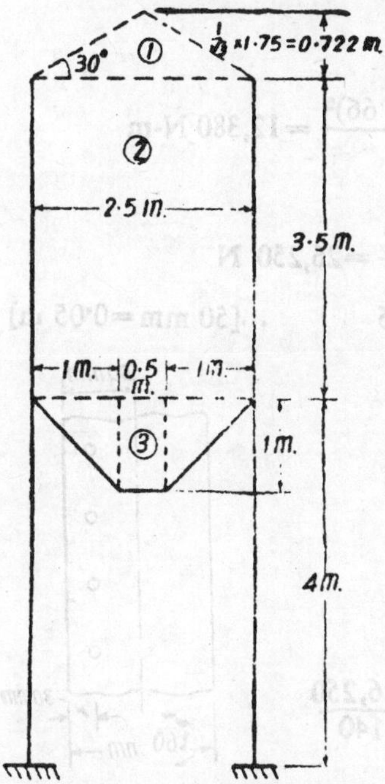

Fig. 19·7.

Volume of coal to be stored

$$=\frac{200{,}000}{8000}$$

$$=25 \text{ m}^3.$$

Capacity provided—

(1) $2{\cdot}5\times2{\cdot}5\times\frac{1}{3}\times0{\cdot}722$

$$=1{\cdot}503 \text{ m}^3.$$

(2) $2{\cdot}5\times2{\cdot}5\times3{\cdot}5$

$$21{\cdot}875 \text{ m}^3.$$

(3) $\frac{1}{3}\left[2{\cdot}5\times2{\cdot}5+0{\cdot}5\times0{\cdot}5+\sqrt{(2{\cdot}5^2\times(0{\cdot}5)^2}\right]$

$$=\frac{1}{3}\,[6{\cdot}25+0{\cdot}25+1{\cdot}25]$$

$$=\frac{1}{3}\times7{\cdot}75$$

$$=2{\cdot}583 \text{ m}^3.$$

Total volume

$$=25{\cdot}961 \text{ m}^3.$$

Design of Side Walls. All the four walls are continuous. These will be designed as continuous slab.

The pressure of coal at depth h will be

$$p=wh\cos\alpha\times\frac{\cos\alpha-\sqrt{\cos^2\alpha-\cos^2\phi}}{\cos\alpha+\sqrt{\cos^2\alpha-\cos^2\phi}}$$

Here $\alpha=\phi$

$\therefore$ $p=wh\cos\phi.$

Horizontal component of this

$$=wh\cos^2\phi$$

At 3·5 m depth horizontal pressure

$$=8000\times3{\cdot}5\times(0{\cdot}866)^2$$

$$=21{,}000 \text{ N/m}^2$$

Concrete mix of M 15 is used.

Assume thickness of wall as 160 mm

Effective span $=2{\cdot}5+0{\cdot}16=2{\cdot}66$ m.

B.M. at corners for square frame will be

$$=\frac{pl^2}{12}$$

$$=\frac{21{,}000\times(2{\cdot}66)^2}{12}=12{,}380 \text{ N-m}$$

Direct tension in the wall

$$=\frac{21{,}000\times2{\cdot}5}{2}=26{,}250 \text{ N}$$

Net B.M. $=12{,}380-26{,}250\times0{\cdot}05$ ...[50 mm=0·05 m]

$=12{,}380-1312{\cdot}5$

$=11067{\cdot}5$ Nm.

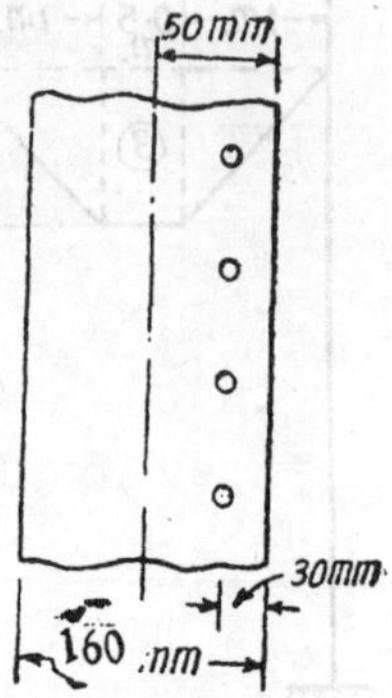

Fig. 19·8

Effective depth

$$=\sqrt{\frac{11067{\cdot}5\times1000}{0{\cdot}87\times1000}}$$

$=112{\cdot}8$ mm

Provide overall depth of 160 mm.

Area of steel required

$$=\frac{11067{\cdot}5\times1000}{0{\cdot}87\times140\times130}+\frac{26{,}250}{140}$$

$=690+187{\cdot}5$

$=877{\cdot}5 \text{ mm}^2$.

Provide 10 mm ϕ bars at 80 mm centres.

B.M. at centre of span

$$=\frac{pl^2}{24}=\frac{21{,}00\times(2{\cdot}66)^2}{24}$$

$=6{,}190$ Nm

Net B.M. at centre of span

$=6{,}190-26{,}250\times0{\cdot}05$

$=6{,}190-1312{\cdot}5$

$=4{,}877{\cdot}5$ N-m

Area of steel required

$$=\frac{4877{\cdot}5\times1000}{0{\cdot}87\times40\times130}+\frac{26{,}250}{140}$$

$=308+187{\cdot}5$

$=495{\cdot}5 \text{ mm}^2$.

Provide 10 mm ϕ bars at 120 mm centres.

Thickness of slab is varied from 120 mm at top to 160 mm at bottom.

Reinforcement at depth of 2·5 m

Pressure at depth of 2·5 m

$$=8000\times 2{\cdot}5\times(0{\cdot}866)^2$$

$$=15{,}000\ \text{N/m}^2$$

B.M. at support $= \dfrac{15{,}000\times(2{\cdot}66)^2}{12}$

$$=8{,}847\ \text{Nm}$$

Direct tension in wall

$$=15{,}000\times\frac{2{\cdot}5}{2}$$

$$=18{,}750\ \text{N per metre}$$

Thickness of slab $=120+\dfrac{2{\cdot}5}{3{\cdot}5}\times 40$

$$=120+28{\cdot}6$$

$$=148{\cdot}6\ \text{mm}$$

Effective depth $=148{\cdot}6-30=118{\cdot}6$ mm

Net B.M. $=8847-18{,}750\times\dfrac{\left(\dfrac{148{\cdot}6}{2}-30\right)}{1000}$

$$=8847-830{\cdot}5$$

$$=8016{\cdot}5\ \text{Nm.}$$

Area of steel required

$$=\frac{8016{\cdot}5\times 1000}{2{\cdot}87\times 118{\cdot}6\times 140}+\frac{18{,}750}{140}$$

$$=555+134$$

$$=689\ \text{mm}^2.$$

Provide 10 mm ϕ bars at 100 mm c. to c.

B.M. at centre of span

$$=\frac{pl^2}{24}=\frac{15{,}000\times(2{\cdot}66)^2}{24}$$

$$=4423{\cdot}5\ \text{Nm}$$

Net B.M. $=4423{\cdot}5-830{\cdot}5$

$$=3593\ \text{Nm.}$$

Area of steel required

$$=\frac{3593\times 1000}{0{\cdot}87\times 118{\cdot}6\times 140}+\frac{18{,}750}{140}$$

$$=249+134$$

$$=383\ \text{mm}^2.$$

Provide 10 mm ϕ bars at 200 mm c. to c.

Reinforcement at depth of 1·5 m.

Pressure at depth of 1·5 m

$$=8000\times1{\cdot}5\times(0{\cdot}866)^2=9000\text{ N/m}^2.$$

B.M. at support $=\dfrac{9000\times(2{\cdot}66)^2}{12}=5308$ N-m

Direct tension in the wall

$$=\frac{9000\times2{\cdot}5}{2}=11{,}250\text{ N.}$$

Thickness of slab $=120+\dfrac{1{\cdot}5}{3.5}\times40=120+17=137$ mm.

Effective depth$=137-30=107$ mm

Net B.M. $=5308-11{,}250\times\dfrac{\left(\dfrac{137}{2}-30\right)}{1000}$

$=5308-11{,}250\times0{\cdot}0385$

$=5308-433{\cdot}2$

$=4874{\cdot}8$ N-m.

Area of steel required

$$=\frac{4874{\cdot}8\times1000}{0{\cdot}87\times107\times140}+\frac{11{,}250}{140}$$

$=374+80{\cdot}4$

$=454{\cdot}4\text{ mm}^2.$

Provide 10 mm ϕ bars at 160 mm c. to c.

B.M. at centre of span

$$=\frac{pl^2}{24}=\frac{9000\times(2{\cdot}66)^2}{24}$$

$=2654$ Nm.

Net B.M. $=2654-11{,}250\times0{\cdot}0385$

$=2220{\cdot}8$ N-m.

Area of steel required

$$=\frac{2220{\cdot}8\times1000}{0{\cdot}87\times107\times140}+\frac{11{,}250}{140}$$

$=170{\cdot}4+80{\cdot}4$

$=250{\cdot}8\text{ mm}^2.$

Provide 10 mm ϕ bars at 240 mm c. to c.

Thus side walls are provided with top rib of 300 mm width by 250 mm depth to allow for attachment of conveyor belts etc.

Design of side walls as girders. Assume thickness of hopper bottom as 140 mm thick with 20 mm thick lining.

The load on the side walls

$$= \tfrac{1}{4}\text{th of weight of coal in bunker}$$

$$= \frac{25{\cdot}961 \times 8000}{4}$$

$$= 51{,}922 \text{ N}$$

$\frac{1}{4}$th of weight of sloping bottom and lining

$$= \left(\frac{2{\cdot}5 + 0{\cdot}5}{2}\right) \sqrt{2} \times \frac{16}{100} \times 24{,}000$$

$$= 8145 \text{ N}$$

Side wall $= 3{\cdot}5 \times \left(\frac{12+16}{2 \times 100}\right) \times 2{\cdot}5 \times 24{,}000$

$$= 29{,}400 \text{ N}$$

Top rib $= 2{\cdot}5 \times 0{\cdot}25 \times (0{\cdot}3 - 0{\cdot}12) \times 24{,}000$

$$= 2700 \text{ N}$$

(rib 400 mm × 400 mm) Bottom rib $= 2{\cdot}5 \times 0{\cdot}4 \times 0{\cdot}4 \times 24{,}000$

$$= 9600 \text{ N}$$

Flat bottom $0{\cdot}2 \times 0{\cdot}2 \times 0{\cdot}6 \times 24{,}000 = 576$ N

$\frac{1}{4}$th weight of gate $= 150$ N

Total $= 102{,}493$ N

Design of Columns

Load of coal bunker on each column $= 102{,}493$ N

Self-weight of columns $= 9{,}000$ N

Total 111,493 N

Say 111,500 N

Wind pressure is taken as 1500 N/m²

Wind pressure on surcharge coal

$$= \tfrac{1}{2} \times 0{\cdot}722 \times 2{\cdot}5 \times 1500 = 1354 \text{ N}$$

acting at $7{\cdot}5 + \frac{0{\cdot}722}{3} = 7{\cdot}74$ m from bottom.

Wind pressure on vertical wall of bunker

$$= 3{\cdot}1 \times 3{\cdot}5 \times 1500$$

$$= 16{,}280 \text{ N acting at } 4 + \frac{3{\cdot}5}{2}$$

$$= 5{\cdot}75 \text{ m from base.}$$

Wind pressure on hooper bottom

$$=\frac{2.82+0.82}{2}\times 1\times 1500$$

$=2730$ N acting at $(4-0.4)$

$=3.6$ m from base.

Wind pressure on columns (300 mm × 300 mm)

$=2\times 0.3\times 4\times 1510$

$=3600$ N acting at 2 m from base.

Total moment at base

$=1354\times 7.74+16{,}280\times 5.75+2730\times 3.6+3600\times 2$

$=10{,}490+93{,}630+9828+7600$

$=121{,}538$ N-m.

Vertical force in each column

$$=\frac{121{,}538}{2.8\times 2}=21{,}700\text{ N.}$$

Maximum force in column

$=111{,}500+21{,}700=133{,}200$ N.

Horizontal shear force per column

$$=\frac{1354+16{,}280+2730+3600}{4}$$

$$=\frac{23{,}964}{4}=5991\text{ N.}$$

B.M. in column $=\dfrac{5991\times 4}{2}=11{,}982$ Nm.

$=11{,}982{,}000$ N-mm.

300 mm × 300 mm column is provided with 4 bars of 20 ϕ.

Area of steel $=1256\text{ mm}^2$.

Eccentricity $=\dfrac{11{,}982{,}000}{133{,}200}=90$ mm.

Equivalent area of cross-section

$=300\times 360+17\times 1256$

$=90{,}000+21{,}352=111{,}352\text{ mm}^2$.

Equivalent moment of inertia

$=\frac{1}{12}\times 300\times (300)^3+17\times 1256\times (100)^2$

$=(67{,}500+21{,}352)\times 10^4$

$=88{,}852\times 10^4\text{ mm}^4$.

Max. and minimum stresses are given by

$$\sigma=\frac{133{,}200}{111{,}352}\pm\frac{11{,}982{,}000\times150}{88{,}852\times10^4}$$

$$=1{\cdot}2\pm2{\cdot}022.$$

Some portion of column will be in tension therefore, the following procedure is followed to check the section

$$\frac{e}{D}=\frac{9}{30}=0{\cdot}3$$

% of steel on each face

$$=\frac{1256}{9900}\times\frac{100}{2}$$

$$=0{\cdot}700$$

Cover $=\frac{5}{30}=\frac{1}{6}$

Referring to graph in appendix, allowable average stress

$$=44\%.$$

Allowable average stress

$$=5\times0{\cdot}44=2{\cdot}2\ \text{N/mm}^2.$$

Actual average stress $=\frac{133{,}200}{9000}$

$$=1{\cdot}48\ \text{N/mm}^2.\quad \text{Safe.}$$

Span of girder $=2{\cdot}66$ m.

Assuming max. B.M. $=\frac{Wl}{8}$

$$=\frac{102{,}493\times2{\cdot}66}{8}$$

$$=34{,}080\ \text{Nm}.$$

Effective depth required

$$=\sqrt{\frac{34{,}080\times1000}{0{\cdot}87\times120}}$$

$$=571{\cdot}4\ \text{mm}.$$

Effective depth provided

$$=3500+200=3700\ \text{mm}.$$

Area of steel required

$$=\frac{34{,}080\times1000}{0{\cdot}87\times3700\times140}$$

$$=80\ \text{mm}^2.$$

Provide 6 bars of 12 mm ϕ.

Shear force $=\frac{102,493}{2}$

$=51,246$ N

Shear stress $=\frac{51,246}{3700\times120}=0{\cdot}115$ N/mm².

No shear reinforcement is required, however nominal stirrups are provided.

Hopper Bottom. Hopper bottom is subjected to direct tension due to weight of coal, self-weight and weight of gate.

Vertical load

$=51,922+8145+576+150$

$=60,793$ N.

Direct tension

$=60,793 \text{ cosec } 45°$

$=85,960$ N.

Area of steel required

$=\frac{85,960}{140}$

$=614$ mm².

Provide 8 bars of 10 mm ϕ.

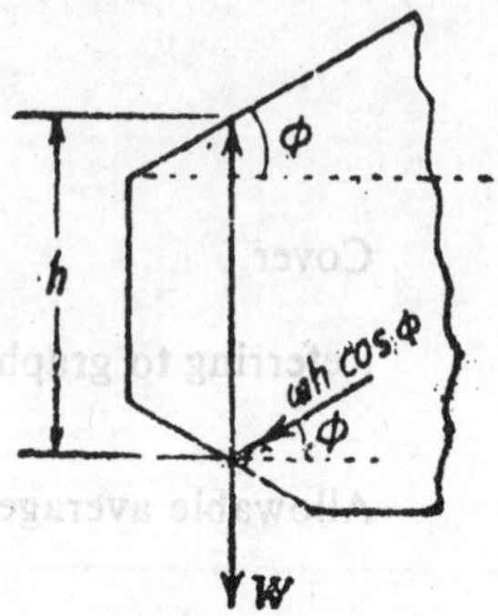

Fig. 19·9

Design of sloping bottom as slab. Hopper bottom has to resist bending due to pressure of coal and weight of slab and lining.

The slab is considered to span horizontally between the intersection of adjacent sloping faces. The span of slab increases and pressure head due to coal decreases. Consider a strip of 1 m width measured parallel to the edge of hopper at average depth of

$$3{\cdot}5+0{\cdot}5+0{\cdot}3=4{\cdot}3 \text{ m}.$$

(This is the depth from centre of hopper to coal surface)

The average normal pressure for this case

$$p_n=wh(\cos^2\alpha+\cos^2\alpha\sin\phi\cos\phi+\cos^2\phi\sin^2\alpha)$$

where,
w = density of coal

ϕ = angle of repose of coal

α = angle made by hopper with the horizontal

$$p_n=8000\times4{\cdot}3(\cos^2 45°+\cos^2 45° \sin 30° \cos 30+\cos^2 30°\times\sin^2 45°)$$

$$=8000\times4{\cdot}3\left(\frac{1}{2}+\frac{1}{2}\times\frac{1}{2}\times\frac{\sqrt{3}}{2}+\frac{3}{4}\times\frac{1}{2}\right)$$

$$=36,540 \text{ N}$$

Normal pressure due to weight of slab and lining

$=1\times1\times0{\cdot}16\times24{,}000\cos45°=2715$ N

Total load $=39{,}255$ N

Span $=1{\cdot}25$ m

$$\text{B.M.}=\frac{wl^2}{12}=\frac{39{,}255\times(1{\cdot}25)^2}{12}=5111 \text{ N-m}$$

$$\text{Effective depth}=\sqrt{\frac{5111\times1000}{0{\cdot}87\times1000}}=77 \text{ mm}$$

Provide overall depth of 140 mm with effective depth of 110 mm

$$\text{Area of steel}=\frac{5111\times1000}{0{\cdot}87\times110\times140}=380 \text{ mm}^2.$$

Provide 8 mm ϕ bars at 130 mm c. to c. at corners.

Provide 8 mm ϕ bars at 260 mm centres at the middle.

Example 19·2. *Design a silo of the dimensions shown in Fig. 19·10 for storing wheat. The inside diameter is 6 m and height of cylindrical portion is 20 m. The conical bottom slopes at 45°. The opening is 500 mm diameter. Use Airy's theory.*

Solution. The mix of concrete used is M 20

Weight of wheat $=8000$ N/m^3

$\mu=0{\cdot}466$

$\mu'=0{\cdot}444$

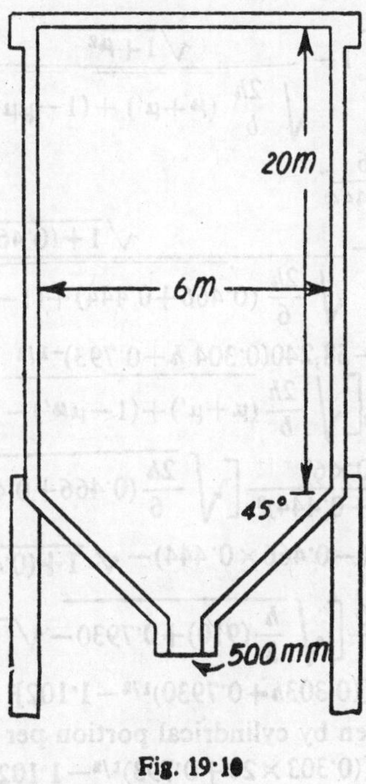

Fig. 19·10

Shallow portion

Height at which rupture plane cuts vertical wall

$$h' = d \tan \theta$$

$$h' = d\left[\mu + \sqrt{\frac{\mu(1+\mu^2)}{\mu+\mu'}}\right]$$

$$= 6\left[0.466 + \sqrt{\frac{0.444\,(1+0.444^2)}{0.466+0.444}}\right]$$

$$= 7.5 \text{ m nearly.}$$

$$p_h = \frac{wh}{\left[\sqrt{\mu'+(\mu+\mu')} + \sqrt{1+\mu^2}\right]}$$

$$= \frac{8000h}{\left[0.466(0.466+0.44) + \sqrt{1+(0.466)^2}\right]}$$

$$= 2598\, h$$

$$P = p_h \times \frac{h}{2} = 1299\, h^2$$

Vertical load taken from 7.5 m length per 1 m perimeter $= p_h \times \frac{h}{2} \times \mu$

$$= 1299 \times (7.5)^2 \times 0.444 = 32450 \text{ N}$$

Deep portion

$$P_h = \frac{wb}{(\mu+\mu')}\left[1 - \frac{\sqrt{1+\mu^2}}{\sqrt{\frac{2h}{b}(\mu+\mu') + (1-\mu\mu')}}\right]$$

$$= \frac{8000 \times 6}{0.466 \times 0.444} \times \left[1 - \frac{\sqrt{1+(0.466)^2}}{\sqrt{\frac{2h}{6}(0.466+0.444) + (1-0.466 \times 0.444)}}\right]$$

$$= 52{,}810 - 58{,}240(0.304\, h + 0.793)^{-1/2}$$

$$P = \frac{wb^2}{2(\mu+\mu')^2}\left[\sqrt{\frac{2h}{b}(\mu+\mu') + (1-\mu\mu')} - \sqrt{1+\mu^2}\right]^2$$

$$= \frac{8000 \times 6^2}{2(0.466+0.444)^2}\left[\sqrt{\frac{2h}{6}(0.466+0.444) + (1-0.466 \times 0.444)} - \sqrt{1+(0.466)^2}\right]^2$$

$$= \frac{144{,}000}{(910)^2}\left[\sqrt{\frac{h}{3}(910) + 0.7930} - \sqrt{1.2172}\right]^2$$

$$= 174{,}000[(0.303h + 0.7930)^{1/2} - 1.102]^2$$

Vertical load taken by cylindrical portion per 1 m perimeter

$$= 174{,}000[(0.303 \times 20 + 0.793)^{1/2} - 1.102]^2 \times 0.444$$

$$= 174{,}000 \times 2.298 \times 0.444 = 177{,}500 \text{ N}$$

Table 19·4

Depth	p_h *N*	*Hoop tension* $p_h \times d/2 = 3p$	*Hoop steel* */140*	*Steel*
2	5196	15,58		at 200 mm c. to c.
4	10,392	31,176		10 mm. ϕ at 200 mm c. to c.
6	15,588	46,764	334·0	10 mm. ϕ at 200 mm c. to c.
8	20,340	6 0	435·9	10 mm. ϕ at 150 mm c. to c.
10	23,020	69,060	491·9	10 mm. ϕ at 150 mm c. to c.
12	25,140	75,420	538·7	10 mm. ϕ at 130 mm c. to c.
14	26,860	80,580	575·6	10 mm. ϕ at 130 mm c. to c.
16	28,290	70	696·2	10 mm. ϕ at 120 mm c. to c.
18	29,510	88,530	632·4	10 mm, ϕ at 120 mm c. to c.
20	30,750	91,710	655·1	10 mm. ϕ at 110 mm c. to c.

For first 6 m., 10 mm ϕ at 200 mm c. to c. are provided.

For next 4 m., 10 mm ϕ at 150 mm c. to c. are provided.

For next 4 m., 10 mm ϕ at 130 mm c. to c. are provided.

For next 4 m., 10 mm ϕ at 120 mm c. to c. are provided.

For next 2 m., 10 mm ϕ at 110 mm c. to c. are provided.

Design of wall section. Hoop tension at bottom of cylindrical portion = 91,710 N. Steel provided as 10 mm ϕ at 10 mm c. to *c., i.e.* 714 mm² in one metre length. For M20 mix allowable tension in concrete is 1·2 N/mm², $m = 13$.

$$T = 1{\cdot}2[1000t_s + (m-1)A_s]$$

where t_c = thickness of wall

$$\therefore \quad 91{,}710 = 1.2[1000\,t_c + 12\times714]$$

$$\therefore \quad t_c = \frac{76{,}425-8568}{1000}$$

$$= \frac{67{,}857}{1000} = 67.857 \text{ mm}$$

Use 120 mm thickness of wall throughout.

Vertical reinforcement is provided at 0·3% of wall section.

Vertical steel $= \dfrac{0.3\times1000\times120}{100} = 360 \text{ mm}^2$

Provide 10 mm ϕ at 200 mm centre to centre.

Load taken by wall. Providing roof on top say 120 mm thick and allowing live load of 1500 N/m², load per 1 metre perimetre of wall

$$= \frac{(2880+1500)\times\frac{\pi}{4}d^2}{\pi d}$$

$$= 4380\times\frac{6}{4} = 6570 \text{ N.}$$

Wall load for 20 m length

$$= 120\times1000\times20\times0.024 = 57{,}600 \text{ N}$$

Grain load carried by wall

$$= 177{,}500 \text{ N}$$

Total load $= 6570 + 57{,}600 + 177{,}500 = 241{,}670$ N

Compressive stress $= \dfrac{241{,}670}{120\times1000} = 2.01 \text{ N/mm}^2$

Stresses due to wind pressure. Take wind pressure of 1500 N/m² and shape factor as 0·7.

B.M. $= 1500\times6.24\times0.7\times20\times10$ N-m
$= 1{,}310{,}000$ N-m $= 131\times10^7$ N-mm

M.I. of silo $= \dfrac{\pi}{64}[(6.24)^4-(6.0)^4]$

$$= \frac{\pi}{64}[1517-1296]$$

$$= \frac{\pi}{64}\times221 = 10.85 \text{ m}^4$$

$$Z = \frac{10.85}{3.12} = 3.477 \text{ m}^3 = 3.477\times10^9 \text{ mm}^3$$

Max. stress $= \dfrac{131\times10^7}{3.477\times10^9} = \dfrac{131}{347.7} = 0.38 \text{ N/mm}^2$

Total compressive stress

$= 2.01 + 0.38 = 2.39$ N/mm². Safe.

Conical Bottom

Surcharge on conical portion

$$=\frac{\frac{\pi}{4}d^2\times 20\times 8000-\pi d\times 177{,}500}{\frac{\pi}{4}d^2}$$

$$=160{,}000-\frac{177{,}500\times 4}{6}=160{,}000-118{,}330$$

$$=41{,}670\ \text{N/m}^2.$$

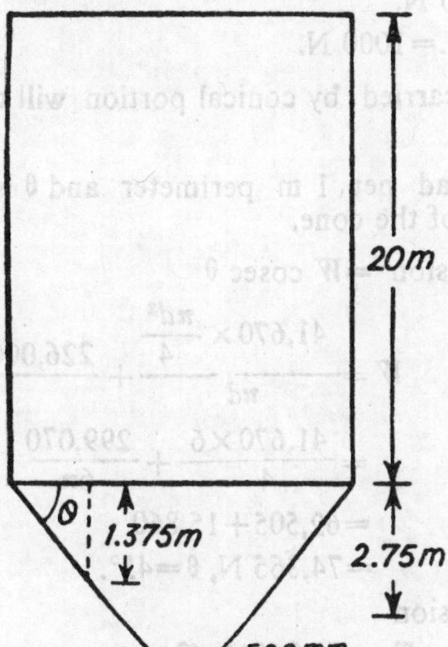

Fig. 19·11

Wt. of grain in conical portion

$$=8000\times\frac{2\cdot75}{3}\left[\frac{\pi}{4}\times 6^2+\frac{\pi}{4}\times(0\cdot5)^2+\sqrt{\frac{\pi}{4}\times 6^2\times\frac{\pi}{4}\times(0\cdot5)^2}\right]$$

$$=\frac{22{,}000\ \pi}{3}\times\frac{\pi}{4}\left[36+0\cdot25+3\right]$$

$$=\frac{22{,}000\ \pi}{3}\times 39\cdot25=226{,}000\ \text{N}.$$

Wt. of concrete in conical portion assuming thickness as 12 cm.

$$=24{,}000\times\frac{2{\cdot}75}{3}\Big[\frac{\pi}{4}\{(6{\cdot}24)^2-(6{\cdot}00)^2\}+\frac{\pi}{4}\{(0{\cdot}74)^2-(0{\cdot}50)^2\}$$

$$+\frac{\pi}{4}\sqrt{\{(6{\cdot}24)^2-(6{\cdot}00)^2\}\times\{(0{\cdot}74)^2-(0{\cdot}50)^2\}}\Big]$$

$$=\frac{22{,}000\times\pi}{4}\Big[12{\cdot}24\times0{\cdot}24+1{\cdot}24\times0{\cdot}24$$

$$+\sqrt{12{\cdot}24\times0{\cdot}24\times1{\cdot}24\times0{\cdot}24}\Big]$$

$$=5500\,\pi\times0{\cdot}24\,[12{\cdot}24+1{\cdot}24+3{\cdot}90]$$
$$=5500\,\pi\times0{\cdot}24\times17{\cdot}38$$
$$=72{,}070\text{ N.}$$

Wt. of gate etc. $=1000$ N.

Total weight carried by conical portion will cause meridional tension in the cone.

If W is the load per 1 m perimeter and θ is the inclination with the horizontal of the cone.

Meridional tension $=W\operatorname{cosec}\theta$

Now
$$W=\frac{41{,}670\times\frac{\pi d^2}{4}}{\pi d}+\frac{226{,}000+72{,}070+1000}{\pi d}$$
$$=\frac{41{,}670\times6}{4}+\frac{299{,}070}{6\pi}$$
$$=62{,}505+15{,}860$$
$$=74{,}365\text{ N},\ \theta=45°.$$

Meridional tension
$$T_m=74.365\times\sqrt{2}$$
$$=105{,}200\text{ N.}$$

Meridional reinforcement
$$=\frac{105{,}200}{140}=751\text{ mm}^2$$

Provide 10 mm ϕ at 100 mm c. to c. The bars go on converging in the bottom of cone. Half the bars are stopped half way.

Hoop tension. Hoop tension will depend on the diameter at various heights of the cone and intensity of pressure. As the diametre is greater at the top and intensity of pressure is greatest at the bottom, the hoop tension will be found at middle of the cone and reinforcement provided throughout.

The section chosen is 1·375 m below the end of cylindrical portion.

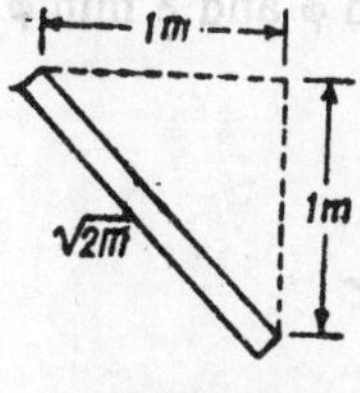

Fig. 19·12 (*a*)

Vertical pressure

$$41{,}670+8000\times 1{\cdot}375$$
$$=41{,}670+11{,}000$$
$$=52{,}670\ \text{N/m}^2.$$

Wt. of concrete

$$=120\times\sqrt{2}\times 1000\times 0{\cdot}024$$
$$=4070\ \text{N}.$$

Grain pressure $=(41{,}670+8000\times 1{\cdot}375)\left(\dfrac{1-\sin\phi}{1+\sin\phi}\right)$

$$=52{,}670\times\left(\frac{1-\sin\tan^{-1}0{\cdot}466}{1+\sin\tan^{-1}0{\cdot}466}\right)$$

$$=52{,}670\times 0{\cdot}4062 - 21{,}390\ \text{Nm}^2.$$

Normal pressure

$$=\frac{(52{,}670+4070)\cos\theta}{\sqrt{2}}+\frac{21{\cdot}390\sin\theta}{\sqrt{2}}$$

$$=\frac{56{,}740}{2}+\frac{21{,}390}{2}$$

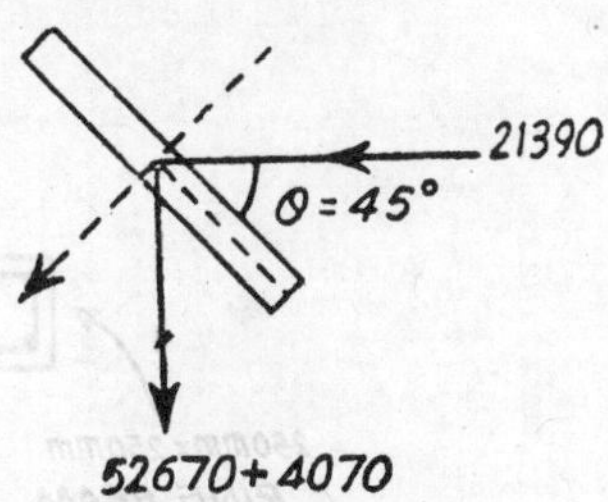

Fig. 19 12 (*b*)

$$=\frac{78{,}130}{2}$$
$$=39{,}065\ \text{N/m}^2$$

$$\therefore\quad \frac{H}{r_1}+\frac{T}{r_2}=p$$

Now $\dfrac{1}{r_2}=0$

$$r_1=1{\cdot}1625\sqrt{2}$$

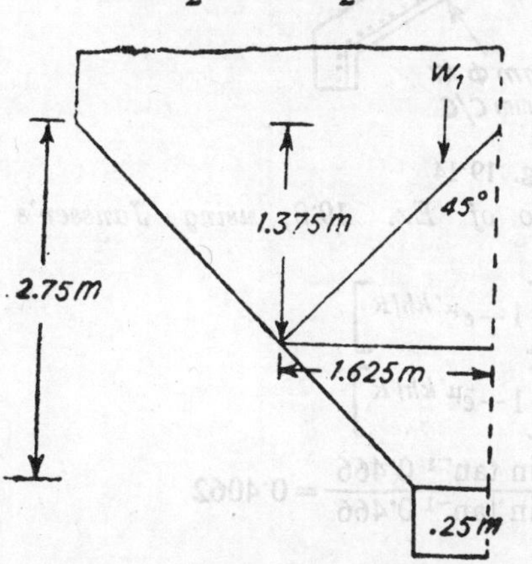

Fig. 19·13

where H is hoop tension and T is meridional tension

$$H=pr=39{,}065\times 1{\cdot}625\sqrt{2}$$
$$=89{,}800\ \text{N}.$$

Hoop reinforcement $=\dfrac{89{,}800}{140}$

$$=640\ \text{mm}^2.$$

Ring beam at the junction of cylindrical portion and conical base

Hoop compression on the ring

= Hoop compression due to pressure from conical base − Hoop tension on cylindrical portion

$$=105{,}200-91{,}710=13{,}490\ \text{N}.$$

250 mm×250 mm section with 4 bars 20 mm ϕ and 8 mm ϕ stirrups at 300 mm c. to c. may be provided

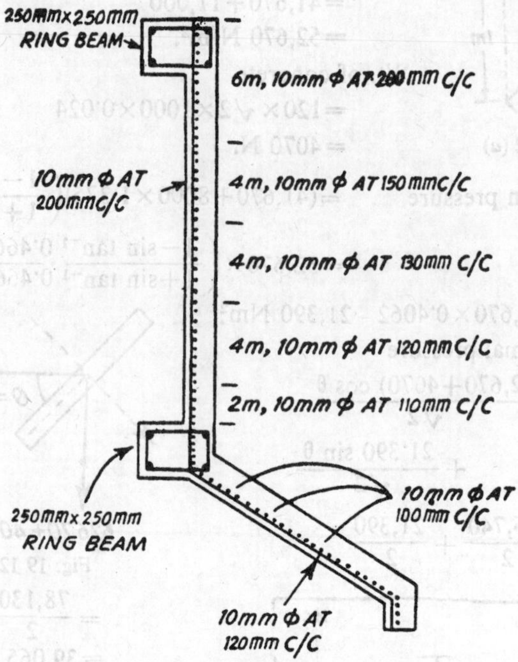

Fig. 19·14

Ex. 19·3. *Design the silo of Ex. 19·2, using Janssen's theory.*

Sol.

$$p_h = \frac{wR}{\mu'}\left[1 - e^{\mu' kh/K}\right]$$

$$p_v = \frac{wR}{\mu' k}\left[1 - e^{\mu' kh/R}\right]$$

$$k = \frac{1 - \sin\tan^{-1} 0{\cdot}466}{1 + \sin\tan^{-1} 0{\cdot}466} = 0{\cdot}4062$$

$$\mu' = 0{\cdot}444$$

$$w = 8000 \text{ N/m}^3.$$

$$R = \frac{A}{U} = \frac{\frac{\pi}{4}d^2}{\pi d} = \frac{d}{4} = \frac{6}{4} = 1{\cdot}5 \text{ m}$$

$$p_h = \frac{8000 \times 1{\cdot}5}{0{\cdot}444}\left[1 - e^{0{\cdot}444 \times 0{\cdot}4062h/1{\cdot}5}\right]$$

$$= 27{,}030\,[1 - e^{-0{\cdot}120\,h}]$$

$$p_s = \frac{27{,}030}{0{\cdot}4062}\,[1 - e^{-{\cdot}120\,h}] = 66{,}550\;[1 - e^{-0{\cdot}120\,h}]$$

Table 19·5

Depth	p_h	*Hoop tension* $p_h \times d/2$ $= 3p_h$	*Area of steel in* mm^2	*Reinforcement*
2	5,763	17,289	123 5	10 mm ϕ 200 mm c. to c.
4	10,540	31,622	226·0	10 mm. ϕ 200 mm c. to c.
6	13,870	41,610	297	10 mm. ϕ 200 mm c. to c.
8	16,670	50,010	375	10 mm. ϕ 200 mm c. to c.
10	18,880	56,640	405	10 mm. ϕ 160 mm c. to c.
12	26,620	61,860	442	10 mm. ϕ 160 mm c. to c.
14	21,990	65,970	472	10 mm. ϕ 160 mm c. to e.
16	23,050	69,150	494	10 mm. ϕ 140 mm c. to c.
18	23,910	71,730	512	10 mm. ϕ 140 mm c. to c.
20	24,580	73,740	527	10 mm. ϕ 140 mm c. to c.

Use 120 mm thickness of wall with vertical reinforcement of 12 mm ϕ at 300 mm c. to c.

For conical portion

Surcharge pressure $p_s = \frac{p_h}{k} = \frac{24{,}580}{0\;4062} = 60{,}520\ \text{N/m}^2$

Taking same data from previous example

$$W=\frac{60{,}520\times\frac{\pi}{4}d^2}{\pi d}+\frac{299{,}070}{\pi d}$$

$$=60{,}520\times 6/4+15{,}860$$

$$=90{,}780+15{,}860=106{,}640 \text{ N}$$

Meridional tension $=106{,}640\times\sqrt{2}=150{,}800$ N

Reinforcement $=\frac{150{,}800}{140}=1080 \text{ mm}^2$

Provide 12 mm ϕ at 100 mm c. to c.

Hoop tension. At middle of conical portion.

Vertical load $=60{,}520+11{,}000=71{,}520 \text{ N/m}^2$

Normal pressure $=\frac{71{,}520}{2}+\frac{22{,}240}{2}$

$$=\frac{93{,}760}{2}$$

$$=46{,}880 \text{ N/m}^2$$

$$H=r_1p=1{\cdot}625\sqrt{2}\times 46{,}880=107{,}700 \text{ N}$$

Hoop reinforcement $=\frac{170{,}700}{140}=770 \text{ mm}^2$

Use 10 mm ϕ at 100 mm c. to c.

Ring beam can be designed in the same way.

Load from grain on wall. At bottom of cylindrical portion load taken

$$=\frac{\pi}{4}d^2\,[20\times 8000-60{,}520] \text{ N.}$$

Load per 1 m perimeter

$$=\frac{\frac{\pi}{4}d^2\,(99{,}480)}{\pi d}$$

$$=\frac{6}{4}\times 99{,}480$$

$$=149{,}200 \text{ N.}$$

Compressive stress in wall due to grain load, self-load, top slab load and wind load is worked out in the same way as in Ex. 19·2.

20

Formwork

20·1. The timber or steel or other materials used to form moulds for pouring concrete in-situ, are described as formwork falsework, centring and shuttering.

20·2. Requirements for Good Formwork

The primary requirements are strength, stiffness, smoothness and good shape at the concrete face.

It is wasteful to cut off timber that are not in the way of pouring concrete. The extra length may prove useful at other times.

Joints should be as simple as possible. Lapped or butted and cleated types of joints are generally used.

All formwork should be erected having in mind that it should be easy to strip after concreting. No piece of formwork should be keyed into the concrete. Shutters forming recesses must be levelled for free withdrawal. As few nails as is practicable should be used. They should be left standing.

Striking of forms follows a definite order and form must be so designed that they can be stripped in the order required.

1. Shutters to vertical (non-load bearing) faces, column boxes, beam sides and wall forms.

2. Shutters forming soffits to slabs, horizontal or inclined which carry only light loads—slabs to roofs floors and canopies.

3. Soffit shutters carrying heavy loads—beam and girder bottoms.

The cost of formwork may be as much as 20 per cent of the total cost. Therefore, the wastage of material and labour should be minimised.

20·3. Precautions Against Timber Wastage

1. Stresses due to fluid concrete should be analysed and the minimum sizes of timber and maximum spacing supports should be aimed at.

2. Timber should be ordered of lengths to avoid wastage in cutting.

3. All long lengths should be cut first, and the small lengths should be obtained from the cut pieces.

4. Shutters should be made strip easy without any damage, so that they may be used often times.

6. All shutters after they are stripped, should be immediately cleaned, loose nails if any removed and these should be stacked nearly for reuse.

6. Small cleats, wedges, bolts and other parts should be stacked in separate boxes. Bolts and other metal parts should be cleaned and oiled for re-use.

Fig. 20·1 shows formwork for beam bottom and column shutter. As the column shutter is stripped off first, the beam bottom should not be placed against the column shutter as in Fig. 20·1 (*a*) but should rest over the column shutter as in Fig. 2·01 (*b*).

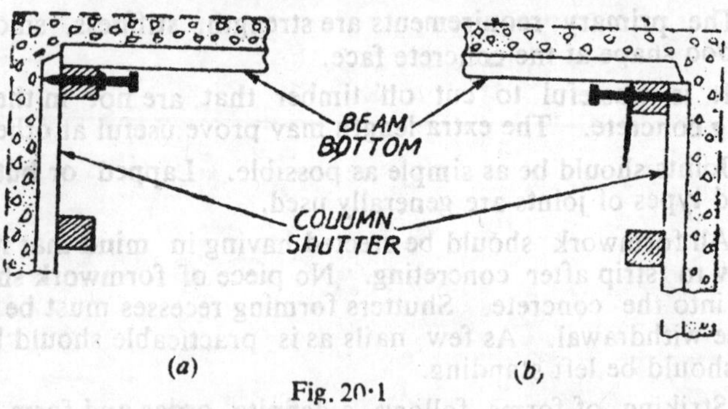

Fig. 20·1

Fig. 20 2 shows form work for slab and beam. As the sides of beam are stripped first and then decking to slab is stripped and

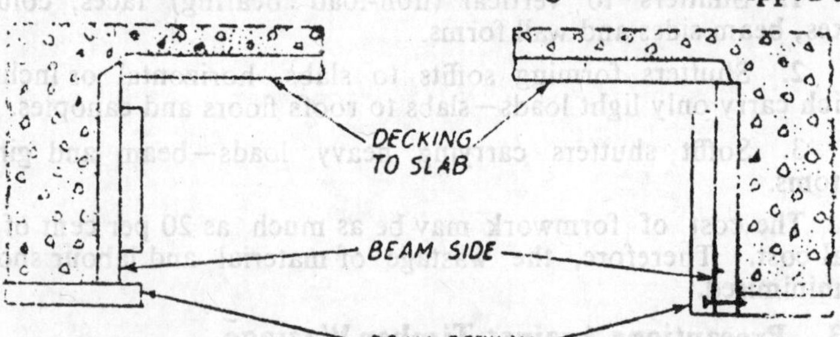

Fig. 20·2

finally the beam bottoms are stripped, the decking to slab should not be placed against the sides of beam and these sides should not be held on top of the beam bottoms as in Fig. 20·2 (*a*), but the

Table 20·1

Permissible stresses in N/mm²

Name	Density N/m³	Modulus of Elasticity E kN/mm²	Bending and tension			Shear		Compersion parallel			Compression perpendicular		
			Inside	Outside	Wet	Horizontal	Along grain	Inside	Outside	Wet	Inside	Outside	Wet
Fir	4500	9·4	7·8	6·6	5·6	0·6	0·8	6·0	5·2	4·2	1·6	1·2	1·0
Deodar	5450	9·5	10·2	8·8	7·0	0·7	1·0	7·8	7·0	5·6	2·6	2·1	1·7
Kail	5150	6·8	6·6	5·6	5·0	0·6	0·8	5·2	4·6	3·8	1·7	1·3	1·0
Chir	5750	9·8	8·4	7·0	6·0	0·6	0·9	6·4	5·6	4·6	2·2	1·7	1·4

decking of slab should rest on the beam sides and the beam sides should be nailed to beam bottoms as shown in Fig. 20·2 (*b*).

A routine inspection should be made of all formwork before concreting is allowed and one carpenter should always be in attendance during the actual pouring of concrete. The carpenter should continually inspect the formwork for any bulges or other signs of failure.

20·4. Timber for Formwork

Timber used for formwork is generally soft wood for which safe values are given in Table 20·1 taken from I.S.I. 883.

The timber should have the following qualities :

(*i*) It should be readily available and economical.

(*ii*) It should be easily worked and should nail easily without splitting.

(*iii*) It should be as light as possible to permit the handling.

(*iv*) It should be stiff to avoid undue deflection when loaded and strong enough to carry safely the load and pressure that may be applied during concreting.

(*v*) It should be reasonably stable and should not be liable to warp when exposed to sun and rain and or wetted during concreting.

(*vi*) The boards used to form the facing to shutters to wall columns are called sheeting and those used for formwork to concrete floor slabs are called decking. The thickness of boards used will depend on the spacing of the supports. Minimum thickness of 25 mm is used. Where shuttering is to be used often times thickness of 30 to 40 mm is used.

20·5. Loads and Pressure on Formwork

The unset concrete applies two types of loads on formwork.

(*i*) Dead and live load imposed by concreting flat roofs and flat slabs.

(*ii*) Hydrostatic pressure of the fluid concrete acting against the faces of vertical and steeply inclined forms used in walls, pitched roofs etc.

Loads on flat slab form. These include—

(*a*) Dead weight of concrete and reinforcing steel resting on the forms. This may be taken as 24,000 N/m^3.

(*b*) The temporary live loads of workmen and equipment during the actual concreting.

(c) Hydrostatic pressure in vertical walls. Unset fluid concrete acts in a similar way as water and exerts pressure in all directions. The actual pressure at any level in a vertical form is the total weight of all the concrete above it. As the concrete is not completely fluid and as the steel reinforcement does not add to the load, it is useful to assess the pressure in a column box as 21,000 N/m². In vertical forms the pressure is due to fluidity of the concrete and after the concrete sets the pressure is zero. Therefore, the pressure on the formwork will depend on the depth of concrete poured before the concrete sets. The setting time may be taken as one hour. Therefore, if hourly output of concreting is known and the cross-section area of column box is known the depth of fluid concrete can be calculated.

20·6. Timber in Joists and Studs

Joists which support decking and studs to vertical sheeting are supported by other timbers at right angles. These are known as leaders and wailings respectively. Any irregularity in the joist or stud widths is transferred from the rigid leaders and wailings to the flexible sheeting and will result in wavy surface on the concrete face. Therefore, joists and studs should be planed to width on their edges at least.

20·7. Formwork to Columns

Concrete in columns is poured quickly and, therefore, considerable pressures are developed at the bottom of the box. The column may have any geometrical shape and may have to

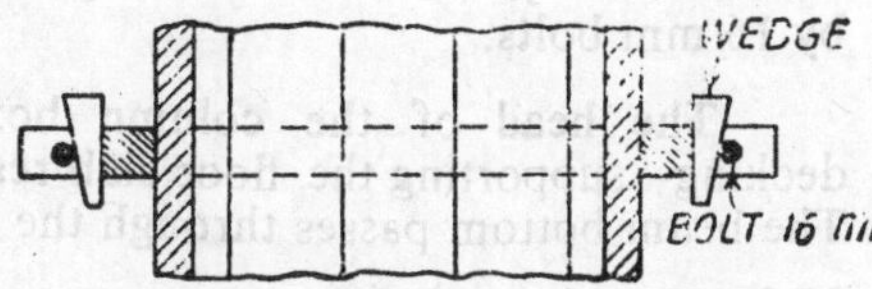

(a)
Fig. 20·3

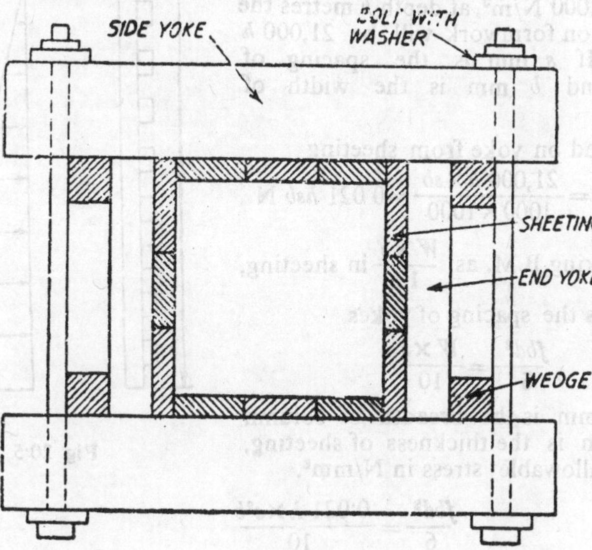

(b)
Fig. 20·3

support a flat slab or carry main and secondary beams.

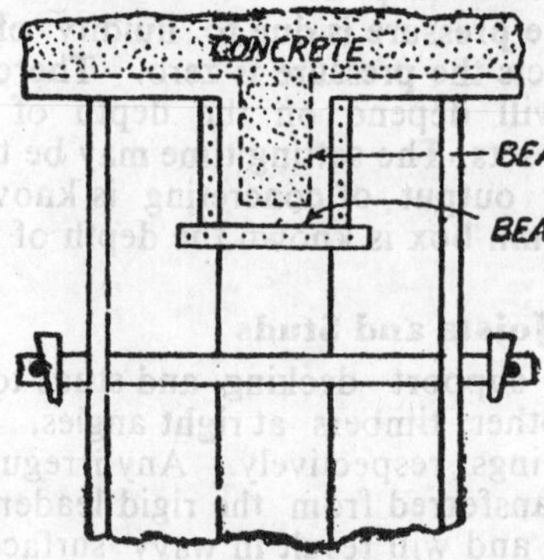

Fig. 20·4

Fig. 20·3 shows formwork for column. The sheeting is nailed to horizontal yokes of stout timber. The yokes are held at the ends by 16 mm bolts.

The head of the column box is shown in Fig. 20·4. The decking supporting the floor slab rests on top of the column box. The beam bottom passes through the column sheeting.

20·8. Spacing of Yokes

If p is the pressure from concrete density 21,000 N/m³, at depth h metres the pressure on formwork will be $21,000\,h$ N/m². If s mm is the spacing of yokes and b mm is the width of column.

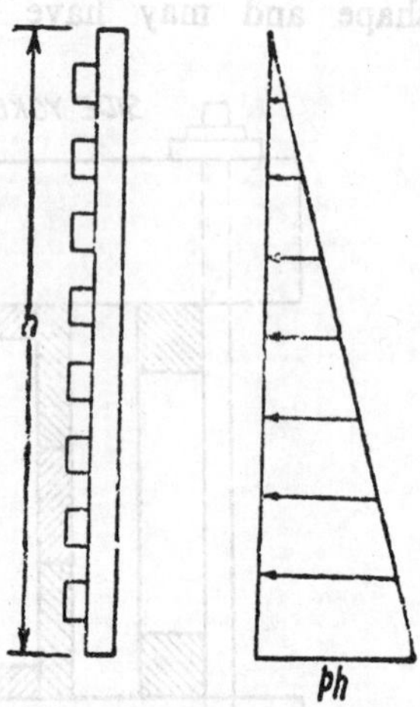

Fig. 20·5

Load on yoke from sheeting

$$W=\frac{21,000\,h\times sb}{1000\times 1000}=0{\cdot}021\,hsb \text{ N}$$

Taking B.M. as $\frac{W\times s}{10}$ in sheeting,

where s is the spacing of yokes

$$\frac{fbd^2}{6}=\frac{W\times s}{10}$$

where b mm is the breadth of column and d mm is the thickness of sheeting, and f is allowable stress in N/mm².

$$\frac{fbd^2}{6}=\frac{0{\cdot}021\,h\times s^2b}{10}$$

$$s^2=\frac{5f\,d^2}{0{\cdot}063\,h}$$

$$\therefore \quad s = 8.9\, d \sqrt{\frac{f}{h}}$$

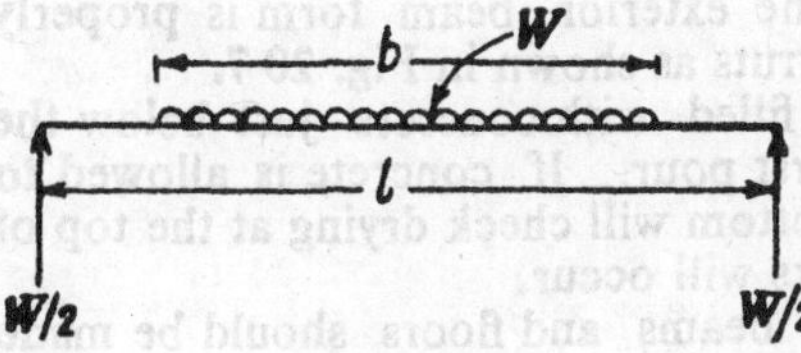

Fig. 20·6

If l mm is the span of yoke centre to centre between the bolts.

Bending moment

$$= \frac{W}{4}\left(l - \frac{b}{2}\right)$$

$$= \frac{W}{8}(2l - b)$$

If Z is the modulus of section of yoke, and f is the allowable stress,

$$\frac{W}{8}(2l - b) = fZ.$$

$$\therefore \quad \frac{0.021\, h\, sb}{8}(2l - b) = fZ.$$

$$\therefore \quad s = \frac{8fZ}{0.021\, hb\,(2l - b)} = \frac{381\, fZ}{hb\,(2l - b)}$$

The spacing of yoke is close at bottom of column.

20·9. Beam and Floor Formwork

Fig. 20·7 shows a section through deck and secondary beams. The cleats 100 mm × 2 mm to 3 mm are fixed to the sides of beams

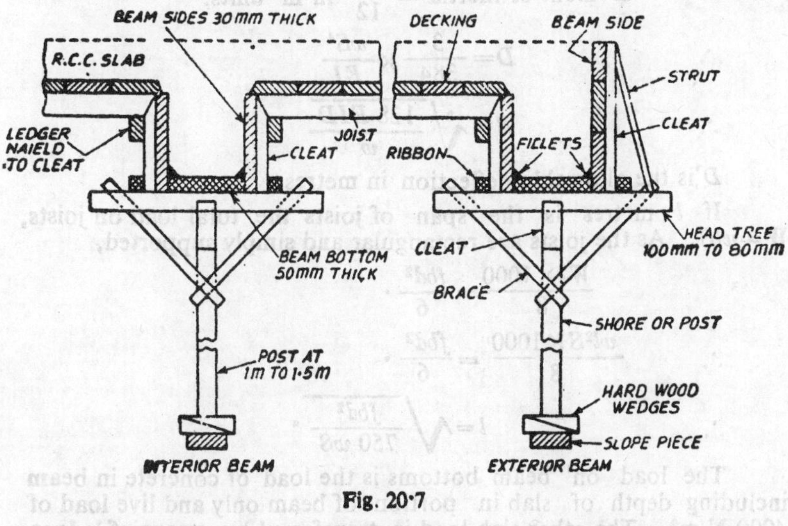

Fig. 20·7

and joists, which support the decking, are supported on ledgers, which are fixed to cleats. The beam form is supported on a head tree. The shore is connected to the head tree by the cleats. At the bottom two hard wood wedges are provided over a sole piece 200 mm×50 mm to distribute the load on floor and provide a means of nailing wedges. The exterior beam form is properly braced by means of cleats and struts as shown in Fig. 20 7.

The columns should be filled with concrete just below the lowest beam bottom at the first pour. If concrete is allowed to flow into the beams the beam bottom will check drying at the top of the column and, therefore, cracks will occur.

All construction joints in beams and floors should be made in the middle third of the span, where the bending moment will be maximum and shear will be minimum. The stop ends should be put in pieces and not around steel.

20·10. Load on Decking

The decking will have to take the immediate impact of live loads, therefore, full value of 4000 N/m² is taken. As the load on decking is comparatively light the design is based on minimum deflection. Therefore, spacing of joists will depend on the allowable deflection. The deflection should not exceed 3 mm.

As the joists are partly continuous the deflection D may be taken as $\frac{3}{384}\times\frac{wl^4}{EI}$,

where w=load on decking N per sq metre
S=spacing of joists in metres
E=modulus of elasticity in N per sq metre
I=moment of inertia $=\frac{bd^3}{12}$ in m⁴ units.

$$\therefore \quad D=\frac{3}{384}\times\frac{wS^4}{EI}$$

$$\therefore \quad S=\sqrt[3]{\frac{128\,EID}{w}}$$

D is the allowable deflection in metres.

If l metres is the span of joists the total load on joists, $W=wlS$. As the joists are rectangular and simply supported,

$$\frac{Wl\times 1000}{8}=\frac{fbd^2}{6}.$$

$$\therefore \quad \frac{wl^2S\times 1000}{8}=\frac{fbd^2}{6}.$$

$$\therefore \quad l=\sqrt{\frac{fbd^2}{750\,wS}}.$$

The load on beam bottoms is the load of concrete in beam including depth of slab in portion of beam only and live load of 4000 N/m². The other slab load is transferred by means of ledges to the head tree and thence to shores.

To minimise the spacing of shores the beam bottoms are made 50 mm thick. If B is the width of beam and if p is the spacing of shores, load on beam bottoms will be wBp, where w is load in N per metre square.

Now deflection will be the criterion of design.

$$\therefore \quad D=\frac{3}{384}\times\frac{wBp\times p^3}{EI}$$

$$\therefore \quad p=\sqrt[4]{\frac{128\,EID}{wB}}$$

where EI=N metre square
w=N per metre square
B is in metres.
D is allowable deflection in metres.

The safe capacity of a timber shore or post carrying a head tree may be limited to the bearing strength on the side grain of the head tree or by the slenderness ratio of the shore. Allowable safe stress$=f\left(1-\frac{h}{50d}\right)$, where f is allowable stress as short column, h is the height and d the least dimension of the column.

The load on the shores is the sum of the total weight of the beam and live load on it, plus the area of slab and live load on an area contained by the spacing of shores and the clear span between beams.

Ex. 20·1. ***Design sheeting and yokes for a column 300 mm × 300 mm having height of 3 m upto bottom of beam.***

Sol. Centre to centre of bolts is kept 600 mm.

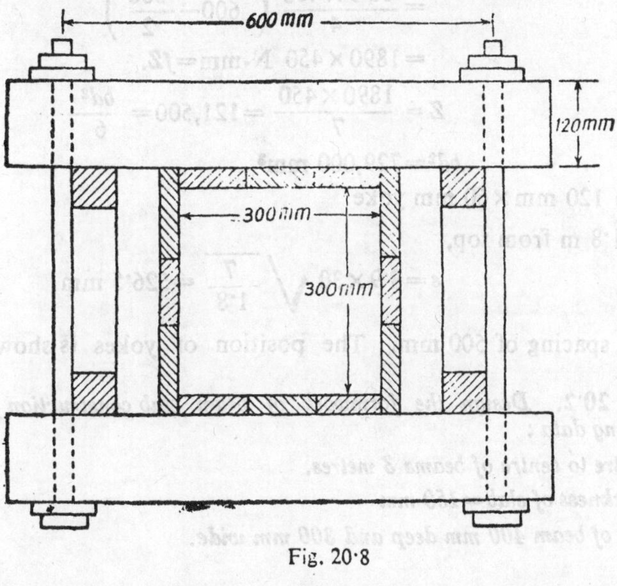

Fig. 20·8

Design of Sheeting

Hydrostatic pressure at base $=3\times21{,}000=63{,}000$ N/m²

If s mm is the spacing of yokes at base,

$$\text{Load taken by yoke}=63{,}000\ \frac{300}{1000}\times\frac{s}{1000}\ \text{N}$$

$$=18{\cdot}9\,s\ \text{N}$$

Using thickness of sheeting 30 mm and timber used Deodar, allowable stress in wet condition in bending will be 7 N/mm².

$$\therefore \quad \frac{fbd^2}{6}=\frac{W\times s}{10}$$

$$\therefore \quad \frac{7\times300\times300^2}{6}=\frac{18{\cdot}2s\times s}{10}$$

$$\therefore \quad s=\sqrt{\frac{7\times300\times900\times10}{6\times18{\cdot}9}}$$

$$=\sqrt{\frac{3{,}150{,}000}{18.9}}=408{\cdot}3\ \text{mm}$$

Use spacing of 400 mm upto 1·2 m.

Design of Yoke. Span $=600$ mm load is acting in 300 mm length.

Bending moment

$$=\frac{W}{4}\left(l-\frac{b}{2}\right)$$

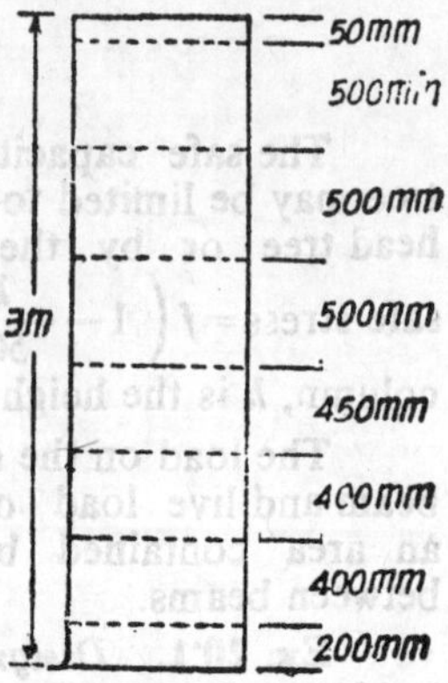

Fig. 20·9

$$=\frac{18{\cdot}9\times400}{4}\left(600-\frac{300}{2}\right)$$

$$=1890\times450\ \text{N-mm}=fZ.$$

$$\therefore \quad Z=\frac{1890\times450}{7}=121{,}500=\frac{bd^2}{6}$$

$$\therefore \quad bd^2=729{,}000\ \text{mm}^3$$

Use 120 mm × 60 mm yoke

At 1·8 m from top,

$$s=8{\cdot}9\times30\sqrt{\frac{7}{1{\cdot}8}}=526{\cdot}5\ \text{mm}$$

Use spacing of 500 mm. The position of yokes is shown in Fig. 20·9.

Ex. 20·2. *Design the formwork of beam slab construction from the following data :*

Centre to centre of beams 3 metres.

Thickness of slab = 150 mm

Rib of beam 400 mm deep and 300 mm wide.

Take live load with impact on decking 4000 N/m² and live load for beam bottoms as 4000 N/m². Deodar is to be used for all formwork.

Sol. Allowable stress in bending wet condition

$$=7 \text{ N/mm}^2$$

Allowable stress in comp. along grain outside

$$=7 \text{ N/mm}^2$$

Allowable stress in comp. perpendicular to grain outside

$$=2{\cdot}1 \text{ N/mm}^2$$

$$E=9{\cdot}5 \text{ kN/mm}^2.$$

Decking. Using 25 mm thick sheeting and allowing max. deflection of 3 mm.

$$\text{Spacing of joists}=\sqrt[4]{\frac{128\,EID}{w}}$$

Taking one metre width of decking,

$$w=4000+24\times 150 \text{ N/m}^2$$

$$=7600 \text{ N/m}^2$$

$$EI=9{\cdot}5\times 1000\times 10^6\times \frac{1}{12}\times\left(\frac{25}{1000}\right)^3 \text{ N-m}$$

$$=\frac{9{\cdot}5\times(25)^3}{12}$$

$$D=\frac{3}{1000}=0{\cdot}003 \text{ metre}$$

∴ Spacing of joists in metres

$$=\sqrt[4]{\frac{\frac{9{\cdot}5\times(25)^3}{12}\times 128\times 0{\cdot}003}{7600}}$$

$$=0{\cdot}79 \text{ m}$$

Use spacing of 0·75 m c/c.

Joists. Span of joists $l=3-0{\cdot}15=2{\cdot}85$ metres

Load on joists per metre run $=7600\times 0{\cdot}75=5700$ N

$$\frac{wl^2}{8}=fZ \quad \therefore \quad \frac{5700\times(2{\cdot}85)^2\times 1000}{8}=7Z$$

$$\therefore \quad Z=\frac{5700\times(2{\cdot}85)^2\times 1000}{56}=826{,}754 \text{ mm}^3$$

$$\therefore \quad bd^2=826{,}754\times 6=4{,}960{,}524 \text{ mm}^3$$

Use 250 mm deep 90 mm wide joists,

$$bd^2=250\times 250\times 90=5{,}625{,}000 \text{ mm}^3$$

Beam sides. Use 25 mm thick sheeting.

Beam bottom. Use 50 mm thick sheeting.

Spacing of poles, $p=\sqrt[3]{\dfrac{128\ EID}{wB}}$

$$EI=9{\cdot}5\times1000\times(10)^6\times\frac{0{\cdot}3}{12}\left(\frac{50}{1000}\right)^3$$

$$=\frac{9{\cdot}5\times0{\cdot}3\times50^3}{12}=29687{\cdot}5\ \text{Nm}^2$$

$$D=0{\cdot}003\ \text{m.}$$

$$w=24\times550+4000=17{,}200\ \text{N/m}^2$$

$$\therefore\quad p=\sqrt[4]{\frac{128\times29{,}687{\cdot}5\times0{\cdot}003}{17{,}200\times0{\cdot}3}}=1{\cdot}219\ \text{m}$$

Use spacing of 1·20 m.

Design of shores

Load from beam $=17{,}200\times0{\cdot}3\times1{\cdot}20=6192$ N

Load from slab $=7600\times2{\cdot}85\times1{\cdot}20=25{,}990$ N

Total load $=32{,}182$ N

Area required in bearing

$$=\frac{32{,}182}{2{\cdot}1}=15{,}320\ \text{mm}^2$$

160 mm × 100 mm shore will be alright.

From buckling point of view,

Allowable stress $=7\left[1-\dfrac{300}{50\times100}\right]=2{\cdot}8\ \text{N/mm}^2$

Allowable load $=160\times100\times2{\cdot}8=44{,}800$ N. Safe

21

Sheet Pile, Pier and Wharf Structures

21·1. Sheet Pile Structures

Sheet piles are used to construct retaining walls, bulkheads and cofferdams. Fig. 21·1 shows various types of sheet pile structures.

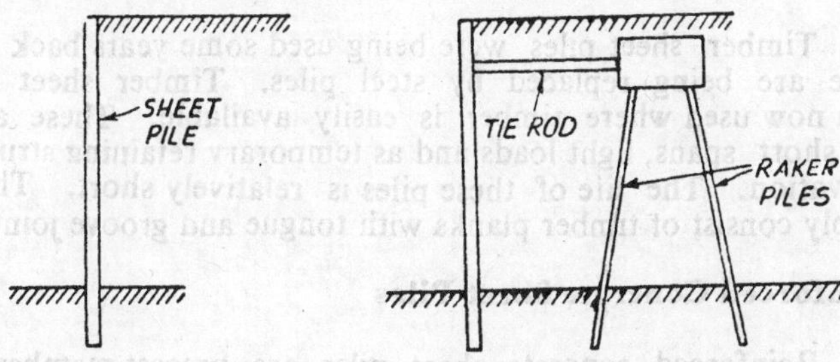

(*a*) Sheet pile (*b*) Anchored sheet piling

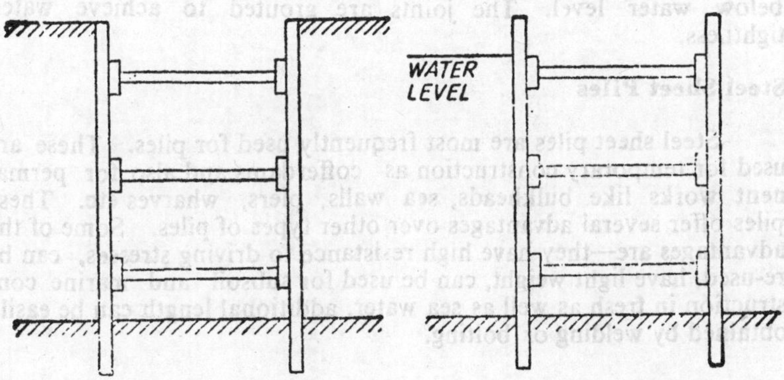

(*c*) Braced sheet piles (*d*) Cofferdam

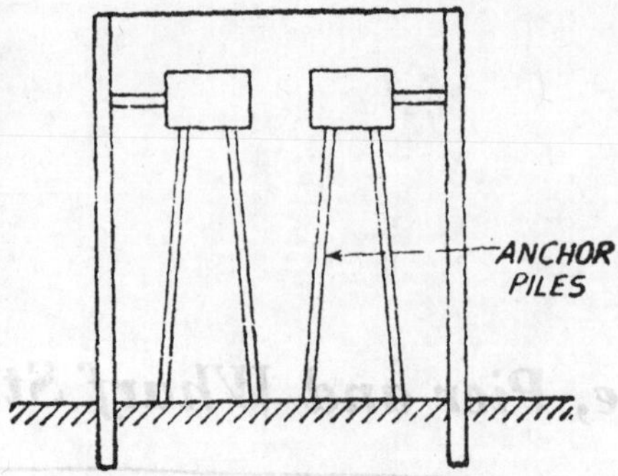

Anchored Bulkheads
Fig. 21·1. Sheet pile structures.

Sheet piles may be made of timber, reinforced concrete or steel.

Timber Sheet Piles

Timber sheet piles were being used some years back but now these are being replaced by steel piles. Timber sheet piles are even now used where timber is easily available. These are used for short spans, light loads and as temporary retaining structures in excavation. The life of these piles is relatively short. These piles simply consist of timber planks with tongue and groove joints.

Reinforced Concrete Sheet Piles

Reinforced concrete sheet piles are precast members having tongue and groove joints as shown in Fig. 21·2. These are used for permanent work where lengths are relatively short. The tongue in the pile may be throughout its length or through the portion below water level. The joints are grouted to achieve watertightness.

Steel Sheet Piles

Steel sheet piles are most frequently used for piles. These are used for temporary construction as cofferdams and also for permanent works like bulkheads, sea walls, piers, wharves etc. These piles offer several advantages over other types of piles. Some of the advantages are—they have high resistance to driving stresses, can be re-used, have light weight, can be used for subsoil and marine construction in fresh as well as sea water, additional length can be easily obtained by welding or bolting.

Fig. 21·3 shows various types of sheet piles in common use.

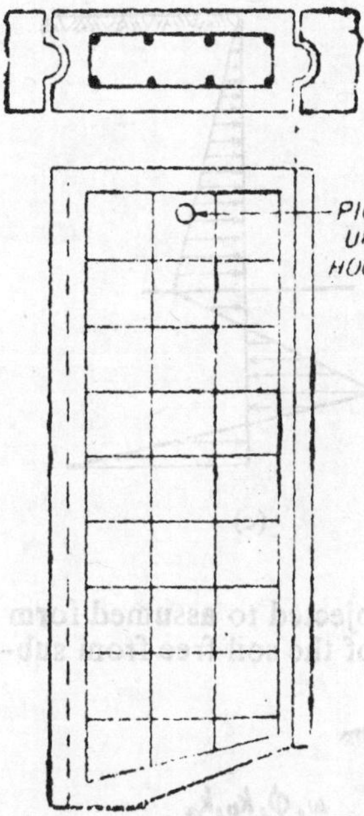

Reinforced sheet pile
Fig. 21·2

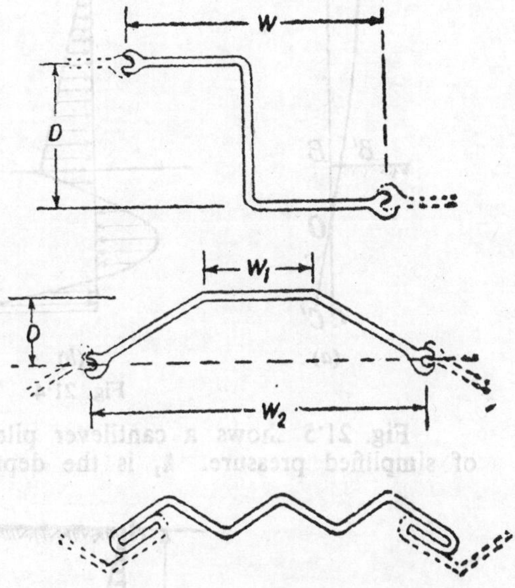

Some common steel sheet piles
Fig. 21·3

21·2. Cantilever Sheet Piling

In these sheet piles, lower portion is embedded in soil below dredge lines and upper portion is unsupported. These types of piles are used only for moderate heights as for large heights heavier sections are required, which prove uneconomical.

Actual pressure distribution due to earth in sheet piles is rather complicated Therefore, certain simplified nature of earth pressure is assumed in design. It is assumed in design that at some point *O*, the sheet pile rotates as shown in Fig. 21·4 (*a*). In portion *AB* the sheet pile will be acted upon by active pressure from right. Below *B* the sheet pile will be acted upon by active pressure from right and passive pressure from left. This is evident from the deflected form of pile. In portion *AB* the soil is stretched. Below *B*, the soil on right gets stretched and the soil on left is pressed. Similarly at *C*, the soil on right is pressed and soil on left is stretched, therefore, the pile will be subjected to passive pressure from right and active pressure from left. Fig. 21·4 (*b*) shows probable stress distribution against the pile. However, for design purposes, simplified distribution as shown in Fig. 21·4 (*c*) is assumed for computing earth pressure. For cohesionless soils either Rankine's theory or Coulomb's theory is applied.

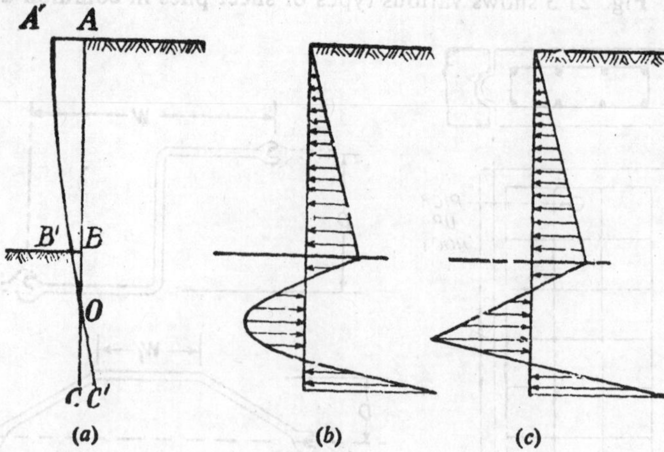

Fig. 21·4

Fig. 21·5 shows a cantilever pile subjected to assumed form of simplified pressure. h_1 is the depth of the soil free from sub-

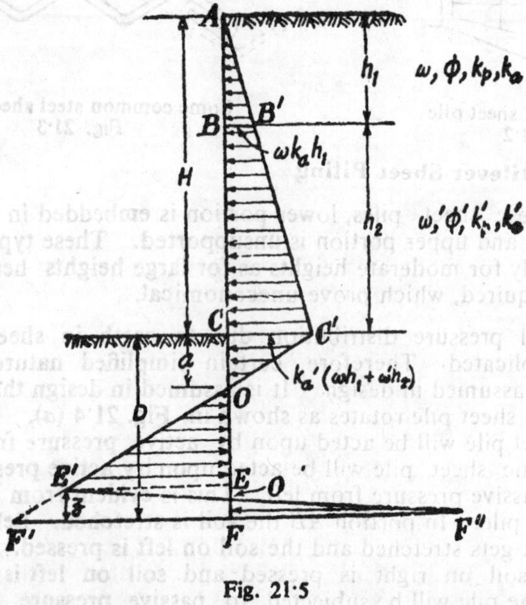

Fig. 21·5

soil water where properties of soil are ω=density, ϕ=angle of repose, k_p=passive pressure constant and k_a active pressure constant, h_2 is the depth of earth with sub-soiled water, where properties of soil are ω'=density, ϕ'=angle of repose, k_p'=passive pressure constant and k_a'=active pressure constant, $h_1+h_2=H$, the height of sheet pile supporting the retained earth. D is the depth of embedment of the pile. The depth D is to be determined. In portion AB there is only active pressure acting. At B pressure will be $\omega k_a h_1$. At C the pressure will be k_a' $(\omega h_1+\omega' h_2)$. Below C the active pressure will act from the right and passive pressure from the left till at O the pressure will be zero. Taking distance CO as a.

$$k_a'(\omega h_1+\omega' h_2)+\omega' k_a' a=\omega' k_p' a$$

$$a=\frac{k_a(\omega h_1+\omega' h_2)}{\omega(k_p'-k_a')}$$

Below O the pressure diagram will be $C'O$ extended till F' z distance above base and then pressure will act as shown by $F'F$. If active pressure were to act all along the pile from right and passive pressure from left, the pressure at base F will be

$$FF'=(\omega' k_p' D-\omega k_a' h_1)-\omega' k_a' h_2-\omega' k_a' D$$

$$=\omega' D(k_p'-k_a')-(\omega k_a' h_1+\omega' k_a' h_2)$$

$$=\omega' D(k_p'-k_a')-k_a'\ (\omega h_1+\omega' h_2)$$

The net pressure at F will be due to passive pressure at F from right and active pressure from left.

$$FF''=\omega k_p' h_1+\omega' k_p' h_2+\omega' k_p' D-\omega' k_a' D$$

$$=\omega' D(k_p'-k_a')+(\omega k_p' h_1+\omega' k_p' h_2)$$

$$=\omega' D(k'_p-k_a'+k_p'(\omega h_1+\omega' h_2)$$

$$F'F''=2\omega' D(k'_p-k_a')+(\omega h_1+\omega' h_2)\ (k_p'-k_a')$$

$$=(k_p'-k_a')[2\omega' D+(\omega h_1+\omega' h_2)]$$

The pressure acting on the pile will be as shown shaded in Fig. 21·5. To simplify calculations $OE'F'F$ is added to $OE'O'$ and OFF''. This will not affect the results. The forces acting are shown in Fig. 21·6. P_1 will be the area of $ABCOC'B'$ say acting at y from O acting towards left. P_2 will be the area of $E'F'F''O'$ acting at $z/3$ from base acting towards left. P_3 will be the area of $OF'F$ acting at $(D-a)/3$ from base acting towards right.

There are two unknowns D and z. The two equations are the total horizontal forces will be zero and total moment about base F will be zero.

$$\therefore \quad P_1+P_2+P_3=0 \qquad \text{...(i)}$$

$$P_1(D-a+y)+P_2\frac{z}{3}-P_3\frac{(D-a)}{3}=0 \qquad \text{...(ii)}$$

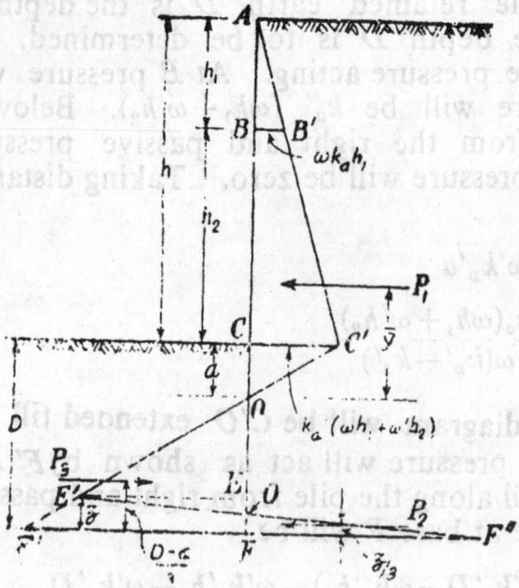

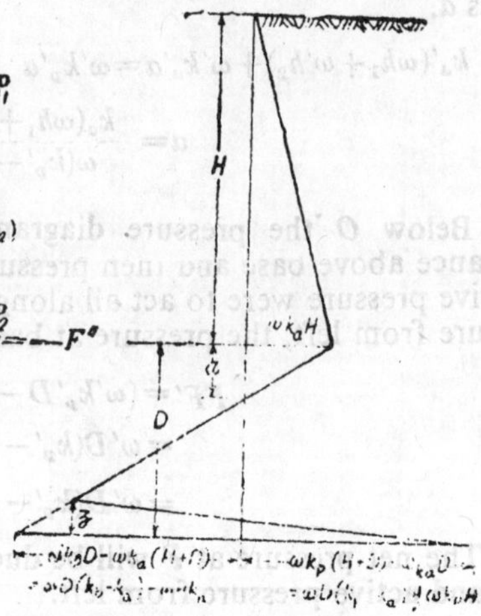

Fig. 21·6

Fig. 21·7

$$P_1 = \frac{\omega k_a h_1^2}{2} + k_a'(2\omega h_1 + \omega' h_2)\frac{h_2}{2} + \frac{k_a'(\omega h_1 + \omega' h_2)a}{2}$$

$$P_2 = \frac{z}{2}(k_p' - k_a')[2\omega' D + (\omega h_1 + \omega' h_2)]$$

$$P_3 = \frac{(D-a)}{2}[\omega' D(k_p' - k_a') - k_a'(\omega h_1 + \omega' h_2)].$$

From equation (*i*) value of z is obtained in terms of D and putting this value in equation (*ii*), value of D can be obtained by trial and error.

In case there is no difference in retained soil and bed soil $\omega = \omega'$, $\phi = \phi'$, $k_p = k_p'$, $k_a = k_a'$.

$$\therefore \quad a = \frac{k_a H}{k_p - k_a}$$

$$y = \frac{2a + H}{3}$$

$$P_1 = \frac{\omega k_a H(H+a)}{2}$$

$$P_2 = \frac{z\omega(k_p - k_a)(2D + H)}{2}$$

$$P_3 = \frac{(D-a)\omega[D(k_p - k_a) - k_a H]}{2}$$

$$\frac{\omega k_a H(H+a)}{2}+\frac{z\omega(k_p-k_a)(2D+H)}{2}$$

$$=\frac{z\omega[D(k_p-k_a)-k_aH]}{2}$$

$$\therefore \quad z=\frac{(D-a)\,[D(k_p-k_a)-k_aH]-k_aH(H+a)}{(k_p-k_a)\,(2D+H)} \qquad \ldots(iii)$$

$$P_1(D-a+y)+P_2\,\frac{z}{3}-P_3\,\frac{(D-a)}{3}=0.$$

$$\frac{\omega k_a H(H+a)\,(D-a+y)}{2}+\frac{z^2\omega(k_p-k_a)\,(2D+H)}{6}$$

$$-\frac{(D-a)^2\omega[D(k_p-k_a)-k_aH]}{6}=0 \qquad \ldots(iv)$$

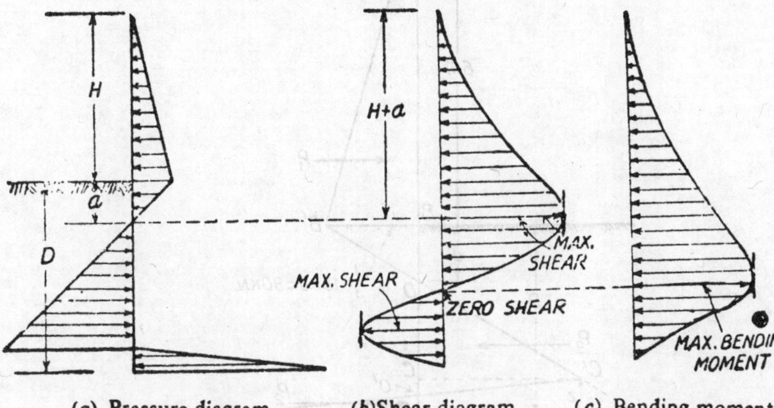

(a) Pressure diagram (b) Shear diagram (c) Bending moment diagram

Fig. 21·8

D will be approximately equal to 0·9 H to H. By trial and error the value of D can be obtained from equation (iv).

Shear force and bending moment diagrams can be drawn and maximum value of shear and bending moment can be obtained for designing the section.

Ex. 21·1. *A cantilever bulkhead is to be designed for height of 6 metres above bed level. If $\omega=15$ kN/m³ and $\phi=30°$, find the minimum depth to which the bulkhead has to be penetrated. Find the maximum shear and bending moment.*

Sol.
$$k_a=\frac{1-\sin\phi}{1+\sin\phi}=\frac{1-\sin 30°}{1+\sin 30°}$$

$$=\frac{1-\frac{1}{2}}{1+\frac{1}{2}}=\frac{1}{3}.$$

$$k_p=\frac{1+\sin\phi}{1-\sin\phi}=\frac{1+\frac{1}{2}}{1-\frac{1}{2}}=3.$$

$$a=\frac{k_aH}{k_p-k_a}=\frac{\frac{1}{3}\times 6}{3-\frac{1}{3}}=0{\cdot}75 \text{ m.}$$

$$EE'=15\times 3\times D-15\times \tfrac{1}{3}(6+D)$$
$$=40D-30$$

$$EE''=15\times 3(6+D)-15\times \tfrac{1}{3}D$$
$$=40D+270$$

$$E'E''=80D+240$$

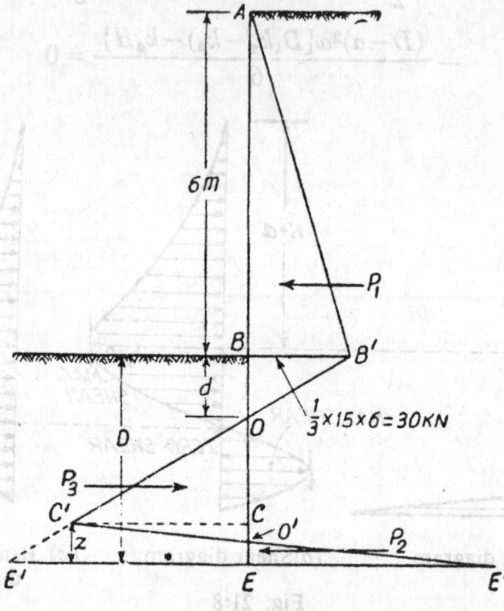

Fig. 21·9.

$$P_1=\frac{30(6+0{\cdot}75)}{2}=\frac{202{\cdot}5}{2}$$

acting at $\frac{2\times 0{\cdot}75+6}{3}=2{\cdot}5$ m from O

$$P_2=\frac{z}{2}\,(80D+240)$$
$=z(40D+120)$ acting at $z/3$ from base

$$P_3=\frac{(D-0{\cdot}75)(40D-30)}{2}$$ acting at $\frac{(D-0{\cdot}75)}{3}$ from base

$$\therefore\quad \frac{20{\cdot}25}{2}+4z(D+3)=\frac{(D-0{\cdot}75)\,(4D-3)}{2}$$

$$z=\frac{(D-0{\cdot}75)\,(4D-3)-20{\cdot}25}{8(D+3)}$$

$$=\frac{4D^2-6D+2{\cdot}25-20{\cdot}25}{8(D+3)}$$

$$=\frac{4D^2-6D-18}{8(D+3)}$$

$$=\frac{2D^2-3D-9}{4(D+3)} \qquad \ldots(i)$$

Taking moments about E,

$$\frac{202{\cdot}5}{2}\times(D-0{\cdot}75+2{\cdot}5)+\frac{10\times 4z^2(D+3)}{3}$$

$$=\frac{10(D-0{\cdot}75)\,(4D-3)\,(D-0{\cdot}75)}{6}$$

$$\therefore \frac{20{\cdot}25}{2}\,(D+1{\cdot}75+)\,\frac{4(D+3)\,(2D^2-3D-9)^2}{3\times 10(D+3)^2}=\frac{(D-0{\cdot}75)^2(4D-3)}{6}$$

$$\therefore \frac{20{\cdot}25(D+1{\cdot}75)}{2}+\frac{(2D^2-3D-9)^2}{12(D+3)}=\frac{(D-0{\cdot}75)^2(4D-3)}{6}$$

$$\therefore \quad 121{\cdot}5(D+1{\cdot}75)\,(D+3)+(2D^2-3D-9)^2=2(D-0{\cdot}75)^2\,(4D-3)(D+3)$$

$$\therefore \quad 4D^4-18D^3-135\,D^2-594D-719{\cdot}2=0$$

$$\therefore \quad D^4+4{\cdot}50D^2-33{\cdot}75D^2-148{\cdot}5D-179{\cdot}8=0.$$

Let $D=6$ m

$\therefore$ L.H.S. $=1296+972{\cdot}1-1214-890-179{\cdot}8=-15{\cdot}7$

Let $D=6{\cdot}1$

$\therefore$ L.H.S. $=1385+1021-1255-905{\cdot}7+179{\cdot}8=+75{\cdot}5$

Let $D=6{\cdot}02$ m

$\therefore$ L.H.S. $=1313+981{\cdot}7-1223-893{\cdot}3-179{\cdot}8=1{\cdot}4$

$\therefore$ Minimum depth will be $6{\cdot}022$ m

$$z=\frac{2\times(6{\cdot}022)^2-3\times 6{\cdot}022-9}{4(6{\cdot}022+3)}$$

$$=\frac{72{\cdot}51-18{\cdot}066-9}{36{\cdot}088}$$

$$=\frac{45{\cdot}444}{36{\cdot}088}$$

$$=1{\cdot}26 \text{ m}$$

$$EE'=40\times 6{\cdot}022-30$$

$$=210{\cdot}88-30$$

$$=210{\cdot}88$$

$$EE''=40D+270$$
$$=40\times 6{\cdot}022+270$$
$$=240{\cdot}88+270$$
$$=570{\cdot}88$$
$$CO'=\frac{EE'\times 4{\cdot}012}{5{\cdot}272}$$
$$=\frac{210{\cdot}88\times 4{\cdot}012}{5{\cdot}272}$$
$$=160{\cdot}4$$
$$CO'=\frac{1{\cdot}26\times 16{\cdot}4}{160{\cdot}4+510{\cdot}88}=0{\cdot}31 \text{ m}$$
$$O'E=1{\cdot}26-0{\cdot}31=0{\cdot}95 \text{ m}$$

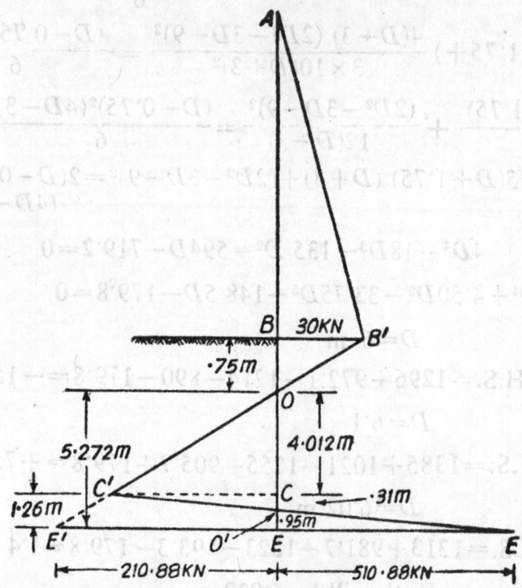

Fig. 21·10

Shear force in AB

$$F_y=\tfrac{1}{3}\times 15\times\frac{y^2}{2}=\frac{y^2}{4}$$

$$F_6=\frac{360}{4}=90 \text{ kN}$$

Bending moment

$$M_y=\tfrac{1}{3}\times 15\times\frac{y^3}{6}=\frac{5y^3}{6}$$

$$M_6=\frac{5\times 6^3}{6}=180 \text{ kN·m}$$

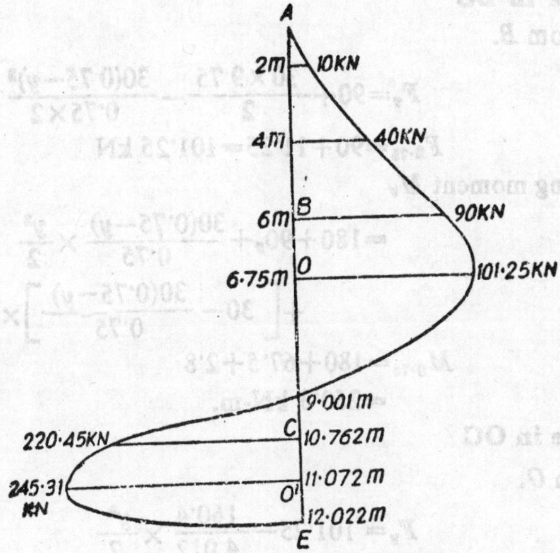

(a) Shear force diagram.

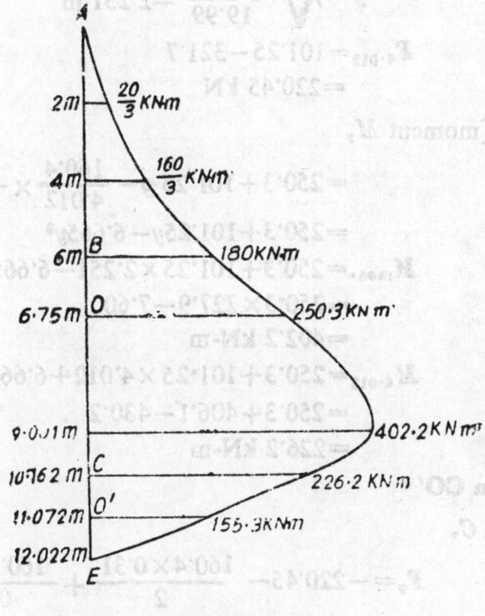

(b) Bending moment diagram.
Fig. 21·11.

Shear force in BO

y from B.

$$F_y = 90 + \frac{30 \times 9.75}{2} - \frac{30(0.75-y)^2}{0.75 \times 2}$$

$$F_{0.75} = 90 + 11.25 = 101.25 \text{ kN}$$

Bending moment M_y

$$= 180 + 90_y + \frac{30(0.75-y)}{0.75} \times \frac{y^2}{2}$$

$$+ \left[30 - \frac{30(0.75-y)}{0.75}\right] \times \frac{y^2}{6}$$

$$M_{0.75} = 180 + 67.5 + 2.8$$

$$= 250.3 \text{ kN-m.}$$

Shear force in OC

y from O.

$$F_y = 101.25 - \frac{160.4}{4.012} \times \frac{y^2}{2}$$

$$= 101.25 - 19.99\, y^2$$

Shear force will be zero at,

$$y = \sqrt{\frac{101.25}{19.99}} = 2.251 \text{ m}$$

$$F_{4.012} = 101.25 - 321.7$$

$$= 220.45 \text{ kN}$$

Bending moment M_y

$$= 250.3 + 101.25\, y - \frac{160.4}{4.012} \times \frac{y^3}{6}$$

$$= 250.3 + 101.25y - 6.665y^3$$

$$M_{max.} = 250.3 + 101.25 \times 2.251 - 6.665 \times (2.251)^3$$

$$= 250.3 \times 227.9 - 7.60$$

$$= 402.2 \text{ kN-m}$$

$$M_{4.012} = 250.3 + 101.25 \times 4.012 + 6.665 \times (4.012)^3$$

$$= 250.3 + 406.1 - 430.2$$

$$= 226.2 \text{ kN-m}$$

Shear force in CO′

y from C.

$$F_y = -220.45 - \frac{160.4 \times 0.31}{2} + \frac{160.4(0.31-y)^2}{0.31 \times 2}$$

$$F_{0.31} = -220.45 - 24.86$$

$$= 245.31 \text{ kN}$$

$$\text{Bending moment} = 226{\cdot}2 - 220{\cdot}45y - \frac{160{\cdot}4(0{\cdot}31-y)}{0{\cdot}31} \times \frac{y^2}{2}$$
$$- \left[160{\cdot}4 - \frac{160{\cdot}4(0{\cdot}31-y)}{0{\cdot}31}\right] \times \frac{y^2}{6}$$
$$M_{0{\cdot}31} = 226{\cdot}2 - 220{\cdot}45 \times 0{\cdot}31 - 160{\cdot}4 \times \frac{(0{\cdot}31)^2}{6}$$
$$= 216{\cdot}2 - 68{\cdot}3 = 2{\cdot}6 = 155{\cdot}3 \text{ kN-m}$$

Shear force in O'E

y from O'.

$$F_y = -245{\cdot}31 + \frac{5\,0{\cdot}88}{0{\cdot}95} \times \frac{y^2}{2}$$
$$= -245{\cdot}31 + 268{\cdot}9\,y^2$$

Bending moment

$$M_y = 155{\cdot}3 - 245{\cdot}31\,y + \frac{510{\cdot}88}{0{\cdot}95} \times \frac{y^3}{6}.$$

21·3. Anchored Sheet Piles

Cantilever sheet piles prove to be uneconomical for long lengths. Therefore, for long sheet piles, anchors are used at top which reduce considerably bending moments in the sheet pile. Anchored sheet piles are used for dock and harbour structures.

Fig. 21·12 shows anchored sheet pile with minimum length embedded. The deflected form is shown. The pile rotates about the

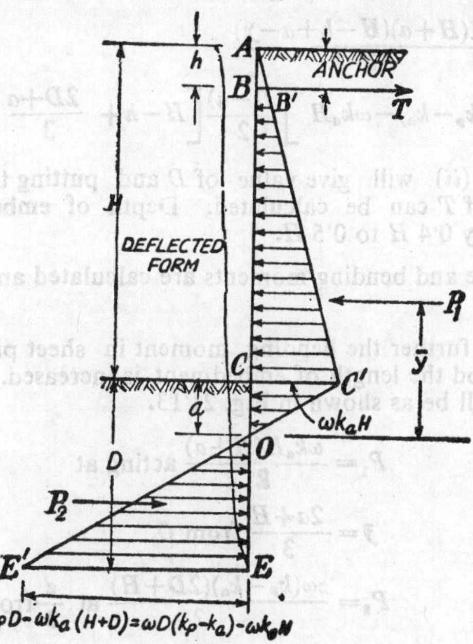

Fig. 21·12

base E. In portion AC there will be active pressure from right. In portion CE there will be active pressure acting from right and passive pressure from left.

$$CC'=\omega k_a H$$

$$EE'=\omega k_p D-\omega k_a\,(H+D)$$

$$=\omega D(k_p-k_a)-\omega k_a H.$$

$$a=\frac{k_a H}{k_p-k_a}$$

$$P_1=\frac{\omega k_a H(H+a)}{2} \text{ acting at}$$

$$y=\frac{2a+H}{3} \text{ from } O.$$

$$P_2=[\omega D(k_p-k_a)-\omega k_a H]\times\frac{(D-a)}{2}$$

$$\text{acting at } \frac{D-a}{3} \text{ from base.}$$

If T is the tension in anchor,

$$P_1-P_2-T=0 \qquad ...(i)$$

Moment about anchor rod will be zero,

$$P_1[H-h+a-y]=P_2\left[H-h+D-\frac{D-a}{3}\right]$$

$$\therefore \frac{\omega k_a H(H+a)(H-h+a-y)}{2}$$

$$=\left[\omega D(k_p-k_a)-\omega k_a H\right]\frac{(D-a)}{2}\left[H-h+\frac{2D+a}{3}\right] \qquad ...(i$$

Equation (ii) will give value of D and putting this value of D in (i), value of T can be calculated. Depth of embedment will be approximately 0·4 H to 0·5 H.

Shear force and bending moments are calculated and the sheet pile designed.

To reduce further the bending moment in sheet pile and tension in anchor rod the length of embedment is increased. The pressure diagram will be as shown in Fig. 21·13.

$$P_1=\frac{\omega k_a H(H+a)}{2} \text{ acting at}$$

$$y=\frac{2a+H}{3} \text{ from } O.$$

$$P_2=\frac{z\omega(k_p-k_a)(2D+H)}{2} \text{ at } \frac{z}{3} \text{ from base.}$$

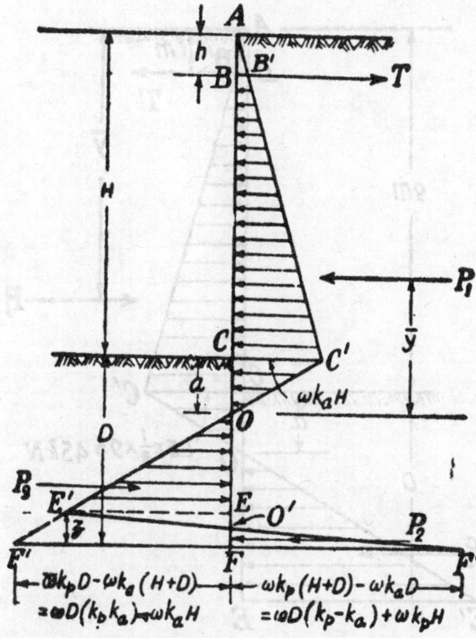

Fig. 21·13

$$P_3=\frac{(D-a)\omega[D(k_p-k_a)-k_aH]}{2}$$

acting at $\frac{D-a}{3}$ from base

$$P_1+P_2-P_3-T=0 \qquad \ldots(i)$$

$$P_1(H-h+a-\bar{y})+P_2\left(H-h+D-\frac{z}{3}\right)-P_3\left[H-h+D-\frac{(D-a)}{3}\right]=0 \qquad \ldots(ii)$$

The value of z can be obtained from equation (*ii*) and putting this value in equation (*i*) value of T can be obtained.

Ex. 21·2. ***Find the minimum penetration for an anchored bulkhead of height 9 metres above bed-level, $\phi=30°$ and $\omega=15$ kN/m³. The anchor is 1 metre from top. Find the max. bending and max. shear in pile and tension in anchor.***

Sol.

$$k_a=\frac{1-\sin\phi}{1+\sin\phi}=\frac{1-\frac{1}{2}}{1+\frac{1}{2}}=\frac{1}{3}.$$

$$k_p=\frac{1+\sin\phi}{1-\sin\phi}=\frac{1+\frac{1}{2}}{1-\frac{1}{2}}=3.$$

$$a=\frac{k_aH}{k_p-k_a}=\frac{9}{8}\text{ metres}$$

$$CC'=\frac{1}{3}\times15\times9=45\text{ kN.}$$

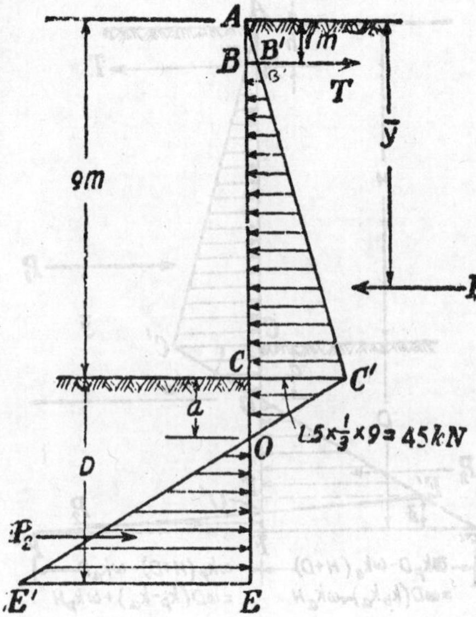

Fig. 21·14

$$EE'=3\times15\,D-\frac{1}{3}\times15(9+D)$$

$$=(40\,D-45)$$

$$P_1=\frac{45(9+\frac{8}{9})}{2}\text{ acting at}$$

$$\bar{y}=\frac{2\times9+\frac{9}{8}}{3}=\frac{51}{8}\text{ metres from top.}$$

$$P_2=(40\,D-45)\times\frac{(D-\frac{9}{8})}{2}\text{ acting at }\frac{(D-\frac{9}{8})}{3}\text{ from base.}$$

Taking moments about anchor,

$$P_1(\bar{y}-1)=P_2\left[8+D-\frac{(D-\frac{9}{8})}{3}\right]$$

$$\therefore\quad \frac{45(81/8)}{2}\times\left(\frac{51}{8}-1\right)=(40\,D-45)\frac{(D-\frac{9}{8})}{2}\times\left[\frac{27+3D-D+\frac{9}{8}}{3}\right]$$

$$\frac{4{\cdot}5\times81\times43}{16\times8}=(4D-4{\cdot}5)\frac{(8D-9)(16D+201)}{16\times24}$$

$$3\times4{\cdot}5\times81\times43=(4D-4{\cdot}5)(8D-9)(16D+201)$$

$$\therefore\quad 3\times4{\cdot}5\times81\times43=512D^3+5280D^2-13{,}824D+8140{\cdot}5$$

$$\therefore\quad D^3+10{\cdot}31D^2-27D+15{\cdot}9-91{\cdot}83=0$$

$\therefore \quad D^3+10{\cdot}31\,D^2-27D-75{\cdot}93=0$

Let $\quad D=4$ metres.

$\therefore \quad$ L.H.S. $=64+164{\cdot}96-108-75{\cdot}93=+45{\cdot}03$

$D=3{\cdot}5$ metres.

$\therefore \quad$ L.H.S. $=42{\cdot}88+126{\cdot}3-94{\cdot}5-75{\cdot}93=-1{\cdot}25$

$D=3{\cdot}51$ metres.

L.H.S. $=43{\cdot}25+127{\cdot}1-94{\cdot}75-75{\cdot}93=-0{\cdot}34$

$$EE'=40\times3{\cdot}51-45$$

$$=140{\cdot}4-45$$

$$=95{\cdot}4 \text{ kN.}$$

$$P_1=\frac{45\times81}{16}=227{\cdot}8 \text{ kN}$$

$$P_2=(40\times3{\cdot}51-45)\ \frac{(3{\cdot}51-\frac{9}{8})}{2}$$

$$=95{\cdot}4\times1{\cdot}1925=113{\cdot}8 \text{ kN}$$

$$T=P_1-P_2$$

$$=227{\cdot}8-113{\cdot}8=114 \text{ kN}$$

Max. shear $\quad =P_1-T=227{\cdot}8-114{\cdot}0=113{\cdot}8$ kN

Shear at anchor $\quad =T-\frac{1}{3}\times15\times\frac{1^2}{2}=114{\cdot}0-2{\cdot}5$

$$=111{\cdot}5 \text{ kN}$$

$$F_y=T-\frac{1}{3}\times15\times\frac{y^2}{2}=114-2{\cdot}5y^2$$

For max. B.M., $\quad F_y=0$

$$y=\sqrt{\frac{114}{2{\cdot}5}}$$

$$=6{\cdot}753 \text{ m}$$

$$M_{max}=T(6{\cdot}753-1)-\frac{1}{3}\times15\times\frac{y^3}{6}$$

$$=114\times5{\cdot}753-\frac{5}{6}\times(6{\cdot}753)^3$$

$$=655{\cdot}8\times256{\cdot}6$$

$$=399{\cdot}2 \text{ kN-m.}$$

Shear force and B.M. in AB

y from *A*.

$$F_y=\frac{1}{3}\times15\times\frac{y^2}{2}=2{\cdot}5\,y^2$$

$$F_1=2{\cdot}5 \text{ kN}$$

$$M_y = \frac{1}{3} \times 15 \times \frac{y^3}{6} = \frac{5y^3}{6}$$

$$M_1 = \frac{5}{6} \text{ kN-m.}$$

Shear force and B.M. in BC

y from A.

$$F_y = \frac{1}{3} \times 15 \times \frac{y^2}{2} - T$$

$$= 2{\cdot}5\, y^2 - 114$$

$$F_9 = 202{\cdot}5 - 114 = 88{\cdot}5 \text{ kN}$$

$$M_y = \frac{1}{3} \times 15 \times \frac{y^3}{6} - T(y-1)$$

$$= \frac{5y^3}{6} - 114(y-1)$$

$$M_3 = \frac{5 \times 27}{6} - 228 = -205{\cdot}5 \text{ kN-m}$$

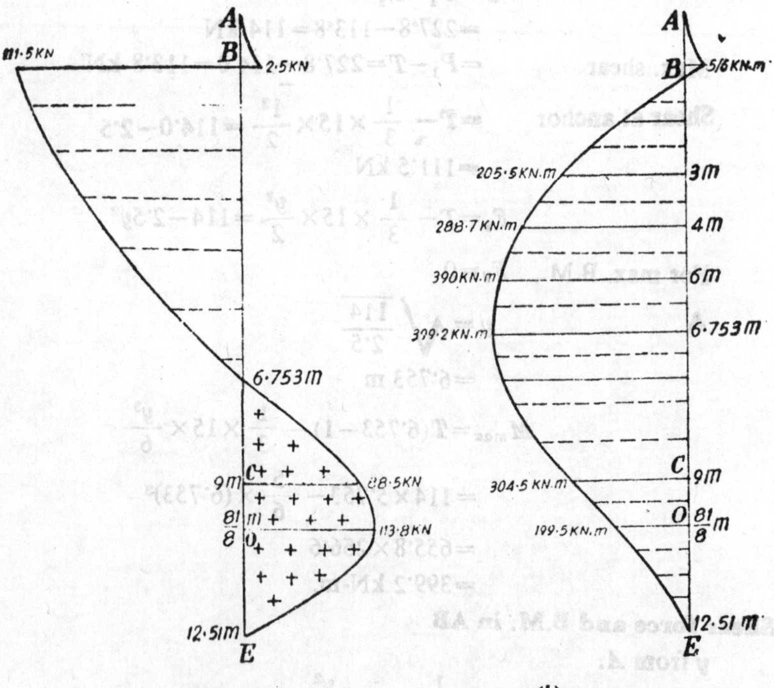

(a) Shear force diagram (b) Bending moment diagram

Fig. 21·15

$$M_4 = -\frac{5\times 64}{6} - 342 = -288{\cdot}7 \text{ kN-m}$$

$$M_6 = 180 - 570 = -390 \text{ kN-m}$$

$$M_{6\cdot 753} = \frac{5\times(6{\cdot}753)^3}{6} - 114\times 5{\cdot}753$$

$$= 256{\cdot}6 - 655{\cdot}8 = -399{\cdot}2 \text{ kN-m}$$

$$M_9 = \frac{5\times 9^3}{6} - 114\times 8$$

$$= 607{\cdot}5 - 912 = -304{\cdot}5 \text{ kN-m.}$$

Shear force and B.M. in CO

y from C.

$$F_y = 88{\cdot}5 + 45\times\frac{9}{8\times 2} - \frac{45}{9/8}\times\frac{(\frac{9}{8}-y)^2}{2}$$

$$= 88{\cdot}5 + 25{\cdot}3 - 20(\tfrac{9}{8}-y)^2$$

$$= 113{\cdot}8 - 20(\tfrac{9}{8}-y)^2$$

$$F_{\frac{9}{8}} = 113{\cdot}8 \text{ kN}$$

$$M_y = -304{\cdot}5 + 88{\cdot}5y + \frac{45}{\frac{9}{8}}(\tfrac{9}{8}-y)\times\frac{y^2}{2}$$

$$+\left[\frac{45\times(\frac{9}{8}-y)}{\frac{9}{8}}\right]\times\frac{y^2}{6}$$

$$M_{\frac{9}{8}} = -304{\cdot}5 + 99{\cdot}5 + 9{\cdot}5$$

$$= -195{\cdot}5 \text{ kN-m.}$$

Shear force and B.M. in OE

y from O.

$$F_y = 113{\cdot}8 - \frac{95{\cdot}4}{(3{\cdot}51-\frac{9}{8})}\times\frac{y^2}{2}$$

$$= 113{\cdot}8 - \frac{(381{\cdot}5y)^2}{19{\cdot}08}$$

$$= 113{\cdot}8 - (20y)^2$$

$$M_y = -195{\cdot}5 + 113{\cdot}8y - \frac{95{\cdot}4}{(3{\cdot}51-\frac{9}{8})}\times\frac{y^3}{6}$$

$$= -195{\cdot}5 + 113{\cdot}8y - \frac{(20y)^3}{3}$$

Ex. 21·3. *If the depth of embedment is increased to 5 m in problem 21·2, find the max. B. M. and max. S.F. and tension in the anchor.*

Sol. $P_1 = \dfrac{45(9+\frac{9}{8})}{2} = \dfrac{45\times 81}{16}$ kN

acting at $\dfrac{2\times 9+\frac{9}{8}}{3} = \dfrac{51}{8}$ m from top

$$P_2 = 755\times\frac{z}{2} = 377{\cdot}5\,z$$

acting at $\dfrac{z}{3}$ from base.

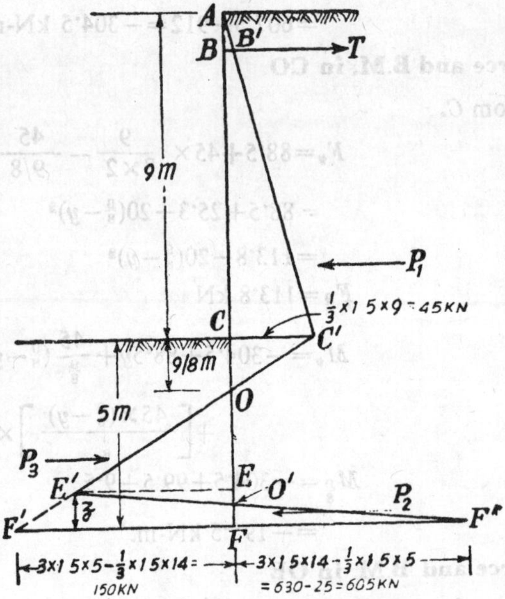

Fig. 21·16

$$P_3 = 150\times\frac{(5-\frac{9}{8})}{2} = \frac{150\times 31}{16}$$

acting at $\dfrac{5-\frac{9}{8}}{3} = \dfrac{31}{24}$ m from base.

Taking moments about anchor,

$$\frac{45\times 81}{16}\times\left(\frac{51}{8}-1\right)+377{\cdot}5z\left(8+5-\frac{z}{3}\right)$$

$$=\frac{150\times 31}{16}\times\left(8+5-\frac{31}{24}\right)$$

$$\therefore\quad \frac{45\times 81\times 43}{16\times 8}+\frac{37{\cdot}75z(39-z)}{3}=\frac{15\times 31\times 281}{16\times 24}$$

$$\therefore \quad -\frac{37 \cdot 75 z^2}{3} + 37 \cdot 75 \times 13z - \frac{15 \times 31 \times 281}{16 \times 24} + \frac{4 \cdot 5 \times 81 \times 43}{8} = 0$$

$$\therefore \quad z^2 - 39z + \frac{45 \times 31 \times 281}{16 \times 24 \times 37 \cdot 75} + \frac{13 \cdot 5 \times 81 \times 43}{128 \times 37 \cdot 75} = 0$$

$$\therefore \quad z^2 - 39z + 27 \cdot 05 - 9 \cdot 73 = 0$$

$$z^2 - 39z + 17 \cdot 32 = 0$$

$$(z - 19 \cdot 5)^2 = (19 \cdot 5)^2 - 17 \cdot 32$$

$$= 380 \cdot 2 - 17 \cdot 32 = 362 \cdot 88$$

$$z = -19 \cdot 05 + 19 \cdot 5 = 0 \cdot 45 \text{ m}$$

$$\therefore \quad P_1 = \frac{45 \times 81}{16} = 227 \cdot 8 \text{ kN}$$

$$P_2 = 37 \cdot 75 \times 0 \cdot 45 = 169 \cdot 8 \text{ kN}$$

$$P_3 = \frac{150 \times 31}{16} = 290 \cdot 6 \text{ kN}$$

$$T = P_1 + P_2 - P_3$$

$$= 227 \cdot 8 + 169 \cdot 8 - 290 \cdot 6$$

$$= 397 \cdot 6 - 290 \cdot 6$$

$$= 107 \text{ kN (tension in anchor)}$$

Max. shear $= 227 \cdot 8 - 107 \cdot 0 = 120 \cdot 8$ kN.

For Max. B.M. shear should be zero,

$$\therefore \quad T - \frac{1}{3} \times 15 \times \frac{y^2}{2} = 0$$

$$\therefore \quad y = \sqrt{\frac{107}{2 \cdot 5}}$$

$$y = \sqrt{42 \cdot 80} = 6 \cdot 542 \text{ m}$$

$$\text{Max. B.M.} = T(6 \cdot 542 - 1) - \frac{1}{3} \times 15 \times \frac{y^3}{6}$$

$$= 107 \times 5 \cdot 542 - \frac{5(6 \cdot 542)^3}{6}$$

$$= 593 \cdot 0 - 232 \cdot 8$$

$$= 360 \cdot 2 \text{ kN-m.}$$

21·4. Anchor for Sheet Piles

Anchors for sheet piles may be in the form of concrete blocks or beams embedded in soil or battered piles to take horizontal pull. Fig. 21·17 shows some anchors commonly used.

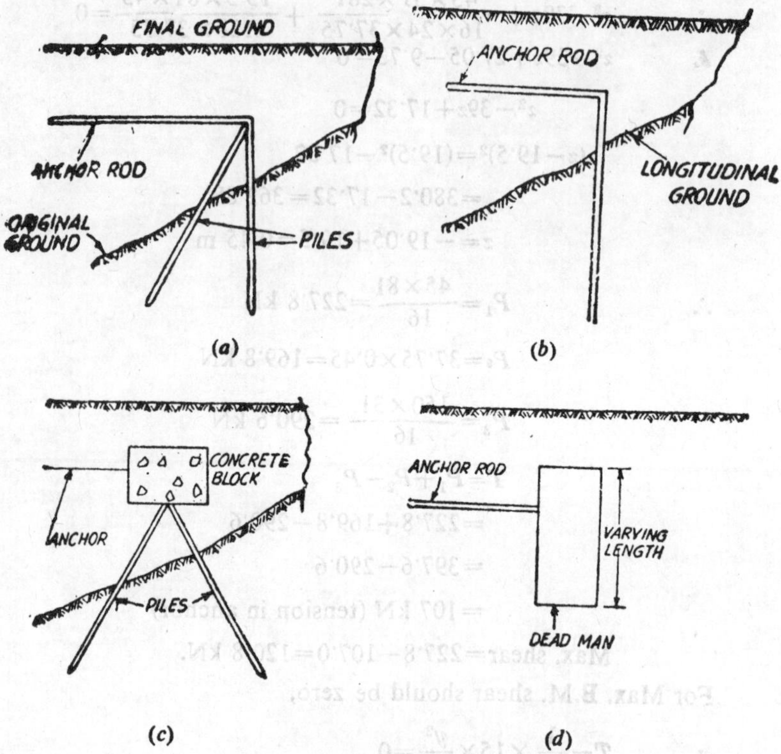

Fig. 21·17. Commonly used anchors.

The anchorage should be located in soil behind the portion of the earth developing active pressure on the wall.

21·5. Caisson Foundation

Caisson foundations are used where heavy loads are to be transferred through deep layer of weak soil to a firm base. These foundations are also used in river or other constructions, where foundations are to be taken below scour depth. The function performed by this foundation is the same as by pile foundation. Caisson foundations are preferred at places where pile driving vibrations are not allowed or at places where very large bearing area is required. Caisson foundations are also used for bridge piers, abutments, docks, wharves, sea walls etc. Fig. 21·18 shows some caisson foundations commonly used.

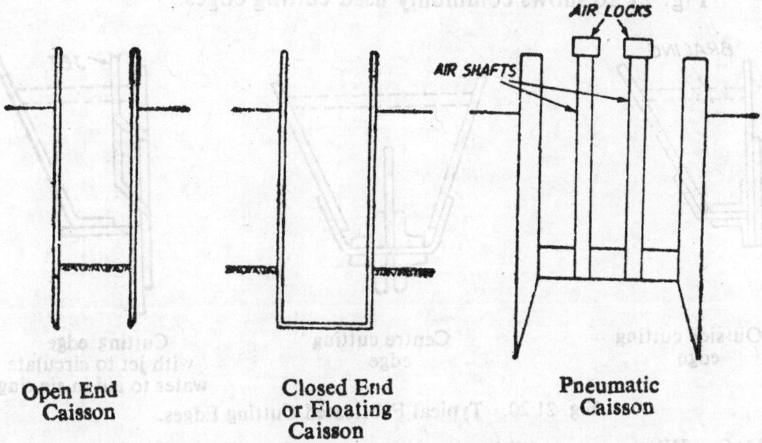

Fig. 21·18. Caisson Foundations.

Design Considerations

In caisson foundations part of the load is transferred to the soil through skin friction. Total capacity is the sum of the resistance at the base and skin friction. Negative skin friction may be developed under certain circumstances. The effect of negative skin friction is generally neglected.

Caisson should be checked at the base against overturning due to horizontal forces. These are also checked against sliding. In cases where caisson is subjected to very heavy loads, the resistance of caisson to lateral loads is sometimes increased by racking piles as shown in Fig. 21·19.

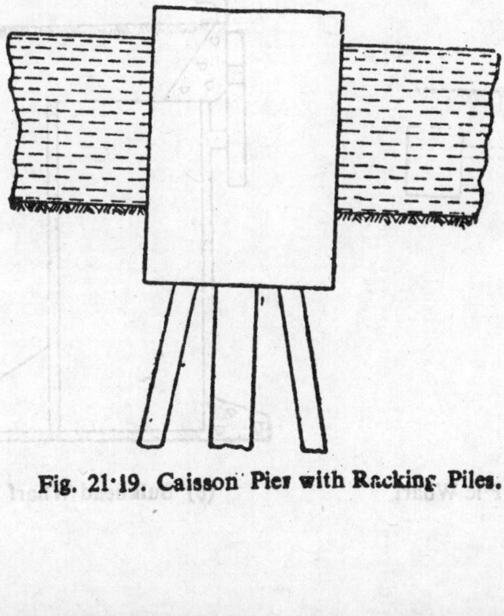

Fig. 21·19. Caisson Pier with Racking Piles.

Fig. 21 20 shows commonly used cutting edges.

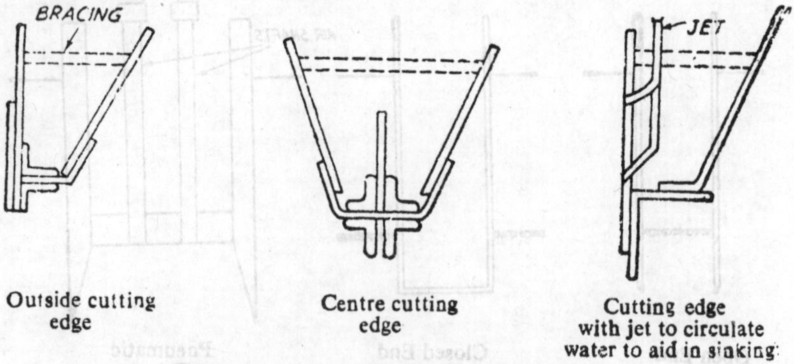

Fig. 21·20. Typical Fabricated Cutting Edges.

21·6. Wharves and Piers

Wharves and piers are structures located at edge or inside water and vessels are tied to these structures and they provide direct connection between water and load carriers. A wharf is a structure to which a vessel is tied and has water only on one side. A pier is a structure which projects into water and has water on both sides.

Fig. 21·21 shows various types of wharf structures commonly used and Fig. 21·22 shows various types of pier structures.

Fig. 21·22 (*a*) shows pier structure consisting of two rows of sheet piles tied together by anchor rods above water level. The portion between parallel rows of sheet piles is filled to form a pier. This type of structure is economical for short depths.

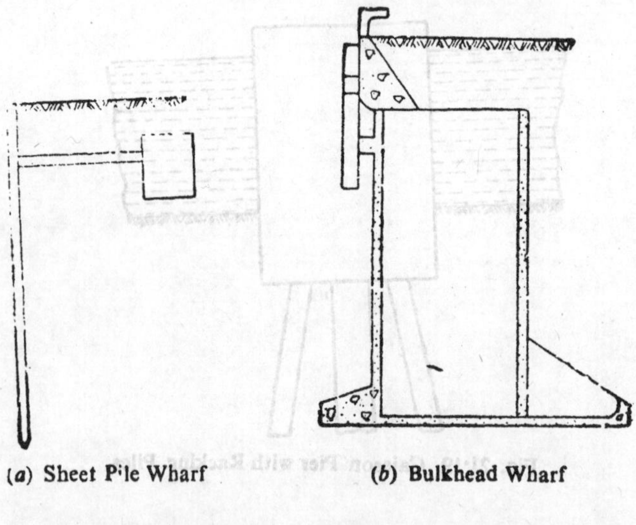

(*a*) Sheet Pile Wharf (*b*) Bulkhead Wharf

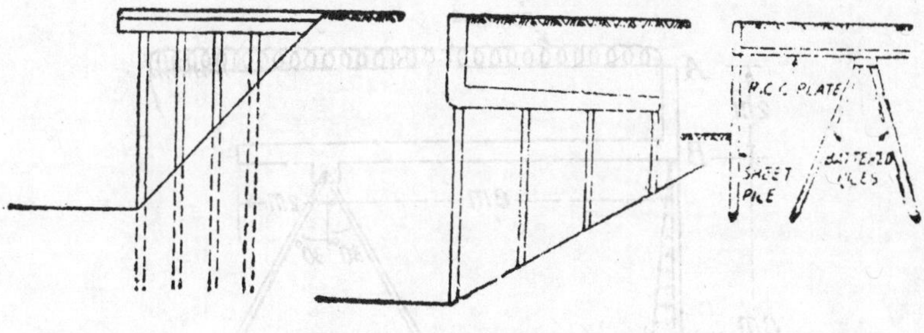

(c) (d) (e)

Fig. 21·21. Wharf structures on piles.

Fig. 21·22 (*a*) shows pier structure on sheet piles. This type of structure is economical for large depths. The pier shown in Fig. 21·22 (*c*) is used for large width and in deep waters. Top of piles are tied together at the top by anchor rods.

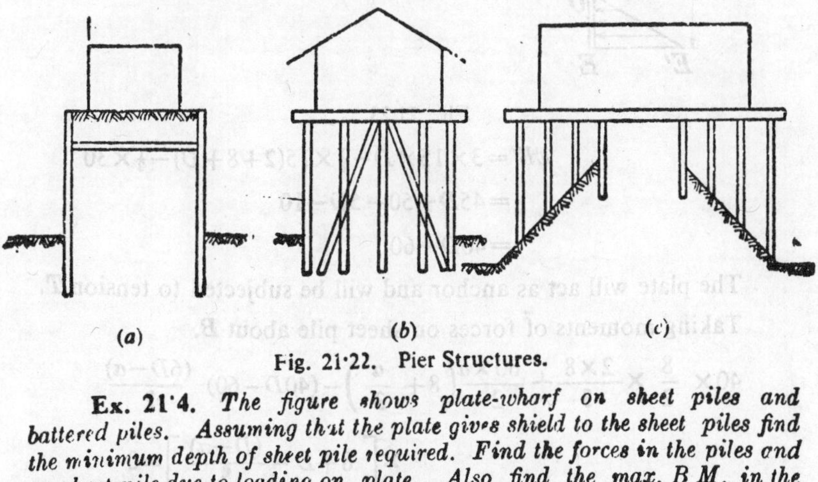

(a) (b) (c)

Fig. 21·22. Pier Structures.

Ex. 21·4. *The figure shows plate-wharf on sheet piles and battered piles. Assuming that the plate gives shield to the sheet piles find the minimum depth of sheet pile required. Find the forces in the piles and on sheet pile due to loading on plate. Also find the max. B M. in the plate. Take density of earth 15 kN/m³ and* $\phi=30°$.

Sol. Let D be the depth of the sheet pile. The pressure on sheet pile is shown in Fig. 21·23. As the sheet pile is protected by the plate, the effect of surcharge will not come on the portion BC of sheet pile but on CE portion the effect of surcharge will be there.

$$k_a=\frac{1-\sin\phi}{1+\sin\phi}=\frac{1}{3},\ k_p=3.$$

$$CC'=\tfrac{1}{3}\times 15\times 8=40\ \text{kN}$$

$$CC''=\tfrac{1}{3}\times 15(8+2)+\tfrac{1}{3}\times 30$$

$$=50+10=60\ \text{kN}$$

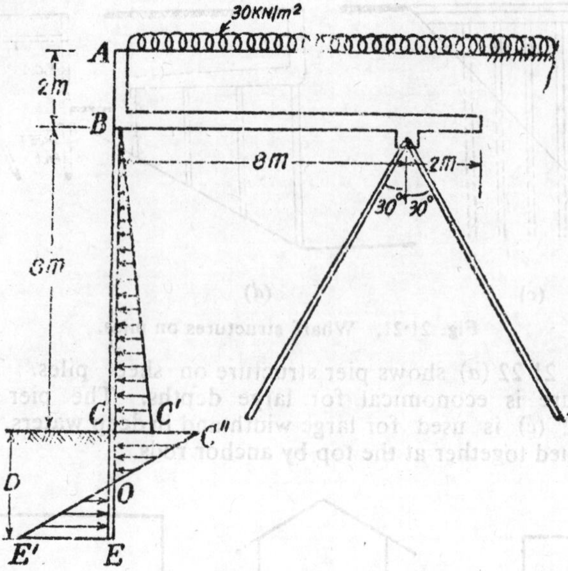

Fig. 21·23

$$EE' = 3\times15\times D - \tfrac{1}{3}\times15(2+8+D) - \tfrac{1}{3}\times30$$
$$= 45D - 50 - 5D - 10$$
$$= 40D - 60$$

The plate will act as anchor and will be subjected to tension T.

Taking moments of forces on sheet pile about B.

$$40\times\frac{8}{2}\times\frac{2\times8}{3} + \frac{60\times a}{2}\left(8+\frac{a}{3}\right) - (40D-60)\,\frac{(6D-a)}{2}\times\left[8+D-\frac{(D-a)}{3}\right] = 0$$

Now $60 + \tfrac{1}{3}\times15\times a = 3\times15\times a$

$$\therefore \quad a = \frac{6}{4} = 1\text{·}5 \text{ m}$$

$$\therefore \quad \frac{160\times16}{2} + \frac{90\times17}{3} - \frac{(40D-60)\,(D-1\text{·}5)\,(24+2D+1\text{·}5)}{2\times3} = 0$$

$$\therefore \quad \frac{1483}{12} = \frac{(2D-3)\,(D-1\text{·}5)\,(2D+25\text{·}5)}{3}$$

$$1483 = 4(2D-3)\,(D-1\text{·}5)\,(2D+25\text{·}5)$$
$$= 16D^3 + 156D^2 - 576D + 549$$

$\therefore \quad 16D^3+156D^2-576D-934=0$

$\therefore \quad D^3+9.75D^2-36D-58.38=0$

Let $\quad D=3.5$ m

$\therefore \quad$ L.H.S. $=42.88+119.5-126-58.38 \;=22,$

$D=3.8$ m

L.H.S. $=54.88+140.8-136.8-58.38=-0.5$

$$T=\frac{40\times 8}{2}+\frac{60\times 1.5}{2}(40\times 3.8-60)\left[\frac{3.8-1.5}{2}\right]$$

$$=160+45-\frac{72\times 2.3}{2}$$

$$=205-82.8$$

$$=122.2 \text{ kN.}$$

Forces on sheet pile and piles

$$P_1=\frac{1}{3}\times 30\times 2$$

$=20$ kN acting at 1 m from B

$$P_2=\frac{1}{3}\times 15\times\frac{(2)^2}{2}$$

$=10$ kN acting at $\frac{2}{3}$ m from B

$R_1+(R_2+R_3)\cos 30=300+300$

$\therefore \quad R_1+\dfrac{\sqrt{3}(R_2+R_3)}{2}=600 \qquad \ldots(i)$

Take moments about F

$$P_1(8\sqrt{3}+1)+P_2\left(8\sqrt{3}+\frac{2}{3}\right)+122.2\times 8\sqrt{3}$$
$$-(300+300)\times 5+R_3\sin 30\,[8\sqrt{3}+8\sqrt{3}]=0$$

$\therefore \quad 20\times 14.86+10\times 14.53+122.2\times 13.86-3000+13.86\,R_3=0$

$297.2+145.3+1691-3000+13.86\,R_3=0$

$$\therefore \quad R_3=\frac{866.6}{13.86}=62.52 \text{ kN}$$

Take moments about G

$$P_1(8\sqrt{3}-1)+P_2\left(8\sqrt{3}-\frac{2}{3}\right)+122.2\times 8\sqrt{3}$$
$$+600\times 5-13.86\,R_2=0$$

$\therefore \quad 20\times 12.86+10\cdot 13.19+169.1+3000-13.86\,R_2=0$

$$\therefore R_2 = \frac{5080.1}{13.86} = 366.5 \text{ kN}$$

$$\therefore R_1 = 600 \frac{\sqrt{3}(62.52+366.5)}{2}$$

$$= 228.5 \text{ kN}$$

$$M_B = -20 \times 1 - 19 \times \frac{2}{3}$$

$$= 26.7 \text{ kN-m}$$

Shear in BH

$$F_x = R_1 - (30 + 15 \times 2)$$
$$x = 228.5 - 60\,x$$

For shear force to be zero

$$x = \frac{228.5}{60} = 3.81 \text{ metres}$$

$$M_{3.81} = 228.5 \times 3.81$$
$$-60 \times \frac{(3.81)^2}{2} - 26.7$$
$$= 870.6 - 435.4 - 26.7$$
$$= 408.5 \text{ kN-m}$$

$$M_c = -60 \times 2 \times 1$$
$$= -120 \text{ kNm}$$

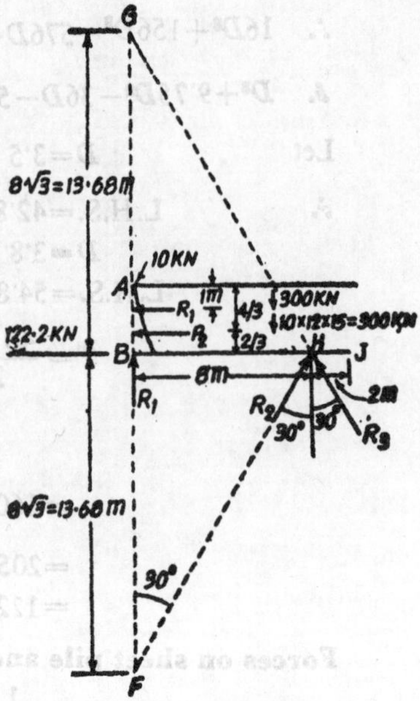

Fig. 21·24

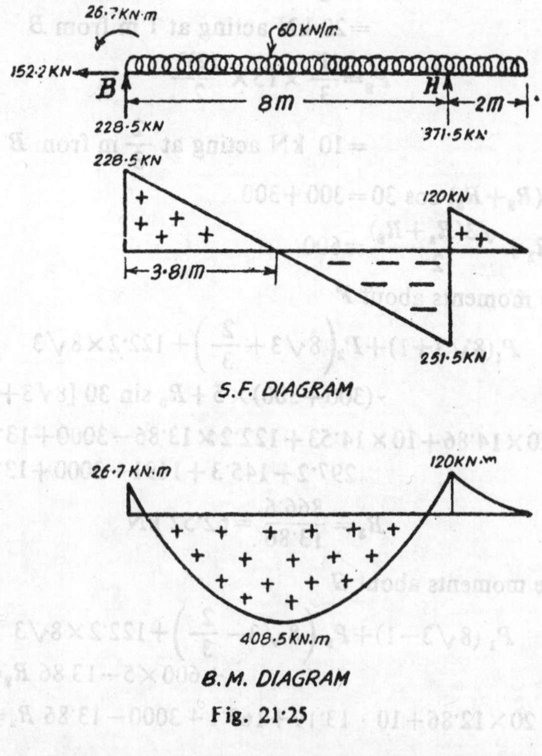

Fig. 21·25

21·7. Design of Wharf and Pier Structures

These structures are designed to resist both vertical and horizontal forces. The dead load consists of load of structure itself, fill and warehouse etc. Live load consists of cargoes and varies from 5000 N/m² to 50,000 N/m². Hydrostatic pressure is not very important as it acts on both sides of the structure. However the uplift forces must be taken into consideration.

Horizontal forces in the structure are caused by the impact of ships and the structure should be designed to resist these forces. The kinetic energy of the moving ships has to be overcome by the structure.

The velocity of ship at impact for design considerations is generally 0·2 m/sec. Generally in design it is assumed that 50% of the energy of the moving ship is absorbed by structure. The energy of absorption is given by

$$W_s V^2/2g$$

where W_s is the weight of the ship and V is the velocity of the ship at the time of impact.

In gravity structures mass and fill help in absorbing energy of impact but cushioning by pads or fenders should be provided to prevent damage to ships. Cushioning system should be such that individual units can be readily replaced.

In design, the structure is assumed to be fixed at top and base. It is assumed that the structure deflects by Δ at the top.

$$\Delta = \frac{PH^3}{12EI}$$

$$\text{Strain Energy} \quad = \tfrac{1}{2} P\Delta = \frac{P^2 H^3}{24EI}$$

where I is moment of inertia and E is modulus of elasticity.

Let N be number of bents sharing impact of ship which is taken as number of bents in half length of the ship. Impact energy per bent

$$= \frac{W_s V^2}{2gN}$$

It is assumed that all the piles in a bent are of the same section.

$$\therefore \qquad \Delta = \frac{P_1 H_1^2}{12EI} = \frac{P_2 H_2^2}{12EI} = \frac{P_3 H_3^3}{12EI} = \ldots\ldots \qquad \ldots(i)$$

Total energy per bent

$$\frac{P_1 H_1^3}{24EI} + \frac{P_2^2 H_2^3}{24EI} + \ldots\ldots$$

$$= \frac{W_s V^3}{2gN}$$

From (*i*)

$$P_2 = P_1\left(\frac{H_1}{H_2}\right)^3$$

$$P_3 = P_2\left(\frac{H_2}{H_3}\right)^3 \qquad \ldots(ii)$$

$$P_1{}^2H_1{}^3 + P_1{}^2\frac{H_1{}^6}{H_2{}^3} + \frac{P_1{}^2H_1{}^6}{H_3{}^3} + \ldots\ldots = \frac{12\,EIW_sV^2}{gN}$$

$$\therefore \quad P_1{}^2H_1{}^6\left[\frac{1}{H_1{}^3} + \frac{1}{H_2{}^3} + \ldots\ldots\right] = \frac{12EIW_sV^2}{gN} \qquad \ldots(iii)$$

From this P_1 can be calculated.

From (*ii*), P_2, P_3...... can be calculated.

Thus horizontal forces in all piles are known and from this maximum B.M. at top and bottom of each pile will be,

$$M_2 = \frac{1}{2}P_1H_1,$$

$$M_2 = \frac{1}{2}P_2H_2$$

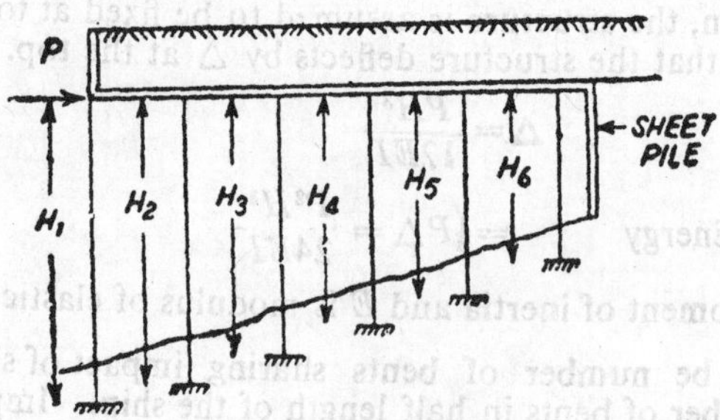

Fig. 21·26

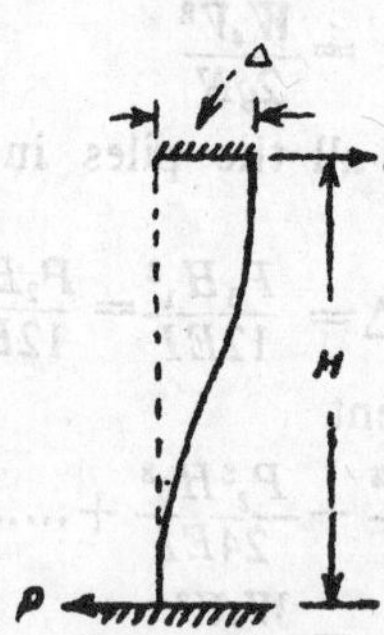

Fig. 21·27

21·8. Wharf Supported on Wells and Piles

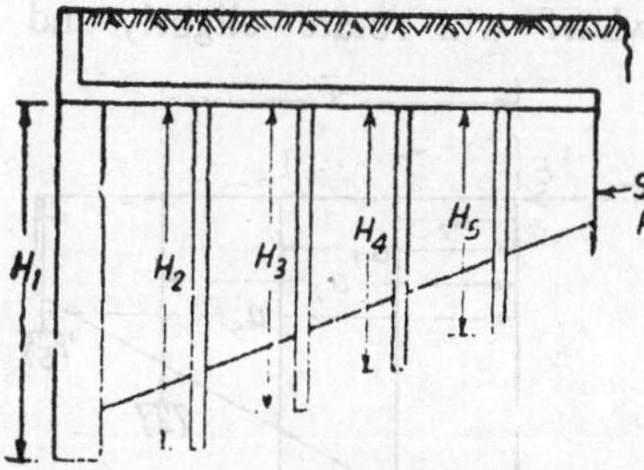

Fig. 21·28

The well is taken as fixed at top and hinged at base. The piles are taken as fixed at top and bottom. Let H_1, I_1, E_1 be the height, moment of inertia and Young's modulus for the well and I and E be the moment of inertia and Young's modulus for the piles with heights H_2, H_3 and so on.

For well $$M=\frac{3\,EI\Delta}{H^2}=PH$$

$$\therefore \quad \Delta=\frac{PH^3}{3EI}$$

For piles $$M=\frac{6EI\Delta}{H^2}=\frac{PH}{2}$$

$$\therefore \quad \Delta=\frac{PH^3}{12EI}$$

$$\therefore \quad \frac{P_1H_1^3}{3E_1I_1}=\frac{P_2H_2^3}{12EI}=\frac{P_3H_3^3}{12EI} \qquad ...(i)$$

(*i*) If the bent is subjected to horizontal force F,

$$F=P_1+P_2+P_3\text{.....} \qquad ...(ii)$$

From equations (*i*) and (*ii*) values of P_1, P_2, P_3 can be calcuted and then moments in well and piles can be calculated as,

$$M_1=P_1H_1,\ M_2=\frac{P_1H_2}{2},\ M_3=\frac{P_3H_3}{2} \text{ and so on.}$$

(*ii*) Impact of ship, with W_s as weight, V as speed and N bents taking the impact.

Kinetic energy per bent

$$=\frac{W_sV^2}{2gN}=\frac{P_1\Delta}{2}+\frac{P_2\Delta}{2}+\frac{P_3\Delta}{2}+\text{......}$$

$$=\frac{P_1^3H_1^3}{6E_1I_1}+\frac{P_2^2H_2^3}{24EI}+\frac{P_3^2H_3^3}{24EI}+ \qquad ...(iii)$$

From equations (*i*) and (*ii*), values of P_1, P_2,..... can be calculated and then moments can be caculated.

21·9. Vertical Forces in Piles due to Horizontal Force on Wharf

Due to horizontal force the wharf will rotate slightly and create vertical reactions in piles. The formation will be as shown in Fig. 21·29.

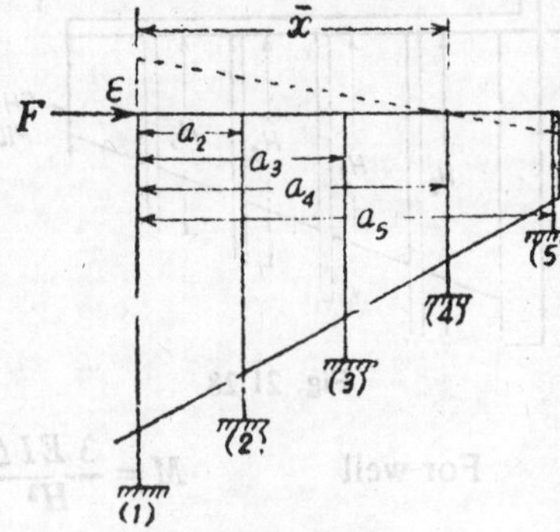

Fig. 21·29.

Let ε be the change in length of the first pile and then change in any other pile is given by $\frac{\varepsilon(\bar{x}-a)}{\bar{x}}$. If R_1, R_2, R_3 are the reactions in piles,

$$R_1+R_2+R_3\ldots\ldots=0$$

$$R_1=\frac{EA_1\varepsilon(\bar{x}-a_1)}{H_1\bar{x}}$$

$$R_2=\frac{EA_2\varepsilon(\bar{x}-a_1)}{H_2\bar{x}} \text{ and so on}$$

If areas are equal,

$$\frac{EA\;\varepsilon(\bar{x}-a)}{H_1\bar{x}}+\frac{EA\;\varepsilon(\bar{x}-a_2)}{H_2\bar{x}}+\ldots\ldots=0 \qquad \ldots (i)$$

From equation (*i*) values of $\bar{x}$ can be obtained.

Cutting the piles at the middle, *i.e.* point of contraflexure and taking moment about the first pile,

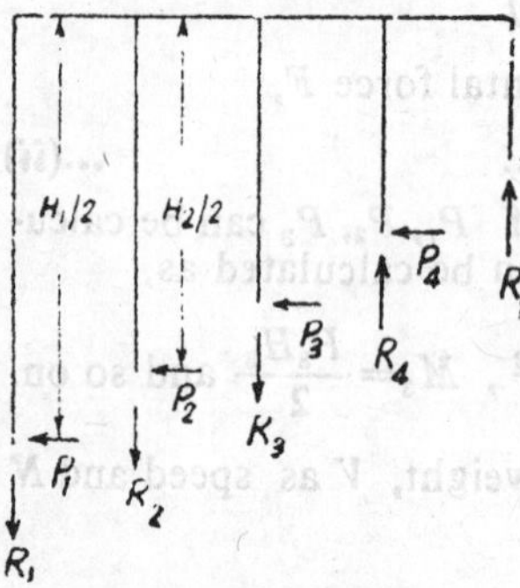

Fig. 21·30

$$\frac{P_1H_1}{2}+\frac{P_2H_2}{2}\ldots\ldots$$

$$=-\left[R_1a_1+R_2a_2\ldots\ldots\right]$$

$$=-\left[\frac{Ea\varepsilon(\bar{x}-a_1)a_1}{H_1\bar{x}}+\frac{EA\varepsilon(\bar{x}-a_2)a_2}{H_2\bar{x}}\right] \qquad \ldots (ii)$$

From equation (*ii*) value of $EA\varepsilon$ can be calculated and then values of reactions can be calculated.

Ex. 21·5. *Find the forces in the piles 300 mm×300 mm of a bent as shown in Fig. 21·31 subjected to horizontal force of 200 kN, $E=2\times10^4$ N/mm². The lengths of piles shown are assumed fixed lengths.*

Sol. $P_1+P_2+P_3+P_4+P_5=200 \qquad \ldots (i)$

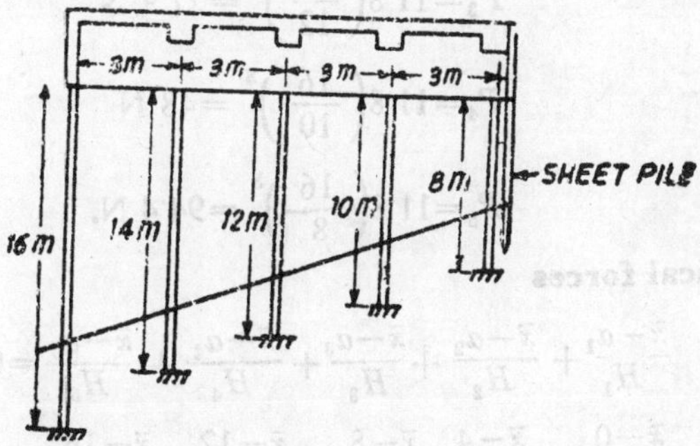

Fig. 21·31

$$\frac{P_1H_1^3}{12EI}=\frac{P_2H_2^3}{12EI}=\frac{P_3H_3^3}{12EI}=\frac{P_4H_4^3}{12EI}=\frac{P_5H_5^3}{12EI} \quad ...(ii)$$

$$P_2=P_1\left(\frac{H_1}{H_2}\right)^3,$$

$$P_3=P_1\left(\frac{H_1}{H_3}\right)^3,$$

$$P_4=P_1\left(\frac{H_1}{H_4}\right)^3,$$

$$P_5=P_1\left(\frac{H_1}{H_5}\right)^3.$$

$$P_1\left(\frac{H_1}{H_1}\right)^2+P_1\left(\frac{H_1}{H_2}\right)^2......=200$$

$$\therefore \quad P_1=\frac{200}{\left(\frac{H_1}{H_1}\right)^3+\left(\frac{H_1}{H_2}\right)^3+\left(\frac{H_1}{H_2}\right)^3......}$$

$$=\frac{200}{1+\left(\frac{16}{14}\right)^3+\left(\frac{16}{12}\right)^2+\left(\frac{16}{10}\right)^2+\left(\frac{16}{8}\right)^2}$$

$$=\frac{200}{1+1{\cdot}493+2{\cdot}370+4{\cdot}096+8{\cdot}000}$$

$$=\frac{200}{16{\cdot}959}$$

$$=11{\cdot}8 \text{ N.}$$

$$P_2=11{\cdot}8\left(\frac{16}{14}\right)^3=17{\cdot}6 \text{ N}$$

$$P_3=11{\cdot}8\left(\frac{16}{12}\right)^3=27{\cdot}9 \text{ N.}$$

$$P_4=11{\cdot}8\left(\frac{16}{10}\right)^3=48 \text{ N}$$

$$P_5=11{\cdot}8\left(\frac{16}{8}\right)^3=94{\cdot}4 \text{ N.}$$

Vertical forces

$$\frac{\bar{x}-a_1}{H_1}+\frac{\bar{x}-a_2}{H_2}+\frac{\bar{x}-a_3}{H_3}+\frac{\bar{x}-a_4}{H_4}+\frac{\bar{x}-a_5}{H_5}=0$$

$$\therefore \quad \frac{\bar{x}-0}{16}+\frac{\bar{x}-4}{14}+\frac{\bar{x}-8}{12}+\frac{\bar{x}-12}{10}+\frac{\bar{x}-16}{8}=0$$

Multiplying by $8\times10\times3\times7=1680$

$$\therefore \quad 105\bar{x}+120\bar{x}-480+140\bar{x}-1120+168\bar{x}-2016+210\bar{x}-3360=0$$

$$\therefore \quad \bar{x}=\frac{6976}{743}=9{\cdot}389 \text{ metres}$$

$$\frac{P_1H_1}{2}+\frac{P_2H_2}{2}+\frac{P_3H_3}{2}+\frac{P_4H_4}{2}+\frac{P_5H_5}{2}$$

$$=-EA\varepsilon\left[\frac{(\bar{x}-a_1)a_1}{H_1\bar{x}}+\frac{(\bar{x}-a_2)a_2}{H_2\bar{x}}+\frac{(\bar{x}-a_3)a_3}{H_3\bar{x}}+\frac{(\bar{x}-a_4)a_4}{H_4\bar{x}}+\frac{(\bar{x}-a_5)a_5}{H_5\bar{x}}\right]$$

$$11{\cdot}8\times8+17{\cdot}6\times7+27{\cdot}9\times6+48{\cdot}3\times5+94{\cdot}4\times3=\frac{-EA\varepsilon}{9{\cdot}389}$$

$$\left[0+\frac{5{\cdot}389\times4}{14}+\frac{1{\cdot}389\times8}{12}+\frac{2{\cdot}611\times12}{10}+\frac{6{\cdot}611\times16}{8}\right]$$

$$\therefore \quad 94{\cdot}4+123{\cdot}2+167{\cdot}4+241{\cdot}5+376{\cdot}7=\frac{-EA\varepsilon}{9{\cdot}389}\times[0+1{\cdot}540+0{\cdot}926-3{\cdot}133-13{\cdot}222]$$

$$\therefore \quad AE\varepsilon=\frac{1004{\cdot}1\times9{\cdot}389}{13{\cdot}889}=678{\cdot}7$$

$$\varepsilon=\frac{678{\cdot}7}{18\times10^8}$$

$$=3{\cdot}86\times10^{-7} \text{ metres.}$$

$$=3{\cdot}86\times10^{-4} \text{ mm.}$$

$$R_1=\frac{EA\varepsilon}{H_1}=\frac{678{\cdot}8}{16}=42{\cdot}4 \text{ kN.}$$

$$R_2 = \frac{EA\varepsilon(9{\cdot}389-4)}{H_2 \times 9{\cdot}389}$$

$$= \frac{678{\cdot}7 \times 5{\cdot}389}{14 \times 9{\cdot}389} = 27{\cdot}8 \text{ kN.}$$

$$R_3 = \frac{EA\varepsilon\ (9{\cdot}389-8)}{H_3 \times 9{\cdot}389}$$

$$= \frac{678{\cdot}7 \times 1{\cdot}389}{12 \times 9{\cdot}389} = 8{\cdot}4 \text{ kN.}$$

$$R_4 = \frac{EA\varepsilon\ (9{\cdot}389-12)}{H_4 \times 9{\cdot}389}$$

$$= \frac{678{\cdot}7(-2{\cdot}611)}{10 \times 9{\cdot}389} = -18{\cdot}9 \text{ kN.}$$

$$R_5 = \frac{EA\varepsilon\ (9{\cdot}389-16)}{H_5 \times 9{\cdot}389}$$

$$= \frac{-678{\cdot}7 \times 6{\cdot}611}{8 \times 9{\cdot}389}$$

$$= -59{\cdot}7 \text{ kN.}$$

21·10. Bridge Piers

Pier is the foundation unit over which super-structure passes. Fig. 21·32 shows different types of piers in common use.

Type (*a*) is used in massive structures and in rapidly running waters. Type (*b*) is used on the ground. It is lighter and much more economical but requires more formwork. Type (*c*) is used for low piers over unstable ground. Outside piles are generally battered. Type (*d*) are used for high bridges. They are generally low piers with steel columns or concrete piers upto bridge level.

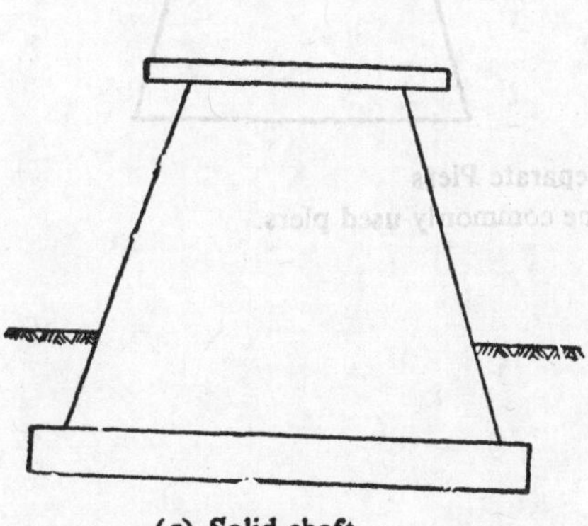

(*a*) Solid shaft.

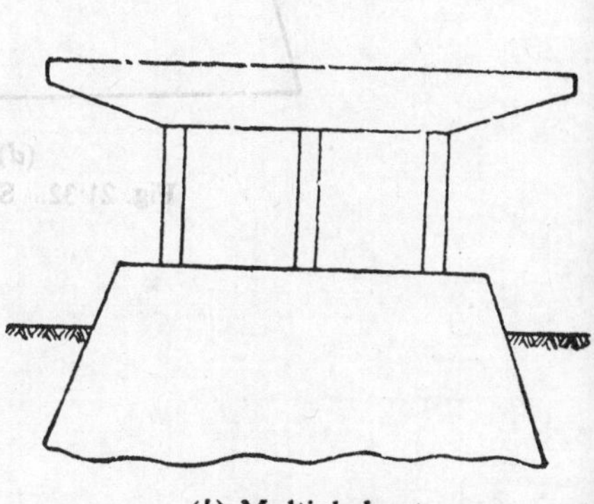

(*b*) Multiple bent.

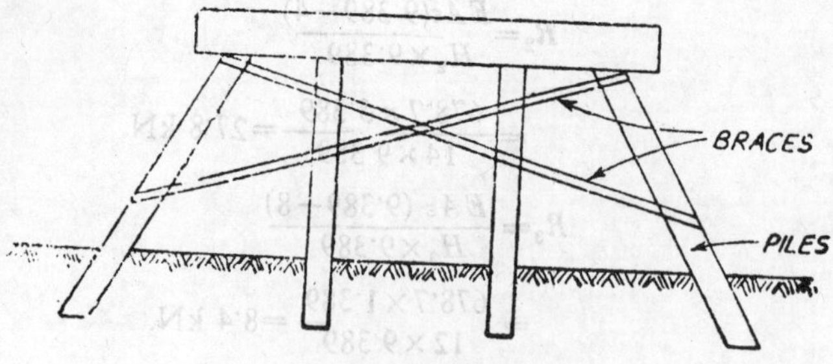

(c)
Fig. 21·32. Pile bent

Piers are designed to take dead loads of super-structure, live loads on super-structure, weight of the pier and uplift, lateral forces due to wind, longitudinal forces, wind pressures and centrifugal forces. Forces exerted by current of water must be taken into consideration. Pressure exerted by current is given by

$$P=KwAV^2/2g$$

where, P=Total water pressure
w=Unit weight of water
A=Area of wetted surface
V=Velocity of current
g=Gravity acceleration.

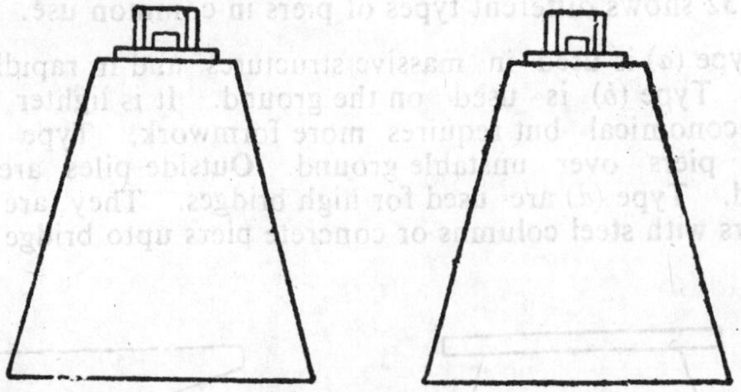

(d) Separate Piers
Fig. 21·32. Some commonly used piers.

22

Detailing of RCC Structures

Detailing of structure consists of preparations of drawings which give the reinforcement details for the correct arrangement of reinforcement in the structure and the bar bending schedule for the correct fabrication of the various reinforcing bars.

A drawing should show the overall size of concrete members, including the sizes of any chamfers and fillets, it should also show the exact position, shape, size and spacing of the reinforcement within the concrete members and the concrete cover to be provided to the reinforcement. The position of holes for pipes and position and details of construction joints should be shown on the drawing.

The details of structures for all the RCC members designed in previous chapters were shown in respective chapters. In this chapter certain reinforcement details not covered in earlier chapters will be covered.

Continuous Beams

Details of reinforcement in continuous beams are shown in Fig. 22·1. In Fig. 22·1 (*a*) the reinforcement of secondary beam is not shown. Steel reinforcement at junction with column is shown in Fig. 22·1 (*b*).

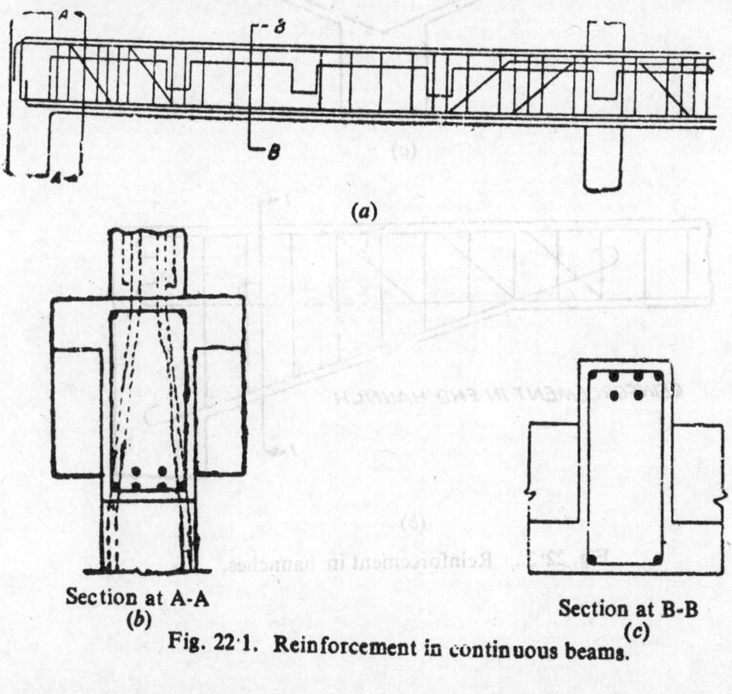

Fig. 22·1. Reinforcement in continuous beams.

Haunched Beams

In heavy structures like warehouses etc. the shear stresses and negative bending moment at supports may be high and in such cases it may be economical to provide beams with haunches as shown in Fig. 22·2.

Fig. 22·3 shows typical main tensile reinforcement in haunches in internal and external columns. Haunch bars are placed parallel to haunch. The stirrups in haunches can be provided vertically as shown in Fig. 22·3 or normal to haunch as shown in Fig. 22·4. Two sets of stirrups are required in haunches, one for main reinforcement and other for haunch reinforcement.

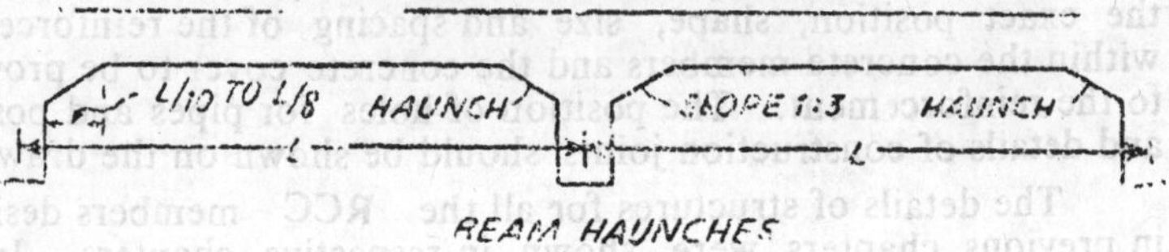

Fig. 22·2. Beam haunches.

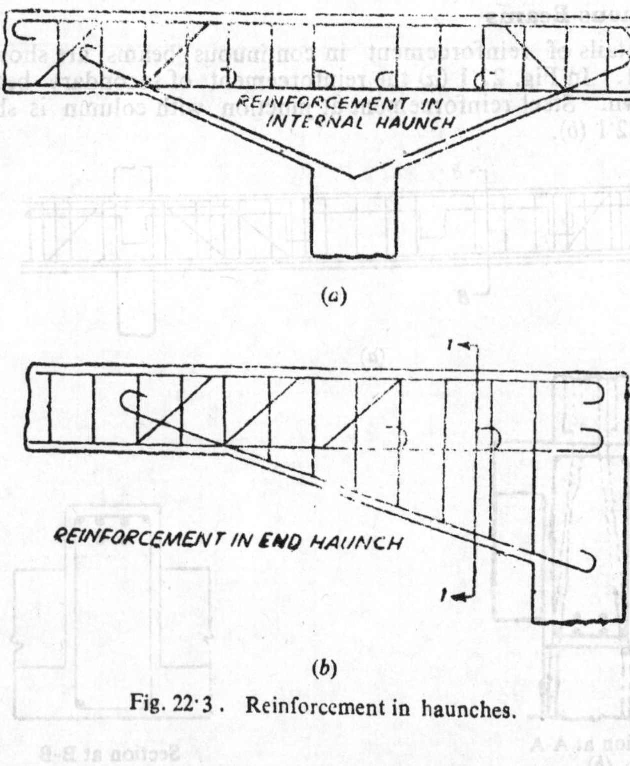

Fig. 22·3. Reinforcement in haunches.

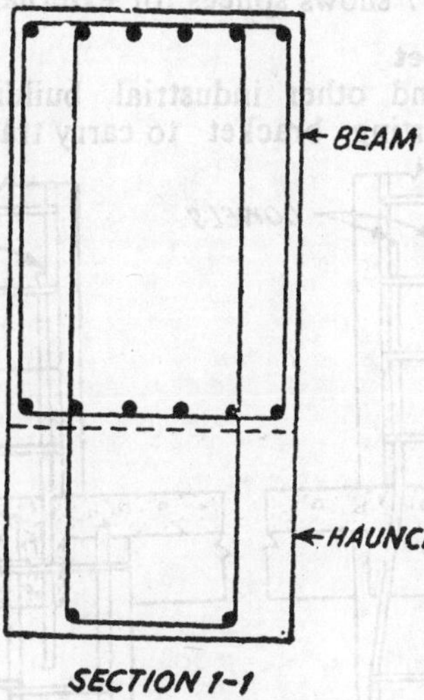

Fig. 22·3 (c)

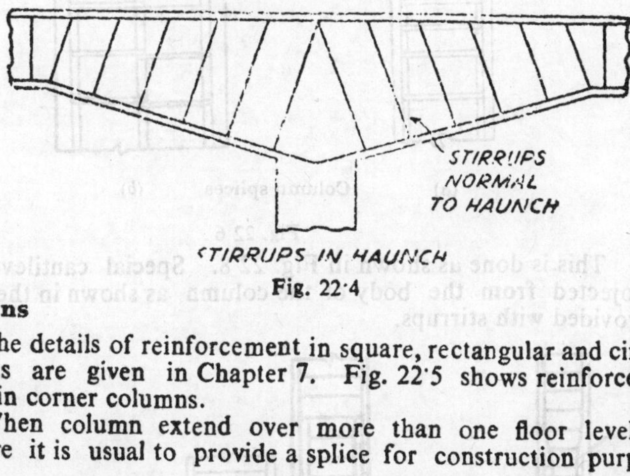

Fig. 22·4

Columns

The details of reinforcement in square, rectangular and circular columns are given in Chapter 7. Fig. 22·5 shows reinforcement details in corner columns.

When column extend over more than one floor level of a structure it is usual to provide a splice for construction purposes.

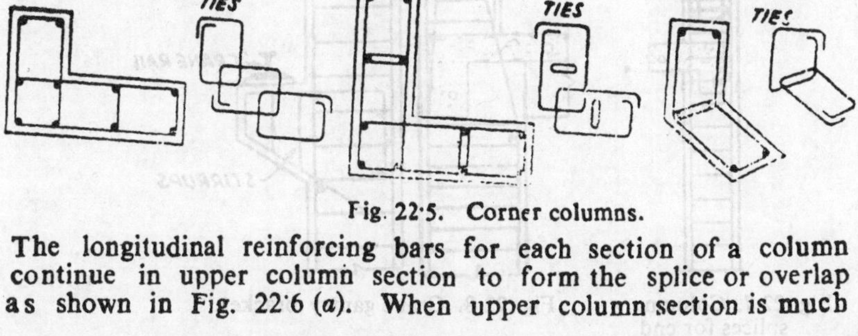

Fig. 22·5. Corner columns.

The longitudinal reinforcing bars for each section of a column continue in upper column section to form the splice or overlap as shown in Fig. 22·6 (*a*). When upper column section is much smaller than lower column section dowel bars are used as shown in Fig. 22·6 (*b*). Fig. 22 7 shows splices for external columns.

Crane Gantry Bracket

In workshops and other industrial buildings, columns are provided with supporting bracket to carry track of an overhead

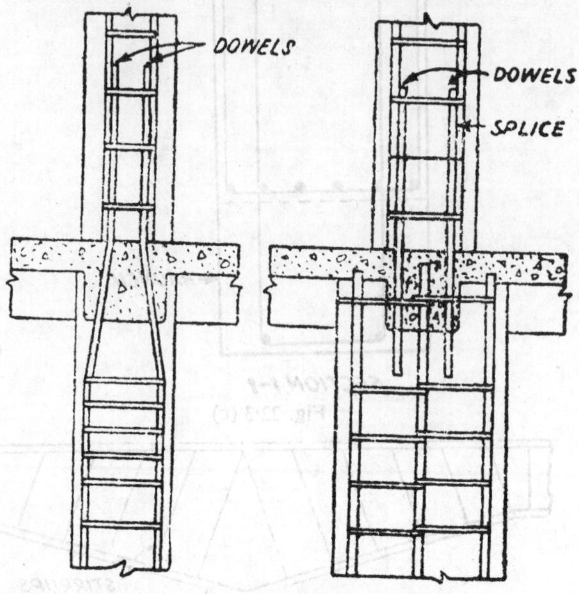

(*a*) Column splices (*b*)

Fig. 22·6

crane. This is done as shown in Fig. 22·8. Special cantilever bars are projected from the body of the column as shown in the figure and provided with stirrups.

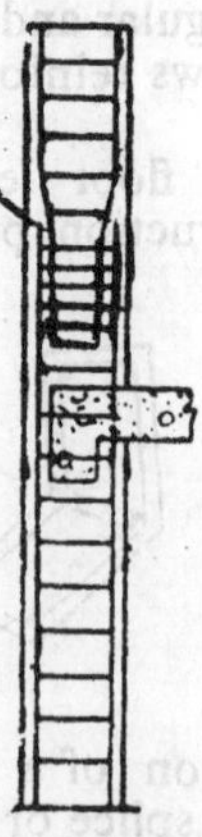

Fig. 22·7. Column splices for end column.

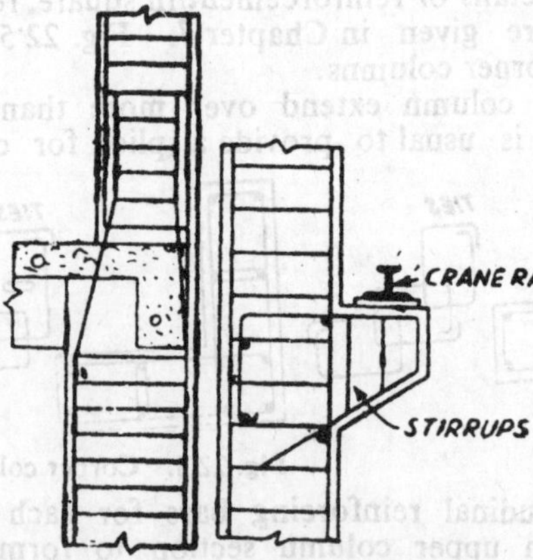

Fig. 22·8. Crane gantry bracket.

Ties

Ties are structural members subjected to axial tension. These are used as diagonal braces for columns such as required in water towers, as horizontal ties for arched roofs and vertical ties (hangers) in bridges. As a tie is a tension member there is no tendency to burst like columns and therefore no binders are required, however, minimum reinforcement is provided as binders to hold main bars. The bars are provided with adequate anchorage and bond lengths. Details of reinforcement in ties for light loading are shown in

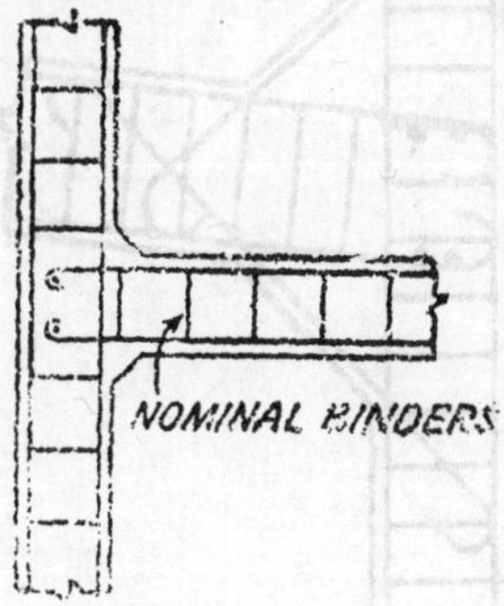

(a)

Fig. 22·9. When ties have to carry heavy loads the ends are more

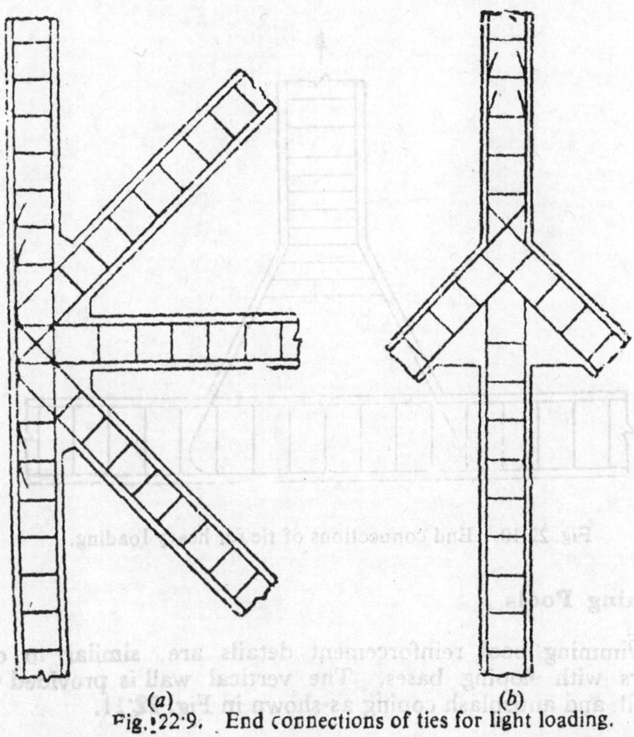

(a) (b)

Fig. 22·9. End connections of ties for light loading.

splayed out to distribute the load adequately as shown in Fig. 22·10.

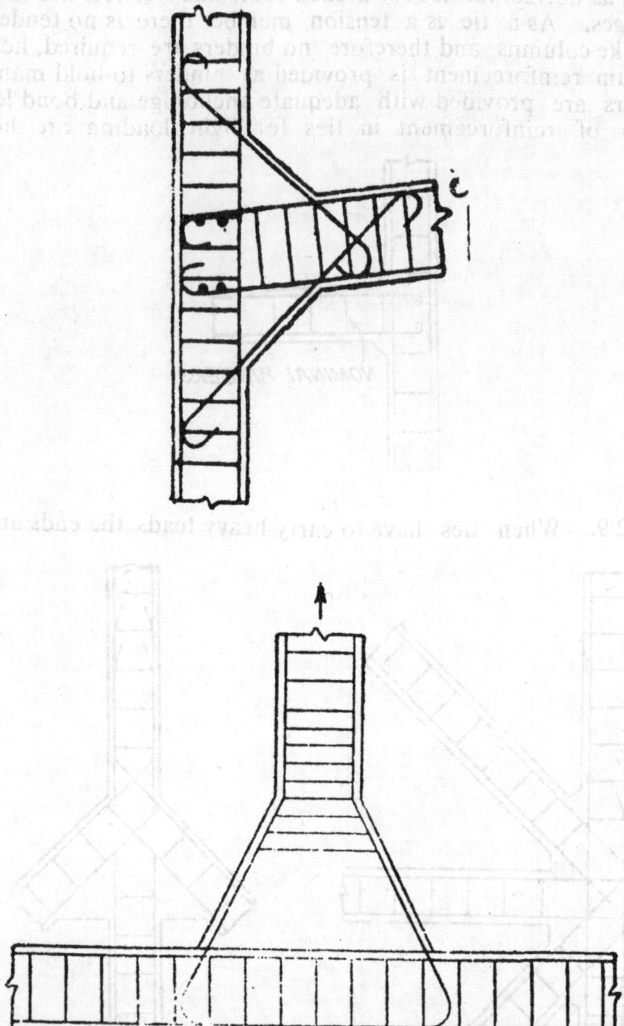

Fig. 22·10. End connections of tie for heavy loading.

Swimming Pools

Swimming pool reinforcement details are similar to open reservoirs with sloping bases. The vertical wall is provided with hand rail and antisplash coping as shown in Fig. 22·11.

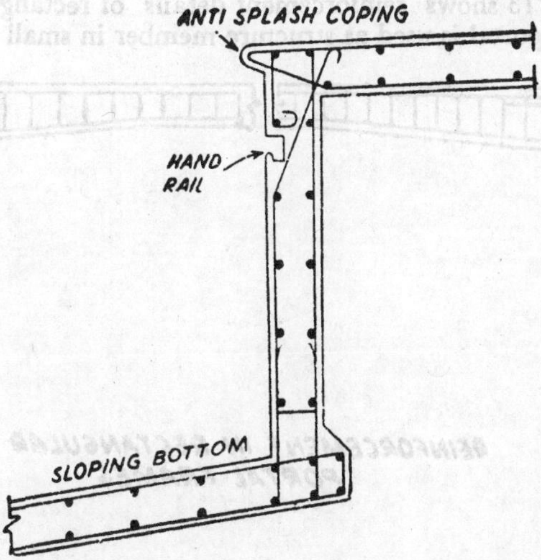

Fig. 22·11. Reinforcement in swimming pool wall.

Portals

Fig. 22·12 shows the basic reinforcement details for a three hinged portal. In this type of frame two halves are pinned at the apex as shown.

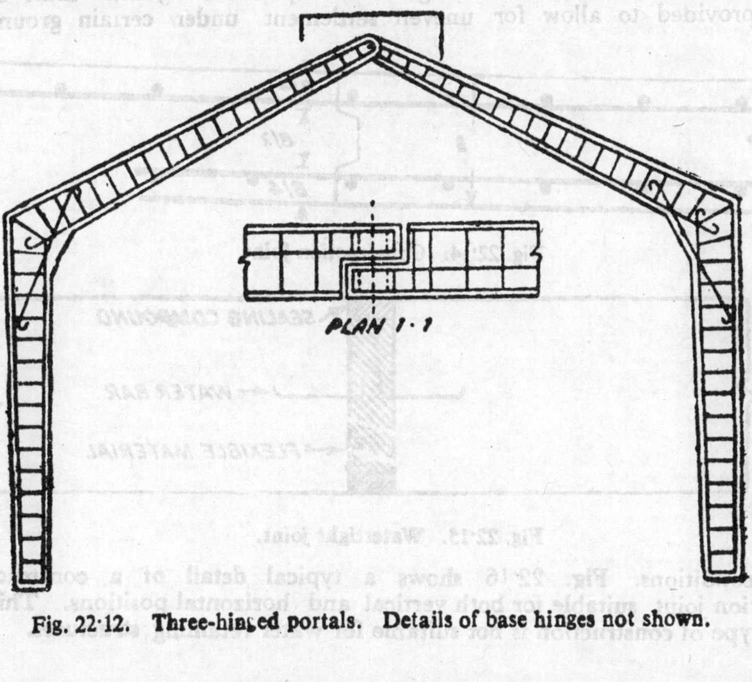

Fig. 22·12. Three-hinged portals. Details of base hinges not shown.

Fig. 22·13 shows reinforcement details of rectangular portals. This type of portal is used as structure member in small bridges.

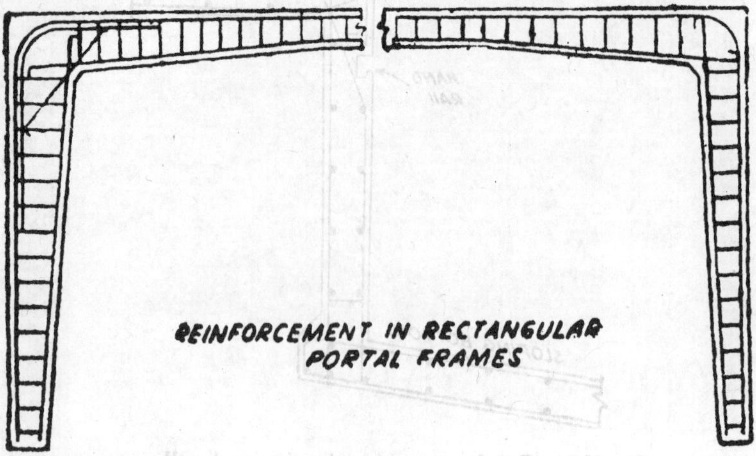

Fig. 22·13

Joints

In reinforced concrete construction two types of joints are provided. One type is temporary construction joint and other type is permanent functional joint which allows for free expansion and contraction with change of temperature. Joints must be provided to allow for uneven settlement under certain ground

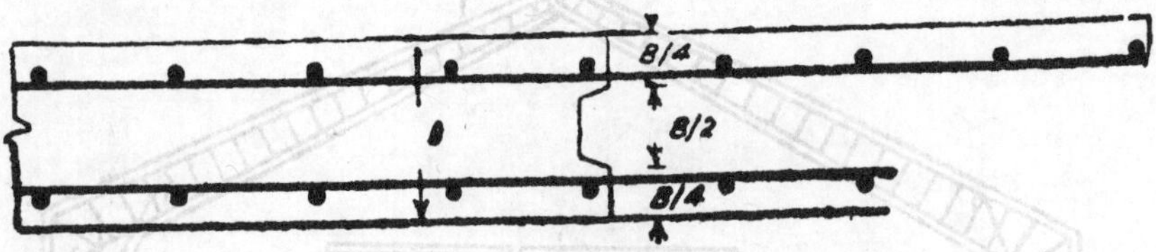

Fig. 22·14. Construction joint.

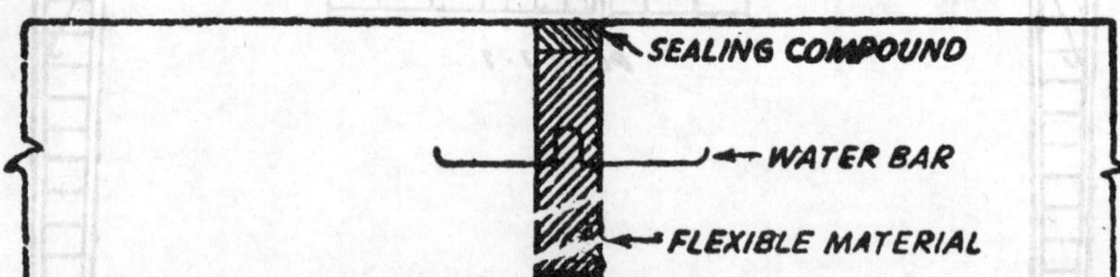

Fig. 22·15. Watertight joint.

conditions. Fig. 22·16 shows a typical detail of a construction joint suitable for both vertical and horizontal positions. This type of construction is not suitable for water retaining structures.

Expansion Joints

This type of joint is provided in long structures like retaining walls, buildings, ducts etc. These joints are also provided

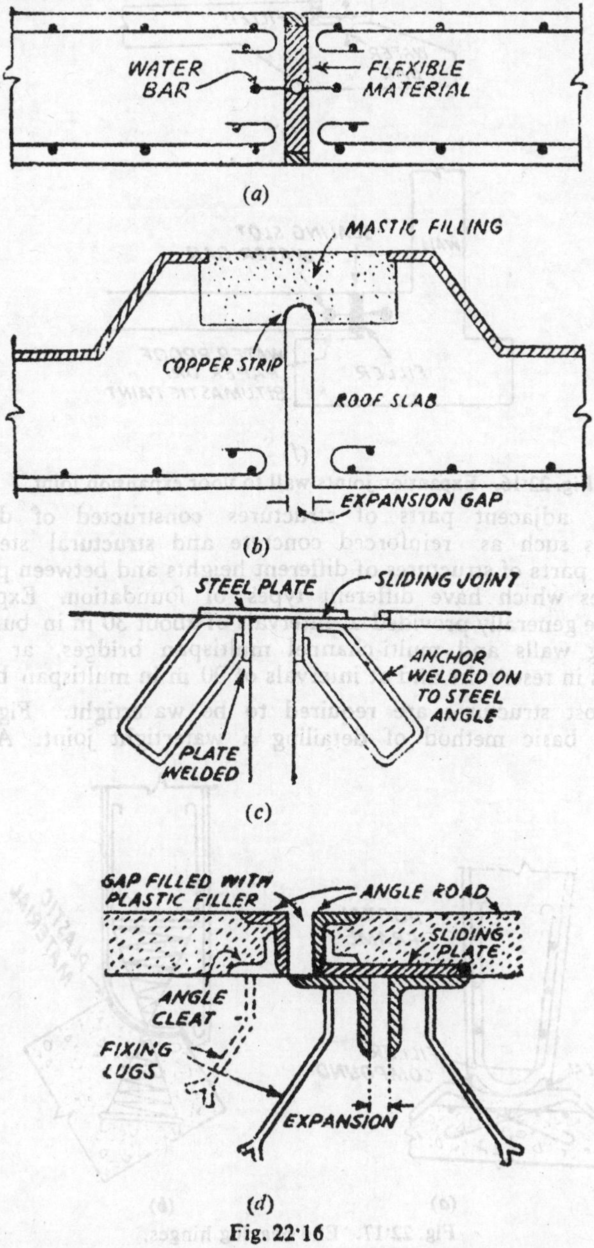

Fig. 22·16

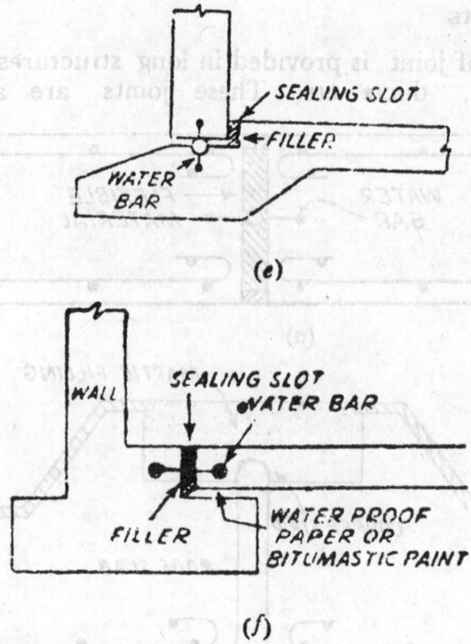

Fig. 22·16. Expansion joints wall to floor expansion joint.

between adjacent parts of structures constructed of different materials such as reinforced concrete and structural steelwork, between parts of structures of different heights and between parts of structures which have different types of foundation. Expansion joints are generally provided at intervals of about 30 m in buildings, retaining walls and multi-channel multispan bridges, at 60 m intervals in reservoirs and at intervals of 90 m in multispan bridges.

Most structures are required to be watertight. Fig. 22·15 shows a basic method of detailing a watertight joint. A water

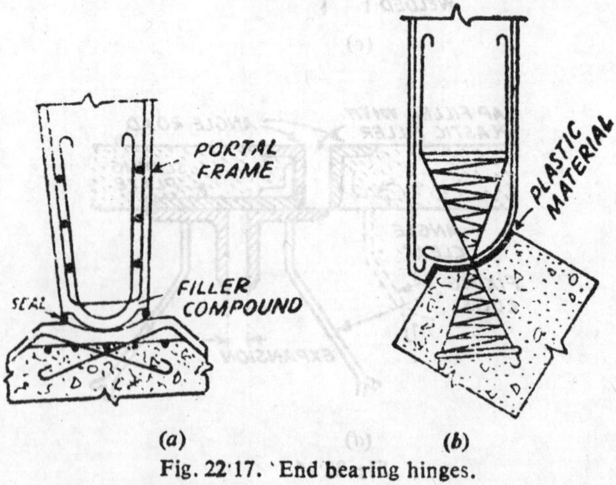

Fig. 22·17. End bearing hinges.

bar of rubber, or plastic is placed across the joints, which can expand or contract without failure because of its shape. On each side of this water bar is placed a flexible material which is sealed with a suitable bitumen or plastic plug. Fig. 22·16 shows some expansion joints used in RCC structures.

Hinges

Many reinforced concrete structures such as arch bridges, portal frames etc. are designed on the condition that the part of the structure act as hinges. In large structures, fitting a normal metal hinge would be very expensive and it may be more economical to

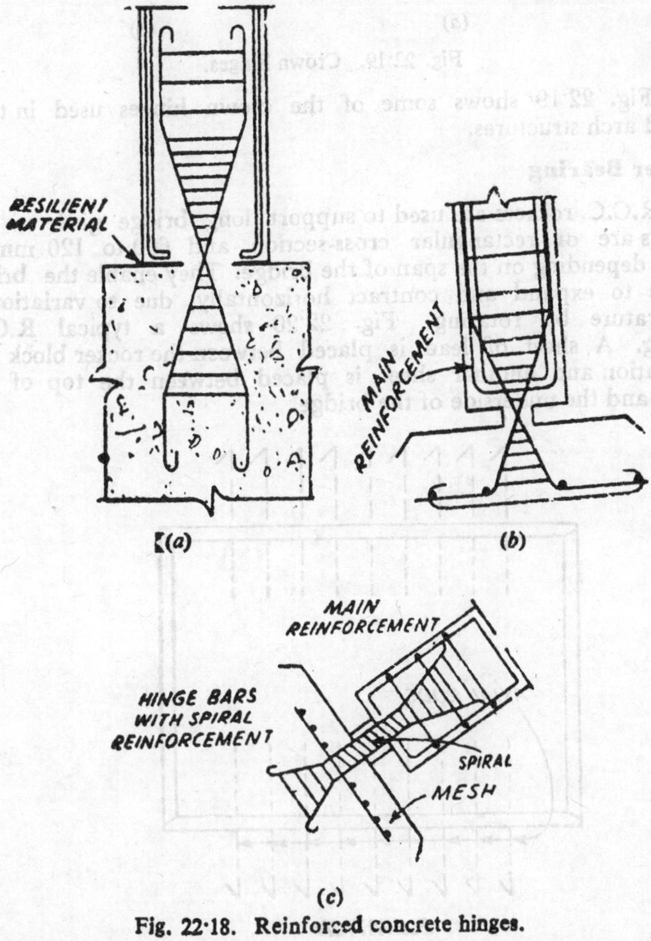

Fig. 22·18. Reinforced concrete hinges.

form hinges using ordinary reinforcing bars. This is possible due to the fact that in most of RCC structures actual rotation required is very small.

Fig. 22·17 shows two simple types of end bearing hinges. Fig 22·18 shows some other type of reinforced concrete hinges suitable for heavy structures.

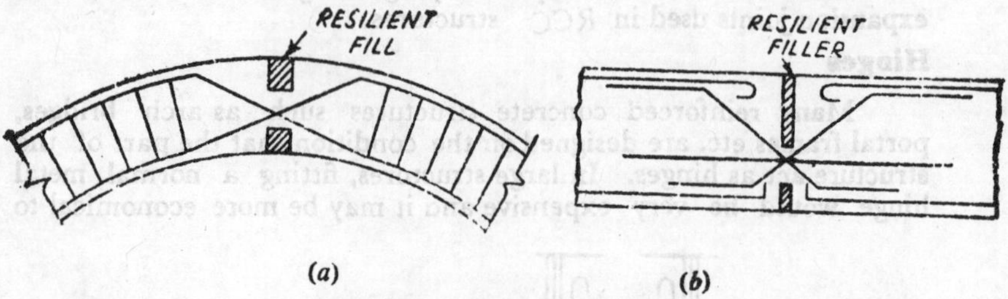

Fig. 22·19. Crown hinges.

Fig. 22·19 shows some of the crown hinges used in three pinned arch structures.

Rocker Bearing

R.C.C. rockers are used to support long bridge girders. The rockers are of rectangular cross-section and 600 to 120 mm in height depending on the span of the bridge. They enable the bridge girders to expand and contract horizontally, due to variation of temperature by rotating. Fig. 22·20 shows a typical R.C.C. bearing. A sheet of lead is placed between the rocker block and foundation and another sheet is placed between the top of the rocker and the underside of the bridge.

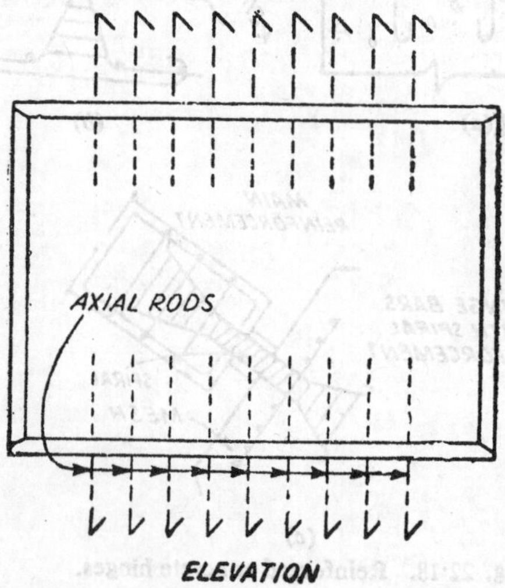

(a)

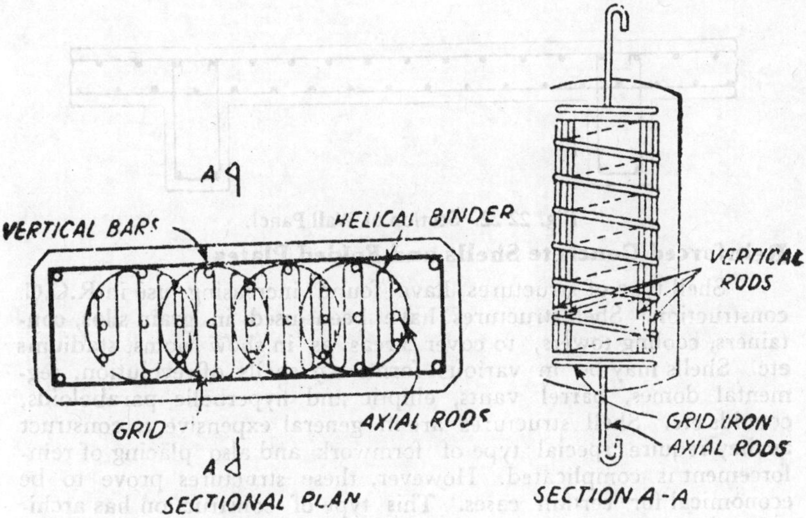

(b) (c)

Fig. 22·20.

Sliding Bearing

Sliding bearings are used in bridges to support bridge girders to allow for expansion and contraction of bridge decks. Fig. 22·21 shows two types of sliding plate bearings.

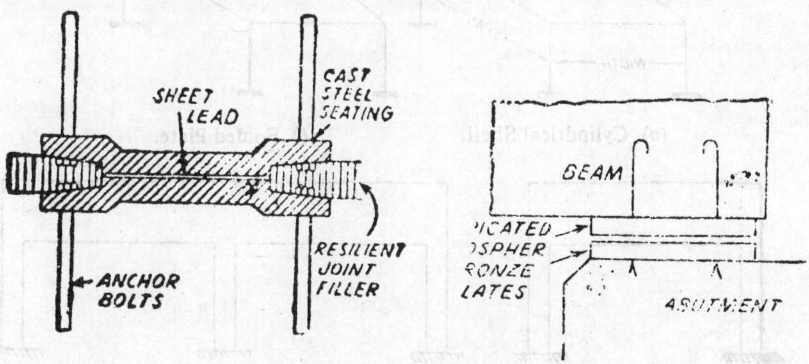

(a) (b)

Fig. 22·21. Cast steel bearing.

Reinforced Concrete Walls

Reinforced concrete walls are sometimes used as curtain or filler walls. These walls are designed to take vertical loads and lateral thrust. In case the walls are provided between piers or columns, the walls are designed to take only lateral thrust and the columns or piers are designed to take vertical loads and side thrust. Fig. 22·22 shows reinforcement details in R.C.C. wall.

Fig. 22·22. Section of Wall Panel.

Reinforced Concrete Shells and Folded Plates

Shell type of structures have found increasing use in R.C.C. construction. Shell structures have been used in grain silos, containers, cooling towers, to cover areas as in show rooms, stadiums etc. Shells may be in various forms *e.g.* shells of revolution, segmental domes, barrel vauts, elliptic and hyperbolic paraboloids, conoids etc. Shell structures are in general expensive to construct as they require special type of formwork and also placing of reinforcement is complicated. However, these structures prove to be economical for certain cases. This type of construction has architectural beauty. Fig 22·23 shows some shell and folded plate structures. Fig. 22·24 shows reinforcement details in shell and plate structures.

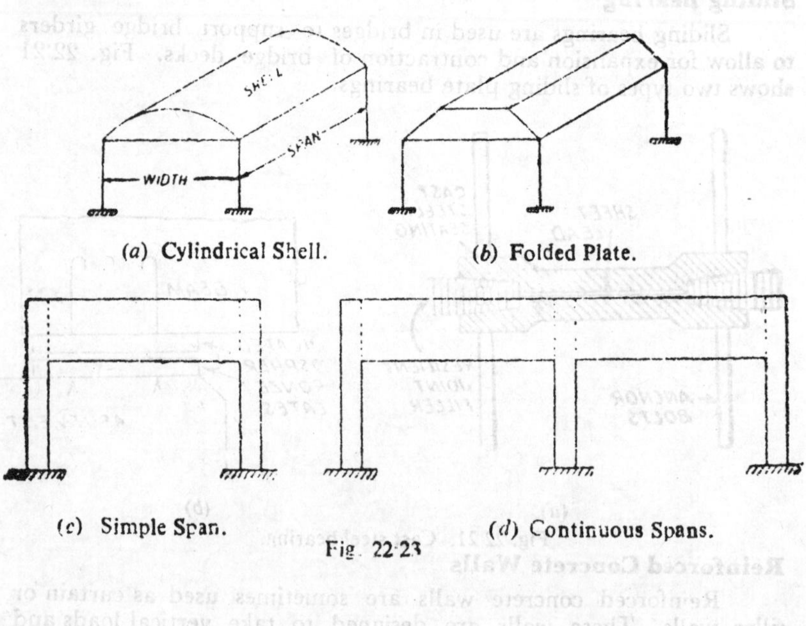

(*a*) Cylindrical Shell. (*b*) Folded Plate.

(*c*) Simple Span. (*d*) Continuous Spans.

Fig. 22·23

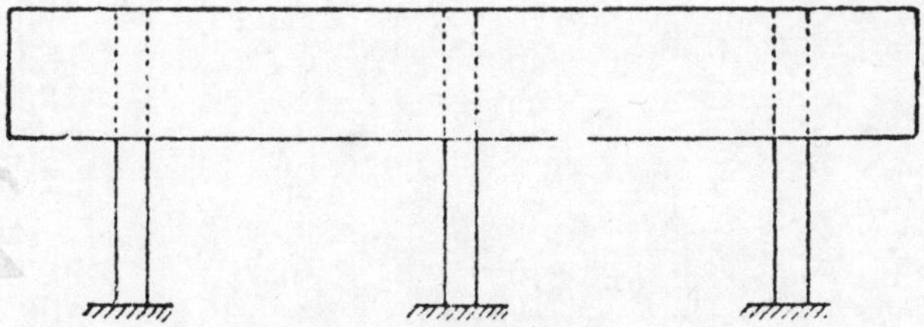

Continuous Spans with Overhangs

Fig. 22·23 (*e*)

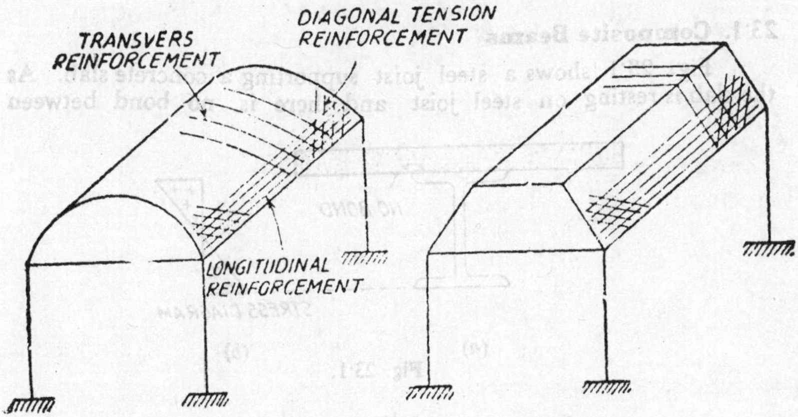

Shell **Folded Plate**

Fig. 22·24. Reinforcement Details.

23

Composite Sections

23·1. Composite Beams

Fig. 23·1 shows a steel joist supporting a concrete slab. As the slab is resting on steel joist and there is no bond between

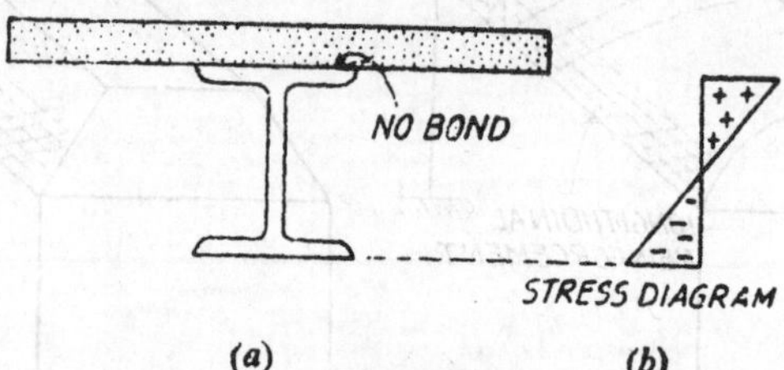

Fig. 23·1.

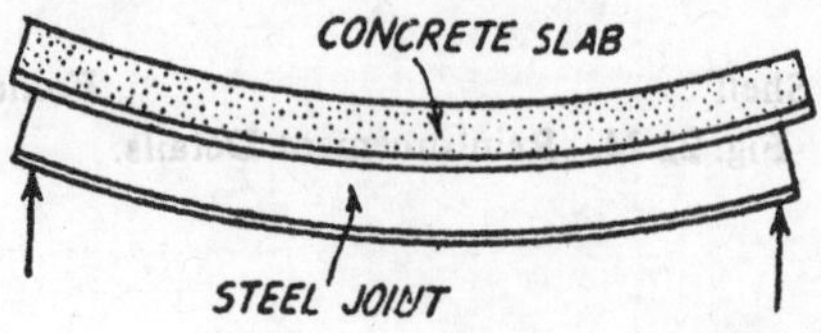

Fig. 23·2.

the two, the two will deflect independent of each other as shown in Fig. 23·2. The steel section will have stress diagram as shown in Fig. 23·1 (*b*).

Had there been proper bond between concrete slab and steel joist, they would have deflected together as one piece and acted as a *T*-beam in which most of the compression would have been taken by the concrete and tension by the joist.

In these composite sections, the greatest shear stress occurs at the neutral axis which is always very near the top flange of the joist. Therefore, composite action can be achieved only by using shear connectors, such as cuttings from plates, angles, channels, spirals or bent and hooked bars which are welded on top flange of the joist as shown in Fig. 23·3.

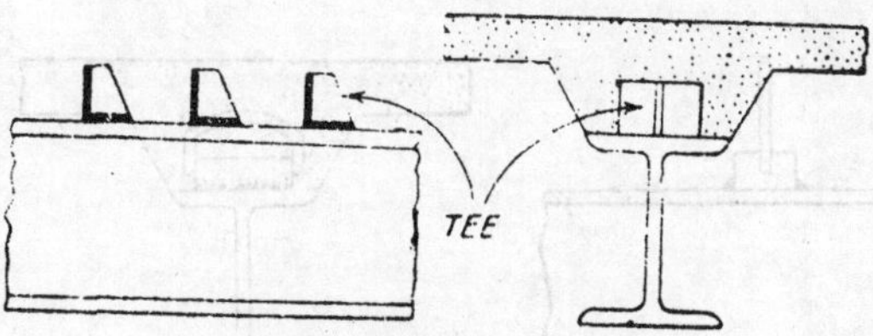

Shear connector (Rigid)
(a)

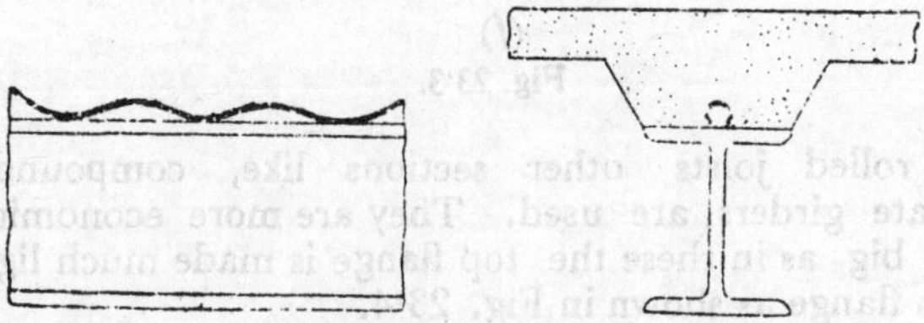

Spiral shear connectors
(b)

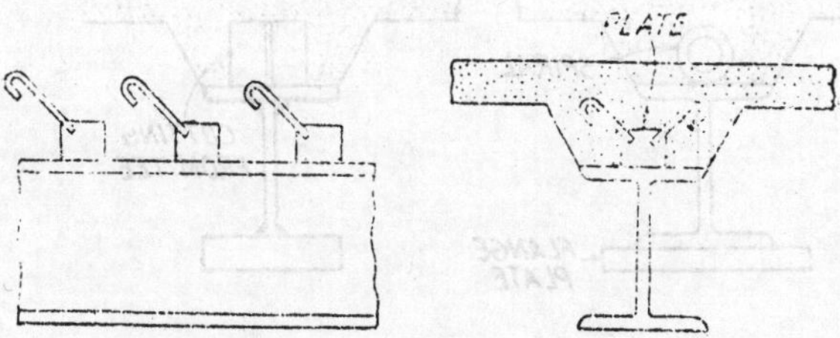

Bent and hooked bar shear connectors (Anchor)
(c)

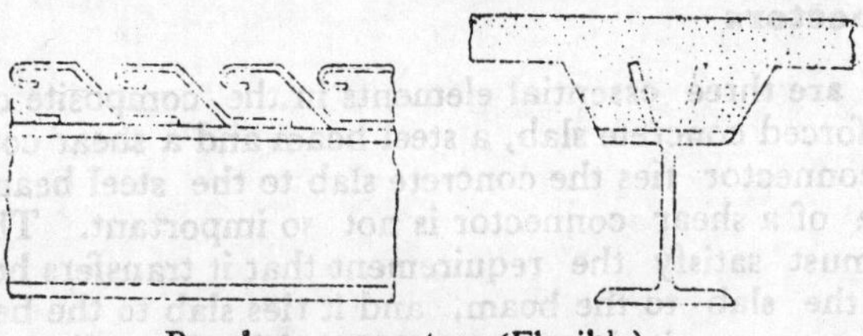

Bar shear connectors (Flexible)
(d)

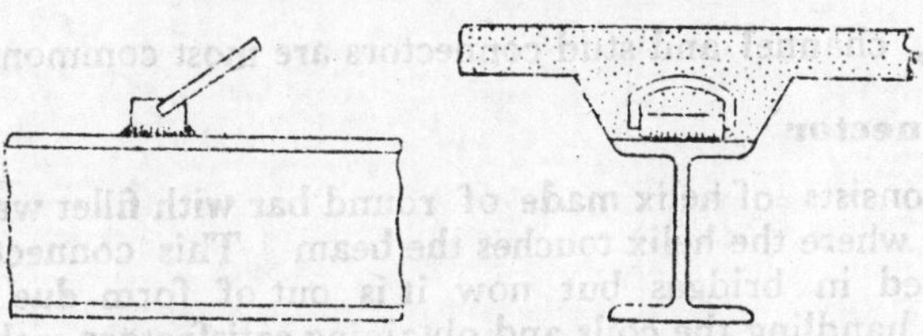

Sloped anchor shear connectors at ends of beam
(e)

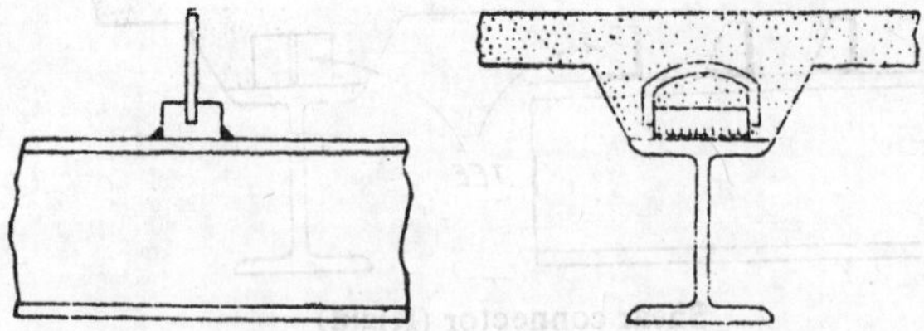

Vertical anchor shear connectors at central portion of beams.

(*f*)

Fig. 23·3.

Beside rolled joists other sections like, compound beams, welded plate girders are used. They are more economical where spans are big as in these the top flange is made much lighter than the bottom flange as shown in Fig. 23·4.

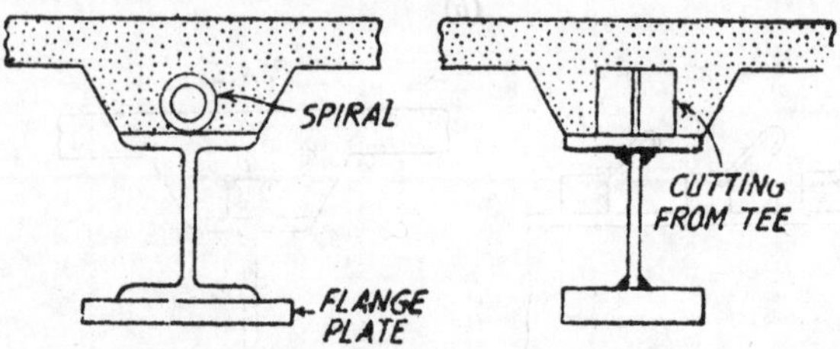

Compound section Welded plate girder

Fig. 23·4.

Shear Connectors

There are three essential elements in the composite construction—a reinforced concrete slab, a steel beam and a shear connector. The shear connector ties the concrete slab to the steel beam. The actual shape of a shear connector is not so important. The shear connector must satisfy the requirement that it transfers horizontal shear from the slab to the beam, and it ties slab to the beam in a satisfactory manner, it is easy to use, quick to install and economical.

Spiral, channel and stud connectors are most commonly used.

Spiral Connector

This consists of helix made of round bar with fillet welded at each point, where the helix touches the beam This connector was formerly used in bridges but now it is out of form due to cost, problems in handling the coils and obtaining satisfactory welds.

Channel Connector

This is a straight forward device. It is short length of rolled channel with one flange welded to the steel beam while the upper flange ties the slab to the beam. Channel connectors are commonly used for bridge construction. Channels are generally installed in shop. Welded channels are used where girder or beam is fabricated. Channels are seldom used on rolled sections or where field installation of connectors is required.

Stud Shear Connectors

These are most commonly used shear connectors throughout the world. This connector is essentially a short round piece of cold drawn steel which is welded to the beam at one end having an upset head at the other. The stud resists the horizontal shear by bending and the anchoring effect of the head prevents the slab from lifting away from the beam.

Welded stud may be placed quickly where required. Shear connector studs are generally welded in place after all other fabrication is over or after the beams are erected. Welded stud connectors are available in diameters from 6 to 20 mm and in lengths from 40 to 200 mm. For bridge construction commonly used size is 20 by 100 mm. From practical considerations maximum possible stud diameter should be used.

23·2. The width of slab which is considered as part of the composite beam is taken as for T-beam, *i.e.* minimum of

(*i*) one-third of the effective span.

(*ii*) distance between the centres of the ribs.

(*iii*) twelve times the thickness of the slab plus breadth of the rib, where the steel member is encased.

The concrete slab is laid over the joists either when the joists are propped throughout its length or when the joists are not propped. In the first case the steel section will have stresses due to its self-load whereas the dead load from slab, and live load will produce stresses in the composite section. In the second case the steel section will have stresses due to self-dead load and dead load of the slab, whereas composite section will have stresses due to live load.

23·3. (*i*) **Propped.** The area of the haunch of concrete over the steel is small and its effect if neglected.

If m is the modular ratio which will depend on the mix used, an equivalent steel section is obtained by taking the flange width B_s as B_s/m. Then total area of the composite section will be

$B_s/m \times d_s + A_s$, where d_s is the thickness of slab. The equivalent composite section is shown in Fig. 23·5 (*b*).

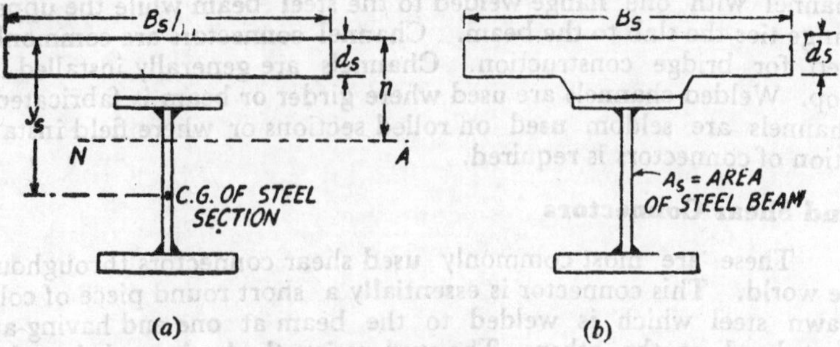

Fig. 23·5

In case neutral axis is not in the slab,

$$n = \frac{\dfrac{B_s d_s^2}{2m} + A_s y_s}{\dfrac{B_s d_s}{m} + A_s}$$

In case neutral axis lies in slab,

$$\frac{B_s}{m} \times \frac{n^2}{2} = A_s\ (y_s - n)$$

$$\therefore \quad \frac{n^2 B_s}{2m} nA_s - A_s y_s = 0$$

$$\therefore \quad n = \frac{-A_s + \sqrt{A_s^2 + \dfrac{2B_s A_s y_s}{m}}}{\dfrac{B_s}{m}}$$

$$= \frac{A_s m}{B_s}\left[\sqrt{1 + \frac{2B_s y_s}{mA_s}} - 1\right]$$

In almost all cases the neutral axis in below the slab.

If I_s is the M.I. of the steel section about its C.G.

$$I_e = I_s + A_s\ (y_s - n)^2 + \frac{B_s}{m} \times \frac{d_s^3}{m} + \frac{B_s d_s}{m}\left(n - \frac{d_s}{2}\right)^2$$

where I_e is the equivalent M.I. of the composite section.

In case N.A. lies in slab,

$$I_e = \frac{B_s n^3}{3m} + I_s + A_s (y_s - n)^2$$

Let M_s be the bending moment due to dead load of steel section, M_C be the bending moment due to dead load of concrete slab and M_A be the applied bending moment.

Let Z_b be the modulus of section at bottom of steel section alone and Z_{bc} and Z_{tc} be the modulus of section of bottom and top of composite section respectively.

$$\text{Max. stress in steel} = \frac{M_s}{Z_b} + \frac{M_C + M_A}{Z_{bc}}$$

Maximum stress in concrete

$$= \left[\frac{M_C + M_A}{Z_{tc}}\right] \times \frac{1}{m}$$

If D is the overall depth of the composite section

$$Z_b = \frac{I_s}{D - y_s}$$

$$Z_{bc} = \frac{I_c}{(D - n)}$$

$$Z_{tc} = \frac{I_c}{n}$$

The horizontal shear F_h at top of the steel section will be given as

$$F_h = \frac{F \dfrac{B_s d_s}{m}\left(n - \dfrac{d_s}{2}\right)}{I_c}$$

where F is the shear force at the section.

If s is the spacing of shear connectors, the force on the connectors and weld will be $s \times F_h$.

The shear connectors will be fairly close together at the ends of the composite members as the shear is maximum there. The maximum spacing for shear connectors is four times the depth of the concrete slab.

(*ii*) **Unpropped**

$$\text{Max. stress in steel} = \frac{M_s}{Z_b} + \frac{M_C}{Z_b} + \frac{M_A}{Z_{bc}}$$

$$\text{Max. stress in concrete} = \left(\frac{M_A}{Z_{tc}}\right)\frac{1}{m}.$$

23·4. Shear Resistance of Connectors

The design requirement for different types of shear connectors are given below.

(*i*) **Rigid connectors.** The safe shearing resistance is given by

$$Q = F_b \times A_b$$

where Q = safe shearing resistance

F_b = permissible bearing pressure on concrete

$$= 0.25\ \sigma_{cu} \sqrt[3]{\frac{A}{A_b}} \not> 0.6\ \sigma_{cu}$$

A = the area to which bearing pressure is transmitted and is equal to product of width of the top flange of steel joists at surface of contact and the depth of concrete slab including the haunch.

A_b = the bearing area of the connector, that is the area of transmitting face of the connector at right angles to the joist flange.

The longitudinal shear stress along the shearing surface between two successive connectors should not be greater than two and a half times the permissible shear stress of concrete. $S_h = 2.5\ bq$, where b is the width of flange of rigid connector at surface of contact and q is permissible shear stress measured as inclined tension in concrete.

The projected area along a slope of 1 to 5 from one rigid connector on to another should be at least three times the area of the face connector.

As far as possible, rigid connectors shall be associated with anchors so that shear is resisted partly by bond of the concrete and partly by the bearing pressure of the concrete against the face of the rigid connectors.

(*ii*) **Anchor connectors.** Anchor connectors are either used along or are used in conjunction with rigid connectors. The safe shear resistance $Q = K\sigma_{st}\ A_t$,

where A_t = cross-sectional area of the anchor bar,

K = coefficient depending on bond length.

(*a*) Anchors with bond length of not less than 40 times the diameter of the bar including the hook in the compression zone of concrete. $K = 1.0$

(*b*) Looped anchors used in connection with rigid connectors with diameter of loop not less than 15 times the diameter of bar. $K = 1.0$

(*c*) Looped anchors used in combination with rigid connectors with diameter of loop less than 15 times the diameter of bar. $K = 0.7$

(*d*) **Hooked anchors used in combination with rigid connectors where the ends of the straight anchors are hooked but the bond length in the compression zone of the concrete is less than 40 times the diameter of the bar. $K=0.5$**

Anchors may be either vertical or inclined at 45°.

Spacing of anchors when used alone shall not be less than 0·7 times the depth of the slab and shall not be greater than two times the depth of slab.

Spacing of anchors when associated with rigid connectors shall not be greater than two and a half times the depth of the slab.

The anchors may be welded to the beam or the ends bent over and well placed completely around. The anchors shall be brought upon the top surface of flange, then bent over and ends hooked. The portion from the upper bent to the hook should be at least ten times the diameter of the bar.

(*iii*) **Spiral connectors.** The shear resistance $Q=315d\times\sqrt[4]{\sigma_{cu}}$ N, where $Q=$ the safe shear resistance of one pitch of a spiral bar and d is the diameter of round bar in mm.

The diameter of the spiral bars shall be between 12 mm and 20 mm. The spirals shall extend at least half way in the slab. The ratio of the pitch of the spiral to the diameter shall be between 0·5 to 0·4. The developed length of spiral pitch shall not be less than 20 times the diameter of the bar. From fabrication consideration, the spiral pitch shall be within the limits 100 to 400 mm.

(*iv*) **End shear connector.** At each end of a simply supported girder an end connector shall be provided to counteract the effect of temperature, shrinkage and creep in addition to the external shear forces. The shear resistance of such an end connector shall have the value $\dfrac{M\times m_s}{I}$, where M is the max. bending moment on the composite beam due to additional loads operating after the composite action has been effected, m_s is the statical moment of the transformed area on the slab side of the contact surface about the neutral axis of the composite section and I is the moment of inertia of the transformed composite section.

The end shear connector shall consist of mechanical device having adequate shear resistance. Such mechanical device may consist of cut piece of rolled steel with the bearing face directed towards the centre of the beam or it may consist of anchor bars spreading out into slab away from the centre of the beam.

23·5. Design of Slab

The slab is designed spanning transversely to the steel girders which support it. In bridges the slab is at least 16 cm thick and in buildings it may be as less as 12 cm. It is important that a sufficient depth of concrete be provided to ensure adequate anchorage of the shear connectors, and consequently, haunching of concrete over the steel girder is preferably done. The steel girders are placed at 1·5 to 3 m centres in bridges and at greater spacings in buildings.

23·6. Design of Steel Section

(i) Propped

Max. stress in concrete$=\left(\dfrac{M_C+M_A}{Z_{tc}}\right)\times\dfrac{1}{m}$. Knowing value of allowable stress in concrete, value of M_C+M_A and m, the modulus of section of composite section Z_{tc} can be obtained. The value of Z_{bc} will be 1·5 to 2·5 times the value of Z_b and from equation of max. tensile stress in steel$=\dfrac{M_s}{Z_b}+\dfrac{M_C+M_A}{Z_{bc}}$, Z_{bc} and Z_b can be approximately obtained. Value of self-load of steel section is assumed and modified if the assumption is widely off the work.

[**Note.** Design of composite columns is given in Article 7·6].

Ex. 23·1. *Design a composite foot bridge having clear width of 2·5 m and effective span of 16 m. The bridge is to be designed for live load of 4000 N/m², concrete mix used is M 15.*

Sol. Kerb of 400 mm × 300 mm is provided. Two girders at 2·1 m centres are provided.

Dead load bending moment

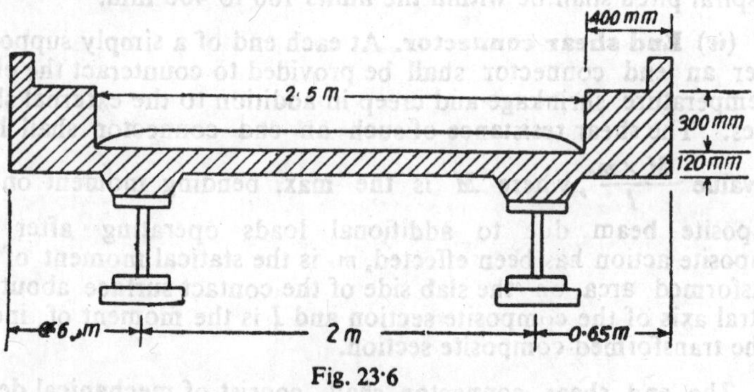

Fig. 23·6

Weight of kerb $=0·4\times0·3\times1\times24{,}000$

$=2880$ N/m

Wearing coat $=1\times1\times0·08\times22{,}000$

Average 80 mm thick) $=1760$

Slab $=12\times24$

$=2880$ N/m²

Parapet $=600$ N/m

$$\text{Reaction} \quad = 600 + 2880 + \frac{(2880+1760)\times 2{\cdot}5}{2}$$

$$= 600 + 2880 + 5800$$

$$= 9280 \text{ N.}$$

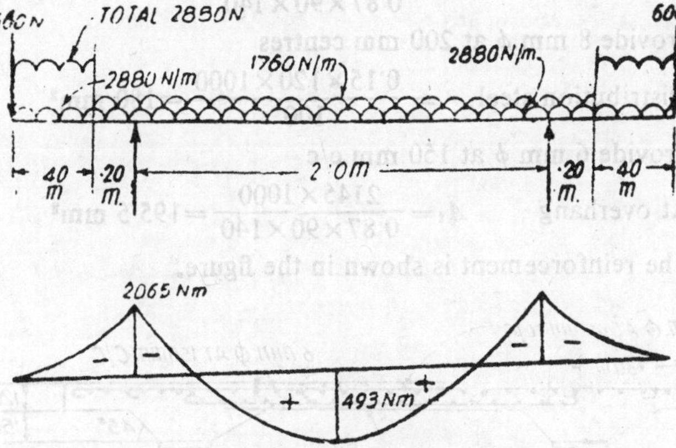

Fig. 23·7

$$\text{Max. −ve moment} = -\left[600\times 0{\cdot}60 + 2880\times 0{\cdot}40 + 2880\times\frac{(60)^2}{2} + 1760 + \frac{(0{\cdot}20)^2}{2}\right]$$

$$= -[360 + 1152 + 518 + 35]$$

$$= -2065 \text{ Nm.}$$

Max. +ve moment at centre of slab

$$= \frac{(2880+1760)\times(2{\cdot}1)^2}{8} - 2065$$

$$= 2558 - 2065 = +493 \text{ N-m.}$$

Live load B.M.

$$\text{−ve moment} \quad = -4000\times\frac{(0{\cdot}02)^2}{2} = -80 \text{ N-m}$$

Moment at centre of mid-span

$$= \frac{4000\times(2{\cdot}1)^2}{8} = +2205 \text{ N-m.}$$

Design moments

In centre of span Max. +ve

$$= 493 + 2205 = 2698 \text{ N-m}$$

$$\text{Overhang moment} \quad = -(2065+80) = -2145 \text{ N-m}$$

Effective depth $d=\sqrt{\frac{2698\times1000}{0.87\times1000}}=56\text{ mm}$

Provided depth of 120 mm will be alright with effective depth of 90 mm

$$A_t=\frac{2698\times1000}{0.87\times90\times140}=246\text{ mm}^2$$

Provide 8 mm ϕ at 200 mm centres

Distribution steel $=\frac{0.15\times120\times1000}{100}=180\text{ mm}^2$

Provide 6 mm ϕ at 150 mm c/c

At overhang $A_t=\frac{2145\times1000}{0.87\times90\times140}=195.5\text{ mm}^2$

The reinforcement is shown in the figure.

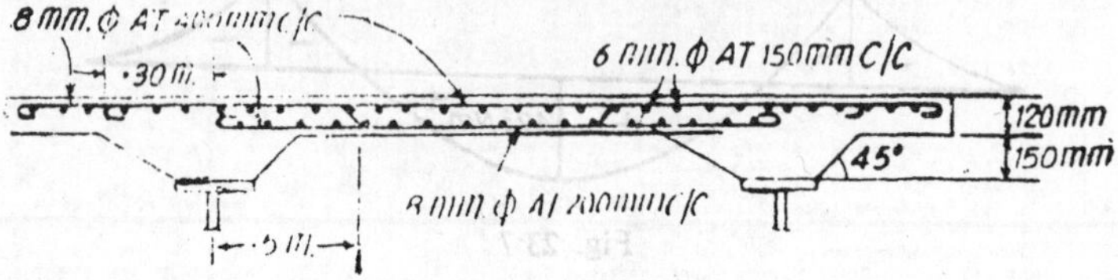

Fig. 23·8

Design of Steel girder—*Propped condition*

Taking girder weight as 1000 N/m

$$M_s=\frac{1000\times(16)^2}{8}=32{,}000\text{ N-m}$$

$$M_c=9280\times\frac{(16)^2}{8}=296{,}960\text{ N-m}$$

$$M_A=4000\times\frac{2.5}{2}\times\frac{(16)^2}{8}$$

$$=160{,}000\text{ N-m}$$

Max. stress in steel $=140=\frac{M_s}{Z_b}+\frac{M_C+M_A}{Z_{bc}}$

$\therefore$ $$140=\frac{32{,}000\times1000}{Z_b}+\frac{(296{,}960+160{,}000)\times1000}{Z_{bc}}$$

Taking $Z_{bc}=1.5\,Z_b$

$\therefore$ $$140=\frac{32{,}000\times1000}{Z_b}+\frac{456{,}960\times1000}{1.5\,Z_b}$$

$\therefore$ $$Z_b=\frac{(32{,}000+304{,}640)\times100}{14}$$

$$=2407\times10^3\text{ mm}^3.$$

Use ISMB $550 \times 103{\cdot}7$ kg/m

$$M.I. = 64{,}893{\cdot}6 \times 10^4 \text{ mm}^4$$

$$A = 13{,}211 \text{ mm}^2$$

$$Z_b = (2359{\cdot}8 \times 10^3) \text{ mm}^3.$$

Composite section

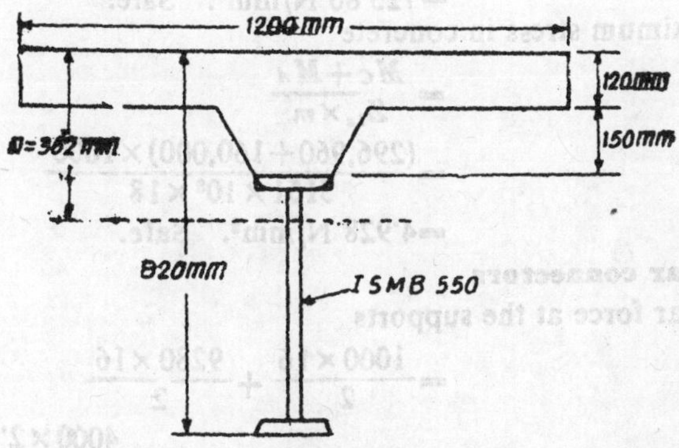

Fig. 23·9

$$n = \frac{\frac{1200}{18} \times 120 \times 60 + 13{,}211 \times (275 + 260)}{\frac{1200}{18} \times 12 + 13{,}211}$$

$$= \frac{7{,}679{,}000}{21{,}211} = 362 \text{ mm}$$

$$I_e = (64{,}893{\cdot}6) \times 10^4 + 13{,}211 \times (820 - 362 - 275)^2 + \frac{1200}{18} \times \frac{(120)^2}{12}$$

$$+ \frac{1200}{18} \times 12(362 - 60)^2$$

$$= 186{,}503{\cdot}67 \times 10^4 \text{ mm}^4$$

$$Z_{bc} = \frac{(1{,}865{,}036) \times 10^4}{820 - 36{\cdot}2}$$

$$= \frac{(1{,}865{,}036) \times 10^4}{458} = 4070 \times 10^3 \text{ mm}^3$$

$$Z_{tc} = \frac{(1{,}865{,}036) \times 10^4}{362} = (5151 \times 10^3) \text{ mm}^3$$

Max. stress in steel $= \frac{M_s}{Z_b} + \frac{M_C + M_A}{Z_{bc}}$

$= \frac{32,000 \times 1000}{(2359.8) \times 10^3}$

$+ \frac{(296,960 + 160,000) \times 1000}{(4070) \times 10^3}$

$= 13.56 + 112.3$

$= 125.86$ N/mm². Safe.

Maximum stress in concrete

$= \frac{M_C + M_A}{Z_{tc} \times m}$

$= \frac{(296,960 + 160,000) \times 1000}{5151 \times 10^3 \times 18}$

$= 4.928$ N/mm². Safe.

Shear connectors

Shear force at the supports

$= \frac{1000 \times 16}{2} + \frac{9280 \times 16}{2}$

$+ \frac{4000 \times 2.5}{2} \times \frac{16}{2}$

$= 8000 + 74,240 + 40,000 = 129,240$ N

$$F_h = \frac{F \frac{B_s d_s}{m} (n + d_s)}{I_s}$$

$$= \frac{122,240 \times \frac{1200 \times 120}{18} \left(362 - \frac{120}{2} \right)}{(1,865,036) \times 10^4}$$

$$= \frac{122,240 \times 800 \times 302}{(1,865,036) \times 10^4}$$

$= 158.3$ N per mm.

Use the shear connectors, ISNT 80, length 200 mm

Using 5 mm weld on both legs,

Length of weld $4 \times 80 = 320$ mm

Strength $= 320 \times 359 = 114,880$ N

Strength of tee in shear

$$= Q = 0.25 \sigma_{cu} \sqrt[2]{\frac{A}{A_b}} \times A_b$$

$$= 0.25 \times 10 \times \sqrt[2]{\frac{190 \times 270}{80 \times 200}} \times 80 \times 200$$

$= 88,490$ N.

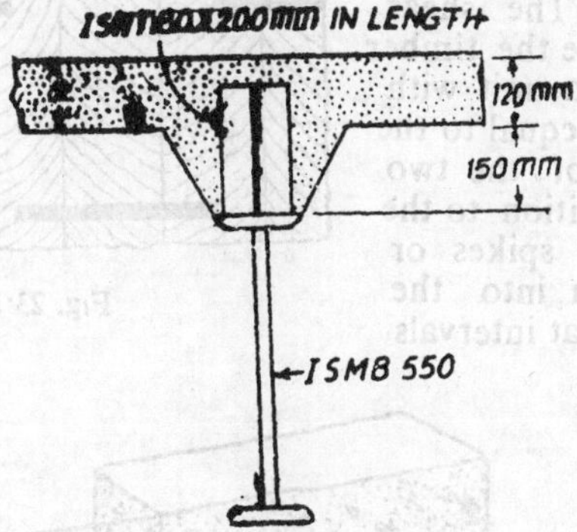

Fig. 23·10

$$\text{Spacing of connectors} = \frac{88,490}{158 \cdot 3} = 559 \text{ mm}$$

Maximum spacing$=2 \cdot 5\ d_s = 300$ mm.

Use the connectors at 300 mm centres.

23·7. Timber Concrete Composite Sections

It consists of a slab consisting of laminated base made of treated timber interlocked with concrete slab. This type should be used for bridge decks, piers docks, warehouses, ramps and other structures requiring heavy floors.

The timber base consists of planks 30 to 40 mm thick laid on edge with alternate pieces raised from 40 to 50 mm to form longitudinal grooves. Steel plates called shear developers as shown in Fig. 23·11 (*a*) are driven into the longitudinal grooves and extend

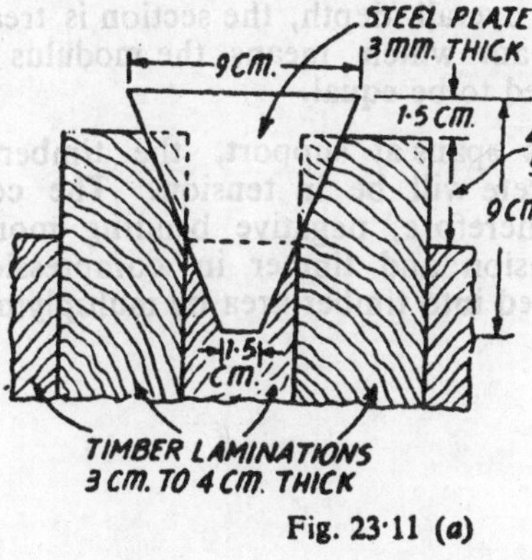

Fig. 23·11 (*a*)

about 15 mm above the high laminators. The shear developers combine the timber and concrete into a unit with an effective depth equal to the overall thickness of the two materials. In addition to the shear developers, spikes or dowels are driven into the raised laminators at intervals ½ to 1 metre.

Fig. 23·11 (*b*)

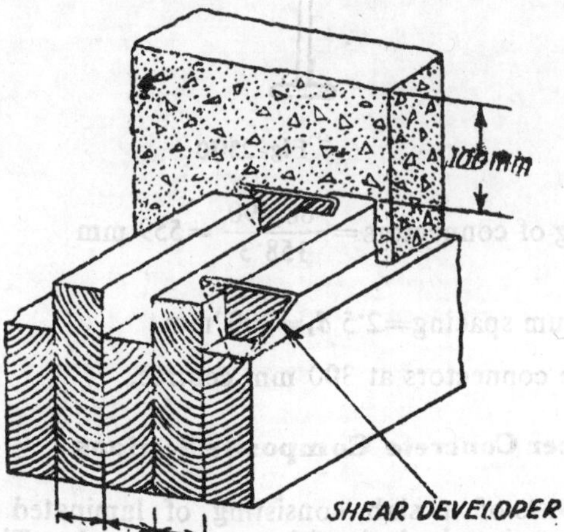

Fig. 23·11 (*c*)

For high single span length all the timbers should be of full length. Since the usual procedure is to put the timber in first place and then add the shear developers, reinforcing bars and concrete the entire floor, therefore, dead load of the composite section is carried by the timber section. The live load is taken by the composite section. For section where the concrete thickness is less than half the overall depth, the section is treated as a homogeneous timber beam which means the modulus of timber and concrete are assumed to be equal.

In continuous spans at support, the timber will be in compression and concrete will be in tension. The concrete cannot take any tension, therefore, negative bending moment is resisted by the steel in tension and timber in compression. Therefore, steel area is converted into timber area by multiplying with modular ratio.

24

Reinforced Concrete Bridges

24·1. Reinforced concrete is almost universally used for bridge construction both for highways and rail-roads. Its durability, rigidity, economy and ease with which pleasing appearance can be obtained makes it suitable for this purpose.

Different types of R.C. bridges in common use are :

1. Solid slab bridges. 2. Simple girder bridges. 3. Continuous girder bridges. 4. Cantilever bridges. 5. Rigid frame bridges both single span and multispan. 6. Arch bridges. 7. Long trusses. 8. Bow string girder bridges.

24·2. Use of Reinforced Concrete Bridges

Reinforced concrete bridges are generally used for highway bridges, their use for rail road bridges is limited. However, reinforced concrete floors are finding increasing use in steel bridges as they are cheaper than other types of floors. Structural steel is generally used for long span rail-road bridges. For short span rail-road bridges reinforced concrete is generally used.

24·3. Selection of Type of Bridge

Cost is generally the governing factor for the selection of type of bridge. The exact arrangement economically suitable for a particular case depends upon the natural condition *e.g.* foundation and specified requirements like width of the bridge, loading, appearance, waterway, time of construction, difficulties in formwork etc. Looking to various factors involved in the selection of type of bridge it is rather difficult to decide upon the type of bridge. However, experience is invaluable in deciding the type of bridge.

For bridges having one span following arrangements should be considered : 1. Simply supported deck. 2. Right angle rigid frame. 3. Right angle frame with concealed cantilevers. 4. Simply supported girder with concealed cantilevers.

Simply supported deck type is statically determinate structure and is simplest to design. However, the cost of this structure is highest. When unyielding foundations are available rigid frame type is most economical. For long spans type 2 and 4 should be considered.

For bridges with several spans the following arrangements should be considered : (1) Number of simply supported girder spans. (2) A combination of girders provided with cantilever and short spans supported by the edge of cantilevers, such a bridge is known as balanced cantilever bridge. (3) Continuous girders. (4) Multi-span rigid frames. (5) Arches.

First two types, *i.e.* number of simply supported girders and balanced cantilever type are statically determinate and easy to design.

These types are used where unyielding foundations are not available. The cost of type 4 is relatively more than the cost of type 2. When unyielding foundations are available continuous girders, multiple rigid frames and arches are best suited. Of these types, cost of multiple rigid frame is lowest. Where heavy piers are required or where bridge is to be supported on already existing piers, type 3 is used.

24·4. Economic Span Length

The cost of a bridge is the total cost of sub-structure and super-structure. For small variation in span, the cost of pier will not be affected. The cost of floorway will be the same for any small variation in span length. The cost of main girders and cross-girders will be directly proportional to the span. If L is the total length of the bridge and n is the number of piers, span $l = L/n+1$. The cost of sub structure will be the cost of abutments which will not vary, plus cost of piers. Let C_A be the cost of two abutments and C_P be the cost of one pier. Let C_T be the cost of main girders and cross-girders per unit length and C_F be the cost of flooring per unit length.

Total cost $\quad C = C_A + nC_P + LC_F + LC_T$

$$C_T \propto l$$

$$\therefore \quad C_T = kl$$

$$n = \frac{L}{l} - l$$

$$\therefore \quad C = C_A + C_P \frac{L}{l} - C_P + LC_F + Lkl$$

For cost to be minimum $\dfrac{dC}{dl} = 0$

$$-C_P \frac{L}{l^2} + Lk = 0$$

$$\therefore \quad \frac{C_P}{l} = kl = C_T$$

$$\therefore \quad C_P = lC_T$$

When the cost of pier is equal to the cost of main girders and cross-girders of one span, the total cost of the bridge will be the minimum.

24·5. Types of Loading

For the purposes of computing maximum stresses in any girder or member of a bridge the following loads shall, where applicable, be taken into account :

(i) Dead load.
(ii) Live load.
(iii) Impact effect.
(iv) Centrifugal force.
(v) Wind load.
(vi) Lateral loads such as nosing effect in rail way bridges and horizontal forces on parapets.
(vii) Longitudinal forces.
(viii) Seismic forces.

In addition to the stresses caused by above loading, the following stresses shall also taken into account where applicable :

(a) Secondary stresses,
(b) Temperature stresses, and
(c) Erection stresses.

24·6. Dead Load

The dead load is the weight of the structure and any permanent loads fixed thereon. The dead load initially assumed shall be checked after the design is completed and the design shall be revised if the actual calculated dead load exceeds the assumed dead load by more than 2½% or if the assumed dead load effect on a member varies from the actual dead load effect to such an extent as to adversely affect the design of such member.

Dead Load on Foot Bridges and Highway Bridges. The weight of floor system for bridges can be calculated. The weight of a component is assumed and checked after design. For example, for a bridge flooring of R C C supported on main girders when designing R C C flooring, its dead weight is assumed which is checked after designing the flooring. The necessary correction in dead weight is made if necessary and the dead weight of cross-beams is assumed when designing these and verified after design. Same procedure is adopted for main girders.

24·7. Live Load

(a) Foot Bridges and Footways

For all parts of bridges, floors acccessible only to pedestrians and for all footways the loading shall be 4000 N/m², where crowd loads are likely to occur the loading shall be 5000 N/m².

Kerbs 600 mm or more in width shall be designed for the above loads. If the kerb width is less than 600 mm no live load shall be applied.

The main girders, trusses, arches or other members supporting the footways shall be designed for the following live loads per square metre of footways area ·

(*i*) For effective spans of 7·5 metres or less, 4000 N/m² and for crowded location 5000 N/m².

(*ii*) For effective spans of over 7·5 metres but not exceeding 30 metres, the intensity of load shall be determined according to the formula—

$$P=P'-\left(\frac{400L-3000}{9}\right)$$

(*iii*) For effective spans of over 30 metres, the intensity of load shall be determined according to the formula—

$$P=\left(P'-2600-\frac{48,000}{L}\right)\left(\frac{165-W}{15}\right)$$

where P'=4000 N/m² or 5000 N/m², as the case may be

P=live load in N/m²

L=effective span of the main girder, truss or arch in metres

W=width of the footway in metres.

Where vehicles mount the footway, each part of the footway shall be capable of carrying a wheel load of 4 tonnes which shall be deemed to include impact, distributed over a contact area of 300 mm in diameter. The working stress shall be increased by 25%.

(*b*) **For Road Bridges**

I.R.C. class AA loading. This loading is to be adopted within certain municipal limits, in certain existing or complicated industrial areas, in other specified areas, and along certain specified highways. Bridges designed for class AA loading should be checked for class A loading also as under certain conditions, heavier stresses may be obtained under class A loading.

I.R.C. class A loading. This loading is to be normally adopted on all roads in which prominent bridge and culverts are constructed.

I.R.C. class B loading. This loading is normally adopted for temporary structures and for bridges in specified areas. Structures with timber span are to be regarded as temporary structures for this clause.

The standard wheeled or tracked vehicle or trains of vehicles are illustrated in Figs. 24·1 and 24·3. The trailers attached to the driving unit are not to be considered as detachable.

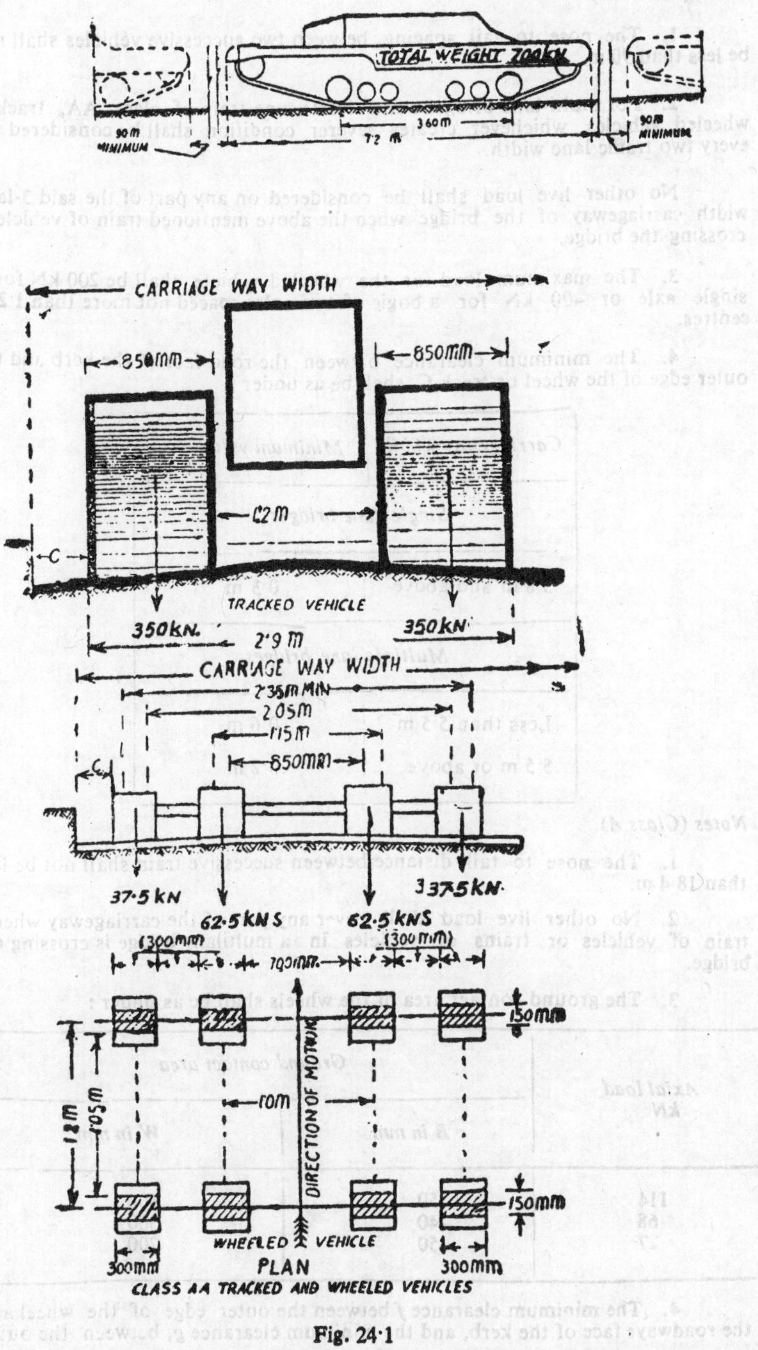

CLASS AA TRACKED AND WHEELED VEHICLES

Fig. 24·1

Notes (Class AA)

1. The nose to tail spacing between two successive vehicles shall not be less than 90 m.

2. For multi-lane bridges and culverts, one train of class AA, tracked wheeled vehicles, whichever creates severer condition shall be considered for every two traffic lane width.

No other live load shall be considered on any part of the said 3-lane width carriageway of the bridge when the above mentioned train of vehicle is crossing the bridge.

3. The maximum load for the wheeled vehicle shall be 200 kN for a single axle or 400 kN for a bogie of two axles spaced not more than 1·2 m centres.

4. The minimum clearance between the road face of the kerb and the outer edge of the wheel or track C, shall be as under :

Carriageway width	*Minimum value of C*
Single lane bridges	
3·8 m and above	0·3 m
Multiple lane bridges	
Less than 5·5 m	0·6 m
5·5 m or above	0·2 m

Notes (Class A)

1. The nose to tail distance between successive train shall not be less than 18·4 m.

2. No other live load shall cover any part of the carriageway when a train of vehicles or trains of vehicles in a multilane bridge is crossing the bridge.

3. The ground contact area of the wheels shall be as under :

Axial load kN	*Ground contact area*	
	B in mm	*W in mm*
114	250	500
68	240	300
27	150	200

4. The minimum clearance *f* between the outer edge of the wheel and the roadways face of the kerb, and the minimum clearance *g*, between the outer

edges of passing or crossing vehicles, on multi-lane bridges shall be as given below :

Clear carriage way width	*g*	*f*
5·5 to 7·5 m	Uniformly increasing from 0·4 to 1·2 m	0·15 m for all carriageway widths
Above 7·5 m	1·2 m	

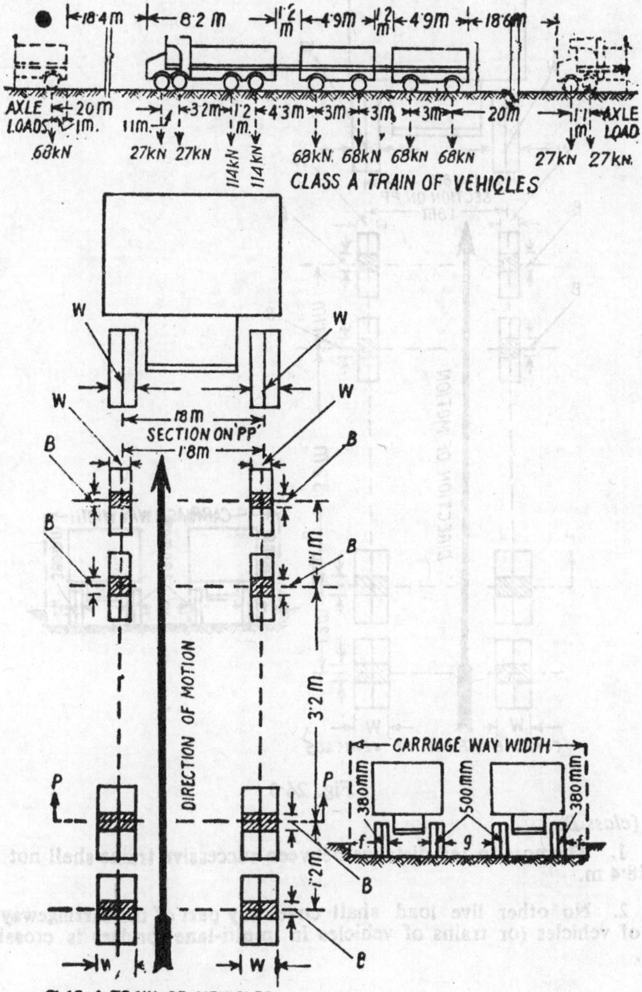

Fig. 24·2

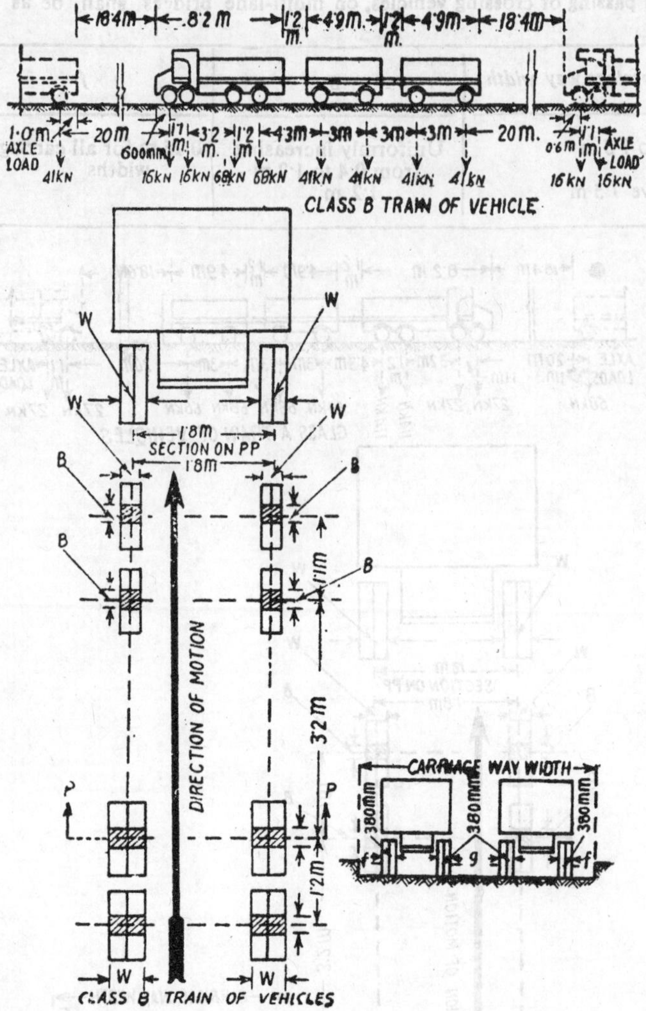

Fig. 24·3

Notes (class B)

1. The nose to tail distance between successive trains shall not be less than 18·4 m.

2. No other live load shall cover any part of the carriageway when a train of vehicles (or trains of vehicles in multi-lane bridge) is crossing the bridge.

3. The ground contact area of the wheels shall be as under :

Axle load kN	*Ground contact area*	
	B in mm	*W in mm*
68	200	380
41	150	300
16	125	175

4 The minimum clearance *f*, between outer edge of the wheel and the roadway face of the kerb, and the minimum clearance *g*, between the outer edges of passing or crossing vehicles on multi-lane bridge shall be as given below :

Clear carriage way width	*f*	*g*
5·5 to 7·5 m	Uniformly increasing from 0·4 to 1·2 m	0 15 m for all carriage way widths
Above 7·5 m	1·2 m	

Within the kerb to kerb width of the roadway, the standard vehicle or train shall be assumed to travel parallel to the length of the bridge and to occupy any position which will produce maximum stresses provided that the minimum clearances between a vehicle and the roadway face of a kerb and between two passing or crossing vehicles are not encroached upon.

For each standard vehicle or train, all the axles of a unit of vehicles shall be considered as acting simultaneously in a position causing maximum stresses.

Vehicles in adjacent lanes shall be taken as headed in the direction producing maximum stresses.

The spaces on the carriageway left uncovered by the standard train or vehicle shall not be assumed as subject to any additional live load.

24·8. Impact Effect

No allowance for impact is to be made for foot bridges.

Road Bridges

(1) For road bridges for class A or B loading. The impact factor shall be determined from the following formula which is applicable for spans between 3 m and 45 m.

For R.C. Bridges $I=\dfrac{4\cdot5}{6\cdot0+L}$

where L is the span in metres.

For spans 3 m and less impact factor will be 50% for concrete bridges. Graph in Fig. 24·4 may be used for determining impact factor.

For road bridges for class AA loading

(*a*) For spans less than 9 m.

(*i*) For tracked vehicle	25% for spans upto 5 m linearly reducing to 10 per cent for spans of 9 m.
(*ii*) For wheeled vehicles	25 per cent.

(*b*) For spans of 9 m or more.

Reinforced concrete bridges.

Tracked vehicle	10 per cent upto a span of 40 m and in accordance with the curve indicated in Fig. 24·4 for spans in excess of 40 m.
Wheeled vehicle	25 per cent for spans upto 12 m and in accordance with the curve indicated in Fig. 24·4 for spans in excess of 12 m.

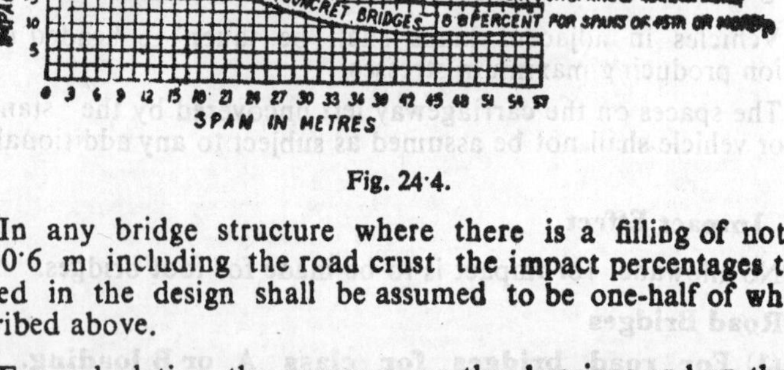

Fig. 24·4.

In any bridge structure where there is a filling of not less than 0·6 m including the road crust the impact percentages to be allowed in the design shall be assumed to be one-half of what is prescribed above.

For calculating the pressure on the bearings and on the top surface of the bed blocks, full value of the appropriate impact percentage shall be allowed, but for the design of piers, abutments and structures generally below the level of the top of the bed block, the appropriate impact percentage shall be multiplied by the factors given below.

(a) For calculating the pressure at the bottom surface of the bed block. ...0·5

(b) For calculating the pressure on the top 3 m of the structure below the bed block. ...0·5 decreasing uniformly

(c) For calculating the pressure on the portion of the structure more than 3 m below the bed block. ...zero

24·9. Centrifugal Force

For road bridges centrifugal force shall be determined from the following formula :

$$C=\frac{WV^2}{127R}$$

where C=Centrifugal force in kN acting normally to traffic (1) at the point of action of the wheel loads (2) uniformly distributed over every metre on which a uniformly distributed load acts.

W=Live load (1) in kN in case of wheel loads each wheel load being considered as acting over the ground contact specified in article 24·7 and (2) in kN per linear metre in case of a uniformly distributed live load.

V=Design speed of the vehicle using the bridge, in km per hour.

R=Radius of curvature in metres.

The centrifugal force shall be considered to act at a height of 1·2 m above the level of the carriages.

24·10. Wind Loads

All structures shall be designed for lateral wind forces given below. These forces shall be considered to act horizontally and in such a direction that the resultant stresses in the member under consideration are maximum.

The wind force on structure shall be assumed as a horizontal force of the intensity specified in Table 24·1 and acting on an area calculated as follows :

(a) *For a deck structure.* The area of the structure as seen in elevation including the floor sy,tem and railings.

(*b*) *For a through or half through structure.* The area of the elevation of the windward truss as specified in (*a*) above plus half the area of elevation above the deck level of all other trusses or girders.

The pressure given in Table 24·1 shall be doubled for bridges situated in area such as Kathiawar Peninsula and the Bengal and Orissa coasts.

Table 24·1. Wind Pressures and Wind Velocities

H	*V*	*P*	*H*	*V*	*P*
0	80	400	30	147	1414
3	98	580	38	154	1510
6	107	730	46	159	1660
9	115	880	54	163	1760
12	123	980	62	168	1850
15	128	1070	76	175	2000
18	133	1120	91	181	2100
21	138	1220	106	186	2240
24	143	1270	122	190	2340
27	145	1360			

where H=Average height in metres of the exposed surface above the mean retarding surface (ground or bed level or water level).

V=Horizontal velocity of wind in kilometres per hour at height H.

P=Horizontal wind pressure in N per sq. metre at height H.

The lateral wind force against any exposed moving load shall be considered as acting at 1·5 m above the roadway and shall be assumed to have the following values :

Highway bridges ordinary ...3000 N per linear metre

Highway bridges carrying tramway...4500 N per linear metre.

The bridges should not be considered to be carrying any live load when the wind velocity at deck level exceeds 130 km per hour.

The total assumed wind force shall not be less than 4500 N per linear metre in the plan of the loaded chord and 2250 N per linear metre in the plane of the unloaded chord on through or half through truss, latticed or other similar spans and not less than 4500 N per linear metre on deck spans.

A wind pressure of 2400 N per square metre of the unloaded structure shall be used if it produces greater stresses than those produced by the combined wind forces as stated above.

24·11. Lateral Loads

Forces on Parapets. Railings on parapets shall have a minimum height above the adjacent roadway or footway surface of **4 m** less on half the horizontal width of the top rail or top of the parapet. They shall be designed to resist a lateral horizontal force and a vertical force each of 1500 N/m applied simultaneously at the top of the railing or parapet. These forces shall also be considered in the design of main structural members of the bridges provided with foot paths ; where, however, foot paths are not provided, these forces need not be considered in the design of the main structural member.

Kerbs. Kerbs shall be designed for lateral loading of 7500 N/m run of the kerb applied horizontally at the top of the kerb. This load need not be taken for the design of supporting structure.

24·12. Longitudinal Forces

Road Bridges

In all road bridges, provision shall be made for longitudinal forces arising from any one or more of the following causes :

(*a*) Tractive effect caused through acceleration of the driving wheels ;

(*b*) Braking effect resulting from the application of brakes to braked wheels ; and

(*c*) Frictional resistance offered to movement of free bearings due to change of temperature or any other cause.

The braking effect on a simply supported span or a continuous unit of spans or any other type of bridge unit shall be assumed to have the following values :

(*a*) In the case of single lane or two lane bridge ; twenty per cent of the first train load plus five per cent of the loads of the succeeding trains or part thereof, the train loads in one lane only being considered for this purpose.

(*b*) In the case of bridges having more than two lanes ; as in (*a*) above for the first two lanes plus five per cent of the loads on the lanes in excess of the two.

The force due to braking effect shall be assumed to act along a line parallel to the roadway and 1·2 m above it. While transferring the force to the bearing, the change in the vertical reaction at the bearings should be accounted for.

The longitudinal force at any free bearing shall be limited to the sum of dead and live load reactions at the bearing multiplied by the appropriate coefficient of friction. The coefficient of friction at the bearings shall be assumed to have the following values :

For roller bearings ...0·13

For sliding bearings of hard copper alloy ...0·15

For sliding bearings of steel on cast iron or steel on steel ...0·25

For sliding bearing of steel on ferro asbestos ...0·20

The longitudinal force at the fixed bearing shall be taken as the algebraic sum of the longitudinal forces at the free bearings in the bridge unit under consideration and the force due to braking effect on the wheels as mentioned above.

The effects of braking force on bridge structures without bearings such as arches, rigid frames, etc., shall be calculated in accordance with approved methods of analysis of indeterminate structures.

The effects of the longitudinal forces and all other horizontal forces should be calculated upto a level where the resultant passive earth resistance of the soil below the deepest scour level (floor level in case of a bridge having pucca floor) balances these forces.

24·13. Seismic Loads

Seismic loads need only be taken in account if the local conditions so require and the allowance for horizontal acceleration shall depend on such local conditions, but shall not exceed 0·12 of gravity.

24·14. Frictional Resistance of Expansion Bearings

Where the frictional resistance of the expansion bearings has to be taken into account, the following coefficients shall be assumed in calculating the amount of friction bearings :

For roller bearings ...0·03

For sliding bearing of steel on cast iron or steel bearings...0·25

For sliding bearings of steel on ferro asbestos ...0·20

24·15. Secondary Stresses

Secondary stresses fall into two groups :

(*a*) Stresses which are the result of eccentricity of connections and off-panel point loading (for example, loads rolling direct on chords, self-weight of member and wind load on member). Secondary stresses due to unknown eccentricities arising out of inaccuracies in fabrication are already allowed for in the factor of safety.

(*b*) Stresses which are the result of the elastic deformation of the structure combined with the rigidity of the joints.

24·16. Temperature Effect

Where any portion of the structure is not free to expand or contract under the variation of temperature, allowance shall be made for the stresses resulting from this condition. The coefficient of expansion shall be taken as $11·7 \times 10^{-6}$ per 1°C.

24·17. Erection Forces and Effects

The weight of all permanent and temporary material together with all other forces and effects which can operate on any part of the structure during erection shall be taken into account.

24·18. Width of Roadway and Footway

(*i*) For high level bridges constructed for the use of road traffic only, the width of roadway shall not be less than 3·8 m for a single lane bridge and shall be increased by a minimum of 3 m for every additional lane of traffic on a multiple lane bridges. Road bridges shall be either one-lane, two line or four-lane. Three-lane bridges shall not be constructed. In the case of four-lane or multiples of two-lane, a central verge of atleast 1·2 metres shall be provided.

(*ii*) For bridges constructed for the use of combined road and tramway traffic, the roadway widths given in (*i*) above shall be increased by 4 m for a single track tramway and by 7·6 m for a double track tramway.

(*iii*) When a footway is provided its width shall not be less than 1·5 m.

24·19 Deck Slab Bridges

This is the simplest type of RCC Bridge and easiest to construct. This type of bridge is suitable as submersible bridge. Upto

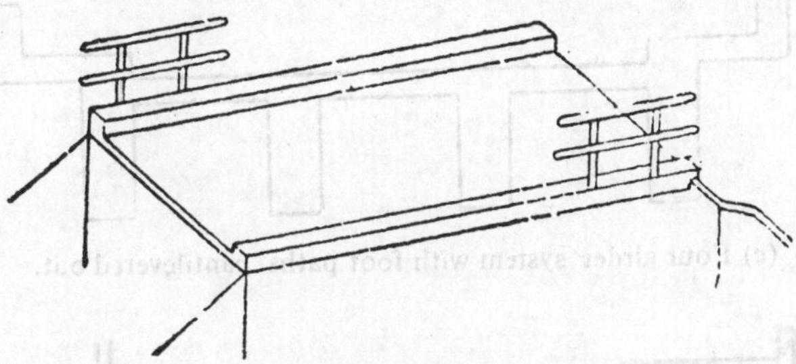

Fig. 24·5

spans of 8 m it is most economical. The thickness of deck slab is quite considerable and the amount of steel and concrete required are generally more than those in a girder bridge. However the construction is much simpler and placement of concrete and steel is easy. The cost of formwork and labour is much less.

24·20. Girder Bridges

This is the most common type of bridge. This type of bridge is economical for spans between 10 to 20 m.

Floor arrangement for deck girder bridges. The floor arrangement for a girder bridge depends on the width of the roadway. For two-lane or wider bridges the roadway is supported on a number of longitudinal girders, generally without any transverse beams. For bridges having narrow width, arrangement used is parapet girders as main supporting members with thick slab

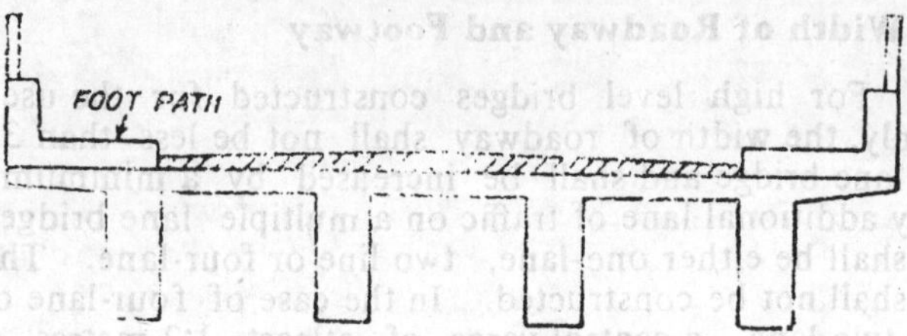

(*a*) Four girder system.

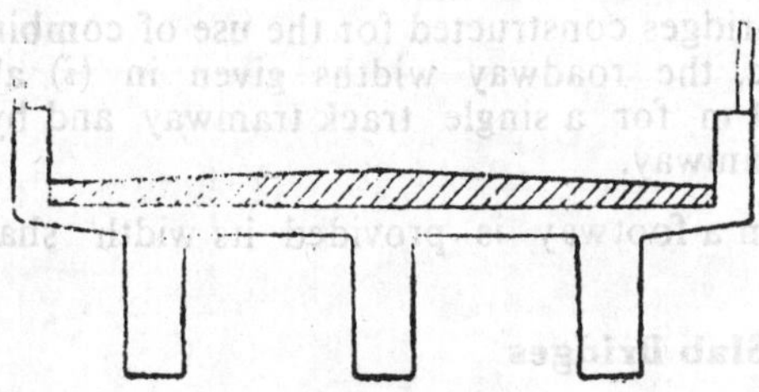

(*b*) Three girder system.

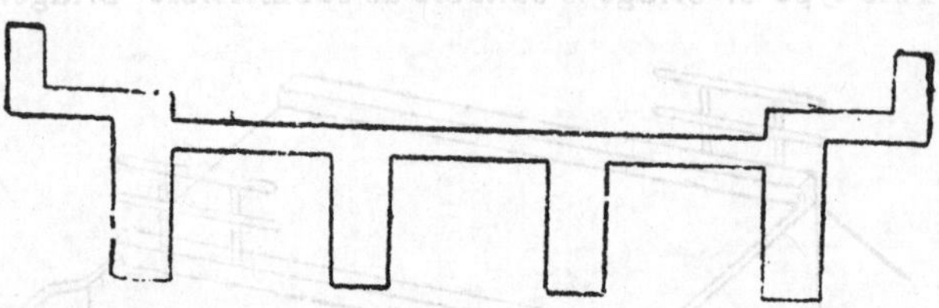

(*c*) Four girder system with foot paths cantilevered out.

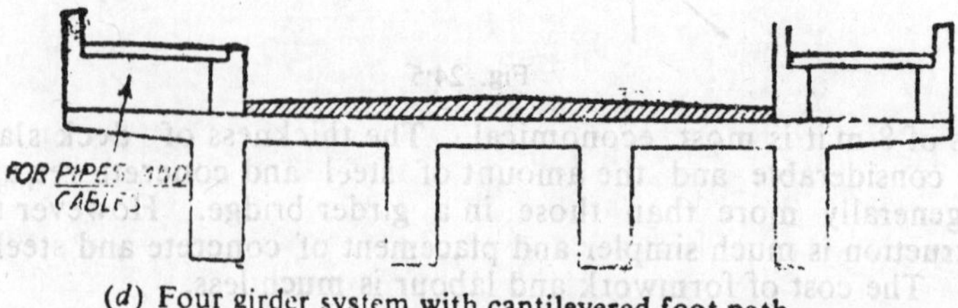

(*d*) Four girder system with cantilevered foot path.

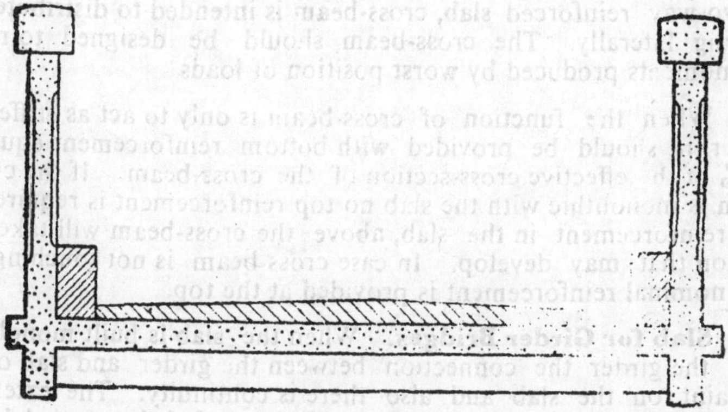

(*e*) Cross-section of parapet girder road bridges.

Fig. 24·6. Typical cross-section of deck girder bridges.

or transverse beams with thinner slab. This type of arrangement is suitable for bridges carrying light traffic.

Spacing of Main Girders. The spacing of girders affects to a considerable extent the cost of a bridge. A comparative statement of different arrangements is to be made before any arrangement is to be adopted. A closer spacing of girders results in a thin slab and lighter section for main girders, however the number of girders is more. As the number of girders is more, formwork is costly and number of bearings required are more, on the other hand, wider spacing between girders gives thicker slab but the number of girders is less, resulting in less cost of formwork and less number of bearings. Where formwork is expensive, cost of bearings is quite considerable and materials are relatively cheap, a wide spacing of girders is economical. On the other hand, where formwork is cheap and materials are costly closer spacing of girders is desirable.

Cross-beams in Girder Bridges. Cross-girders are provided to connect main girders at suitable intervals. Generally maximum spacing of cross-girder is 10 m.

By providing cross-girders at closer intervals the slab can be designed as two way slab thus resulting in thinner section of slab. The cross-girders in that case are to be designed for the live load in the most unfavourable position.

In certain floor design, floor slab is not designed as two-way reinforced slab but only one-way slab spanning between main girders, the cross-girders provided serve to stiffen the main girders. Another function performed by cross-beams is to equalize in a partly loaded bridge the deflection of the girders carrying heavy loading with those of the girders carrying light loads.

Reinforcement of Cross-beams. When the slab is designed as two-way reinforced slab, cross-beam is intended to distribute the loading laterally. The cross-beam should be designed to resist the moments produced by worst position of loads.

When the function of cross-beam is only to act as stiffening girder, it should be provided with bottom reinforcement equal to 0·3% of the effective cross-section of the cross-beam. If the cross-beam is monolithic with the slab no top reinforcement is required as the reinforcement in the slab, above the cross-beam will take any tension that may develop. In case cross-beam is not touching the slab nominal reinforcement is provided at the top.

Slab for Girder Bridges. When the slab is built monolithic with the girder the connection between the girder and slab offers restraint on the slab and also there is continuity. The extent of restraint depends upon the relative stiffnesses of girders and slab and number and spacing of cross-girders. The actual condition is somewhere between a continuous slab fixed at supports.

The bending moments in slabs due to concentrated loads can be found either by Pigeoud's method as discussed in chapter 24 or by effective width formula given below as per I.R.C. code section 3.

1. Solid slabs spanning in one direction

(*i*) For a single concentrated load, the effective width may be calculated in accordance with the following equation :

$$b_e = kx\left(1-\frac{x}{L}\right)+b_w$$

where b_e = the effective width of the slab on which load acts.

L = the effective span.

x = the distance of the centre of gravity of the concentrated load from the nearer support.

b_w = the breadth of concentration area of the load, *i.e.* the dimension of the tyre of track contact area over the load surface of the slab in a direction at right angles to the span plus twice the thickness of the wearing coat or surface finish above the structure slab, and

k = a constant having the values given in Table 24·2, depending on the ratio B/L, where B is the width of the slab.

Provided that the effective width shall not exceed the actual width of the slab. And provided further that in case of a load near the unsupported edge of a slab, the effective width will not exceed the above value nor half the above value plus the distance of the load from the unsupported edge.

Table 24·2

B/L	*k for simply supported slab*	*k for continuous slab*	*B/L*	*k for simply supported slab*	*k for continuous slab*
0·1	0·40	0 40	1·1	2 60	2·28
0·2	0·80	0·80	1·2	2·64	2·36
0·3	1 16	1·16	1 3	2·72	2·40
0.4	1·48	1·44	1·4	2·80	2·44
0·5	1·72	1·68	1·5	2·84	2·48
0·6	1 96	1·84	1·6	2·88	2·52
0·7	2·12	1·96	1·7	2·92	2·56
0·8	2·24	2·08	1·8	2·96	2·60
0·9	2·36	2·16	1·9	3·00	2·60
1·0	2·48	2·24	2 and above	3·00	2·60

(*ii*) For two or more concentrated loads in a line in the direction of the span, the bending moment per unit width of slab shall be calculated separately for each load according to its appropriate effective width of slab as calcaluted in (*i*) above.

(*iii*) For two or more loads not in a line in the direction of the span, if the effective width of a slab for one load overlaps the effective width of slab for an adjacent load, the resultant effective width for the two loads equals the sum of the reactive effective widths for each load minus the width of the overlap, provided that the slab so designed is tested for the two loads acting separately.

2. Solid slab cantilever

(*i*) For a single concentrated load the effective width may be calculated in accordance with the following equations :

$$b_e = 1{\cdot}2x + b_w$$

(**Note** : Slabs designed on the above basis need not be checked for shear.)

where b_e = the effective width ;

x = the distance of the centre of gravity of the concentrated load from the face of the cantilever support ;

b_w = the breadth of concentration area of load, *i.e.* the dimension of the tyre or track contact area over the load surface of the slab in a direction parallel to the supporting edge of the cantilever plus twice the thickness of the wearing coat or surface finish above the structural slab.

Provided that the effective width of the cantilever slab shall not exceed the length of the cantilever slab measured parallel to the support. And provided further that when the concentrated load is placed near one of the two extreme ends of the length of cantilever slab in the direction parallel to the support, the effective width shall not exceed the above value, nor shall it exceed half the above value plus the distance of the concentrated load from the nearer extreme end measured in the direction parallel to the fixed edge.

(*ii*) For two or more concentrated loads, if the effective width of slab for one load overlaps the effective width of slab for an adjacent load, the resultant effective width for the two loads shall be taken as equal to the sum of the respective effective widths for each load minus the width of overlap provided that the slab so designed is tested for the two loads acting separately.

(*iii*) For slabs other than solid slabs, the effective width shall depend on the ratio of the transverse and longitudinal flexural rigidities of the slab. When this ratio is one, that is when the transverse and longitudinal flexural rigidities are approximately equal, the value of effective width as found for solid slabs may be used. But as the ratio decreases a proportionately smaller value shall be taken.

Dispersion of loads along the span

The effective length of wheel track load in the direction of span length shall be taken as equal to the dimension of the tyre contact area over the wearing surface of the slab inclusive of the thickness of wearing surface.

Depth of Slab. The depth of slab in a panel it governed by the maximum positive moment in the panel. Generally same thickness is provided for all panels. If the negative moment at the support is greater than the positive moment, it may be economical to provide slab haunches at junction with girders. The length of the haunch should not be less than 3 times its depth.

Main Reinforcement in slab. Main reinforcement of slab consists of top and bottom reinforcement placed in the direction of span. Reinforcement to resist negative B.M. is also required in central part of each span and the amount of this reinforcement should not be less than one-third of the bottom reinforcement in that span. The negative B.M. is produced in the span when adjacent span is loaded.

Distribution reinforcement

(*i*) For solid slabs spanning in one direction, distributing bars shall be provided at right angles to the main tensile bars to provide for the lateral distribution of the loads. The distribution reinforcement shall be such as to produce a resisting moment in the direction perpendicular to the span equal to 0·3 times the moment due to concentrated live loads plus 0·2 times the moment due to other loads such as dead load shrinkage, temperature, etc.

(*ii*) In cantilever slabs, the distribution steel, to resist a moment equal to that of 0·3 times live-load moment plus 0·2 times dead load moment shall be provided half at top and half at the bottom of the slab.

(*iii*) The pitch of the distributing bars shall be not greater than thrice the effective depth of the slab or 450 mm whichever is less.

Slabs provided with cantilevers. In girder bridges it is generally economical to provide deck cantilever out from the outermost girders. This has the following advantages :

(*i*) Positive B.M. in the exterior span is reduced and thus the distribution of moments in exterior spans is made similar to interior spans.

(*ii*) The loads on all girders are equalised.

(*iii*) The number of girders is reduced without increasing the spacing of the griders.

(*iv*) The length of the sppporting piers is reduced and thus the distance between outermost girders is reduced which governs the length of the pier.

The cantilever loads produce bending moments not only in cantilevers but also in other spans. As the slab is monolithic with heavy girders the cantilever bending moments are partly resisted by the torsional resistance of the outermost girders and partly transferred to the slab. The amount of B.M. transferred to girder depends on the torsional rigidity of the girders and number of cross beams.

24·21. General Design Requirements

1. Bar Sizes. (*i*) The maximum size of the reinforcement shall be 50 mm diameter or a section of equivalent area, unless the particular condition warrants the adoption of special reinforcement design.

(*ii*) Excepting for wiremesh and similar reinforcement the diameter of any reinforcement bar, including transverse ties spirals, stirrups and all secondary reinforcement shall generally be not less than 6 mm.

(*iii*) The diameter of the longitudinal reinforcement bars in column shall be not less than 12 mm.

(*iv*) The diameter of wires under tensile stress in connected mesh and similar reinforcement in slabs shall be not less than 3 mm.

2. Curtailment of Bars. (*i*) To prevent large changes in the moment of resistance the points at which the bars are curtailed shall be suitably spread.

(*ii*) In the case of simply supported spans, at least 20 per cent of the steel required to resist the maximum bending moment shall be carried over to the supports along the tension side of the beam. In case of a span continuous beyond a support, at least 10 per cent of the steel required to resist the maximum positive bending moment shall be so carried over.

3. **Curved or Sloped Reinforcement.** Where the alignment of the reinforcement deviates from the normal to the plane of bending as in the case of a beam with curved or sloped soffit, only the area of the reinforcement effective in the direction normal to the plane of bending shall be considered.

4. **Lateral Support in the Beams.** For T-beams and L-beams with spans over 9 m in length, diaphragms shall be placed at suitable points at the direction of designer. The spacings of such diaphragms shall not be more than 30 times the minimum width of the web. Cross girders shall be provided at the bearings and also at the ends of the cantilevers.

In case of beams if the length between adequate lateral restraints be greater than thirty times the width of the compression flange, the permissible compressive stress in concrete shall be reduced by a factor equal to $(1.5 - L/60B)$, where L is the unsupported length and B is the width of the compression flange of the beam.

5. **Deep Beams.** If the depth of the web of beams exceeds 1.5 m additional longitudinal reinforcement with a geometrical percentage in relation to the web of at least 0.5% shall be provided on each of the faces at a spacing between bars not exceeding 200 mm. It will not be necessary to provide such reinforcement if the principal reinforcement is distributed over the tensile zone of the beam and the exact position of the reinforcement is taken into account in the strength calculations.

6. **Distribution of Live Loads on Longitudinal Beams.** (*i*) In calculating the bending moments on longitudinal beams and stringers, the longitudinal distribution of the wheel loads through the wearing coat, deck slab and filling need not be considered.

(*ii*) When longitudinal beams are connected together by transverse members like deck slab cross girders, diaphragms, soffit slab, etc., the distribution of bending moments between longitudinal girders shall be calculated by one of the following methods, subject to the conditions, if any, mentioned against each :

(*a*) Finding the reactions on the longitudinals assuming the supports of the deck slab as unyielding. This method is applicable where there are only two longitudinal girders.

(*b*) Distributing loads between longitudinals by Courbon's method, strictly with its limitations.

(*c*) Distributing loads between longitudinals by any rational method of grid analysis, *e.g.* the method of harmonic analysis as given by Handry and Jaeger or Morice and Little's version of the isotropic plate theory of Guyon and Massonet.

(*iii*) In calculating the shear force on sections of longitudinal beams, wheel loads or track loads shall be allocated to respective longitudinal beams, by any rational method. Alternatively the following method shall be followed :

(*a*) All loads within a distance of 5·5 m from the supports shall be distributed according to reactions of the longitudinal beams assuming the deck slab either simply supported or continuous as the case may be, the supports being assumed unyielding or by the procedure referred in para (*b*) below, whichever is greater.

(*b*) For loads, distances greater than 5·5 m from the support, distribution of the loads between the longitudinals for purpose of finding shearing forces shall be assumed to be same for bending moment given above.

7. **Distribution of Live Loads on Transverse Beams.** (*i*) In designing the transverse floor beams the wheel loads or track loads shall not be assumed to be spread or distributed through the deck slab, filling and wearing coat along the length of the floor beams.

(*ii*) Distribution of load between longitudinal beam for the purpose of finding bending moments and shears in intermediate transverse floor beams shall be made by one of the methods given in 6 (*ii*) above.

8. **T and L Beams.** (*i*) For the purpose of calculation of stress at any section of a T-beam or L-beam, or in the calculation of its moment or inertia the effective width of the slab, to function as the compression flange of the beam, shall be the least of the following :

In case of T-beams

(*a*) One-fourth the effective span of the beam.

(*b*) The distance between the centres of the ribs of the beam.

(*c*) The breadth of the rib plus 12 times the thickness of the slab plus the width of the haunches, if any. The width of the haunch shall for this purpose, be limited to the actual width or 3 times the depth of the haunch, whichever is less

In case of L-beams

(*a*) One-tenth the effective span of the beam.

(*b*) The breadth of rib plus one-half the clear distance between the ribs.

(*c*) The breadth of the rib plus 4 times the thickness of the slab plus the width of the haunch, if any. The width of the haunch shall for this purpose, be limited to the actual width or 3 times the depth of haunch, whichever is less.

(*ii*) Isolated members in which the T-form or L-form is used only for the purpose of providing additional compression area, shall have a flange thickness not less then one-half the width of the web and a total flange width not more than four times the width of the web. For effective stress transfer it is desirable to splay the junction of the web and the flange so as to form an angle of not less than 110° as shown in Fig. 24·7.

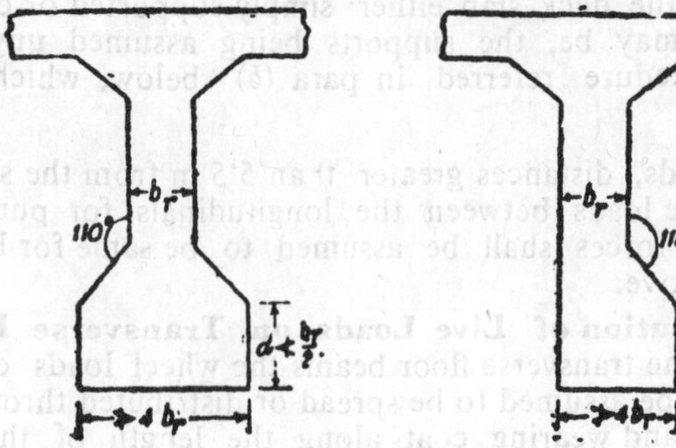

Fig. 24·7

(*iii*) Where the principal reinforcement in a slab which is considered as the flange of a T-beam or L-beam is parallel to the beam, transverse reinforcement shall be provided at the top of the flange. The reinforcement shall be equal to sixty per cent of the main reinforcement of the slab at its centre unless it is specifically calculated. The length of such reinforcing bars shall be as indicated in Fig. 24·8.

(*vi*) The flange shall not ordinarily be considered as effective in computing the shear and diagonal tension resistance of T-beam or L-beam.

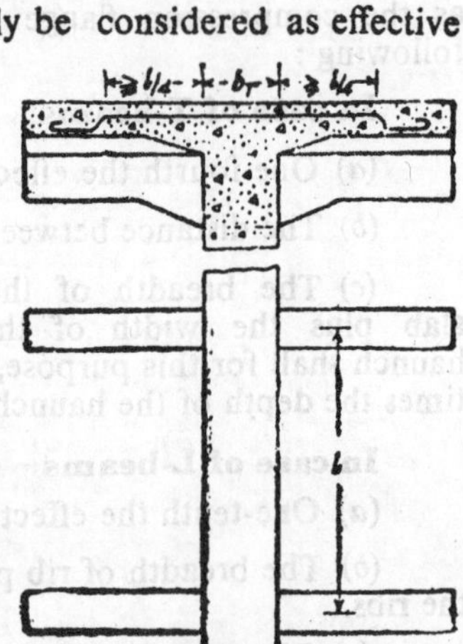

Fig. 24·8

9. Increase in Permissible Stresses. (*i*) When the effect of temperature, shrinkage and creep is taken into consideration, the permissible stresses should not be exceeded in the case of determinate structures but may be exceeded by 15 per cent in the case of indeterminate structures.

(*ii*) When the effect of wind forces is taken into consideration, the permissible stresses may be exceeded by 25%.

(*iii*) When the worst combination of loads met with during erection is considered, the permissible working stresses may be exceeded by 25%.

(*iv*) When the effect of seismic force is also considered the permissible working stresses may be exceeded by 50%.

10. Balanced Cantilver and Continuous Construction

(*i*) **Moments.** The section where the dead load moment is of the sign opposite to that of live-load moment the design bending moment for this condition of loading shall be taken as equal to the live load moment plus 0·7 times dead load moment with proper signs. This shall be in addition to moments expected under normal loads.

(*ii*) **Articulation.** For beams resting on articulations, to prevent damage to articulation the design of the bearings shall be such that it will not induce concentrated edge stresses and will be able to allow angular deflection of the cantilevers and suspended spans.

(*iii*) **Distribution of Reinforcement in the Tensile Flanges.** When the flanges of T-beam are subjected to tensile stresses in longitudinal direction (zone of negative moments), the bars of the principal reinforcements shall be distributed not only above the web but also in a suitable position in the flange, on either side of the web.

11. Cover. (*a*) The thickness of concrete cover (exclusive of plaster or other decorative finish) shall be as follows :

(*i*) At each end of a reinforcing bar, not less than 25 mm nor less than the diameter of such bar.

(*ii*) For a longitudinal reinforcing bar in a column, or beam not less than 40 mm nor less than the diameter of such bar.

(*iii*) In no case shall the cover to any reinforcement including stirrups or binders be less than 35 mm or less than the diameter of such bar.

In case of bars which are not round, the diameter shall be taken as the diameter of a circle giving an equivalent area.

(*b*) The cover of concrete in direct contact with earth faces shall be 25 mm more than that specified above.

(*c*) For reinforced concrete member, totally immersed in sea water, the cover shall be 40 mm and members subjected to sea spray, the cover of concrete shall be 50 mm more than that specified in (*a*).

(*d*) Increased cover thickness, as may be decided by the designer, shall be provided when surfaces of concrete members are exposed to the action of harmful chemicals, acid vapours, saline atmosphere, sulphurous smoke (as in case of bridges over steam operated railways) etc. It may however be noted that dense and strong concrete is a better guarantee of protecting the reinforcement from the harmful effects of chemicals than increasing the thickness of cover.

(*e*) For controlled concrete with specified cube strength greater than 20 N/mm^2 the increase in thickness specified in (*b*), (*c*) and (*d*) may be reduced by half.

12. Web Reinforcement. Where web reinforcement is not required to carry shear, minimum reinforcement of area not less than 0·15% of the area of the web in plan shall be provided.

Example 24·1. *Design a slab bridge for the following data :*

Clear span ...5 m.

Clear width of roadway.........6·8 m.

Live load.........class A loading.

Mix used ....M 20.

Average thickness of wearing coat.........8 cm.

Sol. Assume thickness of slab as 35 cm with effective depth of 32 cm.

Effective span = 5 + 0·32 = 5·32 m.

60 cm wide kerbs are provided.

Overall width of bridge = 6·8 + 2 × 0·6 = 8·0 m.

Dead load B.M.

Dead load of slab per square metre

$$= 1 \times 1 \times \frac{35}{100} \times 24{,}000 = 8400 \text{ N}$$

Dead weight of wearing coat per square metre

$$= 1 \times 1 \times \frac{8}{100} \times 22{,}000 = 1760 \text{ N}$$

Total dead load per square metre

$$= 10{,}160 \text{ N}.$$

Dead load B·M. $= 10{,}160 \times \frac{(5{\cdot}32)^2}{8}$

940 N-m.

Live load B.M.

$$\text{Impact factor} = \frac{4{\cdot}5}{6{\cdot}0 + L} = \frac{4{\cdot}5}{6 + 5{\cdot}32}$$

$$= \frac{4{\cdot}5}{11{\cdot}32} = 0{\cdot}397$$

Clear distance between two trains

$$= 0{\cdot}4 + 0{\cdot}8\,\frac{(6{\cdot}8 - 5{\cdot}5)}{2}$$

$$= 0{\cdot}4 + 0{\cdot}52 = 0{\cdot}92 \text{ m.}$$

Centre to centre distance between wheels of two trains

$$=0{\cdot}92+0{\cdot}5=1{\cdot}42 \text{ m}$$

Distance of centre of wheel from edge of kerb

$$=0{\cdot}15+\frac{0{\cdot}50}{2}=0{\cdot}40 \text{ m.}$$

As the loads are dispersed in direction of span, it is assumed that the maximum B.M. occurs when the loads are symmetrically placed about centre of span. B.M. it centre will be taken for design purposes. The position of loads as shown in Fig. 24·9. Loads with axle load of 11·4 tonnes will be placed to give maximum B.M.

$$\frac{B}{L}=\frac{8}{5{\cdot}32}=1{\cdot}5$$

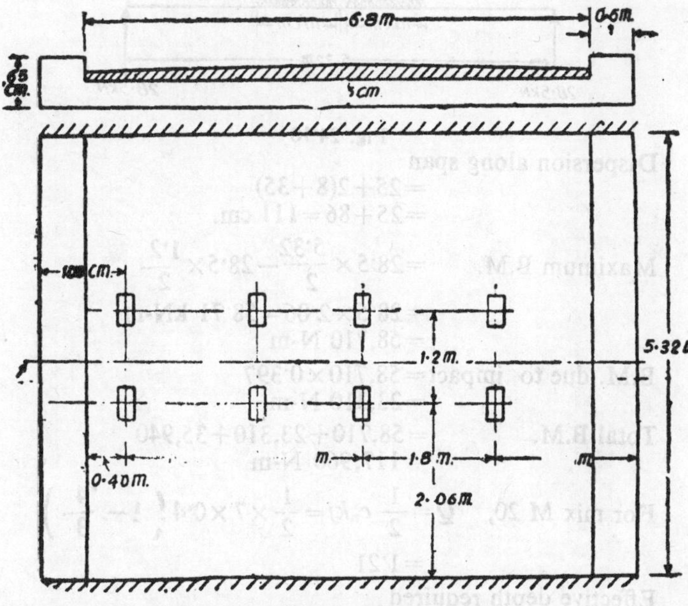

Fig. 24·9

Effective width of slab will be found by IRC code formula

$$k=2{\cdot}84$$

$$x=2{\cdot}06 \text{ m.}$$

Effective width $\quad =kx\left(1-\frac{x}{L}\right)+b_v$

$$=2{\cdot}84\times2{\cdot}06\times\left(1-\frac{2{\cdot}06}{5{\cdot}32}\right)+(0{\cdot}5+2\times0{\cdot}08)$$

$$=3{\cdot}58+0{\cdot}66=4{\cdot}24 \text{ m.}$$

Effective width of different loads overlap.

$\therefore$ Effective width $=1{\cdot}0+1{\cdot}8+1{\cdot}42+1{\cdot}8+1{\cdot}98=8{\cdot}0$ m.

Load per metre width $=\dfrac{2\times114}{8}=28{\cdot}5$ kN

When only one train of load acts, effective width

$$=1{\cdot}0+1{\cdot}8+\frac{4{\cdot}24}{2}=4{\cdot}92 \text{ m.}$$

As effective width is more, load per metre width will be less and thus first case will give maximum B.M.

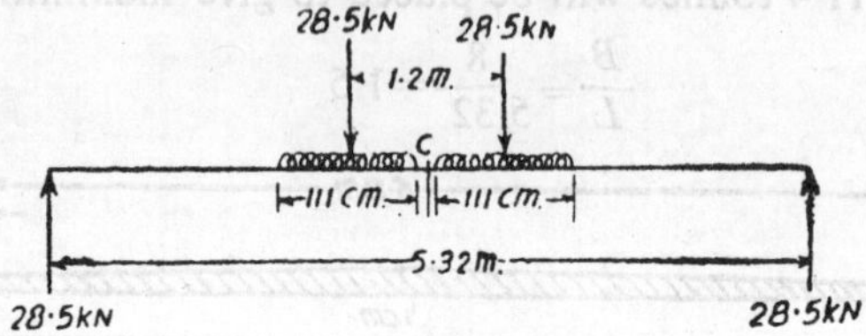

Fig. 24·10

Dispersion along span

$$=25+2(8+35)$$
$$=25+86=111 \text{ cm.}$$

Maximum B.M. $=28{\cdot}5\times\dfrac{5{\cdot}32}{2}-28{\cdot}5\times\dfrac{1{\cdot}2}{2}$

$$=28{\cdot}5\times2{\cdot}06=58{\cdot}71 \text{ kN-m}$$
$$=58{,}710 \text{ N-m}$$

B.M. due to impact $=58{,}710\times0{\cdot}397$

$$=23{,}310 \text{ N-m}$$

Total B.M. $=58{,}710+23{,}310+35{,}940$

$$=117{,}960 \text{ N-m}$$

For mix M 20, $Q=\dfrac{1}{2}\,c_n kj=\dfrac{1}{2}\times7\times0{\cdot}4\left(1-\dfrac{\cdot4}{3}\right)$

$$=1{\cdot}21$$

Effective depth required

$$=\sqrt{\frac{117{,}960\times1000}{1{\cdot}21\times100}}=311{\cdot}8 \text{ mm}$$

Provide overall depth of 350 mm with effective depth of

$$350-28-8=317 \text{ mm}$$

$$A_t=\frac{117{,}960\times1000}{140\times0{\cdot}87\times31}=3050 \text{ mm}^2$$

Using 16 mm ϕ bars,

$$\text{Spacing}=\frac{201\times1000}{3050}=66 \text{ mm}$$

Provide 16 mm ϕ bars at 60 mm centres.

Distribution Reinforcement. Distribution steel is provided to resist 0·3 times live load B.M. and 0·2 times dead load moment.

$$M = 0{\cdot}3(58{,}710+23{,}310)+0{\cdot}2\times 35{,}940$$
$$=24{,}606+7188$$
$$=31{,}794 \text{ N-m}$$

Effective depth $=317-8-6=303$ mm

$$A_s=\frac{31{,}794\times 1000}{140\times 0{\cdot}87\times 303}=857 \text{ mm}^2$$

Using 12 mm ϕ bars,

$$\text{S pacing}=\frac{131\times 1000}{857}=131 \text{ mm.}$$

Provide 12 mm ϕ bars at 130 mm centres.

Shear. Maximum shear force occurs when load is near the support. Dispersion width in direction of span=1·11 m. Fo' maximum shear force wheel load should be at 1·11/2 m from end.

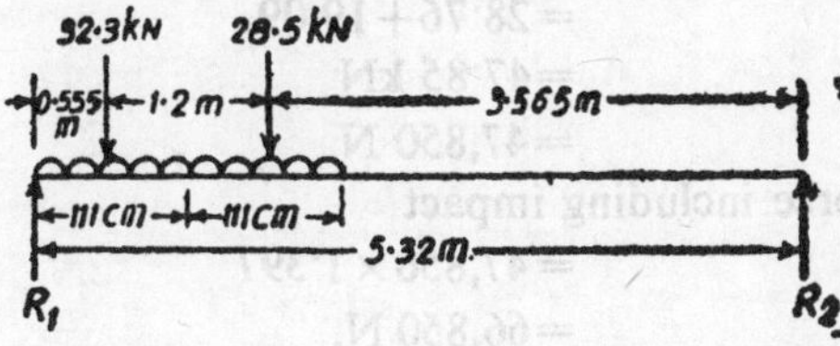

Fig. 24·11

Effective width of slab for first wheel

$$=2{\cdot}84\times\frac{1{\cdot}11}{2}\times\left(1-\frac{0{\cdot}555}{5{\cdot}32}\right)+0{\cdot}66$$
$$=2{\cdot}84\times\frac{1{\cdot}21}{2}\times\frac{4{\cdot}76}{5{\cdot}32}+0{\cdot}66$$
$$=1{\cdot}412+0{\cdot}66=2{\cdot}072 \text{ m.}$$

Effective widths will overlap.

Effective widths for two trains

$$=1+1{\cdot}8+1{\cdot}42+1{\cdot}8+\frac{2{\cdot}072}{2}$$
$$=7{\cdot}056 \text{ m.}$$

Load per metre width

$$=\frac{2\times 114}{7{\cdot}056}=32{\cdot}3 \text{ kN}$$

When only one train of wheels is acting effective width will be much more and that case need not be considered.

Effective width for second wheel

$$=2{\cdot}84\times1{\cdot}755\left(1+\frac{1{\cdot}755}{5{\cdot}32}\right)+0{\cdot}06$$

$$=2{\cdot}84\times1{\cdot}755\times\frac{3{\cdot}565}{5{\cdot}32}+0{\cdot}66$$

$$=3{\cdot}340+0{\cdot}66=4{\cdot}0\text{ m.}$$

Effective widths of two trains of wheel will overlap.

Effective width $=1{\cdot}0+1{\cdot}8+1{\cdot}42+1{\cdot}8+1{\cdot}98=8$ m.

Load per metre width

$$=\frac{2\times114}{8}=28{\cdot}5\text{ kN}$$

Effective width when one train of loads is acting will be much more and hence that case need not be considered.

$$\text{Reaction } R_L=32{\cdot}3\times\frac{(5{\cdot}32-0{\cdot}555)}{5{\cdot}32}+28{\cdot}5\times\frac{3{\cdot}565}{5{\cdot}32}$$

$$=28{\cdot}76+19{\cdot}09$$

$$=47{\cdot}85\text{ kN}$$

$$=47{,}850\text{ N}$$

Shear force including impact

$$=47{,}850\times1{\cdot}397$$

$$=66{,}850\text{ N.}$$

Check for Shear Stress

$$\text{Shear stress}=\frac{66{,}850}{1000\times0{\cdot}87\times317}=0{\cdot}242\text{ N/mm}^2$$

$$\frac{100\,A_s}{bd}=100\times\left(\frac{201\times1000}{60}\right)\times\frac{1}{1000\times317}=1{\cdot}1$$

Allowable shear stress from Table 1·9 is 0·39 N/mm². Safe.

Check for Bond. Half the bars are taken straight and half the bars are bent up.

$$\text{Bond stress}=\frac{66{,}850}{0{\cdot}87\times317\times\frac{100}{12}\times50{\cdot}2}$$

$$=0{\cdot}578\text{ N/mm}^2.$$

Design of Kerb. Kerb is to be designed for a live load of 4000 N/m² and horizontal load of 7500 N/m.

Assume weight of railing as 600 N/m run.

Live load per metre run of kerb $=0{\cdot}6\times400=2400$ N.

Dead load of kerb $=0{\cdot}65\times0{\cdot}6\times1\times24{,}000=9360$ N.

Total load per metre run $= 600 + 2400 + 9360$

$= 12,360$ N.

$$\text{B.M.} = \frac{12,360 \times 5{\cdot}32}{8}$$

$= 43,720,000$ N-mm.

For determining the live load B.M. in slab, width of kerb is also taken into consideration. The kerb should be designed for live load B.M.

Live load B.M. due to class A loading for 600 mm width

$= 0{\cdot}6\,(58,710 + 23,310)$

$= 49,212$ N-m

Total B.M. $= 43,720 + 49,212$

$= 92,932$ N-m

$= 92,932,000$ N-mm.

Effective depth required

$$= \sqrt{\frac{92,932,000}{600 \times 1{\cdot}21}}$$

$= 358$ mm.

Effective depth provided

$= 650 - 25 - 8 = 617$ mm.

Area of steel required

$$= \frac{92,932,000}{0{\cdot}87 \times 617 \times 140}$$

$= 1250\ \text{mm}^2$.

Provide 7 bars of 16 mm ϕ.

$$\text{Maximum S.F.} = \frac{12,360 \times 5{\cdot}32}{2} + 0{\cdot}6 \times 66,850$$

$= 32,880 + 40,110$

$= 72,990$ N.

$$\text{Shear stress} = \frac{72,990}{600 \times 0{\cdot}87 \times 617}$$

$= 0{\cdot}226\ \text{N/mm}^2$.

$$\frac{100\,A_s}{bd} = \frac{100 \times 7 \times 201}{600 \times 617} = 0{\cdot}37$$

Allowable shear stress from Table 1·9 is 0·26.

$$\text{Bond stress} = \frac{72,990}{0{\cdot}87 \times 617 \times 50 \cdot 2 \times \frac{100}{12}}$$

$= 0{\cdot}325\ \text{N/mm}^2$

8 mm ϕ stirrups at 20 cm centres are provided as nominal shear reinforcement.

Design for Horizontal Loads

Horizontal force $= 7500$ N.

B.M. due to this load

$$= 7500 \times 300$$
$$= 2{,}250{,}000 \text{ N-mm.}$$

$$A_t \text{ required} = \frac{2{,}250{,}000}{140 \times 0{\cdot}87 \times (600-30)}$$
$$= \frac{2{,}250{,}000}{140 \times 0{\cdot}87 \times 570}$$
$$= 32{\cdot}4 \text{ mm}^2.$$

This steel is provided by vertical leg of stirrups of 8 mm ϕ at 20 cm c/c.

Horizontal shear force

$$= 7500 \text{ N.}$$

$$\text{Shear stress} = \frac{7500}{1000 \times 0{\cdot}87 \times 617}$$
$$= 0{\cdot}015 \text{ N/mm}^2.$$

Details of reinforcement are shown in Fig. 24·12.

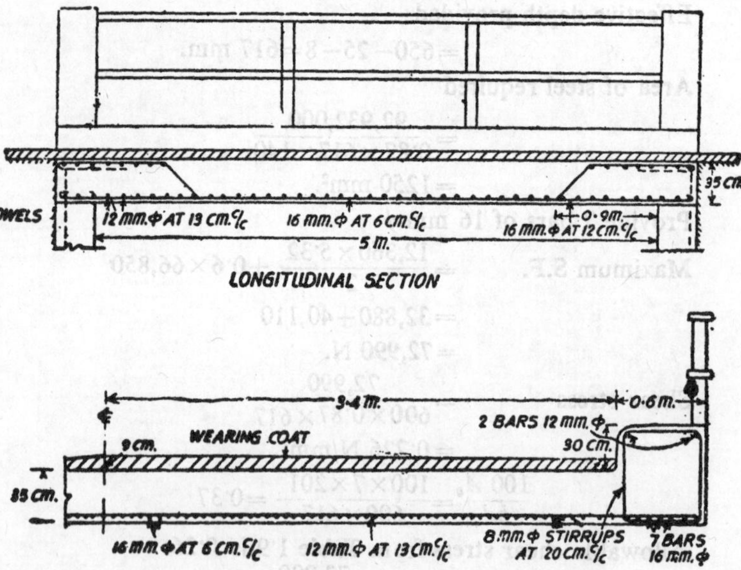

Fig. 24·12.

Ex. 24·2. *Design a reinforced concrete foot bridge having clear road width of 2·5 m and effective span of 12 m. The bridge is to be designed for live load of 4000 N/m². Concrete mix used is M 15.*

Sol. Kerb of 40 cm × 30 cm is provided. Only one central girder is provided and the roadway is cantilevered out from the girder on either side as shown in Fig. 24·13.

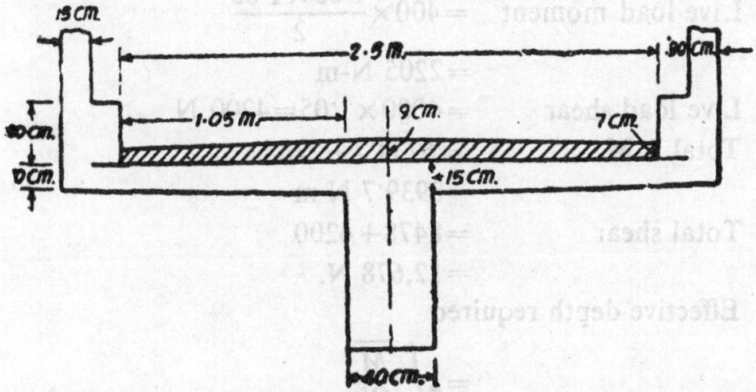

Fig 24·12

Least rib width required

$$=\frac{\text{Span}}{40}=\frac{1200}{30}=40 \text{ cm}.$$

Rib width of 40 cm is provided for the beam. Assume average thickness of 8 cm for the wearing coat. The thickness of slab is assumed as 15 cm at junction with beam and 10 cm at the edge. The weight of the parapet is assumed as 600 N/m run.

Dead load moments at junction with beam

Weight of kerb $=0{\cdot}4\times0{\cdot}3\times1\times24{,}000=2880$ N.

Wearing coat $=1\times1\times0{\cdot}08\times22{,}000\times1{\cdot}05$

$=1848$ N.

Slab rectangular portion

$=0{\cdot}1\times1\times1{\cdot}05\times24{,}000$

$=2520$ N.

Slab triangular portion

$=\frac{1}{2}\times1{\cdot}05\times0{\cdot}5\times24{,}000$

$=630$ N.

Parapet $=600$ N

Total dead load moment at junction with rib

$$=2880\,[1{\cdot}05+0{\cdot}15]+1848\times\frac{1{\cdot}05}{2}+2520\times\frac{1{\cdot}05}{2}+630\times\frac{1}{3}\times1{\cdot}05+600\times(1{\cdot}05+0{\cdot}225)$$

$$=3456+970{\cdot}2+1323+220{\cdot}5+795$$

$$=6734{\cdot}7 \text{ N}.$$

Dead load shear $=2880+1848+2520+630+600$
$=8478$ N.

Live load moment $=400\times\frac{1.05\times1.05}{2}$

$=2205$ N-m

Live load shear $=4000\times1.05=4200$ N

Total B.M. $=6734.7+2205$

$=8939.7$ N-m

Total shear $=8478+4200$

$=12,678$ N.

Effective depth required

$$=\sqrt{\frac{M}{Qb}}$$

$$=\sqrt{\frac{8939.7\times1000}{0.87\times1000}}$$

$=101.3$ mm.

Provide overall depth of 150 mm with effective depth of 120 mm.

Area of steel required

$$=\frac{8939.7\times1000}{0.87\times120\times140}=612\text{ mm}^2.$$

Provide 10 mm ϕ bars at 12 cm centres.

Distribution steel

Average thickness of slab

$$=\frac{100+150}{2}=125\text{ mm.}$$

Area of distribution steel based on average thickness

$$=\frac{0.15}{100}\times1000\times125=187.5\text{ mm}^2$$

Provide 6 mm ϕ bars at 15 cm centres.

Shear stress $=\frac{12,678}{1000\times0.87\times120}=0.121$ N/mm^2. Safe

Load on Main Girder. Dead load from two cantilevers

$=2\times8478=16,956$ N.

Dead weight of wearing coat and slab portion above the rib
$=0.08\times0.4\times22,000+0.4\times0.15\times1\times24,000$

$=724+1440=2164$ N.

Total dead load from slab

$=16,956+2164=9,120$ N .

Assume depth of rib as

$=850$ mm

Dead weight of rib $=1\times0{\cdot}4\times0{\cdot}85\times24{,}000=8160$ N

Live load. For calculating live load moments in girder, live load per square metre of slab is to be taken from the formula

$$P=P'-\frac{400L-3000}{9}$$

where $P'=4000$ N/m², $L=$span$=12$ m

$$\therefore\quad P=4000-\frac{(400\times12-3000)}{9}=4000-200$$

$=3800$ N/m².

Live load per metre run of the beam

$=3800\times2{\cdot}5=9500$ N

Total load per metre run of the beam

$=19{,}120+8160+9500$

$=36{,}780$ N.

Maximum B.M. at centre

$$=\frac{wl^2}{8}$$

$$=\frac{36{,}780\times(12)^2}{8}$$

$=662{,}040$ N-m.

Maximum shear force at support

$$=\frac{36{,}780\times12}{2}=220{,}980 \text{ N}$$

$$\text{Economical depth}=\sqrt{\frac{M}{2{\cdot}1\,b_r}}=\sqrt{\frac{662{,}040\times1000}{2{\cdot}1\times400}}$$

$=888$ mm.

Overall depth of 1 m with rib depth of 0·85 m is provided. Effective depth provided is 920 mm.

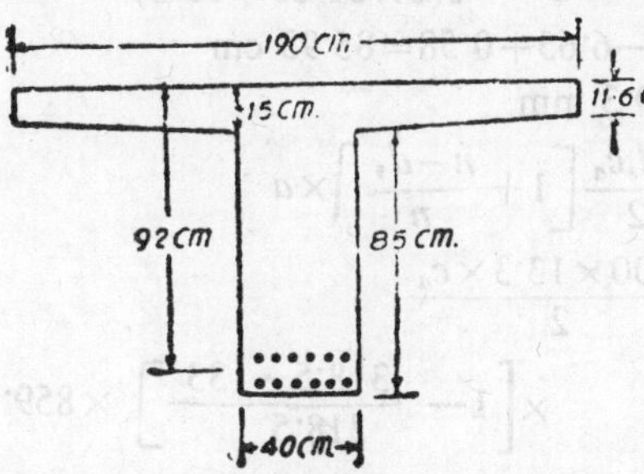

Fig. 24·14.

Average thickness of flange$=125$ mm.

Assume lever arm

$$=d-\frac{d_s}{2}=920-\frac{125}{2}$$

$=857{\cdot}5$ mm.

Area of steel

$$=\frac{662{,}040\times1\,000}{140\times857{\cdot}5}$$

$=5520$ mm².

Provide 12 bars of 25 mm ϕ. Area of steel provided = 5892 mm².

Flange width will be least of

(i) $12d_s + b_r = 12 \times 12.5 + 40 = 150 + 40 = 190$ cm.

(ii) $\frac{\text{Span}}{3} = \frac{1200}{3} = 400$ cm.

Effective width of flange

$$= 190 \text{ cm.}$$

Thickness of flange at edge of effective flange width

$$= 15 - 5 \times \frac{75}{110} = 15 - 3.4$$

$$= 11.6 \text{ cm.}$$

Check for Stresses

Average thickness of flange

$$= \frac{15 + 11.6}{2} = 13.3 \text{ cm.}$$

Assume neutral axis to lie in web.

Taking moments about N.A.

$$190 \times 13.3 \left(n - \frac{13.3}{2}\right) = 18 \times 58.92\,(92 - n)$$

$$\frac{190 \times 13.3}{18 \times 58.92}(n - 6.65) = 92 - n$$

$$2.38n - 15.83 = 92 - n$$

$$3.38\,n = 107.83$$

$$n = \frac{107.83}{3.38} = 31.85 \text{ cm.}$$

$$\text{Lever arm} = d - \frac{d_s}{2} + \frac{d_s^2}{6(2n - d_s)}$$

$$= 92 - \frac{13.3}{2} + \frac{(13.3)^2}{6(2 \times 31.85 - 13.3)}$$

$$= 92 - 6.65 + 0.58 = 85.93 \text{ cm}$$

$$= 859.3 \text{ mm.}$$

$$\text{M.R.} = \frac{Bd_s c_a}{2}\left[1 + \frac{n - d_s}{n}\right] \times a$$

$$662{,}040 \times 1000 = \frac{1900 \times 13.3 \times c_a}{2} \times \left[1 - \frac{318.5 - 133}{318.5}\right] \times 859.3$$

$$c_a = 3.85 \text{ N/mm}^2$$

Stress in teel $= mc_a \dfrac{d-n}{n}$

$$= 18 \times 3{\cdot}85 \times \frac{920-318{\cdot}5}{318{\cdot}5}$$

$$= 131 \text{ N/mm}^2.$$

Check for Bond

4 bars of 25 mm ϕ are taken straight. Sufficient anchorage ength is provided.

Check for Shear

Maximum shear stress

$$= \frac{220{,}680}{859{\cdot}3 \times 400} = 0{\cdot}642 \text{ N/mm}^2$$

$\dfrac{100\,A_s}{bd}$ corresponding to 4 bars taken st gh

$$= \frac{100 \times 4 \times 491}{400 \times 859{\cdot}3} = 0{\cdot}57$$

Permissible shear stress from Table 1·9 for the above value of $\dfrac{100\,A_s}{bd}$ for M 15 concrete is 0·31 N/mm².

Shear force corresponding to shear intensity of 0·31 N/mm²

$= 0{\cdot}31 \times 859{\cdot}3 \times 400$

$= 106{,}553$ N

Bend 4 bars at 1 m from support and 2 bars at 2 m from support and 2 bars at 3 m from support.

Dead load shear $= 27{,}280 \times 6 - 27{,}280 \times 3$

$= 81{,}840$ N.

load shear $= \frac{1}{2} \times \frac{3}{4} \times 9 \times 9500$

$= 32{,}062$

Total shear $= 81{,}840 + 32{,}062$

$= 113{,}902$ N

At 3 m from the support 12 bars are taken straight and

$$\frac{100\,A_s}{bd} = \frac{100 \times 12 \times 491}{400 \times 859{\cdot}3} = 1{\cdot}71$$

Permissible shear stress

$= 0\ 44$ N/mm²

Shear force corresponding to this shear stress

$= 0{\cdot}448 \times 59{\cdot}3 \times 400$

$= 151{,}236 \text{ N} > 113{,}902$ N

No shear reinforcement is required at 3 m from support.

Shear force at 2 m metre support

Dead load shear

$=27,280\times6-27,280\times2$

$=109,120$ N

Live load shear

$=\frac{1}{2}\times\frac{5}{6}\times10\times9500$

$=39,580$

Total shear

$=109,120+39,580$

$=148,700$ N.

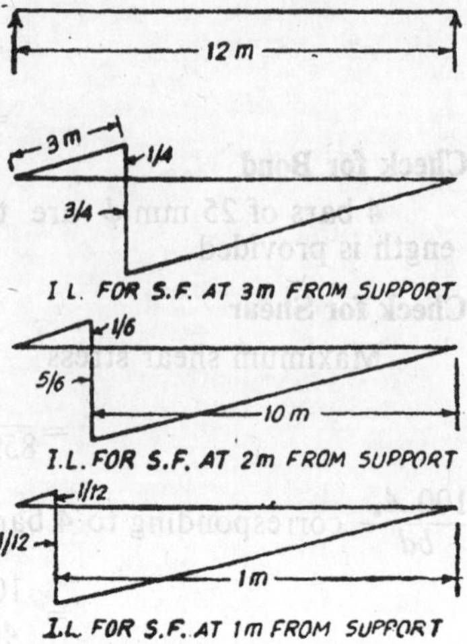

Fig. 24 15.

Shear at 1 metre from support

Dead load shear

$=27,280\times6-27,280\times1$

$=136,400$

Live load shear

$=\frac{1}{2}\times\frac{11}{12}\times11\times9500=47,900$

Total shear force

$=136,400+47,900$

$=184,300.$

Design for shear reinforcement

4 bars are bent up at 1 m from support and 2 bars are bent up at 2 m from support and 2 bars are bent up at 3 m from support.

Shear taken by 4 bent up bars

$=4\times491\times140\times0{\cdot}707$

$=194,360.$

Net shear force at support for which stirrups are to be provided

$=220,680-194,360$

$=26,320$ N

Using 8 mm ϕ 2-legged stirrups

$$\text{Spacing}=\frac{t\times A_t\times a}{F}=\frac{140\times2\times50{\cdot}2\times859{\cdot}3}{26,320}$$

$=450$ mm.

Provide 8 mm ϕ 2-legged stirrups at 13 cm, centres for distance of 1 m from support.

Shear at 1 m from support
$=184{,}300$ N

Shear taken by 2 bent bars
$=2\times491\times140\times0{\cdot}707$
$=97{,}180$ N.

Shear for which reinforcement is required
$=184{,}300-97{,}180$
$=87{,}120$

Use 8 mm ϕ 2-legged stirrups

$$\text{Spacing}=\frac{t\times A_t\times a}{F}$$

$$=\frac{140\times2\times50{\cdot}2\times859{\cdot}3}{87{,}120}$$

$=138{\cdot}6$ mm.

Provide 8 mm ϕ 2-legged stirrups at 13 cm c/c upto distance of 2 m from the support.

S.F. at 2 m from support
$=148{,}700$ N

Shear taken by 2 bent bars
$=97{,}180$ N

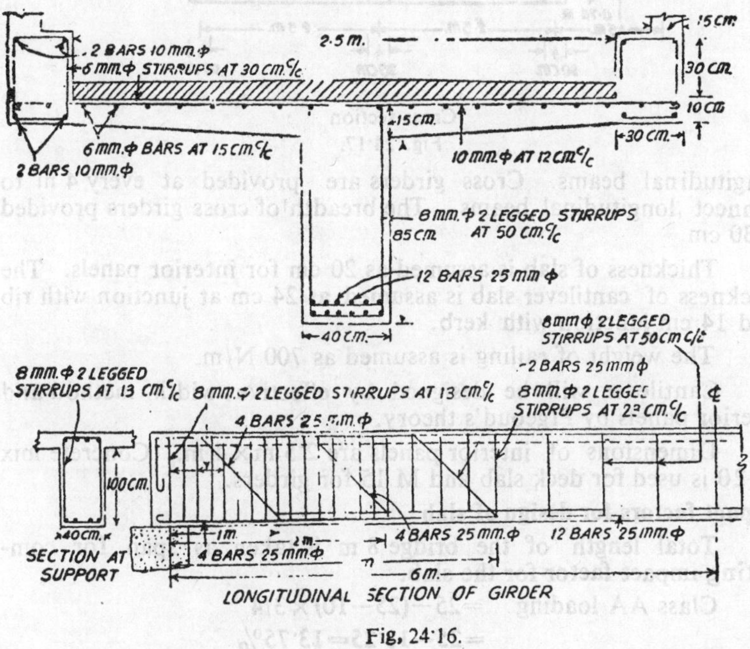

Fig. 24·16.

Shear to be taken by stirrups

$$=148{,}700-97{,}180=51{,}520 \text{ N}$$

Using 8 mm ϕ 2-legged stirrups

$$\text{Spacing}=\frac{140\times2\times50{\cdot}2\times859{\cdot}3}{51{,}520}=234 \text{ mm.}$$

Provide 8 mm ϕ stirrups at 23 cm c/c.

8 mm ϕ 2-legged stirrups at 50 cm centres are provided in remaining portion.

Details of reinforcement in beam and slab are shown in Fig. 24·16.

Ex. 24·3. *Design a girder bridge for the following data :*

Clear width of roadway $=6{\cdot}8$ *m.*
Span centre to centre of bearings $=16$ *m.*
Live load $=$*class A and class AA*
Average thickness of wearing boat $=8$ *cm.*

Sol. Three girders are provided as centre to centre distance of 2·5 m kerbs of 0·5 m width are provided. Cross-section of the bridge is shown in Fig. 24·17. Rib width of 30 cm is provided for

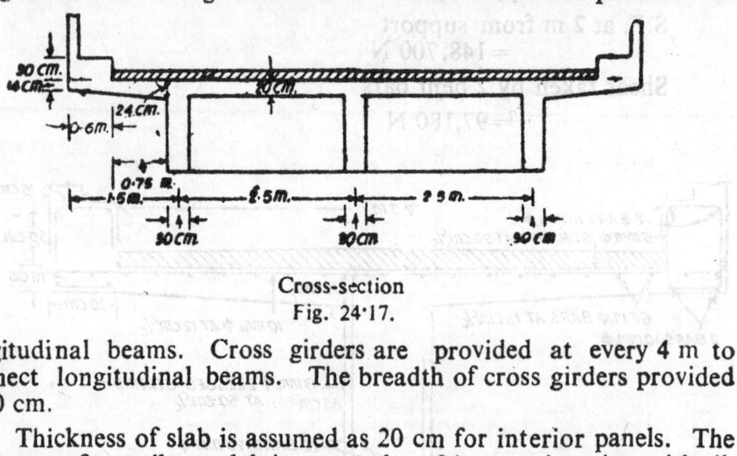

Cross-section
Fig. 24·17.

longitudinal beams. Cross girders are provided at every 4 m to connect longitudinal beams. The breadth of cross girders provided is 30 cm.

Thickness of slab is assumed as 20 cm for interior panels. The thickness of cantilever slab is assumed as 24 cm at junction with rib and 14 cm junction with kerb.

The weight of railing is assumed as 700 N/m.

Cantilever will be designed by effective width method and interior panels by Pigeoud's theory.

Dimensions of interior panels are 2·5 m×4 m. Concrete mix M 20 is used for deck slab and M 15 for girders.

Impact factors for design of slab

Total length of the bridge 8 m is taken as span, for computing impact factor for the slab.

Class AA loading $=25-(25-10)\times3/4$

$$=25-11{\cdot}25=13{\cdot}75\%$$

Class A loading, impact factor $=\dfrac{4\cdot5}{6+8}=0\cdot32$

Design of cantilever portion

Dead load B.M.

Item	*Load on meter run in N*	*Distance of G.C. from edge of cantilever*	*Moment N-m*
Parapet	700	1·275	892·5
Kerb	$0\cdot44\times0\cdot6\times1\times24{,}000=6340$	1·05	665·7
Wearing coat	$0\cdot08\times0\cdot75\times22{,}000=1320$	0·375	495
slabr ectangular	$0\cdot14\times0\cdot75\times24{,}000=2520$	0·375	945
Slab triangular	$\frac{1}{2}\times0\cdot75\times0\cdot10\times24{,}000=900$	0·25	225

Total load = 11,780 N Total D.L.B.M. = 9214·5 N-m

Live Load B.M.

Class AA loading

Maximum clearance of class AA loading from kerb is 1·2 m hence this loading will not come in cantilever portion.

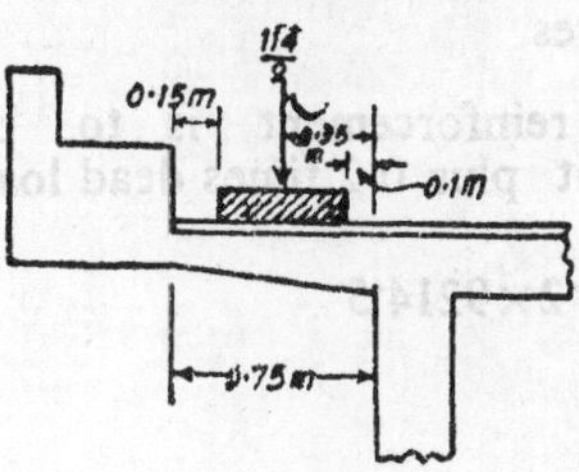

Fig. 24·18

Class A Loading

Average thickness of cantilever slab $=\dfrac{14+24}{2}=19$ cm.

Distance of C.G. of load from edge of cantilever, $x=0\cdot35$ m.

Effective width of slab

$$=1\cdot2x+b_w$$

$$=1\cdot2\times0\cdot35+(0\cdot25+2\times0\cdot08)$$

$$=0\cdot42+0\cdot41=0\cdot83 \text{ m}$$

Load per metre width of slab

$$=\frac{114/2}{0\cdot83}=68\cdot67 \text{ kN}$$

Dispersion width along cantilever

$$=0\cdot5+2\,(0\cdot08+0\cdot19)$$

$$=0\cdot5+0\cdot54=1\cdot04 \text{ m}.$$

Some portion of dispersion width goes behind the edge of cantilever.

Dispersion width in cantilever portion

$$=\frac{0\cdot54}{2}+0\cdot5+0\cdot1$$

$$=0\cdot87 \text{ m}.$$

Portion of load in cantilever

$$=68{\cdot}67\times\frac{0{\cdot}87}{1{\cdot}04}$$
$$=57{\cdot}4\text{ kN}$$

$$\text{B.M.}=57{\cdot}4\times\frac{0{\cdot}87}{2}$$
$$=25\text{ kN-m}$$
$$=25{,}000\text{ N-m.}$$

B.M. including impact

$$=1{\cdot}32\times25{,}000$$
$$=33{,}000\text{ N-m}$$

Total B.M. $=33{,}300+9214{\cdot}5=42{,}214{\cdot}5$ N-m

Effective depth $=\sqrt{\dfrac{42{,}214{\cdot}5\times1000}{1{\cdot}21\times1000}}$

$$=186{\cdot}78\text{ mm.}$$

Provide overall depth of 240 mm with **effective depth of** 209 mm.

Area of steel required

$$=\frac{42{,}214{\cdot}5\times1000}{140\times0{\cdot}87\times209}=166\text{ mm}^2$$

Provide 12 mm ϕ bars at 6 cm centres.

Distribution steel. Distribution reinforcement is to be provided for 0·3 times live load moment plus 0·2 times dead load moment.

$$M=0{\cdot}3\times33000+0{\cdot}2\times9214{\cdot}5$$
$$=9900+1842{\cdot}9$$
$$=117{,}442{\cdot}9$$

Effective depth $=209-6-4$
$$=199\text{ mm}$$

$$A_t=\frac{117{,}442{\cdot}9\times1000}{14\times0{\cdot}87\times199}=484\text{ mm}^2$$

Half reinforcement is to be provided at top and half at bottom.

8 mm ϕ bars at 17 cm centres are provided both at top and bottom.

Shear

Live load shear including impact

$$=57{\cdot}4\times1{\cdot}32$$
$$=75{\cdot}768\text{ kN}$$
$$=75{,}768\text{ N}$$

Dead load shear $=11{,}780$ N

Total shear $=11{,}780+75{,}768=87{,}548$

$$=\frac{87{,}548}{1000\times0{\cdot}87\times209}=0{\cdot}48\text{N/mm}^2$$

$$\frac{100\,A_s}{bd}=\frac{100x}{1000}\frac{(113\times1000)}{60\times209}=0{\cdot}91$$

Average slab thickness $=(14+24)/2=19$

Allowable shear from Table 1·9

$=0{\cdot}32\times k=0{\cdot}39\times1{\cdot}2$

$=0{\cdot}48$ N/mm².

Design Interior Panels

Dead weight of slab per square metre

$=1\times1\times0{\cdot}2\times24{,}000=4800$ N

Dead weight of wearing coat

$=1\times1\times0{\cdot}08\times22{,}000=1760$ N

Total D.L. $=6560$ N/m²

Live Load

Class AA tracked vehicle. One wheel is placed in the centre of the panel.

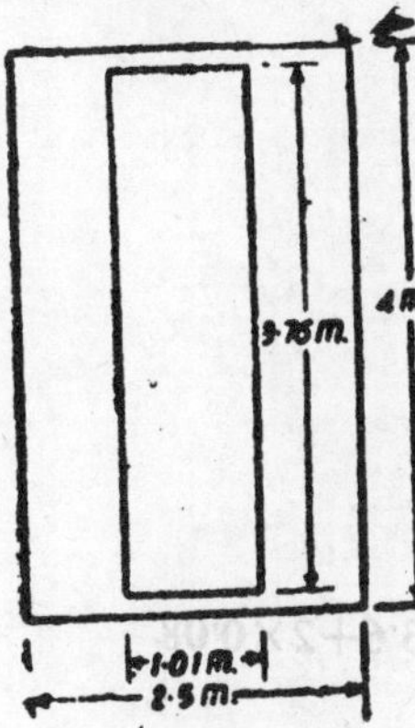

Fig. 24·19 (*a*)

Thickness of wearing coat $=0{\cdot}08$

$u=0{\cdot}85+2\times0{\cdot}08=1{\cdot}01$

$v=3{\cdot}6+2\times0{\cdot}08=3{\cdot}76$

$$\frac{u}{B}=\frac{1{\cdot}05}{2{\cdot}5}=0{\cdot}404$$

$$\frac{v}{L}=\frac{3{\cdot}76}{4}=0{\cdot}94$$

$$\frac{L}{B}=\frac{4}{2{\cdot}5}=1{\cdot}6$$

Referring to Pigeoud's curve

$m_1=8{\cdot}5\times10^{-2}$; $m_2=2{\cdot}4\times10^{-2}$

Moment along $M_1=W\,(m_1+0{\cdot}15\,m_2)$

$$=\frac{700{,}000}{2}\times(8{\cdot}5\times10^{-2}+0{\cdot}15\times2{\cdot}4\times10^{-2})$$

$$=\frac{700{,}000}{2}\times8{\cdot}86\times10^{-2}$$

$=31{,}010$ N-m

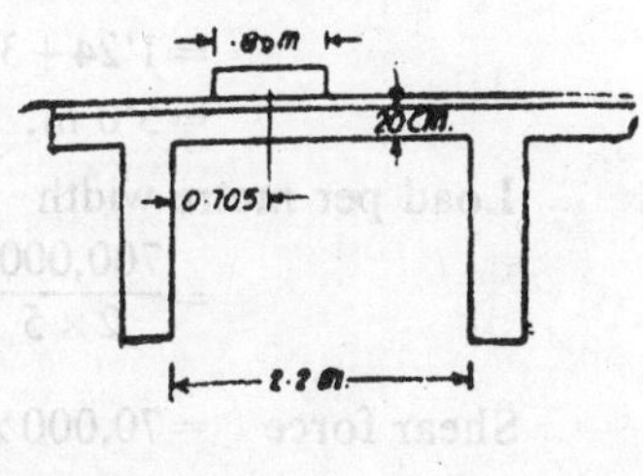

Fig. 24·19 (*b*)

As the slab is continuous B.M taken for design will be 0·8 times M.

B.M. including impact and continuity factor

$M_1=1{\cdot}1375\times31{,}010\times0{\cdot}8=28{,}219$

$$M_2=\frac{700{,}000}{2}\,(2{\cdot}4\times10^{-2}+0{\cdot}15\times8{\cdot}5\times10^{-2})$$

$$= \frac{700,000}{2} \times 3 \cdot 675 \times 10^{-3}$$

$$= 12,862 \cdot 5 \text{ N-m}$$

B.M. including impact and continuity factor

$$M_1 = 1 \cdot 1375 \times 12862 \cdot 5 \times 0 \cdot 8 = 11,705 \text{ N-m}$$

Shear

Dispersion in the direction of span

$$= 0 \cdot 85 + 2\,(0 \cdot 08 + 0 \cdot 20)$$

$$= 0 \cdot 15 + 0 \cdot 56 = 1 \cdot 41 \text{ m.}$$

For maximum shear load is kept such that whole dispersion is in the span. Load is kept at $\frac{1 \cdot 41}{2} = 0 \cdot 705$ m from the edge of beam.

Effective width of slab

$$= kx\left(1 - \frac{x}{l}\right) + b_w$$

Breadth of cross girder

$$= 30 \text{ cm.}$$

∴ Clear length of panel

$$= 4 - 0 \cdot 3 = 3 \cdot 7 \text{ m}$$

$$\frac{B}{L} = \frac{3 \cdot 7}{2 \cdot 2} = 1 \cdot 68$$

k for continuous slab $= 2 \cdot 56$

∴ Effective width of slab

$$= 2 \cdot 56 \times 0 \cdot 70\left(1 - \frac{0 \cdot 705}{2 \cdot 2}\right) + 3 \cdot 6 + 2 \times 0 \cdot 08$$

$$= 2 \cdot 56 \times 0 \cdot 705 \times \frac{1 \cdot 495}{2 \cdot 2} + 3 \cdot 76$$

$$= 1 \cdot 24 + 3 \cdot 76$$

$$= 5 \cdot 0 \text{ m.}$$

Load per metre width

$$= \frac{700,000}{2 \times 5} = 70,000 \text{ N}$$

$$\text{Shear force} = 70,000 \times \frac{(2 \cdot 2 - 0 \cdot 705)}{2 \cdot 2}$$

$$= 70,000 \times \frac{1 \cdot 495}{\cdot} = 47,600 \text{ N}$$

Shear force including impact

$$= 1 \cdot 1375 \times 47,600$$

$$= 54,145 \text{ N}$$

Class AA wheeled vehicle. There are two possibilities which should be investigated for finding the maximum B.M. in the panel. First possibility is when two loads of 37·5 kN and two loads of 62·5 kN each are symmetrically placed about the centre line as shown in Fig. 24·20 (*a*). Second position is when two loads of 37·5 kN each and 4 loads of 62·5 kN each are placed in the panel with central two loads of 62·5 kN placed at the centre line as shown in Fig. 24·20 (*b*).

However, on actual analysis it is seen that first case gives maximum B.M. and thus calculation for first case only are shown.

$$u_1=0·30+2\times0·08=0·46 \text{ m}$$
$$v_1=0·15+2\times0·08=0·31 \text{ m}$$

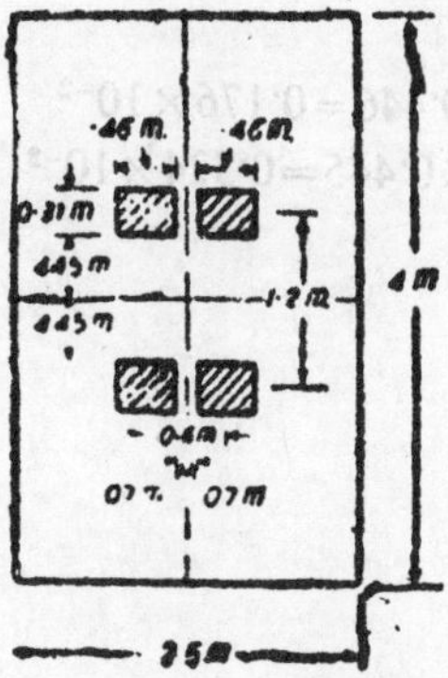

Fig. 24·20 (*a*)

The analysis for this case is done as given below.

For each load

Find m_1 and m_2 for

(*i*) $u=2(u_1+x)$ and $v=2(v_1+y)$

and multiply by $(u_1+x)(v_1+y)$.

(*ii*) $u=2x$ and $v=2y$ and multiply by xy.

(*iii*) $u=2(u_1+x)$ and $v=2y$ and multiply by $y(u_1+x)$.

(*iv*) $u=2x$ and $v=2(v_1+y)$ and multiply by $x(v_1+y)$.

For design B.Ms. subtract [(*iii*)+(*iv*)] from [(*i*)+(*ii*)] and multiply by $W/(u_1v_1)$.

(*i*) $u=2(0·46+0·07)=1·06$

$v=2(0·31+0·445)=1·51$

$$\frac{}{B}=\frac{1·06}{2·50}=0·420$$

$$L=\frac{1·51}{4}=0·38$$

Referring to graphs of $k=1·6$

$m_1=13·5\times12^{-2}$

$m_2=6·3\times10^{-2}$

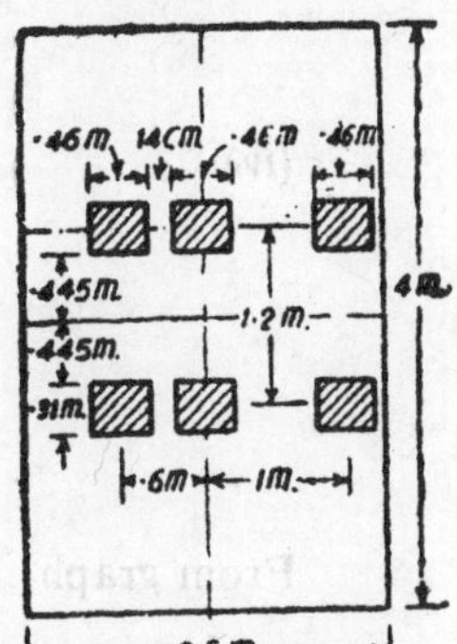

Fig. 24·20 (*b*)

These must be multiplied by $0·53\times0·755$.

Modified values are

$$m_1=13·5\times10^{-2}\times0·53\times0·755=5·4\times10^{-2}$$
$$m_2=6·3\times10^{-2}\times0·53\times0·755=2·52\times10^{-2}$$

(*ii*)

$$u=2\times0{\cdot}07=0{\cdot}14$$

$$v=2\times0{\cdot}445=0{\cdot}89$$

$$\frac{u}{B}=\frac{0{\cdot}04}{2{\cdot}5}=0{\cdot}056$$

$$\frac{v}{L}=\frac{0{\cdot}89}{4}=0{\cdot}22$$

From graph $m_1=23\times10^{-2}$

$m_2=10{\cdot}4\times10^{-2}$.

These are to be multiplied by $0{\cdot}07\times0{\cdot}445$.

Modified values are

$$m_1=23\times10^{-2}\times0{\cdot}07\times0{\cdot}446=0{\cdot}176\times10^{-2}$$

$$M_2=10{\cdot}4\times10^{-2}\times0{\cdot}07\times0{\cdot}445=0{\cdot}324\times10^{-2}$$

(*iii*)

$$u=2(0{\cdot}46+0{\cdot}07)1{\cdot}06$$

$$v=2\times0{\cdot}445=0{\cdot}89$$

$$\frac{u}{B}=\frac{1{\cdot}06}{2{\cdot}50}=0{\cdot}425$$

$$\frac{v}{L}=\frac{0{\cdot}89}{4}=0{\cdot}22.$$

From graph, $m_1=14{\cdot}8\times10^{-2}$

$m_2=9{\cdot}5\times10^{-2}$.

These are multiplied by $0{\cdot}53\times0{\cdot}445$

Modified values are

$$m_1=14{\cdot}8\times10^{-2}\times0{\cdot}53\times0{\cdot}44=3{\cdot}47\times10^{-2}$$

$$m_2=9{\cdot}5\times10^{-2}\times0{\cdot}53\times0{\cdot}44=2{\cdot}23\times10^{-2}$$

(*iv*)

$$u=2\times0{\cdot}07=0{\cdot}14$$

$$v=2(0{\cdot}31+0{\cdot}445)=1{\cdot}51$$

$$\frac{u}{B}=\frac{0{\cdot}14}{2{\cdot}5}=0{\cdot}056$$

$$\frac{v}{L}=\frac{1{\cdot}51}{4{\cdot}0}=0{\cdot}38$$

From graph $m_1=19{\cdot}2\times10^{-2}$

$m_2=6{\cdot}8\times10^{-2}$

These are to be multiplied by $0{\cdot}07\times0{\cdot}755$.

Modified values are

$$m_1=19{\cdot}2\times10^{-2}\times0{\cdot}07\times0{\cdot}755=1{\cdot}038\times10^{-2}$$

$$m^2=6{\cdot}8\times10^{-2}\times0{\cdot}07\times0{\cdot}755=0{\cdot}362\times10^{-2}$$

Design values of m_1 and m_2 are

$$m_1 = 5{\cdot}4\times10^{-2}+0{\cdot}716\times10^{-2}-3{\cdot}47\times10^{-2}-1{\cdot}038\times10^{-2}$$

$$=1{\cdot}608\times10^{-2}$$

$$m_2=2{\cdot}52\times10^{-2}+0{\cdot}324\times10^{-2}-2{\cdot}23\times10^{-2}-0{\cdot}362\times10$$

$$=0{\cdot}252\times10^{-2}$$

B.M. due to load is $\dfrac{W}{u_1 v_1}\,(m_1+0{\cdot}15\,m_2)$

Total M_1 due to 4 loads

$$=\frac{2\times62{,}500}{0{\cdot}31\times0{\cdot}46}(1{\cdot}608\times10^{-2}+0{\cdot}15\times0{\cdot}252\times10^{-2})$$

$$+\frac{2\times37{,}500}{0{\cdot}31\times0{\cdot}46}(1{\cdot}608\times10^{-2}+0{\cdot}15\times0{\cdot}252\times10^{-2})$$

$$=\frac{2\times100{,}000}{0{\cdot}31\times0{\cdot}46}\times1{\cdot}6458\times10^{-2}$$

$$=23{,}100 \text{ N-m}$$

B.M. including impact and continuity factor

$$M_1=23{,}100\times1{\cdot}1375\times0{\cdot}8=21{,}021 \text{ N-m}$$

Total M_2 due to 4 loads

$$=\frac{2\times62{,}500}{0{\cdot}31\times0{\cdot}46}(0{\cdot}252\times10^{-2}+0{\cdot}15\times1{\cdot}608\times10^{-2})$$

$$+\frac{2\times37{,}500}{0{\cdot}31\times0{\cdot}46}(0{\cdot}252\times10^{-2}+0{\cdot}15\times1{\cdot}608\times10^{-2})$$

$$=\frac{2\times1000{,}000}{0{\cdot}31\times0{\cdot}46}\times0{\cdot}4932\times10^{-3}$$

$$=6910 \text{ N-m}$$

B.M. including impact and continuity m_2

$$=1{\cdot}1375\times6910\times0{\cdot}8$$

$$=6288 \text{ N-m}$$

Second position of loads is not investigated as it gives less bending moments.

Shear due to class AA wheeled vehicle. Dispersion width in direction of span

$$=0{\cdot}30+2(0{\cdot}20+0{\cdot}08)$$

$$=0{\cdot}3+0{\cdot}56$$

$$=0{\cdot}86 \text{ m.}$$

Loads are placed such that outermost load is at distance of $\dfrac{0{\cdot}86}{2}=0{\cdot}43$ m from edge of the beam.

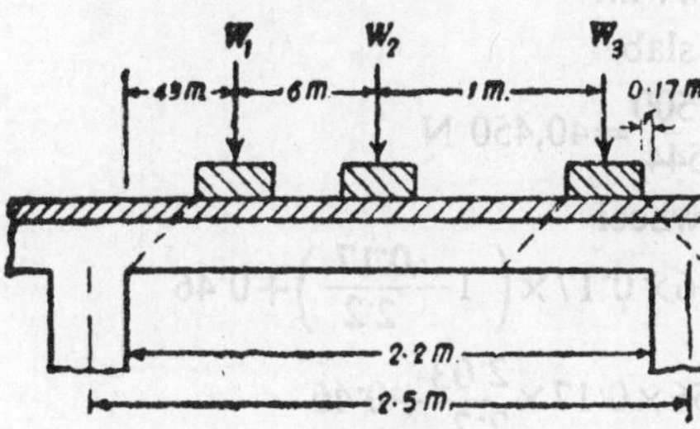

Fig. 24·21

The position of loads for maximum shear is shown in Fig. 24·21.

Effective width for first wheel

$$=kx\left(1-\frac{x}{l}\right)+b_w$$

$$=2{\cdot}56\times0{\cdot}43\times\left(1-\frac{0{\cdot}43}{2{\cdot}2}\right)+0{\cdot}30+2\times0{\cdot}08$$

$$=2{\cdot}56\times0{\cdot}43\times\frac{1{\cdot}77}{2{\cdot}2}+0{\cdot}46$$

$$=0{\cdot}9+0{\cdot}46=1{\cdot}36 \text{ m.}$$

But centre to centre distance of two wheels is 1·2 m and thus effective widths will overlap.

Average effective width for one wheel

$$=\frac{1{\cdot}2+1{\cdot}36}{2}$$

$$=1{\cdot}28 \text{ m.}$$

Load per metre width of slab

$$=\frac{37{,}500}{1{\cdot}28}=29{,}300 \text{ N}$$

Effective width for second wheel

$$=2{\cdot}56\times1{\cdot}03\left(1-\frac{1{\cdot}03}{2{\cdot}21}\right)+0{\cdot}46$$

$$=2{\cdot}56\times1{\cdot}03\times\frac{1{\cdot}17}{2{\cdot}2}+0{\cdot}46$$

$$=1{\cdot}428+0{\cdot}46=1{\cdot}888 \text{ m.}$$

But centre to centre distance of two wheels is 1·2 m and thus, effective widths overlap.

Average effective width for one wheel

$$=\frac{1{\cdot}2+1{\cdot}888}{2}$$

$$=1{\cdot}544 \text{ m.}$$

Load per metre width of slab

$$=\frac{62{,}500}{1{\cdot}544}=40{,}450 \text{ N}$$

Effective width for third wheel

$$=2{\cdot}56\times0{\cdot}17\times\left(1-\frac{0{\cdot}17}{2{\cdot}2}\right)+0{\cdot}46$$

$$=2{\cdot}56\times0{\cdot}17\times\frac{2{\cdot}03}{2{\cdot}2}+0{\cdot}46$$

$$=0{\cdot}404+0{\cdot}46=0{\cdot}864 \text{ m}$$

Load per metre width

$$=\frac{62{,}500}{0{\cdot}864}=72{,}400 \quad \text{N}$$

Portion of W_3 in span

$$=\frac{72{,}400}{0{\cdot}86}\times 0{\cdot}75$$

$$=63{,}139 \text{ N}$$

Shear force at the edge

$$=29{,}300\times\frac{(2{\cdot}2+0{\cdot}43)}{2{\cdot}2}+40{,}450\times\frac{1{\cdot}17}{2{\cdot}2}$$

$$+\frac{72{,}40}{0{\cdot}86}\times 0{\cdot}75\times\frac{0{\cdot}75}{2}\times\frac{1}{2{\cdot}2}$$

$$=23{,}400+21{,}500+10{,}750=55{,}650 \text{ N}$$

Shear force including impact

$$=1{\cdot}1375\times 55{,}650$$

$$=63{,}301 \text{ N}$$

Class A loading. For maximum B.M. one wheel of 57 kN should be placed at the centre of span and other at 1·2 m from it as shown in Fig. 24·22.

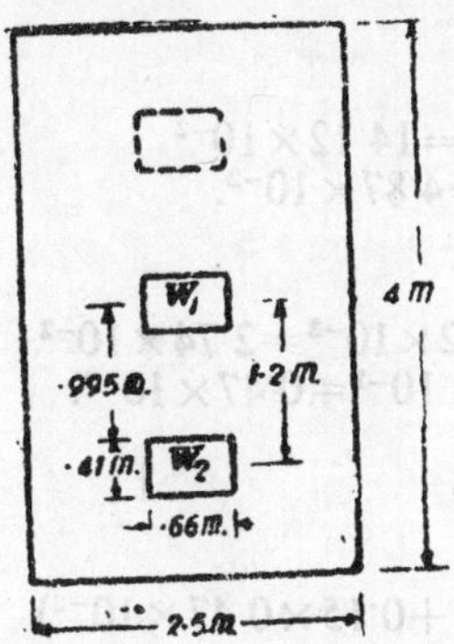

Fig. 24·22

Imaginary load $W_3=W_2$ is placed on the other side of W_1 to make loading symmeterical. B .M. at centre of panel will be that due to W_1 plus half due to loads W_2 and W_3.

B.M. Due to load W_1

$$u=0{\cdot}5+2\times 0{\cdot}08=0{\cdot}66 \text{ m}$$

$$v=0{\cdot}25+2\times 0{\cdot}08=0{\cdot}41 \text{ m}$$

$$\frac{u}{B}=\frac{0{\cdot}66}{2{\cdot}5}=0{\cdot}264$$

$$\frac{v}{L}=\frac{0{\cdot}41}{4}=0{\cdot}102$$

From graph, $m_1=18{\cdot}8\times 10^{-2}$

$$m_2=14{\cdot}8\times 10^{-2}$$

$$M_1'=W\,(m_1+0{\cdot}15\,m_2)$$

$$=57{,}000(18{\cdot}8\times 10^{-2}+0{\cdot}15\times 14{\cdot}8\times 10^{-2})$$

$$=57{,}000\times 21{\cdot}02\times 10^{-2}$$

$$=14{,}980 \text{ N-m}$$

$$M_2'=W(m_2+0{\cdot}15\,m_1)$$

$$=57{,}000\,(14{\cdot}8\times 10^{-2}+0{\cdot}15\times 18{\cdot}8\times 10^{-2})$$

$$=57{,}000\times 17{\cdot}62\times 10^{-2}=10{,}040 \text{ N-m.}$$

B.M. due to W_2

$$u_1=0{\cdot}5+2\times 0{\cdot}08=0{\cdot}66$$

$$v_1=0{\cdot}25+1\times 0{\cdot}08=0{\cdot}41 \text{ m.}$$

(*i*) $u=0.66$; $v=2(0.995+0.41)=2.81$

$$\frac{u}{B}=\frac{0.66}{2.5}=0.264$$

$$\frac{v}{L}=\frac{2.8}{4}=0.702$$

From graph, $m_1=12\times10^{-2}$

$m_2=3.8\times10^{-2}$

These are to be multiplied by 1·405.

Modified values are

$m_1=16.86\times10^{-2}$

$m_2=5.34\times10^{-2}$.

(*ii*) $u=0.66,\quad v=2\times0.995=1.99$

$$\frac{u}{B}=\frac{0.66}{2.5}=0.264$$

$$\frac{v}{L}=\frac{1.99}{4}=0.498$$

From graph, $m_1=14.2\times10^{-2}$

$m_2=4.9\times10^{-2}$.

These ar e to be multiplied by 0·995.

Modified values of m_1 and m_2 are

$m_1=14.2\times0.995\times10^{-2}=14.12\times10^{-2}$

$m_2=4.9\times0.995\times10^{-2}=4.87\times10^{-2}$.

Design values of m_1 and m_2 are

$m_1=16.86\times10^{-2}-14.12\times10^{-2}=2.74\times10^{-2}$

$m_2=5.34\times10^{-2}-4.87\times10^{-2}=0.47\times10^{-2}$.

$$M_1''=\frac{W}{v}(m_1+0.15\,m_2)$$

$$=\frac{57,000}{0.41}(2.74\times10^{-2}+0.15\times0.47\times10^{-2})$$

$$=\frac{57,000}{0.41}\times2.81\times10^{-2}=3910\text{ N-m}$$

$$M_2''=\frac{57,000}{0.41}(0.47\times10^{-2}+0.15\times2.74\times10^{-2})$$

$$=\frac{57,000}{0.41}\times0.881\times10^{-2}=1230\text{ N-m.}$$

Total B.M. due two loads will be

$M_1=11,980+3910=15,890$ N-m

$M_2=10,040+1230$ N-m

$=11,270$ N-m

B.M. including impact and continuity

$$\overline{M}_1 = 1{\cdot}32 \times 15{,}890 \times 0{\cdot}8$$
$$= 16{,}780 \text{ N-m}$$

B.M. including impact and continuity

$$\overline{M}_2 = 1{\cdot}32 \times 11{,}270 \times 0{\cdot}8$$
$$= 11{,}901 \text{ N-m.}$$

Shear due to class A loading

Dispersion width in the direction of span

$$= 0{\cdot}50 + 2\ (0{\cdot}2 + 0{\cdot}08) = 1{\cdot}06 \text{ m.}$$

For maximum shear load should be placed at distance of $1{\cdot}60/2 = 0{\cdot}53$ m from the edge. In this position second load will be on the other beam.

Effective width $= kx\left(1 - \dfrac{x}{l}\right) + b_w$

$$= 2{\cdot}56 \times 0{\cdot}53\left(1 - \frac{0{\cdot}53}{2{\cdot}2}\right) + 0{\cdot}25 + 2 \times 0{\cdot}08$$
$$= 2{\cdot}56 \times 3{\cdot}53 \times \frac{1{\cdot}67}{2{\cdot}2} + 0{\cdot}41$$
$$= 1{\cdot}05 + 0{\cdot}41$$
$$= 1{\cdot}46 \text{ m.}$$

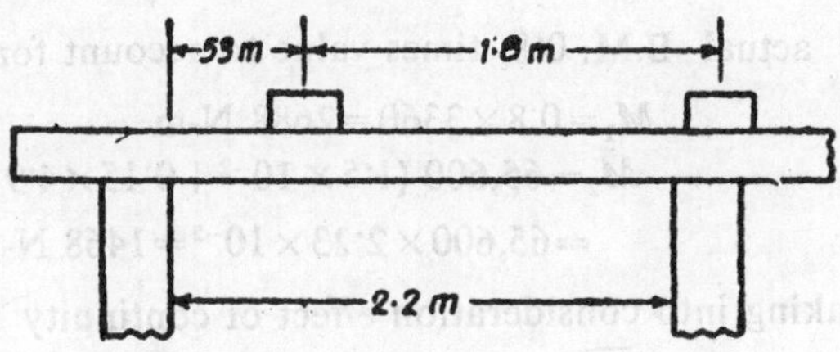

Fig. 24·23

Distance between two wheels is 1·2 m and hence effective width overlap.

Average effective per wheel

$$= \frac{1{\cdot}46 + 1{\cdot}2}{2}$$
$$= 1{\cdot}33 \text{ m.}$$

Load per metre width of the beam

$$= \frac{57{,}000}{1{\cdot}33}$$
$$= 42{,}750 \text{ N}$$

Shear force $=42{,}750\times\dfrac{(2{\cdot}2-0{\cdot}53)}{2{\cdot}2}$

$=32{,}450$ N

S.F. including impact

$=1{\cdot}32\times32{,}450=42{,}834$ N.

Design live load B.M. and S.F.

$M_1=28{,}209$ N-m due to class AA tracked vehicle.

$M_2=11{,}901$ N-m due to class A loading.

S.F.$=63{,}301$ N due to class AA wheeled vehicle.

Dead load

Dead load per square metre $=6560$ N/m^2

Dead load B.M. will be found by Pigeoud's method.

Total load on panel $=4\times2{\cdot}5\times6560=65{,}600$ N

$\dfrac{u}{B}=1,\ \dfrac{v}{L}=1$ as the panel is loaded with uniformly distributed load

$m_1=4{\cdot}9\times10^{-2},\ m_2=1{\cdot}5\times10^{-2}.$

$M_1=65{,}600(4{\cdot}9\times10^{-2}+0{\cdot}15\times1{\cdot}5\times10^{-2})$

$=65{,}600\times5{\cdot}125\times10^{-2}=3360$ N-m

Taking actual B.M. 0·8 times value to account for continuity

$M_1=0{\cdot}8\times3360=2688$ N-m

$M_2=65{,}600\ (1{\cdot}5\times10^{-2}+0{\cdot}15\times4{\cdot}9\times10^{-2})$

$=65{,}600\times2{\cdot}23\times10^{-2}=1468$ N-m

B.M. taking into consideration effect of continuity

$\overline{M_2}=0{\cdot}8\times1468=1174{\cdot}4$ N-m

Dead load shear force is assumed as

$\dfrac{6560\times2{\cdot}2}{2}=7216$ N

Total $\bar{M}_1$ due the dead load and live load

$=28{,}219+2668$

$=39{,}907$ N-m

Total $\bar{M}=11{,}901+1174{\cdot}4=13{,}075{\cdot}4$

Total S.F.$=63{,}301+7216$ N

$=70{,}517$ N

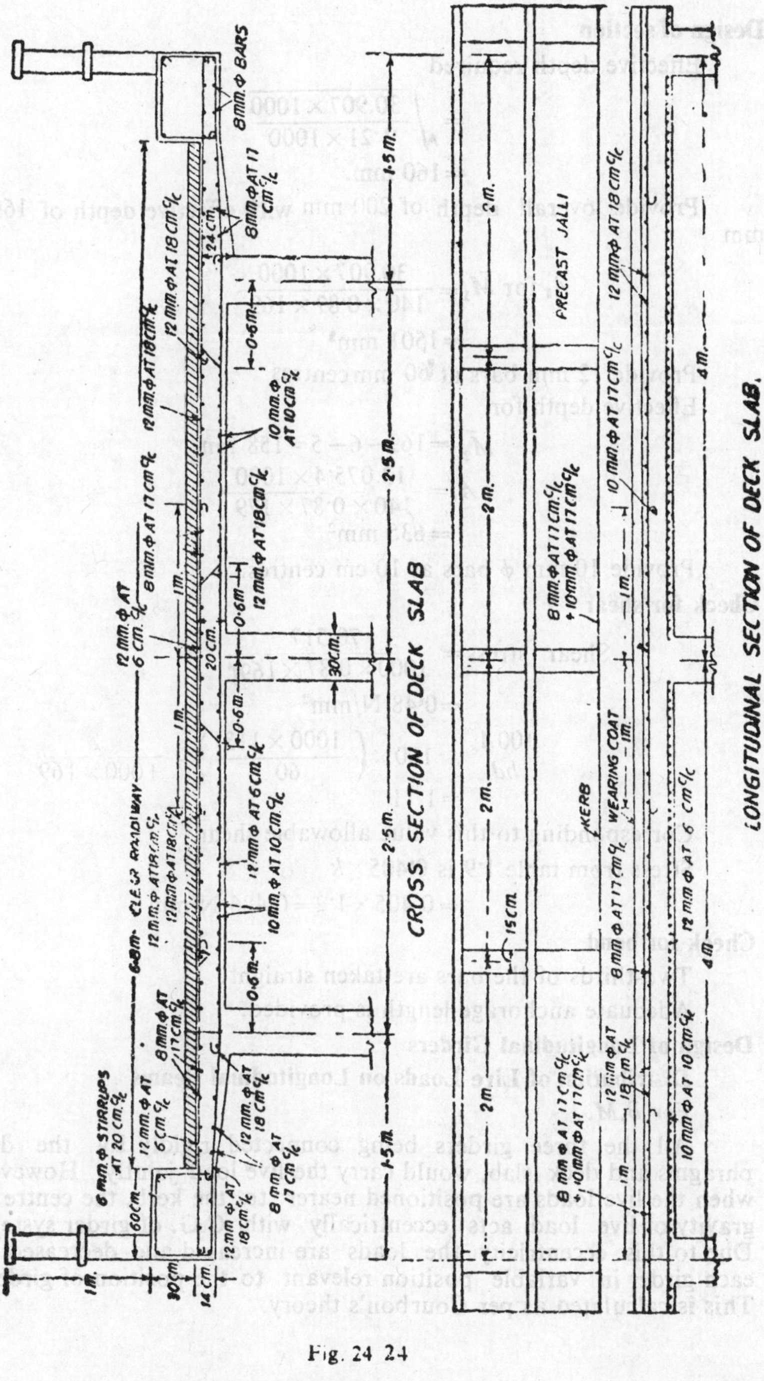
CROSS SECTION OF DECK SLAB
LONGITUDINAL SECTION OF DECK SLAB
PRECAST JALLI
KERB
WEARING COAT
6·8 m. CLEAR ROADWAY
6 mm. Φ STIRRUPS AT 20 CM. %
8 mm. Φ BARS

Fig. 24 24

Design of section

Effective depth required

$$=\sqrt{\frac{30{,}907\times 1000}{1{\cdot}21\times 1000}}$$

$$=160 \text{ mm.}$$

Provide overall depth of 200 mm with effective depth of 169 mm

$$A_t \text{ for } \bar{M}_1=\frac{30{,}907\times 1000}{140\times 0{\cdot}87\times 169}$$

$$=1501 \text{ mm}^2$$

Provide 12 mm bars at 60 mm centres

Effective depth for

$$\bar{M}_2=169-6-5=158 \text{ mm}$$

$$A_t=\frac{13{,}075{\cdot}4\times 1000}{140\times 0{\cdot}87\times 179}$$

$$=635 \text{ mm}^2$$

Provide 10 mm ϕ bars at 10 cm centres.

Check for shear

$$\text{Shear stress}=\frac{70{,}517}{1000\times 0{\cdot}87\times 169}$$

$$=0{\cdot}48 \text{ N/mm}^2$$

$$\frac{100A_s}{bd}=100\times\left(\frac{1000\times 113}{60}\right)\times\frac{1}{1000\times 169}$$

$$=1{\cdot}11$$

Corresponding to this value allowable shear

Stress from table 1·9 is $0{\cdot}405\times k$

$$=0{\cdot}405\times 1{\cdot}2=0{\cdot}486 \text{ N/mm}^2.$$

Check for bond

Two-thirds of the bars are taken straight.

Adequate anchorage length is provided.

Design of Longitudinal Girders

Distribution of Live Loads on Longitudinal Beams

For B.M.

All the three girders being connected rigidly by the diaphragms and deck slab, would carry the live load jointly. However, when the live loads are positioned nearer to the kerb, the centre of gravity of live load acts eccentrically with C.G. of girder system. Due to this eccentricity the loads are increased and decreased on each girder in variable position relevant to the position of girders. This is calculated as per Courbon's theory.

Reaction factor is given by

$$R_x = \frac{\Sigma W}{n}\left[1 + \frac{\Sigma I}{\Sigma p_x^2 I} \times p_x \times e\right]$$

where R_x = Reaction factor for girder under consideration.
I = Moment of inertia of each girder.
p_x = Distance of girder from axis of the bridge.
e = Eccentricity of live load with respect to axis of bridge.
ΣW is total load.
n is number of girders.

Courbon's theory is applicable when following conditions are satisfied :

(*i*) The span width ratio is greater than 2 and less than 4.

(*ii*) There are at least five symmetrical cross-girders or diaphragms integrated with the longitudinals.

(*iii*) The depth of the cross-girder or diaphragms is at least 3/4 of the depth of longitudinal. girder.

Class A Loading. Reaction on the girders will be maximum when the eccentricity is maximum. Eccentricity will be maximum when the loads are very near to kerb. Position of loads for maximum eccentricity is shown in Fig. 24·25.

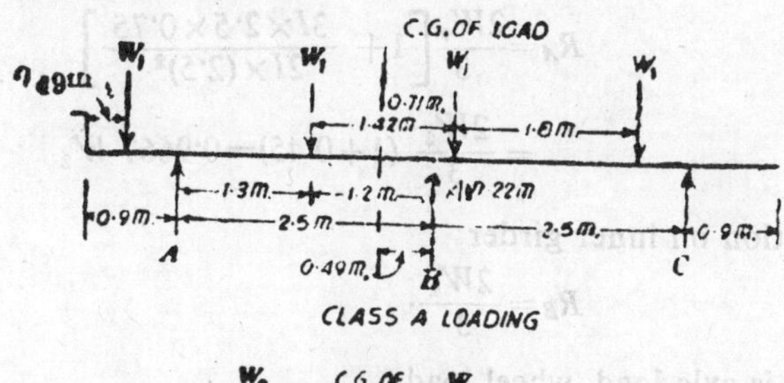

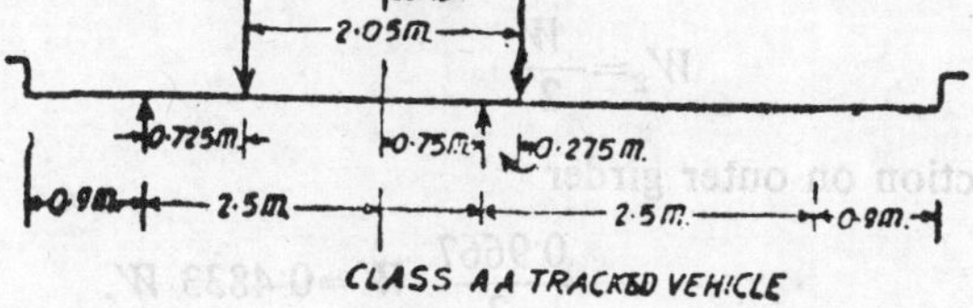

Fig. 24·25

All the girders are assumed to have the same moment of inertia.

Reaction factor for outer girder

$$R_A = \frac{4W_1}{3}\left[1 + \frac{3I}{2I \times (2{\cdot}5)^2} \times 2{\cdot}5 \times 0{\cdot}49\right]$$

$$= \frac{4W_1}{3}\left[1 + \frac{3 \times 0{\cdot}49}{5}\right] = 1{\cdot}725\ W_1.$$

$$R_B = \frac{4W_1}{3}\left[1 + 0\right] = \frac{4W_1}{3}.$$

If W is axle load, wheel load W_1

$$= \frac{W}{2}.$$

Reaction factor for outer girder

$$= \frac{1{\cdot}725}{2} W = 0{\cdot}8625\ W.$$

Reaction factor for inner girder

$$= \frac{4}{3} \times \frac{1}{2} W = 0{\cdot}6667\ W.$$

Class AA Loading. Position of loads for maximum eccentricity is shown in Fig. 24·25.

Reaction on outer girder

$$R_A = \frac{2W_2}{3}\left[1 + \frac{3I \times 2{\cdot}5 \times 0{\cdot}75}{2I \times (2{\cdot}5)^2}\right]$$

$$= \frac{2W_2}{3}(1 + 0{\cdot}45) = 0{\cdot}9667\ W_2$$

Reaction on inner girder

$$R_B = \frac{2W_2}{3}.$$

If W is axle load, wheel load

$$W_2 = \frac{W}{2}.$$

Reaction on outer girder

$$= \frac{0{\cdot}9667}{2} W = 0{\cdot}4833\ W.$$

Reaction factor for inner girder

$$= \frac{2}{3} \times \frac{W}{2} = \frac{W}{3}.$$

Distribution of Loads on Longitudinal Beam for Shear Force. This is divided in two parts first due to the loads placed beyond the diaphragm closest to the support under consideration and the second part due to loads between support and first intermediate diaphragm.

For first case, *i.e.* when loads are placed beyond the diaphragm closest to the support reaction factors are found as for B.M. by Courbon's theory.

For the second case following procedure is used. Consider a load placed in the deck slab between longitudinal beam *A* and *B*

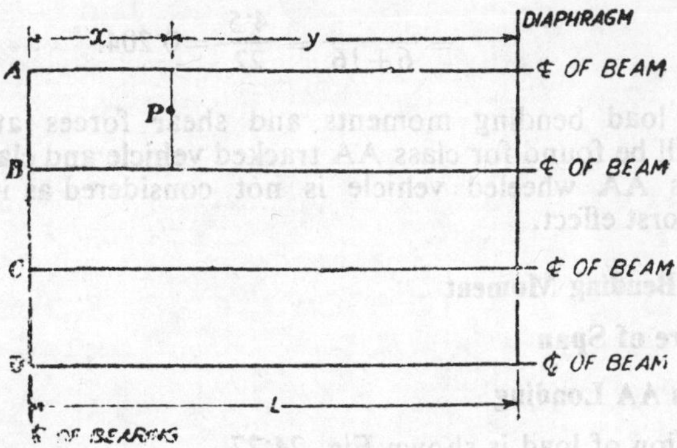

Fig. 24·26

as shown in Fig. 24·26. This load is distributed between beams *A* and *B* as if the slab were simply supported. Let these loads be P_A and P_B on beams *A* and *B* respectively.

There loads P_A and P_B are now distributed to support diaphragm as if beams were simply supported between these points.

That portion of the load going to the supports is a direct shearing force in the beam and that which goes to the diagram is redistributed among all the beams by normal Courbon's method.

Thus the S.F. due to P_A equals to

$$P_A\left[\frac{y}{L}+\frac{x}{nL}\left(1+\frac{\Sigma I}{\Sigma I p_x^2}\times p_x\times e\right)\frac{\text{Span}-L}{\text{Span}}\right]$$

Dead load from slab per girder

Total dead load of deck $=2\times 11{,}780+6\,560\times 5{\cdot}3$

$=23{,}560+34{,}768$

$=58{,}328$ N

It is assumed that dead load is shared equally by all the girders.

Dead load per girder $=\dfrac{58{,}328}{3}=19{,}442$ N

Live load

Impact Factors

Span = 16 m.

Impact factor for class AA loading

$$= 10\%.$$

Impact factor for class A loading

$$= \frac{4.5}{6+L}$$

$$= \frac{4.5}{6+16} = \frac{4.5}{22} = 0.204.$$

Live load bending moments and shear forces at different sections will be found for class AA tracked vehicle and class *A* loading. Class AA wheeled vehicle is not considered as it does not produce worst effect.

Live Load Bending Moment

Centre of Span

Class AA Loading

Position of load is shown Fig. 24·27.

$$\text{B.M.} = \left[\frac{4+3.1}{2}\right] \times 700$$

$$= 2485 \text{ kN-m}$$

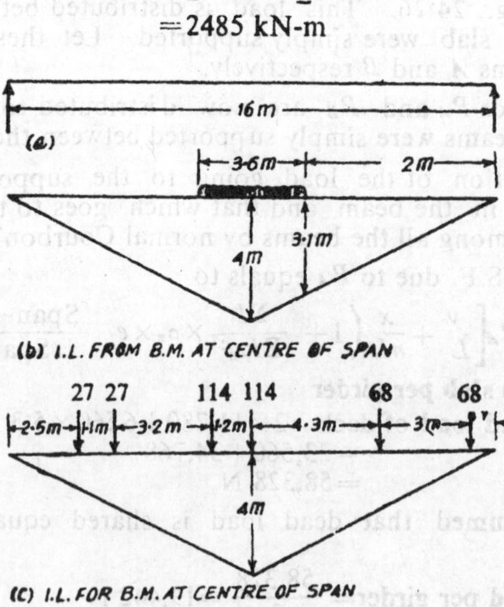

Fig. 24·27

B.M. including impact and reaction factor for outer girder

$$=2485\times1{\cdot}1\times0{\cdot}4833=1321 \text{ kN-m}$$

$$=1{,}321{,}000 \text{ N-m}$$

B.M. including impact and reaction factor for inner girder

$$=2485\times1{\cdot}1\times\tfrac{1}{3}=911{\cdot}13 \text{ N-m}$$

$$=911{,}130 \text{ N-m}$$

Class A loading. Position of loads for maximum B.M. at centre of span is shown in Fig. 24·27.

$$\text{B.M.}=114\times4+114\times\frac{6{\cdot}8}{8}\times4+27\times\frac{2{\cdot}5\times4}{8}$$

$$+27\times\frac{3{\cdot}6\times4}{8}+68\times\frac{0{\cdot}7\times4}{8}+68+\frac{3{\cdot}7\times4}{8}$$

$$=456+387{\cdot}6+33{\cdot}75+48{\cdot}6+23{\cdot}8+125{\cdot}8$$

$$=1075{\cdot}55 \text{ kN-m}$$

B.M. including impact factor and reaction factor for outer girder

$$=1075{\cdot}55\times0{\cdot}8625\times1{\cdot}204=1117 \text{ kN-m}$$

B.M. including impact factor and reaction factor for inner girder

$$=1075{\cdot}55\times0{\cdot}6667\times1{\cdot}204=863{\cdot}4 \text{ kN-m}$$

It is seen that maximum B.M. for inner girder and outer girder is given by class AA tracked vehicle.

Maximum live load moment for outer girder=1,331,000 N-m

Maximum live load moment for inner girder=911,130 N-m.

Quarter Span

Class AA loading. Position of load for maximum B.M. is shown in Fig. 24·28.

$$\text{B.M.}=\frac{(3+2{\cdot}325)}{2}\times700=1863{\cdot}75 \text{ kN-m}$$

B.M. including impact factor for outer girder

$$=1863{\cdot}75\times1{\cdot}1\times0{\cdot}4833=998{\cdot}2 \text{ kN-m}$$

$$=998{,}200 \text{ N-m}$$

B.M. including impact factor for inner girder

$$=1863{\cdot}75\times1{\cdot}1\times\tfrac{1}{3}=683{\cdot}4\text{ kN-m}$$

$$=683{,}400\text{ N-m.}$$

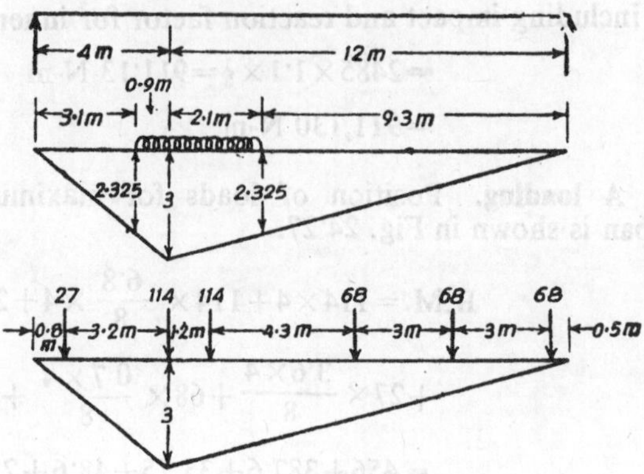

Fig. 24·28

Class A loading. Position of loads for maximum B.M. is shown in Fig. 24·28.

$$\text{B.M.}=114\times3+114\times\frac{3}{12}\times10{\cdot}8+68\times\frac{3}{12}\times(0{\cdot}5+3{\cdot}5+6{\cdot}5)+2{\cdot}7\times\tfrac{3}{4}\times0{\cdot}8$$

$$=342+307{\cdot}8+178{\cdot}5+16{\cdot}2=844{\cdot}5\text{ kN-m}$$

B.M. including impact factor for outer girder

$$=844{\cdot}5\times0{\cdot}8625\times1{\cdot}204=877\text{ kN-m}$$

B.M. including impact factor for inner girder

$$=844{\cdot}5\times0{\cdot}6667\times1{\cdot}24=677{\cdot}9\text{ kN-m}$$

Maximum B.M. is given by class AA loading.

Maximum live load B.M. for outer girder = 998,200 N-m.

Maximum live load B.M. for inner girder = 683,400 N-m.

Live load shear. Maximum shear at all sections is given by class AA tracked vehicle. For maximum shear at support, load should be as near the support as possible. The length of the load is 3·6 m. Whole of load will lie between support and first cross-girder. Live load shear at support is found by the method described earlier. For intermediate section shear will be found by drawing influence lines. The reactions factor as found by Courbon's theory will be used for finding shear at intermediate sections.

The position of centre line of load for maximum shear force is shown in Fig. 24·29. Let P_A, P_B and P_C be the loads coming on the longitudinal considering the slab as simply supported.

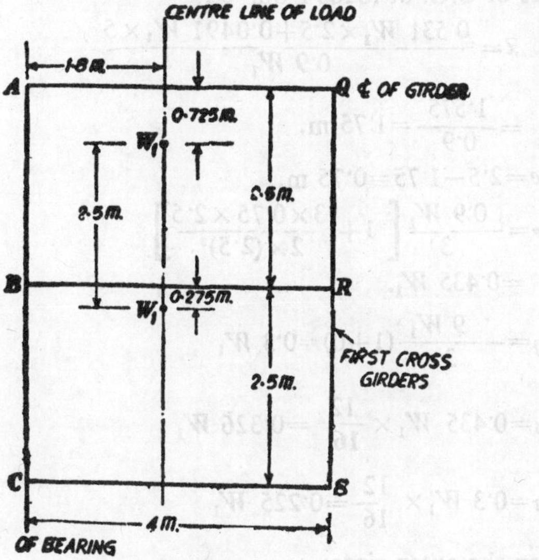

Fig. 24·29 (*a*)

$$P_A = W_1 \frac{(2{\cdot}5 - 0{\cdot}725)}{2{\cdot}5}$$
$$= 0{\cdot}71\ W_1$$
$$P_B = \frac{0{\cdot}725\ W_1}{2{\cdot}5} + \frac{(2{\cdot}5 - 0{\cdot}275)\ W_1}{2{\cdot}5}$$
$$= 0{\cdot}29\ W_1 + 0{\cdot}89\ W_1$$
$$= 1{\cdot}18\ W_1$$
$$P_C = \frac{0{\cdot}275\ W_1}{2{\cdot}5}$$
$$= 0{\cdot}11\ W_1$$

Due to P_A, shea force at $A = R_A'$

$$= 0{\cdot}71 W_1 \times \frac{2}{4}$$
$$= 0{\cdot}3905\ W_1$$
$$R_Q = 0{\cdot}71\ W_1 \times \frac{1{\cdot}8}{2{\cdot}2}$$
$$= 0{\cdot}3195\ W_1$$

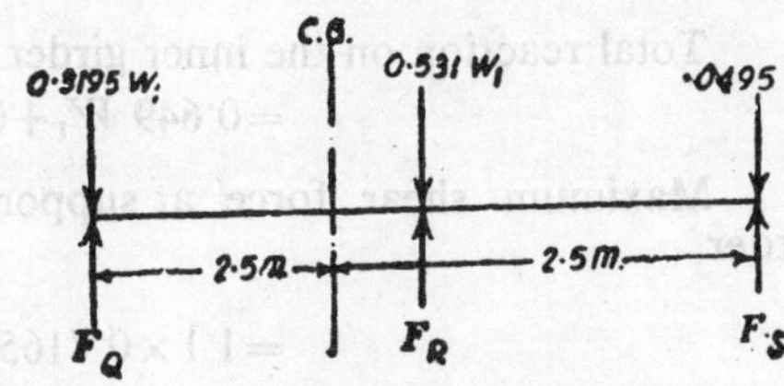

Fig. 24·29 (*b*)

Due to P_B, shear force at

$$B = R_B' = 1{\cdot}18\ W_1 \times \frac{2{\cdot}2}{4} = 0{\cdot}649\ W_1$$

$$R_R = 1{\cdot}18 \times \frac{1{\cdot}8}{2{\cdot}2} = 0{\cdot}531\ W_1$$

Due to P_C. shear force at

$$C = R_C' = 0{\cdot}11\ W_1 \times \frac{2{\cdot}2}{4} = 0{\cdot}0605\ W_1$$

$$R_S = 0{\cdot}11\ W_1 \times \frac{1{\cdot}8}{4} = 0{\cdot}0495\ W_1$$

The loads on the cross-girder, *i.e.* R_Q, R_R and R_S are to be distributed by normal Courbon's theory.

$$\Sigma W = 0{\cdot}3195\ W_1 + 0{\cdot}531\ W_1 + 0{\cdot}0495\ W_1$$
$$= 0{\cdot}9\ W_1.$$

Let $\bar{x}$ be the distance of C.G. of loads from Q.

$$\bar{x}=\frac{0{\cdot}531\ W_1\times2{\cdot}5+0{\cdot}0491\ W_1\times5}{0{\cdot}9\ W_1}$$

$$=\frac{1{\cdot}575}{0{\cdot}9}=1{\cdot}75\text{ m.}$$

Eccentricity $e=2{\cdot}5-1{\cdot}75=0{\cdot}75$ m.

$$F_P=\frac{0{\cdot}9\ W_1}{3}\left[1+\frac{3\times0{\cdot}75\times2{\cdot}5}{2\times(2{\cdot}5)^2}\right]$$

$$=0{\cdot}435\ W_1.$$

$$F_Q=\frac{\cdot\ 9\ W_1}{3}(1+0)=0{\cdot}3\ W_1$$

Due to F_P: $R_A=0{\cdot}435\ W_1\times\frac{12}{16}=0{\cdot}326\ W_1$

Due to F_Q: $R_B=0{\cdot}3\ W_1\times\frac{12}{16}=0{\cdot}225\ W_1$

$\therefore$ Total reaction on the outer girder

$$=0{\cdot}3905\ W_1+0{\cdot}326\ W_1=0{\cdot}7165\ W_1$$

Total reaction on the inner girder

$$=0{\cdot}649\ W_1+0{\cdot}225\ W_1=0{\cdot}874\ W_1$$

Maximum shear force at support including impact for outer girder

$$=1{\cdot}1\times0{\cdot}7165\times\frac{700{,}000}{2}=275{,}850\text{ N}$$

Maximum shear force at support including impact for inner girder

$$=1{\cdot}1\times0{\cdot}874\times\frac{700{,}000}{2}=336{,}490\text{ N}$$

Shear at centre of span. Position of load for maximum shear is shown on influence line diagram in Fig. 24·30.

$$\text{S.F.}=\frac{\left(\frac{1}{2}+\frac{1}{2}\times\frac{5{\cdot}4}{8}\right)}{2}\times700{,}000$$

$$=293{,}300\text{ N}$$

S.F. including impact for outer girder

$$=1{\cdot}1\times0{\cdot}4833\times293{,}300$$

$$=155{,}900\text{ N}$$

S.F. including impact for inner girder
$=1{\cdot}1\times\frac{1}{3}\times293{,}300$
$=107{,}560$ N

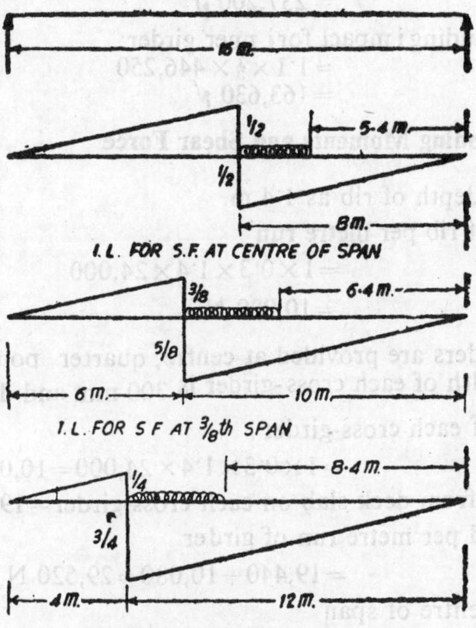

Fig. 24·30

Shear at 3/8th Span

$$\text{S.F.}=\frac{\frac{5}{8}+\frac{5}{8}\times\frac{6{\cdot}4}{10}}{2}\times700{,}000$$
$=358{,}750$ N

S.F. including impact for outer girder
$=358{,}750\times0{\cdot}4833\times1{\cdot}1$
$=209{,}100$ N

S.F. including impact for inner girder
$=358{,}750\times\frac{1}{3}\times1{\cdot}1$
$=131{,}540$ N

Shear at Quarter Span

$$\text{S.F.}=\frac{\left(\frac{3}{4}+\frac{3}{4}\times\frac{8{\cdot}4}{12}\right)}{2}\times700{,}000$$
$=0{\cdot}6375\times700{,}000$
$=446{,}250$ N

S.F. including impact for outer girder

$$=1{\cdot}1\times0{\cdot}4833\times446{,}250$$
$$=237{,}200 \text{ N}$$

S.F. including impact for inner girder

$$=1{\cdot}1\times\tfrac{1}{3}\times446{,}250$$
$$=163{,}630 \text{ N}$$

Dead Load Bending Moments and Shear Force

Assume depth of rib as 1·4 m.

Weight of rib per metre run

$$=1\times0{\cdot}3\times1{\cdot}4\times24{,}000$$
$$=10{,}080 \text{ N}$$

Cross girders are provided at centre, quarter points and at the supports. Width of each cross-girder is 300 mm and depth 1·4 m.

Weight of each cross-girder

$$=1\times0{\cdot}3\times1{\cdot}4\times24{,}000=10{,}080 \text{ N}$$

Reaction from deck slab on each cross-girder = 19,440 N

Total load per metre run of girder

$$=19{,}440+10{,}080=29{,}520 \text{ N}$$

B.M. at centre of span

$$=\frac{29{,}520\times16\times16}{8}+10{,}080\times4+\frac{10{,}080}{2}\times8$$

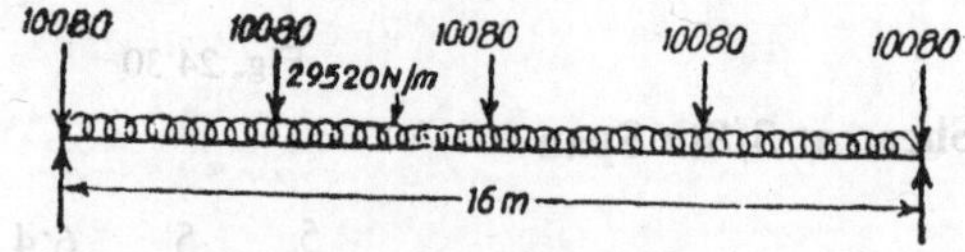

Fig. 24·31.

$$=944{,}640+40{,}320+40{,}320$$
$$=1{,}025{,}280 \text{ N-m}$$

B.M. at quarter span

$$=29{,}520\times8\times4-29{,}520\times4\times2+10{,}080\times4+\frac{10{,}080}{2}\times4$$
$$=708{,}480+40{,}320+20{,}160$$
$$=768{,}960 \text{ N-m}$$

Dead load shear at support

$$=29{,}520\times8+10{,}080+\frac{10{,}080}{2}$$
$$=236{,}160+10{,}080+5040$$
$$=251{,}280 \text{ N}$$

Dead load shear at quarter span

$$=251,280-29,520\times4$$
$$=133,200 \text{ N}$$

Dead load shear at 3/8th span

$$=251,280-29,520\times6-10,080$$
$$=64,080 \text{ N}$$

Dead load shear at centre of span

$$=\frac{10,080}{2}=5040 \text{ N.}$$

Design bending moments and shear forces

Outer Girder

Total B.M. at centre of span

$$=132,100+1,025,280=2,346,280 \text{ N-m}$$

Total B.M. at quarter span

$$=998,200+768,960=1,767,160 \text{ N-m}$$

Total shear force at support

$$=275,850+251,280=527,130 \text{ N}$$

Total shear force at quarter span

$$=237,200+133,200=370,400 \text{ N}$$

Total shear force at 3/8th span

$$=209,100+64,080=273,180 \text{ N}$$

Total shear force at centre of span

$$=155,900+5040=160,940 \text{ N}$$

Inner Girder

Total B.M. at centre of span

$$=911,130+1,025,280=1,936,410 \text{ N-m}$$

Total B.M. at quarter span

$$=683,400+768,960=1,458,360 \text{ N-m}$$

Total shear force at support

$$=336,490+251,280=587,770 \text{ N}$$

Total shear force at quarter span

$$=163,630+133,200=296,830 \text{ N}$$

Total shear force at 3/8th span

$$=131,540+64,080=195,620 \text{ N}$$

Total shear force at mid-span

$$=107,560+5040=112,600 \text{ N}$$

Design of section

Outer Girder. Max. B.M. $=2,346,280$ N-m

The beam is designed as T-beam
Assume effective depth as 145 cm.

Assume lever arm $= 0{\cdot}9 \times 145$

$$A_t = \frac{2{,}346{,}280 \times 1000}{140 \times 0{\cdot}9 \times 1450}$$

$$= 12{,}864 \text{ mm}^2$$

Provide 16 bars of 32 mm ϕ

A_t provided $= 12{,}864 \text{ mm}^2$

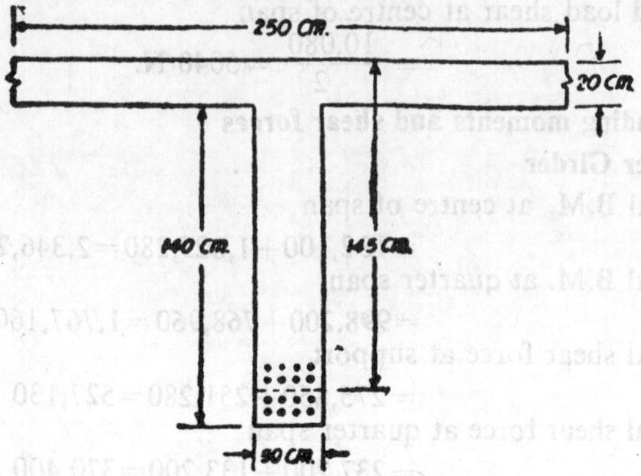

Fig. 24·32.

Check for stresses. Flange width will be least of

(*i*) $12d_s + b_r = 12 \times 20 + 30 = 240 + 30 = 270$ cm.

(*ii*) c/c $= 2{\cdot}5$ m $= 250$ cm.

(*iii*) $\frac{\text{Span}}{4} = \frac{1600}{4} = 400$ cm.

$\therefore$ $B = 250$ cm.

For checking stresses average thickness of flange is assumed as 20 cm.

The concrete mix used for slab is M 20 and for girders M 15. As the whole compression is assumed to be taken by flange value of m assumed is for M 200, *i.e.* $m = 13$.

Assume N.A. to lie in web.

$$250 \times 20\,(n - 10) = 13 \times 128{\cdot}64\,(145 - n)$$

$$\frac{5000}{13 \times 128{\cdot}64}\,(n - 10) = 145 - n$$

$$2{\cdot}99\,n - 29{\cdot}9 = 145 - n$$

$$3{\cdot}99\,n = 174{\cdot}9$$

$$n = \frac{174{\cdot}9}{3{\cdot}99} = 43{\cdot}8 \text{ cm.}$$

Lever arm $a=d-\dfrac{d_s}{2}+\dfrac{d_s^2}{6\,(2n-d_s)}$

$$=145-10+\frac{20\times20}{6(2\times43{\cdot}8-20)}$$

$$=135+0{\cdot}98=135{\cdot}98 \text{ cm}=1359{\cdot}8 \text{ mm}$$

$$\text{B.M.}=Bd_sc_a\left[1+\frac{(n-d_s)}{n}\right]\times a$$

$$2{,}346{,}280\times1000=2500\times200\times c_a\left[1+\frac{438-20}{438}\right]\times1359{\cdot}8$$

$$c_a=\frac{2{,}346{,}280}{50\times1359{\cdot}8\times1{\cdot}544}$$

$$=2{\cdot}244 \text{ N/mm}^2$$

Area of steel required at quarter span

$$=\frac{1{,}767{,}160\times1000}{140\times0{\cdot}9\times1450}=9650 \text{ mm}^2$$

12 bars of 32 mm ϕ are required.

Check for Shear and Bond. Maximum shear at support

$$=527{,}130 \text{ N}$$

4 bars are to be taken straight and adequate anchorage is provided

Shear stress at support

$$=\frac{527{,}130}{300\times0{\cdot}9\times1500}=1{\cdot}3 \text{ N/mm}^2$$

Shear reinforcement is required

Shear stress at centre of s pan

$$=\frac{160{,}940}{300\times1359{\cdot}8}=0{\cdot}393 \text{ N/mm}^2$$

$$\frac{100\,A_s}{bd}=\frac{100\times12{,}864}{300\times1359{\cdot}5}=3$$

From Table 1·9, permissible shear stress is 0·44 N/mm²
No shear reinforcement is required.
Shear stress at 3/8th span

$$=\frac{272{,}180}{300\times0{\cdot}9\times50}=0{\cdot}698 \text{ N/mm}^2$$

Bars are bent in sets, with two bars in each set at every 0·95 m. Five sets of bent up bars are provided. Bent up bars will be effective upto $5\times0{\cdot}95=4{\cdot}75$ m from support. At every section two sets of bars, *i.e.* 4 bars are effective as the bars are bent at spacing of 0·707 *a*. Shear taken by 4 bent up 32 mm ϕ bars

$$=4\times804\times1400\times0{\cdot}707=318{,}000 \text{ N}$$

Net shear force at support for which shear reinforcement is to be provided

$$=527,130-318,000=209,130 \text{ N}$$

Using 10 mm ϕ 2-legged stirrups,

$$\text{Spacing}=\frac{2\times 78{\cdot}5\times 140\times 0{\cdot}87\times 1500}{209,130}$$

$$=137 \text{ mm.}$$

Provide 10 mm ϕ 2-legged stirrups at 13 cm centres upto 3·8 m from support. After 3·8 m from support only two bars are effective.

Shear taken by two bars

$$=159,000 \text{ N}$$

Net shear at quarter point for which stirrups are required

$$=370,400-159,000=211,400 \text{ N}$$

10 mm ϕ stirrups at 10 cm centres are provided between 3·8 m and 4·70 m from support. Beyond 4·70 m from support no bar is effective.

Providing 12 mm ϕ 2-legged stirrups,

$$\text{Spacing}=\frac{2\times 113\times 140\times 0{\cdot}9\times 1450}{370,400}$$

$$=111 \text{ mm.}$$

As the shear at 4·7 m is less than that at quarter span provide 12 mm ϕ 2-legged stirrups at 13 cm centres upto 6 m from support.

Shear force at 3/8th span

$$=273,180 \text{ N}$$

Using 12 mm ϕ 2-legged strriups,

$$\text{Spacing}=\frac{2\times 113\times 140\times 0{\cdot}9\times 145}{273,180}$$

$$=151 \text{ mm}$$

Provide 12 mm ϕ stirrups 5 cm centres up to 7·5 m from support.

In the remaining portion provide 12 mm ϕ stirrups at 50 cm centres.

Inner Girder. Maximum B.M. at centre

$$=911,130+1,025,280=1,936,410 \text{ N-m}$$

Area of steel required at centre

$$=\frac{1,936,410\times 1000}{140\times 0{\cdot}9\times 1450}=10,600 \text{ mm}^2$$

Provide 14 bars of 32 mm ϕ

Area of steel required at quarter span

$$=\frac{1,452,360}{140\times0.9\times1450}=7970\text{ mm}^2$$

10 bars of 32 mm ϕ are required.

Check for Shear and Bond. Maximum shear at support $=587,770$ N

Assume effective depth as 150 cm at support.

6 bars of 32 mm ϕ are taken straight. Adequate anchorage length is provided.

Shear stress at support

$$=\frac{587.770}{300\times0.9\times1500}=1.45\text{ N/mm}$$

Shear stress at centre of span

$$=\frac{112,600}{300\times0.9\times}=0.288\text{ N/mm}^2$$

Shear stress a 3/8th span

$$=\frac{195,620}{300\times0.9\times1450}=0.491\text{ N/mm}^2$$

$$\frac{100A_s}{bd}=\frac{100\times14\times804}{300\times0.9\times1450}=3.2$$

Permissible shear stress from Table 1·9 is 0·44 N/mm^2.

Shear reinforcement is required at 3/8th span.

Shear reinforcement is provided in beyond 3/8th span or 7 m from support.

Bars are bent in sets at every 0·95 m, with two bars in each set. Four sets of bent up bars are provided. Bent up bars will be effective upto $4\times0.95=3.8$ m from support. At every section two sets of bars, *i.e.* 4 bars are effective as the bars are bent at spacing of 0·707 *a*.

Shear taken by 4 bent up bars
$=318,000$

Net shear for which shear reinforcement is to be provided on supports

$587,770-318,000=269,770$ N

Providing 12 mm ϕ 2-legged stirrups,

$$\text{Spacing}=\frac{4\times113\times140\times0.9\times1450}{269,770}$$

$=150$ mm

Provided 12 mm ϕ 2-legged stirrups at 150 mm at centres. This spacing is provided upto 3·8 m from support. After 2·85 m from

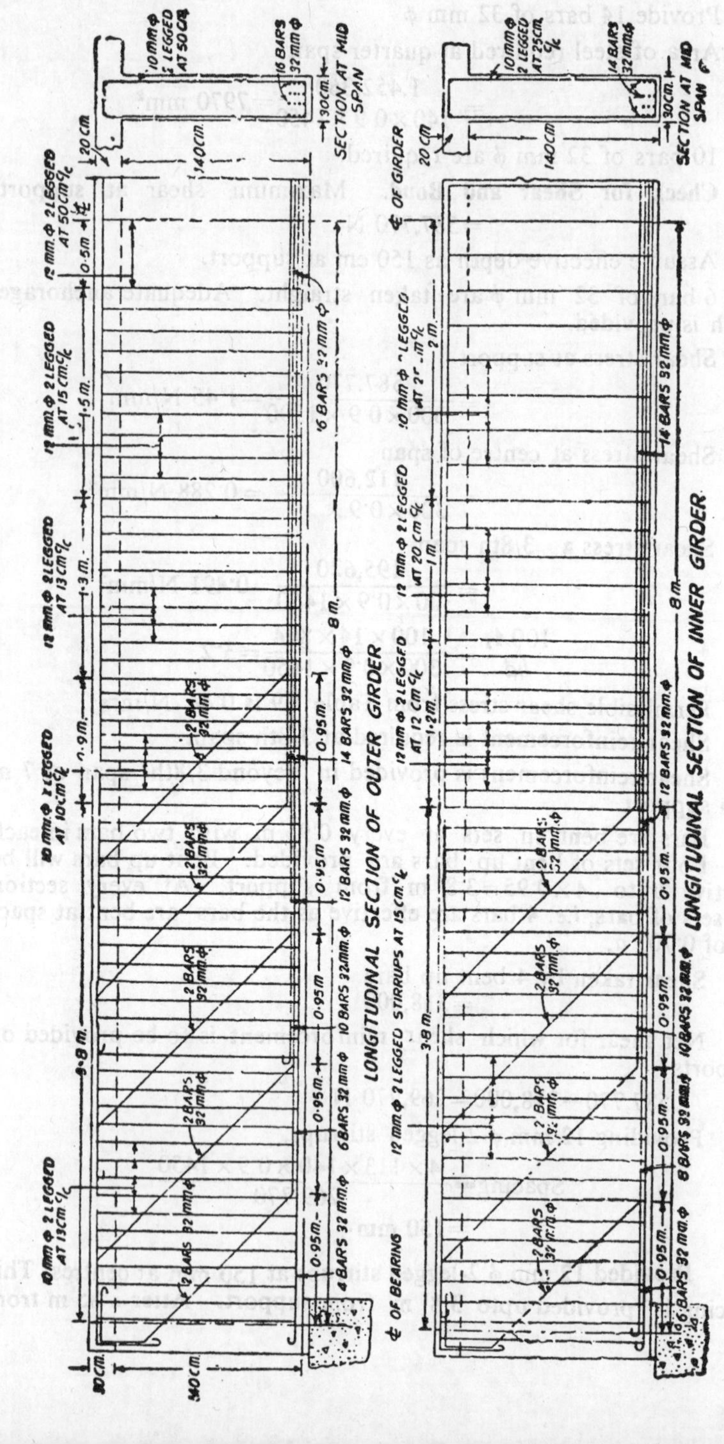

SECTION AT MID SPAN
16 BARS 32 mm φ
10 mm φ 2 LEGGED AT 50 CM
14 BARS 32 mm φ
10 mm φ 2 LEGGED AT 25 CM
SECTION AT MID SPAN
140 CM
20 CM
30 CM
12 mm φ 2 LEGGED AT 50 CM %C
12 mm φ 2 LEGGED AT 15 CM %C
12 mm φ 2 LEGGED AT 13 CM %C
10 mm φ 2 LEGGED AT 10 CM %C
10 mm φ 2 LEGGED AT 13 CM %C
12 mm φ 2 LEGGED STIRRUPS AT 15 CM %C
12 mm φ 2 LEGGED AT 20 CM %C
12 mm φ 2 LEGGED AT 12 CM %C
6 BARS 32 mm φ
14 BARS 32 mm φ
12 BARS 32 mm φ
10 BARS 32 mm φ
8 BARS 32 mm φ
4 BARS 32 mm φ
2 BARS 32 mm φ
0·95 m
3·8 m
8 m
℄ OF GIRDER
℄ OF BEARINGS
LONGITUDINAL SECTION OF OUTER GIRDER
LONGITUDINAL SECTION OF INNER GIRDER

Fig. 24·33

support only two bars are effective but S.F. is less than that at support.

Shear force at quarter span

$$=296{,}830 \text{ N}$$

Providing 12 mm ϕ 2-legged stirrups,

$$\text{Spacing}=\frac{2\times 113\times 140\times 0{\cdot}9\times 1450}{296{,}830}$$

$$=139 \text{ mm.}$$

Provide 12 mm ϕ stirrups at 120 mm centres between 3·8 m from support and 5·0 m from support. Provide 12 mm ϕ 2-legged stirrups at 20 cm centres between 5·0 m and 6·0 m from support.

Shear at 3/8th span = 195,620 N.

Using 12 mm ϕ 2-legged stirrup,

$$\text{Spacing}=\frac{2\times 113\times 140\times 0{\cdot}9\times 1450}{195{,}620}$$

$$=211 \text{ mm}$$

Provide 12 mm ϕ 2-legged stirrups at 200 mm centres upto 1 m from the centre of the span.

Maximum web reinforcement

$$=0{\cdot}15\% \text{ of area of web in plan}$$

$$=\frac{0{\cdot}15}{100}\times 30\times 100=4{\cdot}5 \text{ cm}^2$$

Providing 10 mm ϕ stirrups,

$$\text{Spacing}=\frac{5\times 0{\cdot}785\times 100}{4{\cdot}5}=34{\cdot}8 \text{ cm}$$

Provide 10 mm ϕ stirrups at 25 cm centres

Details of inner and outer girders are shown in Fig. 24·33.

Design cross-girders. Self-weight of cross-girder

$$=10{,}080 \text{ N/m}$$

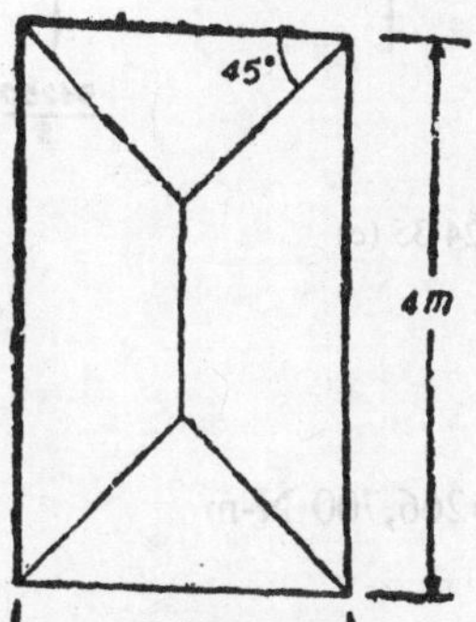

Fig. 24·34 (*a*)

Dead load from slab

$$=2\times\frac{1}{2}\times 2{\cdot}5\times 1{\cdot}25\times 6560$$

$$=20{,}500 \text{ N}$$

This load will be assumed as uniformly distributed load per metre run

$$=\frac{20{,}500}{2{\cdot}5}$$

$$=8200$$

Total dead load per metre run

$$=10{,}080+8200$$

$$=18{,}280 \text{ N}$$

Girder is assumed to be rigid.

$$\text{Reaction} = \frac{18,280 \times 5}{3}$$

$$= 30,466{\cdot}7 \text{ N}$$

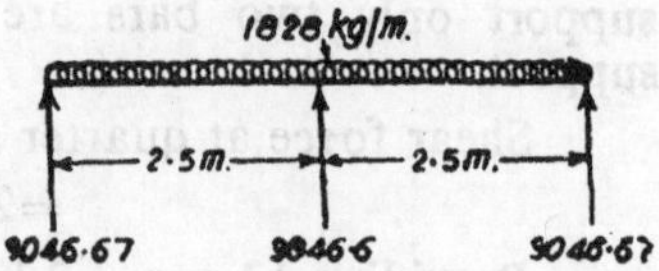

Fig. 24·34 (b)

Live load. Maximum B.M. and S.F. occurs due to class AA tracked vehicle.

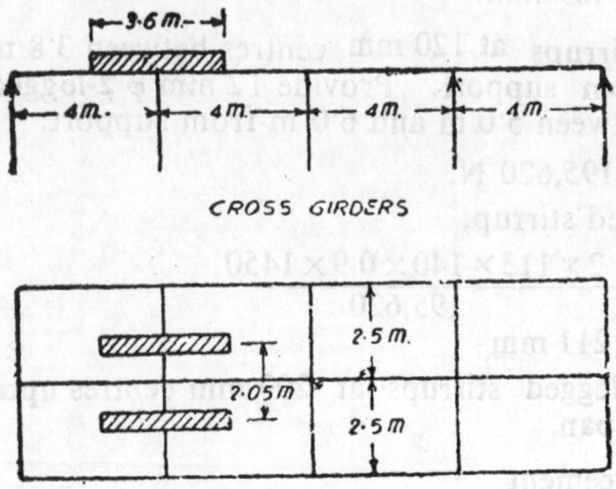

Fig 24·35 (a, b)

For maximum B.M. and S.F. in the cross-girder, load should be on the cross-girder. The position of load for maximum B.M. in the cross-girder is shown in Fig. 24·35.

Slab is assumed to be simply supported between two cross-girders.

Load coming on the cross-girder

$$= 700,000 \times \frac{\left(4 - \frac{1{\cdot}8}{2}\right)}{4}$$

$$= 700,000 \times \frac{3{\cdot}1}{4}$$

$$= 542,500 \text{ N}$$

Cross-girder is assumed to be rigid.

Reaction on each longitudinal girder

$$= \frac{542,500}{3} \text{ N}$$

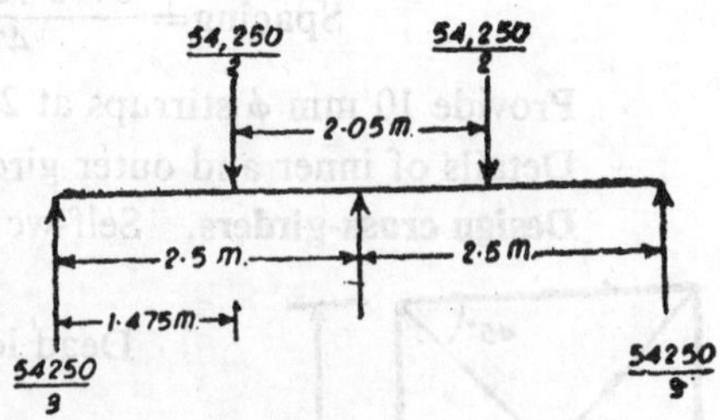

Fig. 24·35 (c)

Maximum B.M. occurs under the load.

Maximum live load B.M.

$$= \frac{542,500}{3} \times 1{\cdot}475 = 266,700 \text{ N-m}$$

B.M. including impact

$$= 1{\cdot}1 \times 266,700 = 293,370 \text{ N-m}$$

Dead load B.M. at 1·475 m from the support

$$=30{,}466{\cdot}7\times1{\cdot}475-18{,}280\times\frac{(1{\cdot}475)^2}{2}$$

$$=45{,}000-19{,}900$$

$$=25{,}100 \text{ N-m}$$

Total B.M. $=293{,}370+25{,}100=318{,}470$ N-m

Live load shear including impact

$$=\frac{542{,}500}{3}\times1{\cdot}1$$

$$=198{,}917 \text{ N}$$

Total shear force $=198{,}917+30{,}466{\cdot}7=229{,}383{\cdot}7$ N

The cross-girder is designed as T-beam.

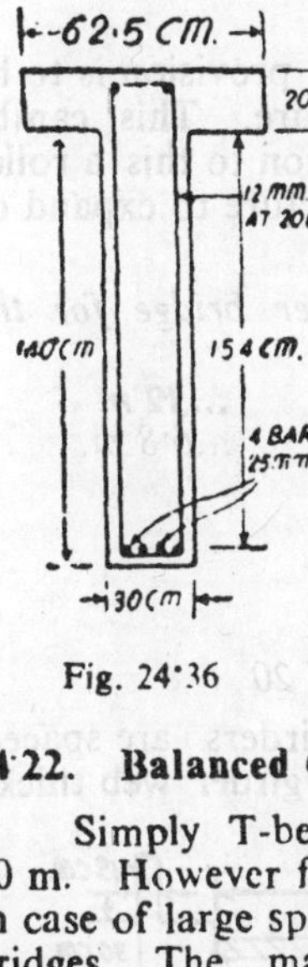

Fig. 24·36

Assume effective depth $=1540$ mm

$$\frac{318{,}470\times1000}{10\times0{\cdot}9\times1540}=1650 \text{ mm}^2$$

Provide 4 bars of 25 mm ϕ.

Shear stress

$$=\frac{229{,}383{\cdot}7}{300\times0{\cdot}9\times1540}=0{\cdot}54 \text{ N/mm}^2$$

Using 12 mm ϕ, 2-legged stirrups.

$$\text{Spacing}=\frac{140\times2\times113\times0{\cdot}9\times154}{229{,}383{\cdot}7}$$

$$=236 \text{ mm.}$$

Provide 12 mm ϕ, 2-legged stirrups at **200 mm** centres.

24·22. Balanced Cantilever Bridges

Simply T-beam slab bridges can be used upto span of about 20 m. However for larger spans these bridges prove uneconomical. In case of large spans it becomes desirable to use continuous girder bridges. The maximum bending moments are considerably reduced and thus resulting in economical design. The continuous girder bridges can be used when there is no possibility of unequal settlement of supports. Any unequal settlement of support will induce bending moments much greater than design moments and thus endanger the life of the structure. In such cases, where possibili of settlement occurs a combination of cantilevers and suspended spans is used. The construction mainly consists of alternate spans with projecting cantilevers, the ends of which are used to support a supended span. The joint between suspended span and edge of cantilever is known as articulation.

Generally a sliding plate bearing is used at this joint. These types of structures are also called articulated structures. Fig. 24·37 shows typical arrangement of balanced cantilever bridge.

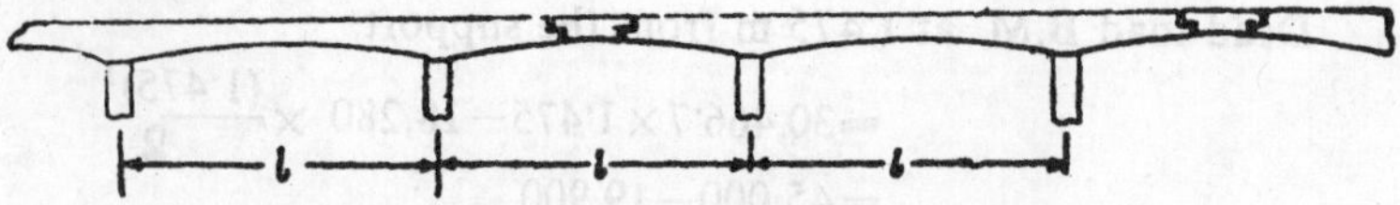

Fig. 24·37.

Cantilever is generally 0·2 to 0·25 of main supported span. This type of bridge has been used upto span of 60 m.

The suspended span is designed as a simply supported span. The cantilever is designed for the dead load and full live load on it and the supported span. For maximum −ve moment in the main supported span only cantilevers and suspended span should be loaded with live load. For maximum +ve moment in the main supported span only main supported span should be loaded with live load. Maximum shear is also found at various sections by plotting influence lines at various sections.

At the supports of main supported span provision is to be made for the angular movement of super-structure. This can be done by providing a rocker bearing. In addition to this a roller bearing is provided at one end to permit the structure to expand or contract.

Example 24·4. *Design a balanced cantilever bridge for the following data* :

Span centre to centre of bearings ...*32 m*
Clear width of roadway ...*6·8 m.*

Loads

2 lanes of class A

1 lane of class AA

Concrete mix used *M 20*

Sol. Two-girder system is adopted. The girders are spaced at centre to centre distance of **4·1 m.** For main girder web thick-

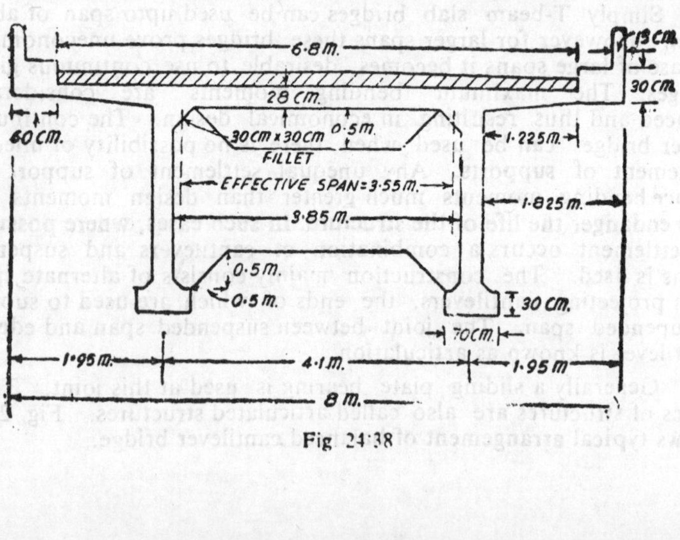

Fig 24·38

ness of 25 cm is adopted. Bottom of girder is provided with the flange. The width of flange provided is 70 cm and thickness 30 cm. General arrangement of deck is shown in Fig. 24·38.

Main supported span is 32 m, cantilevers of 8 m length are provided on either side of the main supported span. Thus suspended span is 16 m length. Stiffeners are provided at every 4 m and articulations. A clearance of 0·5 m is kept between stiffener and deck slab and thus slab is supported by only longitudinal beams. Slab will be designed as one way reinforced.

Design of deck slab. Assume thickness of slab as 28 cm for the supported span. For cantilever thickness will vary from 28 cm at junction with web to 13 cm at junction with kerb.

Average thickness of wearing coat is assumed as 8 cm. Weight of parapet wall is assumed as 1000 N/m. 30 cm × 30 cm fillets are provided at junction of supported span with web.

Bending moments in slab will be found by effective width method.

Design of cantilever

Dead load moments

Item	*Weight* *N*	*Distance of C.G. from edge*	*Moment* *N-m*
Parapet	1000	1·75	1750
Kerb	0·45×0·6×24,000=6480	1·525	9881
Wearing coat slab	0·08×1×1·225×22,000 =2155	$\frac{1·225}{2}$	1320
Slab—Rectangle	0·15×1·225×24,000 =4410	$\frac{1·255}{2}$	2702
Triangle	½×0·13×1·225×24,000 =1911	$\frac{1·225}{3}$	780·3

Total D.L. moment=16,433 N-m

Total load=14,956 N.

Live load moments Class AA. Minimum distance of class AA loading from kerb is 1·1 m and thus very small portion of class AA wheel will lie on cantilever which will give very small B.M.

Class A. Position of load for maximum B.M. is shown in Fig. 24·39.

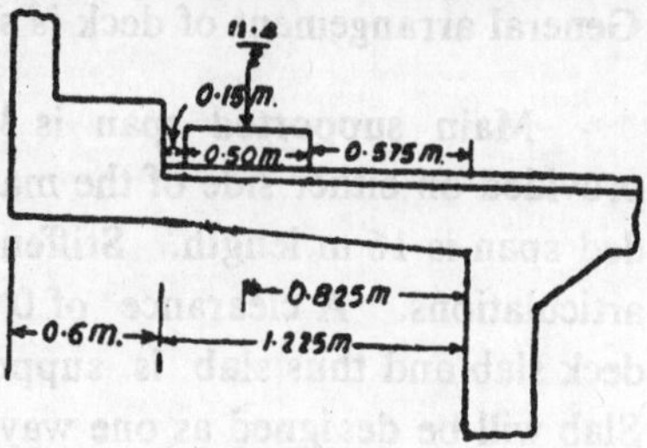

Fig. 24·39

Effective width of slab

$$=2{\cdot}2x+b$$

$$=1{\cdot}2\times0{\cdot}825+0{\cdot}25+2\times0{\cdot}08$$

$$=0{\cdot}99+0{\cdot}41=1{\cdot}40\text{ m}$$

Centre to centre distance of two wheels is 1·2 m and thus effective width overlap. Effective width for two wheel loads

$$=1{\cdot}40+1{\cdot}20=2{\cdot}60\text{ m.}$$

Load per metre width of slab

$$=\frac{2\times\frac{114}{2}}{2{\cdot}6}=43{\cdot}14\text{ kN}$$

$$=43{,}840\text{ N}$$

$$\text{B.M.}=43{,}840\times0{\cdot}825$$

$$=36{,}170\text{ N-m}$$

Live load B.M. including impact

$$=1{\cdot}32\times36{,}170 \quad \text{(span is less than 3 m)}$$

$$=47{,}744\text{ N-m}$$

$$\text{Total B.M.}=47{,}744+16.433$$

$$=64{,}177\text{ N-m}$$

Design of section

Effective depth required

$$\sqrt{\frac{64{,}177\times1000}{1{\cdot}21\times1000}}=230\text{ mm}$$

Provide overall depth of 28 cm with effective depth of 24·5 cm.

Area of steel required

$$=\frac{64{,}177\times1000}{0{\cdot}87\times245\times140}=2150\text{ mm}$$

Using 20 mm ϕ bars,

$$\text{Spacing}=\frac{314{\cdot}2\times1000}{21502}=146\text{ mm}$$

Provide 20 mm ϕ bars at 13 cm centres.

Distribution steel. Distribution steel is to be provided to give moment of resistance of 0·3 times live load moment plus 0·2 times dead load moment.

$$M = 0{\cdot}3 \times 47{,}744 + 0{\cdot}2 \times 16{,}433{\cdot}3$$
$$= 14{,}323 + 3287$$
$$= 17{,}610 \text{ N-m}$$

Using 10 mm ϕ bars, effective depth

$$= 24{\cdot}5 - 1{\cdot}0 - 0{\cdot}5 = 23 \text{ cm}$$

$$A_t = \frac{17{,}610 \times 1000}{140 \times 0{\cdot}87 \times 230} = 628 \text{ mm}^2$$

Half reinforcement is to be provided at top and half at bottom.

Area of steel on each face $= 314 \text{ mm}^2$

Using 10 mm ϕ bars,

$$\text{Spacing} = \frac{78{\cdot}5 \times 1000}{314} = 249 \text{ mm}$$

Provide 10 mm ϕ bars at 22 cm centres and distribution steel both at top and bottom.

Check for shear

Position of load for maximum shear is shown in Fig. 24·39. Maximum live load shear $= 43{,}840$ N.

Shear including impact $= 1{\cdot}32 \times 43{,}840 = 57{,}869$ N

Total D.L.$-$L.L. shear $= 14{,}956 + 57{,}869 = 72{,}824$ N

$$\text{Shear stress} = \frac{72{,}824}{1000 \times 0{\cdot}87 \times 245} = 0{\cdot}36 \text{ N/mm}^2$$

$$\frac{100A_s}{bd} = \frac{100 \times \left(\dfrac{314 \times 1000}{130}\right)}{1000 \times 245} = 1\%$$

Permissible shear from Table 1·9 $= 0{\cdot}39\,k$

$$k = 1{\cdot}20$$

Safe in shear.

Supported span. Effective span is taken as centre to centre of fillets

$$= 3{\cdot}55 \text{ m}$$

Dead weight of slab per square metre

$$= 0{\cdot}28 \times 1 \times 1 \times 24{,}000 = 6720 \text{ N}$$

Dead weight of wearing coat per square metre

$$= 0{\cdot}08 \times 1 \times 22{,}000 = 1760 \text{ N.}$$

Total dead load per square metre $=6720+1760$
$=8480$ N

$$\text{Dead load B.M.} =\frac{wl^2}{8}=\frac{8480\times(3{\cdot}55)^2}{8}=13{,}610 \text{ N-m}$$

—ve B.M. due to cantilever dead load=16,433 N-m

Live Load Moment

Class AA tracked vehicle. Impact factor

$$=25-(25-10)\times\tfrac{3}{4}$$
$$=13{\cdot}75\%$$

Position of loads for maximum B.M. is shown in Fig. 24·40.

$$b_e=\left(1-\frac{x}{l}\right)+b_w$$

Value of k for continuous slabs is 2·6.

$$b_e=2{\cdot}6\times0{\cdot}5\left(1+\frac{0{\cdot}75}{3{\cdot}55}\right)+3{\cdot}6+2\times0{\cdot}08$$

$$=2{\cdot}6\times0{\cdot}75+\frac{2{\cdot}80}{3{\cdot}55}+3{\cdot}76$$

$$=1{\cdot}54+3{\cdot}76$$
$$=5{\cdot}30 \text{ m.}$$

Load per metre width

$$=\frac{700{,}000}{5{\cdot}30}=132{,}000 \text{ N}$$

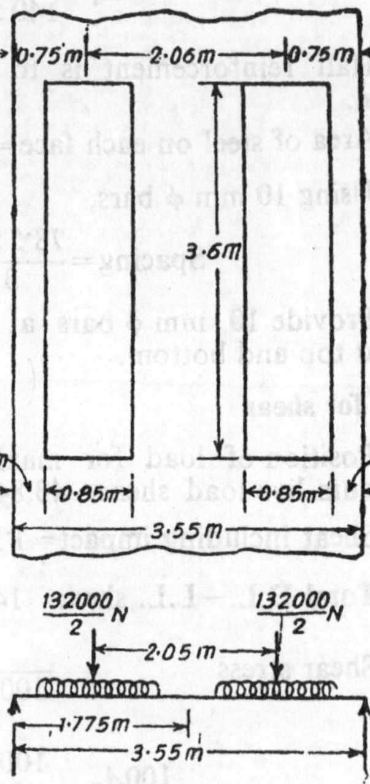

Fig. 24·40

$$\text{Max. B.M.}=\frac{132{,}000}{2}\times1{\cdot}775-\frac{132{,}000}{2}\times\frac{2{\cdot}05}{2}$$

$$=\frac{132{,}000}{2}\times0{\cdot}75=49{,}500 \text{ N-m}$$

B.M. including impact

$$=1{\cdot}1375\times49{,}500=56{,}306 \text{ N-m}$$

Class AA wheeled vehicle. Position of loads for maximum bending moment at centre of span is shown in Fig. 24·41

Dispersion of loads at right angles to span is given by

$$b_e=kx\left(1+\frac{x}{l}\right)+b_w$$
$$k=2{\cdot}6$$

Effective width for first wheel

$$=2{\cdot}6\times0{\cdot}675\left(1+\frac{0{\cdot}675}{3{\cdot}55}\right)+0{\cdot}15+2\times0{\cdot}08$$

$$=1{\cdot}422+0{\cdot}41$$

$$=1{\cdot}832 \text{ m.}$$

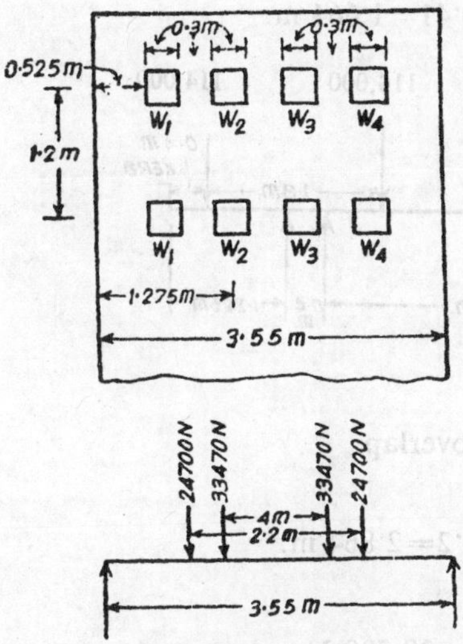

Fig. 24·41

Effective width of two wheels overlap. Effective width for two loads $=1{\cdot}2+1{\cdot}832=3{\cdot}032$ m.

Load per metre width

$$=\frac{2\times37{,}500}{3{\cdot}032}=24{,}700 \text{ N}$$

Effective width for second wheel.

$$b_e=2{\cdot}6\times1{\cdot}275\left(8+\frac{1{\cdot}275}{3{\cdot}55}\right)+0{\cdot}41$$

$$=2{\cdot}124+0{\cdot}41=2{\cdot}534 \text{ m.}$$

Effective widths of two wheels overlap. Total effective width of two loads $=2{\cdot}534+1{\cdot}2$

$$=3{\cdot}734 \text{ m.}$$

Load per metre width of slab

$$=\frac{2\times62{,}500}{3{\cdot}734}=33{,}470 \text{ N}$$

B.M. at centre of span

$$=(24{,}700+33{,}470)\times\frac{3{\cdot}55}{2}$$

$$-24{,}700\times1{\cdot}1-33{,}470\times0{\cdot}5$$

$$=103{,}300-27{,}170-16{,}735$$

$$=59{,}395 \text{ N-m.}$$

B.M. including impact

$$=59{,}395\times1{\cdot}1375=67{,}562 \text{ N-m.}$$

Class AA wheeled vehicle gives maximum live load B.M.

Class A loading does not give maximum live load B.M.

Total B.M. at centre of span

$$=13{,}610+67{,}562-16{,}433{\cdot}3$$

$$=64{,}739 \text{ N-m.}$$

Maximum —ve B.M. at centre of span. Maximum —ve B.M. at the centre of supported span occurs when loads are very near to kerb in cantilever portion. The position of loads for maximum —ve B.M. at centre of span is shown in Fig. 24·42.

Effective width for load in supported span

$$=2·6\times0·575\left(1-\frac{0·575}{3·55}\right)+0·41$$

$$=1·254+0·41=1·664 \text{ m.}$$

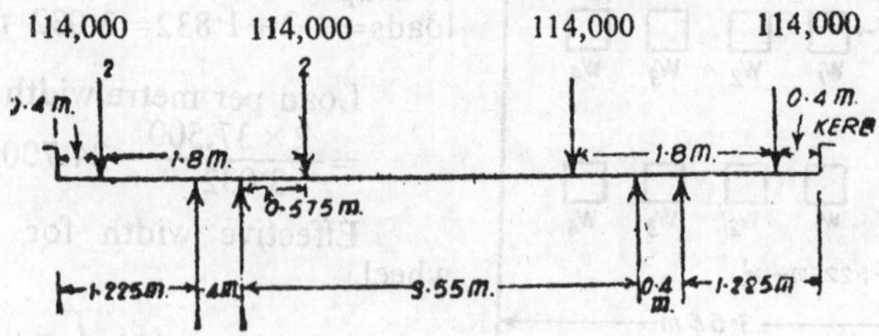

Fig. 24·42

Effective widths of two wheels overlap

Effective width of two loads

$$=1·664+1·2=2·864 \text{ m.}$$

Load per metre width

$$=\frac{114,000}{2·864}=39,780 \text{ N}$$

Positive B.M. at centre of span due to this load

$$=39,780\times0·575 \text{ N-m.}$$

B.M. including impact

$$=39,780\times0·575\times1·32=30,193 \text{ N-m}$$

Total +ve B.M. at centre of span

$$=301,930+13,610=43,803 \text{ N-m.}$$

Maximum cantilever moment due to above position of loads plus due to dead load was found to be 64,177·7 N-m is design of cantilever.

Maximum —ve B.M. at centre of supported span

$$=64.177·7-43,803=20,374·7 \text{ N-m.}$$

Max. +ve B.M. at centre

$$=64,739 \text{ N-m.}$$

Max. —ve B.M. at centre

$$=20,374·7 \text{ N-m.}$$

Effective depth required

$$=\sqrt{\frac{64,739\times1000}{1·21\times1000}}=231 \text{ mm}$$

Provide overall depth of 28 cm with effective depth of 24·5 cm

A_t required for +ve B.M.

$$= \frac{64,739 \times 1000}{140 \times 0·87 \times 245}$$

$$= 2169 \text{ mm}^2.$$

Providing 20 mm ϕ bars,

$$\text{Spacing} = \frac{314 \times 1000}{2169} = 144 \text{ mm}$$

Provide 20 mm ϕ bars at 13 cm centres.

A_t required for −ve B.M.

$$= \frac{20,374·7 \times 1000}{140 \times 0·87 \times 245} = 848 \text{ mm}^2$$

Provide 20 mm ϕ bars at 26 cm centres. Half the bars from cantilever are taken straight.

Distribution steel

0·3 L.L.B.M. + 0·2 D.L. B.M.

$$= 0·3 + 67,562 + 0·2\ (13,610 - 16,433·3)$$
$$= 20,268 - 565 = 20,213 \text{ N-m.}$$

Provide 10 mm ϕ bars, Effective depth

$$= 24·5 - 1·0 - 0·5 = 23 \text{ cm.}$$

$$A_t = \frac{20,213 \times 1000}{0·87 \times 230 \times 840}$$

$$= 553 \text{ mm}^2$$

Provide 10 mm ϕ bars,

$$\text{Spacing} = \frac{78·5 \times 1000}{553 - 1} = 141 \text{ mm}$$

Provide 10 mm ϕ bars at 10 cm centres.

Distribution steel at top

0·3 L.L. B.M. + 0·2 D.L. B.M.

$$= 0·3(47,744 - 30,193) + 0·2(16,433·3 - 13,610)$$
$$= 5265 + 565 = 5830 \text{ N-m.}$$

$$A_t \text{ required} = \frac{5830 \times 1000}{0·87 \times 250 \times 140}$$

$$= 192 \text{ mm}^2$$

Providing 10 mm ϕ bars,

$$\text{Spacing} = \frac{78·5 \times 1000}{192} = 408 \text{ mm}$$

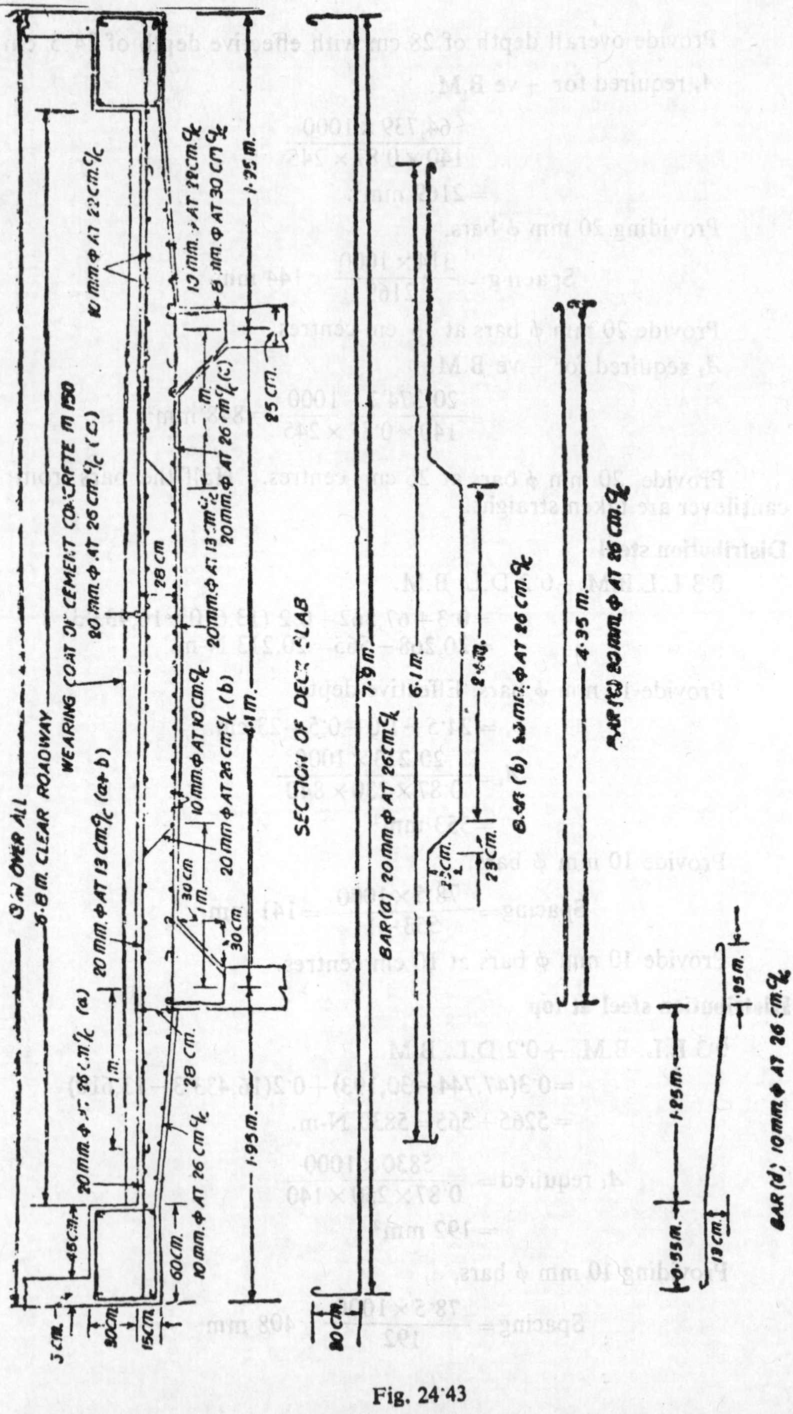

Fig. 24·43

Provide 10 mm ϕ bars at 22 cm. centres as for cantilever.

The section of deck slab is not checked for shear as in T-beam bridge it was found that 20 cm. thick slab was safe in shear. In this case thickness provided is 28 cm.

Cross-section of deck slab is shown in Fig. 24·43.

Main Girder

Profile of girder. Parabolic profile for the main girder is adopted. Thickness of web of girder is kept 25 cm. Depth of web is kept 3·4 m, at the support and 1·7 m at the centre of the span as shown in Fig. 24·44 (*a*). Thickness of bottom flange is kept 30 cm. and width is kept 70 cm. as shown in Fig. 24·44 (*b*). Thickness and width of the bottom flange are kept uniform throughout the span of the girder. The web is widened to width of 70 cm. for distance of 60 cm. at the articulation. The winding is done gradually in a distance of 2·4 m. The web is widened to a width of 70 cm. at the support for distance of 70 cm. The widening is gradually in a distance of 1·5 m.

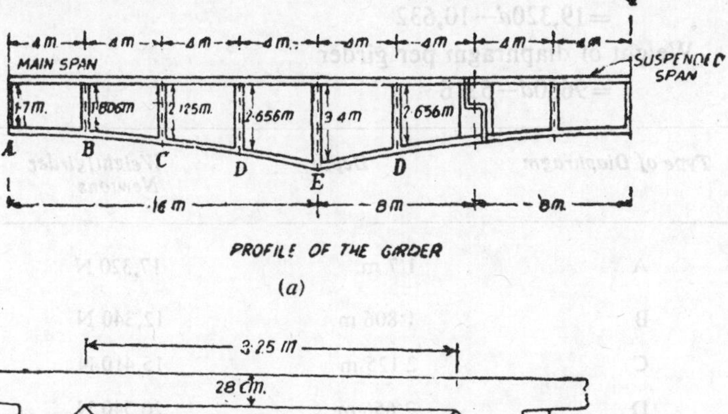

PROFILE OF THE GIRDER

(*a*)

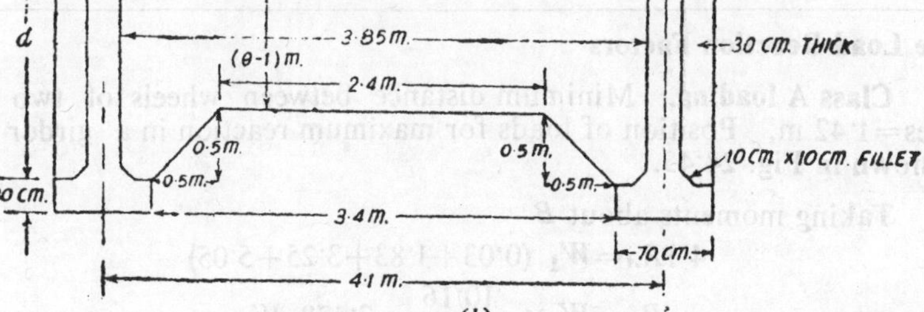

(*b*)

Fig. 24·44

Position of stiffeners and depth of girder at stiffeners is shown in Fig. 24·44 (*a*). The shape of stiffener provided is shown in Fig. 24·44 (*b*).

Load transferred from slab to each girder

1. Parapet $=1000$ N
2. Kerb $=6480$ N
3. Slab $=(4410+1911)-2{\cdot}175\times0{\cdot}28\times24{,}000$
   $=20{,}936$ N
4. Surfacing $0{\cdot}8\times3{\cdot}3\times22{,}000=5984$ N
5. Haunch $\frac{1}{2}\times0{\cdot}3\times0{\cdot}3\times24{,}000=1080$ N

   Total $=35{,}480$ N

Diaphragms. Let d be the depth of web at any section.

Volume of each diaphragm

$$=2\times0{\cdot}3\times0{\cdot}225\times d+3{\cdot}85\times0{\cdot}2\times d-2\times\tfrac{1}{2}\times0{\cdot}3\times0{\cdot}1\times0{\cdot}1$$
$$-2\times\tfrac{1}{2}\times0{\cdot}1\times0{\cdot}1\times0{\cdot}2-\tfrac{1}{2}\,(3{\cdot}25+2{\cdot}25)\times0{\cdot}2\times0{\cdot}5$$
$$-\tfrac{1}{2}\times(3{\cdot}4\times2{\cdot}4)\times0{\cdot}5\times0{\cdot}2-0{\cdot}3\times0{\cdot}3\times\tfrac{1}{2}\times0{\cdot}2\times2$$
$$=0{\cdot}135d+0{\cdot}67d-0{\cdot}003-0{\cdot}002-0{\cdot}275-0{\cdot}145-0{\cdot}018$$
$$=0{\cdot}805d-0{\cdot}443$$

Weight of diaphragm

$$=(0{\cdot}805d-0{\cdot}443)\times24{,}000$$
$$=19{,}320d-10{,}632$$

Weight of diaphragm per girder

$$=9660d-5316$$

Type of Diaphragm	*Depth*	*Weight/girder Newtons*
A	1·7 m.	17,320 N
B	1·806 m.	12,340 N
C	2·125 m	15,410 N
D	2·656 m.	20,540 N
E	3·4 m.	27,740 N

Live Load Reaction Factors

Class A loading. Minimum distance between wheels of two lanes $=1{\cdot}42$ m. Position of loads for maximum reaction in a girder is shown in Fig. 24·45.

Taking moments about B

$$4{\cdot}1R_A=W_1\ (0{\cdot}03+1{\cdot}83+3{\cdot}25+5{\cdot}05)$$

$$R_A=W_1\times\frac{10{\cdot}16}{4{\cdot}1}=2{\cdot}478\ W$$

If W is the axle load $W_1 = \frac{W}{2}$

$$R_A = 1{\cdot}239\ W$$

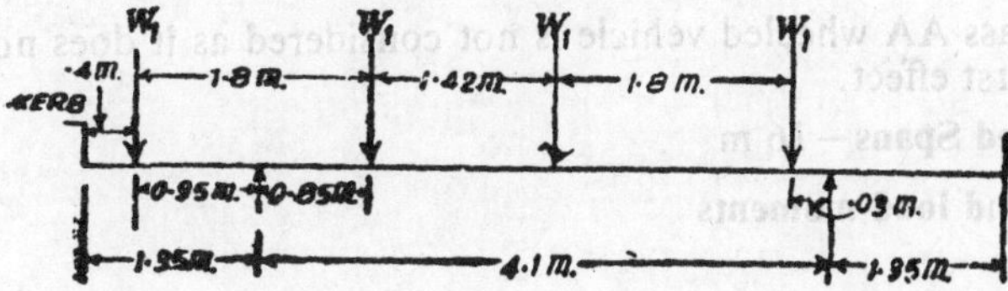

Fig. 24·45

∴ Reaction factor for class A loading is 1·239.

Class AA loading. Minimum clear distance from kerb = 1·2 m

Distance of wheel from kerb = 1·625 m.

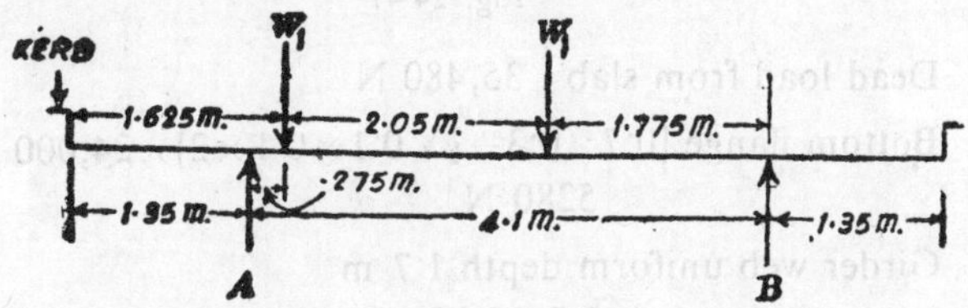

Fig. 24·46

Position of loads for maximum reaction is shown in Fig. 24·46.

Taking moments about B

$$4{\cdot}1\ R_A = W_1(1{\cdot}775 + 3{\cdot}825)$$

$$R_A = \frac{5{\cdot}6}{4{\cdot}1} W_1 = 1{\cdot}366\ W_1$$

If W is axle load, wheel load

$$W_1 = \frac{W}{2}$$

$$R_A = 0{\cdot}683\ W_1$$

Reaction factor for class AA loading is 0·683.

Impact Factors

Class A loading

Suspended span $I = \frac{4{\cdot}5}{6+L} = \frac{4{\cdot}5}{6+16} = 0{\cdot}204$

Cantilever $I = \dfrac{4{\cdot}5}{6 + \text{Cantilever span} + \dfrac{\text{Suspended span}}{2}}$

Main span $I = \frac{4{\cdot}5}{6+32} = 0{\cdot}1185$

Class AA tracked vehicle

Impact factor

0·1 for all spans

Class AA wheeled vehicle is not considered as it does not produce worst effect.

Suspended Spans – 16 m

Dead load moments

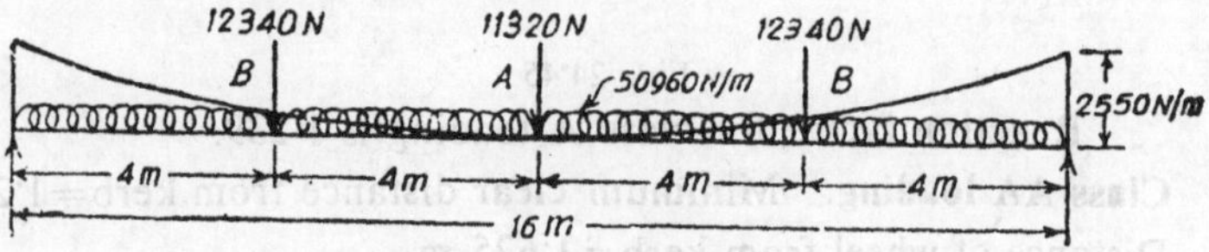

Fig. 24·47

1. Dead load from slab = 35,480 N
2. Bottom flange $[0{\cdot}7 \times 0{\cdot}3 + \frac{1}{2} \times 0{\cdot}1 \times 0{\cdot}1 \times 2] \times 24{,}000$
   $= 5280$ N
3. Girder web uniform depth 1·7 m
   $= 1{\cdot}7 \times 0{\cdot}25 \times 24{,}000 = 10{,}200$ N

Total uniformly distributed load = 50.960 N

Due to parabolic variation in depth of girder dead weight of parabolic portion will vary in parabolic form. Weight at support $= 0{\cdot}425 \times 0{\cdot}25 \times 1 \times 24{,}000 = 2550$ N/m

Dead load **B.M.** at centre of span

$$= \frac{50{,}960 \times (16)^2}{8} + 12{,}340 \times 4 + \frac{11.320}{2} \times 8 + \frac{2550 \times 16}{48} \times 16$$

$$= 1{,}630{,}720 + 49{,}360 + 45{,}280 + 13{,}600$$

$$= 1{,}738\ 960 \text{ N-m.}$$

Dead load **B.M.** at quarter span

$$= \frac{3}{4} \times 1{,}620{,}720 + \frac{11{,}320}{2} \times 4 + 12{,}340 \times 4 + \frac{15}{16} \times 13{,}600$$

$$= 1{,}223{,}040 + 22.640 + 49{,}360 + 12.750$$

$$= 1{,}307{,}790 \text{ N-m}$$

Dead load **B.M.** at 3/8th span

$$= \frac{15}{16} \times 1.630.720 + 1340 \times 4 + \frac{11{,}320}{2} \times 6 + 13.600$$

$=1,528,800+49,360+33,960+13,600$

$=1,625,720$ N-m.

Dead Load Shear Force

Dead load shear at support

$$=50,960\times 8+12,340+\frac{11,320}{2}+\frac{1}{2}\times 2250\times 8$$

$$=407,680+12,340+5660+6800$$

$$=432,480 \text{ N.}$$

Dead load shear at quarter span

$$=50,960\times 4+12,340+5660+\frac{255\times 16}{48}\left(\frac{wl}{48}\right)$$

$$=203,840+12,340+5660+850=222,690 \text{ N}$$

Dead load shear at 3/8th span

$$=50,960\times 2+5660+\frac{255\times 16}{6\times 64}\left(\frac{wl}{6\times 64}\right)$$

$$=101,920+5660+110$$

$$=107,690 \text{ N}$$

Dead load shear at centre = 5660 N

Live Load B.M. In Example 24·3, T-Beam slab bridge was designed for a span of 16 m. Bending moment values taken here are from the influence line diagrams plotted in Ex. 24·3.

Centre of Span

Class AA tracked vehicle

B.M. = 2485 kN-m

Bending moment including impact and reaction factor

$$2485\times 1{\cdot}1\times 0{\cdot}683$$

$$=1867 \text{ kN-m}=1,867,000 \text{ N-m.}$$

Class A loading

B.M. = 1075·55 kN-m.

Bending moment including impact and reaction factor

$$=1075{\cdot}55\times 1{\cdot}239\times 1{\cdot}204$$

$$=1605 \text{ kN-m}=1,605,000 \text{ N-m}$$

Maximum live load B.M. is due to Class AA tracked vehicle is 1,867,000 kN-m.

Total B.M. at centre of span

$$=1,867,000+1,738,960$$

$$=3,605,960 \text{ N-m}$$

Quarter Span. Maximum live load B.M. at quarter span occurs due to class *AA* tracked vehicles.

B.M. = 1863·75 k N-m (See Ex. 24·3)

Bending moment including impact and reaction factor

$$= 1863{\cdot}75 \times 0{\cdot}683 \times 1{\cdot}1$$
$$= 1400 \text{ kN-m}$$
$$= 1{,}400{,}000 \text{ N-m}$$

Total B.M. at quarter span

$$= 1{,}400{,}000 + 1{,}307{,}790 = 2{,}707{,}790 \text{ N-m.}$$

Shear Force. Maximum shear force in the girder occurs due to class *AA* tracked vehicle.

Shear at Support. Position of load for maximum shear force is shown in Fig. 24·48 (*b*).

$$\text{S.F.} = \frac{1 + 0{\cdot}775}{2} \times 700{,}000$$

$$= 621{,}250 \text{ N}$$

Shear force including impact and reaction factor

$$= 621{,}250 \times 0{\cdot}683 \times 1{\cdot}1$$
$$= 466{,}700 \text{ N}$$

Total shear at support

$$= 432{,}480 + 466{,}700$$
$$= 899{,}180 \text{ N}$$

Shear at mid-span

S.F. = 293,300 N

(See Ex. 24·3)

Shear force including impact and reaction factor

$$= 293{,}300 \times 0{\cdot}683 \times 1{\cdot}1$$
$$= 220{,}400$$

Total shear at mid-span

$$= 220{,}400 + 5660$$
$$= 226{,}060 \text{ N.}$$

Shear at 3/8th Span. Position of load for maximum S.F. is shown in Fig. 24·48 (*c*).

$$\text{S.F.} = \frac{\frac{5}{8} + \frac{5}{8} \times \frac{6{\cdot}4}{10}}{2} \times 700{,}000$$

$$= 358{,}750 \text{ N}$$

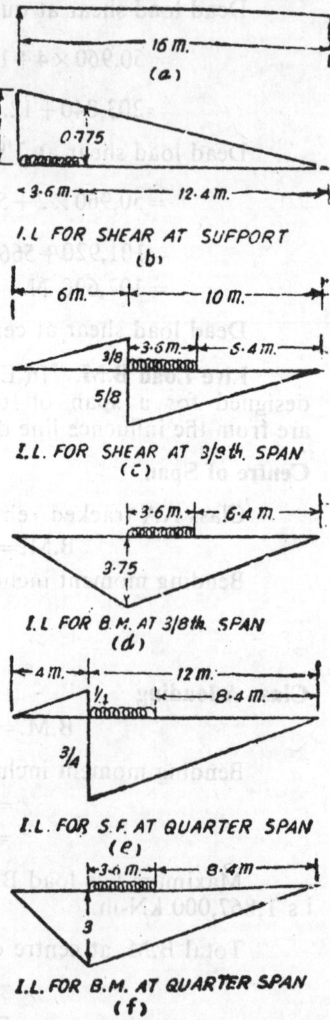

Fig. 24·48

Shear force including impact and reaction factor

$$=358,750\times1{\cdot}1\times0{\cdot}683$$
$$=269,600\text{ N}$$

B.M. due to above position of loads [See Fig. 24·48 (*d*)]

$$=\frac{\left[3{\cdot}75+\frac{6{\cdot}4}{10}\times3{\cdot}75\right]}{2}\times700,000\times0{\cdot}683\times1{\cdot}1$$
$$=3{\cdot}075\times700,000\times0{\cdot}683\times1{\cdot}1$$
$$=1,617,000\text{ N-m}$$

Total B.M. due to D.L. and live load in the above mentioned position

$$=1,625,720+1,617,000=3,242,720\text{ N-m}$$

Effective depth at this section = 216 cm.

$$\tan\beta=\frac{4y_c(l-2x)}{l^2}=\frac{4\times1{\cdot}7\times(32-28)}{32\times32}=\frac{1{\cdot}7}{64}$$

$$\frac{M}{d}\tan\beta=\frac{3,242,720}{2{\cdot}16}\times\frac{1{\cdot}7}{64}=40,000\text{ N}$$

This shear is to be added to dead load and live load shear.

Total S.F. at 3/8th span

$$=269,600+107,690+40,000$$
$$=417,290\text{ N.}$$

Shear at 1/4th Span

Position of load for maximum S.F. is shown in Fig. 24·48 (*e*)

$$\text{S.F.}=\frac{\left(\frac{3}{4}+\frac{3}{4}\times\frac{8{\cdot}4}{12}\right)}{2}\times700,000$$
$$=446,250\text{ N}$$

Shear force including impact and reaction factor

$$=446,250\times0{\cdot}683\times1{\cdot}1$$
$$=335,200\text{ N}$$

B.M. due to above position of loads [See 24·48 (*f*)]

$$=\frac{\left(3+\frac{8{\cdot}4}{12}\times3\right)}{2}\times700,000$$
$$=1,785,000\text{ N-m}$$

B.M. including impact and reaction factor

$$=1,785,000\times0{\cdot}683\times1{\cdot}1$$
$$=1,341,000\text{ N-m}$$

Total B.M. due to dead load and above mentioned position of loads

$$=1,341,000+1,307,790$$
$$=2,648,790 \text{ N-m}$$

Effective depth $=2{\cdot}286$ m.

$$\tan\beta=\frac{1{\cdot}7}{32}$$

$$\frac{M}{d}\tan\beta=\frac{2,648,790}{2{\cdot}286}\times\frac{1{\cdot}7}{32}=61,400 \text{ N}$$

Total shear at quarter span

$$=385,200+222,690+61,400$$
$$=619,290 \text{ N}$$

Design of Section

Centre of Span. Max. +ve B.M.

$$=3,605,960 \text{ N-m}$$
$$=3,605,960,000 \text{ N-mm}$$

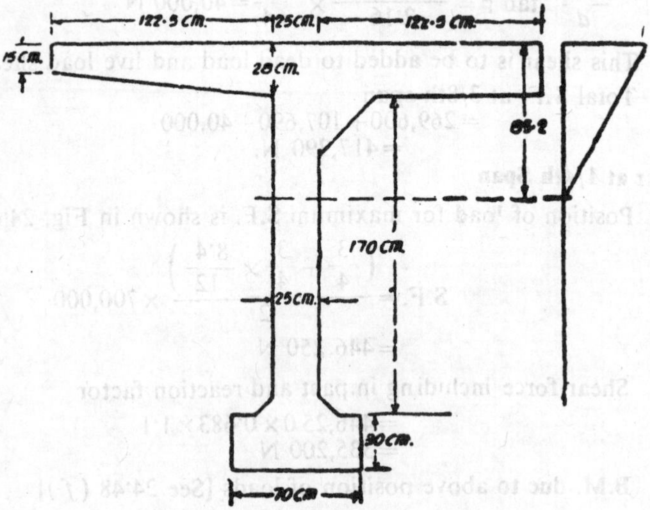

Fig. 24·49

Width of flange will be least of

(*i*) $12d_s+b_r=12\times28+25=336+25=361$ cm.

(*ii*) $\frac{\text{Span}}{4}=\frac{1600}{4}=400$ cm.

(*iii*) Twice maximum cantilever porjection $+b_r$

$$=2\times122{\cdot}5+25=270 \text{ cm.}$$

Effective depth $=170+28+30-15=213$ cm.

Moment of resistance of the section

Depth of critical N.A.

$$=0{\cdot}4d=0{\cdot}4\times 213=85{\cdot}2 \text{ cm}$$

Moment of resistance of rib

$$=Qbd^2=1{\cdot}213\times 250\times(2130)^2$$
$$=1{,}376{,}000{,}000 \text{ N-mm}$$

For moment of resistance of flange, lever arm is given by

$$a=d-\frac{d_s}{2}+\frac{d_s^2}{6(2n-d_s)}$$
$$=213-14+\frac{28\times 28}{6(2\times 85{\cdot}2-28)}$$
$$=199+1=200 \text{ cm.}$$

Moment of resistance

= Area of flange × Average stress × Lever arm

$$=\left[122{\cdot}5\times 28+122{\cdot}5\times\frac{15+28}{2}\right]$$
$$\times\frac{7}{2}\left[1+\frac{85{\cdot}2-28}{85{\cdot}5}\right]\times 200\times 1000$$
$$=[3430+2634]-\tfrac{1}{2}[1+0{\cdot}67]\times 200\times 1000$$
$$=7{,}100{,}000{,}000 \text{ N-mm}$$

Total moment of resistance of the section

$$=7{,}100{,}000{,}000+1{,}376{,}000{,}000$$
$$=8{,}476{,}000{,}000 \text{ N-mm}$$

Actual B.M. is 3,605,860,000

Assume lever arm $=d-\frac{d_s}{2}=199$ cm.

Area of steel $=\frac{3{,}605{,}860{,}000}{140\times 1990}=12{,}950 \text{ mm}^2$

Provide 36 mm ϕ bars, number of bars required

$$=\frac{12{,}950}{1018}=12{\cdot}7$$

Provide 14 bars of 36 mm ϕ.

Section at quarter span

Effective depth $=180{\cdot}6+28+15=228{\cdot}6$ cm.

Assume lever arm $=228{\cdot}6-14$

$=214{\cdot}6$ cm.

Area of steel required

$$=\frac{2,707,790\times 1000}{140\times 2146}$$

$$=9200 \text{ mm}^2$$

9 bars of 36 mm ϕ are required.

Design of Shear Reinforcement

Shear stress at mid-span

$$=\frac{226,060}{250\times 1990}=0{\cdot}455 \text{ N/mm}^2$$

$$\frac{100\,A_s}{bd}=\frac{100\times 12,950}{250\times 1990}=2{\cdot}6$$

Permissible shear from Table 1·9 is 0·49 N/mm².

No shear reinforcement is required

Shear stress at 3/8th span

$$=\frac{417,290}{250\times 2020}=0{\cdot}875$$

Shear reinforcement is required

Shear stress at quarter span

$$=\frac{619,290}{250(2286-15)}=1{\cdot}15 \text{ N/mm}^2$$

Shear stress at support

$$=\frac{899,180}{250\times 0{\cdot}87\times 2505}=1{\cdot}64 \text{ N/mm}^2.$$

Shear reinforcement at 3/8th span. Two bars of 36 mm ϕ are bent at 1·5 m from sapport and the two bars are bent at intervals of every 1·5 m for a distance of 4 m and then at intervals of 2·0 m.

Shear taken by 2 bent bars

$$=2\times 1018\times 140\times 0{\cdot}707$$

$$=202,000 \text{ N}$$

Net shear at 3/8th span for which stirrups are required

$$=417,290-202,000=215,290 \text{ N}.$$

Using 12 mm ϕ 2-legged stirrups,

$$\text{Spacing}=\frac{2\times 113\times 140\times 2020}{215,290}=300 \text{ mm}$$

Provide 12 mm ϕ stirrups at 30 cm centres.

Shear reinforcement at quarter span. Shear taken by two bent up bars at intervals of 2·0 m

$$=\frac{2\times 1018\times 140\times 1{\cdot}414\times 2146}{20,000}$$

$$=476,000 \text{ N}.$$

Net shear for which stirrups are to be provided

$$=619{,}290-76{,}000=143{,}290 \text{ N}$$

Provide 12 mm ϕ stirrups at 30 cm centres.

Shear reinforcement at support. Shear taken by two bent up bars at intervals of 1·5 m

$$=\frac{2\times1018\times140\times1{\cdot}414\times0{\cdot}87\times2605}{1500}$$

$$=615{,}000 \text{ N.}$$

Net shear for which stirrups are to be provided

$$=899{,}180-615{,}000=284{,}180 \text{ N.}$$

Using 12 mm ϕ 2-legged stirrups,

$$\text{Spacing}=\frac{2\times113\times140\times0{\cdot}87\times2605}{284{,}180}=250 \text{ mm}$$

As the web is widened at the articulation and a number of bars are bent there, stirrups at 30 cm, c/c are provided.

Check for Bond

4 bars are straight and a sufficient length of anchorage is provided.

Design of Cantilever. The web width is widened to width of bottom flange at articulation.

Weight of widened portion

$$=[4\times\tfrac{1}{2}\times0{\cdot}225\times2{\cdot}125$$
$$+2\times0{\cdot}6\times0{\cdot}225\times2{\cdot}125]\times24{,}000$$

$$=68{,}850 \text{ N}$$

Reaction from suspended span

$$=432{,}480+15{,}410=447{,}890 \text{ N}$$
$$=447{,}900 \text{ N}$$

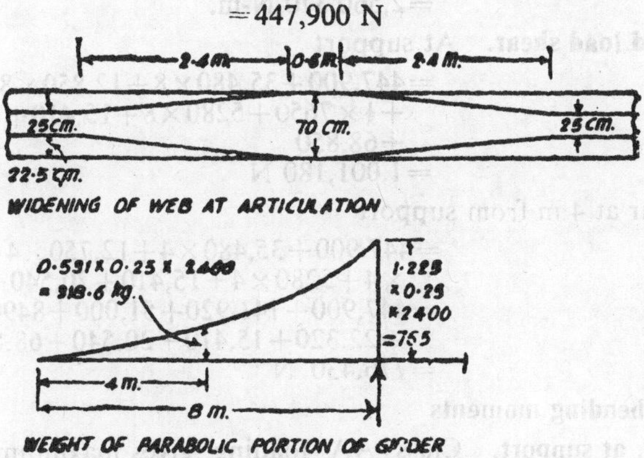

Fig. 24·50

Dead load B.M. at support

Item	Moment	
Reaction from suspended span	$=44{,}790\times 8$	$=3{,}583{,}200$
From slab 35480 N	$=35{,}480\times\frac{8^2}{2}$	$=1{,}055{,}360$
Girder web uniform $0{\cdot}25\times 2{\cdot}125\times 24{,}000$	$=\frac{12{,}758\times 8^2}{2}$	$=408{,}000$
Girder parabola	$=\frac{1}{3}\times 7650\times 8\times\frac{8}{4}$	$=32{,}800$
Bottom flange	$=5280\times\frac{8^2}{2}$	$=168{,}960$
Diaphragms		
C	$=15{,}410\times 8$	$=123{,}280$
D	$=20{,}540\times 4$	$=82{,}160$
Widened portion	$=68{,}850\times 8$	$=550{,}800$
	Total D.L.B.M.	$=6{,}004{,}560$

Dead load B.M. at 4 m from support

$$=\frac{3{,}583{,}200}{2}+35{,}480\times\frac{4^2}{2}+12.750\times\frac{4^2}{2}$$
$$+\frac{1}{3}\times 3186\times 4\times\frac{4}{4}\times 5280\times\frac{4^2}{2}$$
$$+15{,}410\times 4+68{,}850\times 4$$
$$=1{,}791{,}600+283{,}840+102{,}000+4250$$
$$+42{,}243+275{,}400+61{,}640$$
$$=2{,}560{,}370 \text{ N-m.}$$

Dead load shear. At support

$$=447{,}900+35{,}480\times 8+12{,}850\times 8$$
$$+\tfrac{1}{3}\times 7650+5280\times 8+15{,}410+20{,}540$$
$$+68{,}850$$
$$=1{,}001{,}180 \text{ N}$$

Shear at 4 m from support

$$=447{,}900+35{,}480\times 4+12{,}750\times 4+\tfrac{1}{3}\times 3186$$
$$\times 4+5280\times 4+15{,}410+20{,}540+68{,}850$$
$$=447{,}900+141{,}920+51{,}000+8490$$
$$+22{,}320+15{,}410+20{,}540+68.850$$
$$=776{,}430 \text{ N}$$

Live load bending moments

B.M. at support. Class AA loading gives maximum B.M. at support. The position of loads for maximum B.M. at support is shown on influence line diagram in Fig. 24·51 (*b*).

$$\begin{aligned}\text{B.M.} &= 68\times\frac{8}{16}(2{\cdot}7+5{\cdot}7+8{\cdot}7+11{\cdot}7)+114\times 8\\ &\qquad +114\times 6{\cdot}8+27\times 3{\cdot}6+27\times 2{\cdot}5\\ &= 34\times 28{\cdot}8+912+77{\cdot}52+97{\cdot}2+67{\cdot}5\\ &= 979{\cdot}2+1851{\cdot}9=2831{\cdot}1\text{ kN-m}\\ &= 2831{,}100\text{ N-m}\end{aligned}$$

B.M. including impact factor and reaction

$$\begin{aligned}&= 2{,}831{,}100\times 1{\cdot}239\times 1{\cdot}204\\ &= 4{,}224{,}000\text{ N-m.}\end{aligned}$$

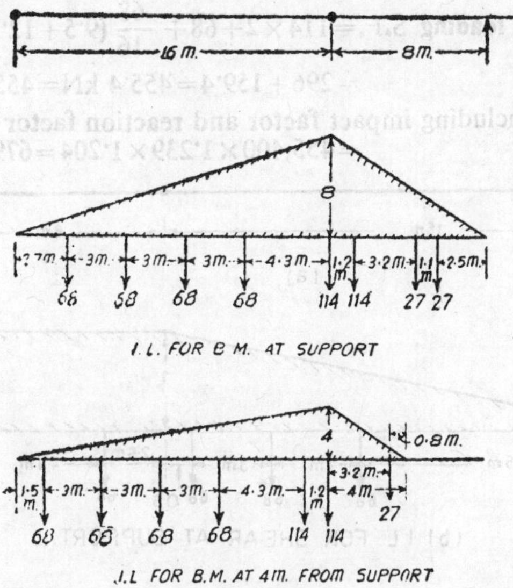

Fig. 24·51

Total B.M. at support

$$\begin{aligned}&= 4{,}224{,}000+6{,}004{,}560\\ &= 10{,}228{,}560\text{ N-m.}\end{aligned}$$

Live load B.M. at 4 m from support. Class A gives maximum B.M. Position of load for maximum B.M. is shown in Fig. 24·51 (*c*).

$$\begin{aligned}\text{B.M.} &= 68\times\frac{4}{16}(1{\cdot}5+4{\cdot}5+7{\cdot}5+10{\cdot}5)+114\times 148\\ &\qquad \times\frac{4}{16}+114\times 4+27\times 0{\cdot}8\\ &= 17\times 24+421{\cdot}8+456+21{\cdot}6\\ &= 408+899{\cdot}4\\ &= 1307{\cdot}4\text{ kN-m}\\ &= 1{,}307{,}400\text{ N-m}\end{aligned}$$

B.M. including impact and reaction factor

$$=1{,}307{,}400\times1{\cdot}239\times1{\cdot}204$$

$$=1{,}950{,}000 \text{ N-m}$$

Total B.M. $=1{,}950{,}000+2{,}560{,}370$

$$=4{,}510{,}370 \text{ N-m.}$$

Shear

Shear at support. Shear due to Class AA tracked vehicle $=700{,}000$ N. Shear including impact and reaction factor

$$=700{,}000\times0{\cdot}683\times1{\cdot}1=525910 \text{ N,}$$

Class A loading S.F. $=114\times2+68+\dfrac{68}{16}(9{\cdot}5+12{\cdot}5+15{\cdot}5)$

$$=296+159{\cdot}4=455{\cdot}4 \text{ kN}=455{,}400 \text{ N.}$$

Shear including impact factor and reaction factor

$$=455{,}400\times1{\cdot}239\times1{\cdot}204=679{,}400 \text{ N}$$

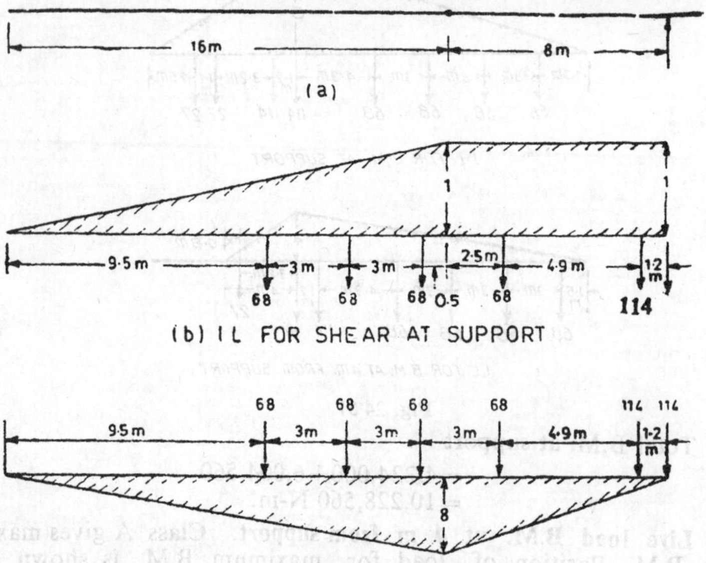

Fig. 24·52

B.M. due to above position of loads [Fig. 24·52 (*c*)]

$$=[114\times1{\cdot}2+68\times5{\cdot}5+68\times\tfrac{1}{2}(9{\cdot}5+12{\cdot}5+15{\cdot}5)]\times1{\cdot}239\times1{\cdot}204$$

$$=2665 \text{ kN-m}=2{,}665{,}000 \text{ N-m.}$$

Total B.M. due to dead load and above position of loads

$$=2{,}665{,}000+6{,}004{,}560$$

$$=8{,}669{,}560 \text{ N-m}$$

Effective depth $=3{\cdot}4+0{\cdot}3+0{\cdot}28+0{\cdot}14=3{\cdot}84$ m.

$$\tan\beta=\frac{1{\cdot}7}{8}$$

$$\frac{M}{d}\tan\beta=\frac{8{,}669{,}560}{3{\cdot}84}\times\frac{1{\cdot}7}{8}$$

$$=480{,}000 \text{ N}$$

Net shear $=1{,}001{,}180+679{,}400-480{,}000$
$=1{,}200{,}580$ N

Shear at 4 m from support. Shear due to class AA loading
$=700{,}000\times0{\cdot}683\times1{\cdot}1=525{,}900$ N

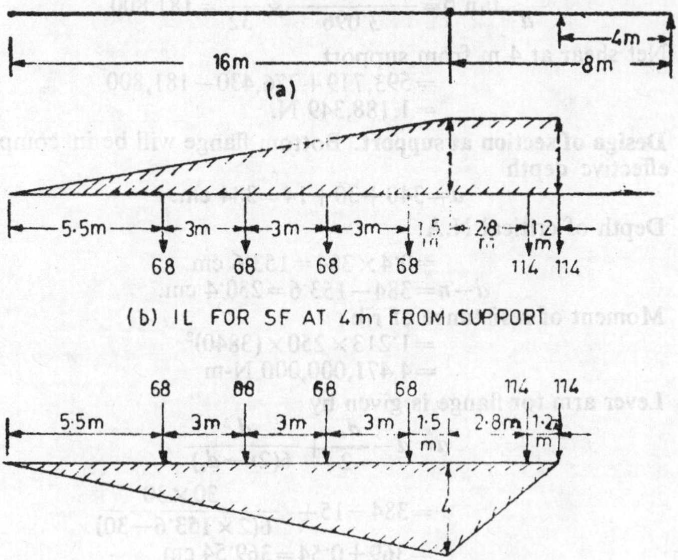

Fig 24·53

Position of loads for maximum S.F. due to class A loading is shown in Fig. 24·53 (*b*).

$$\text{S.F.}=114\times2+\frac{68\times1}{16}(5{\cdot}5+8{\cdot}5+11{\cdot}5+14{\cdot}5)$$

$$=228+\frac{68}{16}\times40$$

$$=228+170=398 \text{ kN}$$

S.F. including impact and reaction factor
$=398{,}000\times1{\cdot}204\times1{\cdot}239=593{,}719$ N

B.M. at 4 m from support due to above position of loads. [Fig. 24·53 (*c*)]

$$=[114\times1{\cdot}2+68\times\frac{1}{4}(5{\cdot}5+8{\cdot}5+11{\cdot}5+14{\cdot}5)]\times1{\cdot}204\times1{\cdot}239$$
$$=[136{\cdot}8+510]\times1{\cdot}204\times1{\cdot}239$$
$$=964{\cdot}7 \text{ kN-m}$$
$$=964{,}700 \text{ N-m.}$$

Total B.M. due to dead load and above position of live load

$$=964{,}700+2{,}560{,}370=3{,}525{,}070$$

Effective depth $=2{\cdot}656+0{\cdot}3+0{\cdot}14=3{\cdot}096$ m.

$$\tan\beta=\frac{5{\cdot}1}{32}$$

$$\frac{M}{d}\tan\beta=\frac{3{,}525{,}070}{3{\cdot}096}\times\frac{5{\cdot}1}{32}=181{,}800$$

Net shear at 4 m from support

$$=593{,}719+776{,}430-181{,}800$$
$$=1{,}188{,}349 \text{ N.}$$

Design of section at support. Bottom flange will be in compression, effective depth

$$d=340+30+14=384 \text{ cm.}$$

Depth of critical N.A.

$$=0{\cdot}4\times384=153{\cdot}6 \text{ cm.}$$
$$d-n=384-153{\cdot}6=230{\cdot}4 \text{ cm.}$$

Moment of resistance of rib

$$=1{\cdot}213\times250\times(3840)^2$$
$$=4{,}471{,}000{,}000 \text{ N-m}$$

Lever arm for flange is given by

$$a=d-\frac{d_s}{2}+\frac{d_s^2}{6(2n-d_s)}$$

$$=384-15+\frac{30\times30}{6(2\times153{\cdot}6-30)}$$

$$=369+0{\cdot}54=369{\cdot}54 \text{ cm.}$$

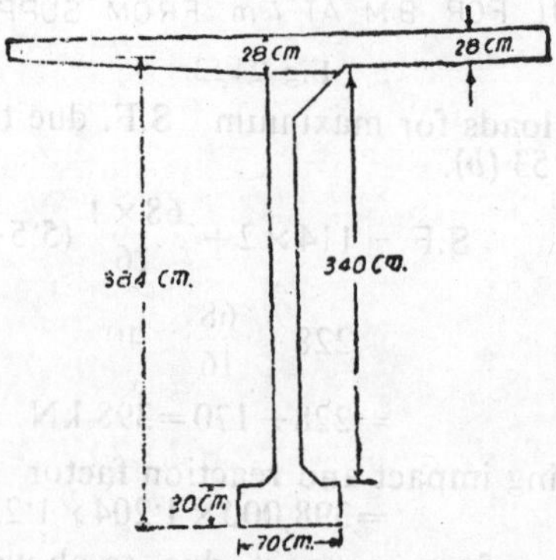

Fig. 24·54

M.R. of Flange $= (70-25)\times 30\times \frac{7}{2}\left[1+\frac{n-d_s}{n}\right] \times a \times 1000$

$= 45\times 30\times \frac{7}{2}\left[1+\frac{153{\cdot}6-30}{153{\cdot}6}\right] \times 369{\cdot}54\times 1000$

$= 45\times 15\times 70\times 1{\cdot}817\times 369{\cdot}54\times 1000$

$= 3{,}174{,}000{,}000$ N-mm

Total moment of resistance

$= 3{,}174{,}000{,}000+4{,}471{,}000{,}000$

$= 7{,}645{,}000{,}000$ N-mm

Actual B.M. is $= 10{,}228{,}560{,}000$ N-mm

Moment of resistance to be provided by compression steel

$= 10{,}228{,}560{,}000-7{,}645{,}000{,}000$

$= 2{,}583{,}560{,}000$ N-m

$$\text{Area of tensile steel} = \frac{4{,}471{,}000{,}000}{140\times 0{\cdot}87\times 3840}+\frac{3{,}174{,}000{,}000}{140\times 3695{\cdot}4}+\frac{2{,}583{,}560{,}000}{140\times 384{,}060}$$

$= 9550+6130+4800 = 20{,}480$ mm^2

Provide 22 bars of 36 mm ϕ.

Area of compression steel is obtained from the formula $(m-1)\,A_c\times(n-f) = mAt_2\,(d-n)$

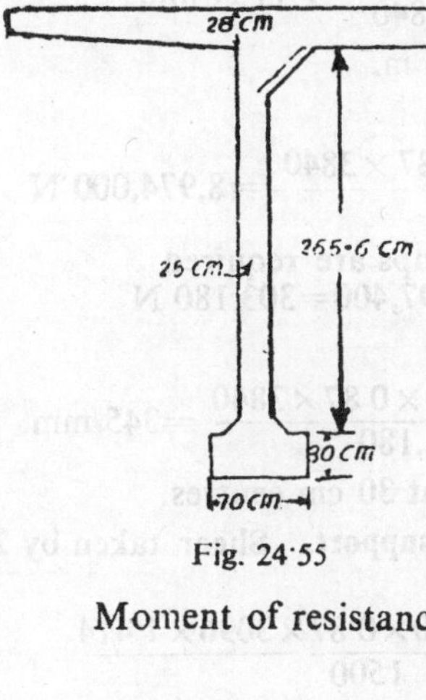

Fig. 24·55

$12\times A_c\times(1536-60)$

$= 13\times 4800\times 2304$

$A_c = \frac{13}{12}\times 4800\times\frac{2304}{1476}$

$= 81{\cdot}4$ cm^2

Provide 8 bars of 36 mm ϕ.

Section at 4 m from support.

Effective depth

$d = 265{\cdot}6+30+14$

$= 309{\cdot}6$ cm.

Depth of critical N.A.

$= 0{\cdot}4\times 309{\cdot}6$

$= 123{\cdot}84$ cm.

Moment of resistance of rib

$= 1{\cdot}213\times 250\times(3096)^2$

$= 2{,}908{,}000{,}000$ N-mm

For moment of resistance of flange lever arms is given by the formula

$$a = d - \frac{d_s}{2} + \frac{d_s^2}{6(2n - d_s)}$$

$$= 309{\cdot}6 - 14 \times \frac{30 \times 30}{6(2 \times 123{\cdot}84 - 30)}$$

$$= 295{\cdot}6 + 0{\cdot}76 = 296{\cdot}36 \text{ cm.}$$

Moment of resistance of flange

$$= (70 - 25) \times 20 \times \frac{7}{2}\left[1 + \frac{123{\cdot}84 - 20}{123{\cdot}94}\right] \times 296{\cdot}36 \times 1000$$

$$= 4 \times 15 \times 7 \times 1{\cdot}757 \times 296{\cdot}36 \times 1000$$

$$= 2{,}460{,}000{,}000 \text{ N-mm.}$$

Total moment of resistance

$$= 2{,}908{,}000{,}000 + 2{,}460{,}000{,}000$$

$$= 5{,}368{,}000{,}000 \text{ N-mm}$$

Actual B.M. $= 4{,}510{,}370$ N-m

$= 4{,}510{,}370{,}000$ N-mm

$$A_t = \frac{4{,}510{,}370{,}000}{140 \times 0{\cdot}87 \times 3096} = 12{,}000 \text{ mm}^2$$

12 bars of 36 mm ϕ are required.

Check for shear. Intensity of shear stress at support

$$= \frac{1{,}200{,}580}{250 \times 0{\cdot}87 \times 3840} = 1{\cdot}44 \text{ N/mm}^2$$

Two bars are bent at intervals of 1·5 m.

Shear taken by bars bent at 1·5 m.

$$= \frac{2 \times 10{\cdot}18 \times 140 \times 1{\cdot}414 \times 0{\cdot}87 \times 3840}{1500} = 8{,}974{,}000 \text{ N}$$

Net shear at support for which stirrups are required

$$= 1{,}204{,}580 - 897{,}400 = 303{,}180 \text{ N}$$

Using 12 mm ϕ 2-legged stirrups,

$$\text{Spacing} = \frac{2 \times 113 \times 140 \times 0{\cdot}87 \times 3840}{303{,}180} = 345 \text{ mm}$$

Provide 12 mm ϕ 2-legged stirrups at 30 cm centres.

Shear reinforcement at 4 m from support. Shear taken by 2 bent bars at 1·5 m interval

$$= \frac{2 \times 1018 \times 140 \times 0{\cdot}87 \times 3096 \times 1{\cdot}414}{1500}$$

$$= 727{,}000 \text{ N}$$

Net shear for which stirrups are required

$$=1,188,349-727,000$$
$$=461,349 \text{ N}$$

Using 12 mm ϕ 2-legged stirrups,

$$\text{Spacing}=\frac{2\times 113\times 140\times 0{\cdot}87\times 3096}{461,349}=200 \text{ mm}$$

Provide 12 mm ϕ 2-legged stirrups at 20 cm centres

Main span 32 m. Dead load B.M.

Load from slab $=35,480$ N/m

Bottom flange $=5280$ N/m

Girder web uniform depth of 1·7 m.

$$=2{\cdot}7\times 0{\cdot}25\times 24,000=10,200 \text{ N}$$

Total uniformly distributed load

$$=35,480+5280+10,200$$
$$=50,960 \text{ N/m}$$

Intensity of parabolic load due to weight of girder

$$=1{\cdot}7\times 0{\cdot}25\times 24,000=10,200 \text{ N}$$

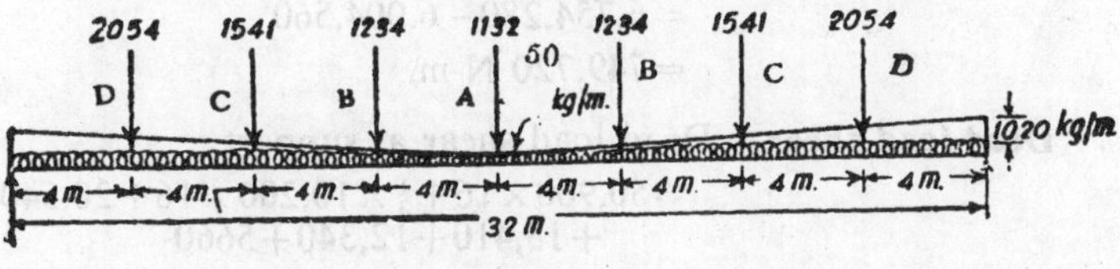

Fig. 24·56

Dead load B.M. at mid-span

$$=50,960\times\frac{32}{8}+20,540\times 4+15,410\times 4+12,340\times 12$$
$$+\frac{11,320}{8}\times 16+\frac{1}{3}\times 16\times 10,200\times\frac{16}{4}$$
$$=6,522,880+82,160+123,280+148,080+90,560+217,600$$
$$=7,184,560 \text{ N-m.}$$

Dead load B.M. at quarter span

$$=\frac{3}{4}\times 50,960\times\frac{(32)^2}{8}+20,540\times 4+15,410\times 8+12,340\times 8$$
$$+\frac{11,320}{2}\times 8+\frac{10,200}{48}\times(32)^2\times\frac{15}{16}$$

$$=4,892,160+82,160+123,280+98,720+45,280+204,000$$

$$=5,445,600 \text{ N-m.}$$

Dead load B.M. at 3/8th span

$$=\frac{15}{16}\times 50,960\times\frac{(32)^2}{8}+20,540\times 4+15,410\times 8+12,340\times 12+\frac{11,320}{20}\times 12+217,600 \text{ (approx)}$$

$$=6,115,240+82,160+123,280+148,080+67,920+217,600$$

$$=6,754,280 \text{ N-m}$$

Cantilever dead load moment $=6,004,560$ N-m

Net dead load B.M. at centre of span

$$=7,184,560-6,004,760$$

$$=1,180,000 \text{ N-m}$$

Net dead load B.M. at quarter span

$$=5,445,600-6,004,560$$

$$=-558,960 \text{ N-m}$$

Net dead load B.M. at 3/8th span

$$=6,754,280-6,004,560$$

$$=749,720 \text{ N-m.}$$

Dead load shear. Dead load shear at support

$$=50,960\times 16+\tfrac{1}{3}\times 10,200\times 16+20,540+15,410+12,340+5660$$

$$=815,360+54,400+53,950$$

$$=932,710 \text{ N}$$

Dead load shear at 1/4th span

$$=50,960\times 8+\frac{10,200\times 32}{48}+15,410+12,340$$

$$=407,680+6,800+33,410$$

$$=447,890 \text{ N}$$

Dead load shear at 3/8th span

$$=50,960\times 4+12.340+5660+\frac{10.300\times 22}{6\times 64}$$

$$=203,840+12.340+5660+850=222,690 \text{ N}$$

Live load moments

Centre of span. Class A loading gives maximum B.M. The position of loads for maximum B.M. is shown in Fig. 24·57 (*b*)

$$\text{B.M.} = 27 \times \tfrac{1}{2}(10{\cdot}5+11{\cdot}6)+114\times\tfrac{1}{2}\times 14{\cdot}8$$
$$+114\times 8+68\times\tfrac{1}{2}(2{\cdot}7+5{\cdot}7+8{\cdot}7+11{\cdot}7)$$
$$=298{\cdot}35+865{\cdot}8+912+979{\cdot}2$$
$$=3055{\cdot}35 \text{ kN-m}$$
$$=3{,}055{,}350 \text{ N-m}$$

B.M. including impact and reaction factor

$$=3{,}055{\cdot}350\times 1{\cdot}1185\times 1{\cdot}239$$
$$=4{,}234{,}000 \text{ N-m}$$

Maximum +ve B.M. at centre of span

$$=\text{Dead load B.M.}+\text{Live load B.M.}$$
$$=118{,}400+4{,}234{,}000$$
$$=5{,}414{,}000 \text{ N-m}$$

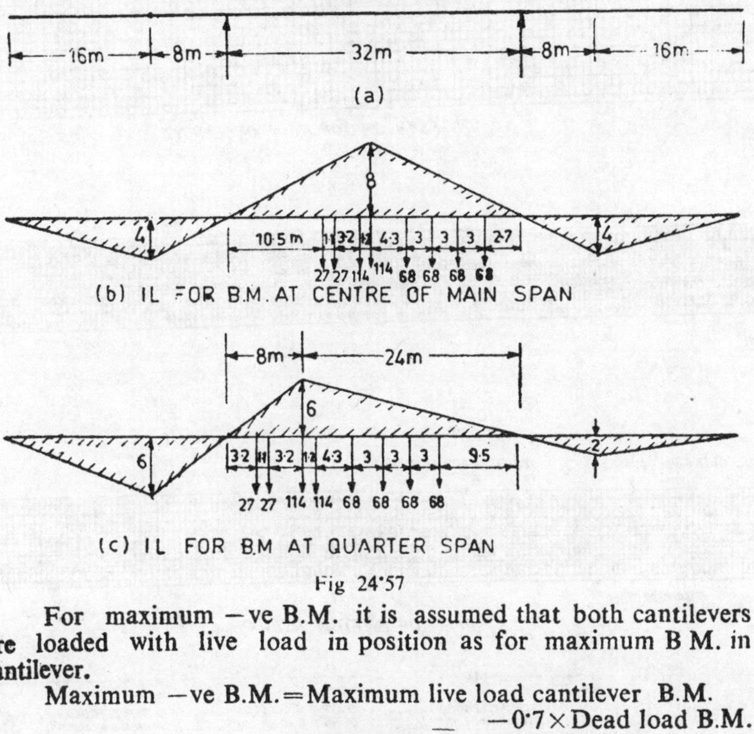

(b) IL FOR B.M. AT CENTRE OF MAIN SPAN

(c) IL FOR BM AT QUARTER SPAN

Fig 24·57

For maximum −ve B.M. it is assumed that both cantilevers are loaded with live load in position as for maximum B M. in cantilever.

Maximum −ve B.M. = Maximum live load cantilever B.M. −0·7 × Dead load B.M.

$$=4{,}224{,}000-0{\cdot}7\times 1{,}180{,}000$$
$$=4{,}224{,}000-826{,}000$$
$$=3{,}398{,}000 \text{ N-m}$$

Quarter span. Class A loading gives maximum B M. Position of loads for maximum B.M. is shown in Fig. 24·57 (*c*).

$$\text{B.M.}=27\times\frac{6}{8}[3{\cdot}7+4{\cdot}8]+68\times\frac{6}{24}(9{\cdot}5+12{\cdot}5+15{\cdot}5+18{\cdot}5)$$
$$+114\times\frac{6}{24}\times 22{\cdot}8+114\times 6$$

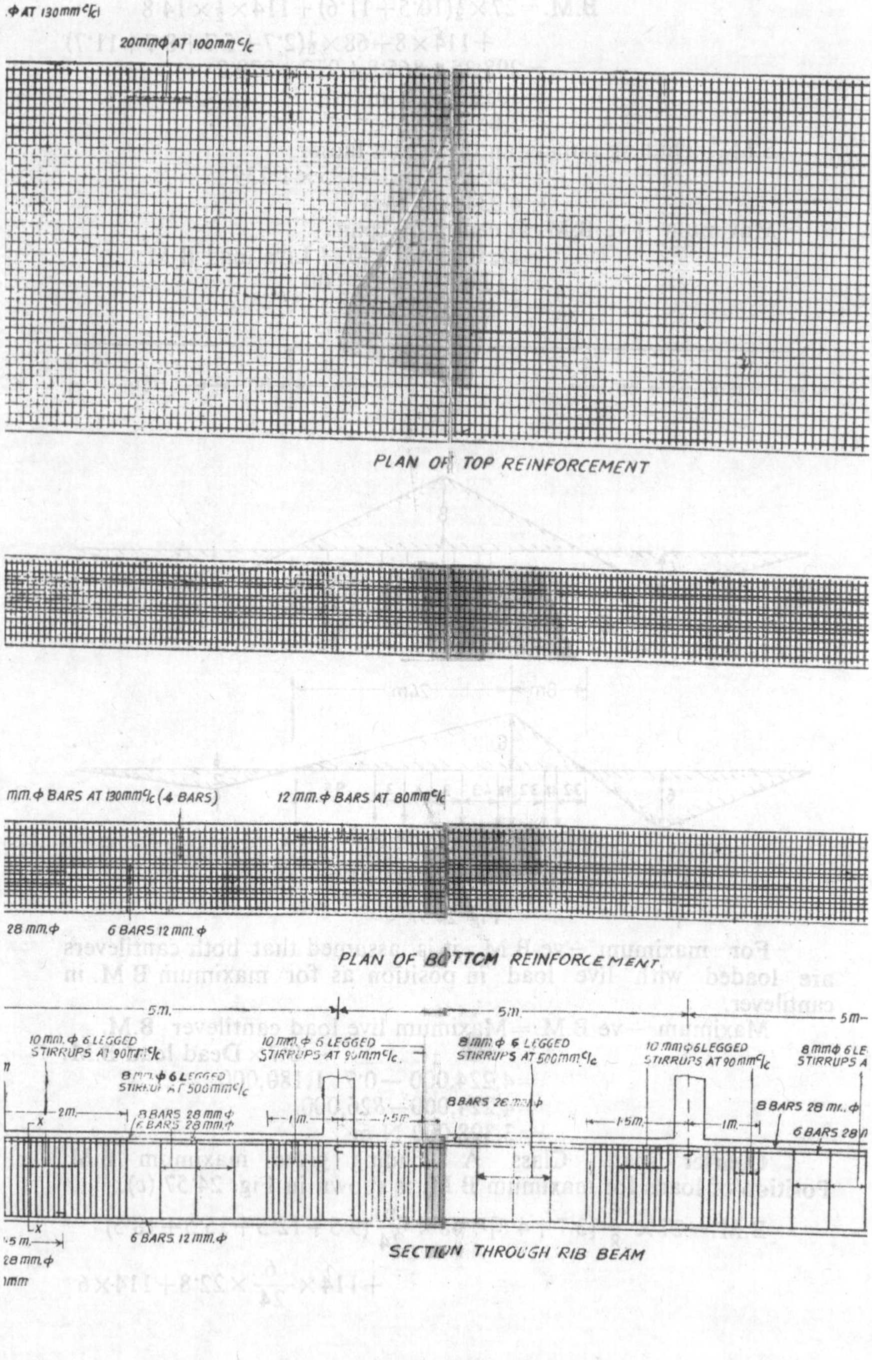

PLAN OF TOP REINFORCEMENT

PLAN OF BOTTOM REINFORCEMENT

SECTION THROUGH RIB BEAM

$$=172{\cdot}2+952+649{\cdot}8+684$$
$$=2458 \text{ kN-m}$$
$$=2{,}458{,}000 \text{ N-m}$$

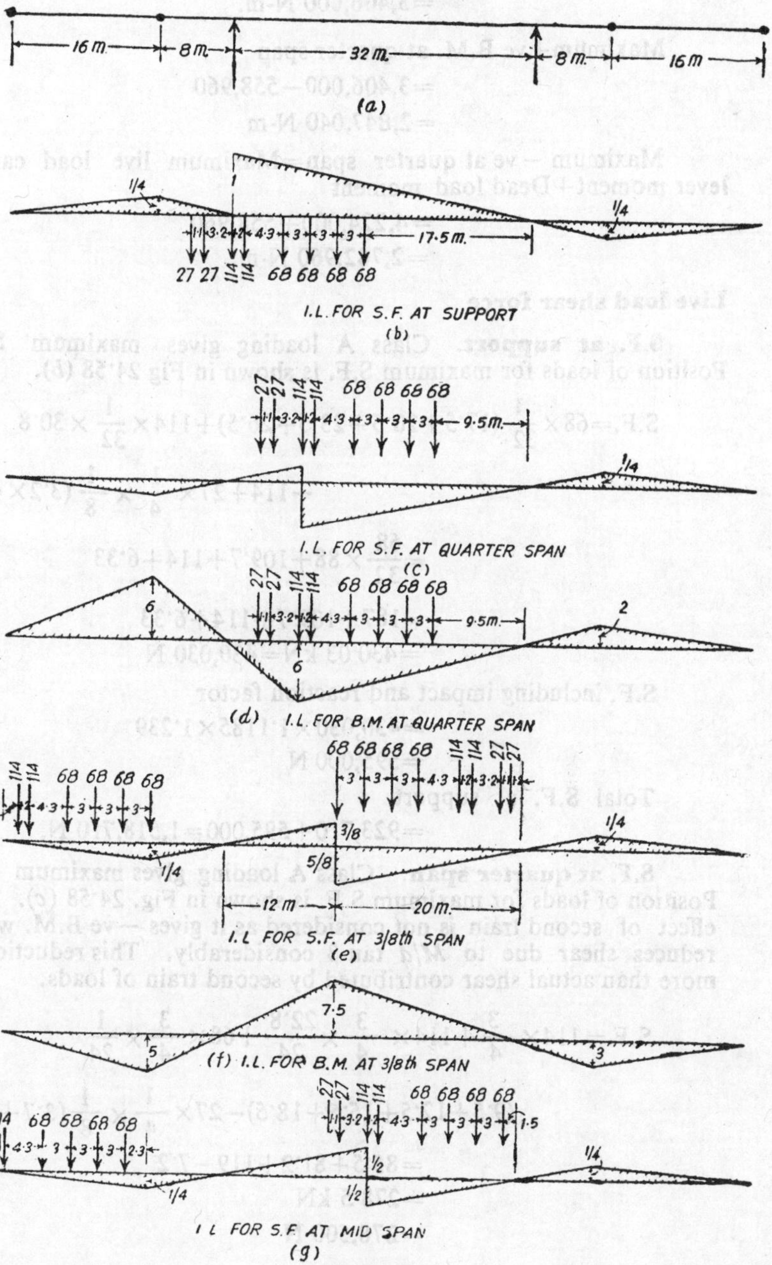

Fig. 24·58

B.M. including impact and reaction factor

$$=2,458,000\times1{\cdot}1185\times1{\cdot}239$$
$$=3,406,000 \text{ N-m.}$$

Maximum+ve B.M. at quarter span

$$=3,406,000-558,960$$
$$=2,847,040 \text{ N-m}$$

Maximum −ve at quarter span=Maximum live load cantilever moment+Dead load moment

$$=4,224,000+558,960$$
$$=2,782,960 \text{ N-m.}$$

Live load shear force

S.F. at support. Class A loading gives maximum S.F. Position of loads for maximum S.F. is shown in Fig 24·58 (*b*).

$$\text{S.F.}=68\times\frac{1}{32}(17{\cdot}5+20{\cdot}5+23{\cdot}5+26{\cdot}5)+114\times\frac{1}{32}\times30{\cdot}8$$
$$+114+27\times\frac{1}{4}\times\frac{1}{8}(3{\cdot}2\times4{\cdot}3)$$
$$=\frac{68}{32}\times88+109{\cdot}7+114+6{\cdot}33$$
$$=187+109{\cdot}7+114+6{\cdot}33$$
$$=430{\cdot}03 \text{ kN}=430,030 \text{ N}$$

S.F. including impact and reaction factor

$$=430,030\times1{\cdot}1185\times1{\cdot}239$$
$$=595,000 \text{ N}$$

Total S.F. at support

$$=923,710+595,000=1,518,710 \text{ N.}$$

S.F. at quarter span. Class A loading gives maximum S.F. Position of loads for maximum S.F. is shown in Fig. 24·58 (*c*). The effect of second train is not considered as it gives −ve B.M. which reduces shear due to $M/d \tan\beta$ considerably. This reduction is more than actual shear contributed by second train of loads.

$$\text{S.F.}=114\times\frac{3}{4}+114\times\frac{3}{4}\times\frac{22{\cdot}8}{24}+68\times\frac{3}{4}\times\frac{1}{24}$$
$$(9{\cdot}5+12{\cdot}5+15{\cdot}5+18{\cdot}5)-27\times\frac{1}{4}\times\frac{1}{8}(3{\cdot}7+4{\cdot}8)$$
$$=85{\cdot}5+81{\cdot}2+119-7{\cdot}2$$
$$=278{\cdot}5 \text{ kN}$$
$$=278,500 \text{ N}$$

S.F. including impact and reaction factor

$$=278{,}500\times1{\cdot}1185\times1{\cdot}239$$

$$=386{,}000 \text{ N}$$

The above position of loads is same as for maximum B.M.

B.M. due to above position of loads

$$=3{,}406{,}000 \text{ N-m}$$

Total B.M. at quarter span

$$=3{,}400{,}000-558{,}960$$

$$=2{,}847{,}040 \text{ N-m}$$

Effective depth $=2{\cdot}125+0{\cdot}28+0{\cdot}15=2{\cdot}555$ m

$$\tan\beta=\frac{1{\cdot}7}{16}$$

$$\frac{M}{d}\tan\beta=\frac{2{,}847{,}040}{2{\cdot}555}\times\frac{1{\cdot}7}{16}=118{,}300 \text{ N}$$

Total S.F. at quarter span

$$=386{,}000+447{,}890+111{,}300$$

$$=952{,}190 \text{ N.}$$

Shear at 3/8th span. Class A loading gives maximum shear. Position of loads for maximum shear is shown in Fig 24·58 (*e*).

$$\text{S.F.}=27\times\frac{5}{8}\times\frac{1}{20}(1{\cdot}2+2{\cdot}3)+114\times\frac{5}{8}\times\frac{1}{20}(5{\cdot}5+6{\cdot}7)$$

$$+68\times\frac{5}{8}\times\frac{1}{20}(11+14+17+20)+\frac{68}{4}$$

$$\times\frac{16}{1}(16+13+10+7)+\frac{1}{4}\times114\times\frac{1}{16}\times(1{\cdot}5+2{\cdot}7)$$

$$=\frac{27}{32}\times3{\cdot}5+\frac{114}{32}\times12{\cdot}2+\frac{68}{32}\times62+\frac{68}{64}\times46+\frac{114}{64}\times4{\cdot}2$$

$$=3+43{\cdot}5+132+48{\cdot}9+7{\cdot}5$$

$$=234{\cdot}9 \text{ kN}$$

$$=234{,}900 \text{ N}$$

S.F. including impact and reaction factor

$$=234{,}900\times1{\cdot}1185\times1{\cdot}239$$

$$=325{,}000 \text{ N}$$

B.M. due to above position of loads [See Fig. 24·58 (*f*)]

$$=\frac{7\cdot5}{20}\times27(1\cdot2+2\cdot3)+\frac{7\cdot5}{20}\times114\times(5\cdot5+6\cdot7)\times\frac{7\cdot5}{20}$$

$$\times68(11+14+17+20)-\frac{5}{16}\times68(16+13+10+7)$$

$$-\frac{5}{16}\times114(1\cdot5+2\cdot7)$$

$$=\frac{7\cdot5}{20}\times27\times3\cdot5+\frac{7\cdot5}{20}\times114\times12\cdot2+\frac{7\cdot5}{20}\times68\times62$$

$$-\frac{5}{16}\times68\times46-\frac{5}{16}\times114\times4\cdot2$$

$$=35\cdot4+521\cdot6+1581-977\cdot6-149\cdot5$$

$$=1011 \text{ kN-m}$$

$$=1,011,000 \text{ N-m}$$

B.M. including impact and reaction factor

$$=1,011,000\times1\cdot1185\times1\cdot239$$

$$=1,400,000 \text{ N-m}$$

Total B.M. $=1,400,000+74,972$

$$=2,149,720 \text{ N-m}$$

Effective depth $=2\cdot236$ m.

$$\tan\beta=\frac{1\cdot7}{32}$$

$$\frac{M}{d}\tan\beta=\frac{2,149,720}{2\cdot236}\times\frac{1\cdot7}{32}=51,100 \text{ N}$$

Total shear at 3/8th span

$$=325,000+51,100+222,690$$

$$=598,790 \text{ N.}$$

Shear at centre of span. Class A loading gives maximum shear. Position of loads for maximum shear is shown in Fig. 24·58 (*g*).

$$\text{S.F.}=\frac{1}{2}\times\frac{68}{16}(1\cdot5+4\cdot5+7\cdot5+10\cdot5)+\frac{1}{2}\times\frac{114}{16}\times14\cdot8$$

$$+\frac{1}{2}\times114+68\times\frac{1}{4}\times\frac{1}{16}(13\cdot7+10\cdot7+7\cdot7+4\cdot7)$$

$$+114\times\frac{1}{4}\times\frac{1}{16}\times0\cdot4$$

$$=51+109\cdot7+39\cdot1+0\cdot7$$

$$=200\cdot5 \text{ kN}$$

$$=200,500 \text{ N}$$

S.F. including impact and reaction factor

$$=200,500\times1\cdot1185\times1\cdot139$$

$$=278,000 \text{ N}$$

Total shear at mid-span

$$=278{,}000+5{,}660$$
$$=283{,}660 \text{ N}$$

Design of Section

Centre of span

Maximum +ve B.M.=5,414,000 N-m

Maximum −ve B.M.=3,398,000 N-m

Moment of resistance of the section (see design of suspended span) for +ve B.M. was found to be 8,476,000,000 N-mm.

Moment of resistance for −ve B.M. will be adequate as steel for +ve B.M. serves as compression reinforcement.

Area of steel at bottom for positive B.M.

$$=\frac{5{,}414{,}000\times1000}{140\times1990}=19{,}450 \text{ mm.}$$

Provide 20 bars of 36 mm ϕ.

$$\text{Area of steel at top}=\frac{3{,}398{,}000\times1000}{140\times0{\cdot}87\times2130}=13{,}100 \text{ mm}^2$$

Provide 13 bars of 36 mm ϕ.

Quarter span

Effective depth d $=212{\cdot}5+28+15=255{\cdot}5$ cm.

Area of steel at bottom for positive B.M.

$$=\frac{2{,}847{,}040\times1000}{140(2555-140)}=8320 \text{ mm}^2$$

10 bars of 36 mm ϕ are provided.

$$\text{Area of steel at top}=\frac{4{,}782{,}960\times1000}{140\times0{\cdot}87\times2555}=15{,}370 \text{ mm}^2$$

16 bars of 36 mm ϕ are provided.

Check for Shear

Shear stress at support

$$=\frac{1{,}518{,}710}{350\times0{\cdot}87\times3880}=1{\cdot}8 \text{ N/mm}^2$$

Shear stress at quarter span

$$=\frac{952{,}190}{250\times0{\cdot}87\times2555}=1{\cdot}71 \text{ N/mm}^2$$

Shear stress at 3/8th span

$$=\frac{598,790}{250\times 2096}=1{\cdot}14\ \text{N/m}^2$$

Shear stress at mid span

$$=\frac{283,660}{250\times 1990}=0{\cdot}57\ \text{N/mm}^2$$

$$\frac{100\,A_s}{bd^*}=\frac{100\times 19,450}{250\times 1990}=3{\cdot}9\%$$

Allowable shear stress from Table 1·9 to 0·51 N/mm²

Provide 12 mm ϕ 2-legged stirrups,

$$\text{Spacing}=\frac{2\times 113\times 140\times 1990}{283,660}=222\ \text{mm}$$

Provide 20 cm spacing.

Shear Reinforcement

3/8th span

Bars are bent in pairs of two bars at interval of 1·5 m.

Shear taken by two bent bars at 1·5 m interval

$$=\frac{2\times 1018\times 140\times 1{\cdot}414\times 2096}{1500}$$

$$=703,000\ \text{N}$$

Actual shear is less than this. Provide 12 mm ϕ stirrups at 30 cm centres.

Quarter span. Shear taken by two bent bars at intervals of 1·2 m

$$=\frac{2\times 1018\times 140\times 1{\cdot}414\times 0{\cdot}87\times 2550}{1200}$$

$$=742,000\ \text{N}$$

Net shear for which stirrups are to be provided

$$=952,190-742,000$$

$$=210,190\ \text{N}$$

Providing 10 mm ϕ 2-legged stirrups,

$$\text{Spacing}=\frac{2\times 113\times 140\times 0{\cdot}87\times 2550}{210,190}=320\ \text{mm}$$

Provide 12 mm ϕ 2-legged stirrups at 30 cm centres.

Support. Shear taken by two bent up bars at 1·2 m intervals

$$=\frac{2\times 1018\times 140\times 1{\cdot}414\times 0{\cdot}87\times 3880}{1200}=1,140,000\ \text{N}$$

Net shear for which stirrups are to be provided

$$=15{,}187{,}100-1{,}140{,}000=378{,}710 \text{ N}$$

Provide 12 mm ϕ 2-legged stirrups,

$$\text{Spacing}=\frac{2\times113\times140\times0{\cdot}87\times3880}{378{,}710}=283 \text{ mm}$$

Provide 12 mm ϕ 2-legged stirrups at 28 cm c/c.

Design of Articulation. Dowel bars are provided at fixed end and roller bearing is provided at free end.

Load on the articulation

Dead load = Reaction from suspended span

$$=432{,}480+15{,}410=447{,}890 \text{ N}$$

Maximum live load reaction

$$=466{,}700 \text{ N}$$

Total reaction $=914{,}590$ N

Bearing plate of 40 cm × 60 cm is provided.

$$\text{Bearing stress}=\frac{941{,}590}{600\times400}=4{\cdot}8 \text{ N/mm}^2. \quad \text{Safe.}$$

Height of bearing is assumed as 40 cm.

Total depth of girder at articulation = 270·5 cm.

Depth of articulation $=\frac{1}{2}(270{\cdot}5-40)=115{\cdot}25$ cm.

Assume effective depth = 150 cm.

$$\text{Shear stress}=\frac{914{,}590}{700\times0{\cdot}87\times1050}$$

$$=1{\cdot}43 \text{ N/mm}^2.$$

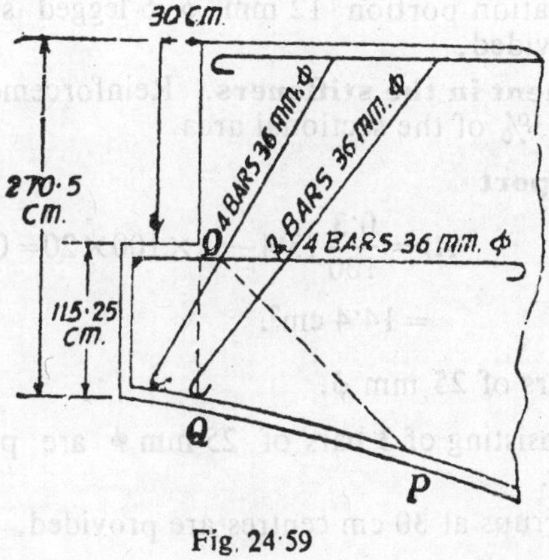

Fig. 24·59

B.M. at section $OQ = 914{,}590 \times 300$

$= 274377{,}000$ N-mm.

Effective depth required

$$= \sqrt{\frac{274{,}377{,}000}{1 \cdot 213 \times 700}} = 570 \text{ mm}$$

Effective depth provided is 105 cm.

Area of steel required

$$= \frac{274{,}377{,}000}{0 \cdot 87 \times 1050 \times 140} = 2140 \text{ mm}^2$$

$$= 21 \cdot 4 \text{ cm}^2$$

4 bars of 26 mm ϕ are provided

Shear force $= 914{,}590$ N

4 bars are bent at 60° and 2 at 45°.

Shear taken by 4 bent bars

$$= 4 \times 1018 \times 140 \times 0 \cdot 866 = 495{,}000 \text{ N}$$

Shear taken by 2 bent bars at 45°

$$= 2 \times 1018 \times 140 \times 0 \cdot 707 = 202{,}000 \text{ N}$$

Total shear taken by bent up bars

$$= 495{,}000 \times 202{,}000 = 697{,}000 \text{ N}$$

Net shear for which stirrups are to be provided

$$= 914{,}590 - 697{,}000 = 217{,}590 \text{ N}$$

Providing 12 mm ϕ 2-legged stirrups,

$$\text{Spacing} = \frac{2 \times 113 \times 140 \times 0 \cdot 87 \times 1050}{217{,}590} = 130 \text{ mm}$$

Provide stirrups at 10 cm centres.

In the articulation portion 12 mm ϕ 6-legged stirrups at 10 cm centres are provided.

Reinforcement in the stiffeners. Reinforcement to be provided should be 0·3% of the sectional area.

Stiffeners at support

$$A_t = \frac{0 \cdot 3}{100}(3 \cdot 4 - 1) \times 100 \times 20 = 0 \cdot 3 \times 2 \cdot 4 \times 20$$

$$= 14 \cdot 4 \text{ cm}^2.$$

Provide 3 bars of 25 mm ϕ.

Stiffeners consisting of 3 bars of 25 mm ϕ are provided both at top and bottom.

10 mm ϕ stirrups at 30 cm centres are provided.

Horizontal reinforcement in web of deep girders. Horizontal reinforcement to be provided on each face should be 0·5% of the area of web.

Area of steel on each face of depth of 1 m.

$$=\frac{0·5}{100}\times 100\times 25=12·5 \text{ cm}^2.$$

Providing 16 mm ϕ bars,

$$\text{Spacing}=\frac{100\times 2·01}{12·5}=16·1 \text{ cm}.$$

16 mm ϕ bars at 16 cm centres are provided on both faces throughout the depth of web.

The details of girder are shown in Fig. 24·59.

24·23. Arch Bridges

For very large spans the use of girder bridges becomes uneconomical. For large span arch bridges can be economically used.

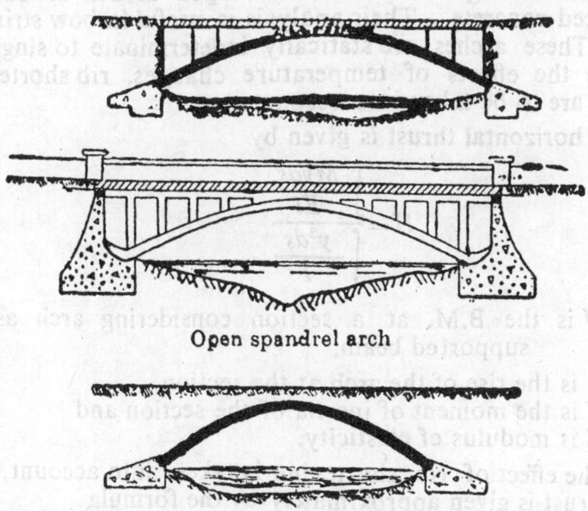

Open spandrel arch

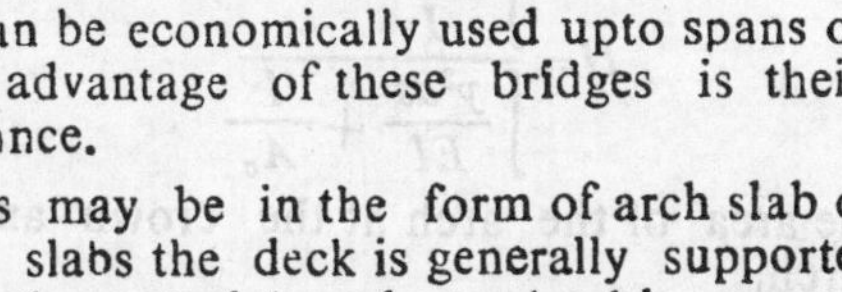

Fig. 24·60. Filled spandrel arch.

These bridges can be economically used upto spans of about 200 m. Another main advantage of these bridges is their graceful and pleasing appearance.

The arches may be in the form of arch slab or arch ribs. In the case of arch slabs the deck is generally supported on the earth filling placed on the arch slab and retained by spandrel walls. This type of arch is also known as filled spandrel arch. In the case of arch rib the deck is supported on the columns which are supported on arch rib. Such an arch is called open spandrel arch.

The types of reinforced concrete arches used are :

1. Three hinged arch, 2. Two hinged arch, 3. Fixed arch.

1. **Three hinged arch.** This type of arch is used for moderate spans and where there is possibility of slight yielding of abutments. This type of structure is a statistically determinate structure and thus there is no effect of temperature changes shrinkage and rib shortening. These effects are not considered in the design.

Most economical shape of the arch is one which coincides with the thrust line. However the thrust line changes with moving loads and hence it is not possible to have shape of arch suited for live loads. The shape of the arch is so selected that there is no B.M. and S.F. in the arch due to dead load. For uniformly distributed loads the shape of the arch will be a parabola.

Certain section for the arch is assumed. Influence lines for B.M., S.F. and normal thrust at various sections are plotted. Moving loads are placed in such a position so as to give maximum values of B.M., S.F. and normal thrust at different sections. The section of the arch is checked for stress at critical points.

2. **Two hinged arch.** Two hinged arches are rarely used in reinforced concrete. Their analysis is useful in bow string girder bridges. These arches are statically indeterminate to single degree and hence the effects of temperature changes, rib shortening and shrinkage are to be taken into account.

The horizontal thrust is given by

$$H=\frac{\int \frac{Myds}{EI}}{\int \frac{y^2ds}{I}}$$

where M is the B.M. at a section considering arch as simply supported beam,

y is the rise of the arch at the section.

I is the moment of inertia of the section and

E is modulus of elasticity.

If the effect of rib shortening is taken into account, the horizontal thrust is given approximately by the formula

$$H=\frac{\int \frac{Myds}{I}}{\int \frac{y^2ds}{EI}+\frac{l}{A_c}}$$

where A_c is the area of the arch at the crown and l is the span of the arch.

Due to the temperature rise of T, the horizontal thrust in the arch is given by

$$H_t=\frac{E\alpha Tl}{\int \frac{y^2ds}{EI}+\frac{l}{A_c}}$$

where α is coefficient of linear expansion.

The effect of shrinkage is to provide negative thrust at the abutments. Generally allowance is made by assuming a fall in temperature of about 8°C.

3. **Fixed arch**. These type of arches are commonly used in reinforced concrete for large spans. These are suitable where rigid abutments are available.

A suitable profile of the arch is assumed so that bending moment and shear force due to dead load are least possible. The dead load includes the weight of filling, arch rib and road surfacing. It is assumed that loads act vertically. The effect of horizontal thrust due to filling is neglected.

A suitable section for the arch rib with certain percentage of steel is assumed. The section is checked for stresses. If there is minor variation in stresses it can be adjusted by changing precentage of steel. If there is major change in stresses another section is assumed and calculations are repeated.

The analysis of arches is done either by method of column analogy or elastic centre. Influence lines for B.M., S.F. and horizontal thrust at crown are plotted. From these, influence lines at other sections can be obtained. Maximum B.M., S.F. and thrust at various points due to moving loads are obtained and sections checked.

Fixed arches are temporarily constructed as three hinged arches for dead loads so that the rib shortening effects due to dead load and some portion of shrinkage stresses are eliminated. Afterwards the hinges are closed making the structure as a fixed arch for live loads.

Proportioning of arch. Site conditions generally determine the ratio of rise to span of the arch. From economy point of view the ratio of rise to span should be between 1/4 to 1/6, depending on the loading and the type of the arch. A large rise to span ratio is desirable as it reduces the thrust on the rib resulting in considerable economy. However, beyond certain limit rise to span ratio is not economical. Rise should never be less than 1/0th to 1/12th of span as with smaller ratio thrust is tremendous and design of abutments becomes uneconomical and sometimes with low bearing capacity of soil, design of abutments becomes impossible.

24·24. Bow String Girder Bridges

Where unyielding foundations for arch abutments are not available or where large head room under the bridge adjacent to the abutments is necessary a bow string girder bridge is used.

A bow string girder is a girder having top curved boom and lower horizontal member. Vertical suspenders are provided between top curved member and horizontal member at suitable intervals. On the suspenders of two girders cross girders are supported on which is supported the roadway. The bow string arch is supported on pin

Fig. 24·61. Bow String Girder Bridge.

bearings to allow rotation at the springs. Roller bearing is provided at one end to allow expansion.

The design of a bow string girder is somewhat similar to that of a two hinged arch. In the case of a two hinged arch the abutments allow the rotation but there is no yielding of the abutments. In the case of a string girder bridge horizontal movements of the arch end is prevented by tie, though due to tension in the tie it will stretch somewhat and thus release horizontal thrust in arch a little. In bow string girder bridge, as the girder is supported on bearings it is free to move due to temperature changes and shrinkage. Thus no appreciable stresses are produced due to temperature effects and shrinkage.

Bow string girder is designed in two ways. One method is to consider the girder as a two hinged arch with elastic tie and it is assumed that no movements are taken by the tie and suspenders. For the purposes of design the curved member is treated as an arch. Generally parabolic shape is used. Horizontal thrust is given by

$$H = \frac{\int \frac{M y d_s}{EI}}{\int \frac{y^2 d_s}{EI} + \frac{l}{A_c} + \frac{l}{m A_s}}$$

where A_s is the area of steel reinforcement, for the purpose of calculation only steel area in the tie is considered. All other terms are already defined.

B.M. and S.F. at various sections can be found as in the case of two hinged arch and the section is checked for stresses.

Second method of design of bow string girder bridges is to consider the moment resisted by the tie. The method is used when inertia of the tie member is quite considerable. Horizontal thrust in this case is found as discussed above Proportion of total B.M. taken by curved boom at section X-X is given by

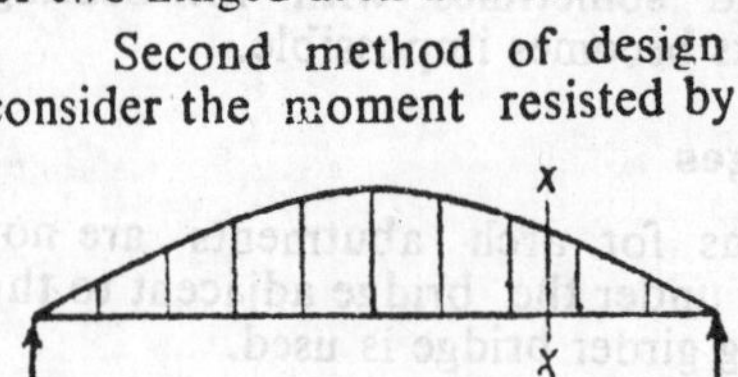

Fig. 24·62

$$M_c = \frac{M_r}{I_c + I_t} \times I$$

where I_c=moment of inertia of rib at section considered.

I_t=moment of inertia of tie (steel alone).

M_e=B.M. at section X-X.

The compressive stresses in the section can be checked.

Tie. Tie is designed for (*i*) uniform tension due to horizontal tension, (*ii*) B.M. dne to vertical load.

$$M_t=\frac{M_x}{I_c-I_t}\times I_t$$

(*iii*) Tension due to B.M. caused by the wind.

Suspenders. Suspenders are designed for the vertical load supported by it and moment caused by wind forces. Moment in the suspender is caused by the wind acting on the suspender and portion for the curved top bottom supported by it acting at respective centres of gravities. In design it is assumed that entire tension is taken by steel.

Effective Wind Pressure

(*a*) **Arch rib without overhead bracing.** In this arrangement wind pressure acting on suspenders and curved boom is transmitted to deck by the suspenders acting as cantilever. The deck acts as plate girder with cross girders acting as stiffeners and ties acting as two booms of the plate girders. The girder is designed with wind pressure acting both on leeward and windward girder plus that acting on moving load. The wind pressure will cause direct tension in leeward tie and direct compression in windward tie.

(*b*) **Arch rib with overhead bracing.** In this arrangement top curved boom with bracing act as one girder and bottom tie with cross-girders act as other girder. Half the wind force is transmitted to top curved truss and half to bottom truss.

Box and Cellular Girder Bridges

Box and cellular girder bridges are commonly used these days because of several advantages they offer. Many highways in Europe and America have been supported on single column and the bridge section has been designed as a box form. The main advantages of this type of construction is high torsion and flexural sitffness with comparatively light weight. This type of construction proves economical for spans between 20 m and 50 m. The section provided may be single cell or multi-cell form. The types shown in Fig. 24·63 (*a*) and (*b*) are also called Spine Beam Bridges.

The types of bridges shown in Figs. 24·63 and 24·64 are monographic slab with voids of some form. The type of construction is generally cast in-situ. Fig. 24·65 shows types of contruction known as multi-beam bridges. The beams usually of precast pretensioned concrete are placed side by side to form the deck. The connection between the beams may be only in the form of shear keys.

In general there are three methods used in the design of these bridges : (*a*) folded plate method (*b*) finite element method and (*c*) finite segment method. First two methods are based on the theory of elasticity and the last one on the ordinary theory.

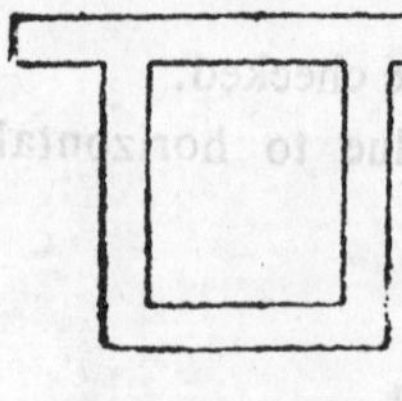

Fig. 21·63 (*b*) Single cell.

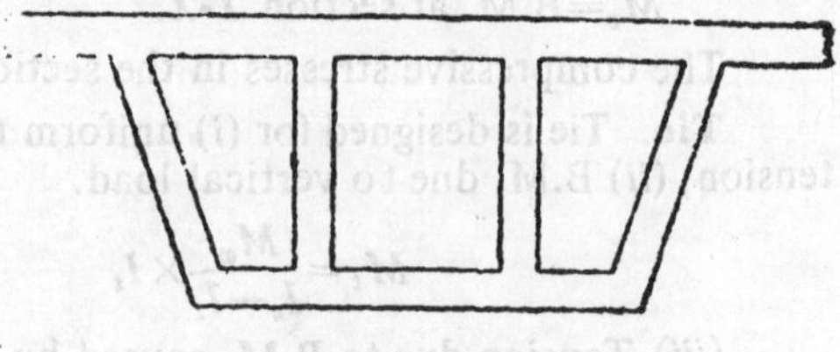

Fig. 24·63 (*b*) Multi-cell.

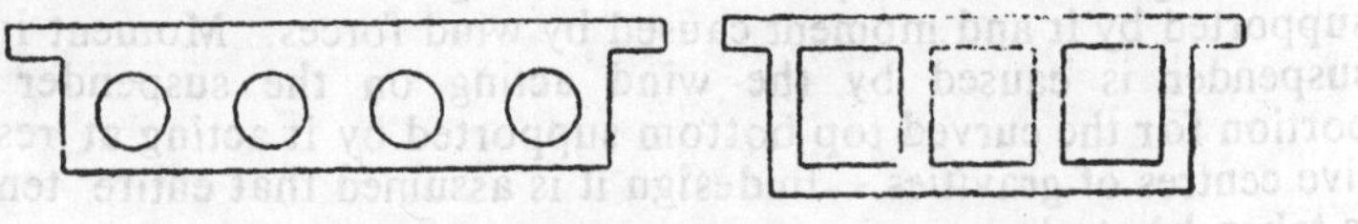

Fig. 24·64. Multi-cell slab bridges construction in-situ.

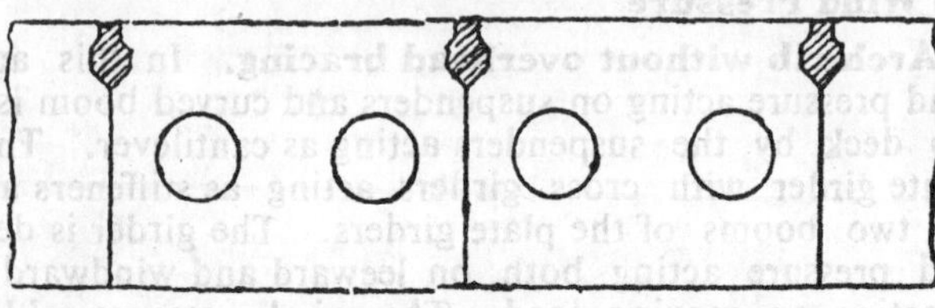

Fig. 24·65 (*a*) Multi-beam with shear key.

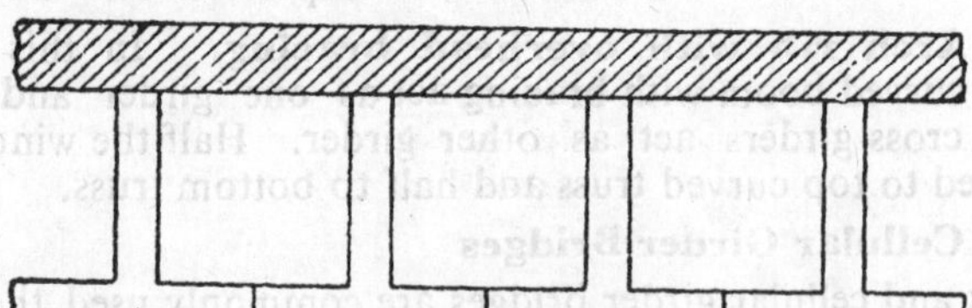

Fig. 24·65 (*b*) Composite multi-span precast construction.

PROBLEMS

1. Design a solid bridge for effective span of 6 m. The clear roadway is 6·8 m. The bridge is to be designed for class A and class AA loading.

2. Design a T-beam slab bridge for effective span of 18 m. The bridge is to be provided with two traffic lanes with foot-paths. The bridge is to be designed for class A and class AA loading.

3. Design a foot bridge having clear width of 3 m. The effective span of bridge is 15 m.

4. Design a balanced cantilever bridge to carry 2 lanes of traffic with foot paths for span of 40 m. The bridge is to be designed for class A and class AA loading.

25

Prestressed Concrete

25·1. Basic Principles. Pretressing is artifically induced compressive stress in a structure before it is loaded so that any tensile stress which might be caused by the external dead and live loads are automatically cancelled, and failure is eliminated.

Following example will clarify the preceding definition of prestressing. Consider a beam consisting of rectangular blocks of concrete 15 cm×40 cm×1·0 metre, ten in number touching end to end, in proper alignment. In order to avoid the effect of gravity this beam is placed on smooth horizontal surface, *see* Fig. 25·1.

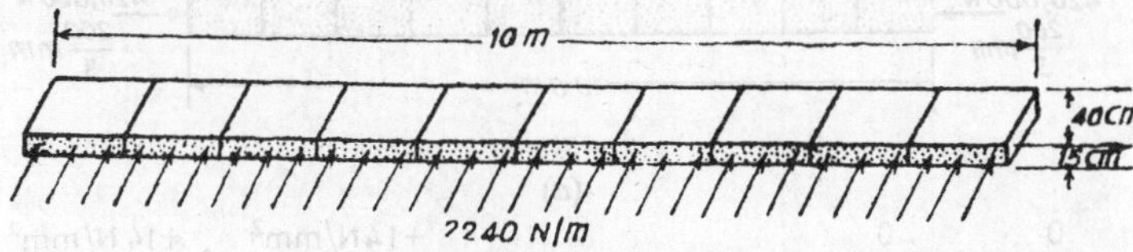

Fig. 25·1

Let it be assumed that the beam has a span of 10 metres and it is desired that it should resist a horizontal load of 2240 N/metre. This loading will cause maximum bending moment of 28,000 N-m and bending stresses of ±7 N/mm². The concrete used for the beam is M 35 and it can safely resist a compressive stress of 14 N/mm². It is obvious, threfore, that the beam which is quite strong in resisting compressive stress is not at all strong to take the tensile stress produced by the applied loading. The beam has to be made strong enough to take up the tensile stresses.

The beam is now subjected to uniform compression of 700 N/mm² or an axial force of 7×150×400=420,000 N is applied on the beam. This stresses resulting from loading, and the axial force are summar sed in Fig. 25·2.

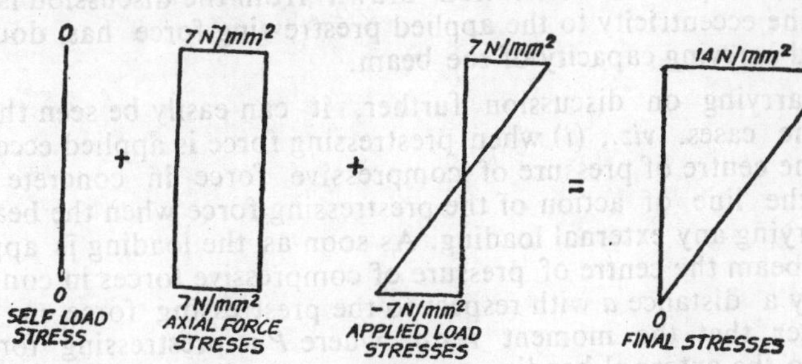

Fig. 25·2

Thus, it is seen that a uniform compression of 420,000 N has enabled the beam, having no tensile resistance, to carry a bending moment of 28,000 N-m safely. This is the basic principle of prestressing. The axial force applied to beam prior to loading is known as prestress force.

From the preceding discussion, it is observed that when the loading is not applied the beam is subjected to a stress of 7 N/mm² whereas it can take a compressive stress of 14 N/mm². In other words, the compressive strength of beam is not fully utilized in unloaded state.

If the prestressing force is now applied at an eccentricity 400/6 mm as shown in Fig. 25·3 (*a*) and the beam is subjected to a load of 4480 N/m the resulting stress at various stages obtained are summarised in Fig. 25·3 (*b*).

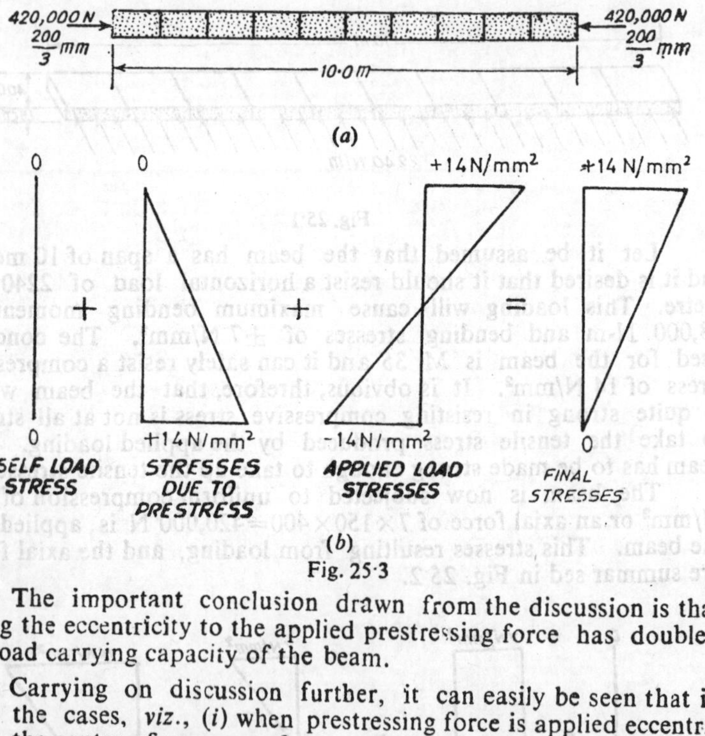

Fig. 25·3

The important conclusion drawn from the discussion is that giving the eccentricity to the applied prestressing force has doubled the load carrying capacity of the beam.

Carrying on discussion further, it can easily be seen that in both the cases, *viz*., (*i*) when prestressing force is applied eccentrically, the centre of pressure of compressive force in concrete was along the line of action of the prestressing force when the beam is not carrying any external loading. As soon as the loading is applied on the beam the centre of pressure of compressive forces in concrete shifts by a distance *a* with respect to the prestressing force in such a manner that the moment $P \times a$, where *P* is prestressing force is equal to the external bending moment.

Case I. Axial prestressing force.

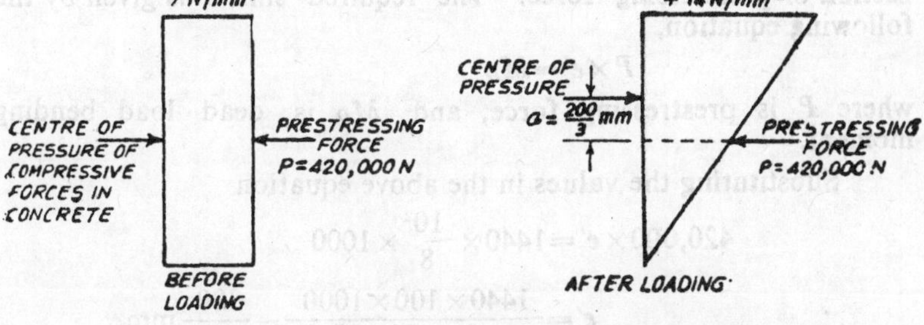

$$M = 420{,}000 \times \frac{200}{3} = 28{,}000{,}000 \text{ N mm}$$
$$= \text{Applied B.M}$$

Fig. 25·4 (*a*)

Case II. Eccentric prestressing force.

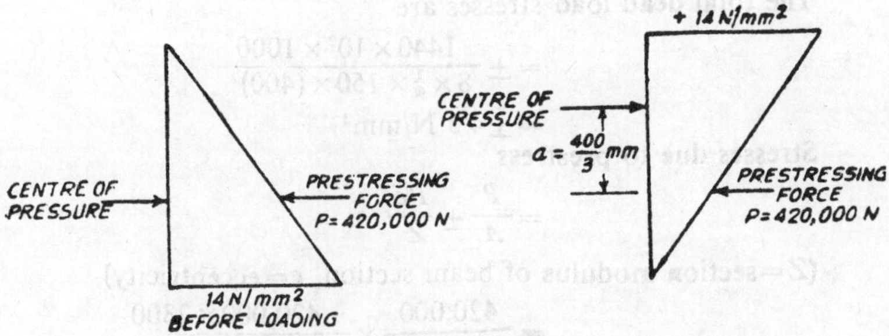

$$M = 420{,}000 \times \frac{40}{3} = 5{,}600{,}000 \text{ N cm.} = \text{Applied B.M.}$$

Fig. 25·4 (*b*)

So far discussion is limited in the sense that the beam is not subjected to gravity or in other words it is not subjected to self-load. Let us examine now the case when in addition to the applied loading the beam is also subjected to self-load as shown in Fig. 25·5. Assuming the density of concrete as 24,000 N/m³, the self-load of beam per metre is $2{\cdot}4 \times 15 \times 40 = 1440$ N. The total load carrying capacity of the beam is 4480 N/m. If we take into account the dead load of beam the useful load carrying capacity of beam is reduced

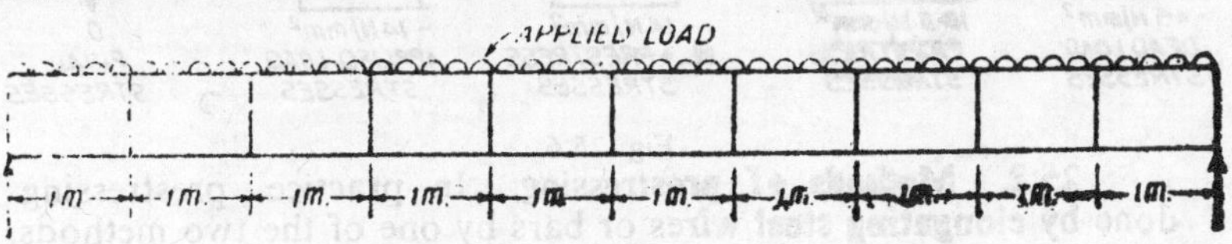

Fig. 25·5.

to 3040 N/m. It will, however, be shown that by slight shift in the prestressing force, the dead load can be carried without increasing section or prestressing force. The required shift e' is given by the following equation,

$$P \times e' = M_D$$

where P is prestressing force, and M_D is dead load bending moment.

Substituting the values in the above equation

$$420{,}000 \times e' = 1440 \times \frac{10^2}{8} \times 1000$$

$$\therefore \quad e' = \frac{1440 \times 100 \times 1000}{8 \times 420{,}000} = \frac{300}{7} \text{ mm}$$

Total eccentricity of prestressing force

$$= \frac{200}{3} + \frac{300}{7}$$

$$= \frac{2300}{21} \text{ mm}$$

The total dead load stresses are

$$= \pm \frac{1440 \times 10^2 \times 1000}{8 \times \frac{1}{6} \times 150 \times (400)^2}$$

$$= \pm 4{\cdot}5 \text{ N/mm}^2$$

Stresses due to prestress

$$= \frac{P}{A} \pm \frac{Pe}{Z}$$

(Z=section modulus of beam section, e=eccentricity)

$$= \frac{420{,}000}{150 \times 400} \times \frac{420{,}000 \times 2300}{21 \times \frac{1}{6} \times 150 \times (400)^2}$$

$$= 7 \pm 11{\cdot}50 = +18{\cdot}50 \text{ and } -4{\cdot}50 \text{ N/mm}^2$$

The stresses at various stages are summarised in Fig 25·6.

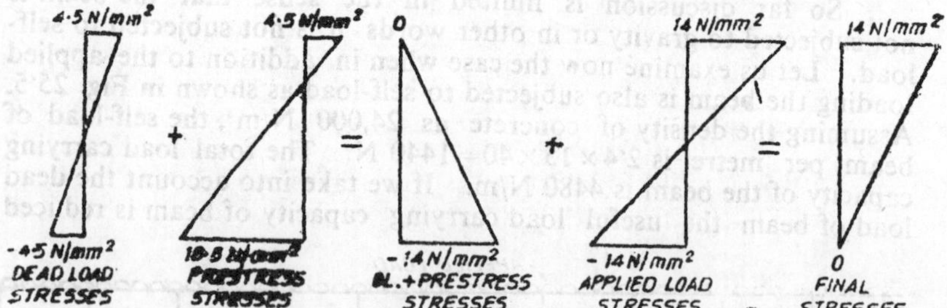

Fig. 25·6

25·2. Methods of prestressing. In practice, prestressing is done by elongating steel wires or bars by one of the two methods.

1. The wires or wire strands are stressed before concrete is placed. After the concrete has attained desired strength, the stress is transferred to the member. The prestress is transferred by bond. The member is said to be pretensioned with the wires or wire strands. The eccentricity of prestress is constant if straight wires are used The methods have been devised to deflect the wires to give variable eccentricity.

2. The second method of applying prestress is by tensioning steel wires, strands, cables or rods after the concrete has attained its strength. The member is then said to be post-tensioned. The stress unit may be straight or curved.

25·3. Pretensioning and Post-tensioning Comparison. In case of pretensioning very heavy moulds or abutments are required to serve as temporary anchorages. It is, therefore, best suited for the mass production of precast members. A single pair of end anchorages can be used for a line of members several metres long.

From the consideration of transporting and handling, the size of precast members is limited. Large sizes are cast-in-situ or near the site and are generally post-tensioned. The other advantage gained by post-tensioning is that 'stage prestressing' can be adopted which is suited for bridge construction. Many systems of prestressing have been devised and some of them are described in brief here so as to give the sizes of cable and end anchorages. The details can be obtained from the literature issued by the makers.

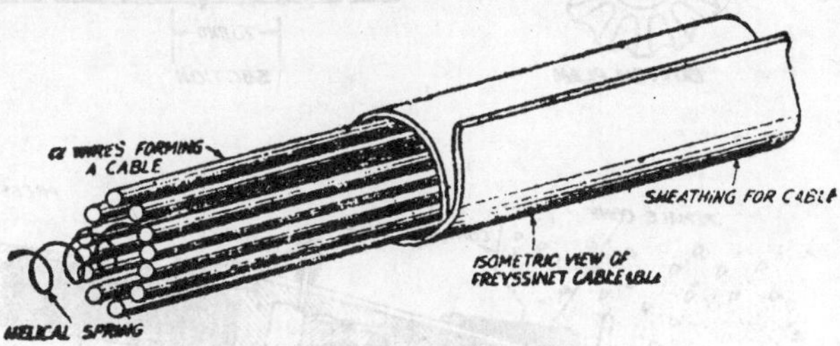

Fig. 25·7 (*a*)

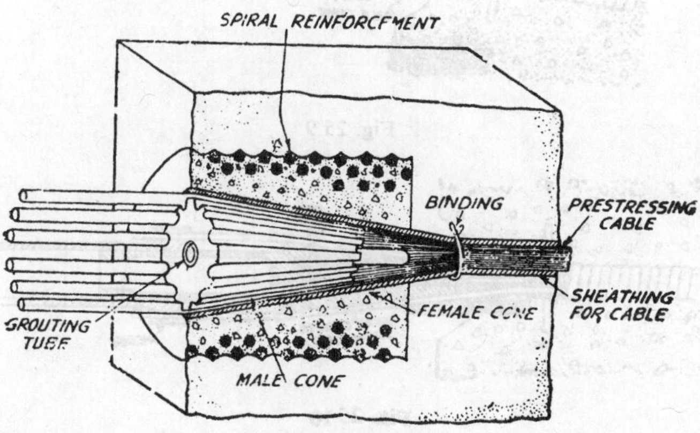

Fig. 25·7 (*b*)

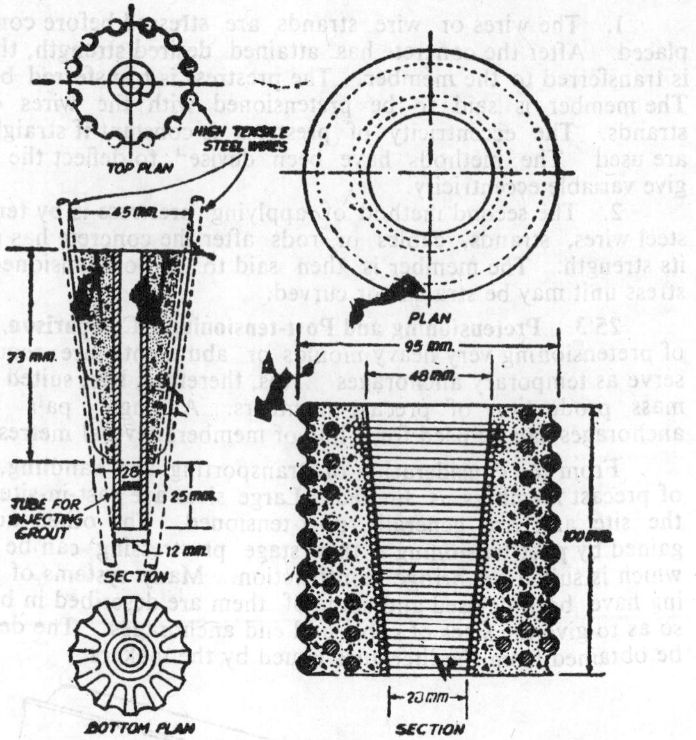

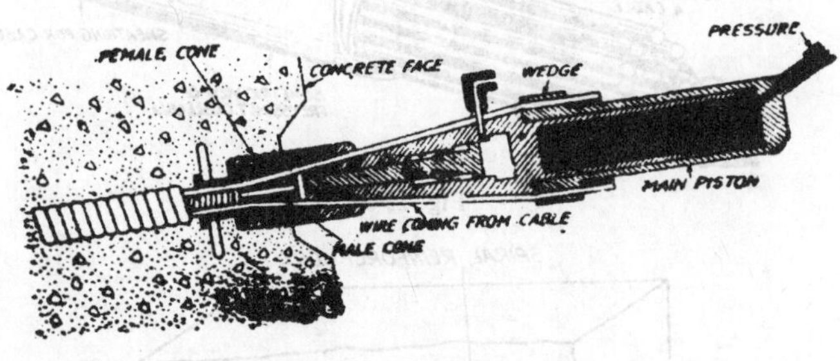

Fig. 25·9

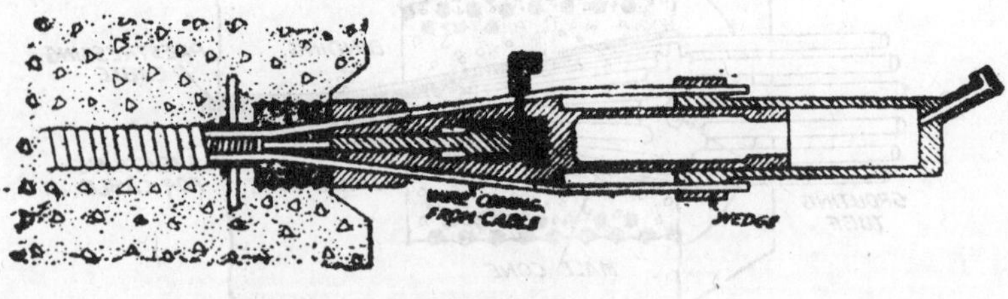

Fig. 25·10

25·4. Freyssinet Method. This method has been developed by a French Engineer Freyssinet. High tensile steel wires of 5 mm dia (0·196″) in units of 8, 10, 12 and 18 and 7 mm dia (0·276″) arranged in 12 wire units are used in this system. The wire units, also known as cables are encased in flexible tube or sheathing. They are provided with an anchorage consisting of reinforced female cone and male cone as shown in Figs. 25·7 and 25·8. The cables are placed in the form in the similar manner as reinforcement. The members are then concreted and after the concrete has attained desired strength the wires are tensioned by means of Freyssinet jack. All the wires in a cable are tensioned simultaneously. When the desired tension is applied the pressure is released and the wires are locked in position. The grout is then pumped through the annular hole in the male cone. Stepwise the procedure is explained in Figs. 25·9 to 25·12.

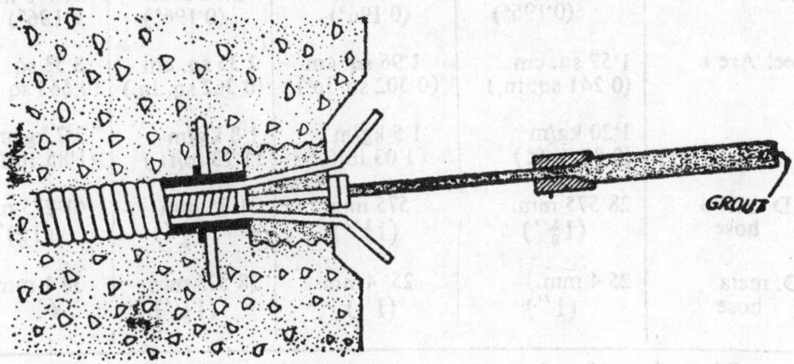

Fig. 25·11

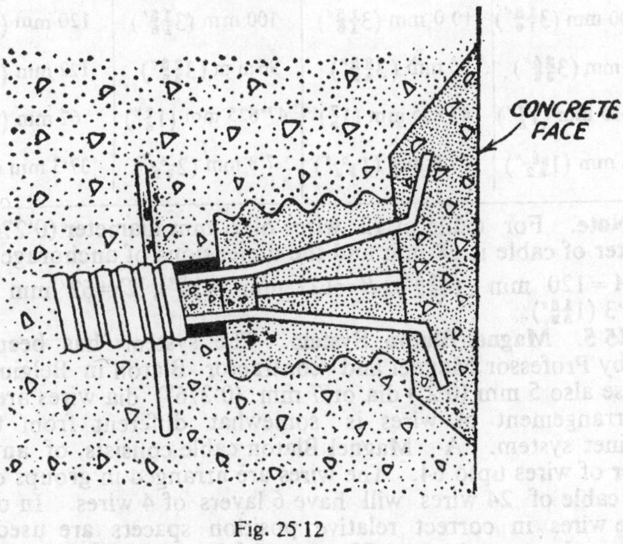

Fig. 25·12

Standard sizes of anchorages and the cable diameters in Freyssinet system are given in Table 25·1.

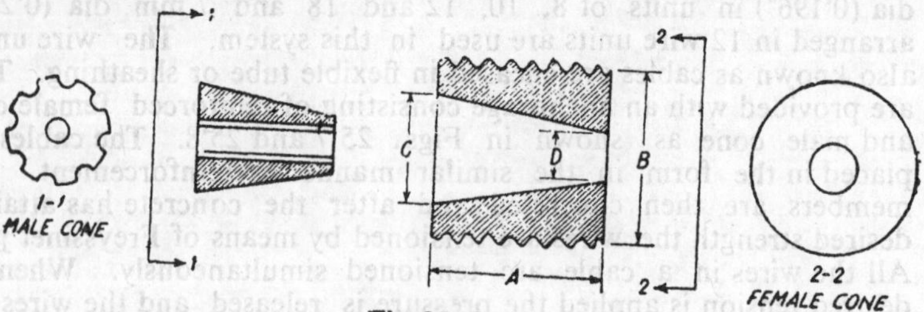

Fig. 25·13

Table 25·1

Type of Cab e	8×5 mm (0·196″)	10×5 mm (0·196″)	12×5 mm (0·196″)	18×5 mm 0·196″)
Steel Are a	1·57 sq. cm (0·241 sq. in,)	1·96 sq. cm (0·302 sq. in)	2·35 sq. cm (0·362 sq. in.)	3·53 sq. cm (0·543 sq. in.)
	1·20 kg/m (0 82 lb/ft.)	1·5 kg/m (1·03 lb/ft	1·8 kg/m (1·23 lb/ft.)	2·7 kg/m (1·85 lb/ft.)
O.D. metal hose	28·575 mm. ($1\frac{1}{8}$″)	·575 mm ($1\frac{1}{8}$″)	31·75 mm ($1\frac{1}{4}$″)	41·275 mm ($1\frac{5}{8}$)″
I.D. meta hose	25·4 mm. (1″)	25· 4 mm (1″)	28 575 mm ($1\frac{1}{8}$″)	38·1 mm (

Dimensions of anchorage

A	100 mm ($3\frac{15}{16}$″)	10 0 mm ($3\frac{15}{16}$″)	100 mm ($3\frac{15}{16}$″)	120 mm ($4\frac{23}{32}$″)
B	96 mm ($3\frac{25}{32}$″)	96 mm ($3\frac{25}{32}$″)	96 m m ($3\frac{25}{32}$″)	122 mm ($4\frac{13}{16}$″)
C	47·625 mm ($1\frac{7}{8}$″)	47·625 mm ($1\frac{7}{8}$)″	47·625 mm ($1\frac{7}{8}$″)	62 mm ($2\frac{7}{16}$″)
D	27·8 mm ($1\frac{3}{32}$″)	27·8 mm ($1\frac{3}{32}$″)	2 7·8 mm ($2\frac{3}{32}$″)	37·3 mm ($1\frac{15}{32}$″)

Note. For cable of 12 wires of 7 mm diameter (0·276″), the diameter of cable is 38 mm and the dimensions of anchorage are :

A=120 mm ($4\frac{23}{32}$″), B=122 mm ($4\frac{13}{16}$″), C=62 mm ($2\frac{7}{16}$″), D=37·3 ($1\frac{15}{32}$″).

25·5. Magnel-Blaton System. The system has been developed by Professor Magnel and contractor Blaton in Belgium. In this case also 5 mm (0·2″) dia or 7 mm (0·276″) dia wires are used. The arrangement of wires is somewhat different from that of Freyssinet system. A Magnel Blaton cable consists of any even number of wires upto 64. The wires are arranged in groups of four, thus a cable of 24 wires will have 6 layers of 4 wires. In order to put the wires in correct relative position spacers are used. The wires are kept at least 4·75 mm ($3\frac{3}{16}$″) apart. The spacers for straight cables are provided at a distance of 1·5 to 2·5 metres and for curved cables the distance is kept upto about 1·5 metres. In order to separate the wires into layers of four, horizontal spacers are provided. They consist of a grid of flat metal strips as shown in Fig. 25·14. The figure also shows the arrangement of wires in cable.

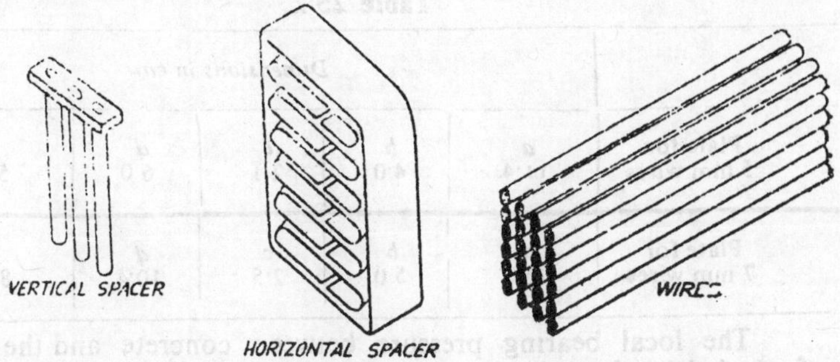

Fig. 25·14

The anchorage in Magnel-Blaton system consists of steel plates known as sandwich plates having two trapezoidal grooves in upper surface and two in the lower surface. A steel wedge properly

25 mm for 5 mm wires

35 mm for 7 mm wires

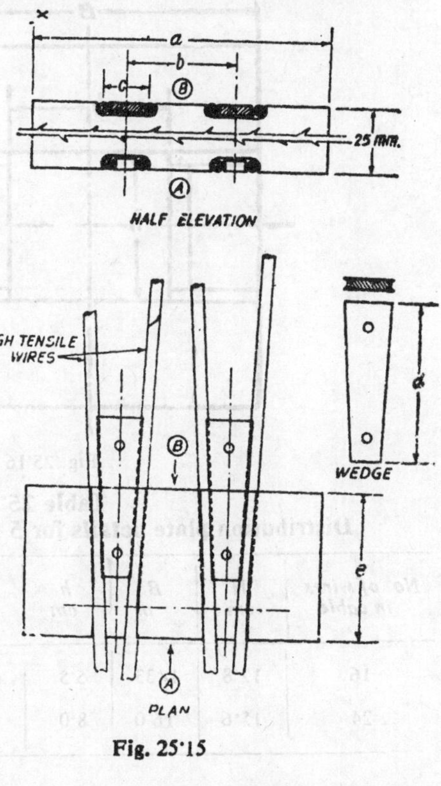

Fig. 25·15

machined fits into each slot and anchors two wires in each groove. Each sandwich plate anchors eight wires. The sandwich plate and the wedge are shown in Fig. 25·15. Their standard sizes are given in Table 25·2.

Table 25·2

	Dimensions in cm				
Plate for 5 mm wires	*a* 11·4	*b* 4·0	*c* 2·1	*d* 6·0	*e* 5·4
Plate for 7 mm wires	*a* 16·0	*b* 5·0	*c* 2·5	*d* 10·4	*e* 8·9

The local bearing pressure between concrete and the edges of sandwich plate is distributed by means of a bearing plate of size dependent on the number of wires in a cable. The distribution plates are of standard size and dimensions are given in Tables 25·3 and 25·4

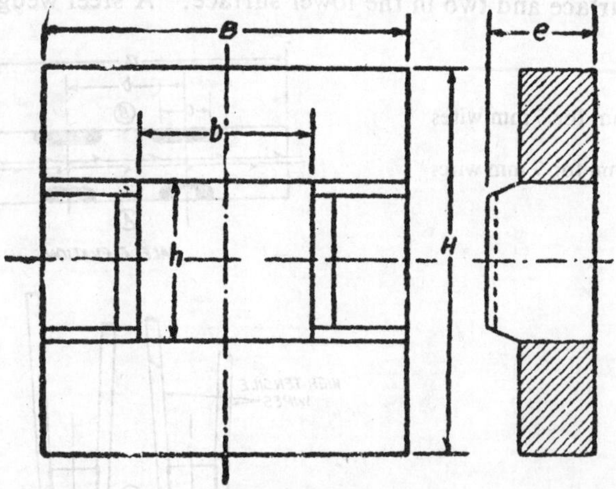

Fig. 25·16 (*a*)

Table 25·3 (A)
Distribution plate details for 5 mm diameter wires

No. of wires in cable	*H cm*	*B m*	*h cm*	*b cm*	*e cm*	*Weight in N (approx.)*
16	12·8	1·33	5·5	6·0	3·8	42
24	15·6	16·0	8·0	6·0	4·0	45

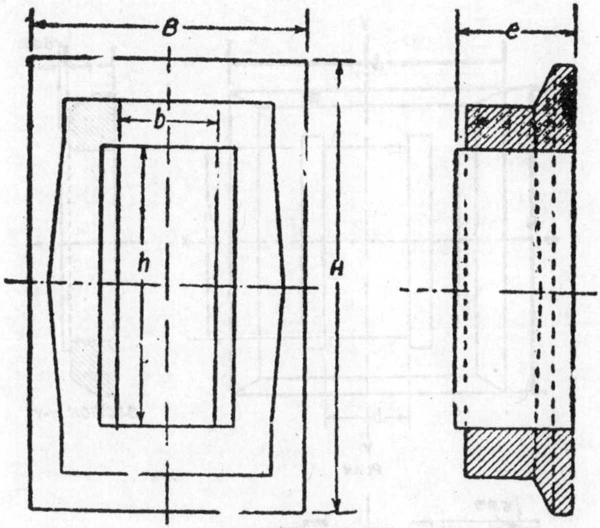

Fig 25·16 (*b*)

Table 25·3 (B)

Distribution plate details for 5 mm dia wires

No. of wires in cables	H cm	B c m	h cm	b c m	e cm	Weight in N approx.
32	19·6	17·0	6	6·0	4·8	68
40	23·0	18·0	13·3	6·0	5·8	91
48	26·2	19·0	15·	6·0	6·4	113
56	29·0	20·0	18·5	·0	7·0	136
64	33·9	21·0	21·1	6·0	7·1	9

For stressing the wires, hydraulically operated stressing jack, grips two wires at a time. When tensioning is completed the wedge for two wires is then driven home. When stressing for one cable is completed, it is invariably grouted. After grouting is completed the anchorage is covered with concrete for protection.

25·6. Assumptions in prestressed concrete design. The following assumptions are made in design of prestressed concrete sections under working load—

(*i*) Plane cross-sections before loading remain plane after loading.

(*ii*) Any change in loading produces change of stress in concrete. The prestressing in the steel is assumed to remain constant.

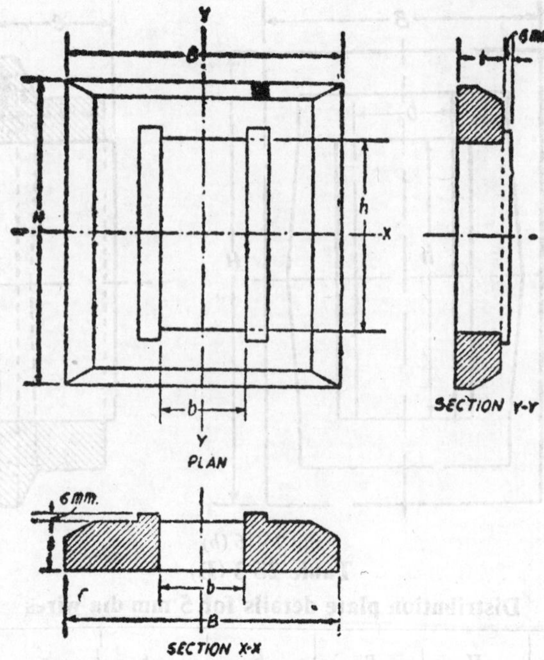

Fig. 25·17

Table 25·4

Distribution plate details for 7 mm dia wires

No. of wires in cables	*cm*	*B cm*	*h cm*	*b cm*	*e cm*	*Approx. wt. in N*
16	14·6	20 3	7·6	7·9	3·2	54
24	18·4	26·0	11·4	7·9	3·8	113
32	24·1	26 0	14·9	7·9	4·4	159
40	29·2	27·1	19·1	7·9	5·1	227
8	35·6	27 9	22·	7·9	6·0	318
56	38·1	29· 2	26· 7	7·9	6 4	363
64	40·6	29· 8	30 2	7·9	7·0	431

25·7. Advantages of prestressed concrete. As already discussed concrete is strong in compression and very weak in tension and hence in RCC structure, steel is used to take up tensile stresses. Though steel is used to take tensile stresses, the concrete below neutral axis will still crack as it cannot sustain the strain as caused in steel. In the case of prestressed concrete structure there are no tensile stresses induced and thus cracking of concrete is avoided. In case of prestressed concrete high strength concrete and high tensile steel are used which prove to be economical. In case of reinforced concrete high strength concrete cannot be used as that will require greater quantity of steel which will be uneconomical. At the same time high strength steel cannot be used as the strain on the tension side would increase resulting in wider cracks in concrete. All these defects are remedied in prestressed concrete. By positioning the prestressing force properly, initial stresses can be induced in such a way that when the member is loaded, the stresses caused in the member will be wholly compressive or slightly tensile which are permissible in concrete. The most effective use of prestressed concrete is obtained when the concrete and the prestressing steel employed are of high quality and strength.

In prestressed concrete structures section of members is considerably reduced thus resulting in reduced dead load of structure. This not only reduces cost of members but also cost of supporting structure and foundations. By properly arranging prestressing force self-load can be carried by member without causing stresses in its section.

25·8. Losses in Prestress The prestressing force which is applied to a member does not remain constant but there is gradual reduction in prestressing force. The amount of force reduced is known as loss of prestress and it generally varies from 15 to 20%.

While assessing the stresses in concrete and steel during tensioning operations and later in service, due regard shall be paid to all losses and variation in stress resulting from creep of concrete and steel the shrinkage of concrete, the shortening of concrete at transfer, friction and slip of anchorage. Unless otherwise determined by actual tests, allowance shall be made in accordance with the values given below

1. Loss of Prestress due to creep of concrete. The loss of prestress due to creep of concrete under load shall be determined for all permanently applied loads including the prestress. The creep due to live loads of intermittent nature and applied for short duration only may be neglected. The loss of prestress is obtained by multiplying the modulus of elasticity of steel E_s, by the creep strain of concrete around the steel. For the purpose of calculating this loss, the concrete strain shall be taken at the centre of gravity of prestressing steel.

The ratio of final creep strain to elastic strain varies with conditions of humidity of the surrounding atmosphere, modulus of elasticity of concrete and its strength at the time of loading. The values of final creep coefficient that is, estimate creep strain/elastic strain at the age of loading can be obtained from Table 25·5

Table 25·5
Creep Coefficient c_e

Conditions of exposures	*Final creep coefficient c_e*
In water	$0{\cdot}5\ k_c$ to $1{\cdot}0\ k_c$
In very humid atmosphere	$1{\cdot}5\ k_c$ to $2{\cdot}0\ k_c$
In average humidity	$2{\cdot}5\ k_c$ to $3{\cdot}0\ k_c$
In dry atmosphere e.g In dry intern : rooms	$2{\cdot}5\ k_c$ to $4{\cdot}0\ k_c$

Where, k_c has the following values depending upon the age of concrete at the time when creep causing stress is applied :

Ratio of F_r/F_c at the time of stressing	*Value of k_c*
0·50	2·20
0·84	1·50
1·00	1·00
1·08	0·75
1·30	0·50

F_r = cube strength of concrete at transfer of prestress on 15 cm cubes.

F_c = cube strength of concrete at 28 days where rapid Portland cement is used and at 7 days where rapid hardening Portland cement is used, on 15 cm cubes.

The values of creep thus computed may be taken as the final values for all practical purposes.

$$\%\ \text{loss of prestress} = \frac{(c_c - 1) \times f \times m}{f_s} \times 100$$

where c_s is creep coefficient
f is stress in concrete at centre of gravity of steel
f_s is stress in steel, and m is modular ratio.

In the absence of experimental data, the ultimate creep strain may be estimated from the following values of creep coefficient.

Table 25·6. Creep Coefficient

Age at Loading	*Creep coefficient*
7 days	2·2
28 days	1·6
1 year	1·1

For the calculation of deformation at some stage before the total creep is reached, it may be assumed that about half the total creep takes place in the first month after loading and that about three-quarters of the total creep takes place in the first six months after loading.

2. **Loss of prestress due to relaxation of steel**. The relaxation losses in steel depend on the type of steel, initial prestress, age and temperature. In the absence of the experimental data. the values given in Table 25·7 may be used.

Table 25·7. Relaxation Losses For Prestressing Steel at 1000 Hours at 27°C

Initial Stress	*Relaxation Loss*
	N/mm²
$0{\cdot}5\,f_p$	0
$0{\cdot}6\,f_p$	35
$0{\cdot}7\,f_p$	70
$0{\cdot}8\,f_p$	90

where f_p is the characteristic strength of prestressing steel.

For tendons at higher temperatures or subjected to large lateral loads, greater relaxation losses may be specified. No reduction in the value of relaxation losses should be made for tendon with a load equal to or greater than relevant jacking force that has been applied for a short time prior to the anchoring of tendon.

3. **Loss of prestress due to shrinkage of concrete.** The loss of prestress due to shrikage of concrete shall be the product of the modulus of elasticity of steel E_s, and the shrinkage of concrete calculated by the following formula:

For pre-tensioning :

Total shrinkage $= 3{\cdot}0 \times 10^{-4}$

For post tensioning :

$$\text{Residual shrinkage} = \frac{2{\cdot}0 \times 10^{-4}}{\log_{10}(T+2)}$$

where T is the age of concrete at transfer in days.

Note : The value of the shrinkage strain for design of post-tensioned concrete may be increased by 50 per cent in dry atmospheric conditions, subject to a maximum value of 0·0003.

For the calculation of deformation of concrete at some stage before the maximum shrinkage is reached, it may be assumed that half of the shrinkage takes place during the first month and that about three-quarters of the shrinkage takes place in first six months after commencement of drying.

4. **Loss of prestress due to shortening of concrete at transfer**. This type of loss occurs when the prestressing tendons

upon release from tensioning devices cause the concrete to be compressed. This loss is proportional to the modular ratio and initial prestress in the concrete and shall be calculated as below, assuming that the tendons are located at their centroid :

(*a*) For pretensioning the loss of prestress in the tendons at transfer shall be calculated on a modular ratio basis using the stress in the adjacent concrete.

(*b*) For members with post-tensioned tendons which are not stressed simultaneously, there is a progressive loss of prestress during transfer due to the gradual application of the prestressing forces. This loss of prestress should be calculated on the basis of half the product of the stress in the concrete adjacent to the tendons averaged along their lengths and the modular ratio.

5. **Loss of prestress due to friction**. The design shall take into consideration all losses in prestress that may occur during tensioning due to friction between the prestressing tendons and the surrounding concrete or any fixture attached to the steel or concrete.

For straight or moderately curved structures with curved or straight cables, the value of prestressing force P_x at distance x from tensioning end shall be calculated by the formula :

$$P_r = P_0 \times e^{-(\mu\alpha + k_f . x)}$$

where

P_0 = prestressing force in the prestressing steel at the tensioning end,

μ = coefficient of friction in curve,

α = the cumulative angle in radians through which the tangent to the cable profile has turned between any points under consideration, and

k_f = friction coefficient for wave effect normally to be taken as 0·15 per 100 m. However where heavy vibration, thin walled ducts and other adverse conditions are involved, this figure may be increased upto 1·5 per 100 m and where the clearance between the duct and cable is sufficiently large to eliminate the wave effect, and the construction of the sheath is of heavy gauge steel tube with water-tight joints and such other arrangements which prevent deformation of the duct profile during vibration of concrete, the coefficient may be reduced to zero.

The values of both P_x and P_0 are those, acting in the direction of the tangent to the curve of the cable.

Unless otherwise proved by tests, the values for the coefficient of friction μ shall be assumed as :

0·55 for steel moving on smooth concrete,

0·35 for steel moving on steel fixed to duct,
0·25 for steel moving on steel fixed to concrete,
0·25 for steel moving on lead.

0·18 to 0·30 for multilayer wire rope cables in rigid rectangular steel sheath and

0·15 to 0·25 for multilayer wire rope cables with spacer plater providing lateral separation.

In circular or similar constructions, where circumferential steel is tensioned by jacks, losses due to friction may be calculated by reducing the above values of μ by 20 per cent.

If the initial tension applied to a prestressed tendon is partially released, the action of friction in the reverse direction caused an alteration in the distribution of stress along the length of the tendon. In such cases, the effect of the reverse friction shall be taken into consideration.

6. Loss of prestress due to slip in anchorage. Any loss of prestress which may occur due to slip of wire during anchoring due to the strain of anchorage shall be allowed for in the design. Loss due to slip in anchorage is of special importance with short members and the necessary additional elongation should be provided for at the time of tensioning to compensate for this loss.

25·9. Initial and Final Stress Conditions. The prestressed beams may not be of separate blocks and, therefore, will be able to take some tensile stress. The following conditions should be satisfied :

(*i*) The stress at the bottom initially should not exceed initial compressive stress C'. If P_i is the initial prestress and e is the eccentricity, A is the area of cross-section and Z_b is the modulus of section equal to I/y_b, where I is the moment of inertia and y_b is the bottom fibre distance from C.G. of the section and M_D is the dead load bending moment.

$$\frac{P_i}{A}+\frac{P_ie}{Z_b}-\frac{M_D}{Z_b}\leqslant C' \quad ...(i)$$

(*ii*) The stress at top initially should not be less than the minimum allowable stress C_t'. C_t' may be compressive or tensile. When tensile its value will be taken negative. If Z_t is the modulus of section equal to I/y_t. where y_t is the top fibre distance from C.G. of the section.

$$\frac{P_i}{A}-\frac{P_ie}{Z_t}+\frac{M_D}{Z_t}\geqslant C_t' \quad ...(ii)$$

(*iii*) Finally after losses have taken place and the remainder prestress is equal to nP_i, where n is the loss factor which is between 0·8 to 0·85, the stress at bottom should not be less than the minimum allowable stress.

$$\frac{nP_i}{A}+\frac{nP_ie}{Z_b}-\frac{M_D}{Z_b}-\frac{M_A}{Z_b}\geq C_t \qquad \text{...(iii)}$$

where M_A is the applied bending moment, C_t is the minimum allowable stress.

(*iv*) The stress at top finally should not be greater than final allowable stress in concrete say C.

$$\frac{nP_i}{A}-\frac{nP_ie}{Z_t}+\frac{M_D}{Z_t}+\frac{M_A}{Z_t}<C \qquad \text{...(iv)}$$

To find possible values of P_i and e, graph may be drawn between $1/P_i$ and e satisfying the above four conditions.

I. From equation (i)

$$e\leq\left(C'+\frac{M_D}{Z_b}-\frac{P_i}{A}\right)\times\frac{Z_b}{P_i}$$

$$\leq\frac{C'Z_b}{P_i}+\frac{M_D}{P_i}-\frac{Z_b}{A}$$

$$\leq\frac{1}{P_i}\left(C'Z_b+M_D\right)-\frac{Z_b}{A}$$

Taking $$e=\frac{1}{P_i}\left(C'Z_b+M_D\right)-\frac{Z_b}{A}$$

i.e. $$\frac{e}{-\dfrac{Z_b}{A}}+\frac{1}{P_i}\times\frac{1}{\dfrac{Z_b}{A(C'Z_b+M_D)}}=1$$

the line representing this equation is shown in Fig. 25·18 and is marked b_i.

For any value of $1/P_i$, e should be less than or equal to value

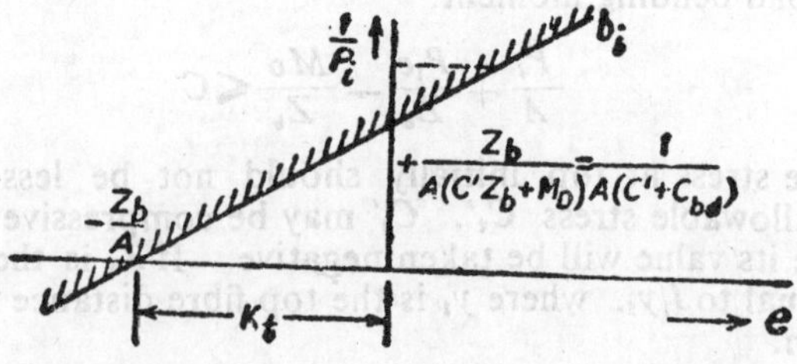

Fig. 25·18

given by this line. Therefore, the inequality is satisfied if the point lies above the line.

II. From equation (ii)

$$e < \left(C'_t - \frac{P_i}{A} - \frac{M_D}{Z_t}\right)\left(\frac{-Z_t}{P_i}\right)$$

$$< -\frac{C'_t Z_t}{P_i} + \frac{Z_t}{A} + \frac{M_D}{P_i}$$

$$< -\frac{P}{P_i}\left(P'_t Z_t + M_D\right) + \frac{Z_t}{A}$$

Taking $$e = +\frac{1}{P_i}\left(M_D - C'_t Z_t\right) + \frac{Z_t}{A}$$

i.e. $$\frac{e}{\frac{Z_t}{A}} + \frac{1}{P_i} \times \frac{1}{\frac{-Z_t}{A(M_D - C'_t Z_t)}} =$$

The line representing this equation is shown in Fig. 25·19 and is marked as t_i.

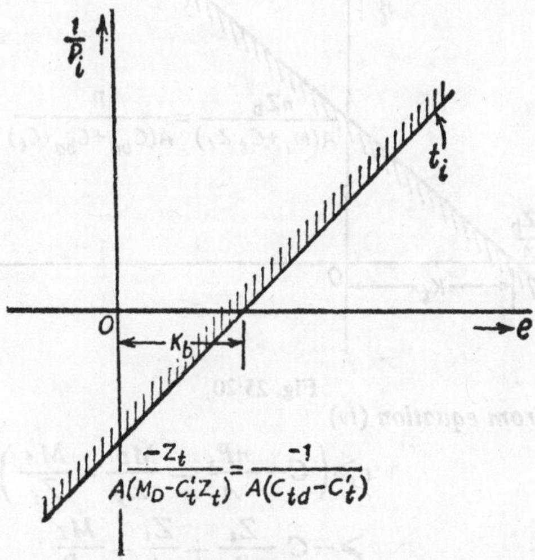

Fig. 25·19

For any value of $\frac{1}{P_i}$ the value of e should be less than the value given by this line. Therefore the inequality is satisfied if the point lies above the line.

III. From equation (iii)

$$e \geqslant \left(C_t - \frac{nP_i}{A} + \frac{M_D}{Z_b} + \frac{M_A}{Z_b}\right) + \frac{Z_b}{nP_i}$$

$$\geqslant \frac{C_t Z_b}{nP_i} - \frac{Z_b}{A} \times \frac{(M_D + M_A)}{nP_i}$$

Let $M_D + M_A = M_T$.

$$e > \frac{1}{P_i}\left(\frac{M_T + C_t Z_b}{n}\right) - \frac{Z_b}{A}.$$

Taking
$$e = \frac{1}{P_i}\left(\frac{M_T + C_t Z_b}{n}\right) - \frac{Z_b}{A}$$

$$\frac{e}{\frac{Z_b}{A}} + \frac{1}{P_i} \times \frac{1}{\frac{nZ_b}{A(M_T + C_t Z_b)}} = 1.$$

The line representing this equation is shown in Fig. 25·20 and is marked as *b*.

For any value of $\frac{1}{P_i}$ the value of *e* should be greater than the value given by this line. Therefore the inequality is satisfied if the point lies below the line.

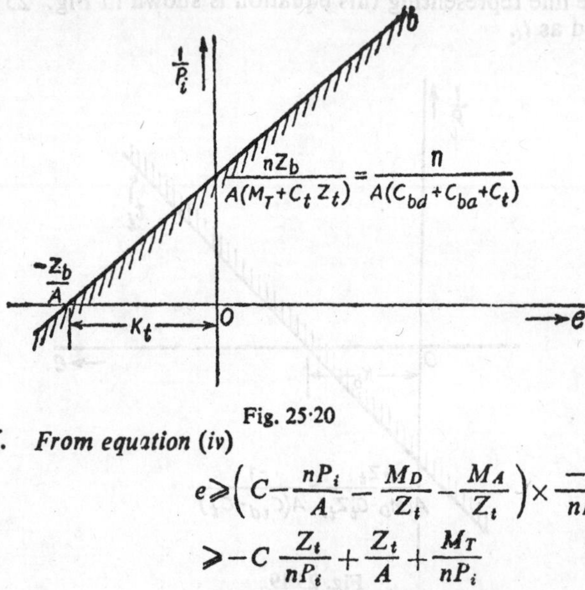

Fig. 25·20

IV. From equation (iv)

$$e \geqslant \left(C - \frac{nP_i}{A} - \frac{M_D}{Z_t} - \frac{M_A}{Z_t}\right) \times \frac{-Z_t}{nP_i}$$

$$\geqslant -C\frac{Z_t}{nP_i} + \frac{Z_t}{A} + \frac{M_T}{nP_i}$$

$$\geqslant \frac{1}{nP_i}\left(M_T - CZ_t\right) + \frac{Z_t}{A}$$

Taking
$$e = \frac{1}{P_i} \times \frac{M_T - C_t Z_t}{n} + \frac{Z_t}{A}$$

$$\frac{e}{\frac{Z_t}{A}} + \frac{1}{P_i} \times \frac{1}{\frac{-Z_t \times n}{A(M_T \times CZ_t)}} = 1.$$

The line representing this equation is shown in Fig. 25·21 and is marked as *t*.

For inequality to be satisfied, for any value of $\frac{1}{P_i}$, the value of e should be greater than the value given by this line. Therefore the point should lie below the line.

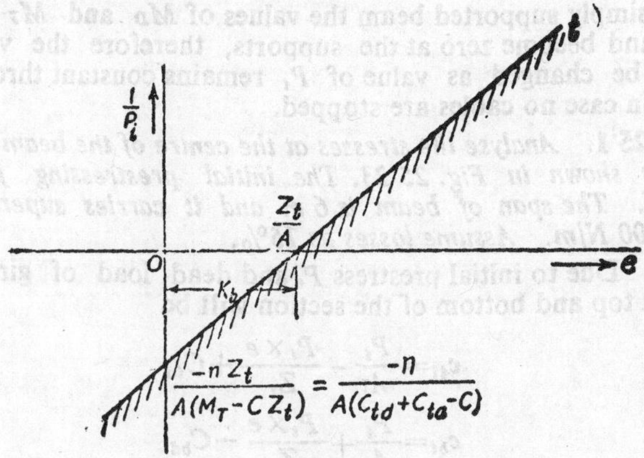

Fig. 25·21

Combining all these inequalities, it will be seen that when line b is above b_i and t is above t_i, the possible value of P_i and e will be given by enclosed area shown hatched.

From this combined diagram, (Fig. 25·22), it is evident, that maximum permissible eccentricity is the inter-section of lines b

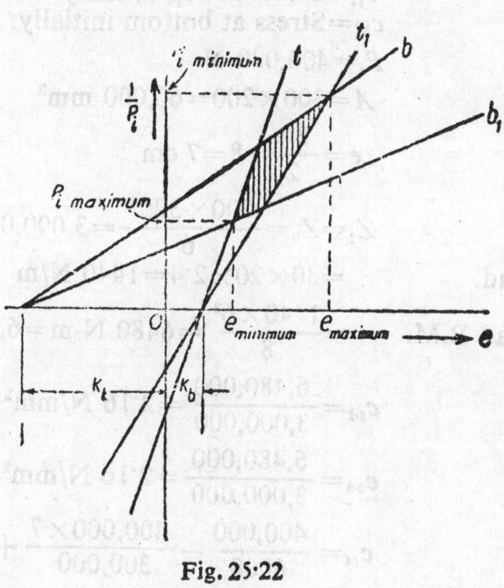

Fig. 25·22

and t_i and then the force P_i is the least. In case this eccentricity is not possible from practical considerations, for economical value of P_i, $\frac{1}{P_i}$ should be taken corresponding to value of e on b line if this is within the shaded zone.

For simply supported beam the values of M_D and M_T go on changing and become zero at the supports, therefore the value of e should be changed as value of P_i remains constant throughout the beam in case no cables are stopped.

Ex. 25·1. *Analyse the stresses at the centre of the beam having section as shown in Fig. 25·23. The initial prestressing force is 400,000 N. The span of beam is 6 m and it carries superimposed load of 8000 N/m. Assume losses as 15%.*

Sol. Due to initial prestress P_i and dead load of girder the stresses at top and bottom of the section will be

$$c_{ti}=\frac{P_i}{A}-\frac{P_i\times e}{Z_t}+C_{td}$$

$$c_{bi}=\frac{P_i}{A}+\frac{P_i\times e}{Z_b}-C_{bd}$$

where,
P_i=Initial prestress.
A=Area of the section.
e=Eccentricity of prestressing force.
Z_t=Section modulus for top fibres.
Z_b=Section modulus for bottom fibres.
c_{td}=Stress at top due to dead load.
c_{bd}=Stress at bottom due to dead load.
c_{ti}=Stress at top initially.
c_{bi}=Stress at bottom initially.

Now $P_i=400{,}000$ N

$$A=300\times 200=60{,}000 \text{ mm}^2$$

$$e=\frac{30}{2}-8=7 \text{ cm}$$

$$Z_t=Z_b=\frac{200\times 300^2}{6}=3{,}000{,}000 \text{ mm}^3$$

Dead load $=30\times 20\times 2{\cdot}4=1440$ N/m

Dead load B.M. $=\frac{1440\times 6^2}{8}=6480$ N-m $=6{,}480{,}000$ N-mm

$$c_{td}=\frac{6{,}480{,}000}{3{,}000{,}000}=2{\cdot}16 \text{ N/mm}^2$$

$$c_{bd}=\frac{6{,}480{,}000}{3{,}000{,}000}=2{\cdot}16 \text{ N/mm}^2$$

$$c_{ti}=\frac{400{,}000}{6000}-\frac{400{,}000\times 7}{300{,}000}+2{\cdot}16$$

$$=6{\cdot}67-9{\cdot}33+2{\cdot}16$$

$$=-0{\cdot}5 \text{ N/mm}^2. \text{ (Tension)}$$

$$c_{bi}=\frac{400{,}000}{60{,}000}+\frac{400{,}000\times 70}{3{,}000{,}000}-2{\cdot}16$$

$$=6{\cdot}67+9{\cdot}33-2{\cdot}16$$

$$=13{\cdot}84\ \text{N/mm}^2.$$

Due to final prestress $P=0{\cdot}85\ P_i$ and dead load and live load, the stresses at top and bottom of the section will be,

$$c_t=\frac{P}{A}-\frac{P\times e}{Z_t}+c_{td}+c_{tl}$$

$$c_b=\frac{P}{A}+\frac{P\times e}{Z_b}-c_{bd}-c_{bl}$$

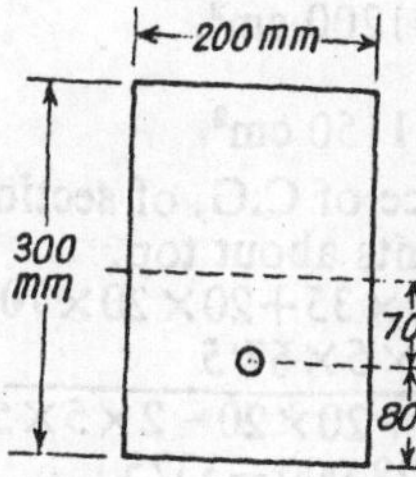

Fig. 25·23

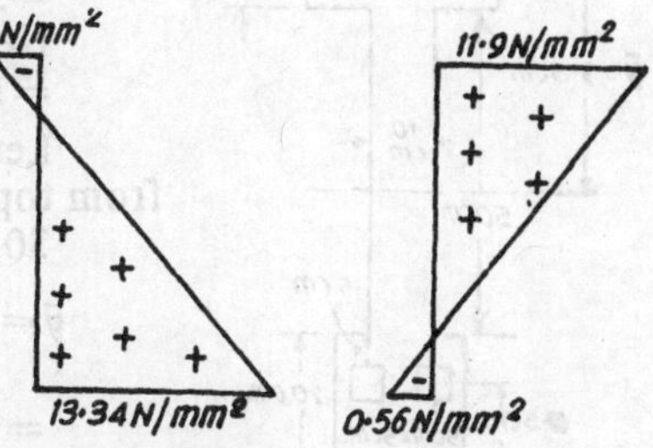

Stress due to initial Pressure+dead load

Final strees

Fig. 25·24

where c_t=Final stress at top

c_b=Final stress at bottom

c_{tl}=Stress at top due to live load

c_{bl}=Stress at bottom due to live load

$$P=0{\cdot}85\times 400{,}000=340{,}000\ \text{N}$$

Live load B.M. $=\dfrac{8000\times 6^2}{8}=36{,}000$ N-m

$=36{,}000{,}000$ N-mm

$$c_{tl}=\frac{36{,}000{,}000}{3{,}000{,}000}=12\ \text{N-mm}^2$$

$$c_{bl}=\frac{36{,}000{,}000}{3{,}000{,}000}=12\ \text{N/mm}^2$$

$$\therefore\quad c_t=\frac{340{,}000}{60{,}000}-\frac{340{,}000\times 70}{3{,}000{,}000}+2{\cdot}16+12{\cdot}0$$

$$=5{\cdot}67-7{\cdot}93+2{\cdot}16+12{\cdot}0=11{\cdot}90\ \text{N/mm}^2$$

$$c_b = \frac{340{,}000}{60{,}000} + \frac{340{,}000 \times 70}{3{,}000{,}000} - 2{\cdot}16 - 12{\cdot}0$$

$$= 5{\cdot}67 + 7{\cdot}93 - 2{\cdot}16 - 12{\cdot}0$$

$$= -0{\cdot}56 \text{ N/mm}^2.$$

Stress diagrams have been shown in Fig. 25·24.

Ex. 25·2. *Analyse the stresses at the centre of beam section shown in Fig. 25·25. The prestressing tendons consist of 2 cables, each having 16 wires of 5 mm diameter. Initial prestress in steel is 1000 N/mm². The span of the beam is 10 m and it has to carry superimposed load of 18,000 N/m. Assume losses as 15%.*

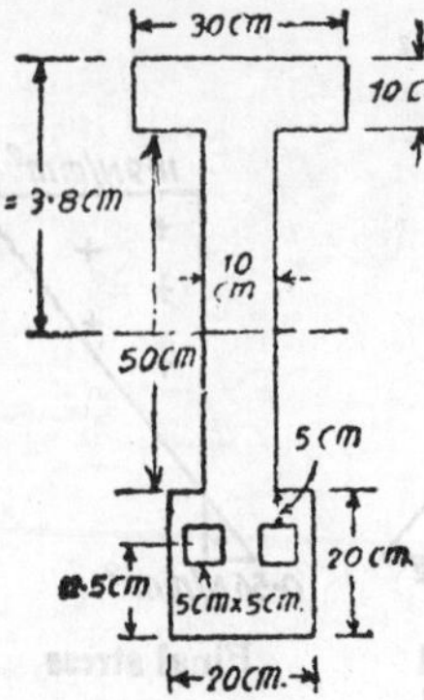

Fig. 25·25

Sol. Gross area of section

$$= 30 \times 10 + 50 \times 10 + 20 \times 20$$

$$= 300 + 500 + 400 = 1200 \text{ cm}^2.$$

Net area

$$= 1200 - 2 \times 5 \times 5 = 1150 \text{ cm}^2.$$

Let $\bar{y}$ be the distance of C.G. of section from top. Taking moments about top.

$$\bar{y} = \frac{30 \times 10 \times 5 + 50 \times 10 \times 35 + 20 \times 20 \times 70 - 2 \times 5 \times 5 \times 67{\cdot}5}{30 \times 10 + 50 \times 10 + 20 \times 20 - 2 \times 5 \times 5}$$

$$= \frac{1500 + 17{,}500 + 28{,}000 - 3375}{300 + 500 + 400 - 50}$$

$$= \frac{43{,}625}{1150} = 38{\cdot}0 \text{ cm}.$$

Moment of inertia about centre of gravity of section

$$= \frac{30 \times 10^3}{12} + 300 \times 33^2 + \frac{10 \times 50^3}{12} + 500 \times 3^2 + \frac{20 \times 20^3}{12} + 400 \times 32^2 - \frac{2 \times 5 \times 5^3}{12} - 50 \times (29{\cdot}5)^2$$

$$= 2500 + 326{,}700 + 104{,}166 + 4500 + 13{,}333 + 409{,}600 - 104 - 43{,}512$$

$$= 870{,}799 \text{ cm}^4$$

$$Z_t = \frac{870{,}799}{38} = 22{,}920 \text{ cm}^3.$$

$$Z_b = \frac{870{,}799}{42} = 20{,}730 \text{ cm}^3.$$

Dead load of girder $= 1200 \times 2{\cdot}4 = 2880$ N/m

$$\text{Dead load B.M.} = \frac{2880 \times 10^2}{8} = 36{,}000 \text{ N-m}$$

$$= 36{,}000{,}000 \text{ N-mm}$$

$$c_{td} = \frac{36{,}000{,}000}{22{,}920{,}000} = 1{\cdot}57 \text{ N/mm}^2$$

$$c_{td}=\frac{36{,}000{,}000}{20{,}730{,}000}=1{\cdot}74 \text{ N/mm}^2$$

Initial prestressing force P_i

$$=\frac{\pi}{4}\times 5^2\times 1000\times 2\times 16=602{,}840 \text{ N}$$

Eccentricity $\quad e=42-12{\cdot}5=29{\cdot}5 \text{ cm}=295 \text{ mm}.$

Initial Stresses

$$c_{ti}=\frac{P_i}{A}-\frac{P_i\times e}{Z_t}+c_{td}$$

$$=\frac{602{,}840}{115{,}000}-\frac{602{,}840\times 295}{22{,}920{,}000}+1{\cdot}57$$

$$=5{\cdot}24-7{\cdot}76+1{\cdot}57=-0{\cdot}95 \text{ N/mm}^2$$

$$c_{bi}=\frac{P_i}{A}+\frac{P_i\times e}{Z_b}-c_{bd}$$

$$=\frac{602{,}840}{115{,}000}+\frac{602{,}840\times 295}{20{,}730{,}000}-1{\cdot}74$$

$$=5{\cdot}24+8{\cdot}58-1{\cdot}74=12{\cdot}08 \text{ N/mm}^2.$$

Final Stresses

Final prestressing force
$=0{\cdot}85\times 602{,}840=512{,}410$ N

Live load B.M.

$$=\frac{18{,}000\times 10^2}{8}=225{,}000 \text{ N-m}$$

$$=225{,}000{,}000 \text{ N-mm}$$

$$c_{tl}=\frac{225{,}000{,}000}{22{,}920{,}000}=9{\cdot}82 \text{ N/mm}^2$$

$$c_{bl}=\frac{225{,}000{,}000}{20{,}730{,}000}=10{\cdot}85 \text{ N/mm}^2$$

$$c_t=\frac{512{,}410}{115000}+\frac{512{,}410\times 295}{20{,}920{,}000}+1{\cdot}57+9{\cdot}80$$

$$=4{\cdot}45-6{\cdot}60+1{\cdot}57+9{\cdot}82$$

$$=9{\cdot}24 \text{ N/mm}^2.$$

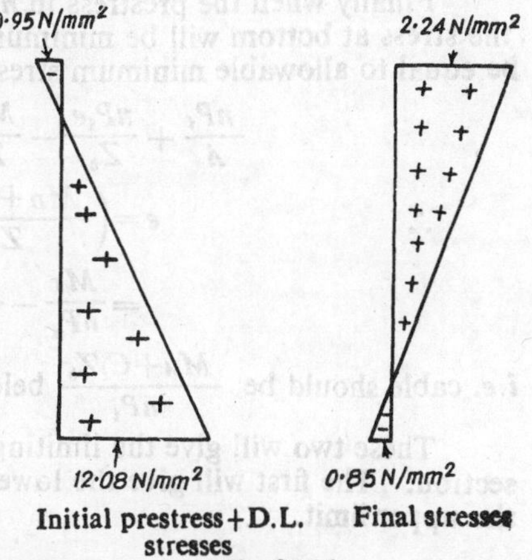

Initial prestress+D.L. stresses $\qquad$ Final stresses

Fig. 25·26

$$c_b=\frac{512{,}410}{115000}-\frac{512{,}410\times 295}{20{,}730{,}000}-1{\cdot}74-10{\cdot}85$$

$$=4{\cdot}45+7{\cdot}29-1{\cdot}74-10{\cdot}85$$

$$=-0{\cdot}85 \text{ N/mm}^2$$

Distribution of stresses is shown in Fig. 25·26.

25·10. Cable Zone. It is seen from the above discussion that the C.G. of cable has to lie within certain limits, so that the stresses are within the allowable limits.

Initially prestressing force is P_i and if it is located at eccentricity e and if M_D is the dead load moment, for minimum stress to be equal to C'_t.

$$\frac{P_i}{A}-\frac{P_i e}{Z_t}+\frac{M_D}{Z_t}=C'_t$$

$$\therefore \qquad e=\left(\frac{P_i}{A}+\frac{M_D}{Z_t}-C'_t\right)\frac{Z_t}{P_i}$$

$$=\frac{Z_t}{A}+\frac{M_D}{P_i}-\frac{C'_t Z_t}{P_i}$$

$$=\frac{Z_t}{A}+\frac{(M_D-C'_t Z_t)}{P_r}$$

$$=k_b+\left(\frac{M_D-C'_t Z_t}{P_i}\right)$$

In case $C'_t=0$,

$$e=k_b+\frac{M_D}{P_i}$$

i.e. at $\frac{M_D}{P_i}$ below the lower kern point.

Finally when the prestress in nP_i, the applied moment is M_A, the stress at bottom will be minimum and in limiting case it should be equal to allowable minimum stress C_t.

$$\frac{nP_i}{A}+\frac{nP_i e}{Z_b}-\frac{M_D}{Z_b}-\frac{M_A}{Z_b}=C_t$$

$$\therefore \qquad e=\left(\frac{M_D+M_A}{Z_b}-\frac{nP_i}{A}+C_t\right)\frac{Z_b}{nP_i}$$

$$=\frac{M_T}{nP_i}-\frac{Z_b}{A}+\frac{C_t Z_b}{nP_i}=\left(\frac{M_T-C_t Z_b}{nP_i}\right)-k_t.$$

i.e. cable should be $\frac{M_A+C_t Z_b}{nP_i}$ below the upper kern point.

These two will give the limiting positions of the cable for the section. The first will give the lower limit and the second will give the upper limit.

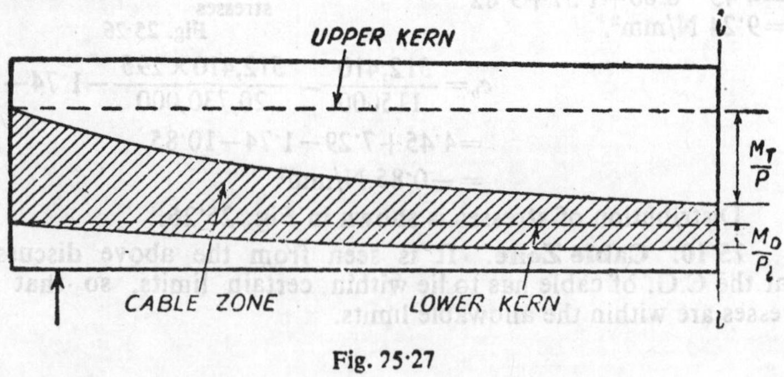

Fig. 25·27

If these limits are found for all sections and curves are drawn showing the lower limit at all sections also the upper limit at all sections, the zone between these is called the cable zone.

For $C_t=0$

The cable zone is shown in Fig. 25·27.

Generally the cable lies on a parabolic curve.

25·11. Design of Sections

(a) *Rectangular section.* The slab thickness will be $\frac{1}{30}$th to $\frac{1}{40}$th of span.

From consideration of buckling, the width of the beam should not be less than $\frac{1}{30}$th of the length between effective restraints or $\frac{1}{4}$th of depth of the beam. The depth will be from $\frac{1}{15}$th to $\frac{1}{20}$th of span.

Taking $C=C'$.

$$\text{(i)}\quad \frac{P_i}{A}+\frac{P_i e}{Z}-\frac{M_D}{Z}=C \qquad \ldots(b_i)$$

$$\text{(ii)}\quad \frac{P_i}{A}-\frac{P_i e}{Z}+\frac{M_D}{Z}=C_t \qquad \ldots(t_i)$$

$$\text{(iii)}\quad \frac{nP_i}{A}+\frac{nP_i e}{Z}-\frac{M_D+M_A}{Z}=C_t. \qquad \ldots(b)$$

$$\text{(iv)}\quad \frac{nP_i}{A}-\frac{nP_i e}{Z}+\frac{M_D+M_A}{Z}=C. \qquad \ldots(t)$$

Subtracting (iii) from (i),

$$\frac{P_i}{A}(1-n)+\frac{P_i e}{Z}(1-n)+\frac{M_A}{A}=C-C_t.$$

Now approximately $P_i=\frac{C}{2}A$,

$$\frac{M_A}{Z}=C-C_t-\frac{C}{2}(1-n)-\frac{CAe}{2Z}(1-n)$$

$$\therefore\quad Z=\frac{M_A}{C-C_t-\frac{C}{2}\left(1-n\right)\left[1+\frac{Ae}{A}\right]} \qquad \ldots(I)$$

Subtracting (iv) from (ii)

$$\frac{P_i}{A}\left(1-n\right)-\frac{P_i e}{Z}\left(1-n\right)-\frac{M_A}{Z}=C_t-C$$

$$\frac{M_A}{Z}=C-C_t+\frac{P_i}{A}\left(1-n\right)\left[1+\frac{Ae}{Z}\right]$$

$$=C-C_t+\frac{C}{2}\left(1-n\right)\left[1-\frac{Ae}{Z}\right]$$

$$Z=\frac{M_A}{C-C_t+\frac{C}{2}\left(1-n\right)\left[1-\frac{Ae}{Z}\right]}\quad ...(II)$$

The first expression will give bigger section.

For trial section,

$$Z=\frac{M_A}{\frac{C}{2}.(3n-1)-C_t}\ \text{approximately}$$

This is obtained by taking

$$\frac{Ae}{Z}=2.$$

(*b*) *I-Section.*

Generally I-sections are used, symmetrical or unsymmetrical. The flanges are splayed for easiness of pouring concrete and to avoid sharp corners.

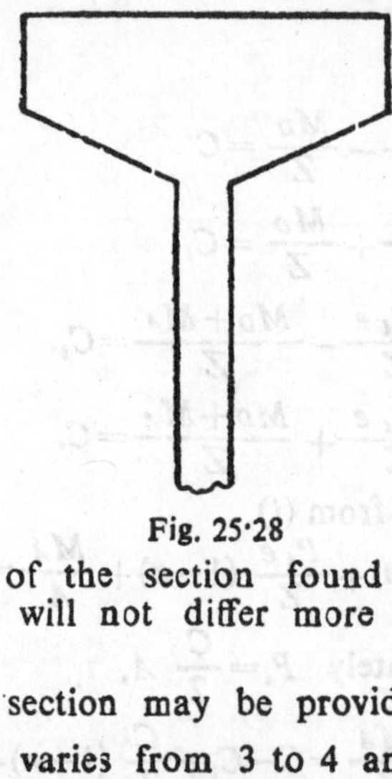

Fig. 25·28

The properties of the section found by taking equivalent rectangle for flanges will not differ more than 1% from actual calculations :

The depth of section may be provided from the equation $h=k\sqrt{M_T}$, where k varies from 3 to 4 and M_T is total moment in kN-m, when dead load moment is 20% or more of total moment. If dead load moment is less than 20%, $h=k\sqrt{M_L}$, where M_L is live load moment.

$$P=\frac{M_T\times 10^5}{0{\cdot}65a}\ \text{or}\ \frac{M_L\times 10^5}{0{\cdot}65h}$$

$$A=\frac{P}{0{\cdot}5C}.$$

where P is the final prestressing force, A is the area of section.

(*i*) The symmetrical I-section can be designed by first finding rectangle section and increasing depth by 10%. The depth of flange

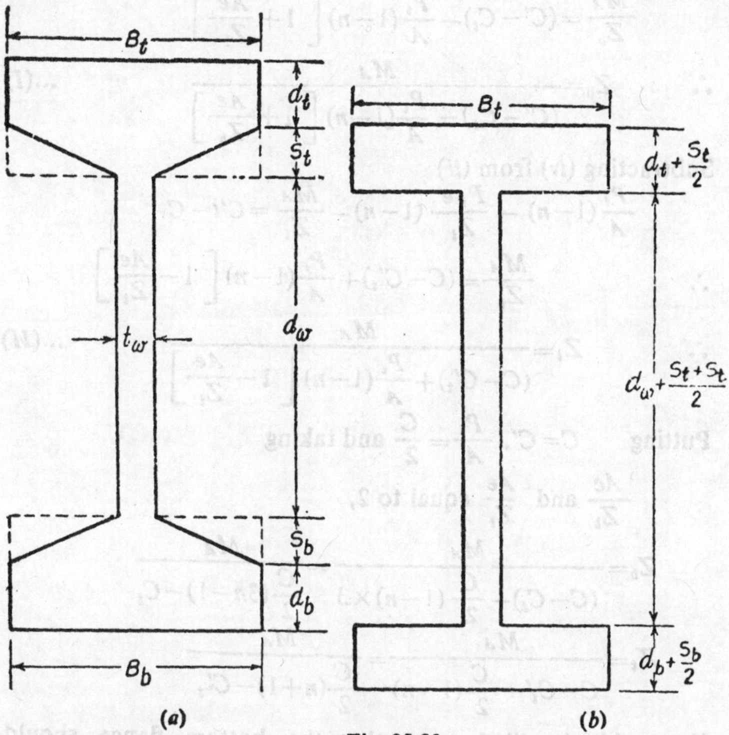

Fig. 25·29

may be taken as 0·1 to 0·2 time overall depth and thickness of web is taken 0·2 to 0·3 times width of top flange or as required for accommodating cables.

(*ii*) *Unsymmetrical I-section*. The following equation are to be satisfied.

$$(i)\quad \frac{P_i}{A}+\frac{P_i e}{Z_b}-\frac{M_D}{Z_b}=C' \qquad \ldots(b_i)$$

$$(ii)\quad \frac{P_i}{A}-\frac{P_i e}{Z_t}+\frac{M_D}{Z_t}=C'_t \qquad \ldots(t_i)$$

$$(iii)\quad \frac{nP_i}{A}+\frac{nP_i e}{Z_b}-\frac{M_D}{Z_b}-\frac{M_A}{Z_b}=C_t \qquad \ldots(b)$$

$$(iv)\quad \frac{nP_i}{A}-\frac{nP_i e}{Z_t}+\frac{M_D}{Z_t}+\frac{M_A}{Z_t}=C \qquad \ldots(t)$$

Subtracting (*iii*) from (*i*)

$$\frac{P_i}{A}(1-n)+\frac{P_i\, e}{Z_b}(1-n)+\frac{M_A}{Z_b}=C'-C_t$$

$$\frac{P_i}{A}(1-n)\left[1+\frac{A_e}{Z_b}\right]+\frac{M_A}{Z_b}=C'-C_t$$

$$\frac{M_A}{Z_b}=(C'-C_t)-\frac{P_i}{A}(1-n)\left[1+\frac{Ae}{Z_b}\right]$$

$$\therefore \quad Z_b=\frac{M_A}{(C'-C_t)-\frac{P_i}{A}(1-n)\left[1+\frac{Ae}{Z_b}\right]} \quad \text{...(I)}$$

Subtracting (*iv*) from (*ii*)

$$\frac{P_i}{A}(1-n)-\frac{P_i e}{Z_t}(1-n)-\frac{M_A}{Z_t}=C't-C$$

$$\therefore \quad \frac{M_A}{Z_t}=(C-C'_t)+\frac{P_i}{A}(1-n)\left[1-\frac{Ae}{Z_t}\right]$$

$$\therefore \quad Z_t=\frac{M_A}{(C-C'_t)+\frac{P_i}{A}(1-n)\left[1-\frac{Ae}{Z_t}\right]} \quad \text{...(II)}$$

Putting $C=C'.\ \frac{P_i}{A}=\frac{C}{2}$ and taking

$\frac{Ae}{Z_b}$ and $\frac{Ae}{Z_t}$ equal to 2,

$$Z_b=\frac{M_A}{(C-C_t)-\frac{C}{2}(1-n)\times 3}=\frac{M_A}{\frac{C}{2}(3n-1)-C_t}$$

$$Z_t=\frac{M_A}{C-C_t'-\frac{C}{2}(1-n)}=\frac{M_A}{\frac{C}{2}(n+1)-C'_t}$$

From this it will be seen that the bottom flange should be made heavier. From buckling point of view the top flange is kept wider than the bottom flange. The bottom flange thickness is provided from consideration of accommodating cables. The prestressing force required will depend on eccentricity available. By increasing available eccentricity, pre-stressing force can be reduced. For this the top flange will have to be made heavier so that C.G. of section is shifted up to provide more eccentricity. All these considerations are taken in designing the section. For longer spans where dead load stresses are not fully counter balanced, heavy top flange is provided to increase eccentricity.

After the section has been assumed, dead load and applied load stresses are worked out. Let C_{td} and C_{ba} be stresses in bottom fibre due to load and applied load respectively, and C_{td} and C_{ta} be stresses in top fibres due to dead load and applied load respectively.

For minimum prestressing force, equations (*b*) and (*t*) should be satisfied. Taking $C'_t=C_t$,

$$\therefore \quad \frac{P_i}{A}-\frac{P_i e}{Z_i}+\frac{M_D}{Z_i}=C$$

$$\frac{nP_i}{A}+\frac{nP_i e}{Z_b}-\frac{M_D}{Z_b}=\frac{M_A}{Z_b}=C_t$$

$$\therefore \quad P_t\left[\frac{1}{A}-\frac{e}{Z_t}\right]=C_t+C_{td}$$

and
$$nP_t\left[\frac{1}{A}+\frac{e}{Z_b}\right]=C_t+C_{bd}+C_{ba}$$

$$\therefore \quad \frac{\frac{1}{A}-\frac{e}{Z_t}}{\frac{1}{A}+\frac{e}{Z_b}}=\frac{n(C_t-C_{td})}{(C_t+C_{bd}+C_{ba})}.$$

From this equation value e can be obtained. If this eccentricity is available the value of P_i can be obtained from equation b. If eccentricity e available is less, then corresponding value of P_t is obtained by putting this value of e in b or t whichever gives more value. The section is designed. If available value of e is more than needed, the section may be revised.

Ex. 25·3. *Design a prestressed slab having a span 16 metres. It carries a load of 10.000 N/m². Take losses at 15%. Allowable initial compressive stress is 1750 N/cm², final compressive stress is 1400 N/cm² and tensile stress is 100 N/cm².*

Sol. Take 1 metre width of slab

$$M_A=\frac{10{,}000\times 16^2}{8}=320{,}000 \text{ N-m}$$

$$Z=\frac{M_A}{\frac{C}{2}(3n-1)-C_t}=\frac{320{,}000\times 100}{\frac{1400}{2}(3\times 0{\cdot}85-1)+100}$$

$$=\frac{32{,}000{,}000}{1185}=27{,}010 \text{ cm}^3=\frac{bd^2}{6}$$

$$\therefore \quad d=\sqrt{\frac{6\times 27{,}010}{100}}=40{\cdot}26 \text{ cm}$$

Use slab thickness of 40 cm.

$$A=40\times 100=4000 \text{ cm}^2$$

$$Z=\frac{100\times 40^2}{6}=26{,}667 \text{ cm}^3$$

$$M_D=\frac{4{,}000\times 2{\cdot}4\times 16^2}{8}=307{,}200 \text{ N-m}$$

$$C_{bd}=C_{td}=\frac{307{,}200\times 100}{26{,}667}=1152 \text{ N/cm}^2$$

$$C_{ba}=C_{ta}=\frac{320{,}000\times 100}{26{,}667}=1199 \text{ N/cm}^2$$

$$(b_i)\ldots=\frac{P_i}{4000}+\frac{P_i\times e}{26{,}667}-1152=1750$$

$$\therefore \quad \frac{1}{P_i} \times \frac{1}{\dfrac{1}{2902 \times 4,000}} - \frac{e}{\dfrac{-26,667}{4000}} = 1$$

$$\therefore \quad \frac{1}{P_i} \times \frac{1}{8{\cdot}614 \times 10^{-8}} + \frac{e}{-6{\cdot}667} = 1 \qquad \ldots(b_i)$$

$$(t_i)\ldots \frac{P_i}{4000} - \frac{P_i \times e}{26,667} + 1152 = -100$$

$$\therefore \quad \frac{1}{P_i} \times \frac{1}{\dfrac{-1}{1252 \times 4,000}} + \frac{e}{\dfrac{26,667}{4,000}} = 1$$

$$\therefore \quad \frac{1}{P_i} \times \frac{1}{-19{\cdot}95 \times 10^{-8}} + \frac{e}{6{\cdot}677} = 1 \qquad \ldots(t_i)$$

$$(b)\ldots \frac{0{\cdot}85\, P_i}{4000} + \frac{0{\cdot}85\, P_i \times e}{26,667} - 1152 - 1199 = -100$$

$$\therefore \quad \frac{1}{P_i} \times \frac{1}{\dfrac{0{\cdot}85}{2251 \times 4,000}} + \frac{e}{\dfrac{-26,667}{4,000}} = 1$$

$$\therefore \quad \frac{1}{P_i} \times \frac{1}{9{\cdot}439 \times 10^{-8}} + \frac{e}{-6{\cdot}667} = 1 \qquad \ldots(b)$$

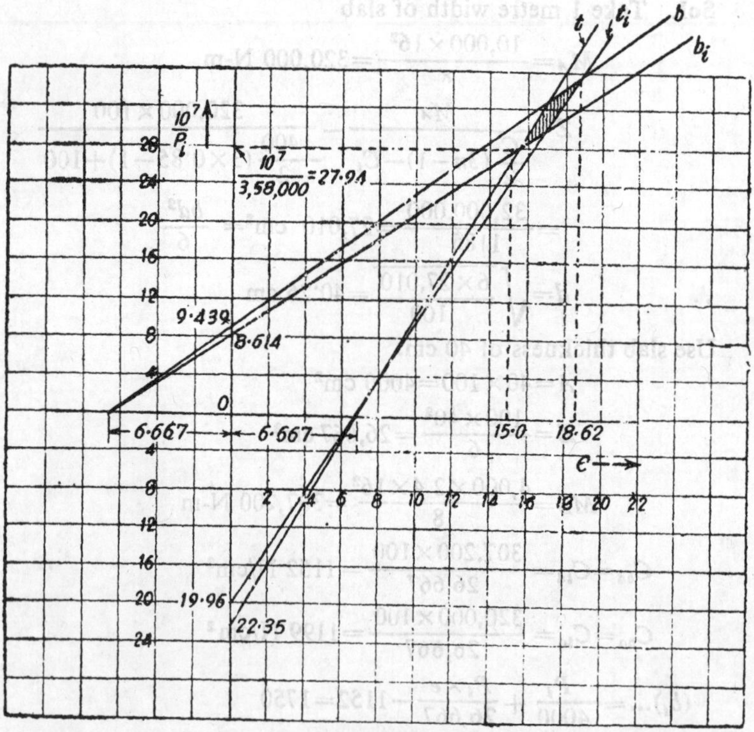

Fig. 25·30

(*t*) $$\frac{0.85\,P_i}{4,000}-\frac{0.85\,P_i\times e}{26,667}+1152+1199=1400$$

$$\therefore \quad \frac{1}{P_i}\times\frac{1}{\frac{-0.85}{951\times 4,000}}+\frac{\frac{e}{26,667}}{4,000}=1$$

$$\therefore \quad \frac{1}{P_i}\times\frac{1}{-22.35\times10^{-8}}+\frac{e}{6.667}=1 \qquad \text{...}(t)$$

For maximum allowable 'e'

$$\frac{\text{Equation } t_i}{\text{Equation } b},\quad \frac{\frac{1}{P_i}\times\frac{10^1}{-19.97}}{\frac{1}{P_i}\times\frac{10^8}{94.39}}=\frac{1-\frac{e}{6.667}}{1+\frac{e}{6.667}}$$

$$\therefore \quad -\frac{9.439}{19.97}=\frac{6.667-e}{6.667+e}$$

$$\therefore \quad -3152-0.4726\,e=6.667-e$$

$$\therefore \quad 0.5274\,e=9.819$$

$$\therefore \quad e=\frac{9.819}{0.5274}=18.62 \text{ cm}$$

Actual eccentricity $=20-3-2=15$ cm.

Putting this eccentricity in equation 'b'

Fig. 25·31

$$\frac{1}{P_i}\times\frac{10^8}{9.439}=1+\frac{15}{6.667}=1+2.25=3.25$$

$$\therefore \quad P_i=\frac{10^8}{9.439\times3.25}=3,260,000 \text{ N.}$$

Putting this eccentricity in equation 't'

$$\frac{1}{P_i}\times\frac{10^8}{-22.35}=1-\frac{15}{6.667}=1-2.25=-1.25$$

$$\therefore \quad P_i=\frac{10^8}{22.35\times1.25}=3,580,000 \text{ N}$$

Use 7 mm ϕ wires.

Allowable stress after losses $=0.6\times1500=900$ N/mm^2

Allowable stress before losses $=\frac{900}{0.85}=1060$ N/mm^2

Max. prestress in cable of 12 wires $=\frac{\pi}{4}\times7^2\times12\times1060$

$=491,100$ N

$$\text{Spacing}=\frac{491,100\times100}{3,580,000}=13{\cdot}72\text{ cm}$$

Adopt spacing of 13 cm

$$\text{Prestressing force in each cable}=\frac{491,100\times13}{13{\cdot}72}=465,200\text{ N}$$

Prestressing force in 100 cm

$$=\frac{100\times465,200}{13}=3,580,000\text{ N}$$

$$\frac{P_i}{4,000}=\frac{3,580,000}{4,000}=895\text{ N/cm}^2$$

$$\frac{0{\cdot}85\times P_i}{4000}=8{\cdot}5\times895=760{\cdot}7\text{ N/cm}^2$$

$$\frac{P_i\times e}{Z}=\frac{3,580,000\times15}{26,667}=2014\text{ N/cm}^2$$

$$\frac{0{\cdot}85P_i\times e}{Z}=0{\cdot}85\times2014=1712\text{ N/cm}^2.$$

Initial Stresses

Bottom...895+2014−1152=2909−1152=1757 N/cm²

Top ..895−2014+1152=2047−2014=33 N/cm².

Final Stresses

Bottom...760·7+1712−1152−1199=2472·7−2351
=121·7 N/cm²

Top ..760·7−1712+1152+1199=3111·7−1712
=1399·7 N/cm².

Initial bottom stress is slightly high, but as some of losses will take place immediately, stress due to prestressing will decrease.

25·12. General Design Requirements

1. **Areas and moments of inertia of sections.** In calculating area, centroid and moment of inertia of cross-section, allowance for the prestressing tendons need not be made in the case of pretensioned members where the prestressing tendons are single wire distributed on the cross-section or strands of wires of relatively small cross-sectional area. Where allowance is made, it shall be on the basis of $(m-1)$ times of the area of the prestressing tendon, m being the modular ratio.

In calculating the area, centroid and moment of inertia of cross-sections of post-tensioned members, deductions shall invariably be made top prestressing tendons, cable ducts or sheaths and such other openings, in the concrete whether they are formed longitudinally or transversely.

These deductions need not, however, be made for loads applied after the duct, sheaths or openings have been grouted or filled with concrete. Where such deductions are not made, a transformed area equivalent to *m* times the area of the prestressing element shall be taken in calculation, *m* being the modular ratio.

2. **Moduli of Elasticity**

(*a*) *Modulus of elasticity of steel*

The value of modulus of elasticity of steel E_s used for the design of prestressed concrete members shall preferably be determined by tests on samples of steel to be used for construction.

Where it is not possible to ascertain the modulus of elasticity by tests or from the manufacturers of the steel, the following values may be adopted.

Type of steel	*Modulus of elasticity* (E_s) kN/mm^2
Plain-drawn wires (cold drawn or cold worked)	210
High tensile steel bars (rolled or heat treated)	200
Strands	195

(*b*) *Modulus of elasticity of concrete* (E_c)

Unless otherwise determined by tests, the modulus of elasticity of concrete E_c shall be assumed to have the values given by the following formulae :

In N/mm^2 : $E_c = 5700\sqrt{F_{ck}}$

F_{ck} being characteristic strength of concrete in N/mm^2.

3. **Permissble Stresses**

Steel

(*a*) At the time of tensioning, the maximum tensiles stress (f_{si}) immediately behind the anchorages shall not exceed (*a*) 80% of the ultimate tensile strength of the wire or bar steel without a definite yield point, and (*b*) the yield stress for steel with a guaranteed yield point.

(*b*) After accounting for all losses due to creep and shrinkage of the concrete, friction effects and relaxation of steel, the maximum tensile stress (f_t) at the point of maximum bending shall not exceed (*a*) 60% of the ultimate strength of the wire or bar for steel without a definite yield point, and (*b*) 80% of the yield stress for steel with guaranteed yield point.

Concrete. The stresses in concrete due to prestressing operations and working loads should not exceed the values given below. Unless otherwise specified, the cube strength refers to that of concrete obtained at 28 days when ordinary Portland cement is used and at 7 days when rapid-hardening cement is used on 15 cm cubes.

(*a*) **Compressive Stress due to initial prestress.** The maximum permissible compressive stress due to bending and direct force at the time of transfer of prestress, c' shall be determined from a graph in which a straight line joins a point given by 0·5 F_r for a concrete with a F_c of 35 N/mm² to a second point giving a permissible stress of 0·43 F_r for concrete with a F_c of 53 N/mm² (Fig. 25·32). The above F_c values apply to post-tensioned work; for pre-tensioned work they shall be increased by 20 per cent.

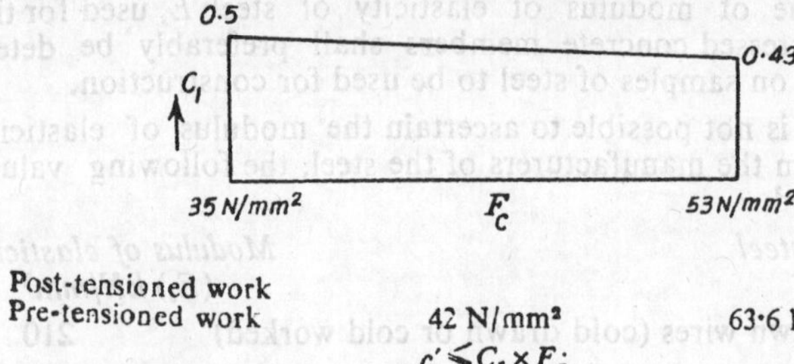

Fig. 25·32. Computation of compressive stress due to initial prestress.

For the purpose of Fig. 25·32, F_r shall be the cube strength of concrete at transfer which in no case shall be less than half the corresponding F_c.

(*b*) **Compressive strength due to final prestress.** The maximum permissible compressive stress (*c*) due to bending and direct force after deduction of full losses in the specified prestress, shall be determined by a straight line relation as in Fig. 25·33, but different stress limits shall apply to the concrete of the structure depending on whether it falls in apart of the section where the compressive stresses are not likely to increase in service (Zone I) or in a part of section where the compressive stresses are likely to increase in service (Zone II) (Fig. 25·33).

(*c*) **Flexural stresses**

(*i*) In structures subjected to dynamic loading under the severest combination of service loads and force applied, all direct force and bending stresses in any section shall remain compressive. In extreme fibre closest to the line of action of the prestressing steel, the maximum compressive stress shall not be less than 5%

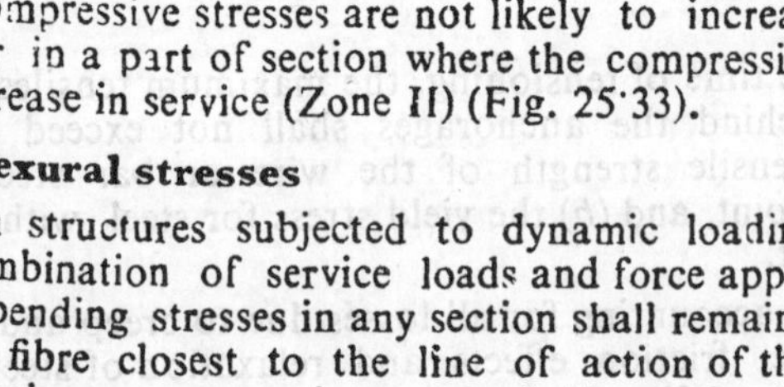

1. Stress in service not likely to increase.
2. Stress in service likely to increase.

Fig. 25·33. Computation of compressive stress due to final prestress.

of the maximum compressive stress permitted on that fibre unless otherwise approved by the appropriate authority. In the latter case, the minimum compressive stress may be reduced upto a limit of zero.

(*ii*) On the other structures under the same circumstances, a tensile stress (f_t) may be permitted, not exceeding one-tenth of the maximum permissible compressive stress or 140 N/cm² for pretensioned and 100 N/cm² for post-tensioned whichever is less.

(*iii*) In case of members assembled out of precast blocks, no tensile stress shall be permitted at any stage of loading upto cracking.

4. **Stages of Loading to be Considered in Design.** The design shall allow for effect of all loads to which the member will be subjected during construction, handling and under service.

In order to determine all the possible extremes or stress, the various loading conditions shall be investigated in same order in which they occur.

The different stages of loading given below shall be investigated and in the case of stages I and III, the various stresses to which the member is subjected shall be maintained within the specified limits. In respect of other stage the resulting factor of safety shall be investigated and shall be brought within the permissible limits.

(*a*) Stage I : At transfer of prestress.

(*b*) Stage II : At handling and erection.

(*c*) Stage III : At design load.

(*d*) Stage IV : At appearance of first crack.

(*e*) Stage V : At failure.

5. **Losses in prestress.** Various losses given in article 25·8 should be considered in design.

25·13. Condition at First Crack. This stage requires the calculation of the load which produces a tensile stress in the member equal to the modulus of rupture of the concrete. The load thus computed shall not be less than 1·1 times the dead load plus 1·25 times the live load placed in such a way as to produce the maximum stresses in the member concerned. The modulus of repture shall be taken as $7\sqrt{F_{ck}}$ where F_{ck} is in N/cm².

The cracking load of a prestressed member assembled out of precast blocks shall be computed on the assumption that the concrete is incapable of taking any tensile stress.

25·14. Ultimate load design. The ultimate loads assumed for purpose of determining the condition at failure shall not be less than the following :

(*a*) For statically determinate structure subjected to dynamic loading, such as bridges. ...1·5 D.L.+3 L.L.

(*b*) For statically determinate structures and subjected to dynamic loading such as buildings. ...1·5 D.L.+2·5 L.L.

2(D.L.+L.L.) whichever is greater

D.L.=design dead load

L.L.=maximum live load including *i*mpact.

Types of Failures

(1) **Fracture of steel in tension.** Failure due to fracture of steel in tension occurs when the percentage of steel provided in a prestressed member is so low that when the concrete in the precompressed tensile zone cracks, the steel is not in a position to take up the additional tensile stress transferred on to it by the cracking of the adjacent fibres of concrete. The percentage of steel p required to avoid such failure may be determined from the formula,

$$p \geqslant \frac{250}{F_s}, \text{ where } p = \frac{A_{ts}}{bD}$$

where F_s=ultimate tensile strength of steel in N/cm^2

A_{ts}=cross-sectional area of the tensioned steel in the tensile zone.

b and D being the width of the bottom and the overall depth of the beam respectively.

(2) **Excessive elongation of steel, followed by crushing of concrete.**

Failure due to excessive elongation of steel consequent crushing of concrete occurs when members are so designed that the percentage of steel provided is such that the increase in load leads to a significant elongation of the steel thus raising the neutral axis closer to the top of compressive fibres. The compressive zone thus gets gradually reduced and ultimately the member will fail through the crushing of the concrete. Failure of this type is preceded by very large cracks of tension zone and cansiderable deflection of the member in bending. These beams are classified as under-reinforced.

(3) **A small elongation of steel, followed by crushing of concrete.**

Failure due to crushing of the concrete occurs when the percentage of steel in a prestressed beam is very high. In such a case with gradual overloading, the tensile stress in steel increases, but is yet within the elastic range, and therefore, the neutral axis remains closer to the centre of gravity of the prestressing steel. In consequence the top compressive fibres undergo excessive plastic deformation under heavy stresses. With further increase in loads, failure takes place by crushing of concrete. These beams are classified as over-reinforced beams.

The upper limit for steel to avoid such a failure may be determined by the considerations given below provided effective bond is ensured between the steel and the concrete.

For rectangular beams— $\dfrac{A_{ts} \times F_s}{b \times d \times F_c} \not> 0{\cdot}240$

For flanged beams— $\dfrac{A_{stw} \times F_s}{b' \times d \times F_c} \not> 0{\cdot}240$

where, $A_{stw} = A_{ts} - A_{tsf}$,

and A_{tsf} = steel area required to develop the ultimate compressive strength of the overhanging portion of the flange

$$= \frac{0{\cdot}68\, F_c\, (b - b')\, t}{F_s}$$

b = breadth of a rectangular member or of flange of a flanged member.

b' = breadth of web of a flanged member,

d = depth from the compression edge to the centre of area of the prestressing steel in the tensile zone, and

t = average thickness of flange of a flanged member.

II. Ultimate Moment

(*a*) for under-reinforced beams the ultimate moment of resistance M_{us} may be calculated as follows—

(*i*) For rectangular beam and flanged beams in which $t \geqslant d/4$

$$M_{us} = K_{us} \times A_{ts} \times F_s \times d \left[1 - 0{\cdot}75 \times \frac{A_{ts} \times F_s}{b \times d \times F_c} \right] \text{ and}$$

(*ii*) For flanged beams in which $t < d/4$

$$M_{us} = K_{us} \times A_{tsw} \times F_s \times d \left[1 - 0{\cdot}75 \times \frac{A_{tsw} \times F_s}{b' \times d \times F_c} \right] + 0{\cdot}7\, F_c \times t (b - b')(d - 0{\cdot}5t)$$

where K_{us} is a constant equal to unity when tensioned steel is effectively bonded to concrete, and equal to 0·7 when the steel is unbonded.

(*b*) For over-reinforced beams, the ultimate moment of resistance M_{us} may be calculated as follows :—

(*i*) For rectangular beams

$$M_{uo}=0\cdot25\times b\times d^2\times0\cdot75\ F_c$$

(*ii*) For flanged beams

$$M_{uc}=0\cdot25\ b'\times d^2\times0\cdot75\ F_c+0\cdot48\times t\times0\cdot75\ F_c\ (b-b')(d+0\cdot5t)$$

Untensioned reinforcement in the tensile zone may be taken into consideration for purposes of ultimate load. The cross-sectional area of the prestressing steel may then be increased by the addition of the effective area of the untensioned reinforcement as follows :

$$\text{Effective area} \quad =\frac{A_{us}\times F_y}{F_s}$$

where, A_{us}= the cross-sectional area of the untensioned reinforcement in the tensile zone.

F_y= the yield stress of the untensioned reinforcement where it is mild steel or 0·2 per cent proof stress for high tensile steel and

F_s= ultimate tensile strength of prestressing steel.

25·15. Shear. The section at any point will be subjected to shear force F which causes shear stress which can be found by usual method $q=\frac{F A\bar{y}}{Ib}$. At this point there will be normal stress due to prestress and applied loading which can be calculated. Let at the point where shear stress is q, the normal stress be f_c, the Mohr circle have been drawn for this case. The diagonal tension will be given by $\sqrt{q^2+\frac{f_c^2}{2}}-\frac{f_c}{2}$. This will be less than q. Thus in case of prestressed sections there is reduction in diagonal tension and, therefore, shear reinforcement.

Where the cable is curved making angle θ with the horizontal, the component $P \sin \theta$, where P is prestress will act contrary to shear force and, therefore, at full load, the net shear force will be much reduced. But when there is only dead load, there might be reversal of direction of shear force if $P \sin\ \theta$ is greater than shear force due to dead load.

From design consideration the slope given to the cable at the ends where shear force is maximum is such that the net shear force when dead load is acting is nearly equal to net shear force when dead load and live load act.

When only dead load acts, net shear force

$$=P\sin\theta-F_d$$

where F_d is the shear due to dead load.

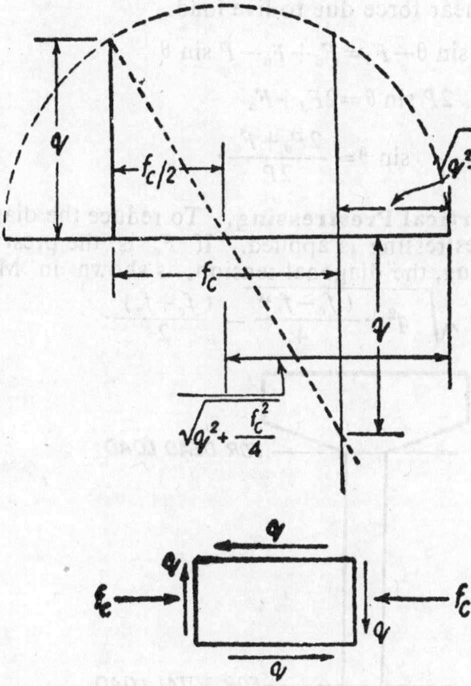

Fig. 22·34

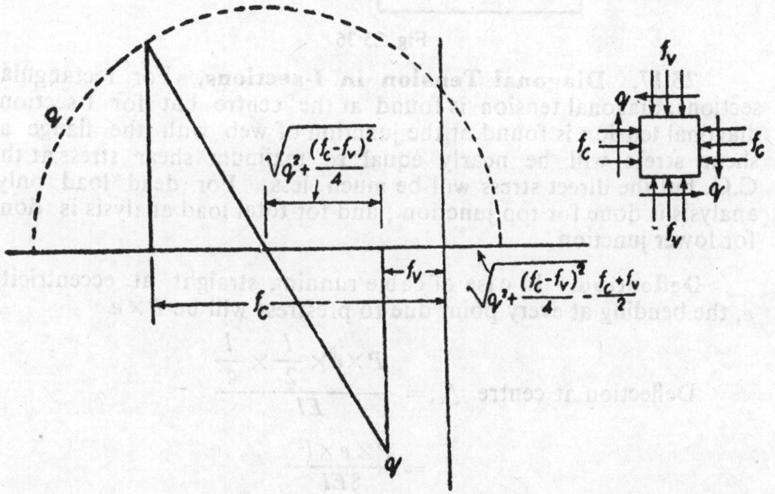

Fig. 25·35

When dead load and live load act,

Net shear force $=F_d+F_a-P\sin\theta$

where F_a is the shear force due to live load.

$$\therefore \quad P\sin\theta-F_d=F_d+F_a-P\sin\theta$$

$$\therefore \quad 2P\sin\theta=2F_d+F_a$$

or
$$\sin\theta=\frac{2F_d+F_a}{2P}.$$

25·16. Vertical Prestressing. To reduce the diagonal tension vertical prestressing is applied. If F_v is the prestress due to vertical prestressing, the diagonal tension, as shown in Mohr circle will be equal to $\sqrt{q^2+\frac{(f_c-f_v)^2}{4}}-\frac{(f_c+f_v)}{2}$.

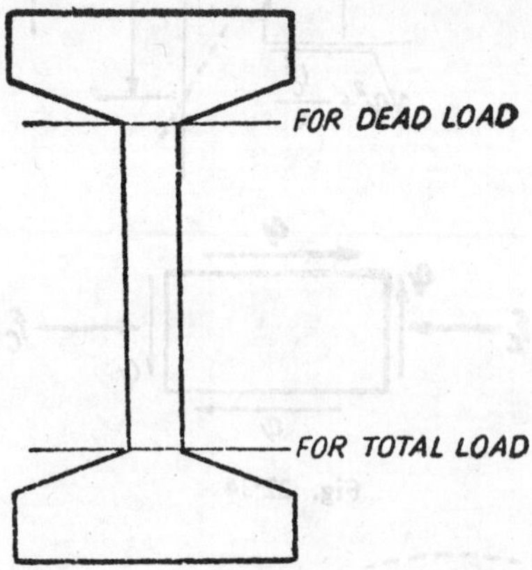

Fig. 25·36

25·17. Diagonal Tension in I-sections. For rectangular sections, diagonal tension is found at the centre but for I-sections diagonal tension is found at the junction of web with the flange as shear stress will be nearly equal to mximum shear stress at the C.G. but the direct stress will be much less. For dead load only, analysis is done for top junction ; and for total load analysis is done for lower junction.

Deflection. In case of cable running straight at eccentricity e, the bending at every point due to prestress will be $P\times e$

$$\text{Deflection at centre } \Delta_s=\frac{P\times e\times\frac{l}{2}\times\frac{l}{4}}{EI}$$

$$=\frac{P\times e\times l^2}{8EI}$$

In this case the cable is curved and has e_x eccentricity at x distance from centre and at this point M.I. is I_x deflection at centre.

$$\Delta c = \int_0^{l/2} \frac{P \times e_x \times \left(\frac{l}{2} - x\right) dx}{EI_x}.$$

25·18. End Block. In post-tensioned beams, the cables are anchored by mechanical means which bear against the end of beams, or are embedded within the concrete.

This results in large concentrated forces on the bearing surfaces. The stress distribution will not be linear in the vicinity of these concentrated forces. Immediately under the bearing surface there will be high compressive stresses. These are bearing stresses for which allowable stress is given as $= f_c \sqrt[3]{\frac{A}{a}}$, where f_c is the compressive stress in concrete, A is area on which the force is transferred, and a is the actual area on which the force is applied.

The stress distribution will be linear at some distance from the anchorage zone. This distance is taken as equal to the depth of

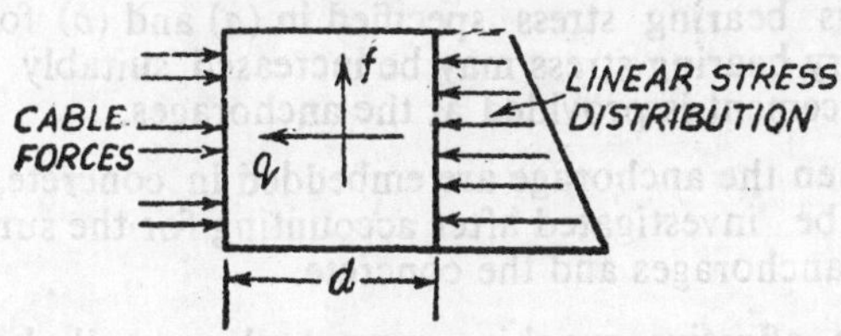

Fig. 25·37

the beam. This distance is called the lead-in-zone in which stresses are discontinuous. The stresses in this zone are transverse stresses acting along the direction of depth and at right angles and these are tensile. Further there will be shear stresses along the planes of tendons.

Besides these there are other transverse tension which cause spalling of conerete.

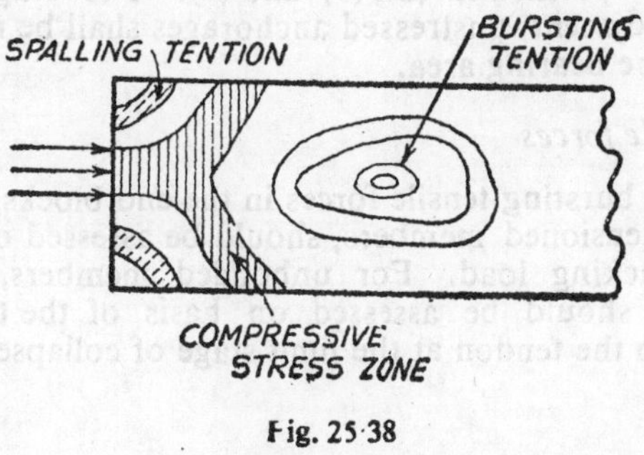

Fig. 25·38

The analysis to get exact stress in the end block is not possible, although theories have been evolved by Magnel and Guyon. They assume stress distribution as two dimensional. The Cement and Concrete Association have suggested some empirical formula from their intensive work done.

General Design Consideration for End Block

Bearing stress

(*a*) On the areas immediately behind external anchorages, the permissible unit bearing stress on the concrete, after accounting for all losses due to relaxation of steel, elastic shorting, creep of concrete, slip and/or seating of anchorage, etc., shall not exceed $0{\cdot}48\, f_{ct} \sqrt{\dfrac{A}{a}}$ or $8{\cdot}0\, f_{ct}$ whichever is smaller, where f_{ct} is the cube strength at transfer, A_b is the bearing area and a is the punching area.

(*b*) During tensioning, the allowable bearing stress specified in (*a*) may be increased by 25 per cent, provided that this temporary value does not exceed f_{ci}.

(*c*) This bearing stress specified in (*a*) and (*b*) for permanent and temporary bearing stress may be increased suitably if adequate hoop reinforcement is provided at the anchorages.

(*d*) When the anchorage are embedded in concrete, the bearing stress shall be investigated after accounting for the surface friction between the anchorages and the concrete.

(*e*) The effective punching area shall generally be the contact area of the anchorage devices which, if circular in shape, shall be replaced by square of equivalent area. The bearing area shall be the maximum area of that portion of the member which is geometrically similar and concentric to the effective punching area.

(*f*) Where a number of anchorages are used, the bearing area A shall not overlap. Where there is already a compressive stress prevailing over the bearing area, as in the case of anchorage placed in the body of a structure, the total stress shall not exceed the limiting values specified in (*a*), (*b*) and (*c*). For stage stressing of cables, the adjacent unstressed anchorages shall be neglected when determining the bearing area.

Bursting tensile forces

(*a*) The bursting tensile forces in the end blocks, or regions of bonded post-tensioned members, should be assessed on the basis of the tension jacking load. For unbonded members, the bursting tensile forces should be assessed on basis of the tendon jacking load or load in the tendon at the limit stage of collapse, whichever is greater.

The bursting tensile force, F_{bst} existing in an individual square end block loaded by a symmetrically placed square anchorage or bearing plate, may be derived from the equation below

$$\frac{F_{bst}}{P_k} = 0{\cdot}32 - 0{\cdot}3\frac{y_{po}}{y_o}$$

where F_{bst} = bursting tensile force,

P_k = load in the tendon assessed as above,

y_{po} = side of loaded area, and

y_o = side of end block.

(*b*) The force F_{bst} will be distributed in a region extending from $0{\cdot}1\ y_o$ to y_o from the loaded face of the end block. Reinforcement provided to sustain the bursting tensile force may be assumed to be acting as its design strength (0·87 times characteristic strength of reinforcement) except that the stress should be limited to a value corresponding to a strain of 0·001 when the concrete cover to the reinforcement is less than 50 mm.

(*c*) In rectangular end blocks, the bursting forces in the two principal directions should be assessed on the basis of above formula. When circular anchorage of bearing plates are used, the side of the equivalent square area should be used. Where groups of anchorages or bearing plates occur, the end blocks should be divided into a series of symmetrically loaded prisms and each prism treated in the above manner. For designing end blocks having a cross-section different in shape from that of the general cross-section of the beam, reference should be made to specialist literature.

(*d*) Compliance with the requirements of (*a*), (*b*) and (*c*) will generally ensure that bursting tensile forces along the load axis are provided for. Alternative methods of design which make allowance for the tensile strength of the concrete may be used, in which case reference should be made to specialist literature.

Consideration should also be given to the spalling tensile stresses that occur in end blocks where the anchorage or bearing plates are highly eccentric, these reach a maximum at the loaded face.

25·19. Magnel's Method. In this method, at distance D from end anchorages, the stress distribution and shear stresses are

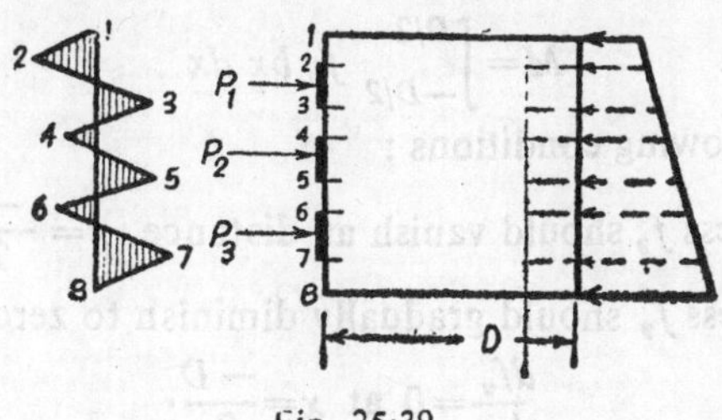

Fig. 25·39

worked out on the section. The stress distribution will depend on the prestress its position and loads acting. The horizontal forces at every section is worked out which will be equal to the sum of forces due to stress distribution above the section and prestressing force action at the anchorages above the section. This is plotted. The section which gives the maximum horizontal shear is investigated for transverse stresses and finally principal tensile stress is found.

The section will be subjected to net horizontal shear force F_h, net vertical force F_v and moment of the forces about the centre of section M.

P_v = total shear above the section

P_h = total horizontal force

$P_1 \cos \alpha$ = horizontal component of prestress

$P_1 \sin \alpha$ = vertical component of prestress.

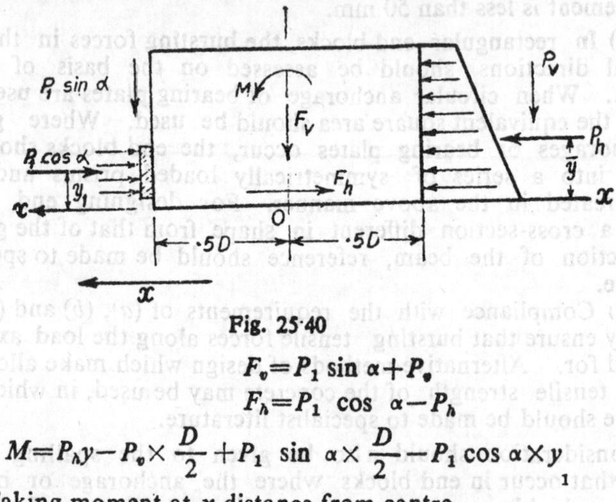

Fig. 25·40

$$F_v = P_1 \sin \alpha + P_v$$

$$F_h = P_1 \cos \alpha - P_h$$

$$M = P_h y - P_v \times \frac{D}{2} + P_1 \sin \alpha \times \frac{D}{2} - P_1 \cos \alpha \times y_1$$

Taking moment at x distance from centre.

from cosideration of equilibrium,

(i) $$F_v = \int_{D/2}^{D/2} f_y \, b \, dx$$

(ii) $$M = \int_{-D/2}^{D/2} f_y \, bx \, dx$$

and with the following conditions :

(a) the stress f_y should vanish at distance $x = \frac{-D}{2}$

(b) the stress f_y should gradually diminish to zero

i.e. $$\frac{df_y}{dx} = 0 \text{ at } x = \frac{-D}{2}.$$

The vertical stress f_y is given as

$$f_y = K_1 \frac{M}{bD^2} + K_2 \frac{F_v}{bD^2}$$

and q shear stress is given as, $q = K_3 \dfrac{F_h}{bD}$.

The values of K_1, K_2 and K_3 are given in Table 25·6.

Table 25·6

$\frac{x}{D}$	K_1	K_2	K_3
0·5	20·00	−2·00	0·000
0·4	9·72	0·00	1·458
0·3	2·56	1·28	2·048
0·2	−1·96	1·96	2·058
0·1	−4·32	2·16	1·728
0·0	−5·00	2·00	1·250
−0·1	−4·48	1·60	0·768
−0·2	−3·24	1·08	0·378
−0·3	−1·76	0·56	0·128
−0·4	−0·52	0·16	0·180
−0·5	0·00	0·00	0·000

There will be compressive stress f_x due to prestressing force. The force is assumed to disperse at 45° till it cuts the outer boundary or the vertical section through the point. The force divided by the intercepted area will give the value of f_x. Principal tensile stress is given by

$$f = \frac{f_y + f_x}{2} + \sqrt{\frac{(f_y - f_x^2)}{4} + q^2}$$

The inclination of the principal tensile stress to the vertical is given by

$$\theta = \frac{1}{2} \tan^{-1} \frac{2q}{f_y - f_x}.$$

In case principal tensile stress is more than allowable, mild steel is provided, generally in the form of a mat or spirals.

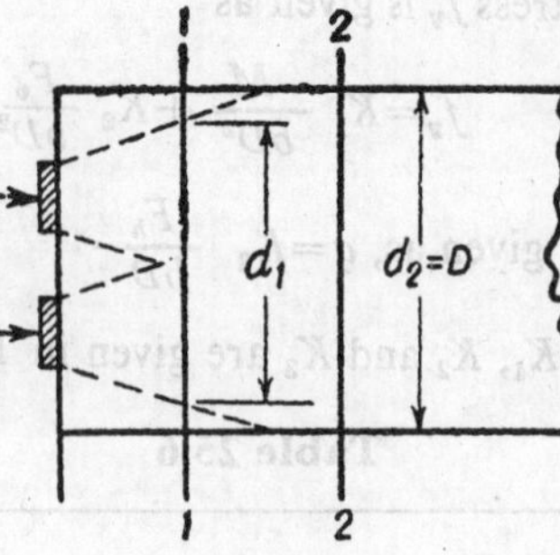

Fig. 25·41

25·20. Empirical Method. In Fig. 25·42, $2a_1$ is the side of the loaded area. If the loaded area is circular, equivalent squsare

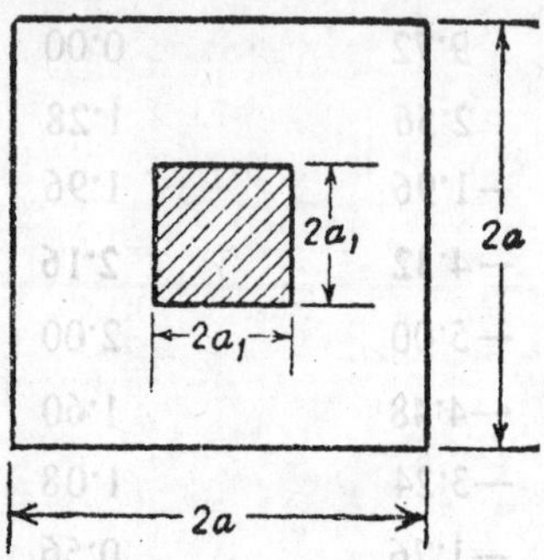

Fig. 25·42

giving the same area is worked out. $2a$ is the side of surrounding prism, it is the smallest dimension either in width direction or in depth direction.

The distribution of transverse stress depends on the ratio a_1/a. The positions of zero transverse stress and maximum transverse stress are not affected by this ratio. The tensile stress increases from zero at distance $0·2a$ to a maximum at $0·5a$ and then decreases to zero at a distance of $2a$.

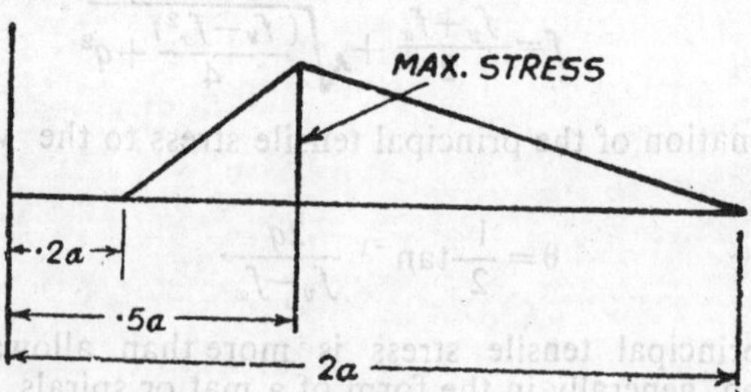

Fig. 25·43

f_c is the uniform compressive stress $= \dfrac{\text{Prestress}}{(2a_1)^2}$

$$f_{y\,max} = f_c \left[\ 0{\cdot}98 - 0{\cdot}825\,\frac{a_1}{a}\ \right]$$

$$\text{Total tensile force, } T = P\left[\ 0{\cdot}48 - 0{\cdot}4\left(\frac{a_1}{a}\right)\right]$$

where P is applied prestressing force.

The reinforcement is in the form of mats, helices or links. The reinforcement is provided in region 0·2a to 1·0a.

$$\text{Reinforcement area} = \frac{T}{14000}\ \text{cm}^2\ \text{(mild steel)}$$

where T is tensile force in N.

In case concrete can take some tensile stress say f_1.

$$T_{corrected} = T\left[\ 1 - \left(\frac{f_1}{f_{y\,max}}\right)\right]$$

The reinforcement is provided in both directions, in depth direction as well as width direction.

A rectangular mesh of bars of diameter 6 to 10 mm is provided. If necessary a number of mats may be provided.

In addition light spiral 6 mm diameter at a pitch of 5 cm is provided round the individual anchorages.

25·21. Mild Steel Reinforcement in Prestressed Concrete Member

It is desirable to provide mild steel reinforcement in the web, where webs are thin and ducts or cables or bars are located in the webs.

Transverse Reinforcement

In case of all members subjected to dynamic loading, web members shall be provided with mild steel reinforcement not less than 0·3 per cent of the sectional area of the web in plan. This reinforcement may be reduced to 0·2 per cent in members where the depth of the web is not more than four times the thickness of the web.

In the case of members not subjected to dynamic loading, mild steel reinforcement shall be provided in such cases where the depth of the web is more than four times the web thickness. Such reinforcement shall not be less than 0·1 per cent of the sectional area of the web in plan. The reinforcement shall be spaced at a distance not greater than the clear depth of the web and the size of such reinforcement shall be as small as possible.

Reinforcement in the form of links, stirrups and helix may require to be provided in planes perpendicular to the line of heavy compression or shock loading to resist the induced tensile stresses. The quantity of such reinforcement shall be determined in each case.

Longitudinal Reinforcement

This inforcement may be reduced to 0·15 per cent in the case of high yield strength deformed reinforcement. The percentage of steel provided, both tensioned and untensioned taken together, should be sufficient so that when the concrete in the precompressed tensile zone cracks, the steel is in a position to take up the additional tensile stress transferred on to it by the cracking of the adjacent fibres of concrete and a sudden failure is avoided.

When the depth of the web exceeds 50 cm, longitudinal distribution reinforcement not less than 0·05 per cent of the area of the web shall be provided on each face. The spacing of the individual bars of such inforcement shall not exceed 20 cm.

All untensioned longitudinal reinforcement shall be restrained in the lateral direction.

Normal longitudinal mild steel reinforcement of not less than 0·2 per cent of the total concrete area shall be provided in all cases except in the case of pretensioned units of small sections.

25·22. Concrete Cover and Spacing of Prestressing Steel

In pretensioned work the cover of concrete measured from the outside of the prestressing tension shall be at least 20 mm.

Where cables and large sized bars are used, the minimum clear cover shall be at least 30 mm or the size of the cable or bar whichever is bigger.

Where prestressed concrete members are exposed to the weather, the cover shall be increased by 12 mm.

The minimum clear spacing between single wires in pretensioned system shall not be less than three times the diameter of the wire used or 1½ times the maximum size of the aggregate whichever is more.

The minimum clear distance between cables or large bars shall not be less than 40 mm or the maximum dimension of the cable or bar, 6·5 mm in excess of the largest size of aggregate used whichever is greater.

Grouping of cables is permissible but not more than two cables should be touching each other in a vertical plane with the spacing between the so grouped cables being not less than 5 cm as shown in Fig. 25·45.

25·23. Slender Beams. The dimensions of the prestressed concrete beams shall be so chosen that the length of the beam between effective lateral restraints does not exceed 30 times the breadth of its compression flange. When the ratio exceeds 30, provided the width is not less than $\frac{1}{2\cdot5}$ of depth, the beam shall be designed as slender beam and the stresses due to bending shall be reduced by multiplying with the slenderness coefficient $1\cdot75-\frac{l_1}{40b}$, where l_1 is the distance between lateral restraints and b the breadth of the compression flange.

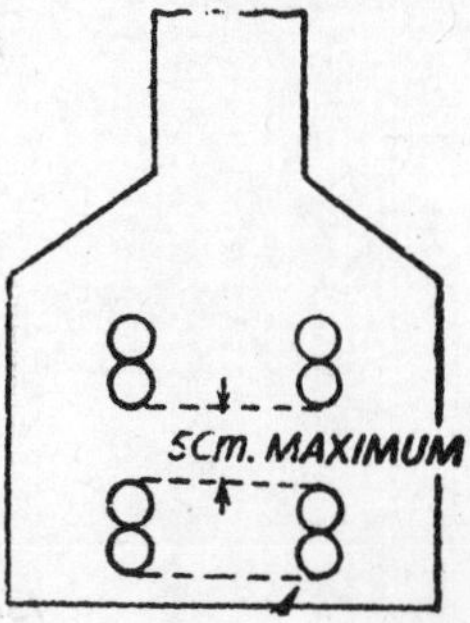

Fig. 25·44

Ex. 25·4. *Design a symmetrical prestressed concrete beam having span of 16 m and carrying superimposed load of 30,000 N/m. The ultimate stress in steel is 1500 N/mm² for 7 mm φ wires. The concrete used has cube strength of 35 N/mm². Assume losses in prestress at 15%.*

Sol. Allowable stresses are

$$c'=0\cdot5\ F_c=\frac{35}{2}=17\cdot5\ \text{N/mm}^2$$

$$c=0\cdot4\ F_c=35\times0\cdot4=14\ \text{N/mm}^2$$

Allowable tensile stress $=1$ N/mm²

Allowable stress in steel after losses

$$=0\cdot6\times1500=900\ \text{N/mm}^2$$

Allowable stress in steel before losses

$$=\frac{900}{0\cdot85}=1059\ \text{N/mm}^2$$

Live load B.M., $M_L=\frac{wl^2}{8}=\frac{30{,}000\times16^2}{8}=960{,}000$ N-m

$$=960\ \text{kN-m}$$

Assume dead load to be less than 20% of live load B.M.

$$h=k\sqrt{M_L}=4\sqrt{960}$$
$$=124\ \text{cm}$$

Provide overall depth as 130 cm

Maximum width of flange $=\frac{16\times100}{30}=53\cdot33$ cm

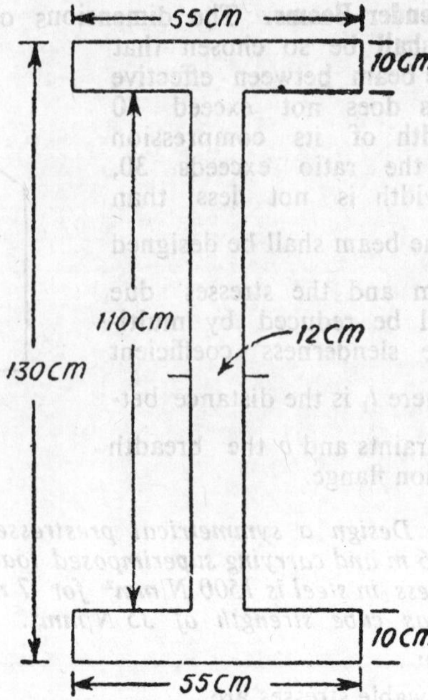

Fig. 25.45

Provide width of 55 cm.

Use cable of 12 wires, 7 mm ϕ, duct hole will be 4·0 cm in diameter.

Therefore, provide 12 cm, thick web.

$$\text{Area } A = 2\times 55\times 10 + 110\times 12$$
$$= 1100 + 1320$$
$$= 2420 \text{ cm}^2$$
$$= 242{,}000 \text{ mm}^2$$

D.L.B.M $\quad M_D = \dfrac{2420\times 2{\cdot}4\times 16^2}{8}$

$$= 186{,}800 \text{ N-m}$$

$$I = \frac{55\times 130^3}{12} - \frac{40\times 110^3}{12}$$
$$= 1007\times 10^4 - 476{\cdot}9\times 10^4$$
$$= 530{\cdot}1\times 10^4 \text{ cm}^4$$
$$= 530{\cdot}1\times 10^8 \text{ mm}^4$$

$$Z = \frac{530{\cdot}1\times 10^4}{65}$$
$$= 81{,}560 \text{ cm}^3$$
$$= 81{,}560{,}000 \text{ mm}^3$$

$$C_{bd}=C_{td}=\frac{185{,}800\times 1000}{81{,}560{,}000}=2.28 \text{ N/mm}^2$$

$$C_{ba}=C_{ta}=\frac{960{,}000\times 1000}{81{,}560{,}000}=11.77 \text{ N/mm}^2$$

$$k_b=k_t=\frac{Z}{A}=\frac{81{,}560}{2420}=33.7 \text{ cm}$$

$$=337 \text{ mm.}$$

Initial conditions (1) $\dfrac{P_i}{2420}+\dfrac{P_i\times e}{81{,}560}-228=1750$

$$\therefore \quad 1+\frac{e}{33.7}=\frac{1978\times 2420}{P_i}$$

$$\therefore \quad \frac{1}{P_i}\times\frac{1}{2.089\times 10^{-7}}+\frac{e}{-33.7}=1 \qquad \ldots(b_i)$$

(2) $\dfrac{P_i}{2420}-\dfrac{P_i\times e}{81{,}560}+228=-100$

$$\therefore \quad 1-\frac{e}{33.7}=\frac{-328\times 2420}{P_i}$$

$$\therefore \quad \frac{1}{P_i}\times\frac{1}{-12.60\times 10^{-7}}+\frac{e}{33.7}=1 \qquad \ldots(t_i)$$

Final conditions (3) $\dfrac{0.85\,P_i}{2420}+\dfrac{0.85\times P_i\times e}{81{,}560}-228-1177=-100$

$$\therefore \quad 1+\frac{e}{33.7}=\frac{1305\times 2420}{0.85\,P_i}$$

$$\therefore \quad \frac{1}{P_i}\times\frac{1}{2.696\times 10^{-7}}+\frac{e}{33.7}=1 \qquad \ldots(b)$$

(4) $\dfrac{0.85\,P_i}{2420}-\dfrac{0.85\,P_i\times e}{81{,}560}+228+1177=1400$

$$\therefore \quad 1-\frac{e}{33.7}=\frac{-5\times 2420}{0.85\,P_i}$$

$$\therefore \quad \frac{1}{P_i}\times\frac{1}{-702.5\times 10^{-7}}+\frac{e}{33.7}=1 \qquad \ldots(t)$$

Maximum allowable eccentricity

$$\frac{\text{Equation 2}}{\text{Equation 3}},\quad \frac{\dfrac{1}{P_i}\times\dfrac{1}{12.60\times 10^{-7}}}{\dfrac{1}{P_i}\times\dfrac{1}{2.696\times 10^{-7}}}=\frac{1-\dfrac{e}{33.7}}{1-\dfrac{e}{-33.7}}$$

$$\therefore \quad -\frac{2{\cdot}696}{12{\cdot}60}=\frac{1-\frac{e}{33{\cdot}7}}{1+\frac{e}{33{\cdot}7}}$$

$$\therefore \quad \frac{9{\cdot}904}{12{\cdot}60}=\frac{2}{1+\frac{e}{33{\cdot}7}}$$

$$9{\cdot}904+0{\cdot}2939e=25{\cdot}2$$

$$\therefore \quad e=\frac{15{\cdot}296}{0{\cdot}2939}$$

$$=52{\cdot}06 \text{ cm.}$$

Putting $e=52$ cm

$$\frac{1}{P_i}\times\frac{10^7}{2{\cdot}696}=1+\frac{e}{33{\cdot}7}=1+\frac{52}{33{\cdot}7}=2{\cdot}543$$

$$\therefore \quad P_i=\frac{10^7}{2{\cdot}696\times 2{\cdot}543}=1{,}458{,}000 \text{ N}$$

Use 7 mm wires.

Initial prestress in one cable of 12 wires

$$=1059\times\frac{\pi}{4}\times 4^2\times 12$$

$$=489{,}000 \text{ N}$$

No. of cables $=\dfrac{1{,}458{,}000}{489{,}000}=2{\cdot}982$ say 3.

Force in each cable$=\dfrac{1{,}458{,}000}{2}=486{,}000$ N

The arrangement of cables is shown in the figure.

The C.G. of cables from base

$$=\frac{2\times 9+1\times 21}{3}=13 \text{ cm}$$

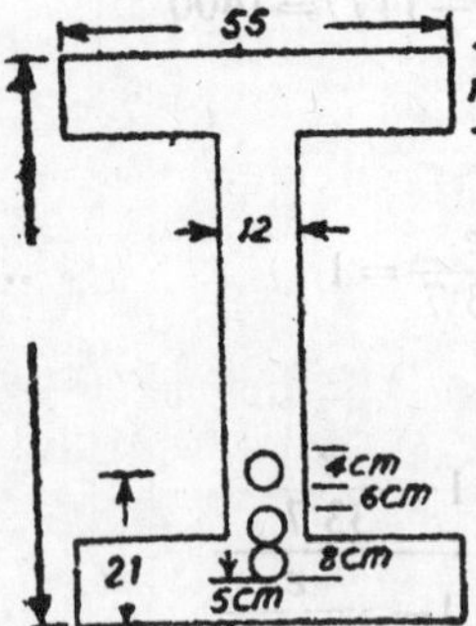

Fig. 25·46.

$$e=65-13=52 \text{ cm}$$

$$d=130-13=117 \text{ cm}$$

Check for ultimate load

$$A_{tot}=0{\cdot}68\,F_c(b-b')\,\frac{t}{F_s}$$

$$=6{\cdot}68\times 3500\times 43\times\frac{10}{150{,}000}$$

$$=6{\cdot}823 \text{ cm}^2$$

$$A_{ts}=\frac{\pi}{4}\times(0{\cdot}7)^2\times12\times3=12{\cdot}86\ \text{cm}^2$$

$$A_{tsw}=A_{ts}-A_{tsf}=13{\cdot}86-6{\cdot}823$$
$$=7{\cdot}037\ \text{cm}^2$$

$$\frac{A_{tsw}\times F_s}{b'\times d\times F_c}=\frac{7{\cdot}037\times150{,}000}{12\times117\times3500}$$

$=0{\cdot}2147<0{\cdot}24$ the section is under reinforced

$$\frac{d}{4}=\frac{117}{4}=29{\cdot}25<t$$

$$M_{us}=K_{us}\times A_{tsw}\times F_s\times d\left[1-0{\cdot}75\times\frac{A_{tsw}\times F_s}{b'\times d\times F_c}\right]+0{\cdot}7\ F_c\times t\ (b-b')\ (d-0{\cdot}5t)$$

$$=1\times7{\cdot}037\times150{,}000\times117\left[1-0{\cdot}75\times\frac{7{\cdot}037\times150{,}000}{12\times117\times3500}\right]+0{\cdot}7\times3500\times10\ (55-12)(11-0{\cdot}5\times170)$$

$$=7{\cdot}037\times150{,}000\times117\times0{\cdot}839+0{\cdot}7\times3500\times10\times43\times112$$
$$=103{,}600{,}000+118{,}000{,}000$$
$$=221{,}600{,}000\ \text{N cm.}$$

Ultimate moment required

$$=1{\cdot}5\ \text{D.L.}+2{\cdot}5\ \text{L.L.}$$
$$=1{\cdot}5\times185{,}800+2{\cdot}5\times960.000$$
$$=278{,}700+2{,}400{,}00$$
$$=2\ 678{,}700\ \text{N-m}$$
$$=267{,}870{,}000\ \text{N-cm}$$

This section is revised as shown in Fig. 25·47.

$$A=2\times55\times10+115\times12$$
$$=1100+1380$$
$$=2480\ \text{cm}^2.$$

$$I=\frac{55\times(135)^3}{12}-\frac{43\times(115)^3}{12}$$
$$=12{,}270{,}000-5{,}450{,}000$$
$$=6{,}820{,}000\ \text{cm}^4$$

$$Z=\frac{6{,}820{,}000}{67{\cdot}5}=101{,}000\ \text{cm}^3$$

$$k_b=k_t=\frac{Z}{A}=\frac{101{,}000}{2480}=40{\cdot}7\ \text{cm}$$

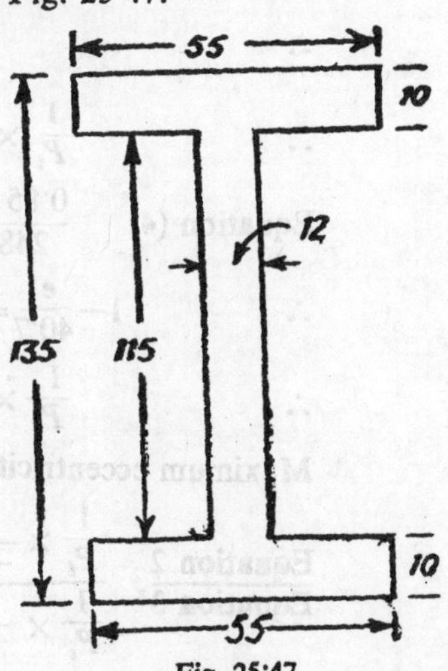

Fig. 25·47.

Dead load B.M., $M_D = \frac{2480 \times 2.4 \times 16^2}{8} = 190{,}400$ N-m

$$C_{bd} = C_{td} = \frac{190{,}400 \times 100}{101{,}000} = 189 \text{ N/cm}^2$$
$$= 1.89 \text{ N/mm}^2$$

$$C_{ba} = C_{ta} = \frac{960{,}000 \times 100}{101{,}000} = 951 \text{ N/cm}^2$$
$$= 9.51 \text{ N/mm}^2$$

Initial condition

Equation (1) $\quad \frac{P_i}{2480} + \frac{P_i \times e}{101{,}000} - 189 = 1750$

$\therefore \quad 1 + \frac{e}{40.7} = \frac{1939 \times 2480}{P_i}$

$\therefore \quad \frac{1}{P_i} \times \frac{1}{2.079 \times 10^{-7}} + \frac{e}{-40.7} = 1 \qquad \ldots(b_i)$

Equation (2) $\quad \frac{P_i}{2480} - \frac{P_i \times e}{101{,}000} + 189 = -100$

$\therefore \quad 1 - \frac{e}{40.7} = -\frac{289 \times 2480}{P_i}$

$\therefore \quad \frac{1}{P_i} \times \frac{1}{-13.95 \times 10^{-7}} + \frac{e}{40.7} = 1 \qquad \ldots(t_i)$

Final condition

Equation (3) $\quad \frac{0.85\, P_i}{2480} + \frac{0.85\, P_i \times e}{101{,}000} - 189 - 951 = -100$

$\therefore \quad 1 + \frac{e}{40.7} = \frac{1040 \times 2480}{0.85 \times P_i}$

$\therefore \quad \frac{1}{P_i} \times \frac{1}{3.296 \times 10^{-7}} + \frac{e}{-40.7} = 1 \qquad \ldots(b)$

Equation (4) $\quad \frac{0.85\, P_i}{2480} - \frac{0.85\, P_i \times e}{101{,}100} + 189 + 951 = 1400$

$\therefore \quad 1 - \frac{e}{40.7} = \frac{260 \times 2480}{0.85\, P_i}$

$\therefore \quad \frac{1}{P_i} \times \frac{1}{13.18 \times 10^{-7}} + \frac{e}{40.7} = 1 \qquad \ldots(t)$

Maximum eccentricity

$$\frac{\text{Equation 2}}{\text{Equation 3}'} \quad \frac{\frac{1}{P_i} \times \frac{1}{-13.95 \times 10^{-7}}}{\frac{1}{P_i} \times \frac{1}{3.296 \times 10^{-7}}} = \frac{1 - \frac{e}{40.7}}{1 - \frac{e}{-40.7}}$$

$$\therefore \quad \frac{-3.296}{13.95}=\frac{1-\dfrac{e}{40.7}}{1+\dfrac{e}{40.7}}$$

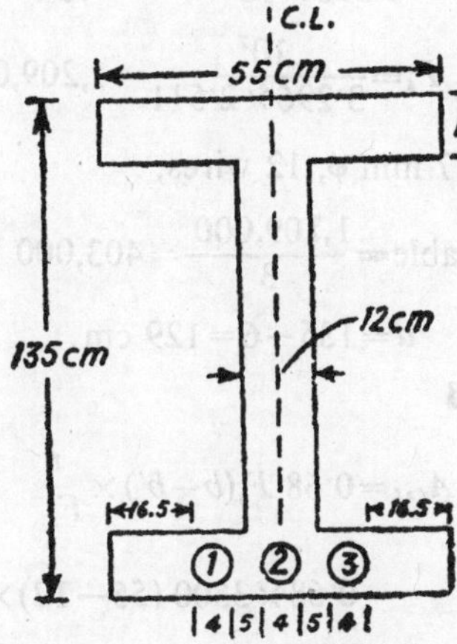

Fig. 25·48

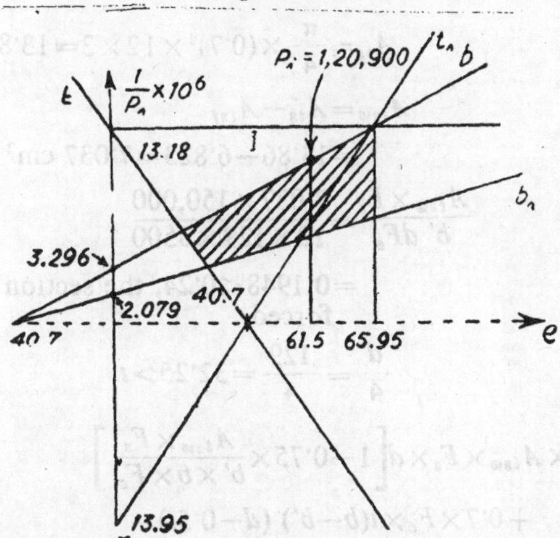

Fig. 25·49

$$\frac{10.654}{13.950}=\frac{2}{1+\dfrac{e}{40.7}}$$

$$\therefore \quad 10.654+0.2616\,e=27.90$$

$$\therefore \quad e=\frac{17.246}{0.2616}$$

$$=65.95 \text{ cm.}$$

Available eccentricity$=67{\cdot}5-6=61{\cdot}5$ cm

Putting this eccentricity in equation 3,

$$\frac{1}{P_i}=\frac{1}{3{\cdot}296\times10^{-7}}=1+\frac{61{\cdot}5}{40{\cdot}7}=2{\cdot}511$$

$$\therefore \qquad P_i=\frac{10^7}{3{\cdot}296\times2{\cdot}511}=1{,}209{,}000 \text{ N}$$

Putting 3 cables, 7 mm ϕ, 12 wires,

$$\text{Force in each cable}=\frac{1{,}209{,}000}{3}=403{,}000 \text{ N}$$

$$d=135-6=129 \text{ cm}$$

Check for ultimate load

$$A_{tsf}=0{\cdot}68\ F_s(b-b')\times\frac{t}{F_s}$$

$$=0{\cdot}68\times3500\,(55-12)\times\frac{10}{150{,}000}$$

$$=6{\cdot}823 \text{ cm}^2$$

$$A_{ts}=\frac{\pi}{4}\times(0{\cdot}7)^2\times12\times3=13{\cdot}86 \text{ cm}^2$$

$$A_{tsw}=A_{ts}-A_{tsf}$$

$$=13{\cdot}86-6{\cdot}823=7{\cdot}037 \text{ cm}^2$$

$$\frac{A_{tsw}\times F_s}{b'\,dF_c}=\frac{7{\cdot}037\times150{,}000}{12\times129\times3500}$$

$=0{\cdot}1948<0{\cdot}24$, the section is under reinforced

$$\frac{d}{4}=\frac{129}{4}=32{\cdot}25>t$$

$$M_{ws}=K_{ws}\times A_{tsw}\times F_s\times d\left[1-0{\cdot}75\times\frac{A_{tsw}\times F_s}{b'\times b\times F_c}\right]$$

$$+0{\cdot}7\times F_c\times t(b-b')\,(d-0{\cdot}5t)$$

$$=1\times7{\cdot}037\times150{,}000\times129\left[1-0{\cdot}75\times\frac{7{\cdot}037\times150{,}000}{12\times129\times3500}\right]$$

$$+0{\cdot}7\times3500\times10(55-12)(129-0{\cdot}5\times10)$$

$$=7{\cdot}037\times150{,}000\times129\times0{\cdot}8539+0{\cdot}7\times3500\times10\times43\times124$$

$$=124{,}000{,}000+130{,}600{,}000$$

$$=254{,}600{,}000 \text{ N-cm}=2{,}546{,}000 \text{ N-m}$$

Ultimate moment required

$=1{\cdot}5$ D.L. $+2{\cdot}5$ L.L.

$=1{\cdot}5\times 190{,}400+2{\cdot}5\times 960{,}000$

$=285{,}600+2{,}400{,}000$

$=2{,}685{,}600$ N-m

Additional u'⁺imate moment required

$=138{,}400$ N-m

Provide mild steel reinforcement 4 bars, 25 mm ϕ, **having** centres at 4 cm from bottom.

$$\text{Equivalent steel} = \frac{19{\cdot}63\times 14{,}000}{150{,}000}=1{\cdot}832 \text{ cm}^2$$

$$\text{Effective depth} = 135-\left(\frac{13{\cdot}86\times 6+1\ 832\times 4}{13{\cdot}86+1{\cdot}832}\right)$$

$$=135-5{\cdot}8$$

$$=129{\cdot}2 \text{ cm}$$

$$A_{tw}=7{\cdot}037+1{\cdot}832=8{\cdot}869$$

$$M_{ut}=8{\cdot}869\times 150{,}000\times 129{\cdot}2\left[1-0{\cdot}75\times\frac{8{\cdot}869\times 150{,}000}{12\times 129{\cdot}2\times 3500}\right]$$

$$+0{\cdot}7\times 3500\times 10\times 43\times(129{\cdot}2-0{\cdot}5\times 10)$$

$$=8{\cdot}869\times 150{,}000\times 129{\cdot}2\times 0{\cdot}8161$$

$$+0{\cdot}7\times 3500\times 10\times 43\times 124{\cdot}2$$

$$=140{,}300{,}000+130{,}800{,}000$$

$$=271{,}100{,}000 \text{ N-m}$$

$=2{,}711{,}000$ N-m $>2{,}685{,}600$. Safe.

Check for stresses

$$\frac{P_i}{A}=\frac{1{,}209{,}000}{2480}=487{\cdot}4 \text{ N/cm}^2$$

$$=4{\cdot}874 \text{ N/mm}^2$$

$$\frac{0{\cdot}85\ P_i}{A}=\frac{0{\cdot}85\times 1{,}209{,}000}{2480}=414{\cdot}3 \text{ N/cm}^2$$

$$=4{\cdot}143 \text{ N/mm}^2$$

$$\frac{P_i\times e}{Z}=\frac{1{,}209{,}000\times 61{\cdot}5}{101{,}000}=736{\cdot}7 \text{ N/cm}^2$$

$$=7{\cdot}367 \text{ N/mm}^2$$

$$\frac{0{\cdot}85\ P_i\times e}{Z}=\frac{0{\cdot}85\times 1{,}209{,}000\times 6}{101{,}000}$$

$$=625{\cdot}8 \text{ N/cm}^2=6{\cdot}258 \text{ N/mm}^2$$

Initial stresses

At bottom $487.4+736.2-189=1034\ N/cm^2=10.34\ N/mm^2$

At top $487.4-736.2+189=-59.8\ N/cm^2=0.598\ N/mm^2$

Final stresses

At bottom $414.3+625.8-189-951=-99.9\ N/cm^2$

$=-0.999\ N/mm^2$

At top $414.3-625.8+189+951=938.5\ N/cm^2$

$=9.385\ N/mm^2$

Mild steel reinforcement

Steel reinforcement

$$=\frac{0.1}{100}\times 12\times 100=1.2\ cm^2$$

Provide 6 mm ϕ bars on both faces at $\dfrac{100\times 0.6}{0.28}$

$=35.56$ cm c/c say at 35 c/c

Longitudinal steel $=\dfrac{0.2}{100}\times 2480=4.96\ cm^2$

Provide 10 bars of 10 mm ϕ

The final section with reinforcement is shown in Fig. 25·50.

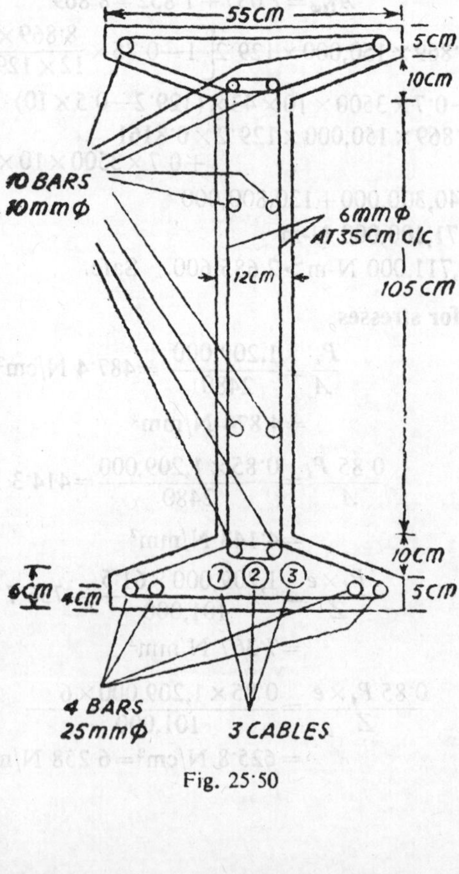

Fig. 25·50

Check for cracking

$$1{\cdot}1 \text{ D.L.}+1{\cdot}25 \text{ L.L.}=1{\cdot}1\times190{,}400+1{\cdot}25\times960{,}000$$
$$=1209{,}440 \text{ N-m}$$

$$\text{Modulus of rupture}=7\sqrt{3500}=414{\cdot}2 \text{ N/cm}^2$$
$$=4{\cdot}142 \text{ N/mm}^2$$

$$\text{Final stress at bottom}=414{\cdot}3+626{\cdot}8-\frac{1{,}209{,}440\times100}{101{,}000}$$
$$=1040{\cdot}1-1197$$
$$=-156{\cdot}9 \text{ N/cm}^2=-1{\cdot}569 \text{ N/mm}^2. \quad \text{Safe.}$$

Cable layout

The cable layout is shown in Fig. 25·51.

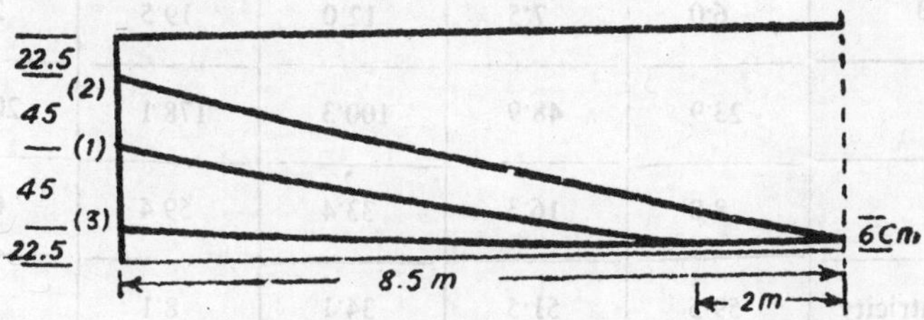

Fig. 25·51

Cable 2 $y=Ax^2+6$

At $x=8{\cdot}5$ m,

$$112{\cdot}5=A\times(8{\cdot}5)^2+6$$

$$\therefore \quad A=\frac{106{\cdot}5}{(8{\cdot}5)^2}$$

$$\therefore \quad y=\frac{106{\cdot}5}{(8{\cdot}5)^2}\times x^2+6 \qquad (0 \text{ to } 8{\cdot}5 \text{ m})$$

Cable 1 $y=A(x-2)^2+6.$

At $x=8{\cdot}5$ m, $67{\cdot}5=A(6{\cdot}5)^2+6.$

$$\therefore \quad A=\frac{61{\cdot}5}{(6{\cdot}5)^2}$$

$$\therefore \quad A=\frac{61{\cdot}5}{(6{\cdot}5)^2}(x-2)^2+6. \qquad (2 \text{ to } 8{\cdot}5 \text{ m})$$

Cable 3

At $x=8{\cdot}5$ m, $22{\cdot}5=A(6{\cdot}5)^2+6.$

$$\therefore \quad A=\frac{16{\cdot}5}{(6{\cdot}5)^2}$$

$$\therefore \quad y=\frac{16{\cdot}5}{(6{\cdot}5)^2}(x-2)^2+6. \qquad (2 \text{ to } 8{\cdot}5 \text{ m})$$

The various ordinates are shown in Table 25·7.

Table 25·7

Cable No.	*Position from bottom*				
	2 m	*4 m*	*6 m*	*8 m*	*8·5 m*
2	1·9	29·6	59·1	100·4	112·5
1	6·0	11·8	29·2	58·2	67·5
3	6·0	7·5	12·0	19·5	22·5
Total	23·9	48·9	100·3	178·1	202·5
C.G.	8·0	16·3	33·4	59·4	67·5
Eccentricity	59·5	51·5	34·1	8·1	0

Check for stresses at quarter section 4 m from centre

$$\frac{P_i}{A}=\frac{1,209,000}{2480}=487·4 \text{ N/cm}^2$$

$$\frac{0·85\,P_i}{A}=\frac{0·85\times 1,209,000}{2480}=414·3 \text{ N/cm}^2$$

$$\frac{P_i\times e}{Z}=\frac{1,209,000\times 51·5}{101,000}=619 \text{ N/cm}^2$$

$$\frac{0·85\times P_i\times e}{Z}=\frac{0·85\times 1,209,000\times 51·5}{101,000}$$

$$=526·1 \text{ N/cm}^2$$

$$C_{bd}=C_{td}=189\left(1-\frac{4^2}{8^2}\right)=94·5 \text{ N/cm}^2$$

$$C_{bl}=C_{td}=951\left(1-\frac{4^2}{8^2}\right)=475·5 \text{ N/cm}^2$$

Initial stresses

At bottom $=487·4+619-94·5=1011·9$ N/cm²

At top $=487·4-619+94·5=-37·1$ N/cm²

Final stresses

At bottom $=414{\cdot}3+526{\cdot}1-94{\cdot}5-475{\cdot}5=370{\cdot}4$ N/cm²

At top $=414{\cdot}3-526{\cdot}1+94{\cdot}5+475{\cdot}5=458{\cdot}7$ N/cm². Safe.

Check can be made for other sections.

End Block. End block is designed by empirical method.

Anchorage diagram = 12·2 cm

Side of equivalent square

$$=\sqrt{\frac{\pi}{4}\times(12{\cdot}2^2)}$$

$$=10{\cdot}8 \text{ cm}=2a_1$$

$$\therefore \quad a_1=5{\cdot}4 \text{ cm}$$

$$2a=45 \text{ cm}$$

$$\therefore \quad a=22{\cdot}5$$

$$\frac{a_1}{a}=\frac{5{\cdot}4}{22{\cdot}5}=0{\cdot}24$$

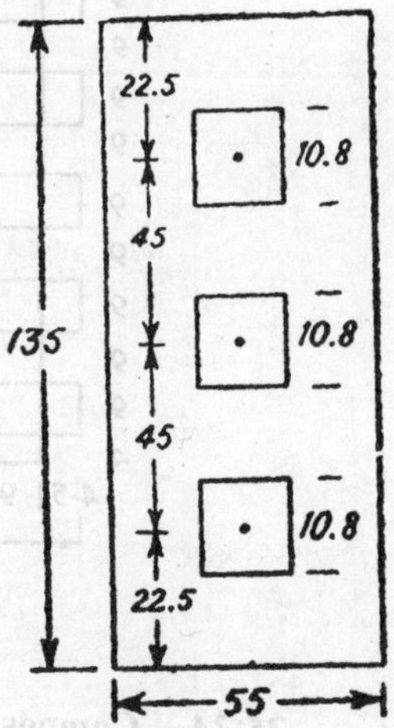

Fig. 25·52

Force per cable

$$=\frac{1{,}209{,}000}{3}=403{,}000 \text{ N}$$

$$f_c=\frac{403{,}000}{(10{\cdot}8)^2}=3455 \text{ N/cm}^2$$

$$f_y=f_c\left[0{\cdot}98-0{\cdot}825\,\frac{a_1}{a}\right]$$

$$=3455\,[0{\cdot}98-0{\cdot}825\times0{\cdot}24]$$

$$=2702 \text{ N/cm}^2$$

Total tensile force $=T=P\times\left[0{\cdot}48-0{\cdot}4\times\frac{a_1}{a}\right]$

$$=403{,}000[0{\cdot}48-0{\cdot}04\times0{\cdot}24]$$

$$=154{,}700 \text{ N}$$

Area of steel $=\frac{154{,}700}{14{,}000}=11{\cdot}07$ cm²

This is to be provided in 45 cm depth

Fig. 25 53

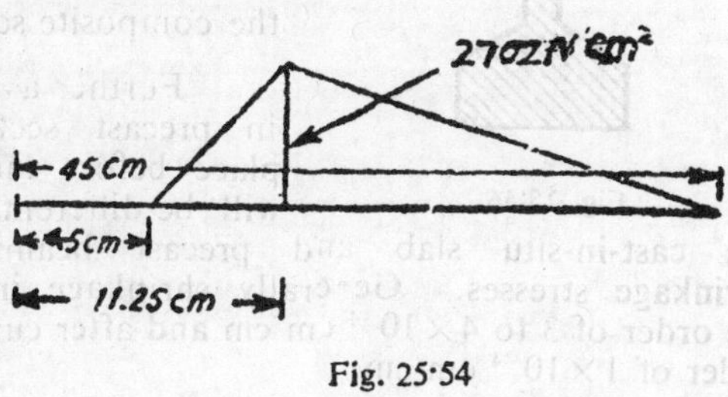

Fig. 25·54

Provide 3 meshes one at 10 cm, 25 cm and 40 cm from end. The mesh consists of 10 mm, ϕ bars, 14 in horizontal direction and 7 bars in vertical direction as shown in Fig. 25·55.

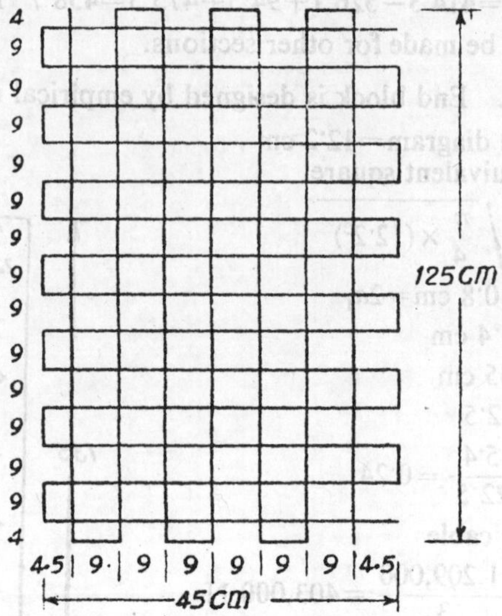

Rectangular Mesh
Fig. 25·55

25·24. Composite Section. Many times, prestressed beams are cast first and launched on site and slab is laid on top of these beams. There are two ways of casting the slab (*i*) when the beams are propped (*ii*) when the beams are unpropped. The action of live load will be same in both cases but in propped beams the load of the slab will not come on the beams while slab is cast, but this dead load of slab and the live load will be taken by the composite section. In case of unpropped beams, the dead load of slab will be borne by the beams as superimposed load and the live load will be taken by the composite section.

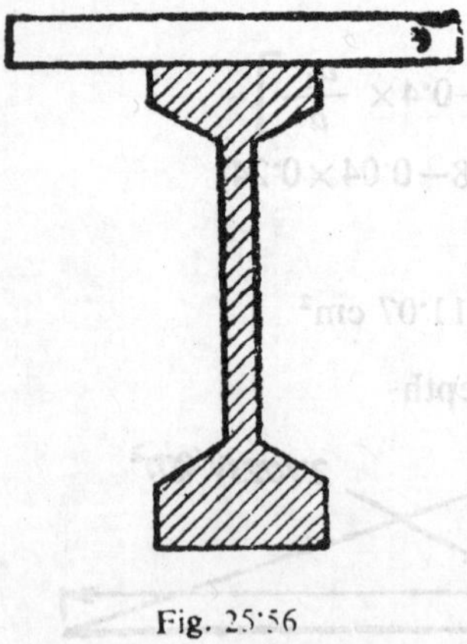

Fig. 25·56

Further as mostly the shrinkage in precast section will have taken place before launching on site there will be differential shrinkage between the cast-in-situ slab and precast beams. This will induce shrinkage stresses. Generally shrinkage in fresh concrete is of the order of 3 to 4×10^{-4} cm/cm and after curing this will be of the order of 1×10^{-4} cm/cm.

Further the prestressed unit will be of a richer mix having higher modulus of elasticity. The in-situ slab is generally M 200 which will have less modulus of elasticity. This fact is taken into account in analysing as a composite section. This section will be equivalent section of precast modulus of elasticity. Let modular ratio be m.

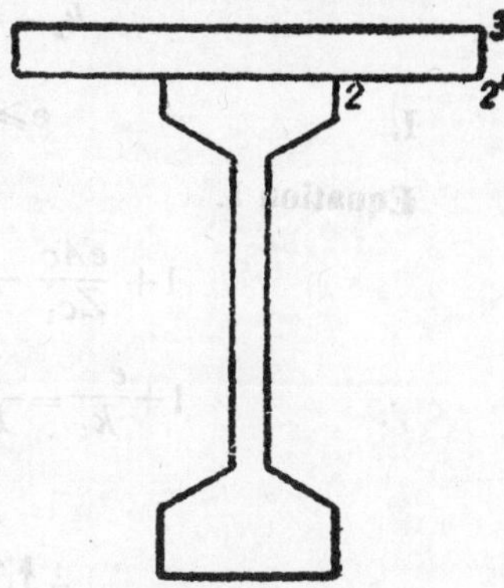

Fig. 25·57

25·25. I. Propped

Let Ic = M.I. of prestressed precast section.

Let Ac = Area of prestressed

I = M.I. of composite section.

P_i = Initial prestress.

η = Loss ratio.

D_{in} = Allowable stress in cast-in-situ concrete.

Zc_1, Zc_2 moduli of section for precast at 1 and 2,

Z_1, Z_2, Z_3 moduli of section for composite at 1, 2 and 3.

Let M_A be the dead load bending moment due to precast unit.

M_B be the dead load bending moment due to in-situ slab and

M_L be the live load bending moment.

$$C_{1A}=\frac{M_A}{Zc_1},\ C_{2A}=\frac{M_A}{Zc_2}$$

$$C_{1B}=\frac{M_B}{Z_1},\ C_{2B}=\frac{M_B}{Z_2},\ C_{3B}=\frac{M_B}{Z_3}m$$

$$C_{1L}=\frac{M_L}{Z_1},\ C_{2L}=\frac{M_L}{Z_2},\ C_{3L}=\frac{M_L}{Z_3}m$$

The following equations will have to be satisfied.

$$(i)\ \frac{P_i}{Ac}+\frac{P_i e}{Zc_1}-C_{1A}\leqslant C'$$

$$(ii)\ \frac{P_i}{Ac}-\frac{P_i e}{Zc_2}+C_{2A}\geqslant C_t'.$$

$$(iii)\ \frac{nP_i}{Ac}+\frac{nP_i e}{Zc_1}-C_{1A}-C_{1B}-C_{1L}\geqslant C_t.$$

$$(iv)\ \frac{nP_i}{Ac}-\frac{nP_i e}{Zc_2}+C_{2A}+C_{2B}+C_{2L}\leqslant C.$$

(v) $C_{3B}+C_{[illegible]}<C_{tn}$.

$$k_b=\frac{Zc_2}{Ac},\ k_t=\frac{Zc_1}{Ac}$$

I. $\qquad e \geqslant k_b$

Equation 1.

$$1+\frac{eAc}{Zc_1}-C_{1A}\times\frac{Ac}{P_i}=C'-\frac{Ac}{P_i}$$

$$\therefore \qquad 1+\frac{e}{k_t}=\frac{1}{P_i}\times Ac[C'+C_{1A}].$$

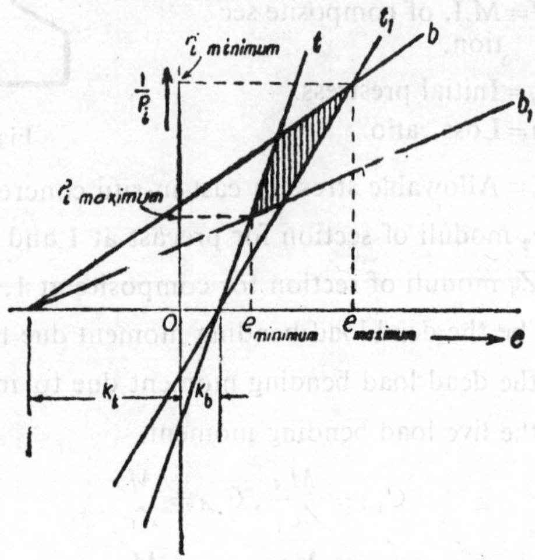

Fig. 25·58

$$\therefore \qquad \frac{e}{-k_t}+\frac{1}{P_i}\times\frac{1}{\dfrac{1}{Ac(C'+C_{1A})}}=1 \qquad \ldots(b_i)$$

Equation 2.

$$1-\frac{e}{k_b}=(C_{2A}-C'_t)\times\frac{Ac}{P_i}$$

$$\therefore \qquad \frac{e}{k_b}+\frac{1}{P_i}\times\frac{1}{\dfrac{-1}{Ac(C_{2A}-C_t')}}=1 \qquad \ldots(t_i)$$

Education 3.

$$1+\frac{e}{k_t}=(C_{1A}+C_{1B}+C_{1L}+C_t)\times\frac{Ac}{nP_i}$$

$$\therefore \quad \frac{e}{-k_t}+\frac{1}{P_i}\times\frac{1}{\dfrac{n}{Ac(C_{1A}+C_{1B}+C_{1L}+C_t)}}=1 \quad ...(b)$$

Equation 4 $1-\frac{e}{k_b}=(C-C_{2A}-C_{2B}-C_{2L})\frac{A}{nP_i}$

$$\therefore \quad \frac{e}{k_b}+\frac{1}{\dfrac{-n}{Ac(C_{2A}+C_{2B}+C_{2L}-C)}}=1 \quad ...(t)$$

The line b should be above b_i and t should be above t_i. Otherwise there will be no solution.

II. $\quad e \leqslant k_b$

The equation number 2 will not be applicable unless $C_i \geqslant C_{2A}$, equation 4 will have always +ve intercept on $1/P_i$ axis as shown in the graph. Therefore, $C \geqslant (C_{2A}+C_{2B}+C_{2L})$.

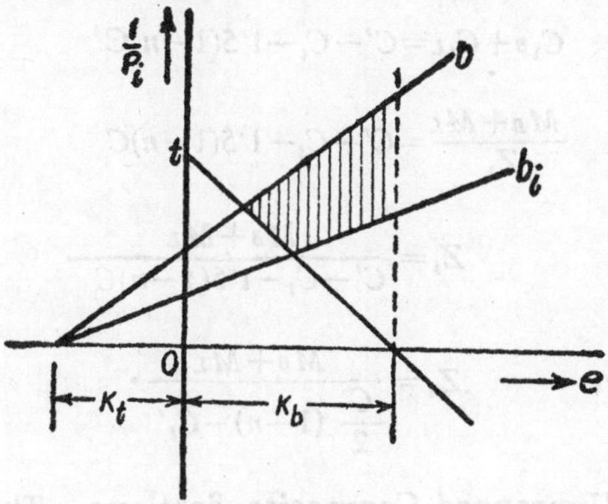

Fig. 25·59

25·26. Design of Propped Composite Sections

$e \geqslant k_b$ and $C_t' = C_t$.

$$\frac{\text{Equation 2}}{\text{Equation 3}}=\frac{\dfrac{P_i}{Ac}\left[1-\dfrac{eAc}{Zc_2}\right]}{\dfrac{nP_i}{Ac}\left[1+\dfrac{eAc}{Zc_1}\right]}=\frac{C_t-C_{2A}}{(C_{1A}+C_{1B}+C_L+C_t)}$$

$$\therefore \quad \frac{1-\dfrac{e\,Ac}{Zc_2}}{1\times\dfrac{e\,Ac}{Zc_1}}=\frac{n(C_t-C_{2A})}{(C_{1A}+C_{1B}+C_{1L}+C_t)}$$

From this, value of e can be obtained ; putting this value of e in equations 2 and 4, the values of P_i are obtained. The greater value will give the prestressing force.

Equation 1 − Equation 3 will give

$$\frac{P_i}{Ac}(1-n)+\frac{P_ie}{Zc_1}(1-n)=C'-C_t-C_{1B}-C_{1L}$$

$$\therefore \quad \frac{P_i}{Ac}\left(1-n\right)\left[1+\frac{e}{k_t}\right]=C'-C_t-C_{1B}-C_{1L}$$

Now $\quad \frac{P_i}{Ac}=\frac{C}{2} \quad$ and $\quad \frac{e}{k_t}=2$

$$\therefore \quad \frac{C}{2}(1-n)\times 3=C'-C_t-C_{1B}-C_{1L}$$

$$\therefore \quad C_{1B}+C_{1L}=C'-C_t-1{\cdot}5(1-n)C$$

$$\therefore \quad \frac{M_B+M_L}{Z_1}=C'-C_t-1{\cdot}5(1-n)C$$

$$\therefore \quad Z_1=\frac{M_B+M_L}{C'-C_t-1{\cdot}5(1-n)C}$$

Similarly, $\quad Z_2=\dfrac{M_B+M_L}{\dfrac{C}{2}(1-n)-C_t'}$.

25·27. Unpropped Composite Sections. The dead load of slab will be taken by the precast section.

$$C_{1A}=\frac{M_A}{Zc_1},\ C_{2A}=\frac{M_A}{Zc_2};\ C_{1B}=\frac{M_B}{Zc_1},\ C_{2B}=\frac{M_L}{Zc_2}$$

$$C_{1L}=\frac{M_L}{Z_1},\ C_{2L}=\frac{M_L}{Z_2};\ C_{3L}=\frac{m\,M_L}{Z_3}.$$

The following equations should be satisfied

(i) $\quad \frac{P_i}{Ac}+\frac{P_ie}{Zc_1}-C_{1A}\leqslant C'$

(ii) $\quad \frac{P_i}{Ac}-\frac{P_ie}{Zc_2}+C_{2A}\geqslant C_t'$

(iii) $\dfrac{nP_i}{Ac}+\dfrac{nP_ie}{Zc_1}-C_{1A}-C_{1B}-C_{1L}\geqslant C_t$

(iv) $\dfrac{nP_i}{Ac}-\dfrac{nP_ie}{Zc_2}+C_{2A}+C_{2B}+C_{2L}\leqslant C_t$

(v) $\dfrac{mM_L}{Z_3}<C_{in}.$

Let $\alpha=\dfrac{Z_1}{Zc_1}$ and $\beta=\dfrac{Z_1}{Zc_2}$.

Equation 1—Equation 3.

$$\frac{P_i}{Ac}(1-n)+\frac{P_ie}{Zc_1}(1-n)=C'-C_t-C_{1B}-C_{1L}$$

$$\frac{P_i}{Ac}(1-n)\left[1+\frac{e}{k_t}\right]=C'-C_t-C_{1B}-C_{1L}$$

$$\frac{P_i}{Ac}=\frac{C}{2} \text{ and } \frac{e}{k_t}=2.$$

$$\frac{C}{2}(1-n)\times 3=C'-C_t-C_{1B}-C_{1L}$$

$$C_{1B}+C_{1L}=C'-C_t-1{\cdot}5\ C(1-n)$$

$$\frac{M_B}{Zc_1}+\frac{M_L}{Zc_1}=C'-C_t-1{\cdot}5\ C(1-n)$$

$$\frac{1}{Z_1}\left(\frac{Z_1}{Zc_1}M_B+M_L\right)=C'-C_t-1{\cdot}5\ C(1-n).$$

$$\therefore \qquad Z_1=\frac{\alpha M_B+M_L}{C'-C_t-1{\cdot}5\ C(1-n)}$$

α varies from 1 to 2.

Equation 2—Equation 4.

$$\frac{P_i}{Ac}(1-n)-\frac{P_ie}{Zc_2}(1-n)=C_{2B}+C_{2L}+C_t'-C$$

$$\therefore \quad \frac{P_i}{Ac}(1-n)\left(1-\frac{e}{k_b}\right)=C_{2B}+C_{2L}-C_t'-C$$

$$\frac{P_i}{Ac}=\frac{C}{2},\ \frac{e}{k_b}=2$$

$$\therefore \quad -\frac{C}{2}(1-n)=C_{2B}+C_{2L}+C_t'-C$$

$$\therefore \quad C_{2B}+C_{2L}=\frac{C}{2}(1+n)-C_t'$$

$$\therefore \quad \frac{M_B}{Zc_2}+\frac{M_L}{Z_2}=\frac{C}{2}(1+n)C_t'$$

$$\therefore \quad \frac{1}{Z_2}\left(\frac{Z_2M_B}{Zc_2}+M_L\right)=\frac{C}{2}(1+n)-C'\ ;\ Z_2=\frac{\beta M_B+M_L}{\frac{C}{2}(1+n)-C_t'}$$

β varies from 2 to 5.

25·28. Two-stage prestressing. The prestressing is carried out in two stages. The first prestressing is applied after the precast section is ready, leaving holes for the second stage prestressing. The precast sections are brought to the site and cast-in-situ slab is laid and allowed to cure. Afterwards the second stage prestress is applied. The second stage prestress is taken up by the composite section. Due to elastic shortening there will be reduction in the first stage prestress.

Let P_1 and P_2 be the first stage prestressing force and second stage prestressing force respectively. Let n_1 be the reduction factor due to elastic shortening in first prestress. Initially when precast section is prestressed.

$$C_1A=\frac{M_A}{Zc_1},\ C_2A=\frac{M_A}{Zc_2}\ ;\ \frac{P_1}{Ac}+\frac{P_1e_1}{Zc_1}\ C_{1A} \geqslant C'$$

$$\frac{P_1}{Ac}-\frac{P_1e_1}{Zc_2}+C_{2A} \geqslant C'_t$$

After the second stage prestress.

$$C_{1B}=\frac{M_B}{Zc_1},\quad C_{2B}=\frac{M_B}{Z_2}$$

$$C_{1L}=\frac{M_L}{Z_1},\quad C_{2L}=\frac{M_L}{Z_2},\quad C_{2L}=\frac{mM_L}{Z_3}$$

(1) $$\frac{nn_1P_1}{Ac}+\frac{nn_1P_1e_1}{Zc_1}+\frac{P_2}{A}+\frac{P_2e_2}{Z_1}-C_{1A}-C_{1B} \leqslant C'$$

(2) $$\frac{nn_1P_1}{Ac}-\frac{nn_1P_1e_1}{Zc_2}+\frac{P_2}{A}-\frac{P_2e_2}{Z_2}+C_{2A}+C_{2B} \geqslant C'_t$$

$$(3) \quad \frac{nn_1P_1}{Ac_1}+\frac{nn_1P_1e_1}{Zc_1}+\frac{nP_2}{A}+\frac{nP_2e_2}{Z_1}-C_{1A}-C_{1B}-C_{1L} \geq C_t$$

$$(4) \quad \frac{nn_1P_1}{Ac}-\frac{nn_1P_1e}{Zc_2}+\frac{nP_2}{A}-\frac{nP_2e_2}{Z_b}+C_{2A}+C_{2B}+C_{2L} \leq C$$

$$(5) \quad \frac{mP_2}{A}-\frac{mP_2e_2}{Z_2} \geq C_{tin}.$$

$$(6) \quad \frac{mnP_2}{A}-\frac{mnP_2e_2}{Z_3}+\frac{mM_L}{Z_3} < C_{tin}.$$

Equation 1—Equation 3

$$\frac{P_2}{A}(1-n)+\frac{P_2e_2}{Z_1}(1-n)=C'-C_t-C_{1L}$$

$$\frac{P_2}{A}(1-n)\times\left(1+\frac{e_2A}{Z_1}\right)=C'-C_t-C_{1L}$$

Assuming $\frac{P_2}{A}=\frac{C}{2}$, $\frac{Ae_2}{Z_1}=2$

$$\therefore \quad \frac{C}{2}(1-n)\times 3=C'-C_t-C_{1L}$$

$$\therefore \quad Z_{1L}=C'-C_t-1{\cdot}5C(1-n)$$

$$\therefore \quad Z_1=\frac{M_L}{C'-C_t-1{\cdot}5C(1-n)}$$

Similarly
$$Z_2=\frac{M_L}{\frac{C}{2}(1-n)-C'_t}$$

$$\frac{Z_1}{ZC_2}=\alpha, \text{ where } \alpha \text{ varies from 1 to } 1{\cdot}5.$$

25·29. Shrinkage Stresses. When two materials undergo different shrinkages, stresses will be induced in them. These are called differential shrinkage stresses. Take the case of R.C.C. column having A_c area of concrete and A_s area of steel with modular ratio as m. Let the shrinkage of concrete be α cm/cm. Steel will not shrink.

To work out stresses, first of all concrete is stopped from having natural shrinkage. This will cause tensile stress in concrete equal to $\alpha\times E_c$. The total tensile force will be $\alpha\times E_c\times A_c$. As there is no external force acting, the combined section is subject to axial compressive force of $\alpha\times E_c\times A_c$. This will cause compressive stress in both concrete and steel.

$$\text{Stress in concrete}=\frac{\alpha\times E_c\times A_c}{A_c+mA_s}$$

$$\text{Net stress in concrete}=\alpha E_c-\frac{\alpha E_cA_c}{A_c+mA_s}$$

$$=\alpha E_c\left[\frac{mA_s}{A_c+mA_s}\right] \text{ tensile}$$

Stress in steel$=\dfrac{m\alpha E_c \times A_c}{A_c+mA_s}$ compressive.

In case of composite sections, the in-situ concrete shrinks more than the precast section. Therefore, there will be differential shrinkage stresses induced.

Let A_{in} be the equivalent area of in-situ concrete and α cm/cm be the differential shrinkage and E_c be the Young's modulus of combined section. Let $\bar{y}$ be the distance of centroid of in-situ concrete from C.G. of the composite section. Let m be the modular ratio and A and I be the area and M.I. of composite section.

If shrinkage is not allowed, stress in-situ concrete$=m\alpha E_c$ tensile.

Total compressive force on composite section will be $A_{in}\times\alpha \times E_c$ acting at $\bar{y}$ from C.G.

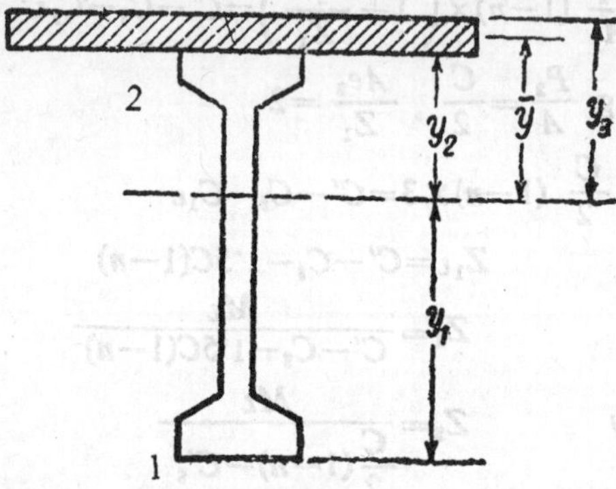

Fig. 25·60

$$\text{Stress at } 1=\frac{A_{in}\alpha E_c}{A}-\frac{A_{in}\alpha E\times\bar{y}\times y_1}{I}$$

$$\text{Stress at } 2=\frac{A_{in}\alpha E_c}{A}+\frac{A_{in}\alpha E_c\times\bar{y}\times y_2}{I}$$

$$\text{Stress at } 3=\frac{mA_{in}\alpha E_c}{A}+\frac{mA_{in}\alpha E_c\times\bar{y}\times y_3}{I}$$

$$\text{Net stress at } 3=m\alpha E_c-\frac{mA_{in}\alpha E_c}{A}-\frac{mA_{in}\alpha E_c\times\bar{y}\times y_3}{I}$$

$$=m\alpha E_c\left[\left(\frac{A-A_{in}}{A}\right)-\left(\frac{A_{in}\times\bar{y}\times y_3}{I}\right)\right] \text{ tensile.}$$

Net stress at 2 in in-situ

$$=m\alpha E_c-\frac{mA_{in}\alpha E_c}{A}-\frac{mA_{in}\alpha E_c\times\bar{y}\times y_2}{I}$$

$$=m\alpha E_c\left[\left(\frac{A-A_{in}}{A}\right)\left(\frac{A_{in}\times\bar{y}\times y_2}{I}\right)\right] \text{ tensile.}$$

Ex. 25 5. *Design a composite beam having span of 10 m. The in-situ slab is 10 cm thick and carries a live load of 20,000 N/m².*

The centre to centre distance of beams is 20 m. The precast sections are propped while laying the slab. Take,

C'=allowable initial compressive stress=1750 N/cm².
C=allowable final ,, ,, =1400 N/cm².
C'_t=initial tensile stress =−100 N/cm².
C_t=final tensile stress =0 N/cm².
C_{in}=allowable stress in in-situ =850 N/cm².
$E_c=3{\cdot}0\times10^6$ N/cm⁶, $m=0{\cdot}8$.
The differential shrinkage is $1{\cdot}5\times10^{-4}$ cm/cm.

Solution.

$$\text{Load from in-situ}=2{\cdot}0\times\frac{10}{100}\times24{,}000=4800\text{ N/m}$$

$$M_B=\frac{4800\times10^2}{8}=60{,}000\text{ N/m}=60\text{ kN-m}$$

Live load from slab=20,000×2=40,000 N

$$M_L=\frac{40{,}000\times10^2}{8}=50{,}000\text{ N-m}=500\text{ kN-m.}$$

As the composite section is subjected to differential shrinkage final stress bottom will be taken as 100 N/cm². Therefore, $C_t=100$ N/cm². The loss factor n will be taken as 0·85.

$$Z_1=\frac{M_B+M_L}{C'-C_t-1{\cdot}5(1-n)C}$$

$$=\frac{(60{,}000+500{,}000)100}{1750-100-1{\cdot}5\times0{\cdot}15\times1400}=\frac{560{,}000\times100}{1335}$$

$$=41{,}950\text{ cm}^3$$

Overall depth$=3\sqrt{M_B+M_L}=3\sqrt{560}=70{\cdot}99$ cm. Adopt overall depth of 75 cm.

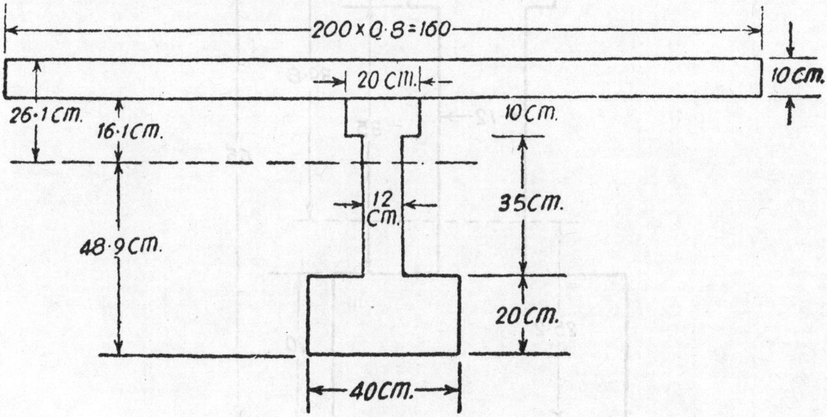

Fig 25.61

All other dimensions are shown in Fig. 25·61.

$$y_1=\frac{160\times10\times70+20\times10\times60+35\times12\times37{\cdot}5+40\times20\times10}{160\times10+20\times10+35\times12+40\times20}$$

$$=\frac{12{,}000+12{,}000+15{,}750+8{,}000}{1{,}600+200+420+800}=\frac{1{,}47{,}750}{3020}=48{\cdot}9\text{ cm.}$$

$$I=\frac{160\times10^3}{12}+1{,}600\times(21{\cdot}1)^2+\frac{20\times(16{\cdot}1)^3}{3}-\frac{8\times(6{\cdot}1)^2}{3}$$

$$+\frac{40\times(48{\cdot}9)^3}{3}-\frac{28\times(28{\cdot}9)^2}{3}$$

$$=13{,}333+712{,}400+27{,}820-605+1{,}559{,}000-225{,}300$$

$$=2{,}086{,}648$$

$$Z_1=\frac{2{,}086{,}648}{48{\cdot}9},\ Z_2=\frac{2{,}086{,}648}{16{\cdot}1},\ Z_3=\frac{2{,}086{,}648}{26{\cdot}1}$$

$$=42{,}690\text{ cm}^3=129{,}600\text{ cm}^3=79{,}980\text{ cm}^3$$

Properties of the precast section

$$yc_1=\frac{20\times10\times60+35\times12\times37{\cdot}5+40\times20\times10}{20\times10+35\times12+40\times20}$$

$$=\frac{12{,}000+15{,}750+8{,}000}{200+420+800}=\frac{35{,}750}{1{,}420}=25{\cdot}2\text{ cm}$$

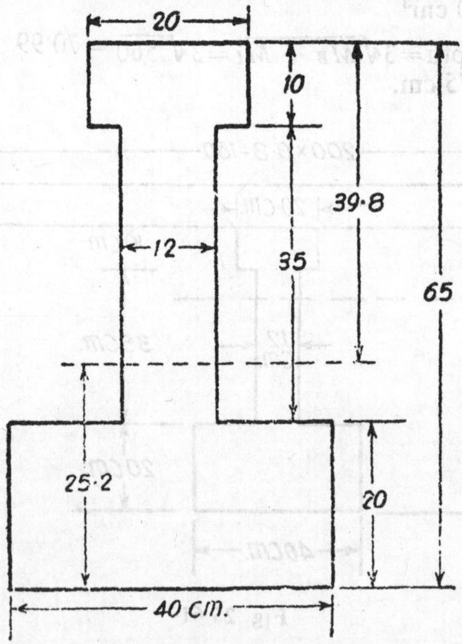

Fig. 25 62

$$yc_2=65-25{\cdot}2=39{\cdot}8 \text{ cm}$$

$$Ic=\frac{20\times(39{\cdot}8)^3}{3}-\frac{8\times(29{\cdot}8)^3}{3}+\frac{40\times(25{\cdot}2)^3}{3}-\frac{28\times(5{\cdot}2)^3}{3}$$

$$=420{,}400-70{,}570+213{,}400-1{,}312=561{,}918 \text{ cm}^4$$

$$Zc_1=\frac{561{,}918}{25{\cdot}2}=22{,}300 \text{ cm}^3$$

$$Zc_2=\frac{561{,}918}{39{\cdot}8}=18{,}860 \text{ cm}^3$$

Dead weight $=1{,}420\times2{\cdot}4=3408$ N/m

$$M_A=\frac{3408\times10^2}{8}=42{,}600 \text{ Nm}$$

$$C_{1A}=\frac{4{,}260{,}000}{22{,}300}=191 \text{ N/cm}^2$$

$$C_{2A}=\frac{4{,}260{,}000}{18{,}860}=225{\cdot}9 \text{ N/cm}^2$$

$$C_{1B}=\frac{6{,}000{,}000}{42{,}590}=140{\cdot}6 \text{ N/cm}^2$$

$$C_{2B}=\frac{6{,}000{,}000}{129{,}000}=46{\cdot}8 \text{ N/cm}^2$$

$$C_{3B}=\frac{0{\cdot}8\times6{,}000{,}000}{79{,}980}=60 \text{ N/cm}^2$$

$$C_{1L}=\frac{50{,}000{,}000}{42{,}690}=1171 \text{ N/cm}^2$$

$$C_{2L}=\frac{50{,}000{,}000}{1{,}29{,}600}=385{\cdot}7 \text{ N/cm}^2$$

$$C_{3L}=\frac{0{\cdot}8\times50{,}000{,}000}{79{,}980}=500{\cdot}1 \text{ N/cm}^2$$

(b_i) .. $$\frac{P_i}{1{,}420}+\frac{P_i\times e}{22{,}300}-191=1750$$

$$\therefore\quad 1+\frac{1{,}420}{22{,}300}\,e=\frac{1941\times1420}{P_i}$$

$$\therefore\quad \frac{1}{P_i}\times\frac{1}{\dfrac{1}{1941\times1{,}420}}+\frac{e}{\dfrac{-22{,}300}{1{,}420}}=1$$

$$\therefore\quad \frac{1}{P_i}\times\frac{1}{3{\cdot}628\times10^{-7}}+\frac{e}{-15{\cdot}7}=1 \qquad ...(b_i)$$

(t_i) .. $$\frac{P_i}{1{,}420}-\frac{P_i\times e}{18{,}860}+225{\cdot}9=-100$$

$$1-\frac{1{,}420}{18{,}860}\,e=\frac{-325{\cdot}9-1{,}420}{P_i}$$

$$-\frac{1}{P_i}\times\frac{1}{\dfrac{1}{325.9-1,420}}+\frac{\dfrac{e}{18,860}}{1,420}=1$$

$$\therefore\quad \frac{1}{P_i}\times\frac{1}{-21.61\times10^{-7}}+\frac{c}{13.28}=1 \qquad \ldots(t_i)$$

$$(b)\ldots\quad \frac{0.85\,P_i}{1,420}+\frac{0.85\,P_i\times e}{22,300}-191-140.6-1171=+100$$

$$\therefore\quad 1+\frac{1,420}{22,300}\,e=\frac{1602.6\times1,420}{0.85\,P_i}$$

$$\therefore\quad \frac{1}{P_i}\times\frac{1}{\dfrac{0.85}{1602.6\times1420}}+\frac{\dfrac{e}{-22,300}}{1,420}=1$$

$$\therefore\quad \frac{1}{P_i}\times\frac{1}{3,735\times10^{-7}}+\frac{e}{-15.7}=1 \qquad \ldots(b)$$

$$(t)\ldots\quad \frac{0.85\,P_i}{1,420}-\frac{0.15\,P_i\times e}{18,860}+225.9+46.3+385.7=1400$$

$$\therefore\quad 1-\frac{1,420}{18,860}\,e=\frac{657.9\times1,420}{0.85\,P_i}$$

$$\therefore\quad \frac{1}{P_i}\times\frac{1}{\dfrac{0.85}{657.9\times1,420}}+\frac{\dfrac{e}{18,860}}{1,420}=1$$

$$\therefore\quad \frac{1}{P_i}\times\frac{3}{9\,097\times10^{-7}}+\frac{e}{13.28}=1 \qquad \ldots(t)$$

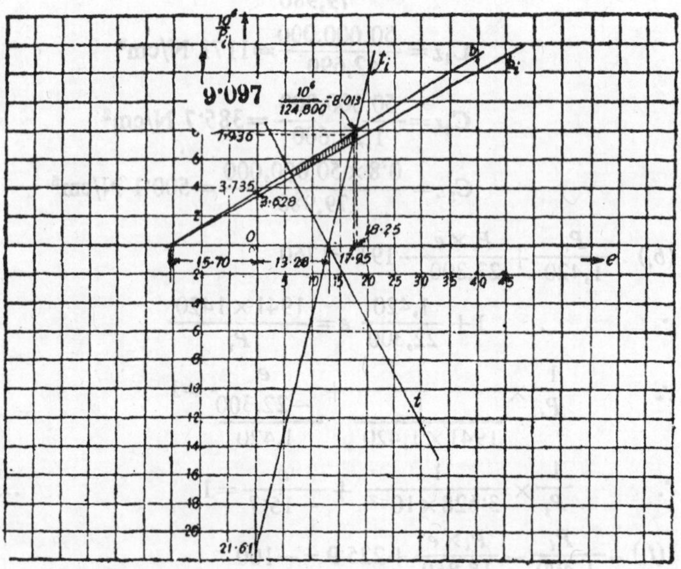

Fig 25 63

For maximum eccentricity

$$\frac{\frac{1}{P_i}\times\frac{10^7}{-21\cdot61}}{\frac{1}{P_i}\times\frac{10^7}{3\cdot735}}=\frac{1+\frac{e}{13\cdot28}}{1+\frac{e}{15\cdot70}}$$

$$\therefore \quad \frac{-3\cdot735}{21\cdot61}=\frac{1-0\cdot0753\,e}{1+0\cdot06369\,e}$$

$$-0\cdot1728=\frac{1-0\cdot0753\,e}{1+0\cdot06369\,e}$$

$$\therefore \quad -0\cdot1728=0\cdot01101\,e=1-0\cdot0753\,e$$

$$\therefore \quad 0\cdot06429\,e=1\cdot1728$$

$$e=\frac{1\cdot1728}{0\cdot06429}=18\cdot25 \text{ cm}$$

Eccentricity provided $=25\cdot2-7\cdot25=17\cdot95$ cm.

For initial prestressing force

(*b*) $\therefore \quad \frac{1}{P_i}\times\frac{10^7}{3\cdot735}=1+\frac{17\cdot95}{15\cdot7}=1+1\cdot144=2\cdot144$

$$\therefore \quad P_i=\frac{10^7}{3\cdot735\times2\cdot144}=1{,}248{,}000 \text{ N}$$

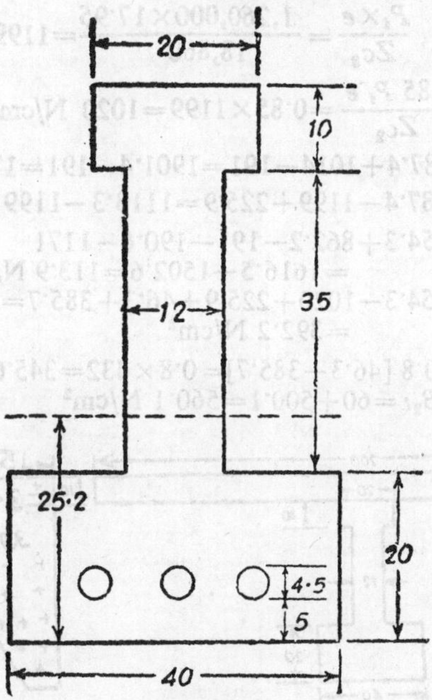

Fig. 25 64

Use 7 mm wires,

Ultimate stress $= 1500 \text{ N/mm}^2$

Working stress $= 0{\cdot}6 \times 1500 = 900 \text{ N/mm}^2$

Initial stress $= \dfrac{900}{0{\cdot}85} = 1059 \text{ N/mm}^2$

Max. force in each cable of 12 wires $= \dfrac{\pi}{4} \times 49 \times 1059 \times 12$

$= 489{,}000$ N

Use 3 cables, force in each $= \dfrac{1{,}248{,}000}{3} = 416{,}000$ N

say 420,000 N

$$P_i = 3 \times 420{,}000 = 1{,}260{,}000 \text{ N}$$

$$\frac{P_i}{Ac} = \frac{1{,}260{,}000}{1{,}420} = 887{\cdot}4 \text{ N/cm}^2$$

$$\frac{0{\cdot}85\, P_i}{Ac} = 0{\cdot}85 \times 887{\cdot}4 = 754{\cdot}3 \text{ N/cm}^2$$

$$\frac{P_i \times e}{Zc_1} = \frac{1{,}260{,}000 \times 17{\cdot}95}{22{,}300} = 1014 \text{ N/cm}^2$$

$$\frac{0{\cdot}85\, P_i\, e}{Zc_1} = 0{\cdot}85 \times 1014 = 862{\cdot}2 \text{ N/cm}^2$$

$$\frac{P_i \times e}{Zc_2} = \frac{1{,}260{,}000 \times 17{\cdot}95}{18{,}860} = 1199 \text{ N/cm}^2$$

$$\frac{0{\cdot}85\, P_i\, e}{Zc_2} = 0{\cdot}85 \times 1199 = 1020 \text{ N/cm}^2$$

Initial stresses
- (*i*) $887{\cdot}4 + 1014 - 191 = 1901{\cdot}4 - 191 = 1710{\cdot}4 \text{ N/cm}^2$
- (*ii*) $887{\cdot}4 - 1199 + 225{\cdot}9 = 1113{\cdot}3 - 1199 = -85{\cdot}7 \text{ N/cm}^2$

Final stresses
- (*iii*) $754{\cdot}3 + 862{\cdot}2 - 191 - 190{\cdot}6 - 1171$
  $= 1616{\cdot}5 - 1502{\cdot}6 = 113{\cdot}9 \text{ N/cm}^2$
- (*iv*) $754{\cdot}3 - 1020 + 225{\cdot}9 + 46{\cdot}3 + 385{\cdot}7 = 1412{\cdot}2 - 1020$
  $= 392{\cdot}2 \text{ N/cm}^2$

$$C_{2B}' + C_{2L}' = 0{\cdot}8\,[46{\cdot}3 + 385{\cdot}7] = 0{\cdot}8 \times 432 = 345{\cdot}6 \text{ N/cm}^2$$

$$C_{3B} + B_{3L} = 60 + 500{\cdot}1 = 560{\cdot}1 \text{ N/cm}^2$$

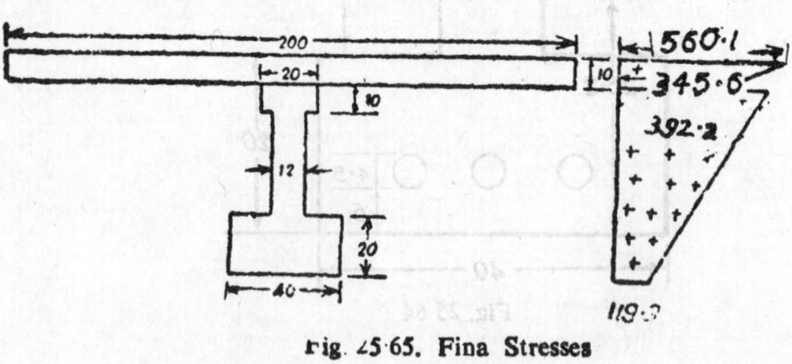

Fig. 25·65. Fina Stresses

Shrinkage stresses

Stress in-situ if shrinkage not allowed

$$=1{\cdot}5\times10^4\times3\times10^6\times(0{\cdot}8)^4=360\ \text{N/cm}^2\ \text{tensile.}$$

Total tensile force $=1{\cdot}5\times10^{-4}\times3\times10^6\times160\times10=720{,}000$ N

Eccentricity $=26{\cdot}1-5=21{\cdot}1$ cm

Force in composite section = 720,000 N compressive acting at 21·1 cm eccentricity.

$$\text{Stress at }1=\frac{72{,}000}{3{,}020}-\frac{72{,}000\times21{\cdot}1}{42{,}690}=23{\cdot}84-35{\cdot}59$$

$$=117{\cdot}5\ \text{kg/cm}^2$$

$$\text{Stress at }2=\frac{720{,}000}{3{,}020}+\frac{720{,}000\times16{\cdot}1}{129{,}600}=238{\cdot}4+117{\cdot}2$$

$$=355{\cdot}6\ \text{N/cm}^2$$

$$\text{Stress at }2'=0{\cdot}8\times\text{Stress at }2=0{\cdot}8-355{\cdot}6=284{\cdot}5\ \text{N/cm}^2$$

$$\text{Stress at }3=0{\cdot}8\left[\frac{720{,}000}{3{,}020}+\frac{720{,}000\times16{\cdot}1}{79{,}980}\right]$$

$$=0{\cdot}8(238{\cdot}4+189{\cdot}9)=0{\cdot}8\times428{\cdot}3=342{\cdot}6\ \text{N/cm}^2$$

Shrinkage stresses not

at $1=-117{\cdot}5$ N/cm²

at $2=+355{\cdot}6$ N/cm²

at $2'=284{\cdot}5-360=-75{\cdot}5$ N/cm²

at $3=342{\cdot}6-360=-17{\cdot}4$ N/cm²

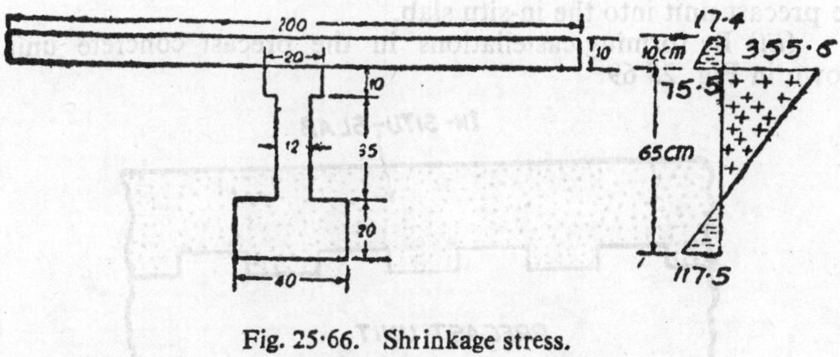

Fig. 25·66. Shrinkage stress.

The Final+Shrinkage stresses are shown in Fig. 25·67. Stress at bottom is −3·3 and is negligible. The design is safe.

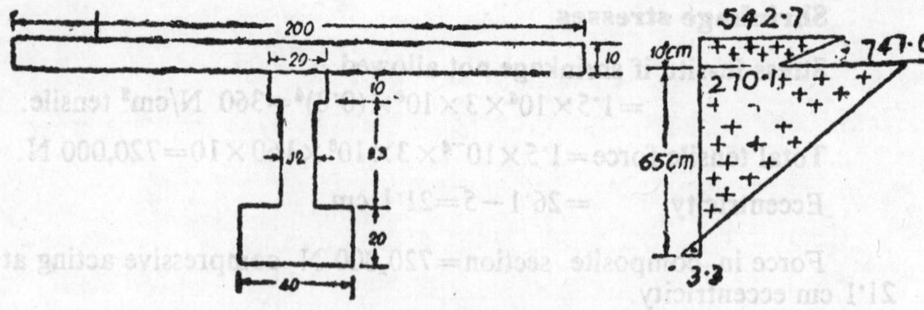

Fig. 25 67. Final+Shrinkage stresses.

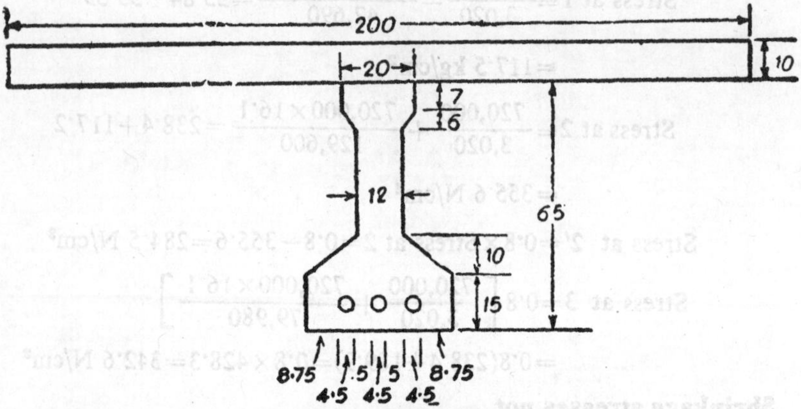

Fig. 25·68. The final section.

25·30. The Joint between Precast and In-situ Concrete. The composite action to be perfect, the sliding between the surface of precast and the cast in-situ due to shear should be considered. The following treatments are given to prevent sliding :

(*i*) The surface of the precast unit is roughened before laying the cast in-situ slab.

(*ii*) By dowel bars, steel bars or stirrups are projected from the precast unit into the in-situ slab.

(*iii*) By forming castellations in the precast concrete unit as shown in Fig. 25·69.

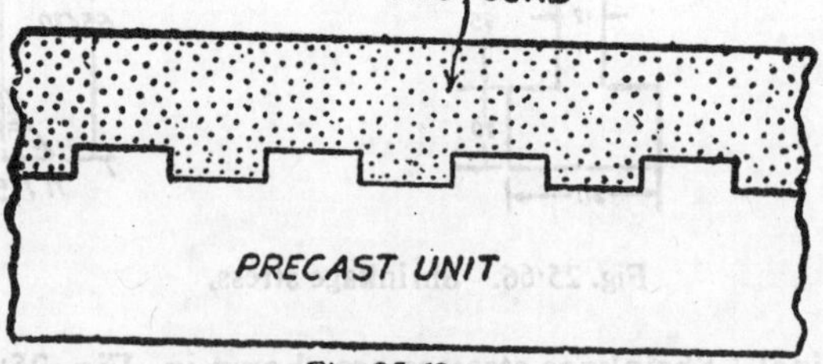

Fig. 25·69

Roughening the surface will be quite sufficient when the top flange of precast unit has sufficient area. Otherwise dowel bars will have to be provided.

25·31. Prestressed Dams. Fig. 25·70 shows a prestressed dam. The top width a is fixed. The other dimensions are designed from consideration that no tension should occur at any stage and stress should not exceed the allowable limit.

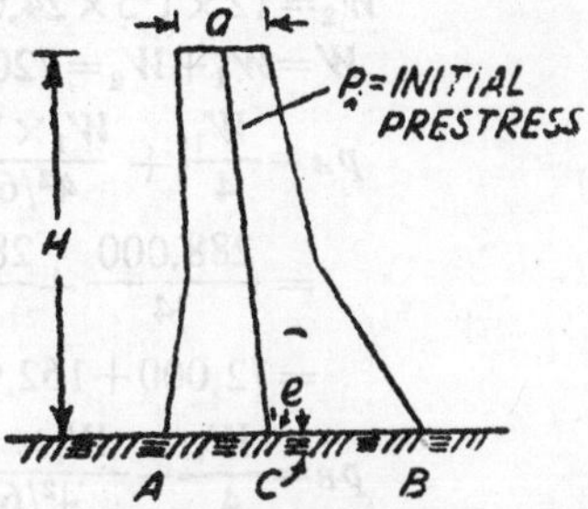

Fig. 25·70

Let p_A and p_B be stresses at A and B due to self-weight of the dam. Let p_A' and p_B' be the stresses due to hydrostatic pressure at A and B respectively. Let f_A and f_B be the stresses due to initial prestress. Let n be the loss in prestress and let c be the max. allowable stress in the material of dam.

At initial condition when there is no water, stress conditions satisfied will be :

At A, $\quad p_A+f_A \leqslant c$ ...(a_i)

At B, $\quad p_B+f_B \geqslant 0$ ...(b_i)

Finally when losses have taken place in prestress and the dam retains water.

At A, $\quad p_A+p_A'+nf_A \geqslant 0$ ...(a)

At B, $\quad p_B+p_B'+nf_B \geqslant c$ ...(b)

As for design of beams, taking equality sign, graph can be drawn for these four equations and suitable values of initial prestress P_i and eccentricity e can be worked out.

The minimum prestress will be obtained by solution of equations b_i and a.

Ex. 25·6. *Find the minimum prestress required and corresponding eccentricity for the dam section shown in Fig. 25·71. Take maximum allowable stress in concrete 300 N/cm², wt. of masonry 24,000 N/m³. Losses in prestress 20%.*

Sol. Take 1 m length of dam.

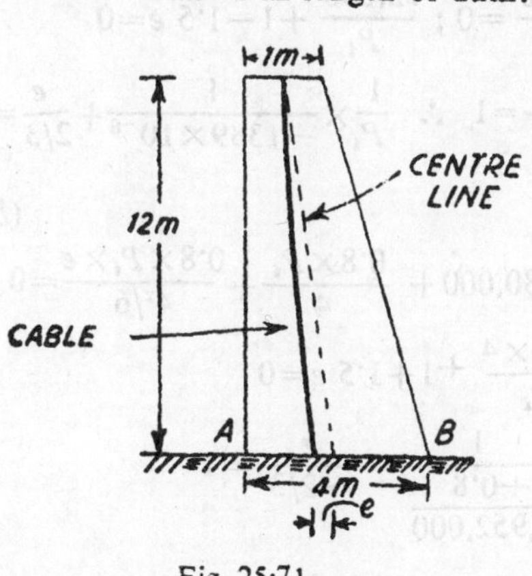

Fig. 25·71

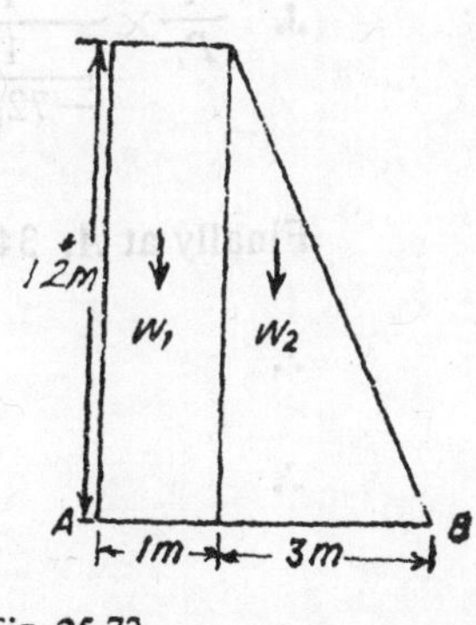

Fig. 25·72

$W_1=12\times1\times24{,}000=288{,}000$ N at 5 m from A

$W_2=12\times1{\cdot}5\times24{,}000=432{,}000$ N at 2 m from A

$W=W_1+W_2=720{,}000$ N.

$$p_A=\frac{W_1}{4}+\frac{W_1\times1{\cdot}5}{4^2/6}+\frac{W^2}{4}+\frac{W_2\times0}{4^2/6}$$

$$=\frac{288{,}000}{4}+\frac{280{,}000\times1{\cdot}5\times6}{16}+\frac{432{,}000}{4}$$

$$=72{,}000+162{,}000+108{,}000=342{,}000\text{ N/m}^2.$$

$$p_B=\frac{W_1}{4}-\frac{W_1\times1{\cdot}5}{4^2/6}+\frac{W_2}{4}$$

$$=72{,}000-162{,}000+108{,}000=+18{,}000\text{ N/m}^2$$

Moment at the base due to hydrostatic pressure

$$=\frac{10^4\times12^3}{6}=2{,}880{,}000\text{ N-m}$$

$$p_A'=-\frac{2{,}880{,}000}{4^2/6}=-1{,}080{,}000\text{ N/m}^2$$

$$p_B'=\frac{2{,}880{,}000}{4^2/6}=1{,}080{,}000\text{ N/m}^2.$$

Initially at A, $342{,}000+\dfrac{P_i}{4}+\dfrac{P_i\times e}{4^2/6}=300\times10^4$

$$\therefore\quad 1+1{\cdot}5\,e=\frac{265{,}800\times4}{P_i}$$

$$\therefore\quad \frac{1}{P_i}\times\frac{1}{\dfrac{1}{10{,}632{,}000}}+\frac{e}{-2/3}=1$$

$$\therefore\quad \frac{1}{P_i}\times\frac{1}{9{\cdot}404\times10^{-8}}+\frac{e}{-2/3}=1 \qquad ...(a_i)$$

At B $18{,}000+\dfrac{P_i}{4}-\dfrac{P_i\times e}{4^2/6}=0$; $\dfrac{72{,}000}{P_i}+1-1{\cdot}5\,e=0$

$$\therefore\quad \frac{1}{P_i}\times\frac{1}{\dfrac{1}{-72{,}000}}+\frac{e}{2/3}=1 \quad\therefore\quad \frac{1}{P_i}\times\frac{1}{-1389\times10^{-8}}+\frac{e}{2/3}=1$$

$$(b_i)$$

Finally at A, $342{,}000-1{,}080{,}000+\dfrac{0{\cdot}8\times P_i}{4}+\dfrac{0{\cdot}8\times P_i\times e}{4^2/6}=0$

$$\therefore\quad -\frac{738{,}000\times4}{0{\cdot}8\,P_i}+1+1{\cdot}5\,e=0$$

$$\therefore\quad \frac{1}{P_i}\times\frac{1}{\dfrac{+0{\cdot}8}{2{,}952{,}000}}+\frac{e}{-2/3}=1$$

$$\therefore \quad \frac{1}{P_i}\times\frac{1}{27{\cdot}10\times10^{-8}}+\frac{e}{-2/3}=1 \qquad \ldots(a)$$

At B $$18{,}000+1{,}080{,}000+\frac{0{\cdot}8P_i}{4}-\frac{0{\cdot}1\ P_i}{4^2/6}=30\times10^4$$

$$\therefore \quad 1-\frac{e}{2/3}=1{,}902{,}000\times\frac{4}{0{\cdot}8P_i}$$

$$\therefore \quad \frac{1}{P_i}\times\frac{1}{\dfrac{0{\cdot}8}{7{,}608{,}000}}+\frac{e}{2/3}=1$$

$$\therefore \quad \frac{1}{P}\times\frac{1}{10{\cdot}52\times10^{-8}}+\frac{e}{2/3}=1 \qquad \ldots(b)$$

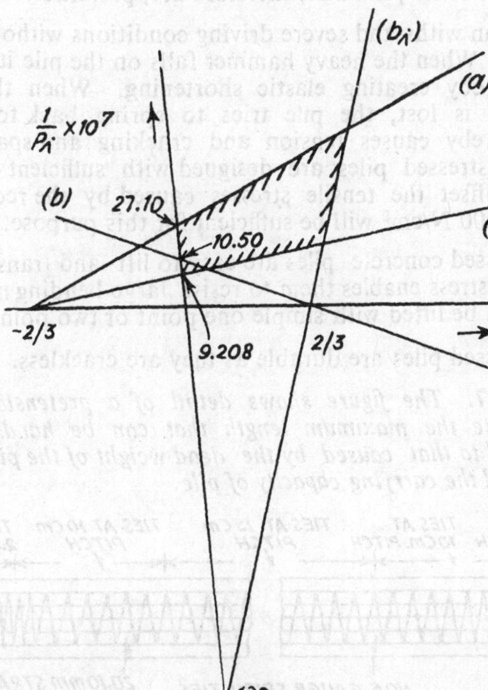

Fig. 25·73.

Solving equations (b_i) and (a),

$$\frac{1}{P_i}\times\frac{1}{-1389\times10^{-8}}+\frac{e}{2/3}=1 \qquad \ldots(b_i)$$

$$\frac{1}{P_i}\times\frac{1}{27{\cdot}10\times10^{-8}}+\frac{e}{-2/3}=1 \qquad \ldots(a)$$

$$\therefore \quad \frac{10^8}{P_i}\left[\frac{1}{27{\cdot}10}-\frac{1}{1389}\right]=2$$

$$P_i=\frac{10^8}{2}\times 0{\cdot}0297=1{,}485{,}000 \text{ N.}$$

Putting this value in equation (b_i),

$$\frac{10^8}{2}\times 0{\cdot}0297\times\frac{1}{-1389\times 10^{-8}}+\frac{e}{2/3}=1$$

$$\therefore \qquad 1{\cdot}5e=1+\frac{0{\cdot}0297}{2778}=1 \text{ (nearly).}$$

$$\therefore \qquad e=0{\cdot}67 \text{ m.}$$

25·32. Prestressed Piles. Prestressed piles have many advantages over the other piles and, therefore are, preferable.

They can withstand severe driving conditions without cracking and spalling. When the heavy hammer falls on the pile it compresses the pile, thereby creating elastic shortening. When the energy of the hammer is lost, the pile tries to spring back to its original position thereby causes tension and cracking and spalling subsequently. Prestressed piles are designed with sufficient compressive stresses to offset the tensile stresses caused by the recoil. Initial prestress of 600 N/cm² will be sufficient for this purpose.

Prestressed concrete piles are easy to lift and transport. Their compressive stress enables them to resist large bending moments, so that they can be lifted with simple one point or two points picks.

Prestressed piles are durable as they are crackless.

Ex. 25·7. *The figure shows detail of a pretensioned square pile. Compute the maximum length that can be handled. Use a moment equal to that caused by the dead weight of the pile plus 50% impact. Find the carrying capacity of pile.*

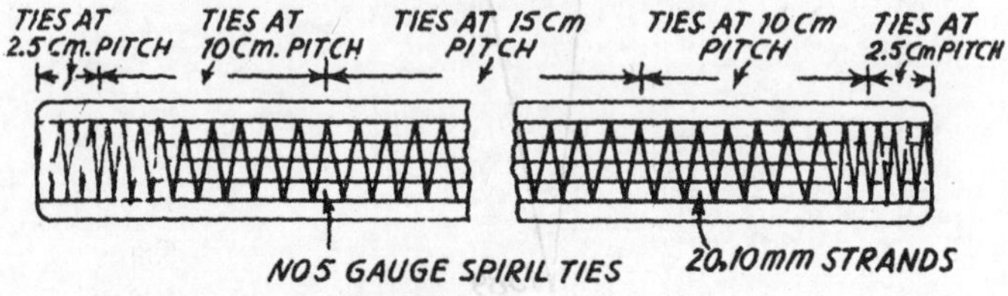

Fig. 25·74.

Size of the pile=45 cm×45 cm

Initial prestress in steel—900 N/mm²

Losses in prestress=20%

Wt. of concrete=24,000 N/m³

Ultimate stresses in concrete=4000 N/cm²

Modulus of rupture=400 N/cm²

Modular ratio=6

Sol. Since handling is temporary a tensile stress near the modulus of rupture will be allowed.

$$\text{Final prestressing force}=0{\cdot}8\times 20\times\frac{\pi}{4}\times 10^2\times 900$$

$$=0{\cdot}8\times 450{,}000\ \pi\ \text{N}=0{\cdot}8\times 1{,}414{,}000\ \text{N}$$

$$\text{Stress in concrete}=\frac{0{\cdot}8\times 1{,}414{,}000}{45\times 45}=554{\cdot}6\ \text{N/cm}^2.$$

Total stress change due to bending will be

$$=554{\cdot}6+400=954{\cdot}6\ \text{N/cm}^2.$$

$$Z=(45)^3/6$$

$$\therefore\quad \text{M.R. of the section}=954{\cdot}6\times\frac{(45)^3}{6}=14{,}500{,}000\ \text{N-cm.}$$

$$=145{,}000\ \text{N-m.}$$

The weight of concrete pile per 1 m $=0{\cdot}45\times 0{\cdot}45\times 24{,}000$

$$=4860\ \text{N/m.}$$

Load on pile including impact $W=7290$ N/m.

(1) **Single pick handling**

Picking is done at a point, such that the max. $-$ve moment is equal to max. $+$ve moment. This can nearly be achieved by picking at 0·3 L.

$$\therefore\ 0{\cdot}045\ WL^2=145{,}000$$

$$\therefore\ L=\sqrt{\frac{145{,}000}{0{\cdot}045\times 7290}}$$

$$=21\ \text{m.}$$

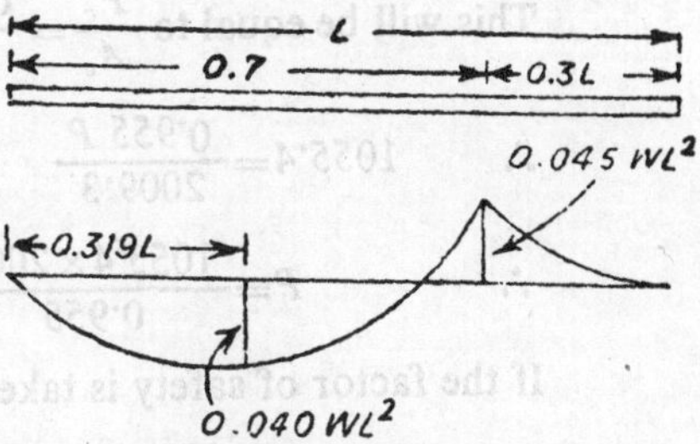

Fig. 25·75

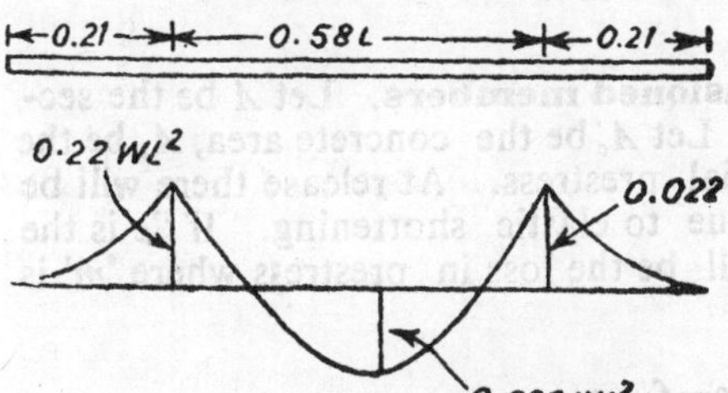

Fig. 25·76.

(2) **Double pick handling**

Picking is done at points, such that the max. $-$ve moment is equal to max. $+$ve moment. This can nearly be achieved by picking at 0·21 L points.

$$\therefore\qquad 0{\cdot}022\ WL^2=145{,}000$$

$$\therefore\qquad L=\sqrt{\frac{145{,}000}{0{\cdot}22\times 7290}}=30\ \text{m.}$$

Let P be the applied load. P_c=load taken by concrete and P_s load taken by steel.

$$P=P_c+P_s,\ \Delta=\frac{P_cL}{A_cE_c}=\frac{P_sL}{A_sE_s}$$

$$\therefore \quad P_s = P_s \times \frac{A_s}{A_c} \times \frac{E_s}{E_c} = m \frac{A_s}{A_c} P_s$$

$$\therefore \quad P = P_c + m \frac{A_s}{A_c} P_c = P_c \left[1 + m \frac{A_s}{A_c} \right]$$

Now $A_s = 20 \times \frac{\pi}{4} \times (10)^2 \text{ mm}^2 = 500\ \pi \text{ mm}^2$

$= 5\pi \text{ cm}^2 = 15{\cdot}70 \text{ cm}^2$

$A_c = 45 \times 45 - 15{\cdot}7 = 2025 - 15{\cdot}7 = 2009{\cdot}3 \text{ cm}^2.$

$$\therefore \quad P_c = \frac{P}{1 + m \frac{A_s}{A_c}} = \frac{P}{1 + 6 \times \frac{15{\cdot}7}{2009{\cdot}3}} = 0{\cdot}955\ P$$

Taking allowable stress $= 0{\cdot}4 \times 4000 = 1600$ N/cm².
Net available stress $= 1600 - 554{\cdot}6 = 1055{\cdot}4$ N/cm².

This will be equal to $\frac{P_c}{A_c} = \frac{0{\cdot}955\ P}{A_c}$

$$\therefore \quad 1055{\cdot}4 = \frac{0{\cdot}955\ P}{2009{\cdot}3}$$

$$\therefore \quad P = \frac{1055{\cdot}4 \times 2009{\cdot}3}{0{\cdot}955} = 2{,}220{,}000 \text{ N} = 2220 \text{ kN}$$

If the factor of safety is taken as 2,

Safe load $= \frac{2220}{2} = 1110$ kN.

25·33. Prestressed tensioned members. Let A be the sectional area of the member. Let A_c be the concrete area, A_s be the steel area. Let f_s be the initial prestress. At release there will be reduction in stress in steel due to elastic shortening. If c is the final stress in concrete, mc will be the loss in prestress where 'm' is the modular ratio.

Therefore, final prestress $= f_s - mc$.

Now $\quad (f_s - mc)\ A_s = c[A - A_s]$

Dividing by A_s and putting $\frac{A_s}{A} = p$

$$(f_s - mc) = c \left[\frac{1}{p} - 1 \right]$$

$$\therefore \quad f_s = c \left(\frac{1}{p} - 1 + m \right) = \frac{c}{p} \left[1 + p\ (m - 1) \right]$$

$$\therefore \qquad c=\frac{pf_s}{1+p(m-1)} \qquad ...(1)$$

Now $\qquad (f_s-mc)=c\left[\frac{1}{p}-1\right]$

$$\therefore \qquad \frac{f_s}{c}-m=\frac{1}{p}-1$$

$$\therefore \qquad \frac{f_s}{c}-m+1=\frac{1}{p}.$$

$$\therefore \qquad p=\frac{1}{\frac{f_s}{c}-m+1}=\frac{1}{\frac{f_s}{c}-(m-1)} \qquad ...(2)$$

Now if P is the tensile force applied to the member, and the final stress in concrete is to be zero, 'n' is the reduction factor in prestress,

$$[A+(m-1)A]nc=P$$

$$\therefore \qquad A[1+(m-1)p]nc=P \qquad ...(3)$$

$$\therefore \qquad A=\frac{P}{[1+(m-1)p]nc}$$

Final stress in steel$=nf_s$.

Ex. 25·8. *Design a pretensioned member to carry a force of 800 kN. Safe compression stress in concrete is 1400 N/cm² and m=6, initial prestress in steel is 100 kN/cm². Take losses in prestress at 20%.*

Sol. $p=\frac{1}{\frac{f_s}{c}-m+1}=\frac{1}{\frac{100{,}000}{1400}-6+1}=\frac{1}{714-5}=\frac{1}{66\cdot4}$

Let 'A' be the area of the section.

$$A=\frac{P}{[1+(m-1)p]nc}=\frac{800{,}000}{\left[1+(6-1)\times\frac{1}{66\cdot4}\right]\times0\cdot8\times1400}$$

$$=\frac{800{,}000}{(1+0\cdot0753)\times0\cdot8\times1400}$$

$$=\frac{800{,}000}{1\cdot0573\times0\cdot8\times1400}=664\cdot5 \text{ sq. cm.}$$

$$A_s=\frac{664\cdot5}{66\cdot4}=10 \text{ cm}^2.$$

25·34. Prestressed Concrete Tanks. The prestressing is used to build large tanks to hold liquids. Preload Corporation of America has developed a system by which continuous prestressing can be done round concrete tank. It consists of a machine called merry-go-round which is supported by a trolley that moves at the top of the tank. The merry-go-round releases wire from a drum, tensions it through a die and wraps it round the tank wall. The wire is anchored at the bottom of the tank and the wrapping is done from bottom to top After prestressing the wall, the tank is filled and steel is covered by guniting.

The concrete is fully hardened in tank walls before the tnnsioned wire is wrapped round to cause hoop compression. The elastic strain in concrete reduces the diameter of the tank and thereby the tension in steel.

The tank is covered with a dome of small rise. The ring beam is provided to support the dome. The dome may be prestressed also.

Let A_s be the steel area per unit height of tank wall and f_s be the initial prestress, Let t_c be the thickness of the tank wall. If 'c' is the allowable compressive stress for concrete, the loss in prestress after wrapping will be mc. The force in steel will be equal to compression in concrete.

$$\therefore \quad (f_s - mc)\, A_s = c \times t_c \times 100$$

Dividing by A_s, and putting $\frac{A_s}{100 t_c} = p.$

$$\left(f_s - mc \right) = c\, \frac{100 t_c}{A_s} = \frac{c}{p}$$

$$\therefore \quad f_s = \frac{c}{p} + mc = c\left[\frac{1}{p} + m\right] = c\left[\frac{1+mp}{p}\right]$$

$$\therefore \quad c = \frac{p f_s}{1+mp} \qquad \ldots(i)$$

$$\text{Also} \quad P = \frac{c}{f_s - mc} = \frac{1}{\frac{f_s}{c} - m} \qquad \ldots(ii)$$

If P is the tensile hoop tension on unit metre, and 'n' is the loss in prestress,

$$(100 t_c + m A_s)\, nc = P$$

$$\therefore \quad 100 t_c [1 + mp]\, nc = P.$$

$$\therefore \quad t_c = \frac{P}{100 nc(1+mp)} \qquad \ldots(iii)$$

Ex. 25·9. *Design the wall of a tank 30 m in diameter to hold water. Its height is 6 m and is free at top and bottom. Take working stress in concrete as 1400 N/cm² and m=6. Take losses due to shrinkage and creep as 20%. Take initial prestress in steel as 100 kN/cm².*

$$p = \frac{1}{\frac{f_s}{c} - m} = \frac{1}{\frac{100{,}000}{1400} - 6} = \frac{1}{71{\cdot}4 - 6} = \frac{1}{65{\cdot}4}$$

$$t_c = \frac{P}{100\, nc(1+mp)}$$

Taking 1 m highest at base

$$\text{Now } P_h = p \times \frac{D}{2} = 6 \times 10^4 \times \frac{30}{2} = 900{,}000 \text{ N}$$

$$\therefore \quad t_c = \frac{900{,}000}{100 \times 0{\cdot}8 \times 1400 \left(1 + \frac{6}{65{\cdot}4}\right)}$$

$$= \frac{9000}{1120 \times 1{\cdot}09175} = 7{\cdot}358 \text{ cm.}$$

Provide thickness of 7·5 cm.

$$A_s = 100\ t_c p = 100 \times 7{\cdot}358 \times \frac{1}{65{\cdot}4} = 11{\cdot}25 \text{ cm}^2.$$

PROBLEMS

1. A prestressed concrete beam, simply supported at its end has its central section with the following properties :

Cross-sectional area=1825 cm²

I_{xx}=1,950,000 cm⁴

Distance of bottom edge from C.G. of the section=49·0 cm
Distance of top edge from the C.G. of the section=41·0 cm
Maximum dead load moment=122,49,000 N-cm
Maximum applied load moment=43,290,000 N-cm

Prestressing force is applied by means of 5 cables of 12 wires each (wire diameter=5 mm) stressed initially to 900 N/mm² with an eccentricity of 40 cm below the C G. of section. If the loss of prestress due to shinkage, creep etc. is assumed 15% determine the stress reached at top and bottom initially and finally.

2. A rectangular prestressed concrete beam 25 cm×40 cm is prestressed with force of 700,000 N applied at eccentricity of 8 cm from the C.G. of the section. The span of the beam is 7 m. The beam carries superimposed load of 10,000 N/m. Find the initial and final stresses at the top and bottom of section. Assume losses at 15%.

3. Design a prestressed concrete beam having span of 20 m carrying superimposed load of 30,000 N/m run. Assume suitable stresses.

4. A prestressed concrete beam having span of 10 m carries superimposed load of 20,000 N/m run. Design a suitable section for the beam.

5. Find the percentage loss of prestress in a post-tensioned beam due to shrinkage and creep for the following data :

$$F_r/F_c = 0{\cdot}5$$

Condition of exposure : Dry atmosphere.
Stress in concrete at centre of steel=500 N/cm²
Stress in steel =900 N/cm²
Modular ratio =6
Age of concrete at transfer =14 days.

6. Design a composite beam having a span of 12 m. The in-situ slab is 12 cm thick The centre to centre of beams is 2·5 m. The slab carries a uniformly distributed load of 30,000 N/m². The precast sections are proposed while laying the slab. Take—

Initially compressive stress in precast concrete 1750 N/cm² and final 1400 N/cm². Allowable tensile stress initial 100 N/cm² and final 0. The modular ratio of the in-situ slab and precast is 8. Allowable compressive stress in in-situ is 700 N/cm².

7. A simply supported rectangular post-tensioned concrete beam is required to span 9 m and carry a load of 7000 N/m in addition to its own weight. Allowable concrete stresses are 1250 N/cm² in compression and 80 N/cm² in tension.

Find the required dimensions of the concrete section, the magnitude of the required prestressing force and the location of the cable centroid. Allow for losses in prestress. *(A.M.I.E. 1968)*

8. A pre-stressed conerete rectangular beam, 25 cm wide × 50 cm deep is pretensioned with 9·7 sq cm of steel, the c.g. of the steel at mid-span is at a depth 42·5 cm from the top of the beam. The dead load and live load bending moments at mid-span are 5400 N-m and 6800 N-m respectively. The initial stress in steel at prestress is 95,000 N/cm². The loss in prestress on account of creep and shrinkage may be taken as 25,000 N/cm². Modular ratio is 5·6. *(A M.I.E. 1969)*

Calculate :

(*a*) Concrete stresses at top and bottom of beam and stress, when pretensioned wires first cut from the units, with the dead load bending moment acting.

(*b*) The effect of loss of prestress due to shrinkage and creep on the stress found in (*a*).

(*c*) Total stresses when the live load bending moment acts.

9. A prestressed concrete beam is required to carry a total distributed load of 27,000 N/m on span of 14 m. Using a rectangular cross-section, determine the suitable dimensions of the beam. Calculate the area of tendons required and also decide their position. The permissible stresses are as under. *(A.M.I E. 1969)*

Concrete =1400 N/cm²
Tendons =105,000 N/cm³

Show the details of the beam and chiefly explain the method you would adopt for prestressing.

10. A floor consists of precast prestressed T-beams placed side by side, with the flanges abutting. A section of a beam is shown in the figure, The beams are spaced 50 cm centres and have a simply supported span of 9 m. The superimposed load is 8200 N/m² of floor area. Determine the magnitude and location of the prestressing force at mid-span on the basis of no tension in concrete. Assume 20% loss in prestress. Find the stress under different stages *viz.* (*i*) prestress only (*ii*) prestress plus D.L. (*iii*) prestress plus D.L. plus L.L. What strength of concrete would you suggest for the beams ? *(A.I.M.E. 1967)*

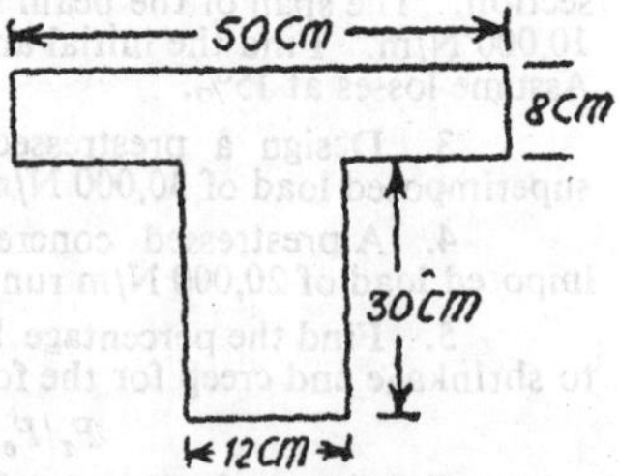

Fig. 25·7

11. An I-shaped prestressed concrete beam spanning 15 m is made up as follows : *(A.M.I.E. 1979)*

Flanges =40 cm × 10 cm
Web =55 cm × 10 cm

Estimate the safe superimposed uniformly distributed loading the beam can carry, after adequate prestressing subject to the following stress limitations at all loadings :

Max. compressive stress in concrete=1400 N/cm²
Max. tensil stress in concrete =0 N/cm²
Max. allowable tension in steel =100,000 N/cm²
Estimate the area of tendons required and their position.
Give cable zoning.

26

Sub-structure of Bridges

The link between superstructure of bridges and its foundation is known as sub structure. In a typical bridge construction the portion of the structure below bearings and above foundation will be sub-structure. A typical bridge arrangement is shown in

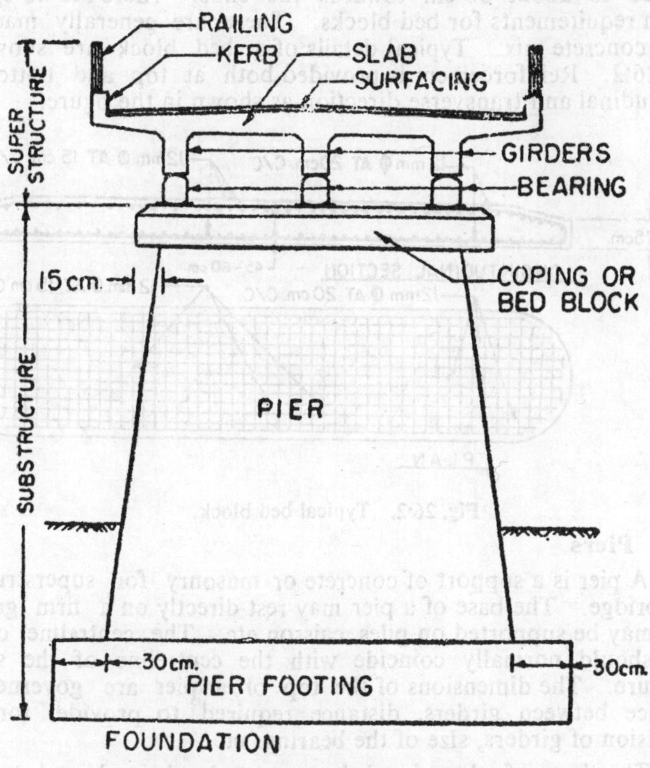

Fig. 26·1. Typical bridge arrangement.

Fig. 26·1. The intermediate support units over superstructure rests are known as piers and the end supports of a superstructure are known as abutments.

The top of the pier and abutments are generally provided with coping or bed blocks shown in Fig. 26·1.

The bottom part of pier known as footing is to transfer the load from pier to the foundation.

26·1. Bed Block

Bed block is the upper part of a pier or abutment. This generally projects 10—30 cm beyond the pier width at the top depending upon the height of the pier. The purpose of the bed block is to prevent pier from weather and hence it is shaped to allow rain water to run off rapidly.

Another purpose of bed block is to distribute the load from the bearings to the pier abutment The thickness of bed block is generally kept between 45 to 60 cm. The thickness is generally tapered to about 30 cm towards the ends. There are no special design requirements for bed blocks. These are generally made of M 15 concrete mix. Typical details of a bed block are shown in Fig. 26·2. Reinforcement is provided both at top and bottom in longitudinal and transverse direction as shown in the figure.

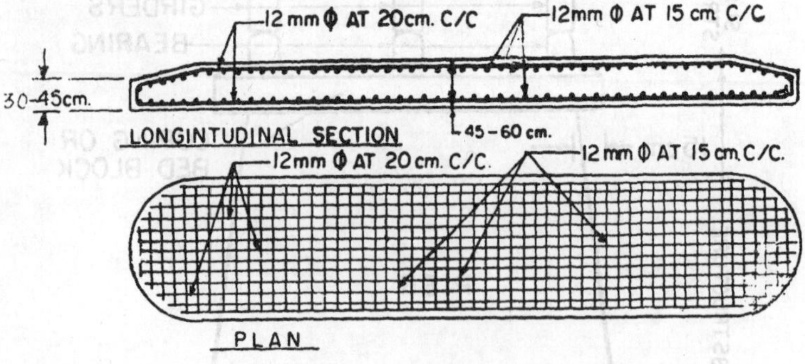

Fig. 26·2. Typical bed block.

26·2. Piers

A pier is a support of concrete or masonry for superstructure of a bridge. The base of a pier may rest directly on a firm ground or it may be supported on piles, caisson etc. The centreline of the pier should normally coincide with the centreline of the superstructure. The dimensions of the top of a pier are governed by distance between girders, distance required to provide for the expansion of girders, size of the bearings etc.

The shape of the pier below water level is shaped to give smooth flow of water to prevent eddy currents and scour.

A typical masonry pier is shown in Fig. 26·1. Typical piers for highway and railway bridges are shown in Fig. 26·3. A solid pier such as shown in Fig. 26·1 is generally used for rail road bridges. The double shaft shown in Fig. 26·3 are used for highway bridges.

The mushroom or hammer head piers are used to avoid skew spans in construction over existing rail-road tracks or highways.

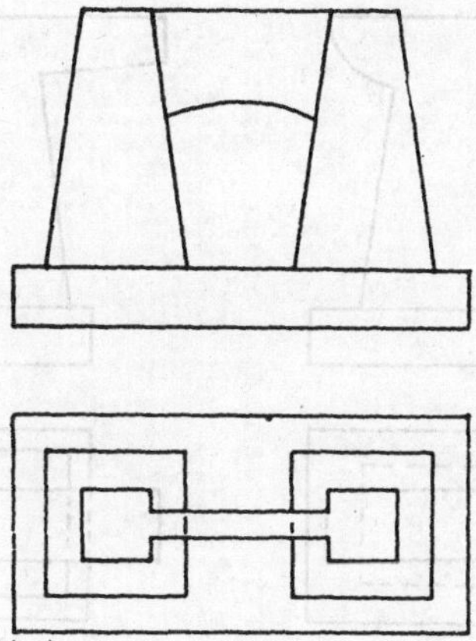

(a.)

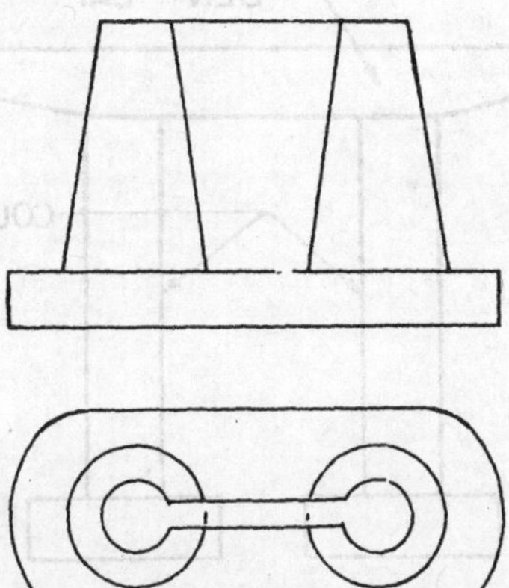

(b.)

Fig. 26·3. Double piers.

Most modern bridge piers are of reinforced concrete construction. R.C.C. piers may be mushroom or hammer head piers shown in Fig. 26·4. Typical R.C.C. trestle pier is shown in Fig. 26·5 and cellular pier is shown in Fig. 26·6.

The concrete used for pier construction should be durable. Large size aggregates are generally used. Nominal reinforcement is provided near the surface of the piers to prevent surface cracks due to temperature and shrinkage.

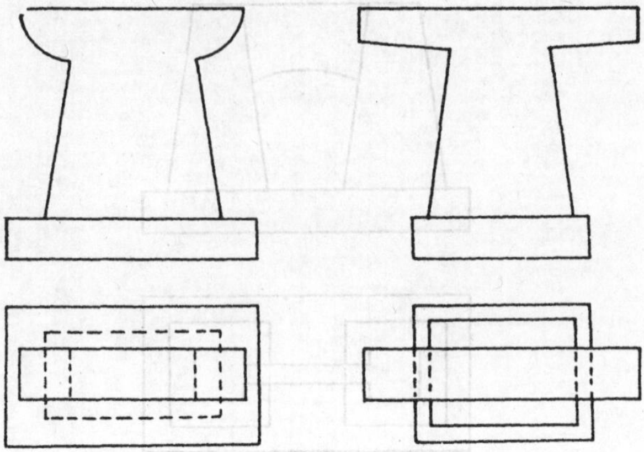

Fig. 26·4. Mushroom or hammerhead piers.

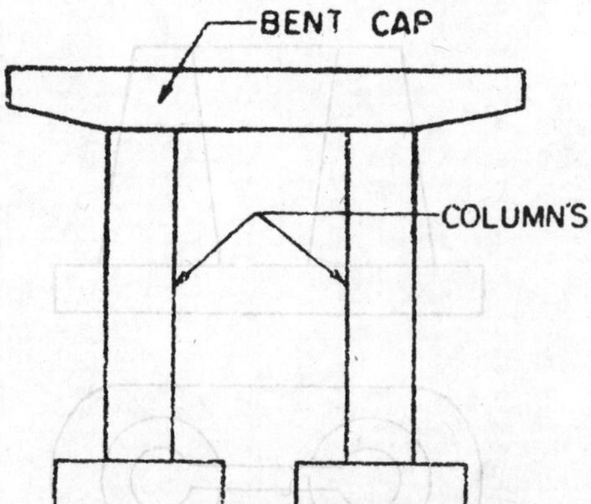

Fig. 26·5. Two-column R.C.C. bent (trestle) for T-beam bridge.

Fig. 26·6. Cellular reinforcement concrete pier.

Sometimes stone masonry is provided to concrete piers. This has the advantages that formwork for concrete is eliminated and the surface cracks in concrete due to temperature changes are avoided. Also stone facing gives good appearance to the piers.

26·3. Pier Dimensions

In bridge over waterway the height of bridge piers is determined by the level of erosion. The presence of bridge piers and abutments causes obstruction in waterway and reduces the effective cross-sectional area of the stream. This results increase in the velocity of the flow and raises water level. The presence of the obstruction due to piers sets up eddy currents around piers cross-section. The effect of eddy currents and increase in velocity is to cause scour. If the scour depth exceeds the foundation depth there is possibility of bridge failure. In order to avoid bridge failures due to erosion, the depth of the pier should be more than the lower limit of scour.

There is no reliable analysis to determine the depth of erosion. A general rule is to take depth of erosion ΔE as four times the height ΔH through which water level rises at the pier due to obstruction. The pier should be at least length ΔE below the stream bed.

The length of the pier at the top is kept about 1·2 m more than the outer dimension of the bearing plates as shown in Fig. 26·7. The width of the pier is kept about 60 cm larger than the length of the bearing plates as shown in Fig. 26·7.

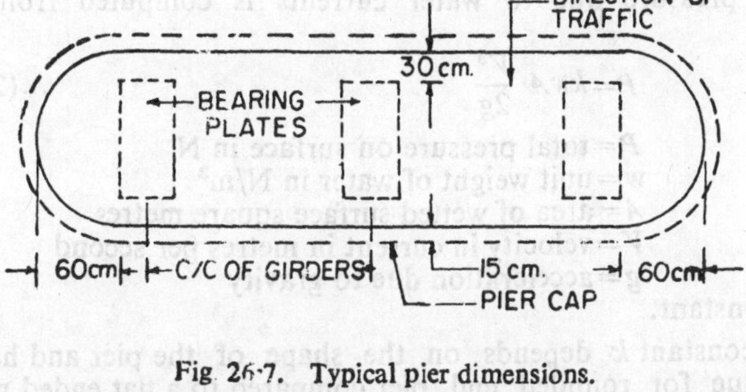

Fig. 26·7. Typical pier dimensions.

The bottom dimensions of the piers are determined from the considerations that no tensile stresses are produced in the section due to all the forces acting on it. The piers are generally provided with batter between 1 in 50 to 1 in 12.

The pier footing is generally rectangular in shape and projects about 30 cm beyond the edge of the pier at the base as shown in Fig. 26·1. The pier footing may rest directly on the firm soil or other suitable foundation.

26·4. Design Load for Piers

The design of piers is relatively simple and does not present any problems. However, the loads on piers may be properly computed and the design verified for the worst combination of loads. Following loads should be taken into consideration while designing bridge piers :

(1) Dead load of the superstructure.

(2) Dead load of the piers less than any hydrostatic uplift caused by surrounding water.

(3) Live load on the superstructure. The effect of load eccentricities due to only some spans loaded should be taken into consideration.

(4) Lateral forces perpendicular to the centreline of the superstructure. These forces include effect of wind on superstructure and moving loads, wind pressure on the portion of pier above water level pressure due to water currents wave action, and ice and the centrifugal force in case of rail-road bridges on curves.

(5) Londitudinal forces acting parallel to the direction of the bridge. These forces are caused by tractive effort of vehicles, braking of vehicles, unequal frictional resistance of bearing and unbalanced thrust in case of arch bridges.

(6) Seismic forces.

The computation of dead load is straight forward. The computations of live load forces and most lateral longitudinal forces is discussed in chapter 1. The wind forces are computed as discussed in Article 1·9, using Table 1·1.

The pressure due to water currents is computed from the equation

$$P=kwA\frac{V^2}{2g} \qquad ...(26·1)$$

where P=total pressure on surface in N
w=unit weight of water in N/m³
A=area of wetted surface square metres
V=velocity in current in metres per second
g=acceleration due to gravity

and k is constant.

The constant k depends on the shape of the pier and has a smaller value for rounded end pier compared to a flat ended pier. The value of k=1·6 will generally be on safe side.

Example 26·1. *Design a suitable pier for the 16 m girder bridge of Example 24·3. Clear width of roadway=6·8 m. Live load on bridge class A and class AA. Height of pier=8 m. Mean velocity of current=3·0 m/sec. High flood level=7 m.*

Solution. Data of Example 24·3 are :

Dead load of superstructure per span

Dead load coming from two outer girders and inner girder

$=2\times 2[2\times 251{,}280]+2\times 251{,}280$
$=1{,}507{,}680$ N.

(It was assumed in the analysis that dead load is taken equally by all girders) $=1{,}507{,}680$ N.

Assume weight of bearings, plates etc. as 100,000 N.

Total dead load $=1{,}607{,}680$ N.

From Fig. 3·9, the distance between the outside of outermost girders

$=5\cdot0+0\cdot3=5\cdot3$ m.

Minimum pier length required at the top

$=5\cdot3+1\cdot2=6\cdot5$ m.

Provide a top length of 7·2 metres in straight portions with semi-circular ends as shown in Fig. 26·8. Use M 10 concrete mix for pier. Assume centre of bearings between two girders as 1·0 m.

Minimum width required at top

$=1\cdot0+2\times0\cdot3=1\cdot6$ m

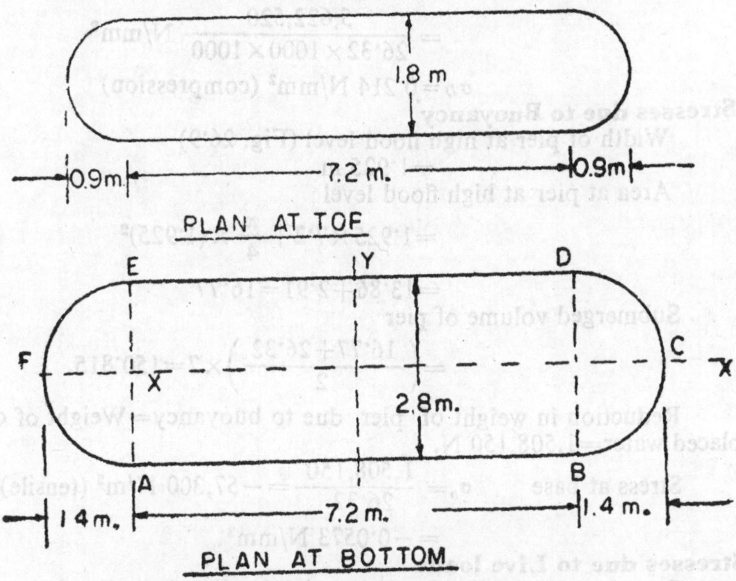

Fig. 26·8. Plan of pier at top and bottom.

Provide a concrete pier with top width of 1·8 m for straight portion and width of 2·8 m at bottom for straight portion.

$$\text{Area of pier at top} = 1\cdot8\times7\cdot2+\frac{\pi}{4}\times(1\cdot8)^2$$
$$=12\cdot96+2\cdot54=15\cdot50.$$

$$\text{Area of pier at bottom} = 2\cdot8\times7\cdot2+\frac{\pi}{4}\times(2\cdot8)^2$$
$$=20\cdot16+6\cdot16=26\cdot32$$

Self-weight of pier $=\left(\frac{15{\cdot}50+26{\cdot}32}{2}\right)\times 8\times 24{,}000$

$=4{,}014{,}720$ N.

Moment of inertia of pier at base about y-axis

$$=\frac{1}{12}\times 2{\cdot}8\times(7{\cdot}2)^3+2\times\frac{\pi}{128}(2{\cdot}8)^4+2\times\frac{\pi}{8}(2{\cdot}8)^2\times(3{\cdot}6)^2$$

$$I_{yy}=87{\cdot}091+3{\cdot}017+79{\cdot}801=169{\cdot}91$$

Moment of inertia about x-axis,

$$I_{xx}=\frac{1}{12}\times 7{\cdot}2\times(2{\cdot}8)^3+\frac{\pi}{64}(2{\cdot}8)^4=13{\cdot}17+3{\cdot}017=16{\cdot}19$$

Design dead load for pier

$=1{,}607{,}800+4{,}014{,}720$ N

$=5{,}622{,}520$ N.

Stresses due to Dead load

Stress at base due to dead load

$$=\frac{5{,}622{,}520}{26{\cdot}32\times 1000\times 1000}\text{ N/mm}^2$$

$\sigma_D=0{\cdot}214$ N/mm² (compression)

Stresses due to Buoyancy

Width of pier at high flood level (Fig. 26·9)

$=1{\cdot}925$ m

Area at pier at high flood level

$$=1{\cdot}925\times 7{\cdot}2+\frac{\pi}{4}\times(1{\cdot}925)^2$$

$=13{\cdot}86+2{\cdot}91=16{\cdot}77$

Submerged volume of pier

$$=\left(\frac{16{\cdot}77+26{\cdot}32}{2}\right)\times 7=150{\cdot}815.$$

Reduction in weight of pier due to buoyancy=Weight of displaced water=1,508,150 N.

Stress at base $\sigma_b=\frac{1{,}508{,}150}{26{\cdot}32}=-57{,}300$ N/m² (tensile)

$=-0{\cdot}0573$ N/mm².

Stresses due to Live loads

Reaction due to live load class AA loading including impact

$=1{\cdot}1\times 700=770$ kN.

Maximum bending caused by eccentricity for this load is about X-axis.

Maximum B.M.$=770\times 0{\cdot}5=385$ kN-m.

Maximum and minimum stress at the base due to this load will be

$$\sigma_l=\frac{770\times 1000}{26{\cdot}32\times 1000\times 1000}\pm\frac{385\times 1000\times 1000\times 1400}{I_{xx}}$$

$$\sigma_t = + \frac{385 \times 1000 \times 1000 \times 1400}{16{\cdot}19 \times (1000)^4}$$

$$= 0{\cdot}029 \pm 0{\cdot}0333$$

$$= 0{\cdot}0623 \text{ N/mm}^2 \text{ (compression)}$$

$$= 0{\cdot}0043 \text{ N/mm}^2 \text{ (tension)}.$$

Stresses Due to Longitudinal Force

Maximum longitudinal force will occur due to class AA loading.

Longitudinal force due to class AA loading

$$= 0{\cdot}2 \times 700 = 140 \text{ kN}.$$

Moment at base of pier due to this force

$$= 140 \times 8 = 1{,}120 \text{ kN-m}$$

$$\text{Stresses at the base} = \pm \frac{1{,}120 \times 1000 \times 1400 \times 1000}{I_{xx}}$$

$$= \pm \frac{1{,}120 \times 1000 \times 1400 \times 1000}{16{\cdot}19 \times (1000)^4}$$

$$= \pm 0{\cdot}968 \text{ N/mm}^2.$$

Stresses Due to Unequal Frictional Resistance at Bearings

The frictional resistance of the bearings on bridge pier may differ and this will cause moment on the pier. Assume coefficient of friction as 0·25 at one bearing and 0·22 at other bearing.

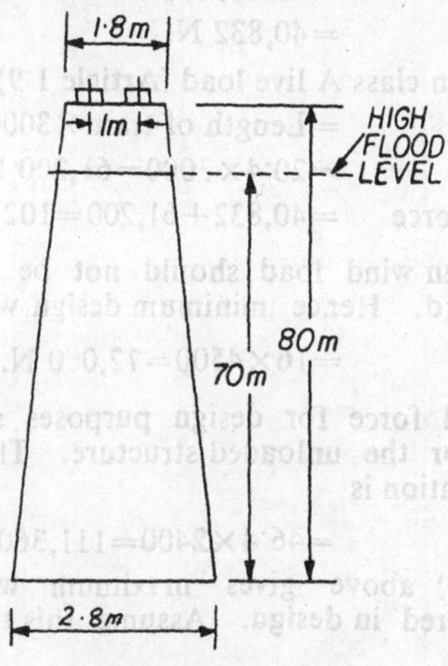

Fig. 26·9. Section of Pier.

Total resistance at one set of bearing with dead and live load

$=0.25(1,607,800+1.1\times700,000)$

$=594,450$ N.

Total resistance at other set of bearings due to dead load only

$=0.22\times1,607,800=353,716$

Unbalancad force at the top of the pier

$=594,450-353,716=240,734$ N

Moment at base $=8\times240,734=1,925,872$ N-m

Stress at the base $=\dfrac{1,925,872\times1000\times1400}{16.19\times(1000)^4}$

$=\pm0.166$ N/mm^2.

Stresses Due to Wind Loads

Exposed height of the structure

=Depth of girder+Thickness of slab +Height of kerb+Height of railing

$=1.4+0.2+0.3+1.0=2.9$ m

Exposed area contributing wind pressure per pier

$=\text{Span}\times\text{Height}=16\times2.9=46.4$ m^2

Assuming average height of about 9 m the wind pressure from Table 24·1 is 880 N/m^2.

(1) Wind force on exposed surface

$=46.4\times880$

$=40,832$ N.

Wind load on class A live load (Article 1·9)

$=\text{Length of train}\times3000$ N/m

$=20.4\times3000=61,200$ N.

Total wind force $=40,832+61,200=102,032$ N.

(2) The design wind load should not be less than 4500 N/m of the loaded chord. Hence minimum design wind force

$=16\times4500=72,000$ N.

(3) The wind force for design purposes should not be less than 2400 N/m^2 for the unloaded structure. The design wind load from this consideration is

$=46.4\times2400=111,360$ N.

Condition (3) above gives maximum wind force. Hence this will be considered in design. Assume this to act at mid height

i.e. $\dfrac{2.9}{2}=1.45$ m from the top of the pier.

Moment at base about y-axis

$$M_w = 111{,}360 \times 9{\cdot}45 = 1{,}052{,}352 \text{ N-m.}$$

The bending takes place about y-axis.

Maximum stress at the base at end of straight portion

$$\sigma_{w1} = \frac{M_w}{I_y} \times 3600$$

$$= \frac{1{,}05{,}2352 \times 3600 \times 1000}{169{\cdot}91 \times (1000)^4}$$

$$= \pm 0{\cdot}0223 \text{ N/mm}^2.$$

Maximum stress at the edge of pier (point C)

$$= \frac{M_w \times 5000}{I_y} = \frac{1{,}052{,}352 \times 5000 \times 1000}{(169{\cdot}91) \times (1000)^4}$$

$$= 0{\cdot}031 \text{ N/mm}^2.$$

Stresses Due to Water Current

Using equation 26·1, the water pressure P is obtained.

Area of wetted surface

$$= 7 \times \frac{(1{\cdot}925 + 2{\cdot}8)}{2} = 16{\cdot}54 \text{ m}^2$$

Water pressure (Eq. 26·1)

$$= \frac{1{\cdot}5 \times 10{,}000 \times 16{\cdot}54 \times 3 \times 3}{2 \times 9{\cdot}81} \text{ N}$$

$$P = 113{,}807 \text{ N.}$$

This force is assumed to act at $\frac{2}{3}$ height

Moment at base $M_w = P \times \frac{2}{3} \times 7$

$$= 113{,}807 \times \frac{14}{3} = 531{,}099 \text{ N-m}$$

Stress at base at edge of straight portion

$$= \frac{M_w}{I_y} \times 3600$$

$$= \frac{531{,}099 \times 3600 \times 1000}{169{\cdot}91 \times (1000)^4}$$

$$= 0{\cdot}0113 \text{ N/mm}^2.$$

Stress at the base at the end of the pier

$$= \frac{M_w \times 5000}{I_y}$$

$$= \frac{531{,}099 \times 5000 \times 1000}{169{\cdot}91 \times (1000)^3}$$

$$= 0{\cdot}0156 \text{ N/mm}^2.$$

Total Stresses

Under wet conditions

Total stress = Dead load + Live load (Eccentric + Longitudinal force + Frictional resistance) + Wind load

At end of straight portion (point A or E) maximum

$$=0{\cdot}214+0{\cdot}0623+0{\cdot}166+0{\cdot}0968+0{\cdot}0223+0{\cdot}031$$

$$=0{\cdot}5943 \text{ N/mm}^2$$

At end of straight portion (point A or C) minimum

$$=0{\cdot}214-0{\cdot}0043-0{\cdot}166-0{\cdot}09685-0{\cdot}0223-0{\cdot}031$$

$$=-0{\cdot}1065 \text{ N/mm}$$

At end of pier (F or C) maximum

$$=0{\cdot}204+0{\cdot}031=0{\cdot}245 \text{ N/mm}^2$$

At end of pier (F or C) minimum

$$=0{\cdot}214-0{\cdot}031=0{\cdot}183 \text{ N/mm}^2$$

Under wet conditions

Total stress = Dead load + Buoyancy + Live load + Wind load + Water currents

At end of straight portion (point A or E) maximum

$$=0{\cdot}214-0{\cdot}0573+0{\cdot}0623+0{\cdot}0968+0{\cdot}166+0{\cdot}0223+0{\cdot}0113$$

$$=0{\cdot}5154 \text{ N/mm}^2$$

At end of straight portion (point A or E) minimum

$$=0{\cdot}214-0{\cdot}0573-0{\cdot}0043-0{\cdot}0968-0{\cdot}166-0{\cdot}0223-0{\cdot}0113$$

$$=-0{\cdot}144 \text{ N/mm}^2$$

At end of pier (F or C) maximum

$$=0{\cdot}214-0{\cdot}0573+0{\cdot}031+0{\cdot}0156$$

$$=0{\cdot}2033 \text{ N/mm}^2$$

At end of pier (F or C) minimum

$$=0{\cdot}214-0{\cdot}0573-0{\cdot}031-0{\cdot}0156$$

$$=0{\cdot}1101 \text{ N/mm}^2$$

Allowable compressive stress in M 10 concrete is 2·5 N/mm² in compression and 0·25 N/mm² in tension. The stresses in pier are within these permissible limits.

26·5. Abutments

An abutment is a structure that supports on end of a bridge span and at the same time supports the embankment upon which the approach roadway to the bridge rests. Thus an abutment serves

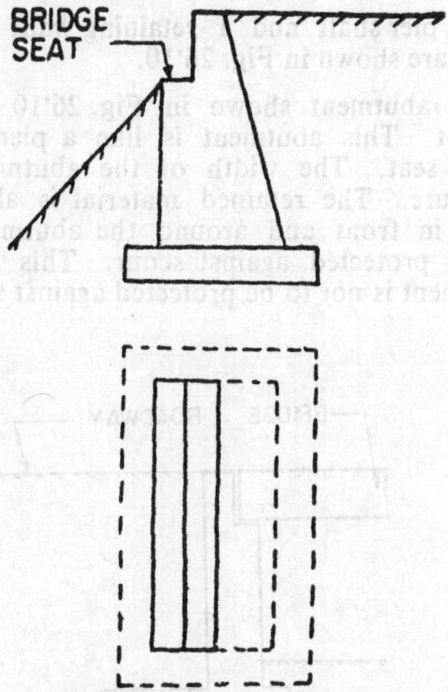

(a.) BREAST OR BURIED ABUTMENT

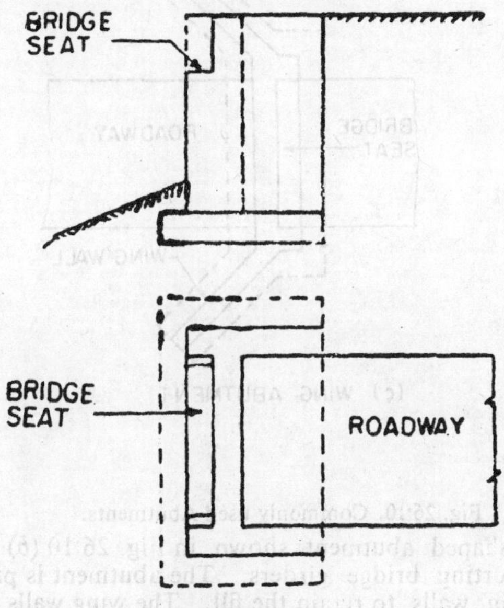

(b) U-ABUTMENT

the purpose of a pier shaft and a retaining wall. Most commonly used abutments are shown in Fig. 26·10.

The breast abutment shown in Fig. 26·10 (*a*) is the simplest type of abutment. This abutment is like a pier with recess at the top for bridge seat. The width of the abutment is designed to resist each pressure. The retained material is allowed to stand on its natural slope in front and around the abutment. The retained material is not protected against scour. This abutment is used where embankment is not to be protected against scour.

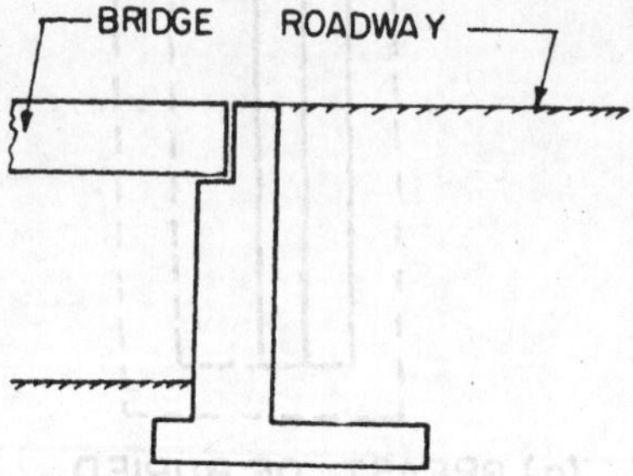

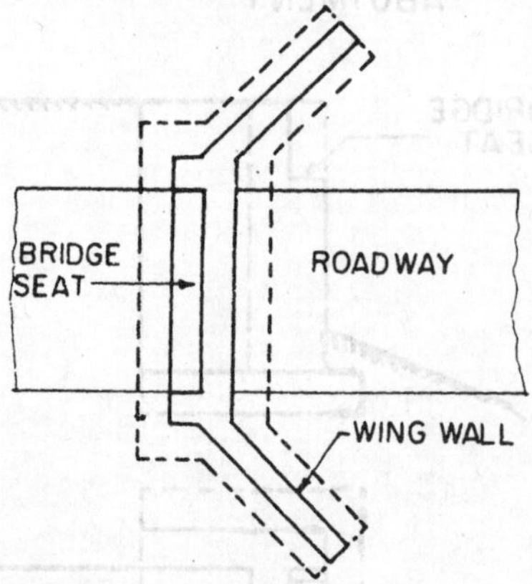

(c) WING ABUTMENT

Fig. 26·10. Commonly used Abutments.

The U-shaped abutment shown in Fig. 26·10 (*b*) has recessed top for supporting bridge girders. The abutment is provided with U shaped wing walls to retain the fill. The wing walls are designed as reinforced concrete walls. They are generally concreted by horizontal tees.

Fig. 26·10 (*c*) shows commonly used wing abutment. This type of abutment has tops of the wings slope to conform to the natural slope of the retained material.

The abutments are provided with drainage system to drain off water and thus reduce hydrostatic pressure at the back of the wall. Commonly used drainage system is to provide weep-holes as shown in Fig. 26·11. These holes consist of pipes 15 to 20 cm in diameter and are spaced about 3 m apart both horizontally and vertically. The weep-holes are prevented from clogging by pockets of gravel in the backfill.

In case if abutments are not provided with adequate drainage, the influence of submerged soil should be taken into consideration.

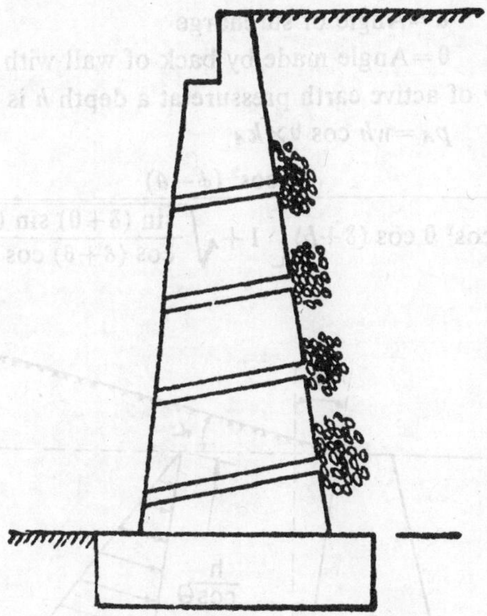

Fig. 26·11. Drainage in an abutment.

26·6. Design Loads for Abutments

Design of abutments is similar to design of retaining walls. In addition to both pressures, the forces from the roadway are to be taken into consideration. An abutment is designed to resist the following forces :

1. Self-weight of the abutment.
2. Thrust on the abutment caused by the retained material.
3. Thrust caused by the live load on the retained material at the back of the abutment.
4. Dead load due to super-structure.

5. Live load on the super-structure.

6. Longitudinal forces caused by tractive effort and braking of vehicles and temperature variation.

Most of the above mentioned forces have been discussed in Chapter 1. The earth pressure can be computed according to any rational theory. Coulomb's theory may be used provided the centre of pressure is assumed at 0·42 of the height from the base. The active and passive earth pressures are given by :

Active Earth Pressure

Notation : ϕ=Angle internal friction of soil.

δ=Angle of friction between retaining wall and earth.

α=Angle of surcharge

θ=Angle made by back of wall with vertical.

Intensity of active earth pressure at a depth h is given by

$$p_A = wh \cos\theta \times k_A$$

where
$$k_A = \frac{\cos^2(\phi-\theta)}{\cos^2\theta \cos(\delta+b)\left[1+\sqrt{\dfrac{\sin(\delta+\theta)\sin(\phi-\alpha)}{\cos(\delta+\theta)\cos(\delta-\alpha)}}\right]^2}$$

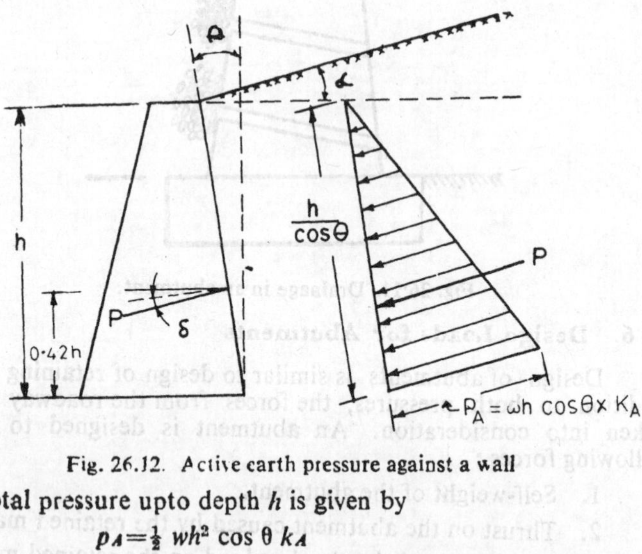

Fig. 26.12. Active earth pressure against a wall

Total pressure upto depth h is given by

$$p_A = \tfrac{1}{2}\, wh^2 \cos\theta\, k_A$$

The resultant earth pressure is inclined at the angle δ to the normal to the face of the wall as shown in Fig. 26·12.

Passive Earth Pressure

Intensity of passive earth pressure at a depth h is given by

$$p_P = wh \cos\theta \, k_P$$

and total earth pressure upto depth h is given by

$$P_P = \frac{1}{2} wh^2 \cos\theta \, k_P$$

where
$$k_P = \frac{\cos^2(\phi+\theta)}{\cos^2\theta \cos(\theta-\delta)\left[1-\sqrt{\dfrac{\sin(\theta+\delta)\sin(\phi-\alpha)}{\cos(\theta-\delta)\cos(\phi-\alpha)}}\right]^2}$$

The resultant earth pressure is inclined at an angle δ to the face of the wall as shown in Fig. 26.13.

The equivalent surcharge for live load acting on the retained material is taken from table 26·1. The surcharge is neglected in design if a reinforced concrete approach slab is provided covering the entire width of the roadway with one end resting on the abutment. The approach slab should have a length of at least 3·5 m.

The figures given in table 26·1 are based on follow ng assumptions :

(1) Length of abutment $L=4·5$ for single lane bridges and 7·6 m for multi-lane bridges.

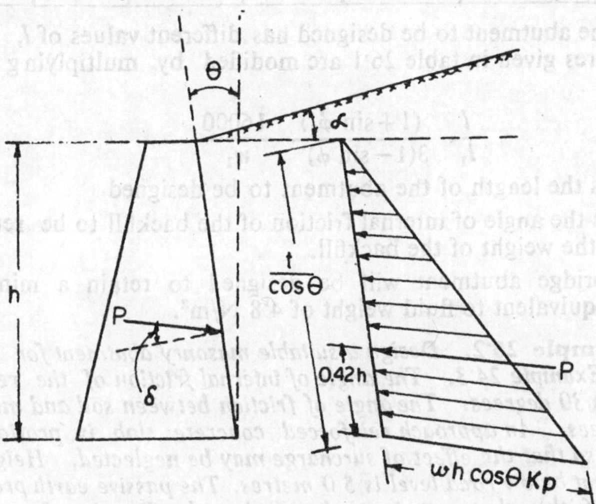

Fig. 26·13. Passive earth pressure against a wall.

(2) Angle of internal friction of the backfill $\phi=30°$

(3) Weight of the backfill, $w=16000$ N/m³

(4) The re ultant earth pressure acts in horizontal direction.

Table 26·1
Equivalent Height of Earth Surcharge in Metres for I.R.C. Standard Loading

Depth of abutment below load level	*Equivalent height of surcharge of earth in metres*					
	Class AA		*Class A*		*Class B*	
	Single lane bridge	*Multi-lane bridge*	*Single lane bridge*	*Multi-lane bridge*	*Single lane bridge*	*Multi-lane bridge*
0·2	26·0	15·4	14·3	17·2	8·3	10·0
1·0	15·0	9·1	8·5	10·0	5·1	5·8
2·0	8·0	5·5	5·1	6·1	3·0	3·7
3·0	6·8	4·1	3·8	4·6	2·3	2·7
4·0	5·5	3·3	3·0	3·5	1·8	2·1
6·0	3·8	2·3	2·2	2·6	1·3	1·5
8·0	3·0	1·8	1·7	2·0	1·0	1·2
10·0 and above	2·6	1·5	1·4	1·7	0·9	1·0

If the abutment to be designed has different values of l, ϕ and w the figures given in table 26·1 are modified by multiplying factor given by

$$\frac{l}{l_1}\ \frac{(1+\sin\phi_1)}{3(1-\sin\phi)}\ \frac{16000}{w_1}$$

where l_1 is the length of the abutment to be designed

ϕ_1 is the angle of internal friction of the backfill to be retained and w_1 is the weight of the backfill.

A bridge abutment will be designed to retain a minimum pressure equivalent to fluid weight of 4·8 N/m³.

Example 26·2. *Design a suitable masonry abutment for girder bridge of Example 24·3. The angle of internal friction of the retained material is 30 degrees. The angle of friction between soil and masonry is 20 degrees. An approach reinforced concrete slab is provided to the bridge so that the effect of surcharge may be neglected. Height of the abutment below road level is 5 0 metres. The passive earth pressure in front of the abutment is to be neglected Density of masonry =20 kN/m³, density of concrete=24 kN/m³ and density of soil=18 kN/m³.*

Solution. Distance between outside of outermost griders

$$=5{\cdot}0+0{\cdot}30=5{\cdot}3$$

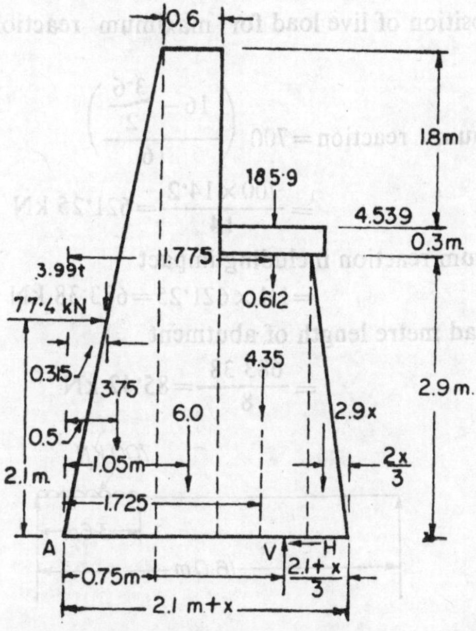

Fig. 26·14. Dimensions of Abutment.

Clear width of road way=6·8 m.

Provide abutment length of 8·0 metres. Provide bed block of 0·30 m depth and top width of 0·75 m. Provide 0·60 m top width for the abutment as shown in Fig. 26·14.

Depth of girder and slab=1·6 m. Provide a height of 1·8 m above the bed block as shown in Fig. 26·14. Provide a projection of 0·75 m in the back of the wall as shown in Fig. 26·14.

Let x be the projection of abutment at the front. The projection x will be determined from the condition that no tension develops at the base.

(1) *Dead load from the temperature.* Total dead load (Example 26·1)

$$=\frac{1,607,680}{2}=803,840 \text{ N}$$

Load per metre length of abutment

$$=\frac{803,840}{8}=100,480 \text{ N}$$

Live Load from Superstructure

Maximum live load reaction is caused by class AA loading.

The position of live load for maximum reaction is shown in Fig. 26·15.

$$\text{Maximum reaction} = 700\left(\frac{16-\frac{3\cdot6}{2}}{16}\right)$$

$$= \frac{700\times14\cdot2}{14} = 621\cdot25 \text{ kN}$$

Maximum reaction including impact

$$= 1\cdot1\times621\cdot25 = 683\cdot38 \text{ kN}$$

Live load metre length of abutment

$$= \frac{683\ 38}{8} = 85\cdot42 \text{ kN}$$

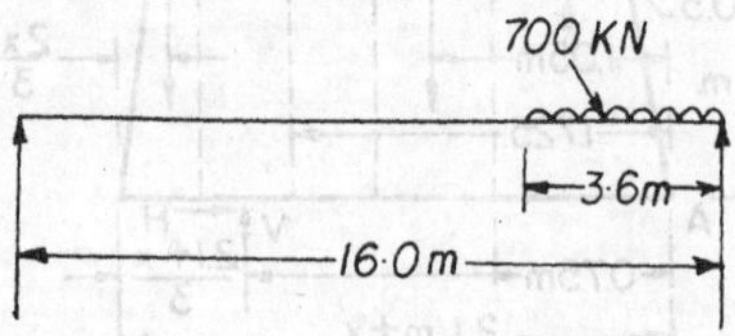

Fig. 26·15. Position of live load for maximum reaction.

Total reaction from superstructure per metre length of abutment. $=100,480+85,420=185,900\text{N}=185\cdot9$ kN.

(2) *Self-weight of the Abutment.* Self-weight of the abutment is divided in five segments as shown in Fig. 26·14. The weight per metre length is given by

Segment $1=0\cdot3\times0\cdot85\times24=6\cdot12$ kN

Segment $2=\frac{1}{2}\times x\times2\cdot9\times20=295$ kN

Segment $3=0\cdot75\times2\cdot9\times20=43\cdot5$ kN

Segment $4=0\cdot6\times5\cdot0\times20=60\cdot0$ kN

Segment $5=\frac{1}{2}\times0\cdot75\times5\cdot0\times20=37\cdot5$ kN.

The centre of gravity of these loads is shown in Fig. 26·16.

(3) *Longitudinal Forces*

Force due to tractive effort$=140$ kN

Force per metre length

$$= \frac{140}{8} = 17\cdot5 \text{ kN}$$

Force due to temperature charges

$$= 0\cdot15\times\text{Load from superstructure}$$

$$= 0\cdot15\times185\cdot9 = 27\cdot89 \text{ kN}$$

Total longitudinal force$=17\cdot5+27\cdot89=45\cdot39$ kN

The force acts at level of bearings.

(4) *Earth Pressure.* Earth pressure is computed from Coulomb's formula.

$\phi=30°$. $\tan\theta=\dfrac{0.75}{5}=0.15$,

$\theta=8.53°$, $\alpha=0$, $\delta=20°$,

Total active earth pressure

$$=\tfrac{1}{2}\times 18\times 5\times 5\times k_A\cos\theta=225\,k_A\cos\theta$$

$$k_A=\frac{\cos^2(\phi-\theta)}{\cos^2\theta\,(\cos\delta+\theta)\left[1+\sqrt{\dfrac{\sin(\delta+\theta)\sin\phi}{\cos(\delta+\theta)\cos\delta}}\right]^2}$$

$=0.3755.$

Total active pressure$=225\times 0.3753\times 0.9889=83.55$ kN

Horizontal component of earth pressure

$$=83.55\cos(\delta+\theta)=83.55\times\cos 28.53$$
$$=83.55\times 0.8786=73.40\text{ kN}$$

Vertical component of earth pressure

$$=83.55\times\sin 28.53=39.90\text{ kN}$$

The earth pressure is assumed to act at $0.42\,h=0.42\times 5=2.1$ m from the base.

The forces acting on the abutment are shown in Fig. 26·16.

The value of x is determined from the condition that no tension develops at base. The resultant of all forces must pass through idmdle third.

Total vertical force

$$V=39.9+37.5+60+43.5+6.12+185.9+29\,x$$
$$=372.92+29\,x.$$

Moments of all forces about A must be zero.

$$73.4\times 2.1+39.9\times 0.315-37.5\times 0.5-60\times 1.05\times 43.5\times 1.725$$
$$+185.9\times 1.725+6.12\times 1.775+295\left(2.1+x-\frac{2x}{3}\right)$$
$$-(372.92+29\,x)\tfrac{2}{3}(2.1+x)=0$$

or $x^2+23.6201\,x-23.7812=0$

Solution of above equation gives

$$x=1.1614$$

Maximum pressure at base

$$=2\times\frac{\text{Total vertical Force}}{\text{Area}}$$
$$=\frac{2\times(39.9+37.5+60+43.5+6.12+185.9+29\times 1.1614)}{2.1+1.1614}$$
$$=244.341\text{ kN/m}^2$$

This is within permissible limits

Provide overall base width of

$$2.1+1.1614=3.2614\text{ m}$$
$$=\text{say } 3.30\text{ metres}$$

Round Bars

Designation	Diameter in mm	Weight per meter in kg.	Perimeter in cm	Cross-sectional area in cm²									
				1	*2*	*3*	*4*	*5*	*6*	*7*	*8*	*9*	*10*
ISRO 5	5	0·15	1·57	0·20	0·39	0·59	0·79	0·98	1·18	1·37	1·57	1·77	1·96
ISRO 6	6	0·22	1·89	0·28	0·57	0·85	1·13	1·42	1·70	1·98	2·26	2·55	1·83
ISRO 8	8	0·40	2·51	0·50	1·01	1·51	2·01	2·51	3·02	3·52	4·02	4·53	5·03
ISRO 10	10	0·62	3·14	0·79	1·57	2·36	3·14	3·93	4·71	5·50	6·28	7·07	7·85
ISRO 12	12	0·89	3·77	1·13	2·26	3·39	4·52	5·65	6·70	7·92	9·05	10·18	11·31
ISRO 16	16	1·58	5·03	2·01	4·02	6·03	8·04	10·05	12·05	12·07	16·08	18·10	20·11
ISRO 20	20	2·47	6·28	3·14	6·28	9·41	12·56	15·71	18·85	21·99	25·14	28·28	31·42
ISRO 25	25	2·85	7·86	4·91	9·82	14·73	19·63	24·54	29·45	34·36	39·27	44·18	49·09
ISRO 28	28	4·80	8·80	6·16	12·32	18·47	24·95	30·79	36·95	43·10	49·26	55·42	61·58
ISRO 32	32	6·31	10·05	8·04	16·48	24·13	32·17	40·21	48·25	56·30	64·34	72·38	80·42
ISRO 36	36	8·00	11·31	10·18	20·36	30·54	40·72	50·90	61·08	71·26	81·44	91·62	101·80
ISRO 40	40	9·87	12·57	12·57	25·13	37·70	50·27	62·83	75·40	87·96	100·53	113·10	125·66

Area of Steel in cm^2 in 1 metre Width of Slab

Diameter of bars in mm	*Area of steel for following spacing in mm*												
	30	*40*	*50*	*60*	*70*	*80*	*90*	*100*	*110*	*120*	*130*	*140*	*150*
5 mm	—	4.90	3.92	3.27	2.80	2.45	2.18	1.96	1.78	1.63	2.51	1.40	1.31
6 mm	—	7.08	5.66	4.72	4.04	4.54	3.14	2.83	2.57	9.36	2.18	2.02	1.89
8 mm	—	12.58	10.06	8.38	7.19	6.29	4.59	5.03	4.57	4.18	3.87	3.59	3.35
10 mm	—	19.63	15.70	13.08	11.21	9.81	8.72	7.85	7.14	6.54	6.04	5.61	5.23
12 mm	—	28.28	22.62	18.85	16.16	14.14	12.57	11.31	10.28	9.43	8.70	8.08	7.54
16 mm	—	—	40.42	33.52	28.73	25.14	22.34	20.11	18.27	16.76	15.47	14.37	13.41
20 mm	--	—	62.84	52.36	44.86	39.28	34.91	31.42	28.56	26.18	24.17	22.44	20.95

Area of Steel in cm² in 1 metre Width of Slab

Area of steel for following spacing in mm

160	*170*	*180*	*200*	*220*	*240*	*260*	*280*	*300*	*320*	*340*	*360*	*380*	*400*
1·23	1·15	—	—	—	—	—	—	—	—	—	—	—	—
1·77	1·66	1·57	1·42	1·29	1·18	—	—	—	—	—	—	—	—
3·14	2·96	2·79	2·52	2·29	2·10	1·93	1·80	1·68	1·57	1·48	1·40	1·33	1·26
4·91	4·62	4·36	3·93	3·51	3·27	3·02	2·80	2·62	2·45	2·31	2·17	2·07	1·96
7·07	6·65	6·28	5·66	5·14	4·17	4·35	4·04	3·77	3·53	3·33	3·09	2·98	2·83
19·57	11·82	11·17	10·06	9·14	8·38	7·74	7·18	6·71	6·29	5·91	5·58	5·29	5·03
19·64	18·48	17·46	15·76	14·28	13·19	12·09	11·22	10·47	9·82	9·24	8·73	8·27	7·88

27

Shell Construction

27·1. Shells

Thin-shell concrete roofs have been extensively used in construction because of impressive appearance and economy in cost. For applications to wide spans, the cost of scaffolding is generally high and for wide spans these roofs are used where spans are similar so that the scaffolding can be reused. Shell roof construction requires special care in laying reinfocement, concrete pouring and compaction ; and hence the labor required for this type of construction is highly skilled.

The curvature in a shell roof is so designed that the load is primarily carried by inplane forces and the bending is negligible. A typical cylindrical shell roofs of single span and multi-spans are shown in Fig. 27·1 (*a*) and (*b*). A typical north-light shell roof is shown in Fig. 27·1 (*c*).

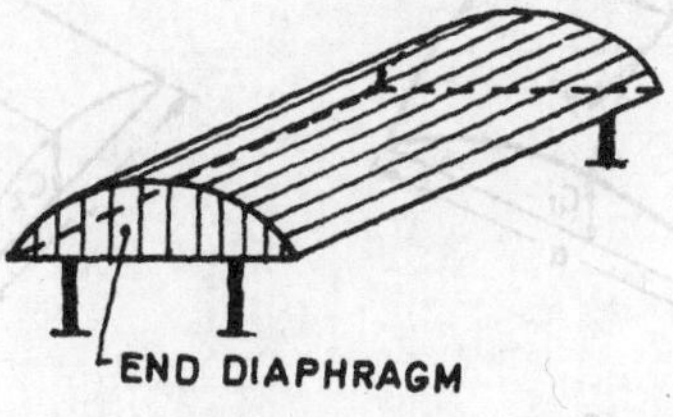

(*a*)

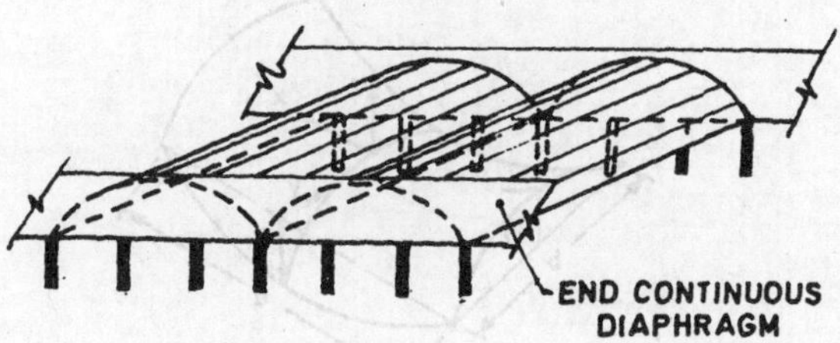

(*b*)

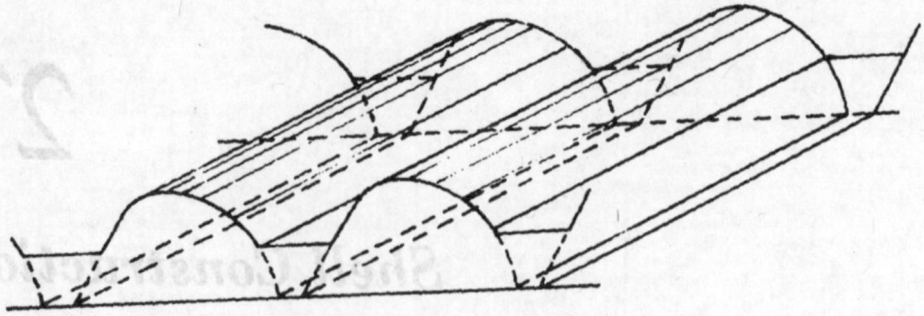

(c)

Fig. 27·1. Typical shell construction.

A shell surface may be generated by rotation or by translation of a plane curve. Consider a thin translational shell rectangular in plan, the middle surface of which is defined by

$$Z(x, y)=\tfrac{1}{2}\, k_2\, (\gamma\, x^2-y^2) \qquad \text{.. (27·1)}$$

where $\gamma=k_1/k_2$ is a shape or curvature parameter of the surface (Fig. 27·2). The constants k_1 and k_2 represent either constant curvature or relative rise to the square of the chord length for parabolic curves. Depending on the sign of γ, the Eqn. (27·1) represents a hyperbolic paraboloid ($\gamma>0$), a parabolic cylinder ($\gamma=0$) and an elleptic paraboloid ($\gamma<0$).

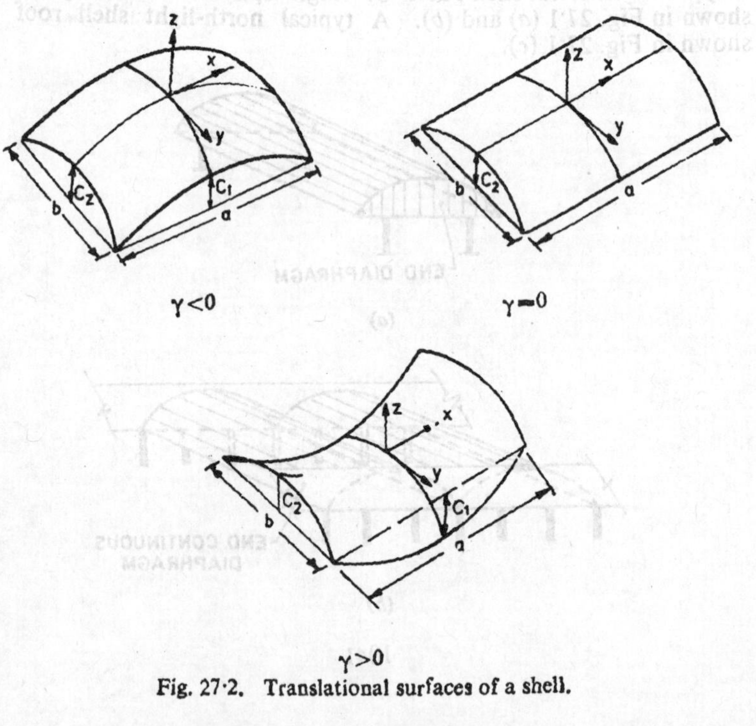

Fig. 27·2. Translational surfaces of a shell.

Two theories namely membrane theory and bending theory have been used to analyze shell structures. In the membrance theory only membrane stresses are considered to be present. These stresses on a small element in cartesian coordinate system are shown in Fig. 27·3 (*a*) and in curvilinear coordinate system these are shown in Fig. 27·3 (*b*).

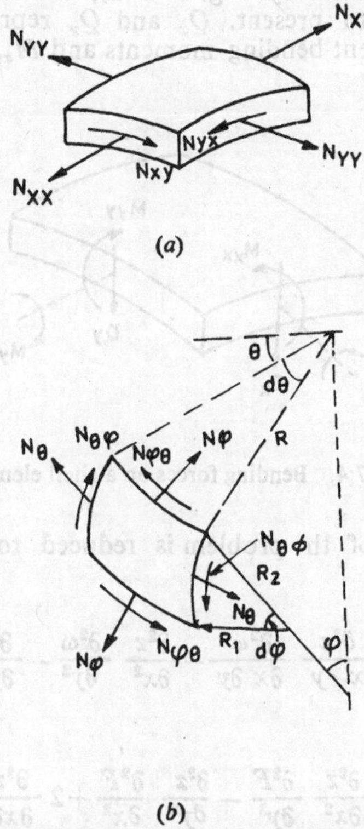

Fig. 27·3. Membrane forces on a shell element.

By considering the equilibrium of these forces it can be shown that in curvilinear coordinate system the problem is reduced to the solution of equations given by

$$\frac{\partial}{\partial\varphi}(RN_\varphi)+R_1\frac{\partial N_{\varphi\theta}}{\partial\theta}-R_1 N_\theta\cos\varphi+p_\varphi RR_1=0$$

$$\frac{\partial}{\partial\varphi}(RN_{\varphi\theta})+R_1\frac{\partial N_\theta}{\partial\theta}+R_1 N_{\varphi\theta}\cos\varphi+p_\theta RR_1=0$$

$$\frac{N_\varphi}{R_1}+\frac{N_\theta}{R_2}=p_r \qquad \text{...(27·2-27·3)}$$

Thus the problem is reduced to solving three differential equations for three unknowns. The solution of this problem has been obtained in literature (see References 43 to 46).

In the bending theory of shell analysis, it is assumed that membrane as well as bending forces are present. Thus in addition to membrane forces given by Fig. 27·3 (*a*), the bending forces given in Fig. 27·4 are also present. Q_x and Q_y represent shear forces. M_{xx} and M_{yy} represent bending moments and M_{xy} and M_{yx} represent torsional moments.

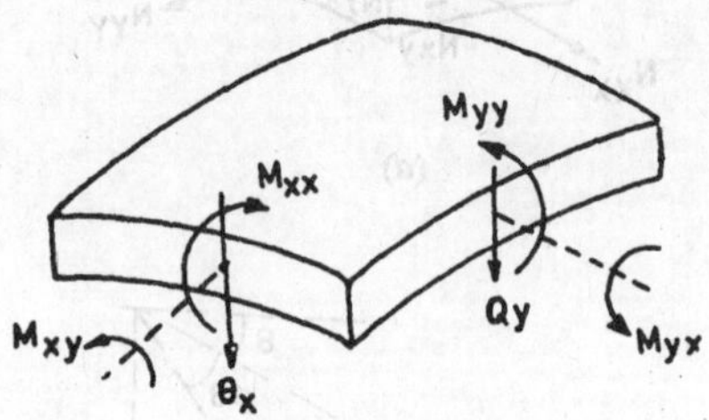

Fig. 27·4. Bending forces on a shell element.

The solution of the problem is reduced to solving differential equations given by

$$\nabla^4 F - Eh\left(2\frac{\partial^2 z}{\partial x\, \partial y}\,\frac{\partial^2 \omega}{\partial x\, \partial y} - \frac{\partial^2 z}{\partial x^2}\,\frac{\partial^2 \omega}{\partial y^2} - \frac{\partial^2 z}{\partial y^2}\,\frac{\partial^2 \omega}{\partial x^2} \right) = 0 \quad \text{...(27·4)}$$

$$D\nabla^4 \omega - \frac{\partial^2 z}{\partial x^2}\,\frac{\partial^2 F}{\partial y^2} - \frac{\partial^2 z}{\partial y^2}\,\frac{\partial^2 F}{\partial x^2} + 2\,\frac{\partial^2 z}{\partial x \partial y}\,\frac{\partial^2 F}{\partial x\, \partial y} = q_n \quad \text{...(27·5)}$$

where F is known as Airy's stress function

ω is the displacement in Z direction

k_2 and γ have been defined in Eqn. (27·1)

q_n is applied surface loading

D is flexural rigidity $= \dfrac{Eh^3}{12(1-\nu^2)}$

h is shell thickness

Membrane and bending forces in terms of stress function and displacement are given by

$$N_{xx}=\frac{\partial^2 F}{\partial y^2},\quad N_{yy}=\frac{\partial^2 F}{\partial x^2}$$

$$N_{xy}=-\frac{\partial^2 F}{\partial x\,\partial y}$$

$$M_{xx}=-D\left[\frac{\partial^2 \omega}{\partial x^2}+v\frac{\partial^2 \omega}{\partial y^2}\right]$$

$$M_{yy}=-D\left[\frac{\partial^2 \omega}{\partial y^2}+v\frac{\partial^2 \omega}{\partial x^2}\right]$$

$$M_{xy}=-D\,(1-v^2)\,\frac{\partial^2 \omega}{\partial x\,\partial y}$$

$$Q_x=-D\frac{\partial}{\partial x}(\nabla^2 w)$$

$$Q_y=-D\frac{\partial}{\partial y}(\nabla^2 w) \qquad ...(27\cdot6)$$

The solution of differential equations given by (27·5) has been obtained in the literature for a variety of cases (*e.g.* see Ref. 43 to 46). A simplified analysis of cylindrical shells is discussed here in subsequent articles.

27·2. Approximate Analysis of Circular Cylindrical Shell

A simplified strength of materials approach to analysis of circular cylindrical shell is presented here. Consider cirular cylindrical shell shown in Fig. 27·5. The shell is considered as simply supported at ends.

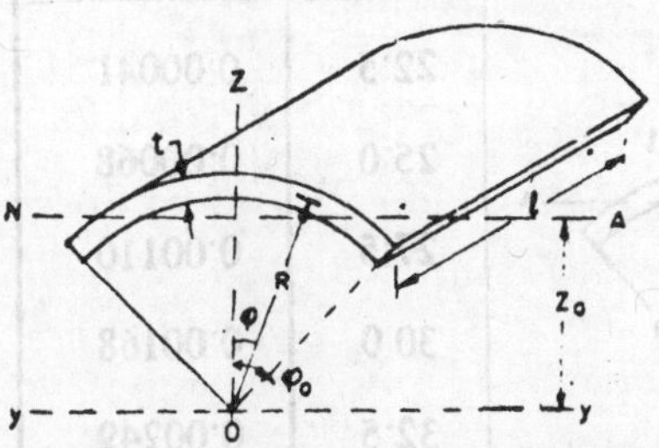

Fig. 27·5. Simply supported circular cylindrical shell.

Let R be the radius of curvature of shell, t the thickness and l the length of the shell. Z axis is axis of symmetry of the shell. Let p be the load per square meter of the shell surface. The load per longitudinal meter run of the shell is given by

$$\omega=2\,Rp\varphi_o \qquad ...(27\cdot7)$$

The longitudinal stresses in the shell are computed assuming it to behave like a beam. Shear force at a section distance x from the support

$$V(x)=\omega\left(\frac{l}{2}-x\right) \qquad \ldots(27{\cdot}8)$$

and bending moment is given by

$$M(x)=\omega\,\frac{l^2}{2}\left(\frac{x}{l}-\frac{x^2}{l^2}\right) \qquad \ldots(27{\cdot}9)$$

Neutral axis of the shell section is computed using strength of materials approach. Distance of neutral axis from origin (Fig. 27·5) is given by

$$Z_0=\frac{R\sin\varphi_0}{\varphi_0} \qquad \ldots(27{\cdot}10)$$

The moment of intertia of the shell section is given by

$$I=R^3 t\left[\varphi_0+\sin\varphi_0\left(\cos\varphi_0-\frac{2\sin\varphi_0}{\varphi_0}\right)\right] \qquad \ldots(27{\cdot}11)$$

The value of I may be written as

$$I=R^3t\alpha \qquad \ldots(27{\cdot}12)$$

The values of α for various values of φ are given in Table 27·1.

Table 27·1. Moment of Inertia Coefficient α For Circular Cylindrical Shell ($I=R^3t\alpha$)

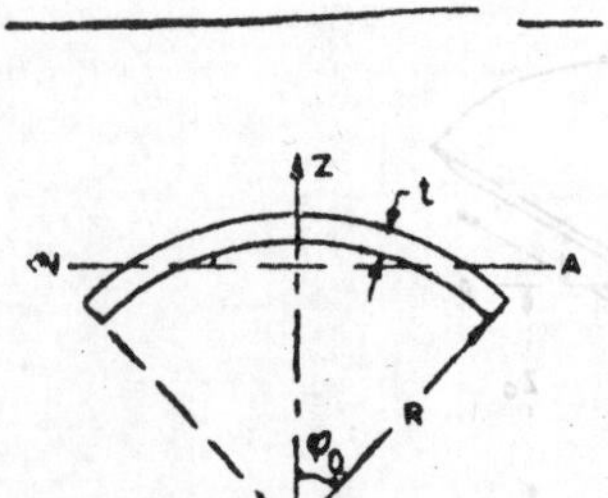

φ_0	α	φ_0	α
22·5	0·00041	37·5	0·00502
25·0	0·00068	40·0	0·00687
27·5	0·00110	45·0	0·01216
30 0	0·00168	50·0	0·02017
32·5	0·00249	55·0	0·03174
35·0	0 00358	60·0	0·04782

Longitudinal stress σ_x is given by

$$\sigma_x=-\frac{MZ}{I} \qquad \ldots(27{\cdot}13)$$

where Z is distance from neutral axes

σ_x is +ve in tension.

To find the transverse stresses in the shell, the shell is considered as an arch and analysis carried out considering a section X-X of the shell as shown in Fig. 27·6.

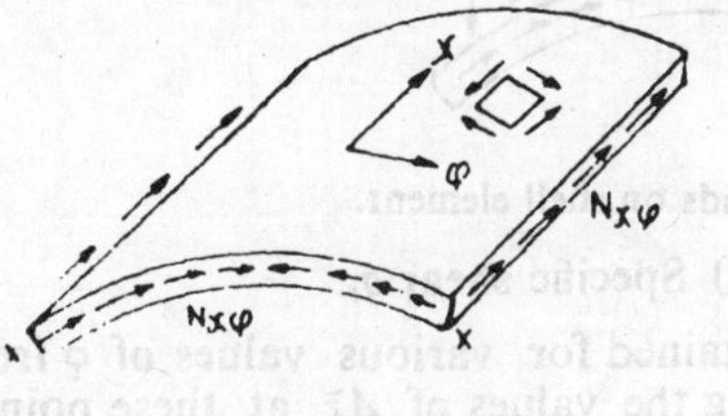

Fig. 27·6

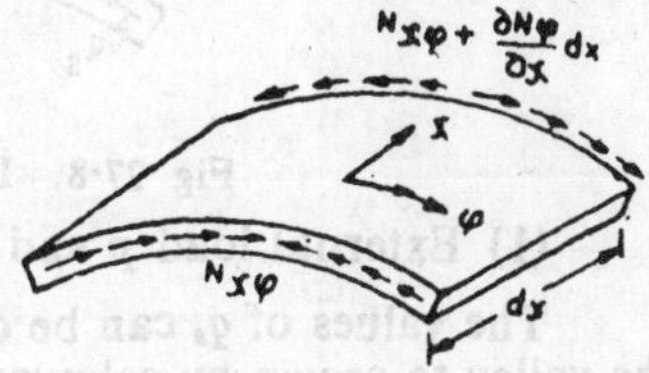

Fig. 27·7

The longitudinal shear flow and complementry transverse shear at distance x from the support can be obtained using strength of materials analysis. Shear is resisted by two edges in section X-X, hence

$$2N_{x\varphi} = \frac{V(x)AZ}{I}$$

or
$$N_{x\varphi} = \frac{V(x)AZ}{2I} \quad ...(27·14)$$

where AZ is the moment of the area about neutral axis of the portion of the cross-section above X-X.

Consider an element of the section between x and $x+dx$ as shown in Fig. 27·7.

The edges of the shell at $\varphi=\varphi_0$ are assumed free, with this assumption, the slice is under equilibrium due to external load ω and the shears $N_{x\varphi}$ and $N_{x\varphi}+\frac{\partial N_{x\varphi}}{\partial x}dx$ acting on the slice.

The difference in two shears is called specific shear and can be obtained from Eqn. (27·14) as

$$\frac{\partial N_{x\varphi}}{\partial x}\, dx = \frac{\partial V(x)}{\partial x}\, dx\, \frac{AZ}{2I} \quad ...(27·15)$$

or specific shear
$$= \frac{\partial N_{x\varphi}}{\partial x} = q_s = \frac{\delta}{dx} V(x) \frac{AZ}{2I} \quad ..(27·16)$$

Substituting value of $V(x)$ from Eqn. (27·8)

$$q_s = -q\, \frac{AZ}{2I} \quad ...(27·17)$$

—ve sign of q_s indicates that it is acting upwards.

A shell slice of unit length at any section is in equilibrium under the following forces (as shown in Fig. 27·8).

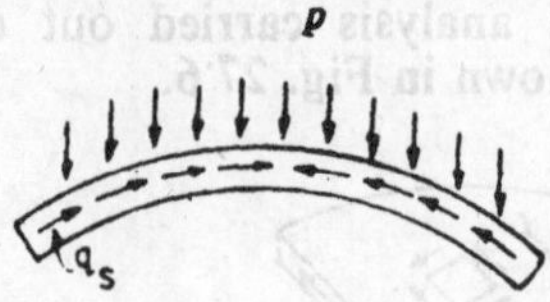

Fig 27·8. Loads on shell element.

(1) External load p and (2) Specific shear q_s.

The values of q_s can be obtained for various values of φ from the valley to crown by calculating the values of $A\bar{z}$ at these points. For a circular cross-section the $A\bar{z}$ can be obtained from

$$A\bar{z}=2R^2t\left(\sin\varphi-\frac{\varphi}{\varphi_0}\sin\varphi_0\right) \quad ...(27\ 18)$$

Knowing ω and q_s the internal stress resultants, *i.e.* moment M_φ, shear Q_φ and normal force N_φ is the arch (shell) can be calculated from statics by starting from the edges. The arch is statically determinate as the edges are considered free.

In the case of internal shell of a multibarrel shell construction, a horizontal thrust and moment are present at the valley of the arch. These are unknown forces, making the arch analysis considered above as indeterminate. These unknown forces are obtained from two additional conditions of zero displacement and rotation at the valley.

Thus, the analysis of shell structure consists of obtaining σ_x from the beam analysis $N_{x\varphi}$ from Eqn. (27·14) and M_φ, Q_φ and N_φ from the arch analysis. The design is governed by $N_{x\varphi}$ at the support and all other values at mid-span.

27·3. Empirical Design of Symmetrical Cylindrical Shells

A design method based on over 250 designs was developed by Bennett and reported in reference 45. All these designs have been actually built. In this method designs of three different shell structures (Fig. 27·9) namely feather edge barrel. barrel with valley beam and barrel with vertically supported edge beam have been discussed.

The main objective of the proposed design method was to obtain the maximum values of all the significant forces in the shell from simple formulae. The forces to be obtained are longitudinal force T_1 (Fig. 27·10) at mid span, shear S at the stiffening beam, transverse force T_2 (Fig. 27·10) at the crown and maximum values of transverse moment M_2. The nomenclature of different terms used in the neutral core given in Table 27·2.

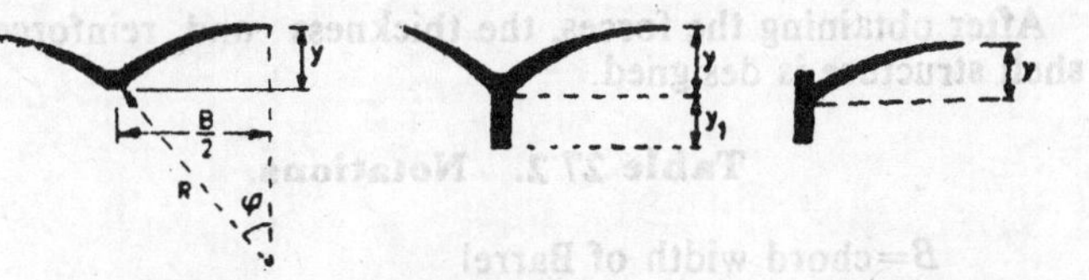

(*a*) Feather edge Barrel (*b*) Barrel with valley beam (*c*) Barrel with vertically s upported edge beam

Fig. 27·9

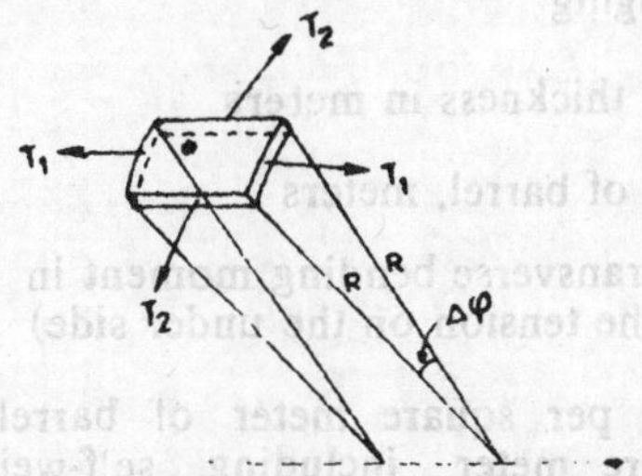

Fig. 27·10. Longitudinal and Transverse forces on shell.

Coefficients for longitudinal and transverse forces, moments and shear are given in Table 27·3. In the Table 27·3, percentage load 'ω' for barrel with vertically supported beam is obtained from Fig. 27·11.

L=Span of barrel

$pR\varphi$=Load per meter run of barrel

ω=Load per meter run carried by barrel

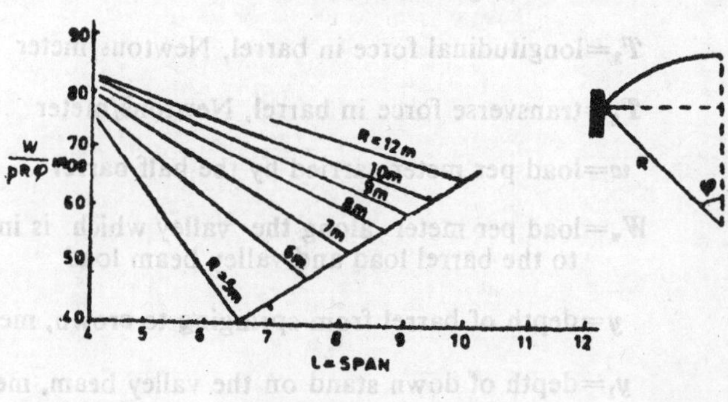

(*a*) (*b*)

Fig. 27·11. Percentage load carried by barrels with supported edge beams.

After obtaining the forces, the thickness and reinforcement of the shell structure is designed.

Table 27.2. Notations

B=chord width of Barrel

θ=angle between the radius and vertical at any point on the barrel

φ=angle between the radius and the vertical at the springing

d=shell thickness in meters

L=span of barrel, meters

M_2=the transverse bending moment in the barrel (positive for the tension on the under side)

p=load per square meter of barrel in Newtons per square meter including self-weight, finishes and superimposed load

q=total load per meter of the barrel and valley beam, including all superimposed loads and variation in thickness due to stiffening.

R=radius of barrel, meters

S=shear force, Newton/meter

ΣS=sum of all the shear forced in the half barrel at the stiffening beam, Newtons

T_1=longitudinal force in barrel, Newtons/meter

T_2=transverse force in barrel, Newtons/meter

w=load per meter carried by the half barrel

W_v=load per meter along the valley which is in addition to the barrel load and valley beam load

y=depth of barrel from springing to crown, meters

y_1=depth of down stand on the valley beam, meters

y_2=total depth from the soffit of the valley beam to the crown of the barrel.

Table 27·3. Coefficients for Designing Symmetrical Cylindrical Shell

	 *Feather edge barrels*	 *Barrels with valley beams*	 *Barrel with vertically supported edge beams*
Weight per metre carried by the barrel ω	$pR\varphi + \frac{W_e}{2}$	$\frac{q}{0{\cdot}84} \; \frac{y}{y_2}$	Determined from Fig. 27·11 after finding $\sqrt{\frac{L}{\varphi}}$
Weight per metre carried by the beam		$q-\omega$	$qR\varphi - \omega$
Tension T_1 in valley of the barrel at mid span	$\frac{\omega L^2}{1{\cdot}76\, y}$	$\frac{qL^2}{1{\cdot}68 y_2}\left(\frac{y}{y_2} - 0{\cdot}48\right)$	$\frac{\omega L^2}{17{\cdot}5\, y}$
Tension T_1 in the beam		$\frac{qL^2}{1{\cdot}68 y_2}\left(1{\cdot}33 - \frac{y}{y_2}\right)$	$\frac{\omega L^2}{2{\cdot}38\, y}$
ΣS	$\frac{\omega L}{1{\cdot}06\,\varphi}$	$\frac{\omega L}{1{\cdot}16\varphi}$	$\frac{\omega L}{1{\cdot}14\,\varphi}$
M_2	$1{\cdot}87 \times 10^{-4} \frac{\omega^2 L}{y}$	$1{\cdot}37 \times 10^{-4} \frac{\omega^2 L}{y_2}$	$1{\cdot}62 \times 10^{-4} \frac{\omega^2 B}{y}$
T_2	$1{\cdot}2\, pR \sqrt[4]{\frac{L}{B}}$	$1{\cdot}04\, pR \sqrt{\frac{L}{B}}$	$1{\cdot}1\, qR$

28

Folded Plate Construction

28·1. Folded Plates

Folded plate construction has been used in reinforced concrete as well as timber construction. Compared to shell construction, this type of construction is economical as the formwork is relatively cheaper and concrete pouring and compaction are easier. Also the inherent stiffness of the folded plate construction permits longer spans. A typical folded plate construction is shown in Fig. 28·1. Folded plate construction is also used as corrugations of floor for ducts and utility troughs as shown in Fig. 28·2.

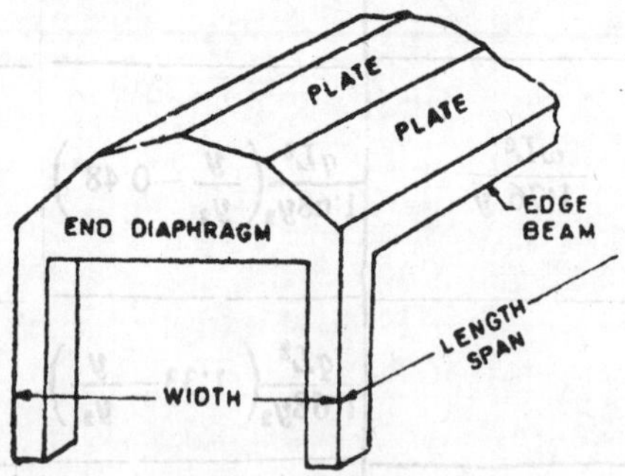

Fig. 28·1. Typical folded plate construction.

Fig. 28·2. Folded plate for corrugate floor construction.

The simplest form of folded plate construction consists of inverted plate in V-shape as shown in Fig. 28·3. The main disadvantage of this construction is that the concrete area to resist compressive forces and the area to place steel may not be adequate. The general form of folded plate construction can be developed by addition of horizontal plates as shown in Fig. 28·4. The application

Fig. 28·3. Simple folded plate construction.

of folded plate construction in north light roof construction as shown in Fig. 28·5. The application of triangular folded plate construction is shown in Fig. 28·6. A folded plate umbrella roof is shown in Fig. 28·7.

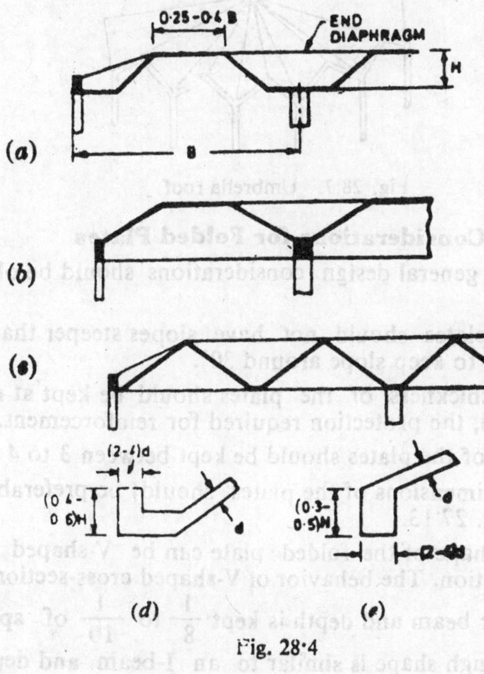

Fig. 28·4

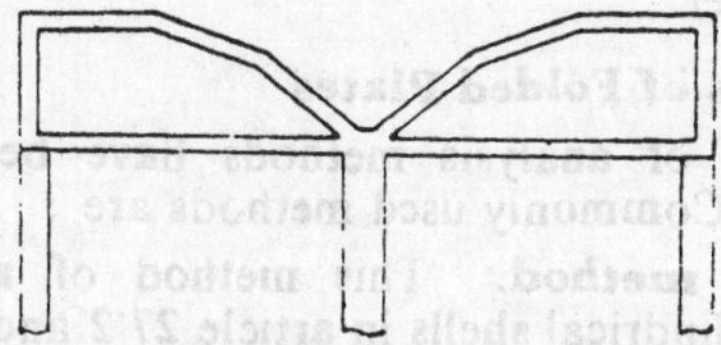

Fig. 28·5. North light folded plate construction.

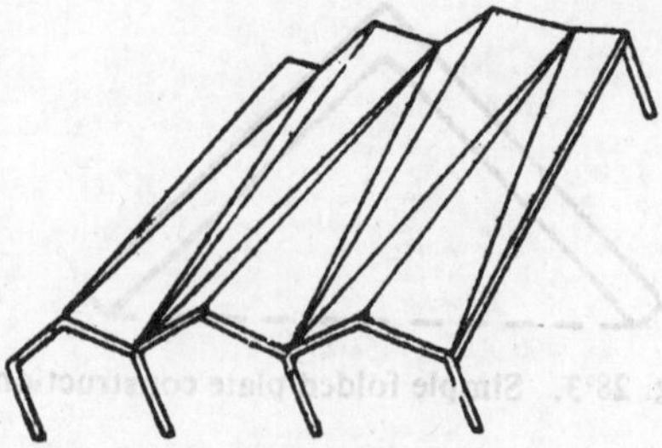

Fig. 28·6. Triangular folded plate construction.

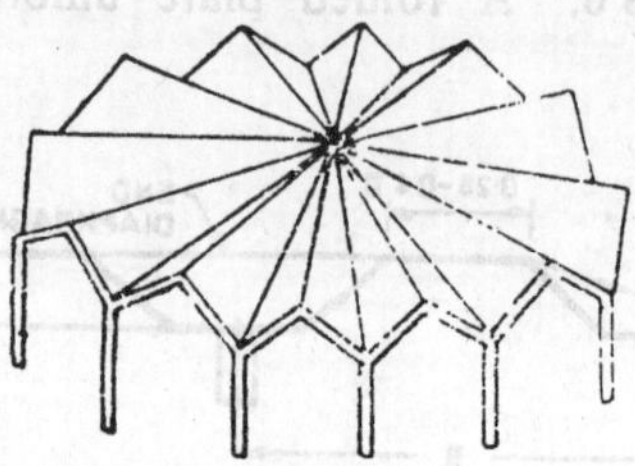

Fig. 28·7. Umbrella roof.

28·2. Design Considerations for Folded Plates

Following general design considerations should be observed as far as possible :

1. The plates should not have slopes steeper than 45°. A good practice is to keep slope around 30°.

2. The thickness of the plates should be kept at minimum, bearing in mind, the protection required for reinforcement.

3. Span of the plates should be kept between 3 to 4 m.

4. The dimensions of the plates should be preferably kept as indicated in Fig. 27·13.

5. The shape of the folded plate can be V-shaped or trough shaped cross section. The behavior of V-shaped cross-section is similar to a rectangular beam and depth is kept $\frac{1}{8}$ to $\frac{1}{10}$ of span. The behavior of trough shape is similar to an I-beam and depth can be somewhat smaller.

28·3. Analysis of Folded Plates

A number of analysis methods have been used to analyse folded plates. Commonly used methods are :

1. **Beam method.** This method of analysis is same as described for cylindrical shells in article 27·2 and consists of combination of beam and arch analysis.

2. Folded plate theory neglecting relative joint displacement.

3. Folded plate theory accounting for relative joint displacement.

4. **Elasticity method.** The following assumptions are made in the analysis :

(*a*) Material is homogeneous and elastic

(*b*) Deflections are small

(*c*) Superposition holds good

(*d*) Longitudinal joints are monolithic with slab

(*e*) Each supporting diaphragm is infinitively stiff parallel to its own plane but is perfectly flexible normal to its plane.

The folded plate theory considering relative joint displacement consists of the following steps :

1. **Elementary transverse slab analysis.** The surface loads are considered as carried transversely by the plates acting as continuous oneway slabs spanning between the unyieldings support at the folds or joints.

2. **Longitudinal plate analysis.** All loads carried transversely to the joints are considered as transferred longitudinally to the end supporting members by the plates acting inclined simple beams. Plate deflections computed from longitudinal analysis have shown that some relative displacement occurs between successive joints. This violates the assumption of unyielding supports made in transverse slab analysis.

3. **Correction for relative joint displacements.** This analysis consists of determining correction resulting from relative joint displacements, both to the transverse one-way slab analysis and to the longitudinal plate analysis.

4. **Superposition.** The analysis of above three steps are combined to give final forces, moments and stresses.

Example 28·1. *Analyse the single span foleded plate roof shown in Fig. 28·8. Span is 15 meter. The load carried by the roof is*

Dead load $=2400\ N/m^2$

Finish load $=700\ N/m^2$

Live load $=700\ N/m^2$.

Sol. Total load per square meter $=3800\ \text{N/m}^2$.

Consider the folded plate to act as a beam.

Load per meter length along the span of the folded plate as a beam

$$\omega = 2(1{\cdot}5)\sqrt{2}\ 3800 \text{ N/m}$$
$$= 16{,}122 \text{ N/m.}$$

As the slope of the folded plate construction is 45″, the neutral axis of the folded plate will be at mid height, *i.e.* 0·75 m from *X-X*.

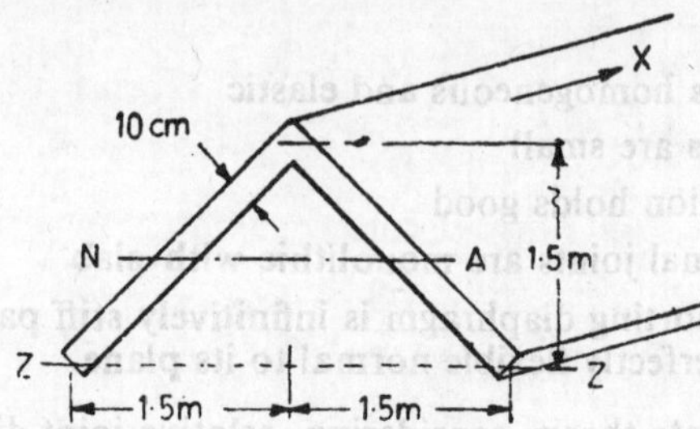

Fig. 28·8

Moment of inertia of the folded plate cross-section

$$I = 2\left(\frac{bd^3}{12}\right)\sin^2 \alpha$$

$$I = 2\left[\frac{10}{100}\ \frac{(1{\cdot}5\sqrt{2})^3}{12}\right]\sin^2 45$$

$$= 0{\cdot}07955 \text{ m}^4.$$

Maximum benting moment considering the folded plate as beam is

$$\frac{wl^2}{8} = \frac{16{,}122 \times 15 \times 15}{8} = 453{,}431 \text{ N-m.}$$

Maximum longitudinal stress

$$\sigma_y = \frac{My}{I} = \frac{453{,}43 \text{ Nm}}{0{\cdot}7955}\left(\frac{1{\cdot}5}{2}\right)$$

$$= 4{,}274{,}964 \text{ N/m}^2$$

or $$= 4{\cdot}275 \text{ N/mm}^2.$$

The find $A\bar{z}$ for various values of *Z*, consider a portion of folded plate above the section where *Z* is constant (Fig. 28·9).

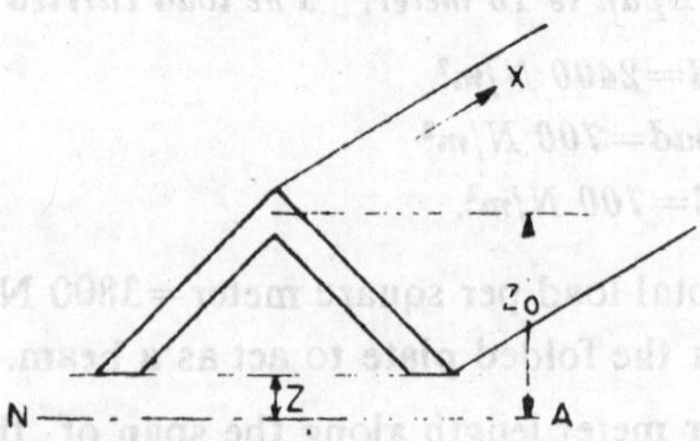

Fig. 28·9

The moment of area above the section about neutral axis

$$A\bar{z}=2\left[\frac{(Z_0-Z)}{\sin\alpha}\,t\right]\left[\frac{(Z_0-Z)}{2}+Z\right]$$

$$A\bar{z}=(Z_0^2-Z^2)\,\frac{t}{\sin\alpha}=(0{\cdot}75^2-Z^2)\,\frac{0{\cdot}1}{\sin\alpha}$$

$$A\bar{z}=0{\cdot}1414\,(0{\cdot}75^2-Z^2)=0{\cdot}1414\,(0{\cdot}5625-Z^2)$$

$N_{x\varphi}$ at support is given by Eqn. (27·14) as

$$N_{x\varphi}=V(x)\,\frac{A\bar{z}}{2I}$$

Maximum value at the support is given by

$$N_{x\varphi}=16{,}122\times\frac{15}{2}\times\frac{0{\cdot}1414\,(0{\cdot}5625-Z^2)}{2\times0{\cdot}07955}$$

$$N_{x\varphi}=107{,}463\,(0{\cdot}5625-Z^2)\text{ N/m}$$

$N_{x\varphi}$ varies parabolically with zero at the top and maximum value occuring at neutral axis ($Z=0$). Maximum value=60,448 N/m.

Specific shear $\quad q_s=\dfrac{(N_{x\varphi})_{max}}{L/2}$

$$q_s=\frac{107{,}463\,(0{\cdot}5625-Z^2)}{(15/2)}$$

$$q_s=14{,}328\,(0{\cdot}5625-Z^2)\text{ N/m.}$$

This also has parabolic variation. The value is zero at the top and maximum at neutral axis.

Next the folded plate is considered to act like an arch. Consider 1 m length of the plate in x direction as shown in

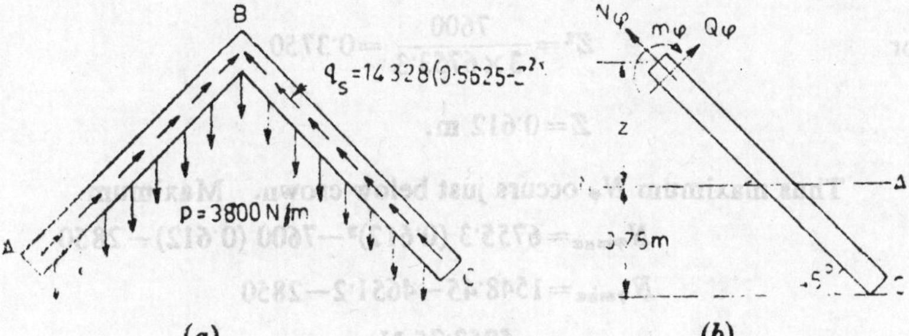

Fig. 28·10

Fig. 28·10 (*a*). Consider a section distance z from neutral axis (Fig. 28·10 (*b*). The forces acting at the section are N_φ, Q_φ and M_φ. The values of these forces can be obtained from statics. Thus

$$N_\varphi=\int_{-0.75}^{Z}-14328\,(0.5625-Z^2)\,\frac{dZ}{\sin 45}$$

$$+3800\sin 45\,\frac{(Z+0.75)}{\sin 45}$$

$$N_\varphi=6755.3\,Z^3-7600\,Z-2850$$

Q_φ is given as

$$Q_\varphi=p\cos 45\,\frac{(Z+0.75)}{\sin 45}=3800\,(Z+0.75)$$

$$Q_\varphi=3800\,Z+2850.$$

Maximum value occurs at crown ($Z=0.75$)

$$Q_{\varphi max}=5700\text{ N}$$

M_φ is given as

$$M_\varphi=-3800\cos 45\left(\frac{Z+0.75}{\sin 45}\right)^2\left(\frac{1}{2}\right)$$

$$M_\varphi=-671.6\,(Z+0.75)^2$$

Maximum value of M_φ occurs at crown ($Z=0.75$)

Maximum $M_{\varphi max}=-1511.1$ Nm.

For maximum value of N_φ, putting $\frac{\partial N\varphi}{\partial Z}=0$ gives

$$6755.3\times 3Z^2-7600=0$$

or $$Z^2=\frac{7600}{3\times 6753.3}=0.3750$$

$$Z=0.612\text{ m.}$$

Thus maximum N_φ occurs just below crown. Maximum

$$N_{\varphi max}=6755.3\,(0.612)^3-7600\,(0.612)-2850$$

$$N_{\varphi max}=1548.45-4651.2-2850$$

$$=-5952.75\text{ N.}$$

Thus all the forces are known.

28·4. Analysis of Multibay Folded Plates

For analyzing the internal bay of a multibay folded plate construction (Fig. 28·11) the method described in article 27·2 and

Fig. 28·11. Multibay folded plate.

illustrated by application to folded plates in Example 28·1 can be used. In this case there will be horizontal thrust and moments acting at the valleys as shown in Fig. 28·12. The horizontal thrust H_0 and moment M_0 will make the arch analysis statically indetermi-

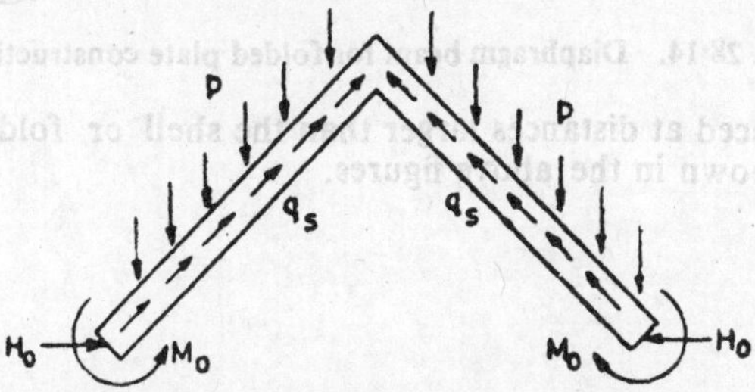

Fig. 28·12. Internal bay of multibay folded plate construction.

nate. Two additional equations are obtained by assumption that the horizontal displacement and rotation at the valley joints are zero.

28·5. Design of Diaphragm Beam

In a barrel shell or folded plate construction, the shell or folded plates transfer the load to diaphragm beams which are supported on columns (Fig. 28·13-14). The load is transfered to diaphragm by

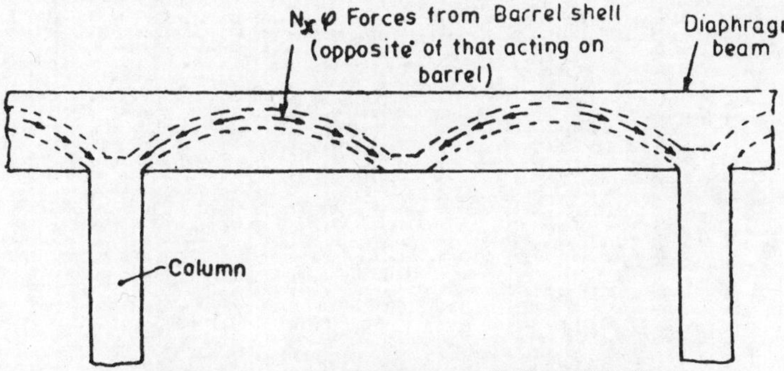

Fig. 28·13. Diaphragm beam for barrel shell.

internal forces $N_{x\varphi}$. The diaphragrms are designed as transverse beams supported on columns as shown in Fig. 28·13-14. Columns

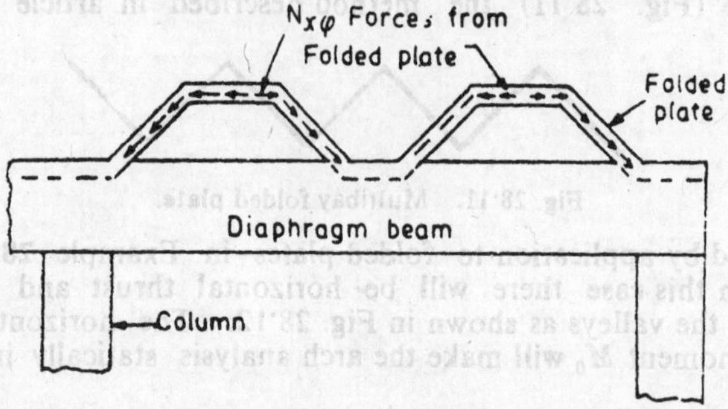

Fig. 28·14. Diaphragm beam for folded plate construction.

may be spaced at distances larger than the shell or folded plate bay width as shown in the above figures.

APPENDIX A

Round Bars

Designation	Diameter in mm	Weight per metre in kg	Perimeter in mm	Cross-sectional area									
				1	2	3	4	5	6	7	8	9	10
ISRO 5	5	0·15	1·57	0·20	0·39	0·59	0·79	0·98	1·18	1·37	1·57	1·77	1·96
ISRO 6	6	0·22	1·89	0·28	0·57	0·85	1·13	1·42	1·70	1·98	2·26	2·55	1·83
ISRO 8	8	0·40	2·51	0·50	1·01	1·51	2·01	2·51	3·02	3·52	4·02	4·53	5·03
ISRO 10	10	0·62	3·14	0·79	1·57	2·36	3·14	3·93	4·71	5·50	6·28	7·07	7·85
ISRO 12	12	0·89	3·77	1·13	2·26	3·39	4·52	5·65	6·70	7·92	9·05	10·18	11·31
ISRO 16	16	1·58	5·03	2·01	4·02	6·03	8·04	10·05	12·05	12·07	16·08	18·10	20·11
ISRO 20	20	2·47	6·28	3·14	6·28	9·41	12·56	15·71	18·85	21·99	25·14	28·28	31·42
ISRO 25	25	3·85	7·86	4·91	9·82	14·73	19·63	24·54	29·45	34·36	39·27	44·18	49·09
ISRO 28	28	4·80	8·80	6·16	12·32	18·47	24·95	30·79	36·95	43·10	49·26	55·42	61·58
ISRO 32	32	6·31	10·05	8·04	16·48	24·13	32·17	40·21	48·25	56·30	64·34	72·38	80·42
ISRO 36	36	8·00	11·31	10·18	20·36	30·54	40·72	50·90	61·08	71·26	81·44	91·62	101·80
ISRO 40	40	9·87	12·57	12·57	25·13	37·70	50·27	62·83	75·40	87·96	100·53	113·10	12566.

APPENDIX B

Area of steel in mm² in 1 metre width of slab

Diametre of bars in mm	Area of steel for following spacing in mm												
	30	40	50	60	70	80	90	100	110	120	130	1·40	150
5 mm	—	4·90	3·92	3·27	2·80	2·45	2·18	1·96	1·78	1·63	2·51	1·40	1·31
68mm	—	7·08	5·65	4·72	4·04	4·54	3·14	2·83	2·57	2·36	2·18	2·02	1·89
mm	—	12·58	10·06	8·38	7·19	6·29	5·59	5·03	4·57	4·18	3·87	3·59	3·35
10 mm	—	19·63	15·70	13·08	11·21	9·81	8·72	7·85	7·14	6·54	6·04	5·61	5·23
12 mm	—	28·28	22·62	18·85	16·16	14·14	12·57	11·31	10·28	9·43	8·70	8·08	7·54
16 mm	—	—	40·42	33·52	28·73	25·14	22·34	20·11	18·27	16·76	15·47	14·37	13·41
20 mm	—	—	62·84	52·36	44·86	39·28	34·91	31·42	28·56	26·18	24·17	22·44	20·95

Area of steel in mm² in 1 metre width of slab

Area of steel for following spacing in mm														
160	*170*	*180*	*200*	*220*	*240*	*260*	*280*	*300*	*320*	*340*	*360*	*380*	*400*
1·23	1·15	—	—	—	—	—	—	—	—	—	—	—	—
1·77	1·66	1·57	1·42	1·29	1·18	—	—	—	—	—	—	—	—
3·14	2·96	2·79	2·52	2·29	2·10	1·93	1·80	1·68	1·57	1·48	1·40	1·33	1·26
4·91	4·62	4·36	3·93	3·51	3·27	3·02	2·80	2·62	2·45	2·31	2·17	2·07	1·96
7·07	6·65	6·28	5·66	5·14	4·17	4·35	4·04	3·77	3·53	3·33	3·09	2·98	2·83
12·57	11·82	11·17	10·06	9·14	8·38	7·74	7·18	6·71	6·29	5·91	5·58	5·29	5·03
19·64	18·48	17·46	15·76	14·28	13·19	12·09	11·22	10·47	9·82	9·24	8·73	8·27	7·88

APPENDIX C

OBJECTIVE TYPE QUESTION

Chapter 1

1·1. The size of fine aggregates does not exceed
(*a*) 2·75 mm (*b*) 3·00 mm
(*c*) 3·75 mm (*d*) 4·75 mm
(*e*) 5·75 mm.

1·2. The strength and quality of concrete, depend upon
(*a*) grading of the aggregates
(*b*) surface area of the aggregates
(*c*) shape of the aggregates
(*d*) surface texture of the aggregates
(*e*) all the above.

1·3. The lower water cement ratio in concrete, introduces
(*a*) smaller creep (*b*) greater density
(*c*) greater strength (*d*) greater wear resistance
(*e*) all the above.

1·4. The entrained air in concrete
(*a*) increases the workability
(*b*) decreases the workability
(*c*) increases the strength of concrete.

1·5. After casting an ordinary cement concrete
(*a*) expands when dried (*b*) shrinks when wetted
(*c*) shrinks when dried.

1·6. For producing impermeable concrete
(*a*) thorough mixing of concrete is required
(*b*) properly graded and non-porous aggregates are required
(*c*) proper compacting of concrete is required
(*d*) proper curing of concrete is required
(*e*) all the above.

1·7. Separation of coarse aggregates from mortar during transportation is known as
(*a*) bleeding (*b*) segregation
(*c*) creeping.

1'8. Separation of water sand cement from a freshly mixed concrete is known as

(a) bleeding (b) segregation

(c) creeping.

1'9. Which statement is correct

(a) water cement paste hardens due to hydration

(b) during hardening cement binds the aggregates together

(c) cement provides strength, durability and water tightness to the concrete

(d) all the above.

1'10. Efflorescence in cement is caused due to an excess of

(a) alumina (b) iron oxide

(c) alkalis (d) magnesium oxide.

1'11. The high strength of rapid hardening cement at an early stage is due to

(a) finer grinding (b) higher content of tricalcium

(c) increased lime content.

1'12. In massive dams the type of cement used is

(a) ordinary portland cement

(b) rapid hardening cement (c) blast furnace slag cement

(d) low heat cement.

1'13. Vicat's apparatus is used to perform the test of

(a) fineness (b) consistency

(c) setting time (d) soundness.

1'14. Inert material of a cement concrete is

(a) water (b) cement

(c) aggregate.

1'15. Sand generally contains salt if it is obtained from

(a) nala beds (b) river beds

(c) sea beds.

1'16. The aggregate containing moisture in pores and having its surface dry is known as

(a) moist-aggregates

(b) saturated surface dry aggregate

(c) dry aggregate.

1'17. Los angeles machine is used to test the aggregate for

(a) crushing strength (b) abrasion resistance

(c) impact value.

1·18. The impurity of mixing water which affects the setting time and strength of concrete is

(a) sodium sulphates (b) sodium chlorides

(c) calcium chlorides

(e) sodium carbonates and bicarbonates.

1·19. Water cement ratio may be defined as

(a) volume of water to that of cement

(b) weight of water to that of cement

(c) weight of concrete to that of water.

1·20. Pick the correct statements

(a) a rich mix of concrete possesses higher strength

(b) the strength of concrete decreases as the water cement ratio increases

(c) good compaction by mechanical vibrations, increases the strength of concrete.

1·21. Pick the correct statements

(a) the concrete gains strength due to hydration of cement

(b) the concrete does not set at freezing point

(c) the strength of concrete increases with age.

1·22. Hardening of cement occurs at

(a) rapid rate during the first few days and afterwards it continues to increase at a reduced rate

(b) slow rate during the first few days and afterwards it continues to increase at a rapid rate

(c) uniform rate throughout the age.

1·23. I.S.I. has specified the full strength of concrete after

(a) 7 days (b) 21 days

(c) 28 days (d) 35 days.

1·24. High temperature

(a) decreases the strength of concrete

(b) increases the strength of concrete

(c) has no effect on the strength of concrete.

1·25. Placing of concrete should preferably be done at a temperature of

(a) 27±2°C (b) 20°C

(c) 0°C.

1·26. Pick up the correct statements

(a) water enables chemical reaction to take place with cement (*b*) water lubricates the mixture of gravel, sand and cement

(*c*) only a small quantity of water is required for hydration of cement

(*d*) strength of concrete largely depends upon its workability.

1·27. Seregation is responsible for

(*a*) honey-combed concrete

(*b*) porous layers in concrete

(*c*) surface scaling in concrete

(*d*) sand streaks in concrete.

1·28. The factors which affect workability are

(*a*) water content and its temperature

(*b*) shape and size of the aggregates

(*c*) grading and surface textures of the aggregates

(*d*) air entraining agents.

1·29. Pick up the correct statements

(*a*) an increase in water content must be accompaired by an increase in cement content

(*b*) angular and rough aggregates reduce the workability of the concrete

(*c*) the slump of the concrete mix decreases due to an increase in temperature

(*d*) large size aggregates increase the workability due to lesser surface area.

1·30. Workability of concrete mix having very low water cement ratio should be ascretained by

(*a*) tensile strength test (*b*) slump test

(*c*) compaction factor test.

1·31. Pick up the correct statements

(*a*) segregation is necessary for a workable concrete

(*b*) consistency does not affect the workability of concrete

(*c*) if the slump increases, workability decreases

(*d*) if the concrete mix is dry, the slump is maximum.

1·32. Proper proportioning of concrete, ensures

(*a*) desired strength and workability

(*b*) desired durability

(*c*) water tightness of the structure

(*d*) resistance to wear.

1·33. The grade of concrete M 15 means that the compressive strength of a 15 cm cube after 28 days is

(*a*) 10 N/mm² (*b*) 15 N/mm²

(*c*) 25 N/mm².

1·34. Shrinkage in concrete can be reduced by using

(*a*) low water cement ratio (*b*) low cement in the concrete

(*c*) proper concrete mix (*d*) presaturated aggregates.

1·35. Pick up the correct statements

(*a*) with passage of time, the strength of cement decreases

(*b*) with passage of time, the strength of cement increases

(*c*) the concrete made with storage deteriorated cement gains strength with work time.

1·36. The process of proper and accurate measurement of concrete ingredients for uniformity of proportions is known

(*a*) grading (*b*) curing

(*c*) mixing (*d*) batching.

1·37. Proper batching ensures

(*a*) economy (*b*) durability

(*c*) workability (*d*) strength.

1·38. The process of mixing, transporting, placing and compacting concrete using ordinary Portland cement should not take more than

(*a*) 60 minutes (*b*) 90 minutes

(*c*) 120 minutes.

1·39. To prevent segregation, the maximum height for placing concrete is

(*a*) 50 cm (*b*) 100 cm

(*c*) 150 cm.

1·40. The process of hardening the concrete mixes by keeping its surface moist for a certain period after compaction is known

(*a*) wetting (*b*) placing

(*c*) curing.

1·41. Curing of pavement, floors, roofs and slabs is done by

(*a*) membrane method (*b*) ponding method

(*c*) covering surface with bags

(*d*) sprinkling water method.

1·42. Longer period of curing of concrete, ensures better

(*a*) volume stability (*b*) strength

(c) wear resistance

(d) water tightness and durability.

1·43. Under normal conditions using an ordinary cement, the period of removal of form work is

(a) 7 days for beam soffits

(b) 14 days for bottom of slabs of spans 4·6 m and above

(c) 21 days for bottom of beam, over 6 m spans

(d) 2 days for vertical sides of columns.

1·44. Joints in concrete structures are provided

(a) to reduce the tensile stresses likely to be developed due to evaporation of water

(b) to minimise the change in the dimensions of the slab

(c) to minimise the necessary cracking.

1·45. The main objects of compaction of concrete are

(a) to eliminate air holes

(b) to achieve maximum density

(c) to provide intimate contact between the concrete and embedded materials.

1·46. Too wet concrete may cause

(a) weakness (b) excessive laitance

(c) segregation (d) lower density.

1·47. Construction joints are provided where

(a) where B.M. and S.F. are small

(b) where it is supported by other member

(c) at 18 metres apart in huge structures

(d) in concrete wall at sill level of windows.

1·48. The surface where two successive placements of concrete meet is known as

(a) construction joint (b) contraction joint

(c) expansion joint.

1·49. Horizontal construction joints in concrete walls are generally provided at

(a) soffit level (b) window sill level

(c) floor level.

1·50. Expansion joints are provided if the length of concrete structures exceeds

(a) 15 m (b) 45 m

(c) 60 m.

1·51. According to the recommendations of I.S.I., expansion joints

(a) are provided where plan changes abraptly

(b) are provided to ensure minimum resistance

(c) are supported on separate columns

(d) do not carry reinforcement across them.

1·52. Construction joints are generally provided in concrete structures of

(a) roads (b) retaining walls

(c) lining of canals (d) lining of tunnels.

1·53. Pick up the correct statements

(a) construction joints are necessarily planned for their locations

(b) expansion joints are provided to accomodate thermal expansion

(c) construction joints are provided to control shrinkage cracks

(d) expansion joints need not be provided in foundation concrete.

1·54. For the construction of cement concrete dams, the maximum permissible size of the aggregates is

(a) 30 mm (b) 40 mm

(c) 50 mm (d) 60 mm.

1·55. For the construction of cement concrete floor, the maximum size of the aggregate is

(a) 8 mm (b) 12 mm

(c) 20 mm.

1·56. The most useless aggregate is one whose surface texture is

(a) smooth (b) granular

(c) glassy (d) honey combed and porous.

1·57. The natural light weight aggregates are obtained from

(a) sedimentary rocks (b) metamorphic rocks

(c) igneous rocks (d) volcanic scoure.

1·58. Pick the correct statements

(a) sands obtained from pits should usually be washed to remove clay and silt

(b) sands obtained from flooded pits, need not be washed before use

(c) sea shore sands contain chlorides which cause effloresance

(d) the chloride in sea shore sand and shingle may cause corrosion of reinforcement if the concrete is porous.

1·59. Pick the correct statements

(*a*) sand stones may be divided into calcareous, siliceous and ferrugineous sand stones

(*b*) concrete using sand stones, cracks due to excessive shrinkage

(*c*) very hard and close grained crystallined lime stones are suitable aggregates but provide low strength

(*d*) broken bricks produee a concrete having good fire resisting qualities.

1·60. Pick up correct statements

(*a*) in properly graded aggregates, bulk density is more

(*b*) in single size aggregates, bulk density is least

(*c*) in single size aggregates, bulk density is maximum.

1·61. The bulk density depends on

(*a*) shape (*b*) grading

(*c*) compaction.

1·62. Pick up the correct statement

(*a*) the free water is the amount of water added when mixing plus the amount of water held on the surface of the aggregates prior to mixing

(*b*) the total water is the free water plus the amount actually absorbed by the aggregates.

1·63. Addition of pozzolana to ordinary portland cement, causes

(*a*) decrease in early strength

(*b*) incsease in strength

(*c*) increase in heat of hydration

(*d*) decrease in curing time

(*e*) increase in shrinkage

(*f*) reduction in bleeding.

1·64. Setting time of cement increases by adding

(*a*) gypsum (*b*) calcium chloride

(*c*) sodium oxide (*d*) calcium sulphate.

1·65. Setting time of cement decrease by adding

(*a*) gypsum (*b*) calcium chloride

(*c*) sodium oxide (*d*) calcium sulphate.

1·66. Bulking of sand is

(*a*) mixing of different sizes of sand particles

(*b*) mixing water with sand

(*c*) swelling of sand when wetted.

1·67. Slump test of concrete is a measure of its

(a) consistancy (b) compressive strength

(c) tensile strength (d) impact value.

1·68 For ensuring quality of concrete, we should use

(a) single sized aggregate (b) two sized aggregate

(c) graded aggregates (d) coarse aggregates.

1·69. The commonly used material in cement manufacture is

(a) sand stone (b) shale

(c) lime stone (d) graphite.

1·70. Portland cement is manufactured from

(a) lime stone and clay (b) gypsum and lime

(c) pozzolana (d) lime, pozzolana and clay.

1·71. For construction of structures in sea water. the cement generally preferred is

(a) portland pozzolana cement

(b) quick setting cement

(c) low heat portland cement

(d) rapid hardening cement.

1·72. The size of fine aggregates does not exceed

(a) 3·75 mm (b) 4·75 mm

(c) 5·75 mm.

1·73. M_{10} grade of concrete approximates

(a) 1 : 3 : 6 mix (b) 1 : 2 : 4 mix

(c) 1 : 1 : 2 mix.

1·74. M_{15} grade of concrete approximates

(a) 1 : 3 : 6 mix (b) 1 : 2 : 4 mix

(c) 1 : 1 : 2 mix.

1·75. Strength of concrete with passage of time

(a) increases (b) decreases

(c) remains constant.

1·76. Concrete gains strength due to

(a) chemical reaction of cement with sand and coarse aggregates

(b) evaporation of water from concrete

(c) hydration of cement.

Chapter 2

2·1. The advantages of reinforced concrete are

(a) more rigidity due to monolithic character
(b) moulding in any desired shape
(c) economy because of less maintenance cost
(d) fire resisting and durability.

2·2. Pick up the correct statements from the following

(a) compression is borne entirely by concrete
(b) plane sections transverse to the centre line of the beam before bending remain plane sections after bending
(c) elastic modulii for concrete and steel have different values within the limits of deformation of the beam
(d) steel possesses initial stresses when embedded in concrete.

2·3. The effective depth is measured from compression edge to

(a) tensile edge (b) tensile reinforcement centroid
(c) neutral axis of the beam
(d) longitudinal central axis.

2·4. If the modular ratio is *m*, stress ratio is '*r*' and effective depth of the beam is '*d*', the depth of the critical neutral axis of the beam is

(a) $\left(\frac{m}{m-r}\right)d$ (b) $\left(\frac{m}{m+r}\right)d$
(c) $\left(\frac{m+r}{m}\right)d$ (d) $\left(\frac{r-m}{m}\right)d$.

2·5. If '*d*' and '*n*' are the effective depth and depth of neutral axis respectively, the lever arm of the beam is

(a) d (b) n
(c) $\left(d+\frac{n}{3}\right)$ (d) $d-\frac{n}{2}$
(e) $d-\frac{n}{3}$.

2·6. A singly reinforced beam has breadth '*b*', effective depth '*d*', depth of neutral axis '*n*' and critical neutral axis 'n_c'. If '*c*' and '*t*' are permissible compressive stress in concrete and permissible tensile stress in steel, then the moment of resistance of the beam if A_t is the steel area is

(a) $bn\frac{c}{2}\left(d-\frac{n}{3}\right)$ (b) $A_t\times t\left(d-\frac{n}{3}\right)$
(c) $\frac{n_c}{2}\left(1-\frac{n_c}{3}\right)cbd^2$.

2·7. The section of a reinforced beam whose most distant concrete fibre in the compression zone and tensile stress in steel reach the permissible stress simultaneously is called

(a) balanced section (b) critical section

(c) economic section.

2·8. If the permissible compressive stress in concrete is 5 N/mm² and permissible stress in steel is 140 N/mm² and $m=18$, the percentage area A_t of the steel required for balanced section is

(a) 0·44 (b) 0·72

(c) 1·00.

2·9. In a singly reinforced beam, if the stress in concrete reaches its allowable limit earlier than that in the steel, the beam section is called

(a) critical section (b) under-reinforced section

(c) over-reinforced section.

2·10. In case 'c' and 't' are stresses in concrete and steel respectively and 'm' is the modular ratio, 'd' is the effective depth and 'n' is the neutral axis depth, then

(a) $\dfrac{mc}{t}=\dfrac{d-n}{n}$ (b) $\dfrac{t}{mc}=\dfrac{d-n}{n}$

(c) $\dfrac{t}{mc}=\dfrac{d+n}{n}$ (d) $\dfrac{mc}{t}=\dfrac{n}{d-n}$.

2·11. Pick up the correct statements

As the percentage of steel increases

(a) depth of neutral axis increases

(b) depth of neutral axis decreases

(c) lever arm increases

(d) lever arm decreases.

2·12. Minimum spacing between horizontal parallel reinforcements of different sizes, should not be less than

(a) one diameter of thicker bar

(b) one diameter of thinner bar

(c) twice the diameter of thinner bar.

2·13. The minimum thickness of the cover at the end of a reinforcing bar should not be less than twice the diameter of the bar subject to a minimum of

(a) 20 mm (b) 25 mm

(c) 30 mm.

2·14. For M_{15} grade concrete and steel of allowable stress 140 N/mm^2, the moment of resistance factor is equal to

(a) 0·85 (b) 0·87

(c) 0·75.

2·15. If the permissible compressive stress for concrete in bending is σ_{cbc}, the modular ratio is

(a) $\dfrac{2800}{\sigma_{cbc}}$ (b) $\dfrac{2800}{2\sigma_{cbc}}$

(c) $\dfrac{2800}{3\sigma_{cbc}}$

2·16. For a continuous beam simply supported at ends

(a) max. sagging B.M. for the end spans is $= +\dfrac{\omega l^2}{10}$

(b) max. hogging B.M. over penultimate supports $= -\dfrac{\omega l^2}{10}$

(c) max. sagging B.M. for the interior spans $= +\dfrac{\omega l^2}{12}$

(d) max. hogging B.M. over interior supports other than penultimate supports $= -\dfrac{\omega l^2}{12}$.

2·17. For normal cases, stiffness of a simply supported beam is satisfied if the ratio of its span to overall depth does not exceed

(a) 10 (b) 15

(c) 20 (d) 25

(e) 30.

2·18. The ratio of the breadth and effective depth of a beam is generally kept as

(a) 0·25 (b) 0·5

(c) between 0·25 to 0·5.

2·19. If permissible compressive stress in concrete is 5 N/mm^2 and tensile stress in steel is 140 N/mm^2 and modular ratio is 18, the depth of the beam is given by

(a) $d=\sqrt{\dfrac{1·15\,M}{b}}$ (b) $d=\sqrt{\dfrac{0·83\,M}{b}}$

(c) $d=\sqrt{\dfrac{0·68\,M}{b}}$

2·20. In the case the weight of triangular brickwork formed by drawing 60° at the ends of span 'L' is W the maximum B.M. at centre of lintel will be

(a) $\frac{WL}{8}$ (b) $\frac{WL}{12}$

(c) $\frac{WL}{6}$ (d) $\frac{WL}{4}$.

2·21. A beam shall be deemed to be a deep beam when the ratio of effective span to overall depth is

(a) 2·5 (b) 2·0

(c) 1·5 (d) 3·0.

2·22. A continuous beam shall be deemed to be a deep beam when the ratio of effective span to overall depth is

(a) 2·5 (b) 2·0

(c) 1·5 (d) 3·0.

2·23. In a beam of width 'b' and effective depth 'd' if 'A_t' is the area of steel and 'm' is the modular ratio then

(a) $n=\frac{mA_t+\sqrt{(mA_t)^2+2bmA_td}}{2}$

(b) $n=-mA_t+\sqrt{(mA_t)^2-2bmA_td}$

(c) $n=-mA_t+\sqrt{(mA_t)^2-2bmA_td}$.

2·24. A simply supported or continuous beam with width 'b' and effective depth 'd' shall be so proportioned that the clear distance between the lateral restraints does not exceed

(a) $50\,b$ or $\frac{200\,b^2}{d}$ (b) $60\,b$ or $\frac{240\,b^2}{d}$

(c) $70\,b$ or $\frac{250\,b^2}{d}$.

2·25. For a cantilever, the clear distance from the free end of the cantilever to the lateral restraint if 'b' is the breadth and 'd' is the effective depth shall not exceed

(a) $20\,b$ or $\frac{80\,b^2}{d}$ (b) $30\,b$ or $\frac{120\,b^2}{d}$

(c) $25\,b$ or $\frac{100\,b^2}{d}$.

2·26. For curtailment of tension reinforcement in flexural members the reinforcement shall extend beyond the point at which it is no longer required to resist flexure excepting at simple support or end of cantilever by

(a) effective depth of the member or 8 times the bar diameter (*b*) effective depth of the member or 10 times the bar diameter

(*c*) effective depth of the member or 12 times the bar diameter.

2·27. The minimum area of tension reinforcement 'A_t' if 'b' is the breadth and 'd' is the depth and 'f_y' is the characteristic strength of reinforcement in N/mm², shall not be less than

(*a*) $A_t = \frac{1 \cdot 0\, bd}{f_y}$ (*b*) $A_t = \frac{0 \cdot 85\, bd}{f_y}$

(*c*) $A_t = \frac{0 \cdot 75\, bd}{f_y}$.

2·28. Maximum area 'A_t' of tension reinforcement 'A_t' if 'b' is the breadth and 'D' is the overall depth shall not exceed

(*a*) 0·05 *bD* (*b*) 0·03 *bD*

(*c*) 0·04 *bD*.

2·29. Where the depth of the web in a beam exceeds 750 mm, side face reinforcement shall be provided along the two faces not less than

(*a*) 0·1 percent if web area

(*b*) 0·15 percent of web area

(*c*) 0·2 percent of web area.

2·30. Side face reinforcement shall be provided in the beam when depth of the web exceeds

(*a*) 500 mm (*b*) 750 mm

(*c*) 1000 mm (*d*) 1200 mm.

Chapter 3

3·1. The pitch of the main bars in a simply supported slab should not exceed its effective depth by

(*a*) three times (*b*) four times

(*c*) five times (*d*) six times.

3·2. The transverse reinforcement provided at right angles to the main reinforcement

(*a*) distributes the load (*b*) resists the temperature stresses

(*c*) resists the shrinkage stresses.

3·3. If the diameter of the main reinforcement provided in a slab is 16 mm, the concrete cover to main reinforcement should be

(*a*) 1 mm (*b*) 14 mm

(*c*) 16 mm.

3·4. For a single concentrated load if 'k' is constant having the value depending upon the ratio of width of the slab to the effective span, 'x' is the distance of the centroid of the concentrated load from the nearer support 'l_{ef}' is the effective span and 'a' is width of the contact area of the concentrated load measured parallel to the supported edge, then the effective width 'b_{ef}' is defined as

(a) $\frac{k}{x}\left(1+\frac{x}{l_{ef}}\right)+a$ (b) $kx\left(1-\frac{x}{l_{ef}}\right)+a$

(c) $kx\left(1+\frac{x}{l_{ef}}\right)+a$.

3·5. For cantilever slab, if 'a', is the distance of the concentrated load from the face of the centilever support and 'a' is width of contact area of the concentrated load measured parellel to the supporting edge, effective width 'b_{ef}' is defined as

(a) $b_{ef}=1·2\ a_1-a$ (b) $2a_1-a$

(c) $1·2\ a_1+a$.

3·6. For simply supported slabs, the vertical deflection limits may generally be assumed to be satisfied provided that the span to depth ratios are not greater than

(a) 15 (b) 20

(c) 25.

3·7. The final deflection due to all loads including the effects of temperature, creep and shrinkage and measured from the as cast level should not normally exceed

(a) $\frac{\text{span}}{250}$ (b) $\frac{\text{span}}{200}$

(c) $\frac{\text{span}}{300}$.

3·8. The deflection including the effects of temperature, creep and shrinkage occurring after erection of partitions and the application of finishes should not normally exceed

(a) $\frac{\text{span}}{250}$ (b) $\frac{\text{span}}{300}$

(c) $\frac{\text{span}}{350}$.

3·9. In a slab, the minimum reinforcement provided is

(a) 0·10 percent of gross area

(b) 0·15 percent of gross area

(c) 0·20 percent of gross area

(d) 0·25 percent of gross area.

3·10. **In a simply supported slab the maximum spacing of distribution steel should be smaller than 5 times the effective depth and**

(*a*) 400 mm (*b*) 450 mm

(*c*) 500 mm (*d*) 550 mm.

3·11. **The amount of reinforcement for main bars in a slab is based upon**

(*a*) a minimum bending moment

(*b*) maximum shear force (*c*) maximum bending moment

(*d*) minimum shear force.

Chapter 4

4·1. **It '*F*' is the shear force, $A\bar{y}$, is the moment of area above the section, about N.A., '*b*' is the breadth of the section, '*I*' is the moment of inertia, shear stress '*q*' at the section of rectangular homogeneous beam is given by**

(*a*) $\frac{F\,A\bar{y}}{I\,b}$ (*b*) $\frac{F\,Ib}{A\bar{y}}$

(*c*) $\frac{FA\bar{y}\,b}{I}$.

4·2. **The maximum shear stress in a beam of rectangular homogeneous beam is**

(*a*) 1·25 times the average (*b*) 1·75 times the average

(*c*) 2·00 times the average (*d*) 1·50 times the average.

4·3. **Distribution of shear intensity of a rectangular homogeneous section of a beam follows**

(*a*) a circular curve (*b*) a straight line

(*c*) a parabolic curve.

4·4. **Spacing of stirrups in a rectangular R.C.C. beam is**

(*a*) kept constant throughout the length

(*b*) decreased towards the centre of the beam

(*c*) increased at the ends

(*d*) increased at the centre of the beam.

4·5. **If '*F*' is the shear force, '*b*' is the width of R.C.C. beam and '*a*' is the lever arm then the maximum shear stress '*q*' is given by**

(*a*) $\frac{F}{ab}$ (*b*) $\frac{ab}{F}$

(*c*) $\frac{Fa}{b}$.

4·6. If 'ϕ' is nominal diameter of the bar 'σ_s' is stress in the bar at the section considered, 'τ_{bd}' is the design bond stress, then the development length L_d is given by

(*a*) $\frac{\phi\sigma_s}{\tau_{bd}}$ (*b*) $\frac{\phi\sigma_s}{2\,\tau_{bd}}$

(*c*) $\frac{\phi\sigma_s}{4\,\tau_{bd}}$ (*d*) $\frac{\phi\sigma_s}{3\,\tau_{bd}}$.

4·7. The length of embedment 'l' for M_{15} concrete when stress in reinforcement is 140 N/mm² is nearly equal to

(*a*) $l=48\,d$ (*b*) $l=58\,d$

(*c*) $l=68\,d$.

4·8. For M_{15} concrete, average bond stress for plain bars in tension is

(*a*) 0·5 N/mm² (*b*) 0·6 N/mm²

(*c*) 1·0 N/mm².

4·9. The standard 'U' type hook for mild steel has diameter of bend as

(*a*) 3 ϕ (*b*) 5 ϕ

(*c*) 4 ϕ.

4·10. If 'ϕ' is the diameter of bar, the anchorage value of the standard hook is equal to

(*a*) 12 ϕ (*b*) 16 ϕ

(*c*) 20 ϕ.

4·11. Lapped splices in tensile reinforcement are generally not used for bars of size larger than

(*a*) 18 mm (*b*) 24 mm

(*c*) 30 mm (*d*) 36 mm.

4·12. An R.C.C. beam not provided with shear reinforcement may develop cracks in its bottom, inclined roughly to the horizontal at

(*a*) 125° (*b*) 35°

(*c*) 45° (*d*) 55°.

4·13. The shear reinforcement in R.C.C. is provided to resist

(*a*) vertical shear (*b*) diagonal tension

(*c*) horizontal shear.

4·14. Minimum shear reinforcement A_{sv} should be equal to or greater than, in case 'b' is breadth of beam, S_v is the stirrup spacing and f_y is the characteristic strength of the stirrup reinforcement in N/mm^2 which shall not be taken greater than 415 N/mm^2 is

(a) $\frac{0.2\, b\, s_v}{f_y}$ (b) $\frac{0.3\, b\, s_v}{f_y}$

(c) $\frac{0.4\, b\, s_v}{f_y}$.

4·15. The maximum spacing of vertical stirrups should not exceed 450 mm or

(a) 0·50 effective depth (b) 0·75 effective depth

(c) 1·00 effective depth.

4·16. The maximum spacing of inclined stirrups inclined at 45° should not exceed 450 mm or

(a) 0·50 effective depth (b) 0·75 effective depth

(c) 1·00 effective depth.

4·17. For R.C.C. beams maximum shear $\tau_{c\,max}$ calculated as average shear stress $\frac{F}{bd}$ should not exceed for M_{15}

(a) 1·0 N/mm^2 (b) 1·6 N/mm^2

(c) 1·8 N/mm^2 (d) 2·0 N/mm^2.

4·18. The allowable tensile stress in mild steel bars provided in R.C.C. is

(a) 140 N/mm^2 (b) 190 N/mm^2

(c) 230 N/mm^2

Chapter 5

5·1. The percentage of steel in compression in doubly reinfrced beams should not exceed

(a) 3 percent (b) 4 percent

(c) 4·5 percent (d) 5·0 percent.

5·2. Steel beam theory is used for

(a) design of simple steel beams

(b) steel beams encased in concrete

(c) doubly reinforced beams ignoring compressive stress in concrete.

5·3. In analysis of doubly reinforced beams by revised I.S. method, the stress in compression steel is to be lower of σ_{sc} and

(a) m × compressive stress in the surrounding concrete

(b) 1·25 m × compressive stress in the surrounding concrete

(c) 1·50 m × compressive stress in the surrounding concrete.

5·4. In doubly reinforced beams by revised I.S. method, the total M.R. is sum of M_1=M.R. as balanced simple beam and M_2=M.R. provided by compression steel. If 'f' is the distance of centroid of compression steel from top edge of compression and A_c is the area of compression steel, M_2 will be equal to

(a) $(m-1)\,A_c\left(\frac{n-f}{n}\right)(d-f)$

(b) $(1{\cdot}5\,m-1)\,A_c\left(\frac{n+f}{n}\right)(d+f)$

(c) $(1{\cdot}5\,m-1)\,A_c\left(\frac{n+f}{n}\right)(d-f)$

(d) $(1{\cdot}5\,m-1)\,A_c\left(\frac{n-f}{n}\right)(d-f)$.

5·5. Lever arm 'a' of doubly reinforced beam by revised method with 'A_c' as steel on compression side and 'f' distance of centroid of compression steel from top edge of compression is given by

(a) $d-\dfrac{\frac{bn^2}{6}+(m-1)\,A_c\left(\frac{n-f}{n}\right)\times f}{\frac{bn}{2}+(m-1)\,A_c\left(\frac{n-f}{n}\right)}$

(b) $d-\dfrac{\frac{bn^2}{6}+(m-1)\,A_c\left(\frac{n+f}{n}\right)\times f}{\frac{bn}{2}+(m-1)\,A_c\left(\frac{n+f}{n}\right)}$

(c) $d-\dfrac{\frac{bn^2}{6}+(1{\cdot}5\,m-1)\,A_c\left(\frac{n-f}{n}\right)\times f}{\frac{bn}{2}+(1{\cdot}5\,m-1)\,A_c\left(\frac{n-f}{n}\right)}$.

5·6. The compression reinforcement shall be effectively anchored in two directions at points not further apart centre to centre than

(a) 24 times the diameter of anchored bar (compression bar)
(b) 16 times the diameter of anchored bar
(c) 24 times the diameter of anchoring bar
(d) 48 times the diameter of anchoring bar.

Chapter 6

6·1. The effective width of the flange of T-beam, with 'b' as the rib width, 'D_f' as the depth of flange and 'l_0' as the distance between point of zero moments in the beam is given by

(a) $b_f = l_0 + b_r + D_f$ (b) $b_f = \frac{l_0}{4} + b_r + 4D_r$

(c) $b_f = \frac{l_0}{12} + b_r + 3D_f$ (d) $b_f = \frac{l_0}{6} + b_r + 6D_f$.

6·2. For continuous beams and frames 'l_0' distance between points of zero moments may be taken as

(a) $l_0 = 0·5$ times the effective span

(b) $l_0 = 0·6$ times the effective span

(c) $l_0 = 0·7$ times the effective span.

6·3. The effective width of T-beams is not to exceed the width of rib 'b_r' plus

(a) the sum of the clear distances to the adjacent beams on either side

(b) half the sum of the clear distances to the adjacent beams on either side

(c) the greater of the clear distances to the adjacent beams on either side.

6·4. The effective width of the flange of L-beam with 'b_r', as the width of rib, 'D_f' as the depth of flange and 'l_0' as the distance between point of zero moments in the beam is given by

(a) $b_f = l_0 + b_r + D_f$ (b) $b_f = \frac{l_0}{4} + b_r + 4D_f$

(c) $b_f = \frac{l_0}{12} + b_r + 3D_f$ (d) $b_f = \frac{l_0}{6} + b_r + 6D_f$.

6·5. For isolated beams, the effective width of the flange of T-beam with 'b' as the actual width of flange, 'b_r' as the width of rib and 'l_0' as the distance between points of zero moments in the beam is given by

(a) $b_f = \frac{l_0}{\left(\frac{l_0}{b}\right) + 6} + b_r$ (b) $b_f = \frac{0·5\, l_0}{\left(\frac{l_0}{b}\right) + 4} + b_r$

(c) $b_f = \frac{l_0}{\left(\frac{l_0}{b}\right) + 4} + b_r$.

6·6. For isolated beams, the effective width of the flange of L-beam with 'b' as the actual width of flange, 'b_r' as the width of rib and 'l_0' as the distance between points of zero moments in the beam is given by

(a) $b_f = \dfrac{l_0}{\left(\dfrac{l_0}{b}\right)+6} + b_r$ (b) $b_f = \dfrac{0.5\, l_0}{\left(\dfrac{l_0}{b}\right)+4} + b_r$

(c) $b_f = \dfrac{l_0}{\left(\dfrac{l_0}{b}\right)+4} + b_r$.

6·7. For simply supported T-beams and L-beams 'l_0' distance between points of zero moments in the beam may be taken as

(a) 1·0 times the effective span

(b) 0·9 times the effective span

(c) 0·7 times the effective span.

6·8. A part of the slab may be considered as the flange of the T-beam or L-beam if

(a) the flange has adequate reinforcement transverse to beam

(b) it is built integrally with the beam

(c) it is effectively bonded together with the beam.

6·9. A T-beam behaves as a rectangular beam of a width equal to its effective flange if its neutral axis

(a) remains within the flange

(b) remains below the slab

(c) coincides with the centroid of the beam.

6·10. If the neutral axis of a T-beam is below the slab, the relationship between the flange width 'B', depth of neutral axis 'n', thickness of slab d_s, effective depth of the beam 'd' gross area of tensile steel 'A_t' and the modular ratio 'm' may be stated as

(a) $Bd_s\left(n-\dfrac{d_s}{2}\right)=m\,A_t\,(d-n)$

(b) $Bd_s\left(n+\dfrac{d_s}{2}\right)=m\,A_t\,(d-n)$

(c) $Bd_s\left(n-\dfrac{d_s}{2}\right)=m\,A_t\,(d+n)$

(d) $Bd_s(n-d_s)=m\,A_t\,(d-n)$.

6·11. The lever arm of the T-beam is given by

(a) $d-\dfrac{d_s}{3}\left(\dfrac{3n+2d_s}{2n+d_s}\right)$ (b) $d-\dfrac{d_s}{3}\left(\dfrac{3n-2d_s}{2n+d_s}\right)$

(c) $d-\dfrac{d_s}{3}\left(\dfrac{3n-2d_s}{2n-d_s}\right)$.

6·12. For *T*-beam, if '*d*' is economical effective depth, *M* is the B.M. '*r*' is the ratio of cost of steel to concrete and '*b*' is the width of the rib, then

(*a*) $d=\sqrt{\dfrac{M_r}{126\ b_r}}$ (*b*) $d=\sqrt{\dfrac{M_r}{100\ b_r}}$

(*c*) $d=\sqrt{\dfrac{M_r}{90\ b_r}}$.

6·13. By revised theory in doubly reinforced *T*-beams if 'c_a' is the max. compressive stress, '*n*' is the depth of neutral axis, and '*f*' is the distance of compression steel from compressive edge, the stress in compressive stress is

(*a*) $mc_a\left(\dfrac{n-f}{n}\right)$ (*b*) $2\ mc_a\left(\dfrac{n-f}{n}\right)$

(*c*) $1{\cdot}5\ mc_a\left(\dfrac{n-f}{n}\right)$.

6·14. In case of *L*-beams if '*V*' is the shear and '*T*' is the torsion, the equivalent shear V_e, for breadth of beam '*b*' is given by

(*a*) $V_e=V+\dfrac{T}{b}$ (*b*) $V_e=V+\dfrac{1{\cdot}6\ T}{b}$

(*c*) $V_e=V+\dfrac{2\ T}{b}$.

6·15. In *L*-beams equivalent bending moment M_{e1}, when '*M*' is B.M., '*T*' is torsion, '*b*' is breadth of beam, '*D*' is overall depth of *L*-beam, is given by

(*a*) $M_{e1}=M+T\left(1+\dfrac{D}{b}\right)$

(*b*) $M_{e1}=M+\dfrac{T\left(1+\dfrac{D}{b}\right)}{2}$

(*c*) $M_{e1}=M+\dfrac{T\left(1+\dfrac{D}{b}\right)}{1{\cdot}7}$.

Chapter 7

7·1. Compression member can be called pedastal, if its effective length does not exceed its least lateral dimension by

(*a*) two times (*b*) four times

(*c*) three times (*d*) six times.

7·2. The reinforcement in a column should not be less than the following, of cross-sectional area

(*a*) 0·5 percent (*b*) 0·6 percent

(*c*) 0·8 percent (*d*) 1·0 percent.

7·3. In case of pedastals, nominal reinforcement should be provided not less than

(*a*) 0·10 percent of the cross-sectional area

(*b*) 0·15 percent of the cross-sectional area

(*c*) 0·20 percent of the cross-sectional area.

7·4. The reinforcement in a column should not be more than the following of cross-sectional area

(*a*) 4 percent (*b*) 5 percent

(*c*) 8 percent (*d*) 6 percent.

7·5. Minimum number of longitudinal bars in a rectangular column should not be less than

(*a*) four (*b*) six

(*c*) eight

7·6. Minimum number of longitudinal bars in a circular column should not be less than

(*a*) four (*b*) six

(*c*) eight.

7·7. The diameter of longitudinal bars of a column should not be less than

(*a*) 8 mm (*b*) 10 mm

(*c*) 12 mm (*d*) 15 mm.

7·8. Spacing of longitudinal bars measured along the periphery of the column should not exceed

(*a*) 200 mm (*b*) 300 mm

(*c*) 350 mm (*d*) 400 mm.

7·9. In a helically reinforced column having helical reinforcement, longitudinal bars within the helical reinforcement shall be atleast

(*a*) six (*b*) eight

(*c*) twelve.

7·10. The diameter of transverse reinforcement of columns should be not less than one-fourth of the largest diameter of the main steel rods and in no case less than

(*a*) 4 mm (*b*) 5 mm

(*c*) 6 mm (*d*) 8 mm.

7·11. Pick up the correct statements. The spacing of transverse reinforcement of column is decided by

(a) the least lateral dimension of the column

(b) sixteen times the diameter of the smallest longitudinal reinforcing rod in the column

(c) forty eight times the diameter of transverse reinforcement.

7·12. Where an increased load on the column having helical reinforcement is allowed for, the pitch of helical turns shall be

(a) not more than 75 mm

(b) not more than one-sixth of the core diameter of the column

(c) not less than 25 mm

(d) not less than three times the diameter of the steel bar forming the helix.

7·13. For heavy structural loads, the type of column used is

(a) column with longitudinal steel and lateral ties

(b) column with longitudinal steel and spirals

(c) composite column in which structural section is encased in concrete.

7·14. A short column is that in which effective length does not exceed

(a) 8 times the least lateral dimension

(b) 12 times the least lateral dimension

(c) 16 times the least lateral dimension.

7·15. For more exact calculations, short column is that in which effective length does not exceed

(a) 40 times least radius of gyration

(b) 45 times least radius of gyration

(c) 50 times least radius of gyration

(d) 60 times least radius of gyration.

7·16. In case of long column, reduction factor 'C_r' if 'l_{ef}' is the effective length and 'b' is the least lateral dimension of column is given by

(a) $1{\cdot}00-\dfrac{l_{ef}}{40\,b}$ (b) $1{\cdot}25-\dfrac{l_{ef}}{48\,b}$

(c) $1{\cdot}50-\dfrac{l_{ef}}{48\,b}$.

7·17. In case of long columns, reduction factor, 'C_r', if 'l_{ef}' is the effective length and 'i_{min}' is the least radius of gyration is given by

(a) $1{\cdot}00-\frac{l_{ef}}{100\ i_{min}}$ (b) $1{\cdot}25-\frac{l_{ef}}{100\ i_{min}}$

(c) $1{\cdot}25-\frac{l_{ef}}{160\ i_{min}}$.

7·18. According to load factor method, the permissible load 'P' on a short column, in which 'A_c' is the cross-sectional area of concrete alone, 'σ_{sc}' is the permissible compressive stress for column bars, 'A_{sc}', is the cross-sectional area of the longitudinal steel, 'σ_{cc}' is the permissible stress in concrete in direct compression, and 'm' is the modular ratio is given by

(a) $A_c\ \sigma_{cc}+mA_s\ \sigma_{cc}$ (b) $A_c\ \sigma_{cc}+1{\cdot}5mA_s\ \sigma_{cc}$

(c) $A_c\ \sigma_{cc}+A_{sc}\ \sigma_{sc}$.

7·19. The cross-sectional area of metal core in composite column shall not exceed

(a) 10 per cent of gross area of the column
(b) 20 per cent of gross area of the column
(c) 30 per cent of gross area of the column.

7·20. The permissible load for short columns with helical reinforcement satisfying condition that the ratio of volume of helical reinforcement to the volume of the core shall not be less than $0{\cdot}36\left[\frac{A_g}{A_c}-1\right]\frac{f_{ck}}{f_y}$ where 'A_g' is the gross area of section, 'A_c' is the core area, 'f_{ck}' is the characteristic compressive strength of the concrete and f_y is the characteristic strength of the helical reinforcement but not exceeding 415 N/mm^2 shall be

(a) 1·05 times the permissible load for similar member with lateral ties or rings

(b) 1·10 times the permissible load for similar member with lateral ties or rings

(c) 1·20 times the permissible load for similar member with lateral ties or rings

7·21. The unsupported length for slender columns shall not exceed

(a) 50 times the least lateral dimension of a column
(b) 60 times the least lateral dimension of a column
(c) 80 times the least lateral dimension of a column.

7·22. All columns shall be designed for minimum eccentricity equal to the unsupported length of column by 500 plus lateral dimension by 30 subject to a minimum of

(a) 10 mm (b) 15 mm
(c) 20 mm (d) 25 mm.

7·23. The longitudinal reinforcing bar in a column of dimension greater than 200 mm shall have concrete cover not less than 40 mm nor less than

(a) 2 diameter of such bar

(b) 1·5 diameter of such bar

(c) 1·0 diameter of such bar.

7·24. The longitudinal reinforcing bar in a column of minimum dimension 200 mm or less using 12 mm diameter or less than 12 mm bars cover may be

(a) 20 mm (b) 25 mm

(c) 30 mm.

7·25. The lap length in compression shall be equal to the development length in compression, but not less than if ϕ is the diameter of the bar

(a) 18 bar diameters (b) 24 bar diameters

(c) 30 bar diameters (d) 36 bar diameters.

Chapter 8

8·1. The maximum ratio of span to depth of a slab simply supported and spanning in two directions is

(a) 20 (b) 25

(c) 30 (d) 35

(e) 40.

8·2. If 'L' and 'B' are the lengths of long and short spans of a two way slab simply supported on all its four edges and carrying a load of 'ω' per unit area, the ratio if the loads 'ω_L' and 'ω_B' carried by slab in 'L' direction and carried by slab in 'B' direction respectively is given by

(a) $\frac{\omega_L}{\omega_B}=\left(\frac{B}{L}\right)^2$ (b) $\frac{\omega_L}{\omega_B}=\left(\frac{B}{L}\right)^3$

(c) $\frac{\omega_L}{\omega_B}=\left(\frac{B}{L}\right)^4$.

8·3. If the ratio of long and short spans of a two way slab with corners held down is 'k', the reduction factor of B.M., 'C' is given by

(a) $C=1-\frac{5}{6}\left(\frac{k^2}{1+k^2}\right)$ (b) $C=1-\frac{5}{6}\left(\frac{k^3}{1+k^3}\right)$

(c) $C=1-\frac{5}{6}\left(\frac{k^2}{1+k^4}\right)$.

8·4. According to I.S. 456 slabs which span in two directions with corners held down, are assumed to be divided in each direction into middle and edge strips such that the width of the middle strip is

(a) half of the width of slab

(b) three-fourth of the width of slab

(c) two-thirds of the width of slab.

8·5. If the sides of a slab simply supported on its edges and spanning in two directions are equal, the maximum bending moment as one way is multiplied by

(a) 0·2 (b) 0·4

(c) 0·5 (d) 0·6.

8·6. Design of a two-way slab simply supported on all its four edges and having no provision to prevent the corners from lifting, is made by

(a) Rankine formula (b) Rankine Grashoff formula

(c) Grashoff formula (d) Rankine-Marcus formula.

8·7. If the ratio of $\frac{L}{B}$ exceeds 2, the slab is considered to span across shorter span only and the B.M. in the slab is

(a) $\frac{WB^2}{8}$ (b) $\frac{WB^2}{4}$

(c) $\frac{WB^2}{16}$.

8·8. For slabs fixed on all the four edges the bending moments may be calculated by Dr. Marcus's method by multiplying by factor '*C*' the B.M. found by Grashoff-Rankine formula. The value of '*C*' is

(a) $1-\frac{5}{18}\left(\frac{k}{1+k^2}\right)$ (b) $1-\frac{5}{18}\left(\frac{k^2}{1+k^2}\right)$

(c) $1-\frac{5}{18}\left(\frac{k^2}{1+k^4}\right)$.

8·9. Tension reinfrcement provided at mid-span in the strip as by I.S I. method, shall extend in the lower part of the slab to within

(a) 0·15 l of a continuous edge

(b) 0·20 l of a continuous edge

(c) 0·25 l of a continuous edge.

8·10. Tension reinforcement provided at mid-span in the strip by I.S I. method shall extend in the lower part of the slab to within

(a) 0·15 l of a discontinuous edge

(*b*) 0·20 l of a discontinuous edge

(*c*) 0·25 l of a discontinuous edge.

8·11. Torsion reinforcement shall be provided at any corner where the slab is simply supported on both edges meeting at that corner. It shall consist of top and bottom reinforcement, each with layers of bars parallel to the sides of the slabs and extending from the edges of a minimum distance of one-fifth of the shorter span. The area of reinforcement in each of these four layers shall be

(*a*) half of the area required for the maximum mid-span moment in the slab

(*b*) three quaters of the area required for the maximum mid-span moment in the slab

(*c*) three-eighths of the area required for the maximum mid-span moment in the slab.

8·12. Torsion reinforcement shall be provided at any corner where the slab is continuous on one edge and simply supported on other edge. It shall consist of top and bottom reinforcement, each with layers of bars placed paralled to the sides of the slab and extending from the edge of a minimum distance of one-fifth of the shorter span. The area of reinforcement in each of these four layers shall be

(*a*) half the area required for the maximum mid-span moment in the slab

(*b*) three quarters of the area required for the maximum mid-span moment in the slab

(*c*) three-eighths of the area required for the maximum mid-span moment in the slab.

8·13. A flat slab is supported

(*a*) on beams (*b*) on columns

(*c*) on beams and columns

(*d*) on columns monolithically built with slab.

8·14. Enlarged head of a supporting column of a flat slab is known as

(*a*) top of the column (*b*) column head

(*c*) drop panel.

8·15. Thickened part of a flat slab over its supporting column is known as

(*a*) top of the column (*b*) column head

(*c*) drop panel.

8·16. The maximum diameter of the column head supporting a flat slab is

(a) 0·20 times the span length

(b) 0·25 times the span length

(c) 0·30 times the span length.

8·17. For slabs with drops, span to effective depth ratios shall be 20 for simply supported edge and 26 for continuous edge. The minimum thickness of slab shall be

(a) 100 mm **(b) 125 mm**

(c) 150 mm.

8·18. For slabs without drops, span to effective depth ratios shall be (20×0·9) for simply supported edge and (26×0·9) for continuous edge. The miximum thickness of slab shall be

(a) 100 mm **(b) 125 mm**

(c) 150 mm.

8·19. The effective width of a column strip of a flat slab, if 'l_1' is the span in the direction moments are being determined centre to centre of supports and 'l_2' is the span transverse to 'l_1', measured centre to centre of supports, is equal to

(a) 0·3 l_2 **(b) 0·25 l_2**

(c) 0·20 l_2.

8·20. The absolute of the positive and average negative bending moments in each direction, if 'l_2' is the span in transverse direction, 'l_n' is the clear span extending from face to face of columns capitals, brackets or walls, but not less than '0·65 l_1' where 'l_1' is the the span in the direction moments are being determined, and 'W' is the load coming on area $l_2 \times l_n$, is given by

(a) $\frac{Wl_n}{12}$ **(b)** $\frac{Wl_n}{10}$

(c) $\frac{Wl_n}{8}$.

8·21. In an interior span, the design moment M_0 shall be distributed in the following proportions

(a) negative design moment$=0·7\ M_0$, positive design moment $=0·30\ M_0$

(b) negative design moment$=0·65 M_0$, positive design moment $=0·35\ M_0$

(c) negative design moment$=0·60\ M_0$, positive design moment $=0·40\ M_0$

8·22. In the end span, if α_e is the ratio of flexural stiffness of the exterior columns to the flexural stiffness of the slab at a joint taken in the direction moments are being determined and is given by $\alpha_e = \frac{\Sigma k_c}{\Sigma k_s}$ where Σk_c is the sum of the flexural stiffnesses of the columns meeting at the joint and Σk_s is the stiffness of the slab, expressed as moment per unit rotation, the total moment 'M_0' is distributed as

(a) Interior negative design moment

$$= M_0 \left[0{\cdot}75 - \frac{0{\cdot}10}{\left(1 + \frac{1}{\alpha_e}\right)} \right]$$

Positive design moment

$$= M_0 \left[0{\cdot}63 - \frac{0{\cdot}28}{\left(1 + \frac{1}{\alpha_e}\right)} \right]$$

Exterior negative design moment

$$= M_0 \left[\frac{0{\cdot}65}{\left(1 + \frac{1}{\alpha_e}\right)} \right]$$

(b) Interior negative design moment

$$= M_0 \left[0{\cdot}65 - \frac{0{\cdot}10}{\left(1 + \frac{1}{\alpha_e}\right)} \right]$$

Positive design moment

$$= M_0 \left[0{\cdot}63 - \frac{0{\cdot}10}{\left(1 + \frac{1}{\alpha_e}\right)} \right]$$

Exterior negative design moment

$$= M_0 \left[\frac{0{\cdot}75}{\left(1 + \frac{1}{\alpha_e}\right)} \right]$$

(c) Interior negative design moment

$$= M_0 \left[0{\cdot}75 - \frac{0{\cdot}10}{\left(1 + \frac{1}{\alpha_e}\right)} \right]$$

Positive design moment

$$= M_0\left[0{\cdot}63 - \frac{0{\cdot}10}{\left(1+\frac{1}{\alpha_c}\right)}\right]$$

Exterior negative design moment

$$= M_0\left[\frac{0{\cdot}65}{\left(1+\frac{1}{\alpha_c}\right)}\right]$$

8·23. The spacing of bars in a flat slab, except where slab is of cellular or ribbed construction, shall not exceed

(*a*) 2 times the thickness of slab

(*b*) 2·5 times the thickness of slab

(*c*) 3·0 times the thickness of slab.

8·24. When drop panels are used the thickness of drop panel for determination of area of reinforcement shall be lesser of thickness of drop and

(*a*) one-half of the distance between edge of drop and edge of capital

(*b*) three-eighths of the distance between edge of drop and edge of capital

(*c*) one-querter the distance between edge of drop and edge of capital.

8·25. Critical section for shear in drop is at a distance from capital of column

(*a*) 1·0 × effective depth of drop

(*b*) 0·75 × effective depth of drop

(*c*) 0·50 × effective depth of drop.

8·26. Critical section for flat slab is at a distance as given below from edge of drop for flat slabs with drops and edge of capital of column for flat slabs without drops

(*a*) 1·0 × effective depth of flat slab

(*b*) 0·75 × effective depth of flat slab

(*c*) 0 50 × effective depth of flat slab.

8·27. In case '*V*' is the shear force at critical section 'b_0' is the perimeter of critical section '*d*' is the effective depth at critical section, nominal shear stress is equal to

(*a*) $\frac{V}{4\,b_0 d}$ (*b*) $\frac{V}{b_0 d}$

(*c*) $\frac{V}{2\,b_0 d}$.

8·28. A circular slab subjected to loading deflects in the form of
(a) semi-hemisphere (b) ellipsoid
(c) parabolloid.

8·29. If 'q' is the uniformly distributed load on a circular slab of radius 'a', simply supported at its edges, the maximum radial moment at the centre is

(a) $\frac{q\,a^2}{16}$ (b) $\frac{3q\,a^2}{16}$

(c) $\frac{2q\,a^2}{16}$.

8·30. If 'q' is the uniformally distributed load on a circular slab of radius 'a', simply supported on its edges circumferential moment at the edges is

(a) $\frac{q\,a^2}{16}$ (b) $\frac{3q\,a^2}{16}$

(c) $\frac{2q\,a^2}{16}$.

8·31. If 'q' is the uniformly distributed load on a circular slab of radius 'a', simply supported on its edges, circumferential moment at the centre is

(a) $\frac{q\,a^2}{16}$ (b) $\frac{3q\,a^2}{16}$

(c) $\frac{2q\,a^2}{16}$.

8·32. If 'q' is the uniformly distributed load on a circular slab of radius 'a' fixed at its edges, the maximum positive radial moment at its centre is

(a) $\frac{q\,a^2}{16}$ (b) $\frac{2q\,a^2}{16}$

(c) $\frac{3q\,a^2}{15}$.

8·33. If 'q' is the uniformly distributed load on a circular slab of radius 'a', fixed at its edges, the maximum negative radial moment at its edges is

(a) $\frac{-q\,a^2}{16}$ (b) $\frac{-2q\,a^2}{16}$

(c) $\frac{-3q\,a^2}{16}$.

8·34. If 'q' is the uniformly distributed load on a circular slab of radius 'a', fixed at its edges, the maximum positive circumferential moment at the centre is

(a) $\frac{q a^2}{16}$ (b) $\frac{2q a^2}{16}$

(c) $\frac{3q a^2}{16}$.

8·35. If 'q' is the uniformly distributed load on a circular slab of radius 'a', fixed at its edges the positive circumferential moment at the edges is

(a) $\frac{q a^2}{16}$ (b) $\frac{2q a^2}{16}$

(c) $\frac{3q a^2}{16}$ (d) 0·0.

8·36. If 'q' is the uniformly distributed load on a circular slab of radius 'a', simply supported at its edges, the positive radial moment at the edges is

(a) $\frac{a^2}{16}$ (b) $\frac{2q a^2}{16}$

(c) $\frac{3q a^2}{16}$ (d) 0·0.

8·37. For a circular slab carrying a uniformly distributed load and fixed at edges, the ratio of maximum negative radial moment and maximum positive radial moment is

(a) 1 (b) 2

(c) 3 (d) 1·5.

8·38. For hexagonal slabs, the design of slab is based on circular slab of equivalent diameter 'D_e' as given below

(a) D_e=1·04 D (b) D_e=1·08 D

(c) D_e=1·10 D.

8·39. For octagonal slabs, the design of slab is based on circular slab of equivalent diameter 'D_e' as given below

(a) D_e=1·04 D (b) D_e=1·08 D

(c) D_e=1·10 D.

8·40. In triangular panels, the design is based on

(a) Rectangle constructed on base passing through the third vertex (*b*) as circular slab of equivalent diameter given by ascribed circle

(*c*) as circular slab of equivalent diameter given by inscribed circle.

8·41. The thickness of the topping of a ribbed slab, varies between

(*a*) 60 mm to 100 mm (*b*) 80 mm to 120 mm

(*c*) 100 mm to 150 mm.

8·42. The breadth of a ribbed slab containing two bars must be

(*a*) 65 mm to 80 mm (*b*) 80 mm to 100 mm

(*c*) 100 mm to 120 mm.

8·43. Ribs should not be spaced centre to centre greater than

(*a*) 1·0 metre (*b*) 1·5 meters

(*c*) 2·0 metres.

8·44. The depth of rib excluding topping should be not more than

(*a*) 2 times their width (*b*) 3 times their width

(*c*) 4 times their width (*d*) 5 times their width.

Chapter 9

9·1. A member subjected to axial load and bending shall be considered safe if $\frac{\sigma_{cc,\, cal}}{\sigma_{cc}}+\frac{\sigma_{cbc,\, cal}}{\sigma_{cbc}} \leq 1$, where $\sigma_{cc,\, cal}$ is calculated direct compressive stress in concrete σ_{cc} is permissible axial compressive stress in concrete, $\sigma_{cbc,\, cal}$ is caluculated bending compressive stress in concrete and σ_{cbc} is permissible bending compressive stress in concrete, and the resultant tension in concrete for uniaxial bending does not exceed three-fourth, the 7-day modulus of rupture of concrete and is

(*a*) not greater than 25 percent of the resultant compression

(*b*) not greater than 30 percent of the resultant compression

(*c*) not greater than 35 percent of the resultant compression.

9·2. In case of columns with ties, if $\sigma_{cc,\, cal}$ is the calculated direct compressive stress in concrete, 'A_c' is the cross-sectional area of concrete excluding any finishing material and reinforcing steel, 'A_{sc}' is the cross-sectional area of the longitudinal steel, 'm' is the modular ratio, then

(*a*) $\sigma_{cc,\, cal}=\frac{P}{A_c+mA_{sc}}$ (*b*) $\sigma_{cc,\, cal}=\frac{P}{A_c+1{\cdot}5\, m\, A_{sc}}$

(*c*) $\sigma_{cc,\, cal}=\frac{P}{A_c+2m\, A_{sc}}$.

9·3. In case of columns with bending 'M_1' and 'Z_1' about axis 1-1 and 'M_2' and 'Z_2' about axis 2-2 and $\sigma_{cbc-cal}$ is the calculated bending stress, then

(a) $\sigma_{cbc,\,cal}=\frac{M_1}{Z_1}$ (b) $\sigma_{cbc,\,cal}=\frac{M_2}{Z_2}$

(c) $\sigma_{cbc,\,cal}$ is greater of $\frac{M_1}{Z_1}$ and $\frac{M_2}{Z_2}$

(d) $\sigma_{cbc,\,cal}=\frac{M_1}{Z_1}+\frac{M_2}{Z_2}$.

9·4. In case column subjected to bending also, has tensile stress greater than three-fourths, the 7-day modulus of rupture of concrete then

(a) the tensile resistance in concrete is ignored

(b) the stresses in concrete and steel shall be calculated by the theory of cracked section and if the stresses are within permissible limits, the section is safe

(c) the direct load should be equal to the algebraic sum of forces on concrete and steel

(d) the moment of the external loads about any two reference lines should be equal to the algebraic sum of the moment of the forces in concrete and steel, about the same lines.

Chapter 10

10·1. A foundation rests on
(a) foundation soil (b) subgrade.

10·2. On an asbsolutely rigid foundation base, the pressure will
(a) be uniform
(b) be more at the edges of the foundation
(c) be zero at the centre of the foundation.

10·3. The thickness of reinforced concrete footing on soil, at its edges, is kept not less than
(a) 10 cm (b) 15 cm
(c) 20 cm (d) 30 cm
(e) 40 cm.

10·4. The thickness of reinforced concrete footing on piles, at its edges is kept not less than
(a) 10 cm (b) 15 cm
(c) 20 cm (d) 30 cm
(e) 40 cm.

10·5. In the case of plain concrete pedestals, the angle 'α' between the plane passing through the bottom edge of the pedestal and the corresponding junction edge of the column with pedestal and the horizontal plane if 'q_0' is the calculated maximum bearing pressure at the base of the pedestal in N/mm² and 'f_{ck}' is characteristic strength of concrete at 28 days in N/mm² will be

(a) $\tan \alpha = 0.9\sqrt{\frac{100\, q_0}{f_{ck}} + 1.0}$

(b) $\tan \alpha \nless \sqrt{\frac{100\, q_0}{f_{ck}} + 1.0}$

(c) $\tan \alpha \nless 0.9\sqrt{\frac{100\, q_0}{f_{ck}} + 1.0}$.

10·6. The weight of a foundation is assumed as
(a) 5 percent of load (b) 10 percent of load
(c) 15 percent of load (d) 20 percent of load.

10·7 The self-weight of the footing, is
(a) not considered for calculating the upward pressure on footing
(b) considered for calculating the upward pressure on footing
(c) not considered for calculating the area of the footing
(d) considered for calculating the area of the footing.

10·8. If 'p' is the net upward pressure in N/m² on footing of a wall whose width is 'b' metres and if 'B' metres is the width of the footing max. B.M. M per metre is given by

(a) $M = \frac{p}{8}(4B - b)(B - b)$ Nm/m

(b) $M = \frac{p}{16}(4B - b)(B - b)$ Nm/m

(c) $M = \frac{p}{32}(4B - b)(B - b)$ Nm/m.

10·9. If 'p' is the net upward pressure in a square footing of side 'B' for a square column of side 'a', the maximum bending is given by

(a) $\frac{pB\,(B-a)^2}{8}$ (b) $\frac{pB\,(B-a)^2}{4}$

(c) $\frac{pB\,(B+a)}{8}$ (d) $\frac{pB\,(B-a)}{4}$.

10·10. If $\sigma_{s,p}$ is allowable punching shear stress, 'B' is the side of a square footing for a square column, of side 'a', carrying a weight 'W', the effective depth 'd_p' of the footing from punching shear condition is

(a) $\dfrac{W(B^2-a^2)}{8B^2 a\sigma_{s,p}}$ (b) $\dfrac{W(B^2-a^2)}{4B^2 a\sigma_{s,p}}$

(c) $\dfrac{W(B^2-a^2)}{8Ba\sigma_{s,p}}$ (b) $\dfrac{W(B-a)}{4B^2 a\sigma_{s,p}}$.

10·11. If the width of the foundation for two equal columns is restricted, the shape of the footing generally adopted is

(a) trapezoidal (b) rectangular

(c) square.

10·12. If the length of a combined rectangular footing for two columns 'l' metres apart is 'L' metres and the projection on the left side of the exterior column is 'x' metres, then the projection 'y' metres on the right side of the exterior column, in order to have a uniformly distributed load, if '$\bar{x}$' is the distance of C.G. of loads from left column, then

(a) $y=\dfrac{L}{2}-(l-\bar{x})$ (b) $y=\dfrac{L}{2}+(l+\bar{x})$

(c) $y=\dfrac{L}{2}+(l-\bar{x})$ (d) $y=L-(l-\bar{x})$.

10·13. In a combined footing for two columns carrying unequal loads, the maximum hogging bending moment (tension on top) will occur at

(a) the more loaded column

(b) mid-way between columns

(c) the point of maximum shear force

(d) the point of zero shear force.

10·14. In a combined footing, the nominal stirrups provided are

(a) 6 legged (b) 8 legged

(c) 12 legged.

10·15. The zone in which reinforcement for transverse bending moment is provided is width of the column plus

(a) effective depth of the footing

(b) twice effective depth of the footing

(c) four times effective depth of the footing.

10·16. The minimum pile spacing in bearing piles in hard stratum is 2 to $2\frac{1}{2}$ times diameter of pile but not less than

(a) 0·6 m (b) 0·8 m

(c) 1·0 m.

10·17. The minimum pile spacing in bearing piles in hard bed rock is 2 times diameter of pile but not less than

(a) 0·6 m (b) 0·8 m

(c) 1·0 m.

10·18. The minimum pile spacing in friction piles is 3 to 5 times diameter of pile but not less than

(a) 0·6 m (b) 0·8 m

(c) 1·0 m.

10·19. In two way rectangular footing the reinforcement in the long direction shall be distributed uniformly across the full width of the footing. For the reinforcement in the short direction, a central band equal to the width of the footing shall be marked along the length of the footing and the portion of reinforcement, if 'β' is the ratio of the long side to the short side of the footing, is equal to

(a) $\frac{1}{(\beta+1)}$ (b) $\frac{1\cdot5!}{(\beta+1)}$

(c) $\frac{2}{(\beta+1)}$.

10·20. If 'R_2' is the radius of circular footing for a column of 'R_1', radius and 'p' is the upward pressure from soil, B.M. at face of column is

(a) $\frac{p\pi}{20}(R_2-R_1)(3R_2^2+2R_1R_2+2R_1^2)$

(b) $\frac{p\pi}{20}(R_2-R_1)(3R_2^2+2R_1R_2-2R_1^2)$

(c) $\frac{p\pi}{20}(R_2-R_1)(3R_2^2-3R_1R_2-2R_1^2)$.

10·21. Effective depth 'd' for circular footing required for B.M., 'M' for column of radius 'R_1' is

(a) $d=\sqrt{\frac{M}{QR_1}}$ (b) $d=\sqrt{\frac{M}{Q\pi R_1}}$

(c) $d=\sqrt{\frac{M}{Q\pi\, R_1/2}}$.

10·22. Effective depth 'd_p' required for circular footing if σ_{sp} is the allowable punching shear stress, 'p' is the upward pressure from soil, 'R_2' is the radius of circular footing and 'R_1' is the radius of the circular column is given by

(a) $d_p=\dfrac{p(R_2^2-R_1^2)}{2R_1\,\sigma_{sp}}$ (b) $d_p=\dfrac{p(R_2^2+R_1^2)}{2R_1\,\sigma_{sp}}$

(c) $d_p=\dfrac{p(R_2^2-R_1^2)}{R_1\,\sigma_{sp}}$.

10·23. The critical section for shear in footings is located as a vertical section located from the face of the column, pedestal or wall at a distance

(a) half the effective depth of footing
(b) three-fourths the effective depth of footing
(c) the effective depth of footing.

10·24. The critical section for shear in footings in piles is located as vertical section located from the face of the column at a distance

(a) half the effective depth of footing
(b) three-fourths the effective depth of footing
(c) the effective depth of footing.

10·25. A mat foundation is provided where bearing capacity of soil is low so that quite considerable area of individual footings required is more than

(a) 40 percent of the area of the building
(b) 50 percent of the area of the building
(c) 60 percent of the area of the building.

10·26. To avoid cracks due to shrinkage in mat foundation small areas as given below are constructed at a time

(a) 25 m × 25 m (b) 15 m × 15 m
(c) 10 m × 10 m.

10·27. The adjacent strips in mat fundations are filled after the already constructed one after

(a) 24 hours (b) 36 hours
(c) 48 hours.

10·28. The construction joints are located at mat foundation at

(a) places of low shear stress
(b) at the column points
(c) at places of least bending moment.

10·29. A pre-cast concrete pile generally used is
(a) square with chamfered corners
(b) circular
(c) octagonal with chamfered corners.

10·30. The maximum permissible stress in timber piles is taken as
(a) 6 N/mm² (b) 5 N/mm²
(c) 4 N/mm².

10·31. The timber piles cannot be subjected to heavy blows, therefore carrying carrying capacity of these piles is restricted to
(a) 250 kN (b) 300 kN
(c) 350 kN.

10·32. The advantage of a concrete pile over a timber pile is
(a) no decay due to termites
(b) no restriction on length
(c) higher bearing capacity
(d) not necessary to cut below the water mark.

10·33. The *H* beams, pipe sections and box type steel piles have certain advantages over other types of piles. Pick up these points
(a) greater capacity per pile
(b) hard driving without shuttering
(c) facility of splicing on the site
(d) can withstand rough handling
(e) good resistance to lateral forces and buckling
(f) can stand sea water
(g) not susceptible to corrosion.

10·34. The piles are supported at one point and the other resting on the ground during handling upto length of
(a) 12 m (b) 10 m
(c) 8 m.

10·35. Piles supported at two or more points while handling when the lengths are greater than
(a) 4 m (b) 6 m
(c) 8 m.

10·36. For piles supported on single point, the distance from butt end should be
(a) 0·25 *l* (b) 0·207 *l*
(c) 0·293 *l*.

10·37. For piles supported at two points, during lifting, the supports should be from the ends for maximum B.M. to be minimum possible

(a) 0·25 l (b) 0·207 l

(c) 0·293 l.

10·38. The longitudinal reinforcement in concrete piles varies from

(a) 1 to 2% (b) 2 to 3%

(c) 3 to 4%.

10·39. The amount of lateral reinforcement in the form of links or loops not less than 5 mm in diameter, should be not lsss than

(a) 0·1 per cent of the gross volume of the pile

(b) 0·2 per cent of the gross volume of the pile

(c) 0·3 per cent of the gross volume of the pile.

10·40. The amount of lateral reinforcement at both ends of piles equal to 3 times the least width of pile should be not less than

(a) 0·5 per cent of the gross volume of the pile

(b) 0·6 per cent of the gross volume of the pile

(c) 0·8 per cent of the gross volume of the pile.

10·41. A pile supported at two points having wt. 'W' and length 'l' supported at 0·207 l from the ends has max. B.M.

(a) 0·043 Wl (b) 0·030 Wl

(c) 0·0214 Wl.

10·42. A pile supported at one point having wt. 'W' and length 'l' supported at 0·293 l from butt end, has max. B.M.

(a) 0·043 Wl (b) 0·030 Wl

(c) 0·0214 Wl.

10·43. The pile caps should be extended beyond exterior piles by

(a) 100 mm to 150 mm (b) 200 mm to 250 mm

(c) 300 mm to 350 mm.

10·44. The piles are embedded in the cap by atleast

(a) 50 mm (b) 100 mm

(c) 150 mm (d) 200 mm.

10·45. The pile cap reinforcement is kept above the head of piles about

(a) 50 mm (b) 100 mm

(c) 150 mm (d) 200 mm.

10·46. The column is centred over the centroid of the pile group. In case of cap on three piles, the column rests on beam B_1 resting on one pile and centre of beam B_2 connecting the other two beams. The maximum $B.M_s$ in two beams are

(a) $\frac{2Wl_1}{9}$ in B_1 (b) $\frac{Wl_1}{9}$ in B_1

(c) $\frac{Wl_2}{3}$ in B_2 (d) $\frac{Wl_2}{6}$ in B_2.

where 'l_1' is span of B_1 and l_2 is span of B_2

Chapter 11

11·1. The horizontal portion of a step in a staircase is known as
(a) riser (b) tread
(c) flight.

11·2. The vertical distance between two steps is known as
(a) riser (b) tread
(c) flight.

11·3. In public buildings rise is about
(a) 10 cm (b) 12 cm
(c) 15 cm.

11·4. The tread in public buildings is about
(a) 30 cm (b) 25 cm
(c) 20 cm.

11·5. The sum of tread plus twice the rise is kept about
(a) 55 cm (b) 60 cm
(c) 70 cm.

11·6. The product of tread and rise is kept between
(a) 300 to 350 (b) 400 to 450
(c) 500 to 550.

11·7. A very comfortable type of stairs is
(a) dog-legged (b) straight
(c) open newel.

Chapter 12

12·1. By Rankine's theory if 'p_1', is the vertical pressure at a depth 'h' on a block of earth weighing 'ω' per unit volume and the angle of repose is 'ϕ', the lateral intensity of pressure p_2 is given by

(a) $p_2 = \frac{\omega h(1-\cos\phi)}{(1+\cos\phi)}$ (b) $p_2 = \frac{\omega h(1-\tan\phi)}{(1+\tan\phi)}$

(c) $p_2 = \frac{\omega h(1-\sin\phi)}{(1+\sin\phi)}$.

12·2. By Rankine's theory total horizontal force 'P_H' on the vertical face of a retaining wall of height 'h', and earth of density 'ω' per unit volume, for a unit run is

(a) $P_H = \frac{\omega h^2(1-\cos\phi)}{2(1+\cos\phi)}$ (b) $P_H = \frac{\omega h^2(1-\sin\phi)}{2(1+\sin\phi)}$

(c) $P_H = \frac{\omega h^2(1-\tan\phi)}{2(1+\tan\phi)}$.

12·3. By Rankine's theory total lateral force 'P' on the vertical face of a retaining wall of height 'h' exerted by the retained earth weighing 'ω' per unit volume having an angle of surcharge $\alpha°_1$ parallel to the profile of soil is

(a) $P = \frac{\omega h^2}{2} \cos\alpha \times \frac{\cos\alpha - \sqrt{\cos^2\alpha - \cos^2\phi}}{\cos\alpha + \sqrt{\cos^2\alpha - \cos^2\phi}}$

(b) $P = \frac{\omega h^2}{2} \cos\alpha \times \frac{\cos\alpha + \sqrt{\cos^2\alpha - \cos^2\phi}}{\cos\alpha - \sqrt{\cos^2\alpha + \cos^2\phi}}$

(c) $P = \frac{\omega h^2}{2} \cos\alpha \times \frac{\cos\alpha - \sqrt{\cos^2\alpha - \cos^2\phi}}{\cos\alpha - \sqrt{\cos^2\alpha + \cos^2\phi}}$.

12·4. By Rankine's theory total lateral pressure on the vertical face of a retaining wall of height 'h' acts parallel to free surface and acts at a distance from base at

(a) $\frac{h}{4}$ (b) $\frac{h}{2}$

(c) $\frac{h}{3}$.

12·5. Graphical representation of Coulomb's theory is known as

(a) Otto Mohr (b) Culmann's method

(c) Rebhann's construction.

12·6. To avoid failture of retaining wall, the factor of safety against sliding should be

(a) 1·25 (b) 1·50

(c) 2·0.

12·7. Expansion joint should be provided in retaining wall if the length is more than

(a) 40 metres (b) 45 metres

(c) 50 metres.

12·8. Cantilever retaining wall can safely be used for heights not more than

(a) 5 m (b) 6 m

(c) 8 m (d) 10 m.

12·9. If 'H' is the overall height of a retaining wall the width of base slab generally lies between

(a) 0·3 H to 0·4 H (b) 0·4 H to 0·5 H

(c) 0·5 H to 0·6 H (d) 0·6 H to 0·7 H.

12·10. Pick up the correct statement

(a) In the stem of cantilever retaining wall, reinforcement is provided on the earth side

(b) In the toe slab of a cantilever retaining wall, reinforcement is provided at the bottom of slab

(c) In the heel slab of a cantilever retaining wall, reinforcement is provided at the top of the slab.

12·11. In case factor of safety against sliding is less than 1·5, a slab is constructed downward generally below stem which is known

(a) a cut-off wall (b) a rib

(c) a key.

12·12. The thickness of stem in a cantilever retaining wall varies. The minimum thickness at the top of stem is

(a) 10 cm (b) 15 cm

(c) 20 cm.

12·13. If the pressure from earth on foundation is 'p' and 'ω' is the density of earth, the minimum depth 'h' of the foundation from the free surface of the earth is

(a) $h=\frac{p}{\omega}\left(\frac{1-\sin\phi}{1+\sin\phi}\right)$ (b) $h=\frac{p}{\omega}\left(\frac{1-\sin\phi}{1+\sin\phi}\right)$

(c) $h=\frac{p}{\omega}\sqrt{\frac{p}{\omega}\times\left(\frac{1-\sin\phi}{1+\sin\phi}\right)^2}$.

12·14. To have pressure wholly compressive under the base of a retaining wall of width 'b', the resultant of the weight of the wall and the pressuse exerted by the earth retained and any other loads should not have excentrity more than

(a) $\frac{b}{3}$ (b) $\frac{b}{4}$

(c) $\frac{b}{6}$.

12·15. When the earth to be retained is of considerable height, anchor rods are provided at the back of the retaining wall approximately at

(a) ⅓rd the height of retaining wall

(b) ¼th the height of retaining wall

(c) ½ the height of retaining wall.

Chapter 13

13·1. The domes are designed for vertical loads only. To account for accidental loads the shell is stressed to

(a) $\frac{2}{5}$-th of allowable stress under vertical loads

(b) $\frac{1}{2}$ of allowable stress under vertical loads

(c) $\frac{3}{5}$ th of allowable stress under vertical loads.

13·2. Temperature and shrinkage stress are accounted for in domes by providing additional reinforcement of

(a) 0·2 percent of cross-sectional area in both directions

(b) 0·25 percent of cross-sectional area in both directions

(c) 0·30 percent of cross-sectional area in both directions.

13·3. A dome of radius 'r' is loaded with uniformly distributed load of 'ω' per unit area. The thrust 'T' per unit length of the circle of latitude acting tangential to surface downwards at an angle 'θ' to horizontal is given by

(a) $T=\frac{\omega r\,(1-\cos\theta)^2}{\sin\theta}$ (b) $T=\frac{\omega r\,(1-\cos\theta)}{\sin^2\theta}$

(c) $T=\frac{\omega r\,(1-\cos\theta)^2}{\sin^2\theta}$.

13·4. At the crown if 'ω' is the load per unit area and 'r' is the radius of the dome and 't' is the thickness of shell of dome, the hoof stress 'H' is given by

(a) $H=\frac{\omega r}{t}$ (b) $H=\frac{\omega r}{2t}$

(c) $H=\frac{\omega r}{4t}$.

13·5. At the crown if 'ω' is the load per unit area and 'r' is the radius of the dome and 't' is the thickness of shell of dome, the meridional stress 'T' is given by

(a) $T=\frac{\omega r}{t}$ (b) $T=\frac{\omega r}{2t}$

(c) $T\,\frac{\omega t}{4t}$.

Chapter 14

14·1. Pick up the correct statements

Reinforced concrete chimneys are used in preference to masonry chimneys as they are

(*a*) low cost (*b*) they are light

(*c*) they have longer life (*d*) foundation is cheaper.

14·2. Pick up the correct statements

(*a*) the chimneys are provided with a brick lining

(*b*) there is air-gap provided between R.C.C. shell and the lining

(*c*) atleast rich mix of M_{20} is used in chimneys.

Chapter 15

15·1. Circular beam loaded uniformly supported on symmetrically placed columns, the angle subtended at the centre by two consecutive columns being 'θ'; the centre of gravity of load on the beam between two consecutive columns will lie on the bisecter of angle subtended by the columns at a distance from centre

(*a*) $\frac{R \sin \theta}{\theta}$ (*b*) $\frac{R \cos \theta}{\theta}$

(*c*) $\frac{R \sin \theta/2}{\theta/2}$.

15·2. Circular beam loaded uniformly supported on symmetrically placed columns, the angle subtended at the centre by two consecutive columns being 'θ', the max. shear at the column is

(*a*) $\omega R\theta$ (*b*) $\frac{\omega R\theta}{2}$

(*c*) $2\omega R\theta$.

15·3. Circular beam loaded uniformly supported on symmetrically placed columns, the angle subtended at the centre by two consecutive columns being 'θ' the B.M. at supports is given by

(*a*) $M_0 = \omega R^2 \, [1 - \theta \cot \theta]$ (*b*) $M_0 = \omega R^2 \, \theta \cot \theta$

(*c*) $M_0 = \omega R^2 \left[1 - \frac{\theta \cot \theta/2}{2} \right]$.

15·4. Semicircular beam supported on three supports equally spaced loaded with uniformly distributed load, the centre of gravity of the total load will lie on the line joining the centre to the centre column at a distance from the centre (*a*) $\frac{R}{\pi}$ (*b*) $\frac{2R}{\pi}$

(*c*) $\frac{R}{2\pi}$.

15·5. Semicircular beam supported on three columns equally spaced loaded with uniformly distributed load 'ω' per unit length, the central column reaction will be

(*a*) ωR (*b*) $\frac{\omega R}{2}$

(*c*) $2\omega R$.

Chapter 16

16·1. Pick up the correct statements. In water tanks

(*a*) the proportion of fine and coarse aggregates to cement should be such as to give high quality concrete to make a dense unpermeable concrete

(*b*) the minimum quantity of cement in the concrete mix shall not be less than 300 N/m^3

(*c*) concrete mix weaker than M_{20} is not used

(*d*) efficient compaction by vibration is preferable.

16·2. Pick up the joints which come under movement joints

(*a*) construction joint (*b*) contraction joint

(*c*) expansion joint (*d*) sliding joint.

16·3. The minimum reinforcement in walls, floors and roofs in each of two directions at right angles for sections upto 100 mm thickness shall have an area of

(*a*) 0·2 per cent of concrete section

(*b*) 0·25 per cent of concrete section

(*c*) 0·30 per cent of concrete section.

16·4. For liquid faces of parts of members, the minimum cover to all reinforcement should be greater of the main bar or

(*a*) 15 mm (*b*) 20 mm

(*c*) 25 mm.

16·5. In circular tanks, if 'D' is the diameter of tank and 'H' is the height a tank and if 't' is the thickness of shell of tank and H^2/Dt is between 6 to 12, the cantilever effect is assumed as $H/3$ or

(*a*) 0·5 m from base (*b*) 1·0 m from base

(*c*) 1·5 m from base.

16·6. In water tanks, for concrete mix of M_{20}, the permissible stress in tension due to bending is 1·5 N/mm², the permissible stress in direct tension is

(a) 1·1 N/mm² (b) 1·2 N/mm²

(c) 1·3 N/mm² (d) 1·4 N/mm².

16·7. In water tanks permissible stress in steel in bending on liquid retaining face or face away from liquid for members less than 215 mm thick is

(a) 100 N/mm² (b) 112·5 N/mm²

(c) 125 N/mm².

16·8. In water tanks for members 225 mm or more in thickness the bending stress allowable on face away from liquid is

(a) 100 N/mm² (b) 112·5 N/mm²

(c) 125 N/mm².

16·9. Tensile stress in shear reinforcement for members of tank less than 225 mm is

(a) 100 N/mm² (b) 112·5 N/mm²

(c) 125 N/mm².

16·10. Tensile stress in shear reinforcement for members of tank 250 mm or more is

(a) 100 N/mm² (b) 112·5 N/mm²

(c) 125 N/mm².

16·11. If 'H' is the depth of water in circular tank in metres 'D' is the diameter of tank in metres, 't' is the thickness of wall in mm and 'A_t' is the area of steel and 'm' is the modular ratio, tensile stress in concrete is given by

(a) tensile stress $= \dfrac{\omega HD}{1000\, t+(m-1)\, A_t}$

(b) tensile stress $= \dfrac{2\omega\, HD}{1000\, t+(m-1)\, A_t}$

(c) tensile stress $= \dfrac{\omega HD/2}{1000\, t+(m-1)\, A_t}$.

Chapter 17

17·1. If 'h' is the depth of water in aqueduct and 'l' is the span of the horizontal slab, for max bending moment in base slab, the relationship of span and depth is given by

(a) $l=h$ (b) $l=2h$

(c) $l=1·5\, h$.

17·2. If 'ω' is the density of water, 'l' is the span of base slab of aqueduct, max. B.M. at centre for metre of slab is

(a) $\frac{\omega l^3}{8}$ (b) $\frac{\omega l^3}{12}$

(c) $\frac{\omega l^3}{24}$.

17·3. If 'h' is the depth of water in aqueduct and 'l' is the span of slab, the B.M. at centre of slab will be negative if

(a) $h > 0{\cdot}6\ l$ (b) $h > 0{\cdot}7\ l$

(c) $h > l\ \sqrt{0{\cdot}075}$.

17·4. Pick the correct statements. The design of box culvert is based on B.M. worked out for the following conditions of loading

(a) no water in the box culvert. The side walls are subjected to earth pressure from outside

(b) water in the box culvert. The side walls are subjected to pressure from inside and earth pressure from outside.

Chapter 18

18·1. Pick up correct statements. For large span roof construction, concrete construction is preferred to roofs of steel trusses

(a) it is economical (b) it is fire resistant

(c) its maintenance cost is negligible.

18·2. The analysis of frames of a multi storeyed buildings is rather cumbersome. Pick from the following, the assumptions which are made which will give results quite satisfactory

(a) little portion of the frame called substitute frame is taken

(b) the moments carried from floor to floor through columns are very small compared with the beam moment. Therefore the analysis of multistoreyed frames is carried out by taking one floor at a time

(a) each floor is taken with columns above and below fixed at far ends

(d) one method of analysis is known as two-cycle method of moment distribution.

18·3. Analysis of building frames under wind loads and seismic forces is analysed by one of the approximate methods known as portal method. State which assumptions are correct from the following for this method

(a) it is assumed that points of contraflexure lie at centres of members, columns as well beams

(b) the shear taken by interior columns is double that taken by external columns

(c) the direct stresses in columns are proportional to their distances from the C.G. of the frame.

18·4. Analysis of building frames under wind loads and seismic forces is analysed by one of the approximate methods, known as cantilever method. State which assumptions are correct from the following for this method

(a) it is assumed that points of contraflexure lie at centres of members, columns as well as beams

(b) the shear taken by interior columns is double that taken by external columns

(c) the direct shear in columns are proportional to their distances from the C.G. of the frame.

18·5. In the factor method of analysis of frames, the girder factor 'g' is defined as follows

(a) $g=\frac{\Sigma k_c}{\Sigma k}$, where Σk_c is the sum of stiffnesses of columns and Σk is the sum of stiffnesses of all members meeting at the joint

(b) $g=\frac{\Sigma k_b}{\Sigma k}$, where Σk_b is the sum of stiffnesses of beams and Σk is the sum of stiffnesses of all members meeting at the joint.

18·6. In the factor method of analysis of frames, the column factor 'c' is defined as

(a) $c=\frac{\Sigma k_c}{\Sigma k}$, where Σk_c is the sum of stiffnesses of columns and Σk is the sum of stiffnesses of all members meeting at the joint.

(b) $c=\frac{\Sigma k_b}{\Sigma k}$, where Σk_b is the sum of stiffnesses of beams and Σk is the sum of stiffnesses of all members meeting at the joint

(c) $c=1-\frac{\Sigma k_c}{\Sigma k}$, where Σk_c is the sum of stiffnesses of columns and Σk is the sum of stiffnesses of all members meeting at the joint.

18·7. In Kani's method for analysis of frames, if k_{AB} is the stiffness of member AB and Σk is the sum of stiffnesses of all members meeting at joint A, then rotation factor for AB at A, γ_{AB} is defined as

(a) $\frac{k_{AB}}{\Sigma k}$ (b) $\frac{k_{AB}}{2\Sigma k}$

(c) $\frac{-k_{AB}}{2\Sigma k}$ (d) $\frac{-3\,k_{AB}}{2\Sigma k}$.

18·8. The displacement factor in Kani's method at A is given by, if k_{AB} is the stiffness of AB and Σk is the stiffnesses of all the members meeting at joint A

(a) $\frac{k_{AB}}{\Sigma k}$ (b) $\frac{k_{AB}}{2\Sigma k}$

(c) $\frac{-k_{AB}}{2\,\Sigma k}$ (d) $\frac{-3\,k_{AB}}{2\Sigma k}$

Chapter 19

19·1. The bunker is that container in which

(a) the diameter is large as compared to height, so that the plane of rupture between the wedge which contains maximum pressure and the remaining fill cuts the top horizontal surface

(b) the height of the container is large compared to its diameter so that the plane of rupture cuts the apposite side.

19·2. The silo is that container in which

(a) The diameter is large as compared to height so that the plane of rupture between the wedge which contains maximum pressure and the remaining fill cuts the top horizontal surface

(b) the height of the container is large compared to its diameter so that the plane of rupture cuts the apposite side.

19·3. In the case of bunkers

(a) the vertical weight is supported by the bottom which is generally hopper shaped sloping in one or more directions

(b) a small portion of vertical weight of stored material is supported by the hopper bottom and large amount of vertical weight is supported by the side walls and the stored material.

19·4. In the case of silos

(a) the vertical weight is supported by the bottom which is generally hopper shaped sloping in one or more directions

(b) a small portion of vertical weight of stored material is supported by the hopper bottom and large amount of vertical weight is supported by the side walls and the stored material.

19·5. In bunker of uniform thickness if 'l' is the length of bunker and 'B' is the breadth of the bunker and if 'p' is the pressure at any depth, the corner bending moment is given by

(a) $\frac{-p}{12}(B^2-Bl+l^2)$ (b) $\frac{-p}{12}(B^2+Bl-l^2)$

(c) $\frac{-p}{24}(l^2+2Bl-2B^2)$ (d) $\frac{-p}{24}(B^2+2Bl-2l^2)$.

19·6. In bunker of uniform thickness if 'l' is the length of bunker and 'B' is the breadth of the bunker and if 'p' is the pressure at any depth, the +ve B.M. at centre of width 'B' is

(a) $\frac{p}{12}(B^2-Bl+l^2)$ (b) $\frac{p}{12}(B^2+Bl-l^2)$

(c) $\frac{p}{24}(B^2+2Bl-2l^2)$ (d) $\frac{p}{24}(l^2+2Bl-2B^2)$.

Chapter 20

20·1. If density of fresh concrete is 21000 N/m³ at depth 'h' metres if 's' mm is the spacing of column formwork yokes and 'd' mm is the thickness of sheeting of formwork and if 'f' N/mm² is the allowable stress in timber, the spacing is given as

(a) $s=8{\cdot}9\,d\left(\frac{f}{h}\right)$ (b) $s=8{\cdot}9\,d\sqrt{\frac{f}{h}}$.

Chapter 21

21·1. Pick up correct statements. Caisson foundations are used

(a) where heavy loads are to be transferred through deep layer of weak soil to a firm base

(b) where foundations are to be taken below scour depth

(c) where pile driving vibration are not allowed

(d) where very large bearing area is required.

21·2. If 'V' is the velocity of ship, and W_s is the weight of the ship the energy of impact with the wharf or pier is given as

(a) $\frac{W_s V^2}{g}$ (b) $\frac{2W_s V^2}{g}$

(c) $\frac{W_s V^2}{2g}$.

Chapter 22

22·1. In heavy structures like warehouses, etc. the shear stresses and negative bending moments may be high and therefore haunches are provided. The main tensile reinforcement in haunches is provided

(a) straight (b) parallel to haunch.

22·2. In heavy structures like warehouses etc., the shear stresses and negative bending moments may be high and therefore haunches are provided. The stirrups in haunches can be provided

(a) vertically (b) normal to haunch.

Chapter 23

23·1. **In composite section when concrete slab is laid over the steel joists when they are propped throughout its length, called propped composite sections**

(a) the steel section will have stresses due to its self load, whereas the dead load of slab and live load will produce stresses in the composite section

(b) the steel section will have stresses due to self load and dead load of the slab, whereas composite section will have stresses due to live load.

23·2. **In composite section when concrete slab is laid over the steel joists, when they are not propped called unpropped composite section**

(a) the steel section will have stresses due to its self load, whereas the dead load of slab and live load will produce stresses in the composite section

(b) the steel section will have stresses due to self load and dead load of the slab, whereas composite section will have stresses due to live load.

23·3. **In composite section, if 'B_s' is the width of flange, 'd_s' is the thickness of R.C.C. slab, 'A_s' is the area of steel joist, 'y_s' is the distance C.G. of equivalent steel section of composite beam from top, 'm' is the modular ratio, neutral axis when not in slab is given by**

(a) $$n=\frac{\dfrac{B_s d_s^2}{2m}+A_s y_s}{\dfrac{B_s A_s}{m}+A_s}$$

(b) $$n=\frac{mA_s}{B_s}\left[-1+\sqrt{1-\frac{2B_s y_s}{mA_s}}\right]$$

23·4. **In composite section, if 'B_s' is the width of flange, 'd_s' is the thickness of slab, 'A_s' is the area of steel joist, 'y_s' is the distance of C.G. of equivalent steel section of composite beam from top, 'm' is the modular ratio, neutral axis when in slab, is given by**

(a) $$n=\frac{\dfrac{B_s d_s^2}{2m}+A_s y_s}{\dfrac{B_s d_s}{m}+A_s}$$

(b) $$n=\frac{mA_s}{B_s}\left[-1+\sqrt{1-\frac{2B_s y_s}{mA_s}}\right]$$

Chapter 24

24·1. For road bridges for class A or B loading, if 'L' is the span in metres, the impact factor 'i' is given by

(a) 25 percent (b) $i=\dfrac{6\cdot0}{4\cdot5+L}$

(c) $i=\dfrac{4\cdot5}{6\cdot0+L}$.

24·2. For road bridges centrifugal force 'C' in kN, if 'W' is live load in kN, 'V' is the speed of vehicle in km/h and 'R' is the radius of curvature in metres

(a) $C=\dfrac{WV^2}{100\ R}$ (b) $C=\dfrac{WV^2}{150\ R}$

(c) $C=\dfrac{WV^2}{127\ R}$.

24·3. Kerbs in bridges shall be designed for lateral loading applied horizontally at the top of kerb of value

(a) 5000 N/m (b) 7500 N/m

(c) 10,000 N/m.

24·4. Solid slabs spanning in one direction, the effective width 'b_e', if 'k' is a constant depending on the ratio 'B/L', where 'B' is the width and 'L' is the span of the slab, 'x' is the distance of the centre of gravity of the concentrated load from the nearer support, 'b_w' is the breadth of concentration area of the load, *i.e.* the dimension of the tyre or track contact area over the load surface of the slab in a direction at right angles to the span plus twice the thickness of the wearing coat or surface finish, is given by

(a) $b_e=kx\left(1+\dfrac{x}{L}\right)+b_w$ (b) $b_e=kx\left(1-\dfrac{x}{L}\right)+b_w$

(c) $b_e=kx\left(1-\dfrac{x}{2L}\right)+b_w$.

24·5. For solid slab cantilever, if 'x' is the distance of the centre of gravity of the concentrated load from the face of the cantilever support, 'b_w' is the breadth of concentration area of load *i.e.* the dimension of the tyre or track contact area over the load surface of the slab in a direction parallel to the supporting edge of the cantilever plus twice the thickness of the wearing coat or surface finish, the effective 'b_e', is given by

(a) $2x+b_w$ (b) $1\cdot5\ x+b_w$

(c) $1\cdot2\ x+b_w$.

24·6. When the effect of wind forces is taken into consideration of design of bridge members, the permissible stresses may be exceeded by

(a) 15 percent (b) 25 percent
(c) 50 percent.

24·7. When the worst combination of loads met with during erection is considered, the permissible working stresses may be exceeded by

(a) 15 percent (b) 25 percent
(c) 50 percent.

24·8. For a longitudinal reinforcement bar, cover in a column or beam in bridges shall not be less than the diameter of bar or

(a) 25 mm (b) 35 mm
(c) 40 mm.

Chapter 25

25·1. In a prestressed member, it is advisable to use

(a) low strength concrete only
(b) high strength concrete only
(c) low strength concrete but high tensile steel
(d) high strength concrete and high tensile steel
(e) high strength cencrete and low tensile steel.

25·2. The minimum cube strength of concrete used for a prestressed member is

(a) 15 N/mm^2 (b) 25 N/mm^2
(c) 30 N/mm^2 (d) 35 N/mm^2.

25·3. The following assumption is made for the design of a prestressed concrete member

(a) plane cross-section before loading remains plane after loading
(a) any change in loading produces change of stress in concrete
(c) the prestressing in the steel is assumed to remain constant.

25·4. If 'P' is the prestressing force, 'A' the area of cross-section, if the prestressing force is applied at the lower kern point, the stress at the top fibres will be

(a) $\frac{P}{A}$ (b) $\frac{2P}{A}$

(c) 0·0.

25·5. If 'P' is the prestressing force, 'A' is the area of cross-section, if the prestressing force is applied at the lower kern point, the stress at the bottom fibres will be

(a) $\frac{P}{A}$ (b) $\frac{2P}{A}$

(c) 0·0.

25·6. If the tendon is placed at an eccentricity 'e' below the centroidal axis of the longitudinal axis of a rectangular prestressed beam of sectional area 'A' and sectional modulus 'Z' and if 'P' is the prestressing force the compressive stress in top fibres will be

(a) $\frac{P}{A}+\frac{Pe}{Z}$ (b) $\frac{P}{A}$

(c) $\frac{P}{A}-\frac{Pe}{Z}$.

25·7. If the tendon is placed at an eccentricity 'e' below the centroidal axis of the longitudinal axis of a rectangular prestressed beam of sectional area 'A' and sectional modulus 'Z' and if 'P' is the prestressing force, the compressive stress in bottom fibres will be

(a) $\frac{P}{A}+\frac{Pe}{Z}$ (b) $\frac{P}{A}$

(c) $\frac{P}{A}-\frac{Pe}{Z}$.

25·8. Modulus of elasticity of concrete E_c shall be assumed if F_{ck} is characteristic strength of concrete in N/mm^2 as

(a) $E_c=2000\ F_{ck}$ (b) $2000\sqrt{F_{ck}}$

(c) $5700\sqrt{F_{ck}}$.

25·9. The modulus of rupture of concrete in N/cm^2 shall be taken if F_{ck} is the characteristic strength of concrete in N/mm^2 as

(a) $5\sqrt{F_{ck}}$ (b) $7\sqrt{F_{ck}}$

(c) $10\sqrt{F_{ck}}$.

25·10. Failure due to fracture of steel occurs when the percentage of steel, $p=\frac{A_{ts}}{bD}$ provided in a prestressed member, is lower than, if F_s is ultimate tensile strength in N/mm^2

(a) $p \leqslant \frac{150}{F_s}$ (b) $p \leqslant \frac{200}{F_s}$

(c) $p < \frac{250}{F_s}$ (d) $p \leqslant \frac{300}{F_s}$.

25·11. After prestressing process is completed, a loss of stress is due to

(*a*) shrinkage of concrete

(*b*) elastic shortening of concrete

(*c*) creep of concrete (*d*) creep of steel.

25·12. The total shrinkage in a pretensioned beam allowed is

(*a*) $3·0 \times 10^{-4}$ (*b*) $4·0 \times 10^{-4}$

(*c*) $3·0 \times 10^{-3}$ (*d*) $3·0 \times 10^{-5}$.

ANSWERS TO OBJECTIVE TYPE QUESTIONS

Chapter 1

1·1.	(*d*)	1·2.	(*e*)	1·3.	(*e*)	1·4.	(*a*)
1·5.	(*c*)	1·6.	(*e*)	1·7.	(*b*)	1·8.	(*a*)
1·9.	(*d*)	1·10.	(*c*)	1·11.	(*c*)	1·12.	(*d*)
1·13.	(*b*)	1·14.	(*c*)	1·15.	(*c*)	1·16.	(*b*)
1·17.	(*b*)	1·18.	(*d*)	1·19.	(*a, b*)	1·20.	(*a, b, c*)
1·21.	(*a, b, c*)	1·22.	(*a*)	1·23.	(*c*)	1·24.	(*a*)
1·25.	(*a*)	1·26.	(*a, b, c, d*)	1·27.	(*a, b, c, d*)	1·28.	(*a, b, c, d*)
1·29.	(*a, b, c, d*)	1·30.	(*c*)	1·31.	(None)	1·32.	(*a, b, c, d*)
1·33.	(*b*)	1·34.	(*a, b, c, d*)	1·35.	(*a, c*)	1·36.	(*d*)
1·37.	(*a, b, c, d*)	1·38.	(*b*)	1·39.	(*b*)	1·40.	(*c*)
1·41.	(*b*)	1·42.	(*a, b, c, d*)	1·43.	(*a, b, c, d*)	1·44.	(*a, b, c*)
1·45.	(*a, b, c*)	1·46.	(*a, b, c, d*)	1·47.	(*a, b, c, d*)	1·48.	(*a*)
1·49.	(*a, b, c*)	1·50.	(*b*)	1·51.	(*a, b, c, d*)	1·52.	(*a, b, c, d*)
1·53.	(*a, b, c, d*)	1·54.	(*b*)	1·55.	(*b*)	1·56.	(*c*)
1·57.	(*d*)	1·58.	(*a, b, c, d*)	1·59.	(*a, b, c, d*)	1·60.	(*a, b*)
1·61.	(*a, b, c*)	1·62.	(*a, b*)	1·63.	(*a, e, f*)	1·64.	(*b*)
1·65.	(*a*)	1·66.	(*c*)	1·67.	(*a*)	1·68.	(*c*)
1·69.	(*c*)	1·70.	(*b*)	1·71.	(*a*)	1·72.	(*b*)
173.	(*a*)	1·74.	(*b*)	1·75.	(*a*)	1·76.	(*c*).

Chapter 2

2·1.	(*a, b, c, d*)	2·2.	(*a, b, c*)	2·3.	(*b*)	2·4.	(*b*)
2·5.	(*e*)	2·6.	(*a, b, c*)	2·7.	(*a, b, c*)	2·8.	(*b*)
2·9.	(*c*)	2·10.	(*b, d*)	2·11.	(*a, d*)	2·12.	(*a*)
2·13.	(*b*)	2·14.	(*b*)	2·15.	(*c*)	2·16.	(*a, b, c, d*)
2·17.	(*c*)	2·18.	(*c*)	2·19.	(*a*)	2·20.	(*c*)
2·21.	(*b*)	2·22.	(*a*)	2·23.	(*c*)	2·24.	(*b*)
2·25.	(*c*)	2·26.	(*c*)	2·27.	(*b*)	2·28.	(*c*)
2·29.	(*a*)	2·30.	(*b*).				

Chapter 3

3·1.	(*a*)	3·2.	(*a, b, c*)	3·3.	(*c*)	3·4.	(*b*)
3·5.	(*c*)	3·6.	(*b*)	3·7.	(*a*)	3·8.	(*c*)
3·9.	(*b*)	3·10.	(*b*)	3·11.	(*c*).		

Chapter 4

4·1. (*a*) 4·2. (*d*) 4·3. (*c*) 4·4. (*b, c*)
4·5. (*a*) 4·6. (*c*) 4·7. (*b*) 4·8. (*b*)
4·9. (*c*) 4·10. (*b*) 4·11. (*c*) 4·12. (*c*)
4·13. (*b*) 4·14. (*c*) 4·15. (*b*) 4·16. (*c*)
4·17. (*b*) 4·18. (*a*).

Chapter 5

5·1. (*b*) 5·2. (*c*) 5·3. (*c*) 5·4. (*d*)
5·5. (*c*) 5·6. (*b, d*).

Chapter 6

6·1. (*d*) 6·2. (*c*) 6·3. (*b*) 6·4. (*c*)
6·5. (*c*) 6·6. (*b*) 6·7. (*a*) 6·8. (*a, b, c*)
6·9. (*a*) 6·10. (*a*) 6·11. (*c*) 6·12. (*a*)
6·13. (*c*) 6·14. (*b*) 6·15. (*c*).

Chapter 7

7·1. (*c*) 7·2. (*c*) 7·3. (*b*) 7·4. (*d*)
7·5. (*a*) 7·6. (*b*) 7·7. (*c*) 7·8. (*b*)
7·9. (*a*) 7·10. (*b*) 7·11. (*a, b, c*) 7·12. (*a, b, c, d*)
7·13. (*c*) 7·14. (*b*) 7·15. (*c*) 7·16. (*b*)
7·17. (*c*) 7·18. (*c*) 7·19. (*b*) 7·20. (*a*)
7·21. (*b*) 7·22. (*c*) 7·23. (*c*) 7·24. (*b*)
7·25. (*b*).

Chapter 8

8·1. (*d*) 8·2. (*c*) 8·3. (*c*) 8·4. (*b*)
8·5. (*c*) 8·6. (*b*) 8·7. (*a*) 8·8. (*c*)
8·9. (*c*) 8·10. (*a*) 8·11. (*b*) 8·12. (*c*)
8·13. (*d*) 8·14. (*b*) 8·15. (*c*) 8·16. (*b*)
8·17. (*b*) 8·18. (*b*) 8·19. (*b*) 8·20. (*c*)
8·21. (*b*) 8·22. (*a*) 8·23. (*a*) 8·24. (*c*)
8·25. (*c*) 8·26. (*c*) 8·27. (*b*) 8·28. (*c*)
8·29. (*b*) 8·30. (*c*) 8·31. (*b*) 8·32. (*a*)
8·33. (*b*) 8·34. (*a*) 8·35. (*d*) 8·36. (*d*)
8·37. (*b*) 8·38. (*b*) 8·39. (*a*) 8·40. (*c*)
8·41. (*a*) 8·42. (*b*) 8·43. (*b*) 8·44. (*d*).

Chapter 9

9·1. (*a*) 9·2. (*b*) 9·3. (*d*) 9·4. (*a, b, c, d*).

Chapter 10

10·1. (*a, b*)	10·2. (*a*)	10·3. (*b*)	10·4. (*d*)
10·5. (*c*)	10·6. (*b*)	10·7. (*a, d*)	10·8. (*c*)
10·9. (*a*)	10·10. (*b*)	10·11. (*b*)	10·12. (*a*)
10·13. (*d*)	10·14. (*b*)	10·15. (*b*)	10·16. (*b*)
10·17. (*a*)	10·18. (*c*)	10·19. (*c*)	10·20. (*c*)
10·21. (*c*)	10·22. (*a*)	10·23. (*c*)	10·24. (*a*)
10·25. (*b*)	10·26. (*c*)	10·27. (*a*)	10·28. (*a*)
10·29. (*a, c*)	10·30. (*c*)	10·31. (*a*)	10·32. (*a, b, c, d*)
10·33. (*a,b,c,d,e*)	10·34. (*c*)	10·35. (*c*)	10·36. (*c*)
10·37. (*b*)	10·38. (*a*)	10·39. (*b*)	10·40. (*b*)
10·41. (*c*)	10·42. (*a*)	10·43. (*a*)	10·44. (*c*)
10·45. (*b*)	10·46. (*a, d*).		

Chapter 11

11·1. (*b*)	11·2. (*a*)	11·3. (*c*)	11·4. (*a*)
11·5. (*b*)	11·6. (*b*)	11·7. (*a*).	

Chapter 12

12·1. (*c*)	12·2. (*b*)	12·3. (*a*)	12·4. (*c*)
12·5. (*c*)	12·6. (*b*)	12·7. (*b*)	12·8. (*b*)
12·9. (*c*)	12·10. (*a, b, c*)	12·11. (*c*)	12·12. (*b*)
12·13. (*b*)	12·14. (*c*)	12·15. (*a*).	

Chapter 13

13·1. (*a*)	13·2. (*a*)	13·3. (*b*)	13·4. (*b*)
13·5. (*b*).			

Chapter 14

14·1. (*a, b, c, d*)	14·2. (*a, b, c*).

Chapter 15

15·1. (*c*)	15·2. (*b*)	15·3. (*c*)	15·4. (*b*)
15·5. (*c*).			

Chapter 16

16·1. (*a, b, c, d*)	16·2. (*b, c, d*)	16·3. (*c*)	16·4. (*c*)
16·5. (*b*)	16·6. (*b*)	16·7. (*a*)	16·8. (*c*)
16·9. (*a*)	16·10. (*c*)	16·11. (*c*).	

Chapter 17

17·1. (*b*)	17·2. (*c*)	17·3. (*c*)	17·4. (*a, b*).

Chapter 18

18·1. (*b, c*) 18·2. (*a, b, c, d*) 18·3. (*a, b*) 18·4. (*a, c*)
18·5. (*a*) 18·6. (*b, c*) 18·7. (*c*) 18·8. (*d*).

Chapter 19

19·1. (*a*) 19·2. (*b*) 19·3. (*a*) 19·4. (*b*)
19·5. (*a*) 19·6. (*c*).

Chapter 20

20·1. (*a*).

Chapter 21

21·1. (*a, b, c, d*) 21·2. (*c*)

Chapter 22

22·1. (*a*) 22·2. (*a, b*).

Chapter 23

23·1. (*a*) 23·2. (*b*) 23·3. (*a*) 23·4. (*b*)

Chapter 24

24·1. (*c*) 24·2. (*c*) 24·3. (*b*) 24·4. (*b*)
24·5. (*c*) 24·6. (*b*) 24·7. (*c*) 24·8. (*c*)

Chapter 25

25·1. (*b, d*) 25·2. (*d*) 25·3. (*a, b, c*) 25·4. (*c*)
25·5. (*b*) 25·6. (*c*) 25·7. (*a*) 25·8. (*c*)
25·9. (*b*) 25·10. (*c*) 25·11. (*a, b, c, d*) 25·12. (*a*)

Appendix D

DESIGN PROBLEMS

A-D. Design reinforced concrete slab panels S1, S2, S3 and S4 of the building having floor plan shown in Fig. A-D 1. The live load on the floor is 4000 N/m².

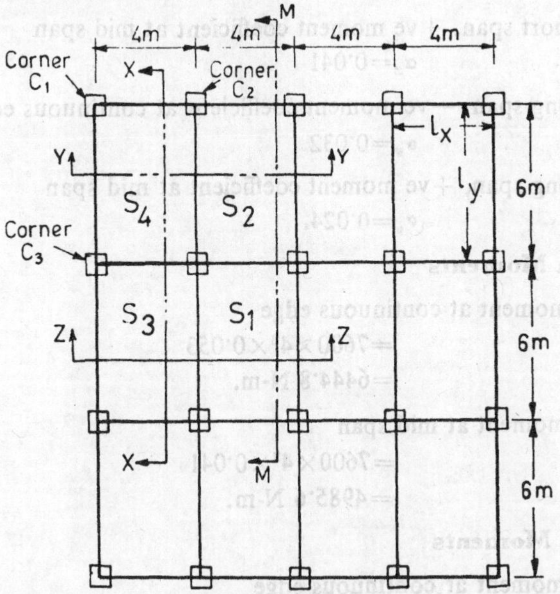

Fig. A-D 1. Building floor plan.

Sol. Assume thickness of concrete slab as 150 mm for all panels.

Assuming density of concrete as 24000 N/m³, self weight of slab per square meter

$$=1\times1\times\frac{150}{1000}\times24000$$

$$=3600 \text{ N/m}^2.$$

Total load on slab $=4000+3600$

$$=7600 \text{ N/m}^2.$$

Table 8·5 is used to obtain bending moment coefficients as per ISI code.

Bending Moments in Slab Panels

$$\frac{l_y}{l_x}=\frac{6}{4}=1\cdot5 \text{ for slab panels.}$$

Bending moment $M_x=\alpha_x\, wl_x^2$

Bending moment $M_y=\alpha_y\, wl_y^2$.

Slab S1. (All edges continuous).

For short span, −ve moment coefficient at continuous edge

$$\alpha_x=0{\cdot}053$$

For short span, +ve moment coefficient at mid span

$$\alpha_x=0{\cdot}041$$

For long span, −ve moment coefficient at continuous edge

$$\alpha_y=0{\cdot}032$$

For long span, +ve moment coefficient at mid span

$$\alpha_y=0{\cdot}024.$$

Short Span Moments

−ve moment at continuous edge

$$=7600\times4^2\times0{\cdot}053$$
$$=6444{\cdot}8 \text{ N-m.}$$

+ve moment at mid span

$$=7600\times4^2\times0{\cdot}041$$
$$=4985{\cdot}6 \text{ N-m.}$$

Long Span Moments

−ve moment at continuous edge

$$=7600\times6^2\times0{\cdot}032$$
$$=8755{\cdot}2 \text{ N-m}$$

+ve moment at mid span

$$=7600\times6^2\times0{\cdot}024$$
$$=6566{\cdot}4 \text{ N-m.}$$

Slab S2. (One short edge discontinuous).

For short span, −ve moment coefficient at continuous edge

$$\alpha_x=0{\cdot}057$$

For short span, +ve moment coefficient at mid span

$$\alpha_x=0{\cdot}044$$

For long span, −ve moment coefficient at continuous edge

$$\alpha_y=0{\cdot}037$$

For long span, +ve moment coefficient at mid span

$$\alpha_y=0{\cdot}028$$

Short Span Moments

$-$ve moment at continuous edge

$=7600\times 4^2\times 0{\cdot}057$

$=6931{\cdot}2$ N-m

$+$ve moment at mid span

$=7600\times 4^2\times 0{\cdot}044$

$=5350{\cdot}4$ N-m.

Long Span Moments

$-$ve moment at continuous edge

$=7600\times 6^2\times 0{\cdot}037$

$=10{,}123{\cdot}2$ N-m

$+$ve moment at mid span

$=7600\times 6^2\times 0{\cdot}028$

$=7660{\cdot}8$ N-m.

Slab S3. (One long edge discontinuous).

For short span, $-$ve moment coefficient at continuous edge

$\alpha_x=0{\cdot}067$

For short span, $+$ve moment coefficient at mid span

$\alpha_x=0{\cdot}051$

For long span, $-$ve moment coefficient at continuous edge

$\alpha_y=0{\cdot}037$

For long span, $+$ve moment coefficient at mid span

$\alpha_y=0{\cdot}028$.

Short Span Moments

$-$ve moment at continuous edge

$=7600\times 4^2\times 0{\cdot}067$

$=8147{\cdot}2$ N-m

$+$ve moment at mid span

$=7600\times 4^2\times 0{\cdot}051$

$=6201{\cdot}6$ N-m.

Long Span Moments

$-$ve moment at continuous edge

$=7600\times 6^2\times 0{\cdot}037$

$=10{,}123$ N-m

+ve moment at mid span

$$=7600\times6^2\times0{\cdot}028$$
$$=7660{\cdot}8 \text{ N-m.}$$

Slab S4. (Two adjacent edges discontinuous).

For short span, −ve moment coefficient at continuous edge

$$\alpha_x=0{\cdot}075$$

For short span, +ve moment coefficient at mid span

$$\alpha_x=0{\cdot}056$$

For long span, −ve moment coefficient at continuous edge

$$\alpha_y=0{\cdot}047$$

For long span, +ve moment coefficient at mid span

$$\alpha_y=0{\cdot}035.$$

Short Span Moments

−ve moment at continuous edge

$$=7600\times4^2\times0{\cdot}075$$
$$=9{,}120 \text{ N·m}$$

+ve moment at mid span

$$=7600\times4^2\times0{\cdot}056$$
$$=6809{\cdot}6 \text{ N·m.}$$

Long Span Moments

−ve moment at continuous edge

$$=7600\times6^2\times0{\cdot}047$$
$$=12{,}859{\cdot}2 \text{ N-m}$$

+ve moment at mid span

$$=7600\times6^2\times0{\cdot}035$$
$$=9576 \text{ N·m.}$$

Maximum moment in slab

$$=12{,}859{\cdot}2 \text{ N-m.}$$

Use M15 concrete mix.

Effective depth is given by

$$d=\sqrt{\frac{M}{Qb}}$$

$$=\sqrt{\frac{12,859{\cdot}2\times1000}{0{\cdot}87\times1000}}$$

$$=121{\cdot}58 \text{ mm.}$$

Provide overall depth of 150 mm.
Using 10 mm diameter bars,
Effective depth $=150-15-5=130$ mm, for long span moments,
Effective depth for short span using 8 mm φ bars is

$$=130-8$$
$$=122 \text{ mm.}$$

Reinforcement in Slab S1

Short Span

Reinforcement for −ve moment at continuous edge

$$=\frac{M}{bjd}=\frac{6444{\cdot}8\times1000}{140\times0{\cdot}87\times122}$$

$$=430{\cdot}2 \text{ mm } \varphi$$

Provide 8 mm φ at 110 mm c/c, area provided

$$=457 \text{ mm}^2$$

Reinforcement for +ve moment at mid span

$$=\frac{4985{\cdot}6\times1000}{140\times0{\cdot}87\times122}$$

$$=335{\cdot}5.$$

Provide 8 mm φ at 110 mm c/c.

Long Span

Reinforcement for +ve moment at mid span

$$=\frac{6566{\cdot}4\times1000}{140\times0{\cdot}87\times130}$$

$$=414{\cdot}7 \text{ mm}^2.$$

Provide 8 mm φ at 100 mm c/c. Area provided

$$=503 \text{ mm}^2.$$

Reinforcement for −ve moment at support

$$=\frac{8755{\cdot}2\times1000}{140\times0{\cdot}87\times130}$$

$$=552{\cdot}9 \text{ mm}^2.$$

Half the bars from mid span are bent near the support, providing 8 mm φ at 200 mm c/c, area provided 252 mm².

Provide additional bars of 10 mm φ at 200 mm c/c at the support, area provided

$$=393 \text{ mm}^2.$$

Total area provided $=252+393$

$$=645 \text{ mm}^2.$$

Reinforcement in Slab S2

Short Span

Effective depth $=122$ mm

Reinforcement for −ve moment at continuous edge

$$=\frac{6931 \cdot 2 \times 1000}{140 \times 0 \cdot 87 \times 122}$$

$$=466 \cdot 4 \text{ mm}^2.$$

Provide 8 mm φ at 100 mm c/c, area provided

$$=503 \text{ mm}^2.$$

Reinforcement for +ve moment at mid span

$$=\frac{5350 \cdot 4 \times 1000}{140 \times 0 \cdot 87 \times 122}$$

$$=360 \cdot 1 \text{ mm}^2.$$

Provide 8 mm φ at 100 mm c/c.

Long Span

Reinforcement for +ve moment at mid span

$$=\frac{7660 \cdot 8 \times 1000}{140 \times 0 \cdot 87 \times 130}$$

$$=483 \cdot 8 \text{ mm}.$$

Provide 8 mm φ at 100 mm c/c, area provided

$$=503 \text{ mm}^2$$

Reinforcement for −ve moment at continuous edge.

$$=\frac{10{,}123 \cdot 2 \times 1000}{140 \times 0 \cdot 87 \times 130}$$

$$=639 \cdot 3 \text{ mm}^2.$$

Half the bars from mid span are bent near the support, providing 8 mm φ at 200 mm c/c, area provided

$$=252\ mm^2$$

Provide additional bars of 10 mm φ at 200 mm c/c, area prov ded=393.

Total area provided=252+393

$$=645\ mm^2.$$

Reinforcement in Slab S3

Short Span

Reinforcement for −ve moment at continuous edge

$$=\frac{8147{\cdot}2\times1000}{140\times0{\cdot}87\times122}$$

$$=548{\cdot}3\ mm^2.$$

Reinforcement for +ve moment at mid span

$$=\frac{6201{\cdot}6\times1000}{140\times0{\cdot}87\times122}$$

$$=417{\cdot}3\ mm^2.$$

Provide 8 mm φ at 110 mm c/c, area provided

$$=457\ mm^2.$$

Half the bars are bent at the support providing 8 mm φ at 220 m c/c for −ve moment, providing area

$$=229\ mm^2.$$

Provide additional reinforcement of 10 mm φ at 220 mm c/c, providing area

$$=351\ mm^2$$

Total area provided=220+351

$$=571\ mm^2.$$

Long Span

Reinforcement for +ve moment at mid span

$$=\frac{7660{\cdot}8\times1000}{140\times0{\cdot}87\times130}$$

$$=483{\cdot}8\ mm.$$

Provide 8 mm φ at 80 mm c/c, area provided

$=629 \text{ mm}^2$.

Reinforcement for −ve moment at continuous edge

$$=\frac{10{,}123\times 1000}{140\times 0{\cdot}87\times 130}$$

$=639{\cdot}3 \text{ mm}^2$.

Provide 8 mm φ bars at 70 mm c/c.

Reinforcement in Slab S4

Short Span

Reinforcement for +ve moment at mid span

$$=\frac{6809{\cdot}6\times 1000}{140\times 0{\cdot}87\times 122}$$

$=458{\cdot}3$

Provide 8 mm φ at 100 mm c/c, area provided

$=503 \text{ mm}^2$

Reinforcement for negative moment at continuous edge

$$=\frac{9120\times 1000}{140\times 0{\cdot}87\times 122}$$

$=613{\cdot}7 \text{ mm}^2$.

Half the bars from mid span are bent near the support providing 8 mm φ at 200 mm c/c, area provided

$=252 \text{ mm}^2$

Additional reinforcement of 10 mm φ a 200 mm c/c is provided giving area

$=393 \text{ mm}^2$

Total area provided$=252+393$

$=645 \text{ mm}^2$

Long Span

Reinforcement for +ve moment at mid span

$$=\frac{9576\times 1000}{140\times 0{\cdot}87\times 130}$$

$=604{\cdot}8$

Provide 8 mm φ at 80 mm c/c, area provided

$=629 \text{ mm}^2$

Reinforcement for $-$ve moment at continuous support

$$=\frac{12,859 \cdot 2\times 1000}{140\times 0 \cdot 87\times 130}$$

$=806\ mm^2$.

Half the bars from mid span are bent up near the support, providing 8 mm φ at 160 mm, area provided

$=315\ mm^2$

Additional reinforcement of 10 mm φ at 160 mm is provided, giving area$=491$.

Area provided $=491+315$

$=806\ mm^2$

The details of reinforcement are shown in Fig. A-D 2.

Torsion Reinforcement

Corner C_1

Torsion reinforcement

$=0 \cdot 75$ of reinforcement for maximum +ve moment

$=0 \cdot 75\times 604 \cdot 8$

$=453 \cdot 6\ mm^2$

Provide 8 mm φ at 110 mm c/c.

Corner C_2

Torsion reinforcement

$=\frac{1}{2}$ of reinforcement at corner C_1

$=\frac{1}{2}\times 453 \cdot 6$

$=226 \cdot 8\ mm^2$.

Provide 8 mm φ at 200 mm c/c.

Corner C_3

Torsion reinforcement

$=\frac{1}{2}$ of reinforcement at corner C_1

$=\frac{1}{2}\times 453 \cdot 6$

$=226 \cdot 8\ mm^2$

Provide 8 mm φ at 200 mm c/c.

Distribution Steel

Steel area $=0 \cdot 15$ % of cross-sectional area

$$=\frac{0 \cdot 15}{100}\times 150\times 1000$$

$=225\ mm^2$

Provide 8 mm φ at 200 mm c/c.

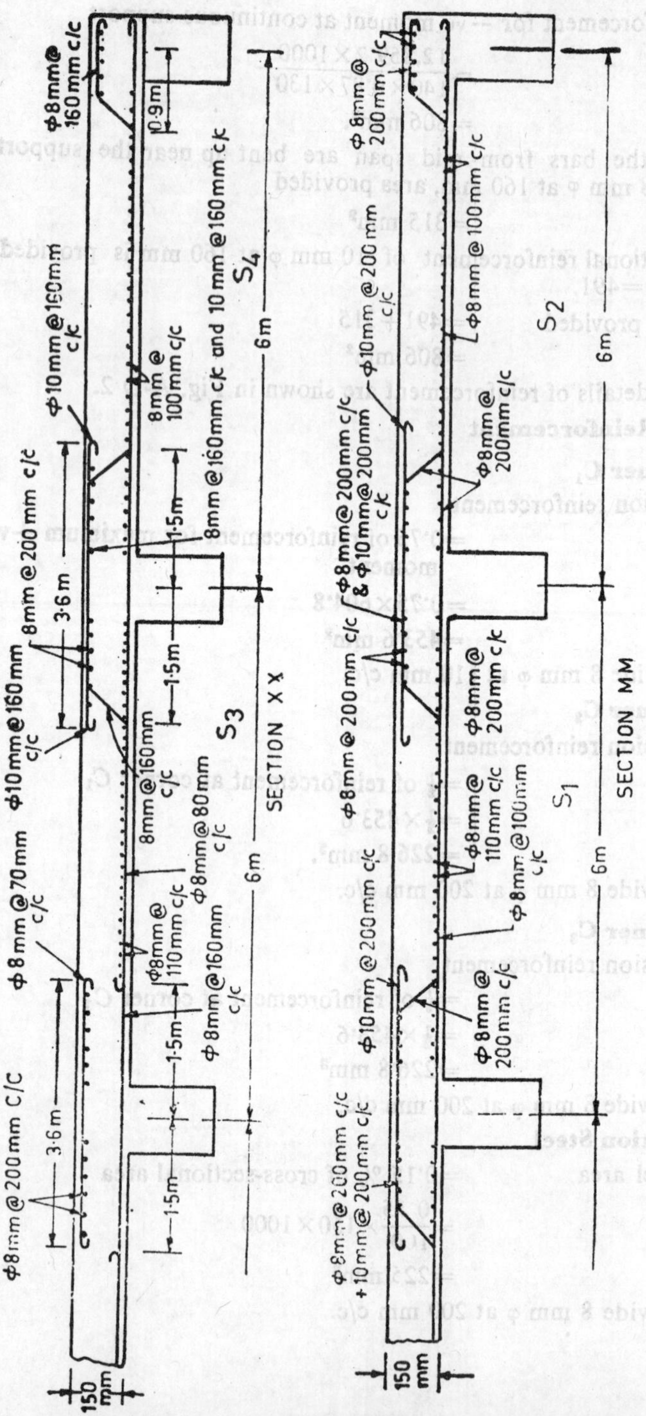
150 mm
φ8 mm @ 200 mm C/C
3·6 m
1·5 m
φ8 mm @ 70 mm c/c
φ10 mm @ 160 mm c/c
8 mm @ 200 mm c/c
φ10 mm @ 160 mm c/c
φ8 mm @ 160 mm c/c
φ8 mm @ 160 mm c/c
8 mm @ 100 mm c/c
8 mm @ 160 mm c/c and 10 mm @ 160 mm c/c
0·9 m
S3
S4
SECTION XX
6 m
φ8 mm @ 200 mm c/c + 10 mm @ 200 mm c/c
φ10 mm @ 200 mm c/c
φ8 mm @ 200 mm c/c
φ8 mm @ 200 mm c/c
φ8 mm @ 200 mm c/c
φ8 mm @ 200 mm c/c
φ8 mm @ 110 mm c/c
φ8 mm @ 100 mm c/c
φ8 mm @ 200 mm c/c & φ10 mm @ 200 mm c/c
φ10 mm @ 200 mm c/c
φ8 mm @ 100 mm c/c
φ8 mm @ 200 mm c/c
S1
S2
SECTION MM

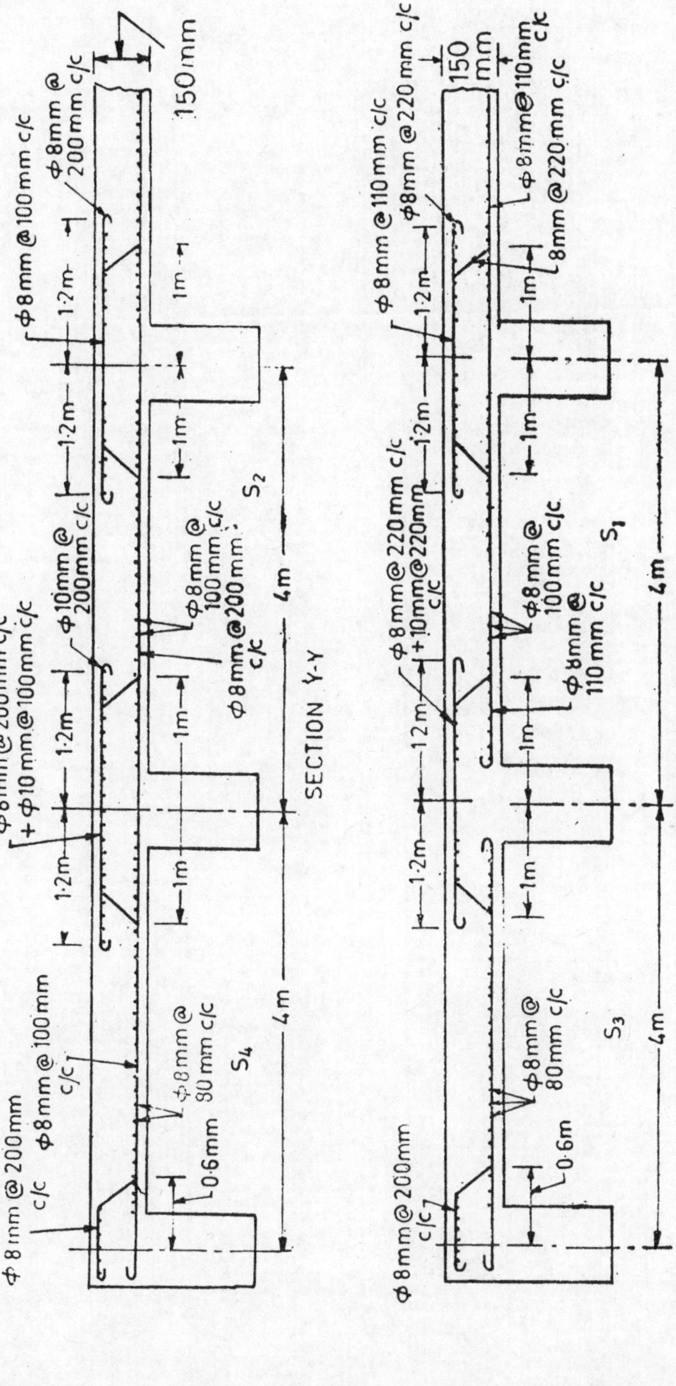

Fig. A-D-2. Details of siab reinforcement.

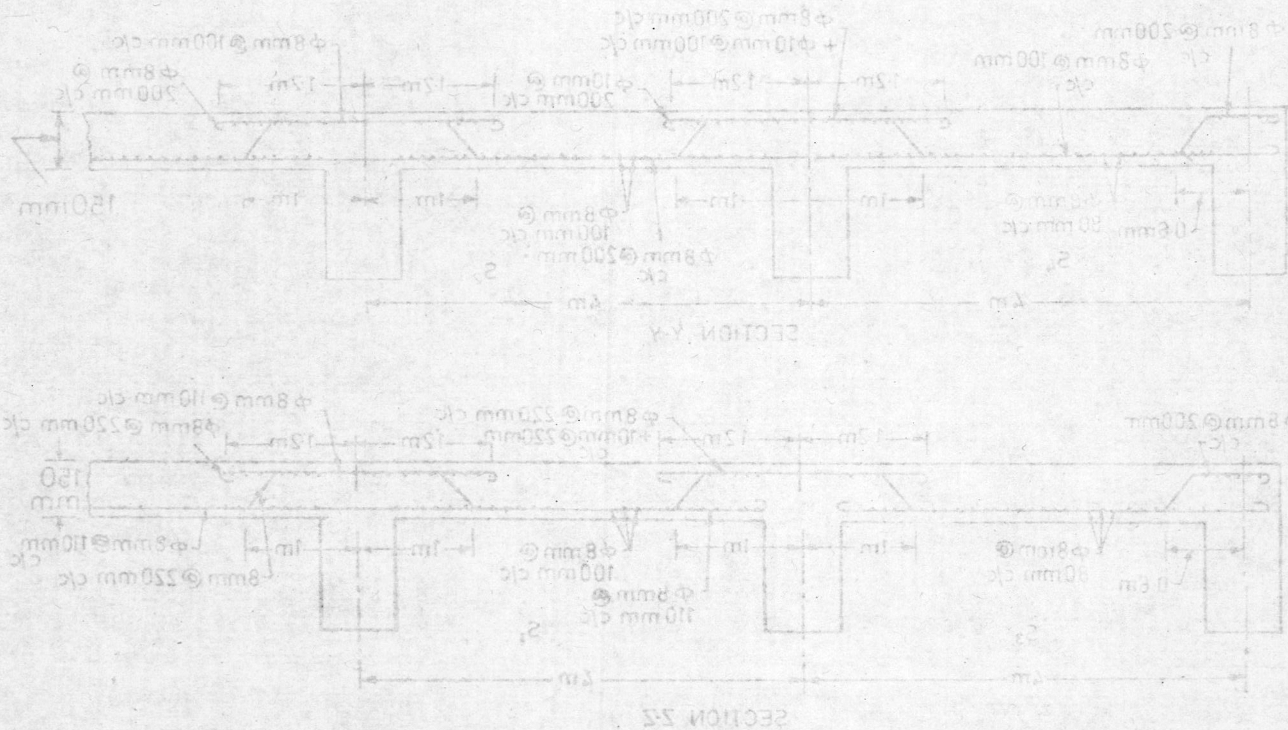

Fig. A-D-2. Details of slab reinforcement.

Appendix A

Graph for Compression with Bending—Rectangular Section—Reinforcement Distributed Equally on Two Sides

Graphs for compression with bending are developed using equations 6·4, 6·6, 6·8 and 6·9. The graphs shown in Fig. A-1 to A-12 are obtained for rectangular sections with equal reinforcement on two sides. These charts are shown for d'/D values of 0·05, 0·10, 0·15 and 0·2, and f_y values of 250 N/mm^2, 415 N/mm^2 and 500 N/mm^2.

The dotted lines in these charts show the stress in the steel reinforcement close to the tension face of the member. The $f_{st}=0$ line represents the case when the neutral axis lies along the outermost row of reinforcement. The function of the graph above line $f_{st}=0$ represents the cases when all the reinforcement is in compression.

The line $f_{st}=f_y$ in these charts represents the case when the outermost tension reinforcement reaches yield stress. For the points below this line, the outermost tension reinforcement undergoes inelastic deformation while inner rows may reach yield stress.

GRAPH-A-1

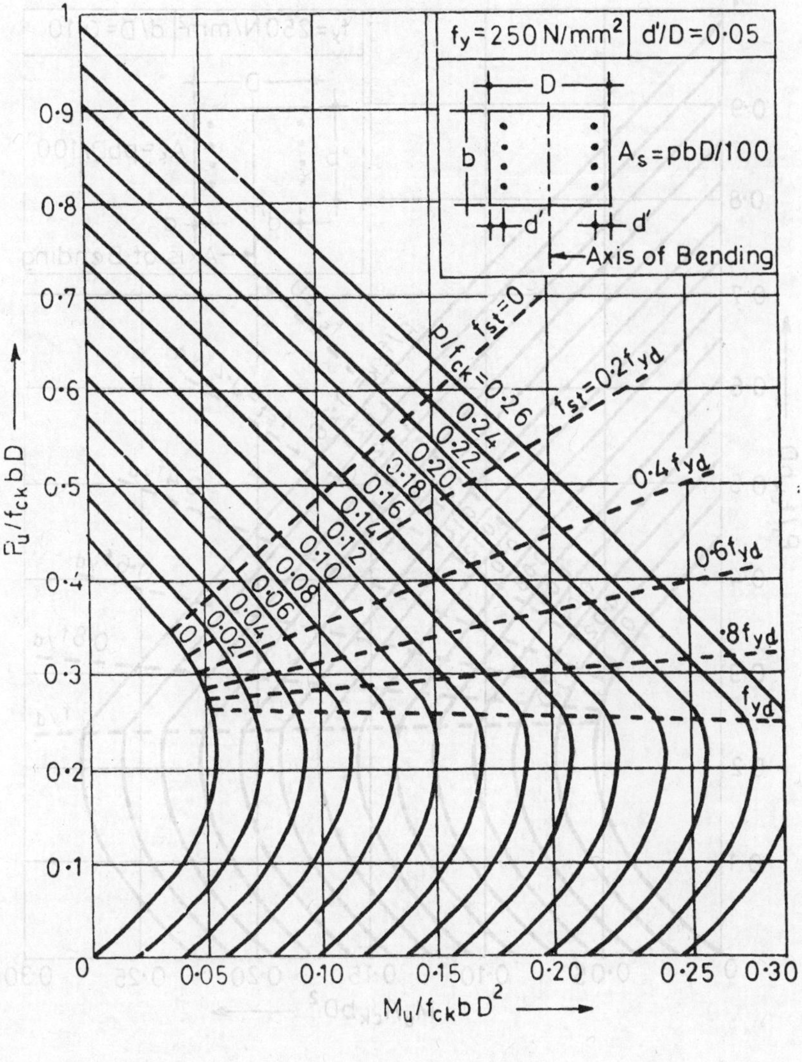

$f_y = 250\ N/mm^2$
$d'/D = 0{\cdot}05$
D
b
d'
$A_s = pbD/100$
Axis of Bending
$f_{st} = 0$
$f_{st} = 0{\cdot}2 f_{yd}$
$0{\cdot}4 f_{yd}$
$0{\cdot}6 f_{yd}$
$\cdot 8 f_{yd}$
f_{yd}
$p/f_{ck} = 0{\cdot}26$
0·24
0·22
0·20
0·18
0·16
0·14
0·12
0·10
0·08
0·06
0·04
0·02
0
$P_u / f_{ck} b D$
$M_u / f_{ck} b D^2$
0
0·05
0·10
0·15
0·20
0·25
0·30
0·1
0·2
0·3
0·4
0·5
0·6
0·7
0·8
0·9
1

GRAPH-A-2

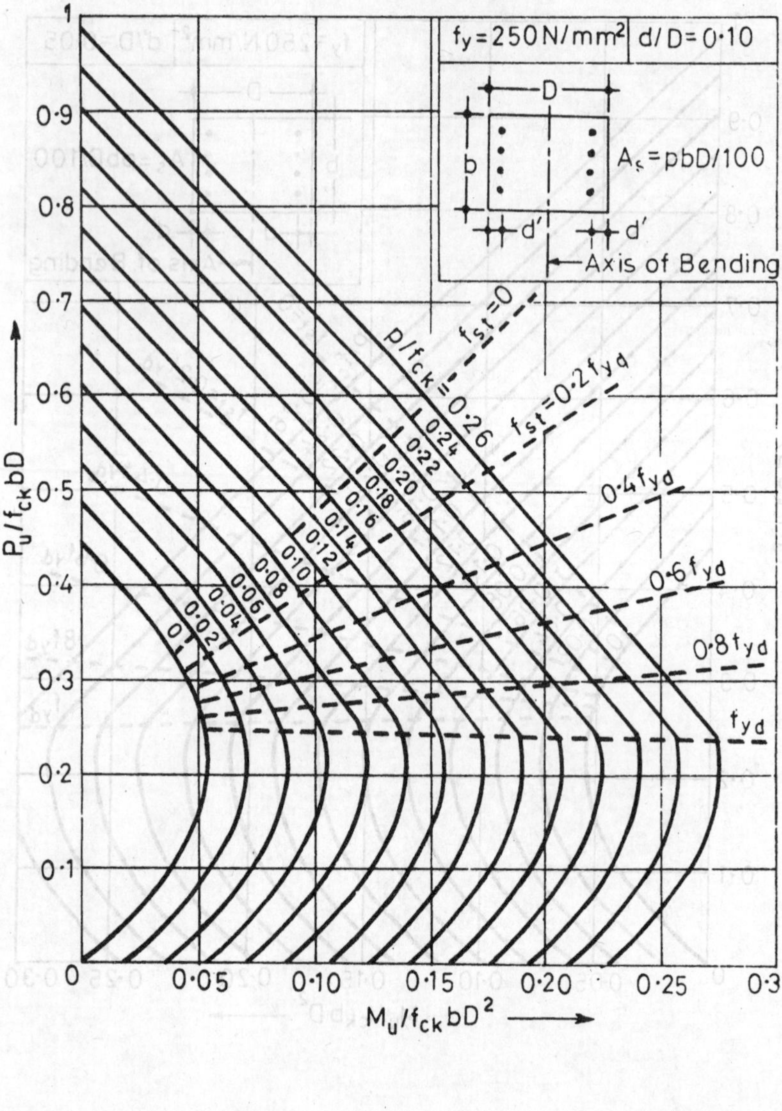

$f_y = 250\ N/mm^2$
$d'/D = 0.10$
D
b
$A_s = pbD/100$
d'
d'
Axis of Bending
$f_{st} = 0$
$f_{st} = 0.2 f_{yd}$
$0.4 f_{yd}$
$0.6 f_{yd}$
$0.8 f_{yd}$
f_{yd}
$p/f_{ck} = 0.26$
0.24
0.22
0.20
0.18
0.16
0.14
0.12
0.10
0.08
0.06
0.04
0.02
0
$P_u/f_{ck}bD$
$M_u/f_{ck}bD^2$
0
0·05
0·10
0·15
0·20
0·25
0·30
0·1
0·2
0·3
0·4
0·5
0·6
0·7
0·8
0·9
1

GRAPH-A-3

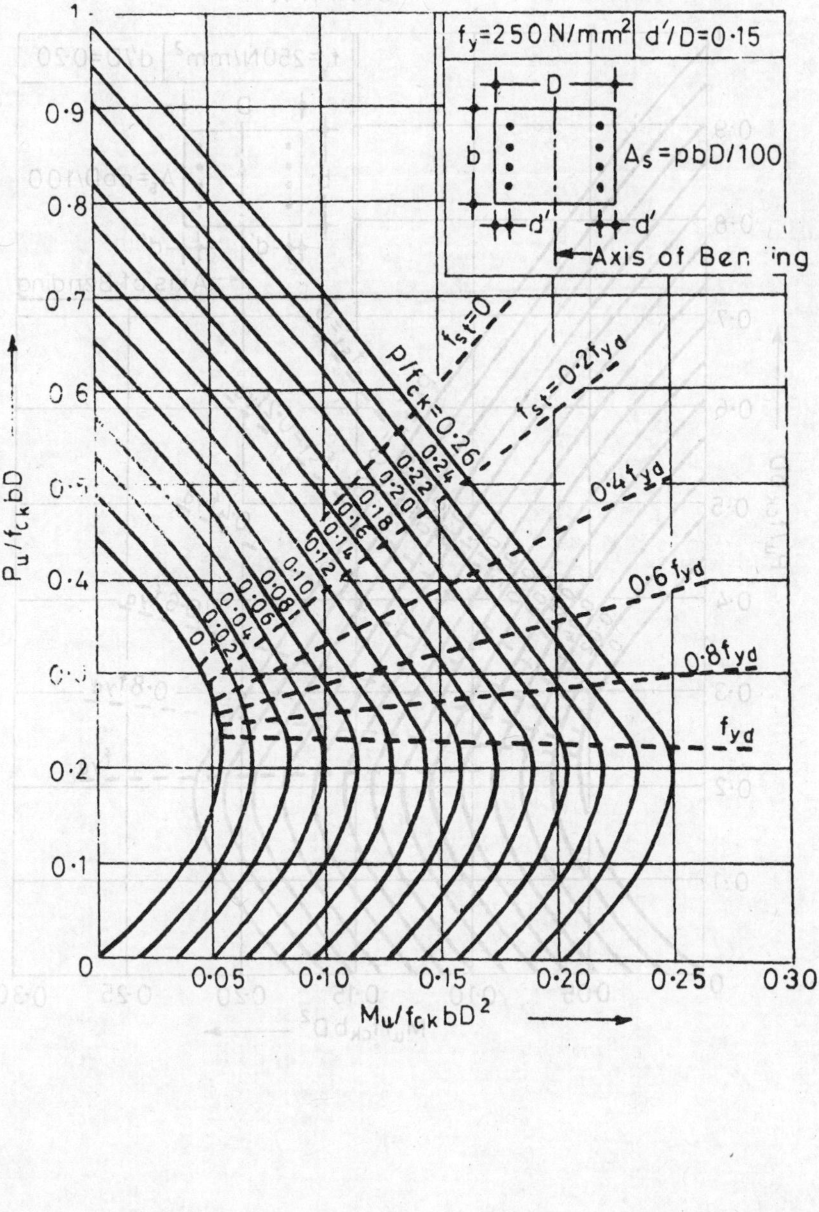

$f_y = 250\ N/mm^2$
$d'/D = 0.15$
D
b
d'
d'
$A_s = pbD/100$
Axis of Bending
$f_{st} = 0$
$f_{st} = 0.2 f_{yd}$
$0.4 f_{yd}$
$0.6 f_{yd}$
$0.8 f_{yd}$
f_{yd}
$p/f_{ck} = 0.26$
0.24
0.22
0.20
0.18
0.16
0.14
0.12
0.10
0.08
0.06
0.04
0.02
0
$P_u/f_{ck}bD$
$M_u/f_{ck}bD^2$
0
0.1
0.2
0.3
0.4
0.5
0.6
0.7
0.8
0.9
1
0.05
0.10
0.15
0.20
0.25
0.30

GRAPH-A-4

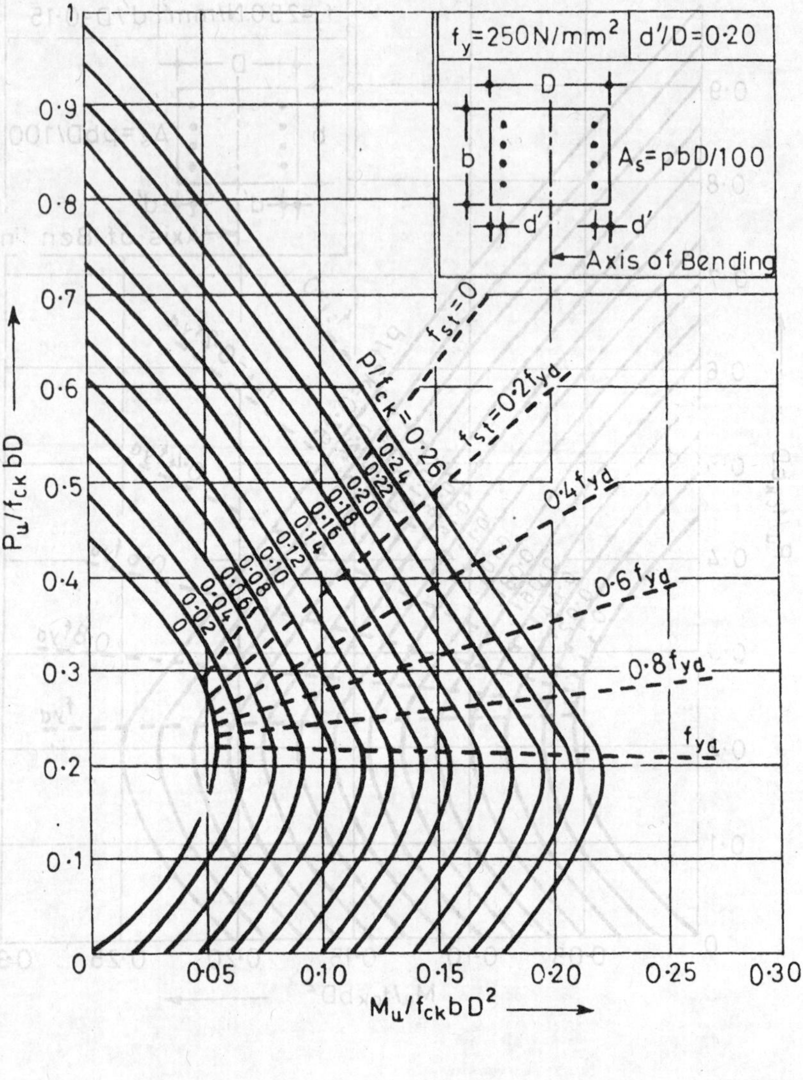

$f_y = 250 N/mm^2$
$d'/D = 0.20$
D
b
$A_s = pbD/100$
d'
d'
Axis of Bending
$f_{st} = 0$
$f_{st} = 0.2 f_{yd}$
$0.4 f_{yd}$
$0.6 f_{yd}$
$0.8 f_{yd}$
f_{yd}
$p/f_{ck} = 0.26$
0.24
0.22
0.20
0.18
0.16
0.14
0.12
0.10
0.08
0.06
0.04
0.02
0
$P_u/f_{ck}bD$
$M_u/f_{ck}bD^2$
0
0.05
0.10
0.15
0.20
0.25
0.30
0.1
0.2
0.3
0.4
0.5
0.6
0.7
0.8
0.9
1

GRAPH-A-5

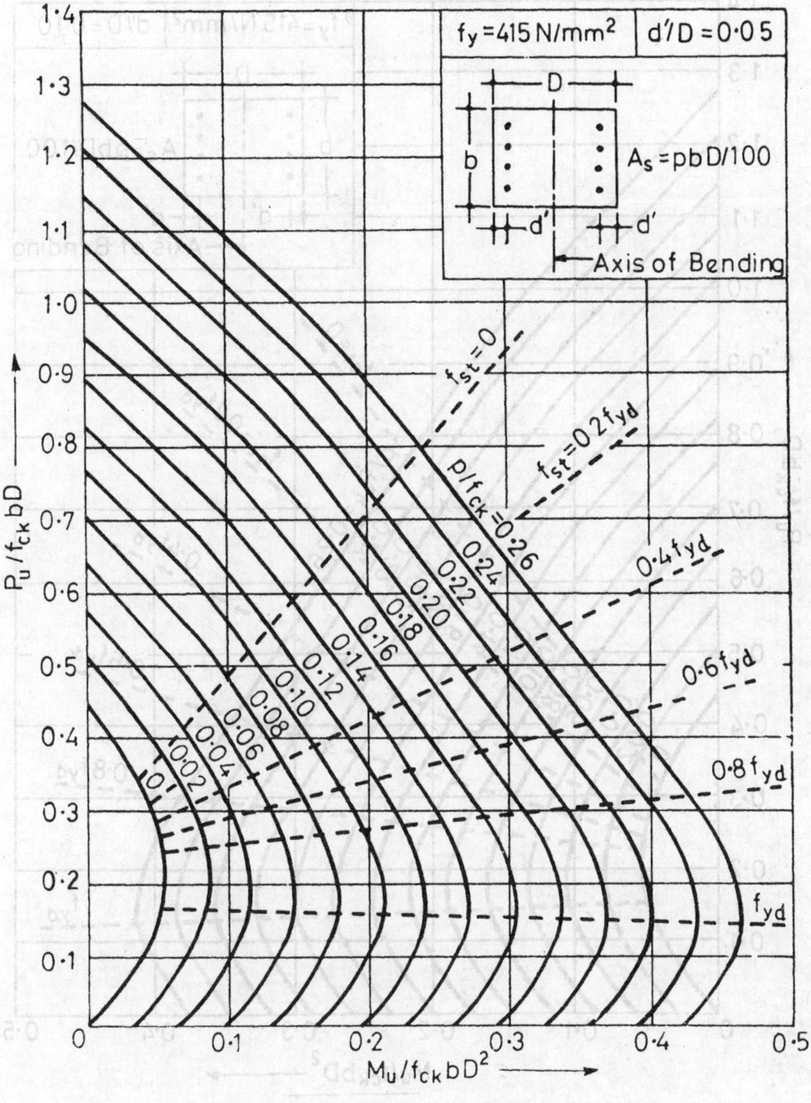

$f_y = 415\ N/mm^2$
$d'/D = 0{\cdot}05$
D
b
d'
d'
$A_s = pbD/100$
Axis of Bending
$f_{st} = 0$
$f_{st} = 0{\cdot}2 f_{yd}$
$0{\cdot}4 f_{yd}$
$0{\cdot}6 f_{yd}$
$0{\cdot}8 f_{yd}$
f_{yd}
$p/f_{ck} = 0{\cdot}26$
0·24
0·22
0·20
0·18
0·16
0·14
0·12
0·10
0·08
0·06
0·04
0·02
0
$P_u/f_{ck}bD$
$M_u/f_{ck}bD^2$
0
0·1
0·2
0·3
0·4
0·5
0·6
0·7
0·8
0·9
1·0
1·1
1·2
1·3
1·4
0·5

GRAPH-A-6

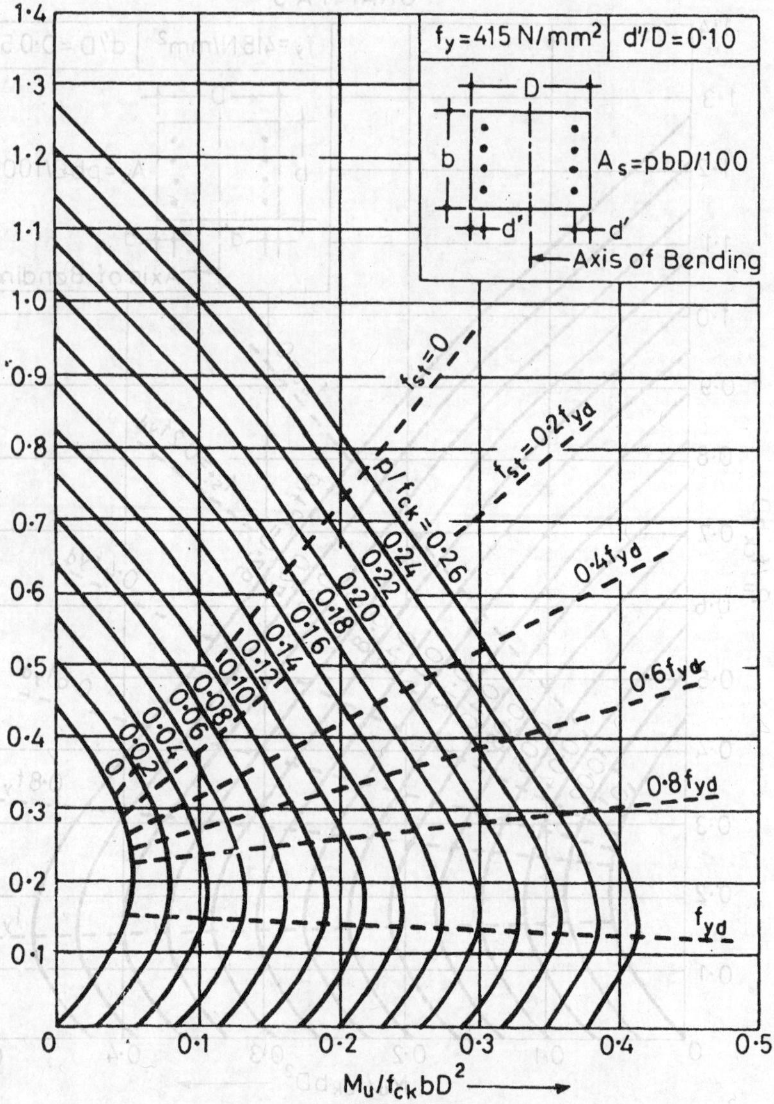

$f_y = 415\ N/mm^2$
$d'/D = 0{\cdot}10$
D
b
$A_s = pbD/100$
d'
d'
Axis of Bending
$f_{st} = 0$
$f_{st} = 0{\cdot}2 f_{yd}$
$0{\cdot}4 f_{yd}$
$0{\cdot}6 f_{yd}$
$0{\cdot}8 f_{yd}$
f_{yd}
$p/f_{ck} = 0{\cdot}26$
0·24
0·22
0·20
0·18
0·16
0·14
0·12
0·10
0·08
0·06
0·04
0·02
0
$P_u/f_{ck}bD$
$M_u/f_{ck}bD^2$
0
0·1
0·2
0·3
0·4
0·5
0·6
0·7
0·8
0·9
1·0
1·1
1·2
1·3
1·4
0·5

GRAPH-A-7

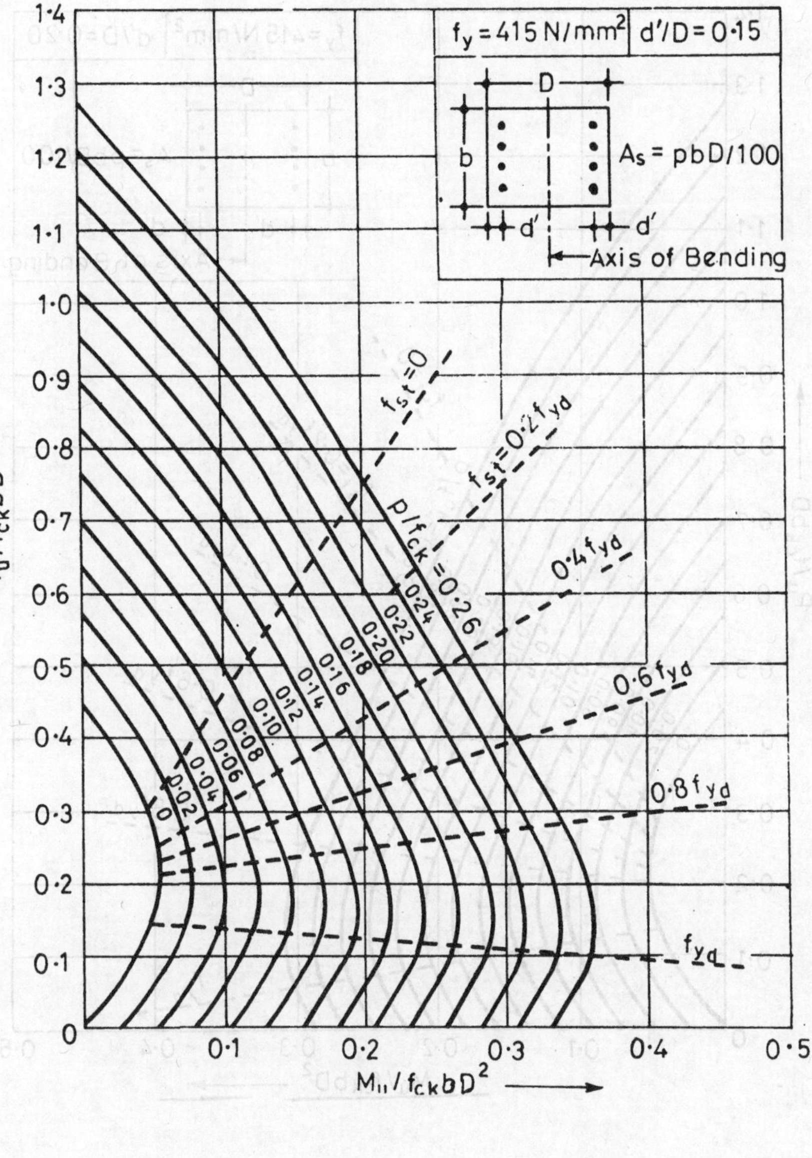

$f_y = 415\ N/mm^2$
$d'/D = 0.15$
D
b
d'
$A_s = pbD/100$
Axis of Bending
$f_{st} = 0$
$f_{st} = 0.2 f_{yd}$
$0.4 f_{yd}$
$0.6 f_{yd}$
$0.8 f_{yd}$
f_{yd}
$p/f_{ck} = 0.26$
0.24
0.22
0.20
0.18
0.16
0.14
0.12
0.10
0.08
0.06
0.04
0.02
0
$P_u/f_{ck}bD$
$M_u/f_{ck}bD^2$
0
0.1
0.2
0.3
0.4
0.5
1.4
1.3
1.2
1.1
1.0
0.9
0.8
0.7
0.6
0.5
0.4
0.3
0.2
0.1

GRAPH-A-8

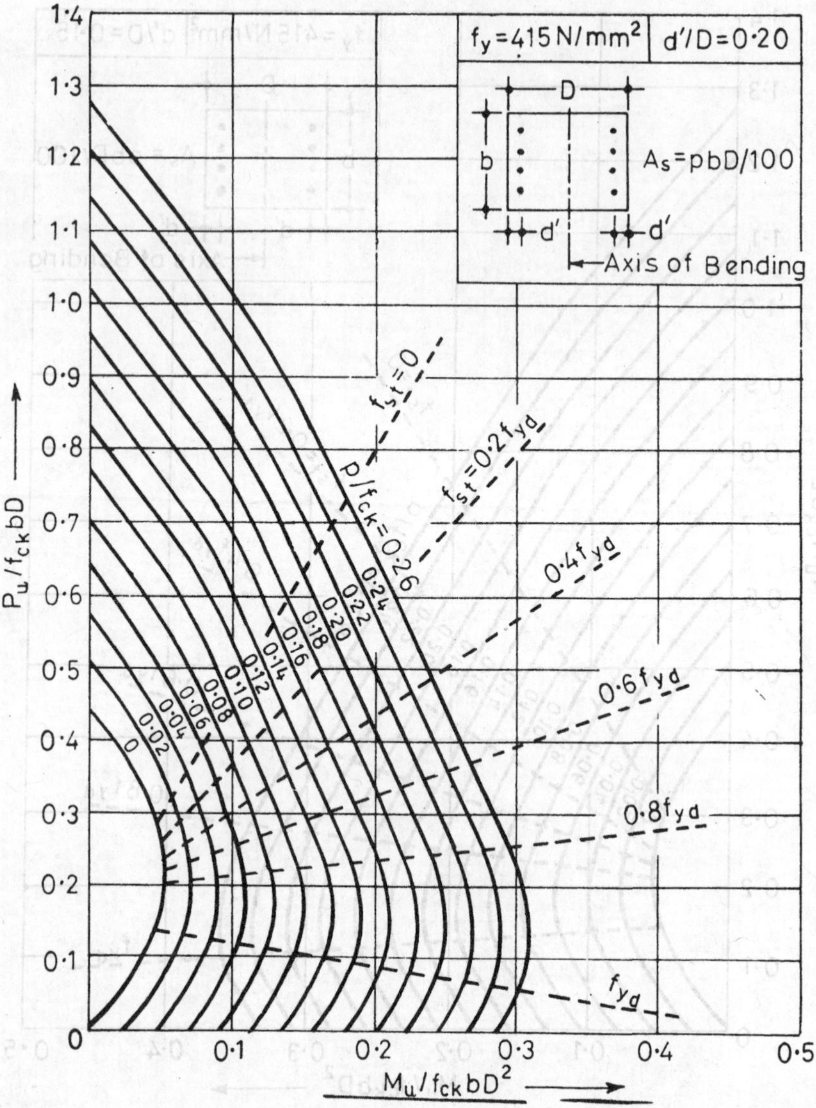

$f_y = 415\ N/mm^2$
$d'/D = 0.20$
D
b
$A_s = pbD/100$
d'
d'
Axis of Bending
$P_u/f_{ck}bD$
$M_u/f_{ck}bD^2$
$f_{st} = 0$
$f_{st} = 0.2 f_{yd}$
$0.4 f_{yd}$
$0.6 f_{yd}$
$0.8 f_{yd}$
f_{yd}
$p/f_{ck} = 0.26$
0.24
0.22
0.20
0.18
0.16
0.14
0.12
0.10
0.08
0.06
0.04
0.02
0

GRAPH-A-9

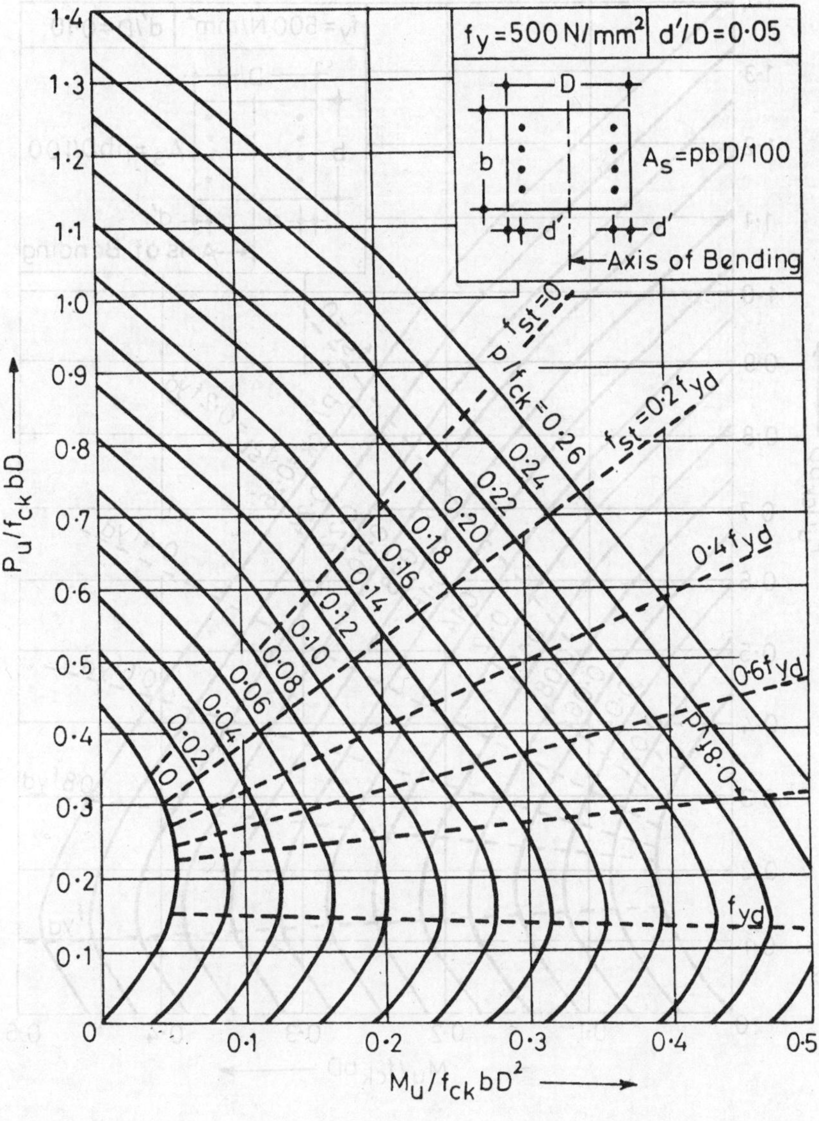

$f_y = 500\ N/mm^2$
$d'/D = 0.05$
D
b
$A_s = pbD/100$
d'
d'
Axis of Bending
$f_{st} = 0$
$p/f_{ck} = 0.26$
0.24
0.22
0.20
0.18
0.16
0.14
0.12
0.10
0.08
0.06
0.04
0.02
0
$f_{st} = 0.2 f_{yd}$
$0.4 f_{yd}$
$0.6 f_{yd}$
$0.8 f_{yd}$
f_{yd}
$P_u/f_{ck}bD$
$M_u/f_{ck}bD^2$
0
0·1
0·2
0·3
0·4
0·5
0·6
0·7
0·8
0·9
1·0
1·1
1·2
1·3
1·4

GRAPH-A-10

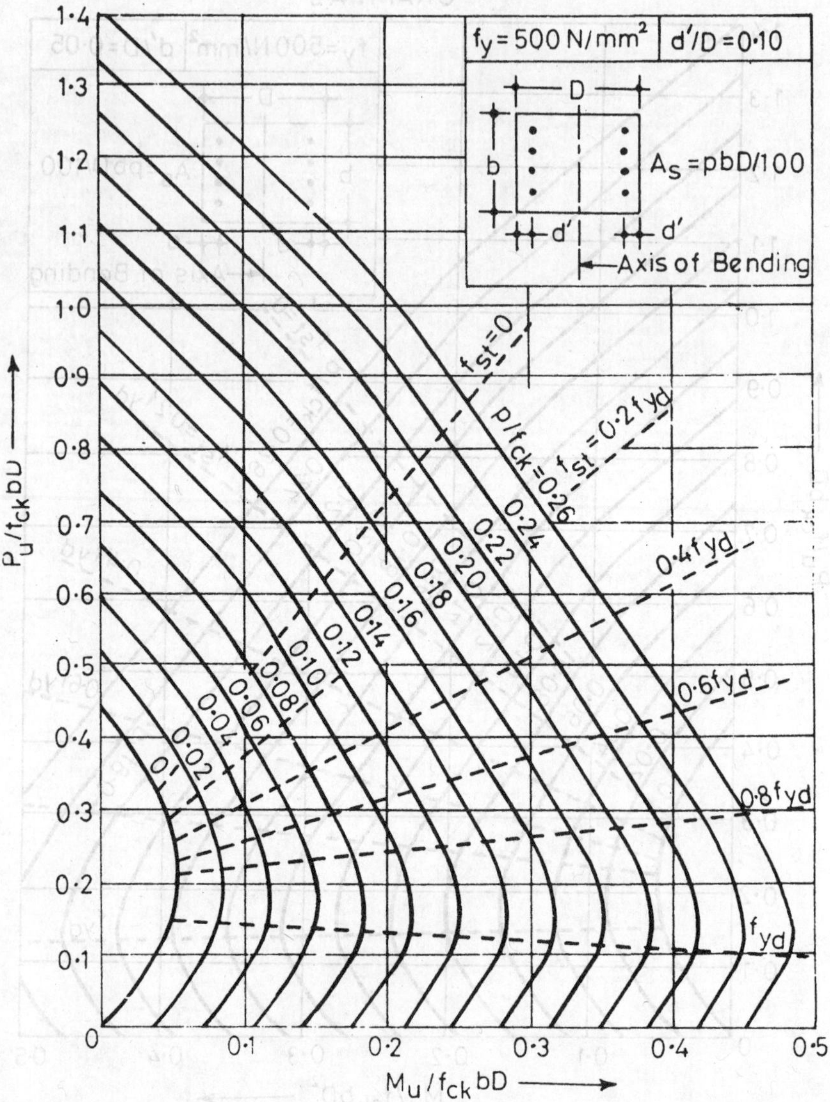

$f_y = 500\ N/mm^2$
$d'/D = 0{\cdot}10$
D
b
$A_s = pbD/100$
d'
d'
Axis of Bending
$f_{st} = 0$
$f_{st} = 0{\cdot}2 f_{yd}$
$0{\cdot}4 f_{yd}$
$0{\cdot}6 f_{yd}$
$0{\cdot}8 f_{yd}$
f_{yd}
$p/f_{ck} = 0{\cdot}26$
0·24
0·22
0·20
0·18
0·16
0·14
0·12
0·10
0·08
0·06
0·04
0·02
0
$P_u/f_{ck}bD$
$M_u/f_{ck}bD$
0
0·1
0·2
0·3
0·4
0·5
0·6
0·7
0·8
0·9
1·0
1·1
1·2
1·3
1·4

GRAPH-A-11

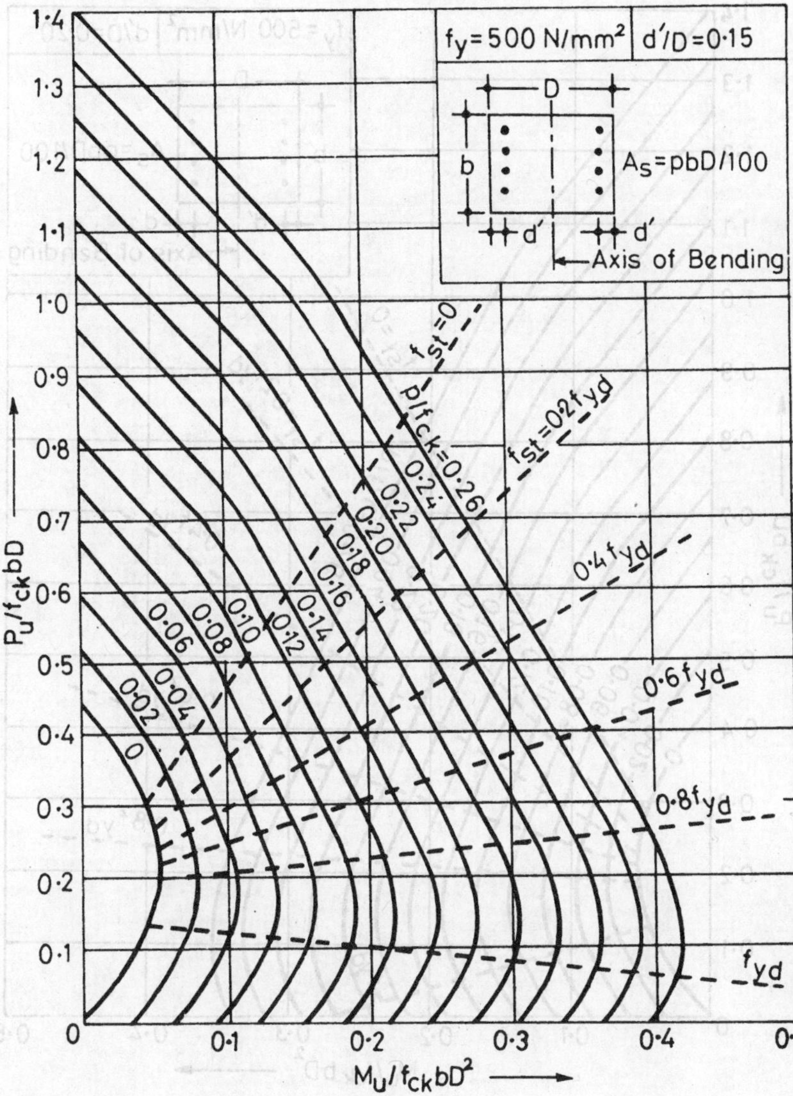

fy = 500 N/mm²
d'/D = 0·15
D
b
d'
d'
As = pbD/100
Axis of Bending
fst = 0
fst = 0·2 fyd
0·4 fyd
0·6 fyd
0·8 fyd
fyd
p/fck = 0·26
0·24
0·22
0·20
0·18
0·16
0·14
0·12
0·10
0·08
0·06
0·04
0·02
0
Pu/fck bD
Mu/fck bD²

GRAPH-A-12

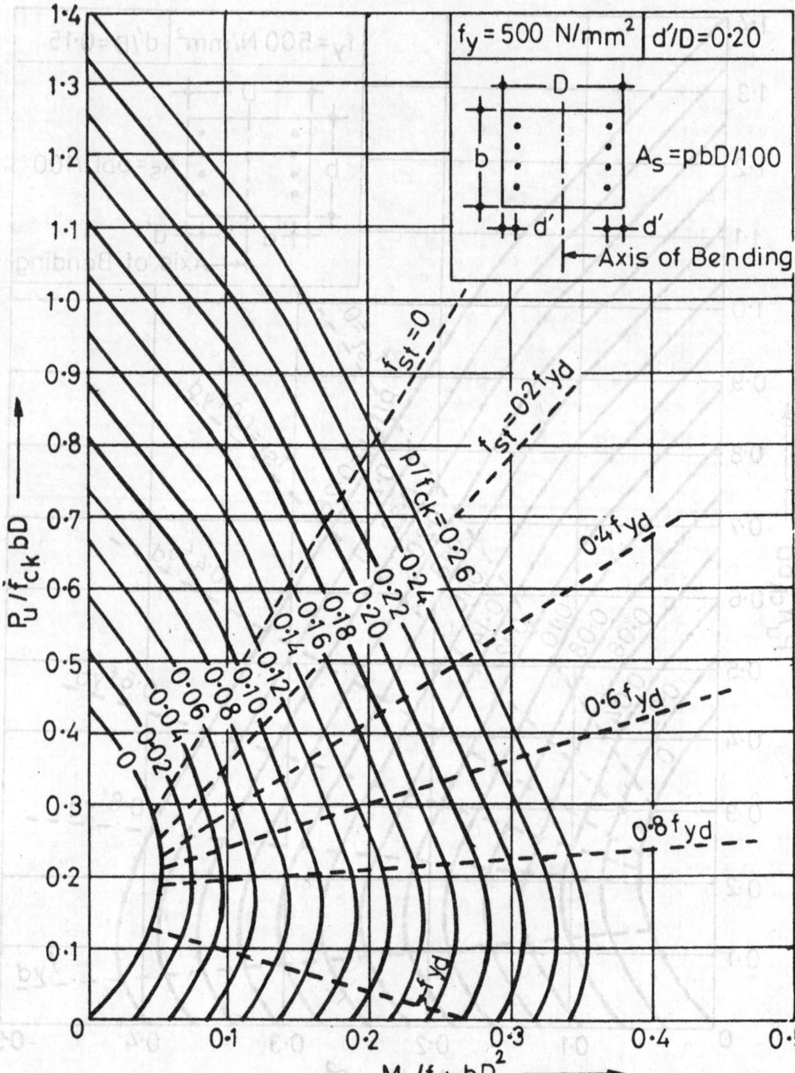

$f_y = 500$ N/mm^2
d'/D=0·20
D
b
A_s =pbD/100
d'
d'
Axis of Bending
$f_{st}=0$
$f_{st}=0·2 f_{yd}$
$0·4 f_{yd}$
$0·6 f_{yd}$
$0·8 f_{yd}$
f_{yd}
$p/f_{ck}=0·26$
0·24
0·22
0·20
0·18
0·16
0·14
0·12
0·10
0·08
0·06
0·04
0·02
0
$P_u/f_{ck}\,bD$
$M_u/f_{ck}\,bD^2$
1·4
1·3
1·2
1·1
1·0
0·9
0·8
0·7
0·6
0·5
0·4
0·3
0·2
0·1
0
0·1
0·2
0·3
0·4
0·5

GRAPHS

SINGLY REINFORCED BEAMS

GRAPH 1

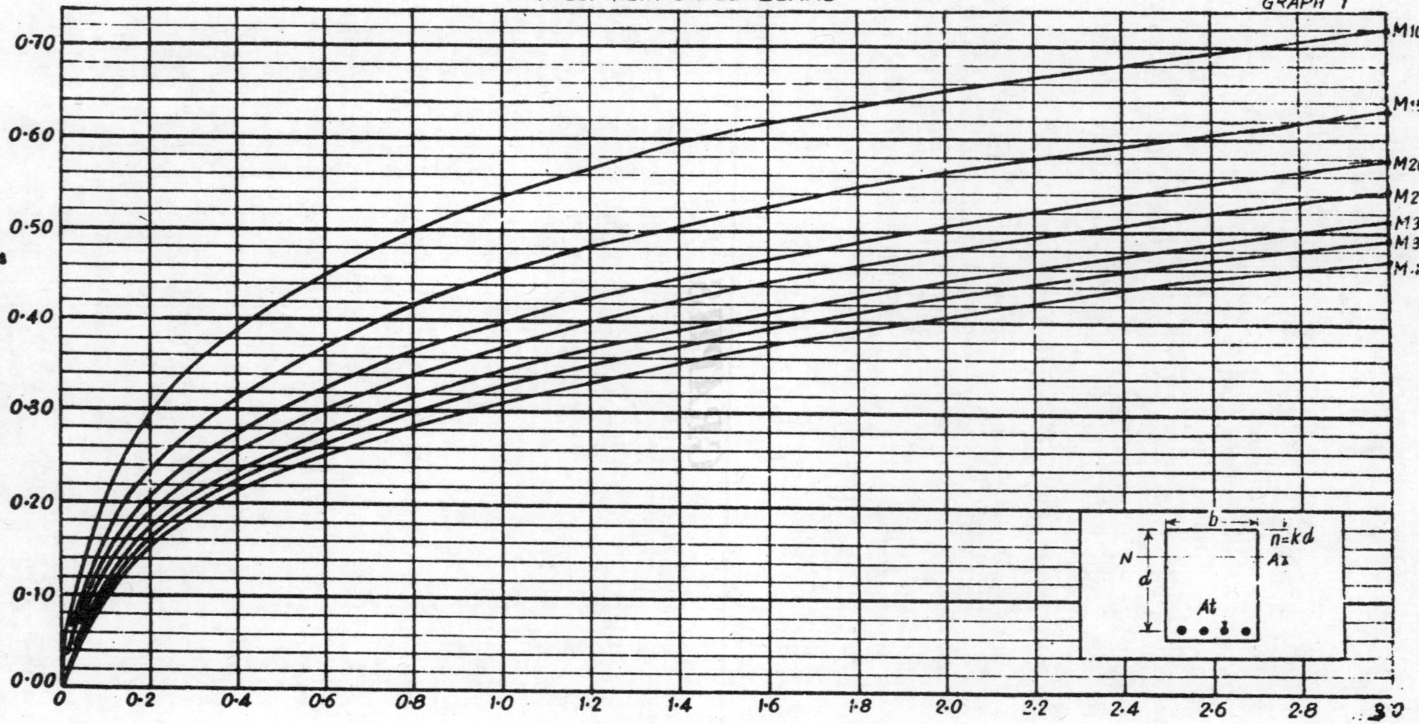

NEUTRAL AXIS CONSTANT $k = \frac{n}{d}$

PERCENTAGE OF STEEL $p = \frac{At}{bd} \times 100$

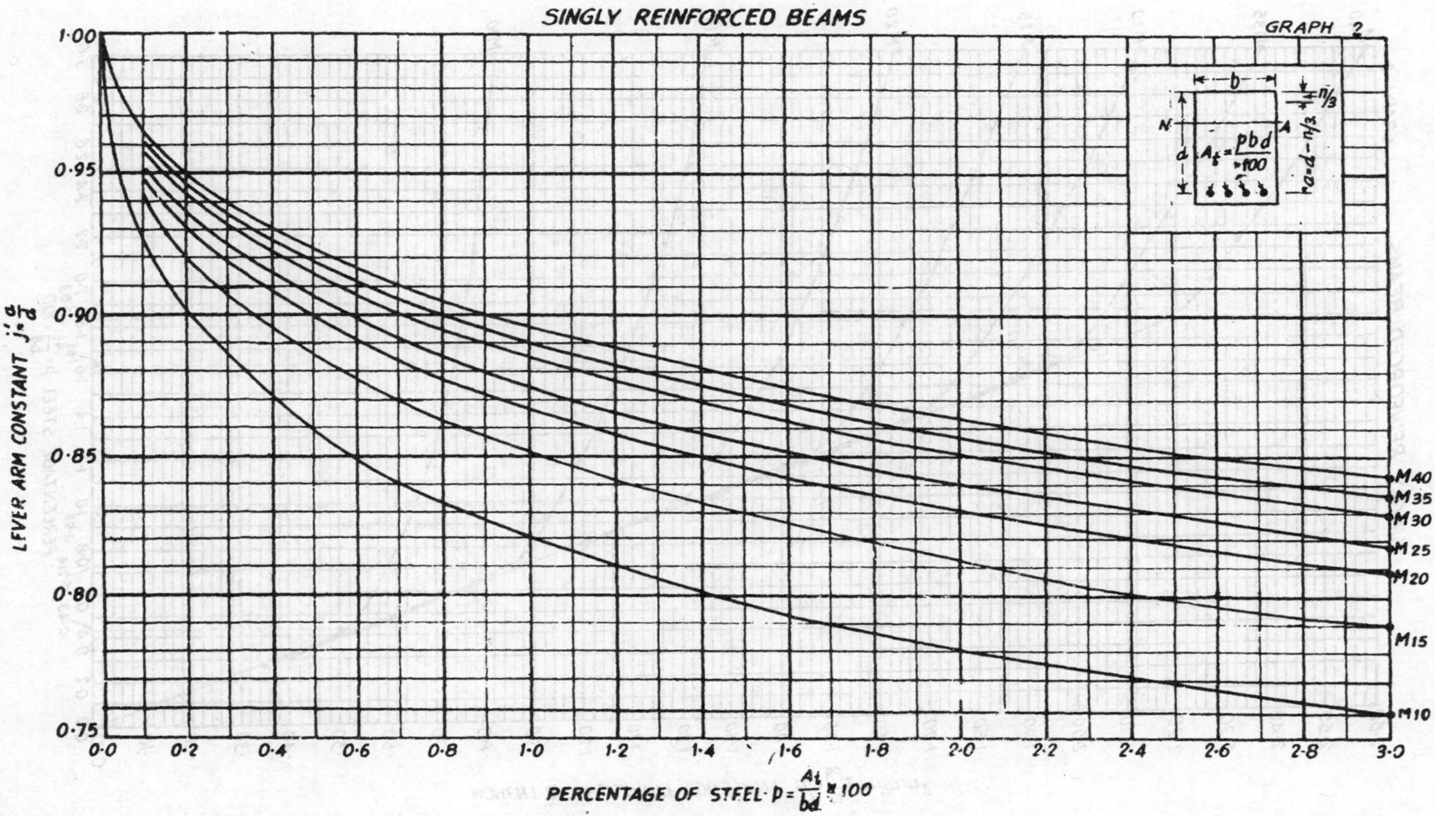

SINGLY REINFORCED BEAMS
GRAPH 2
LEVER ARM CONSTANT j = a/d
1·00
0·95
0·90
0·85
0·80
0·75
0·0
0·2
0·4
0·6
0·8
1·0
1·2
1·4
1·6
1·8
2·0
2·2
2·4
2·6
2·8
3·0
PERCENTAGE OF STEEL p = At/bd × 100
M40
M35
M30
M25
M20
M15
M10
b
N
A
d
At = pbd/100
a = d − n/3
n/3

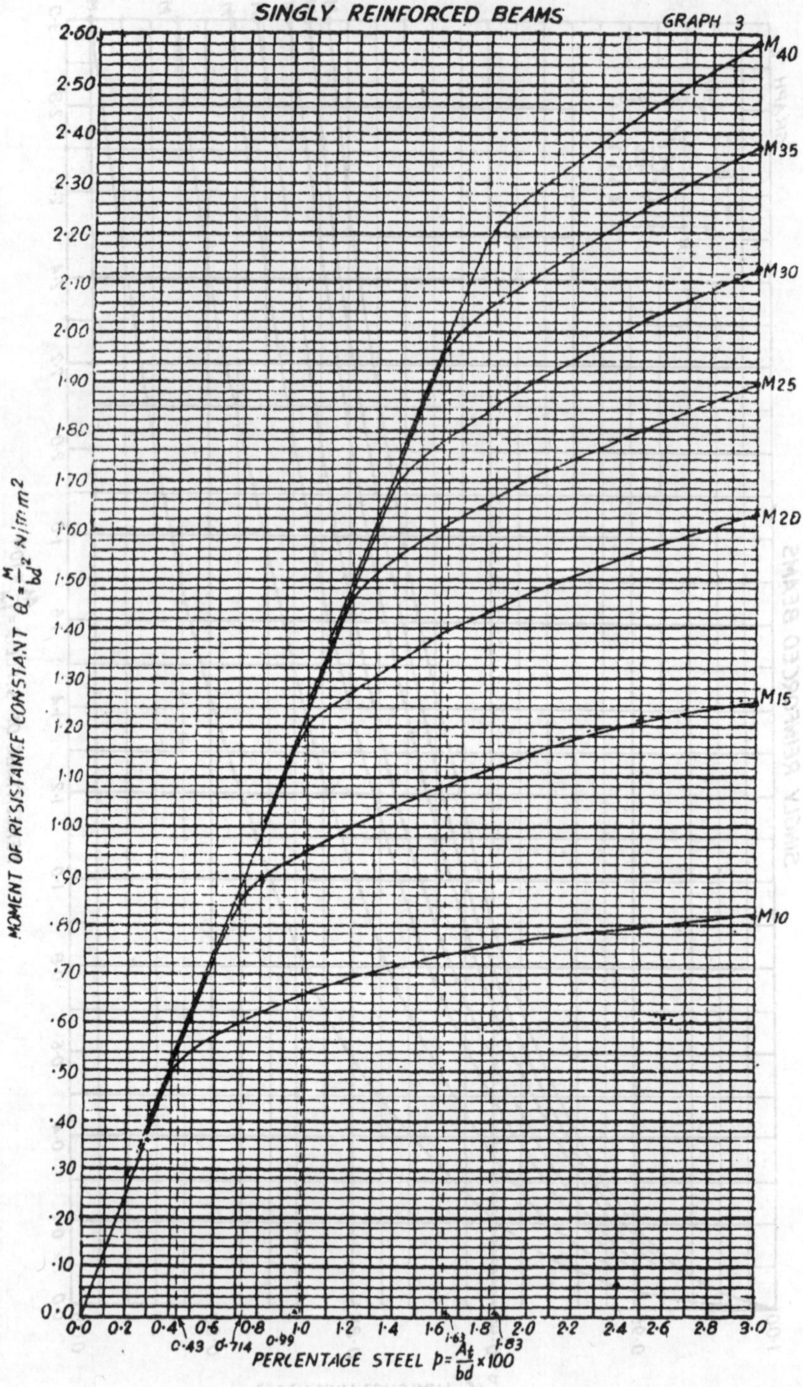
SINGLY REINFORCED BEAMS
GRAPH 3
MOMENT OF RESISTANCE CONSTANT $Q' = \frac{M}{bd^2}$ N/mm^2
PERCENTAGE STEEL $p = \frac{A_t}{bd} \times 100$
M40
M35
M30
M25
M20
M15
M10
0·43
0·714
0·99
1·63
1·83

GRAPH 4

SINGLY REINFORCED BEAMS

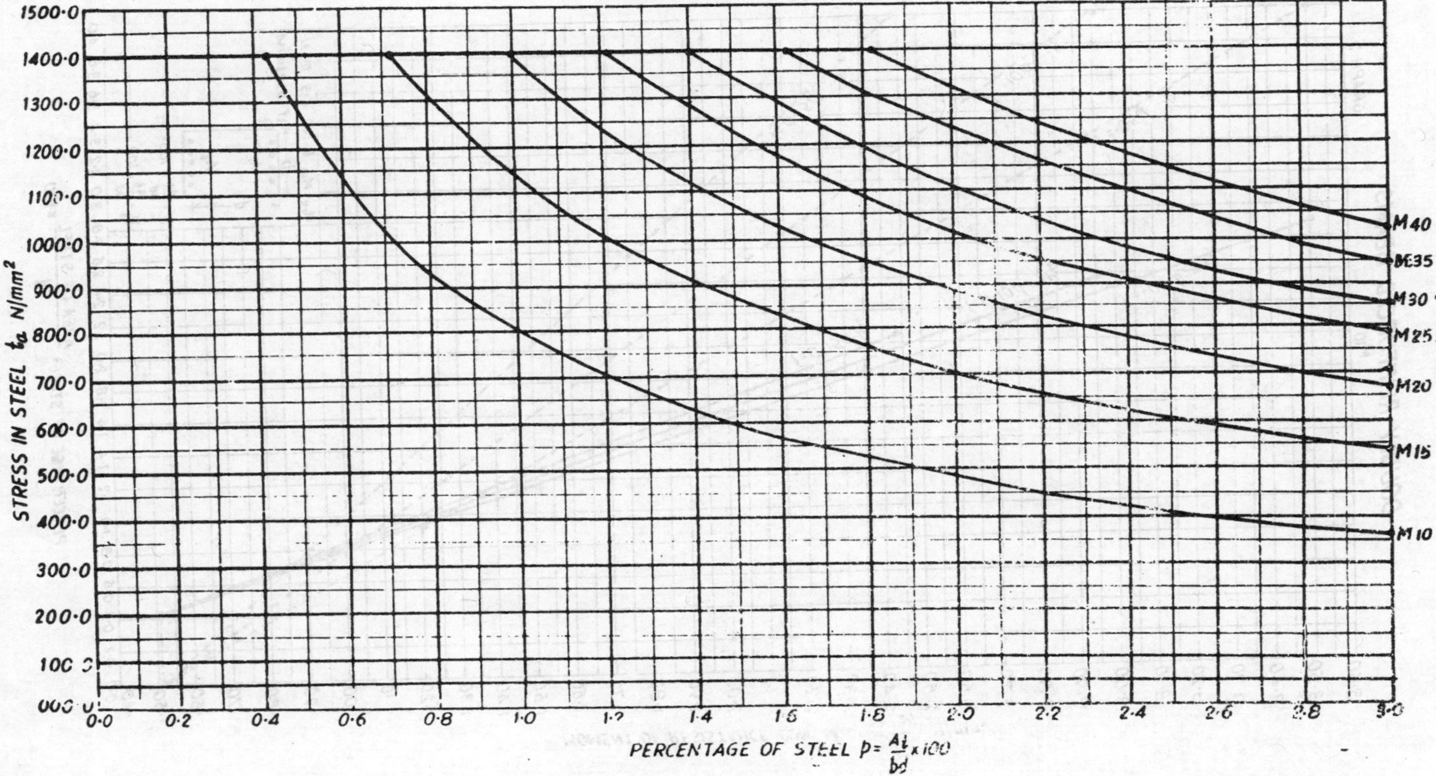

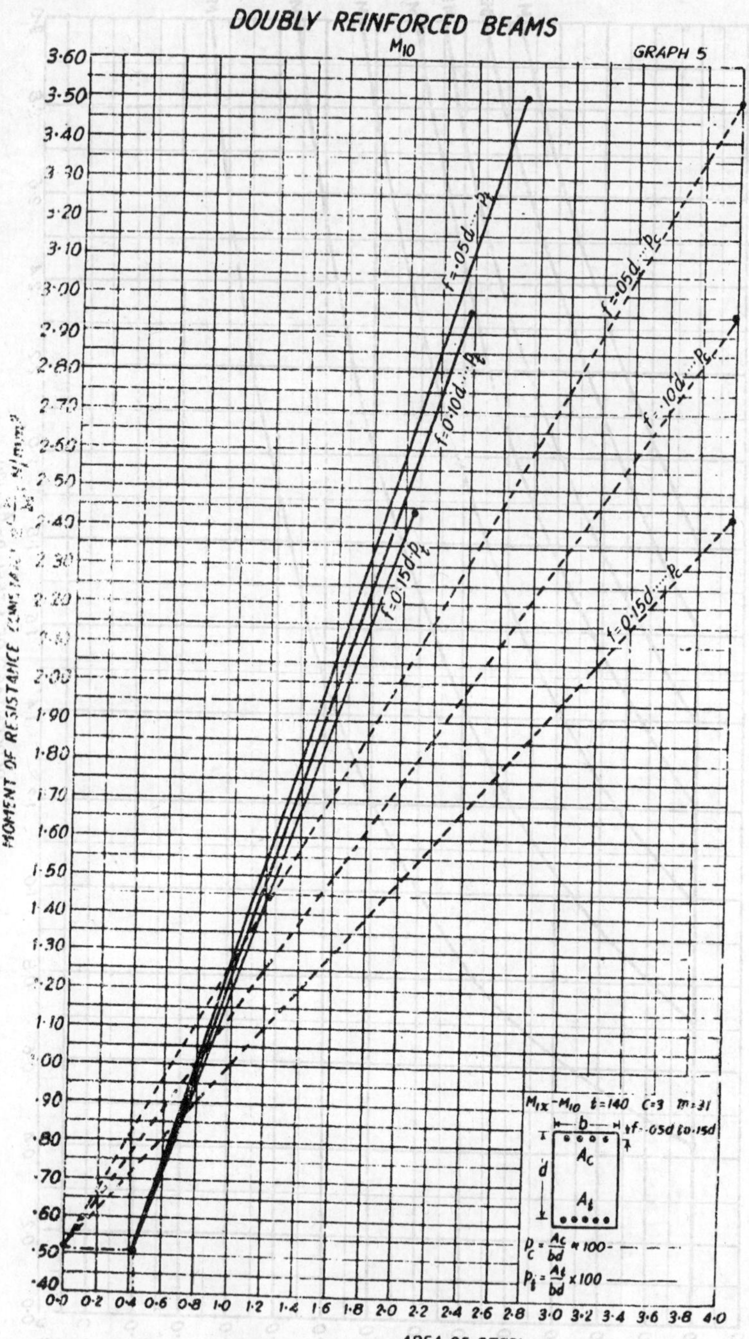
DOUBLY REINFORCED BEAMS
M10
GRAPH 5
MOMENT OF RESISTANCE CONSTANT
f = ·05d ..... Pt
f = 0·10d ..... Pt
f = 0·15d·Pt
f = ·05d ..... Pc
f = ·10d ..... Pc
f = 0·15d ..... Pc
M1x – M10 t = 140 C = 3 m = 31
b
f – ·05d to ·15d
d
Ac
At
Pc = Ac/bd × 100
Pt = At/bd × 100
PERCENTAGE OF STEEL = AREA OF STEEL / bd × 100

DOUBLY REINFORCED BEAMS

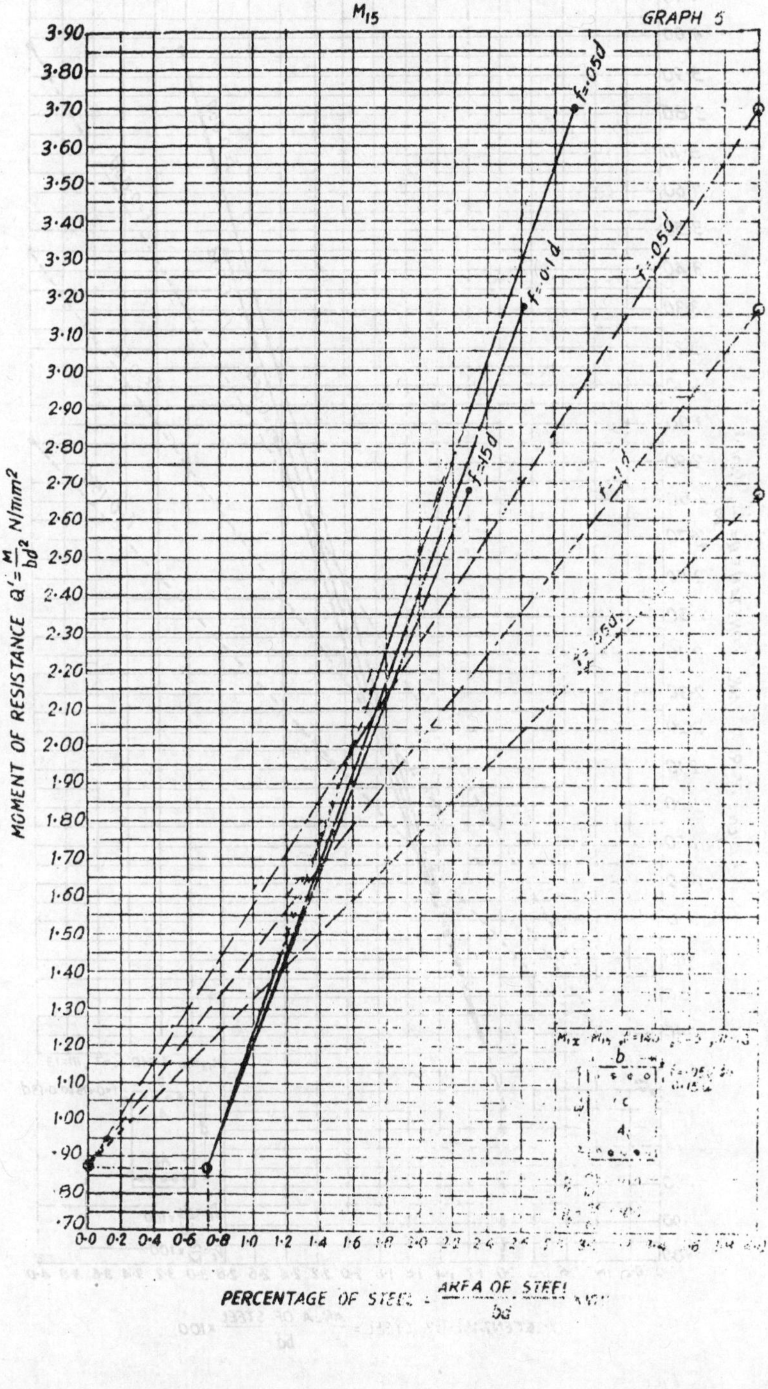

DOUBLY REINFORCED BEAMS

M_{20}

GRAPH 7

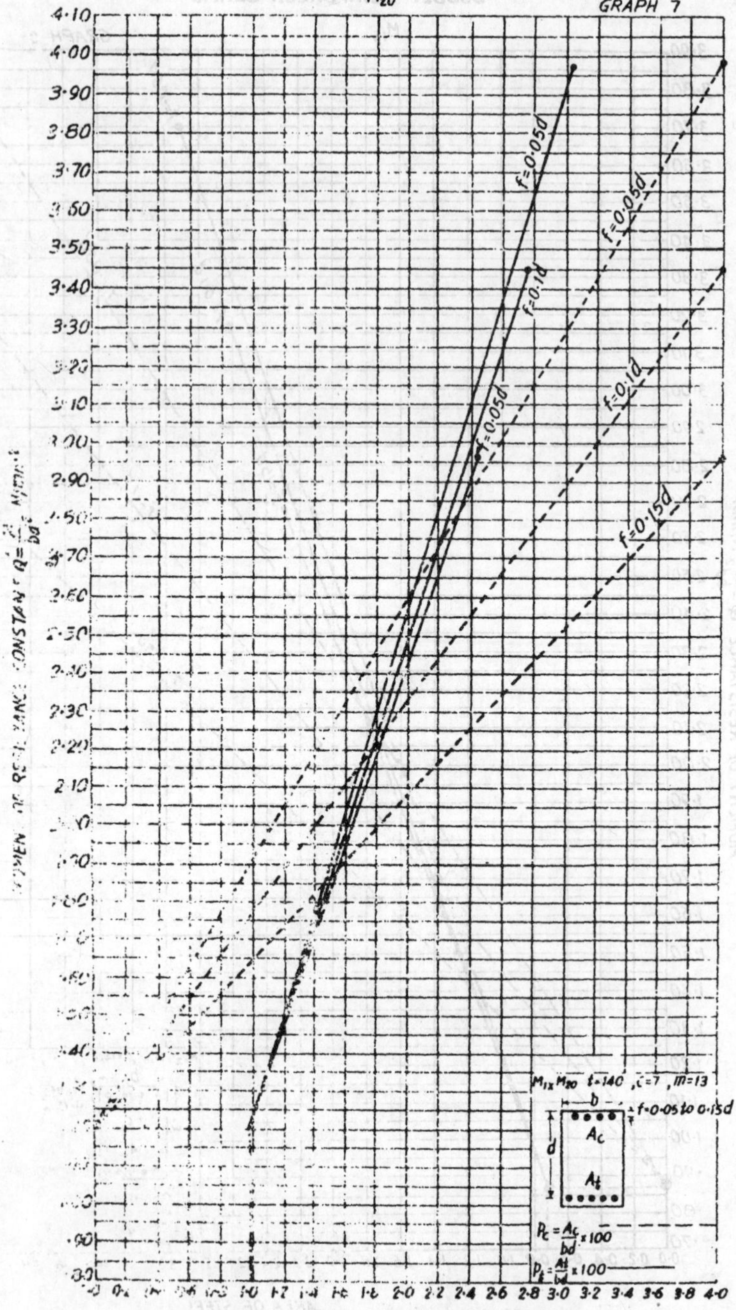

DOUBLY REINFORCED BEAMS

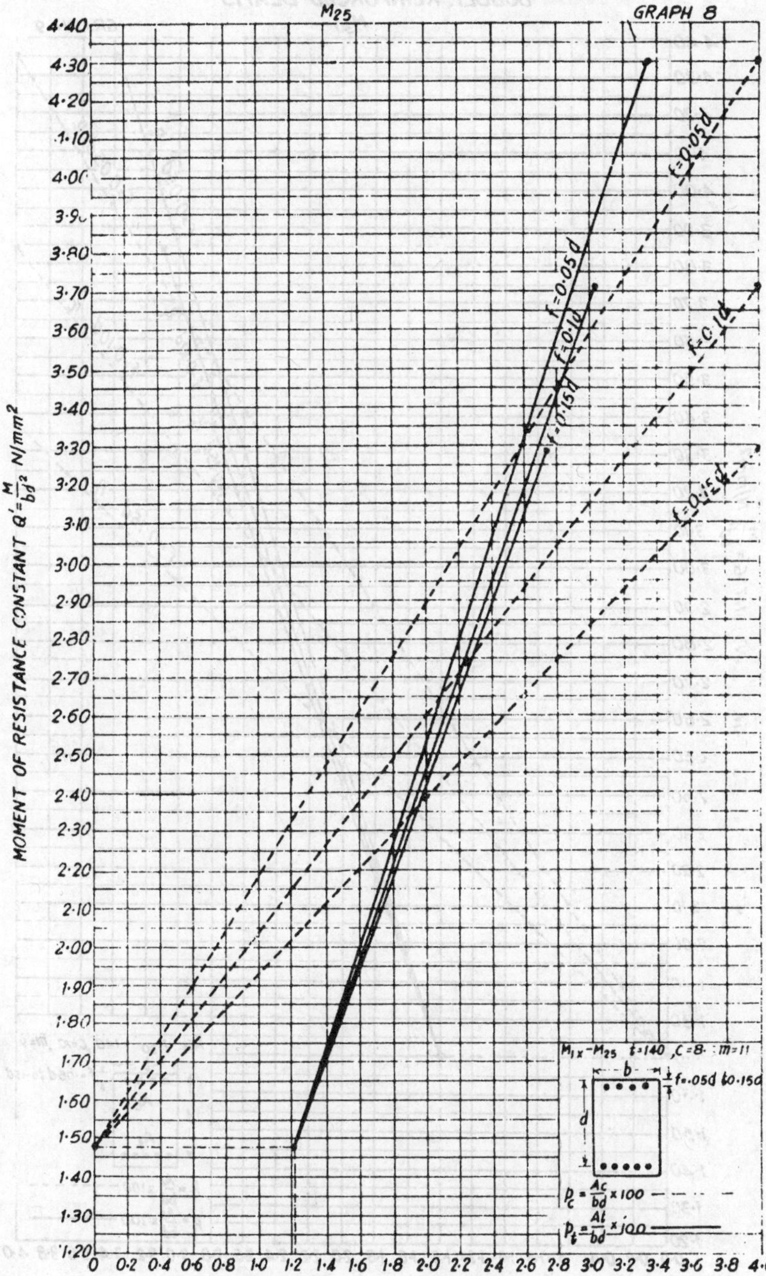

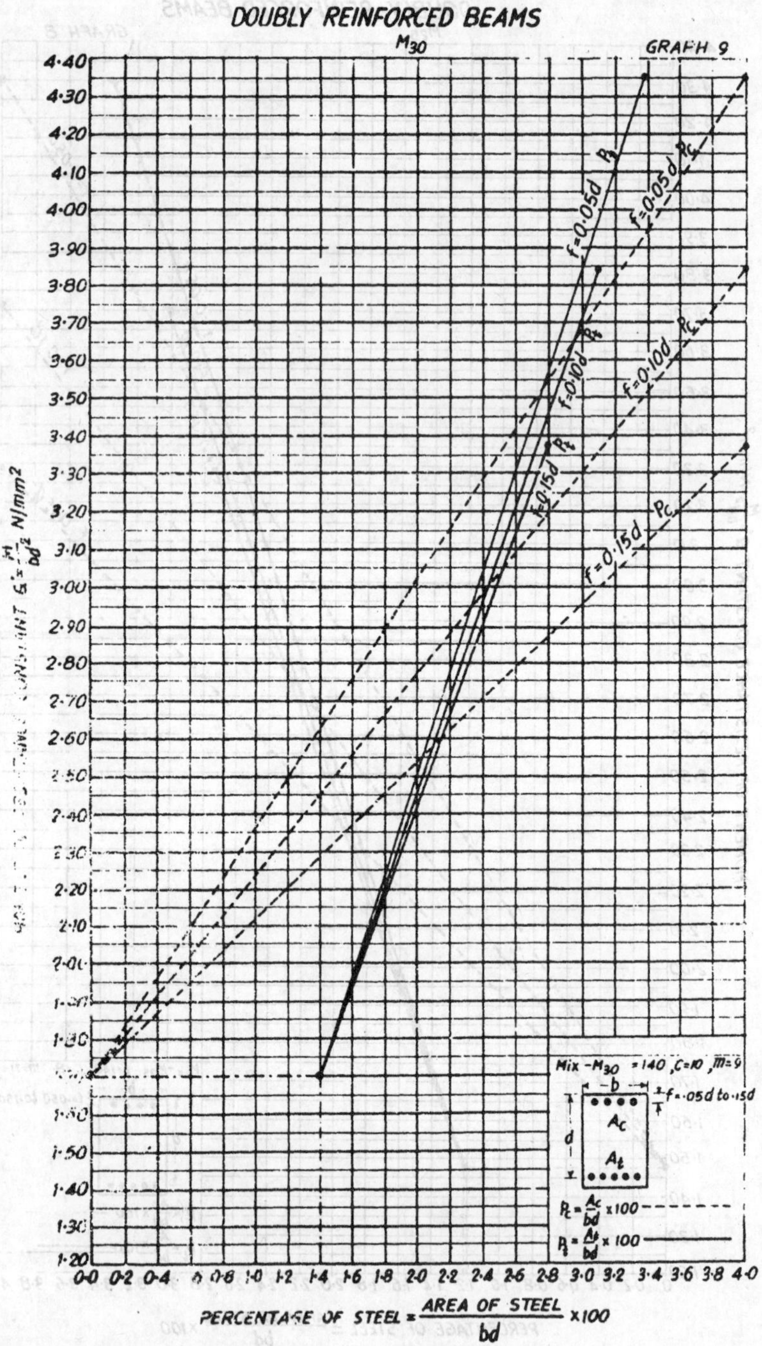

DOUBLY REINFORCED BEAMS
M_{30}
GRAPH 9
f = 0·05d
P_t
P_c
f = 0·10d
f = 0·15d
Mix – M_{30} = 140, C = 10, m = 9
b
f = ·05d to ·15d
d
A_c
A_t
$P_c = \frac{A_c}{bd} \times 100$
$P_t = \frac{A_t}{bd} \times 100$
N/mm^2
4·40
4·30
4·20
4·10
4·00
3·90
3·80
3·70
3·60
3·50
3·40
3·30
3·20
3·10
3·00
2·90
2·80
2·70
2·60
2·50
2·40
2·30
2·20
2·10
2·00
1·90
1·80
1·70
1·60
1·50
1·40
1·30
1·20
0·0
0·2
0·4
0·6
0·8
1·0
1·2
1·4
1·6
1·8
2·0
2·2
2·4
2·6
2·8
3·0
3·2
3·4
3·6
3·8
4·0
PERCENTAGE OF STEEL = $\frac{\text{AREA OF STEEL}}{bd} \times 100$

DOUBLY REINFORCED BEAMS

M_{35}

GRAPH 10

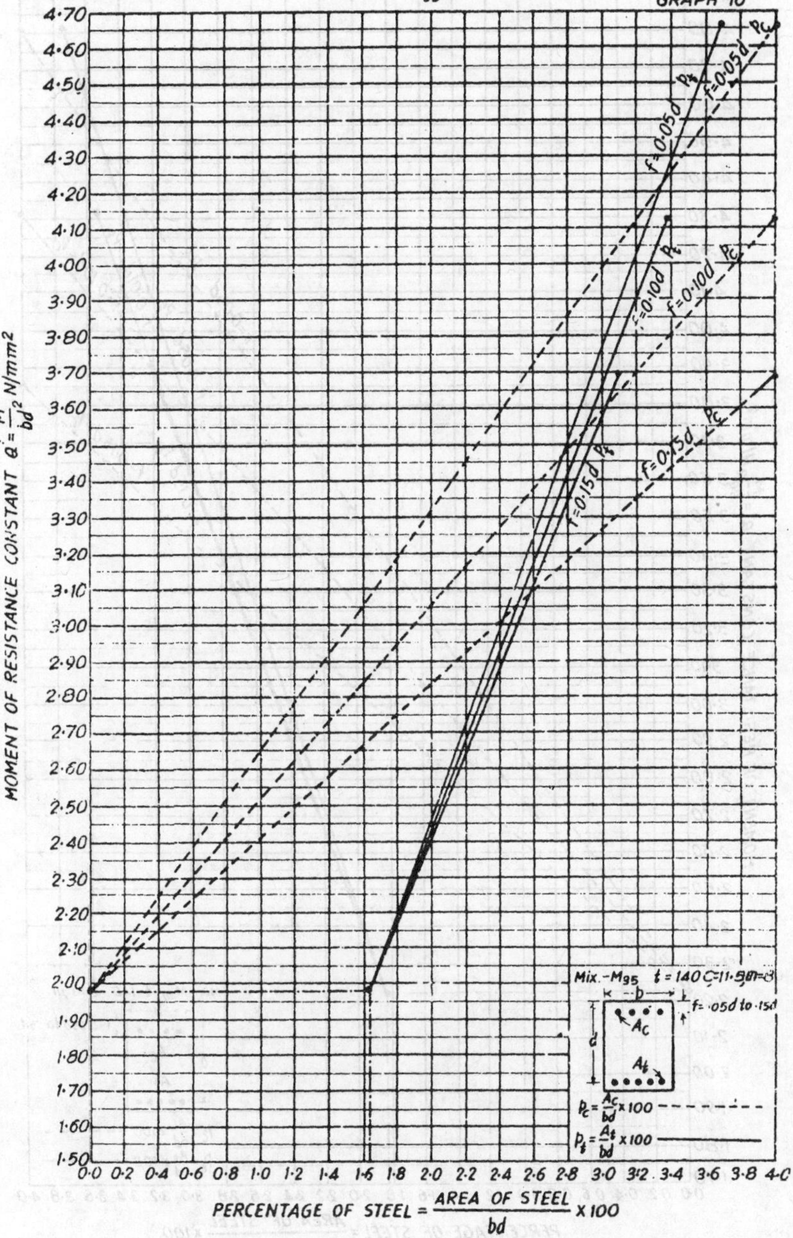

DOUBLY REINFORCED BEAMS

M_{40}

GRAPH 11

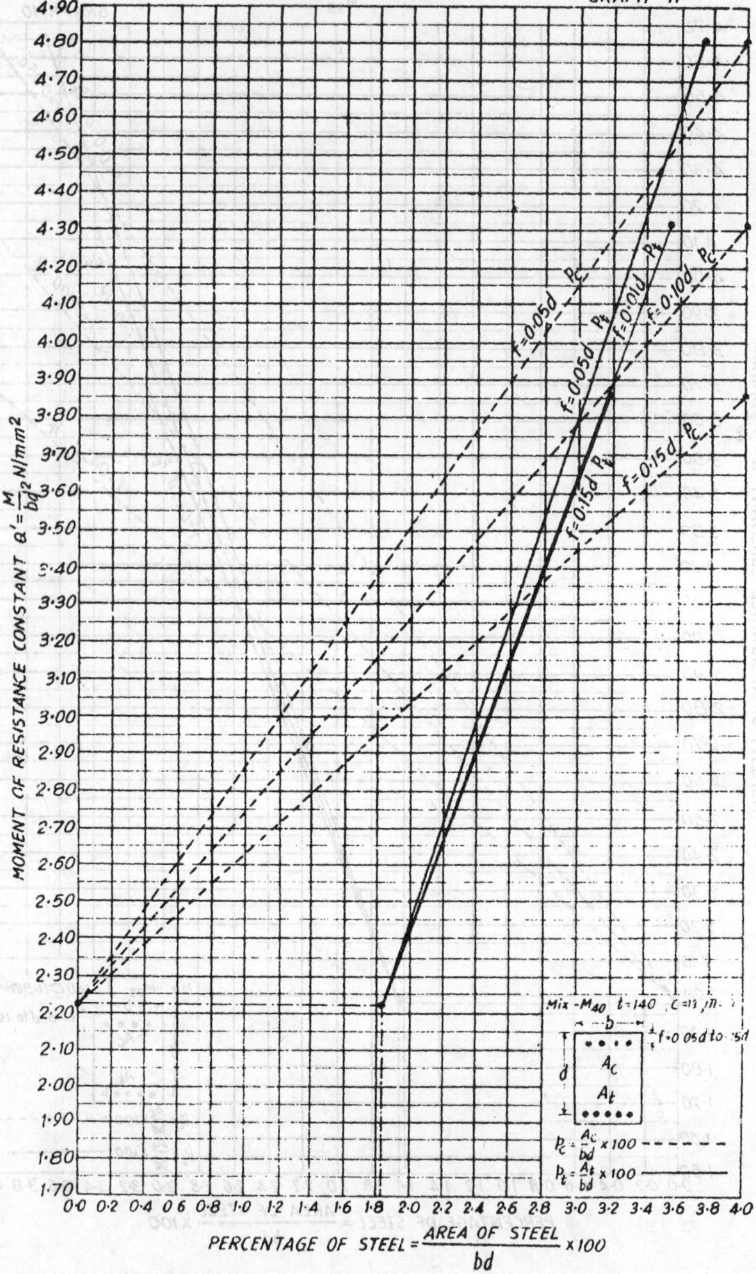

TEE BEAMS

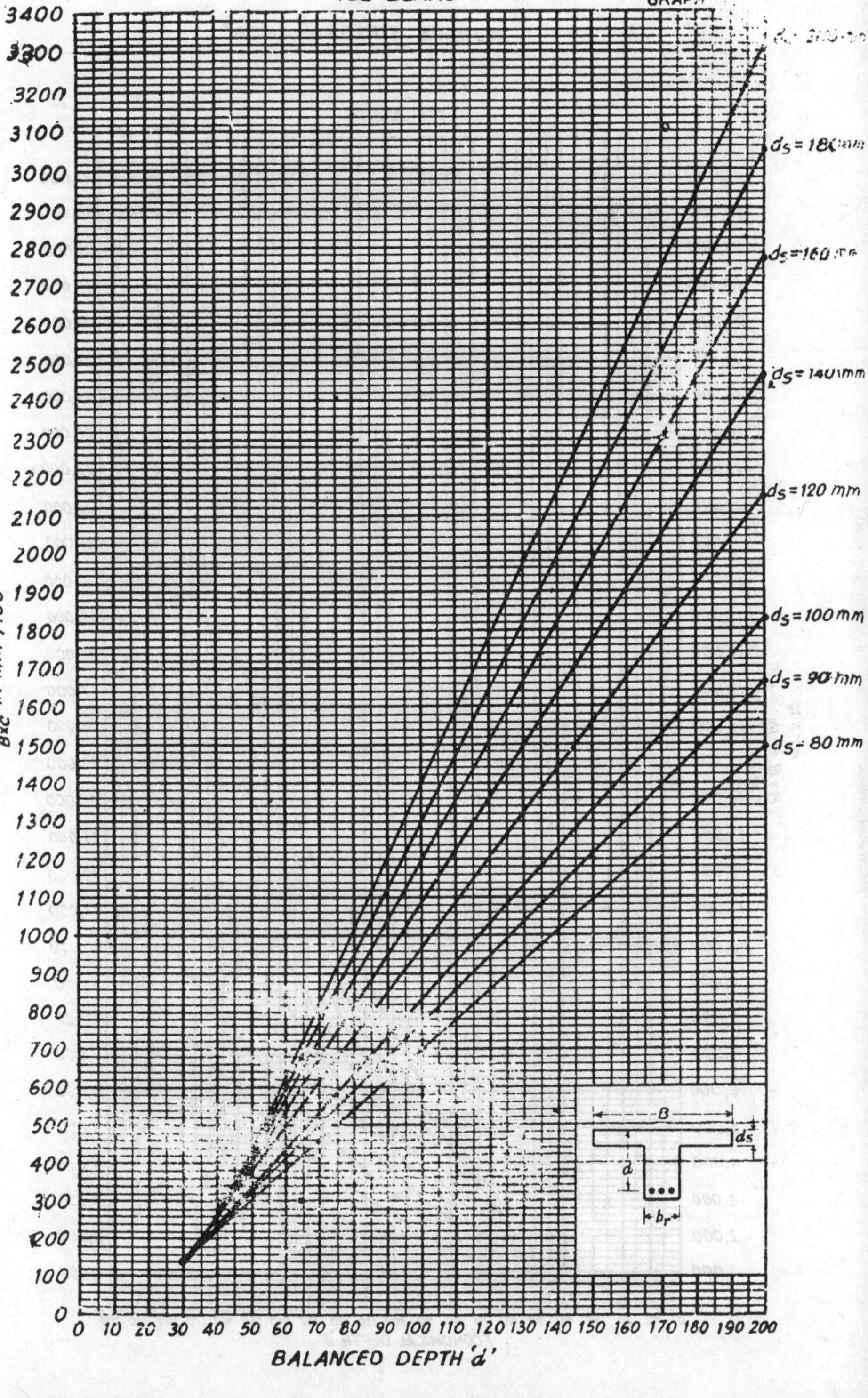

GRAPH
d_s = 200 mm
d_s = 180 mm
d_s = 160 mm
d_s = 140 mm
d_s = 120 mm
d_s = 100 mm
d_s = 90 mm
d_s = 80 mm
$\frac{M}{B \times C}$ IN mm²/100
BALANCED DEPTH 'd'
0 10 20 30 40 50 60 70 80 90 100 110 120 130 140 150 160 170 180 190 200
0 100 200 300 400 500 600 700 800 900 1000 1100 1200 1300 1400 1500 1600 1700 1800 1900 2000 2100 2200 2300 2400 2500 2600 2700 2800 2900 3000 3100 3200 3300 3400
B
d_s
d
b_r

AVERAGE STRESS $\frac{P}{br} = N/mm^2$

TEE BEAMS

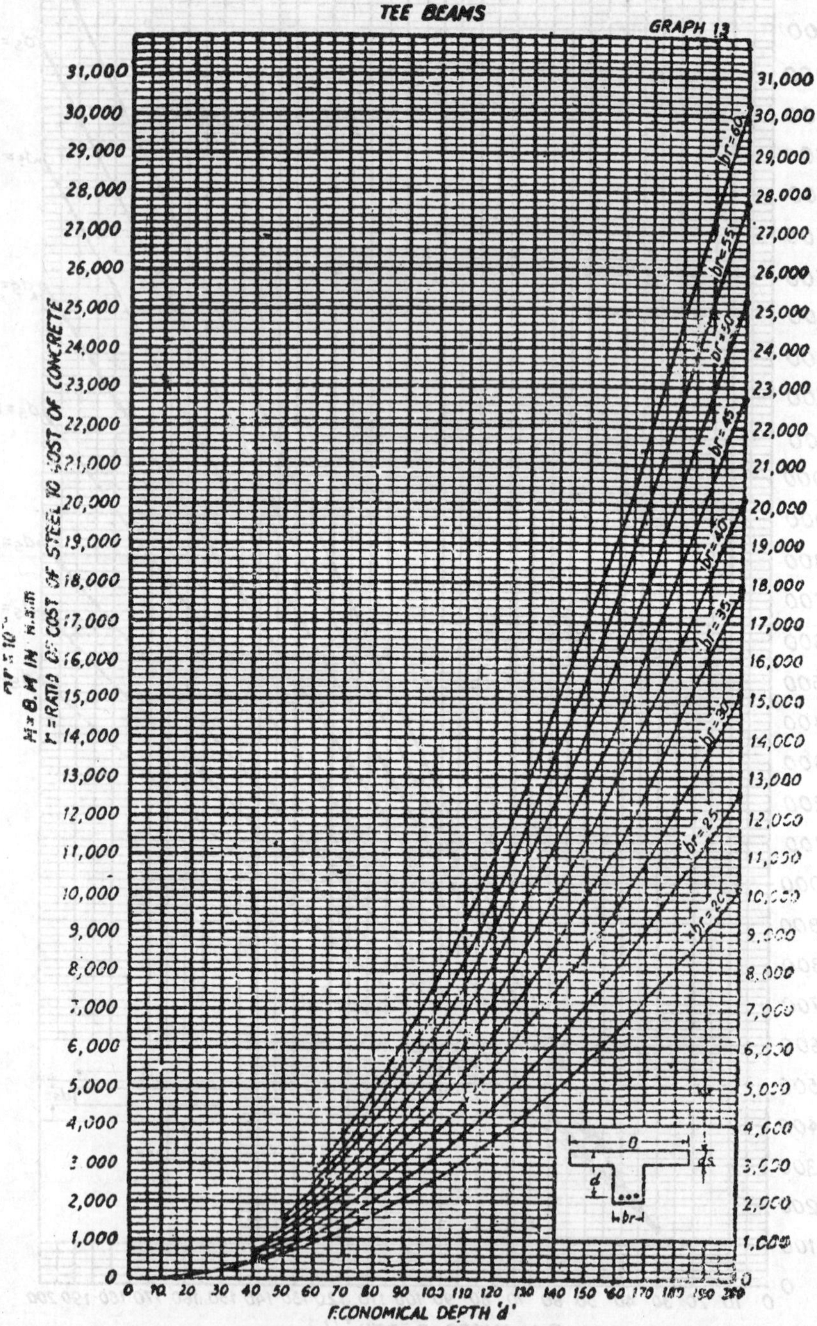

SECTIONS SUBJECTED TO COMBINED BENDING AND DIRECT STRESSES

GRAPH 14

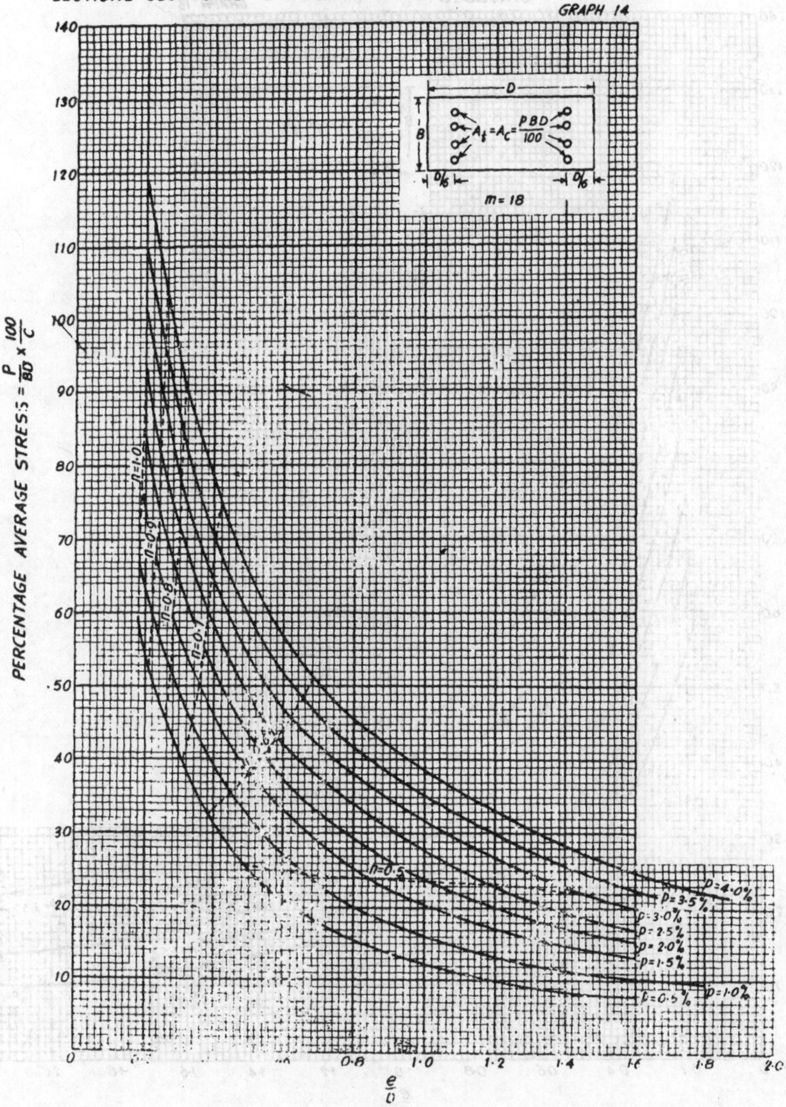

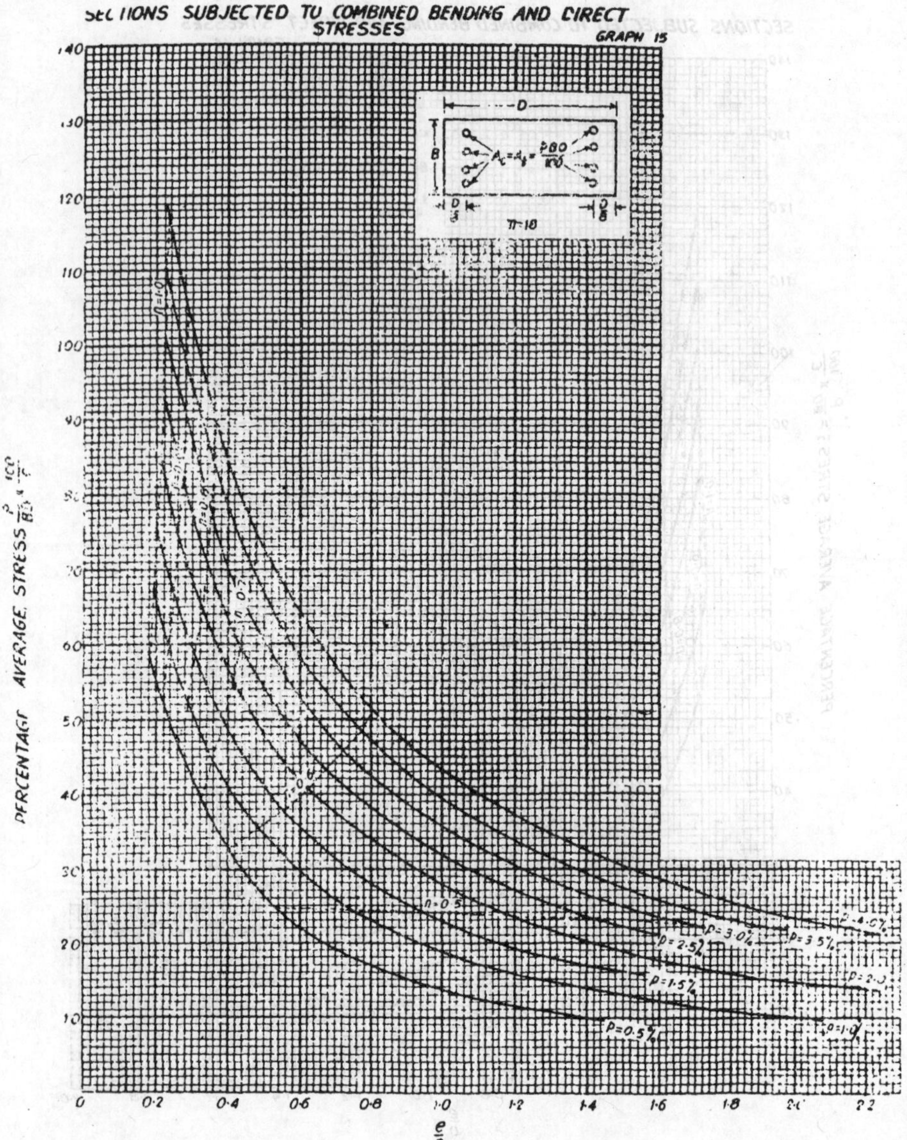
SECTIONS SUBJECTED TO COMBINED BENDING AND DIRECT STRESSES
GRAPH 15
D
B
$A_c = A_t = \frac{pBD}{100}$
$\frac{D}{8}$
$\frac{D}{8}$
m=18
PERCENTAGE AVERAGE STRESS $\frac{P}{BD} \times \frac{100}{c}$
$\frac{e}{D}$
p=0.5%
p=1.0%
p=1.5%
p=2.0
p=2.5%
p=3.0%
p=3.5%
p=4.0%
n=0.5
n=0.6
n=0.7
n=1.0
0
0.2
0.4
0.6
0.8
1.0
1.2
1.4
1.6
1.8
2.0
2.2
10
20
30
40
50
60
70
80
90
100
110
120
130
140

SECTIONS SUBJECTED TO COMBINED BENDING AND DIRECT STRESSES

GRAPH 16

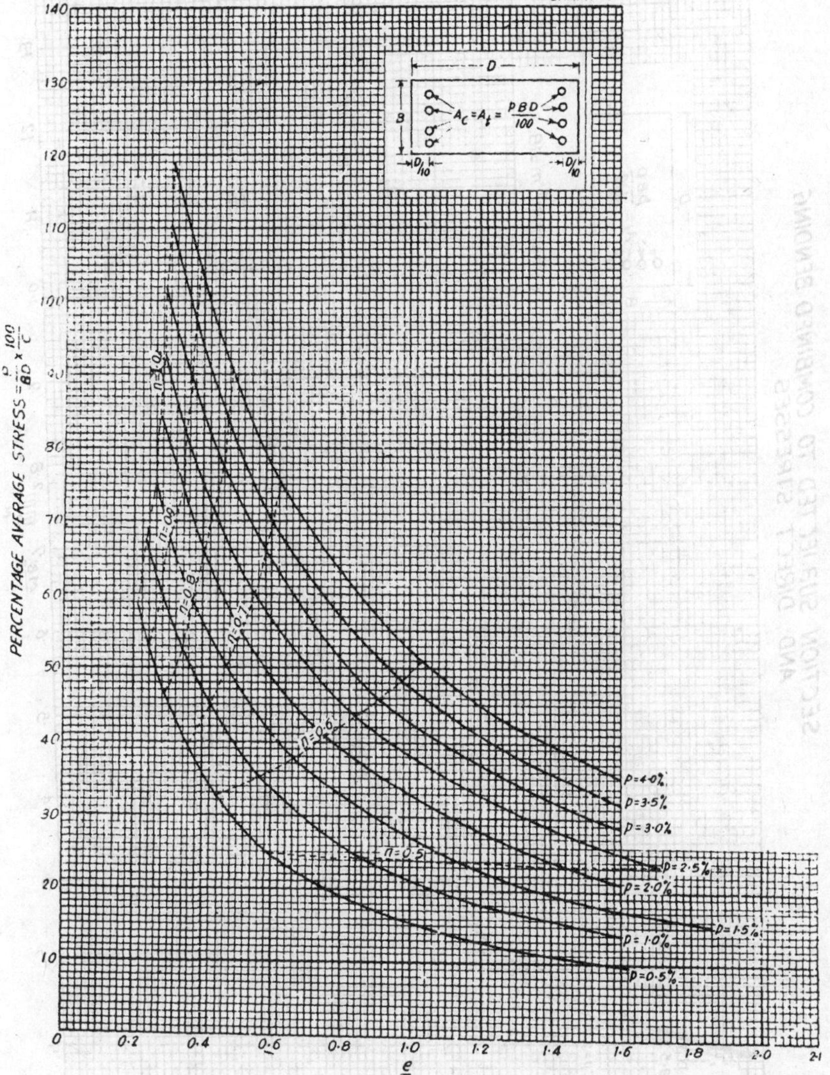

SECTION SUBJECTED TO COMBINED BENDING AND DIRECT STRESSES

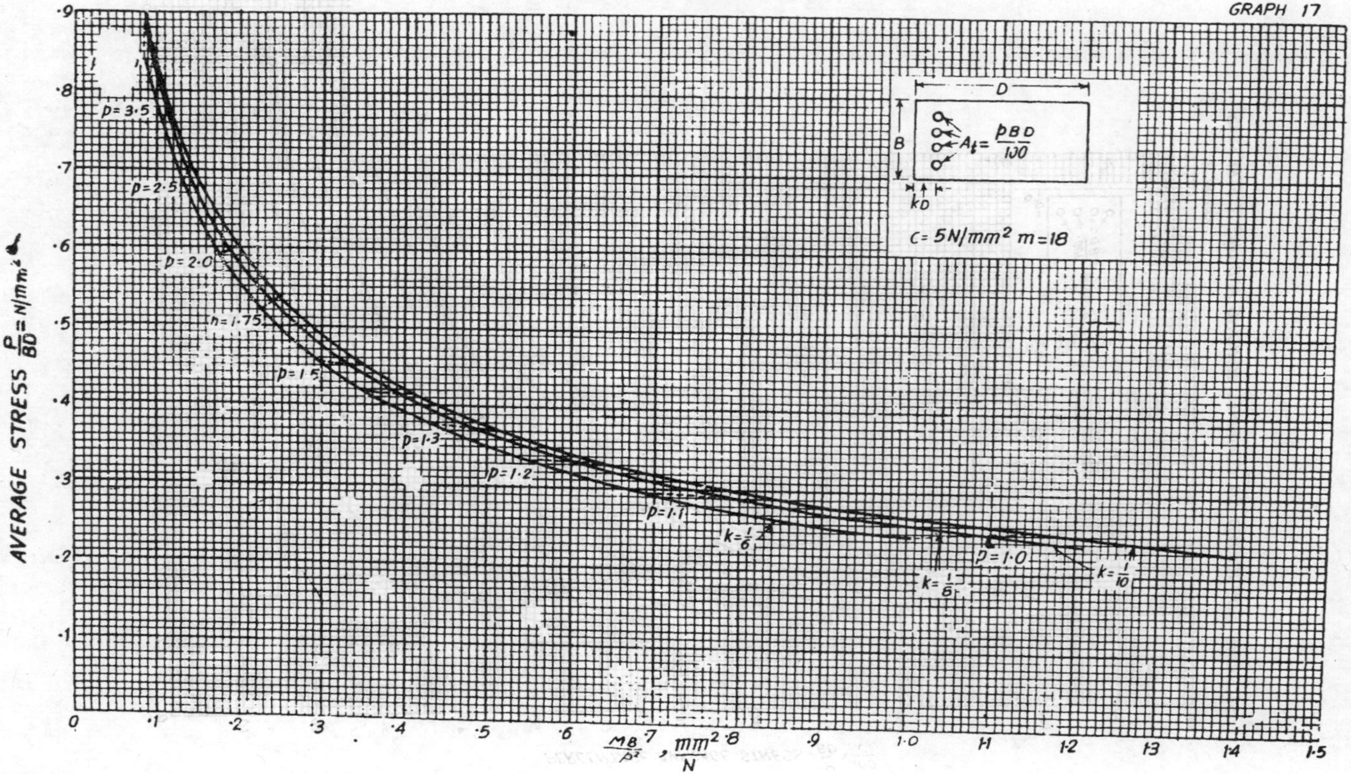

CANTILEVER RETAINING WALL

GRAPH 18

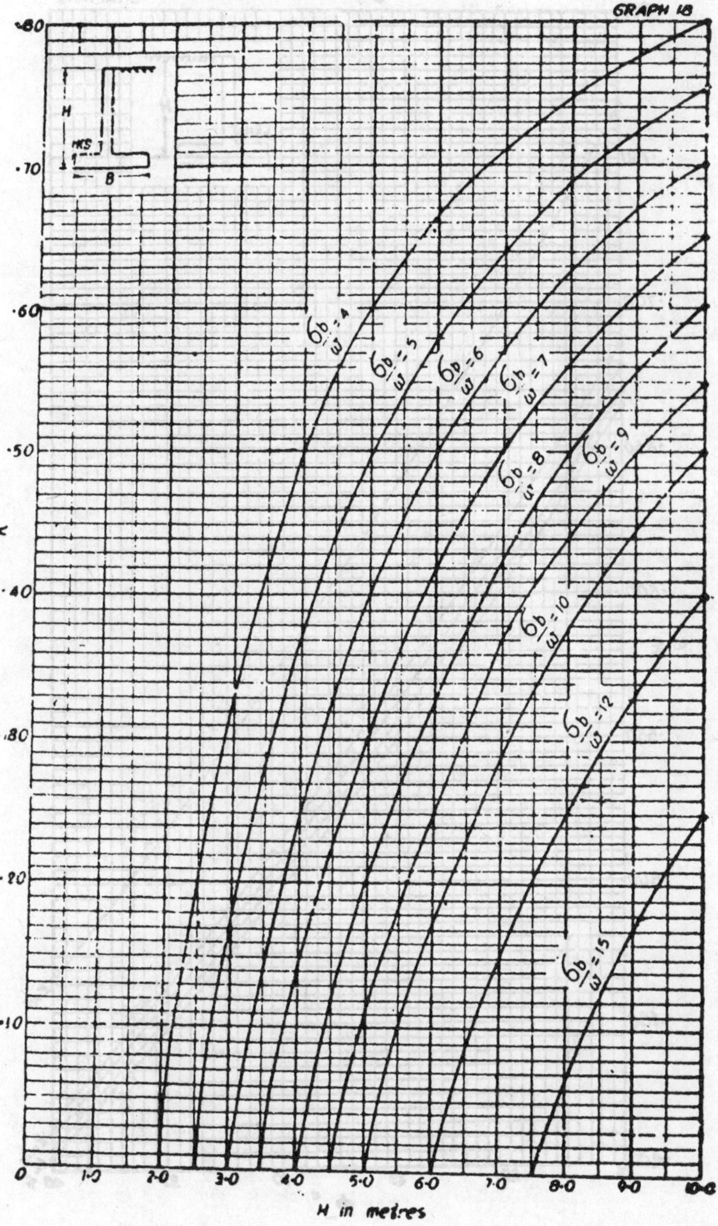

CANTILEVER RETAINING WALL

GRAPH 19

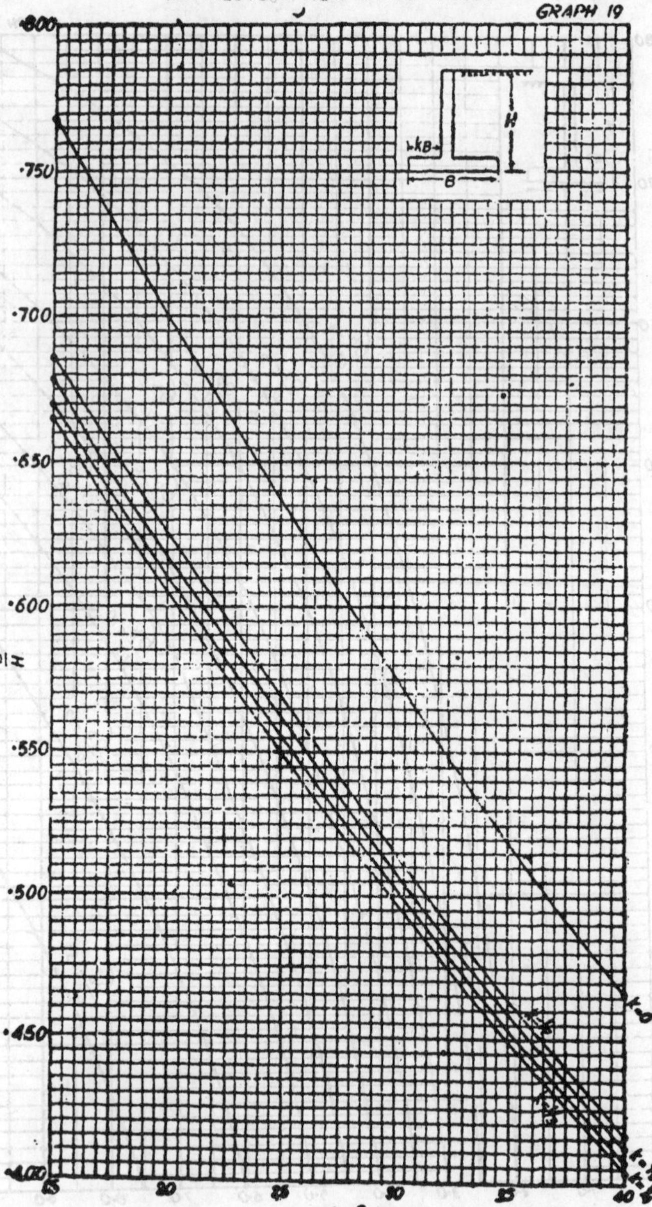

RETAINING WALL BASE WIDTH FROM CONSIDERATIONS OF SLIDING

GRAPH 20

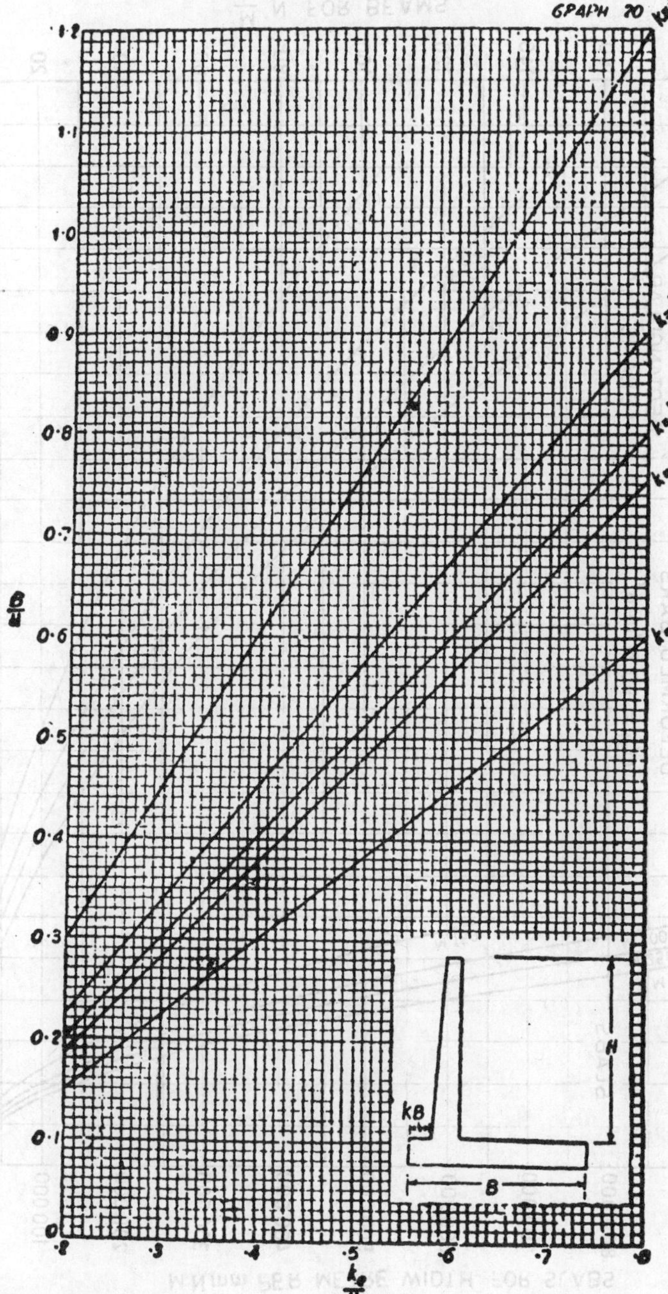

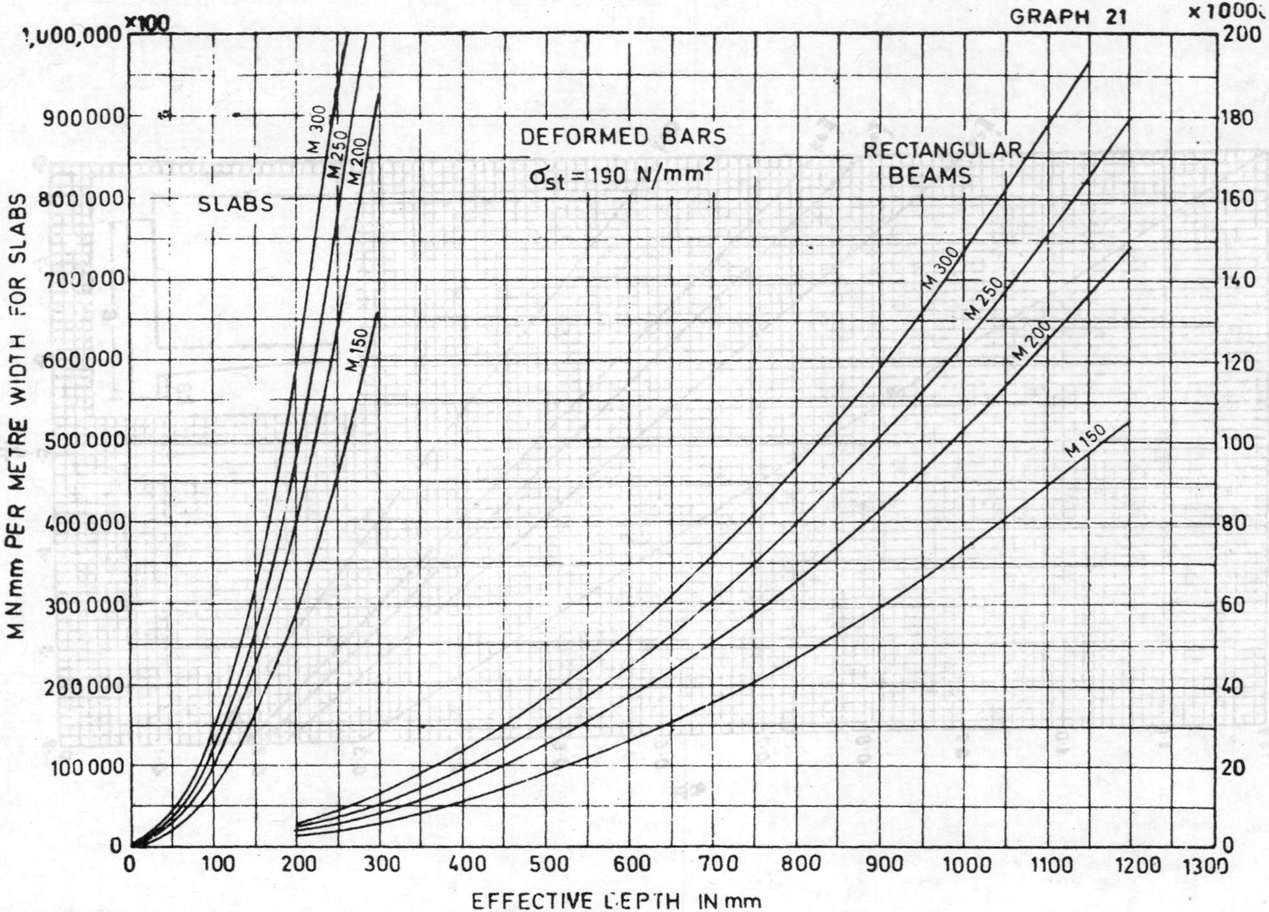
GRAPH 21
DEFORMED BARS
$\sigma_{st} = 190\ N/mm^2$
SLABS
RECTANGULAR BEAMS
M 300
M 250
M 200
M 150
M 300
M 250
M 200
M 150
×100
×1000
1,000,000
900 000
800 000
700 000
600 000
500 000
400 000
300 000
200 000
100 000
0
200
180
160
140
120
100
80
60
40
20
0
0
100
200
300
400
500
600
700
800
900
1000
1100
1200
1300
M N mm PER METRE WIDTH FOR SLABS
$\frac{M}{b}$ N FOR BEAMS
EFFECTIVE DEPTH IN mm

GRAPH 22

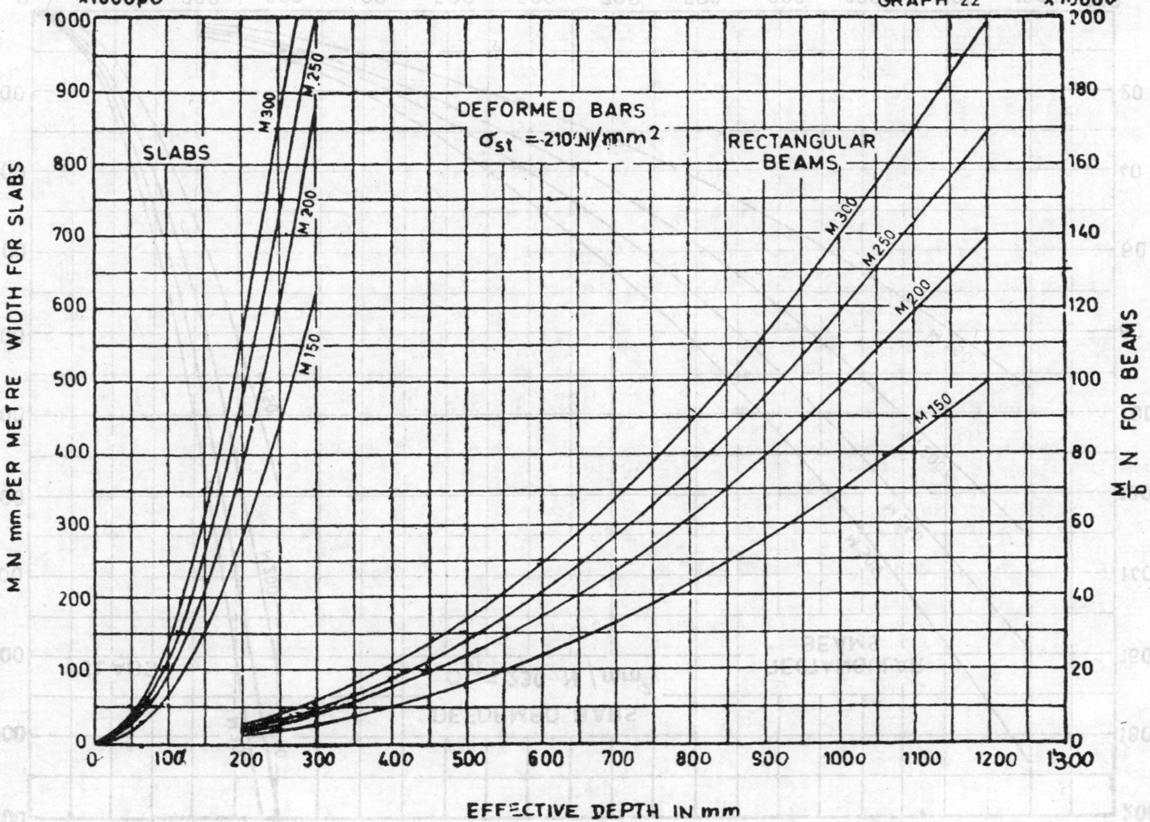

x100000
x10000
SLABS
DEFORMED BARS
σ_{st} = 210 N/mm^2
RECTANGULAR BEAMS
M 300
M 250
M 200
M 150
M N mm PER METRE WIDTH FOR SLABS
M/b N FOR BEAMS
EFFECTIVE DEPTH IN mm

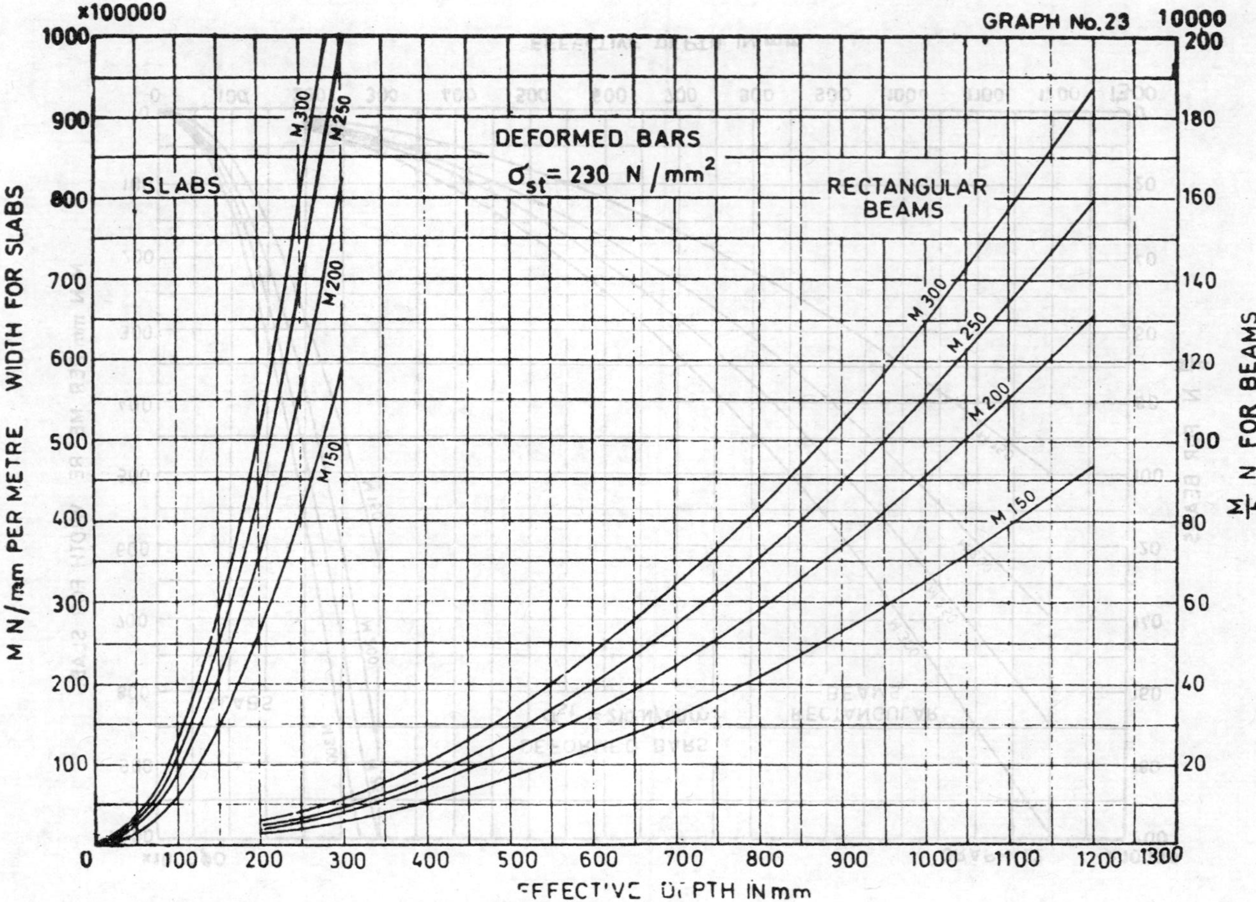

GRAPH No. 23
DEFORMED BARS
$\sigma_{st} = 230\ N/mm^2$
SLABS
RECTANGULAR BEAMS
M 300
M 250
M 200
M 150
x100000
x10000
M N/mm PER METRE WIDTH FOR SLABS
$\frac{M}{b}$ N FOR BEAMS
EFFECTIVE DEPTH IN mm
0
100
200
300
400
500
600
700
800
900
1000
1100
1200
1300
20
40
60
80
100
120
140
160
180
200

GRAPH 24

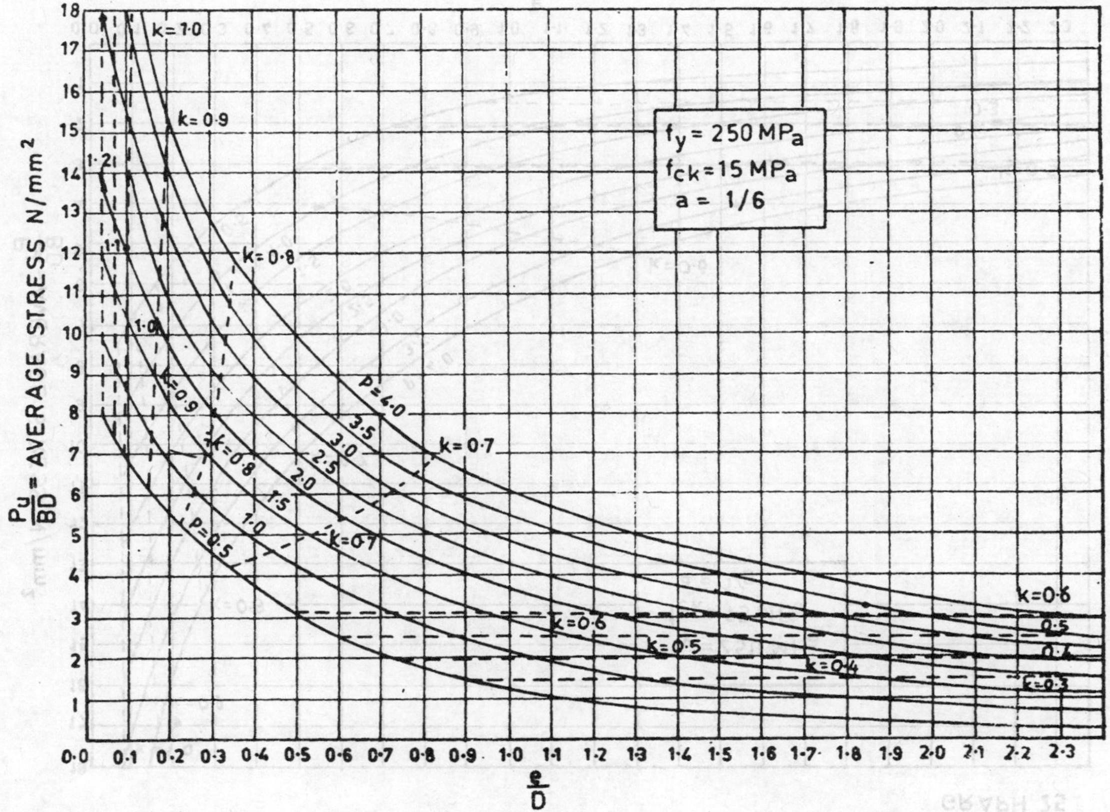

f_y = 250 MPa
f_{ck} = 15 MPa
a = 1/6
$\frac{P_u}{BD}$ = AVERAGE STRESS N/mm²
$\frac{e}{D}$
P = 4·0
3·5
3·0
2·5
2·0
1·5
1·0
P = 0·5
k = 1·0
k = 0·9
k = 0·8
k = 0·7
k = 0·6
k = 0·5
k = 0·4
k = 0·3
1·2
1·1
1·0

GRAPH 25

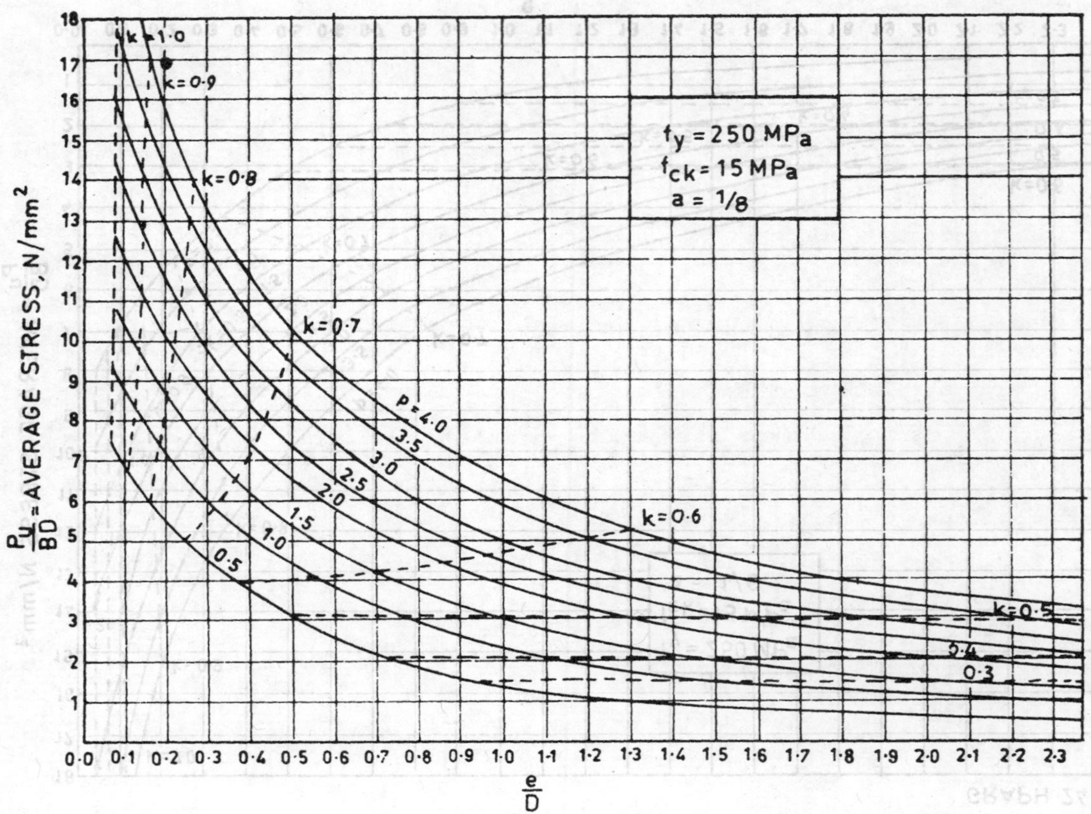

$f_y = 250$ MPa
$f_{ck} = 15$ MPa
$a = 1/8$
k=1·0
k=0·9
k=0·8
k=0·7
k=0·6
k=0·5
0·4
0·3
p=4·0
3·5
3·0
2·5
2·0
1·5
1·0
0·5
$\frac{P_u}{BD}$ = AVERAGE STRESS, N/mm²
$\frac{e}{D}$

GRAPH 26

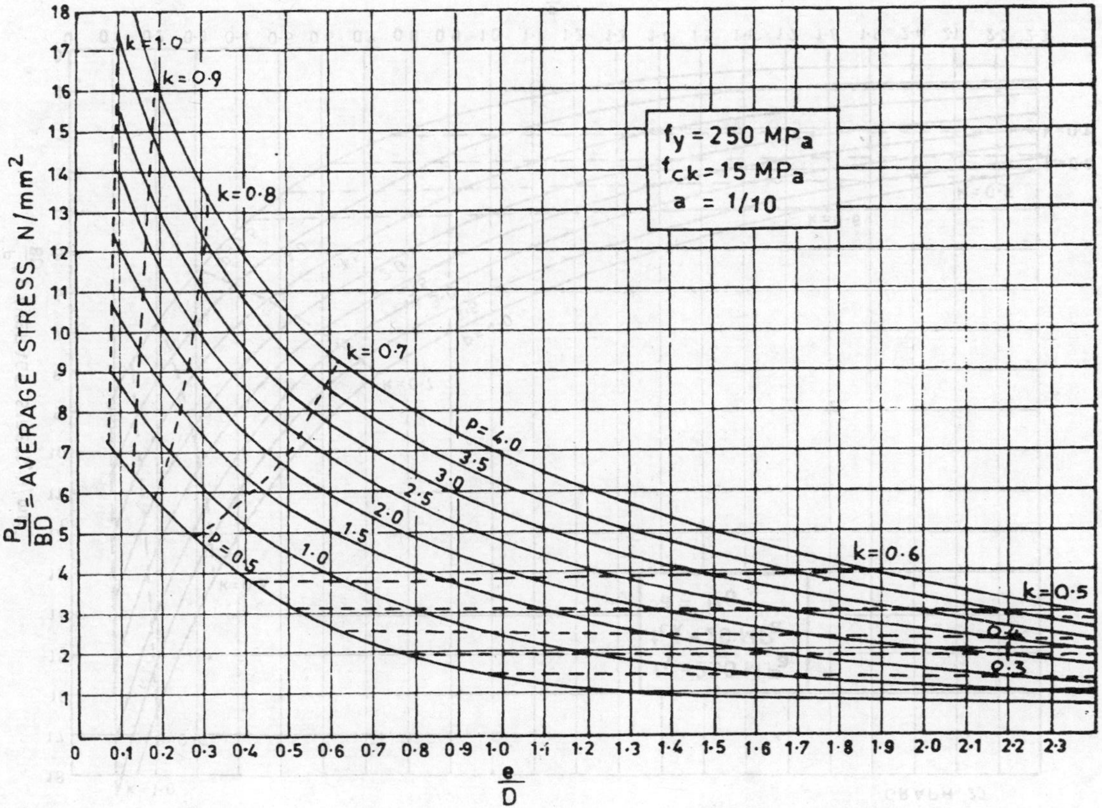

f_y = 250 MPa
f_{ck} = 15 MPa
a = 1/10
$\frac{P_u}{BD}$ = AVERAGE STRESS N/mm^2
$\frac{e}{D}$
k = 1.0
k = 0.9
k = 0.8
k = 0.7
k = 0.6
k = 0.5
0.4
0.3
p = 4.0
3.5
3.0
2.5
2.0
1.5
1.0
p = 0.5

GRAPH 27

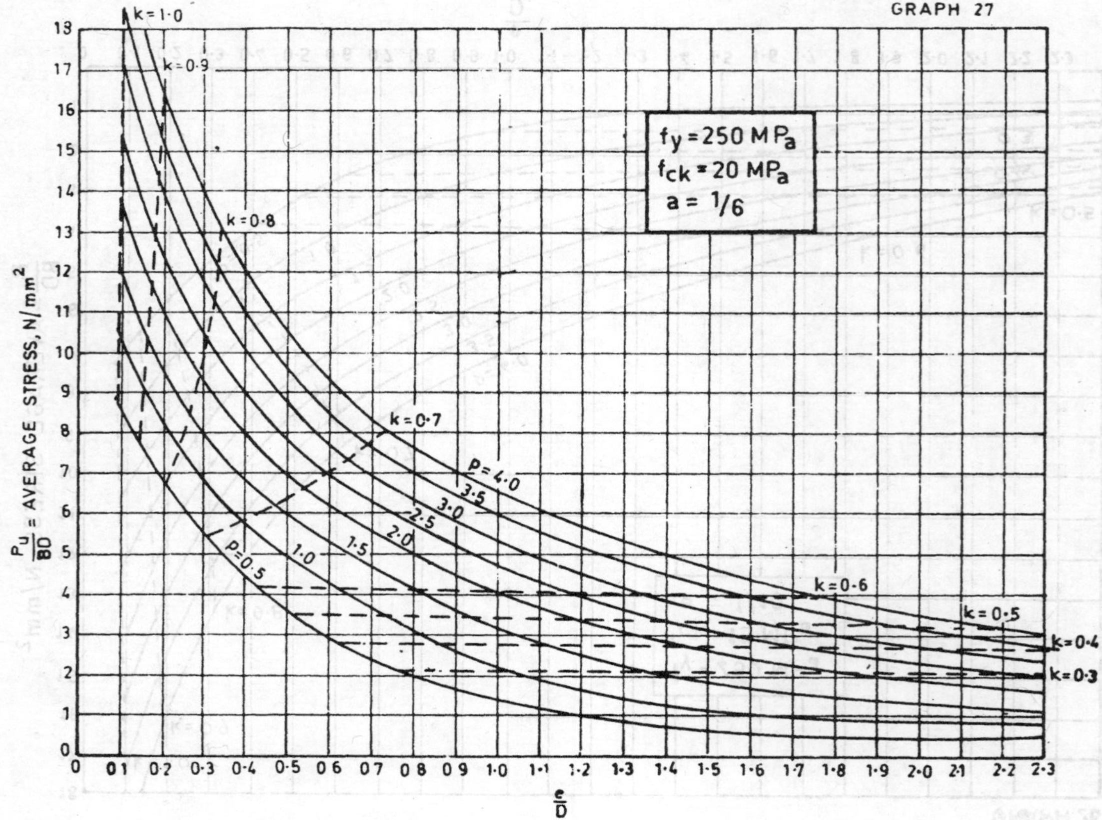

$f_y = 250\ MP_a$
$f_{ck} = 20\ MP_a$
$a = 1/6$
k = 1·0
k = 0·9
k = 0·8
k = 0·7
k = 0·6
k = 0·5
k = 0·4
k = 0·3
P = 4·0
3·5
3·0
2·5
2·0
1·5
1·0
P = 0·5
$\frac{P_u}{BD}$ = AVERAGE STRESS, N/mm²
$\frac{e}{D}$

GRAPH 28

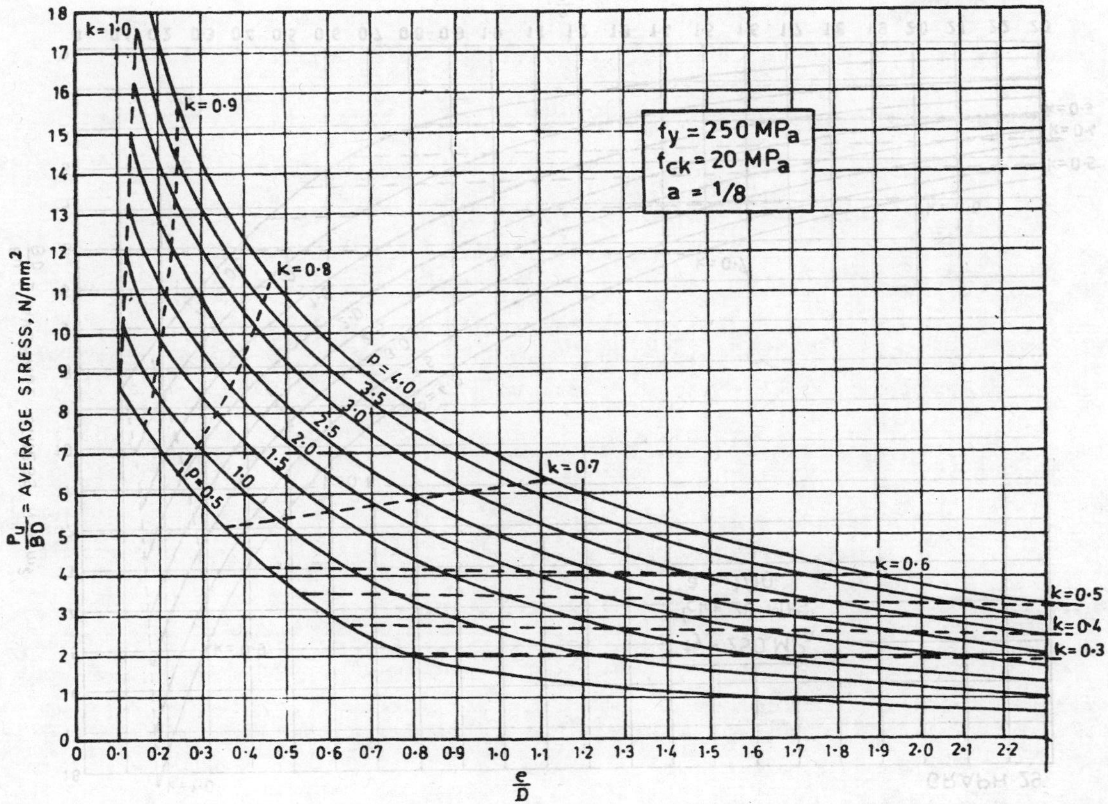

$f_y = 250\ MP_a$
$f_{ck} = 20\ MP_a$
$a = 1/8$
k=1·0
k=0·9
k=0·8
k=0·7
k=0·6
k=0·5
k=0·4
k=0·3
p=4·0
3·5
3·0
2·5
2·0
1·5
1·0
p=0·5
$\frac{P_u}{BD}$ = AVERAGE STRESS, N/mm²
$\frac{e}{D}$
0
0·1
0·2
0·3
0·4
0·5
0·6
0·7
0·8
0·9
1·0
1·1
1·2
1·3
1·4
1·5
1·6
1·7
1·8
1·9
2·0
2·1
2·2
1
2
3
4
5
6
7
8
9
10
11
12
13
14
15
16
17
18

GRAPH 29

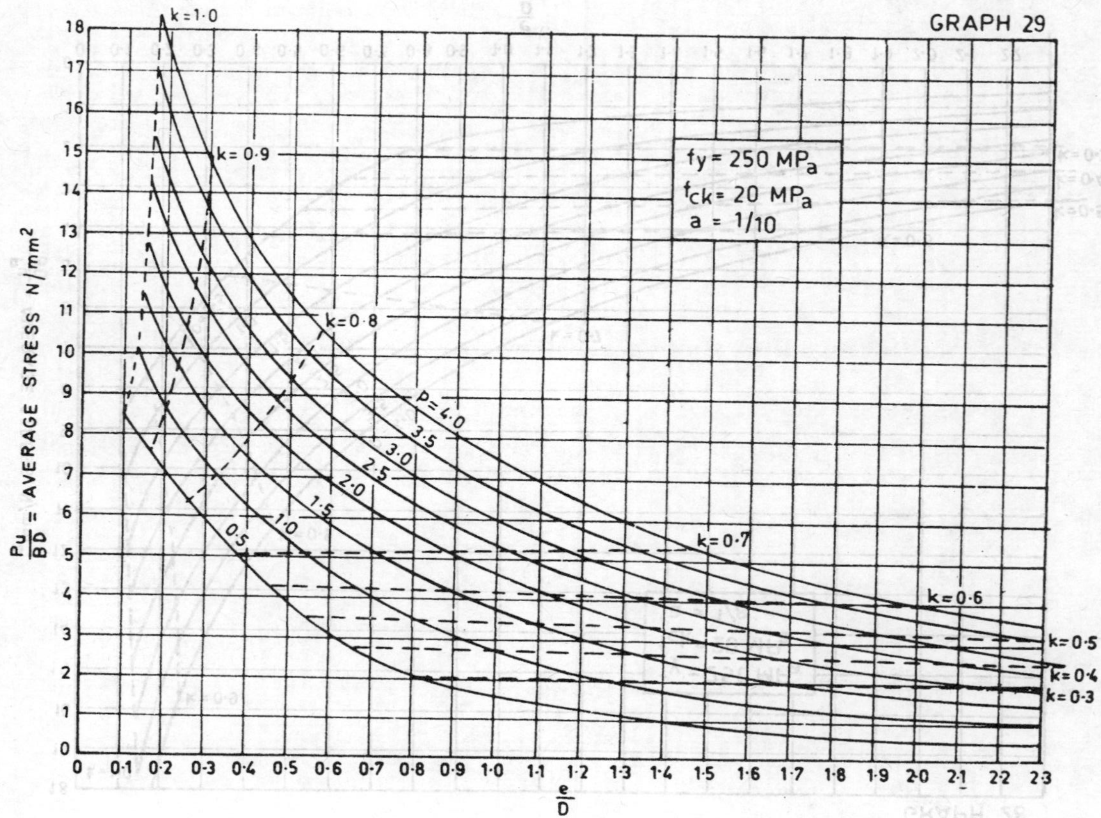

$f_y = 250$ MPa
$f_{ck} = 20$ MPa
a = 1/10
$\frac{P_u}{BD}$ = AVERAGE STRESS N/mm²
$\frac{e}{D}$
P = 4·0
3·5
3·0
2·5
2·0
1·5
1·0
0·5
k = 1·0
k = 0·9
k = 0·8
k = 0·7
k = 0·6
k = 0·5
k = 0·4
k = 0·3

GRAPH 30

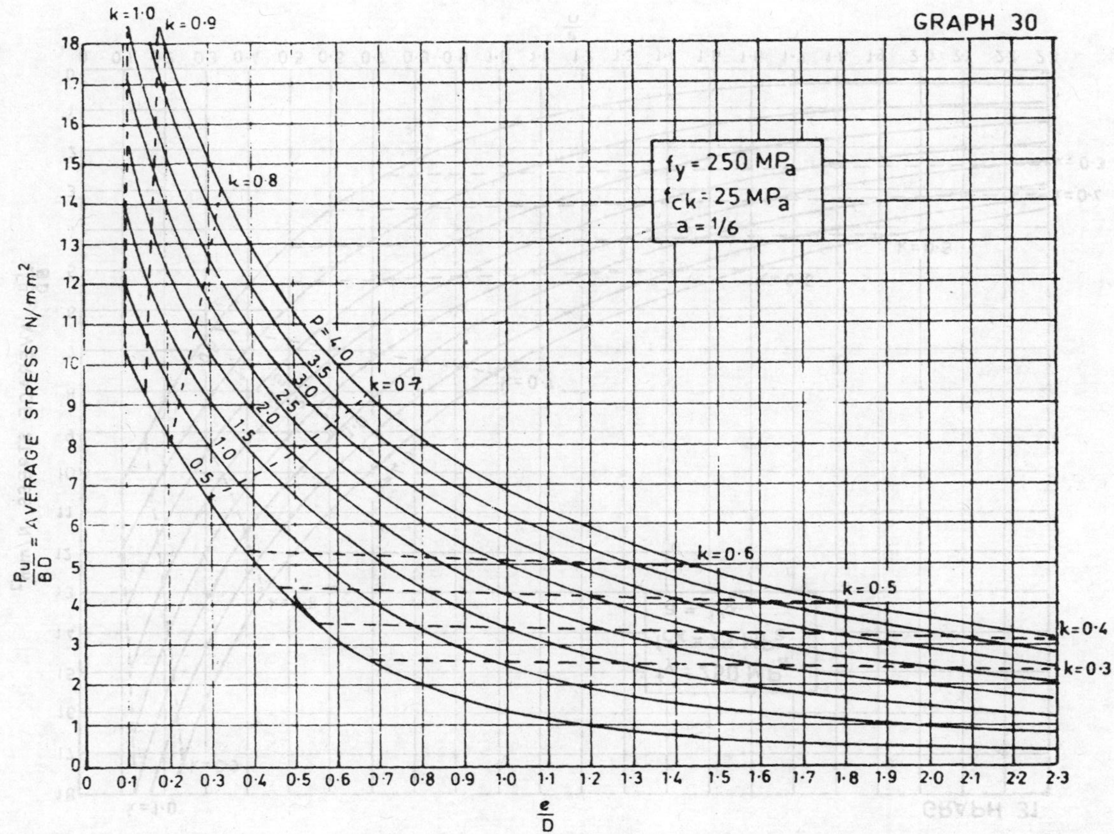

$f_y = 250\ MP_a$
$f_{ck} = 25\ MP_a$
$a = 1/6$
k = 1·0
k = 0·9
k = 0·8
k = 0·7
k = 0·6
k = 0·5
k = 0·4
k = 0·3
p = 4·0
3·5
3·0
2·5
2·0
1·5
1·0
0·5
$\frac{P_u}{BD}$ = AVERAGE STRESS N/mm²
$\frac{e}{D}$
0 1 2 3 4 5 6 7 8 9 10 11 12 13 14 15 16 17 18
0 0·1 0·2 0·3 0·4 0·5 0·6 0·7 0·8 0·9 1·0 1·1 1·2 1·3 1·4 1·5 1·6 1·7 1·8 1·9 2·0 2·1 2·2 2·3

GRAPH 31

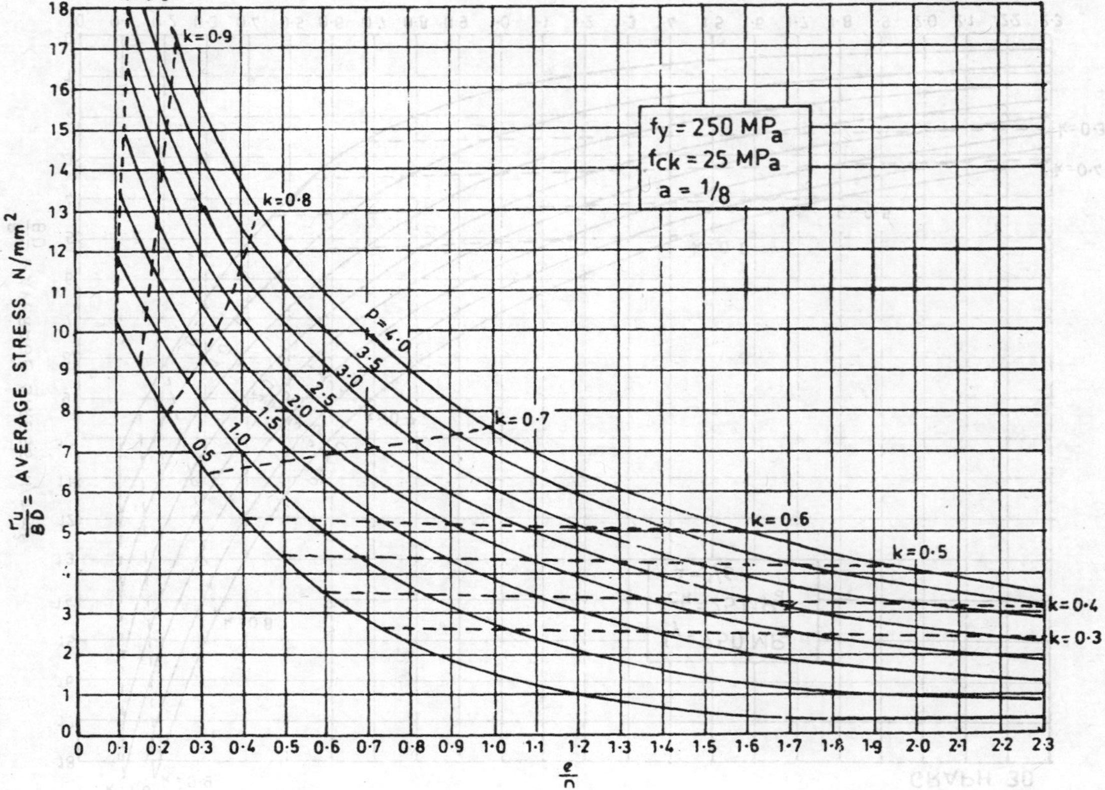

f_y = 250 MPa
f_{ck} = 25 MPa
a = 1/8
$\frac{P_u}{BD}$ = AVERAGE STRESS N/mm²
$\frac{e}{D}$
p = 4·0
3·5
3·0
2·5
2·0
1·5
1·0
0·5
k=1·0
k=0·9
k=0·8
k=0·7
k=0·6
k=0·5
k=0·4
k=0·3
18
17
16
15
14
13
12
11
10
9
8
7
6
5
3
2
1
0
0·1
0·2
0·3
0·4
0·5
0·6
0·7
0·8
0·9
1·0
1·1
1·2
1·3
1·4
1·5
1·6
1·7
1·8
1·9
2·0
2·1
2·2
2·3

GRAPH 32

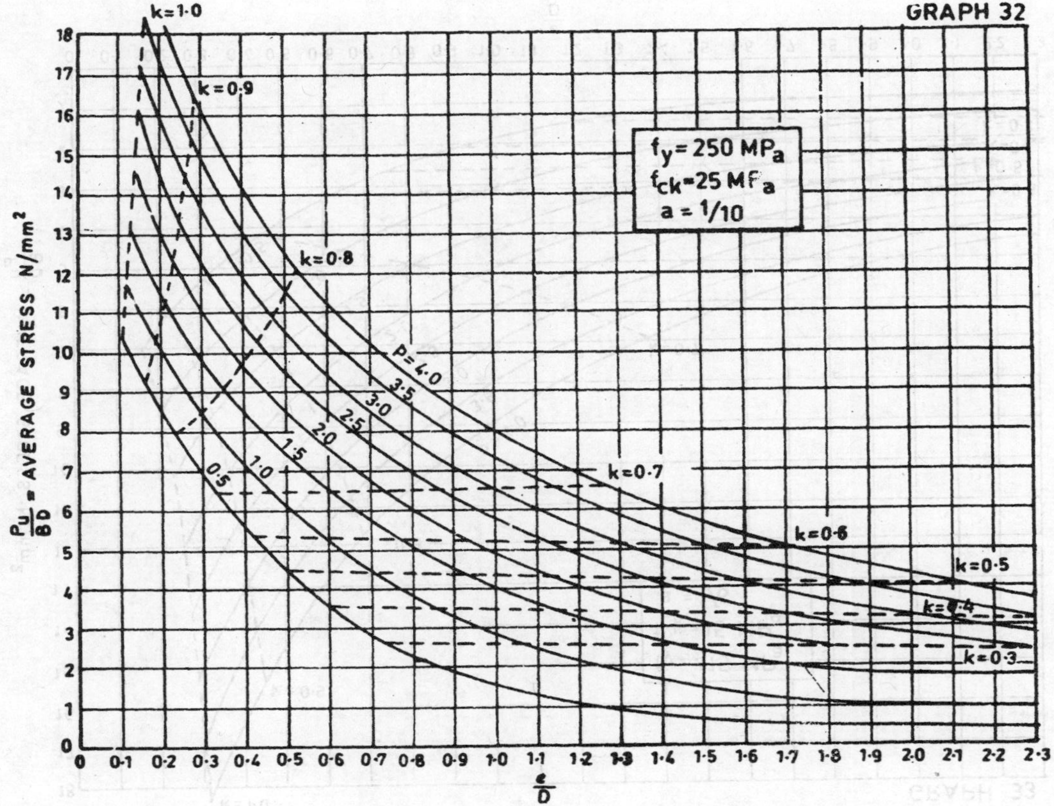

f_y = 250 MPa
f_{ck} = 25 MPa
a = 1/10
k=1·0
k=0·9
k=0·8
k=0·7
k=0·6
k=0·5
k=0·4
k=0·3
p=4·0
3·5
3·0
2·5
2·0
1·5
1·0
0·5
$\frac{P_u}{BD}$ = AVERAGE STRESS N/mm²
$\frac{e}{D}$
0
1
2
3
4
5
6
7
8
9
10
11
12
13
14
15
16
17
18
0·1
0·2
0·3
0·4
0·5
0·6
0·7
0·8
0·9
1·0
1·1
1·2
1·3
1·4
1·5
1·6
1·7
1·8
1·9
2·0
2·1
2·2
2·3

GRAPH 33

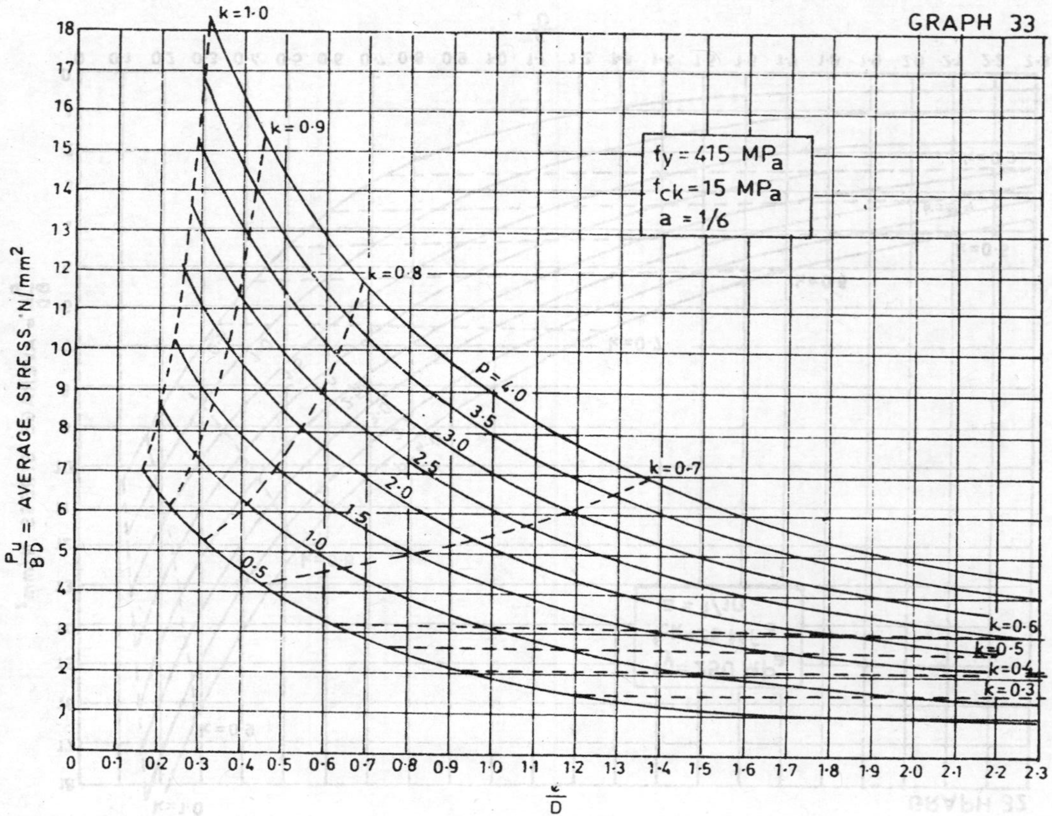

$f_y = 415$ MPa
$f_{ck} = 15$ MPa
a = 1/6
k = 1·0
k = 0·9
k = 0·8
k = 0·7
k = 0·6
k = 0·5
k = 0·4
k = 0·3
p = 4·0
3·5
3·0
2·5
2·0
1·5
1·0
0·5
$\frac{P_u}{BD}$ = AVERAGE STRESS N/mm^2
$\frac{e}{D}$

GRAPH 34

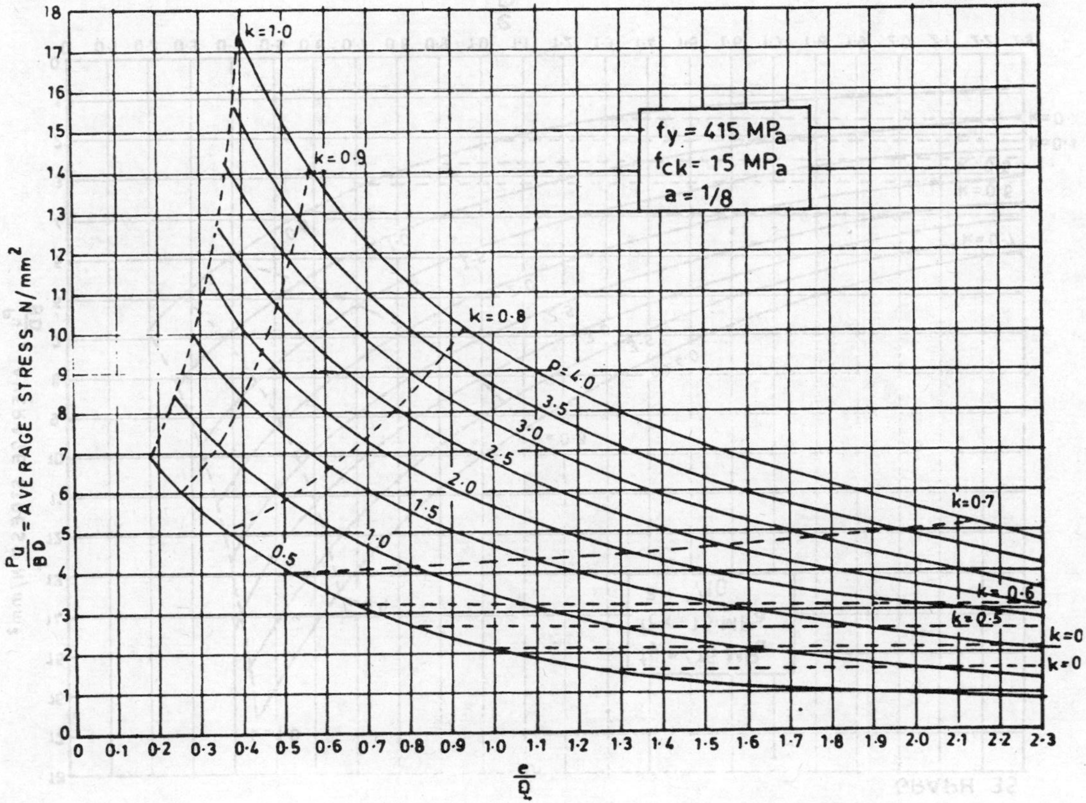

$f_y = 415$ MPa
$f_{ck} = 15$ MPa
a = 1/8
$\frac{P_u}{BD}$ = AVERAGE STRESS N/mm²
$\frac{e}{D}$
k = 1·0
k = 0·9
k = 0·8
k = 0·7
k = 0·6
k = 0·5
k = 0
k = 0
p = 4·0
3·5
3·0
2·5
2·0
1·5
1·0
0·5

GRAPH 35

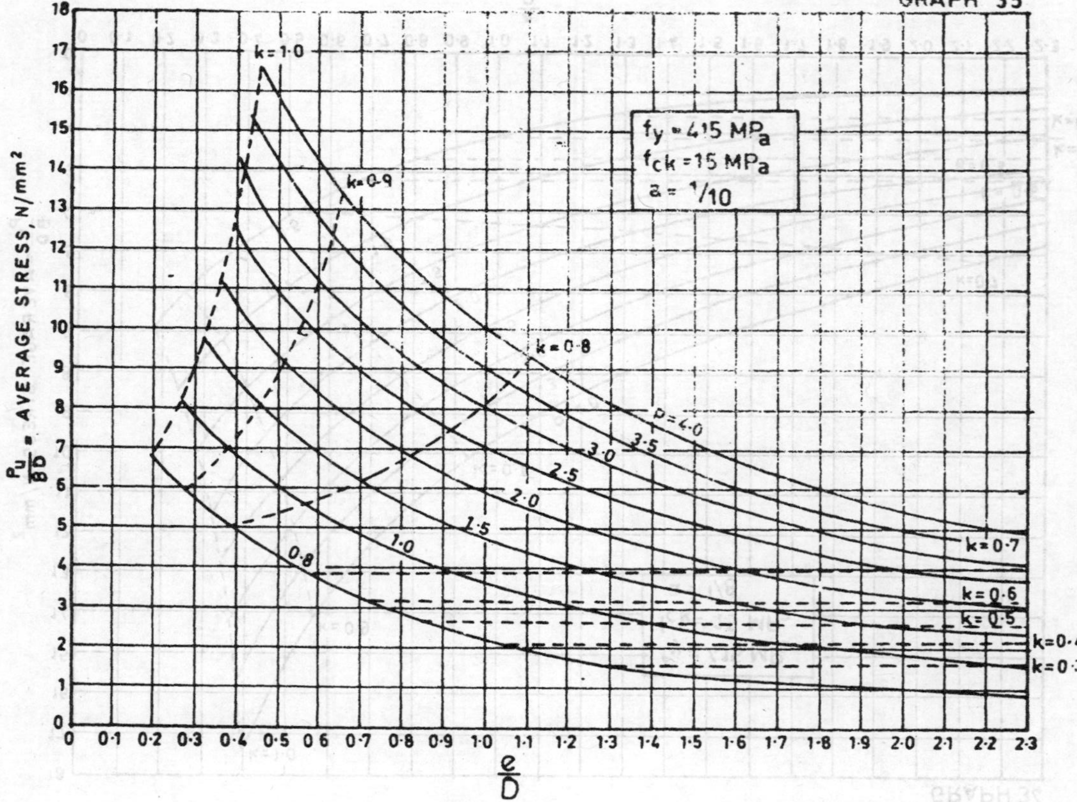

$f_y = 415$ MPa
$f_{ck} = 15$ MPa
$a = 1/10$
k=1·0
k=0·9
k=0·8
k=0·7
k=0·6
k=0·5
p=4·0
3·5
3·0
2·5
2·0
1·5
1·0
0·8
$\frac{P_u}{BD}$ = AVERAGE STRESS, N/mm²
$\frac{e}{D}$
0 0·1 0·2 0·3 0·4 0·5 0·6 0·7 0·8 0·9 1·0 1·1 1·2 1·3 1·4 1·5 1·6 1·7 1·8 1·9 2·0 2·1 2·2 2·3
0 1 2 3 4 5 6 7 8 9 10 11 12 13 14 15 16 17 18

GRAPH 36

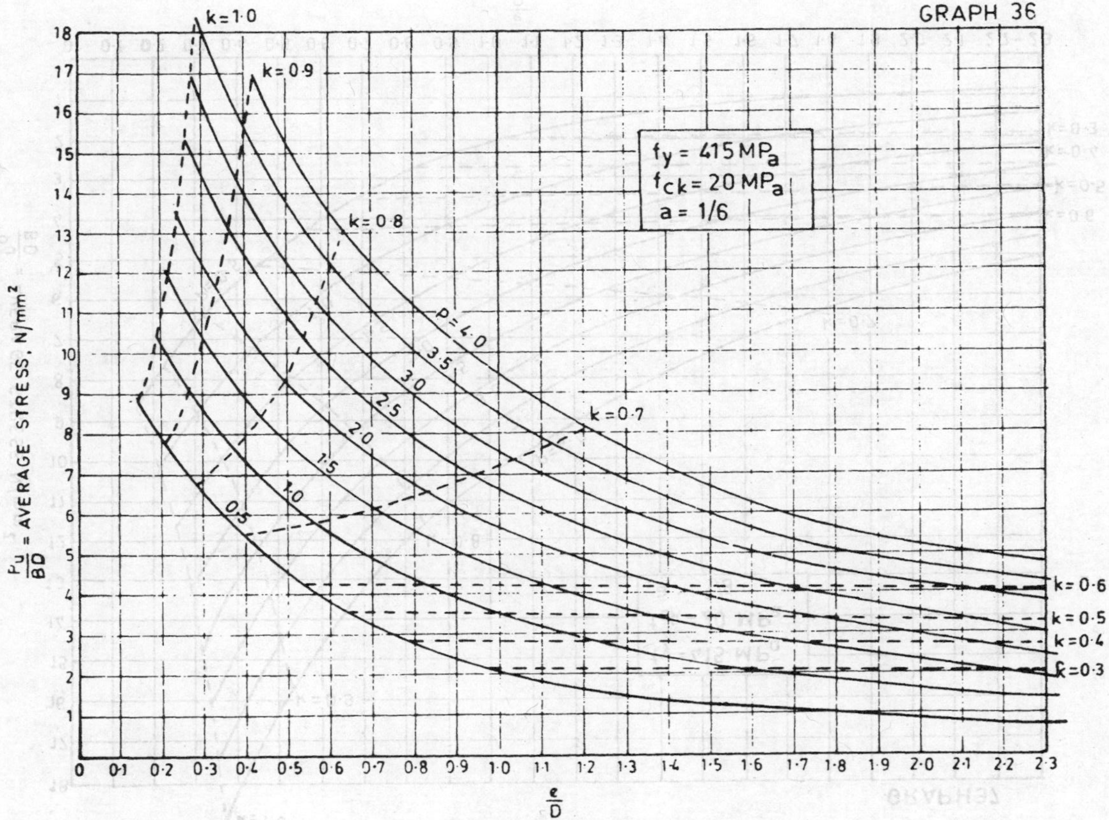

$f_y = 415\ MP_a$
$f_{ck} = 20\ MP_a$
$a = 1/6$
k=1·0
k=0·9
k=0·8
k=0·7
k=0·6
k=0·5
k=0·4
k=0·3
P=4·0
3·5
3·0
2·5
2·0
1·5
1·0
0·5
$\frac{P_u}{BD}$ = AVERAGE STRESS, N/mm²
$\frac{e}{D}$

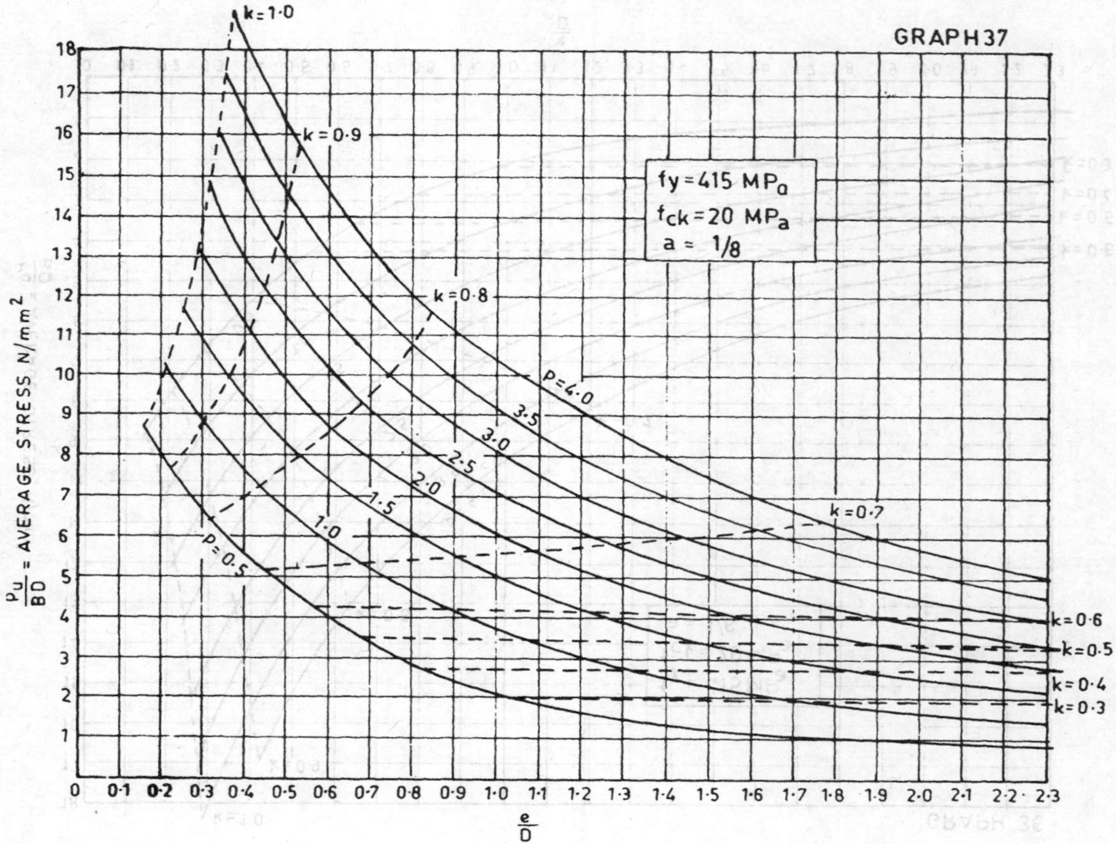
GRAPH 37
fy = 415 MPa
fck = 20 MPa
a = 1/8
k = 1·0
k = 0·9
k = 0·8
k = 0·7
k = 0·6
k = 0·5
k = 0·4
k = 0·3
p = 4·0
3·5
3·0
2·5
2·0
1·5
1·0
p = 0·5
Pu/BD = AVERAGE STRESS N/mm²
e/D

GRAPH 38

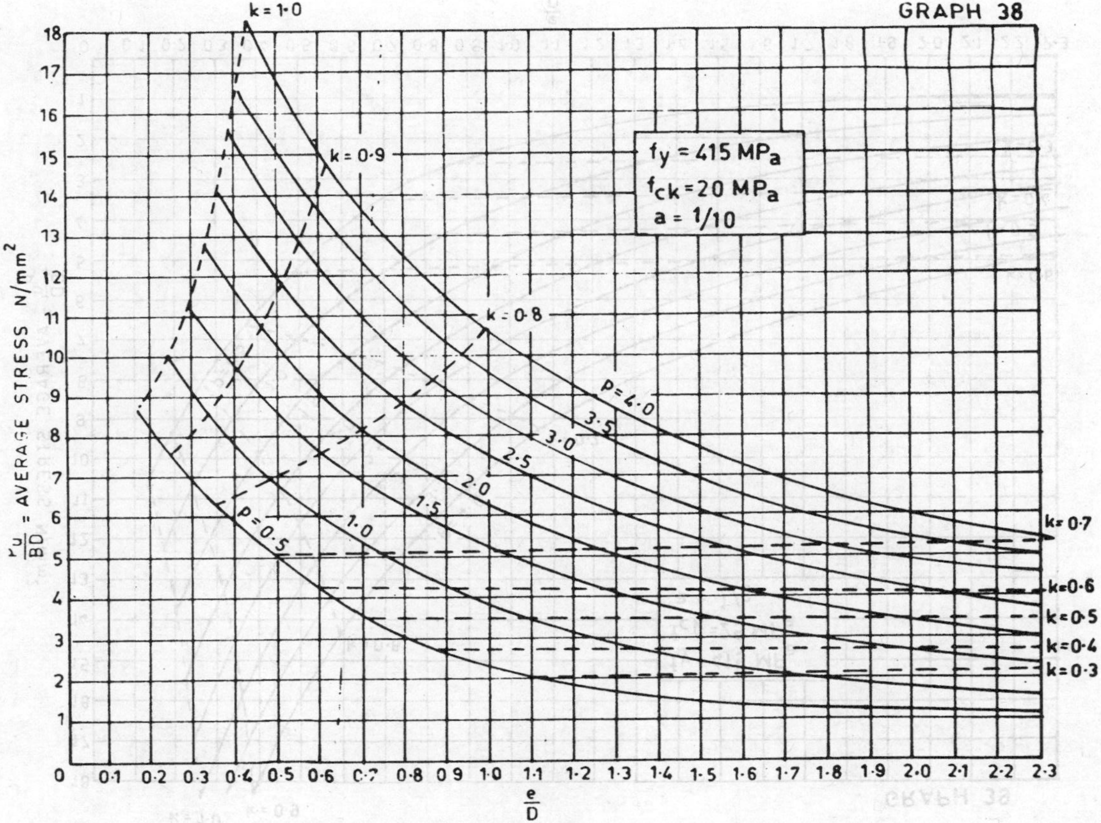

f_y = 415 MPa
f_{ck} = 20 MPa
a = 1/10
k = 1·0
k = 0·9
k = 0·8
k = 0·7
k = 0·6
k = 0·5
k = 0·4
k = 0·3
p = 4·0
3·5
3·0
2·5
2·0
1·5
1·0
p = 0·5
$\frac{P_u}{BD}$ = AVERAGE STRESS N/mm²
$\frac{e}{D}$
0 0·1 0·2 0·3 0·4 0·5 0·6 0·7 0·8 0·9 1·0 1·1 1·2 1·3 1·4 1·5 1·6 1·7 1·8 1·9 2·0 2·1 2·2 2·3
1 2 3 4 5 6 7 8 9 10 11 12 13 14 15 16 17 18

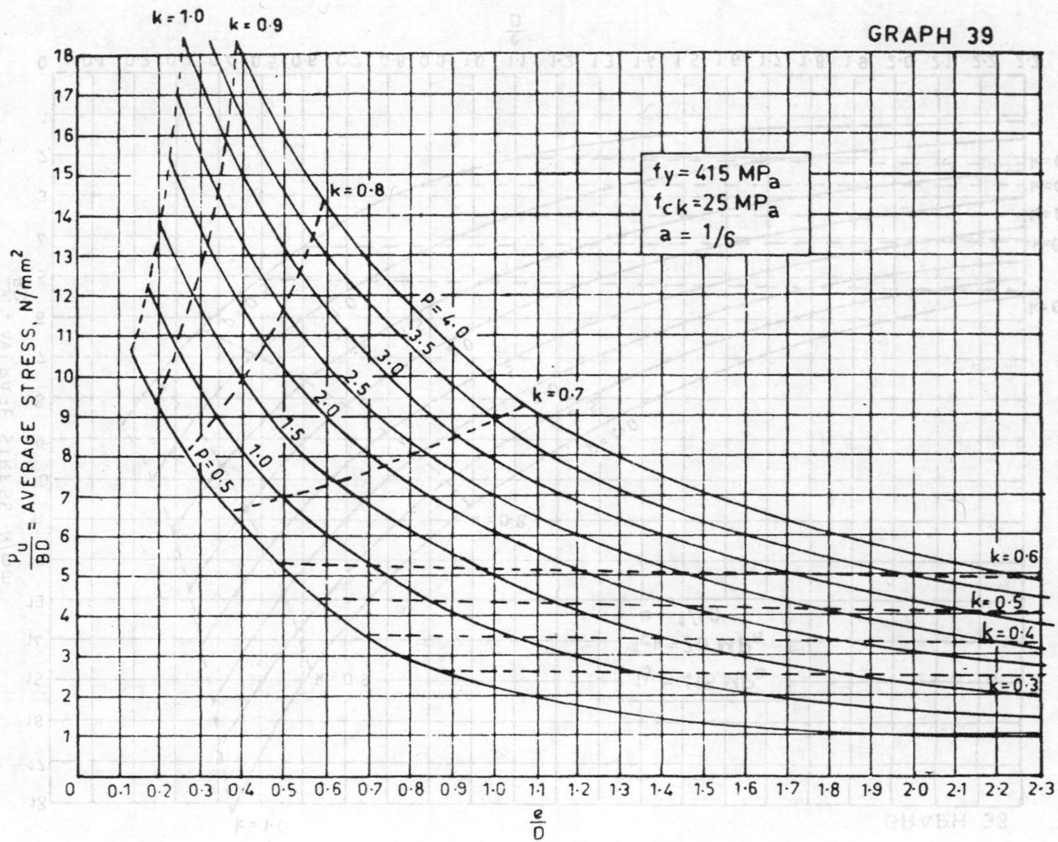
GRAPH 39
$f_y = 415$ MPa
$f_{ck} = 25$ MPa
a = 1/6
$\frac{P_u}{BD}$ = AVERAGE STRESS, N/mm²
$\frac{e}{D}$
k = 1·0
k = 0·9
k = 0·8
k = 0·7
k = 0·6
k = 0·5
k = 0·4
k = 0·3
P = 4·0
3·5
3·0
2·5
2·0
1·5
1·0
P = 0·5

GRAPH 40

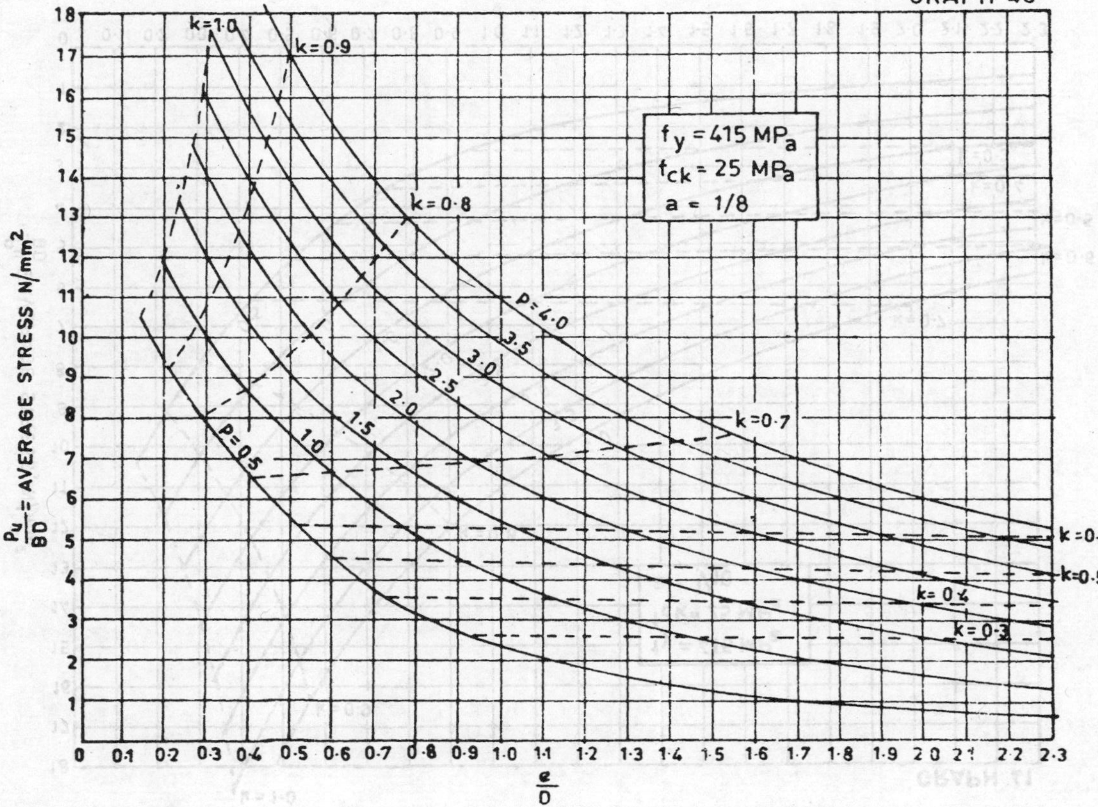

f_y = 415 MPa
f_{ck} = 25 MPa
a = 1/8
k=1·0
k=0·9
k=0·8
k=0·7
k=0·6
k=0·5
k=0·4
k=0·3
p=4·0
3·5
3·0
2·5
2·0
1·5
1·0
p=0·5
$\frac{P_u}{BD}$ = AVERAGE STRESS N/mm²
$\frac{e}{D}$

GRAPH 41

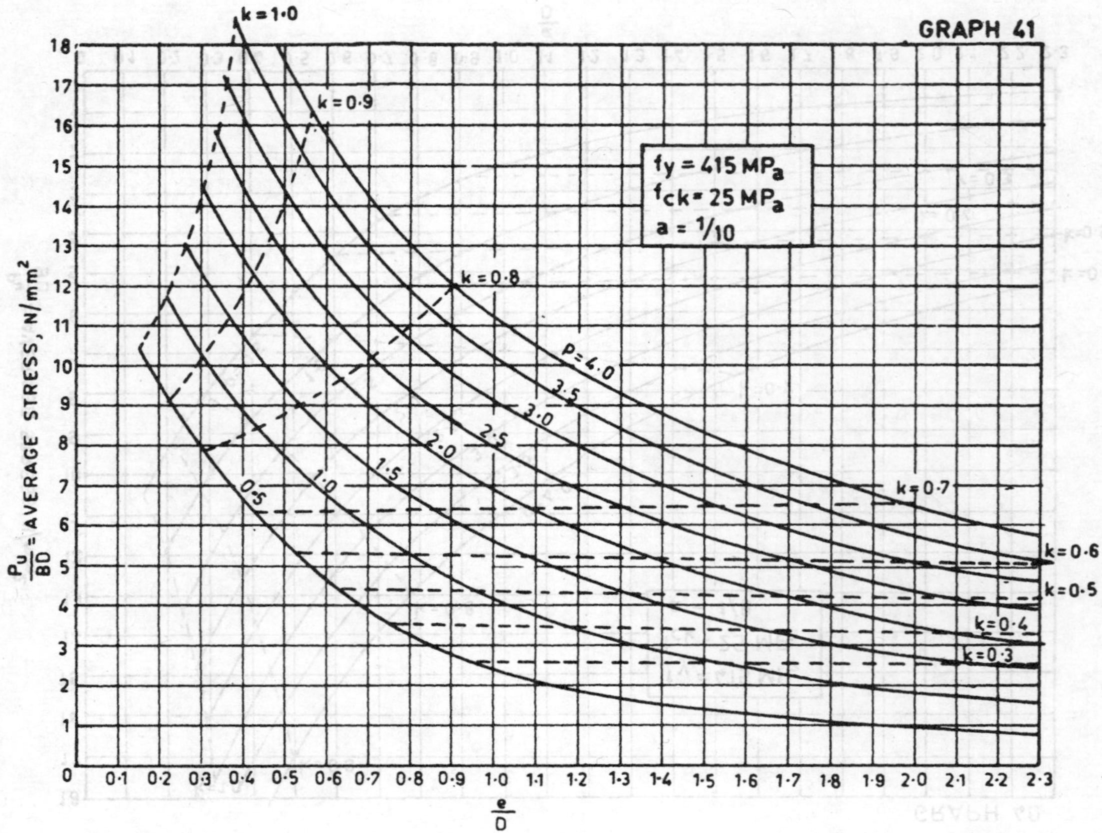

fy = 415 MPa
fck = 25 MPa
a = 1/10
k = 1·0
k = 0·9
k = 0·8
k = 0·7
k = 0·6
k = 0·5
k = 0·4
k = 0·3
p = 4·0
3·5
3·0
2·5
2·0
1·5
1·0
0·5
Pu/BD = AVERAGE STRESS, N/mm²
e/D

GRAPH 42

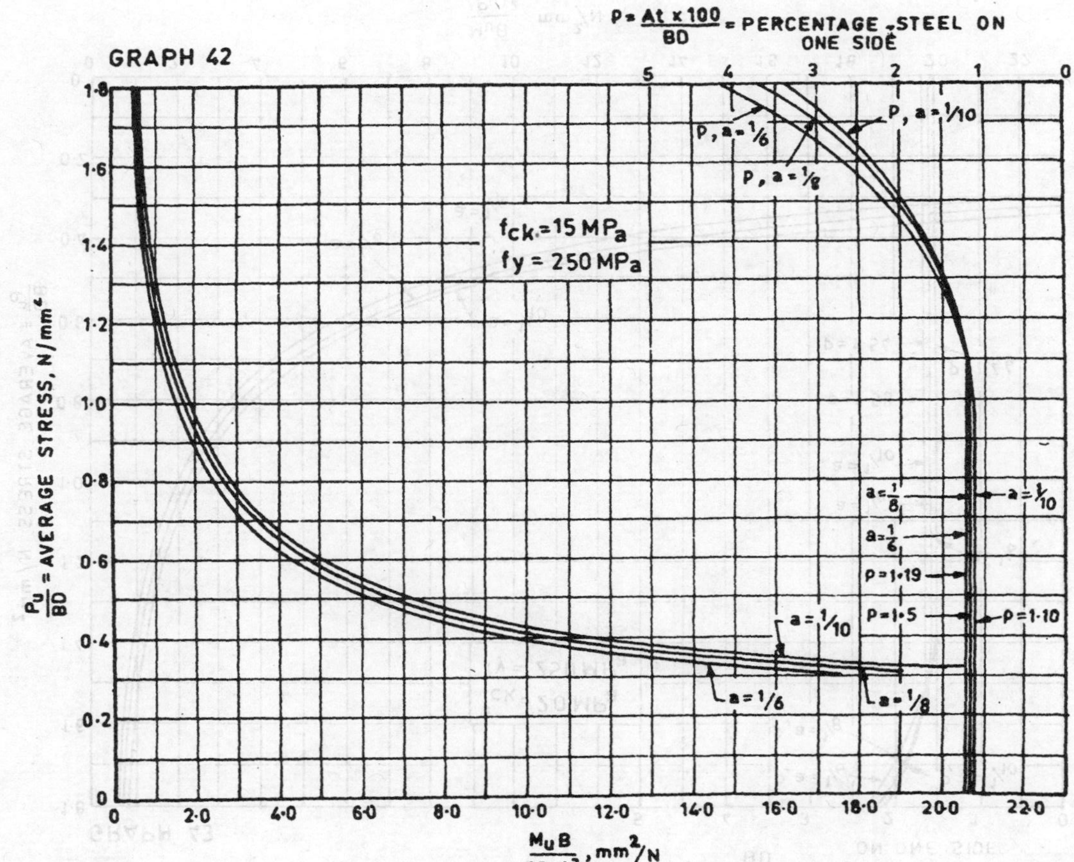

p = At x 100 / BD = PERCENTAGE STEEL ON ONE SIDE
5
4
3
2
1
0
P, a = 1/10
P, a = 1/6
P, a = 1/8
fck = 15 MPa
fy = 250 MPa
a = 1/8
a = 1/10
a = 1/6
p = 1·19
p = 1·5
p = 1·10
a = 1/10
a = 1/6
a = 1/8
Pu/BD = AVERAGE STRESS, N/mm²
1·8
1·6
1·4
1·2
1·0
0·8
0·6
0·4
0·2
0
2·0
4·0
6·0
8·0
10·0
12·0
14·0
16·0
18·0
20·0
22·0
MuB/(Pu)², mm²/N

GRAPH 43

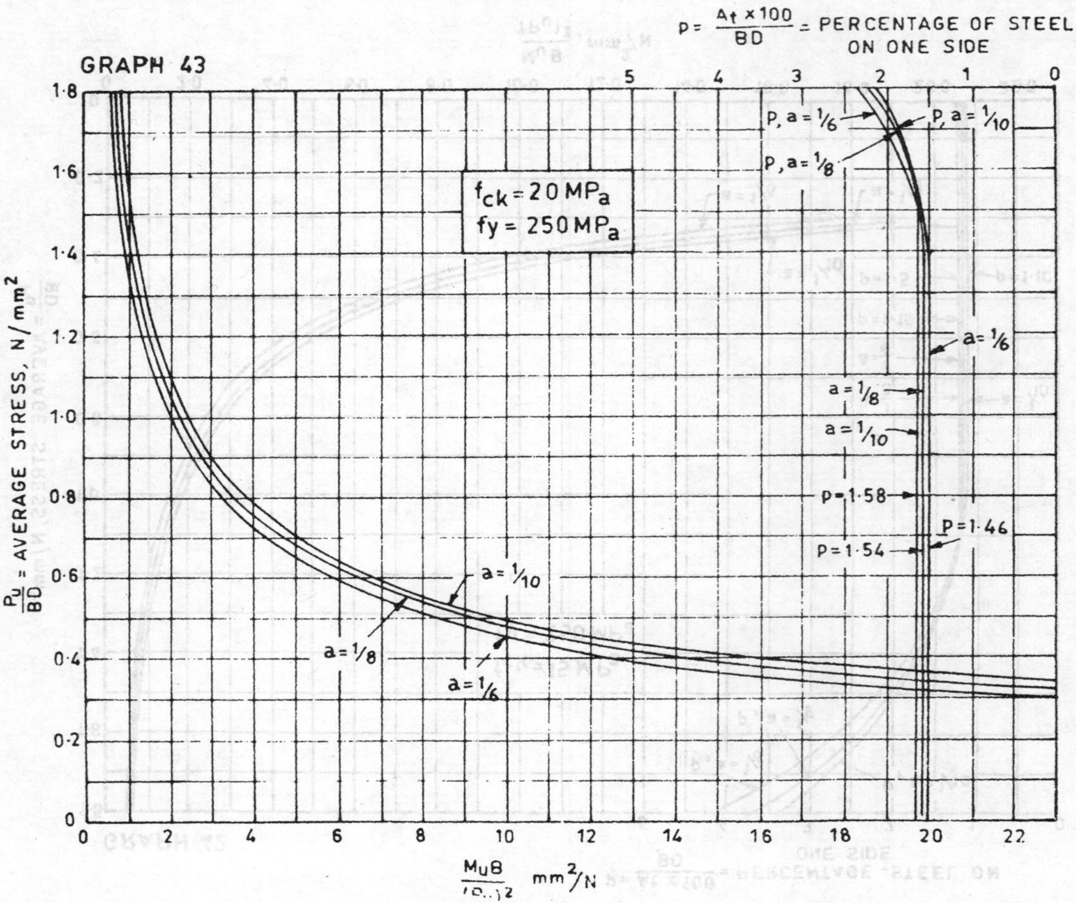

p = At x 100 / BD = PERCENTAGE OF STEEL ON ONE SIDE
fck = 20 MPa
fy = 250 MPa
P, a = 1/6
P, a = 1/10
P, a = 1/8
a = 1/6
a = 1/8
a = 1/10
p = 1·58
P = 1·46
p = 1·54
a = 1/10
a = 1/8
a = 1/6
Pu/BD = AVERAGE STRESS, N/mm2
MuB/(D)2 mm2/N

GRAPH 44

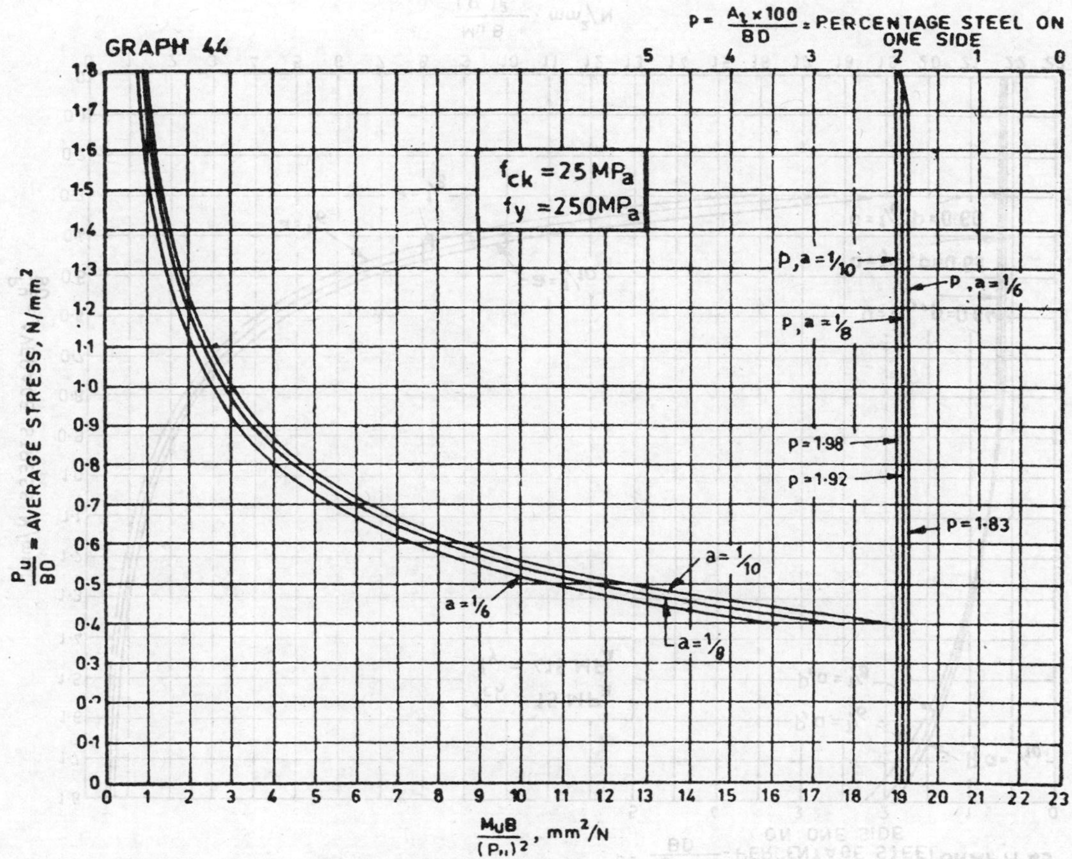

$p = \frac{A_t \times 100}{BD}$ = PERCENTAGE STEEL ON ONE SIDE
$f_{ck} = 25\,MP_a$
$f_y = 250\,MP_a$
p, a = 1/10
p, a = 1/6
p, a = 1/8
p = 1·98
p = 1·92
p = 1·83
a = 1/10
a = 1/6
a = 1/8
$\frac{P_u}{BD}$ = AVERAGE STRESS, N/mm²
$\frac{M_u B}{(P_u)^2}$, mm²/N

GRAPH 45

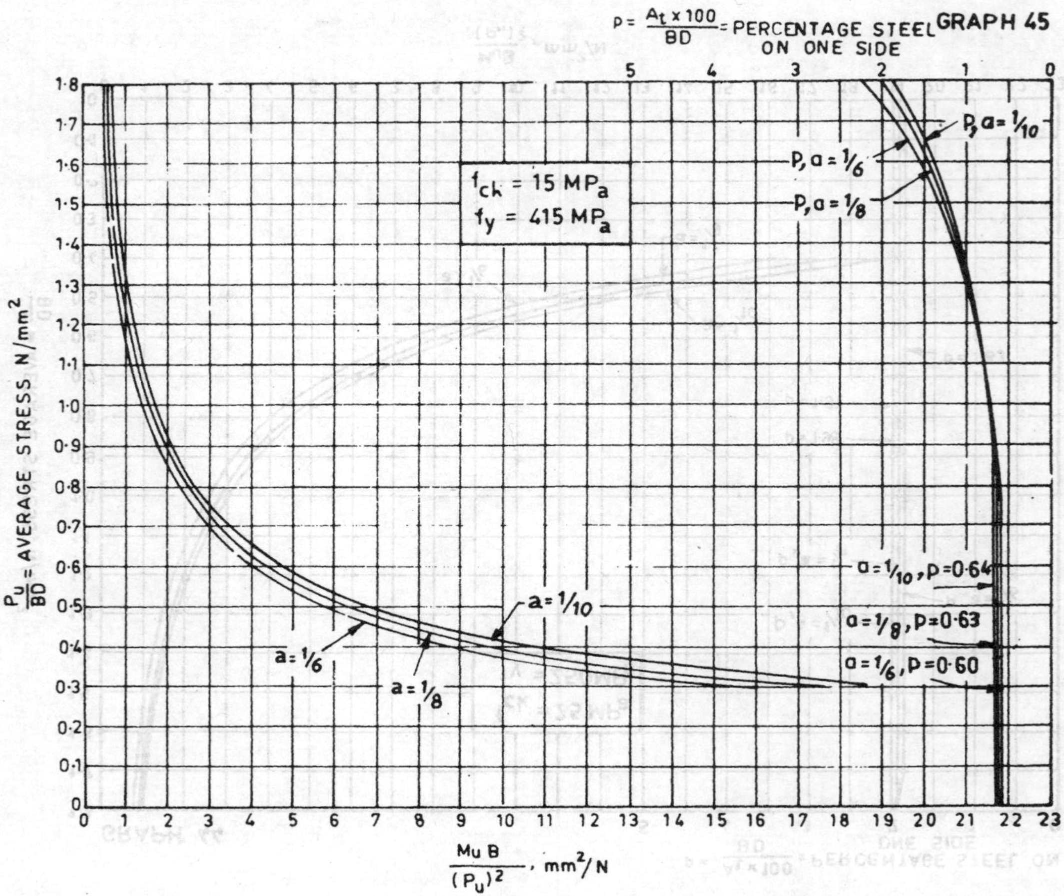

p = At x 100 / BD = PERCENTAGE STEEL ON ONE SIDE
fck = 15 MPa
fy = 415 MPa
P, a = 1/10
P, a = 1/6
P, a = 1/8
a = 1/10, P = 0·64
a = 1/8, P = 0·63
a = 1/6, P = 0·60
a = 1/10
a = 1/6
a = 1/8
Pu/BD = AVERAGE STRESS N/mm²
Mu B/(Pu)², mm²/N

GRAPH 46

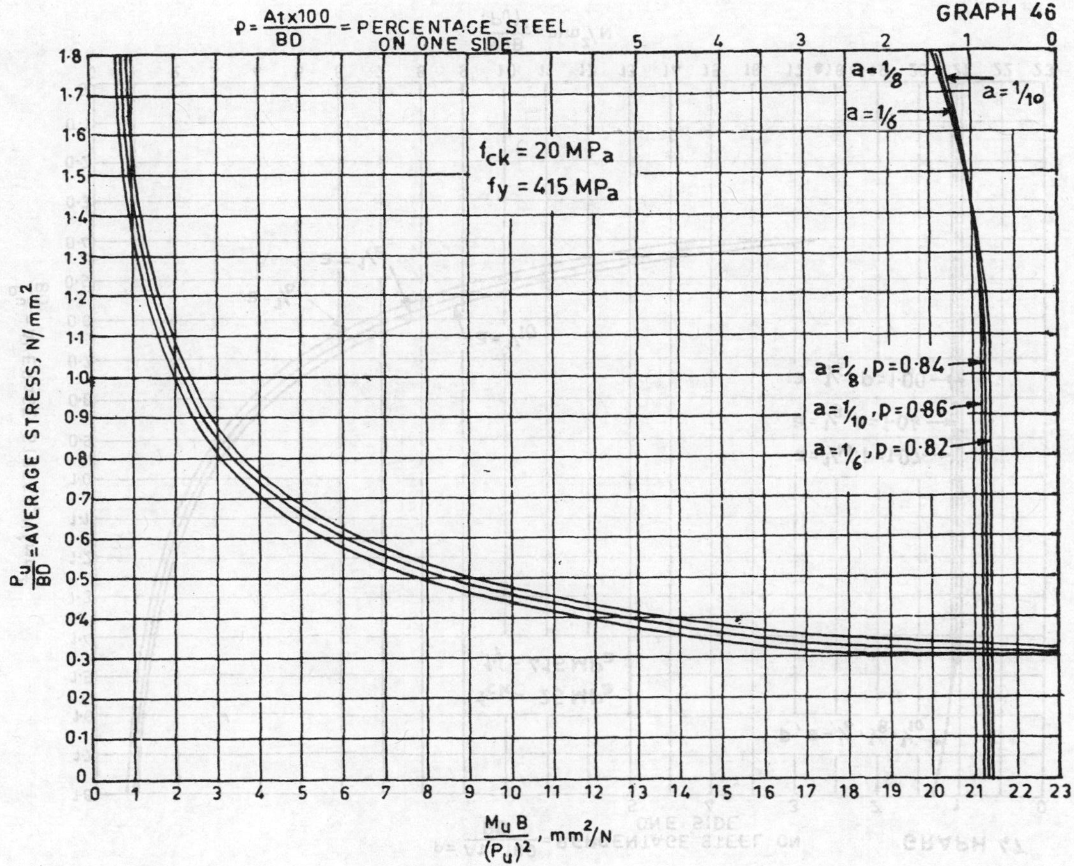

p = At×100 / BD = PERCENTAGE STEEL ON ONE SIDE
fck = 20 MPa
fy = 415 MPa
a = 1/8
a = 1/10
a = 1/6
a = 1/8, p = 0·84
a = 1/10, p = 0·86
a = 1/6, p = 0·82
Pu/BD = AVERAGE STRESS, N/mm²
MuB/(Pu)², mm²/N

GRAPH 47

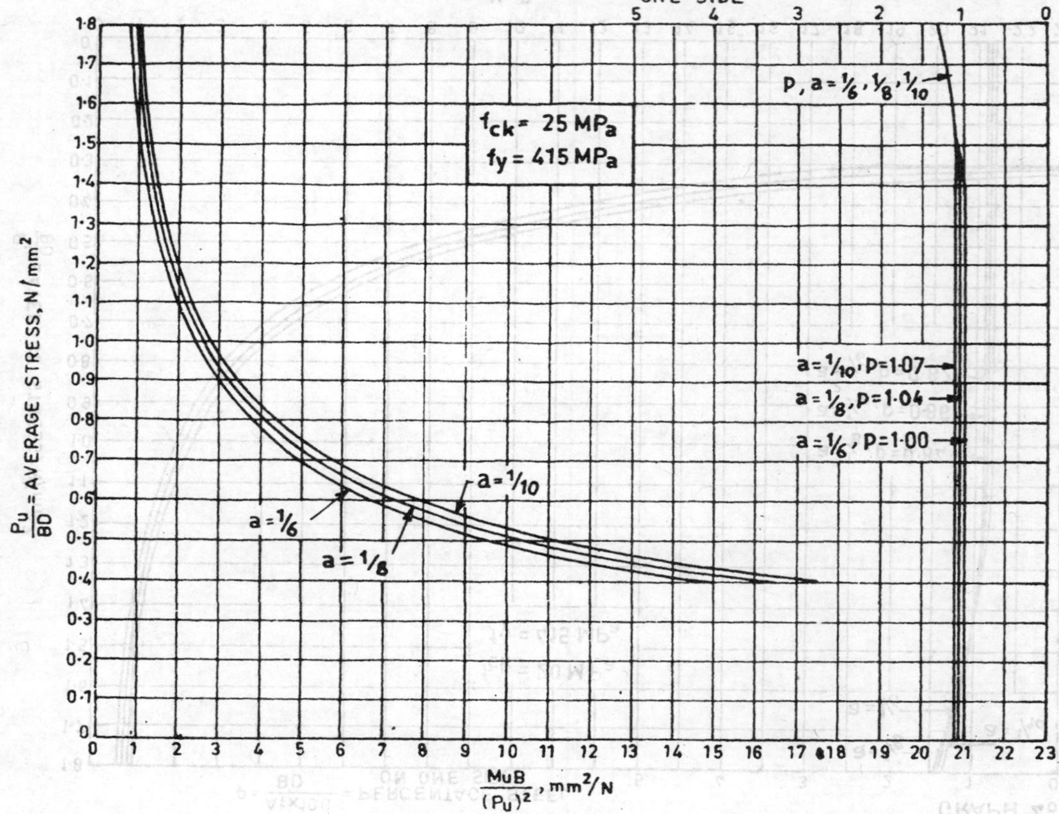

p = At x 100 / BD = PERCENTAGE STEEL ON ONE SIDE
f_ck = 25 MPa
f_y = 415 MPa
p, a = 1/6, 1/8, 1/10
a = 1/10, P = 1·07
a = 1/8, P = 1·04
a = 1/6, P = 1·00
a = 1/10
a = 1/6
a = 1/8
Pu/BD = AVERAGE STRESS, N/mm²
MuB/(Pu)², mm²/N

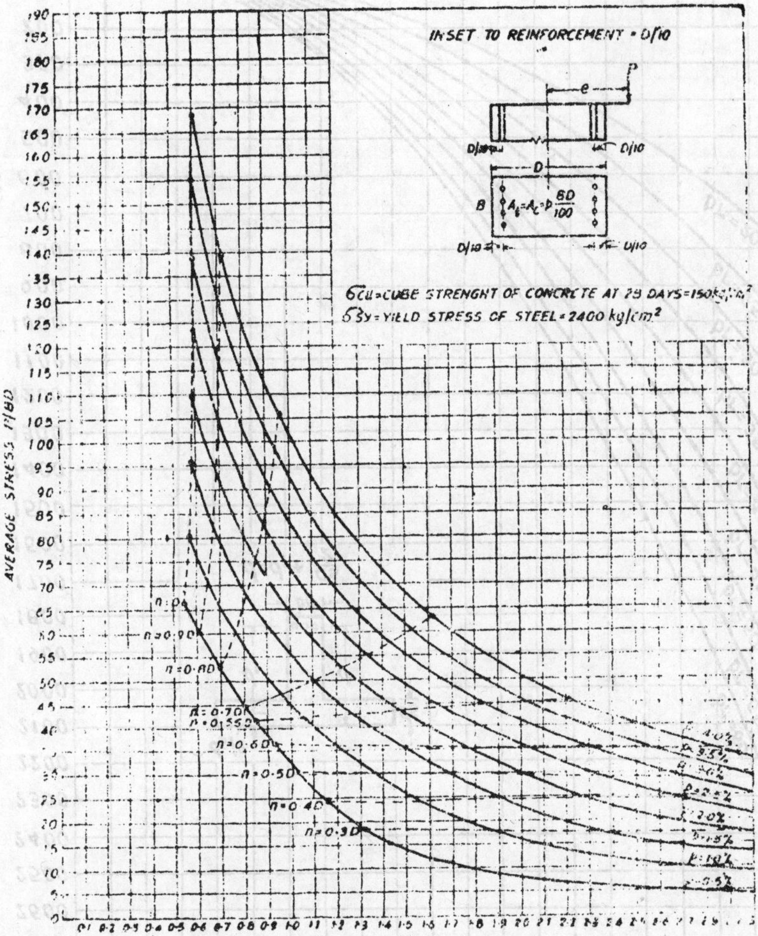

INSET TO REINFORCEMENT = D/10
e
P
D/10
D
B
$A_s = A_c = p\frac{BD}{100}$
σcu = CUBE STRENGHT OF CONCRETE AT 28 DAYS = 150 kg/cm²
σsy = YIELD STRESS OF STEEL = 2400 kg/cm²
AVERAGE STRESS P/BD
e/D
n = 0.90
n = 0.80
n = 0.70
n = 0.60
n = 0.50
n = 0.40
n = 0.30
p = 4.0%
p = 3.5%
p = 3.0%
p = 2.5%
p = 2.0%
p = 1.5%
p = 1.0%
p = 0.5%

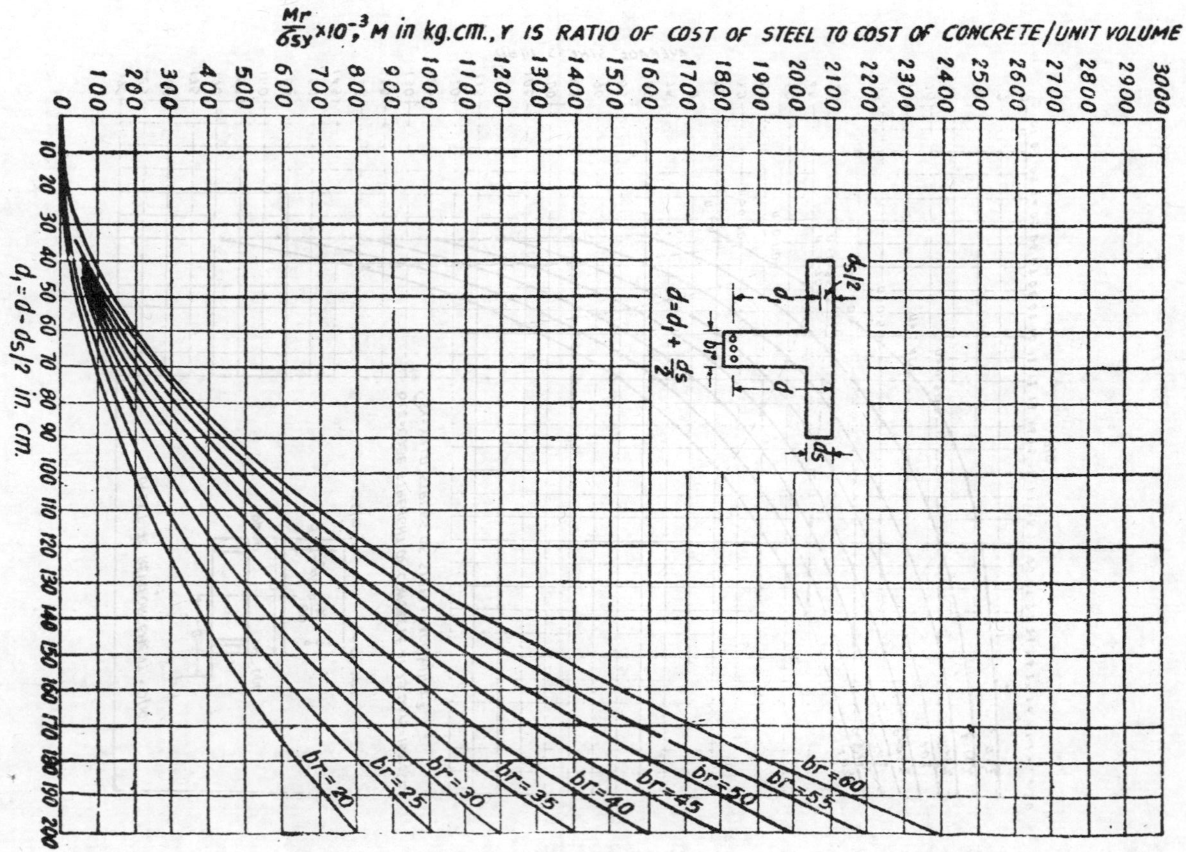

$\frac{Mr}{\sigma sy} \times 10^{-3}$, M in kg.cm., r IS RATIO OF COST OF STEEL TO COST OF CONCRETE/UNIT VOLUME
0
100
200
300
400
500
600
700
800
900
1000
1100
1200
1300
1400
1500
1600
1700
1800
1900
2000
2100
2200
2300
2400
2500
2600
2700
2800
2900
3000
10 20 30 40 50 60 70 80 90 100 110 120 130 140 150 160 170 180 190 200
$d_1 = d - d_s/2$ in. cm.
$d_s/2$
d_1
b_r
d
d_s
$d = d_1 + \frac{d_s}{2}$
br = 20
br = 25
br = 30
br = 35
br = 40
br = 45
br = 50
br = 55
br = 60

BIBLIOGRAPHY

1. Reinforced Concrete Design *by* Sutherland and Reese.
2. Design of Concrete Structures *by* Winter, Ur uhart, Rourke and Nilson.
3. Design of Reinforced Concrete *by* Boris W. Bogulavsky.
4. Reinforced Concrete Structures *by* Dean Peabody.
5. Reinforced Concrete Design *by* G.P. Manning.
6. Reinforced Concrete *by* E. Sigalov and S. Straingin.
7. Reinforced Concrete Designers Handbook *by* Chase E. Reynolds.
8. Advanced Structural Design *by* Cyril S. Benson.
9. Theory and Practice of Reinforced Concrete *by* Clarence W. Dunham.
10. Basic Reinforced Concrete Design *by* George Elwyn Large.
11. Concrete Plain and Reinforced *by* Taylor, Thomson and Smulski.
12. Elementary Reinforced Concrete *by* Morgan.
13. Structural Design Reinforced Concrete *by* Clifford D. Williams and Charles E. Cutts.
14. Examples of Design of Reinforced Concrete Buildings *by* Chase E. Reynolds.
15. Reinforced Concretes Bridges *by* Taylor, Thomson and Smulski.
16. Reinforced Concrete Bridges *by* W.L. Scott.
17. Reinforced Concrete Bridge Design *by* C.S. Chettoe and Haddon C. Adams.
18. Design and Construction of Reinforced Concrete Bridges *by* A.W. Legat, G. Dunn and W.H. Faihurst.
19. Reinforced Concrete Reservoirs and Tanks *by* W.S. Gray.

20. Reinforced Concrete Water, Towers, Bunkers, Silos and Gantries *by* G.S. Gray.

21. The Design of Prismatic Structures *by* A.J. Ashdown.

22. Non-Planar Concrete Roofs *by* J.S. Terrington and F.H. Turner.

23. Reinforced Concrete Chimneys *by* Percy Taylor and Leslie Turner.

24. Theory of Plates and Shells *by* S. Timoshanko and S. Woinowsky Krieger.

25. The Ultimate Load Theory Applied to Design of Reinforced Concrete and Prestressed Concrete Beams and Frames *by* A.L.S. Baker.

26. Prestressed Concrete *by* Gustave Magnel.

27. Basic Design and Construction in Prestressed Concrete *by* J.D. Harris.

28. Design of Prestressed Concrete Structures by T.Y. Lin.

29. Foundation Design *by* Wayne C. Teng.

30. The Design and Constructions Engineering Foundations *by* F.D.G. Henry.

31. Foundations of Bridges and Buildings *by* Jacoby and Devis.

32. Earth Pressures and Retaining Walls *by* Whitney Clark Huntington.

33. Soil Mechanics in Engineeriag Practice *by* Terzaghi and Peak.

34. The Mechanics of Engineering Soil *by* Capper and Cassie.

35. Prestressed Concrete *by* Guyon.

36. Introduction to Prestressed Concrete *by* Abeles.

37. Basic Design and Constructions in Prestressed Concrete *by* J.D. Harries and I.C. Smith.

38. Prestressed Concrete *by* Valley.

39. Prestressed Concrete by Evans and Bennet.

40. Sub-structure Analysis and Design *by* Paul Andersen.

41. Ultimate Load Analysis of Reinforced and Prestressed Concrete Structures *by* L.L. Jones.

42. Ultimate Load Design *by* Dr. Ramakrishanan and Arthur.

43. W. Fluggi's "Stresses in Shells" Springer Verleg, Berlin.

44. Structural Engineering Handbook, edited *by* Gaylord, Mc-Graw Hill Book Company.

45. Bennel J.D. "Emperical Design of Symmetrical Cylinderical Shells" Proceedings of Symposium on Shell Structures, Netherlands 1965.

INDEX

A

B

C